Foundations *of* Ecology

Foundations

Classic Papers

PUBLISHED IN ASSOCIATION WITH
The Ecological Society of America

# of Ecology

with Commentaries

EDITORS

Leslie A. Real and James H. Brown

THE UNIVERSITY OF CHICAGO PRESS

Chicago & London

The University of Chicago Press, Chicago 60637
The University of Chicago Press, Ltd., London

Printed in the United States of America
00 99 98 97 96 95 94 93 92 91 5 4 3 2 1

Library of Congress Cataloging-in-Publication Data

Foundations of ecology : classic papers with commentaries / Leslie A. Real and James H. Brown, editors.
p. cm.
"Published in association with the Ecological Society of America."
Includes bibliographical references.
ISBN 0-226-70593-5 (alk. paper).—
ISBN 0-226-70594-3 (pbk. : alk. paper)
1. Ecology. I. Real, Leslie. II. Brown, James H. III. Ecological Society of America.
QH541.145.F68 1991
574.5—dc20 91-3052
CIP

∞ The paper used in this publication meets the minimum requirements of the American National Standard for Information Sciences—Permanence of Paper for Printed Library Materials, ANSI Z39.48-1984.

Contents

Preface

What is the value of a book of classics? In his often anthologized essay, "Why Read the Classics?" the Italian novelist Italo Calvino suggests that we always return to the roots of modern literature to refresh our senses and perfect our vision. We continue to learn from the classics every time we reread them. What is it we learn each time we reread the classic works in any field of science? Both of us have been struck by the diversity of approaches and contributions that have shaped our field, ranging from the elegance of Connell's barnacle experiments and Simberloff and Wilson's island biogeography experiments to the encyclopedic natural history of Watt and Gleason. Beyond the beauty of individual accounts and studies, we have learned to recognize how seminal contributions critically reshape our thinking about natural order and the processes that account for natural pattern. Often a seminal paper alters irretrievably our perception of the world; for example, Hutchinson's essay on diversity or Lindeman's concept of trophic organization. Sometimes the importance of a contribution is only apparent in retrospect, as is true for Grinnell's concept of niche and Skellam's analysis of spatial structure in populations.

Surveying the historical development of critical concepts informed our own perspective on the development of successful research programs. Why are some ideas or approaches more successful than others? Where do new ideas come from? How are they expressed in ways that have immediate intellectual appeal? What is the relationship between the intellectual fate of an idea and its empirical validation? How are ideas refined and extended to new systems and broader sets of interactions? By rereading the classics, students and researchers will be in a superior position to map out research strategies that confer higher probabilities of success. Finally, rereading the classics is fun!

For the purposes of this volume, a classic is a paper that has made a substantial contribution to our thinking about ecological processes. Every paper reprinted here has had a lasting impact on our field either by posing new problems, demonstrating important effects, or stimulating new dimensions to research. Often the impact of a classic is not immediate but delayed through the major contribution of its progeny. In every case, the papers are exciting to read for their insights and hints, oversights and errors.

Even before the conception of this project, we shared a love of the history of ecology and a study of the development of critical concepts in the field. During 1987, Real taught a course on the classics in ecology in which Brown gave a lecture on the role of the classics in our current understanding of ecological process. Recognizing that much of the organization and structure for a book on this subject could be borrowed from that course, we decided to collaborate on the production of this collection.

Development of this book was given impetus by the seventy-fifth anniversary of The Ecological Society of America, which led us to think carefully about the conceptual history of ecology as a science. During our earliest discussions on the book's scope, we realized that our interests and expertise could not reflect those of the broader ecological community. As a consequence, we asked The Ecological Society of America (ESA) to establish an ad-hoc committee to serve as a board of editors for this volume. The board consisted of representatives from a variety of subdisciplines: James Brown, University of New Mexico (Terrestrial Community Ecology); Linda Brubaker, University of Washington (Paleoecology); Sharon Kingsland, Johns Hopkins University (History of Ecology); Joel Kingsolver, University of Washington (Physiological Ecology); Jane Lubchenco, Oregon State University (Aquatic Community Ecology); Robert Peet, University of North Carolina (Plant Ecology); Leslie Real, University of North Carolina (Theoretical Ecology); and Peter Vitousek, Stanford University (Ecosystem Studies).

Following extensive discussion among the committee members, a tentative outline for this volume was presented to the executive committee of the ESA, which agreed to sponsor the project. That original outline included over 65 papers and book chapters, a total of over 1,700 printed pages! In order to produce a single volume of reasonable size, we eliminated over half of the original suggestions. We decided to exclude all papers published after 1975 and all book excerpts. After reviewing the remaining articles, we soon realized that no two ecologists have identical opinions about what is or is not a classic. The current list obviously represents a compromise. No one will be completely satisfied with our selection; nonetheless, a large proportion of these papers will appear on every ecologist's list. We have tried to reprint those papers that are essential to understanding the origins of contemporary ecology.

Having established the classics list, members of the editorial board assumed responsibility for introducing designated sections. Some elected to invite a collaborator. The individuals in charge of writing introductions were given some license in the final composition of their section, and they made the final decisions on the papers that were included. As the principal editors, we were impressed by the dynamic interaction among the members of the editorial board and by the ever-changing nature of the classics list. The fact that a final list was only determined just prior to publication indicates a vigorous and healthy debate over what constitutes an important idea in our field.

We have consciously broken from the traditional partitioning of subject matter by level of organization, for example, populations, communities, and ecosystems. Instead, we focus on the common intellectual structures that emerge across levels and across taxonomic groups. The book is divided into six sections: foundational papers, theoretical advances, synthetic statements, methodological developments, field studies, and experiments in ecology.

The introductions to each section attempt to place the papers in their broader conceptual and historical context. These introductions are not restricted to discussing the targeted classic papers. They also explore the intellectual antecedents of these seminal contributions and consider the impact of these ideas on subsequent research. These introductions, along with the literature cited, provide an overview of the historical foundations and the current status of ecological science. They also attempt to identify promising directions for future research.

The completion of this kind of project depended upon the efforts of many individuals. Most importantly, we wish to thank the other members of the editorial board. They have shown great patience, insight, and an ability to resolve differences in opinion. On behalf of the board, we would like to thank all of our students and colleagues for their suggestions and advice. We hope this volume represents the combined interests of the larger ecological community. We thank The Ecological Society of America for its encouragement and for including this project in its seventy-fifth anniversary celebration.

Our greatest hope is that this book helps the students who will write the classics of tomorrow. To promote the training of young ecologists, all royalties from the sale of this book will go to a special Ecological Society of America fund supporting graduate and posdoctoral research fellowships.

Leslie A. Real
James H. Brown

PART 1 Foundational Papers

Defining Ecology as a Science

Sharon E. Kingsland

The origins of ecology as a science began with the application of experimental and mathematical methods to the analysis of organism-environment relations, community structure and succession, and population dynamics. The word "oecology" was first coined in the 1860s by a German zoologist, Ernst Haeckel. As a convert to Charles Darwin's theory of evolution, Haeckel believed that a term was needed to refer to the study of the multifaceted struggle for existence that Darwin had discussed in his 1859 treatise *On the Origin of Species* (McIntosh 1985, pp. 7–8).

In his book, Darwin had reviewed the diverse meanings of the struggle for existence, a metaphor standing for all the factors that affected the organism's survival and reproduction. His argument was influenced by his reading of Thomas Robert Malthus's controversial *Essay on Population*, first published in 1798, which pointed out that populations would, if left unchecked, tend to increase geometrically and would soon outstrip their food supply. Malthus had asked his readers to consider the consequences of the struggle that would inevitably follow from this population pressure. He concluded that because this population pressure could never be eliminated, we would never be able to create a truly utopian society where war, famine, and vice were absent. He was advising the utopian social thinkers of his time to adopt a more realistic and conservative estimate of the potential for human progress.

Darwin read the sixth edition of the *Essay on Population* (1826) in 1838, soon after returning from a five-year voyage around the world. In England, when he had had time to reflect on the observations he had made, he realized that species had probably originated from preexisting species and had not been specially created by God. Reading Malthus gave him a crucial insight that eventually led to his theory of natural selection. By turning Malthus's logic back to the natural world, Darwin deduced that the tendency to overproduction would lead to intense competition, heavy mortality, and therefore to an unconscious selection process. Given a long enough period of time one species might split into several new species, each with adaptations shaped by the selective pressures of climate, food supply, predation, and competition. Darwin described his theory as the law of Malthus applied to the natural world, but in fact he had brilliantly reversed Malthus's conclusions by showing that the struggle for existence was the mechanism for open-ended evolutionary change.

The struggle for existence included all forms of competition, direct and indirect, between organisms. One of the most important arguments in *Origin of Species* was the emphasis Darwin placed on competition between individuals of the same species as the chief mechanism of evolutionary change. Ernst Mayr (1982) has argued that the reading of Malthus was so important because it made Darwin realize that the struggle for existence mainly occurred *within* the species, that the members of a population were competing with each other. It was common in natural history writing before Darwin to describe a struggle for existence between different species, for instance between predator and prey. Naturalists accepted the idea that predators would kill off the weak, old, or diseased members of a prey population, in other words, that predation was a form of selection of the unfit. This kind of selection, they thought, actually served to preserve the character of the prey species, because only the deviant individuals were picked off.

But competition within the population could produce real changes, as Darwin recognized. If closely similar individuals were competing intensely for resources, then those with traits that gave them superior competitive ability would eventually replace the inferior and less prolific members of the population. Therefore natural selection was a genuinely creative force, not merely a mechanism for the preservation of the species type. Competition *within* the species had quite a different significance from competition *between* species. Recognition of the intensity of competition between closely related forms was also crucial to the understanding of the subtlety of adaptation and the patterns of species' replacement over a large geographical area.

The work of Darwin and other naturalists of the late-nineteenth century stimulated a more rigorous approach to natural history. In the United States, naturalists such as Stephen Alfred Forbes (1844–1930), Henry Chandler Cowles (1869–1939), and Frederic Edward Clements (1874–1945), all working in the Midwest, began to develop new quantitative methods and theoretical principles that would eventually lay the foundations for a new science called ecology. The term "ecology" was first used in America by a group of professional botanists, who in the 1890s began openly to reject traditional descriptive methods of natural history and concentrated instead on physiological studies of the relationship between organisms and their environment (Cittadino 1980). These botanists were greatly influenced by German and Danish studies in plant geography, themselves an outgrowth of the European Darwinian tradition of research on adaptation and environment. In the creation of the discipline of ecology botanists led the way, but their zoological colleagues followed closely behind and by the 1920s had taken the lead in population studies, which became a major field of research in the 1930s.

The science of ecology, as it was understood both by botanists and zoologists at the turn of the century, signified a dynamic, experimental approach to the study of adaptation, community succession, and population interactions. For botanists in particular, the ecologist was a kind of "outdoor physiologist," someone who studied adaptation and community evolution in the field using the same rigorous methods that the physiologist employed in the laboratory. These botanists were interested in whether the evolution of species could be controlled experimentally by altering the environment, a goal that had obvious agricultural applications. Most of the early biological surveys and other ecological studies were done in connection with the agricultural experiment stations and land-grant colleges being established by the government in each state at the end of the nineteenth century. In the early-twentieth century, much ecological research on evolution, adaptation, and community succession was funded privately by the Carnegie Institution, which established a Desert Botanical Laboratory near

Tucson, Arizona in 1902. In 1904, the Carnegie Institution also opened an experimental station for studies of evolution at Cold Spring Harbor, New York, which became a center of genetics research.

As a high profile and trendy scientific field, ecology attracted many practitioners, but the standards of the new science were not always very rigorous. Hence the complaint voiced by one botanist of the old school, Charles E. Bessey, who bemoaned the popularity of this latest "fad," which seemed only to distract from serious botanical research (Cittadino 1980, p. 171). But the professional interest in ecology was also serious and the field flourished, despite heavy competition from that other great "fad" of the twentieth century, genetics. The Ecological Society of America held its first organizational meeting at the end of 1914 and was officially constituted in 1915 with 284 charter members (Burgess 1977).

Stephen Forbes, who began his ecological career in Illinois in the 1870s, was greatly influenced by Darwinian biology and by the efforts in America to raise the standards in natural history research by such eminent scientific figures as Louis Agassiz, who in 1859 founded the Museum of Comparative Zoology at Harvard University. Stimulated to pursue ecological work by the research ideals of men such as Agassiz, Darwin, and Thomas Henry Huxley, Forbes undertook detailed analyses of the food relations of insects, birds, and fish within the community, believing that exact information was needed before the value of a species to society could be assessed. The immediate justification for his research was its practical importance: his analyses of food webs were pioneering attempts to give agriculture a scientific basis (Forbes 1880). But Forbes also appreciated the importance of setting his data into the broader theoretical framework of evolutionary biology. The enormous labor involved in ecological research was justified partly by its potential contribution to general questions involving the nature of adaptation, the causes of variation, and the origin and extinction of species.

Forbes accepted Darwin's argument for evolution by natural selection, but missing from Darwin's discussion was an explanation of how the struggle for existence produced what seemed to be a well-regulated world where population densities remained fairly stable from year to year. This idea of a balance in nature was commonly accepted by natural historians well before Darwin. Forbes integrated this traditional belief, which harkened back to an earlier teleological view of nature as harmoniously regulated for the benefit of all in accordance with divine wisdom, with the new theoretical writing on evolution. To achieve this integration, he had to go beyond Darwin's own discussion, which did not really address the problem of how balance was achieved between populations.

The theoretical support for the common sense idea that nature was self-regulating came largely from Darwin's compatriot Herbert Spencer (1820–1903), a philosopher and evolutionist who coined the phrase which we now associate with Darwinian evolution, "survival of the fittest." Spencer was trained as an engineer and thought of nature as a moving equilibrium between opposing forces, in this case the forces of population increase and decrease. Spencer believed that there was a necessary adjustment of fertility to mortality, a balance that produced rhythmical population changes about a stable equilibrium value. As evolution progressed, these forces came into perfect balance; Spencer believed that eventually evolution itself would come to a halt (Kingsland 1985; Spencer 1874). Spencer's analysis of evolution, though based almost entirely on armchair reasoning rather than direct observation, was attractive to his contemporaries partly because it supported the prevalent Victorian belief in a harmonious, progressive world. Let the struggle for existence work itself out unfettered, he thought, and improvement would inevitably result (Kingsland 1988).

Spencer himself was not really a Darwinian evolutionist, though he accepted Darwin's main argument for natural selection. He believed that a more important mechanism of change was direct, hereditary adaptation to the environment.

Adaptive change in direct response to the environment was known as "neo-Lamarckian" evolution, named after the French zoologist Jean-Baptiste Lamarck, who in 1809 proposed a theory of evolution based partly on the idea that characteristics acquired during an organism's lifetime could become hereditary (Lamarck 1984). In the late-nineteenth century most evolutionists viewed both Darwinian selection theory and the neo-Lamarckian theory of direct adaptation to the environment as complementary mechanisms of evolutionary change. Spencer's writings did a great deal to popularize Lamarckian evolution. Darwin himself was more accepting of the inheritance of acquired characteristics in later editions of the *Origin*, especially when he was faced with the problem of explaining how habits evolved into hereditary instincts.

Influenced by both Darwin and Spencer, Forbes was concerned to show that despite the intensity of the struggle for existence, natural selection was a beneficial force because it tended to restore a healthy equilibrium to the community. Darwin's struggle for existence and Spencer's balance of forces combined to create an image of nature which was both benign and thrifty. By combining a Spencerian outlook with Darwin's theory of evolution, Forbes developed the theme of a "common interest" between a species and its enemies, the idea being that natural selection would adjust reproductive rates so that they balanced mortality.

His classic description of the community along these lines was put forth in his essay of 1887, "The Lake as a Microcosm," which set out the main goal of ecological research: to analyze how harmony is maintained through the complex predatory and competitive relations of the community. By using the metaphorical language of the "sensibility" of the organic complex, the essay drew attention to the way all species were bound up with others within the community. This concept of the ecological community had European precedents. Karl Möbius's pioneering study of oyster culture had been published in America in English translation in 1883. In this essay Möbius had proposed the term "biocoenosis" for a community of species inhabiting a definite territory (Möbius 1883). Forbes's discussion developed the concept of the community in more detail and his essay exemplifies the ecological viewpoint at that time. His assumption of balance between reproduction and mortality was remarkably long-lived in the ecological literature. As late as the 1950s David Lack had occasion to criticize this still-prevalent assumption; he proposed instead that the reproductive rate in birds was adjusted not to mortality but to the food supply available to the young (Lack 1954).

While animal ecologists focused on community structure and population dynamics, plant ecologists concentrated on ecological succession. Henry Chandler Cowles, whose research concentrated on the sand dunes in the Chicago region, developed a dynamic perspective which he called physiographic ecology (Engel 1983). Trained as a geologist, Cowles stressed the constant dynamic interaction between plant formations and the underlying geological formations. The physiographic viewpoint saw the flora of a landscape as an ever-changing panorama; the ecologist had to discover the laws governing these changes. Cowles's studies of the Indiana dunes yielded the first thorough working out of a complete successional series. Cowles was able to accomplish this task by assuming that vegetational changes in space paralleled successional changes in time. Therefore, as one walked inland from Lake Michigan, one also walked backwards in time. By putting together the spatial sequences of plant formations, Cowles reconstructed the temporal development of plant associations (Cowles 1899). We have included here excerpts from his first major study of the dunes, published in 1899, which illustrates his view of ecology as a study of process. The references accompanying his article reveal the important precedents in European ecological research that served as a foundation for the American school.

Cowles's account of succession was never dogmatic. Though he believed that succession tended toward a stable equilibrium, he did not believe that this equilibrium state was ever reached. Moreover, successional stages leading to the climax community—the final stage of succession—were never in a straight line, but could even regress in the normal course of events. His concept of the community included not only the idea of a continuous, never-ending process of change, but also the idea that all the organisms were connected in a vast, complicated symbiosis.

While Cowles was tracking the ever-shifting landscape of the dunes, his colleague Frederic Clements was studying the more stable grasslands and conifer forests of the western prairie. He developed a theory of the plant community that differed in significant ways from that of Cowles (Tobey 1981). Clements was important also for publishing the first American textbook in ecology, *Research Methods in Ecology* (1905), which discussed the statistical and graphical analytical methods he and other Nebraskan ecologists developed from 1897 to 1905. His ecological theory rested on two ideas, the concept of ecological succession of plant formations, and the treatment of the plant community as a "complex organism" undergoing a life cycle and evolutionary history analogous to the individual organism. The formal presentation of his theory appeared in 1916 in his monumental study *Plant Succession*.

Clements was also influenced by Herbert Spencer and believed that organisms evolved by direct adaptation to changes in the environment. In America the neo-Lamarckian school was prominent during the 1890s, the decade when Clements was developing his ideas (Bowler 1983). He held fast to his Lamarckian ideas, even believing that body cells could modify the germ plasm, long after the theory had been roundly challenged and mostly discredited in America by the 1920s. In Clements's theory, the plant community could be analyzed as a complex organism which grew, matured, and died like an individual organism. Following the Lamarckian model of evolutionary change, the process of plant succession entailed a continual interaction between the habitat and the life forms of the community. The habitat and the populations acted upon one another in a reciprocal way, until finally a stable state, the climax, was reached. If the climate remained stable and no humans intervened, the climax might persist for millions of years.

Clements believed that one of the important processes directing succession was competition between similar plants (Clements, Weaver, and Hanson 1929). Species of trees, for instance, were described as competing sharply when together, whereas the relation of shrubs to trees was thought to be one of subordination and dominance rather than competition. Clements thought that the process of succession reduced the amount of competition within the community as a whole by setting up stable dominance hierarchies among the species of the community as it moved toward the climax stage. The overall character of the climax formation was defined by its dominant plant forms. Though Clements did come to appreciate the importance of animal populations in succession, largely through his collaboration with Victor Shelford, a leading animal ecologist and former student of Cowles's, he continued to think of the community as structured mainly by its plant formations.

Clements also used the idea of the climax to develop a system of classification for the units of vegetation. He subdivided the units of the climax formation into various categories with parallel categories of successional change. His use of esoteric Greek and Latin terms, a penchant that others found tedious, as well as his emphasis on quantitative methods reveal his desire to build ecology into a rigorous discipline. His classification system and theoretical framework dominated American plant ecology in the first decades of the twentieth century.

Clementsian doctrine did not go unchallenged (McIntosh 1985, pp. 76–85). On the agricultural front, wheat farmers in the midwestern prairie criticized the climax theory largely because it advocated a more cautious use of these marginal lands and therefore threatened their livelihood (Worster 1979). The creation of the Dust Bowl in the 1930s, however, showed the wisdom of the kind of ecological awareness that Clements had advocated.

But Clements's fellow ecologists also criticized aspects of his theory. British ecologist Arthur G. Tansley disagreed with the organismic metaphor, though he defended Clements's idea of the plant formation as a natural unit. He preferred to call the community a "quasi-organism" and in 1935 coined the word "ecosystem" as a more accurate characterization of the vegetational unit (Tansley 1935; Tobey 1981). The shift from a biological to a physical model for ecology also opened the way to a mathematical analysis of the system. Fittingly, the volume of *Ecology* in which Tansley's 1935 critique appeared was dedicated to Cowles, who also had no use for the organismic metaphor. Forrest Shreve, an ecologist at the Desert Laboratory of the Carnegie Institution, with which Clements was also affiliated, criticized the idea of the climax as a stable formation controlled by the environment.

These arguments were echoed by another critic, Henry Allen Gleason (1882–1975), who proposed an "Individualistic" concept of the plant association in place of the organismic metaphor (McIntosh 1977). Gleason argued that fixed and definite vegetational structures did not exist. To be sure, it was possible to identify communities that were uniform and fairly stable over a given region, but he denied that all vegetation could be segregated into such communities. He was very conscious of the way that short-term environmental changes in time and space could have profound effects on the abundance of species in an area. Taking the opposite view from Clements, Gleason argued that every plant association was the unique product of the fluctuating environmental conditions of a particular time and place. It was not meaningful, in his view, to compare the plant community to an organism, nor was it possible to create a precisely logical classification of communities, as Clements had tried to do.

These criticisms did not persuade Clements to drop the organismic metaphor or to change his classification system in any way. The only major change in Clements's ideas in later years was to recognize the existence of animal populations as an integral part of the community. However, by the 1950s plant ecologists had abandoned many of the central principles of Clementsian dogma, as well as the more cumbersome features of his classification system, as inappropriate or unproductive. Instead of the "complex-organism" analogy, postwar ecology emphasized the functional system formed by community and environment, the ecosystem (Whittaker 1977).

In the 1920s, animal ecologists who were analyzing community structure began to develop the concept of the ecological niche. Charles Elton's text of 1927, *Animal Ecology,* which is still an excellent introduction to the subject for students, defined the niche concept in the context of his discussion of the food chain (Elton 1966). Before Elton, the niche was a nontechnical term referring to an abstract space in the environment which could be full or empty. Elton employed the idea of the niche, meaning an animal's "place," to redirect attention to the food relationships within the community. He believed that ecologists had to pay more attention to what the animal was actually doing in the community. He used the term niche to refer to the animal's place in the food chain, which also defined the animal's economic role in the community. Elton considered the niche to be a smaller subdivision of the traditional groupings of herbivore, carnivore, insectivore, and so on. By studying how different species occupied the niches within the community, that is, how they took on the major economic roles in the food cycle, one could perceive the basic similarity in structure between communities that appeared quite different in their species composition.

As a general economic category, therefore, a single niche could be occupied by different species. Elton's discussion of the niche did not include what we now call "the competitive exclusion principle," or the principle that two different species cannot occupy the same niche (Hardin 1960). Even before the term niche was used to designate the organism's habitat, this basic principle was well known to naturalists in the nineteenth century. Stephen Forbes, for instance, described in 1884 how species of insects feeding on the strawberry plant avoided direct competition by feeding at different times (Winsor 1972). One of the central problems that field naturalists studied in the early twentieth century was the coexistence of closely allied species and the ways that similar species evaded competition.

Joseph Grinnell (1877–1939), an expert on North American birds and mammals, connected the idea of competitive exclusion to the term *niche* in 1917 when he asserted that "no two species regularly established in a single fauna have precisely the same niche relationships." In his usage the niche was roughly synonymous with habitat. As a natural historian in the twentieth century, Grinnell was part of a shrinking population of field naturalists who published descriptive rather than experimental studies (Hutchinson 1978, pp. 152–54). Descriptive field studies of this kind were often disparaged by experimental biologists who wanted to establish the experimental approach to evolutionary biology as the more rigorous method. But Grinnell's brilliant research in natural history was recognized by contemporary naturalists who resisted the increasing specialization of biology. We have included here the article in which Grinnell refers to competitive exclusion as an axiomatic principle.

Elton also recognized the essential idea of competitive exclusion. Elsewhere he used the term niche in its older meaning of habitat rather than of a position in the food chain, and in that context he adopted Grinnell's conclusion that the limited number of niches and competitive exclusion determined the number of species in an association (Elton 1933, p. 28). But Elton did not raise the principle of competitive exclusion to the status of a key organizing idea of ecological theory. The competitive exclusion principle framed in terms of the niche concept became a more central and much debated ecological idea following the work of a young Russian ecologist, Georgii F. Gause (1910–1986), who designed his doctoral dissertation around an experimental analysis of predatory and competitive interactions between species of yeast and protozoa. These results appeared in 1935 as a small book called *The Struggle for Existence.* Unlike Elton's discussion of community structure, Gause's analysis made use of new mathematical modeling techniques that were being developed in American and European ecology in the 1920s. In fact, his book drew many ecologists' attention to this mathematical work for the first time.

The starting point for mathematical ecology was the analysis of population growth within a single species. Raymond Pearl, a statistician studying human population change after the First World War, discovered in 1920 that human population growth over time seemed to follow a regular, S-shaped curve which he called the "logistic curve" (Pearl and Reed 1920). The differential equation of the curve was $dN/dt = rN(K - N)/K$, where N represents the number of individuals, t the time, r the maximum rate of increase of the population, and K the upper limit of population growth. In the course of his research, Pearl also discovered that a Belgian mathematician named Pierre-François Verhulst had analyzed the curve nearly a century earlier, as part of a larger attempt to determine the law of population growth. The term "logistic" was Verhulst's term. This research had been entirely forgotten until Pearl resurrected it (Kingsland 1985). Determined not to suffer Verhulst's fate, Pearl made strenuous efforts to publicize his discovery of this curve, which (unlike Verhulst) he regarded as an actual law of population growth, comparable to Boyle's law in chemistry. Pearl was responsible for having Gause's book published by an American publisher, no doubt realizing that it would also help to advance his own reputation through Gause's application of the logistic curve to ecological theory.

Pearl's claims were controversial. He carried on an active debate about population growth with social scientists and biologists throughout the 1920s and 1930s. All of this discussion generated a large literature that helped to publicize the curve. Most of Pearl's opponents agreed that although Pearl was incorrect to regard the curve as a law of growth, it was a convenient description of growth because the equation describing it was easy to derive by verbal logic and easy to translate into biological terms. As such, it came to be used by ecologists in the 1930s who wanted to study population fluctuations mathematically. Fisheries ecologists in particular used and developed mathematical models based on the logistic equation (Cushing 1975).

The logistic equation had also been used independently by Italian physicist Vito Volterra (1860–1940) to construct a basic model of competition between two species (Kingsland 1985). Volterra had become interested in ecological problems in the 1920s when his daughter's fiance asked him to analyze certain changes in fish populations in the Adriatic during and after the First World War. Intrigued by the possibility of creating a mathematical science of the "struggle for existence," Volterra took up this challenge and devoted roughly fifteen years to exploring mathematical ecology. His formulations were so technical that they were beyond the reach of most ecologists; even today only his simpler models are well known. Apart from a model of competition, he also developed a model of predation in a two-species system, which coincidentally had been anticipated by Alfred James Lotka (1880–1949), an American mathematician and demographer. The two-species model of predation is now known as the "Lotka-Volterra equations" and is the starting point for most modern discussions of predation.

In 1925 Lotka published an unusual treatise which included a mathematical study of energy transformations within the biosphere, forming the basis for a new science which he called "physical biology" (Lotka 1925). Lotka, trained as a physical chemist, hit on the idea that one could apply thermodynamic principles to biology along the lines of physical chemistry in order to create a new science that focused on energy transformations within the biosphere. Over twenty years later his ideas on the subject were gathered into the book *Elements of Physical Biology*. Though he was not an ecologist, his wide-ranging discussion touched on many ecological problems, including such topics as food webs, the water cycle, and the carbon dioxide, nitrogen, and phosphorus cycles. Lotka's efforts to analyze the earth as a single undivided system, a kind of giant engine or energy transformer, were unprecedented in ecological theory, although his book coincided with a similar biogeochemical approach to ecology advanced in the 1920s by Vladimir I. Vernadsky in Russia. Lotka's clear exposition of the systems approach later influenced Eugene P. Odum and Howard T. Odum, who developed ecosystem ecology in the 1950s.

Lotka's desire to make his new science mathematical led him into the study of population growth and predator-prey interactions. His demographic analysis of stable populations influenced Patrick H. Leslie, who developed a method of analyzing populations using matrix algebra in the 1940s. Lotka was the first to explore the analysis of population interactions using sets of simultaneous differential equations, a method similar to Ludwig von Bertalanffy's "general system theory" of the 1950s, though Bertalanffy did not give Lotka credit for his prior work. As far as population dynamics were concerned, Lotka was not especially interested in competition; the mathematical analysis of competitive interactions was therefore carried forward mainly by Volterra.

Gause had read the publications of Pearl, Lotka, and Volterra and decided to test some of these simple models in a laboratory setting. In his book he referred to the principle of competitive exclusion in passing, regarding it as a natural extension of previous work and not worthy of special attention. A second and more sophisticated series of experiments was published in 1935 in a French monograph that unfortunately was not well known to ecologists (Gause 1935). In this second monograph he developed the idea that competition would force two species into separate ecological niches, thereby enabling them to coexist in the confined space of the test-tube environment.

Subsequent fieldwork on ecological succession that buttressed his laboratory experiments led Gause to appreciate the idea that competitive exclusion was the key to the structure of whole communities (Gause 1936, 1937). He began to think of the niche as a unit structure over which species fought for possession; each species' niche was the place where it alone enjoyed full advantage as a competitor. This meant that at the basis of the structure of the community lay the niche structure and that the stable, regulated community was actually a result of the competition between similar species. The principle of competitive exclusion therefore provided a way of relating community structure and the process of succession directly to the ongoing competitive interactions between the populations within the community. This explanation of community succession was compatible with Clements's ideas about competition, though Gause rejected the organismic metaphor as superfluous. In 1939 he asserted the centrality of the competitive exclusion principle in a commentary to a review of population ecology by Thomas Park (1939).

Gause's historical importance is not that he invented a new principle, but that he drew attention to what had been considered an axiomatic principle by making it a focus of the ecological theory of community structure. As soon as competitive exclusion began to be used as a central organizing idea, it generated controversy mainly because of the structure of the argument. Critics suggested that the principle was tautological and therefore was not useful because all it really told us was that no two species have identical requirements, which is true but trivial. We do not know the nature of these ecological requirements in advance, however. The principle focuses our attention on the fact that species must find ways to partition limited resources in order to coexist. We can discover how they do this by looking for cases where the principle seems not to apply and asking how species manage to live together. What we find most often is that the species have partitioned their resources in subtler ways than we had suspected. Therefore, the principle of competitive exclusion is useful in steering us toward a more profound analysis of ecological relationships. David L. Lack, G. Evelyn Hutchinson, and Robert H. MacArthur developed Gause's hypothesis along these lines in their studies of competition. In the 1960s studies of competition and the niche grew into major fields of research in evolutionary ecology, despite continuous controversy over the meaning of competition and the difficulty of measuring competitive interactions in the field.

A rather different approach to competition, population regulation and mathematical modeling was advanced by Alexander John Nicholson (1895–1965), an Australian entomologist, who teamed up with Victor Albert Bailey, a physicist, in a theoretical analysis of host-parasite interactions. Nicholson devised the original arguments behind the models and Bailey converted them into mathematical form. Reasoning by physical analogy, Bailey considered the movement of parasites in search of hosts to be analogous to Maxwell's theory of the mean free path of a particle in a gas. He assumed that density was uniform and that search proceeded randomly in the population as a whole. Volterra had used a similar analogy in his own models, but where Volterra and Lotka both used continuous-time models, Bailey used more realistic discrete-time models.

Nicholson and Bailey tried to improve on the Lotka-Volterra predation model by taking into account the effects of competition from members of the same species, as well as delays caused by the age distribution of the populations. Nicholson believed that any factor controlling populations had to act with increasing severity as density increased (Nicholson 1933). Only competition seemed to fulfill this requirement; therefore, he considered competition the chief mechanism of population regulation. In their models, Nicholson and Bailey tried to take into account the competition occurring when animals were engaged in a search for essential resources, though they could not directly measure the effects of such competition. With these and other adjustments, they found that instead of the steady-state oscillations of the Lotka-Volterra model, their models predicted an unstable system of increasing oscillations in their theoretical populations. In general, they hoped for a more exact treatment of population regulation with more detailed consideration of the alternative outcomes that would result from making different biological assumptions. The article which we have excerpted here was intended to be the first part in a series on population regulation, as indicated in the title, but the other parts were never completed. Harry S. Smith and Paul DeBach published some of the first experiments based on Nicholson's work in 1941 (DeBach and Smith 1941), and George Varley published the first field test of Nicholson's predictions in 1947 (Varley 1947). David Lack (1954) was also influenced by Nicholson's arguments. The question of density-dependent versus density-independent population regulation erupted in controversy in the 1950s and 1960s, with the main opposition to Nicholson's ideas and to mathematical approaches in general being voiced by H. G. Andrewartha and L. C. Birch (Andrewartha and Birch 1954).

With all the mathematical modeling of the 1930s, there was almost no attempt to incorporate population genetics into ecological models. Just as population geneticists tended to assume that the environment was constant, population ecologists assumed that their populations were genetically uniform. Although ecologists were aware of the mathematical work of J. B. S. Haldane, R. A. Fisher, and Sewall Wright in the 1920s and 1930s, there was little overlap between ecology and genetics until the 1960s. Two notable exceptions in these early decades were E. B. Ford at Oxford University, who collaborated with R. A. Fisher and developed ecological genetics starting in the late 1920s (Ford 1980), and V. A. Kostitzin, a Russian geophysicist living in Paris in the 1930s, who combined the Lotka-Volterra model with an evolutionary, genetic perspective (Scudo and Ziegler 1978). Kostitzin's work, which ended during the Second World War, has made virtually no impact on ecology.

While the new methods and concepts of population ecology were being brought to bear on problems of community structure and succession, another young ecologist, Raymond Lindeman (1915–1942) tackled the problems of succession from a physiological perspective. He fashioned a synthesis of physiological ecology and community ecology which he called the "trophic-dynamic aspect" of ecology. The 1942 article that set out this viewpoint was part of a series of articles on the ecology of a senescent lake, Cedar Creek Bog in Minnesota, which was the subject of his doctoral dissertation. This final article, published posthumously, was path-breaking in its general analysis of ecological succession in terms of energy flow through the ecosystem.

The trophic-dynamic viewpoint was an attempt to demonstrate how the day-to-day processes within a lake affected the long-term changes of ecological succession. From our perspective, using short-term cumulative changes to explain long-term dynamical changes may seem obvious, but ecology had tended to develop along two separate paths that made this kind of integrated analysis difficult. One branch of ecology, termed *autecology,* focused on physiological relations between organisms and environment, while the other branch, termed *synecology,* mapped out the long-term patterns of community succession. Both branches dealt in different time scales and focused on different sets of problems appropriate to the different time scales. Lindeman brought them together again.

Instead of looking at competitive interactions, as Gause had done, Lindeman focused on the trophic or nutritional relationships within the lake. His decision to group the inhabitants of the lake according to their position in the food cycle was stimulated by the work of Charles Elton, who was the first to describe the relationship between food habits and community structure. Lindeman found that there were two parallel food cycles operating in the lake which merged in their common link with bacterial decomposers. While studying the cycling of nutrients through the lake, it became clear to Lindeman that any attempt to separate the organisms from their abiotic habitat was highly artificial. Instead, the lake was an integrated system of the biotic and the abiotic, to which he gave the name "ecosystem." The remaining problem was to understand how all of the processes involving the influx of nutrients and removal of nutrients from the lake, acting in combination with the food cycle, could affect the rate of succession of the whole ecosystem.

Lindeman's mature analysis of the problem was worked out in 1941 while he was visiting G. Evelyn Hutchinson at Yale University. In a 1940 article Hutchinson and A. Wollack discussed the relationship between trophic processes and succession. In a footnote they called attention to the limnological method, as epitomized by American ecologists Edward A. Birge and Chancey Juday, of isolating a suitable volume of space, such as a lake, and studying the transfers of matter and energy across the boundaries of this volume (Hutchinson and Wollack 1940). This approach, though useful for its emphasis on physical and chemical processes, treated the lake as a "black box" without the detailed mapping of the trophic structure that Lindeman was attempting. In a major treatise on biogeochemical processes that was still in progress while Lindeman was visiting Yale, Hutchinson had independently begun to think of classifying biological formations and their developmental stages in terms of photosynthetic and consumer group efficiencies. With access to Hutchinson's research and lecture notes, Lindeman developed a theory of successional change which emphasized the ecological efficiency of energy transfer over the long term (Cook 1977).

One idea that Hutchinson himself did not use greatly, but which Lindeman made the centerpiece of his study, was the ecosystem concept. Though Arthur Tansley had introduced the term in 1935 to describe the climax community, his use of the ecosystem concept had little impact on ecology. By reviving the concept in the context of a dynamic analysis of succession, Lindeman showed how it could be used to organize ecological ideas. The ecosystem unit was a significant departure from traditional interpretations of succession. Frederic Clements and Victor Shelford had organized their 1939 text *Bio-Ecology* around the unit of the plant-animal formation, known as the biome. Their discussion of change within the biome drew on the complex-organism analogy, so that succession was the result of the reciprocal interaction between the living community (biome) and its habitat, moving through stages comparable to the developmental stages of the individual organism (Clements and Shelford 1939).

German limnologist August Thienemann (1918, 1926) had also discussed the relationship between the community (known in Europe as the *biocoenosis*) and the habitat (or biotype) in the context of succession. His work was important for showing that the union of community and habitat need not result in a static, perfectly cycling system. However, he also used an organismic analogy, with succession being the product of the reciprocal interactions between "biocoenosis" and "biotype." By 1939, he combined the biocoenosis with the biotype to form a higher unity called the "biosystem" (Thienemann 1939). His approach differed from Lindeman's in lacking any attempt to work out precisely the relationship between the development of the unit and the internal trophic processes of the system.

It is interesting to note that when Lindeman sent his essay to *Ecology* for publication, two eminent limnologists, Chancey Juday and Paul Welch, recommended that it not be published, largely because it was too theoretical and went far beyond the data available at the time (Cook 1977). Welch recommended that Lindeman set the manuscript aside for ten years in the hope of accumulating a better data base. With Hutchinson's strong support of the manuscript, the editor of the journal, Thomas Park, decided to publish a revised manuscript despite the negative reviews. In fact, Lindeman never saw the article in print because he died in 1942 after a long illness; he was twenty-seven years old. Though published without the desired supporting data, his article opened up new directions for the analysis of the functioning of ecosystems.

These are some of the major publications in ecology which formed the basis for a continuing research tradition in community studies, succession, ecosystem analysis, population dynamics, and organism-environment relations. Many of the ideas and methods presented in these papers were controversial, especially the complex-organism analogy and the use of mathematical techniques for modeling population interactions. Mathematical modeling was controversial because it seemed oversimplified and generated conclusions that went beyond the available data. Many of the laboratory studies that grew from these theoretical forays were criticized for being inapplicable to the more complex field environment, while many field studies were in turn criticized for being too descriptive or not being related to theoretical principles. Ecology is such a heterogeneous science that arguments about methods, approaches, and definitions of central terms are nearly impossible to avoid. In the following sections we consider in detail the elaborations of ecological science that stemmed from these sometimes crude but often imaginative and optimistic beginnings.

Literature Cited

Andrewartha, H. G., and L. C. Birch. 1954. *The distribution and abundance of animals.* Chicago: University of Chicago Press.

Bowler, Peter J. 1983. *The eclipse of Darwinism: Anti-Darwinian evolution theories in the decades around 1900.* Baltimore: Johns Hopkins University Press.

Burgess, Robert L. 1977. The Ecological Society of America: Historical data and some preliminary analyses. In Frank N. Egerton and R. P. McIntosh, eds. *History of American Ecology.* New York: Arno Press.

Cittadino, Eugene. 1980. Ecology and the professionalization of botany in America, 1890–1905. *Stud. Hist. Biol.* 4:171–98.

Clements, Frederic E. 1905. *Research methods in ecology.* Lincoln, Neb.: University Publishing.

——— 1916. *Plant succession, An analysis of the development of vegetation.* Publication no. 242. Washington, D.C.: Carnegie Institution.

Clements, F. E., J. E. Weaver, and H. C. Hanson, 1929. *Plant competition: An analysis of community functions.* Washington, D.C.: Carnegie Institution.

Clements, F. E., and V. E. Shelford. 1939. *Bio-Ecology.* New York: John Wiley and Sons.

Cook, Robert E. 1977. Raymond Lindeman and the trophic-dynamic concept in ecology. *Science* 198: 22–26.

Cowles, Henry C. 1899. The ecological relations of the vegetation on the sand dunes of Lake Michigan. *Bot. Gaz.* 27:95–117, 167–202, 281–308, 361–91.

Cushing, D. H. 1975. *Marine ecology and fisheries.* Cambridge: Cambridge University Press.

DeBach, Paul, and H. S. Smith. 1941. Are population oscillations inherent in nature? *Ecology* 22:363–69.

Elton, Charles. 1933. *The ecology of animals.* London: Methuen.

——— 1966. *Animal ecology.* London: Sidgwick and Jackson.

Engel, J. Ronald. 1983. *Sacred sands: The struggle for community in the Indiana dunes.* Middletown, Conn.: Wesleyan University Press.

Forbes, Steven A. 1880. On some interactions of organisms. *Bull. Ill. St. Lab. Nat. Hist.* 1:1–17.

Ford, E. B. 1980. Some recollections pertaining to the evolutionary synthesis. In Ernst Mayr and W. B. Provine, eds. *The evolutionary synthesis: Perspectives on the unification of biology.* Cambridge, Mass.: Harvard University Press.

Gause, G. F. 1934. *The struggle for existence.* Baltimore: Williams and Wilkins. Reprint 1971. New York: Dover.

——— 1935. *Vérifications expérimentales de la théorie mathématique de la lutte pour la vie.* Paris: Hermann.

——— 1936. The principles of biocoenology. *Quart. Rev. Bio.* 11:320–36.

——— 1937. Experimental populations of microscopic organisms. *Ecology* 18:173–79.

Hardin, Garrett. 1960. The competitive exclusion principle. *Science* 131:1292–97.

Hutchinson, G. Evelyn. 1978. *An introduction to population ecology.* New Haven: Yale University Press.

Hutchinson, G. E., and A. Wollack, 1940. Studies on Connecticut lake sediments, II. Chemical analysis of a core from Linsley Pond, North Branford. *Amer. J. Sci.* 238:483–517.

Kingsland, Sharon E. 1985. *Modeling nature: Episodes in the history of population ecology.* Chicago: University of Chicago Press.

——— 1988. Evolution and debates over human progress from Darwin to sociobiology. *Pop. Devel. Rev.* 14(Suppl.):167–98.

Lack, David L. 1954. *The natural regulation of animal numbers.* Oxford: Clarendon.

Lamarck, Jean-Baptiste. 1984. *Zoological philosophy.* Chicago: University of Chicago Press.

Lotka, Alfred J. 1925. *Elements of Physical Biology.*

Baltimore: Williams and Wilkins. Reprinted in 1956 with revisions as *Elements of Mathematical Biology*. New York: Dover.

McIntosh, Robert P. 1977. H. A. Gleason, individualistic ecologist, 1882–1975. In Frank N. Egerton and R. P. McIntosh, eds. *History of American Ecology*. New York: Arno Press.

——— 1985. *The background of ecology: Concept and theory*. Cambridge: Cambridge University Press.

Mayr, Ernst. 1982. *The growth of biological thought: Diversity, evolution, and inheritance*. Cambridge, Mass.: Harvard University Press.

Möbius, Karl. 1983. The oyster and oyster-culture. *U.S. commission of fish and fisheries, Report for 1880*. Washington, D.C.: U.S. Government Printing Office.

Nicholson, Alexander J. 1933. The balance of animal populations. *J. Anim. Ecol.* 2:132–78.

Park, Thomas. 1939. Analytical population studies in relation to general ecology. *Amer. Midland Nat.* 21:235–55.

Pearl, Raymond, and L. J. Reed. 1920. On the rate of growth of the population of the United States since 1790 and its mathematical representation. *Proc. Nat. Acad. Sci. U.S.A.* 6:275–88.

Scudo, F. M., and J. R. Ziegler, 1978. *The golden age of theoretical ecology, 1923–1940*. Berlin: Springer-Verlag.

Spencer, Herbert. 1874. *Principles of Biology*. 2 vols. Reprint of 1864–67 first edition. New York: D. Appleton.

Tansley, Arthur G. 1935. The use and abuse of vegetational concepts and terms. *Ecology* 16:284–307.

Thienemann, August. 1918. Lebengemeinschaft und Lebensraum. *Naturwissenschaftliche Wochenschrift* 17:281–90; 297–303.

——— 1926. Der Nahrungskreislauf im Wasser. *Verh. dtsch. zool. Ges.* 31:29–79.

——— 1939. Grundzüge einer allgemeinen Ökologie. *Arch. Hydrobiol.* 35:267–85.

Tobey, Ronald C. 1981. *Saving the prairies: The life cycle of the founding school of American plant ecology, 1895–1955*. Berkeley: University of California Press.

Varley, George C. 1947. The natural control of population balance in the Knapweed gall-fly. *J. Anim. Ecol.* 16:139–86.

Whittaker, Robert H. 1977. Recent evolution of ecological concepts in relation to the eastern forests of North America. In Frank N. Egerton and R. P. McIntosh, eds., *History of American ecology*. New York: Arno Press.

Winsor, Mary P. 1972. Forbes, Stephen Alfred. In *Dictionary of scientific biography*, vol. 5. New York: Charles Scribner's Sons.

Worster, Donald. 1979. *Nature's economy: The roots of ecology*. Garden City, N.Y.: Anchor Press.

PAPER 1

ARTICLE IX.—*The Lake as a Microcosm**. BY STEPHEN A. FORBES.

A lake is to the naturalist a chapter out of the history of a primeval time, for the conditions of life there are primitive, the forms of life are, as a whole, relatively low and ancient, and the system of organic interactions by which they influence and control each other has remained substantially unchanged from a remote geological period.

The animals of such a body of water are, as a whole, remarkably isolated—closely related among themselves in all their interests, but so far independent of the land about them that if every terrestrial animal were suddenly annihilated it would doubtless be long before the general multitude of the inhabitants of the lake would feel the effects of this event in any important way. It is an islet of older, lower life in the midst of the higher, more recent life of the surounding region. It forms a little world within itself—a microcosm within which all the elemental forces are at work and the play of life goes on in full, but on so small a scale as to bring it easily within the mental grasp.

Nowhere can one see more clearly illustrated what may be called the *sensibility* of such an organic complex, expressed by the fact that whatever affects any species belonging to it, must have its influence of some sort upon the whole assemblage. He will thus be made to see the impossibility of studying completely any form out of relation to the other forms; the necessity for taking a comprehensive survey of the whole as a condition to a satisfactory understanding of any part. If one wishes to become acquainted with the black bass, for example, he will learn but little if he limits himself to that species. He must evidently study also the species upon which it depends for its existence, and the various conditions upon which *these* depend. He must likewise study the species with which it comes in competition, and the entire system of conditions affecting their prosperity; and by the time he has studied all these sufficiently he will find that he has run through the whole complicated mechanism of the aquatic life of the locality, both animal and vegetable, of which his species forms but a single element.

It is under the influence of these general ideas that I propose to examine briefly to-night the lacustrine life of Illinois, drawing my data

*This paper, originally read February 25, 1887, to the Peoria Scientific Association (now extinct), and published in their Bulletin, was reprinted many years ago by the Illinois State Laboratory of Natural History in an edition which has long been out of print. A single copy remaining in the library of the Natural History Survey is used every year by classes in the University of Illinois, and a professor of zoology in a Canadian university borrows a copy regularly from a Peoria library for use in his own classes. In view of this long-continued demand and in the hope that the paper may still be found useful elsewhere, it is again reprinted. with trivial emendations, and with no attempt to supply its deficiencies or to bring it down to date.

from collections and observations made during recent years by myself and my assistants of the State Laboratory of Natural History.

The lakes of Illinois are of two kinds, fluviatile and water-shed. The fluviatile lakes, which are much the more numerous and important, are appendages of the river systems of the state, being situated in the river bottoms and connected with the adjacent streams by periodical overflows. Their fauna is therefore substantially that of the rivers themselves, and the two should, of course, be studied together.

They are probably in all cases either parts of former river channels, which have been cut off and abandoned by the current as the river changed its course, or else are tracts of the high-water beds of streams over which, for one reason or another, the periodical deposit of sediment has gone on less rapidly than over the surrounding area, and which have thus come to form depressions in the surface which retain the waters of overflow longer than the higher lands adjacent. Most of the numerous "horseshoe lakes" belong to the first of these varieties, and the "bluff-lakes," situated along the borders of the bottoms, are many of them examples of the second.

These fluviatile lakes are most important breeding grounds and reservoirs of life, especially as they are protected from the filth and poison of towns and manufactories by which the running waters of the state are yearly more deeply defiled.

The amount and variety of animal life contained in them as well as in the streams related to them is extremely variable, depending chiefly on the frequency, extent, and duration of the spring and summer overflows. This is, in fact, the characteristic and peculiar feature of life in these waters. There is perhaps no better illustration of the methods by which the flexible system of organic life adapts itself, without injury, to widely and rapidly fluctuating conditions. Whenever the waters of the river remain for a long time far beyond their banks, the breeding grounds of fishes and other animals are immensely extended, and their food supplies increased to a corresponding degree. The slow or stagnant backwaters of such an overflow afford the best situations possible for the development of myriads of Entomostraca, which furnish, in turn, abundant food for young fishes of all descriptions. There thus results an outpouring of life—an extraordinary multiplication of nearly every species, most prompt and rapid, generally speaking, in such as have the highest reproductive rate, that is to say, in those which produce the largest average number of eggs and young for each adult.

The first to feel this tremendous impulse are the protophytes and Protozoa, upon which most of the Entomostraca and certain minute insect larvæ depend for food. This sudden development of their food resources causes, of course, a corresponding increase in the numbers of the latter classes, and, through them, of all sorts of fishes. The first fishes to feel the force of this tidal wave of life are the rapidly-breeding, non-predaceous kinds; and the last, the game fishes, which derive from the others their principal food supplies. Evidently each of these classes

539

must act as a check upon the one preceding it. The development of animalcules is arrested and soon sent back below its highest point by the consequent development of Entomostraca; the latter, again, are met, checked, and reduced in number by the innumerable shoals of fishes with which the water speedily swarms. In this way a general adjustment of numbers to the new conditions would finally be reached spontaneously; but long before any such settled balance can be established, often of course before the full effect of this upward influence has been exhibited, a new cause of disturbance intervenes in the *disappearance of the overflow.* As the waters retire, the lakes are again defined; the teeming life which they contain is restricted within daily narrower bounds, and a fearful slaughter follows; the lower and more defenceless animals are penned up more and more closely with their predaceous enemies, and these thrive for a time to an extraordinary degree. To trace the further consequences of this oscillation would take me too far. Enough has been said to illustrate the general idea that the life of waters subject to periodical expansions of considerable duration, is peculiarly unstable and fluctuating; that each species swings, pendulum-like but irregularly, between a highest and a lowest point, and that this fluctuation affects the different classes successively, in the order of their dependence upon each other for food.

Where a water-shed is a nearly level plateau with slight irregularities of the surface many of these will probably be imperfectly drained, and the accumulating waters will form either marshes or lakes according to the depth of the depressions. Highland marshes of this character are seen in Ford, Livingston, and adjacent counties,* between the headwaters of the Illinois and Wabash systems; and an area of water-shed lakes occurs in Lake and McHenry counties, in northern Illinois.

The latter region is everywhere broken by low, irregular ridges of glacial drift, with no rock but boulders anywhere in sight. The intervening hollows are of every variety, from mere sink-holes, either dry or occupied by ponds, to expanses of several square miles, forming marshes or lakes.

This is, in fact, the southern end of a broad lake belt which borders Lakes Michigan and Superior on the west and south, extending through eastern and northern Wisconsin and northwestern Minnesota, and occupying the plateau which separates the headwaters of the St. Lawrence from those of the Mississippi. These lakes are of glacial origin, some filling beds excavated in the solid rock, and others collecting the surface waters in hollows of the drift. The latter class, to which all the Illinois lakes belong, may lie either parallel to the line of glacial action, occupying valleys between adjacent lateral moraines, or transverse to that line and bounded by terminal moraines. Those of our own state

*All now drained and brought under cultivation.

all drain at present into the Illinois through the Des Plaines and Fox; but as the terraces around their borders indicate a former water-level considerably higher than the present one it is likely that some of them once emptied eastward into Lake Michigan. Several of these lakes are clear and beautiful sheets of water, with sandy or gravelly beaches, and shores bold and broken enough to relieve them from monotony. Sportsmen long ago discovered their advantages and club-houses and places of summer resort are numerous on the borders of the most attractive and easily accessible. They offer also an unusually rich field to the naturalist, and their zoology and botany should be better known.

The conditions of aquatic life are here in marked contrast to those afforded by the fluviatile lakes already mentioned. Connected with each other or with adjacent streams only by slender rivulets, varying but little in level with the change of the season and scarcely at all from year to year, they are characterized by an isolation, independence, and uniformity which can be found nowhere else within our limits.

Among these Illinois lakes I did considerable work during October of two successive years, using the sounding line, deep-sea thermometer, towing net, dredge, and trawl in six lakes of northern Illinois, and in Geneva Lake, Wisconsin, just across the line. Upon one of these Illinois lakes I spent a week in October, and an assistant, Prof. H. Garman, now of the University, spent two more, making as thorough a physical and zoölogical survey of this lake as was possible at that season of the year.

I now propose to give you in this paper a brief general account of the physical characters and the fauna of these lakes, and of the relations of the one to the other; to compare, in a general way, the animal assemblages which they contain with those of Lake Michigan —where also I did some weeks of active aquatic work in 1881—and with those of the fluviatile lakes of central Illinois; to make some similar comparisons with the lakes of Europe; and, finally, to reach the subject which has given the title to this paper—to study the system of natural interactions by which this mere collocation of plants and animals has been organized as a stable and prosperous community.

First let us endeavor to form the mental picture. To make this more graphic and true to the facts, I will describe to you some typical lakes among those in which we worked; and will then do what I can to furnish you the materials for a picture of the life that swims and creeps and crawls and burrows and climbs through the water, in and on the bottom, and among the feathery water-plants with which large areas of these lakes are filled.

Fox Lake, in the western border of Lake county, lies in the form of a broad irregular crescent, truncate at the ends, and with the concavity of the crescent to the northwest. The northern end is broadest and communicates with Petite Lake. Two points projecting inward from the southern shore form three broad bays. The western end opens into Nippisink Lake, Crab Island separating the two. Fox River enters the lake from the north, just eastward of this island, and flows directly through the Nippisink. The length of a curved line extending through the central part of this lake, from end to end, is very nearly three miles, and the width of the widest part is about a mile and a quarter. The shores are bold, broken, and wooded, except to the north, where they are marshy and flat. All the northern and eastern part of the lake was visibly shallow—covered with weeds and feeding waterfowl, and I made no soundings there. The water there was probably nowhere more than two fathoms in depth, and over most of that area was doubtless under one and a half. In the western part, five lines of soundings were run, four of them radiating from Lippincott's Point, and the fifth crossing three of these nearly at right angles. The deepest water was found in the middle of the mouth of the western bay, where a small area of five fathoms occurs. On the line running northeast from the Point, not more than one and three fourths fathoms is found. The bottom at a short distance from the shores was everywhere a soft, deep mud. Four hauls of the dredge were made in the western bay, and the surface net was dragged about a mile.

Long Lake differs from this especially in its isolation, and in its smaller size. It is about a mile and a half in length by a mile in breadth. Its banks are all bold except at the western end, where a marshy valley traversed by a small creek connects it with Fox Lake, at a distance of about two miles. The deepest sounding made was six and a half fathoms, while the average depth of the deepest part of the bed was about five fathoms.

Cedar Lake, upon which we spent a fortnight, is a pretty sheet of water, the head of a chain of six lakes which open finally into the Fox. It is about a mile in greatest diameter in each direction, with a small but charming island bank near the center, covered with bushes and vines—a favorite home of birds and wild flowers. The shores vary from rolling to bluffy except for a narrow strip of marsh through which the outlet passes, and the bottoms and margins are gravel, sand, and mud in different parts of its area. Much of the lake is shallow and full of water plants; but the southern part reaches a depth of fifty feet a short distance from the eastern bluff.

Deep Lake, the second of this chain, is of similar character, with a greatest depth of fifty-seven feet—the deepest sounding we made in these smaller lakes of Illinois. In these two lakes several temperatures were taken with a differential thermometer. In Deep Lake, for example, at fifty-seven feet I found the bottom temperature 53½°—about that of ordinary well-water—when the air was 63°; and in Cedar Lake, at forty-eight feet, the bottom was 58° when the air was 61°.

Geneva Lake, Wisconsin, is a clear and beautiful body of water about eight miles long by one and a quarter in greatest width. The banks are all high, rolling, and wooded, except at the eastern end, where its outlet rises. Its deepest water is found in its western third, where it reaches a depth of twenty-three fathoms. I made here, early in November, twelve hauls of the dredge and three of the trawl, aggregating about three miles in length, so distributed in distance and depth as to give a good idea of the invertebrate life of the lake at that season.

And now if you will kindly let this suffice for the background or setting of the picture of lacustrine life which I have undertaken to give you, I will next endeavor—not to paint in the picture; for that I have not the artistic skill. I will confine myself to the humble and safer task of supplying you the pigments, leaving it to your own constructive imaginations to put them on the canvas.

When one sees acres of the shallower water black with water-fowl, and so clogged with weeds that a boat can scarcely be pushed through the mass; when, lifting a handful of the latter, he finds them covered with shells and alive with small crustaceans; and then, dragging a towing net for a few minutes, finds it lined with myriads of diatoms and other microscopic algæ, and with multitudes of Entomostraca, he is likely to infer that these waters are everywhere swarming with life, from top to bottom and from shore to shore. If, however, he will haul a dredge for an hour or so in the deepest water he can find, he will invariably discover an area singularly barren of both plant and animal life, yielding scarcely anything but a small bivalve mollusk, a few low worms, and red larvæ of gnats. These inhabit a black, deep, and almost impalpable mud or ooze, too soft and unstable to afford foothold to plants even if the lake is shallow enough to admit a sufficient quantity of light to its bottom to support vegetation. It is doubtless to this character of the bottom that the barrenness of the interior parts of these lakes is due; and this again is caused by the selective influence of gravity upon the mud and detritus washed down by rains. The heaviest and coarsest of this material necessarily settles nearest the margin, and only the finest silt reaches the remotest parts of the lakes, which, filling most slowly, remain, of course, the deepest. This ooze consists very largely, also, of a fine organic *debris*. The superficial part of it contains scarcely any sand, but has a greasy feel and rubs away, almost to nothing, between the fingers. The largest lakes are not therefore, as a rule, by any means the most prolific of life, but this shades inward rapidly from the shore, and becomes at no great distance almost as simple and scanty as that of a desert.

Among the weeds and lily-pads upon the shallows and around the margin—the Potamogeton, Myriophyllum, Ceratophyllum, Anacharis, and Chara, and the common Nelumbium,—among these the fishes chiefly swim or lurk, by far the commonest being the barbaric bream[1] or "pumpkin-seed" of northern Illinois, splendid with its green and scarlet and purple and orange. Little less abundant is the common perch (*Perca lutea*) in the larger lakes—in the largest out-numbering the bream itself. The whole sunfish family, to which the latter belongs, is in fact the dominant group in these lakes. Of the one hundred and thirty-two fishes of Illinois only thirty-seven are found in these waters—about twenty-

[1]Lepomis gibbosus.

543

eight per cent.—while eight out of our seventeen sunfishes (*Centrarchinae*) have been taken there. Next, perhaps, one searching the pebbly beaches or scanning the weedy tracts will be struck by the small number of minnows or cyprinoids which catch the eye or come out in the net. Of our thirty-three Illinois cyprinoids, only six occur there—about eighteen per cent.—and only three of these are common. These are in part replaced by shoals of the beautiful little silversides (*Labidesthes sicculus*), a spiny-finned fish, bright, slender, active, and voracious—as well suppliel with teeth as a perch, and far better equipped for self-defense than the soft-bodied and toothless cyprinoids. Next we note that of our twelve catfishes (*Siluridae*) only two have been taken in these lakes—one the common bullhead (*Ictalurus nebulosus*), which occurs everywhere, and the other an insignificant stone cat, not as long as one's thumb. The suckers, also, are much less abundant in this region than farther south, the buffalo fishes[1] not appearing at all in our collections. Their family is represented by worthless carp[2] by two redhorse[3], by the chub sucker[4] and the common sucker (*Catostomus teres*), and by one other species. Even the hickory shad[5]—an ichthyological weed in the Illinois—we have not found in these lakes at all. The sheepshead[6], so common here, is also conspicuous there by its absence. The yellow bass[7], not rare in this river, we should not expect in these lakes because it is, rather, a southern species; but why the white bass[8], abundant here, in Lake Michigan, and in the Wisconsin lakes, should be wholly absent from the lakes of the Illinois plateau, I am unable to imagine. If it occurs there at all, it must be rare, as I could neither find nor hear of it.

A characteristic, abundant, and attractive little fish is the log perch (*Percina caprodes*)—the largest of the darters, slender, active, barred like a zebra, spending much of its time in chase of Entomostraca among the water plants, or prying curiously about among the stones for minute insect larvæ. Six darters in all (*Etheostomatinae*), out of the eighteen from the state, are on our list from these lakes. The two black bass[9] are the most popular game fishes—the large-mouthed species being much the most abundant. The pickerels[10], gar[11], and dogfish[12] are there about as here; but the shovel-fish[13] does not occur.

Of the peculiar fish fauna of Lake Michigan—the burbot[14], white fish,[15] trout,[16] lake herring or cisco,[17] etc., not one species occurs in these smaller lakes, and all attempts to transfer any of them have failed completely. The cisco is a notable fish of Geneva Lake, Wisconsin, but does not reach Illinois except in Lake Michigan. It is useless to attempt to introduce it, because the deeper areas of the interior lakes are too limited to give it sufficient range of cool water in midsummer.

In short, the fishes of these lakes are substantially those of their

---

[1]Ictiobus bubalus. [2]Ictiobus cyprinus. [3]Moxostoma aureolum and M. macrolepidotum. [4]Erimyzon sucetta. [5]Dorosoma cepedianum. [6]Haploidonotus. [7]Roccus interruptus. [8]Roccus chrysops. [9]Micropterus. [10]Esox. [11]Lepidosteus. [12]Amia. [13]Polyodon. [14]Lota. [15]Coregonus clupeiformis. [16]Salvelinus namaycush. [17]Coregonus artedi.

region—excluding the Lake Michigan series (for which the lakes are too small and warm) and those peculiar to creeks and rivers. Possibly the relative scarcity of catfishes (*Siluridae*) is due to the comparative clearness and cleanness of these waters. I see no good reason why minnows should be so few, unless it be the abundance of pike and Chicago sportsmen.

Concerning the molluscan fauna, I will only say that it is poor in bivalves—as far as our observations go—and rich in univalves. Our collections have been but partly determined, but they give us three species of Valvata, seven of Planorbis, four Amnicolas, a Melantho, two Physas, six Limnæas, and an Ancylus among the Gastropoda, and two Unios, an Anodonta, a Sphærium, and a Pisidium among the Lamellibranchiates. *Pisiduim variabile* is by far the most abundant mollusk in the oozy bottom in the deeper parts of the lakes; and crawling over the weeds are multitudes of small Amnicolas and Valvatas.

The entomology of these lakes I can merely touch upon, mentioning only the most important and abundant insect larvæ. Hiding under stones and driftwood, well aware, no doubt, what enticing morsels they are to a great variety of fishes, we find a number of species of ephemerid larvæ whose specific determination we have not yet attempted. Among the weeds are the usual larvæ of dragon-flies—Agrionina and Libellulina, familiar to every one; swimming in open water the predaceous larvæ of Corethra; wriggling through the water or buried in the mud the larvæ of Chironomus—the shallow water species white, and those from the deeper ooze of the central parts of the lakes blood-red and larger. Among Chara on the sandy bottom are a great number and variety of interesting case-worms—larvæ of Phryganeidæ—most of them inhabiting tubes of a slender conical form made of a viscid secretion exuded from the mouth and strengthened and thickened by grains of sand, fine or coarse. One of these cases, nearly naked, but usually thinly covered with diatoms, is especially worthy of note, as it has been reported nowhere in this country except in our collections, and was indeed recently described from Brazil as new. Its generic name is Lagenopsyche, but its species undetermined. These larvæ are also eaten by fishes.

Among the worms we have of course a number of species of leeches and of planarians,—in the mud minute Anguillulidæ, like vinegar eels, and a slender Lumbriculus which makes a tubular mud burrow for itself in the deepest water, and also the curious *Nais probiscidea,* notable for its capacity of multiplication by transverse division.

The crustacean fauna of these lakes is more varied than any other group. About forty species were noted in all. Crawfishes were not especially abundant, and most belonged to a single species, *Cambarus virilis.* Two amphipods occurred frequently in our collections; one, less common here but very abundant farther south—*Crangonyx gracilis*—and one, *Allorchestes dentata,* probably the commonest animal in these waters, crawling and swimming everywhere in myriads among the submerged water-plants. An occasional *Gammarus fasciatus* was also taken in the dredge. A few isopod Crustacea occur, belonging to *Mancasellus tenax*—a species not previously found in the state.

I have reserved for the last the Entomostraca—minute crustaceans of a surprising number and variety, and of a beauty often truly exquisite. They belong wholly, in our waters, to the three orders, Copedoda, Ostracoda, and Cladocera—the first two predaceous upon still smaller organisms and upon each other, and the last chiefly vegetarian. Twenty-one species of Cladocera have been recognized in our collections, representing sixteen genera. It is an interesting fact that twelve of these species are found also in the fresh waters of Europe. Five cyprids have been detected, two of them common to Europe, and also an abundant Diaptomus, a variety of a European species. Several Cyclops species were collected which have not yet been determined.

These Entomostraca swarm in microscopic myriads among the weeds along the shore, some swimming freely, and others creeping in the mud or climbing over the leaves of plants. Some prefer the open water, in which they throng locally like shoals of fishes, coming to the surface preferably by night, or on dark days, and sinking to the bottom usually by day to avoid the sunshine. These pelagic forms, as they are called, are often exquisitely transparent, and hence almost invisible in their native element—a charming device of Nature to protect them against their enemies in the open lake, where there is no chance of shelter or escape. Then with an ingenuity in which one may almost detect the flavor of sarcastic humor, Nature has turned upon these favored children and endowed their most deadly enemies with a like transparency, so tnat wherever the towing net brings to light a host of these crystalline Cladocera, there it discovers also swimming, invisible, among them, a lovely pair of robbers and beasts of prey—the delicate Leptodora and the Corethra larva.

These slight, transparent, pelagic forms are much more numerous in Lake Michigan than in any of the smaller lakes, and peculiar forms occur there commonly which are rare in the larger lakes of Illinois and entirely wanting in the smallest. The transparent species are also much more abundant in the isolated smaller lakes than in those more directly connected with the rivers.

The vertical range of the animals of Geneva Lake showed clearly that the barrenness of the interiors of these small bodies of water was not due to the greater depth alone. While there were a few species of crustaceans and case-worms which occurred there abundantly near shore but rarely or not at all at depths greater than four fathoms, and may hence be called littoral species, there was, on the whole, little diminution either in quantity or variety of animal life until about fifteen fathoms had been reached. Dredging at four or five fathoms were nearly or quite as fruitful as any made. On the other hand, the barrenness of the bottom at twenty to twenty-three fathoms was very remarkable. The total product of four hauls of the dredge and one of the trawl at that depth, aggregating fully a mile and a half of continuous dragging, would easily go into a two-dram vial, and represents only nine animal species—not counting dead shells, and fragments which had probably floated in from shallower waters. The greater part of this little collection was composed of specimens of Lumbriculus and larvæ of Chironomus. There were a few Corethra larvæ, a single Gammarus, three small leeches, and some sixteen mollusks, all but four of which belonged to Pisidium. The others were two Sphæriums, a *Valvata carinata,* and a *V. sincera.* None of the species taken here are peculiar, but all were of the kinds found in the smaller lakes, and all occurred also in shallower water. It is evident that these interior regions of the lakes must be as destitute of fishes as they are of plants and lower animals.

While none of the deep-water animals of the Great Lakes were found in Geneva Lake, other evidences of zoölogical affinity were detected. The towing net yielded almost precisely the assemblage of species of Entomostraca found in Lake Michigan, including many specimens of *Limnocalanus macrurus* Sars; and peculiar long, smooth leeches, common in Lake Michigan but not occurring in the small, Illinois lakes, were also found in Geneva. Many *Valvata tri-carinata* lacked the middle carina, as in Long Lake and other *isolated* lakes of this region.

Comparing the Daphnias of Lake Michigan with those of Geneva Lake, Wis. (nine miles long and twenty-three fathoms in depth), those of Long Lake, Ill. (one and a half miles long and six fathoms deep), and those of other, still smaller, lakes of that region, and the swamps and smaller ponds as well, we shall be struck by the inferior development of the Entomostraca of the larger bodies of water in numbers, in size and robustness, and in reproductive power. Their smaller numbers and size are doubtless due to the relative scarcity of food. The system of aquatic animal life rests essentially upon the vegetable world, although perhaps less strictly than does the terrestrial system, and in a large and deep lake vegetation is much less abundant than in a narrower and shallower one, not only relatively to the amount of water but also to the area of the bottom. From this deficiency of plant life results a deficiency of food for Entomostraca, whether of algæ, of Protozoa, or of higher forms, and hence, of course, a smaller number of the Entomostraca themselves, and these with more slender bodies, suitable for more rapid locomotion and wider range.

The difference of reproductive energy, as shown by the much smaller egg-masses borne by the species of the larger lakes, depends upon the vastly greater destruction to which the paludal Crustacea are subjected. Many of the latter occupy waters liable to be exhausted by drought, with a consequent enormous waste of entomostracan life. The opportunity for reproduction is here greatly limited—in some situations to early spring alone—and the chances for destruction of the summer eggs in the dry and often dusty soil are so numerous that only the most prolific species can maintain themselves.

547

Further, the marshes and shallower lakes are the favorite breeding grounds of fishes, which migrate to them in spawning time if possible, and it is from the Entomostraca found here that most young fishes get their earliest food supplies—a danger from which the deep-water species are measurably free. Not only is a high reproductive rate rendered unnecessary among the latter by their freedom from many dangers to which the shallow-water species are exposed, but in view of the relatively small amount of food available for them, a high rate of multiplication would be a positive injury, and could result only in wholesale starvation.

All these lakes of Illinois and Wisconsin, together with the much larger Lake Mendota at Madison (in which also I have done much work with dredge, trawl, and seine), differ in one notable particular both from Lake Michigan and from the larger lakes of Europe. In the latter the bottoms in the deeper parts yield a peculiar assemblage of animal forms which range but rarely into the littoral region, while in our inland lakes no such deep water fauna occurs, with the exception of the cisco and the large red Chironomus larva. At Grand Traverse Bay, in Lake Michigan, I found at a depth of one hundred fathoms a very odd fish of the sculpin family (*Triglopsis thompsoni* Gir.) which, until I collected it, had been known only from the stomachs of fishes; and there also was an abundant crustacean, Mysis—the "opossum shrimp", as it is sometimes called—the principal food of these deep lake sculpins. Two remarkable amphipod crustaceans also belong in a peculiar way to this deep water. In the European lakes the same Mysis occurs in the deepest part, with several other forms not represented in our collections, two of these being blind crustaceans related to those which in this country occur in caves and wells.

Comparing the other features of our lake fauna with that of Europe, we find a surprising number of Entomostraca identical; but this is a general phenomenon, as many of the more abundant Cladocera and Copepoda of our small wayside pools are either European species, or differ from them so slightly that it is doubtful if they ought to be called distinct.

It would be quite impossible, within reasonable limits, to go into details respecting the organic relations of the animals of these waters, and I will content myself with two or three illustrations. As one example of the varied and far-reaching relations into which the animals of a lake are brought in the general struggle for life, I take the common black bass. In the dietary of this fish I find, at different ages of the individual, fishes of great variety, representing all the important orders of that class; insects in considerable number, especially the various water-bugs and larvæ of day-flies; fresh-water shrimps; and a great multitude of Entomostraca of many species and genera. The fish is therefore directly dependent upon all these classes for its existence. Next, looking to the food of the species which the bass has eaten, and upon which it is therefore indirectly dependent, I find that one kind of the fishes taken feeds upon mud, algæ, and Entomostraca, and another upon nearly every animal substance in the water, including mollusks and decomposing organic matter. The insects taken by the bass, themselves take other insects and small Crustacea. The crawfishes are nearly omnivorous, and of the other crustaceans some eat Entomostraca and some algæ and Protoza. At only the second step, therefore, we find our bass brought into dependence upon nearly every class of animals in the water.

And now, if we search for its competitors we shall find these also extremely numerous. In the first place, I have found that all our young fishes except the Catostomidæ feed at first almost wholly on Entomostraca, so that the little bass finds himself at the very beginning of his life engaged in a scramble for food with all the other little fishes in the lake. In fact, not only young fishes but a multitude of other animals as well, especially insects and the larger Crustacea, feed upon these Entomostraca, so that the competitors of the bass are not confined to members of its own class. Even mollusks, while they do not directly compete with it do so indirectly, for they appropriate myriads of the microscopic forms upon which the Entomostraca largely depend for food. But the enemies of the bass do not all attack it by appropriating its food supplies, for many devour the little fish itself. A great variety of predaceous fishes, turtles, water-snakes, wading and diving birds, and even bugs of gigantic dimensions destroy it on the slightest opportunity. It is in fact hardly too much to say that fishes which reach maturity are relatively as rare as centenarians among human kind.

As an illustration of the remote and unsuspected rivalries which reveal themselves on a careful study of such a situation, we may take the relations of fishes to the bladderwort[1]—a flowering plant which fills many acres of the water in the shallow lakes of northern Illinois. Upon the leaves of this species are found little bladders—several hundred to each plant—which when closely examined are seen to be tiny traps for the capture of Entomostraca and other minute animals. The plant usually has no roots, but lives entirely upon the animal food obtained through these little bladders. Ten of these sacs which I took at random from a mature plant contained no less than ninety-three animals (more than nine to a bladder), belonging to twenty-eight different species. Seventy-six of these were Entomostraca, and eight others were minute insect larvæ. When we estimate the myriads of small insects and Crustacea which these plants must appropriate during a year to their own support, and consider the fact that these are of the kinds most useful as food for young fishes of nearly all descriptions, we must conclude that the bladderworts compete with fishes for food, and tend to keep down their number by diminishing the food resources of the young. The plants even have a certain advantage in this competition, since they are not strictly dependent on Entomostraca, as the fishes are, but sometimes take root, developing then but very few leaves and bladders. This probably happens under conditions unfavorable to their support by the other

[1]Utricularia.

549

method. These simple instances will suffice to illustrate the intimate way in which the living forms of a lake are united.

Perhaps no phenomenon of life in such a situation is more remarkable than the steady balance of organic nature, which holds each species within the limits of a uniform average number, year after year, although each one is always doing its best to break across boundaries on every side. The reproductive rate is usually enormous and the struggle for existence is correspondingly severe. Every animal within these bounds has its enemies, and Nature seems to have taxed her skill and ingenuity to the utmost to furnish these enemies with contrivances for the destruction of their prey in myriads. For every defensive device with which she has armed an animal, she has invented a still more effective apparatus of destruction and bestowed it upon some foe, thus striving with unending pertinacity to outwit herself; and yet life does not perish in the lake, nor even oscillate to any considerable degree, but on the contrary the little community secluded here is as prosperous as if its state were one of profound and perpetual peace. Although every species has to fight its way inch by inch from the egg to maturity, yet no species is exterminated, but each is maintained at a regular average number which we shall find good reason to believe is the greatest for which there is, year after year, a sufficient supply of food.

I will bring this paper to a close, already too long postponed, by endeavoring to show how this beneficent order is maintained in the midst of a conflict seemingly so lawless.

It is a self-evident proposition that a species can not maintain itself continuously, year after year, unless its birth-rate at least equals its death-rate. If it is preyed upon by another species, it must produce regularly an excess of individuals for destruction, or else it must certainly dwindle and disappear. On the other hand, the dependent species evidently must not appropriate, on an average, any more than the surplus and excess of individuals upon which it preys, for if it does so it will continuously diminish its own food supply, and thus indirectly but surely exterminate itself. The interests of both parties will therefore be best served by an adjustment of their respective rates of multiplication such that the species devoured shall furnish an excess of numbers to supply the wants of the devourer, and that the latter shall confine its appropriations to the excess thus furnished. We thus see that there is really a close *community of interest* between these two seemingly deadly foes.

And next we note that this common interest is promoted by the process of natural selection; for it is the great office of this process to eliminate the unfit. If two species standing to each other in the relation of hunter and prey are or become badly adjusted in respect to their rates of increase, so that the one preyed upon is kept very far below the normal number which might find food, even if they do not presently obliterate each other the pair are placed at a disadvantage in the battle for life, and must suffer accordingly. Just as certainly as the thrifty business man who lives within his income will finally dispossess his shiftless competitor who can never pay his debts, the well-adjusted aquatic animal will in time crowd out its poorly-adjusted competitors for food and for the various goods of life. Consequently we may believe that in the long run and as a general rule those species which have survived, are those which have reached a fairly close adjustment in this particular.[1]

Two ideas are thus seen to be sufficient to explain the order evolved from this seeming chaos; the first that of a general community of interests among all the classes of organic beings here assembled, and the second that of the beneficent power of natural selection which compels such adjustments of the rates of destruction and of multiplication of the various species as shall best promote this common interest.

Have these facts and ideas, derived from a study of our aquatic microcosm, any general application on a higher plane? We have here an example of the triumphant beneficence of the laws of life applied to conditions seemingly the most unfavorable possible for any mutually helpful adjustment. In this lake, where competitions are fierce and continuous beyond any parallel in the worst periods of human history; where they take hold, not on goods of life merely, but always upon life itself; where mercy and charity and sympathy and magnanimity and all the virtues are utterly unknown; where robbery and murder and the deadly tyranny of strength over weakness are the unvarying rule; where what we call wrong-doing is always triumphant, and what we call goodness would be immediately fatal to its possessor,—even here, out of these hard conditions, an order has been evolved which is the best conceivable without a total change in the conditions themselves; an equilibrium has been reached and is steadily maintained that actually accomplishes for all the parties involved the greatest good which the circumstances will at all permit. In a system where life is the universal good, but the destruction of life the well-nigh universal occupation, an order has spontaneously arisen which constantly tends to maintain life at the highest limit—a limit far higher, in fact, with respect to both quality and quantity, than would be possible in the absence of this destructive conflict. Is there not, in this reflection, solid ground for a belief in the final beneficence of the laws of organic nature? If the system of life is such that a harmonious balance of conflicting interests has been reached where every element is either hostile or indifferent to every other, may we not trust much to the outcome where, as in human affairs, the spontaneous adjustments of nature are aided by intelligent effort, by sympathy, and by self-sacrifice?

---

[1]For a fuller statement of this argument, see Bul. Ill. State Lab. Nat. Hist., Vol. I, No. 3, pages 5 to 10.

PAPER 2

THE ECOLOGICAL RELATIONS OF THE VEGETATION ON THE SAND DUNES OF LAKE MICHIGAN.

PART I.—GEOGRAPHICAL RELATIONS OF THE DUNE FLORAS.

CONTRIBUTIONS FROM THE HULL BOTANICAL LABORATORY. XIII.

HENRY CHANDLER COWLES.

(WITH FIGURES 1–26)

I. Introduction.

THE province of ecology is to consider the mutual relations between plants and their environment. Such a study is to structural botany what dynamical geology is to structural geology. Just as modern geologists interpret the structure of the rocks by seeking to find how and under what conditions similar rocks are formed today, so ecologists seek to study those plant structures which are changing at the present time, and thus to throw light on the origin of plant structures themselves.

Again, ecology is comparable to physiography. The surface of the earth is composed of a myriad of topographic forms, not at all distinct, but passing into one another by a series of almost perfect gradations; the physiographer studies landscapes in their making, and writes on the origin and relationships of topographic forms. The ecologist employs the methods of physiography, regarding the flora of a pond or swamp or hillside not as a changeless landscape feature, but rather as a panorama, never twice alike. The ecologist, then, must study the order of succession of the plant societies in the development of a region, and he must endeavor to discover the laws which govern the panoramic changes. Ecology, therefore, is a study in dynamics. For its most ready application, plants should be found whose tissues and organs are actually changing at the present time in response to varying conditions. Plant formations should be found which are rapidly passing into other types by reason of a changing environment.

These requirements are met *par excellence* in a region of sand dunes. Perhaps no topographic form is more unstable than a dune. Because of this instability plant societies, plant organs, and plant tissues are obliged to adapt themselves to a new mode of life within years rather than centuries, the penalty for lack of adaptation being certain death. The sand dunes furnish a favorable region for the pursuit of ecological investigations because of the comparative absence of the perplexing problems arising from previous vegetation. Any plant society is the joint product of present and past environmental conditions, and perhaps the latter are much more potent than most ecologists have thought. As will be shown in another place, even the sand dune floras are often highly modified by preexisting conditions, but on the whole the physical forces of the present shape the floras as we find them. The advancing dune buries the old plant societies of a region, and with their death there pass away the influences which contributed so largely to their making. In place of the rich soil which had been accumulating by centuries of plant and animal decay, and in place of the complex reciprocal relations between the plants, as worked out by a struggle of centuries, the advance of a dune makes all things new. By burying the past, the dune offers to plant life a world for conquest, subject almost entirely to existing physical conditions. The primary motive, then, which prompted this present study was the feeling that nowhere else could many of the living problems of ecology be solved more clearly; that nowhere else could ecological principles be subjected to a more rigid test.

This particular investigation was also prompted by the fact that the previous ecological studies of sand-dune floras have been carried on chiefly in European countries, and almost exclusively along marine coasts. There has been considerable difference of opinion as to the influence of salty soils and atmospheres upon the vegetation. It would seem that a comparison of dunes along an inland fresh water lake with those along the sea should yield instructive results.

An ecological study of this character has a natural twofold division. In the first place the plant formations are to be investigated. The species characteristic of each formation must be discovered, together with the facts and laws of their distribution. The progressive changes that take place and the factors in the environment which cause these changes must be discussed. This phase of the subject is largely geographic, and will be the special feature of the present paper. In another paper it is the author's purpose to discuss the adaptations of the plants to their dune environment, paying especial attention to those species which show a large degree of plasticity, and which are found growing under widely divergent conditions. This second phase again has a natural twofold division, one part treating of gross adaptations such as are shown in plant organs and plant bodies, the other dealing with the anatomical structures of the plant tissues.

The material for the present paper has been gathered chiefly from the study of the dunes in northwestern Indiana in the vicinity of Chicago. These studies were carried on in the seasons of 1896, 1897, and 1898, frequent visits being made to various points at all seasons of the year. A portion of the summer seasons of 1897 and 1898 was spent in a more rapid reconnoissance along the entire eastern shore of Lake Michigan, including the group of islands toward the north end of the lake.

The work resulting in this paper has been carried on in connection with the Hull Botanical Laboratory of the University of Chicago, and the author gratefully acknowledges the kindly interest and cooperation shown by his associates among the faculty and students of the botanical department, especially Head Professor John M. Coulter, through whose influence the author was directed along lines of ecological research. The author further desires to express his great indebtedness to Dr. Eugen Warming, professor of botany at Copenhagen: his textbook on ecology and his treatises on the sand-dune floras of

Fox, and High islands. The dunes on North Manitou and Beaver islands were visited. On North Manitou there are prominent areas of dune activity along the southwest coast, the dunes being superposed on bluffs of clay or gravel. There is a flat-topped terrace here, like that at Glen Haven, but in miniature, the height being only 15 meters and the area scarcely half a hectare; the dune perched on this bluff has been rejuvenated and carried inland a few meters, the greatest altitude being 45 meters above the lake. There are also small wandering dunes superposed directly upon the beach. On the west coast the bluffs are steeper and much higher, at times perhaps 60 meters above the lake; the summits are occasionally crowned by established dunes. On Beaver island the southwest coast was not visited, but there are rejuvenated dunes at various points along the west coast, sometimes 45 meters in height; these dunes are superposed upon the beach. The beach dunes here are exceedingly varied and extensive. As previously stated, there are low beach dunes along the east coast of Beaver island. On Mackinac island there are steep clay bluffs, but no dunes. Thus, the islands plainly show that westerly winds, and especially winds from the southwest, are the chief dune formers.

Surveying the lake region as a whole, the dunes are created and shaped almost entirely by westerly winds. In the southern portions of the lake, the northwest winds have the greater sweep and are the chief dune-formers. Northward the southwest winds are the chief factors in determining the location of dunes. In intermediate localities all westerly winds contribute about equally to dune formation, and there is progressive movement of active dunes to the eastward

III. The ecological factors.

The distribution of the plants in the various dune associations is governed by physical and biotic agencies which will be considered somewhat in detail in another place. At this point it seems advisable to give a general survey of these factors, especially in so far as they affect the distribution of plant societies in the region as a whole.

LIGHT AND HEAT.

Nearly all of the dune societies are characterized by a high degree of exposure to *light*. Particularly is this true of the beach and the active dunes. The intensity of direct illumination is greatly increased by reflection; the glare of the white sand is almost intolerable on a bright summer day. The *temperature* relation is even more marked in its influence upon plant life. Because of the absence of vegetation and the general exposure of sand dunes the temperature is higher in summer and lower in winter than in most localities. This great divergence between the temperature extremes is still further increased by the low specific heat of sand. On sandy slopes protected from cold winds the vegetation renews its activity very early in the spring, because of the strong sunlight and the ease with which the surface layers of sand are heated. Willow shoots half-buried in the sand frequently develop fully a week in advance of similar shoots a few centimeters above the surface. Similarly in the autumn the activity of plant life ceases early largely because of the rapid cooling of the superficial layers of sand, as well as because of direct exposure to the cold.

WIND.

The wind is one of the most potent of all factors in determining the character of the dune vegetation. The winds constantly gather force as they sweep across the lake, and when they reach the shore quantities of sand are frequently picked up and carried on. The force with which this sand is hurled against all obstacles in its path may be realized if one stoops down and faces it. The carving of the dead and living trees which are exposed to these natural sand-blasts is another evidence of their power. Fleshy fungi have been found growing on the windward side of logs and stumps completely petrified, as it were, by sand-blast action; sand grains are imbedded in the soft plant body and as it grows the imbedding is continued, so that finally the structure appears like a mass of sand cemented firmly together by the fungus. The bark of the common osier dogwood is red on the leeward side, but white to the windward because the colored outer layers of the bark have been wholly worn away. On the windward side of basswood limbs the softer portions are carved away while the tougher fibers remain as a reticulated network. On the leeward side of these same limbs, the outer bark is intact and even covered more or less with lichens.

The indirect action of the wind produces effects that are considerably more far-reaching than any other factor, for it is the wind which is primarily responsible for sand dunes and hence for their floras. But more directly than this, the wind plays a prominent part in modifying the plant societies of the dunes. The wind is the chief destroyer of plant societies. Its methods of destruction are twofold. Single trees or entire groups of plants frequently have the soil blown away from under them, leaving the roots exposed high above the surface; as will be shown later this process is sometimes continued until entire forests are undermined, the débris being strewn about in great abundance. Again, swamps, forests, and even low hills may be buried by the onward advance of a dune impelled by the winds; in place of a diversified landscape there results from this an all but barren waste of sand.

SOIL.

The soil of the dunes is chiefly quartz sand, since quartz is so resistent to the processes of disintegration. The quartz particles are commonly so light colored that the sand as a whole appears whitish; closer examination reveals many grains that are not white, especially those that are colored by iron oxide. With the quartz there are conspicuous grains of black sand, largely hornblende and magnetite. These black grains often accumulate in streaks, persistent for considerable depths and apparently sifted by the wind; large quartz grains are mingled with these grains of magnetite and hornblende so that it would seem as if grains of higher specific gravity are sifted out together with those of greater absolute weight. The sand of the dunes is remarkably uniform in the size of the particles as compared with beach sand; this feature is due to the selective action of the wind, since the latter agent is unable to pick up and carry for any distance the gravel or large sand particles of the beach.

As is well known, soil made up chiefly of quartz sand has certain marked peculiarities which strongly influence the vegetation. The particles are relatively very large; hence the soil is extremely porous and almost devoid of cohesion between the grains. These features are of especial importance in their effect upon the water and heat relations as shown elsewhere. As a rule, sandy soils are poor in nutrient food materials, nor do they rapidly develop a rich humus soil because of the rapid oxidation of the organic matter.

WATER.

A factor of great importance, here as everywhere, is the water relation. Nothing need be said of atmospheric moisture, since that is sufficient to develop a rich vegetation if properly conserved, as is shown by the luxuriance of neighboring floras. Because of the peculiar physical properties of quartz sand, precipitated water quickly percolates to the water level and becomes unavailable to plants with short roots. The water capacity of sand is also slight, nor is there such pronounced capillarity as is characteristic of many other soils. Again, the evaporation from a sandy surface is commonly quite rapid. All of these features combine to furnish a scanty supply of water to the tenants of sandy soil. The rapid cooling of sand on summer nights may, however, result in a considerable condensation of dew, and thus, in a small way, compensate for the other disadvantages.

The ecological factors thus far mentioned act together harmoniously and produce a striking composite effect upon the vegetation. A flora which is subjected to periods of drought is called a xerophytic flora and its component species have commonly worked out various xerophytic structural adaptations of one sort or another. Again, a flora which is subjected to extreme cold, especially when accompanied by severe winds, takes on various structural adaptations similar to those that are characteristic of alpine and arctic floras. The dune flora is a composite flora, showing both xerophytic and arctic structures. In those situations which are most exposed to cold winds, one finds the best illustrations of the arctic type, while the desert or xerophilous type is shown in its purest expression on protected inland sandy hills. The discussion of the various arctic and desert structures and their relations to each other will be deferred to the second part of this paper.

OTHER FACTORS.

Certain other factors are of minor importance in determining the character of the dune flora. *Forest fires* occur occasionally, and, as will be shown later, they may considerably shorten the lifetime of a coniferous plant society.

Near cities the vegetation is unfavorably influenced by *smoke* and other products issuing from chimneys. In the neighborhood of the oil refineries at Whiting, Ind., the pine trees especially have been injured or destroyed. A careful study would probably show many plant species that have suffered a similar fate.

The *topography* is often a factor of considerable importance. Dune areas are conspicuous for their diversified topography. This factor determines to a great extent the water relation which has been previously considered, the hills and slopes being of course much drier than the depressions. The topography indirectly affects the soil, since it is mainly in the depressions that humus can rapidly accumulate. The direction of slope is a matter of importance, as will be shown in discussing the oak dunes; the greater exposure of the southern slopes to the sun results in a drier soil and a more xerophytic flora on that side.

Animals do not appear to exert any dominating influences on the dune floras. The dispersal of pollen and fruits by their agency is common here as elsewhere; so, too, the changes that animal activities produce in the soil. Near the cities the influence of man is seen, although such influences are slight unless the sand is removed bodily for railroad grading and other purposes.

The influence of *plants*, which so often becomes the dominant factor, is relatively inconspicuous on the dunes. The most important function which dune plants perform for other plants is in the contribution of organic food materials to the soil. The oxidation or removal of decaying vegetation is so complete on the newer dunes that the accumulation of humus is not important. On the more established dunes the mold becomes deeper and deeper, and, after the lapse of centuries, the sandy soil beneath may become buried so deeply that a mesophytic flora is able to establish itself where once there lived the tenants of an active dune. The advance of a wandering dune often results in the burial of a large amount of organic matter; when this matter becomes unburied years afterward it may again furnish a soil for plants. Many fossil soil lines have thus been uncovered on the Sleeping Bear dunes at Glen Haven, Mich.

IV. The plant societies.

A plant society is defined as a group of plants living together in a common habitat and subjected to similar life conditions. The term is taken to be the English equivalent of Warming's *Plantesamfund*, translated into the German as *Pflanzenverein*. The term formation, as used by Drude and others, is more comprehensive, in so far as it is not synonymous. It may be well to consider the individual habitat groups in a given locality as plant societies, while all of these groups taken together comprise a formation of that type, thus giving to the word formation a value similar to its familiar geological application. For example, one might refer to particular sedge swamp societies near Chicago, or, on the other hand, to the sedge swamp formation as a whole; by this application formation becomes a term of generic value, plant society of specific value.

Plant societies may be still further subdivided into patches or zones; the former more or less irregular, the latter more or less radially symmetrical. Patches are to be found in any plant society, where one or another constituent becomes locally dominant; zones are conspicuously developed on the beach and in sphagnous swamps. The term patch or zone has a value like that of variety in taxonomy. Authors disagree, here as everywhere, upon the content and values of the terms employed; this disagreement is but an expression of the fact that there are few if any sharp lines in nature. The above, or any other terminology, is largely arbitrary and adopted only as a matter of convenience.

In the following pages an attempt is made to arrange the plant societies in the order of development, the author's belief being that this order more faithfully expresses genetic relationships than any other. In the historical development of a region the primitive plant societies pass rapidly or slowly into others; at first the changes are likely to be rapid, but as the plant assemblage more and more approaches the climax type of the region, the changes become more slow. In the dune region of Lake Michigan the normal primitive formation is the beach; then, in order, the stationary beach dunes, the active or wandering dunes, the arrested or transitional dunes, and the passive or established dunes. The established dunes pass through several stages, finally culminating in a deciduous mesophytic forest, the normal climax type in the lake region. Speaking broadly, the conditions for plant life become less and less severe through all these stages, until there is reached the most genial of all conditions in our climate, that which results in the production of a diversified deciduous forest. On the beach there are to be found the most extreme of all xerophytic adaptations in this latitude, and, as one passes through the above dune series in the order of genetic succession, these xerophytic structures become less and less pronounced, finally culminating in the typical mesophytic structures of a deciduous forest.

A. THE BEACH.

As the author hopes to show in a subsequent paper, the beach formations of Lake Michigan are of two distinct types. One may be called the xerophytic beach, the other the hydrophytic beach. The conditions that determine these two types are not altogether clear, though their distribution suggests some factors which will contribute to the solution of the problem. Dunes are invariably absent from an area occupied by the situation. If there is a sufficient amount of sand still remaining, the once stationary dune begins to move, not bodily, of course, but none the less steadily and surely. The sand is swept up the low gradient of the windward side, deposited at the crest, and carried down the steep leeward slope by gravitation. In this manner successive parallel layers of the windward slope are carried over the crest, and the dune as a whole advances inland. The simple life-history just outlined is the exception, not the rule. Much more commonly the sand is scattered in many directions, collecting wherever new lodging places can be found. These processes of deposition and removal, dune formation and dune destruction, are constantly going on with seeming lawlessness. However, in the district as a whole the sand is constantly increasing in quantity, whatever may be true of the individual dunes here and there. The outcome is certain to be a wandering dune in the process of time, unless the actions of the wind and wave are checked. Because of the complexity of the conditions when the movement across the country becomes a conspicuous fact, it seems well to apply the term dune-complex to the totality of topographic forms which make up the moving landscape as a whole.

2. *Physical and biological features of the dune-complex.*

It will not be necessary to trace farther the changes involved in the transformation of simple beach dunes into a dune-complex, although the coast of Lake Michigan shows all of the intervening stages. Inasmuch as a single dune-complex illustrates almost all conceivable conditions of a dune's life-history, a careful description of a typical dune-complex will involve all of the essential points. The dune-complex is best developed at Glen Haven, Mich., and Dune Park, Ind. All of the essential features are present in both areas, though developed on a grander scale at Glen Haven. The Dune Park area has been most carefully studied, and most of my photographs were taken there.

The dune-complex is a restless maze. It is a maze because all things that a dune ever does are accomplished there. While there is a general advance of the complex as a whole in the direction of the prevailing winds, individual portions are advancing in all directions in which winds ever blow. It is not at all uncommon to find small dunes advancing over the dune-complex back toward the lake. At Dune Park the main line of advance is southeast, yet some small dunes advance toward the northwest, because taller dunes are situated between them and the lake. These little dunes in the lee of large ones are protected from the westerly and northerly winds but feel the full force of the easterly and southerly winds, and hence advance contrary to the prevailing direction. It is thus a common sight to see two dunes advancing to meet each other; when they come together, of course the two dunes become one and move in the direction of the prevailing winds. From this account it is easy to see that small dunes on the complex may advance in any direction, provided only that they are protected from winds blowing in other directions.

The dune-complex, however, is much more than a maze of little dunes wandering in all directions. At many points there are to be found the stationary embryonic dunes that have been previously described. All stages of their life-history may be seen; the beginning, the climax, the destruction. Here and there the wind sweeps out great hollows, which reach down almost to the water level. Great troughs are carved out by the wind, chiefly at right angles to the lake, but also at all other angles. Here and there vegetation has obtained a foothold on the complex, thus converting portions of it into an established dune. These established dunes may become rejuvenated, or the vegetation may spread until it covers large portions of the complex. The most striking feature of the dune-complex, then, is its topographic diversity.

To one who visits a dune-complex season after season, another feature comes to be as striking as its diversity, and that is its restlessness. From a distance the complex seems always the same, a barren scene of monotony, but the details are never twice alike. A little dune arising on the complex has become enlarged, another has passed from existence without leaving a trace behind. Where a dune was advancing last year, there is now, perhaps, a hollow swept out by the wind. Where last year was a hollow there may now be seen the beginnings of a flora, or again the flora of a former year may have been buried out of sight. The dune-complex, then, is not only a maze, but also a restless maze.

FIG. 7.—Trough-shaped wind-sweep at Beaver island. Dead roots and branches of plants that have been torn up. Embryonic dunes in the background formed of sand brought from the foreground. Sparse annual vegetation in the wind-sweep.

It might seem impossible to unravel the tangled threads of the dune-complex; it is, indeed, impossible to write the details of its history. There is, however, a simplicity in the complexity. While little dunes advance in all directions, the complex as a whole advances in the direction of the prevailing wind. While there are troughs at all angles, the main troughs are likewise in the direction of the wind. The complex is like a river with its side currents and eddies at many points, but with the main current in one direction.

It has already been stated that the windward slope of an advancing dune is very gentle, averaging perhaps about 5°. That portion of the windward slope up which the main wind currents pass is also trough-shaped and may be called a wind-sweep. *Fig. 7* shows a small trough-shaped wind-sweep at Beaver island; the direction of advance is from the foreground to the background. In the path of the wind there may be seen dead branches, the remnants of a vegetation that has been swept away. In the background are small dunes which have been formed by sand carried along the trough by the wind. A wind-sweep more characteristic of the dune-complex is shown in *fig. 17*. Here, likewise, the prevailing wind direction is up the gentle slope away from the foreground. At this particular place there is in reality much more of a trough than is shown in the photograph, since there is a conspicuous rise both at the right and at the left. Just beyond the pines in the background there is a steep pitch downward, the advancing lee slope.

The most remarkable wind-sweep at Dune Park reaches down almost to the water level, appearing like a cañon, by reason of its steep sides from ten to twenty meters in height. This sweep, unlike most of the troughs, is curved so that a wind entering it as a northwest wind becomes a west and finally a southwest wind, and actually contributes to the advance of a dune toward the lake, as will be discussed more fully at another place in connection with *fig. 10*. The concentration of the wind energy which these gorge-like wind-sweeps permit is something remarkable. At no place is the destructive power of the wind upon the vegetation felt more keenly than along the sides of these deeper wind-sweeps. The foreground of *fig. 22* shows the upper part of one of these troughs, and gives a vivid impression of the wind's destructiveness.

The advancing lee slopes, as has been previously mentioned, have a gradient in the neighborhood of 30°. The slope is exactly that at which sand, whose grains have the size and cohesiveness there folund, will ie. *Figs. 9, 13, 14,* and *15* give some conception of the striking features presented by a landscape of which an advancing dune forms a part. Nowhere can there be a sharper line in nature, nowhere a more abrupt transition. The height of these slopes above the country on which they are advancing varies from almost nothing up to thirty meters at Dune Park. The Glen Haven dunes are far more imposing, since there is an almost unbroken line of advance for four kilometers, while the average height is from thirty to sixty meters above the territory on which they are encroaching.

The vegetation of the complex proper is exceedingly sparse. In the winter it appears almost a barren waste. The one plant which seems to be at home in all locations, whether wind-sweeps, exposed summits, or protected lees, is the bugseed, *Corispermum hyssopifolium.* This plant is an annual, and has been previously mentioned as a tenant of the beach and the beach dunes. The bugseed is shown in several of the photographs, but best in the left foreground of *fig. 12*. The seeds are winged and readily dispersed by the wind. Furthermore they germinate rapidly during wet spring weather. This power of rapid germination is a necessary condition of success, since the surface layers of sand dry off very quickly after the wet weather has ceased. The plants are obliged not only to germinate rapidly but also to send roots deep enough to reach beyond the surface desiccation. Even this perfectly successful plant species is often absent from large areas on the dune-complex, probably because of the difficulty which the seeds meet in finding lodgment. It is only an exceptional seed which is allowed to remain stranded on the complex, and many of the seeds which succeed in finding lodgment are likely to be buried too far below the surface to permit germination.

Another plant which deserves especial mention is the cottonwood, *Populus monilifera.* The plasticity of this species is remarkable. Normally at home along protected river bottoms, it is yet able to endure almost all of the severe conditions of the dune-complex. Mention has been made of its importance as a evergreen series, and is in all probability a climax type, at least in certain situations.

A very distinct type of coniferous forest is especially well developed at the south end of the lake. Since it is not developed in exposed situations, or even on old dunes, but in low depressions between dunes, it may be called a pine bottom. These societies are developed where the soil is almost hydrophytic. A common location for these miniature pine forests is about the gently sloping margin of an undrained swamp. *Figs. 9, 13, 14*, and *20* show them in such a situation. The line of demarcation between the sedge swamp and the pines is usually quite sharp. The surface of the soil where the pines grow may be less than a meter above the water level.

The character tree of the pine bottoms is always *Pinus Banksiana*. This species is, perhaps, less common than the white pine at the higher levels, but the white pine is rarely, if ever, present on the bottoms. No growth of trees anywhere in the dune region is so pure as the pine growth here. The most common shrubs in these locations are *Hypericum Kalmianum, Salix glaucophylla, Arctostaphylos Uva-ursi*, and *Juniperus communis. Linnaea borealis, Arabis lyrata, Fragaria Virginiana*, and species of Pyrola are frequent. The development of the pine bottom floras was seen at several points. One of the most interesting cases was in a region of oak dunes, where a railroad company had removed considerable sand and lowered the level several meters. Although surrounded on all sides by oaks and at some distance from a pine flora, the new flora at the lower level is developing into that of a pine bottom.

c. The rejuvenated dunes.—The instability of dune conditions is not confined to the dune-complex. The capture or establishment of a dune is liable to be stopped at any point and retrogression toward the active dune conditions instituted. Even a dune that has long been completely established may have its vegetation destroyed and pass again into a state of activity. This process may be called rejuvenation. Any dune may become rejuvenated if the physical conditions are favorable, but the great majority of rejuvenated dunes are developed from established coniferous dunes ; hence this type is discussed in connection with the evergreen series. The coniferous forests that develop on the windward slopes near the lake are peculiarly subject to destruction. The slightest change in the physical conditions is often sufficient to bring about the destruction of a

FIG. 24.—Beginnings of a heath on a fossil beach at North Manitou island. Grasses, sand cherries, and patches of Hudsonia in the foreground. Low Ammophila dune at the center. Patches of bearberry heath forming farther back, Advance of the coniferous forest out upon the heath.

coniferous society. The removal of a comparatively slight barrier may be enough to direct the entire wind energy against a pine forest.

The formation of a wind-sweep is, perhaps, the most common way for rejuvenation to begin. *Fig. 6* shows a windward slope tenanted by conifers that has become rejuvenated at three points. One of these wind-sweeps is seen at closer range in *fig. 21*. This latter sweep is forty-five meters in height, and the angle of slope varies from twenty to thirty degrees. When once a sweep is formed the tendency to self-perpetuation becomes greater and greater, since the wind becomes more and more concentrated as the sweep grows deeper. The destruction of the forest vegetation is very soon accomplished at such a place. The desiccating influence of the wind becomes increased and

FIG. 25.—Juniper heath on North Manitou island. Young heath patches at the left background. Fully developed heath in the foreground. Advancing coniferous forest at the right background.

makes it difficult even for the xerophytic conifers to survive. At no place is the destructive action of the sand-blast seen so well as in these rejuvenated sweeps. The branches and even the trunks of the trees have the softer parts carved away, while the more resistant portions stand out in conspicuous relief. The leaves, especially of deciduous trees, are torn or withered or even altogether destroyed.

These destructive agencies are aided by another force that is altogether irresistible when the sweeps grow deeper, the force of gravity. *Fig. 22* shows a plant society that is being destroyed mainly by gravity. The view is taken looking at the side of a deep gorge-like wind-sweep which the wind has cut. As the wind blows along, its energy increased by concentration, a large amount of sand is picked up along the base of the steep sides. The sand is as steep as it will lie, so that each removal causes a movement of the sand down the slope. The fallen trees shown in the photograph have been overturned and carried down the slope in just this way. That the direct action of the wind is also powerful enough to destroy without the assistance of gravity is proven by the dead but standing trees at the left, where the action of gravity happens to be much less.

Many plant species resist the process of dune rejuvenation to a surprisingly successful extent. *Fig. 23* shows the last remnant of a plant society that may have been somewhat extensive. The tree at the center is a basswood, a tree which could never develop in such an exposed situation. In all probability this mound is a fragment of a protected lee slope, on which the basswood grew and flourished for a time. The grass at the left is Calamagrostis; the tenacity with which it holds its ground has already been mentioned. Sometimes a group of cedars, *Juniperus Virginiana*, remain at the apex of a conical mound of sand, their associates having been swept away with the sand in which they grew. On the beach at Charlevoix there is a stranded clump of stunted trees of Thuya; they are probably the remnant of a society which has been otherwise destroyed.

As a wind-sweep is developed, and the evergreen vegetation destroyed, many plants that have been previously mentioned as characteristic of bare and exposed situations again make their appearance. The most prominent of these are *Artemisia Canadensis* (or *A. caudata*), *Elymus Canadensis*, *Solidago humilis Gillmani*, *Asclepias Cornuti*, *Œnothera biennis*, *Rosa Engelmanni*, *Calamagrostis longifolia*, *Prunus Virginiana*. In addition to these there come in, of course, the annuals and biennials mentioned in connection with the wind-sweeps on the dune-complex.

While rejuvenated dunes are to be found along the entire

coast, they reach their highest development northward, especially at the summit of the terraces and bluffs. Perched dunes, it would seem, are favorably located for destruction by the wind. At Frankfort and Empire the perched dunes are in the earlier stages of rejuvenation. At Glen Haven these dunes have been rejuvenated, the vegetation entirely destroyed, and the sand

FIG. 26.—Development of a juniper heath in a pasture at Beaver island.

removed inland to form the gigantic moving dunes previously mentioned. The substratum on which the dunes rested remains as a bare gravel mesa, with only the Sleeping Bear left to tell the tale of its former occupation by coniferous dunes. It is barely possible that some of these so-called rejuvenated dunes have never been established, and that they have grown slowly to their present height *pari passu* with the vegetation. This is purely a theory without any facts whatever to support it. The evidence seems to point unmistakably to an establishment followed by rejuvenation. Evergreen vegetation is very poorly adapted for any *pari passu* growth, such as is found on the embryonic dunes.

3. *The oak dunes.*

At the south end of the lake, and as far up the eastern shore as Manistee, there may be seen old dunes covered over with rather open and scrubby oak forests. These dunes have long been established and are entirely free from the destructive sand-laden winds which are so influential in determining the character of the other dune societies. As a rule the oak dunes are low and are separated from the lake by several series of dunes on which the vegetation is less stable.

The dominant tree on the oak dunes is the black oak, *Quercus coccinea tinctoria.*[5] This tree is far more abundant than all others combined. The only other tree that may be called characteristic in the Dune Park region is *Quercus alba.* On some oak dunes there are low trees or tall shrubs of *Sassafras officinale, Cornus florida, Amelanchier Canadensis,* and *Hamamelis Virginiana.* The characteristic shrubs are comparatively few except along the lower margins toward the swamp level, or on shaded northern slopes. The most abundant shrubs are *Vaccinium vacillans* and *V. Pennsylvanicum, Salix humilis, Viburnum acerifolium, Rosa blanda* and *R. humilis,* and *Rhus copallina.*

The herbaceous vegetation of the oak dunes is very diversified and interesting. The trees are always far enough apart to permit an extensive undergrowth of relatively light-loving plants. On the southern slopes, where there is considerable exposure to the sun, there is rarely a continuous vegetation carpet, but a more or less tufted vegetation with intervening patches of naked sand. A large number of herbs are characteristic of such places, for example: *Pteris aquilina, Koeleria cristata, Cyperus Schweinitzii, Carex Pennsylvanica, C. umbellata, C. Muhlenbergii, Tradescantia Virginica, Arabis lyrata, Lupinus perennis, Tephrosia Virginiana, Les-*

[5] The closely related *Quercus rubra* and *Q. coccinea* occur commonly in neighboring plant societies and may be present on the dunes, as may hybrids between any of the three forms here mentioned.

shown by studying the floras of the oak and pine dunes. The former has a flora related to those farther south, containing Opuntia, Euphorbia, and many other plants of southern range. The pine dunes, on the other hand, show the farthest southern limits of many northern plants—for example, the scrub pine itself. Linnaea, the bearberry, and many others have a northern range.

V. Conclusion.

No attempt will be made to summarize the results of this study, but a few of the more striking phenomena of the Lake Michigan dunes and their vegetation will be given. The dunes have been determined in the main by westerly winds. The great majority of the dunes are established, and many of them are perched high up on bluffs. The vegetation is xerophytic, belonging either to the arctic or desert type.

The xerophytic beaches are subdivided into three zones: the lower beach which is washed by summer waves and is essentially devoid of life; the middle beach which is washed by winter waves and is inhabited only by succulent annuals; the upper beach which is beyond present wave action and is inhabited also by biennials and perennials. There are also fossil beaches and gravel terraces with a flora resembling that of the upper beach, but less xerophytic.

Perennial plants are necessary for any extensive dune formation on the beach, since they alone furnish growing obstacles. Such plants must be pronounced xerophytes and be able to endure covering or uncovering. The most successful dune-formers are *Ammophila arundinacea*, *Agropyrum dasystachyum*, *Elymus Canadensis*, *Salix glaucophylla* and *S. adenophylla*, *Prunus pumila*, *Populus monilifera*. Ammophila and Agropyrum form low dunes that have a large area, because of their extensive rhizome propagation. The Elymus dunes do not increase in area since rhizome propagation is absent. The Salix dunes increase both in area and height, because of extensive horizontal and vertical growth. The Populus dunes are the highest and steepest, since the cottonwoods grow quite tall, but do not spread horizontally.

Small dunes are formed in more protected places by plants that are unable to exist on the beach, or where there is rapid dune formation. Among these secondary dune-formers are Andropogon, Arctostaphylos, Juniperus. Primary embryonic dunes may pass gradually into this second type, as this latter passes into the heath.

The stationary embryonic dunes on the beach begin to wander as soon as the conditions become too severe for the dune-forming plants. The first result of this change is seen in the reshaping of the dune to correspond with the contour of a purely wind-made form. The rapidity of this process is largely determined by the success or failure of the dune-formers as dune-holders. The best dune-holders are Calamagrostis, Ammophila, and Prunus.

There are all gradations between a simple moving dune and a moving landscape; the latter may be called a dune-complex. The complex is a restless maze, advancing as a whole in one direction, but with individual portions advancing in all directions. It shows all stages of dune development and is forever changing. The windward slopes are gentle and are furrowed by the wind, as it sweeps along; the lee slopes are much steeper. The only plant that flourishes everywhere on the complex is the succulent annual, *Corispermum hyssopifolium*, although *Populus monilifera* is frequent. The scanty flora is not due to the lack of water in the soil, but to the instability of the soil and to the xerophytic air.

The influence of an encroaching dune upon a preexisting flora varies with the rate of advance, the height of the dune above the country on which it encroaches, and the nature of the vegetation. The burial of forests is a common phenomenon. The dominant forest trees in the path of advancing dunes are *Pinus Banksiana*, *Quercus coccinea tinctoria*, and *Acer saccharinum*. All of these trees are destroyed long before they are completely buried. The dead trees may be uncovered later, as the dune passes on beyond.

In the Dune Park region there are a number of swamps upon

which dunes are advancing. While most of the vegetation is destroyed at once, *Salix glaucophylla*, *S. adenophylla*, and *Cornus stolonifera* are able to adapt themselves to the new conditions, by elongating their stems and sending out roots from the buried portions. Thus hydrophytic shrubs are better able to meet the dune's advance successfully than any other plants. The water relations of these plants, however, are not rapidly altered in the new conditions. It may be, too, that these shrubs have adapted themselves to an essentially xerophytic life through living in undrained swamps. Again it may be true that inhabitants of undrained swamps are better able to withstand a partial burial than are other plants.

Vegetation appears to be unable to capture a rapidly moving dune. While many plants can grow even on rapidly advancing slopes, they do not succeed in stopping the dune. The movement of a dune is checked chiefly by a decrease in the available wind energy, due to increasing distance from the lake or to barriers. A slowly advancing slope is soon captured by plants, because they have a power of vertical growth greater than the vertical component of advance. Vegetation commonly gets its first foothold at the base of lee slopes about the outer margin of the complex, because of soil moisture and protection from the wind. The plants tend to creep up the slopes by vegetative propagation. Antecedent and subsequent vegetation work together toward the common end. Where there is no antecedent vegetation, Ammophila and other herbs first appear, and then a dense shrub growth of Cornus, Salix, *Vitis cordifolia*, and *Prunus Virginiana*. Capture may also begin within the complex, especially in protected depressions, where *Salix longifolia* is often abundant.

Tilia Americana develops rapidly on the captured lee slopes, and the thicket is transformed into a forest. The trees grow densely, and there is little or no vegetation carpet. Associated with Tilia is a remarkable collection of river bottom plants, so that the flora as a whole has a decided mesophytic cast. These plants have developed xerophytic structures that are not present in the river bottoms. Acer and Fagus succeed Tilia and represent the normal climax type of the lake region, the deciduous forest.

On the established windward slopes the development is quite different from that described above. There is a dominance of evergreens instead of deciduous vegetation. The soil conditions are nearly alike on the two slopes, but the air is more xerophytic on the windward slopes. The evergreen flora starts as a heath formed of Arctostaphylos, *Juniperus communis*, and *J. Sabina procumbens*. The heath arises on fossil beaches, secondary embryonic dunes, or wherever the wind is relatively inactive and where the conditions are too xerophytic for the development of a deciduous flora. Before long the heath passes into a coniferous forest, in which *Pinus Banksiana*, *P. Strobus*, or *P. resinosa* dominate. Coniferous forests also occur on sterile barrens and in bottoms, where the conditions are also unfavorable for deciduous forests. A slight change in the physical conditions may bring about the rejuvenation of the coniferous dunes, because of their exposed situation. Rejuvenation commonly begins by the formation of a wind-sweep; the vegetation on either hand is forced to succumb to sand-blast action and gravity.

The evergreen floras are more and more common northward, while to the south there are developed forests in which *Quercus coccinea tinctoria* prevails. The oak forests are more common on inland dunes and on southern slopes. The oaks may follow the pines, when the areas occupied by pines become sufficiently protected from cold winds. The pines have a much wider range of life conditions than the oaks, since they appear at lower levels, higher levels, and on northern or windward slopes. The oaks flourish best on southern slopes. The flora of the oak dunes is xerophytic, but of the desert type, while that of the pine dunes is of the arctic xerophytic type. The pine dunes have a northern flora, the oak dunes a southern flora.

VI. Previous studies of sand dune floras.

A great deal of physiographic work has been done in sand dune areas in total disregard of the plant life, although the results obtained from this study show that the vegetation profoundly modifies the topography. In like manner the flora has often been studied from a purely taxonomic standpoint, little attention being paid to the striking effects of the environment upon plant structures. More recently the ecological standpoint has been taken by a number of investigators, particularly to show the influence of the extreme environment upon plant organs and tissues. The second part of this paper will treat this phase of the subject in some detail. Very little previous work has been done on the geographic phase of the subject from the standpoint of historical development and the order of genetic succession of the various dune types. Still less has there been any adequate study of the modifying influence of vegetation upon topography. These latter phases of the subject have given color to the work which has resulted in this paper.

Warming's work on the sand dune vegetation of Denmark stands in the front rank. In his separate publications and in his text-book of ecology, the conditions on the Danish dunes are quite fully stated. The order of succession, speaking broadly, seems to be quite similar to that along Lake Michigan, but there appears to be less diversity of conditions, and the features appear to be developed on a smaller scale. The strand is succeeded by the wandering or white dunes, and these by the established or gray dunes. Beyond these are sandy fields. Just as along Lake Michigan, the dune floras may pass into the heath and these latter into coniferous forests.

There is a remarkable similarity in the flora of the Danish and Lake Michigan dunes. The same genera and often the same species occur in the two regions. *Cakile maritima* and *Lathyrus maritimus* grow on the strand. *Ammophila arundinacea* (=*Psamma arenaria*), *Elymus arenarius*, and *Agropyrum junceum* grow on the wandering dunes. Where the genera are not common or even nearly related, there are to be found in the two regions plants that have the same life habits. There is thus a striking similarity in the two regions in almost every respect, and that too in spite of the marine conditions in Denmark, as contrasted with the inland fresh-water area in the United States. The life conditions appear to be essentially alike on all dunes, whether marine or not, and there are found not only identical life habits, but even identical plant species.

Warming reports Chlamydomonas on the strand in the same relations as along Lake Michigan. Among the sand-binding plants, Warming and Graebner give an important place to mosses. Along the Lake Michigan dunes, mosses do not appear to any great extent until establishment is nearly complete. On the Denmark coast, the Agropyrum dunes are lower than those formed by Ammophila, just as along Lake Michigan. The Danish dunes have also been studied by Raunkiaer, Paulsen, and Feilberg. Erikson has studied the similar dunes of southern Sweden, Giltay and Massart those of Holland and Belgium.

The dunes on the islands along the German coast have been carefully studied by Buchenau and to some extent by Knuth. Graebner, in his exhaustive work on the North German heath, discusses the origin of the heath on naked dune sand. He gives an important place to algæ and moss protonema, since they precede other vegetation, forming the first humus and causing the sand grains to cohere. It is doubtful if these lower plants are so important as sand-binders along Lake Michigan. Rothert and Klinge have studied the coast vegetation of Russia.

The French dunes have been very carefully studied by Flahault alone and also in association with Combres. Some work has also been done in France by Constantin and Masclef. Willkomm's work in Spain and Portugal, covering a period of nearly fifty years, is very complete and satisfactory. Daveau has worked out the conditions along the coast of Portugal. On these more southern dunes, the plant species resemble those along Lake Michigan less than do those in northern Europe, but the life habits are the same.

The dune flora of South Africa has been touched upon by Thode, that of Chile by Kurtz and Reiche, that of northern Siberia by Kjellman, that of New Zealand by Diels. The tropical dunes of Indo-Malaysia have been studied in detail by Schimper, and are fully discussed in his work on the Indo-Malay strand flora and also in his recent Plant Geography. In the latter work there are several excellent discussions of sand dune vegetation, accompanied by photographs from a number of regions. The tropical dunes have totally different species, but even there the dominant dune-formers are grasses with the same life habits as Ammophila.

Dunes may be formed in deserts and inland regions apart from large bodies of water. Those in the Sahara and in the deserts to the northeastward have been more or less studied. Brackebusch has described dunes in Argentina.

In the United States dunes are common along the Atlantic coast, especially in Massachusetts, New Jersey, North Carolina, and Florida. On the Pacific coast they also occur extensively. None of these marine dunes have been exhaustively studied from the ecological standpoint. One of the best works that has ever appeared on strand floras is that by MacMillan on the shores at the Lake of the Woods. The dune formation is not extensive there, but is most admirably treated. As would be expected, there are many species common to Lake Michigan and the Lake of the Woods. The sand hills in the interior have been studied by Rydberg, Hitchcock, and Pound and Clements. Hill has studied the dune floras about Lake Michigan for many years, and although he has not written a great deal along ecological lines, he has had the ecological standpoint thoroughly in mind and the author has received from him a number of valuable suggestions.

THE UNIVERSITY OF CHICAGO.

BIBLIOGRAPHY.

BOERGESEN, F.: Beretning om et Par Exkursioner i Sydspanien. Bot. Tid. **21** : 139. 1897.

BOERGESEN, F., and PAULSEN, O.: Om Vegetationen paa de dansk vestindiske Öer. Copenhagen. 1898.

BORBÁS, V.: Die Vegetation der ungarischen Sandpuszten mit Rücksicht auf die Bindung des Sandes. Abstract in Bot. Cent. **19** : 92. 1884.

BRACKEBUSCH, L.: Ueber die Bodenverhältnisse des nordwestlichen Theiles der Argentinischen Republik mit Bezugnahme auf die Vegetation. Petermann's Mittheilungen **39** : 153. 1893.

BUCHENAU, F.: Vergleichung der nordfriesischen Inseln mit den ostfriesischen in floristischer Beziehung. Abhandl. Naturw. Ver. Bremen **9**. 1887.

——— Ueber die Vegetations-verhältnisse des "Helms" (Psamma arenaria Röm. et Schultes) und der verwandten Dünengräser. Abhandl. Naturw. Ver. Bremen **10** : 397. 1889.

——— Die Pflanzenwelt der ostfriesischen Inseln. Abhandl. Naturw. Ver. Bremen **11**. 1890.

——— Flora der ostfriesischen Inseln. Norden und Norderney. 1881, 1891. Also various articles on the floras of these islands.

——— Ueber die ostfriesischen Inseln und ihre Flora. Verhandl. des XI. deutschen Geographentages in Bremen. **1895–6** : 129. (Bot. Cent. **66** : 318.)

CLEGHORN, H.: On the sand-binding plants of the Madras beach. Jour. of Bot. **8** : 1858.

CONSTANTIN, J.: Observations sur la flore du Littoral. Jour. de Bot. **1** : 5. 1887.

DAVEAU, J.: La flore littorale du Portugal. Bull. Herb. Boiss. **4** : 209. 1896.

DIELS, L.: Vegetations-Biologie von Neuseeland. Eng. Bot. Jahrb. **22** : 202. 1897.

DRUDE, O.: Handbuch der Pflanzengeographie. Stuttgart. 1890.

——— Ueber die Principien in der Unterscheidung von Vegetationsformationen, erläutert an der centraleuropäischen Flora. Eng. Bot. Jahrb. **11** : 21. 1890.

——— Deutschlands Pflanzengeographie, I Teil. Stuttgart. 1896.

ERIKSON, JOHAN: Studier öfver sandfloran i östra Skåne. Bihang till Kongl. Svenska Vetenskaps-Akademiens Handlingar. **22**. 1896.

FEILBERG, P.: Om Gräskultur paa Klitsletterne ved Gammel Skagen Söborg. 1890.

FLAHAULT, C. La distribution géographique des végétaux dans un coin du Languedoc. Montpellier. 1893.

FLAHAULT, C., et COMBRES, P.: Sur la flore de la Camargue et des alluvions du Rhône. Bull. Soc. Bot. France **41** : 37. 1894.

GIFFORD, J.: The control and fixation of shifting sands. The Engineering Magazine, January 1898. Also various articles in the Forester.

GILTAY, E.: Anatomische Eigenthümlichkeiten in Beziehung auf klimatische Umstände. Nederlandsch Kruidkundig Archief. II. **4**:413. 1886. (Bot. Cent. **36** : 42.)

GRAEBNER, P.: Studien über die norddeutsche Heide. Eng. Bot. Jahrb. **20**: 500. 1895.

HILL, E. J.: The sand dunes of northern Indiana and their flora. Garden and Forest **9** : 353. September 1896.

HITCHCOCK, A. S.: Ecological Plant Geography of Kansas. Trans. Acad. Sci. St. Louis **8** : 55. 1898.

KJELLMAN, F. R.: Om växtligheten på Sibiriens nordkust. Ur Vega-Expeditionens vetensk. iakttagelser **1** : 233. 1882. (Bot. Cent. **13** : 305.)

KLINGE, J.: Die vegetativen und topographischen Verhältnisse der Nordküste der Kurischen Halbinsel. Sitzber. der Dorpater Naturforscher-Gesellschaft **1884** : 76. (Bot. Cent. **21** : 77.)

——— Die topographischen Verhältnisse der Westküste Kurlands. Sitzber. der Dorpater Naturforscher-Gesellschaft **1884** : 603. (Bot. Cent. **21** : 203.)

KNUTH, P.: Botanische Beobachtungen auf der Insel Sylt. Humboldt **1888** : 104.

——— Flora der Insel Helgoland. Kiel. 1896.

KRASSNOFF, A.: Geobotanical studies in the Kalmuck steppes. Trans. Imp. Russ. Geograph. Soc. **22** : 1. 1888. (Russian.) (Eng. Bot. Jahrb. **10**: Litteraturbericht 53.)

KURTZ, F.: Bericht über zwei Reisen zum Gebiet des oberen Rio Salado (Cordillera de Mendoza), ausgeführt in den Jahren 1891–1893. Verhandl. Bot. Ver. Prov. Brandenburg **35** : 95. 1893.

LAMB, F. H.: The sand dunes of the Pacific coast. The Forester **3** : 94. 1897.

LITWINOW, D. J.: Ob okskoi florje w Moskowskoi gubernii. Materialy k posnaniju fauny i flory Rossijskoi Imperii. Moscow. 1895. (Bot. Cent. **66** : 248.)

MACMILLAN, C.: Observations on the distribution of plants along shore at Lake of the Woods. Minnesota Botanical Studies **1** : 949. 1897.

MASCLEF, A.: Études sur la géographie botanique du Nord de la France. Jour. de Bot. **2** : 177. 1888.

MASSART, J.: La biologie de la végétation sur le littoral Belge. Mém. Soc. Roy. Bot. Belgique **32** : 1. 1893.

PAULSEN, O.: Om Vegetationen paa Anholt. Bot. Tid. **21** : 264. 1898.

POUND, R. and CLEMENTS, F. E.: The Phytogeography of Nebraska. I. General Survey. Lincoln. 1898.

RAUNKIAER, C.: Vesterhavets Öst- og Sydkysts Vegetation. Copenhagen. 1889.

REICHE, K.: Die Vegetations-Verhältnisse am Unterlaufe des Rio Maule (Chile). Eng. Bot. Jahrb. **21**: 1. 1896.

ROTHERT, W.: Ueber die Vegetation des Seestrandes im Sommer 1889. Korrespondenzblatt des Naturf.-Ver. zu Riga **32**. (Bot. Cent. **46**: 52.)

RYDBERG, P. A.: Flora of the Sand Hills of Nebraska. Contrib. U. S. Nat. Herb. **3**: 133. 1895.

SCHIMPER, A. F. W.: Die indo-malayische Strandflora. Jena. 1891.

——— Pflanzen-geographie auf physiologischer Grundlage. Jena. 1898.

THODE, J.: Die Küstenvegetation von Britisch-Kaffrarien und ihr Verhältniss zu den Nachbarfloren. Eng. Bot. Jahrb. **12**: 589. 1890.

WARMING, E.: De psammophile Formationer i Danmark. Vidensk. Meddel. fra den naturhist. Foren. **1891**: 153.

——— Exkursionen til Fano og Blaavand i Juli 1893. Bot. Tid. **19**: 52. 1894.

——— Plantesamfund. Copenhagen. 1895. German edition, translated by Knoblauch. 1896.

——— Exkursionen til Skagen i Juli 1896. Bot. Tid. **21**: 59. 1897.

WEBBER, H. J.: Notes on the strand flora of Florida. Abstract in Science, **8**: 658. 1898.

WILLKOMM, M.: Die Strand- und Steppengebiete der iberischen Halbinsel und deren Vegetation. Leipzig. 1852.

——— Statistik der Strand- und Steppenvegetation der iberischen Halbinsel. Eng. Bot. Jahrb. **19**: 279. 1895.

——— Grundzüge der Pflanzenverbreitung auf der iberischen Halbinsel Leipzig. 1896.

NATURE AND STRUCTURE OF THE CLIMAX

By FREDERIC E. CLEMENTS

(Carnegie Institution of Washington, Santa Barbara, California)

(With Plates VI—XI)

CONTENTS

INTRODUCTION

MORE than a century ago when Lewis and Clark set out upon their memorable journey across the continent of North America (1803–6), they were the first to traverse the great climaxes from deciduous woods in the east through the vast expanse of prairie and plain to the majestic coniferous forest of the north-west. At this time the oak-hickory woodland beyond the Appalachians was almost untouched by the ax except in the neighborhood of a few straggling pioneer settlements, and west of the Mississippi hardly an acre of prairie had known the plow. A few years later (1809), Bradbury states that the boundless prairies are covered with the finest verdure imaginable and will become one of the most beautiful countries in the world, while the plains are of such extent and fertility as to maintain an immense number of animals. It appears probable that at this time no other grassland in the world exhibited such myriads of large mammals belonging to but a few species.

The natural inference has been that the prairies were much modified by the grazing of animals and the fires of primitive man, and this has been reinforced by estimates of the population of each. Seton (1929) concludes that the original number of bison was about 60 million with a probable reduction to 40 million by 1800, and that both the antelope and white-tailed deer were equally abundant, while elk and mule-deer each amounted to not more than 10 million at the maximum. However, these were distributed over a billion or two acres, and the average density was probably never more than a score to the square mile. Estimates of the Indian tribes show the greatest divergence, but it seems improbable that the total population within the grassland ever exceeded a half million. The general habit of migration among the animals further insured that serious effects from overgrazing and trampling were but local or transitory, while the influence of fires set by the Indians was even less significant in modifying the plant cover. As to the forests, those of the north-west were still primeval and in the east they were yet to be changed over wide areas by lumbering and burning on a large scale.

THE CLIMAX CONCEPT

The idea of a climax in the development of vegetation was first suggested by Hult in 1885 and then was advanced more or less independently by several investigators during the next decade or so (cf. Clements, 1916; Phillips, 1935). It was applied to a more or less permanent and final stage of a particular succession and hence one characteristic of a restricted area. The concept of the climax as a complex organism inseparably connected with its climate and often continental in extent was introduced by Clements (1916). According to this view, the climax constitutes the major unit of vegetation and as such forms the basis for the natural classification of plant communities. The relation between climate and climax is considered to be the paramount one, while the intimate bond between the two is emphasized by the derivation of the terms from the same Greek root. In consequence, under this concept climax is invariably employed with reference to the climatic community alone, namely, the formation or its major divisions.

At the outset it was recognized that animals must also be considered members of the climax, and the word *biome* was proposed for the purpose of laying stress upon the mutual roles of plants and animals (Clements, 1916*b*; Clements and Shelford, 1936). With this went the realization that the primary relations to the habitat or ece were necessarily different by virtue of the fact that plants are producents and animals consuments. On land, moreover, plants constitute the fixed matrix of the biome in direct connection with the climate, while the animals bear a dual relation, to plants as well as to climate. The outstanding effect of the one is displayed in reaction upon the ece, of the other in coaction upon plants, which constitutes the primary bond of the biotic community.

Because of its emphasis upon the climatic relation, the term climax has come more and more to replace the word formation, which is regarded as an exact synonym, and this process may have been favored by a tendency to avoid confusion with the geological use. The designation "climatic formation" has now and then been employed, but this is merely to accentuate its nature and to distinguish it from less definite usages. Furthermore, climax and biome are complete synonyms when the biotic community is to be indicated, though climax will necessarily continue to be employed for the matrix when plants alone are considered.

Nature of the climax

This theme has been developed in considerable detail in earlier works (Clements, 1916, 1920, 1928; Weaver and Clements, 1929), as well as in a recent comprehensive treatment by Phillips (1935), and hence a summary account of the major features will suffice in the present place. These may be conveniently grouped under the following four captions, i.e. unity, stabilization and change, origin and relationship, and objective tests.

Unity of the climax

The inherent unity of the climax rests upon the fact that it is not merely the response to a particular climate, but is at the same time the expression and the indicator of it. Because of extent, variation in space and time, and the usually gradual transition into adjacent climates, to say nothing of the human equation, neither physical nor human measures of a climate are adequately satisfactory. By contrast, the visibility, continuity, and sessile nature of the plant community are peculiarly helpful in indicating the fluctuating limits of a climate, while its direct response in terms of food-making, growth and life-form provides the fullest possible integration of physical factors. Naturally, both physical and human values have a part in analyzing and in interpreting the climate as outlined by the climax, but these can only supplement and not replace the biotic indicators.

It may seem logical to infer that the unity of both climax and climate should be matched by a similar uniformity, but reflection will make clear that such is not the case. This is due in the first place to the gradual but marked shift in rainfall or temperature from one boundary to the other, probably best illustrated by the climate of the prairie. In terms of precipitation, the latter may range along the parallel of 40° from nearly 40 in. at the eastern edge of the true prairie to approximately 10 in. at the western border of the mixed grassland, or even to 6 in. in the desert plains and the Great Valley of California. Such a change is roughly 1 in. for 50 miles and is regionally all but imperceptible. The temperature change along the 100th meridian from the mixed prairie in Texas to that of Manitoba and Saskatchewan is even more striking, since only one association is concerned. At the south the average period without killing frost is about 9 months, but at the north it is less than 3, while the mean annual temperatures are 70 and 33° F. respectively. The variation of the two major factors at the extremes of the climatic cycle is likewise great, the maximum rainfall not infrequently amounting to three to four times that of the minimum.

The visible unity of the climax is due primarily to the life-form of the dominants, which is the concrete expression of the climate. In prairie and steppe, this is the grass form, with which must be reckoned the sedges, especially in the tundra. The shrub characterizes the three scrub climaxes of North America, namely, desert, sagebrush, and chaparral, while the tree appears in three subforms, coniferous, deciduous, and broad-leaved evergreen, to typify the corresponding boreal, temperate, and tropical climaxes. The life-form is naturally reflected in the genus, though not without exceptions, since two or more forms or subforms, herb or shrub, deciduous or evergreen, annual or perennial, may occur in the same genus. Hence, the essential unity of a climax is to be sought in its dominant species, since these embody not only the life-form and the genus, but also denote in themselves a definite relation to the climate. Their reactions and coactions are the most controlling both in kind and amount, and thus they determine the conditions under which all the remaining species are associated with them. This is true to a less degree of the animal influents, though their coactions may often be more significant than those of plants.

Stabilization and change

Under the growing tendency to abandon static concepts, it is comprehensible that the pendulum should swing too far and change be overstressed. This consequence is fostered by the fact that most ecological studies are carried out in settled regions where disturbance is the ruling process. As a result, the climax is badly fragmented or even absent over wide areas and subseres are legion. In all such instances it is exceedingly difficult or entirely impossible to strike a balance between stability and change, and it becomes imperative to turn to regions much less disturbed by man, where climatic control is still paramount. It is likewise essential to employ a conceivable measure of time, such as can be expressed in human terms of millennia rather than in eons. No student of past vegetation entertains a doubt that climaxes have evolved, migrated and disappeared under the compulsion of great climatic changes from the Paleozoic onward, but he is also insistent that they persist through millions of years in the absence of such changes and of destructive disturbances by man. There is good and even conclusive evidence within the limitations of fossil materials that the prairie climax has been in existence for several millions of years at least and with most of the dominant species of to-day. This is even more certainly true of forests on the Pacific Coast, owing to the wealth of fossil evidence (Chaney, 1925, 1935), while the generic dominants of the deciduous forests of the Dakota Cretaceous and of to-day are strikingly similar.

It can still be confidently affirmed that stabilization is the universal tendency of all vegetation under the ruling climate, and that climaxes are characterized by a high degree of stability when reckoned in thousands or even millions of years. No one realizes more clearly than the devotee of succession that change is constantly and universally at work, but in the absence of civilized man this is within the fabric of the climax and not destructive of it. Even in a country as intensively developed as the Middle West, the prairie relicts exhibit almost complete stability of dominants and subdominants in spite of being surrounded by cultivation (cf. Weaver and Flory, 1934). It is obvious that climaxes display superficial changes with the season, year or cycle, as in aspection and annuation, but these modify the matrix itself little or not at all. The annuals of the desert may be present in millions one year and absent the next, or one dominant grass may seem prevailing one season and a different one the following year, but these changes are merely recurrent or indeed only apparent. While the modifications represented by bare areas and by seres in every stage are more striking, these are all in the irresistible process of being stabilized as rapidly as the controlling climate and the interference of man permit.

In brief, the changes due to aspection, annuation or natural coaction are superficial, fleeting or periodic and leave no permanent impress, while those of succession are an intrinsic part of the stabilizing process. Man alone can destroy the stability of the climax during the long period of control by its climate, and he accomplishes this by fragments in consequence of a destruction that is selective, partial or complete, and continually renewed.

Origin and relationship

Like other but simpler organisms, each climax not only has its own growth and development in terms of primary and secondary succession, but it has also evolved out of a preceding climax. In other words, it possesses an ontogeny and phylogeny that can be quantitatively and experimentally studied, much as with the individuals and species of plants and animals (*Plant Succession*, 1916, pp. 181, 342). Out of the one has come widespread activity in the investigation of succession, while interest in the other lingers on the threshold, chiefly because it demands a knowledge of the climaxes of more than one continent. With increasing research in these, especially in Europe and Asia, it will be possible to test critically the panclimaxes already suggested (Clements, 1916, 1924, 1929), as well as to determine the origin and relationships of the constituent formations.

This task will also require the services of paleo-ecology for the reconstruction of each eoclimax, which has been differentiated by worldwide climatic changes into the existing units of the panclimax or panformation. As it is, there can be no serious question of the existence of a great hemispheric clisere constituted by the arctic, boreal, deciduous, grassland, subtropical and tropical panclimaxes. Desert formations for the most part constitute an exception and may well be regarded as endemic climaxes evolved in response to regional changes of climate (Clements, 1935).

It is a significant fact that the boreal formations of North America and Eurasia are more closely related than the coniferous ones of the former, but this seeming anomaly is explained by the greater climatic differences that have produced the forests of the Petran and Sierran systems. The five climaxes concerned are relatively well known, and it is possible to indicate their relationships with some assurance, and all the more because of their parallel development on the two great mountain chains. In the case of deciduous forest and grassland, only a single formation of each is present in North America, and the problem of differentiation resolves itself into tracing the origin and relationship of the several associations. It has been suggested that the mixed prairie by virtue of its position, extent and common dominants represents the original formation in Tertiary times, an assumption reinforced by its close resemblance to the steppe climax (Clements, 1935). It is not improbable that the mixed hardwoods of the southern Appalachians bear a similar relation to the associations of the modern deciduous forest (Braun, 1935).

Tests of a climax

As has been previously indicated, the major climaxes of North America, such as tundra, boreal and deciduous forest, and prairie, stand forth clearly as distinct units, in spite of the fact that the prairie was first regarded as comprising two formations, as a consequence of the changes produced by overgrazing. The other coniferous and the scrub climaxes emerge less distinctly because of the greater similarity of life-form within each group, and hence it is necessary to appeal to criteria derived from the major formations just mentioned. This insures uniformity of basis and a high degree of objectivity, both of which are qualities of paramount importance for the natural classification of biomes. In fact, entire consistency in the application of criteria is the best warrant of objective results, though this is obviously a procedure that demands a first-hand acquaintance with most if not all the units concerned and over a large portion of their respective areas.

The primary criterion is that afforded by the vegetation-form, as is illustrated by the four major climaxes. The others of each group, such as coniferous forest or scrub, are characterized also by the same form in the dominants, but this is not decisive as between related climaxes and hence recourse must be taken to the other tests. The value of the life-form is most evident where two climaxes of different physiognomy are in contact, as in the case of the lake forest of pine-hemlock and the deciduous forest of hardwoods. The static view would make the hemlock in particular a dominant of the deciduous formation, but the evidence derived from the vegetation-form is supported by that of phylogeny and by early records of composition and timber-cut to show that two different climaxes are concerned. Secondary forms or subforms rarely if ever mark distinctions between climaxes, but do aid in the recognition of associations. This is well exemplified by the tall, mid and short grasses of the prairie and somewhat less definitely by the generally deciduous character of the Petran chaparral and the typically evergreen nature of the Sierran.

As would be expected, the most significant test of the unity of a formation is afforded by the presence of certain dominant species through all or nearly all of the associations, but often not in the role of dominants by reason of reduced abundance. Here again perhaps the best examples are furnished by prairie and tundra, though the rule applies almost equally well to deciduous forest and only less so to coniferous ones because of a usually smaller number of dominants. For the prairie, the number of such species, or *perdominants*, found in all or all but one or two of the five associations is eight, namely, *Stipa comata, Agropyrum smithi, Bouteloua gracilis, Sporobolus cryptandrus, Koeleria cristata, Elymus sitanion, Poa scabrella* and *Festuca ovina*. Even when a species is lacking over most of an association, as in the case of *Stipa comata* and the true prairie, it may be represented by a close relative, such as *S. spartea*, which is probably no more than a mesic variety of it. As to the three associations of the deciduous forest, a still larger number of dominant species occur to some degree in all; the oaks comprise *Quercus borealis, velutina, alba, macrocarpa, coccinea, muhlenbergi, stellata* and *marilandica*, and the hickories, *Carya ovata, glabra, alba* and *cordiformis*.

It was the application of this test by specific dominants that led to the recognition of the two climaxes in the coniferous mantle of the Petran and

Sierran cordilleras. The natural assumption was that such a narrow belt could not contain more than one climax, especially in view of its physiognomic uniformity, but this failed to reckon with the great climatic differences of the two portions and the corresponding response of the dominants. The effect of altitude proved to be much more decisive than that of region, dominants common to the montane and subalpine zones being practically absent, though the rule for the same zone in each of the two separate mountain systems. Long after the presence of two climaxes had been established, it was found that Sargent had anticipated this conclusion, though in other terms (1884, p. 8).

As would be inferred, the dominants of related associations belong to a few common genera for the most part. Thus, there are a dozen species of *Stipa* variously distributed as dominants through the grassland, nearly as many of *Sporobolus*, *Bouteloua* and *Aristida*, and several each of *Poa*, *Agropyrum*, *Elymus*, *Andropogon*, *Festuca* and *Muhlenbergia*. In the deciduous forest, *Quercus*, *Carya* and *Acer* are the great genera, and for the various coniferous ones, *Pinus*, *Abies* and *Picea*, with species of *Tsuga*, *Thuja*, *Larix* and *Juniperus* hardly less numerous.

The perennial forbs that play the part of subdominants also possess considerable value in linking associations together, and to a higher degree in the deciduous forest than in the prairie, owing chiefly to the factors of shade and protection. Over a hundred subdominants belonging to two score or more of genera, such as *Erythronium*, *Dicentra*, *Trillium*, *Aquilegia*, *Arisaema*, *Phlox*, *Uvularia*, *Viola*, *Impatiens*, *Desmodium*, *Helianthus*, *Aster* and *Solidago*, range from Nova Scotia or New England beyond the borders of the actual climax to Nebraska and Kansas. Across the wide expanse of the prairie climax, species in common are only exceptional, these few belonging mostly to the composites, notably *Grindelia squarrosa*, *Gutierrezia sarothrae*, *Artemisia dracunculus* and *vulgaris*. On the other hand, the number of genera of subdominants found throughout the grassland is very large.

The greater mobility of the larger mammals in particular renders animal influents less significant than plants as a criterion, but several of these possess definite value and the less mobile rodents even more. The antelope and bison are typical of the grassland climax, the first being practically restricted to it, while jack-rabbits, ground-squirrels and kangaroo-rats are characteristic dwellers in the prairie, as is their chief foe, the coyote.

The remaining criteria are derived from development directly or indirectly, though this is less evident in the case of the ecotone between two associations. Here the mixing of dominants and subdominants indicates their general similarity in terms of the formation, within which range their preferences assign them to different associations. The evidence from primary succession is of value only in the later stages as a rule, since initial associes like the reed-swamp may occur in several climaxes. With subseres, however, all or nearly all the stages are related to the particular climax and such seres denote a corresponding unity in development. This is especially true of all subclimaxes and most evidently in the case of those due to fire. More significant still are postclimaxes in both grassland and forest. For example, the associes of species of *Andropogon*, which is subclimax to the oak-hickory forest, constitutes a postclimax to five out of the six associations of the prairie. On the other hand, the community of *Ulmus*, *Juglans*, *Fraxinus*, etc., found on flood-plains through the region of deciduous forest, forms a common subclimax to the three associations.

In addition to such ontogenetic criteria, phylogeny supplies tests of even greater value. This is notably the case with the two associations of the montane and subalpine coniferous forests of the west, though perhaps the most striking application of this criterion is in connection with the lake forest of pine-hemlock. Though the concrete evidence for such a climax recurs constantly through the region of the Great Lakes to the Atlantic, it is fragmentary and there is no evident related association to the westward. However, the four genera are represented by related species in the two regions, namely, *Pinus strobus* by *P. monticola*, *P. banksiana* by *P. contorta*, *Tsuga canadensis* by *T. heterophylla*, *Larix laricina* by *L. occidentalis*, and *Thuja occidentalis* by *T. plicata*, though the last two genera have changed from a subclimax role in the east to a climax one in the west. As suggested earlier, phylogenetic evidence of still more direct nature is supplied by the mixed prairie with the other enclosing associations and by the remnants of a virgin deciduous forest that exhibits a similar genetic and spatial relation to the associations of this climax (cf. Braun, 1935).

Finally, it is clear that any test will gain in definiteness and accuracy of application whenever dependable records are available with respect to earlier composition and structure. These may belong entirely to the historical period, as in the case of scientific reports or land surveys, they may bridge the gap between the present and the past as with pollen statistics, or they may reach further back into the geological record, as with leaf-impressions or other fossils (Chaney, 1925, 1933; Clements, 1936). Two instances of the scientific record that are of the first importance may be given as examples. The first is the essential recognition by Sargent of the pine-hemlock climax under the name of the northern pine belt (1884), at a time when relatively little of this had been logged, by contrast with 90 per cent. or more at present (cf. also Bromley, 1935). The second is an account, discovered and communicated by Dr Vestal, of the prairies of Illinois as seen by Short in *ca.* 1840. This is of heightened interest since its discovery followed little more than a year after repeated field trips had led to the conclusion that all of Iowa, northern Missouri and most of Illinois were to be assigned to the true prairie,[1] a decision

[1] The true prairie is characterized by the three eudominants, *Stipa spartea*, *Sporobolus asper* and *S. heterolepis*. The presence of tall-grasses in it to-day, particularly *Andropogon furcatus* and *nutans*, is the mark of the disclimax due to the varied disturbances associated with settlement.

Phot. 1. Sierran subalpine climax: Consociation of *Tsuga mertensiana*; Crater Lake, Oregon.

Phot. 2. Proclimax of Sagebrush (*Artemisia tridentata*), a disclimax due to overgrazing of mixed prairie; climatically a preclimax.

confirmed for Illinois by Short's description, and supported by the more general accounts of Bradbury (*ca.* 1815) and Greeley (1860).

CLIMAX AND PROCLIMAX

Essential relations

In accordance with the view that development regularly terminates in the community capable of maintaining itself under a particular climate, except when disturbance enters, there is but one kind of climax, namely, that controlled by climate. This essential relation is regarded as not only inherent in all natural vegetation, but also as implicit in the cognate nature of the two terms. While it is fully recognized that succession may be halted in practically any stage, such communities are invariably subordinate to the true climax as determined by climate alone. From the very meaning of the word, there can not be climaxes scattered along the developmental route with a genuine climax at the end. There is no intention to question the reality of such pauses, but only to emphasize the fact that they are of a different order from the climax.

While it is natural to express new ideas by qualifying an old term, this does not conduce to the clearest thinking or the most accurate usage. Even more undesirable is the fact that the meaning of the original word is gradually shifted until it becomes either quite vague or hopelessly inclusive. At the hands of some, climax has already suffered this fate, and fire, disease, insects, and human disturbances of all sorts are assumed to produce corresponding climaxes (cf. Chapman, 1932). On such an assumption corn would constitute one climax, wheat another, and cotton a third, and it would then become imperative to begin anew the task of properly analyzing and classifying vegetation.

In the light of two decades of continued analysis of the vegetation of North America, as well as the application of the twin concepts of climax and complex organism by workers in other portions of the globe and the strong support brought to them by the rise of emergent evolution and holism (Phillips, 1935), the characterization of the climax as given in *Plant Succession,* in 1916, still appears to be both complete and accurate. "The unit of vegetation, the climax formation, is an organic entity. As an organism, the formation arises, grows, matures and dies. Its response to the habitat is shown in processes or functions and in structures that are the record as well as the result of these functions. Furthermore, each climax formation is able to reproduce itself, repeating with essential fidelity the stages of its development. The life-history of a formation is a complex but definite process, comparable in its chief features with the life-history of an individual plant. The climax formation is the adult organism, of which all initial and medial stages are but stages of development. . . . A formation, in short, is the final stage of vegetational development in a climatic unit. It is the climax community of a succession that terminates in the highest life-form possible in the climate concerned."

To-day this statement would need modification only to the extent of substituting "biome" for climax or formation and "biotic" for vegetational. This characterization has recently been annotated and confirmed by Phillips' masterly discussion of climax and complex organism, as cited above, a treatise that should be read and digested by everyone interested in the field of dynamic ecology and its wide applications.

Proclimaxes

As a general term, proclimax includes all the communities that simulate the climax to some extent in terms of stability or permanence but lack the proper sanction of the existing climate. Certain communities of this type were called potential climaxes in *Plant Succession* (p. 108; 1928, p. 109), and two kinds were distinguished, namely, preclimax and postclimax. To avoid proposing a new term in advance of its need, subclimax was made to do double duty, denoting both the subfinal stage of succession, as well as apparent climaxes of other kinds. This dual usage was criticized by Godwin (1929, p. 144) and partially justified by Tansley in an appended note on the ground just given. However, this discussion made it evident that a new term was desirable and proclimax was accordingly suggested (Clements, 1934). While this takes care of the use of subclimax in the second sense noted above, it is better adapted by reason of its significance to apply to all kinds of subpermanent communities other than the climax proper. However, there is still an important residuum after subclimax, preclimax and postclimax have been recognized, and it is proposed to call these *disclimaxes*, as indicated later.

The proclimax may be defined as any more or less permanent community resembling the climax in one or more respects, but gradually replaceable by the latter when the control of climate is not inhibited by disturbance. Besides its general function, it may be used as a synonym for any one of its divisions, as well as in cases of doubt pending further investigation, such as in water climaxes. The four types to be considered are subclimax, disclimax, preclimax and postclimax.

Subclimax

As the stage preceding the climax in all complete seres, primary and secondary, the subclimax is as universal as it is generally well understood. The great majority of such communities belong to the subsere, especially that following fire, owing to the fact that disturbance is to-day a practically constant feature of most climaxes. Fire and fallow are recurrent processes in cultivated regions generally and they serve to maintain the corresponding subsere until protection or conversion terminates the disturbance. Though the subclimax is just as regular a feature of priseres, these have long ago ended in the climax over most of the climatic area and the related subclimax communities are consequently much restricted in size and widely scattered. Smallness is

Phot. 3. Fire subclimax of dwarf *Pinus* and *Quercus*; the "Plains", New Jersey Pine-barrens.

Phot. 4. Subclimax of consocies of *Aristida purpurea* in field abandoned 15 years; Great Plains.

naturally a characteristic of nearly all subclimaxes, the chief exceptions being due to fire or to fire and logging combined, but by contrast they are often exceedingly numerous.

Because of its position in the succession, the subclimax resembles the preclimax in some respects and in a few instances either term may be properly applied. The distinction between subclimax and disclimax presents some difficulty now and then, as the amount of change necessary to produce the latter may be a matter of judgment. This arises in part also from the structural diversity of formation and association, as a consequence of which the dominants of a particular type of subsere vary in different areas. When there is but a single dominant, as in many burn subclimaxes, no question ensues, but if two or more are present, the decision between subclimax and disclimax may be less simple, as is not infrequent in scrub and grassland.

Examples of the subclimax are legion, the outstanding cases being mostly due to fire, alone or after lumbering or clearing. Most typical are those composed of "jack-pines" or species with closed cones that open most readily after fire. Each great region has at least one of these, e.g. *Pinus rigida, virginiana* and *echinata* in the east, *P. banksiana* in the north, *P. murrayana* in the Rocky Mountains, and *P. tuberculata, muricata* and *radiata* on the Pacific Slope. *Pinus palustris* and *taeda* play a similar role in the "piney" woods of the Atlantic Gulf region, as does *Pseudotsuga taxifolia* in the north-west. The characteristic subclimaxes of the boreal forest are composed of aspen (*Populus tremuloides*), balsam-poplar (*P. balsamifera*), and paper-birch (*Betula papyrifera*), either singly or in various combinations. Aspen also forms a notable subclimax in the Rocky Mountains, for the most part in the subalpine zone. Prisere subclimaxes are regular features of bogs and muskeags throughout much or all of the boreal and lake forests, the three dominants being *Larix laricina, Picea mariana* and *Thuja occidentalis,* often associated as zonal consocies. Where pines are absent in the region of the deciduous forest, two xeric oaks, *Quercus stellata* and *marilandica,* may constitute a subclimax, and this role is sometimes assumed by small trees, *Sassafras, Diospyrus* and *Hamamelis* being especially important.

Subclimaxes in the grassland are composed largely of tall-grasses, usually in the form of a consocies. In the true prairie, this part is taken by *Spartina cynosuroides* and in the desert plains by *Sporobolus wrighti,* while *Elymus condensatus* plays a similar role in the mixed prairie and in portions of the bunch-grass prairies. The function of the tall Andropogons is more varied; they are typically postclimax rather than subclimax, though they maintain the latter relation along the fringe of the oak-hickory forest and in oak openings. They occupy a similar position at the margin of the pine subclimax in Texas especially, and hence they are what might be termed "sub-subclimax" in such situations. Beyond the forest and in association with *Elionurus, Trachypogon,* etc., they appear to constitute a faciation of the coastal prairie. Chaparral proper is to be regarded as a climax, but with a change of species it extends into the montane and even the subalpine zone and there constitutes a fire subclimax. In the foothills of southern California, the coastal sagebrush behaves in like manner where it lies in contact with the chaparral.

The disposition of seral stages below the subclimax that exhibit a distinct retardation or halt for a longer or shorter period is a debatable matter. It is entirely possible to include them among subclimaxes, but this would again fail in accuracy and definiteness and hence lead to confusion. The decision may well be left to usage by providing a term for such seral or sub-subclimax communities as persist for a long or indefinite period because of continued or recurrent edaphic control or human disturbance. By virtue of its significance, brevity and accord with related terms, the designation "serclimax" is suggested, with the meaning of a seral community usually one or two stages before the subclimax, which persists for such a period as to resemble the climax in this one respect. For reasons of brevity and agreement, the connecting vowel is omitted, but the *e* remains long as in sere.

For the most part, serclimaxes are found in standing water or in saturated soils as a consequence of imperfect drainage. The universal example is the reed-swamp with one or more of several consocies, such as *Scirpus*, *Typha*, *Zizania*, *Phragmites* and *Glyceria*: this is typical of the lower reaches of rivers, of deltas and of certain kinds of lakes, the great tule swamps of California affording outstanding instances. Another type occurs in coastal marshes in which *Spartina* is often the sole or major dominant, while sedge-swamps have a wider climatic range but are especially characteristic of northern latitudes and high altitudes. The Everglades of Florida dominated by *Cladium* constitute perhaps the most extensive example of the general group, though *Carex* swamps often cover great areas and the grass *Arundinaria* forms jungle-like cane-brakes through the south. Among woody species, *Salix longifolia* is an omnipresent consocies of sand-bars and river-sides, but the most unique exemplar is the cypress-swamp of the south, typified by *Taxodium*. In boreal and subalpine districts the distinctive serclimax is the peat-bog, moor or muskeag, more or less regularly associated with other seral communities of *Carex* and usually of *Larix* or *Picea* also in the proper region.

Frequent burning may retard or prevent the development of the normal fire subclimax and cause it to be replaced by a preceding stage. This may be a scrub community or one kept in the shrub form by repeated fires, but along the Atlantic and Gulf Coasts it is usually one of *Andropogon virginicus*, owing to its sufferance of burning. The so-called "balds" of the southern Appalachians are seral communities of heaths or grasses initiated and maintained primarily by fire. Finally, there are serclimaxes of weeds, especially annuals, in cultivated districts, and a somewhat similar community of native annuals is characteristic of wide stretches in the desert region.

Phot. 5. Serclimax of *Taxodium, Nyssa* and *Quercus* forming a cypress swamp; Paris, Arkansas.

Phot. 6. Disclimax of *Bouteloua, Muhlenbergia* and *Opuntia*, due to overgrazing of mixed prairie, Great Plains.

Disclimax

As with the related concepts, the significance of this term is indicated by a prefix, *dis-*, denoting separation, unlikeness or derogation, much as in the Greek *dys*, poor, bad. The most frequent examples of this community result from the modification or replacement of the true climax, either as a whole or in part, or from a change in the direction of succession. These ensue chiefly in consequence of a disturbance by man or domesticated animals, but they are also occasionally produced by mass migration. In some cases, disturbance and the introduction of alien species act together through destruction and competition to constitute a quasi-permanent community with the general character of the climax. This type is best illustrated by the *Avena-Bromus* disclimax of California, which has all but completely replaced the bunch-grass prairie.[1] A similar replacement by *Bromus tectorum* has more recently taken place over lárge areas of the Great Basin, while *Poa pratensis* has during the last half-century steadily invaded the native hay-fields and pastures of the true prairie, an advance first noted by Bradbury in 1809. An even more striking phenomenon is the steadily increasing dominance of *Salsola* over range and crop land in the west, and this is imitated by *Sisymbrium* and *Lepidium* in the north-west. It is obvious that all cultivated crops belong in the same general category, but this point hardly requires consideration.

Probably the example most cited in North America is that of the short-grass plains, which actually represent a reduction of the mixed prairie due to overgrazing, supplemented by periodic drouth. Over most of this association, the mid-grasses, *Stipa*, *Agropyrum*, etc., are still in evidence, though often reduced in abundance and stature, but in some areas they have been practically eliminated. Similar though less extensive partial climaxes of short-grasses characterise pastures in the true and both pasture and range in the coastal prairie, the dominants regularly belonging to *Bouteloua*, *Buchloe* or *Hilaria*. Of essentially the same nature is the substitution of annual species of *Bouteloua*

[1] The grassland climax of North America comprises six well-marked associations (*Plant Indicators*, 1920; *Plant Ecology*, 1929). The mixed prairie, so-called because it is composed of both mid-grasses and short-grasses, is more or less central to the other five and is regarded as ancestral to them. To the east along the Missouri and Mississippi Rivers, it has become differentiated into the true prairie formed by other species of mid-grasses pertaining mostly to the same genera, and this unit is flanked along the western margin of the deciduous forest by a proclimax of tall grasses, chiefly *Andropogon*. Southward the true prairie is replaced by coastal prairie, which in the main occupies the Gulf region of Texas and Mexico and is constituted by similar dominants but of different species. The desert plains are characterized primarily by species of *Bouteloua* and *Aristida*, which range from western Texas to the edge of the deserts of Mexico and Arizona. In the north-west the short-grasses disappear and the Palouse prairie of eastern Washington and adjacent regions is formed by mid-grasses of the bunch-grass life-form, among which *Agropyrum spicatum* is the eudominant. The same life-form signalizes the California prairie, found from the northern part of the state southward into Lower California, but its especial character is derived from endemic species of *Stipa*. As indicated in the discussion, the short-grass plains, composed of *Bouteloua*, *Buchloe*, and *Carex*, are not climatic in nature, and this statement applies likewise to the tall-grass meadows of *Andropogon* mentioned above.

and *Aristida* in the desert plains for perennial ones of the same genera, which is a case of short-grasses being followed by still shorter ones.

In other instances, the effect of disturbance is to produce a community with the appearance of a postclimax, when the life-form concerned is that of an undershrub or tall grass. This is notably the case in the mixed prairie when overgrazing is carried to the point of breaking up the short-grass sod and permitting the dominance of *Artemisia frigida* or *Gutierrezia sarothrae*. In essence, the wide extension of sagebrush (*Artemisia tridentata*) and of creosote-bush (*Larrea tridentata*) is the same phenomenon, though each of these is a climax dominant in its own region. In the case of *Opuntia*, the peculiar life-form suggests an important difference, but the numerous species behave in all significant respects like other shrubs, though with the two advantages of spines and ready propagation.

The communities of tall-grasses formed by species of *Andropogon* originally presented some difficulty, since these naturally have all the appearance of a postclimax to the prairie. Probably the greater number are to be assigned to this type, but the evidence from reconnaissance and record indicates that in the true prairie and especially the eastern portion, *Andropogon furcatus* in particular now constitutes a disclimax due to pasturing, mowing and in some measure to fire also (Clements, 1933). A characteristic disclimax in miniature is to be found in the "gopher gardens" of the alpine tundra, where coaction and reaction have removed the climax dominants of sedges and grasses to make place for flower gardens of perennial forbs. "Towns" of prairie-dogs and kangaroo-rats often produce similar but much more extensive communities.

Selective cutting not infrequently initiates disclimaxes, as may likewise the similar action of other agents such as fire or epidemic disease. The most dramatic example is the elimination of the chestnut (*Castanea dentata*) from the oak-chestnut canopy, but of even greater importance has been the extreme reduction and fragmentation of the lake forest through the overcutting of white pine. Finally, what is essentially a disclimax may result from climatic mass migration, such as in the Black Hills of South Dakota has brought together *Pinus ponderosa* from the montane climax of the Rocky Mountains and *Picea canadensis* from the boreal forest.

Preclimax and postclimax

These related concepts were first advanced in *Plant Succession* (1916, 1928) and have since been discussed in *Plant Ecology* (1929) and in the organisation of the relict method (1934). They are both direct corollaries of the principle of the clisere, the spatial series of climaxes that are set in motion by a major climatic shift, such as that of the glacial epoch with its opposite phases. The clisere is most readily comprehended in the case of high ranges or summits, such as Pikes Peak where the entire series of climaxes is readily visible, and is what Tournefort described on Mount Ararat in his famous journey of 1700.

However, this is but an expression of the continental clisere in latitude, which achieves perhaps its greatest regularity in North America. A similar relation is characteristic of the longitudinal disposition of climaxes in the temperate zone between the two oceans, the portion from deciduous forest through prairie to desert being the most uniform.

With the exception of the two extremes, arctalpine and tropical, each climax has a dual role, being preclimax to the contiguous community of so-called higher life-form and postclimax to that of lower life-form. This may be illustrated by the woodland climax, which is postclimax to grassland and preclimax to montane forest. The arctic and alpine tundras exhibit only the preclimax relation, to boreal and subalpine forest respectively, since a potential lichen climax attains but incomplete expression northward or upward. While the general primary relation is one of water in terms of rainfall and evaporation, temperature constantly enters the situation and at the extremes may be largely controlling, as in the tundra especially. However, in our present imperfect knowledge of causal factors it is simpler and more definite to determine rank by position in the cliseral sequence, each community higher in altitude or latitude being successively preclimax to the preceding one. This relation is likewise entirely consistent in the clisere from deciduous forest to desert, as it is among the associations of the same climax, though in both these cases the zonal grouping may be more or less obscured.

Wherever concrete preclimaxes or postclimaxes occur, either between climaxes or within a single one, they are due to the compensation afforded by edaphic situations. The major examples of the latter are provided by valleys, especially gorges and canyons, long and steep slope-exposures, and by extreme soil-types such as sand and alkali. The seration is a series of communities produced by a graduated compensation across a valley and operating within a formation or through adjacent ones, while the ecocline embraces the differentiation brought about by shifting slope-exposures around a mountain or on the two sides of a high ridge. In the case of such soils as sand or gravel at one extreme and stiff clay at the other, the edaphic adjustment may sometimes appear contradictory. Thus, sand affords a haven for postclimax relicts in the dry prairie and for preclimax ones in the humid forest region, while the effect of heavy soils is just the reverse. However, this is readily intelligible when one recalls the peculiar properties of such soils in terms of absorption, chresard and evaporation (Clements, 1933).

Preclimax

Since they occupy the same general antecedent position with respect to the climax, it is necessary to distinguish with some care between subclimax and preclimax, especially in view of the fact that they often exhibit the same life-form. However, this is not difficult when the priseres and subseres have been investigated in detail, as the actual composition and behavior of the two communities are usually quite different. Moreover, in the first, reaction leads to the entry of the climax dominants with ultimate conversion, while in the second the compensation by local factors is rarely if ever to be overcome within the existing climate, short of man-made disturbance.

Preclimaxes are most clearly marked where two adjacent formations are concerned, either prairie and forest or desert and prairie. Examples of the first kind are found in the grassy "openings" and oak savannahs of the deciduous forest and in the so-called "natural parks" along the margin of the montane and boreal forests. They are also well developed on warm dry slope-exposures or xeroclines in the Rocky Mountains. In the one, compensation is usually afforded by a sandy or rocky soil, in the other by a local climate due to insolation. Desert climaxes regularly bear the proper relation to circumjacent grassland, but this is somewhat obscured by the shrub life-form, which would be expected to characterize the less xeric formation. This may be explained, however, by the wide capacity for adaptation shown by such major dominants as *Larrea tridentata* and *Artemisia tridentata*, a quality that is lacking in most of their associates. Left stranded as relict communities in desert plains and mixed prairie by the recession of the last dry phase, they have profited by the overgrazing of grasses to extend across a territory much larger than that in which they are climax. Here they have all the appearance of a postclimax, especially in the case of *Larrea*, which commonly attains a stature several times that found in the desert. However, since this is the direct outcome of disturbance in terms of grazing, it is better regarded as a disclimax, particularly since the climax grasses still persist in it to some degree.

Within the same formation, the more xeric associations or consociations are preclimax to the less xeric ones. This is the general relation between the oak-hickory and beech-maple associations of the deciduous forest, the former occupying in the latter the warmer drier sites produced by insolation or type of soil. A similar relation may obtain in the case of faciations, the *Quercus stellata-marilandica* community often being a border of marginal preclimax to the more mesic oak-hickory faciations. Such preclimaxes naturally persist beyond the limits of the association proper as relicts in valleys or sandy soils and then assume the role of postclimaxes to the surrounding grassland, a situation strikingly exemplified in the "Cross Timbers" of Texas. In the montane forest of the Rockies, the consociation of *Pinus ponderosa* is preclimax to that of *Pseudotsuga taxifolia*, and a similar condition recurs in all forests where there is more or less segregation of consociations.

In the mixed prairie, fragments of the desert plains occur all along the margin as preclimaxes, the most extensive one confronting the Colorado Valley, where it is at the same time postclimax to the desert. The mixed prairie constitutes relicts of this type where it meets the true prairie. The most frequent examples are provided by *Bouteloua gracilis* and *Sporobolus cryptandrus*, though as with all the short-grasses in this role, grazing has played some part.

Phot. 7. Postclimax of *Quercus, Juglans, Ulmus, Fraxinus*, etc., Canadian River, near Oklahoma City.

Phot. 8. Postclimax of tall-grasses, *Andropogon, Calamovilfa* and *Panicium* in sandhills; Thedford, Nebraska.

Postclimax

As a general rule, postclimax relicts are much more abundant than those that represent preclimaxes, owing in the first place to the secular trend toward desiccation in climate and in the second to the large number of valleys, sandhills and sandy plains, and escarpments in the grassland especially. Postclimaxes of oak-hickory and of their flood-plain associates, elm, ash, walnut, etc., are characteristic features of the true and mixed prairies, holding their own far westward in major valleys but limited as outliers on ridges and sandy stretches to the eastern edge. However, the compensation afforded by the last two is incomplete as a rule and the postclimax is typically reduced to the savannah type. The latter is an almost universal feature where forest, woodland or chaparral touches grassland, owing to the fact that shrinkage under slow desiccation operates gradually upon the density and size of individuals. Savannah is derived from the reduction of deciduous forest along the eastern edge of the prairie, of the aspen subclimax of the boreal forest along the northern, and of the montane pine consociation, woodland or chaparral on the western and southern borders, recurring again on the flanks of the Sierras and Coast Ranges in California. On the south, the unique ability of the mesquite (*Prosopis juliflora*) to produce root-sprouts after fire, its thorniness, palatable pods and resistant seeds have permitted it to produce an extensive savannah that often closely simulates a true woodland climax.

As would be expected, a point is reached in the reduction of rainfall westward in the prairie where sand no longer affords compensation adequate for trees. In general this is along the isohyet of 30 in. in the center and south, and of about 20 in. in the north. Southward from the parallel of 37° the further shrinkage of the oak savannah may be traced in the "shinry", which dwindles from four or five feet to dwarfs only "shin" high. With these are associated tall-grasses, principally *Andropogon* and *Calamovilfa* in the form *gigantea*. To the north of this line, the shin oaks are absent and the tall-grasses make a typical postclimax that extends into Canada, though the compensatory influence of sand is still sufficient to permit an abundance of such low bushes as *Amorpha*, *Ceanothus*, *Artemisia filifolia* and *Yucca*, as well as depauperate hackberry and aspen. In the vast sandhill area of central Nebraska, the tall-grass postclimax attains its best development, which is assumed to reflect the climate when the prairies were occupied by the Andropogons and their associates some millions of years ago. The gradual decrease to the rainfall of the present has led to the tall-grasses finding refuge in all areas of edaphic compensation, not only in sand but likewise on foothills and in valleys, and in addition along the front of the deciduous forest.

Structure of the climax

Community functions

The nature of community functions and their relation to the structure of climax and sere have been discussed in considerable detail elsewhere (*Plant Succession*, 1916, 1928; *Plant Ecology*, 1929; *Bio-ecology*, 1936), and for the present purpose it may well suffice to emphasize the difference in significance between major or primary and minor or secondary functions. The former comprise aggregation, migration, ecesis, reaction, competition, cooperation, disoperation, and coaction, together with the resulting complexes, invasion and succession. Any one of these may have a profound effect upon community structure, but the driving force in the selection and grouping of life-forms and species resides chiefly in reaction, competition, and coaction. Migration deals for the most part with the movement and evolution of units under climatic compulsion, and succession with the development and regeneration of the climax in bare or denuded areas.

In contrast to these stands the group of minor functions that are concerned with numbers and appearance or visibility as it may be termed. The first is annuation, in accordance with which the abundance of any species may fluctuate from dry to wet phases of the various climatic cycles or the growth differ in terms of prominence, the two effects not infrequently being combined. For the grassland, a season of rainfall more or less extreme in either direction often emphasizes one dominant at the expense of others, though the balance is usually redressed by the following year, while in the desert in particular the swing in number of annuals may be from almost complete absence to seasonal dominance, again with one or few species taking the major role. Aspection is mainly the orderly procession of societies through each growing season, more or less modified by changes in number ensuing from annuation. Hibernation and estivation merely affect seasonal appearance and are forms of aspection, with the temporary suspension of coaction effects. While usually applied to the animal members of the biome, it is obvious that plants exhibit certain responses of similar nature. Diurnation is likewise best known in the case of animals, especially nocturnal ones, but it is exhibited also by the vertical movement of plankton and in different form by the opening and closing of flowers and the "sleep" movements of leaves.

Roles of constituent species: dominants

The abundant and controlling species of characteristic life-form were long ago termed dominants (Clements, 1907, 1916), this property being chiefly determined by the degree of reaction and effective competition. In harmony with the concept of the biome, it has become desirable to consider the role of animals likewise; since their influence is seen chiefly in coaction by contrast to the reaction of plants, the term *influent* has been applied to the important species of land biomes (cf. Clements and Shelford, 1936). It is an axiom that the life-form of the dominant trees stamps its character upon forest and woodland, that of the shrub upon chaparral and desert, and the grass form on prairie, steppe and tundra. There are seral dominants as well as climax ones, and these give the respective impresses to the stages of prisere and subsere. Finally, there are considerable differences in rank or territory even among the dominants of each formation. The most important are those of wide range that bind together the associations of a climax; to these the term *perdominant* (*per*, throughout) may well be applied. In contrast to these stand the dominants more or less peculiar to each association, such as beech or chestnut in their respective communities and *Sporobolus asper* in the true and *Stipa comata* in the mixed prairie, for which *eudominant* may be employed.

Subdominants regularly belong to a life-form different from that of the dominants and are subject to the control of the latter in a high degree, as the name indicates. They are best exemplified by the perennial forbs, though biennials and annuals may serve as seral subdominants; all three may be actual dominants in the initial stages of succession and especially in the subsere. The term *codominant* has so far had no very definite status; it is hardly needed to call attention to the presence of two or more dominants, since this is the rule in all cases with the exception of consociation and consocies. In contrast to the types mentioned stands a large number of secondary or accessory species that exhibit no dominance, which may be conveniently referred to as *edominants*, pending more detailed analysis.

Influents

As indicated previously, the designation of *influent* is applied to the animal members of the biome by virtue of the influence or coaction they exert in the community. The significance of this effect depends much upon the life-form and to a large degree upon the size and abundance of the species as well, and is seen chiefly in the coactions involved in food, material, and shelter. Influents may be grouped in accordance with these properties, or they may be arranged with respect to distribution and role in climax or sere, or to time of appearance (Clements and Shelford, 1936). For general purposes it is perhaps most convenient to recognize subdivisions similar to those for dominants and with corresponding terms and significance. Thus, a *perfluent* would occur more or less throughout the formation, while the *eufluent* would be more or less typical or peculiar to an association. *Subfluents* would mark the next lower degree of importance, roughly comparable to that of subdominant, while minute or microscopic influents of still less significance might well be known as *vefluents*.

Climax and seral units

No adequate analysis of vegetation or of the biome is possible without taking full account of development. As the first step, this involves a distinction between climax communities proper and those that constitute the successional movement toward the final stages. The two groups differ in composition, stability, and type of control, but they agree in the possession of dominants, subdominants, and influents. These primary differences made it desirable to recognize two series of communities, viz. climax and seral and to propose corresponding terms, distinguished by the respective suffixes, *-ation* and *-ies* (Clements, 1916). These have gradually come into use as the feeling for dynamic ecology has grown and bid fair to constitute a permanent basis for all such studies. It is not supposed that they embrace all the units finally necessary for a complete system, but their constant application to the great climaxes of North America for nearly two decades indicates that they meet present needs in the matter of analysis.

Not all communities can be certainly placed in the proper category at the outset, but the number of doubtful cases is relatively small and few of these present serious difficulty under combined extensive and intensive research. This statement, however, presupposes an experience sufficiently wide and long to permit distinguishing between climaxes and the various types of proclimax, as well as recognizing the characteristic features of subclimaxes in particular. Comparative studies over a wide region are indispensable and the difficulties will disappear to the degree that this is achieved. While ecotones and mictia necessarily give rise to some questions in this connection, these in turn are resolved by investigations as extensive as they are detailed.

Climax units

In the organization of these, four types of descending rank and importance were distinguished within the formation, namely, association, consociation, society, and clan. Like the formation itself, the first two were based upon the dominant and its life-form, while the last were established upon the subdominant and its different life-form. It was recognized at the time that the association contained within itself other units formed by the dominants (cf. *Plant Indicators*, 1920, pp. 107, 276), and two further divisions, *faciation* and *lociation*, with corresponding seral ones, *facies* and *locies*, were suggested and submitted to Prof. Tansley for his opinion as to their desirability. These have been tested in the course of further field studies and have now and then been used in print, though the complete series was not published until 1932 (cf. Shelford, 1932). The climax group now comprises the following units, viz. association, consociation, faciation, lociation, society, and clan. At the beginning, it was intended to replace society by *sociation* for the sake of greater uniformity in terms, but the former had attained such usage that the idea was relinquished. However, the use of society in quite a different sense by students of social relations, especially among insects, again raises the question of the desirability of such a substitution, in view of the growing emphasis upon bioecology (cf. also Du Rietz, 1930; Rübel, 1930).

Phot. 9. Association of mixed prairie, *Stipa*, *Agropyrum*, *Bouteloua*, etc.: Monument, Colorado.

Phot. 10. Foothill faciation of the desert-plains association, *Bouteloua eriopoda*, *B. gracilis*, *B. hirsuta*, *B. filiformis*, etc.: Safford, Arizona.

Association.

Under the climax concept this represents the primary division of the biome or formation, and hence differs entirely from the generalized unit of the plant sociologists, for which the term *community* is to be preferred. Each biome consists regularly of two or more associations, though the lake forest and the desert scrub embody two apparent exceptions, each seeming to consist of one association only. However, these are readily explained by the fact that the western member of the former has been obscured by the expansion of montane and coast forests in the north-west, while one or more additional associations of the desert climax occur to the southward in Mexico, and apparently in South America also.

The number of associations in a particular formation is naturally determined by the number of primary differences and these in turn depend upon the presence of eudominants. Just as the unity of the formation rests upon the wide distribution of several major dominants or perdominants, so the association is also marked by one or more dominants peculiar to it, and often as well by differences in the rank and grouping of dominants held in common. Thus, in the true prairie association, the eudominants are *Stipa spartea*, *Sporobolus asper* and *heterolepis*; for the desert plains, *Bouteloua eriopoda*, *rothrocki* and *radicosa* and *Aristida californica*, while *Stipa comata* and *Buchloe* take a similar part in the mixed prairie. In the deciduous climax, the characteristic dominants of one association are supplied by the beech and hard maple, of a second by chestnut and chestnut-oak, though the oak-hickory association, of wider range and greater complexity, is comparatively poor in eudominants by contrast with the number of species.

The structural and phyletic relations of the associations of a climax are best illustrated by the grassland, which is the most highly differentiated of all North American formations, largely as an outcome of its great extent. The most extensive and varied unit is the mixed prairie, which occupies a generally median position with respect to the other five associations of this climax. Originally, it derived its dominants from three separate regions, *Stipa*, *Agropyrum* and *Koeleria* coming from Holarctica, *Sporobolus* from the south, and the short-grasses from the Mexican plateaux, and it still exhibits the closest kinship with the Eurasian steppe. It contains nearly all the genera that serve as dominants in the related associations, while many of the eudominants of these have all the appearance of direct derivatives from its species, as is shown by *Stipa*, *Sporobolus*, *Poa*, and *Agropyrum*. The evolution of both species and communities is evidently in response to the various subclimates, that of the true prairie being moister, of the coastal warmer as well; the desert plains are hotter and drier, the California prairie marked by winter rainfall and the Palouse by snowfall.

Consociation.

In its typical form the consociation is constituted by a single dominant, but as a matter of convenience the term is also applied to cases in which other dominants are but sparingly present and hence have no real share in the control of the community. It has likewise been convenient to refer in the abstract to each major dominant of the association as a consociation, though with the realization that it occurs more frequently in mixture than by itself. In this sense it may be considered a unit of the association, though the actual area of the latter is to be regarded as divided into definite faciations. Consociation dominants fall into a more or less regular series with respect to factor requirements, especially water content, and often exhibit zonation in consequence. This is a general feature of mixed prairie where *Agropyrum smithi* and *Stipa comata* are the chief mid-grasses, the former occupying swales and lower slopes, the latter upper slopes and ridges.

The consociation achieves definite expression over a considerable area only when the factors concerned fluctuate within the limits set by the requirements of the dominant or when the other dominants are not found in the region. The first case may be illustrated by *Pinus ponderosa* in the lower part of the montane forest and by *Adenostoma fasciculatum* in the Sierran chaparral, while the second is exemplified by *Picea engelmanni* in the Front Range of Colorado, its usual associate, *Abies lasiocarpa*, being absent from the district. In rolling terrain like that of the prairie, each consociation will recur constantly in the proper situation but is necessarily fragmentary in nature. Such behavior is characteristic of dominants with a postclimax tendency, as with *Stipa minor* and *Elymus condensatus* in swales and lower levels of the mixed prairie.

Faciation.

This is the concrete subdivision of the association, the entire area of the latter being made up of the various faciations, except for seral stages or fragments of the several consociations. Each faciation corresponds to a particular regional climate of real but smaller differences in rainfall/evaporation and temperature. It may be characterized by one or two eudominants, such as *Hilaria jamesi* and *Stipa pennata* in the southern mixed prairie, but more often it derives its individuality from a sorting out or a recombination of the dominants of the association. As is evident, the term is formed from the stem *fac-*, show, appear, as seen in *face* and *facies*, and the suffix *-ation*, which denotes a climax unit.

During the past decade, much attention has been given to the recognition and limitation of faciations on the basis of the presence or absence of a eudominant, such as *Hilaria*, *Buchloe*, or *Carex*, or a change in the rank or grouping of common dominants, like *Stipa*, *Agropyrum*, *Sporobolus* or *Bouteloua*. In the prairie this task has been complicated by overgrazing, cultivation and related disturbances, while selective lumbering and fire have added to the difficulties, in the deciduous forest especially. In general, temperature appears to play the leading part in the differentiation of faciations, since they usually fall into a sequence determined by latitude or altitude, though rainfall/evaporation is naturally concerned also. The mixed prairie exhibits the largest number, but it is approached in this respect by the deciduous forest as a consequence of wide extent and numerous dominants. Over the Great Plains from north to south, the successive faciations are *Stipa-Bouteloua*, *Bouteloua-Carex*, *Stipa-Agropyrum-Buchloe*, *Bouteloua-Buchloe*, *Hilaria-Stipa-Bouteloua*, and *Agropyrum-Bouteloua*. However, the short-grass communities are to be regarded as disclimaxes wherever the mid-grasses have been eliminated or nearly so, a condition that fluctuates in relation to dry and wet phases of the climatic cycle.

Lociation.

In its turn, the lociation is the subdivision of the faciation, the term being derived from *locus*, place, as indicating a general locality rather than a large region. Nevertheless, a lociation may occupy a relatively extensive territory up to a hundred miles or more in extent, by comparison with several hundred for the faciation. It is characterized by more or less local differences in the abundance and grouping of two or more dominants of the faciation. These correspond to considerable variations in soil, contour, slope-exposure or altitude, but all within the limits of the faciation concerned. As a consequence, lociations are very often fragmented, recurring here and there as alternes with each other, and frequently with proclimaxes of various types. Like most climax units, they have been modified by disturbance in some degree, and this fact must be constantly kept in mind in the task of distinguishing them from subclimax or disclimax.

A detailed knowledge of the faciation is prerequisite to the recognition of the various lociations in it. The number for a particular faciation naturally depends upon the extent of the latter and the number of dominants concerned. Consequently, lociations are more numerous in the faciations of the mixed prairie and desert plains, of the chaparral and the oak-hickory forest. As would be expected, they are often most distinct in ecotones and in districts where there is local intrusion of another dominant. In correspondence with their local character, it is important to eliminate or diminish superimposed differences through restoration of the original cover by means of protection enclosures and thus render it possible to disclose the true composition.

Society.

This term has had a wide range of application, but by dynamic ecologists it has generally been employed for various groupings of subdominants, of which those constituted by aspects or by layers are the most important. In addition, there is a host of minor communities formed by cryptogams in the ground layer or on host-plants and other matrices. The soil itself represents a major layer, divisible into more or less definite sublayers. Animals regularly assume roles of varying importance in all of these, especially the insects, arachnids and crustacea, and hence most if not all societies comprise both subdominant plants and subinfluent animals. It is doubtful whether animals form true societies independently of their food-plants or those used for materials or shelter, but this is a question that can be answered only after the simplest units, namely, family and colony, have been recognized and coordinated in terms of their coactions.

In view of what has been said previously, it seems desirable to employ society as the general term for all communities of subdominants and subinfluents above the rank of family and colony, much as community is the inclusive term for all groupings of whatsoever rank. This then permits carrying out the suggestion made two decades ago that the major types of societies be set apart by distinctive names. In accordance, it is here proposed to call the aspect society a *sociation* and the layer society a *lamiation*, while the corresponding seral terms would be *socies* and *lamies*. Many of the societies of cryptogams and minute animals would find their place in these, particularly so for those of the surface and soil layers, but many others take part in a miniature sere or *serule*, such as that of a moldering log, and may best receive designations that suggest this relation.

Sociation. Wherever societies are well developed, they regularly manifest a fairly definite seasonal sequence, producing what have long been known as aspects (Pound and Clements, 1898). As phenomena of the growing season, these were first distinguished as early spring or prevernal, vernal proper, estival, and serotinal or autumnal, but there may also be a hiemal aspect, especially for animals, in correspondence with an actual and not merely a calendar winter as in California.

Sociations are determined primarily by the relation between the life cycle of the subdominants and the seasonal march of direct factors, temperature in particular. So far as the matrix of plants is concerned, the constituent species may be in evidence throughout the season, but they give character to it only during the period of flowering, or fruiting in the case of cryptogams. They are present largely or wholly by sufferance of the dominants, and they are to be related to the reactions of these and competition among themselves rather more than to the habitat factors as such. In grassland and desert, they are often more striking than the dominants themselves, sometimes owing to stature but chiefly as an effect of color and abundance, and they may also attain much prominence in woods with the canopy not too dense.

Sociations are usually most conspicuous and best developed in grassland, four or even five distinct aspects occurring in the true prairie from early spring to autumn. In the mixed prairie these are usually reduced to three, and in the desert plains and desert proper, to two major ones, summer and winter, in which however there may be subaspects marked by *sations*, as indicated

Phot. 11. Sociation of *Erigeron* and *Psoralea*, estival aspect; Belmont Prairie, Lincoln, Nebraska.

Phot. 12. Lamiation of mid-herbs, *Laportea*, *Physostegia*, *Impatiens*, etc., in Oak-Hickory forest.

later. Short seasons due to increasing latitude or altitude afford less opportunity, and the tundra, both alpine and arctic, usually exhibits but two aspects, the sociations however taking a conspicuous role as in the prairie. In woodland the number and character of the sociations depend largely upon the nature of the canopy, and for deciduous forest the flowery sociations regularly belong to the spring and autumn aspects, when the foliage is either developing or disappearing.

It is convenient to distinguish sociations as simple or mixed with respect to the plant matrix in accordance with the presence of a single subdominant or of two or more. However, when animals are included in the grouping, such a distinction appears misleading and may well be dropped. The word "mixed" would be more properly applied to plant-animal societies were it not for the fact that this appears to be the universal condition. Since seasonal insects are legion, many of the societies in which they take part are best denoted as sations.

Lamiation. The term for a layer society is derived from the stem *lam-*, seen in lamina and lamella. As is well known, layers are best developed in forests with a canopy of medium density, so that under the most favorable conditions as many as five or six may be recognized above the soil. In such instances, there are usually two shrub stories, an upper and lower, often much interrupted, followed by tall, medium and low forb layers, and a ground community of mosses, lichens and other fungi, and usually some delicate annuals (cf. Hult, 1881; Grevillius, 1894). The soil population is perhaps best treated as a single unit, though it may exhibit more or less definite sublayers. When the various layers beneath the dominants are distinct, each is regarded as a lamiation, but in many cases only one or two are sufficiently organized to warrant designation, e.g. shrub or tall-herb lamiation.

Layers are often reduced to a single lamiation of low herbs in the climax forest, especially of conifers, and even this may be entirely lacking in dense chaparral. Two or three layers of forbs may be present in true prairie in particular, the upper lamiation being much the most definite and often concealing the grass dominants in the estival aspect, but the structure of grassland generally reflects the greater importance of sociations. Climaxes of sagebrush and desert scrub exhibit no proper lamiations, owing to the interval between the individual dominants, but the herbaceous societies of the interspaces show something of this nature.

As a rule, well-developed lamiations also manifest a seasonal rhythm, corresponding to aspects or to subaspects. These constitute recognizable groupings in the lamiation and for the sake of determination and analysis may be termed *sations*. This word is a doublet of season, both being derived from the root *sa-*, sow, hence grow or appear. Because of the frequent interplay of aspects and layers, the sation may for the present be employed for the subdivision of both sociation and lamiation, and especially where seasonal species of invertebrates play a conspicuous role.

Clan. This is a small community of subordinate importance but commonly of distinctive character. It is marked by a density that excludes all or nearly all competing species, in consequence of types of propagation that agree in the possession of short offshoots. Extension is usually by bulb, corm, tuber, stolon or short rootstock, each of which produces a more or less definite family grouping; in fact, most clans are families developed in the climax matrix and sometimes with a blurred outline in consequence. Clans of a particular species such as *Delphinium azureum* or *Solidago mollis* are dotted throughout the respective sociation, often in large numbers, and contribute a distinctive impress much beyond their abundance. Like all units, but small ones especially, they are subject to much fluctuation with the climatic cycle, as a result of which they may pass into societies or be formed by the shrinkage of the latter.

Seral units.

The concepts of dominance and subdominance apply to the sere as they do to the climax, as does that of influence also, and the corresponding sets of units bear the same general relation to each other. Each of the four major units is the developmental equivalent of a similar community in the climax series and this is likewise true of the various kinds of societies. They constitute the successive stages of each sere, both primary and secondary, including the subclimax, where they often achieve their best expression. It has also been customary to employ seral terms for preclimax and postclimax, and this appears to be the better usage for disclimax and proclimaxes in general. From the fragmentary nature of bare areas and suitable water bodies in particular, seral communities are often but partially developed and one or more units will be lacking in consequence. Thus, the reed-swamp associes is frequently represented by a single one of its several consocies and the minor units are even more commonly absent.

The associes is the major unit of every sere, the number being relatively large in the prisere and small in the subsere. The universal and best understood examples are those of the hydrosere, in which *Lemna, Potamogeton, Nuphar, Nymphaea, Nelumbo* and others form the consocies of the floating stage, and *Scirpus, Typha, Phragmites,* etc., are the dominants of the reed-swamp or amphibious associes. As already indicated, every consocies may occur singly and often does when the habitat offers just the proper conditions for it or the others have failed to reach the particular spot. When the ecial range is wider, various combinations of two or three dominants will appear, to constitute corresponding facies. Locies are less definitely marked as a rule, except in swamps of vast extent, but are to be recognized by the abundance of reed-like dominants of lower stature, belonging to other species of *Scirpus,* to *Heleocharis, Juncus,* etc. Both facies and locies seem to be better developed in sedge-swamp with its larger number of dominants, though the Everglades with the single consocies of *Cladium* form a striking exception.

The tree-swamps of the south-eastern United States contain a considerable number of consocies, such as *Taxodium distichum*, *Nyssa aquatica*, *biflora* and *ogeche*, *Carya aquatica*, *Planera aquatica*, *Persea palustris* and *borbonia*, *Magnolia virginiana*, *Fraxinus pauciflora*, *profunda* and *caroliniana*, and *Quercus nigra*. These are variously combined in several different facies, though a more detailed and exact study of the swamp sere may show the presence of two woody associes, distinguished by the depth or duration of the water. As with other scrub communities, the heath associes of peat-bog and muskeag comprises a large number of dominants and presents a corresponding wealth of facies and locies.

In the hydrosere of the deciduous forest, the typical subclimax is that of the flood-plain associes, composed of species of *Quercus*, *Ulmus*, *Fraxinus*, *Acer*, *Betula*, *Juglans*, *Celtis*, *Platanus*, *Liquidambar*, *Populus* and *Salix* for the most part. There are at least three well-marked facies, namely, northern, central and southern, each with a number of more or less distinct locies. The swamp associes or subclimax of the lake and boreal forests consists of *Larix laricina*, *Picea mariana* and *Thuja occidentalis*, occurring often as consocies but generally in the form of zoned facies. A large number of fire subclimaxes appear in the form of consocies, as with many of the pines, but associes are frequent along the Atlantic Coast, as they are in the boreal climax, where aspens and birches are chiefly concerned. The number of shrubs and small trees that play the part of seral dominants in the deciduous climax is much larger, producing not merely a wide range of associes but of facies and locies as well. More than a dozen genera and a score or so of species are involved, chief among them being *Sassafras*, *Diospyrus*, *Asimina*, *Hamamelis*, *Prunus*, *Ilex*, *Crataegus* and *Robinia*. The subclimax of the xerosere is constituted for the most part by species of *Quercus*, forming an eastern, a south-eastern and a western associes, the last with two well-marked facies, one of *stellata* and *marilandica*, the other of *macrocarpa* and *Carya ovata*.

Among postclimax associes, those of grassland and scrub possess a large number of dominants and exhibit a corresponding variety of facies and locies, together with fairly definite consocies. In the sandhills of Nebraska, the tall-grasses concerned are *Andropogon halli*, *furcatus* and *nutans*, *Calamovilfa longifolia*, *Eragrostis trichodes*, *Elymus canadensis* and *Panicum virgatum*; some of these drop out to the northward and others to the south, thus producing at least three regional facies. The mesquite-acacia associes of the south-west possesses a larger number of dominants and manifests a greater variety of facies through its wide area, and this is likewise true of the coastal sagebrush of California.

With reference to seral societies, it must suffice to point out that these are of necessity poorly developed in the initial stages of both hydrosere and xerosere, as the dominants are relatively few. Even in the reed-swamp, true layers are the exception, being largely restricted to such subdominants as

Alisma, Pontederia, Hydrocotyle and *Sagittaria,* which are found mostly in borders and intervals. However, in extensive subclimaxes and postclimaxes the situation is quite different. The tall-grass associes of sandhills is often quite as rich in saties and lamies as the true prairie, while the various subclimaxes of the several great forest types may equal the latter in the wealth of subdominants for each season and layer, the actual communities being very much the same.

Serule.

This term, a diminutive of *sere,* has been employed for a great variety of miniature successions that run their short but somewhat complex course within the control of a major community, especially the climax and subclimax. They resemble ordinary seres in arising in bare spots or on matrices of different sorts, such as earth, duff, litter, rocks, logs, cadavers, etc. Parasites and saprophytes play a prominent and often exclusive role in them, and plants and animals may alternate in the dominant parts. The organisms range from microscopic bacteria and worms to mites, larvae and imagoes on the one hand and large fleshy and shelf fungi on the other. The most important of these in terms of coaction and abundance are known as *dominules* (Clements and Shelford, 1936), with *subdominule* and *edominule* as terms for the two degrees of lesser importance. On the same model are formed *associule, consociule,* and *sociule* in general correspondence with the units of the sere itself. In addition there are families and colonies of these minute organisms, which are essentially similar to those of the initial stages of the major succession. Up to the present, little attention has been devoted to the development and structure of serules, but they are coming to receive adequate consideration in connection with bio-ecological problems. Many of the coactions, however, have long been the subject of detailed research in the conversion of organic materials.

Rank and correspondence of units

The following table exhibits the actual units of climax and sere, as well as their correspondence with each other. However, for the complete and accurate analysis of a great climax and especially the continental mass of vegetation, it is necessary to invoke other concepts, chiefly that of the proclimax and of communities mixed in space or in time. The several proclimaxes have been characterized (pp. 262–8), and the ecocline and seration briefly defined (p. 267). To these are to be added the *ecotone* and *mictium,* both terms of long standing, the former applied to the mixing of dominants between two units, the latter to the mixed community that intervenes between two seral stages or associes. Finally, there will be the several types of seres in all possible stages of development, the prisere in the form of hydrosere, xerosere, halosere or psammosere in regions less disturbed and a myriad of subseres in those long settled.

TABLE OF CLIMAX AND SERAL UNITS

Eoclimax
Panclimax

Climax (formation)

Climax	*Sere*
Association	Associes
Consociation	Consocies
Faciation	Facies
Lociation	Locies
Sociation	Socies
Lamiation	Lamies
Sation	Saties
Clan	Colony
	Family

Serule

Associule
Consociule
Sociule

As indicated previously, the word *community* is employed as a general term to designate any or all of the preceding units, while *society* may well be used to include those of the second division, i.e. sociation, etc. These are characterized by subdominants in contrast to the dominants that mark the first group. It has also been pointed out that the entire area of the association is divided into faciations and that the consociation is the relatively local expression of complete or nearly complete dominance on the part of a single species. The clan corresponds to the family as a rule, but in some cases resembles the colony in being formed by two species.

Families and colonies may also appear in climax communities, but this is regularly in connection with the serule.

Panclimax and eoclimax

The comprehensive treatment of these concepts is reserved for the succeeding paper in the present series, but it is desirable to characterize them meanwhile. The *panclimax* (παν, all, whole) comprises the two or more related climaxes or formations of the same general climatic features, the same life-form and common genera of dominants. The relationship is regarded as due to their origin from an ancestral climax or *eoclimax* (ἠώς, dawn), of Tertiary or even earlier time, as a consequence of continental emergence and climatic differentiation. In the past, eoclimaxes formed a series of great biotic zones in the northern hemisphere with the pole as a focus, and this zonal disposition or clisere is still largely evident in the arrangement of panclimaxes at the present. It is striking in the case of the arctic tundra and taiga or boreal forest, fairly evident for deciduous forest and prairie-steppe, and somewhat obscure for woodland and chaparral-macchia, while the position of deserts is largely determined by intervening mountain ranges. This is true likewise of grassland in some degree, and taken with the former broad land connection between North and South America explains why both prairie and desert panclimax contain at least one austral formation.

In the light of what has been said earlier, it is readily understood that panclimax and panformation are exact synonyms, as are eoclimax and eoformation. Panbiome and eobiome are the corresponding terms when the biotic community is taken as the basis for research.

Prerequisites to research in climaxes

It would be entirely superfluous to state that the major difficulty in the analysis of vegetation is its complexity, were it not for the fact that it is too often taken as the warrant for the static viewpoint. This was embodied in the original idea of the formation as a unit in which communities were assembled on a physiognomic basis, quite irrespective of generic composition and phyletic relationship. It is not strange that this view and its corollaries should have persisted long past its period of usefulness, since this is exactly what happened with the artificial system of Linnaeus, but the time has come to recognize fully that a natural system of communities must be built just as certainly upon development and consequent relationship as must that of plant families. Complexity is an argument for this rather than against it, and especially in view of the fact that the complexity discloses a definite pattern when the touchstone of development is applied to it.

Though the mosaic of vegetation may appear to be a veritable kaleidoscope in countries long occupied by man, the changes wrought upon it are readily intelligible in terms of the processes concerned. As emphasized previously, the primary control is that of climate, in a descending scale of units that correspond to formation, association, and faciation. Upon this general pattern are wrought the more circumscribed effects of physiography and soil, and both climatic and edaphic figures are overlaid and often more or less completely obscured with a veneer applied by disturbance of all possible kinds. Even above this may be discerned the effect, transient but nonetheless apparent, of such recurrent changes as annuation and aspection. Moreover, the orderly pattern of climate is complicated by great mountain ranges so that such climaxes as tundra and taiga occur far beyond their proper zone, and the effect is further varied by the relative position of the axis.

The migrations of climaxes in the past are a prolific source of fragmentary relicts, the interpretation of which is impossible except in terms of dynamics. This is likewise true of savannah, which represents the shrinkage of forest and scrub under a drying climate and is then usually further modified by fire or grazing. Fragmentation from this and other causes is characteristic of every diversified terrain and reaches its maximum when human utilization enters the scene upon a large scale. Somewhat similar in effect though not in process is the reduction of number of dominants by distance, with the consequence that an association of several may be converted into a consociation of one. Such a shrinkage naturally bears some relation also to climate and physiography, especially as seen in the glacial period, and finds its best illustration in the general poverty of dominants in the coniferous and deciduous climaxes of Europe, by contrast with those of eastern Asia and North America. A similar contrast obtains between the grassland of Asia and of North America, the latter being much richer in dominants, while South America approximates it closely in this respect.

On the part of the investigator, the difficulties in the way of an extensive and thoroughgoing study of climaxes are usually more serious. They arise partly from the handicap too often set by state or national boundaries and partly from the limitations of funds and time. They are also not unrelated to the fact that it is easiest to know a small district well and to assume that it reflects larger ones with much fidelity. As a consequence, it is impossible to lay too much stress upon the need for combining intensive and extensive methods in the research upon climaxes, insofar as their nature, limits and structure are concerned. The detailed development in terms of primary and secondary succession lends itself much more readily to local or regional investigation, but even here a wider perspective is essential to accurate generalization.

REFERENCES

Bradbury, J. "Travels in the Interior of North America, *ca.* 1815." In **Thwaites'** *Early Western Travels*, **5**, 1904.

Bromley, S. W. "The original forest types of southern New England." *Ecol. Mon.* **5**, 61–89, 1935,

Chaney, R. W. "A comparative study of the Bridge Creek flora and the modern redwood forest." *Publ. Carneg. Instn*, No. 349, 1925.

Chaney, R. W. and **E. I. Sanborn.** "The Goshen Flora of West-central Oregon." *Publ. Carneg. Instn*, No. 439, 1933.

Chapman, H. H. "Is the longleaf type a climax?" *Ecology*, **13**, 328–34, 1932.

Clements, F. E. *Plant Physiology and Ecology*, New York, 1907.

Clements, F. E. *Plant Succession*, Washington, 1916.

Clements, F. E. "Development and structure of the biome." *Ecol. Soc. Abs.* 1916.

Clements, F. E. *Plant Indicators*, Washington, 1920.

Clements, F. E. "Phylogeny and classification of climaxes." *Yearb. Carneg. Instn*, **24**, 334–5, 1925.

Clements, F. E. *Plant Succession and Indicators*, New York, 1928.

Clements, F. E. "The relict method in dynamic ecology." This JOURN. **22**, 39–68, 1934.

Clements, F. E. "Origin of the desert climate and climax in North America." This JOURN. **24**, 1936.

Clements, F. E. and **E. S. Clements.** "Climate and climax." *Yearb. Carneg. Instn*, **32**, 203, 1933.

Clements, F. E. and **V. E. Shelford.** *Bio-ecology*, 1936.

Du Rietz, G. E. "Classification and nomenclature of vegetation." *Svensk. Bot. Tid.* **24**, 489, 1930.

Godwin, H. "The subclimax and deflected succession." This JOURN. **17**, 144, 1929.

Greeley, H. *An Overland Journey to California in* 1859, New York, 1860.

Grevillius, A. Y. "Biologisch-physiologische Untersuchungen einiger Schwedischen Hainthälchen." *Bot. Z.* **52**, 147–68, 1894.

Hult, R. "Forsök til analytisk behandling af växformationerna." *Medd. Soc. Faun. Flor. Fenn.* **8**, 1881.

Hult, R. "Blekinges vegetation. Ett bidrag till växformationernas utvecklingshistorie." *Medd. Soc. Faun. Flor. Fenn.* **12**, 161, 1885. *Bot. Zbl.* **27**, 192, 1888.

Lewis, M. and **W. Clark.** Journal, 1803–1806. In **Thwaites'** *The Original Journals of the Lewis and Clark Expedition*, 1904–05.

Phillips, J. "The biotic community." This JOURN. **19**, 1–24, 1931.

Phillips, J. "Succession, development, the climax and the complex organism: an analysis of concepts. Part I." This Journ. **22**, 554–71, 1934.

Phillips, J. "Succession, development, the climax and the complex organism: an analysis of concepts. Part II." This Journ. **23**, 210–46, 1935.

Phillips, J. "Succession, development, the climax and the complex organism: an analysis of concepts. Part III." This Journ. **23**, 488–508, 1935.

Rübel, E. *Pflanzengesellschaften der Erde*, Bern-Berlin, 1930.

Sargent, C. S. *Report on the Forests of North America (exclusive of Mexico)*, Washington, 1884.

Seton, E. T. *Lives of Game Animals*, **3**, New York, 1929.

Shelford, V. E. "Basic principles of the classification of communities and habitats and the use of terms." *Ecology*, **13**, 105–20, 1932.

Short, C. W. "Observations on the botany of Illinois." *West. J. Med. Surg.* **3**, 185, 1845.

Tansley, A. G. "Editorial note." This Journ. **17**, 146–7, 1929.

Tansley, A. G. "The use and abuse of vegetational concepts and terms." *Ecology*, **16**, 284–307, 1935.

Tournefort, J. P. *Relation d'un Voyage du Levant*, Paris, 1717.

Weaver, J. E. and **F. E. Clements.** *Plant Ecology*, New York, 1929.

Weaver, J. E. and **E. L. Flory.** "Stability of climax prairie and some environmental changes resulting from breaking." *Ecology*, **15**, 333–47, 1934.

The individualistic concept of the plant association*

H. A. Gleason

The continued activity of European ecologists, and to a somewhat smaller extent of American ecologists as well, in discussing the fundamental nature, structure, and classification of plant associations, and their apparently chronic inability to come to any general agreement on these matters, make it evident that the last word has not yet been said on the subject. Indeed, the constant disagreement of ecologists, the readiness with which flaws are found by one in the proposals of another, and the wide range of opinions which have been ably presented by careful observers, lead one to the suspicion that possibly many of them are somewhat mistaken in their concepts, or are attacking the problem from the wrong angle.

It is not proposed to cite any of the extensive recent literature on these general subjects, since it is well known to all working ecologists. Neither is it necessary to single out particular contributions for special criticism, nor to point out what may appear to us as errors in methods or conclusions.

It is a fact, as Dr. W. S. Cooper has brought out so clearly in a manuscript which he has allowed me to read, and which will doubtless be in print before this, that the tendency of the human species is to crystallize and to classify his knowledge; to arrange it in pigeon-holes, if I may borrow Dr. Cooper's metaphor. As accumulation of knowledge continues, we eventually find facts that will not fit properly into any established pigeon-hole. This should at once be the sign that possibly our original arrangement of pigeon-holes was insufficient and should lead us to a careful examination of our accumulated data. Then we may conclude that we would better demolish our whole system of arrangement and classification and start anew with hope of better success.

Is it not possible that the study of synecology suffers at the present time from this sort of trouble? Is it not conceivable that, as the study of plant associations has progressed from its originally simple condition into its present highly organized and complex state, we have attempted to arrange all our facts in ac-

* Contributions from The New York Botanical Garden, No. 279.

cordance with older ideas, and have come as a result into a tangle of conflicting ideas and theories?

No one can doubt for a moment that there is a solid basis of fact on which to build our study of synecology, or that the study is well worth building. It is the duty of the botanist to translate into intelligible words the various phenomena of plant life, and there are few phenomena more apparent than those of their spatial relations. Plant associations exist; we can walk over them, we can measure their extent, we can describe their structure in terms of their component species, we can correlate them with their environment, we can frequently discover their past history and make inferences about their future. For more than a century a general progress in these features of synecology can be traced.

It has been, and still is, the duty of the plant ecologist to furnish clear and accurate descriptions of these plant communities, so that by them the nature of the world's vegetation may be understood. Whether such a description places its emphasis chiefly on the general appearance of the association, on a list of its component species, on its broader successional relations, or on its gross environment, or whether it enters into far greater detail by use of the quadrat method, statistical analysis,[1] or exact environometry, it nevertheless contributes in every case to the advancement of our understanding of each association in detail and of vegetation in all its aspects in general.

It is only natural that we should tend to depart from the various conclusions which we have reached by direct observation or experiment, and to attempt other more general deductions as well. So we invent special terms and methods for indicating the differences between associations and the variation of the plant life within a single community. We draw conclusions for ourselves, and attempt to lay down rules for others as to ways and means of defining single associations, by character species, by statistical studies, by environmental relations, or by successional history. We attempt to classify associations, as individual

[1] Pavillard has cast serious doubt on the efficiency of the statistical method in answering questions of synecology. His argument, based solely on European conditions, needs of course no reply from America, but it may properly be pointed out that the intimate knowledge of vegetational structure obtained in this way may easily lead to a much fuller appreciation of synecological structure, entirely aside from any merits of the actual statistical results.

examples of vegetation, into broader groups, again basing our methods on various observable features and arriving accordingly at various results. We even enter the domain of philosophy, and speculate on the fundamental nature of the association, regard it as the basic unit of vegetation, call it an organism, and compare different areas of the same sort of vegetation to a species.

The numerous conclusions in synecology which depend directly upon observation or experiment are in the vast majority of cases entirely dependable. Ecologists are trained to be accurate in their observations, and it is highly improbable that any have erred purposely in order to substantiate a conclusion not entirely supported by facts. But our various theories on the fundamental nature, definition, and classification of associations extend largely beyond the bounds of experiment and observation and represent merely abstract extrapolations of the ecologist's mind. They are not based on a pure and rigid logic, and suffer regularly from the vagaries and errors of human reason. A geneticist can base a whole system of evolution on his observations of a single species: ecologists are certainly equally gifted with imagination, and their theories are prone to surpass by far the extent warranted by observation.

Let us then throw aside for the moment all our pre-conceived ideas as to the definition, fundamental nature, structure, and classification of plant associations, and examine step by step some of the various facts pertinent to the subject which we actually know. It will not be necessary to illustrate them by reference to definite vegetational conditions, although a few instances will be cited merely to make our meaning clear. Other illustrations will doubtless occur to every reader from his own field experience.

We all readily grant that there are areas of vegetation, having a measurable extent, in each of which there is a high degree of structural uniformity throughout, so that any two small portions of one of them look reasonably alike. Such an area is a plant association, but different ecologists may disagree on a number of matters connected with such an apparently simple condition. More careful examination of one of these areas, especially when conducted by some statistical method, will show that the uniformity is only a matter of degree, and that two sample quadrats with precisely the same structure can scarcely be discovered.

Consequently an area of vegetation which one ecologist regards as a single association may by another be considered as a mosaic or mixture of several, depending on their individual differences in definition. Some of these variations in structure (if one takes the broader view of the association) or smaller associations (if one prefers the narrower view) may be correlated with differences in the environment. For example, the lichens on a tree-trunk enjoy a different environment from the adjacent herbs growing in the forest floor. A prostrate decaying log is covered with herbs which differ from the ground flora in species or in relative numbers of individuals of each species. A shallow depression in the forest, occupied by the same species of trees as the surroundings, may support several species of moisture-loving herbs in the lower stratum of vegetation. In other cases, the variations in vegetational structure may show no relation whatever to the environment, as in the case of a dense patch of some species which spreads by rhizomes and accordingly comes to dominate its own small area. The essential point is that precise structural uniformity[2] of vegetation does not exist, and that we have no general agreement of opinion as to how much variation may be permitted within the scope of a single association.

In our attempts to define the limits of the association, we have but two actually observable features which may be used as a basis, the environment and the vegetation. Logically enough, most ecologists prefer the latter, and have developed a system based on character-species. In northern latitudes, and particularly in glaciated regions, where most of this work has been done, there is a wide diversity in environment and a comparatively limited number of species in the flora. A single association is therefore occupied by few species, with large numbers of individuals of each, and it has not been difficult to select from most associations a set of species which are not only fairly common and abundant, but which are strictly limited to the one association. But in many parts of the tropics, where diversity of environment has been reduced to a minimum by the practical completion of most physiographic processes and by the long-continued cumulation of plant reactions, and where the flora is

[2] It has often occurred to the writer that much of the structural variation in an association would disappear if those taxonomic units which have the same vegetational form and behavior could be considered as a single ecological unit.

extraordinarily rich in species, such a procedure is impracticable or even impossible. Where a single hectare may contain a hundred species of trees, not one of which can be found in an adjacent hectare, where a hundred quadrats may never exhibit the same herbaceous species twice, it is obvious that the method of characteristic species is difficult or impracticable.

It is also apparent that different areas of what are generally called the same association do not always have precisely the same environment. A grove of *Pinus Strobus* on soil formed from decomposed rocks in the eastern states, a second on the loose glacial sands of northern Michigan, and a third on the sandstone cliffs of northern Illinois are certainly subject to different environmental conditions of soil. An association of prairie grass in Illinois and another in Nebraska undoubtedly have considerable differences in rainfall and available water. A cypress swamp in Indiana has a different temperature environment from one in Florida.

Two environments which are identical in regard to physiography and climate may be occupied by entirely different associations. It is perfectly possible to duplicate environments in the Andes of southern Chile and in the Cascade Mountains of Oregon, yet the plant life is entirely different. Duplicate environments may be found in the deserts of Australia and of Arizona, and again have an entirely different assemblage of species. Alpine summits have essentially the same environment at equal altitudes and latitudes throughout the world, apart from local variations in the component rock, and again have different floras. It seems apparent, then, that environment can not be used as a means of defining associations with any better success than the vegetation.

At the margin of an association, it comes in contact with another, and there is a transition line or belt between them. In many instances, particularly where there is an abrupt change in the environment, this transition line is very narrow and sharply defined, so that a single step may sometimes be sufficient to take the observer from one into the other. In other places, especially where there is a very gradual transition in the environment, there is a correspondingly wide transition in the vegetation. Examples of the latter condition are easily found in any arid mountain region. The oak forests of the southern Coast Range in California in many places descend upon the grass-covered foothills by a wide transition zone in which the trees become very gradually fewer and farther apart until they ultimately disappear completely. In Utah, it may be miles from the association of desert shrubs on the lower elevations across a mixture of shrubs and juniper before the pure stands of juniper are reached on the higher altitudes. It is obvious, therefore, that it is not always possible to define with accuracy the geographical boundaries of an association and that actual mixtures of associations occur.

Such transition zones, whether broad or narrow, are usually populated by species of the two associations concerned, but instances are not lacking of situations in which a number of species seem to colonize in the transition zone more freely than in either of the contiguous associations. Such is the case along the contact between prairie and forest, where many species of this type occur, probably because their optimum light requirements are better satisfied in the thin shade of the forest border than in the full sun of the prairie or the dense shade of the forest. Measured by component species such a transition zone rises almost to the dignity of an independent association.

Species of plants usually associated by an ecologist with a particular plant community are frequently found within many other types of vegetation. A single boulder, partly exposed above the ground at the foot of the Rocky Mountains in Colorado, in the short-grass prairie association, may be marked by a single plant of the mountain shrub *Cercocarpus*. In northern Michigan, scattered plants of the moisture-loving *Viburnum cassinoides* occur in the xerophytic upland thickets of birch and aspen. Every ecologist has seen these fragmentary associations, or instances of sporadic distribution, but they are generally passed by as negligible exceptions to what is considered a general rule.

There are always variations in vegetational structure from year to year within every plant association. This is exclusive of mere periodic variations from season to season, or aspects, caused by the periodicity of the component species. Slight differences in temperature or rainfall or other environmental factors may cause certain species to increase or decrease conspicuously in number of individuals, or others to vary in their vigor or luxuriance. Coville describes, in this connection, the remarkable variation in size of an *Amaranthus* in the Death Valley, which was three meters high in a year of abundant rainfall, and its progeny only a decimeter high in the following year of drought.

The duration of an association is in general limited. Sooner or later each plant community gives way to a different type of vegetation, constituting the phenomenon known as succession. The existence of an association may be short or long, just as its superficial extent may be great or small. And just as it is often difficult and sometimes impossible to locate satisfactorily the boundaries of an association in space, so is it frequently impossible to distinguish accurately the beginning or the end of an association in time. It is only at the center of the association, both geographical and historical, that its distinctive character is easily recognizable. Fortunately for ecology, it commonly happens that associations of long duration are also wide in extent. But there are others, mostly following fires or other unusual disturbances of the original vegetation, whose existence is so limited, whose disappearance follows so closely on their origin, that they scarcely seem to reach at any time a condition of stable equilibrium, and their treatment in any ecological study is difficult. The short-lived communities bear somewhat the same relation to time-distribution as the fragmentary associations bear to space-distribution. If our ecological terminology were not already nearly saturated, they might be termed ephemeral associations.

Now, when all these features of the plant community are considered, it seems that we are treading upon rather dangerous ground when we define an association as an area of uniform vegetation, or, in fact, when we attempt any definition of it. A community is frequently so heterogeneous as to lead observers to conflicting ideas as to its associational identity, its boundaries may be so poorly marked that they can not be located with any degree of accuracy, its origin and disappearance may be so gradual that its time-boundaries can not be located; small fragments of associations with only a small proportion of their normal components of species are often observed; the duration of a community may be so short that it fails to show a period of equilibrium in its structure.

A great deal has been said of the repetition of associations on different stations over a considerable area. This phenomenon is striking, indeed, and upon it depend our numerous attempts to classify associations into larger groups. In a region of numerous glacial lakes, as in parts of our northeastern states, we find lake after lake surrounded by apparently the same communities, each of them with essentially the same array of species in about the same numerical proportions. If an ecologist had crossed Illinois from east to west prior to civilization, he would have found each stream bordered by the same types of forest, various species of oaks and hickories on the upland, and ash, maple, and sycamore in the alluvial soil nearer the water. But even this idea, if carried too far afield, is found to be far from universal. If our study of glacial lakes is extended to a long series, stretching from Maine past the Great Lakes and far west into Saskatchewan, a very gradual but nevertheless apparent geographical diversity becomes evident, so that the westernmost and easternmost members of the series, while still containing some species in common, are so different floristically that they would scarcely be regarded as members of the same association. If one examines the forests of the alluvial floodplain of the Mississippi River in southeastern Minnesota, that of one mile seems to be precisely like that of the next. As the observer continues his studies farther down stream, additional species very gradually appear, and many of the original ones likewise very gradually disappear. In any short distance these differences are so minute as to be negligible, but they are cumulative and result in an almost complete change in the flora after several hundred miles.

No ecologist would refer the alluvial forests of the upper and lower Mississippi to the same associations, yet there is no place along their whole range where one can logically mark the boundary between them. One association merges gradually into the next without any apparent transition zone. Nor is it necessary to extend our observations over such a wide area to discover this spatial variation in ecological structure. I believe no one has ever doubted that the beech-maple forest of northern Michigan constitutes a single association-type. Yet every detached area of it exhibits easily discoverable floristic peculiarities, and even adjacent square miles of a single area differ notably among themselves, not in the broader features, to be sure, but in the details of floristic composition which a simple statistical analysis brings out. In other words, the local variation in structure of any association merges gradually into the broader geographical variation of the association-type.

This diversity in space is commonly overlooked by ecologists, most of whom of necessity limit their work to a comparatively small area, not extensive enough to indicate that the small observed floristic differences between associations may be of much significance or that this wide geographical variation is actually in operation. Yet it makes difficult the exact definition of any association-type, except as developed in a restricted locality, renders it almost impossible to select for study a typical or average example of a type, and in general introduces complexities into any attempt to classify plant associations.

What have we now as a basis for consideration in our attempts to define and classify associations? In the northeastern states, we can find many sharply marked communities, capable of fairly exact location on a map. But not all of that region can be thus divided into associations, and there are other regions where associations, if they exist at all in the ordinary sense of the word, are so vaguely defined that one does not know where their limits lie and can locate only arbitrary geographic boundaries. We know that associations vary internally from year to year, so that any definition we may make of a particular community, based on the most careful analysis of the vegetation, may be wrong next year. We know that the origin and disappearance of some are rapid, of others slow, but we do not always know whether a particular type of vegetation is really an association in itself or represents merely the slow transition stage between two others. We know that no two areas, supposed to represent the same association-type, are exactly the same, and we do not know which one to accept as typical and which to assume as showing the effects of geographical variation. We find fragmentary associations, and usually have no solid basis for deciding whether they are mere accidental intruders or embryonic stages in a developing association which may become typical after a lapse of years. We find variation of environment within the association, similar associations occupying different environments, and different associations in the same environment. It is small wonder that there is conflict and confusion in the definition and classification of plant communities. Surely our belief in the integrity of the association and the sanctity of the association-concept must be severely shaken. Are we not justified in coming to the general conclusion, far removed from the prevailing opinion, that an association is not an organism, scarcely even a vegetational unit, but merely a *coincidence?*

This question has been raised on what might well be termed negative evidence. It has been shown that the extraordinary variability of the areas termed associations interferes seriously with their description, their delimitation, and their classification. Can we find some more positive evidence to substantiate the same idea? To do this, we must revert to the individualistic concept of the development of plant communities, as suggested by me in an earlier paper.[3]

As a basis for the presentation of the individualistic concept of the plant association, the reader may assume for illustration any plant of his acquaintance, growing in any sort of environment or location. During its life it produces one or more crops of seeds, either unaided or with the assistance of another plant in pollination. These seeds are endowed with some means of migration by which they ultimately come to rest on the ground at a distance from the parent plant. Some seeds are poorly fitted for migration and normally travel but a short distance; others are better adapted and may cover a long distance before coming to rest. All species of plants occasionally profit by accidental means of dispersal, by means of which they traverse

[3] I may frankly admit that my earlier ideas of the plant association were by no means similar to the concept here discussed. Ideas are subject to modification and change as additional facts accumulate and the observer's geographical experience is broadened. An inkling of the effect of migration on the plant community appeared as early as 1903 and 1904 (Bull. Illinois State Lab. Nat. Hist. **7:** 189.) My field work of 1908 covered a single general type of environment over a wide area, and was responsible for still more of my present opinions (Bull. Illinois State Lab. Nat. Hist. **9:** 35–42). Thus we find such statements as the following: "No two areas of vegetation are exactly similar, either in species, the relative numbers of individuals of each, or their spatial arrangement" (l. c. 37), and again: "The more widely the different areas of an association are separated, the greater are the floral discrepancies. . . . Many of these are the results of selective migration from neighboring associations, so that a variation in the general nature of the vegetation of an area affects the specific structure of each association" (l. c. 41). Still further experience led to my summary of vegetational structure in 1917 (Bull. Torrey Club **44:** 463–481), and the careful quantitative study of certain associations from 1911 to 1923 produced the unexpected information that the distribution of species and individuals within a community followed the mathematical laws of probability and chance (Ecology **6:** 66–74).

distances far in excess of their average journey. Sometimes these longer trips may be of such a nature that the seed is rendered incapable of germination, as in dispersal by currents of salt water, but in many cases they will remain viable. A majority of the seeds reach their final stopping-point not far from the parent, comparatively speaking, and only progressively smaller numbers of them are distributed over a wider circle. The actual number of seeds produced is generally large, or a small number may be compensated by repeated crops in successive years. The actual methods of dispersal are too well known to demand attention at this place.

For the growth of these seeds a certain environment is necessary. They will germinate between folds of paper, if given the proper conditions of light, moisture, oxygen, and heat. They will germinate in the soil if they find a favorable environment, irrespective of its geographical location or the nature of the surrounding vegetation. Herein we find the *crux* of the question. The plant individual shows no physiological response to geographical location or to surrounding vegetation *per se*, but is limited to a particular complex of environmental conditions, which may be correlated with location, or controlled, modified, or supplied by vegetation. If a viable seed migrates to a suitable environment, it germinates. If the environment remains favorable, the young plants will come to maturity, bear seeds in their turn, and serve as further centers of distribution for the species. Seeds which fall in unfavorable environments do not germinate, eventually lose their viability and their history closes.

As a result of this constant seed-migration, every plant association is regularly sowed with seeds of numerous extra-limital species, as well as with seeds of its own normal plant population. The latter will be in the majority, since most seeds fall close to the parent plant. The seeds of extra-limital species will be most numerous near the margin of the association, where they have the advantage of proximity to their parent plants. Smaller numbers of fewer species will be scattered throughout the association, the actual number depending on the distance to be covered, and the species represented depending on their means of migration, including the various accidents of dispersal. This thesis needs no argument in its support. The practical universality of seed dispersal is known to every botanist as a matter of common experience.

An exact physiological analysis of the various species in a single association would certainly show that their optimal environments are not precisely identical, but cover a considerable range. At the same time, the available environment tends to fluctuate from year to year with the annual variations of climate and with the accumulated reactionary effects of the plant population. The average environment may be near the optimum for some species, near the physiological limit of others, and for a third group may occasionally lie completely outside the necessary requirements. In the latter case there will result a group of evanescent species, variable in number and kind, depending on the accidents of dispersal, which may occasionally be found in the association and then be missing for a period of years. This has already been suggested by the writer as a probable explanation of certain phenomena of plant life on mountains, and was also clearly demonstrated by Dodds, Ramaley, and Robbins in their studies of vegetation in Colorado. In the first and second cases, the effect of environmental variation toward or away from the optimum will be reflected in the number of individual plants and their general luxuriance. On the other hand, those species which are limited to a single type of plant association must find in that and in that only the environmental conditions necessary to their life, since they have certainly dispersed their seeds many times into other communities, or else be so far removed from other associations of similar environment that their migration thence is impossible.

Nor are plants in general, apart from these few restricted species, limited to a very narrow range of environmental demands. Probably those species which are parasitic or which require the presence of a certain soil-organism for their successful germination and growth are the most highly restricted, but for the same reason they are generally among the rarest and most localized in their range. Most plants can and do endure a considerable range in their environment.

With the continuance of this dispersal of seeds over a period of years, every plant association tends to contain every species of the vicinity which can grow in the available environment. Once a species is established, even by a single seed-bearing plant, its further spread through the association is hastened, since it no longer needs to depend on a long or accidental migration, and this spread is continued until the species is eventually distributed throughout the area of the association. In general, it may be considered that, other things being equal, those species of wide extent through an association are those of early introduction which have had ample time to complete their spread, while those of localized or sporadic distribution are the recent arrivals which have not yet become completely established.

This individualistic standpoint therefore furnishes us with an explanation of several of the difficulties which confront us in our attempts to diagnose or classify associations. Heterogeneity in the structure of an association may be explained by the accidents of seed dispersal and by the lack of time for complete establishment. Minor differences between neighboring associations of the same general type may be due to irregularities in immigration and minor variations in environment. Geographical variation in the floristics of an association depends not alone on the geographical variation of the environment, but also on differences in the surrounding floras, which furnish the immigrants into the association. Two widely distant but essentially similar environments have different plant associations because of the completely different plant population from which immigrants may be drawn.

But it must be noted that an appreciation of these conditions still leaves us unable to recognize any one example of an association-type as the normal or typical. Every association of the same general type has come into existence and had its structure determined by the same sort of causes; each is independent of the other, except as it has derived immigrants from the other; each is fully entitled to be recognized as an association and there is no more reason for regarding one as typical than another. Neither are we given any method for the classification of associations into any broader groups.

Similar conditions obtain for the development of vegetation in a new habitat. Let us assume a dozen miniature dunes, heaped up behind fragments of driftwood on the shore of Lake Michigan. Seeds are heaped up with the sand by the same propelling power of the wind, but they are never very numerous and usually of various species. Some of them germinate, and the dozen embryonic dunes may thenceforth be held by as many different species of plants. Originally the environment of the dunes was identical and their floristic difference is due solely to the chances of seed dispersal. As soon as the plants have developed, the environment is subject to the modifying action of the plant, and small differences between the different dunes appear. These are so slight that they are evidenced more by the size and shape of the dune than by its flora, but nevertheless they exist. Additional species gradually appear, but that is a slow process, involving not only the chance migration of the seed to the exact spot but also its covering upon arrival. It is not strange that individuals are few and that species vary from one dune to another, and it is not until much later in the history of each dune, when the ground cover has become so dense that it affects conditions of light and soil moisture, and when decaying vegetable matter is adding humus to the sand in appreciable quantities, that a true selective action of the environment becomes possible. After that time permanent differences in the vegetation may appear, but the early stages of dune communities are due to chance alone. Under such circumstances, how can an ecologist select character species or how can he define the boundaries of an association? As a matter of fact, in such a location the association, in the ordinary sense of the term, scarcely exists.

Assume again a series of artificial excavations in an agricultural region, deep enough to catch and retain water for most or all of the summer, but considerably removed from the nearest areas of natural aquatic vegetation. Annually the surrounding fields have been ineffectively planted with seeds of *Typha* and other wind-distributed hydrophytes, and in some of the new pools *Typha* seeds germinate at once. Water-loving birds bring various species to other pools. Various sorts of accidents conspire to the planting of all of them. The result is that certain pools soon have a vegetation of *Typha latifolia*, others of *Typha angustifolia*, others of *Scirpus validus;* plants of *Iris versicolor* appear in one, of *Sagittaria* in another, of *Alisma* in a third, of *Juncus effusus* in a fourth. Only the chances of seed dispersal have determined the allocation of species to different pools, but in the course of three or four years each pool has a different appearance, although the environment, aside from the reaction of the various species, is precisely the same for each. Are we dealing here with several different associations, or with a single association, or with merely embryonic stages of some future association? Under our view, these become merely academic questions, and any answer which may be suggested is equally academic.

But it must again be emphasized that these small areas of vegetation are component parts of the vegetative mantle of the land, and as such are fully worthy of description, of discussion, and of inquiry into the causes which have produced them and into their probable future. It must be emphasized that in citing the foregoing examples, the existence of associations or of successions is not denied, and that the purpose of the two paragraphs is to point out the fact that such communities introduce many difficulties into any attempt to define or classify association-types and successional series.

A plant association therefore, using the term in its ordinarily accepted meaning, represents the result of an environmental sorting of a population, but there are other communities which have existed such a short time that a reasonably large population has not yet been available for sorting.

Let us consider next the relation of migration and environmental selection to succession. We realize that all habitats are marked by continuous environmental fluctuation, accompanied or followed by a resulting vegetational fluctuation, but, in the common usage of the term, this is hardly to be regarded as an example of succession. But if the environmental change proceeds steadily and progressively in one direction, the vegetation ultimately shows a permanent change. Old species find it increasingly difficult or impossible to reproduce, as the environment approaches and finally passes their physiological demands. Some of the migrants find establishment progressively easier, as the environment passes the limit and approaches the optimum of their requirements. These are represented by more and more individuals, until they finally become the most conspicuous element of the association, and we say that a second stage of a successional series has been reached.

It has sometimes been assumed that the various stages in a successional series follow each other in a regular and fixed sequence, but that is frequently not the case. The next vegetation will depend entirely on the nature of the immigration which takes place in the particular period when environmental change reaches the critical stage. Who can predict the future for any one of the little ponds considered above? In one, as the bottom silts up, the chance migration of willow seeds will produce a willow thicket, in a second a thicket of *Cephalanthus* may develop, while a third, which happens to get no shrubby immigrants, may be converted into a miniature meadow of *Calamagrostis canadensis*. A glance at the diagram of observed successions in the Beach Area, Illinois, as published by Gates, will show at once how extraordinarily complicated the matter may become, and how far vegetation may fail to follow simple, pre-supposed successional series.

It is a fact, of course, that adjacent vegetation, because of its mere proximity, has the best chance in migration, and it is equally true that in many cases the tendency is for an environment, during its process of change, to approximate the conditions of adjacent areas. Such an environmental change becomes effective at the margin of an association, and we have as a result the apparent advance of one association upon another, so that their present distribution in space portrays their succession in time. The conspicuousness of this phenomenon has probably been the cause of the undue emphasis laid on the idea of successional series. But even here the individualistic nature of succession is often apparent. Commonly the vegetation of the advancing edge differs from that of the older established portion of the association in the numerical proportion of individuals of the component species due to the sorting of immigrants by an environment which has not yet reached the optimum, and, when the rate of succession is very rapid, the pioneer species are frequently limited to those of the greatest mobility. It also happens that the change in environment may become effective throughout the whole area of the association simultaneously, or may begin somewhere near the center. In such cases the pioneers of the succeeding association are dependent on their high mobility or on accidental dispersal, as well as environmental selection.

It is well known that the duration of the different stages in succession varies greatly. Some are superseded in a very short time, others persist for long or even indefinite periods. This again introduces difficulties into any scheme for defining and classifying associations.

A forest of beech and maple in northern Michigan is lumbered, and as a result of exposure to light and wind most of the usual herbaceous species also die. Brush fires sweep over the clearing and aid in the destruction of the original vegetation. Very soon the area grows up to a tangle of other herbaceous and shrubby species, notably *Epilobium angustifolium*, *Rubus strigosus*, and *Sambucus racemosa*. This persists but a few years before it is overtopped by saplings of the original hardwoods which eventually restore the forest. Is this early stage of fire-weeds and shrubs a distinct association or merely an embryonic phase of the forest? Since it has such a short duration, it is frequently regarded as the latter, but since it is caused by an entirely different type of environmental sorting and lacks most of the characteristic species of the forest, it might as well be called distinct. If it lasted for a long period of years it would certainly be called an association, and if all the forest near enough to provide seeds for immigration were lumbered, that might be the case. Again we are confronted with a purely arbitrary decision as to the associational identity of the vegetation.

Similarly, in the broad transition zone between the oak-covered mountains and the grass-covered foothills in the Coast Range of California, we are forced to deal arbitrarily in any matter of classification. Shall we call such a zone a mere transition, describe the forests above and the grasslands below and neglect the transition as a mere mixture? Or shall we regard it as a successional or time transition, evidencing the advance of the grasslands up the mountain or of the oaks down toward the foothills? If we choose the latter, we must decide whether the future trend of rainfall is to increase, thereby bringing the oaks to lower elevations, or to decrease, thereby encouraging the grasslands to grow at higher altitudes. If we adopt the former alternative, we either neglect or do a scientific injustice to a great strip of vegetation, in which numerous species are "associated" just as surely as in any recognized plant association.

The sole conclusion we can draw from all the foregoing considerations is that the vegetation of an area is merely the resultant of two factors, the fluctuating and fortuitous immigration of plants and an equally fluctuating and variable environment. As a result, there is no inherent reason why any two areas of the earth's surface should bear precisely the same vegetation, nor any reason for adhering to our old ideas of the definiteness and distinctness of plant associations. As a matter of fact, no two areas of the earth's surface do bear precisely the same vegetation, except as a matter of chance, and that chance may be broken in another year by a continuance of the same variable migration and fluctuating environment which produced it. Again, experience has shown that it is impossible for ecologists to agree on the scope of the plant association or on the method of classifying plant communities. Furthermore, it seems that the vegetation of a region is not capable of complete segregation into definite communities, but that there is a considerable development of vegetational mixtures.

Why then should there be any representation at all of these characteristic areas of relatively similar vegetation which are generally recognized by plant ecologists under the name of associations, the existence of which is indisputable as shown by our field studies in many parts of the world, and whose frequent repetition in similar areas of the same general region has led us to attempt their classification into vegetational groups of superior rank?

It has been shown that vegetation is the resultant of migration and environmental selection. In any general region there is a large flora and it has furnished migrating seeds for all parts of the region alike. Every environment has therefore had, in general, similar material of species for the sorting process. Environments are determined principally by climate and soil, and are altered by climatic changes, physiographic processes, and reaction of the plant population. Essentially the same environments are repeated in the same region, their selective action upon the plant immigrants leads to an essentially similar flora in each, and a similar flora produces similar reactions. These conditions produce the well known phenomena of plant associations of recognizable extent and their repetition with great fidelity in many areas of the same region, but they also produce the variable vegetation of our sand dunes and small pools, the fragmentary associations of areas of small size, and the broad transition zones where different types of vegetation are mixed. Climatic changes are always slow, physiographic processes frequently reach stages where further change is greatly retarded, and the accumulated effects of plant reaction often reach a condition beyond which they have relatively little effect on plant life. All of these conspire to give to certain areas a comparatively uniform environment for a considerable period of time, during which continued migration of plants leads to a smoothing out of original vegetational differences and to the establishment of a relatively uniform and static vegetational structure. But other physiographic processes are rapid and soon develop an entirely different environment, and some plant reactions are rapid in their operation and profound in their effects. These lead to the short duration of some plant communities, to the development, through the prevention of complete migration by lack of sufficient time, of associations of few species and of different species in the same environment, and to mixtures of vegetation which seem to baffle all attempts to resolve them into distinct associations.

Under the usual concept, the plant association is an area of vegetation in which spatial extent, describable structure, and distinctness from other areas are the essential features. Under extensions of this concept it has been regarded as a unit of vegetation, signifying or implying that vegetation in general is composed of a multiplicity of such units, as an individual representation of a general group, bearing a general similarity to the relation of an individual to a species, or even as an organism, which is merely a more striking manner of expressing its unit nature and uniformity of structure. In every case spatial extent is an indispensable part of the definition. Under the individualistic concept, the fundamental idea is neither extent, unit character, permanence, nor definiteness of structure. It is rather the visible expression, through the juxtaposition of individuals, of the same or different species and either with or without mutual influence, of the result of causes in continuous operation. These primary causes, migration and environmental selection, operate independently on each area, no matter how small, and have no relation to the process on any other area. Nor are they related to the vegetation of any other area, except as the latter may serve as a source of migrants or control the environment of the former. The effect of these primary causes is therefore not to produce large areas of similar vegetation, but to determine the plant life on every minimum area. The recurrence of a similar juxtaposition over tracts of measurable extent, producing an association in the ordinary use of the term, is due to a similarity in the contributing causes over the whole area involved.

Where one or both of the primary causes changes abruptly, sharply delimited areas of vegetation ensue. Since such a condition is of common occurrence, the distinctness of associations is in many regions obvious, and has led first to the recognition of communities and later to their common acceptance as vegetational units. Where the variation of the causes is gradual, the apparent distinctness of associations is lost. The continuation in time of these primary causes unchanged produces associational stability, and the alteration of either or both leads to succession. If the nature and sequence of these changes are identical for all the associations of one general type (although they need not be synchronous), similar successions ensue, producing successional series. Climax vegetation represents a stage at which effective changes have ceased, although their resumption at any future time may again initiate a new series of successions.

In conclusion, it may be said that every species of plant is a law unto itself, the distribution of which in space depends upon its individual peculiarities of migration and environmental requirements. Its disseminules migrate everywhere, and grow wherever they find favorable conditions. The species disappears from areas where the environment is no longer endurable. It grows in company with any other species of similar environmental requirements, irrespective of their normal associational affiliations. The behavior of the plant offers in itself no reason at all for the segregation of definite communities. Plant associations, the most conspicuous illustration of the space relation of plants, depend solely on the coincidence of environmental selection and migration over an area of recognizable extent and usually for a time of considerable duration. A rigid definition of the scope or extent of the association is impossible, and a logical classification of associations into larger groups, or into successional series, has not yet been achieved.

The writer expresses his thanks to Dr. W. S. Cooper, Dr. Frank C. Gates, Major Barrington Moore, Mr. Norman Taylor, and Dr. A. G. Vestal for kindly criticism and suggestion during the preparation of this paper.

into the body of the partly eaten bantam and replaced it in the same spot where he found it. Next morning the seemingly impossible was made a practical certainty, for he found the body of a screech owl with the claws of one foot firmly imbedded in the body of the bantam. He very kindly presented me with the owl which, upon dissection, proved to be a female, its stomach containing a very considerable amount of bantam flesh and feathers, together with a great deal of wheat. (It seems probable that the wheat was accidentally swallowed with the crop of the bantam during the feast, but there was so much that it seems strange the owl did not discard it while eating). How a bird only 9.12 inches in length could have dealt out such havoc in so short a time is almost incredible, but, although purely circumstantial, the evidence against the owl appeared altogether too strong for even a reasonable doubt. The doctor and I wished to make as certain as possible, however, so the poisoned bantam was replaced and left for several days, but without any further results. For the above mentioned reasons I am rather doubtful as to the net value of this owl from an economic standpoint, although birds in a wild state would not give them such opportunities for such wanton killing as birds enclosed in pens.

THE NICHE-RELATIONSHIPS OF THE CALIFORNIA THRASHER.[1]

BY JOSEPH GRINNELL.

THE California Thrasher (*Toxostoma redivivum*) is one of the several distinct bird types which characterize the so-called "Californian Fauna." Its range is notably restricted, even more so than that of the Wren-Tit. Only at the south does the California Thrasher occur beyond the limits of the state of California, and in that direction only as far as the San Pedro Martir Mountains and

[1] Contribution from the Museum of Vertebrate Zoölogy of the University of California.

San Quintin, not more than one hundred and sixty miles below the Mexican line in Lower California.

An explanation of this restricted distribution is probably to be found in the close adjustment of the bird in various physiological and psychological respects to a narrow range of environmental conditions. The nature of these critical conditions is to be learned through an examination of the bird's habitat. It is desirable to make such examination at as many points in the general range of the species as possible with the object of determining the elements common to all these points, and of these the ones not in evidence beyond the limits of the bird's range. The following statements in this regard are summarized from the writer's personal experience combined with all the pertinent information afforded in literature.

The distribution of the California Thrasher as regards life-zone is unmistakable. Both as observed locally and over its entire range the species shows close adherence to the Upper Sonoran division of the Austral zone. Especially upwards, is it always sharply defined. For example, in approaching the sea-coast north of San Francisco Bay, in Sonoma County, where the vegetation is prevailingly Transition, thrashers are found only in the Sonoran "islands," namely southerly-facing hill slopes, where the maximum insolation manifests its effects in a distinctive chaparral containing such lower zone plants as Adenostoma. Again, around Monterey, to find thrashers one must seek the warm hill-slopes back from the coastal belt of conifers. Everywhere I have been, the thrashers seem to be very particular not to venture even a few rods into Transition, whether the latter consist of conifers or of high-zone species of manzanita and deer brush, though the latter growth resembles closely in density and general appearance the Upper Sonoran chaparral adjacent.

While sharply delimited, as an invariable rule, at the upper edge of Upper Sonoran, the California Thrasher is not so closely restricted at the *lower* edge of this zone. Locally, individuals occur, and numbers may do so where associational factors favor, down well into Lower Sonoran. Instances of this are particularly numerous in the San Diegan district; for example, in the Lower Sonoran "washes" at the mouths of the canyons along the south base of the San Gabriel Mountains, as near San Fernando, Pasadena, and

Azusa. A noticeable thing in this connection, however, is that, on the desert slopes of the mountains, where *Toxostoma lecontei* occurs on the desert floor as an associational homologue of *T. redivivum* in the Lower Sonoran zone, the latter "stays put" far

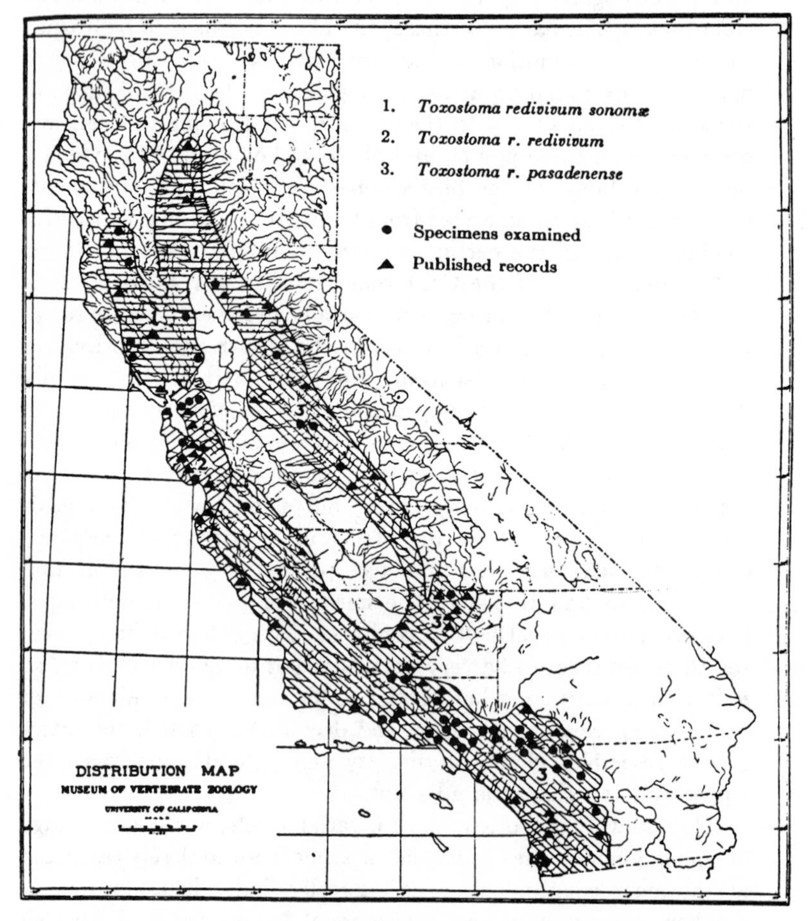

Figure 1.

more closely; that is, it strays but little or not at all below the typical confines of its own zone, namely *Upper* Sonoran. The writer's field work in the vicinity of Walker Pass, Kern County, provides good illustrations of this. A tongue or belt of Lower Sonoran extends from the Mohave Desert over the low axial mountain ridge at the head of Kelso Creek and thence down along the valley of the South Fork of the Kern River nearly to Isabella. Leconte's Thrasher is a conspicuous element in this Lower Sonoran invasion, but no California Thrashers were met with in this region below the belt of good Upper Sonoran on the flanking mountain sides, as marked by the presence of digger pine, blue oak, sumach, silk-tassel bush, and other good zone-plants. Similar zonal relationships are on record from San Gorgonio Pass, Riverside County, as well as elsewhere.

Reference now to the general range of the bird under consideration (see p. 429), as compared with a life-zone map of California (Pacific Coast Avifauna No. 11, Pls. I, II), will show to a remarkable degree how closely the former coincides with the Upper Sonoran zone. The thrasher is, to be sure, one of the elements upon the presence of which this zone was marked on the map; but it was only one of many, both plant and animal; and it is concordance with the aggregate that is significant. Diagnosis of zonation similarly is possible in scores of places where change in altitude (which as a rule means change in temperature) is the obvious factor, as up the west flank of the Sierra Nevada, or the north wall of the South Fork valley, already referred to, in Kern County, or on the north wall of the San Jacinto Mountains. The California Thrasher is unquestionably delimited in its range in ultimate analysis by temperature conditions. The isothermic area it occupies is in zonal parlance, Upper Sonoran.

The second order of restriction is faunal, using this term in its narrowed sense, indicating dependence upon atmospheric humidity. The California Thrasher does not range interiorly into excessively arid country, although the Upper Sonoran zone may, as around the southern end of the Sierra Nevada, continue uninterruptedly towards the interior in a generally latitudinal direction. This is true where extensive areas are considered, but locally, as with zones, individuals or descent-lines may have invaded short distances beyond the normally preferred conditions. An example of this situation is to be found on the north and west slopes of the San Jacinto Mountains, where California Thrashers range around onto arid chaparral slopes, intermingling with such arid Upper Sonoran birds as Scott's Oriole and the Gray Vireo. It is questionable, however, as to what extent faunal restriction really operates in this case; for reference to the zone map, again, shows that a vast tract of Lower Sonoran, lying to the east of the desert divides, extends continuously north to the head of Owens Valley. Really the only unbroken bridge of Upper Sonoran towards the east from the west-Sierran habitat of *Toxostoma redivivum* is around the southern end of the Sierra Nevada — a very narrow and long route of possible emigration, with consequent factors unfavorable to invasion, irrespective of either temperature or humidity, such as interrupted associational requirements and small aggregate area. In this particular bird, therefore, faunal restriction may be of minor importance, as compared with zonal and associational controls.

That faunal conditions have had their influence on the species, however, is shown by the fact of geographic variation within its range. The thrasher throughout its habitat-as-a-whole, is subjected to different degrees of humidity. Amount of rainfall is, in a general way, an index of atmospheric humidity, though not without conspicuous exceptions. Comparing the map of the ranges of the subspecies of *T. redivivum* (p. 429) with a climatic map of the State, direct concordance is observed between areas of stated rainfall on the latter and the ranges of the respective subspecies. It will be seen that the race *T. r. pasadenense* occupies an area of relatively low humidity, the race *T. r. sonomæ* of higher humidity and the race *T. r. redivivum* of highest humidity, in fact a portion of California's fog-belt. The distinctive color-tones developed are, respectively, of gray, slate and brown casts. In the thrasher, therefore, we may look to faunal influences as having most to do with differentiation within the species. In this case it is the faunal variation over the occupied country which is apparently responsible for the intra-specific budding, or, in other words, the origination of new specific divarications.

Wherever it occurs, and in whichever of the three subspecies it is represented, the California Thrasher evinces strong associational predilections. It is a characteristic element in California's famous chaparral belt. Where this belt is broadest and best developed, as in the San Diegan district and in the foothill regions bordering the great interior valleys, there the Thrasher abounds. The writer's personal field acquaintance with this bird gives basis for the following analysis of habitat relations.

The California Thrasher is a habitual forager beneath dense and continuous cover. Furthermore, probably two-thirds of its foraging is done on the ground. In seeking food above ground, as when patronizing cascara bushes, the thrasher rarely mounts to an exposed position, but only goes as high as is essential to securing the coveted fruits. The bird may be characterized as semi-terrestrial, but always dependent upon vegetational cover; and this cover must be of the chaparral type, open next to the ground, with strongly interlacing branch-work and evergreen leafy canopy close above — not forest under-growth, or close-set, upright stems as in new-growth willow, or matted leafage as in rank-growing annual herbage.

The Thrasher is relatively omnivorous in its diet. Beal (Biological Survey Bulletin no. 30, p. 55) examined 82 stomachs of *Toxostoma redivivum* and found that 59 percent of the food was of a vegetable nature and 41 animal. A large part of this food consisted of ground-beetles, ants, and seeds, such as are undoubtedly obtained by working over the litter beneath chaparral. The bird's most conspicuous structural feature, the long curved bill, is used to whisk aside the litter, and also to dig, pick-fashion, into soft earth where insects lie concealed. Ground much frequented by Thrashers shows numerous little pits in the soil surface, less than an inch deep, steep on one side and with a little heap of earth piled up on the opposite side. As already intimated, the Thrasher at times ascends to the foliage above, for fruit and doubtless some insects. Much in the way of berries and seeds may also be recovered from the ground in what is evidently the Thrasher's own specialized method of food-getting. Even granting this specialization, I do not see why the chaparral, alone, should afford the exclusive forage-ground; for the same mode of food-getting ought to be just as useful on the forest floor, or even on the meadow. The further fact, of widely omnivorous diet, leads one to conclude that it is *not* any peculiarity of food-source, or way of getting at it, that alone limits the Thrasher associationally. We must look farther.

The amateur observer, or collector of specimens, is struck by the apparent "shyness" of the Thrasher — by the ease with which it eludes close observation, or, if thoroughly alarmed, escapes detection altogether. For this protective effect the bird is dependent upon appropriate cover, the chaparral, and upon its ability to co-operate in making use of this cover. The Thrasher has strong feet and legs, and muscular thighs, an equipment which betokens powers of running; the tail is conspicuously long, as in many running birds; and correlatively the wings are short, rounded, and soft-feathered, indicating little use of the flight function. The colors of the bird are non-conspicuous — blended, dark and light browns. The nests of the Thrasher are located in dense masses of foliage, from two to six feet above the ground, in bushes which are usually a part of its typical chaparral habitat. In only exceptional cases is the chosen nesting site located in a bush or scrubby tree, isolated more or less from the main body of the chaparral.

These various circumstances, which emphasize dependence upon cover, and adaptation in physical structure and temperament thereto, go to demonstrate the nature of the ultimate associational niche occupied by the California Thrasher. This is one of the minor niches which with their occupants all together make up the chaparral association. It is, of course, axiomatic that no two species regularly established in a single fauna have precisely the same niche relationships.

As a final statement with regard to the California Thrasher, we may conclude, then, that its range is determined by a narrow phase of conditions obtaining in the Chaparral association, within the California fauna, and within the Upper Sonoran life-zone.

36. The Balance of Animal Populations.—Part I.

By A. J. NICHOLSON, D.Sc., Division of Economic Entomology, Commonwealth Council for Scientific and Industrial Research, Canberra, and V. A. BAILEY, M.A., D.Phil. (Oxon.), F.Inst.P., Associate Professor of Physics, University of Sydney *.

[Received November 9, 1934 : Read April 2, 1935.]

(Text-figures 1–15.)

INTRODUCTION.

Although our work has been developed quite independently of others, it is desirable to preface it with a few remarks about other theoretical investigations of population problems.

Since 1907 Lotka has published a large number of mathematical articles † on the population problem, and in 1925 he produced 'The Elements of Physical Biology' ‡, a book containing much interesting information concerning animal populations. His fundamental equations (Chapter VI.) for an association of any number of species state the fact that the rate of change in mass of each species is a function of the masses of all the species in the same association, of the masses of their food materials, and of certain parameters. These equations, in the form given, do not appear to be capable of leading to *definite* general conclusions §, as they seem to be consistent with all possible conclusions. Moreover, we have not been able to derive our theory from Lotka's fundamental equations. Competition does not appear explicitly in any of his equations, and few, if any, indicate the existence of this factor.

In 1910 Fisk ‖ published a curve, derived arithmetically, which is nearly the same as the "competition curve" (see Nicholson ¶, fig. 1), and later Thompson ** made use of a formula (supplied to him by Professor Deltheil of Toulouse) which is essentially the same as that upon which the "competition curve" is based. Both these authors, however, used this curve for the investigation of the problem of superparasitism, and did not apply it to the general problem of competition. Thompson also published a number of articles †† which are mainly concerned with the investigation of the special problem of the overtaking of a geometrically increasing population of hosts by parasites that are initially introduced in very small numbers. Thus he deals neither with the steady densities of animals, nor definitely with interspecific oscillation.

* Communicated by THE SECRETARY.

† The following require particular mention :—Proc. Nat. Acad. Sci. vol. vi. no. 7, pp. 410–415 (1920) ; Journ. Washington Acad. Sci. vol. xiii. no. 8 (1923) ; Proc. Nat. Acad. Sci. vol. xviii. no. 2, pp. 172–178 (1932).

‡ Williams & Wilkins Co., Baltimore.

§ In particular situations, however, he is able to obtain definite results ; for example, he finds that oscillations of population densities occur when an entomophagous parasite and a host-species alone interact.

‖ Journ. Econ. Ent. vol. iii. pp. 88–97 (1910).

¶ Journ. Animal Ecology, vol. ii. no. 1, pp. 132–178 (1933).

** Ann. Fac. Sci. Marseille, (2) vol. ii. pp. 69–89 (1924).

†† *Vide* Thompson, W. R., E. M. B., 29, H.M. Stationery Office (1930).

An extensive mathematical investigation of the population problem is given by Volterra *. His work and the present investigation have something in common, but there are important differences in the fundamental formulation of the problem, as has already been indicated †, and these naturally have led to differences in the conclusions reached. Volterra concludes that the interaction of animals produces fluctuation of the population-densities, and he also shows that for each species there is a steady value of the density (dependent upon the properties of the interacting animals and the nature of the interaction) about which the fluctuations take place. It may be remarked that, although this work is extensive, it is by no means comprehensive ‡, a fact which Volterra himself makes quite clear. Moreover, it is evident from his equations that he supposes the number of prey eaten, and the number of offspring produced by predators, to be *proportional* to the number of encounters with prey. Consequently, although Volterra speaks of the "eating species" ("*specie mangiante*") and of the "eaten species" ("*specie mangiata*"), he actually considers predators of the special kind here referred to as *parasites* (*i. e.*, entomophagous parasites), for, unlike true predators, such parasites may attack hosts in proportion to the number met, and thus produce offspring in the same proportion.

Stanley § has published a mathematical investigation of the problem of the natural control of populations of *Tribolium confusum* Duval. This problem is of a very special kind, for populations of *T. confusum* are limited by individuals attacking other individuals of the same species. Stanley, like Lotka and Bailey, makes use of ideas derived from the kinetic theory of gases.

Bremer ‖ and Zwölfer ¶ use mathematical expressions to represent the changes in numbers that take place during several successive generations of animals. These expressions, however, merely amount to representing symbolically the obvious fact that the number of animals existing at any given time is equal to the number produced less the number that has been destroyed.

Janisch ** has made much use of mathematics in his investigation of the influence of physical factors on the duration of life and the mortality of insects, and he uses the conclusions so reached to explain the initiation and termination of outbreaks of insects. His mathematical work, however, is entirely devoted to problems concerning individuals, and so has no direct relation to the present investigation. Moreover, his conclusions about the effect of physical factors on populations refer only to *changes* in population-densities; they do not show what determines the densities to be changed, nor why populations are limited.

Section I.—Fundamental Hypotheses and Method of Investigation.

In order to exist and produce offspring animals must obtain food, mates, and suitable places in which to live.

This hypothesis is clearly axiomatic, from which it follows that the problem

* Mem. Accad. Lincei, Cl. Sci. fis. e nat. cccxxiii. (6) ii. fasc. 3 (1926); Rend. R. A. dei Lincei, vol. vi. ser. 6 (1927); 'Leçons sur la théorie mathématique de la lutte pour la vie,' (Gauthier-Villars et Cie, Paris, 1931).

† Bailey, V. A., Q. J. Math. vol. ii. no. 5, pp. 68–77 (Oxford 1931).

‡ This point is not appreciated by some biologists. Thus Thompson says, "Professor Volterra has succeeded in elaborating an absolutely general theory of biological associations" (E.M.B. 29, H.M. Stationery Office, 1930, p. 56).

§ Canadian Journ. Research, vol. vi. pp. 632–671 (1932); *ibid.* vol. vii. pp. 426–433 (1932).

‖ Z. angew. Ent. vol. xiv. no. 2, pp. 254–272 (1928).

¶ Z. angew. Ent. vol. xix. no. 1, pp. 1–21 (1932).

** Trans. Ent. Soc. London, vol. lxxx. pt. 2, pp. 137–168 (1932); 'Das Exponentialgesetz,' (Julius Springer, Berlin, 1927).

of finding things is of outstanding importance to animals. Even when animals begin life in favourable situations in which food and mates are readily available, such situations must have been found for them by their parents or ancestors.

The searching of animal populations is always random.

This is the second fundamental hypothesis upon which the present investigation is based. By definition random searching is completely unorganized searching. Consequently, as organized searching by whole populations is unknown amongst animals *, the second hypothesis represents a true fact of nature.

It is important to make a sharp distinction between the searching of individuals and that of populations. Systematic searching by each of a large number of independent individuals does not make the searching of a population systematic, any more than the organized but independent running of each of a large number of ships makes the operations of a fleet organized. So long as individuals, or groups of individuals, search independently, the searching within a population is completely unorganized, and so is wholly random.

Systematic but independent searching by animals merely increases their individual efficiencies in making contact with the things they seek, and so a smaller population of animals than would otherwise be necessary can make a given number of contacts. It will shortly be shown that the probability of making random contacts can be expressed by a simple formula (and curve), which is independent of the individual efficiencies of the animals. It is therefore clear that the problem of finding things remains fundamentally the same whether animals move (or are scattered) in a truly random way, or individually search with system and intelligence.

The remainder of the present investigation † consists of an examination of the problem of the random searching of animals for the things they require for existence. Many different kinds of situations are investigated in which animals interact in various ways and are subject to various kinds of environmental factors. Necessarily, attention has been confined to comparatively simple types of situations, and so the conclusions cannot generally be applied *directly* to situations observed in nature. The conclusions, however, show the *kinds of effects* that are produced by competition due to searching under various circumstances, and so should help observers in their analyses of complex situations occurring in nature. Moreover, the competition that exists between searching animals is shown to produce certain effects, such as balance and interspecific oscillation, which are independent of other environmental factors in their essential features, and so must actually exist—unless the fundamental hypotheses given are wrong, or there is some factor capable of nullifying the effects of competition, which seems most improbable. The conclusions reached are known to be consistent in a general way with what has been observed to occur in nature ; but further observations and experiments need to be made in order to discover whether this is also true in detail.

Most of the investigations which follow are concerned with the problem of *discontinuous interaction*, i.e., interaction which occurs only during particular periods of time that succeed each other at intervals equal to the length of a generation. This simplification was introduced principally for arithmetical convenience in the earlier stages of the investigation, but it approximates

* With the possible exception of those animals which lay claim to territories, *e. g.*, some species of birds, see Howard, H. E., " Territory in Bird Life " (John Murray, London, 1920), and Nicholson, Journ. Animal Ecology, vol. ii. no. 1, p. 171 (1933).

† A brief outline of the whole investigation has already been given by Nicholson, *loc. cit.*

to what actually happens with many species of animals. The same simplification was naturally adopted in the mathematical investigation used to verify the arithmetical work, but subsequently the investigation was extended to the problem of continuous interaction.

The form of argument in the present series of communications is almost entirely mathematical, but, to assist readers whose mathematical attainments are not large, non-mathematical argument is also used when this can be done concisely.

Definition of certain Properties of Animals.

When an animal has completed its search it has traversed, with respect to objects of a given type, an area equal to the distance it has moved multiplied by twice the extreme distance at which it can perceive objects of the given type. This will be called the *areal range* of the animal for objects of the given type. Clearly an animal has a different areal range for each kind of object it seeks, for the extremes at which these can be perceived differ from each other.

Similarly we may speak of the *volumetric range* of an animal, this being the volume of the environment examined for objects of a given type. If, however, volumetric ranges be considered instead of areal ranges, the problems and calculations remain unaltered, and, as most animals search over surfaces, it is convenient in general to represent their efficiency in searching by their areal ranges.

The efficiency of an animal in finding what it seeks depends on many things besides its areal range. For example, it depends upon the movement, the inconspicuousness, and the defensive properties of the objects sought, the system followed by the animal when searching, and the density of the objects sought. It is convenient to have a measure of the efficiency of an animal when searching which includes the effects of all factors, other than the density of objects sought, that influence this efficiency. When searching for objects of a given type under given conditions an animal has a certain efficiency, and it is possible to imagine an equivalent animal that would have the same efficiency if it found all the objects of the given type occurring in its areal range and all these objects were stationary. The areal range of this equivalent animal may be used as a measure of the efficiency of the given animal, and will be called the *area of discovery* of the given animal. Thus, if an animal fails to perceive half the objects that occur in its areal range, its area of discovery has half the value of its areal range.

The efficiency in searching of a species will be represented by the area of discovery of an average individual, both sexes being included when determining the average.

It is important to note that the area of discovery of an animal is dependent not only upon the properties of this animal, but also upon those of the objects it seeks, as has just been shown. Consequently, any given animal has a different area of discovery for each kind of object it seeks. Furthermore, the value of the area of discovery is influenced by general environmental conditions, such as climate. However, to avoid undue complexity in the earlier parts of this investigation, it is assumed that environmental conditions remain constant, so that the areas of discovery of given species of animals for given kinds of objects are constant.

The number of times a population would be multiplied in each generation if freed from the destructive action of given factors will be called the *power of increase* of the species with respect to the given factors and under the given conditions. Except where otherwise specifically stated, the phrase "power of increase" will always be used with respect to the whole of the destructive factors considered in any given instance. Clearly this property determines the fraction of animals that must be destroyed by the given factors in order to prevent indefinite increase in density.

Intra-specific Competition.

If few animals search in a very large area it is clear that the chance of their crossing each other's tracks is small, whereas if there are large numbers of animals in the area the chance of the tracks crossing is greatly increased. Consequently, with progressively increasing density of population the animals spend a progressively greater fraction of their time in searching over areas that have already been searched, so the chance of an individual finding the things it requires progressively diminishes. It is shown elsewhere * that the approximate character of this diminution of success with increasing density can easily be determined by simple arithmetical calculation, and that it is the same in all problems in which the searching of a population is unorganized. An exact investigation of this problem is the following :—

Let u_1 be the number of objects originally present in a unit of area, and let u be the number undiscovered after an area s has been traversed.

The number of previously undiscovered objects found in traversing ds is $u\,ds$.

This must be equal to the decrease $-du$ of the number of undiscovered objects.

$$\therefore \quad -du = u\,ds,$$

and since

$$u = u_1 \quad \text{when} \quad s = 0,$$

$$\therefore \quad u = u_1 e^{-s}, \qquad (1)$$

i. e.,

$$s = \log (u_1/u). \qquad (2)$$

The fraction of the original objects which is discovered is $(u_1 - u)/u_1$, *i. e.*, $1 - e^{-s}$. This is represented by a curve published elsewhere †.

Such competition appears implicitly in Volterra's arguments and equations, but in the later stages of his investigation he introduces what appears to be an additional competition factor (*vide* 'Leçons sur la théorie mathématique de la lutte pour la vie,' pp. 78 and 199). In certain situations in the earlier parts of his investigation he finds that the density of a species may increase indefinitely with time, and, as this does not occur in nature, he introduces "*termes d'amortissement*" to represent resistance to increase in numbers. Such a term is then used for each of the species in a biological association, so there is reason to believe that Volterra has superimposed terms representing competition on other terms, some of which already imply this competition.

Simplifications used.

In order to avoid undue complexity the factors about to be mentioned are neglected in the earlier parts of this investigation, but they are later taken into account when there is reason to believe that they may influence the conclusions reached.

The investigation will be limited to random searching for food and suitable places in which to live, the problem of finding mates being neglected because it is believed that animals experience little difficulty in finding mates except

* Nicholson, Journ. Animal Ecology, vol. ii. no. 1 (1933).

† Nicholson, *loc. cit.* fig. 1.

when the population densities are exceedingly small compared with their steady values.

In spite of the fact that we assume all objects or animals of a given kind or species to have identical properties, it is probable that some of the objects are more difficult to find than others and that some of the individual animals are more efficient searchers than others. Consequently, when most of the objects have been found, the remainder is likely to contain more than the average proportion of objects that are difficult to find, so that the area of discovery of the animals for the objects is reduced. Similarly, when competition for objects is the factor that controls the density of the searching animals, it is probable that the less efficient animals will be amongst those destroyed, so that the area of discovery of the survivors is above the average for the population. Thus the two factors tend to counteract one another, though, of course, they will not necessarily do so exactly.

The special difficulty introduced by the two factors just mentioned is that they may cause the areas of discovery of the animals to change progressively with increasing competition. They do not, however, alter the major fact with which we are concerned, namely, that with increasing competition the chance of survival of an average individual progressively diminishes—they merely influence the rate at which the chance of survival diminishes. The investigations which follow indicate that if we were to introduce these complicating factors the steady densities of the animals would be slightly different from those calculated, but in all major features the conclusions would be uninfluenced. The only important effect of these factors appears to be a tendency to damp interspecific oscillations.

The effects of hunger and satiety greatly complicate the problem of competition for food. Thus animals that are replete, or weak from hunger, are likely to search over smaller areas than are animals that succeed in finding all the food they require by thorough search. So the effect of hunger and satiety is to cause the areas of discovery of animals to vary in relation to the density of food. This effect is clearly of no importance in the steady state, for in this state only those animals that are not surplus can find sufficient food for survival, and so the degree of hunger is constant. As will later be shown, however, hunger and satiety have some influence on interspecific oscillation.

Entomophagous parasites hunt for the food (viz., the hosts) required by their offspring. Consequently the offspring do not need to search, for they are born only in places where there is an ample supply of food for their full development, and so there is here no problem about the effects of hunger and satiety. Moreover, when a mature parasite meets a suitable host it lays one or more eggs, the number of eggs laid depending generally upon the species of parasite and the species of host attacked. Thus the number of offspring produced by a parasite is at least approximately proportional to the number of encounters of mature parasites with suitable hosts. For these reasons the problem of the interaction of entomophagous parasites and their hosts is a relatively simple one with which to begin our investigation of the interaction of animals, as we shall now do.

Section II.—The Interaction of one Host- and one Parasite-Species.

We are here concerned with the steady densities of the interacting animals when one species of host and one species of parasite interact, these animals being uninfluenced by the activities of other animals, and the general environmental conditions remaining constant. We assume, for the reasons just given, that the number of hosts destroyed and the number of offspring produced by parasites are proportional to the number of encounters between mature parasites and hosts in a suitable condition to be attacked. This includes the assumption that mature parasites have an unlimited supply of eggs, but we will shortly see that, in the steady state, the chance of a mature parasite exhausting its egg-supply is small unless this egg-supply is very small. Since the egg-supplies of parasites are seldom, if ever, very small, the above assumption is justified.

For a parasite-species with a given area of discovery there is clearly a particular density at which it effectively searches a fraction of the environment equal to the fraction of hosts that is surplus, and at this density exactly the surplus of hosts is destroyed. Also there is a particular density of a host-species which is just sufficient to maintain the above density of parasites from generation to generation. When the densities of the interacting animals have these particular values it is clear that they must remain constant indefinitely in a constant environment. This situation will be referred to as the *steady state*, and the densities of the animals when in this state as their *steady densities*.

If the density of a parasite is above its steady value more than the surplus of hosts is destroyed, so that the density of hosts is reduced in the following generation. This process must continue at least until such time as the scarcity of hosts reduces the density of parasites to the steady value. Similarly if the density of parasites is below the steady value, or if the host-species has a density above or below its steady value, a reaction is set up which *tends* to cause the densities of the animals to return to their respective steady values. But, as will be shown later, the reaction to displacement causes the densities of interacting animals to swing to the opposite side of their steady values, so causing oscillation in density. For the present, however, discussion will be limited to the steady state.

1. *Determination of the steady state.*

Let h be the density of hosts which survive attack by parasites (*i. e.*, final density of hosts);

p the density of parasites;

F the power of increase of the hosts (so that, with equal numbers of each sex, a female host lays 2F eggs); and

a the area of discovery of a parasite individual (*i. e.*, with equal numbers of each sex the area of discovery of a female parasite is $2a$).

For convenience we introduce the functional symbol $\mathscr{L}$, defined by

$$\mathscr{L}(x) \equiv \frac{\log x}{x-1}. \qquad (3)$$

This has the property that

$$x\mathscr{L}(x) = \mathscr{L}(x^{-1}). \qquad (4)$$

The number of host eggs laid in a unit area, *i. e.*, the initial host-density, is Fh ($=u_1$ in the notation of p. 555), and to maintain the steady density of the hosts the number h ($=u$) of the hosts which develop from these eggs must remain unattacked. So, by equation (2) the parasites must collectively traverse an area

$$s = \log(\mathrm{F}h/h) = \log \mathrm{F}.$$

Since each traverses an area a, the p traverse an area $s=pa$.

$$\therefore \quad pa=\log F. \qquad (5)$$

To maintain the steady density of the parasites in the succeeding generation, which comes from the $Fh-h$ host-offspring attacked (one parasite from each host), the latter quantity must be equal to p.

$$\therefore \quad p=h(F-1),$$

$$\therefore \quad ha=(\log F)/(F-1),$$

i. e., by (3)

$$ha=\mathscr{L}(F). \qquad (6)$$

The initial host-density Fh is then given by

$$Fha=\mathscr{L}(F^{-1}). \qquad (7)$$

So, if the specific characteristics F and a be known, the equations (5) and (6) give the densities p and h respectively. When a is equal to 1 the dependence of p, h and Fh on F is illustrated by the curves in text-fig. 1. We thus conclude :—

C. 1. The steady density of parasites and that of hosts vary inversely as the area of discovery, but otherwise depend on the properties of the host (from equations) (5) and (6).

C. 2. Raising the power of increase of the host *increases* the steady density of parasites and the initial steady density of hosts, but *reduces* the final steady density of hosts (see text-fig. 1).

Text-figure 1.

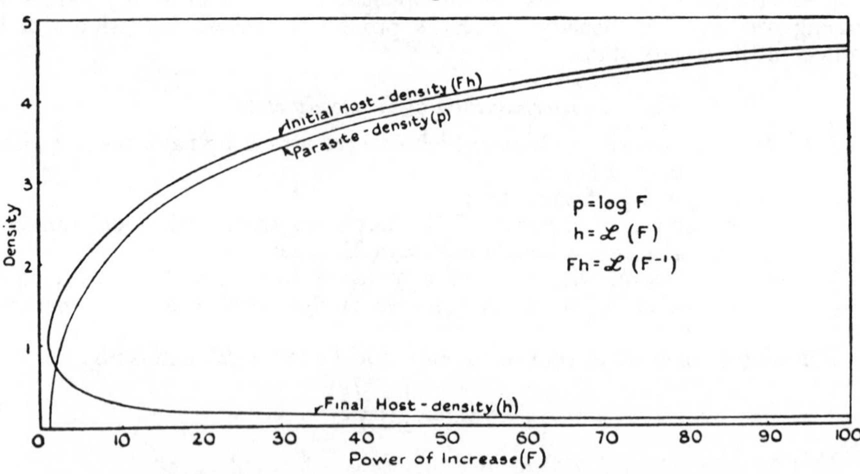

Relation between population density and power of increase.
The area of discovery of the parasite is taken as 1.

One may particularly notice the curious conclusion that a parasite species becomes less successful as its efficiency (*i. e.*, its area of discovery) increases (C. 1). Paradoxical as this conclusion appears, it is easy to see that it is entirely reasonable, for in the steady state a parasite destroys just the surplus of host-offspring, and an efficient parasite maintains its host at a low density. Consequently only a low density of an efficient parasite can be supported by its host.

It follows from C. 2 that great fecundity does not necessarily cause a species to be abundant, but rather tends to cause it to be scarce in the adult state.

2. *Each parasite requires either more or less than one host individual for its full development.*

Let n be the number of parasites for which one host suffices as food;

$$\therefore \quad p=n(\mathrm{F}h-h)=nh(\mathrm{F}-1).$$

Equation (5) must still hold true, so

$$nha=\mathscr{L}(\mathrm{F}). \quad \ldots \ldots \ldots \ldots \quad (8)$$

This replaces (6), and so :—

C. 3. The final steady density of hosts (h) varies inversely as the number (n) of parasites that develop at the expense of one host.

When each parasite requires m hosts for its development the host-density is also given by the above calculation, but with n set equal to $1/m$. Therefore

$$ha=m\mathscr{L}(\mathrm{F}). \quad \ldots \ldots \ldots \ldots \quad (9)$$

Hence :—

C. 4. The final steady density of hosts (h) varies directly as the number (m) of hosts required by one parasite for its full development.

Thus, by C. 3, the efficiency of a parasite in controlling its host is increased when a greater number of parasites develop at the expense of a single host—a conclusion diametrically opposite to the view generally held. It is clear, however, that, when several parasites can develop at the expense of a single host individual, the particular density of parasites necessary to find the surplus fraction of hosts can be supported by a relatively low density of hosts. C. 4 may be confirmed by a similar argument.

Wasps which collect caterpillars, spiders, etc., and store these in cells as provision for their offspring, provide examples of "parasites" which require more than one host for their full development. Such insects may produce offspring in proportion to the number of hosts they find; so for the purposes of this investigation they can be considered as parasites, although normally they are not given this name.

3. *A host individual can be attacked more than once.*

It has so far been assumed that once a host individual has been attacked it is no longer available for attack; but it is known that parasites sometimes attack hosts that have already been attacked. When host individuals are attacked more than once there are two possible situations, viz. :—

(1) owing to the limited food-supply the number of eggs laid in a host that can give rise to mature parasites is limited; and

(2) interference prevents the development of any parasites from eggs laid in the same host.

The first situation may be investigated as follows :—

In a very large area A the parasites collectively meet $\mathrm{N}=\mathrm{F}h\mathrm{A}s$ hosts, where $s=pa$. As the number of hosts in this area is $\mathrm{F}h\mathrm{A}$, the probability that an egg which is being laid in this area would be received by a selected individual is $(FhA)^{-1}$, i. e., s/N. Hence, by a well-known argument *, the probability that n eggs are laid in this individual in the N encounters is

$$P_n=\frac{N!}{(N-n)!\,n!}\left(\frac{s}{N}\right)^n\left(1-\frac{s}{N}\right)^{N-n}.$$

In the limit when A, and so N, is made very large this becomes

$$P_n=\frac{s^n}{n!}e^{-s}. \qquad (10)$$

This also gives the proportion of the hosts in a unit area which have n eggs laid in them †.

So of the Fh hosts initially present in a unit area FhP_n ultimately receive n eggs. If M be the maximum number of eggs in a host which can survive to become adult parasites, then the total number which survive is

$$\sum_1^M FhP_n n+\sum_{M+1}^{\infty} FhP_n M.$$

In the steady state this must be equal to p, so using (10)

$$p=Fh\left[\sum_1^M\frac{s^n}{(n-1)!}e^{-s}+\sum_{M+1}^{\infty}M\frac{s^n}{n!}e^{-s}\right].$$

$$\therefore\quad Fha=se^s\left[\sum_1^M\frac{s^n}{(n-1)!}+M\sum_{M+1}^{\infty}\frac{s^n}{n!}\right]^{-1} \qquad (11)$$

In order to preserve the parasite-density constant equation (5) must still hold true.

Hence we may set $s=\log F$ in (11), and so obtain an expression for determining h.

When M is very large it becomes

$$Fha\doteq 1. \qquad (12)$$

The average number α of eggs laid in an attacked host is

$$\frac{\sum_1^{\infty}FhP_n n}{\sum_1^{\infty}FhP_n}=\frac{s\sum_1^{\infty}\frac{s^{n-1}}{(n-1)!}}{\sum_1\frac{s^n}{n!}}=\frac{s}{1-e^{-s}};$$

$$\therefore\quad \alpha=\mathscr{L}(F^{-1}). \qquad (13)$$

But the average number of eggs laid in *any* one of the hosts is s, i. e., log F. We thus conclude :—

C. 5. The steady density of mature parasites is the same as before, and is given by equation (5).

C. 6. The steady density of hosts is given accurately by equation (11), and approximately by equation (12) when the maximum capacity (M) of a host individual for a parasite is large. The latter equation shows that when the maximum capacity is large, the final density of hosts is inversely proportional to the power of increase (F) of the host-species and the area of discovery (a) of the parasite-species, and that it is less than in the original situation (p. 558).

* *Cf.* Borel and Deltheil, ' Probabilités, Erreurs ' (Armand Colin, Paris), p. 25.

† Equation (1) may be regarded as a special example of equation (10) when $n=0$.

C. 7. The average number of parasite eggs per attacked host is shown by equation (13) to be derivable from the uppermost curve in text-fig. 1.

Situation (2) (p. 589) results in mature parasites being produced only from those hosts in which just one egg is laid. The proportion P_1 of the Fh hosts, present in unit area, which have just one egg laid in them is given by setting $n=1$ in the formula (10). Hence the number of mature parasites produced is FhP_1, where $P_1=se^{-s}$.

Hence to preserve the parasite-density constant requires that

$$p=Fhpae^{-pa}.$$

To preserve the host-density constant equation (5) must still hold true; so the above expression leads to

$$ha=1. \quad . \quad . \quad . \quad . \quad . \quad . \quad . \quad . \quad . \quad . \quad . \quad (14)$$

So we conclude :—

C. 8. The steady density of mature parasites is again the same.

C. 9. The steady density of hosts is shown by equation (14) to be equal to the reciprocal of the area of discovery (a) and to be greater than in the original simple situation (equation (6)).

C. 5 and C. 8 follow immediately from the fact that only at one particular density can a given species of parasite find and destroy the surplus fraction of hosts having a given power of increase.

That the steady density of hosts is less than in the simple situation (p. 558), when more than one parasite can develop from an attacked host (*vide* C. 6), may be shown by the same verbal argument that is used in support of C. 3.

When no mature parasites develop from eggs laid in hosts that have been attacked more than once it is clear that, not only must sufficient hosts be found to support the steady density of mature parasites, but, in addition, other hosts must be found which, though attacked, do not give rise to mature parasites. This is possible only when the steady density of hosts is greater than in the simple situation (*vide* C. 9).

4. *The parasites are unable to attack all previously unattacked hosts met.*

a. *Partial immunity of hosts.*

When, for example, two-thirds of the hosts met are not attacked, owing to their inconspicuousness, or to some other defensive property, the area of discovery of a parasite is, by definition (p. 554), one-third of what it would be if the parasite attacked all the hosts met. Thus the area of discovery of a parasite is proportional to the attacked fraction of the hosts that are met. Consequently, by C 1. :—

C. 10. The steady density of hosts increases with the efficiency of their special protective properties.

When some hosts live in protected situations this may either increase the steady densities of hosts and parasites, or prevent the parasites from controlling the hosts. Thus, if hosts protected from attack constitute a fraction greater than that which must survive in the steady state, it is clear that under no circumstances can the parasites destroy all the surplus hosts This is essentially the situation examined in more detail later (p. 565).

b. *The parasites have limited egg-supplies.*

Let N be the large number of parasites in a large area A, and let N′ be the number of eggs laid by them. The probability that an egg which is laid comes from a selected individual is 1/N; hence by a well-known argument the probability that n are laid by this individual is

$$P_n = \frac{N'!}{n!(N'-n)!}\left(\frac{1}{N}\right)^n\left(1-\frac{1}{N}\right)^{N'-n}.$$

Since each parasite lays on the average one egg, we have N'=N; so, when N is made indefinitely large this expression becomes

$$P_n = \frac{e^{-1}}{n!} = \frac{\cdot 37}{n!}. \qquad (15)$$

Hence the parasites which lay n eggs each are a fraction $\cdot 37/n!$ of the total number.

Text-fig. 2 has been constructed by plotting $\cdot 37/n!$ against n.

So we conclude :

C. 11. The parasites which lay n eggs each are a fraction $\cdot 37/n!$ of the total number, when each parasite on the average succeeds in laying one egg.

C. 12. Unless the maximum number of eggs which a parasite may lay is small, the steady density of hosts is but little increased by the limitation of the egg-supply of the parasites.

Text-fig. 2 illustrates C. 12, and is independent of the properties of the host and parasite except in one respect, namely, that only one egg is laid each time a parasite meets a host (a simplification which has little influence on the

Text-figure 2.

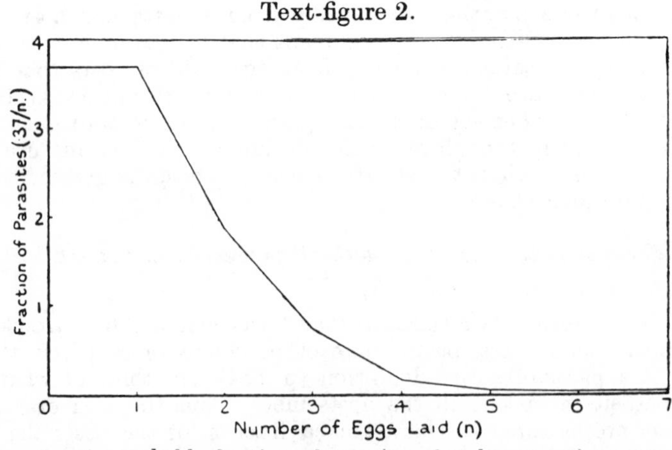

Diagram showing the probable fraction of parasites that lay any given number of eggs. The simple situation is taken in which a parasite attacks only the previously unattacked hosts it finds, and lays only one egg in each host it attacks.

result). It will be observed that the fraction of parasites which lay more than five eggs each in the steady state is negligible. So, if we assume equal numbers of the sexes, it is clear that the limitation of the egg-supplies of parasites can have but a negligible influence on the steady densities of hosts and parasites, unless the maximum number of eggs which can be laid by each female is less than ten. As most parasites can produce many more than ten eggs each, it is evident that we may, without serious error, assume parasites to have unlimited egg-supplies when investigating problems in which the steady state is approached.

5. *Some of the parasite eggs laid do not give rise to mature adults.*

We are here concerned only with normal factors, such as unfavourable weather conditions, which destroy an approximately constant fraction of the parasite offspring, and do so independently of the densities of the animals. Abnormal destructive factors initiate fluctuations but cannot influence the steady state. The more complex problem of the destruction of parasites by natural enemies and competitors, which is influenced by the densities of the interacting animals, will be examined later.

Let f be the fraction of the parasite offspring which are not destroyed by the environmental factors.

$$\therefore \quad p=f\,(\mathrm{F}h-h).$$

Then equation (6) is replaced by

$$fha=\mathscr{L}(\mathrm{F}),$$

for equation (5) still remains true. So :—

C. 13. The steady density of mature parasites is unaltered.

C. 14. The steady density of hosts varies inversely as the fraction of the parasite offspring that survives.

C. 13 and C. 14 are supported respectively by the verbal arguments given for C. 8 and C. 9.

6. *Some hosts are destroyed by factors other than parasites.*

We are here concerned only with the effects of an environmental factor which normally destroys a particular fraction of hosts and does so independently of their density (*cf.* section 5).

There are three possible situations that need to be examined separately, namely, that the destructive environmental factor operates (1) after, (2) before, and (3) at the same time as the parasites attack the hosts.

Situation (1).

Let f be the fraction of hosts, available to the destructive environmental factor, which survives its action.

Of the Fh initial hosts the number $h'=h/f$ survive after the action of the parasites, where h is the number of hosts surviving finally.

So, by equation (2),

$$pa=\log\,(\mathrm{F}h/h')\,;$$

$$\therefore \quad pa=\log\,(f\,\mathrm{F}). \qquad (16)$$

Also Fh—h' new parasites are produced;

$$\therefore \quad p=h(\mathrm{F}-f^{-1})\,;$$

$$\therefore \quad \mathrm{F}ha=\mathscr{L}(f^{-1}\mathrm{F}^{-1}). \qquad (17)$$

Thus p and Fh depend on f F in the same way as p and Fh depend on F in the simple situation (p. 558). So f F may be called the effective power of increase of the host under the given environmental conditions.

The number h' of hosts surviving after the attack of the parasites is then given by

$$h'=\mathscr{L}(f\,\mathrm{F}). \qquad (18)$$

Situation (2).

With f defined as in situation (1), let $h'=f\,\mathrm{F}h$ be the number of hosts surviving the action of the destructive environmental factor.

Then by equation (2)

$$pa=\log(h'/h);$$

$$\therefore\quad pa=\log(f\,\mathrm{F}). \qquad (19)$$

Since $h'-h$ new parasites are produced,

$$\therefore\quad p=h(f\,\mathrm{F}-1);$$

$$\therefore\quad ha=\mathscr{L}(f\,\mathrm{F}). \qquad (20)$$

Hence p is diminished and h is increased as if F in the simple situation were diminished to $f\,\mathrm{F}$.

Situation (3).

Let h_s be the density of surviving hosts when the parasites have collectively traversed an area s.

$$\therefore\quad -dh_s=h_s ds+\lambda h_s ds$$

$$=(1+\lambda)h_s ds,$$

where λ is a positive constant representing the destructive influence of the environment.

$$\therefore\quad (1+\lambda)s=\log(\mathrm{F}h/h_s),$$

since the initial host density is $\mathrm{F}h$.

When

$$s=pa,\quad h_s=h;$$

$$\therefore\quad pa=(\log \mathrm{F})/(1+\lambda). \qquad (21)$$

The number of hosts parasitized is

$$\int_0^{pa} h_s ds=(\mathrm{F}h-h)/(1+\lambda).$$

This must reproduce p parasites,

$$\therefore\quad p=h(\mathrm{F}-1)/(1+\lambda).$$

Hence by equation (21)

$$ha=\mathscr{L}(\mathrm{F}). \qquad (22)$$

We thus conclude :—

C. 15. The steady density of mature parasites is decreased in all situations (by equations (16), (19), (21)).

C. 16. When the destructive environmental factor operates after the parasites the initial and final steady densities of hosts are decreased, but not by as much as the fraction destroyed by the environmental factor (by equation (17)).

C. 17. When the destructive environmental factor operates before the parasites the initial and final steady densities of hosts are increased (by equation (20)).

C. 18. When the destructive environmental factor operates at the same time as the parasites the steady density of hosts is unchanged (by equation (22)).

The outstanding features of these conclusions may readily be established by verbal reasoning. Thus, when a destructive environmental factor destroys, say, half of the available hosts, the power of increase of the host with reference to the given parasites has only half the value it otherwise would have (see p. 554). Consequently for the steady state the fraction of available hosts that the parasites must leave unattacked is doubled, and this is possible only if the steady density of parasites is decreased (*cf.* C. 15). When the power of increase of its host is reduced, in the steady state a parasite commences to operate in a lower density, and leaves a higher density (C. 2 and text-fig. 1). If, then, the destructive environmental factor operates after the parasites, the initial steady density is decreased (*cf.* C. 16), the density of hosts left after the operation of the parasites is increased (*cf.* equation (18)), while the density of hosts left after the operation of the destructive environmental factor is decreased (*cf.* C. 16), for this clearly must be equal to the initial steady density divided by the actual power of increase. C. 17 may be supported by means of a similar verbal argument. When the parasites and the destructive environmental factor operate simultaneously the situation is intermediate between the two already discussed, so it may reasonably be expected that the steady densities of the hosts would be little, if at all, influenced by the action of the destructive environmental factor (*cf.* C. 18).

That an additional destructive factor may produce no change in the steady densities of the hosts against which it operates, or may even increase these densities, should be particularly noted. This illustrates a fact which will become increasingly prominent as this investigation proceeds, namely, that the indirect effects produced when a factor modifies an existing equilibrium system are often very different from the direct effects which the same factor tends to produce.

7. *The effective period of the parasite is longer than the vulnerable period of the host.*

So long as the hosts are vulnerable during the whole period when the parasites hunt for them the parasites exert their maximum possible influence upon the density of hosts, and, whether the hosts are also vulnerable at other times or not, their steady density clearly remains unaltered.

On the other hand, if the effective period of the parasite is such, for example, that only half the area which can be traversed by the parasites is searched during the vulnerable period of the host, it is clear that the area of discovery of the parasite has only half the value it otherwise would have, and so the steady densities of the parasites and the hosts are both doubled (by C. 1). Thus :—

C. 19. The steady density of hosts varies inversely as that fraction of the total area traversed by the parasites which is searched while the hosts are vulnerable ; the steady density of the parasites varies similarly.

8. *An imperfect correspondence exists between the effective period of the parasite and the vulnerable period of the host.*

It may happen (1) that some host individuals pass through their vulnerable stages when there are no parasites to attack them, or (2) that some parasites are in a condition to attack only at times when there are no hosts available. In situation (1) it is clear that the parasites must attack a larger fraction of the hosts available to them in order that the surplus fraction of hosts, taken as a whole, may be attacked, and this is possible only if the density of parasites is increased. In situation (2) the same steady density of parasites as in the simple situation (p. 558) is required to operate when hosts are available, so the total density of parasites must be increased. As in both situations an increased density of parasites is supported by the surplus fraction of hosts, it is clear that the steady densities of hosts must be increased. It is also clear that if the fraction of hosts available to the parasites is less than the surplus fraction of hosts, under no circumstances can the parasites destroy all the surplus hosts and consequently the hosts are left uncontrolled. So :—

C. 20. Lack of complete correspondence of the life-cycles increases the steady densities of hosts and parasites, and may cause the hosts to be completely out of control although attacked by the parasites.

9. *Conclusion.*

We have now examined the influence of a number of factors on the simple situation of the interaction of one host and one parasite-species. All of these either modify the efficiency of parasites in finding and attacking hosts, or modify the rate at which hosts would multiply in successive generations if free from attack by parasites. Consequently, when we assign values to the areas of discovery of parasites and the powers of increase of hosts in subsequent problems these values will be considered to take into account the influence of any or all of the modifying factors that have been investigated here, even though they are not explicitly mentioned.

Section III.—More Complex Types of Interaction between Parasites and Hosts.

It is necessary first to define the sense in which certain convenient terms are here used. A parasite-species which attacks only one species of host is called a *specific parasite* ; one which attacks hosts of two or more species is called a *general parasite* ; and one which attacks hosts of two or more given species is called a *common parasite* of the given hosts. Similarly we speak of *specific*, *general*, and *common hosts* according as they are attacked by one, by several, or by several given species of parasites. The inter-relationships mentioned in these definitions are those existing under the conditions to which any given investigation applies. Thus a parasite known to attack many species of hosts is a specific parasite in a given environment if it attacks only one species of host there.

A. Several Specific Parasites attack a Common Host.

When several species of specific parasites that have different areas of discovery simultaneously attack the one-host-species, that having the greatest area of discovery must eventually displace the others completely, for individuals of this species always have greater opportunities of finding hosts, and so of leaving offsprings, than have individuals of the other species. Therefore :—

C. 21. There can be no steady state when several species of specific parasites simultaneously attack a common host-species, unless the areas of discovery of the parasites have identical values (which is infinitely improbable).

Nevertheless balance between a common host-species and its parasites may occur when each of several parasite-species that attack the host attacks it at a different stage of its life-cycle. This is so because each parasite operates in a different density of hosts, viz., that left by the preceding parasite. Consequently, intra-specific competition may so adjust the density of each parasite to the available density of hosts that the steady state is possible. This is

C. 43. The effectiveness of a species of parasite in controlling the density of a particular host-species at a low value is increased if the parasite also attacks other species of hosts.

In the example just given it is clear that three times as many parasites search for the given host-species as are supported by it. The efficiency of the common parasite in finding hosts of the given species is therefore equal to that of a *specific parasite* with an area of discovery three times as great. The area of discovery of such an equivalent specific parasite will be called the *equivalent area of discovery* of a general parasite for a given host under given conditions. Consequently:

C. 44. Of the hosts attacked by a general parasite-species a certain fraction belongs to any given one of its host-species, and the equivalent area of discovery of the parasite for the given host-species varies inversely as this fraction.

The equivalent area of discovery is thus a measure of the effectiveness of a parasite in controlling the density of a given host when the parasite also attacks other species of hosts, and so it is a convenient property to use when investigating complex interactions between species, as we propose to do later.

IV.—Interspecific Oscillation.

In previous sections we confined attention to the steady state, *i. e.*, to situations in which the densities of interacting animals were so adjusted that the animals would maintain these densities unaltered from generation to generation in an unvarying environment. If the conclusions reached by the study of the steady state are to correspond to reality it is necessary that animals should at least tend to reach the steady state under natural conditions; so here we investigate what happens in subsequent generations when populations of interacting animals are not at their steady densities in any given generation.

1. *A specific parasite and a specific host interact.*

The reasoning involved in this problem is similar to that used when establishing equations (5) and (6) for determining the steady densities p and h.

Thus, if p_n and h_n represent the densities of the parasites and hosts respectively at the nth generation, then the parasites traverse an area $s=ap_n$ among host-offspring whose initial density is Fh_n and final density is h_{n+1}, and so by equation (1):

$$h_{n+1}=Fh_ne^{-ap_n}. \quad . \quad . \quad . \quad . \quad . \quad . \quad . \quad . \quad . \quad . \quad (80)$$

The number of hosts attacked by the p_n parasites is Fh_n-h_{n+1}, and since these produce an equal number of mature parasites in the next generation it follows that

$$p_{n+1}=Fh_n-h_{n+1}. \quad . \quad . \quad . \quad . \quad . \quad . \quad . \quad . \quad . \quad . \quad (81)$$

The curves in text-figs. 10 and 11, which were constructed by means of arithmetical reasoning, are found to be in good numerical agreement with these equations.

These diagrams show that the densities of the two species oscillate perpetually about their respective steady densities, and that the oscillations successively increase in magnitude.

That these results are generally true is established in a recent publication *.

* V. A. Bailey, Proc. Camb. Phil. Soc. vol. xxix. pt. iv. (1933). It has not yet, however, been found possible to prove that *large* oscillations perpetually increase in magnitude.

As a convenient introduction to a more complicated problem we give here an outline of the argument.

From (80) and (81) it is easily seen that

$$h_{n+1}=p_{n+1}/(e^{ap_n}-1), \quad . \quad . \quad . \quad . \quad . \quad . \quad . \quad . \quad . \quad (82)$$

and

$$Fh_n=p_{n+1}/(1-e^{-ap_n}).$$

Therefore

$$p_{n+2}=p_{n+1}F(1-e^{-ap_{n+1}})/(e^{ap_n}-1). \quad . \quad . \quad . \quad . \quad . \quad . \quad . \quad (83)$$

Setting $p_n=p+P_n$, where $p=a^{-1}\log F$ (the steady density), we obtain

$$P_{n+2}-P_{n+1}=p_{n+1}\left(\frac{1-e^{-aP_{n+1}}+F(1-e^{aP_n})}{Fe^{aP_n}-1}\right). \quad . \quad . \quad . \quad . \quad (84)$$

Elimination of p^n between (80) and (81) leads to

$$h_{n+1}=h_{n+2}Fe^{a(h_{n+1}-Fh_n)}, \quad . \quad . \quad . \quad . \quad . \quad . \quad . \quad . \quad (85)$$

from which, on setting $h_n=h+H_n$, where $h=a^{-1}\mathscr{L}(F)$ * (the steady density), we obtain

$$H_{n+2}-H_{n+1}=h_{n+1}(e^{a(H_{n+1}-FH_n)}-1). \quad . \quad . \quad . \quad . \quad . \quad . \quad . \quad (86)$$

We first prove that neither p_n nor h_n can tend asymptotically to any limit.

From (83) it is easily seen that if an asymptotic limit of p_n exists it must have the value p of the steady density. But the following argument shows that even the value p cannot be attained asymptotically by p_n.

Let us suppose that asymptotic approach has commenced before the rth generation, and that p_n is very close in value to p when $n>r$.

Then P_n, P_{n-1}, etc., are of the first order of smallness; so neglecting their mutual products and their powers higher than the first in the expansion of (84) this becomes

$$P_{n+2}-P_{n+1}=\mathscr{L}(F)(P_{n+1}-FP_n). \quad . \quad . \quad . \quad . \quad . \quad . \quad . \quad (87)$$

This is an equation in Finite Differences which may be written, with u_n replacing P_n, in the form

$$(E^2-2bE+c^2)u_n=0,$$

where E is the well-known operator, and

$$2b=1+\mathscr{L}(F), \quad c^2=\mathscr{L}(F^{-1}).$$

The general solution of this equation is

$$u_n=A\mid c\mid^n \sin(\phi n+B), \quad . \quad . \quad . \quad . \quad . \quad . \quad . \quad . \quad (88)$$

where

$$c=\sqrt{\mathscr{L}(F^{-1})}, \quad \phi=\cos^{-1}\left(\frac{1+\mathscr{L}(F)}{2\sqrt{\mathscr{L}(F^{-1})}}\right)$$

and A, B are arbitrary, but necessarily real, constants.

We thus find that the behaviour of u_n, *i. e.*, of small P_n, is non-asymptotic, and so the value p cannot be asymptotically attained by p_n.

Similarly (85) shows that h_n can have no asymptotic limit other than the steady value h, and with h_n in the neighbourhood of h (86) reduces to the approximate form

$$H_{n+2}-H_{n+1}=\mathscr{L}(F)(H_{n+1}-FH_n), \quad . \quad . \quad . \quad . \quad . \quad . \quad . \quad (89)$$

which is identical with that of (87).

* $\mathscr{L}(F)=\log(F)/(F-1)$.

Hence even the value h cannot be asymptotically attained by h_n.

It can now be shown that, even when the magnitudes of p_n and h_n are unrestricted, p_n cannot perpetually exceed p. The argument is in the form of a "reductio ad absurdum" based on a relation deduced from (80) and (81).

Similarly it can be shown that p_n cannot perpetually be less than p. Hence

p_n oscillates perpetually about the steady value p. . . . (90)

From this result, with the help of (82), we can now deduce that

h_n oscillates perpetually about the steady value h. . . . (91)

It is easy, moreover, to show that when the parasite-density passes through its steady value the host-density attains a maximum or minimum value. For when $p_{n-1}<p<p_n$ (80) requires that $h_{n-1}<h_n>h_{n+1}$; hence h_n is a maximum. Thus, in agreement with text-figs. 10–13,

the parasite-oscillations lag behind the host-oscillations by about a quarter of a period . . . . (92)

On multiplying (80) and (81) throughout by a, they become

$$\eta_{n+1}=\mathrm{F}\eta_n e^{-\pi_n},$$

and

$$\pi_{n+1}=\mathrm{F}\eta_n-\eta_{n+1},$$

where

$$\eta_n=ah_n, \quad \pi_n=ap_n.$$

The curves representing η_n and π_n in terms of n are evidently independent of a. By a change of the *vertical scale* alone these curves can be made to represent h_n and p_n in terms of n. So

the characters of the curves for h_n and p_n are independent of the area of discovery. . . . . (93)

To examine the principal features of the oscillations of p_n and h_n when small we return to (88). The period T of such an oscillation, in generations, is given by

$$\mathrm{T}=2\pi/\cos^{-1}\left(\frac{1+\mathscr{L}(\mathrm{F})}{2\sqrt{\mathscr{L}(\mathrm{F}^{-1})}}\right). \quad . \; . \; . \; . \; . \; . \; . \quad (94)$$

Since $|\,c\,|>1$ the factor $|\,c\,|^n$ increases with n and so with time. Hence the incremental factor I is equal to $|\,c\,|^{\mathrm{T}}$, *i. e.*,

$$\mathrm{I}=\{\mathscr{L}(\mathrm{F}^{-1})\}^{\mathrm{T}/2}. \quad . \; . \; . \; . \; . \; . \; . \; . \; . \quad (95)$$

Thus:

the small oscillations ultimately become large ones . . . . . . (96)

The following table shows numbers calculated by means of these two formulæ :—

F.	T.	I.
2	8·2	3·8
4	6·3	6·8
∞	4·0	∞

These numbers are in excellent agreement with the corresponding results for small oscillations shown in text-figs. 10 and 11.

We thus conclude :

C. 45. The densities of the interacting species oscillate about their respective steady values (statements (90) and (91)).

C. 46. The small oscillations ultimately become large ones (statement (96)).

C. 47. The small oscillations of both species are identical in character (equations (87) and (89)).

C. 48. The parasite-oscillations lag behind the host-oscillations by about a quarter of a period (statement (92)).

C. 49. For small oscillations the greater the power of increase the greater is the rate of increase of the amplitude (equations (95) and (94)).

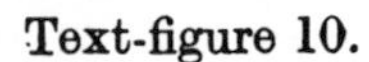
Text-figure 10.

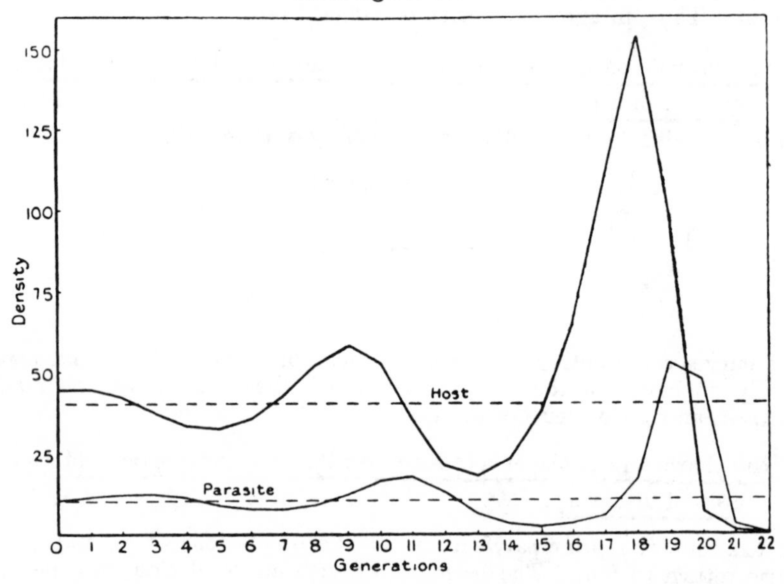

A specific parasite and a specific host interact. Power of increase of host 2. Area of discovery of parasite ·035. Original arbitrary displacement of initial host-density from 40 to 44. Host-curve shows initial density. Parasite-curve drawn to half vertical scale of host-curve. Dotted lines show steady densities.

C. 50. The characters of the density-curves for both species are independent of the area of discovery (statement (93)).

Conclusions C. 45 to C. 49 are illustrated by the curves in text-figs. 10 and 11. These curves were derived arithmetically by assuming an initial arbitrary displacement of the density of one of the species from its steady value and then, by the use of the " competition curve " *, calculating the densities of the animals at the end of the first generation. With these new data the densities of the animals in the second generation were calculated, and so on.

Verbal argument is given elsewhere † which shows that the oscillation of the population-densities is due to the fact that, although any displacement

* Nicholson, Journ. Animal Ecology, vol. ii. no. 1, fig. 1 (1933).

† Nicholson, *loc. cit.*

of a species from its steady density sets up a reaction tending to cause a return to this density, there is always a delay in the occurrence of the reaction. Consequently the density passes the steady value, and this new displacement sets up a new reaction of opposite sign to the preceding one. As this oscillation is wholly due to the interaction of animals it will be called *interspecific oscillation*, in order to distinguish it from oscillation due to external causes.

In addition to illustrating C. 45 to C. 49, text-figs. 10 and 11 show that the period of oscillation increases as the amplitude becomes large.

Text-figure 11.

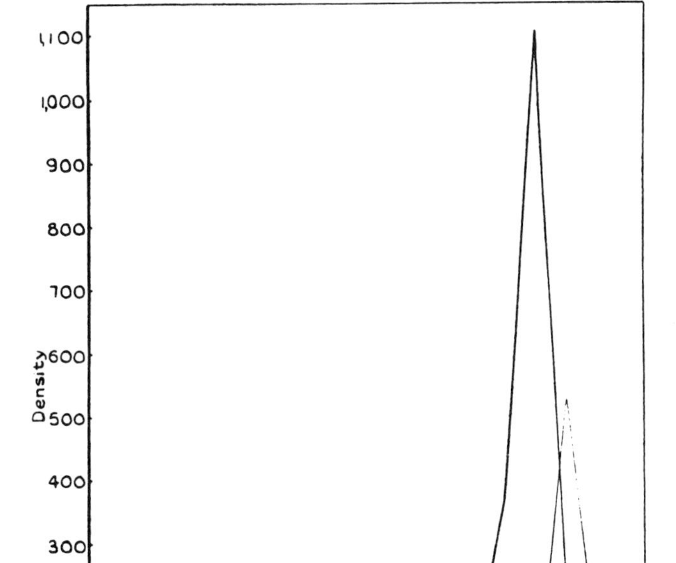

A specific parasite and a specific host interact. Power of increase of host 4. Area of discovery of parasite ·035. Original arbitrary displacement of initial host-density from 53·3 to 58. Host-curve shows initial density. Parasite-curve drawn to half vertical scale of host-curve.

C. 50 has been confirmed by the construction of curves for other numerical examples, but its truth is at once evident from the fact that the unit of area may have any value. If, for example, the absolute value of a unit of area in text-fig. 10 were one square yard, we might take the unit of area as two square yards, so halving the numerical value of the area of discovery of the parasite, and the only effect of this would be to double the values of the densities of the animals. Clearly the character of the oscillation could not be influenced by merely changing the arbitrary unit of measurement.

2. *Several specific parasites and a specific host interact.*

The curves in text-fig. 12 show the character of interspecific oscillation when a common host is attacked by two successive species of specific parasites.

Text-figure 12.

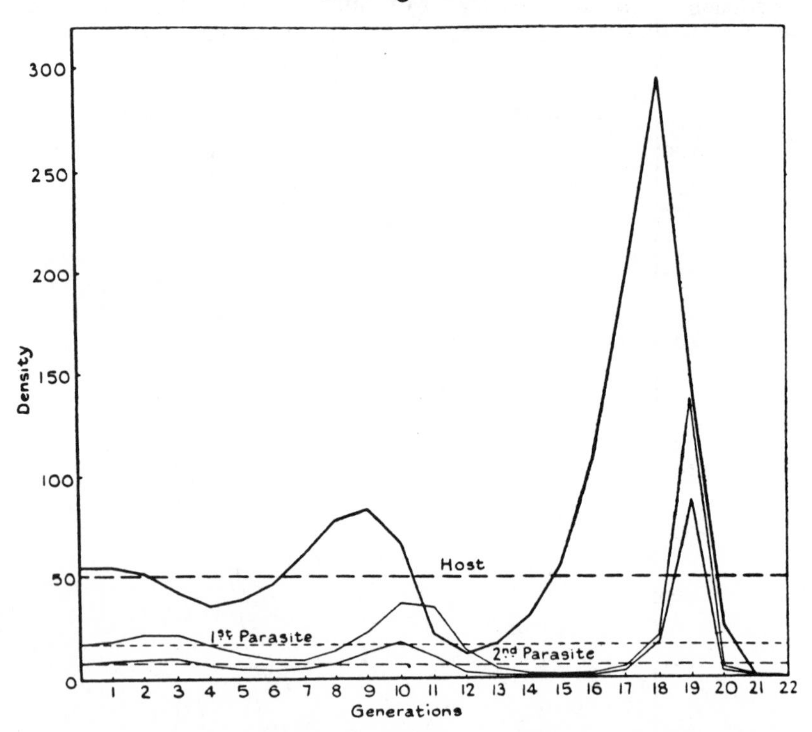

Two specific parasites and a common host interact. Power of increase of host 2. Area of discovery of first parasite ·025, and of second parasite ·035. Original arbitrary displacement of initial host-density from 50 to 55. Host-curve shows initial density.

These and similar curves, based on other numerical examples, show that C. 45 to C. 50 are true whether a host is attacked by one or by several specific parasites.

3. *A specific host, a specific parasite, and a specific hyperparasite interact.*

The curves in text-fig. 13 show the influence of a hyperparasite on interspecific oscillation. Except for the presence of the hyperparasite the data are the same as those on which text-fig. 10 is based. It will be observed that the presence of the hyperparasite has shortened the period of oscillation and that successive oscillations grow less rapidly. The hyperparasite curve lags behind the parasite curve by about a quarter of a period, just as the parasite curve lags behind the host curve. Consequently the hyperparasite attacks the parasite most severely when the host-density is at a minimum and least severely when the host-density is at a maximum—that is to say the violence of the reaction between the host and parasite is lessened, and this

Text-figure 13.

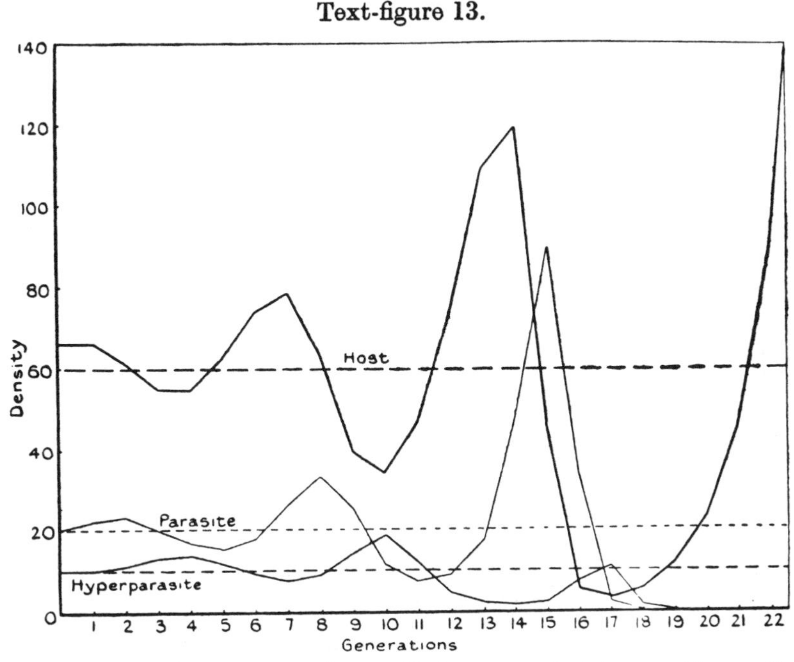

A specific host, a specific parasite, and a specific hyperparasite interact. Power of increase of host 2. Area of discovery of parasite ·035, and of hyperparasite ·04. Original arbitrary displacement of initial host-density from 60 to 66. Host-curve shows initial density. Parasite-curve is based on the density of parasites that remain unattacked.

appears sufficient to account for the above-mentioned influence of a hyperparasite on interspecific oscillation.

4. *Ultimate effects of increasing oscillation.*

The curves in text-figs. 10 to 13 suggest that in general successive large oscillations perpetually increase in magnitude, though it has not yet been found possible to prove this mathematically (see p. 583). It is inconceivable that such perpetual increase takes place in nature, for this is certainly not compatible with what we observe to happen. When considering large fluctuations, however, account must be taken of other factors besides those already used in this investigation.

Each female is the centre of diffusion of her offspring, so when the density of a species becomes very low as a result of violent oscillation the zones of diffusion from the various females can seldom overlap and may frequently be separated from one another by great intervals. Thus the interacting animals exist in numerous disconnected small groups, within each of which interspecific oscillation follows its course independently of that in the other groups. Since the numbers of animals in these groups are very small compared with species-populations, the great reduction in numbers soon produced by increasing oscillation frequently exterminates the groups, but meanwhile some hosts will have migrated into the surrounding previously unoccupied country and have established new groups there. It follows that :

C. 51. A probable ultimate effect of increasing oscillation is the breaking up of the species-population into numerous small widely separated groups which wax and wane and then disappear, to be replaced by new groups in previously unoccupied situations.

Parasites necessarily can develop only in situations where hosts exist, and so they at least commence their search in such situations. Consequently the regions containing hosts are searched more efficiently than the intervening regions which are devoid of hosts. Thus, when the hosts exist in widely separated groups, the density of parasites necessary to find the surplus hosts is much less than it would be if the parasites searched evenly over the whole environment, and the density of hosts necessary to support this low density of parasites is similarly low. Hence :

C. 52. The densities maintained by animals which occur in such scattered groups are lower than their calculated steady densities.

It should be noticed, however, that the *relative* values of the densities such animals maintain should generally be similar to the relative values of their calculated steady densities, for the densities tend to oscillate about their calculated steady values *within those areas in which the animals interact,* and these constitute a certain fraction of the environment. But differences in the powers of diffusion of different species will modify this relation, for they influence the abundance and size of the groups.

When the density of a species becomes very great as a result of increasing oscillation the retarding influence of such factors as scarcity of food or of suitable places in which to live is bound to be felt. Clearly these factors will prevent unlimited increase in density, so :

C. 53. Large fluctuations produced by increasing oscillation may ultimately be limited by factors other than parasites, so that the oscillation is perpetually maintained at a large constant amplitude in a constant environment.

It is possible that a similar result might be produced by great differences in the properties of different individuals of the same species. Thus if some host-individuals are far more difficult to find than others most of these would remain unattacked even when the parasites delivered an exceptionally severe attack. This would limit the degree to which the host-population could be reduced, and so would cause the oscillations to be maintained at a constant amplitude in a constant environment.

Whether C. 51 and C. 52, or C. 53 apply in any particular instance depends on whether the limiting effects of very low or of very high densities first come into operation.

5. *A general parasite, a host not controlled by it, and a specific host interact.*

Verbal arguments are given elsewhere * which show that, when some of the host-species that a given parasite attacks are not under its control, this lessens the violence of the reactions of the given parasite to variations in the density of the host it does control, and may lead to interspecific oscillation, which decreases progressively with time. A quantitative investigation of this problem is as follows :—

Let us suppose a parasite-species to interact with a specific host-species and with a general host-species whose density, available to attack by this parasite, remains constant.

* Nicholson, Journ. Animal Ecology, vol. ii. no. 1, p. 164.

Let p_n, h_n, and h' be their respective densities at the nth generation, F and F′ the respective powers of increase of the hosts, and a and a' the areas of discovery of the parasite for the respective hosts.

Then by reasoning similar to that in Sub-Section I., the following two recurrence relations are easily obtained :

$$\left.\begin{aligned} h_{n+1}&=\mathrm{F}h_n e^{-ap_n}, \\ p_{n+1}&=\mathrm{F}h_n-h_{n+1}+\mathrm{F}'h'(1-e^{-a'p_n}). \end{aligned}\right\} \qquad (97)$$

Eliminating p_n and p_{n+1} we obtain the following difference-equation for h_n:—

$$a^{-1}\log\left(\frac{h_{n+2}}{\mathrm{F}h_{n+1}}\right)+\mathrm{F}h_n-h_{n+1}+\mathrm{F}'h'\left\{1-\left(\frac{h_{n+1}}{\mathrm{F}h_n}\right)^{a'/a}\right\}=0. \quad (98)$$

The steady density h of the specific host is therefore

$$h=a^{-1}\mathscr{L}(\mathrm{F})-h'\mathrm{F}'\left(\frac{1-\mathrm{F}^{-a'/a}}{\mathrm{F}-1}\right), \qquad (99)$$

i. e., the introduction of the general host reduces the steady density of the specific host.

The necessary condition for the steady state to be possible is therefore

$$a'h'\mathrm{F}'/\mathscr{L}(\mathrm{F}^{-a'/a})<1. \qquad (100)$$

The variation of p_n and h_n with increasing n may be investigated by methods similar to those used in Section I., but the work is necessarily more complicated, and so we limit the investigation to the behaviour near the steady state.

Accordingly in (98) we set

$$h_n=h+\mathrm{H}_n,$$

where H_n, H_{n+1}, etc. are taken as quantities of the first order of smallness, and after expansion of the first and last terms in (98) we retain only quantities of the first order.

This results in the following linear difference equation :—

$$\mathrm{H}_{n+2}-2b\mathrm{H}_{n+1}+c^2\mathrm{H}_n=0, \qquad (101)$$

where

$$2b=1+ah+a'h'\mathrm{F}'\mathrm{F}^{-a'/a},$$

$$c^2=ah\mathrm{F}+a'h'\mathrm{F}'\mathrm{F}^{-a'/a}.$$

In virtue of (99) therefore

$$2b=1+\mathscr{L}(\mathrm{F})+g'\left\{1-\frac{\mathscr{L}(\mathrm{F})}{\mathscr{L}(\mathrm{F}^{a'/a})}\right\}, \qquad (102)$$

$$c^2=\mathscr{L}(\mathrm{F}^{-1})-g'\left\{\frac{\mathscr{L}(\mathrm{F}^{-1})}{\mathscr{L}(\mathrm{F}^{a'/a})}-1\right\}, \qquad (103)$$

where

$$g'=a'h'\mathrm{F}'\mathrm{F}^{-a'/a}.$$

The general solution of (101) is

$$\mathrm{H}_n=\mathrm{A}(b+\sqrt{b^2-c^2})^n+\mathrm{B}(b-\sqrt{b^2-c^2})^n, \qquad (104)$$

where A and B are arbitrary constants.

We now examine the properties of this expression under different possible circumstances.

With the help of (100) it is easily seen that

$$0<g'/\mathscr{L}(F^{a'/a})<1. \qquad (105)$$

Let us suppose that g' is increased monotonically from 0 to $\mathscr{L}(F^{a'/a})$, and determine the effect of this on the quantities b and c^2.

Then the quantity $2b$ will change from $1+\mathscr{L}(F)$ to $1+\mathscr{L}(F^{a'/a})$.

It follows that always

$$\tfrac{1}{2}<b<1, \qquad (106)$$

and that b^2 will change from $\tfrac{1}{4}\{1+\mathscr{L}(F)\}^2$ to $\tfrac{1}{4}\{1+\mathscr{L}(F^{a'/a})\}^2$.

The cofactor of g' in (103) is always positive because $\mathscr{L}(F^{-1})>1$ and $\mathscr{L}(F^{a'/a})<1$. Hence c^2 will decrease from $\mathscr{L}(F^{-1})$ to $\mathscr{L}(F^{a'/a})$. Thus c^2 will pass from a value which is greater than 1 to a value less than 1 but positive.

It is easy to see for a real quantity $x\neq 1$ that $(1+x)^2>4x$. Hence

$$\tfrac{1}{4}\{1+\mathscr{L}(F^{a'/a})\}^2>\mathscr{L}(F^{a'/a}).$$

Also since $\tfrac{1}{4}\{1+\mathscr{L}(F)\}^2<1$ and $\mathscr{L}(F^{-1})>1$ it follows that

$$\tfrac{1}{4}\{1+\mathscr{L}(F)\}^2<\mathscr{L}(F^{-1}).$$

Hence a value G′ of g' exists below which $b^2<c^2$ and above which $b^2>c^2$. This is clearly seen if b^2 and c^2 are represented graphically as ordinates with g' as the abscissa.

There are thus two situations to be examined in connection with (104), namely :—

I. $b^2<c^2$.

Introducing the real angle $\phi=\cos^{-1}(b/|c|)$, the equation (104) becomes

$$H_n=A|c|^n \sin(\phi n+B),$$

where A and B are arbitrary constants.

This represents an oscillation whose amplitude increases or decreases with time according as $|c|\gtrless 1$.

The period T, expressed in generations, is

$$T=2\pi/\cos^{-1}(b/|c|).$$

As g' increases from 0 to G′, ϕ decreases from $\phi_0=\cos^{-1}\{1+\mathscr{L}(F)/2\sqrt{\mathscr{L}(F^{-1})}\}$ to 0, and $|c|$ decreases to a value less than 1. Hence, the effect of introducing the general host up to a certain density is to increase the period ultimately to an infinite value and to convert the increasing oscillations into damped oscillations.

II. $b^2>c^2$.

From the formulæ (102) and (103) we obtain this :—

$$1-2b+c^2=ah(F-1).$$

The right-hand side is positive, so

$$1-2b+c^2>0,$$

$$\therefore\ (1-b)^2>b^2-c^2>0.$$

From (106) $1>b$, so

$$1>b+\sqrt{b^2-c^2}.$$

Also $b^2>b^2-c^2>0$, and from (106) $b>0$; hence $b-\sqrt{b^2-c^2}>0$.

We thus conclude that

$$1>b+\sqrt{b^2-c^2}>b-\sqrt{b^2-c^2}>0.$$

Equation (104) then shows that H_n converges on the value 0, and does so ultimately like the quantity $A(b+\sqrt{b^2-c^2})^n$. So the host-density converges on the steady value from above or below according as A is positive or negative.

The special situations arising when $b^2=c^2$ or $c^2=1$ need not be discussed, since their existence is infinitely improbable.

In regard to the parasite-density we may now proceed as follows :—

Setting in the first of equations (97) $h_n=h+H_n$ and $p_n=p+P_n$, where P_n, P_{n+1}, etc., are of the first order of smallness, we find that

$$ahP_n=H_n-H_{n+1}.$$

Using (104) this leads to

$$P_n=C(b+\sqrt{b^2-c^2})^n+D(b-\sqrt{b^2-c^2})^n,$$

where

$$C=A(1-b-\sqrt{b^2-c^2}) \text{ and } D=B(1-b+\sqrt{b^2-c^2}).$$

Thus the parasite-densities vary with time similarly to the host-densities.

We can investigate in a similar way the more general problem in which there are several species of general hosts present. It is easy to verify that the resultant equations corresponding to equations (97) to (105) are those obtained by prefixing the summation symbol Σ to all the terms in (97) to (105) which contain h', either explicitly or implicitly as in g'.

Thus, corresponding respectively to (102), (103), and (105), we now have

$$2b=1+\mathscr{L}(F)+\Sigma g'-\mathscr{L}(F)\Sigma\frac{g'}{\mathscr{L}(F^{a'/a})},$$

$$c^2=\mathscr{L}(F^{-1})+\Sigma g'-\mathscr{L}(F^{-1})\Sigma\frac{g'}{\mathscr{L}(F^{a'/a})},$$

and

$$0<\Sigma\frac{g'}{\mathscr{L}(F^{a'/a})}<1.$$

Since $\mathscr{L}(F^{a'/a})<1$, and since $g'>0$,

therefore $$0<\Sigma g'<\Sigma\frac{g'}{\mathscr{L}(F^{a'/a})}<1.$$

We now determine the effect on b and c^2 of increasing $\Sigma\frac{g'}{\mathscr{L}(F^{a'/a})}$ from 0 to 1.

It is clear that $2b$ will change from $1+\mathscr{L}(F)$ to $1+(\Sigma g')_m$, where $(\Sigma g')_m$ is the maximum value of $\Sigma g'$. Hence (106) is also valid for the present situation, and b^2 will change from $\frac{1}{4}\{1+\mathscr{L}(F)\}^2$ to $\frac{1}{4}\{1+(\Sigma g')_m\}^2$.

Also c^2 will decrease from $\mathscr{L}(F^{-1})$ to $(\Sigma g')_m$.

But $$\tfrac{1}{4}\{1+(\Sigma g')_m\}^2>(\Sigma g')_m \text{ and } \tfrac{1}{4}\{1+\mathscr{L}(F)\}^2<\mathscr{L}(F^{-1})\,;$$

therefore the two situations $b^2\lesseqgtr c^2$ can occur.

We now proceed exactly as in I. and II. above, and thus conclude that h_n and p_n behave in ways similar to those found to exist for the type of interaction in which only one species of general host is present.

The foregoing investigation leads to these conclusions :—

C. 54. When a parasite attacks other host-species in addition to the one it controls there are three possible situations, namely (1) increasing oscillation, (2) decreasing oscillation, and (3) non-oscillatory convergence

Text-figure 14.

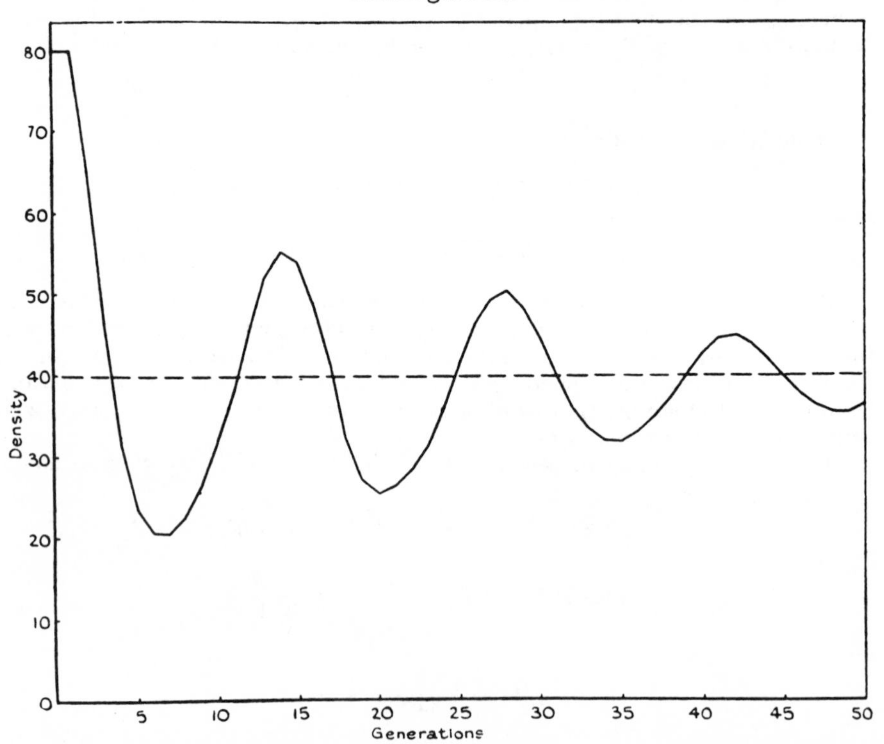

A specific host and a general parasite interact. Power of increase of host 2. Area of discovery of parasite ·01. In the steady state the specific host has an initial density of 40, and some other host of the given parasite has an initial density of 100. The initial density of the latter host is unaffected by the activity of the given parasite and remains indefinitely at 100. Original arbitrary displacement of initial density of specific host from 40 to 80. Curve shows initial density of specific host.

of the density on the steady value. If increasing oscillation is produced it grows less rapidly than when the specific host and specific parasite interact alone.

C. 55. The period of oscillation progressively lengthens with increasing density of one or more of those host-species which are not controlled by the given parasite.

The curves in text-figs. 14 and 15, which have been determined arithmetically, illustrate the above conclusions. From text-fig. 14 we find that for the small oscillations the period T is 13·9 and the decremental factor I is ·46. The formulæ (102) and (103) give respectively $b=\cdot846$, $|c|=\cdot940$, and therefore $\phi=\cdot448$. Hence T=14·0 and I=·42, in satisfactory agreement with the values given by text-fig. 14.

Text-figure 15.

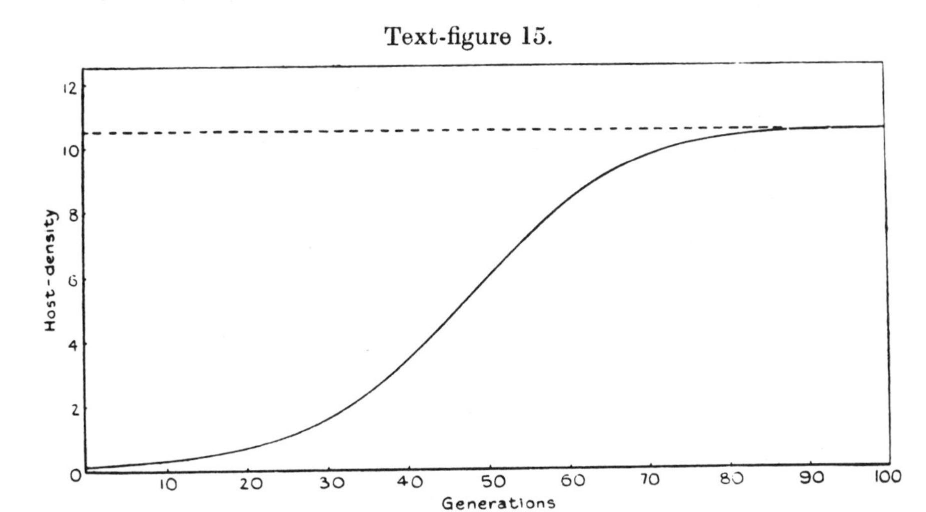

A specific host and a general parasite interact. Power of increase of host 10. Area of discovery of parasite ·01. In the steady state the specific host has an initial density of 10·5, and some other host of the given parasite has an initial density of 245, which is maintained unchanged in all generations. Originally the parasite is maintained by the host it does not control, and the specific host is introduced at a density of 0·1. Curve shows initial density of specific host.

For the situation illustrated in text-fig. 15 the formulæ (102) and (103) give respectively $b^2=\cdot395$, $c^2=\cdot351$, and so $b^2>c^2$. This corresponds to non-oscillatory convergence, in agreement with the curve in text-fig. 15.

Section V. Continuous Interaction.

The determination of the densities of animal populations when their interaction is continuous has been investigated by Lotka *, Volterra †, and Bailey ‡.

The earliest discussion, due to Lotka, is limited to two interacting species and leads in general to damped oscillations of the densities. It neglects the consideration of two factors in the problem, namely (1) the delay in the action of the food eaten by the parasites on the latter's production of offspring and (2) the effects of age-distribution in the animal populations. Lotka has also discussed the general problem involving any number of species, but so broadly that the subject is left somewhat indeterminate.

Volterra's discussion of the same simple problem attempts to take (1)

* Proc. Nat. Acad. Sci. vol. vi. no. 7, pp. 410–415 (1920); Journ. Washington Acad. Sci. vol. xiii. no. 8 (1923); 'Elements of Physical Biology' (Williams & Wilkins Co., Baltimore, 1925).

† 'Léçons sur la théorie mathématique de la lutte pour la vie' (Gauthier-Villars et Cie, Paris, 1931).

‡ Q. J. Maths. vol. ii. no. 5, pp. 68–77 (Oxford, 1931); Proc. Roy. Soc. N.S.W. vol. lxvi. pp. 387–393 (Sydney, 1932); Proc. Roy. Soc. A, vol. cxliii. pp. 75–88 (1933).

into account but neglects (2). His investigation of the problem where any number of species interact neglects both (1) and (2). Moreover, the introduction into his equations of "*termes d'amortissement*" is, as we have already indicated in Section I., a source of error. He concludes in general that where the steady state is possible, bounded fluctuations occur.

In the investigation of Bailey the factors * (1) and (2) are an integral part of the theory, so his conclusions are less open to criticism. This work also presents the convenience of including the problem of discontinuous interaction as a special case.

The further details of the investigation of the problem of continuous interaction between two species are given in a recent paper †, so we give here only such a brief summary ‡ as is needed for the discussion.

For simplicity the probability that a host-individual is able to interact with any parasites it meets is assumed to be independent of the host's age.

Let $j(\eta)$ represent the product $v(\eta) f(\eta)$, where $f(\eta)$ is the probability at birth that a parasite will survive environmental causes of death till the age η, and $v(\eta)$ is the effective areal velocity of parasites of age η in regard to their ability to find hosts.

It is easy to see that as η increases from 0 to ∞ the quantity $j(\eta)$ must, for any real parasite-species, increase from 0 to a maximum value and then decrease again to 0, but always remain positive.

Following Prony's method. we can express $j(\eta)$ as a sum of exponential terms; as a suitable first approximation, taking only two such terms, we express $j(\eta)$ as follows:

$$j(\eta)=a(e^{-b\eta}-e^{-c\eta})$$

where a, b, c are positive constants and $c>b$.

The calculations then show that:—

C. 56. Near the steady state the host- and parasite-densities both oscillate with increasing violence about their respective steady values.

This behaviour is similar to that found when the interaction between the two species is discontinuous (*cf.* C. 46).

More recently it has been found possible to establish § this result also for a closer approximation to the form of $j(\eta)$, namely

$$j(\eta)=a\sum_{n=1}^{m}\frac{e^{-\alpha_n\eta}}{(\alpha_1-\alpha_n)(\alpha_2-\alpha_n)(\alpha_3-\alpha_n)\ldots},$$

where the factor $(\alpha_n-\alpha_n)$ is omitted from the nth denominator and all $\alpha_n>0$.

Apart from a constant factor, this formula is the same as Bateman's expression || for the amount at the time η of the mth product in a series of successive radioactive transformations, when initially only one substance is present. It therefore follows that $j(\eta)$ is positive and bounded for all positive values of η; since also $j(0)=0$ and $j(\infty)=0$, it is clear that the above formula is a suitable approximation.

The fundamental equations of the problem of interaction between any number of animal species are also derived in the paper referred to, and may serve as the basis of future investigations of this problem.

* It is clear that a thorough use of (2) must inevitably include (1) as a special part of it.

† Bailey, Proc. Roy. Soc. A, vol. cxliii. pp. 75–88 (1933).

‡ See also Bailey, Proc. Roy. Soc. N.S.W. vol. lxvi. p. 391 (1932).

§ This investigation will be published later.

|| See E. Rutherford, 'Radioactive Substances and their Radiations,' p. 422 (Cambridge University Press, 1913).

These equations lead to the following important conclusions about the steady state when P species of parasites interact with Q species of hosts :

C. 57, if $P<Q$ a steady state cannot exist.

C. 58, if $P=Q$ a steady state in general is possible.

C. 59, if $P>Q$ a steady state may or may not be possible according to the special properties of the interacting animals. A numerical example has already been given (p. 571) of a steady state with $P=2$, $Q=1$. Examples of interaction for which no steady state can exist have been mentioned elsewhere *, namely those in which the interaction-functions of the host-species are similar functions.

A study of Volterra's equations for the interaction of several species shows that they imply interactions included in the special type just mentioned. It therefore follows that they do not allow steady states to exist unless $P=Q$, which necessarily means that only with an even number of interacting species can a steady state exist ; this is precisely the result obtained by Volterra himself. It is therefore evident that the neglect of the effect of age-distribution in his equations results in the restriction of Volterra's theory to very special types of interaction, and leads to a conclusion about the steady state which does not appear to be generally true.

We thus conclude that in the formulation of the problem of interaction between several species of animals it is essential to consider the effects of age-distribution. The resulting complexity introduced into the fundamental equations makes their full investigation a task to be undertaken only by professional mathematicians, so it is to be hoped that some of them will, like Volterra, interest themselves in this question, which is so important to Biology.

Summary.

Section I.

1. The problem of competition between animals when searching for the things they require for existence is examined.

2. It is shown that, even when individual animals search systematically, the searching of populations is always random, and that the competition resulting from random searching can be expresssed by a simple formula or curve of general application.

Section II.

3. The interaction in the steady state of a specific entomophagous parasite and its specific host is examined, the detailed conclusions being expressed by C. 1 to C. 20.

Section III.

4. The steady state of interacting entomophagous parasites and hosts, having discontinuous interaction, is examined in the following situations :— Several specific parasites attack a common host ; specific parasites are themselves attacked ; a parasite attacks two or more specific hosts. The effects produced by the modification of the area of discovery of a parasite, by the introduction of an additional specific parasite, and by the modification of the power of increase of a common host are also investigated.

5. It is shown that the steady state may exist in all the situations just mentioned provided the animals have properties that lie between certain

* V. A. Bailey, Proc. Roy. Soc. N.S.W. vol. lxvi. p. 390 (1932).

limits, and that Volterra's conclusion, that only an even number of interacting species can be in a steady state, is incorrect.

6. The detailed conclusions are expressed by C. 21 to C. 44.

Section IV.

7. The problem of the interaction of entomophagous parasites and their hosts when not in the steady state is examined for certain relatively simple types of interaction.

8. It is concluded that the interaction of animals may by itself cause the densities of the animals to oscillate about their steady values. According to the circumstances the magnitude of the oscillations may decrease, remain constant, or increase with time. In the latter situation the oscillation may ultimately split the species-population into numerous small groups of individuals which fluctuate in density independently of one another.

9. The detailed conclusions are expressed by C. 45 to C. 55.

Section V.

10. The problem involving continuous interaction between entomophagous parasites and their hosts is discussed.

11. It is concluded that it is essential to consider the effects of age-distribution. When this is done it is found, as in Section IV., that the interaction of two species leads to oscillations of the animal densities about their steady values, and that the magnitude of these oscillations increases with time.

*12. The discussions by Lotka and Volterra of the problem involving continuous interaction are criticised, and are shown to lead to results inconsistent with some of the conclusions of the more rigorous theory, due to Bailey, in which the effects of age-distribution are explicitly considered.

13. The detailed conclusions are expressed by C. 56 to C. 59.

* While this communication was in the press two papers by G. F. Gause (' Science,' vol. lxxix. p. 16 (1934), Journ. Expt. Biol. (London), vol. xii. no. 1, pp. 44–48) (1935)), came to our notice. In these papers experimental studies of the interaction of certain micro-organisms are described, and the results are discussed in the light of Lotka's and Volterra's equations.

THE TROPHIC-DYNAMIC ASPECT OF ECOLOGY

RAYMOND L. LINDEMAN
Osborn Zoological Laboratory, Yale University

Recent progress in the study of aquatic food-cycle relationships invites a reappraisal of certain ecological tenets. Quantitative productivity data provide a basis for enunciating certain trophic principles, which, when applied to a series of successional stages, shed new light on the dynamics of ecological succession.

"COMMUNITY" CONCEPTS

A chronological review of the major viewpoints guiding synecological thought indicates the following stages: (1) the static species-distributional viewpoint; (2) the dynamic species-distributional viewpoint, with emphasis on successional phenomena; and (3) the trophic-dynamic viewpoint. From either species-distributional viewpoint, a lake, for example, might be considered by a botanist as containing several distinct plant aggregations, such as marginal emergent, floating-leafed, submerged, benthic, or phytoplankton communities, some of which might even be considered as "climax" (cf. Tutin, '41). The associated animals would be "biotic factors" of the plant environment, tending to limit or modify the development of the aquatic plant communities. To a strict zoologist, on the other hand, a lake would seem to contain animal communities roughly coincident with the plant communities, although the "associated vegetation" would be considered merely as a part of the environment[1] of the animal community. A more "bio-ecological" species-distributional approach would recognize both the plants and animals as co-constituents of restricted "biotic" communities, such as "plankton communities," "benthic communities," etc., in which members of the living community "co-act" with each other and "react" with the non-living environment (Clements and Shelford, '39; Carpenter, '39, '40; T. Park, '41). Coactions and reactions are considered by bio-ecologists to be the dynamic effectors of succession.

The trophic-dynamic viewpoint, as adopted in this paper, emphasizes the relationship of trophic or "energy-availing" relationships within the community-unit to the process of succession. From this viewpoint, which is closely allied to Vernadsky's "biogeochemical" approach (cf. Hutchinson and Wollack, '40) and to the "oekologische Sicht" of Friederichs ('30), a lake is considered as a primary ecological unit in its own right, since all the lesser "communities" mentioned above are dependent upon other components of the lacustrine food cycle (cf. figure 1) for their very existence. Upon further consideration of the trophic cycle, the discrimination between living organisms as parts of the "biotic community" and dead organisms and inorganic nutritives as parts of the "environment" seems arbitrary and unnatural. The difficulty of drawing clear-cut lines between the living *community* and the non-living *environment* is illustrated by the difficulty of determining the status of a slowly dying pondweed covered with periphytes, some of which are also continually dying. As indicated in figure 1, much of the non-living nascent ooze is rapidly reincorporated through "dis-

---

[1] The term *habitat* is used by certain ecologists (Clements and Shelford, '39; Haskell, '40; T. Park, '41) as a synonym for *environment* in the usual sense and as here used, although Park points out that most biologists understand "habitat" to mean "simply the place or niche that an animal or plant occupies in nature" in a species-distributional sense. On the other hand, Haskell, and apparently also Park, use "environment" as synonymous with the *cosmos*. It is to be hoped that ecologists will shortly be able to reach some sort of agreement on the meanings of these basic terms.

solved nutrients" back into the living "biotic community." This constant organic-inorganic cycle of nutritive substance is so completely integrated that to consider even such a unit as a lake primarily as a biotic community appears to force a "biological" emphasis upon a more basic functional organization.

This concept was perhaps first expressed by Thienemann ('18), as a result of his extensive limnological studies on the lakes of North Germany. Allee ('34) expressed a similar view, stating: "The picture than finally emerges . . . is of a sort of superorganismic unity not alone between the plants and animals to form biotic communities, but also between the biota and the environment." Such a concept is inherent in the term *ecosystem*, proposed by Tansley ('35) for the fundamental ecological unit.[2] Rejecting the terms "complex organism" and "biotic community," Tansley writes, "But the more fundamental conception is, as it seems to me, the whole *system* (in the sense of physics), including not only the organism-complex, but also the whole complex of physical factors forming what we call the environment of the biome. . . . It is the systems so formed which, from the point of view of the ecologist, are the basic units of nature on the face of the earth. . . . These *ecosystems*, as we may call them, are of the most various kinds and sizes. They form one category of the multitudinous physical systems of the universe, which range from the universe as a whole down to the atom." Tansley goes on to discuss the ecosystem as a category of rank equal to the "biome" (Clements, '16), but points out that the term can also be used in a general sense, as is the word "community." The *ecosystem* may be formally defined as the system composed of physical-chemical-biological processes active within a space-time unit of any magnitude, i.e., the biotic community *plus* its abiotic environment. The concept of the ecosystem is believed by the writer to be of fundamental importance in interpreting the data of dynamic ecology.

[2] The ecological system composed of the "biocoenosis + biotop" has been termed the *holocoen* by Friederichs ('30) and the *biosystem* by Thienemann ('39).

Trophic Dynamics

Qualitative food-cycle relationships

Although certain aspects of food relations have been known for centuries, many processes within ecosystems are still very incompletely understood. The basic process in trophic dynamics is the transfer of energy from one part of the ecosystem to another. All function, and indeed all life, within an ecosystem depends upon the utilization of an external source of energy, solar radiation. A portion of this incident energy is transformed by the process of photosynthesis into the structure of living organisms. In the language of community economics introduced by Thienemann ('26), autotrophic plants are *producer* organisms, employing the energy obtained by photosynthesis to synthesize complex organic substances from simple inorganic substances. Although plants again release a portion of this potential energy in catabolic processes, a great surplus of organic substance is accumulated. Animals and heterotrophic plants, as *consumer* organisms, feed upon this surplus of potential energy, oxidizing a considerable portion of the consumed substance to release kinetic energy for metabolism, but transforming the remainder into the complex chemical substances of their own bodies. Following death, every organism is a potential source of energy for saprophagous organisms (feeding directly on dead tissues), which again may act as energy sources for successive categories of consumers. Heterotrophic bacteria and fungi, representing the most important saprophagous consumption of energy, may be conveniently differentiated from animal consumers as special-

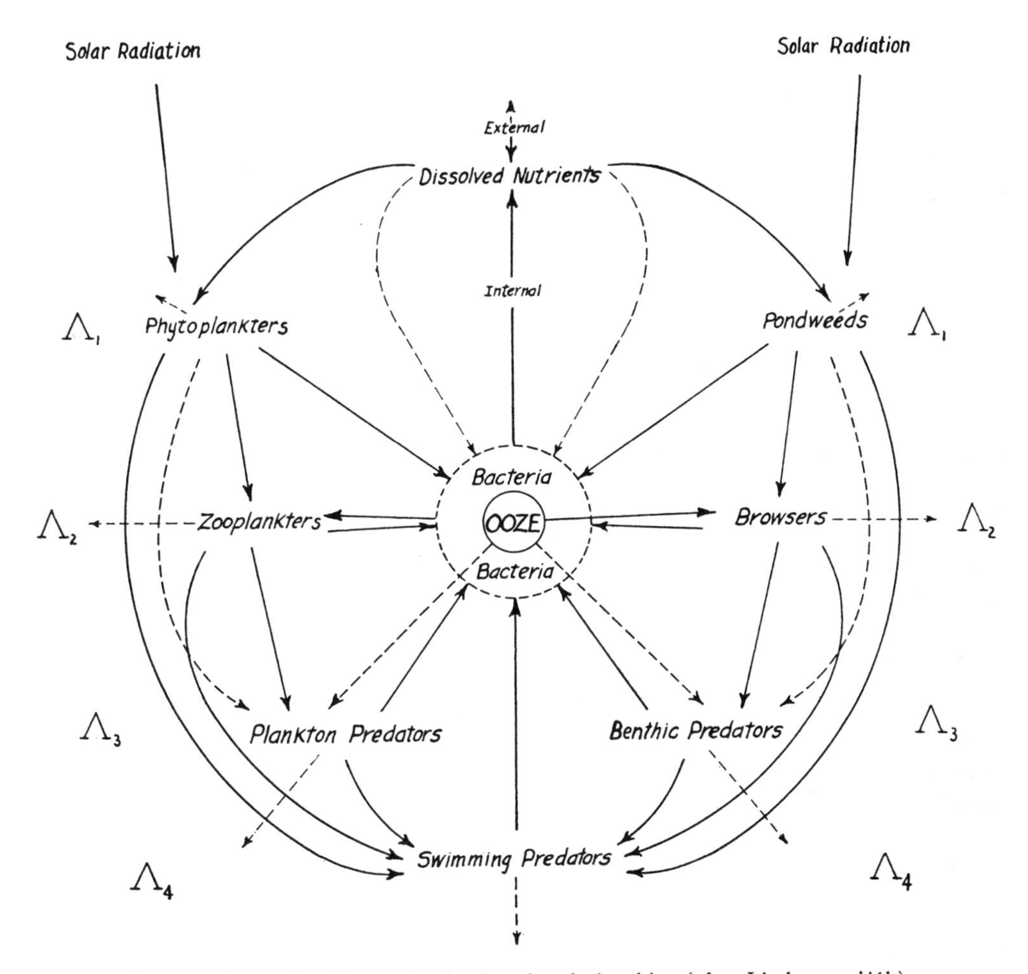

FIG. 1. Generalized lacustrine food-cycle relationships (after Lindeman, '41b).

ized *decomposers*[3] of organic substance. Waksman ('41) has suggested that certain of these bacteria be further differentiated as *transformers* of organic and inorganic compounds. The combined action of animal consumers and bacterial decomposers tends to dissipate the potential energy of organic substances, again transforming them to the inorganic state. From this inorganic state the autotrophic plants may utilize the dissolved nutrients once more in resynthesizing complex organic substance, thus completing the food cycle.

The careful study of food cycles reveals an intricate pattern of trophic predilections and substitutions underlain by certain basic dependencies; food-cycle diagrams, such as figure 1, attempt to portray these underlying relationships. In general, predators are less specialized in food habits than are their prey. The ecological importance of this statement seems to have been first recognized by Elton ('27), who discussed its effect on the survival of prey species when predators are numerous and its effect in enabling predators to survive when their

[3] Thienemann ('26) proposed the term *reducers* for the heterotrophic bacteria and fungi, but this term suggests that decomposition is produced solely by chemical reduction rather than oxidation, which is certainly not the case. The term *decomposers* is suggested as being more appropriate.

usual prey are only periodically abundant. This ability on the part of predators, which tends to make the higher trophic levels of a food cycle less discrete than the lower, increases the difficulties of analyzing the energy relationships in this portion of the food cycle, and may also tend to "shorten the food-chain."

Fundamental food-cycle variations in different ecosystems may be observed by comparing lacustrine and terrestrial cycles. Although dissolved nutrients in the lake water and in the ooze correspond directly with those in the soil, the autotrophic producers differ considerably in form. Lacustrine producers include macrophytic pondweeds, in which massive supporting tissues are at a minimum, and microphytic phytoplankters, which in larger lakes definitely dominate the production of organic substance. Terrestrial producers are predominantly multicellular plants containing much cellulose and lignin in various types of supporting tissues. Terrestrial herbivores, belonging to a great number of specialized food groups, act as *primary consumers* (sensu Jacot, '40) of organic substance; these groups correspond to the "browsers" of aquatic ecosystems. Terrestrial predators may be classified as more remote (secondary, tertiary, quaternary, etc.) consumers, according to whether they prey upon herbivores or upon other predators; these correspond roughly to the benthic predators and swimming predators, respectively, of a lake. Bacterial and fungal decomposers in terrestrial systems are concentrated in the humus layer of the soil; in lakes, where the "soil" is overlain by water, decomposition takes place both in the water, as organic particles slowly settle, and in the benthic "soil." Nutrient salts are thus freed to be reutilized by the autotrophic plants of both ecosystems.

The striking absence of terrestrial "life-forms" analogous to plankters[4] (cf. figure 1) indicates that the terrestrial food cycle is essentially "mono-cyclic" with macrophytic producers, while the lacustrine cycle, with two "life-forms" of producers, may be considered as "bicyclic." The marine cycle, in which plankters are the only producers of any consequence, may be considered as "mono-cyclic" with microphytic producers. The relative absence of massive supporting tissues in plankters and the very rapid completion of their life cycle exert a great influence on the differential productivities of terrestrial and aquatic systems. The general convexity of terrestrial systems as contrasted with the concavity of aquatic substrata results in striking trophic and successional differences, which will be discussed in a later section.

Productivity

Definitions.—The quantitative aspects of trophic ecology have been commonly expressed in terms of the productivity of the food groups concerned. Productivity has been rather broadly defined as the general rate of production (Riley, '40, and others), a term which may be applied to any or every food group in a given ecosystem. The problem of productivity as related to biotic dynamics has been critically analyzed by G. E. Hutchinson ('42) in his recent book on limnological principles. The two following paragraphs are quoted from Hutchinson's chapter on "The Dynamics of Lake Biota":

> The dynamics of lake biota is here treated as primarily a problem of energy transfer . . . the biotic utilization of solar energy entering the lake surface. Some of this energy is transformed by photosynthesis into the structure of phytoplankton organisms, representing an energy content which may be expressed as Λ_1 (first level). Some of the phytoplankters will be eaten by

[4] Francé ('13) developed the concept of the *edaphon*, in which the soil microbiota was represented as the terrestrial equivalent of aquatic plankton. This concept appears to have a number of adherents in this country. The author feels that this analogy is misleading, as the edaphon, which has almost no producers, represents only a dependent side-chain of the terrestrial cycle, and is much more comparable to the lacustrine microbenthos than to the plankton.

zooplankters (energy content Λ_2), which again will be eaten by plankton predators (energy content Λ_3). The various successive levels (i.e., stages[5]) of the food cycle are thus seen to have successively different energy contents (Λ_1, Λ_2, Λ_3, etc.).

Considering any food-cycle level Λ_n, energy is entering the level and is leaving it. The rate of change of the energy content Λ_n therefore may be divided into a positive and a negative part:

$$\frac{d\Lambda_n}{dt} = \lambda_n + \lambda_n',$$

where λ_n is by definition positive and represents the rate of contribution of energy from Λ_{n-1} (the previous level) to Λ_n, while λ_n' is negative and represents the sum of the rate of energy dissipated from Λ_n and the rate of energy content handed on to the following level Λ_{n+1}. The more interesting quantity is λ_n which is defined as the true *productivity* of level Λ_n. In practice, it is necessary to use mean rates over finite periods of time as approximations to the mean rates $\lambda_0, \lambda_1, \lambda_2$. . . .

In the following pages we shall consider the quantitative relationships of the following productivities: λ_0 (rate of incident solar radiation), λ_1 (rate of photosynthetic production), λ_2 (rate of primary or herbivorous consumption), λ_3 (rate of secondary consumption or primary predation), and λ_4 (rate of tertiary consumption). The total amount of organic structure formed per year for any level Λ_n, which is commonly expressed as the annual "yield," actually represents a value uncorrected for dissipation of energy by (1) respiration, (2) predation, and (3) post-mortem decomposition. Let us now consider the quantitative aspects of these losses.

Respiratory corrections.—The amount of energy lost from food levels by catabolic processes (respiration) varies considerably for the different stages in the life histories of individuals, for different levels in the food cycle and for different seasonal temperatures. In terms of annual production, however, individual deviates cancel out and respiratory differences between food groups may be observed.

[5] The term *stage,* in some respects preferable to the term *level,* cannot be used in this trophic sense because of its long-established usage as a successional term (cf. p. 23).

Numerous estimates of average respiration for photosynthetic producers may be obtained from the literature. For terrestrial plants, estimates range from 15 per cent (Pütter, re Spoehr, '26) to 43 per cent (Lundegårdh, '24) under various types of natural conditions. For aquatic producers, Hicks ('34) reported a coefficient of about 15 per cent for Lemna under certain conditions. Wimpenny ('41) has indicated that the respiratory coefficient of marine producers in polar regions (diatoms) is probably much less than that of the more "animal-like" producers (peridinians and coccolithophorids) of warmer seas, and that temperature may be an important factor in determining respiratory coefficients in general. Juday ('40), after conducting numerous experiments with Manning and others on the respiration of phytoplankters in Trout Lake, Wisconsin, concluded that under average conditions these producers respire about ⅓ of the organic matter which they synthesize. This latter value, 33 per cent, is probably the best available respiratory coefficient for lacustrine producers.

Information on the respiration of aquatic primary consumers is obtained from an illuminating study by Ivlev ('39a) on the energy relationships of *Tubifex.* By means of ingenious techniques, he determined calorific values for assimilation and growth in eleven series of experiments. Using the averages of his calorific values, we can make the following simple calculations: *assimilation* (16.77 cal.) − *growth* (10.33 cal.) = *respiration* (6.44 cal.), so that respiration in terms of growth $= \frac{6.44}{10.33} = 62.30$ per cent. As a check on the growth stage of these worms, we find that $\frac{\text{growth}}{\text{assimilation}} = 61.7$ per cent, a value in good agreement with the classical conclusions of Needham ('31, III, p. 1655) with respect to embryos: the efficiency of all developing embryos is numerically similar, between 60 and 70 per cent, and independent of temperature within the range of biological tolerance. We may therefore conclude that the worms were growing at nearly maximal efficiency, so that the above respiratory coefficient is nearly minimal. In the absence of further data, we shall tentatively accept 62 per cent as the best available respiratory coefficient for aquatic herbivores.

The respiratory coefficient for aquatic predators can be approximated from data of another important study by Ivlev ('39b), on the energy transformations in predatory yearling carp. Treating his calorific values as in the preceding paragraph, we find that *ingestion* (1829 cal.) − *defecation* (454 cal.) = *assimilation* (1375 cal.), and *assimilation* − *growth* (573 cal.) = *respiration* (802 cal.), so that respiration in terms of growth $= \frac{802}{573} = 140$ per cent, a much higher coefficient than that found for the primary consumer, *Tubifex*. A rough check on this coefficient was obtained by calorific analysis of data on the growth of yearling green sunfishes (*Lepomis cyanellus*) published by W. G. Moore ('41), which indicate a respiratory coefficient of 120 per cent with respect to growth, suggesting that these fishes were growing more efficiently than those studied by Ivlev. Since Moore's fishes were fed on a highly concentrated food (liver), this greater growth efficiency is not surprising. If the maximum growth efficiency would occur when $\frac{\text{growth}}{\text{assimilation}}$ = 60–70 per cent (AEE of Needham, '31), the AEE of Moore's data (about 50 per cent) indicates that the minimum respiratory coefficient with respect to growth might be as low as 100 per cent for certain fishes. Food-conversion data from Thompson ('41) indicate a minimum respiratory coefficient of less than 150 per cent for young black bass (*Huro salmoides*) at 70° F., the exact percentage depending upon how much of the ingested food (minnows) was assimilated. Krogh ('41) showed that predatory fishes have a higher rate of respiration than the more sluggish herbivorous species; the respiratory rate of *Esox* under resting conditions at 20° C. was 3½ times that of *Cyprinus*. The form of piscian growth curves (cf. Hile, '41) suggests that the respiratory coefficient is much higher for fishes towards the end of their normal life-span. Since the value obtained from Ivlev (above) is based on more extensive data than those of Moore, we shall tentatively accept 140 per cent as an average respiratory coefficient for aquatic predators.

Considering that predators are usually more active than their herbivorous prey, which are in turn more active than the plants upon which they feed, it is not surprising to find that respiration with respect to growth in producers (33 per cent), in primary consumers (62 per cent) and in secondary consumers (>100 per cent) increases progressively. These differences probably reflect a trophic principle of wide application: the percentage loss of energy due to respiration is progressively greater for higher levels in the food cycle.

Predation corrections.—In considering the predation losses from each level, it is most convenient to begin with the highest level, Λ_n. In a mechanically perfect food cycle composed of organically discrete levels, this loss by predation obviously would be zero. Since no natural food cycle is so mechanically constituted, some "cannibalism" within such an arbitrary level can be expected, so that the actual value for predation loss from Λ_n probably will be somewhat above zero. The predation loss from level Λ_{n-1} will represent the total amount of assimilable energy passed on into the higher level (i.e., the true productivity, λ_n), plus a quantity representing the average content of substance killed but not assimilated by the predator, as will be discussed in the following section. The predation loss from level Λ_{n-2} will likewise represent the total amount of assimilable energy passed on to the next level (i.e., λ_{n-1}), plus a similar factor for unassimilated material, as illustrated by the data of tables II and III. The various categories of parasites are somewhat comparable to those of predators, but the details of their energy relationships have not yet been clarified, and cannot be included in this preliminary account.

Decomposition corrections. — In conformity with the principle of Le Chatelier, the energy of no food level can be completely extracted by the organisms which feed upon it. In addition to the energy represented by organisms which survive to be included in the "annual yield," much energy is contained in "killed" tissues which the predators are unable to digest and assimilate. Average coefficients of indigestible tissues, based largely of the calorific equivalents of the "crude fiber" fractions in the chemical analyses of Birge and Juday ('22), are as follows:

Nannoplankters	ca. 5%
Algal mesoplankters	5–35%
Mature pondweeds	ca. 20%
Primary consumers	ca. 10%
Secondary consumers	ca. 8%
Predatory fishes	ca. 5%

Corrections for terrestrial producers would certainly be much higher. Although the data are insufficient to warrant a generalization, these values suggest increasing digestibility of the higher food levels, particularly for the benthic components of aquatic cycles.

The loss of energy due to premature death from non-predatory causes usually must be neglected, since such losses are exceedingly difficult to evaluate and under normal conditions probably represent relatively small components of the annual production. However, considering that these losses may assume considerable proportions at any time, the above "decomposition coefficients" must be regarded as correspondingly minimal.

Following non-predated death, every organism is a potential source of energy for myriads of bacterial and fungal saprophages, whose metabolic products provide simple inorganic and organic solutes reavailable to photosynthetic producers. These saprophages may also serve as energy sources for successive levels of consumers, often considerably supplementing the normal diet of herbivores (ZoBell and Feltham, '38). Jacot ('40) considered saprophage-feeding or coprophagous animals as "low" primary consumers, but the writer believes that in the present state of our knowledge a quantitative subdivision of primary consumers is unwarranted.

Application.—The value of these theoretical energy relationships can be illustrated by analyzing data of the three ecosystems for which relatively comprehensive productivity values have been published (table I). The summary account of Brujewicz ('39) on "the dynamics of living matter in the Caspian Sea" leaves much to be desired, as bottom animals are not differentiated into their relative food levels, and the basis for determining the annual production of phytoplankters (which on theoretical grounds appears to be much too low) is not clearly explained. Furthermore, his values are stated in terms of thousands of tons of dry weight for the Caspian Sea as a whole, and must be roughly transformed to calories per square centimeter of surface area. The data for Lake Mendota, Wisconsin, are

TABLE I. *Productivities of food-groups in three aquatic ecosystems, as g-cal/cm²/year, uncorrected for losses due to respiration, predation and decomposition. Data from Brujewicz ('39), Juday ('40) and Lindeman ('41b).*

	Caspian Sea	Lake Mendota	Cedar Bog Lake
Phytoplankters: Λ_1	59.5	299	25.8
Phytobenthos: Λ_1	0.3	22	44.6
Zooplankters: Λ_2	20.0	22	6.1
Benthic browsers: Λ_2	}	1.8*	0.8
Benthic predators: Λ_3	} 20.6	} 0.9*	0.2
Plankton predators: Λ_3	}	}	0.8
"Forage" fishes: $\Lambda_3(+\Lambda_2?)$	0.6	?	0.3
Carp: $\Lambda_3(+\Lambda_2?)$	0.0	0.2	0.0
"Game" fishes: $\Lambda_4(+\Lambda_3?)$	0.6	0.1	0.0
Seals: Λ_5	0.01	0.0	0.0

* Roughly assuming that ⅔ of the bottom fauna is herbivorous (cf. Juday, '22).

taken directly from a general summary (Juday, '40) of the many productivity studies made on that eutrophic lake. The data for Cedar Bog Lake, Minnesota, are taken from the author's four-year analysis (Lindeman, '41b) of its food-cycle dynamics. The calorific values in table I, representing annual production of organic matter, are uncorrected for energy losses.

TABLE II. *Productivity values for the Cedar Bog Lake food cycle, in g-cal/cm²/year, as corrected by using the coefficients derived in the preceding sections.*

Trophic level	Uncorrected productivity	Respiration	Predation	Decomposition	Corrected productivity
Producers: Λ_1	70.4±10.14	23.4	14.8	2.8	111.3
Primary consumers: Λ_2	7.0±1.07	4.4	3.1	0.3	14.8
Secondary consumers: Λ_3	1.3±0.43*	1.8	0.0	0.0	3.1

* This value includes the productivity of the small cyprinoid fishes found in the lake.

Correcting for the energy losses due to respiration, predation and decomposition, as discussed in the preceding sections, casts a very different light on the relative productivities of food levels. The calculation of corrections for the Cedar Bog Lake values for producers, primary consumers and secondary consumers are given in table II. The application of similar corrections to the energy values for the food levels of the Lake Mendota food cycle given by Juday ('40), as shown in table III, indicates that Lake Mendota is much more productive of producers and primary consumers than is Cedar Bog Lake, while the production of secondary consumers is of the same order of magnitude in the two lakes.

In calculating total productivity for Lake Mendota, Juday ('40) used a blanket correction of 500 per cent of the annual production of all consumer levels for "metabolism," which presumably includes both respiration and predation. Thompson ('41) found that the "carry-

TABLE III. *Productivity values for the Lake Mendota food cycle, in g-cal/cm²/year, as corrected by using coefficients derived in the preceding sections, and as given by Juday ('40).*

Trophic Level	Uncorrected productivity	Respiration	Predation	Decomposition	Corrected productivity	Juday's corrected productivity
Producers: Λ_1	321*	107	42	10	480	428
Primary consumers: Λ_2	24	15	2.3	0.3	41.6	144
Secondary consumers: Λ_3	1†	1	0.3	0.0	2.3	6
Tertiary consumers: Λ_4	0.12	0.2	0.0	0.0	0.3	0.7

* Hutchinson ('42) gives evidence that this value is probably too high and may actually be as low as 250.

† Apparently such organisms as small "forage" fishes are not included in any part of Juday's balance sheet. The inclusion of these forms might be expected to increase considerably the productivity of secondary consumption.

ing-capacity" of lakes containing mostly carp and other "coarse" fishes (primarily Λ_3), was about 500 per cent that of lakes containing mostly "game" fishes (primarily Λ_4), and concluded that "this difference must be about one complete link in the food chain, since it usually requires about five pounds of food to produce one pound of fish." While such high "metabolic losses" may hold for tertiary and quaternary predators under certain field conditions, the physiological experiments previously cited indicate much lower respiratory coefficients. Even when predation and decomposition corrections are included, the resultant productivity values are less than half those obtained by using Juday's coefficient. Since we have shown that the necessary corrections vary progressively with the different food levels, it seems probable that Juday's "coefficient of metabolism" is much too high for primary and secondary consumers.

Biological efficiency

The quantitative relationships of any food-cycle level may be expressed in terms of its efficiency with respect to lower levels. Quoting Hutchinson's ('42) definition, "the efficiency of the productivity of any level (Λ_n) relative to the productivity of any previous level (Λ_m) is defined as $\frac{\lambda_n}{\lambda_m} 100$. If the rate of solar energy entering the ecosystem is denoted as λ_0, the efficiencies of all levels may be referred back to this quantity λ_0." In general, however, the most interesting efficiencies are those referred to the previous level's productivity (λ_{n-1}), or those expressed as $\frac{\lambda_n}{\lambda_{n-1}} 100$. These latter may be termed the *progressive efficiencies* of the various food-cycle levels, indicating for each level the degree of utilization of its potential food supply or energy source. All efficiencies discussed in the following pages are progressive efficiencies, expressed in terms of relative productivities $\left(\frac{\lambda_n}{\lambda_{n-1}} 100\right)$. It is important to remember that efficiency and productivity are not synonymous. Productivity is a rate (i.e., in the units here used, cal/cm²/year), while efficiency, being a ratio, is a dimensionless number. The points of reference for any efficiency value should always be clearly stated.

The progressive efficiencies $\left(\frac{\lambda_n}{\lambda_{n-1}} 100\right)$ for the trophic levels of Cedar Bog Lake and Lake Mendota, as obtained from the productivities derived in tables II and III, are presented in table IV. In view of the uncertainties concerning some of the Lake Mendota productivities, no definite conclusions can be drawn from their relative efficiencies. The Cedar Bog Lake ratios, however, indicate that the progressive efficiencies increase from about 0.10 per cent for production, to 13.3 per cent for primary consumption, and to 22.3 per cent for secondary consumption. An uncorrected efficiency of tertiary consumption of 37.5 per cent ± 3.0 per cent (for the weight ratios of "carnivorous" to "forage" fishes in Alabama ponds) is indicated in data published by Swingle and Smith ('40). These progressively increasing efficiencies may well represent a fundamental trophic principle, namely, that the consumers at progressively higher levels in the food cycle are progressively more efficient in the use of their food supply.

TABLE IV. *Productivities and progressive efficiencies in the Cedar Bog Lake and Lake Mendota food cycles, as g-cal/cm²/year*

	Cedar Bog Lake		Lake Mendota	
	Productivity	Efficiency	Productivity	Efficiency
Radiation	≤118,872		118,872	
Producers: Λ_1	111.3	0.10%	480*	0.40%
Primary consumers: Λ_2	14.8	13.3%	41.6	8.7%
Secondary consumers: Λ_3	3.1	22.3%	2.3†	5.5%
Tertiary consumers: Λ_4	—	—	0.3	13.0%

* Probably too high; see footnote of table III.
† Probably too low; see footnote of table III.

At first sight, this generalization of increasing efficiency in higher consumer groups would appear to contradict the previous generalization that the loss of energy due to respiration is progressively greater for higher levels in the food cycle. These can be reconciled by remembering that increased activity of predators considerably increases the chances of encountering suitable prey. The ultimate effect of such antagonistic principles would present a picture of a predator completely wearing itself out in the process of completely exterminating its prey, a very improbable situation. However, Elton ('27) pointed out that food-cycles rarely have more than five trophic levels. Among the several factors involved, increasing respiration of successive levels of predators contrasted with their successively increasing efficiency of predation appears to be important in restricting the number of trophic levels in a food cycle.

The effect of increasing temperature is alleged by Wimpenny ('41) to cause a decreasing consumer/producer ratio, presumably because he believes that the "acceleration of vital velocities" of consumers at increasing temperatures is more rapid than that of producers. He cites as evidence Lohmann's ('12) data for relative *numbers* (not biomass) of Protophyta, Protozoa and Metazoa in the centrifuge plankton of "cool" seas (741:73:1) as contrasted with tropical areas (458:24:1). Since Wimpenny himself emphasizes that many metazoan plankters are larger in size toward the poles, these data do not furnish convincing proof of the allegation. The data given in table IV, since Cedar Bog Lake has a much higher mean annual water temperature than Lake Mendota, appear to contradict Wimpenny's generalization that consumer/producer ratios fall as the temperature increases.

The Eltonian pyramid

The general relationships of higher food-cycle levels to one another and to community structure were greatly clarified following recognition (Elton, '27) of the importance of size and of numbers in the animals of an ecosystem. Beginning with primary consumers of various sizes, there are as a rule a number of food-chains radiating outwards in which the probability is that predators will become successively larger, while parasites and hyper-parasites will be progressively smaller than their hosts. Since small primary consumers can increase faster than larger secondary consumers and are so able to support the latter, the animals at the base of a food-chain are relatively abundant while those toward the end are progressively fewer in number. The resulting arrangement of sizes and numbers of animals, termed the pyramid of Numbers by Elton, is now commonly known as the Eltonian Pyramid. Williams ('41), reporting on the "floor fauna" of the Panama rain forest, has recently published an interesting example of such a pyramid, which is reproduced in figure 2.

The Eltonian Pyramid may also be expressed in terms of biomass. The weight of all predators must always be much lower than that of all food animals, and the total weight of the latter much lower than the plant production (Bodenheimer, '38). To the human ecologist, it is noteworthy that the population density of the essentially vegetarian Chinese, for example, is much greater than that of the more carnivorous English.

The principle of the Eltonian Pyramid has been redefined in terms of productivity by Hutchinson (unpublished) in the following formalized terms: the rate of production cannot be less and will almost certainly be greater than the rate of primary consumption, which in turn cannot be less and will almost certainly be greater than the rate of secondary consumption, which in turn . . . , etc. The energy-relationships of this principle may be epitomized by means

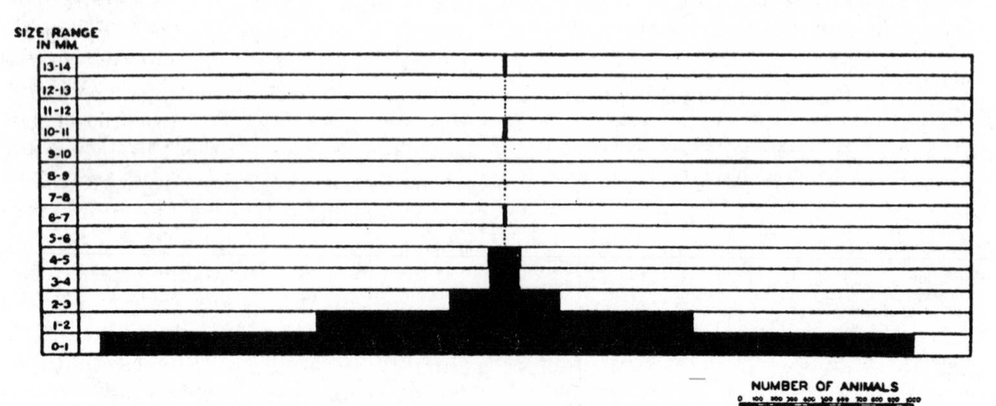

FIG. 2. Eltonian pyramid of numbers, for floor-fauna invertebrates of the Panama rain forest (from Williams, '41).

of the productivity symbol λ, as follows:

$$\lambda_0 > \lambda_1 > \lambda_2 \ldots > \lambda_n.$$

This rather obvious generalization is confirmed by the data of all ecosystems analyzed to date.

Trophic-Dynamics in Succession

Dynamic processes within an ecosystem, over a period of time, tend to produce certain obvious changes in its species-composition, soil characteristics and productivity. Change, according to Cooper ('26), is the essential criterion of succession. From the trophic-dynamic viewpoint, succession is the process of development in an ecosystem, brought about primarily by the effects of the organisms on the environment and upon each other, towards a relatively stable condition of equilibrium.

It is well known that in the initial phases of hydrarch succession (oligotrophy → eutrophy) productivity increases rapidly; it is equally apparent that the colonization of a bare terrestrial area represents a similar acceleration in productivity. In the later phases of succession, productivity increases much more slowly. As Hutchinson and Wollack ('40) pointed out, these generalized changes in the rate of production may be expressed as a sigmoid curve showing a rough resemblance to the growth curve of an organism or of a homogeneous population.

Such smooth logistic growth, of course, is seldom found in natural succession, except possibly in such cases as bare areas developing directly to the climax vegetation type in the wake of a retreating glacier. Most successional seres consist of a number of *stages* ("recognizable, clearly-marked subdivisions of a given sere"—W. S. Cooper), so that their productivity growth-curves will contain undulations corresponding in distinctness to the distinctness of the stages. The presence of stages in a successional sere apparently represents the persistent influence of some combination of limiting factors, which, until they are overcome by species-substitution, etc., tend to decrease the acceleration of productivity and maintain it at a more constant rate. This tendency towards *stage-equilibrium* of productivity will be discussed in the following pages.

Productivity in hydrarch succession

The descriptive dynamics of hydrarch succession is well known. Due to the essentially concave nature of the substratum, lake succession is internally complicated by a rather considerable influx of nutritive substances from the drainage basin surrounding the lake. The basins of lakes are gradually filled with sediments, largely organogenic, upon which a series of vascular plant stages successively replace one another until a more or less stable (climax) stage is attained. We are concerned here, however, primarily with the productivity aspects of the successional process.

Eutrophication. — Thienemann ('26) presented a comprehensive theoretical discussion of the relation between lake succession and productivity, as follows: In oligotrophy, the pioneer phase, productivity is rather low, limited by the amount of dissolved nutrients in the lake water. Oxygen is abundant at all times, almost all of the synthesized organic matter is available for animal food; bacteria release dissolved nutrients from the remainder. Oligotrophy thus has a very "thrifty" food cycle, with a relatively high "efficiency" of the consumer populations. With increasing influx of nutritives from the surrounding drainage basin and increasing primary productivity (λ_1), oligotrophy is gradually changed through mesotrophy to eutrophy, in which condition the production of organic matter (λ_1) exceeds that which can be oxidized (λ_1') by respiration, predation and bacterial decomposition. The oxygen supply of the hypolimnion becomes depleted, with disastrous effects on the oligotrophconditioned bottom fauna. Organisms especially adapted to endure temporary anaerobiosis replace the oligotrophic species, accompanied by anaerobic bacteria which during the stagnation period cause reduction rather than oxidation of the organic detritus. As a result of this process, semi-reduced organic ooze, or *gyttja*, accumulates on the bottom. As oxygen supply thus becomes a limiting factor of productivity, relative efficiency of the consumer groups in utilizing the synthesized organic substance becomes correspondingly lower.

The validity of Thienemann's interpretation, particularly regarding the trophic mechanisms, has recently been challenged by Hutchinson ('41, '42), who states that three distinct factors are involved: (1) the edaphic factor, representing the potential nutrient supply (primarily phosphorus) in the surrounding drainage basin; (2) the age of the lake at any stage, indicating the degree of utilization of the nutrient supply; and (3) the morphometric character at any stage, dependent on both the original morphometry of the lake basin and the age of the lake, and presumably influencing the oxygen concentration in the hypolimnion. He holds that true eutrophication takes place only in regions well supplied with nutrients, lakes in other regions developing into "ideotrophic types." The influx of phosphorus is probably very great in the earliest phases, much greater than the supply of nitrogen, as indicated by very low N/P ratios in the earliest sediments (Hutchinson and Wollack, '40). A large portion of this phosphorus is believed to be insoluble, as a component of such mineral particles as apatite, etc., although certainly some of it is soluble. The supply of available nitrogen increases somewhat more slowly, being largely dependent upon the fixation of nitrogen by microorganisms either in the lake or in the surrounding soils. The photosynthetic productivity (λ_1) of lakes thus increases rather rapidly in the early phases, its quantitative value for lakes with comparable edaphic nutrient supplies being dependent on the morphometry (mean depth). Since deep lakes have a greater depth range for plankton photosynthesis, abundant oxygen and more chance for decomposition of the plankton detritus before reaching the bottom, such deep lakes may be potentially as productive as shallower lakes, in terms of unit surface area. Factors tending to lessen the comparative productivity of deep lakes are (1) lower temperature for the lake as a whole, and (2) greater dilution of nutrients in terms of volume of the illuminated "trophogenic zone" of the lake. During eutrophication in a deep lake, the phosphorus content of the sediment falls and nitrogen rises, until a N/P ratio of about 40/1 is established. "The decomposition of organic matter presumably is always liberating some of this phosphorus and nitrogen. Within limits, the more organic matter present the easier will be such regeneration. It is probable that benthic animals and anion exchange play a part in such processes" (Hutchinson, '42). The progressive filling of the lake basin with sediments makes the lake shallower, so that the oxygen supply of the hypolimnion is increasingly, and finally completely, exhausted during summer stagnation. Oxidation of the sediments is less complete, but sufficient phosphorus is believed to be regenerated from the ooze surface so that productivity in terms of surface area remains relatively constant. The nascent ooze acts as a trophic buffer, in the chemical sense, tending to maintain the productivity of a lake in stage-equilibrium (*typological equilibrium* of Hutchinson) during the eutrophic stage of its succession.

The concept of eutrophic stage-equilibrium seems to be partially confused (cf. Thienemann, '26; Hutchinson and Wollack, '40) with the theoretically ideal condition of complete *trophic equilibrium*, which may be roughly defined as the dynamic state of continuous, complete utilization and regeneration of chemical nutrients in an ecosystem, without loss or gain from the outside, under a periodically constant energy source—such as might be found in a perfectly balanced aquarium or terrarium. Natural ecosystems may tend to approach a state of trophic equilibrium under certain conditions, but it is doubtful if any are sufficiently autochthonous to attain, or maintain, true trophic equilibrium for any length of time. The biosphere as a whole, however, as Vernadsky ('29, '39) so vigorously asserts, may exhibit a high degree of true trophic equilibrium.

The existence of prolonged eutrophic stage-equilibrium was first suggested as a result of a study on the sediments of Grosser Plöner See in Germany (Groschopf, '36). Its significance was recognized by Hutchinson and Wollack ('40), who based their critical discussion on chemical analyses (ibid.) and pollen analyses (Deevey, '39) of the sediments of Linsley Pond, Connecticut. They reported a gradual transition from oligotrophy to eutrophy (first attained in the oak-hemlock pollen period), in which stage the lake has remained for a very long time, perhaps more than 4000 years. They report indications of a comparable eutrophic stage-equilibrium in the sediments of nearby Upper Linsley Pond (Hutchinson and Wollack, unpublished). Similar attainment of stage-equilibrium is indicated in a preliminary report on the sediments of Lake Windermere in England (Jenkin, Mortimer and Pennington, '41). Every stage of a sere is believed to possess a similar stage-equilibrium of variable duration, although terrestrial stages have not yet been defined in terms of productivity.

The trophic aspects of eutrophication cannot be determined easily from the sediments. In general, however, the ratio of organic matter to the silt washed into the lake from the margins affords an approximation of the photosynthetic productivity. This undecomposed organic matter, representing the amount of energy which is lost from the food cycle, is derived largely from level Λ_1, as plant structures in general are decomposed less easily than animal structures. The quantity of energy passed on into consumer levels can only be surmised from undecomposed fragments of organisms which are believed to occupy those levels. Several types of animal "microfossils" occur rather consistently in lake sediments, such as the carapaces and postabdomens of certain cladocerans, chironomid head-capsules, fragments of the phantom-midge larva *Chaoborus*, snail shells, polyzoan statoblasts, sponge spicules and rhizopod shells. Deevey ('42), after making comprehensive microfossil and chemical analyses of the sediments of Linsley Pond, suggested that the abundant half-carapaces of the planktonic browser *Bosmina* afford "a reasonable estimate of the quantity of zooplankton produced" and that "the total organic matter of the sediment is a reasonable estimate of the organic matter produced by phytoplankton and littoral vegetation." He found a striking similarity in the shape of the curves representing *Bosmina* content and total organic matter plotted against depth, which, when plotted logarithmically against each other, showed a linear relationship expressed by an empirical power equation. Citing Hutchinson and Wollack ('40) to the effect that the developmental curve for organic matter was analogous to that for the development of an organism, he pressed the analogy further by suggesting that the increase of zooplankton (*Bosmina*) with reference to the increase of organic matter (λ_1) fitted the formula $y = bx^k$ for allometric growth (Huxley, '32), "where y = *Bosmina*, x = total organic matter, b = a constant giving the value of y when $x = 1$, and k = the 'allometry constant,' or the slope of the line when a double log plot is made." If we represent the organic matter produced as λ_1 and further assume that *Bosmina* represents the primary consumers (λ_2), neglecting benthic browsers, the formula becomes $\lambda_2 = b\lambda_1{}^k$. Whether this formula would express the relationship found in other levels of the food cycle, the development of other stages, or other ecosystems, remains to be demonstrated.[6] Stratigraphic analyses in Cedar Bog Lake (Lindeman and Lindeman, unpublished) suggest a roughly similar increase of both organic matter and *Bosmina* carapaces in the earliest sediments. In the modern senescent lake, however, double logarithmic plottings of the calorific values for λ_1 against λ_2, and λ_2 against λ_3, for the four years studied, show no semblance of linear relationship, i.e., do not fit any power equation. If Deevey is correct in his interpretation of the Linsley Pond microfossils, allometric growth would appear to characterize the phases of pre-equilibrium succession as the term "growth" indeed implies.

The relative duration of eutrophic stage-equilibrium is not yet completely understood. As exemplified by Linsley Pond, the relation of stage-equilibrium to succession is intimately concerned with the trophic processes of (1) external influx and efflux (partly controlled by climate), (2) photosynthetic productivity, (3) sedimentation (partly by physiographic silting) and (4) regeneration of nutritives from the sediments. These processes apparently maintain a relatively constant ratio to each other during the extended equilibrium period. Yet the food cycle is not in true trophic equilibrium, and continues to fill the lake with organic sediments. *Succession* is continuing, at a rate corresponding to the rate of sediment accumulation. In the words of Hutchinson and Wollack ('40), "this means that during the equilibrium period the lake, through the internal activities of its biocoenosis, is continually approaching a condition when it ceases to be a lake."

Senescence.—As a result of long-continued sedimentation, eutrophic lakes attain senescence, first manifested in bays and wind-protected areas. Senescence is usually characterized by such pond-like conditions as (1) tremendous increase in shallow littoral area populated with pondweeds and (2) increased marginal invasion of terrestrial stages. Cedar Bog Lake, which the author has studied for several years, is in late senescence, rapidly changing to the terrestrial stages of its succession. On casual inspection, the massed verdure of pondweeds and epiphytes, together with sporadic algal blooms, appears to indicate great photosynthetic productivity. As pointed out by Wesenberg-Lund ('12), littoral areas of lakes are virtual hothouses, absorbing more radiant energy per unit volume than deeper areas. At the present time the entire aquatic area of Cedar Bog Lake is essentially littoral in nature, and its productivity per cubic meter of water is probably greater than at any time in its history. However, since radiant energy (λ_0) enters a lake only from the surface, productivity must be defined in terms of surface area. In these terms, the present photosynthetic productivity pales into insignificance when compared with less advanced lakes in similar edaphic regions; for instance, λ_1 is less than ⅓ that of Lake Mendota, Wisconsin (cf. table IV). These facts attest the essential accuracy of Welch's ('35) generalization that productivity declines greatly during senescence. An interesting principle demonstrated in Cedar Bog Lake (Lindeman, '41b) is that during late lake senescence general productivity (λ_n) is increasingly influenced by climatic factors, acting through

---

[6] It should be mentioned in this connection that Meschkat ('37) found that the relationship of population density of tubificids to organic matter in the bottom of a polluted "Buhnenfeld" could be expressed by the formula $y = a^x$, where y represents the population density, x is the "determining environmental factor," and a is a constant. He pointed out that for such an expression to hold the population density must be maximal. Hentschel ('36), on less secure grounds, suggested applying a similar expression to the relationship between populations of marine plankton and the "controlling factor" of their environment.

water level changes, drainage, duration of winter ice, snow cover, etc., to affect the presence and abundance of practically all food groups in the lake.

Terrestrial stages.—As an aquatic ecosystem passes into terrestrial phases, fluctuations in atmospheric factors increasingly affect its productivity. As succession proceeds, both the species-composition and the productivity of an ecosystem increasingly reflect the effects of the regional climate. Qualitatively, these climatic effects are known for soil morphology (Joffe, '36), autotrophic vegetation (Clements, '16), fauna (Clements and Shelford, '39) and soil microbiota (Braun-Blanquet, '32), in fact for every important component of the food cycle. Quantitatively, these effects have been so little studied that generalizations are most hazardous. It seems probable, however, that productivity tends to increase until the system approaches maturity. Clements and Shelford ('39, p. 116) assert that both plant and animal productivity is generally greatest in the subclimax, except possibly in the case of grasslands. Terrestrial ecosystems are primarily convex topographically and thus subject to a certain nutrient loss by erosion, which may or may not be made up by increased availability of such nutrients as can be extracted from the "C" soil horizon.

Successional productivity curves.—In recapitulating the probable photosynthetic productivity relationships in hydrarch succession, we shall venture to diagram (figure 3) a hypothetical hydrosere, developing from a moderately deep lake in a fertile cold temperate region under relatively constant climatic conditions. The initial period of oligotrophy is believed to be relatively short (Hutchinson and Wollack, '40; Lindeman '41a), with productivity rapidly increasing until eutrophic stage-equilibrium is attained. The duration of high eutrophic productivity depends upon the mean depth of the basin and upon the rate of sedimentation, and productivity fluctuates about a high eutrophic mean until the lake becomes too shallow for maximum growth of phytoplankton or regeneration of nutrients from the ooze. As the lake becomes shallower and more senescent, productivity is increasingly influenced by climatic fluctuations and

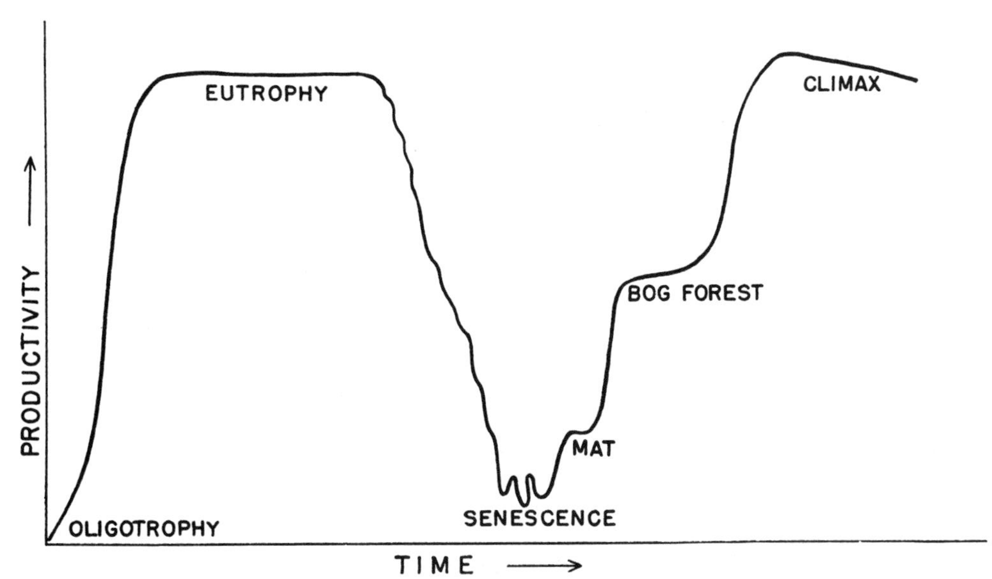

FIG. 3. Hypothetical productivity growth-curve of a hydrosere, developing from a deep lake to climax in a fertile, cold-temperate region.

gradually declines to a minimum as the lake is completely filled with sediments.

The terrestrial aspects of hydrarch succession in cold temperate regions usually follow sharply defined, distinctive stages. In lake basins which are poorly drained, the first stage consists of a mat, often partly floating, made up primarily of sedges and grasses or (in more coastal regions) such heaths as *Chamaedaphne* and *Kalmia* with certain species of sphagnum moss (cf. Rigg, '40). The mat stage is usually followed by a bog forest stage, in which the dominant species is *Larix laricina*, *Picea mariana* or *Thuja occidentalis*. The bog forest stage may be relatively permanent ("edaphic" climax) or succeeded to a greater or lesser degree by the regional climax vegetation. The stage-productivities indicated in figure 3 represent only crude relative estimates, as practically no quantitative data are available.

Efficiency relationships in succession

The successional changes of photosynthetic efficiency in natural areas (with respect to solar radiation, i.e., $\left(\frac{\lambda_1}{\lambda_0} 100\right)$ have not been intensively studied. In lake succession, photosynthetic efficiency would be expected to follow the same course deduced for productivity, rising to a more or less constant value during eutrophic stage-equilibrium, and declining during senescence, as suggested by a photosynthetic efficiency of at least 0.27 per cent for eutrophic Lake Mendota (Juday, '40) and of 0.10 per cent for senescent Cedar Bog Lake. For the terrestrial hydrosere, efficiency would likewise follow a curve similar to that postulated for productivity.

Rough estimates of photosynthetic efficiency for various climatic regions of the earth have been summarized from the literature by Hutchinson (unpublished). These estimates, corrected for respiration, do not appear to be very reliable because of imperfections in the original observations, but are probably of the correct order of magnitude. The mean photosynthetic efficiency for the sea is given as 0.31 per cent (after Riley, '41). The mean photosynthetic efficiency for terrestrial areas of the earth is given as 0.09 per cent ± 0.02 per cent (after Noddack, '37), for forests as 0.16 per cent, for cultivated lands as 0.13 per cent, for steppes as 0.05 per cent, and for deserts as 0.004 per cent. The mean photosynthetic efficiency for the earth as a whole is given as 0.25 per cent. Hutchinson has suggested (cf. Hutchinson and Lindeman, '41) that numerical efficiency values may provide "the most fundamental possible classification of biological formations and of their developmental stages."

Almost nothing is known concerning the efficiencies of consumer groups in succession. The general chronological increase in numbers of *Bosmina* carapaces with respect to organic matter and of *Chaoborus* fragments with respect to *Bosmina* carapaces in the sediments of Linsley Pond (Deevey, '42) suggests progressively increasing efficiencies of zooplankters and plankton predators. On the other hand, Hutchinson ('42) concludes from a comparison of the P : Z (phytoplankton : zooplankton) biomass ratios of several oligotrophic alpine lakes, ca 1 : 2 (Ruttner, '37), as compared with the ratios for Linsley Pond, 1 : 0.22 (Riley, '40) and three eutrophic Bavarian lakes, 1 : 0.25 (Heinrich, '34), that "as the phytoplankton crop is increased the zooplankton by no means keeps pace with the increase." Data compiled by Deevey ('41) for lakes in both mesotrophic (Connecticut) and eutrophic regions (southern Wisconsin), indicate that the deeper or morphometrically "younger" lakes have a lower ratio of bottom fauna to the standing crop of plankton (10–15 per cent) than the shallower lakes which have attained eutrophic equilibrium (22–27 per cent). The ratios for senescent Cedar Bog Lake, while not directly comparable because of its essentially littoral nature, are even higher. These meager data suggest that the efficiencies of consumer groups may increase throughout the aquatic phases of succession.

For terrestrial stages, no consumer efficiency data are available. A suggestive series of species-frequencies in mesarch succession was published by Vera Smith-Davidson ('32), which indicated greatly increasing numbers of arthropods in successive stages approaching the regional climax. Since the photosynthetic productivity of the stages probably also increased, it is impossible to determine progressive efficiency relationships. The problems of biological efficiencies present a practically virgin field, which appears to offer abundant rewards for studies guided by a trophic-dynamic viewpoint.

In conclusion, it should be emphasized that the trophic-dynamic principles indicated in the following summary cannot be expected to hold for every single case, in accord with the known facts of biological variability. *à priori*, however, these principles appear to be valid for the vast majority of cases, and may be expected to possess a statistically significant probability of validity for any case selected at random. Since the available data summarized in this paper are far too meager to establish such generalizations on a statistical basis, it is highly important that further studies be initiated to test the validity of these and other trophic-dynamic principles.

Summary

1. Analyses of food-cycle relationships indicate that a biotic community cannot be clearly differentiated from its abiotic environment; the *ecosystem* is hence regarded as the more fundamental ecological unit.

2. The organisms within an ecosystem may be grouped into a series of more or less discrete trophic levels (Λ_1, Λ_2, Λ_3, . . . Λ_n) as producers, primary consumers, secondary consumers, etc., each successively dependent upon the preceding level as a source of energy, with the producers (Λ_1) directly dependent upon the rate of incident solar radiation (productivity λ_0) as a source of energy.

3. The more remote an organism is from the initial source of energy (solar radiation), the less probable that it will be dependent solely upon the preceding trophic level as a source of energy.

4. The progressive energy relationships of the food levels of an "Eltonian Pyramid" may be epitomized in terms of the productivity symbol λ, as follows:

$$\lambda_0 > \lambda_1 > \lambda_2 \ldots > \lambda_n.$$

5. The percentage loss of energy due to respiration is progressively greater for higher levels in the food cycle. Respiration with respect to growth is about 33 per cent for producers, 62 per cent for primary consumers, and more than 100 per cent for secondary consumers.

6. The consumers at progressively higher levels in the food cycle appear to be progressively more efficient in the use of their food supply. This generalization can be reconciled with the preceding one by remembering that increased activity of predators considerably increases the chances of encountering suitable prey.

7. Productivity and efficiency increase during the early phases of successional development. In lake succession, productivity and photosynthetic efficiency increase from oligotrophy to a prolonged eutrophic stage-equilibrium and decline with lake senescence, rising again in the terrestrial stages of hydrarch succession.

8. The progressive efficiencies of consumer levels, on the basis of very meager data, apparently tend to increase throughout the aquatic phases of succession.

Acknowledgments

The author is deeply indebted to Professor G. E. Hutchinson of Yale University, who has stimulated many of the trophic concepts developed here, generously placed at the author's disposal several unpublished manuscripts, given valuable counsel, and aided the final development of this paper in every way possible. Many of the concepts embodied in the successional sections of this paper were developed independently by Professor Hutchinson at Yale and by the author as a graduate student at the University of Minnesota. Subsequent to an exchange of notes, a joint preliminary abstract was published (Hutchinson and Lindeman, '41). The author wishes to express gratitude to the mentors and friends at the University of Minnesota who encouraged and helpfully criticized the initial development of these concepts, particularly Drs. W. S. Cooper, Samuel Eddy, A. C. Hodson, D. B. Lawrence and J. B. Moyle, as well as other members of the local Ecological Discussion Group. The author is also indebted to Drs. J. R. Carpenter, E. S. Deevey, H. J. Lutz, A. E. Parr, G. A. Riley and V. E. Shelford, as well as the persons mentioned above, for critical reading of preliminary manuscripts. Grateful acknowledgment is made to the Graduate School, Yale University, for the award of a Sterling Fellowship in Biology during 1941–1942.

Literature Cited

Allee, W. C. 1934. Concerning the organization of marine coastal communities. Ecol. Monogr., **4**: 541–554.

Birge, E. A., and C. Juday. 1922. The inland lakes of Wisconsin. The plankton. Part I. Its quantity and chemical composition. Bull. Wisconsin Geol. Nat. Hist. Surv., **64**: 1–222.

Bodenheimer, F. S. 1938. Problems of Animal Ecology. London. Oxford University Press.

Braun-Blanquet, J. 1932. Plant Sociology. N. Y. McGraw-Hill Co.

Brujewicz, S. W. 1939. Distribution and dynamics of living matter in the Caspian Sea. Compt. Rend. Acad. Sci. URSS, **25**: 138–141.

Carpenter, J. R. 1939. The biome. Amer. Midl. Nat., **21**: 75–91.

——. 1940. The grassland biome. Ecol. Monogr., **10**: 617–687.

Clements, F. E. 1916. Plant Succession. Carnegie Inst. Washington Publ., No. 242.

—— **and V. E. Shelford.** 1939. Bio-Ecology. N. Y. John Wiley & Co.

Cooper, W. S. 1926. The fundamentals of vegetational change. Ecology, **7**: 391–413.

Cowles, H. C. 1899. The ecological relations of the vegetation of the sand dunes of Lake Michigan. Bot. Gaz., **27**: 95–391.

Davidson, V. S. 1932. The effect of seasonal variability upon animal species in a deciduous forest succession. Ecol. Monogr., **2**: 305–334.

Deevey, E. S. 1939. Studies on Connecticut lake sediments: I. A postglacial climatic chronology for southern New England. Amer. Jour. Sci., **237**: 691–724.

——. 1941. Limnological studies in Connecticut: VI. The quantity and composition of the bottom fauna. Ecol. Monogr., **11**: 413–455.

——. 1942. Studies on Connecticut lake sediments: III. The biostratonomy of Linsley Pond. Amer. Jour. Sci., **240**: 233–264, 313–338.

Elton, C. 1927. Animal Ecology. N. Y. Macmillan Co.

Francé, R. H. 1913. Das Edaphon, Untersuchungen zur Oekologie der bodenbewohnenden Mikroorganismen. Deutsch. Mikrolog. Gesellsch., Arbeit. aus d. Biol. Inst., No. 2. Munich.

Friederichs, K. 1930. Die Grundfragen und Gesetzmässigkeiten der land- und forstwirtschaftlichen Zoologie. 2 vols. Berlin. Verschlag. Paul Parey.

Groschopf, P. 1936. Die postglaziale Entwicklung des Grosser Plöner Sees in Ostholstein auf Grund pollenanalytischer Sedimentuntersuchungen. Arch. Hydrobiol., **30**: 1–84.

Heinrich, K. 1934. Atmung und Assimilation im freien Wasser. Internat. Rev. ges. Hydrobiol. u. Hydrogr., **30**: 387–410.

Hentschel, E. 1933–1936. Allgemeine Biologie des Südatlantischen Ozeans. Wiss. Ergebn. Deutsch. Atlant. Exped. a. d. Forschungs- u. Vermessungsschiff "Meteor" 1925–1927. Bd. **XI.**

Hicks, P. A. 1934. Interaction of factors in the growth of *Lemna:* V. Some preliminary observations upon the interaction of temperature and light on the growth of *Lemna.* Ann. Bot., **48**: 515–523.

Hile, R. 1941. Age and growth of the rock bass *Ambloplites rupestris* (Rafinesque) in Nebish Lake, Wisconsin. Trans. Wisconsin Acad. Sci., Arts, Lett., **33**: 189–337.

Hutchinson, G. E. 1941. Limnological studies in Connecticut: IV. Mechanism of intermediary metabolism in stratified lakes. Ecol. Monogr., **11**: 21–60.

——. 1942. Recent Advances in Limnology (*in manuscript*).

—— **and R. L. Lindeman.** 1941. Biological efficiency in succession (Abstract). Bull. Ecol. Soc. Amer., **22**: 44.

—— **and Anne Wollack.** 1940. Studies on Connecticut lake sediments: II. Chemical analyses of a core from Linsley Pond, North Branford. Amer. Jour. Sci., **238**: 493–517.

Huxley, J. S. 1932. Problems of Relative Growth. N. Y. Dial Press.

Ivlev, V. S. 1939a. Transformation of energy by aquatic animals. Internat. Rev. ges. Hydrobiol. u. Hydrogr., **38**: 449–458.

——. 1939b. Balance of energy in carps. Zool. Zhurn. Moscow, **18**: 303–316.

Jacot, A. P. 1940. The fauna of the soil. Quart. Rev. Biol., **15**: 28–58.

Jenkin, B. M., C. H. Mortimer, and W. Pennington. 1941. The study of lake deposits. Nature, **147**: 496–500.

Joffe, J. S. 1936. Pedology. New Brunswick, New Jersey. Rutgers Univ. Press.

Juday, C. 1922. Quantitative studies of the bottom fauna in the deeper waters of Lake Mendota. Trans. Wisconsin Acad. Sci., Arts, Lett., **20**: 461–493.

——. 1940. The annual energy budget of an inland lake. Ecology, **21**: 438–450.

Krogh, A. 1941. The Comparative Physiology of Respiratory Mechanisms. Philadelphia. Univ. Pennsylvania Press.

Lindeman, R. L. 1941a. The developmental history of Cedar Creek Bog, Minnesota. Amer. Midl. Nat., **25**: 101–112.

——. 1941b. Seasonal food-cycle dynamics in a senescent lake. Amer. Midl. Nat., **26**: 636–673.

Lohmann, H. 1912. Untersuchungen über das Pflanzen- und Tierleben der Hochsee, zugleich ein Bericht über die biologischen Arbeiten auf der Fahrt der "Deutschland" von Bremerhaven nach Buenos Aires. Veröffentl. d. Inst. f. Meereskunde, N.F., A. Geogr.-naturwissen. Reihe, Heft 1, 92 pp.

Lundegårdh, H. 1924. Kreislauf der Kohlensäure in der Natur. Jena. G. Fischer.

Meschkat, A. 1937. Abwasserbiologische Untersuchungen in einem Buhnenfeld unterhalb Hamburgs. Arch. Hydrobiol., **31**: 399–432.

Moore, W. G. 1941. Studies on the feeding habits of fishes. Ecology, **22**: 91–95.

Needham, J. 1931. Chemical Embryology. 3 vols. N. Y. Cambridge University Press.

Noddack, W. 1937. Der Kohlenstoff im Haushalt der Natur. Zeitschr. angew. Chemie, **50**: 505–510.

Park, Thomas. 1941. The laboratory population as a test of a comprehensive ecological system. Quart. Rev. Biol., **16**: 274–293, 440–461.

Rigg, G. B. 1940. Comparisons of the development of some Sphagnum bogs of the Atlantic coast, the interior, and the Pacific coast. Amer. Jour. Bot., **27**: 1–14.

Riley, G. A. 1940. Limnological studies in Connecticut. III. The plankton of Linsley Pond. Ecol. Monogr., **10**: 279–306.

——. 1941. Plankton studies. III. Long Island Sound. Bull. Bingham Oceanogr. Coll. **7** (3): 1–93.

Ruttner, F. 1937. Limnologische Studien an einigen Seen der Ostalpen. Arch. Hydrobiol., **32**: 167–319.

Smith-Davidson, Vera. 1932. The effect of seasonal variability upon animal species in a deciduous forest succession. Ecol. Monogr., **2**: 305–334.

Spoehr, H. A. 1926. Photosynthesis. N. Y. Chemical Catalogue Co.

Swingle, H. S., and E. V. Smith. 1940. Experiments on the stocking of fish ponds. Trans. North Amer. Wildlife Conf., **5**: 267–276.

Tansley, A. G. 1935. The use and abuse of vegetational concepts and terms. Ecology, **16**: 284–307.

Thienemann, A. 1918. Lebensgemeinschaft und Lebensraum. Naturw. Wochenschrift, N.F., **17**: 282–290, 297–303.

——. 1926. Der Nahrungskreislauf im Wasser. Verh. deutsch. Zool. Ges., **31**: 29–79. (or) Zool. Anz. Suppl., **2**: 29–79.

——. 1939. Grundzüge einen allgemeinen Oekologie. Arch. Hydrobiol., **35**: 267–285.

Thompson, D. H. 1941. The fish production of inland lakes and streams. Symposium on Hydrobiology, pp. 206–217. Madison. Univ. Wisconsin Press.

Tutin, T. G. 1941. The hydrosere and current concepts of the climax. Jour. Ecol. **29**: 268–279.

Vernadsky, V. I. 1929. La biosphere. Paris. Librairie Felix Alcan.

——. 1939. On some fundamental problems of biogeochemistry. Trav. Lab. Biogeochem. Acad. Sci. URSS, **5**: 5–17.

Waksman, S. A. 1941. Aquatic bacteria in relation to the cycle of organic matter in lakes. Symposium on Hydrobiology, pp. 86–105. Madison. Univ. Wisconsin Press.

Welch, P. S. 1935. Limnology. N. Y. McGraw-Hill Co.

Wesenberg-Lund, C. 1912. Über einige eigentümliche Temperaturverhaltnisse in der Litoralregion. . . . Internat. Rev. ges. Hydrobiol. u. Hydrogr., **5**: 287–316.

Williams, E. C. 1941. An ecological study of the floor fauna of the Panama rain forest. Bull. Chicago Acad. Sci., **6**: 63–124.

Wimpenny, R. S. 1941. Organic polarity: some ecological and physiological aspects. Quart. Rev. Biol., **16**: 389–425.

ZoBell, C. E., and C. B. Feltham. 1938. Bacteria as food for certain marine invertebrates. Jour. Marine Research, **1**: 312–327.

Addendum

While this, his sixth completed paper, was in the press, Raymond Lindeman died after a long illness on 29 June, 1942, in his twenty-seventh year. While his loss is grievous to all who knew him, it is more fitting here to dwell on the achievements of his brief working life. The present paper represents a synthesis of Lindeman's work on the modern ecology and past history of a small senescent lake in Minnesota. In studying this locality he came to realize, as others before him had done, that the most profitable method of analysis lay in reduction of all the interrelated biological events to energetic terms. The attempt to do this led him far beyond the immediate problem in hand, and in stating his conclusions he felt that he was providing a program for further studies. Knowing that one man's life at best is too short for intensive studies of more than a few localities, and before the manuscript was completed, that he might never return again to the field, he wanted others to think in the same terms as he had found so stimulating, and for them to collect material that would confirm, extend, or correct his theoretical conclusions. The present contribution does far more than this, as here for the first time, we have the interrelated dynamics of a biocoenosis presented in a form that is amenable to a productive abstract analysis. The question, for instance, arises, "What determines the length of a food chain?"; the answer given is admittedly imperfect, but it is far more important to have seen that there is a real problem of this kind to be solved. That the final statement of the structure of a biocoenosis consists of pairs of numbers, one an integer determining the level, one a fraction determining the efficiency, may even give some hint of an undiscovered type of mathematical treatment of biological communities. Though Lindeman's work on the ecology and history of Cedar Bog Lake is of more than local interest, and will, it is hoped, appear of even greater significance when the notes made in the last few months of his life can be coordinated and published, it is to the present paper that we must turn as the major contribution of one of the most creative and generous minds yet to devote itself to ecological science.

G. EVELYN HUTCHINSON.

YALE UNIVERSITY.

PART 2 Theoretical Advances

The Role of Theory in the Rise of Modern Ecology

Leslie A. Real and Simon A. Levin

Interest in the mathematical description of ecological processes is quite old. The twelfth century Italian mathematician Leonardo Fibonacci derived the famous series that bears his name by contemplating a population of reproducing rabbits. Fibonacci's interests were entirely aesthetic; whether or not he intended to understand rabbits is unclear. Still, we can claim Fibonacci as our first theoretical ecologist. That Fibonacci would concentrate on aspects of population growth to illustrate certain mathematical properties illustrates the natural tie between ecological processes and their numerical descriptions.

Since the mid 1950s, contemporary ecological theory has developed around a core of issues that have shaped the current spectrum of laboratory and field experiments. Theoreticians have focused on six general problems: (1) the number and relative abundance of species in a community; (2) niche theory; (3) the demography and population dynamics of single species; (4) multi-species interactions; (5) population structure, including the effect of spatial structure in populations along with the problem of scale in ecological processes; and (6) the implications of individual behavior on population phenomena.

In this collection of reprinted articles, we have included representative classic papers from each of these six categories. These classics either provided the seminal formulation of a particularly important ecological problem or crystallized an already existing problem into a form which all contemporary theory takes as its starting point. In all cases, the papers contain fruitful insights and unsolved problems that still occupy the talents of contemporary investigators.

The Number of Species in a Community

Charles Darwin provided a solution to the problem of the origin of individual species. But he did not tell us why a community consists of 100, 1,000 or 10,000 species. A major intellectual challenge facing modern ecologists, suggested by May (1986), is to discover those forces that determine the magnitude of species found in a given community or geographic location. To achieve this end, we must first have accurate methods for estimating the number of species in a community.

Attempts at characterizing the species composition of a community are rather old. Compilations of species lists by locale were a major preoccupation of the French Encyclopedists and continue today as cataloges of flora and fauna. Rigorous quantitative general models of species composition and relative abundance, however, were first proposed in the mid 1940s. Extensive sampling of a given community yields information on both the number of species and the number of individuals in each of these species. Fisher, Corbet, and Williams (1943) combined these two forms of information to reveal a common pattern in community structure. A plot of the frequency of species in the sample represented by a given number of individuals follows a "hollow curve"—that is, few species are very common, while many species are rare. Fisher, Corbet, and Williams concluded that the data were best modelled by means of a logarithmic series:

$$S = \sum_{i=1}^{\infty} \frac{\alpha\lambda^i}{i} \qquad (1)$$

where S = the number of species in the community and $\alpha\lambda^i/i$ = the number of species represented by i individuals in the community. The parameters α and λ can be estimated empirically, and are related through the relation $\lambda = N/(N+\alpha)$, where N is the total number of individuals in the sample. The series (1) may be seen to converge to the finite sum

$$S = \alpha \ln\left(1 + \frac{N}{\alpha}\right) \qquad (2)$$

where N = the total number of individuals in the sample.

While the Fisher et al. model was a seminal advance, using the logarithmic series has many problems. Most notably, the series predicts that the mode of the distribution is at one individual. Available data suggests that the mode is usually substantially larger than one. Typically, the distribution is humped and not uniformly declining. For instance, a plot of the relative abundance of breeding bird species in Quaker Run, New York indicates that the distribution has its peak at ~10 breeding pairs, not at a point corresponding to single individuals (Williams 1964).

Preston (1948, 1962), in a series of elegant and very important papers, provided an alternative model that laid the foundation for all contemporary discussions of species abundance patterns. Preston perceptively noted that when we compare species abundances, we usually employ relative rather than absolute terms: one species is twice as abundant as some other species—or four times, or eight times. Characterizing species abundances in relative terms leads naturally to using a geometric or logarithmic scale for abundance (as opposed to the arithmetic scale used by Fisher). In his 1948 paper, Preston transformed the ordinate axis into "octaves," where species were grouped into geometric classes of abundance based on log to the base 2, that is, class A corresponds to species represented by single individuals, class B corresponds to species consisting of 2 − 4, C = 4 − 8, etc. Preston found that data on species abundance from many diverse communities when plotted on the octave scale appeared to be normally distributed. Hence, log transformation leads to a log-normal distribution of species relative abundance.

The log-normal distribution can be characterized by the following relation:

$$S_R = S_0 \exp\left\{-\frac{R}{2\sigma^2}\right\},$$

Where S_R = the number of species occurring in the Rth octave to the right or left of the modal octave, S_0 = the number of species in the modal octave, and σ^2 = the variance of the distribution. The modal octave is that octave containing the largest number of species in the distribution, that is, the peak of the distribution.

The second important observation in the 1948 paper concerned the total number of individuals *regardless of species* in each octave. A plot of this total number against the octave scale generates an "individuals curve." Preston noted that often the empirical data suggest that the peak of the individuals curve corresponds to the terminal octave of the species abundance curve. In his 1962 paper, Preston termed such distributions "canonical." Technically, we can define a parameter $\gamma = R_N/R_{max}$, where R_N = the octave for the maximum of the individuals curve, and R_{max} = the octave that terminates the species abundance curve. When $\gamma = 1$, the species relative abundance distribution is canonical.

We have reprinted Preston's 1962 paper in this section since it represents his fullest treatment of the implications of the canonical distribution. The shape of a log-normal distribution is uniquely characterized by: (1) the total area under the curve (that is, the total number of species in the community); and (2) the variance in the distribution. In most cases, these two parameters are independent. However, Preston showed that when the distribution is canonical the species number S and the variance in the distribution are not independent. The two terms can be related by the expression

$$S = \sigma \sqrt{\pi/2}\ \exp\left\{\frac{(\sigma \ln 2)^2}{2}\right\}$$

(see Sugihara 1980). After establishing the general properties of the canonical distribution, Preston proceeds to use these properties to deduce the species area relationship

$$S = c\,A^z$$

where A = the area, and c and z are empirical constants. The large number of examples compiled by Preston and presented in his 1962 paper suggested that $z \approx 1/4$. This general empirical result became the focus of considerable speculation and conjecture over the next 20 years.

In his 1948 paper, Preston gave no account of why species relative abundances should be log-normally distributed. It remained a purely phenomenological description. MacArthur (1957, 1960) presented his now famous "broken-stick" model in an attempt to account for the log-normal distribution. MacArthur imagined the community's niche space could be represented as a stick or line of specified length. If there are S species in the community, then we can imagine the stick being randomly and simultaneously broken at S − 1 places. The relative lengths of the S fragments could then represent each specie's relative abundance. Unfortunately, the broken-stick model does not generate a log-normal distribution, and Preston (1962) demonstrated the general inappropriateness of the MacArthur model. Several other authors pointed out weaknesses in the broken-stick model, and MacArthur eventually abandoned it.

MacArthur had hoped for some biological model based on evolutionary principles to explain the log-normal distribution. May (1975) sought an explanation based strictly on statistical sampling properties. May argued that relative abundances are likely to be governed by the multiplicative interaction of multiple independent factors. The statistical Central Limit Theorem applied to the product of such factors would imply a log-normal distribution of abundance.

May further suggested that the canonical distribution is merely a mathematical property of all log-normal distributions for large numbers of species S. May conjectured that γ, under not too restrictive biological assumptions, would range normally from ≈ 0.2 to ≈ 1.8. Consequently, the canonical property is also a mathematical artifact of the Central Limit Theorem.

However, Sugihara (1980) has pointed out that (1) there are no mathematical or statistical limits on γ except that it must be positive, and that (2) the variance in the relationship between S and σ in natural communities is far less than that predicted from May's conjecture. Sugihara returned to a biological rather than a statistical explanation and produced a modified form of the MacArthur broken-stick model. Sugihara imagines the community niche space as a multidimensional hypervolume rather than a one-dimensional space. This hypervolume is sequentially (rather than simultaneously) broken into S fragments, where the volume represents relative abundance. Sugihara shows that the model generates the canonical lognormal distribution ($\gamma = 1$) as well as the species area constant $z \approx 1/4$.

Niche Theory

No concept in ecology has been more variously defined or more universally confused than "niche." Nonetheless, the concept has become symbolic of the whole field of ecology, and most of the papers reprinted in this volume, in one way or another, have something to do with the organism's niche (Whittaker and Levin 1975). "Niche" has become part of the common parlance of environmental policy and issues, and no term in ecology has been more appropriated into common language (it has even become immortalized in poetry by Dr. Seuss). Canonization of the concept has not been restricted to the public. Pianka feels the concept is the cornerstone of ecology and that a "Periodic Table of Niches" would provide a Rosetta Stone for all ecological interactions in which ecological processes could be understood as some combination of elemental niches. While few ecologists hold such high hope, most would agree that niche is a central concept of ecology, even though we do not know exactly what it means.

The Oxford English Dictionary attributes the origin of the word "niche" to the French word *niche* meaning "a shallow ornamental recess or hollow formed in a wall of a building, usually for the purpose of containing a statue or other decorative object." The English form was used as early as 1616 by Ben Johnson in his *Prince Henrie's Barrier:* "There Porticos were built . . . the nieces fild with statues." Modern ecological definitions of niche are invariably linked to four preeminent ecologists: Grinnell, Elton, Gause, and Hutchinson. The first use of niche in ecology appears in Grinnell's (1917) "the niche-relationships of the California Thrasher" (reprinted in part 1 of this volume). Grinnell stuck rather closely to the French definition and associated the specie's niche with its place or location in the environment, emphasizing adaptation to the physical structure of the habitat. Grinnell (1928) subsequently identified the niche as the "ultimate distributional unit within which each species is held by its structural and functional limitations." The Grinnellian niche concept has been most frequently interpreted as a place concept of niche; that is, where the organism lives and is indistinguishable from its habitat. Elton (1927) redefined the specie's niche by emphasizing its functional role in the community: the niche is "the status of an animal in its community." Elton was particularly concerned with the specie's relationship to food and enemies. Elton's concept idealized what the species does rather than where it lives—the functional niche concept. Gause (1934) adapted this Eltonian concept of niche and extended the notion to competitive interactions among different species. For Gause, "a niche indicates what place the given species occupies in a community, i.e. what are its habits, food, and mode of life." Gause used this niche concept to formulate his now-famous exclusion principle: "as a result of competition two similar species scarcely ever occupy similar niches, but displace each other in such a manner that each takes possession of certain peculiar kinds of food and modes of life in which it has an advantage over its competition." The competitive exclusion principle, based upon the Eltonian niche concept, has been widely acclaimed. Hutchinson and Deevey (1949) believed the principle "the most important development in theoretical ecology," and "one of the chief foundations of modern ecology." In 1944, the British Ecological Society held a symposium on the ecology of closely related species that "centered about Gause's contention (1934) that two species with similar ecology cannot live together in the same place . . ." (Lack 1944).

The niche concept and the competitive exclusion principle came to dominate most ecological thinking over the next 30 years and to some degree continues as a major focus of ecological research. While the concept and principle were well entrenched by the late 1940s and early 1950s, it was not until Hutchinson's (1957) "Concluding remarks" that the niche concept was rigorously defined and its relationship to competition and species diversity was rigorously explored. The importance of Hutchinson's paper cannot be overestimated. Hutchinson succeeded in combining both the Eltonian and Grinnellian concepts of niche into one model. Consider some environmental variable that can be ordered along some axis. Along this axis, the specie's upper and lower limits that allow reproduction and survival can be defined by the values x' and x''. The line segment between x' and x'' constitutes the specie's niche along this one axis. Each species will be subject to a set of important environmental variables $x_1, \ldots, x_n$ that are assumed independent (and therefore orthogonal), and the upper and lower limits for survival and reproduction are definable for each of the n-axes. The specie's requirements can then be defined by a n-dimensional hypervolume N which Hutchinson terms the "fundamental niche." Different species $S_1, \ldots, S_m$ would then have different fundamental niches $N_1, \ldots, N_m$ with respect to these n-environmental variables. The species will only occur in locations or points of the physical space or "biotop space" B, where conditions in the fundamental niche are realized. Then the Grinnellian niche and the Eltonian niche are united through correspondence between points in N and points in B.

Hutchinson then noted, rather ingeniously, that the fundamental niche so defined constitutes a set, and that the algebra of sets could be used to describe species interactions—particularly competition. Competition between two species, S_1 and S_2, for Hutchinson could essentially be described as the intersection of their fundamental niches ($N_1 \cap N_2$). The competitive exclusion principle could be reformulated by the condition: when N_1 and N_2 are identical, there are no points in B where S_1 and S_2 can cooccur. Rarely will N_1 and N_2 be identical, instead, they may show different degrees of overlap. Consequently, there are regions of niche space unique to S_1 and other regions unique to S_2 where survival and reproduction can occur. If there are points in biotop space B that correspond to points in these unique regions, then both species can occur in the community. Hutchinson termed this portion of niche space occupied by a species in the presence of competitors the "realized niche." The formal algebra of set theory could then be used to describe the equilibrium community structure as the partitioning of niche space into a set of realized niches with their corresponding points in biotop space. Hutchinson's belief that set theory could adequately describe ecological processes echoes Joseph Woodger's (1937) earlier efforts to establish a set-theoretic foundation for the whole of biology.

Hutchinson also realized that a catalog of the relative proportion of biotop space occupied by the realized niche of the species in the community would generate a distribution of species relative abundance. Unfortunately, he exclusively relied on MacArthur's broken-stick calculations as a model of niche partitioning. The inadequacies of the broken-stick analogy escaped Hutchinson, and the restriction to partitioning along a single dimension were lightly glossed over. However, Sugihara's (1980) multidimensional niche-partitioning model seems to be both more realistic and predictive of species abundance patterns. Hutchinson's early vision of a community ecology based on set theory may have been fulfilled after all.

Demography and Single-species Population Growth

The two papers reprinted here, Cole (1954) and May (1974), represent two different perspectives on single-species population phenomena. May begins with a general equation that encapsulates population growth and examines the variety of dynamical behaviors that can emerge over a range of parameter values. May's goal is to uncover the dynamic complexities of simple equations and their phenomenological relationship to the kinds of temporal complexity observed in nature. Cole, on the other hand, takes the phenomenological dynamics of populations as an epiphenomenon and, instead, focuses on the evolution of the particular parameter values underlying the pattern of growth. These two concerns are not mutually exclusive, but complementary. May provides a methodology that enables the ecologist to appreciate fully the dynamic properties implied by a particular evolutionary scenario. Similarly, given the variety of temporal growth patterns, what kinds of variation in selection regime are necessary to account for this diversity? Unfortunately, these two approaches have not been fully integrated, and often ecologists working in one area are not particularly concerned with questions and developments in the other.

Demography

Hutchinson traced the dichotomy between phenomenological population dynamics and the evolutionary approach to the "Oxford School," which he associated principally with David Lack. The Oxford approach is best illustrated in Lack's treatment of clutch size evolution in birds. Prior to Lack, ornithologists had not thought much about the number of eggs laid by the females of different species. Lack was struck by two features: first, the number seemed very low, and second, when an egg was destroyed or removed, another egg was often produced to replace the missing one. These features obviously implied that birds were not producing as many eggs as possible. To Lack, this automatically implied an evolutionary problem. Why should any bird reduce its fecundity, since fecundity is a direct component of fitness? What is important is not that Lack posed a certain answer to this problem, but that he raised the problem in the first place. The focus is not on the population-dynamic consequences of a fecundity pattern, but on the evolutionary basis for the pattern. Lack suggested that a species reduces its clutch to maximize the number of young that survive and fledge. If too many eggs are laid, there will be insufficient resources available to fledge them all. Overall total fitness to the parent (measured as young fledged), then, is maximized at a reduced clutch size, and the evolutionary problem is resolved. In his 1954 book, *The Natural Regulation of Animal Numbers,* Lack summarized his arguments and amassed a considerable body of evidence relating geographic trends in resource availability and clutch size. Lack's explanation remains the predominant view among avian ecologists.

The themes outlined in Lack (1954) were echoed in Cole's (1954) classic paper, which explicitly explored the link between life-history phenomena and population growth. Cole, like Lack, was impressed by the evolutionary problem of why animals show reduced or delayed reproduction. Cole considered two categories: species that reproduce a single time and then die (annuals or semelparous organisms) versus organisms that reproduce multiple times over many seasons (perennials or iteroparous organisms). Using simple population models, Cole compared the fitness of annual versus perennial organisms and arrived at his now-infamous paradox:

> For an annual species, the absolute gain in intrinsic population growth which could be achieved by changing to the perennial reproductive habit would be exactly equivalent to adding one individual to the average litter size.

It certainly seems easier in terms of evolution to add *one* individual to the litter than to undergo all the adaptations necessary for long life. Therein lies the paradox.

Almost all contemporary research on the evolution of life-history phenomena is rooted in the resolution of this paradox. Cole appears to have ignored two important forces that either singly or together can resolve this apparent paradox. Murdoch (1966) suggested that there are trade-offs between fecundity and survivorship, and that these trade-offs, when coupled with uncertainty in juvenile survivorship, could promote iteroparity. Charnov and Schaffer (1973) formalized this resolution using the same kinds of mathematical arguments employed in the original Cole paper. They point out that the number of offspring which must be added to the litter depends on the ratio of adult to juvenile survival. Cole's result critically depends on this ratio being equal to one.

All modern treatments of the evolution of life-history phenomena begin with the structure of the trade-offs between fecundity and survivorship, and with the relationship between these trade-offs and variability in different stages of the life-history. While not explicitly addressing these trade-offs, they are implicit in Lack's thinking and in his arguments on the evolution of clutch size. It is this implicit incorporation of trade-offs that gives Lack's writing its contemporary tone and value. Cole's contribution to modern ecological thought rests on his construction of a mathematical form for stating the critical issues and their possible resolution. The modern treatment of life-history phenomena emerges as the union of Lack's evolutionary thinking with Cole's mathematical rigor.

Population Dynamics

May's (1974) paper, which explores the complex, dynamic behavior implicit in some simple forms of population growth models, represents both an end and a beginning to an era of theoretical ecology. Before the 1974 paper and beginning with Pearl's introduction of the logistic equation, most ecologists characterized population growth as basically simple in pattern. From some small initial size, populations would grow until they reached the upper limit of their resources; density-dependent effects on fecundity and survivorship would then bring them to an equilibrium static size. Without intervening forces, populations would smoothly approach this limit and then stay there.

Unfortunately, populations rarely exhibit such simple behavior. Instead, they crash and fall, often showing fluctuations in size over several orders of magnitude. Modifications to the simple logistic equation to account for fluctuations incorporate time lags in density effects, age structure and demographic stochasticity, interactions with other species (for example, predation and parasitism), and environmental stochasticity. A huge literature has been built around these complicating influences, the belief being that in order to account for complex population dynamics, we need to add complicating factors to the simpler models of density-dependent regulation. May's (1974, 1976) articles challenged this basic assumption.

First, May noted that in many biologically important scenarios population growth is discrete with nonoverlapping generations. Populations subject to this kind of growth are best represented by systems of difference equations rather than the usual continuous differential equation. For example, for nonoverlapping generations a discrete analogue to the logistic equation can be represented by

$$N_{t+1} = N_t \exp\left\{r\left(1 - \frac{N_t}{k}\right)\right\}$$

where N_t is the population size at time t, N_{t+1} the size one generation hence, and r, k are defined as usual. The use of difference equations was in itself not new. Rather, May's second important point (and by far the more important) was to demonstrate the full implications of such simple, nonlinear discrete time models.

Unlike the continuous time-logistic equation of Pearl, the discrete time equation generates a wide variety of dynamical behaviors depending on parameter values. Pearl's continuous time logistic equation implies only two possible behaviors. If $r < 0$, then the population crashes to zero. If $r > 0$, then N equilibrates at k. In the discrete case, stable point equilibria only occur for values of $2 > r > 0$. As the growth rate r increases, the equilibrium loses stability, giving way to cycles of different periods. As r increases, the period of the fluctuation increases through a series of bifurcations: 2-point cycles give way to 4-point cycles, 4-point to 8-point, and so on. At a critical value r_c, a region of "chaos" arises, where (formally) "there are cycles of period 2, 3, 4, 5, . . . , n, . . . , where n is any positive integer, along with an uncountable number of initial points for which the system does not eventually settle into any finite cycle; whether the system converges on a cycle, and if so, which cycle, depends on the initial population point." In regions of chaos there is an infinite number of population trajectories that are aperiodic and would have the appearance of a randomly chosen segment of a stochastic process. However, population behavior is not stochastic but instead is absolutely determined. If we know the initial population size, then the size at any time in the future is unique and can be calculated from the model. Yet slightly different initial conditions can generate quite different and often complicated dynamics. As May states, "What is remarkable, and disturbing, is that the simplest, purely deterministic, single-species models give essentially arbitrary dynamic behavior once r is big enough."

The implications of this complexity in behavior are devastating for traditional approaches to understanding population fluctuations. Are temporal fluctuations the consequence of complicating outside influences or are they inherent to the system and generated as a *consequence* of density-dependence? The question has turned upside-down. Instead of density-dependence acting to stabilize populations, now strong density-dependence generates wild swings in population size. Devising techniques to disentangle this intellectual knot has been the preoccupation of mathematicians and theoretical ecologists since May's papers. In this regard, May's work championed a new beginning for theory.

Original attempts relied upon assuming a particular model for population growth, determining the parameter values that make the model chaotic, and checking to see if the parameter values estimated from natural populations fall in this chaotic region. This kind of investigation led Hassell et al. (1976) to believe that chaos was rare in nature. More direct model-independent methods for detecting chaos were developed by Packard et al. (1980) and Takens (1981), and have been used by Schaffer (1984; Schaffer and Kot 1985) to argue for the existence of chaotic dynamics in natural populations.

Sensitive dependence on initial conditions, aperiodicity, and emergent complexity in the behavior of simple, nonlinear difference equations had been detected in other fields. Lorenz (1963) constructed a simple model for weather that was the first model to show the features of chaotic dynamics, and mathematicians, physicists, and biologists all credit Lorenz with the first discoveries in nonlinear chaotic systems. He constructed the first "strange attractor"—the Lorenz attractor—and applied this simple model toward explanations of turbulent flows and other apparently stochastic phenomena. While Lorenz was the first to discover the existence of complicated behavior from simple nonlinear models, his contributions were mostly restricted to specialty journals and did not have much impact on the wider scientific community. His paper, published in the *Journal of the Atmospheric Sciences,* went largely unnoticed for almost 20 years (Gleick 1987). By the 1980s, however, it was cited on average over 100 times a year.

The transformation of chaos theory into a dominant intellectual movement was in large part due to the emergence of a general mathematical theory of nonlinear dynamics that embraced chaos (Smale 1967; Li and Yorke 1975; Packard et al. 1980; Takens 1981) and the demonstration that these simple nonlinear models could influence our interpretation of real world phenomena. May's papers, published in *Science* and then *Nature,* explored the full implication of chaotic dynamics in population models and provided the impetus to begin looking for chaos in other fields that use temporal sequences—physics, meteorology, economics, sociology, and even history. The discovery of chaotic systems has revolutionized our thinking about dynamical systems (Gleick 1987). Ecologists should take pride in recognizing that the origins of this revolution can be traced directly to issues in population biology.

Individual Behavior

A mechanistic understanding of community organization must incorporate the differential ability of individual organisms to acquire and exploit limited resources (for example, mates, nutrients, energy, nesting sites, etc.). Adaptations in individuals that increase the efficiency in and likelihood of acquiring resources will be favored by natural selection. Often these adaptations are behavioral.

MacArthur and Pianka (1966) were the first to model explicitly the strategies by which organisms exploit resources most efficiently. Theirs is strictly an adaptationist view borrowing heavily from an economic analogy: "In this paper we undertake to determine in which patches a species would feed and which items would form its diet if the species acted in the most economical fashion. Hopefully, natural selection will often have achieved such optimal allocation of time and energy expenditures." MacArthur and Pianka noticed that the appropriate mathematical tools for analyzing such problems involve optimization, and that specific optimization models pose specific evolutionary and behavioral hypotheses. Dominating almost all models in behavioral ecology is the hypothesis that the organism's rate of energy acquisition can stand as a proxy measure of evolutionary fitness. In some unspecified manner, rate of energy gain translates directly into fitness components. Consequently, for empirical purposes we need only examine energy acquisition and unit time to understand selection on foraging behavior.

This simplistic view has given way to more complicated models. Nonetheless, their basic optimization approach is still followed. The simplest form of optimization analysis and the one advanced by MacArthur and Pianka tracks the marginal changes in the benefits and costs of alternative activities. According to MacArthur and Pianka, "an activity should be enlarged as long as the resulting gain in time spent per unit food exceeds the loss. . . . The problem is to find which components of a time or energy budget increase and which decrease as certain activities are enlarged."

If one starts with the natural selection assumption and the belief that efficiency in energy acquisition is a fitness component, then this simple method seems to lead to an obvious truism. However, their simple procedure, when examined closely, leads to considerable difficulties in interpretation. Part of the continuing project of behavioral ecology, and particularly of foraging ecology, has been to clarify its "simple procedure." What exactly do we mean by gain and loss? How are they to be measured? What is the exact connection between traditional fitness (measured in terms of reproduction and survival) and energy acquisition? What aspects of the organism's behavioral repertoire contribute to foraging efficiency? Can selection improve performance in foraging independently from other aspects of the phenotype? And so the questions go.

MacArthur and Pianka do not attempt to answer these questions. Instead they present a simple graphical model that leads to some general qualitative predictions about directions of change in diet breadth with changes in food availability. Their conclusions are much too general to lead to specific predictions amenable to experimental investigation. Like much of MacArthur's work, the contribution lies not so much in the solution, but in posing the problem and pointing to a general method for attacking the problem.

In an ingenious treatment of the optimization analysis of animal foraging, Schoener (1971) expanded upon the original contribution of MacArthur and Pianka. He challenged the simplicity of the MacArthur-Pianka approach and noted that the optimization criterion can be different for different organisms. Specificially, he suggested that organisms might be selected for either of two strategies: (1) minimizing the time spent acquiring a given amount of energy; or (2) maximizing total energy acquisition per unit time. MacArthur and Pianka considered only the latter criterion. Evidence assessed by Belovsky (1984) seems to support energy maximization as the appropriate criterion. Nonetheless, the idea that there may be alternative objectives for the organism remains very important.

MacArthur and Pianka focus on two problems faced by individual foragers: the number of different food types to include in the diet, and the types of patches to include in a foraging excursion. They apply essentially the same analyses to both problems. Charnov (1976) suggested an alternative formulation of the patch-choice problem where the organism must choose the time spent in each patch before giving up and moving to a new patch. The currency in the Charnov model is the same as in MacArthur and Pianka; organisms maximize the expected rate of energy gain. The decision variables are different: time versus number.

Charnov showed in a simple result that he termed the Marginal Value Theorem that the organism should remain in a patch until the rate of energy gain in the patch falls below the expected rate of energy gain among all remaining patches. In other words, if it is on average better for the organism to move than to stay, then it moves. This theorem has had considerable influence principally because it makes quite specific quantitative predictions which can be measured in a variety of foraging situations. Still, the basic idea of a marginal condition specifying switches in behavior or the optimal value of decision variables was suggested by MacArthur and Pianka—indeed, it was their basic principle.

Since the MacArthur and Pianka (1966), Schoener (1971), and Charnov (1976) papers, foraging ecology and theory have grown into a rich subdiscipline (reviewed by Stephens and Krebs 1986). The simplest energy maximization models have been extended to include nutritional constraints (Belovsky 1984; Pulliam 1975), effects of environmental stochasticity and risk-taking behavior (Real and Caraco 1986; Ellner and Real 1989) and a variety of psychological constraints such as learning (Kacelnik and Krebs 1985), memory (Sherry and Schacter 1987), time perception (Kacelnik et al. 1990), and probability and frequency perception (Real 1987, 1990). The link between energy acquisition and/or other kinds of foraging behavior has been more closely linked to traditional fitness (McNamara and Houston 1986; Mangel and Clark 1988); and there is growing interest in the genetic base of ecologically important behaviors.

It was clear from the early papers that this new field could develop in two direction: either into a deeper investigation of the psychological and economic foundations of individual behavior or into an examination of the consequences of individual behavior for population growth and community organization. By far, most attention has been directed toward the first concern. Most of the developments listed above involve refinements in characterizing individual choice behavior, often borrowing concepts from psychology, neurophysiology, and/or economics. Linking behavior to population phenomena has received little attention. The importance of this link and possible theoretical methods for establishing connections between individuals and populations have been suggested (Real 1983; Lomnicki 1988; Hassell and May 1985). Establishing this link is fundamental since it is only in this translation to higher-level phenomena that we can understand the selection consequences of behavior.

Multispecies Interactions

Populations do not exist in isolation but rather interact with other species populations through a variety of mechanisms—competition, predation, parasitism, mutualism, for example. Extension of simple mathematical models incorporating population interactions originated with the pioneering efforts of Sir Ronald Ross and W. R. Thompson (Kingsland 1985). Their work, produced mostly between 1910 and 1920, focused on models of host-parasite epidemiology. Thompson employed simple algebraic expressions to represent cyclical patterns of host-parasite abundance, while Ross recognized the advantage in modeling interactions through coupled differential equations. A full and mature treatment of species interactions, however, is most often associated with Alfred Lotka and Vito Volterra.

Both Lotka and Volterra used sets of coupled differential equations to describe species interactions. While producing models of almost identical structure, they used quite different logic to arrive at the system's description. In his 1925 book *Elements of Physical Biology,* Lotka articulated what can be described as an early version of general systems theory. The community can be decomposed into components $x_1, \ldots, x_n$ (in this case different species) and the change in any one component is some function of all the other components in the system; that is:

$$\frac{dx_i}{dt} = f(x_1, \ldots, x_n), \quad i = 1, \ldots, n$$

After describing the basic system, Lotka used the Taylor Series to approximate $f(x_1, \ldots, x_n)$. The stability properties of this dynamical system could then be determined using the phase-plane analysis developed by Poincairé. Lotka used this basic method to demonstrate the now-familiar neutral, oscillatory predator-prey dynamics, as well as to describe competition.

On the other hand, Volterra, a distinguished mathematician known especially for his work on integral equations (which he applied to ecology), took a different approach based on purely physical analogies. In the simplest "mass action" models, the interaction between two species (predator and prey, for example) is determined by the likelihood of their encounter, which is in turn proportional to their densities. If P represents the predator population density, and V represents the prey population density, then the population dynamics of each species follows the set of equations:

$$\frac{dV}{dt} = rV - \alpha VP$$

$$\frac{dP}{dt} = \beta VP - dP$$

where r is the growth rate of prey, d is the death rate of predators, and α and β are proportionality constants. In the paper we have reprinted here, Volterra (1926) showed that this kind of system would show neutrally stable oscillations.

The specific models developed by Lotka and Volterra underwent successive changes. Volterra showed, for instance, that by the addition of competitive interactions among members of the same species, the stability changed from neutral oscillations to damped oscillations. The general method of qualitative analysis employed by Lotka and Volterra—cataloge the set of differential equations and solve for the stability properties around equilibrium—has remained a mainstay of theoretical ecology since it was first introduced. The Lotka-Volterra approach has had a substantial impact in other fields as well. Paul Samuelson, for instance, attributes the ideas in his pathbreaking work *The Foundations of Economic Analysis* to the original comparative approach in Lotka.

At the same time Lotka and Volterra were developing their differential equation approach to predator-prey interactions and competition, Nicholson and Bailey introduced an alternative difference equation approach to modeling parasitoid-host interactions. May (1986) has argued that the Nicholson-Bailey modeling approach has been more successfully applied—especially for small, short-lived species that clearly have nonoverlapping generations, like insects and their parasites.

The formal approach of Lotka and Volterra is instructive regarding the sorts of dynamics that might occur in species interactions but should not be taken literally as a model for actual populations. Although mathematicians have remained intrigued with them because of their straightforward character, no effort has ever been successful at identifying the parameters needed to apply these equations to real settings. Their value lies primarily in their use as starting points for theoretical investigation, for example in examining the implications of deviations from the assumptions of the models such as the addition of spatial structure (to be discussed later).

A more static and less successful approach was developed in the late 1950s and 1960s. Borrowing heavily from information theory, several ecologists argued for a functional relationship between the diversity of a natural community and stability. The argument went something like this: if there are many species in a system, then a disturbance to any one species will have little effect on the whole assemblage. Similarly, a general disturbance will have little effect since the disturbance will be distributed over the whole system. This argument was crystallized in an influential paper by Margalef (1969), in which he summarized the diversity-stability relation in a specific equation. This phenomenological belief in a relation between diversity and stability went unchallenged for over a decade. The argument, however, was not rigorously grounded in any way, and Margalef's equation is simply a short-lived expression of a commonly shared belief. May (1975) undertook a rigorous investigation of the diversity-stability relationship using the identical mathematical techniques introduced by Lotka 50 years earlier. May found that when the number of interacting species in a community increases, then the requirements necessary to guarantee mathematical stability become more restrictive. Increasing diversity of species per se decreases rather than increases community stability. It can be argued that such an approach is too artificial, since the kinds of interactions among species (competition, predation, etc.) are assumed to be random—and communities are not random assemblages. Nonetheless, the value of May's analysis was to recast the fuzzy verbal arguments of the earlier investigators into a rigorous form where the terms had precise meaning. He used techniques that were already in the ecological literature but had been overlooked, and showed that factors other than diversity per se must be included in the theory if a general diversity-stability relationship was to hold true. These were enormously valuable contributions that forced theoreticians to undertake a rigor which had essentially disappeared after the pioneering work of Lotka and Volterra.

Today we recognize that the whole debate about diversity and stability is flawed to a large extent by imprecision in defining these terms. What do we mean when we say "diversity," "stability," or the related term "complexity"? Depending on the definition one chooses for these terms, one can obtain diametrically opposed results. We will not review the topic here but will simply point out that the absence of hard results in this area of research is the principal reason why a once intensely discussed topic receives little theoretical or empirical attention today.

Movement and Scale

The earlier discussion of foraging theory relates to one of the fundamental problems in ecology, as well as in evolutionary biology: how do organisms relate to environmental heterogeneity in space and time? The notion of a patch is a simplistic one, since more typically one finds a spectrum of variation in environmental quality. The evolutionary responses of species to this variation determine the patterns of coexistence within a community. To some extent, dispersal is an evolutionary response to spatially uncorrelated temporal variation in environmental quality; conversely, the result of dispersal may be to spread offspring or members of a genotype across a range of environments, thereby reducing the temporal variability they face. Therefore, the selection regime for dispersal is shaped by, and in turn shapes, the scale of environmental variation facing the species.

The issue of scale is a central theoretical problem in ecology, from evolutionary ecology (where the definition of the effective population involves spatial scale, the concept of punctuated equilibria raises problems of temporal scale, and the problem of levels of selection raises issues of organizational scale that are central to the concept of evolutionary change) to ecosystem science (where fundamental challenges involve the hierarchical organization of food webs, the interface between local population change and global climate change, and nutrient and element dynamics). No papers on scale are included in this volume (but see discussion in Levin 1989), however, the closely related issue of patch dynamics and the effects of disturbance, one of the most exciting theoretical topics of the past two decades, owes its origins to the seminal paper by A. S. Watt (1947, reprinted below in Part 5). Watt's paper has stimulated a wide range of papers on patch dynamics in a diversity of ecosystems (Levin and Paine 1974; Paine and Levin 1981; Pickett and White 1985).

The response of a species to scale, as already discussed, is interwoven with the species characteristics of dispersal, dormancy, and foraging behavior. The biology of dispersal and its theoretical characterization has been one of the most actively pursued topics in contemporary population biology (Levin 1974, 1976; Okubo 1980). Certainly the key paper in triggering this interest was Skellam's (1951), which we have chosen to reprint in this section. Although the methods for describing the movement of organisms through the use of diffusion equations were available for a century, as Skellam points out, they were not exploited. Numerous papers (for example, Brownlee 1911, and especially those by Dobzhansky and Wright 1943, and Fisher 1937 exploring different aspects of population genetics) had used these methods with considerable success. Fisher (1937) introduced the diffusion equation to describe the rate of spread of an advantageous allele, and his insightful conjectures concerning the asymptotic speed of advance were confirmed by Kolmogorov, et al. (1937). Dobzhansky and Wright (1943) applied these approaches to quantify the movement of *Drosophila* but did not carry them much further. It was left to Skellam to place the whole subject within an ecological context, which he did in his 1951 paper.

Skellam's paper contains a number of important advances. First, he provided a rigorous derivation and motivation for the diffusion model. Whatever its drawbacks, the diffusion model has proved remarkably robust. These simple models are extremely flexible and can be modified to accommodate different biological scenarios. The usefulness of the approach is not in the details of its assumptions; surely organisms do not move randomly. However, neither do molecules; yet the theory of heat has been built successfully on the identical model. The strength in both cases, instead, resides in the demonstration that such a simple model adequately captures many of the prominent features of population spread, and hence that the details of individual movements are irrelevant to explaining these patterns. Karieva (1982, 1983) very effectively demonstrated the adequacy of simple diffusion models for explaining the macroscopic features of the foraging behavior of phytophagous insects. Furthermore, he showed that where the model is inadequate, it is not wrong by much and the predictive power of the model can be improved significantly by simply incorporating aspects of habitat heterogeneity.

Skellam's second contribution was his application of the problem of asymptotic speeds of propagation to the spread of introduced species—a problem of considerable importance in contemporary conservation biology. He examined a number of specific cases, including oaks and muskrats, and provided many insights that encouraged both mathematicians and ecologists to tackle this problem. Recent papers (Lubina and Levin 1988; Andow et al. 1991) demonstrate the continuing usefulness of Skellam's approach for understanding the problem of spread.

The third contribution by Skellam concerns his development of the concept of critical patch size (see also Okubo 1980; Platt and Denman 1975; Kareiva 1985). The major influence of this idea has been in oceanography, due principally to the independent application of identical methods to the problem of red tides (Kierstead and Slobodkin 1953). The basic idea is that diffusive losses through a boundary will be roughly at a rate proportional to the length of the perimeter of a patch, which is of decreasing importance as patch area increases. Thus, if a favorable region for growth is small, losses exceed growth, and the population will go extinct. Beyond a certain threshold, however, the size of which can be computed, a region of favorable conditions (for example, a burst of nutrients) will be sufficient to support growth. Thus, in nature one should not see patches smaller than the critical size. This has been an extremely stimulating concept in oceanography (see for example, Steele 1978 and the contributions therein) and has recently been applied to agricultural systems (Kareiva 1985).

Literature Cited

Andow, D. A., P. M., Karieva, S. A. Levin, and A. Okubo. 1991. Spread of invading organisms. *Landscape Ecology*.

Belovsky, G. 1984. Herbivore optimal foraging: A comparative test of three models. *Amer. Nat.* 124:97–115.

Brownlee, J. 1911. The mathematical theory of random migration and epidemic distribution. *Proc. Roy. Soc. Edinburgh* 31:262–89.

Charnov, E. L. 1976. Optimal foraging: The marginal value theorem. *Theor. Pop. Biol.* 9:129–36.

Charnov, E. L., and W. M. Schaffer. 1973. Life-history consequences of natural selection: Cole's result revisited. *Amer. Nat.* 107:791–93.

Cole, L. C. 1954. The population consequences of life history phenomena. *Quart. Rev. Biol.* 29:103–37.

Dobzhansky, T., and S. Wright. 1943. Genetics of natural populations. X. Dispersion rates in *Drosophila pseudoobscura*. *Genetics* 28:304–40.

Ellner, S., and L. A. Real. 1989. Optimal foraging models for stochastic environments: Are we missing the point? *Comm. Theor. Biol.* 1:129–58.

Elton, C. 1927. *Animal ecology*. London: Sedgwick and Jackson.

Fisher, R. A. 1937. The wave of advance of advantageous genes. *Ann. Eugen.* London 7:355–69.

Fisher, R. A., A. S. Corbet, and C. B. Williams. 1943. The relation between the number of species and the number of individuals in a random sample of an animal population. *J. Anim. Ecol.* 12:42–58.

Gause, G. F. 1934. *The struggle for existence*. New York: Hafner.

Gleick, J. 1987. *Chaos: Making a new science*. New York: Viking.

Grinnell, J. 1917. The niche-relationships of the California Thrasher. *Auk* 34:427–33.

——— 1928. Presence and absence of animals. *Univ. Calif. Chron.* 30:429–50.

Hassell, M. P., J. H. Lawton, and R. M. May. 1976. Patterns of dynamical behavior in single species populations. *J. Anim. Ecol.* 45:471–86.

Hassell, M. P., and R. M. May. 1985. From individual behavior to population dynamics. Pp. 3–32 in R. Sibley and R. Smith, eds. *Behavioural ecology*. Oxford: Blackwell.

Hutchinson, G. E. 1957. Concluding remarks. *Cold Spring Harbor Symp. Quant. Biol.* 22:415–27.

Hutchinson, G. E., and E. S. Deevey, Jr. 1949. Ecological studies on populations. In G. S. Avery, Jr., ed., *Survey of biological progress*, vol. 1. New York: Academic Press.

Kacelnik, A., D. Brunner, and J. Gibbon. 1990. Timing mechanisms in optimal foraging: Some applications of scalar expectancy theory. Pp. 61–82 in R. N. Hughes, ed., *Behavioural mechanisms of food selection*. NATO ASI Series G, vol. 20. Heidelberg: Springer-Verlag.

Kacelnik, A., and J. R. Krebs. 1985. Learning to exploit patchily distributed food. Pp.–189–206 in R. M. Sibley and R. H. Smith, eds., *Behavioural ecology*. Oxford: Blackwell.

Karieva, P. 1982. Experimental and mathematical analyses of herbivore movement: Quantifying the influence of plant spacing and quality on foraging discrimination. *Ecol. Monogr.* 52:261–82.

——— 1983. Local movement in herbivorous insects: Applying a passive diffusion model to mark-recapture field experiments. *Oecologia* 57:322–27.

——— 1985. Finding and losing host plants by flea beetles: Patch size and surrounding habitat. *Ecology* 66:1809–16.

Kierstead, H., and L. B. Slobodkin. 1953. The size of water masses containing plankton blooms. *J. Mar. Res.* 12:141–47.

Kingsland, S. E. 1985. *Modeling nature: Episodes in the history of population ecology*. Chicago: University of Chicago Press.

Kolmogorov, A., I. Petrovsky, and N. Piscounov. 1937. Étude de l'équations de la diffusion avec croissance de la quantité de la matrière et son application à un problème biologique. *Bull. Univ. Moscou. Ser. Internation.* Sec. A 1(6):1–25.

Lack, D. 1944. Symposium on the ecology of closely allied species. *J. Anim. Ecol.* 13:176–77.

——— 1954. *The natural regulation of animal numbers*. Oxford: Oxford University Press.

Levin, S. A. 1974. Dispersion and population interactions. *Amer. Nat.* 108:207–8.

——— 1976. Population dynamic models in heterogeneous environments. *Ann. Rev. Ecol. Syst.* 7:287–311.

——— 1989. Challenges in the development of a theory of community and ecosystem structure and function. Pp. 242–55 in J. Roughgarden, R. M. May, and S. A. Levin, eds., *Perspectives in ecological theory*. Princeton: Princeton University Press.

Levin, S. A., and R. T. Paine. 1974. Disturbance, patch formation, and community structure. *Proc. Nat. Acad. Sci. U.S.A.* 71:2744–47.

Li, T. Y., and J. A. Yorke. 1975. Period three implies chaos. *Amer. Math. Monthly* 82:985–92.

Lomnicki, A. 1988. *Population ecology of individuals*. Princeton: Princeton University Press.

Lorenz, E. N. 1963. Deterministic nonperiodic flow. *J. Atmosph. Sci.* 20:130–41.

Lotka, A. J. 1925. *Elements of physical biology*. Reprinted 1956. New York: Dover.

Lubina, J. A., and S. A. Levin. 1988. The spread of a reinvading species: Range expansion in the California sea otter. *Amer. Nat.* 131:526–43.

MacArthur, R. H. 1957. On the relative abundance of bird species. *Proc. Nat. Acad. Sci. U.S.A.* 43: 293–95.

——— 1960. On the relative abundance of species. *Amer. Nat.* 94:25–36.

MacArthur, R. H., and E. R. Pianka. 1966. On optimal use of a patchy environment. *Amer. Nat.* 100: 603–9.

Mangel, M., and C. W. Clark. 1988. *Dynamic modeling in behavioral ecology.* Princeton: Princeton University Press.

Margalef, R. 1969. Diversity and stability: A practical proposal and a model of interdependence. *Brookhaven Symp. Biol.* 22:25–37.

May, R. M. 1973. *Stability and complexity in model ecosystems.* Princeton: Princeton University Press.

——— 1974. Biological populations with nonoverlapping generations: Stable points, stable cycles, and chaos. *Science* 186:645–47.

——— 1975. Patterns of species abundance and diversity. Pp. 81–120 in M. L. Cody and J. M. Diamond, eds., *Ecology and evolution of communities.* Cambridge, Mass.: Harvard University Press.

——— 1976. Simple mathematical models with very complicated dynamics. *Nature* 261:459–67.

——— 1986. The search for patterns in the balance of nature: Advances and retreats. *Ecology* 67: 1115–26.

McNamara, J. M., and A. I. Houston. 1986. The common currency for behavioral decisions. *Amer. Nat.* 127:358–78.

Murdoch, W. W. 1966. Population stability and life history phenomena. *Amer. Nat.* 100:5–11.

Okubo, A. 1980. *Diffusion and ecological problems: Mathematical models.* Lecture Notes in Biomathematics 10. Heidelberg: Springer-Verlag.

Packard, N., J. P. Crutchfield, J. D. Farmer, and R. S. Shaw, 1980. Geometry from a time series. *Phys. Rev. Letters* 45:712–16.

Paine, R. T., and S. A. Levin. 1981. Intertidal landscapes: Disturbance and the dynamics of pattern. *Ecol. Monogr.* 51:145–78.

Pickett, S. T. A., and P. S. White. 1985. *The ecology of natural disturbance and patch dynamics.* Orlando: Academic Press.

Platt, T. and K. L. Denman. 1975. A general equation for the mesoscale distribution of phytoplankton in the sea. *Mém. Soc. Roy. Sci. Liège,* 6ᵉ série, tome VII:31–42.

Preston, F. W. 1948. The commonness and rarity of species. *Ecology* 29:254–83.

——— 1962. The canonical distribution of commonness and rarity. Parts I and II. *Ecology* 43:185–215, 410–32.

Pulliam, H. R. 1975. Diet optimization with nutrient constraints. *Amer. Nat.* 109:765–68.

Real, L. A. 1983. Microbehavior and macrostructure in pollinator-plant interactions. Pp. 287–304 in L. A. Real, ed., *Pollination biology.* New York: Academic Press.

——— 1987. Objective benefit versus subjective perception in the theory of risk-sensitive foraging. *Amer. Nat.* 130:399–411.

——— 1990. Predator switching and the interpretation of animal choice behavior: The case for constrained optimization. Pp. 1–21 in R. N. Hughes, ed., *Behavioral mechanisms of food selection.* NATO ASI Series, Series A: Life Sciences. Heidelberg: Springer-Verlag.

Real, L. A., and T. Caraco. 1986. Risk and foraging in stochastic environments. *Ann. Rev. Ecol. Syst.* 17:371–90.

Schaffer, W. M. 1984. Stretching and folding in lynx fur returns: Evidence for a strange attractor in nature. *Amer. Nat.* 124:798–820.

Schaffer, W. M., and M. Kot. 1985. Nearly one dimensional dynamics in an epidemic. *Theor. Biol.* 112: 403–27.

Schoener, T. W. 1971. Theory of feeding strategies. *Ann. Rev. Ecol. Syst.* 2:369–404.

Scudo, F., and J. Ziegler. 1978. *The golden age of theoretical ecology: 1923–1940.* Lecture Notes in Biomathematics 22. Heidelberg; Springer-Verlag.

Sherry, D. F., and D. L. Schacter. 1987. The evolution of multiple memory systems. *Psychol. Rev.* 94: 439–54.

Skellam, J. G. 1951. Random dispersal in theoretical populations. *Biometrika* 38:196–216.

Smale, S. 1967. Differentiable dynamical systems. *Bull. Amer. Math. Soc.* 73:747–817.

Steele, J. H., ed. 1978. *Spatial pattern in plankton communities.* NATO Conference Series IV: Marine Sciences, vol. 3. New York: Plenum.

Stephens, D. W., and J. R. Krebs. 1986. *Foraging theory.* Princeton: Princeton University Press.

Sugihara, G. 1980. Minimal community structure: An explanation of species abundance patterns. *Amer. Nat.* 116:770–87.

Takens, F. 1981. Detecting strange attractors in turbulance. vol. 898. *Dynamic systems and turbulance: Warwick 1980.* Heidelberg: Springer-Verlag.

Volterra, V. 1926. Fluctuations in the abundance of a species considered mathematically. *Nature* 118: 558–60.

Whittaker, R. H., and S. A. Levin, eds. 1975. *Niche: Theory and application.* Stroudsburg, Pa.: Dowden, Hutchinson, and Ross.

Williams, C. B. 1964. *Patterns in the balance of nature and related problems in quantitative ecology.* New York: Academic Press.

Woodger, J. 1937. *Axiomatic method in biology.* Cambridge: Cambridge University Press.

PAPER 8

ECOLOGY

VOL. 43 SPRING 1962 No. 2

THE CANONICAL DISTRIBUTION OF COMMONNESS AND RARITY: PART I

F. W. PRESTON

Preston Laboratories, Butler, Pennsylvania

POSTULATES AND THEORY

Introduction

In an earlier paper (Preston 1948) we found that, in a sufficiently large aggregation of individuals of many species, the individuals often tended to be distributed among the species according to a lognormal law. We plotted as abscissa equal increments in the logarithms of the number of individuals representing a species, and as ordinate the number of species falling into each of these increments. We found it convenient to use as such increments the "octave," that is the interval in which representation doubled, so that our abscissae became simply a scale of "octaves," but this choice of unit is arbitrary. Whatever logarithmic unit is used, the graph tended to take the form of a normal or Gaussian curve, so that the distribution was "lognormal." We called this the "Species Curve."

In the present paper we take up a point merely mentioned in 1948 that not only is the distribution lognormal, but the constants or parameters seem to be restricted in a peculiar way. They are not fixed, but they are interlocked. The nature of this restriction and interlocking is the main theme of the present paper.

In the earlier paper we graduated the experimental results with curves of the form

$$y = y_0 e^{-(aR)^2} \qquad (1)$$

where y is the number of species falling into the Rth "octave" to the right or left of the mode, y_0 is the number in the modal octave, and a was treated as an arbitrary constant, to be found from the experimental evidence. This constant is related to the logarithmic standard deviation σ by the formula

$$a^2 = \tfrac{1}{2}\sigma^2 \qquad (2)$$

and we noted that it had a pronounced tendency to come out at a figure not far from 0.2, so that σ would have a value of about 3.5 octaves, and, since there are 3.3 octaves to an "order of magnitude," σ would be a little more than an order of magnitude.

If we now make a 2nd graph in which, using the same abscissae, we plot as ordinate not the number of species (y) that fall in each interval but the number of individuals which those y species comprise, we get another lognormal curve with the same standard deviation as the first graph, but with its mode or peak displaced to the right. This we call the "Individuals Curve."

Though we can use a Gaussian curve to "graduate" the observed points of the species curve, the curve extends infinitely far to left and right, while the number of observed points is necessarily finite. This is a common situation in statistical work and usually causes no complications, but in our problem there results an additional piece of information. The Individuals Curve necessarily lacks at least part of the descending limb. It terminates over the last observed point of the Species Curve, and this is long before the Individuals Curve begins to become asymptotic to the horizontal axis. In the earlier paper we noted that, as a matter of observation, it seems to terminate at its crest, so that, in effect, only half of the curve is present.

What I failed to observe in 1948 was that when the Individuals Curve terminates at its crest or very close to it, the value of "a" in equation (1) and of the standard deviation "σ," is fixed within narrow limits, and this value is in fact the one actually observed. This does not mean that "a" is a true constant, but only that it is not independent of y_0 or of the total number of species, N. It may be said that "a" is a function of y_0, so that given one, the other is settled. Thus we are reduced from 2 seemingly disposable parameters or constants to one. More generally, given any one piece of information about our collection, for instance given either the total number of species or the total number of individuals, everything else is fixed.

The word "Canonical"

I have ventured to call such an equation "canonical." It appears likely that this term was introduced into mathematical physics by J. Willard Gibbs: I quote from the preface to Volume 2 of his Collected Works (1931): "We return to the consideration of statistical equilibrium . . . we consider especially ensembles of systems in which the logarithm of probability of phase is a linear function of the energy. This distribution, on account of its unique importance in the theory of statistical equilibrium, I have ventured to call *canonical*" (italics his).

By a sort of rough analogy, I have designated as "canonical," for ecological purposes, that particular lognormal distribution of the abundances of the various species (or genera, families, etc.) whose "Individuals Curve" terminates at its crest. This way of describing it is probably imperfect, and, as shown later, it apparently corresponds to a situation in space or time where the individuals, or pairs, are distributed at random, not clumped on the one hand nor over-regularized on the other. Thus a better definition may be possible, and in that case preferable; but, however defined, it is a distribution that seems to have special importance in the general theory of ecological ensembles, and so I have ventured to call it "canonical."

In the present paper we trace the consequences of assuming the distribution canonical. We examine how nearly the experimental results fit the purely theoretical curves. These experimental results, though some of them involve "collections" having many millions of individuals, take us only as far as a few hundred species. We attempt to estimate what would happen if we had thousands or scores of thousands of species, where we have no actual counts of individuals but may sometimes be able to estimate them roughly, and we draw such other tentative conclusions as occur to us.

As the scale of abscissae we may use octaves, which is equivalent to taking "logarithms to the base 2," as we did in 1948, or we may use "orders of magnitude" which is more convenient when dealing with very large numbers. This is equivalent to taking logarithms to base 10 as James Fisher (1952) did. For theoretical purposes a scale of natural logarithms (i.e to base 2.718) would be most convenient. Williams (1953) found it convenient to work with logarithms to base 3.

Consequences of assuming that the individuals curve terminates at its crest

This matter may be stated briefly, anticipating the more detailed statement of the next heading, as follows: As shown in Preston (1948) the distance between the crests of the Species and Individuals Curves is $\ln 2/(2a^2)$ or $(\ln 2)\sigma^2$, where σ is the logarithmic standard deviation in octaves.

Though we describe our distribution as lognormal, it actually is finite, and species and individuals are not found infinitely distant from the mode either to the right or the left. In industrial "quality control" work it is customary to say that not more than one specimen out of a thousand should be expected beyond the 3-sigma limit, but in none of the biological examples we have yet encountered do we have as many as a thousand species to work with. We should therefore expect the finite distribution to end short of the 3-sigma limit. In fact, with the number of species in our examples to date we should expect it to terminate at about 2.5 to 2.8 sigma. If we take the latter value, the assumption that the distributions terminate where the Individuals Curve reaches its crest is equivalent to setting

$(\ln 2)\sigma^2 = 2.8\sigma$, or $\sigma = 4.0$ octaves approximately.

But this is just about what we find in our observations; it corresponds to an "a" value of 0.175.

Thus by an appeal to observation, but not by pure theory, we can reach the conclusion that very often the finite distribution does in fact end just about where the Individuals Curve reaches its crest. This seems to make it advisable to restate the matter more formally, in order to cover the complete range of possible values of the number of species involved.

The Individuals Curve

In the Species Curve each octave contains a certain number of species and each of these is represented by roughly the same number of individuals. Multiplying the one figure by the other gives the total number of individuals that have, in effect, been assigned to that octave. By making this computation for each octave we can construct the "Individuals Curve." This can be done for the observed points, but here we are concerned with its theoretical form.

For the Species Curve we have, as in equation (1) above

$$y = y_o e^{-a^2 R^2}.$$

Let the number of individuals per species, the

"representation" of the species, be n_o at the modal octave. Then at the Rth octave from the mode it is $n_o 2^R$ individuals per species, and the octave holds $yn_o 2^R$ individuals, or:

$$Y = (n_o y_o)[2^R e^{-a^2R^2}] = (n_o y_o)[e^{R \ln 2} e^{-a^2R^2}]$$
$$= n_o y_o \, e^{\left(\frac{\ln 2}{2a}\right)^2} e^{-a^2\left(R - \frac{\ln 2}{2a^2}\right)^2} \qquad (3)$$

This is a lognormal curve with the mode displaced by an amount $\frac{\ln 2}{2a^2}$ octaves: it has the same dispersion constant as the Species Curve, and it has the modal height $Y_o = n_o y_o \, e^{\left(\frac{\ln 2}{2a}\right)^2}$.

(See Figure 1.)

Equations like (3) become somewhat simpler in appearance if we use "natural orders of magnitude" in place of "octaves," i.e. if we use intervals in which the frequency or abundance of a species increases in the ratio 2.718 instead of 2.0, and if we use σ, the standard deviation in orders of magnitude, instead of the coefficient a.

This method of working is convenient if we have available adequate tables of natural logarithms, for then the observed frequencies are easily classified into their natural orders of magnitude. I think, however, that it will be more convenient if we continue, as we have begun, by using "octaves," which do not depend on the availability of such tables, or on the alternative method of converting ordinary logarithms to natural ones.

The effects of a finite number of species

Referring to the Species Curve in Figure 1, we note that in theory this curve extends infinitely far both to left and to right, but the long "tails" are exceedingly close to the R axis as asymptote. The area under the curve, or the integral of the curve, represents the number of species we have accumulated, as we go from minus infinity to any given point. This area is at first so small that not until we are within about 9 octaves of the mode (for this particular case, where we have a total of 178 species) have we accumulated enough area to correspond to a single species. This is the beginning of the real, finite, distribution. As we continue to the right we accumulate species rapidly; then we pass the mode and accumulate them increasingly slowly. Finally we reach a point some 9 octaves to the right of the mode where the remaining area is scarcely enough to hold one more species. In practice,

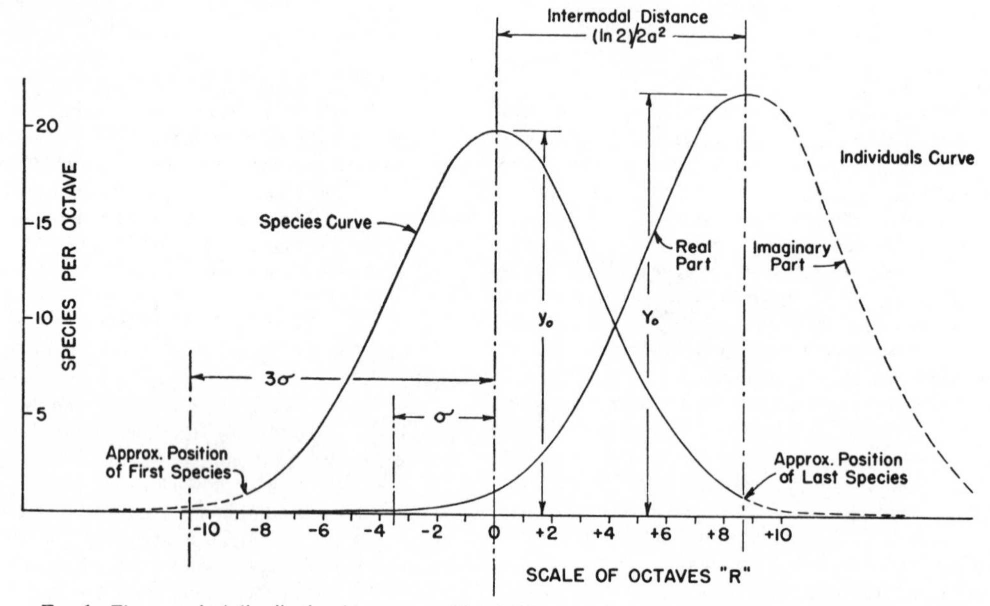

FIG. 1. The canonical distribution for an ensemble of 178 species. For this number of species, the coefficient $a = 0.200$ and the standard deviation σ is 3.53 octaves for both "Species" and "Individuals" curves. The modal height of the species curve is $y_o = 20$ species. The ordinate scale for the individuals curve is arbitrary: see text for explanation. The intermodal distance is 8.68 octaves. The real part of each curve is drawn solid: the first (rarest) and last (commonest) species ought to lie at about this distance (8.68 octaves or 2.45 σ) from the mode of the species curve, or perhaps a little, but only a little, more.

the finite distribution ends at, or near, this point.

The numerical value of the integral between any specified limits can be found from published tables, but the integral itself cannot be expressed readily in analytical form, suitable for finding a general solution to our problems. Nor, if it could, would it define the end of the distribution with real precision; at best it can only give an idea of the most probable position of the end of the distribution. We have taken as the most probable position that point where the remaining area under the tail of the curve corresponds to half a species. This mean that there is a 1:1 chance that we have not quite reached the end or that we have just passed it.

In Table I we have given, for various values of the total number, N, of species involved, that value of x, or R_{max}/σ,[1] which makes

$$\frac{1}{\sqrt{2\pi}}\int_{-x}^{x} e^{-q^2/2}\, dq = \frac{N-1}{N} \qquad (4)$$

TABLE I. The parameters of the canonical ensemble

N	x	σ	a	y_o	R_{max}	r	r^2	I/m
100...	2.576	3.72	0.190	10.7	9.58	765	5.8×10^5	2.69×10^6
200...	2.807	4.05	0.175	19.7	11.4	2,720	7.40×10^6	3.75×10^7
400...	3.024	4.36	0.162	36.6	13.2	9,410	8.85×10^7	4.82×10^8
800...	3.227	4.66	0.152	68.5	15.0	32,770	1.07×10^8	6.23×10^9
1,000...	3.291	4.75	0.149	84.0	15.6	49,670	2.47×10^9	1.47×10^{10}
2,000...	3.481	5.02	0.141	158.9	17.5	185,400	3.44×10^{10}	2.16×10^{11}
4,000...	3.662	5.28	0.134	302	19.3	645,500	4.17×10^{11}	2.75×10^{12}
8,000...	3.836	5.54	0.128	576	21.2	2,409,000	5.80×10^{12}	4.02×10^{13}
10,000...	3.891	5.61	0.126	711	21.8	3,651,000	1.33×10^{13}	0.93×10^{14}
100,000...	4.418	6.38	0.111	6,130	28.2	2.88×10^8	8.29×10^{16}	6.61×10^{17}
1,000,000...	4.892	7.06	0.100	56,500	34.5	2.43×10^{10}	5.90×10^{20}	5.21×10^{21}

(Note. ln 2 = 0.69315 and the reciprocal of this is 1.443) σ is to be ascertained by multiplying x by 1.443.
Then a is to be ascertained by dividing 0.707 by σ.
Then y_o is to be ascertained from the formula $y_o = 0.3989\ N/\sigma$.
R_{max} is to be ascertained by multiplying x by σ.
$r = 2\ R_{max}$
$I/m = 1.25\ r^2\ \sigma$

That is to say, we have left one species out of N to be divided between the two tails, to left of $-x$ and to right of $+x$, or half a species per tail. The values are taken from Lowan (1942).

It can be shown that this point is very nearly the same as would be obtained by setting $y = 0.4$ (species per octave) in equation (1), which would give

$$y = 0.4 = y_o\, e^{-(aR)^2} \text{ or } (aR)^2 = \ln\,(2.5\, y_o) \qquad (5)$$

We now have to trace the consequences of assuming that the crest of the Individuals Curve coincides with the finite end of the Species Curve.

Estimating the canonical constants

We have seen that the distance between the modes of the species and individuals curves is $\ln 2)/2a^2 = \sigma^2 \ln 2$, and from Table I we have the values of R_{max}/σ or "x." Equating the two we have

$$R_{max} = x\sigma = \sigma^2 \ln 2 \text{ or } \sigma = x/\ln 2 = 1.44\ x \qquad (6)$$

whence

$$R_{max} = 1.44\ x^2. \qquad (7)$$

Since we already have the values of x for various values of N, the total number of species, we are now in a position to add to Table I the values of σ and of R_{max}, and this has been done. Further, since $a^2 = 1/2\sigma^2$ or $a = 0.49/x$, we can also add the value of a. Again, the number of species (y_o) in the modal octave of the Species Curve can also be computed, for

$$N = y_o\, \sigma\sqrt{2\pi}$$

so that

$$y_o = 0.399\ N/\sigma = 0.277\ N/x. \qquad (8)$$

This value, also, is therefore added to Table I and this completes all the unknowns. Given the total number of species, the first column of Table I, we can ascertain all the constants of equation (1). The basic equation of the distribution has therefore become canonical, in the sense that nothing is left to chance; once N is specified, everything is determined.

The relation between total individuals and total species

The canonical equation implies that there is a definite relationship between these 2 quantities, and if "m," as defined below, is known, it is easily calculated. When we permit ourselves the liberty of making this theoretical estimate, we can expect only rough agreement with it in practice in any particular instance. The actual termination of the Species Curve may be some distance from where our estimates place its "most probable" position and, as shown later, the crest of the Individuals Curve is not precisely at its termination if "contagion" is present. The computation however is worth making; it may lead to some understanding of the problem.

We have denoted by n_o the number of individuals in the modal octave representing a single species. Let us denote by R_{max} the "range," in octaves over which the finite distribution of species extends to left or right of the mode of the Species Curve. In a complete "universe" or lognormal curve, the range should be the same to left or right, though samples usually show a truncation at the left end.

[1] Note that this defines x as the half-range in terms of the logarithmic standard deviation, σ, as the unit.

The commonest species, located at $+R_{max}$, ought to have $n_o\,2^{R_{max}}$ individuals. Similarly, the rarest, located at $-R_{max}$, ought to have $n_o/2^{R_{max}}$. If we denote the number $2^{R_{max}}$ by r, the commonest species has $n_o r$ individuals, and the rarest n_o/r. The ratio of the numbers of individuals of the commonest to those of the rarest is then r^2. Since we know R_{max}, approximately, we can add to Table I the values of r and of r^2.

Now the rarest species must include one individual or one pair, and if we suppose the species to be viable, we probably must assume that it is somewhat more than this. We can call its number of individuals (or pairs), "m," for minimum, where m is probably a rather small number, without committing ourselves immediately to an estimate of m. Then the modal species will have mr individuals per species, and the commonest will have about mr^2. In equation (3) we have denoted mr^2 by the symbol Y_o.

The total number of Individuals in the whole ensemble is

$$I = \tfrac{1}{2}\sqrt{\pi}\, Y_o/a = \tfrac{1}{2}\sqrt{2\pi}\, Y_o\, \sigma$$

whence

$$I/m = \tfrac{1}{2}\sqrt{2\pi}\, r^2\, \sigma = 1.25\, r^2\, \sigma \qquad (9)$$

This value of I/m, as a function of the total number of species N, is given in the last column of Table I.

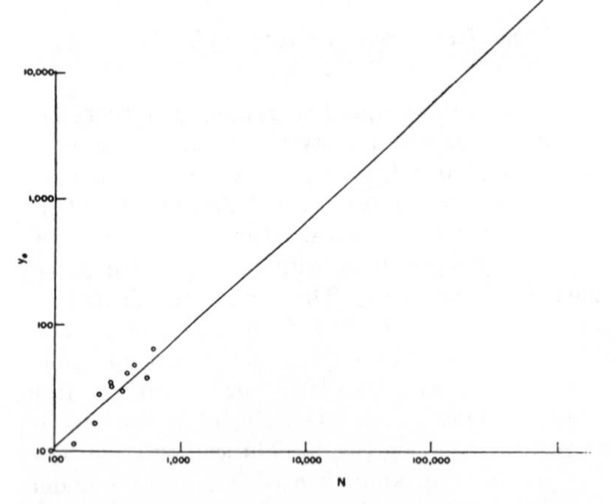

FIG. 2. The relation between the number of species (y_o) in the modal octave and the total number of species (N) in the ensemble. The solid line is the computed relationship from equation 18, and is pure theory. The observed points are taken from Table II.

In Figure 2 we have plotted y_o as a function of N, using "log-log" plotting. The curve is substantially a straight line. Observed points from earlier work, and from computations made on the estimates of Fisher and of Merikallio, are plotted at the lower end of the curve, which seems to lie reasonably well among them.

Note that the line we have drawn is a purely theoretical one. It is not drawn to fit the observed points which were added after the curve had been drawn. Unless the theory bore some relation to the facts we might very well find all the points to one side of the curve and not indicating much the same slope as that of the theoretical line. The fact that in position and in slope the theoretical line, at its lower end, lies close to where we should place an empirical line drawn among the points, is encouraging.

In Figure 3 we plot the logarithmic standard deviation σ and the coefficient "a," against N, using semi-logarithmic plotting. In Figure 4 we plot the value of R_{max}, or half-ranges against N. It appears that R_{max} is nearly a rectilinear function of log N. In Figure 5 we plot r, r^2, and I/m as functions of N, log-log ploting. All 3 lines are nearly straight.

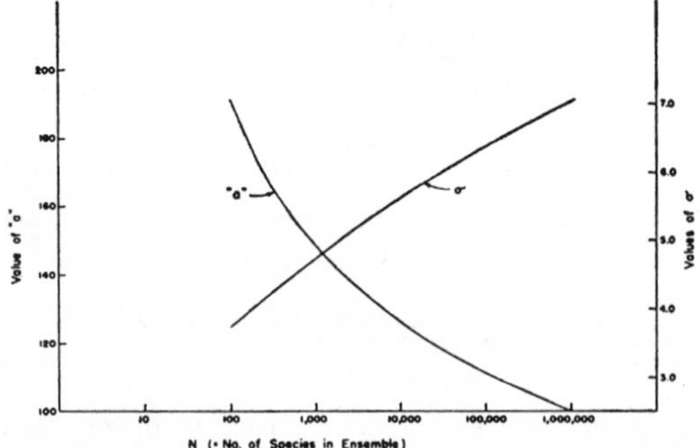

FIG. 3. The relation between the coefficient "a" in equation 7, or the standard deviation σ, and the total number of species (N) in the ensemble. (Computed relationship)

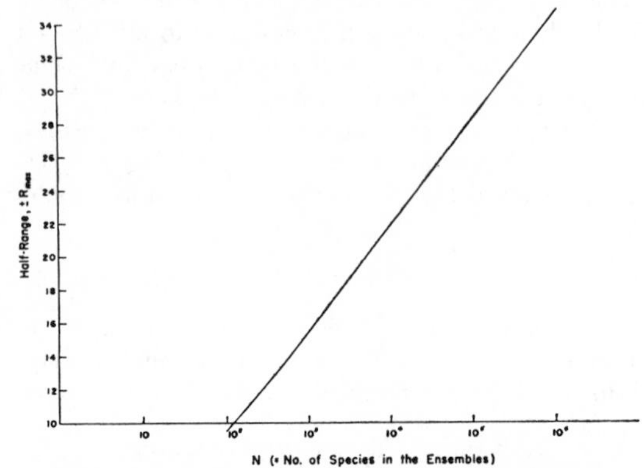

FIG. 4. Half-range (± R_{max}) as a function of N, the total number of species. The "Range" is the number of octaves which the finite distribution may be expected to cover.

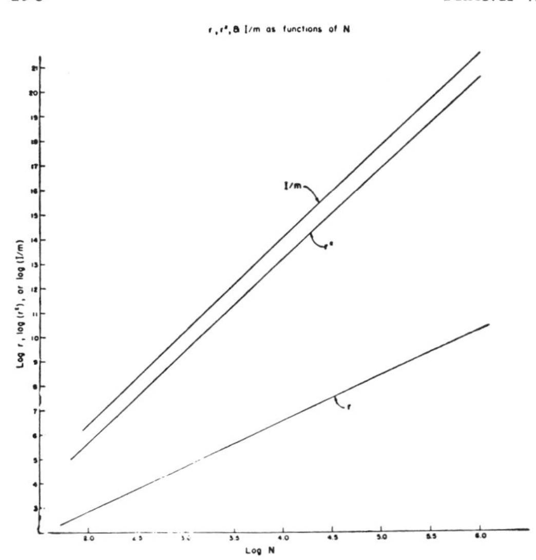

FIG. 5. The relation of r, r^2, and I/m to N. Here r is the value of $2 R_{max}$, I/m is the total number of individuals divided by that minimum number of individuals (m) that may be assumed necessary to keep a species in existence, and N is the total number of species in the ensemble.

The equations of these straight lines, as determined by a least-squares fitting to the data of Table I are

$$\log r = 1.872 \log N - 0.875 \quad (10)$$

$$\log r^2 = 3.745 \log N - 1.751 \quad (11)$$

$$\log I/m = 3.821 \log N - 1.21 \quad (12).$$

These equations are, as we have seen, all intimately connected, and if one has to be modified, all must.

Note that these equations are only approximations, because the lines are not quite straight. Equations (10) and (11) are probably close enough for any purposes that I can foresee; equation (12), particularly in its inverted form where it relates log N with log (I/m), may usefully be modified slightly for different ranges of the value of N (See below, under "The Species-Area Curve.").

The relation between I and N in a complete ensemble

We have, in equation (12) a relation between I and N if we can make a reasonable estimate of the value of m. We have tentatively identified m as the number of individuals or pairs in the rarest species and, since species are all the time being exterminated, we may expect that m will be a number not far from unity. In practice we often find it so as shown below. But a full discussion of this would involve many biological considerations. For instance we find in practice that m is frequently less, even appreciably less, than unity, and the temporary interpretation we have given then has no meaning. Another, related, meaning can be given to m but the simplest interpretation for the present is a purely mathematical one. Any 2 of the 3 quantities N, y_o, and σ (or "a"), define the size and shape of the Species-Curve, but they do not define its position along the axis of x, i.e. of R. The quantity "m" specifies this position. It thus determines not merely the number of individuals theoretically representing the rarest species, but also the number of individuals representing any of the other species, including the commonest. In consequence, the relation is really between (I/m) and N, not between I and N directly, but if we know, or can estimate, I and N, we can get an estimate of m.

The Species-Area equation

There is one other formula that may be useful to us. In some cases we may regard individuals (or pairs of birds), as being distributed uniformly, statistically speaking, over wide areas. Let the density of individuals (or pairs) be ρ per acre. Then the formula

$$I = \rho A \quad (13)$$

gives the number of individuals or pairs to be expected on an area of A acres.

We can substitute this in equation (12) and obtain

$$\log N = 0.262 \log (\rho A/m) + 0.316 \quad (14)$$

$$N = 2.07 (\rho/m)^{.0262} A^{.0262} \quad (15)$$

This is the Species-Area Equation under "ideal" conditions such that the area we consider is populated with a complete, not a truncated, lognormal ensemble, and that the density of the population (ρ) does not change substantially over the range of areas we are considering. The first stipulation is important because in contiguous areas, for which Species-Area Curves are often propounded (and this includes my own companion paper on "Time and Space and the Variation of Species"), the smaller areas act much like "samples" of larger ones.

If we are dealing with isolates that take the form of complete canonical ensembles, and if ρ and m are substantially unchanged from isolate to isolate, then equation (15) may be written.

$$N \propto A^{0.262} \quad (16)$$

and since, 0.262 is not far from 1/4, this may be written

$$N \propto \sqrt[4]{A} \text{ approximately} \qquad (17)$$

which I have referred to as "the fourth-root law." This is useful for quick calculations.

As a matter of fact, as mentioned earlier, the exponent 0.262 is an average index over the whole range shown in Fig. 5 or Table I. It was obtained by comptuing a theoretical regression line from that table. Over the range for which we have some observational data, i.e. when the number of species is between 100 and 1000, the index as computed from theory is more nearly 0.270, and our equation becomes

$$N = 1.83\ (\rho/m)^{0.270}\ A^{0.270} \qquad (18)$$

At the present time, therefore, equation (18) is more useful in practice than (17), and this is graphed in Figure 6 for various values of ρ/m, and tabulated in Table II.

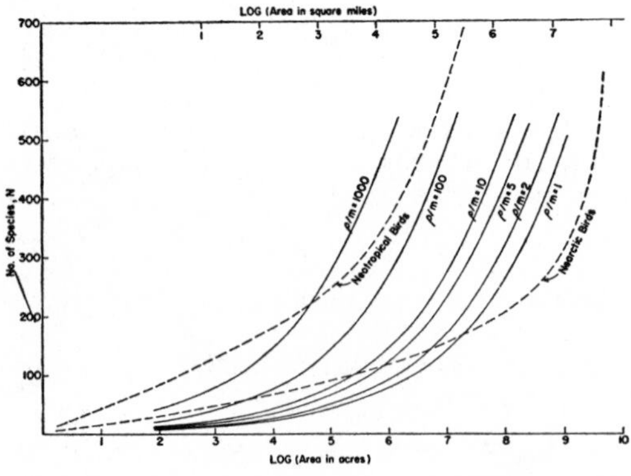

FIG. 6. The Species Area Curve for Isolates (solid line) and for Samples (broken line). Here ρ is the areal density (e.g. number of birds per acre) and m is the minimum number of individuals assumed necessary to keep a species in existence.

TABLE II. Number of species to be expected on area (A) of different sizes when given the density (ρ) of individuals per unit area and the minimum number (m) of individuals needed to keep a species in existence

A in acres ρ/m	10	10^2	10^3	10^4	10^5	10^6	10^7	10^8	10^9	10^{10}
1....	3.4	6.3	11.8	22	41	76	142	265	493	917
2....	4.1	7.6	14.2	27	49	92	171	320	595	1107
5....	5.5	10.3	19.2	36	67	124	230	430	796	1490
10....	6.3	11.8	22	41	76	142	265	492	917	1710
100....	11.8	22	41	76	142	265	492	917	1710	3180
1,000....	22	41	76	142	265	492	917	1710	3180	5920
10,000....	41	76	142	265	492	917	1710	3180	5920	11×10^3
100,000....	76	142	265	492	917	1710	3180	5920	11×10^3	20.5×10^3
1,000,000....	142	265	492	917	1710	3180	5920	11×10^3	20.5×10^3	38.3×10^3

In all probability, for the range from 50 to 200 species, the index is slightly higher still, perhaps 0.280. A number of our counts are within this range, and we have to consider them also when we deal with families rather than species. However the differences between 0.262 and 0.270 or even 0.280 are small compared with the uncertainties of our experimental or observational information. When we are dealing with areas that are merely samples of larger areas the index will tend, except for very small samples, to fall well below 0.262.

Fictive areas

Not all areas of equal size are equal in carrying capacity. A thousand square miles of the Greenland ice-cap are not equivalent to a thousand square miles of Neotropical forest, for instance. The treatment given above assumes that the areas we consider are reasonably comparable in fertility or, more generally, in "carrying capacity." The problems presented when this is not the case are discussed below where the concept of equivalent or "Fictive" areas is introduced.

Theoretical Species-Area Curves (for complete lognormal ensembles)

The equation is $N = 1.83\ (\rho/m)^{0.270}\ A^{0.270}$ or $\log N = 0.262 + 0.270 \log (\rho/m) + 0.270 \log A$. Table II gives the values of N over the range $10 < A < 10^{10}$, or $1 < \log A < 10$, and for $\rho/m = 1, 2, 5, 10, 100, 1{,}000, 10{,}000, 100{,}000$, and 1,000,000. In Fig. 6 we graph the entries of Table II. The solid curves are all alike, merely displaced to left or right. For comparison there are added the observed curves for Nearctic and Neotropical birds, taken from the companion paper (Preston 1960). These 2 curves, in broken lines, represent the behavior of "samples," not of complete canonical ensembles, and their form is discussed later in the present paper.

COMPARISON OF THEORY AND OBSERVATION, AND THE SPECIES-AREA CURVE FOR ISOLATES

In this section and the next we shall be dealing with the properties of isolated "universes" or isolated "populations," hereinafter called simply "isolates," as contrasted with the properties of "samples." In general the isolate tends to give a distribution-curve that is symmetrical and looks like a fair approximation to a complete lognormal. On the other hand, if we have a "sample," which is a small fraction of an isolate and is a random sample thereof, the distribution tends to look like a truncated or decapitated lognormal, an unsymmetrical distribution, of which we showed examples in Preston (1948).

We have shown above the mathematical consequences of 2 assumptions; viz, that abundance is typically distributed lognormally among species, and that this distribtuion is Canonical in the sense that not all lognormals meet the requirement, but only those in which a definite relation exists between the number of species, (N), the number of species in the modal octave (y_o), and the logarithmic standard deviation (σ). We now consider what degree of confirmation or refutation is available from observation.

The Canonical Parameters

The relative constancy of σ and "a"

In any Gaussian distribution or, in our case, any lognormal distribution, we have from equation (8),

$$N = \sqrt{2\pi}\,(y_o\,\sigma) \qquad (19)$$

That is, the total number of species in the "universe" is proportional to the product of the number of species in the modal octave and the logarithmic standard deviation. As the number of species increases, y_o or σ, or both, must increase. When the distribution is canonical both increase, but σ increases only slowly while y_o increases rapidly. In fact if we double N, y_o increases by about 85%, but σ by only 8% in the range of most interest to us.

Similarly, since "a," the "modulus of precision," is related to σ by the formula $\sigma^2 = 1/(2a^2)$, "a" also is relatively constant. This is what we found in Preston (1948).

The numerical values of "a" and σ

Not only are these values relatively constant over the accessible range of values of N, but Table III shows that, in this range, where N averages perhaps two or three hundred species, "a" is about 0.175 and σ is about 4 octaves, or something like one and a fifth orders of magnitude.

Now, referring to Table I, we see that for 314 species we should theoretically have an average value of "a" of about 0.169. This agreement is close, perhaps fortuitously so (see below on contagious distributions), but it warrants a few comments.

The only attempts to get a picture of the complete ensemble by direct observation are those of Fisher (1952) and Merikallio (1958). There are considerable difficulties with the experimental work and some uncertainties, some of which the authors have indicated. The other results come from estimates of what the "universe" is like as a result of studying a sample. Indeed neither Fisher nor Merikallio was studying a perfect "isolate," though they approximated it. The sample theoretically has the same modal height and the same dispersion as the universe, and it has also a 3rd variable, the position of the "Veil-line," or what is the same thing in the end, the abscissa of the mode. This 3rd disposable variable makes our estimates of the other 2 more uncertain than they would otherwise be, and therefore agreement in our estimate of "a" or σ within about 6% seems in part fortuitous. For a discussion of the fitting of truncated Gaussian distributions see Hald (1952). Furthemore, when we are dealing with truncated ensembles, we do not directly observe the value of N, the total number of species, but have to estimate it from the sample. This throws a further strain upon the interpretation of our observations.

TABLE III. (Observed Relationships). N is the number of species estimated, on the basis of the sample, to be present in the total "universe" or "population": y_0 is the number of species in the modal octave, and "a" is the "modulus of precision" of the lognormal distribution

Instance	N (estimated)	y_o	a	Reference
Saunders (birds)	91	10	.194	Preston 1948
Dirks (moths)	410	48	.207	Preston 1948
Dirks (female moths)	363	42	.205	Preston 1948
Williams (moths)	273	35	.227	Preston 1948
King (moths)	277	33	.152	Preston 1948
Seamans (moths)	332	30	.160	Preston 1948
Maryland birds	233	28	.213	Preston 1957
Nation-wide bird count	530	38	.129	Preston 1958
Nearctic estimate	600	65	.19	Preston 1948
Land Birds of England and Wales	142	11.2	.14	Fisher 1952
Breeding Birds of Finland	204	16.5	.146	Merikallio 1958

The relation between y_o and N

To the extent that we find the correct relationship between σ and N we must necessarily find a correspondingly correct relationship between y_o and N but, since σ is so nearly constant over the observable range of N while y_o varies rather rapidly, y_o may throw some further light on the matter. Table III gives the observed values of y_o for various values of N, most of which are estimated from incomplete or truncated distributions: Figure 7 shows the data in graphical form. It should be once more emphasized that the line is purely theoretical. The small circles represent observed points from Table III. The line is not "fitted" to the points and then extrapolated; the line is from theory, the points from observation. But it will be observed that the line passes neatly among the observed points which lie in a narrow band straddling the line so that if we did "fit" a line to the observations it would not be very different from the theoretical one. The observed points not only lie reasonably near the line, but they parallel its course. This gives us an additional point of agreement between theory and observation, beyond what we get from considering the relationship of σ and N.

The relation between N and I in a complete ensemble

Here we exclude discussion of the relation between N and I in "samples." The best way to get complete ensembles is probably to deal with "isolates" such as the fauna and flora of islands which are in internal equilibrium, but not necessarily in equilibrium with, or even appreciably affected by, the populations of other land masses. However, so far as I know, we have no counts of individuals for such islands, nor even good estimates of the numbers of individuals. Fortunately when our areas become reasonably large, of the order perhaps of 50,000 to 100,000 square miles, the distribution begins to approximate closely to that of a complete ensemble (in the case of birds) even though the area is in intimate contact with neighboring land areas. We have at least 2 fairly good estimates of the number of species and of individuals on such areas: James Fisher's (1952) estimate of the *land* birds of England and Wales, and Merikallio's (1958) estimate of the breeding birds of Finland.

The land birds of England and Wales

Fisher lists some 142 land birds as regular breeding species, and the distribution looks reasonably complete and symmetrical (see his Figure I). From our equation (12), we compute that I/m = 10,000,000 very closely. Fisher estimates that I is 63,000,000 *individuals,* hence m should be approximately 6 individuals or 3 pairs. This is, I think, a reasonable estimate of the number of pairs of the rarest regularly-breeding species.

The value of r^2 may be computed from formula (11) and comes out very close to 2,000,000. Then $mr^2 = 6.3 \times 2 \times 10^6 = 12.6$ million individuals. Fisher gives a value of 10 million approximately, so again the agreement is good. The value of r may be computed from formula (10) or taken as the square root of 2 million. Then we get mr = 8,900 individuals per species in the modal octave. Our computation from his data (which are not given in great detail) is 8,500. The agreement is almost too good.

The breeding birds of Finland

Merikallio (1958) lists about 204 species. On this basis we should expect I/m to be about 41 million. Merikallio gives a value of I of 31 million *pairs,* hence our computation suggests in this case that m is about one pair (0.76 pairs). We can also compute that r = 2,810, whence $mr = 2.1 \times 10^3$ for the modal species and $mr^2 = 6.0 \times 10^6$ for the commonest. The figure Merikallio gives for his commonest species is 5.7×10^6 pairs, where once more the agreement is accidentally too good, and our computation from his figure gives the modal representation as 4.26×10^3, which is fair.

The relation between N and A: the Species-Area Curve for complete Canonical ensembles

In general there can be no complete count of individuals on areas large enough to approximate complete ensembles. Both Fisher and Merikallio use methods of estimating, not counting, individuals except for the rarest species. Their methods could of course be extended to other areas and other taxonomic groups and it might then be seen how nearly the theory fits the observations.

We can however assume that in certain parts of the world we have isolates, or near-isolates, of some taxonomic groups on islands of different sizes between which there is very limited floral or faunal interchange, and that on each of these islands there may be a fair approximation to a canonical distribution. These islands ought to be numerous enough for us to strike a reasonably good statistical correlation; they should be similar climatically and not too dissimilar in vegetation cover and soil. There are 2 obvious groups of such islands, the East Indies and the West, and there are taxonomic groups such as land mammals, land reptiles, and amphibia that do not readily cross stretches of sea. Birds qualify partially, and are usually better known. Unfortunately, the number of species on the smaller islands is usually far below what we like for statistical work. Elsewhere (Preston 1957) I have suggested that we ought to have something like 200 species, and this is out of the question for the groups mentioned. We must accordingly do the best we can with smaller numbers.

The mammals of the East Indies

Darlington (1957, p. 480) discusses the mammals of Sumatra, Borneo and Java. These islands are comparable in latitude and climate (a matter that cannot be neglected) and their areas are given as 167,000; 290,000; and 49,000 square miles respectively, while their "units" of mammals are 55, 47, and 33. "Units" in this case are a mixture of species and genera. Striking an average of the 2 larger islands gives us an imaginary island 228,000 mi² and 51 species. Comparing this with the smaller island of Java by means of the formula

$$z = \log (N_1/N_2)/\log (A_1/A_2) \text{ (from eq. (18)) } \quad (20)$$

where the suffix 1 applies to the one island and the suffix 2 to the other, we find that

$$z = 0.28$$

which is very close to what is called for by the theory of Equation (20).

Amphibia and reptiles of the West Indies

Darlington (1957, p. 483) in discussing the Anura and the lizards and snakes of Cuba, Hispaniola, Jamaica and Puerto Rico, with some support from the small islands of Montserrat and Saba in the Lesser Antilles, concludes that "division of area by ten divides the fauna by two." (The "herps" on 40,000 mi² total about 80, and on 4,000 mi² about 40.) Darlington's statement implies that

$$z = \log 2/\log 10 = \log 2 = 0.301$$

which is quite close to the theoretical value of 0.28. The smaller islands support Darlington's views, but since the number of species is very small, neither agreement nor disagreement would carry much weight.

It may be noted that if we take frogs and toads alone, since the legitimacy of lumping these with lizards and snakes may be questioned, we have an average of 25½ species on the 2 larger islands, with an average of 35,000 mi², and an average of 14½ species on the two smaller islands, with an average of 4000 mi². The computation then gives $z = 0.26$, which is also the theoretical value. The reptiles (snakes and lizards combined) give a value of 0.318; the lizards alone give a value of 0.294. The snakes are too few to be used safely for a caculation.

Birds of the West Indies

The populations of birds, particularly seafowl, on the various islands of the West Indies are probably somewhat less isolated from one another than the frogs and lizards. On the other hand there is quite likely not much gene-flow among a large proportion of the land-birds. The populations of each island may therefore approximate to canonical lognormals and the exponent z may be expected to approach 0.26-0.28, the theoretical value, or to be a little lower.

Bond (1936 and 1956, with supplements) has given detailed accounts of these birds, from which I attempted to compile a list of the breeding species of all the major islands and a number of the smaller ones. The areas I took from the Encyclopedia Britannica, edition of 1949. In the case of a few species, while Bond leaves no doubt that they breed within the West Indies, there may be some uncertainty whether they breed regularly on some of the islands. This is probably unimportant in the case of the major islands, where the birds are probably better known, but I may be in error in connection with some of the smaller ones. In order to permit others to correct me, I give in Table IV a list of the islands I have used, their areas and my estimates of the numbers of breeding species of birds, and in Figure 7 I have graphed the results. Note that I have excluded introduced species. Note also that the Isle of Pines has an abnormally rich fauna for its size. This probably comes about from its not being an "isolate," but rather a "sample" of Cuba, probably a truncated distribution.

The line is not a calculated regression line, but

TABLE IV. Birds of the West Indies

Island	Area in Square Miles	Breeding Species of Birds
Cuba	43,000	124
(Isle of Pines)	(11,000)	(89)
Hispaniola	47,000	106
Jamaica	4,470	99
Puerto Rico	3,435	79
Bahamas	5,450	c74
Virgin Islands	465	35
Guadalupe	600	37
Dominica	304	36
St. Lucia	233	35
St. Vincent	150	35
Grenada	120	29

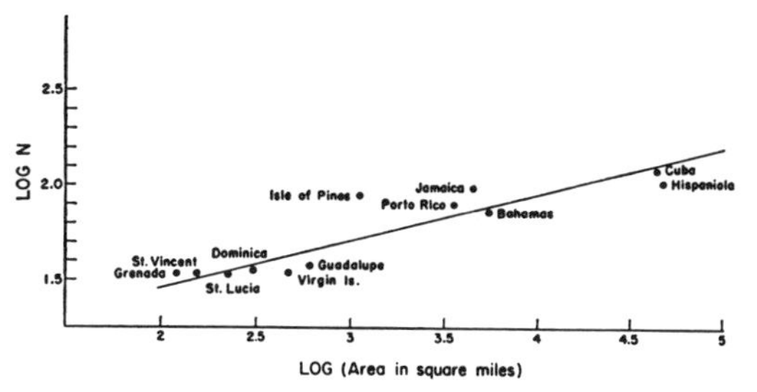

FIG. 7. Birds of the West Indies: Species-Area Curve. The abscissa gives the areas of the various islands; the ordinate is the total number of bird species breeding on each island.

is drawn "by eye" and seems reasonable; it is scarcely worth more refinement in view of the errors I may have made in my estimates or the general question of how legitimate it may be to include seafowl in a study of "isolates." Its slope corresponds to $z = 0.24$, which seems to agree well with expectations.

Birds of the East Indies

The "East Indies" here includes 2 or 3 islands of the Sunda Shelf, and several other islands not strictly on the shelf or even part of the same zoogeographical province, and one island, Ceylon, far removed to the west. However, the islands are all essentially tropical and climatically not too dissimilar. I have depended heavily on information supplied by Dr. Kenneth Parkes who has used, besides his own knowledge, the findings of Delacour and Mayr. The 1949 Edition of the Encyclopedia Britannica gives somewhat different counts of species.

A list of the islands with their areas and breeding species as given in Table V, and the results are graphed in Figure 8. The slope of the curve is $z = 0.288$, close to the theoretical value which in this case should be about 0.27. The curve is drawn "by eye," but seems likely to be as close

TABLE V. Breeding birds of some East Indian islands

Island	Area in Square Miles	Breeding Species of Birds
New Guinea	312,000	540
Borneo	290,000	420
Phillipines (excluding Palawan)	144,000	368
Celebes	70,000	220
Java	48,000	337
Ceylon	25,000	232
Palawan	4,500	111
Flores	8,870	143
Timor	300 × 60 = 18,000	137
Sumba	4,600	108 (or 103)

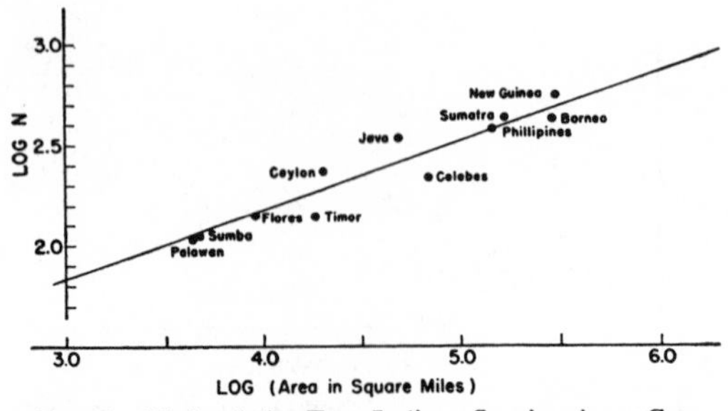

FIG. 8. Birds of the East Indies: Species-Area Curve. The abscissa gives the areas of the various islands; the ordinate is the total number of bird species breeding on each island.

as the observational evidence warrants. Good counts on some smaller islands might help but, in view of differences of opinion among taxonomists as to what constitutes a valid species, and in view also of the fact that with small islands statistical fluctuations or "errors" are likely to be important, I am not sure that much would be gained.

It may be noted that Java is rich for its size, while Celebes is poor. Ceylon is relatively "rich," but it may be acting to some extent as a sample of India and thus be somewhat enriched. New Guinea is famous for its wealth of species but some fraction of these may be an enrichment from the Australian mainland. Mayr (1944) comments that Timor is poor in birds, being both peripheral and arid.

The birds of Madagascar and the Comoro Islands

Madagascar is one of the large islands of the world, with an area of 229,000 mi^2. The Comoro Islands lie off its northeast coast, in the Mozambique Channel, in Latitude 12°S. They are 4 small islands, averaging about 200 mi^2 apiece, and have received their bird fauna predominantly from Madagascar (Benson 1960). Madagascar has apparently been an isolated island since early Mesozoic (Triassic) times while the Comoros date back to the mid-Tertiary (Miocene) times. The 1949 edition of the Encylopedia Britannica gives about 260 species of birds for Madagascar. Rand (1936) gives his estimate as 237 breeding species. I have comprised on 250. Benson (1960, p. 18) gives the breeding bird tally as follows: on Grand Comoro (366 square miles), 35 species; on Moheli (83 square miles), 34 species; on Anjouan (178 square miles), 35 species; on Mayotte (170 square miles), 27 species.

Madagascar is roughly a thousand times as large as the average Comoro Island and there are no islands of intermediate size in the immediate neighborhood. We can, however, make the assumption that over the ages, many species of birds from Madagascar, Africa, and occasionally elsewhere have made landfalls on the Comoros, and that the islands have as large a number of species as they can support. On that basis we can plot them on a graph along with the avifauna of Madagascar, and this has been done in Figure 9. The line has been drawn by eye, and its equation is

$$N = 7.95\ A^{0.28}$$

The exponent, 0.28, is very slightly above the theoretical value, and the coefficient, 7.95, only a little below the theoretical value of 10, appropriate

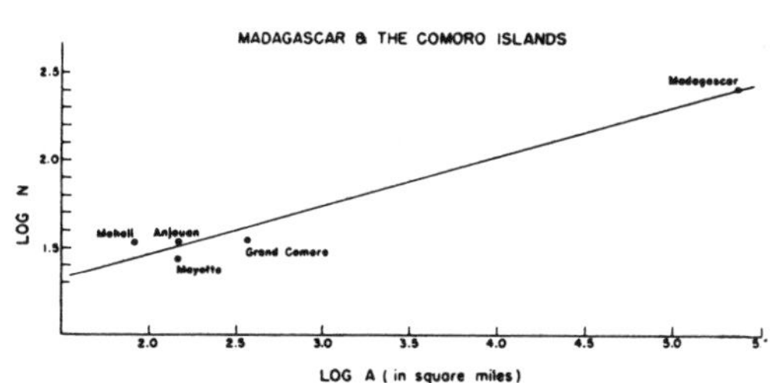

FIG. 9. Birds of Madagascar and the Comoro Islands: Species-Area Curve. The abscissa gives the areas of the various islands; the ordinate is the total number of bird species breeding on each island.

when A is in square miles and ρ/m is assumed to be unity.

The land vertebrates of islands in Lake Michigan

Hatt *et al.* (1948, p. 149) list 12 islands in Lake Michigan ranging in area from 2.5 acres to 37,400 acres, and in species-count of land vertebrates from 4 to 120. The 2 smallest islands with 8 and 4 species respectively are of dubious value for our purposes, but the remaining 10 islands, with a minimum of 19 species, seem satisfactory. Land Vertebrates are defined as amphibians, reptiles, birds and mammals. Table VI gives the information.

Figure 10 presents the data in graphical form for the 10 larger islands and a log-log plot yields a fair approximation to a straight line. The curve as drawn has the equation $N = 10\ A^{0.24}$ (where A is in acres). This implies a high density of individual "vertebrate animals" compared with what we encounter among birds. The slope, or exponent 0.24, is not far below the theoretical 0.27. If we include the 2 lowest points, a very questionable procedure, and give them equal weight with the other points, the slope of the curve increases to about 0.30. Giving them even a little weight brings the exponent close to the theoretical value. I am inclined to suspect that most islands have achieved a rough approximation to internal equilibrium, i.e. to a canonical distribution, but that there is a modest exchange of individuals among the islands or with the mainland, perhaps by water in summer or ice in winter. This tends to depress the index slightly below the value for strict isolates, and makes the islands to some extent "samples" of the mainland.

TABLE VI. Land vertebrates of islands in Lake Michigan

Area in Acres	Species	Area	Species	Area	Species	Area	Species
37,400	120	3,400	68	130	27	16	19
13,000	88	895	38	115	36	3	4
5,000	77	270	37	75	42	2.5	8

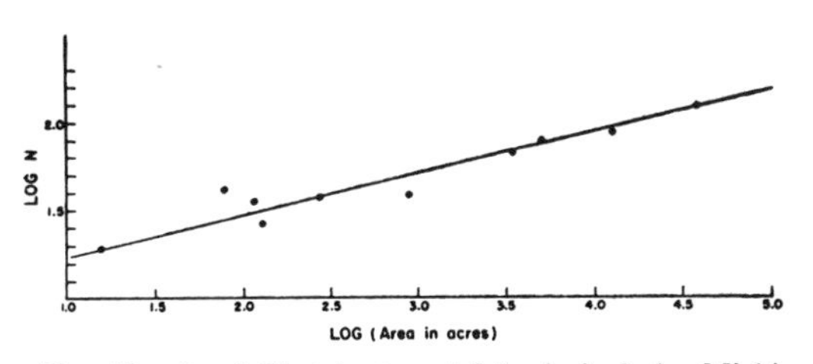

FIG. 10. Land Vertebrates of Islands in Lake Michigan: Species-Area Curve. The abscissa gives the areas of the various islands; the ordinate is the total number of species on each island.

The land plants of the Galapagos Islands

So far we have been considering faunas, and it may be well to consider a flora. Kroeber (1916) has made a detailed study of 18 of the Galapagos Islands. He does not give the areas of any of the islands, and I have obtained these from other sources, which are not always in very exact agreement, or by estimates from the U.S. Hydrographic Survey maps. One small island, which he calls Brattle, I have not been able to identify with assurance, and this has therefore been omitted, reducing our count to 17 islands. Since we shall have occasion to analyse Kroeber's results further, I have here included an outline map of the islands as Fig. 11.

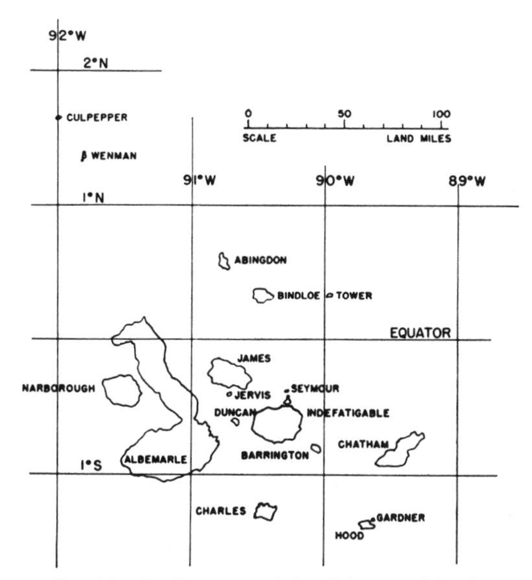

FIG. 11. Outline map of the Galapagos Islands.

We shall work principally with Kroeber's Table VI, from which our Table VII has been prepared,

TABLE VII. The land plants of the Galapagos Islands

No.	Name	Area (Sq. mi.)	Species
1..............	Albemarle	2249	325
2..............	Charles	64	319
3..............	Chatham	195	306
4..............	James	203	224
5..............	Indefatigable	389	193
6..............	Abingdon	20	119
7..............	Duncan	7.1	103
8..............	Narborough	245	80
9..............	Hood	18	79
10..............	Seymour	1	52
11..............	Barrington	7.5	48
12..............	Gardner	0.2	48
13..............	Bindloe	45	47
14..............	Jervis	1.87	42
15..............	Tower	4.4	22
(16)..............	(Brattle)	—	(16)
17..............	Wenman	1.8	14
18..............	Culpepper	0.9	7

with an identifying number added for convenience, and with my estimate of the area of each island also added. There is a good deal of scatter, because area is not the only factor that decides the richness of faunas and floras, and accordingly we have computed 2 regression lines, one for the 12 larger islands and one for all 17. The two are scarcely distinguishable but this is a coincidence; too much attention should not be paid, as a rule, to very small populations. The regression line drawn in Figure 12 is the one for the 12 larger islands, but the "points" for the others are indicated. The slope of this line is z = 0.325, a little higher than the theoretical value of 0.27 or thereabouts.

Summary of the z-values for isolates

We now have 7 independent estimates of the index z of Species-Area curves for "isolates," all of them relating to islands. Some of these are oceanic islands, some are islands in a fresh-water lake, some are in tropical, some in temperate latitudes. We have floras and faunas, and among the faunas we have mammals, birds, reptiles, amphibians, and "land vertebrates."

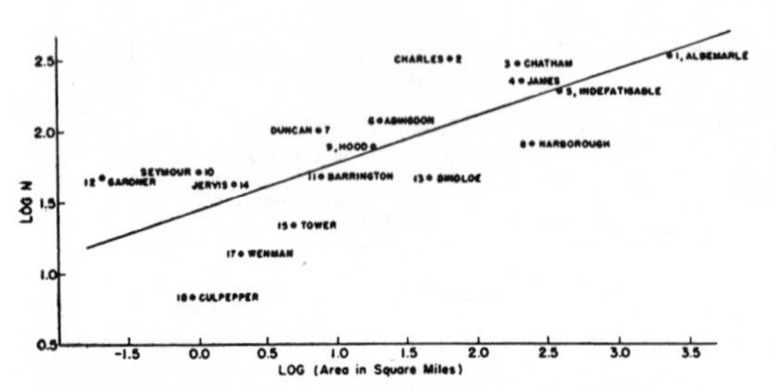

FIG. 12. Plants of the Galapagos Islands. Species-Area Curve. The abscissa gives the areas of the various islands; the ordinate is the total number of species on each island.

Table VIII summarizes our results. It is clear that the average of the 7 z-values comes very close to the theoretical figure. The individual departures from this average figure are in general modest, and the agreement may be regarded as satisfactory. Departures may be expected because of several factors. For instance, large islands are more likely to have high mountains, and therefore special habitats, than small ones, and then area alone would not adequately describe the opportunities for faunas. Other factors are discussed below.

TABLE VIII. The Species-Area curve for isolates (The coefficient "z" of the Species Area Curve-Summary

Fauna or Flora	Locality	z-value	Fig. No.
Mammals.........	East Indies	0.280	none
Amphibian and Reptiles........	West Indies	0.301	none
Birds.............	West Indies	0.240	7
Birds.............	East Indies	0.333	8
Birds.............	Madagascar and Comoros	0.280	9
Land Vertebrates...	Islands in Lake Michigan	0.239	10
Land Plants.......	Galapagos Islands	0.325	12
	Average (of 7)...	0.285	
Land Plants.......	Many areas	0.222	13

The flowering plants of the world

Williams (1943b) has plotted on a log-log basis all the data available to him for flowering plants of adequately defined areas, and these areas range from a few square centimeters to the land area of the globe. He has divided the information into 6 categories, the flora of arctic, temperate, subtropical, tropical, and desert regions and of oceanic islands, and he comments that over the linear portion of the graph, which extends from about 0.1 km^2 to 10^2 km^2, "to double the number of species, the area must be increased by thirty-two times." This may be compared with Darlington's statement that the area must be increased about ten-fold. Williams' statement amounts to saying that our exponent z = 0.20 approximately. Williams is working with the upper boundary of his scatter plot, the boundary between the area where there are observed points and the area where there are none. It happens that this boundary is quite well defined and Williams marks it with a broken line. The lower boundary on the other hand is nebulous. Williams treats the upper boundary as a statement of optimum conditions, i.e. as indicating the maximum number of species that can be accommodated on various sizes of "quadrat."

The slope is actually a little steeper than Williams' estimate; I make it 0.222. In Figure 13 I have reproduced all of Williams' points for

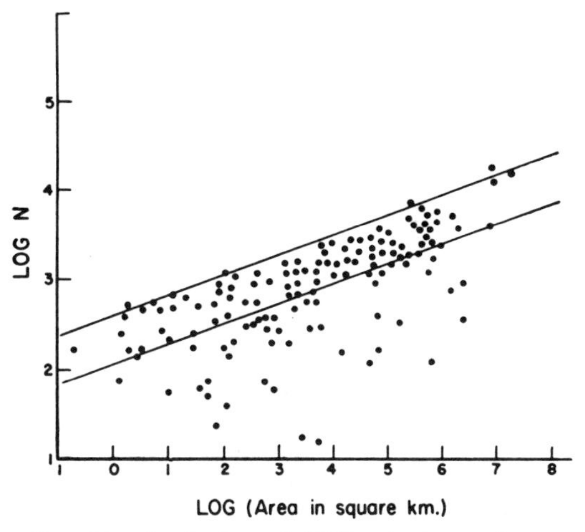

FIG. 13. Re-plotting of Williams' plants of many islands and other areas. Species-Area relationship. The upper curve is Williams' upper limit of maximum richness. The lower is the regression line for all plotted points. N is the number of species reported from the various areas.

areas between 0.1 and 10^8 km². Below this range many or most of the points must refer to "samples," i.e. to truncated distributions; it is virtually certain that a large proportion of the remainder do likewise. This gives us 135 points with which to work. I have drawn 2 lines on the graph, one marking the boundary as estimated by eye, and the other the regression line of "y on x" (log N on log A) computed by orthodox methods. The 2 lines turn out to have identical slopes, z = 0.222. The equation of the regression line is in fact

$$N = 118\, A^{0.222} \qquad (21)$$

$$\log N = 2.07 + 0.222 \log A. \qquad (22)$$

By separating the categories into which Williams divides his data, it is possible to compute or estimate values of z for the tropical, temperate, and desert floras. I have merely estimated the slopes of the upper boundary lines. For tropical and desert regions, the slope seems to be about 0.25, for temperate regions, 0.22. The regression lines may be a little flatter. In any case they are all roughly similar, and all a little below the theoretical value, which over this range, which extends to more than 10,000 species, is about 0.265. We may speculate that though, on unit area we have approximately complete canonical distributions, yet some samples are truncated, thus depressing the index below its theoretical value, from 0.26 down to 0.22. It may be worth noticing that the depression is by no means so great as we found in Preston (1960) for the bird faunas of the Neotropical region (0.16+) or Nearctic (0.12+) where the truncation is presumably more pronounced.

We may also note that while the statements of Darlington and Williams, that it takes 10 times, or 32 times, the area to double the number of species, sound considerably at variance, yet when expressed as an index the difference is only between 0.22 and 0.30. It will not escape notice that the average of these two is 0.26, almost exactly the theoretical value, but I think this is accidental.

Our justification for ignoring those areas in Williams' graph below 0.1 km² lies less in the fact that Williams himself did so than in the fact that we here have many counts involving less than 50 species, and scarcely any with more than a hundred, and this brings us to about the limit for satisfactory statistical work. (This restriction makes it difficult to work with the extensive tabulations of Vestal 1949. However see below.)

There is another calculation we can make, of a somewhat more risky kind, since the index 0.222 differs appreciably from the theoretical one of 0.262. We can compute the theoretical number of individuals per square meter and see how it agrees with what Williams has to say on the subject.

The regression line not far from the middle of the range we have studied gives 1000 species at about log A = 4.2 or A = 1.6×10^4 km². From Table I we should have $I/m = 1.47 \times 10^{10}$ individuals when N = 1000 species. Then if we assume tentatively that m = 1 or thereabouts, this involves approximately one flowering plant to the square meter, averaged over tropical, temperate, and arctic areas. The boundary line would give as an "optimum" or maximum concentration something like 30 cm² to a flowering plant.

Williams says, on p. 260, "Available information indicates that there is about one plant per square centimeter in temperate grassland, about one per square meter in woodland, and about one per hundred square meters upwards in semi-desert areas." Thus our estimates fall well within the range of concentrations given by Williams, and since we have evidence that m can vary at least over the range from 10 to 1/10 or rather more, the agreement can only be regarded as satisfactory.

Summarizing this section, we might say that there are a number of cases where the theory of isolated canonical ensembles seems to be confirmed by the exprimental, or observational, evidence; or at least the observations seem to accord with the theory.

The Properties of "Samples"

Introduction

In the previous two sections we examined first the theory of complete canonical ensembles and then the available observations on ensembles that seemed likely to be substantially complete. Under ideal conditions we think of these as "isolates." Information of this sort is rare, and in general we have to content ourselves with "samples" drawn from a presumably complete canonical universe. The relationship between sample and universe was discussed briefly in Preston (1948). What we get, if the sample is a random one, is a truncated lognormal distribution. To a close approximation the sample has the same modal height (y_0) as the universe, and it indicates correctly the logarithmic standard deviation (σ). From these figures the total number of species (N) in the unknown universe may be computed, though we may have observed or collected, in the sample, only 70 or 80% of them.

A large part of the field work of an ecologist consists of taking samples, so the properties of samples have great practical importance. The purpose of the sampling is in nearly all cases to deduce something about the "population," "universe," or "stand" or "community" that is being sampled, so the relation between sample and universe is a matter of prime concern. Some of the problems that arise are of broad scientific interest. In large samples, which typically should contain at least a couple of hundred species and at least forty or fifty thousand individuals, the problems are largely analytical in nature; in small or very small samples, biological peculiarities, especially "contagion," positive or negative (See Cole 1946), become very important; with larger samples the "contagion" tends to smooth out and disappear.

A "sample" may be obtained by catching moths in a light trap. Obviously we do not catch all the moths that are on the wing, and it is clear that we are dealing with a sample, though it is far from clear what "universe" we are sampling. It is important not to assume that we know this from general considerations; in particular, in this instance, we must not assume that it is the population of a definite geographical area. Some of the noctuid and hawk moths we catch may be very strong fliers and may have crossed the English Channel or some larger body of water from foreign parts; some of the geometrids may be very weak fliers and originate entirely close at hand. Similar considerations apply to birds and even to plants, and even more obviously to wind-blown pollen. The universe being sampled is simply what the sample says it is; it must be ascertained from the internal evidence, not from assumptions. This applies to "quadrats" of plants, and hence to the Species-Area curves thereof, just as it applies to moths in a light trap.

Fundamental differences between sample and universe

There are 2 very obvious differences between a sample and a universe: first, the ratio of species to individuals is vastly higher in the sample than in the universe, and second, there are vastly more species represented by a single specimen ("singletons"), or by a few specimens, in the sample if it is a "random" one.

Let us suppose that we have a "fixed" universe which will not expand as we collect our random sample, and let us examine our collection as we collect it.

The first moth into a light trap may be one of the commonest species, but it is more likely to be one of some other, since the commonest species is not as plentiful as all the other species put together; hence it may very well be that at first we collect half a dozen moths all of different species. Thus we accumulate species, at first, as rapidly as we accumulate individuals. As we continue collecting this situation ceases to obtain, and after a time the addition of new individuals piles up indefinitely, while the addition of a new species is a rare event. We have reached the point of diminishing returns.

In Preston (1948) we showed that doubling the catch of individuals was approximately equivalent to withdrawing a graph of the lognormal distribution from under a Veil-line to a distance of one more octave. This is a very close approximation for most of the cases of practical importance, or at any rate those cases we were considering in 1948. But actually withdrawing the graph by one octave rather more than doubles the count of individuals. It doubles the count for all those octaves that were previously withdrawn and it adds in addition one individual for each species in the new octave now unveiled. This modification of our 1948 statement is of significance only at the beginning of collecting.

The index "z" in the Arrhenius Equation, when collecting at random from a fixed universe

The Arrhenius equation, in ecological work, is the statement that the number of species (N) is connected with the area (A) by the equation

$$N = KA^z$$

where K and z are constants. This equation is not given by Arrhenius but is implied by his work. (See Preston 1960). This is of the same form as our equation (16). Note that we are not referring to that other Arrhenius equation that concerns the rate of chemical reactions and their variation with temperature.

Suppose in our collecting we have begun at the octave R_{max} and have reached octave R, going from right to left across our Species-Curve.

The number of species we have collected is

$$Q = y_R + y_{R+1} + y_{R+2} + \ldots + y_{R_{max}} = \sum_{y_R}^{y_{R_{max}}} (y) \qquad (23)$$

The number of individuals collected is

$$I = y_R + 2y_{R+1} + 2^2 y_{R+2} + 2^3 y_{R+3} + \ldots + 2^{(R_{max}-R)} y_{R_{max}} \qquad (24)$$

Let us now collect till the next octave is fully exposed. The additional number of species collected is

$$\Delta Q = y_{R-1} \qquad (25)$$

The additional number of individuals collected is

$$\Delta I = I + y_{R-1} \qquad (26)$$

For a given number of species (N) in the fixed universe, we can, provided it is canonical, obtain the value of y at any octave, and also the value of I/m.

Assuming that m does not change as we collect, and there is no reason why in a fixed universe it should, we can compute the index z in the Arrhenius equation as

$$z = (\Delta \log N / \Delta \log I) \text{ (See Preston 1960)} \qquad (27)$$

This has been done for the case where N = 200 species in the universe, and where in consequence $\sigma = 4.07$ octaves if the universe is canonical. The tabulation is not reproduced here, but in Figure 14 we give a graph of the Index z as ordinate against the percentage of species collected. It

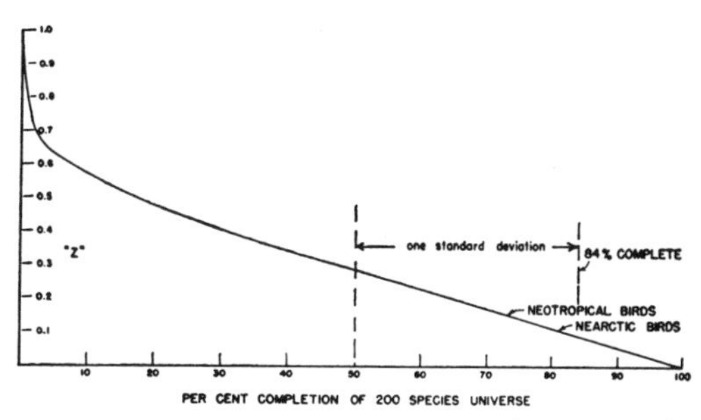

FIG. 14. The index "z" of the Arrhenius equation for a truncated distribution.

will be seen that at first we collect species as fast as we collect individuals (z = 1), but by the time we have collected 15 or 16 species (7 or 8% of the total of 200 species available), z has fallen to 0.6, and from this point on it continues to fall steadily till it reaches z = 0.0 as the last species is collected.

The inconstancy of z

It follows that "z" in the Arrhenius equation is not by any means a constant under the conditions we have stipulated, and since the argument applies to increasing sizes of quadrat, i.e. to a species-area curve, as well as to a sample of moths in a light trap, it follows that a log-log plot of a species-area curve will not give a straight line, at least not under the stipulated conditions.

The reason we sometimes find a fair approximation to a straight line in the log-log plot is that, after we reach a certain degree of completeness, the universe begins to expand as fast as the sample. This seems to happen typically when we have collected some 75 or 85% of the initially-available species. The remaining species are very rare, and it requires a prodigious effort and the capturing of innumerable individuals to get these rarities by strictly random collecting. We are likely to obtain a lot of "strays," "casuals," and "accidentals" before we succeed in netting all the rare species that are legitimately present.

In Figure 14, Williams index of 0.22 would be reached at 64% completion, but his data probably relate to a mixture of plant-isolates (with a z index of 0.27) and of "samples" with an index well below 0.22. The index we found (Preston 1960) for the neotropical avifauna, z = 0.16, is reached at 73% completion, and the figure of z = 0.12 (nearctic avifauna) is reached at 83% completion. This last point is at just about one standard deviation beyond the mode. Collecting by random methods beyond this point is a rather unprofitable procedure; deliberate searching now pays dividends.

The Species-Area Curve for samples

Even when the sample-area seems to be well defined and our methods systematic rather than random, as in determining the breeding birds of a county or some such area, if we succeed in locating every pair and in counting them, we shall, in general, have a sample rather than a universe. Usually it will approximate to being a sample of a much larger area, perhaps of an area as large as a state.

From the same data as before we can construct a Species-Area curve for such samples. Let us hide the canonical curve completely behind the veil line, and then withdraw it octave by octave. At each step of the withdrawal we add a number of individuals, slightly more than doubling, in fact, the number we have already counted, as given by equation (26). A number of individuals may be taken as proportional to an area. At the same time we add a number of species given by equation (25). We plot the accumulated area logarithmically as abscissa, and the accumulated species either arithmetically (Gleason) or logarithmically (Arrhenius) as ordinate. In Figure 15 we have used the latter method of plotting.

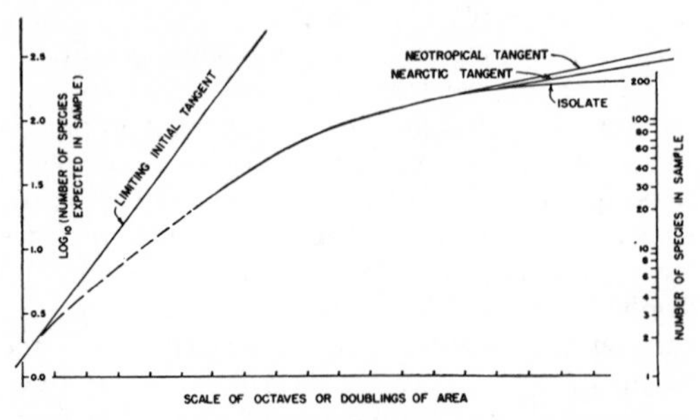

FIG. 15. Theoretical Species-Area curve for samples.

In this figure, as in the previous one, we assume that the universe initially sampled consists of 200 species. The left-hand end of the curve is shown as a broken line, till we have captured or observed 20 species; below this level we must in practice expect all sorts of erratic results, and indeed for some little distance beyond it. The initial part of the curve is fairly straight and quite steep, but not so steep as the limiting tangent, except at the very origin. This tangent is indicated; it represents the start where every new specimen is likely to be of a new species. See Preston (1960) for a discussion of such tangents.

At the right-hand end of the curve, we have 3 lines. The lowest is the theoretical behavior of an isolate or fixed universe. There are only 200 species all told, so the curve flattens to horizontal at this level. In practice, however, before we reach this point the effort to collect the few remaining rare species by random methods results, as we have seen, in causing the universe itself to expand. The middle line, therefore, is drawn as a straight line, tangent to the curve at about 80% completion (i.e. when we have collected 160 species) and having a shape corresponding to $z = 0.12$. This resembles the situation we encountered in our study of the Species-Area Curve of Nearctic breeding birds (Preston 1960). In that case it seemed as though initially our universe consisted of somewhere around 75 or 100 species and by the time we had collected 80% of them the universe was expanding as fast as the sample.

The upper line is the corresponding tangent for the neotropical fauna, with a slope of 0.16. As we go farther to the right (we are not now limited to the right hand end of the graph at 200 species) both of these tangents will ultimately curl upwards in practice, as discussed in Preston (1960).

The general appearance of the curves, with the double-logarithmic or Arrhenius type of plotting, is that of a pair of more-or-less straight lines meeting at an obtuse angle, but rounded off one into the other, the steeper of the lines being toward the left. The graph resembles fairly well the 2 curves found in practice in the previous paper.

The size of the universe in terms of the properties of the sample

Hidden beyond the Veil-line of our truncated distribution, which is our sample, are a number of octaves which are needed to complete our universe. So far as species are concerned, this may amount to only a modest addition, but for individuals each octave represents a doubling of the population. The complete universe is therefore usually many times as large as the sample.

In Table I the quantity x is the half-range in terms of the standard deviation as unit, i.e. $x = R_{max}/\sigma$. To a first approximation, for the cases we have encountered and are likely to encounter in practice, with N between 200 and 800, x is close to 3.0.

When we have collected as far as one standard deviation beyond the mode, we have accumulated 84% of the species in that universe, but we still lack 2 σ or thereabouts of having the complete curve. Since σ is usually around 4 or 5 octaves, we still lack some 8 or 9 octaves. The size of the universe is therefore 2^8 or 2^9 times the size of the sample, or 250 to 500 times as large.

This statement is better put in this form: we should have to collect 250 to 500 times as many specimens before we had a reasonable chance of collecting one specimen (or pair) of the rarest species. Better yet, the statement is true in its original form if $m = 1$, and it frequently seems to be near this.

This would permit us, for instance, to estimate the overwintering population of the birds of the nearctic, using the Audubon Christmas Bird Counts (Preston 1958), if we consider only those species that were found by random methods. Suppose for the sake of illustration, that in a typical single year about 10^7 individuals were seen and that these represented 84% of the complete distribution and about 500 species. Then σ should be about 4.5 octaves and x about 3.1. The end of the distribution would be about 9 octaves away, and the total number of overwintering individuals would be about 500 times the size of the sample, or 5×10^9 birds. (Again we are assuming that $m = 1$.) This is roughly equivalent to saying that, since the nearctic is roughly 4 or 5×10^9 acres, we have one bird per acre wintering in the nearctic. They will of course be far from uniformly distributed, most of them being along the seacoasts or the more southern parts of the country, where some acres will be black with redwings and starlings. Whether the estimate is fairly close or somewhat high may depend on whether different parties of observers saw the same large flocks at different places, and on various other problematical matters. The computation merely illustrates the possibility that we may be able, by more refined methods and especially, perhaps, with a surer knowledge of the value of "m," to estimate the total wintering population.

We can also compute what fraction of a population must be observed or collected in order to reach the mode, i.e. to collect 50% of the species and get at least a rough idea of the height of the mode. With $m = 1$, this fraction is simply $1/r$, which may be ascertained from Table I. So, with a population of 200 species, we reach the mode when we have captured or observed 1/2720 of the individuals. With 600 species we need to observe or capture only 1/17,000 of the total.

Graphical construction

Suppose we have collected beyond the mode and the truncated distribution appears as shown in Figure 16. From the point A where the curve intersects the Veil-line, draw a horizontal line to intersect the curve again at B. From B draw the vertical line BB_1 intersecting the R axis in B_1. Let C_1 be the end of the finite distribution actually observed. B_1C_1 is the number (R) of octaves theoretically missing to the left of A, the curve being assumed quite symmetrical. Then 2^R is the ratio of size of universe to size of sample.

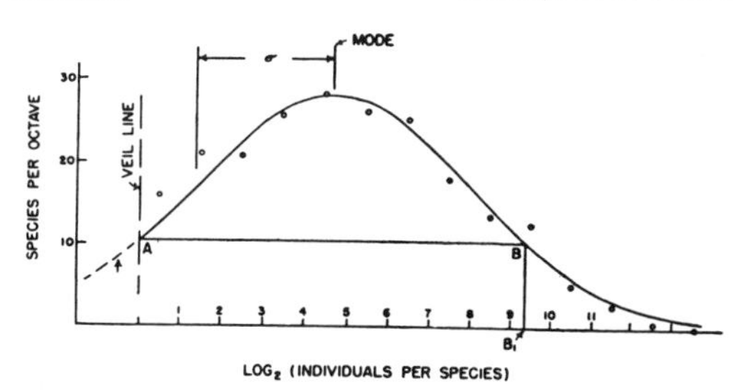

FIG. 16. Maryland State-wide Bird Count: graphical estimate of degree of completeness.

This particular plot is taken from Preston (1957) and represents the State-wide bird counts of Maryland.

Small samples

For our purposes most samples are to be regarded as small if the number of species involved is less than about a hundred. The number of individuals may be extraordinarily diverse. If our ensembles were canonical, the number of individuals for any given number of species would be ascertainable from Table I. But small biological "universes," are greatly distorted by "contagion." This is stated succinctly by Cain and Castro (1959): Most plants "are more or less clumped or contagiously distributed" and by Hopkins (1955): "Results support the view that most plant individuals are aggregated." A recent paper by Hairston (1959) may almost be regarded as a warning that contagion or clumping is so general and widespread that it is scarcely worth while constructing theories about noncontagious distributions. The warning is sound, and was given earlier by Cole (1946), who emphasizes the fact that biological material, whether animal or vegetable, is rarely distributed at random, but is nearly always more or less "contagiously distributed," and he examines the possibility of handling such distributions adequately by mathematical methods. But whereas Hairston's emphasis is on the "clumped" distribution, which he believes is characteristic more particularly of the rarer species, Cole emphasizes that "contagion" can be "positive" (clumped) or "negative" (over-regularized).

A great deal depends on the nature of the biological material with which one is working, and even upon the season of the year. Thus in the breeding season most passerine birds are, owing to their "territorial" propensities, over-regularized spatially, and therefore on a small area they are over-regularized in abundance. (See below, under Thomas, Hicks, Williams, Walkinshaw, etc.). But in the winter they "flock" and become "clumped," and it is not solely the rarer forms that do this.

Plants are perhaps the most obviously clumped material, and sometimes approximate to pure cultures, as with Hopkins' (1955) *Zostera* community. However, this can be matched among colonially nesting birds, like Beebe's (1924) "pure culture" of boobies in the Galapagos Islands, or the Emperor Penguin (*Aptenodytes forsteri*) in Antarctica. Therefore it seems to me that we are justified in examining the properties of randomly distributed populations, and treating them as a "norm," and we may treat contagion, whether positive or negative, as introducing a disturbance, modification, or perturbation into our calculations.

Notwithstanding Hairston's (1959) comment that, so long as we are dealing with a single community, "departure from randomness increases with sample size," I think we are justified in assuming that in this paper we shall rarely be dealing with a single community, but with what he calls heterogenous material, in which departure from randomness decreases with increasing sample size. Thus in order to study contagion in our present context we must examine the properties of rather small samples.

Since in the present paper we are concerned with abundance-distributions and want to know whether they conform to the lognormal type, and in particular to the canonical lognormal, a small sample for our purposes is primarily one that has rather few species, say a hundred or less. For with less than about one hundred we cannot plot the distribution graphically with much success. It is true we might use analytical methods rather than graphical ones, as being more powerful tools, but statistical fluctuations are not thereby prevented from confusing the issue, and so in Preston (1957) I suggested that an adequate sample called for something like a minimum of 200 species and something like a minimum of 40,000 individuals.

Very little in the way of plant material meets the requirement of 200 species, or even 100, and much of it, in fact most of it, appears to have less than 50, often much less. The individuals may be very numerous, especially in our northern latitudes where the floras are poor compared with the tropics (Cain and Castro, 1959) but, because the majority of plant distributions are "positively" contagious, we have an abnormally high number of individuals without accumulating many species. Thus, in order to deal at all with the published literature on plant distributions, i.e., their relative abundances in a "community" or their Species-Area curves, we have to compile a tabulation and make a graph of what to expect of samples containing less than 100 species.

Criteria for small samples

With small samples in this sense, it is easier and perhaps more accurate to reduce them to graphs, and the easiest computation to make concerns the (logarithmic) standard deviation σ. Therefore in Table IX below we use the methods

TABLE IX. Properties of small canonical ensembles

Number of species N	Logarithmic standard deviation σ (octaves)	Ratio of individuals to species I/N or more properly I/mN
100	3.72	26,600
80	3.62	12,100
60	3.48	7,500
40	3.26	3,300
20	2.84	390
10	2.37	65
6	2.00	20
3	1.40	3.7
2	0.97	1.5

of our first section to get estimates of the value of σ of I/mN for canonical ensembles ("universes") with less than 100 species.

These values should be regarded as approximate only. They are graphed in Figure 17. On that same graph are shown a number of experimental points exhibiting contagion, positive and negative, which will be discussed later. It should be understood that these figures relate to complete, non-truncated, canonical ensembles; but where small samples seem not to be too severely truncated we can compute σ as if the distribution were normal in order to get a rough picture of the effects of contagion and the "distortion" produced thereby. We shall see later that the observed departure of σ from its expected or canonical value gives an estimate of the degree of contagion, as indeed is otherwise obvious, and that contagious distributions do not end at the crest of the "Individual Curve."

Skewness as a criterion

It is very likely that contagious distributions are somewhat skewed, the mode lying to the right of the mean, i.e. at a higher abundance of individuals per species than the mean for negative contagion. This criterion, being subject to com-

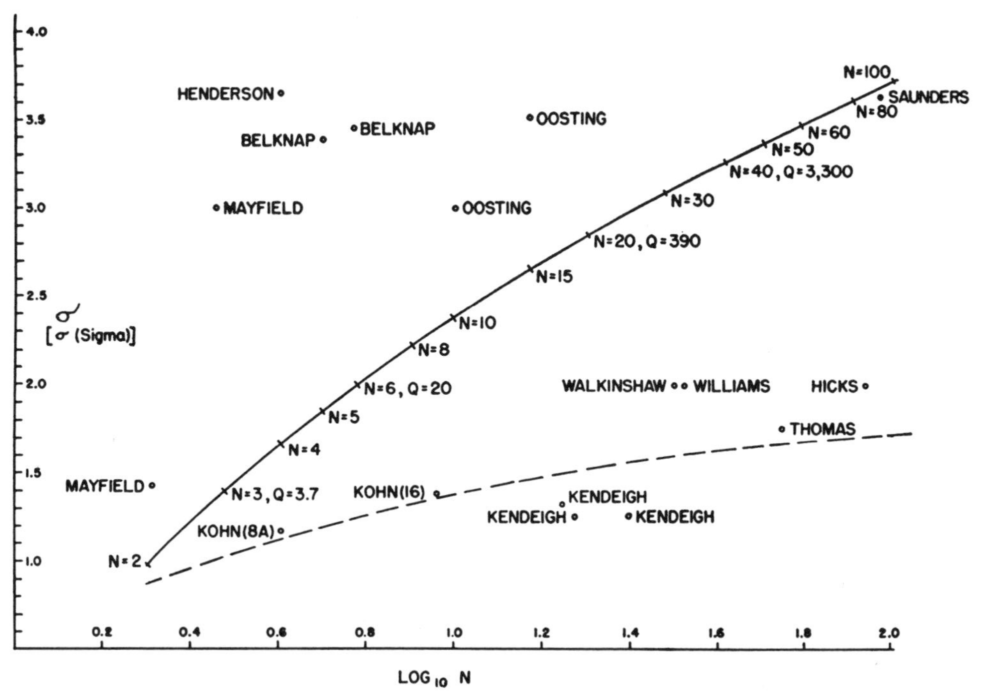

FIG. 17. The logarithmic standard deviation for small samples and universes, as a function of the number of species: N less than 100. The full line is the Canonical expectation, the broken line is the MacArthur "broken-stick" expectation. The Canonical line seems to divide positively contagious aggregates (above) from over-regularized aggregates (below).

putation, might sometimes be useful, but it is subject to greater statistical uncertainty than the standard deviation, which for the present seems sufficient for our purpose.

Examples of negative contagion

There are available a number of useful counts of breeding birds, which, being disposed to "defend a territory," tend to be distributed more uniformly over the countryside than we should otherwise expect. There is more resistance to the incoming of an additional pair of a species that is already common in the area than there is to the coming of a species not yet well represented. Thus on a given area, the species are more nearly of equal abundance than they would otherwise be. This amounts to saying that the standard deviation in practice falls below its canonical value, and the points should lie below the solid line of Fig. 17.

Thomas. The breeding birds of Neotoma, south-central Ohio

This report, given in Preston (1960), is valuable because we have the counts for 10 separate years on the same 65 acres. If we take each year and list the numbers of singletons, doubletons, and so on, discarding the names of the species, and then strike an average for the 10 years, gather the results into "octaves" and graph the outcome, we obtain Figure 18. Such a curve is somewhat typical of one-year counts and other small samples (Preston 1948), but in this case it does not make use of all the available informa-

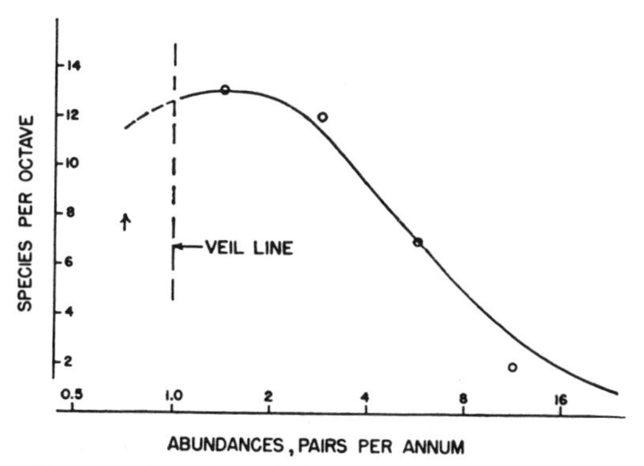

FIG. 18. A typical year's count of the breeding birds of "Neotoma," by E. S. Thomas. (Species Curve)

tion. In particular, it does not discriminate between singletons (say) of species that occur as singletons every year and of species that occur only once in 10 years.

If we therefore retain the names of the birds and make this distinction, we can distinguish between species that have a 1:1 chance of appearing in any one year and those that have only one chance in 4 or one in 8. This produces Figure 19, where, so to speak, we get a peep behind the

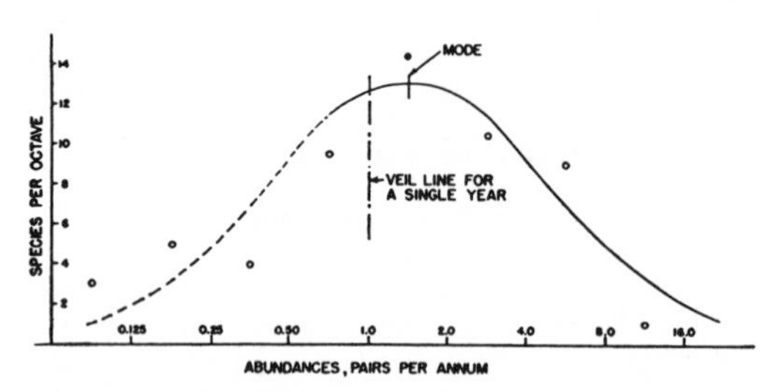

FIG. 19. Birds of "Neotoma," a typical year reconstructed from a knowledge of what species were present in each of 10 years. (Species Curve)

veil. Although the curve as drawn lies reasonably well among the observed points, which generally alternate above and below it, there could be some argument that a better curve might be one skewed to the right, descending abruptly and terminating at abundance 16. We shall not debate the point, but merely note that the number of species involved (N) is 56 excluding the cowbird, that σ is 1.73 octaves as against the canonical expectation of 3.37, and I/N is less than 3. The canonical expectation in a complete universe would give $I/mN = 5000$ or thereabouts. We may also note that the individuals curve Figure 20 continues past its own crest into its descending limb for an

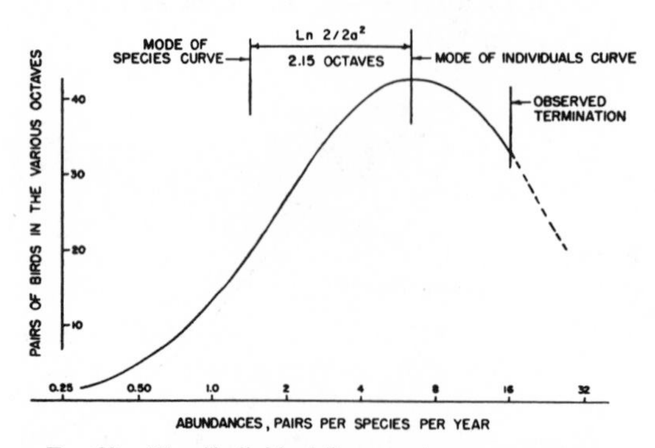

FIG 20. The "individuals" curve for an average year at "Neotoma." The observed termination is well beyond the crest, as it should be for a negatively-contagious or over-regularized distribution.

octave or so, a necessary consequence of the low σ value of the Species-Abundance curve.

Hicks (1935). Breeding birds near Westerville, Ohio

This also is a 10 year count, in the course of which 86 species were observed, though the average for a single year was 63 or less. In Figure 21 we give the graph, on the same basis as Fig.

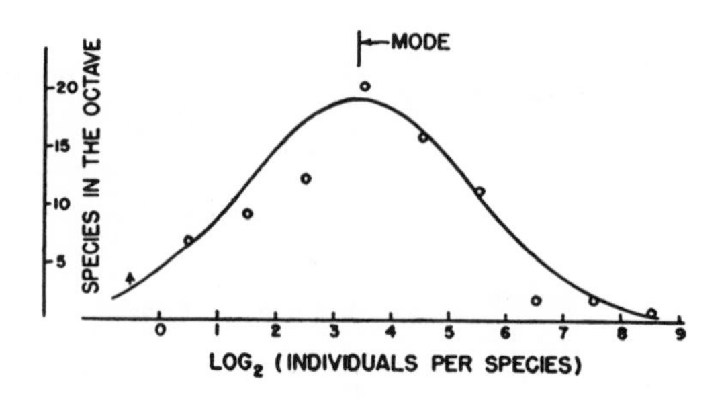

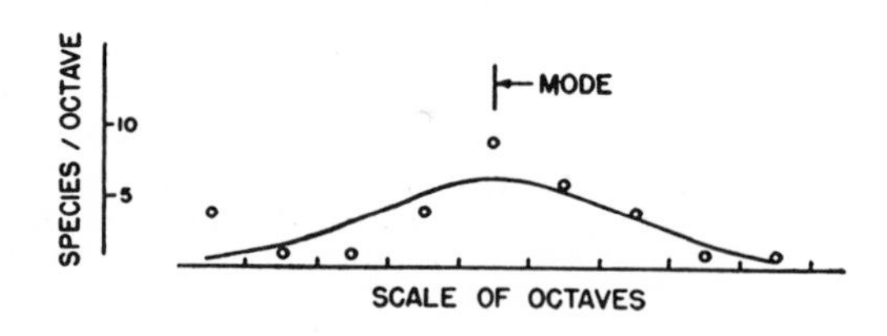

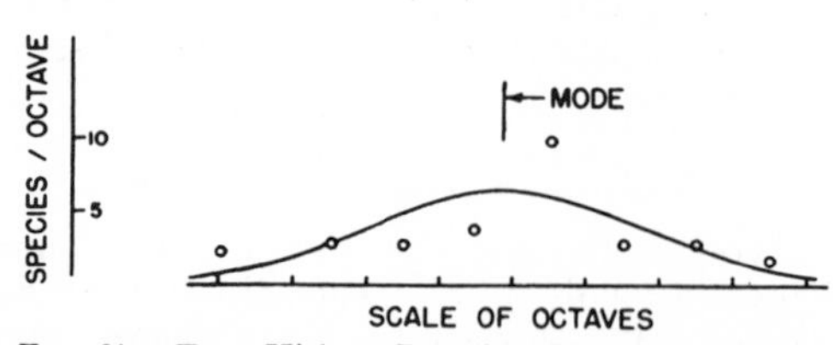

FIG. 21. Top, Hicks. Breeding Birds near Westerville, Central Ohio. Ten year count. Center, Walkinshaw. Breeding Birds near Battle Creek, Michigan. Ten year count. Bottom, Williams. Breeding Birds near Cleveland, Ohio. Fifteen year count.

19, except that I have used the accumulated totals for the 10 years and have not struck an average by dividing this result by 10. The value of σ is almost exactly 2.0 octaves, the average value of I/N is about 3.5 for a single year, and there is a suggestion of skewness, as expected.

Walkinshaw (1947). Breeding birds on 83 acres of brushy fields

This is another 10 year count. Some 31 species were accumulated. The distribution is graphed in Fig. 21. The standard deviation is almost exactly 2.0.

Williams (1947). Breeding birds of a beech-maple forest near Cleveland, Ohio

This is a 15 year count and 33 species, including the Cowbird, were accumulated on 65 acres. The value of σ is approximately 2.0, and I/N is about 4.2 pairs per species per annum. (Bottom Fig. 21.)

The counts of the individuals curves for Hicks, Walkinshaw and Williams all seem to lie about 1½ to 2 octaves to the left of the termination of the curve, which again is what is expected.

Kendeigh (1946). Breeding bird counts in New York State

These counts extended, in form usable by us, over 3 years in the early 1940's. The results from Kendeigh's tables I, II, and III, may be condensed for our purposes into the form of Table X below, with Thomas' results for "Neotoma" added for comparison.

TABLE X. Kendeigh's breeding bird counts in New York state

Kendeigh's Table Number	I	II	III	"Neotoma"
Character of Woodland	Beech Maple Hemlock	Hemlock Beech	Beech Maple Hemlock	Mixed Deciduous
Acreage involved	8	21	62	69
Number of species known breeding	24	17	18	56 in 10 years
Av. pairs per Species per Annum (I/N)	1.22	2.48	4.0	—
Standard Deviation (σ) observed σ	1.26	1.32	1.22	1.76
Standard Deviation for a canonical ensemble with same number of species σ^1	3.0	2.75	2.80	3.45

So far we have dealt solely with birds, and have tentatively ascribed the departure from the canonical results to "negative contagion," or, in biological language, to the territorial demands of the birds. Let us now turn to an entirely different fauna, marine gastropods.

Kohn (1959). Gastropods (Conus) in Hawaii

Kohn examines the abundance-distribution of cone-shell species from several collecting localities, and plots them in terms of the MacArthur hypothesis discussed below. I have chosen to examine here two instances, Kohn's Figure 8A, because it is the first he gives, and his Fig. 16, which is the one that agrees most closely with the MacArthur distribution.

For his Fig. 8A there are 4 species, and 136 individuals, so that I/N (Kohn's m/n) is 34 individuals per species. This is more than we expect in a canonical distribution. Yet the standard deviation σ appears to be only 1.11 octaves, far below the 1.67 octaves expected for a canonical distribution. For his Fig. 16 there are 9 species and 182 individuals, giving I/N = 20.2. This is much less than the canonical expectation, unless our "m" (not his) is not unity but 2 or 3, as it may perhaps be. The standard deviation is about 1.38 octaves, far below the canonical 2.3 octaves.

These results suggest, but do not prove, that species of *Conus* may be over-regularized in distribution, so that the various species in a quadrat are of more uniform commonness than a random distribution would give. In correspondence Dr. Kohn says that *Conus* is not known to "defend a territory," but not much is known about it in this respect. We do know however that the genus is carnivorous, but the different species have different food-preferences, some eating marine worms, some fishes, and some other animals. Thus there is a possibility that food supply of itself might induce a non-random distribution, either directly, or indirectly by causing quasi-territorial behavior. All these examples of presumed negative contagion (over-regularized distribution) are plotted on Fig. 17, and it will be noted that they all fall below the line, often a very long way below it.

Examples of positive contagion

Oosting (1942). Plants of the Carolina piedmont

From Oosting's Table #1, dealing with fields abandoned for one year, I took Field #4, which has 15 species reported. The abundances are given in terms of "densities," and the standard deviation works out at about $\sigma = 3.52$ octaves, as against the 2.65 expected for a canonical ensemble.

Oosting (1942). Shrubs and vines in a 15 year old stand of the Carolina piedmont

There are 10 species, and the standard deviation is $\sigma = 3.0$ octaves, as against the canonical 2.37.

I think these examples from Oosting are sufficient. He provides many other tables, and it seems, from a rather casual inspection, that most of the other tabulations might corroborate these 2.

The Herons of West Sister Island, in Lake Erie off Toledo

This information comes from Mr. Harold Mayfield. American Egrets first nested, so far as is known, in 1946. In that year a count showed that the heronry consisted of about 1500 Black Crowned Night Herons, 200 Great Blue Herons, and 10 American Egrets. This gives a σ-value of 2.95 octaves, far above the canonical value (1.40) for 3 species. The next year the number of Egrets doubled, and if everything else remained constant, the σ-value would increase a little. In 1945 there were no Egrets, presumably, and so only two species nested. The σ-value may presumably have been 1.44, which is still well above the canonical expectation of 0.97 for 2 species.

Belknap (1951). Breeding birds on an Island in Lake Ontario.

Belknap censused a small island of one acre from 1948 to 1951. He gives his count for 1951 and comments briefly on earlier years. This island, like West Sister Island above, presumably behaves as an isolate, and this feature may be valuable. There were, in 1951, six species and a comparatively large number of individuals or pairs (965 occupied nests), giving a value of $I/N = 160.8$, far above the canonical expectation of 20 for $N = 6$. In Table XI we give the tabulation, with the figures predicted by the MacArthur hypothesis for comparison; the latter will be discussed later in this section.

TABLE XI. Belknap's breeding birds on an island in Lake Ontario in 1951

Species	Observed Nests	MacArthur Hypothesis Prediction
Ring-billed Gull	793	394
Common Tern	117	233
Herring Gull	34	153
Double-crested Cormorant	19	99
Black Crowned Night Heron	1	59
Black Duck	1	27
	965	965

The standard deviation is $\sigma = 3.47$ octaves, against the 2.0 of canonical expectation and a still lower figure for the McArthur distribution.

Belknap. The same island in 1950

Belknap does not give this count, but states that there were no Night Herons in 1950, that there were less than half as many terns, that there were twice as many Ring-billed Gulls, and that the Herring Gulls were "relatively stable." This leaves us with 5 species and a rough estimate of the 1950 population, given in Table XII.

TABLE XII. Belknap's Island in 1950

Species	Estimated Nests	MacArthur Prediction
Ring-billed Gull	1600	780
Common Tern	55	439
Herring Gull	34	267
Double-crested Cormorant	19	154
Black Duck	1	69
	1709	1709

The logarithmic Standard Deviation for the first column is 3.4 octaves, compared with 1.85 for the canonical ensemble of 5 species. Belknap's description of the 1948 and 1949 situations would lead to a similar conclusion.

Henderson. A communal roost of passerines near Oberlin, Ohio

Henderson describes this as a roost of "blackbirds," though it includes 500 Robins (*T. migratorius*), and the Redwinged Blackbirds were not positively identified. The pattern appears to be approximately as shown in Table XIII.

TABLE XIII. Henderson's blackbird roost

Redwings	5 (not positively identified)
Cowbirds	50
Robins	500
Grackles	5,000 to 10,000
Starlings	50,000

If we omit the Redwings which "were suspected but never surely identified," the logarithmic standard deviation comes out at about 1.1 orders of magnitude or 3.65 octaves. If we include the Redwings, the standard deviation will be greater, but it is almost off the map (Fig. 17) anyway. In this connection we may note that, while by ordinary standards the roost is a communal one, there is often in such cases a partial segregation of the species, just as there may be a negro quarter in a town.

These half dozen values of "clumped" or "colonial" distributions are plotted as points on Fig. 17. They all lie far above the line, and this is presumably true of "positively contagious" distributions generally.

Seemingly contagion-free examples

It seems only fair to note here that in looking for examples of contagious distributions, I found

3 that did not act quite as expected, but instead fell close to the line. One was Saunders' tally of the birds of Quaker Run Valley (Fig. 17). The explanation may well be the one advanced by MacArthur, that there are really several communities involved and mixing them tends to produce such a result. Hairston has come close to saying the same thing.

Another example was a count of plants by Buell & Cantlon (1951). The same reason is possibly valid, though the catagion to be removed is of the opposite sign.

Finally I took the counts made by Mr. H. H. Mills and myself of the day-flying herons of the heronry of Stone Harbor, New Jersey, in 1959 (unpublished). Six species were involved and the count was of the homing herons and egrets at sundown, not of their nests. The difficulty here is the impracticability of distinguishing in the gloaming between Snowy Egrets and immature Little Blue Herons, for it was late in the season and the immatures were largely on the wing. One estimate of the partition gave a point nearly on the line. Another estimate would place it distinctly above the line.

In any case the general picture seems clear. Gregarious ensembles tend to fall above the line, territory-guarding ones tend to fall below it. Thus the canonical distribution seems to justify itself as a "norm" corresponding to randomly-distributed species devoid of both positive and negative contagion, which can be regarded as perturbations of the canonical.

Possible restatement of the canonical hypothesis

We are now in a position to make a surmise that the canonical hypothesis is a statement, perhaps only an approximate one, of the behavior of lognormal ensembles when the individuals in the ensemble act completely independently and neither attract nor repel others of their own species. If I can interpret Hairston's views in the light of my own, it is the situation that obtains when the "community" is completely devoid of "organization." When "organization" appears, it results in "contagion" and not solely the "clumping" or positive contagion that is Hairston's primary concern, but also the "negative contagion" of Cole and the "regularity" of Hopkins (1955 and 1957).

This suggests that our graphical description of a canonical ensemble as one whose Individuals Curve crests at its termination may very well be replaced by an analytical description of a lognormal ensemble whose individuals (or pairs) completely "ignore" others of their own species, that is, are unaffected by their proximity or distance. Such a reformulation might be very useful, as well as more satisfying esthetically than the one we have used, but I do not know whether it would prove so simple to handle mathematically.

The MacArthur distribution

A recent suggestion by MacArthur (1957), originally based on the analogy of the probable lengths of the fragments of a randomly broken stick, proposes that a population of m individuals may be apportioned among n species according to the law

$$m_r = (m/n) \sum_{i=1}^{r} [1/(n - i + 1)] \qquad (28)$$

where m_r is the number of individuals assigned to the rth rarest species and m/n is our I/N.

The actual operation of computing the abundance of the various species may be carried out as follows: suppose, for example, that n = 57 species. The rarest species will have 1/57 (m/n): the next rarest (1/57 + 1/56)(m/n): the next (1/57 +1/56 + 1/55)(m/n) and so on. Now for comparison with the lognormal distribution, having obtained the MacArthur figures for each species we take the logarithms thereof and gather them into octaves for plotting, or compute the logarithmic standard deviation by orthodox methods.

In Fig. 22 we graph the distribution for 4 instances, where N = 5, 10, 100, and 1000 species respectively. The scale of octaves at the bottom applies to all 4 graphs as abscissa, but the scale of ordinates varies, and is given in each case. With this method of plotting, the curve is single-humped, tangent to the x-axis on the left but cutting it abruptly on the right, skewed somewhat to the right and therefore only a rough approximation to a Gaussian or "normal" curve. We may, however, calculate the standard deviation "as if" the curves were normal, and we find that for 10 species σ = 1.49 octaves, and for 100 species σ = 1.72 approximately. We may also easily compute σ for 2 species as 0.79 octave and for 3 species as 1.00 octave. This permits us to draw, as a broken line in Fig. 17, the approximate position of the standard deviation as a function of the number of species in the "community," as predicted by the MacArthur hypothesis. It lies far below the value predicted by the canonical lognormal hypothesis. It passes very close to Kohn's 2 points, and lies among the points for anti-gregarious or territorially-minded breeding birds.

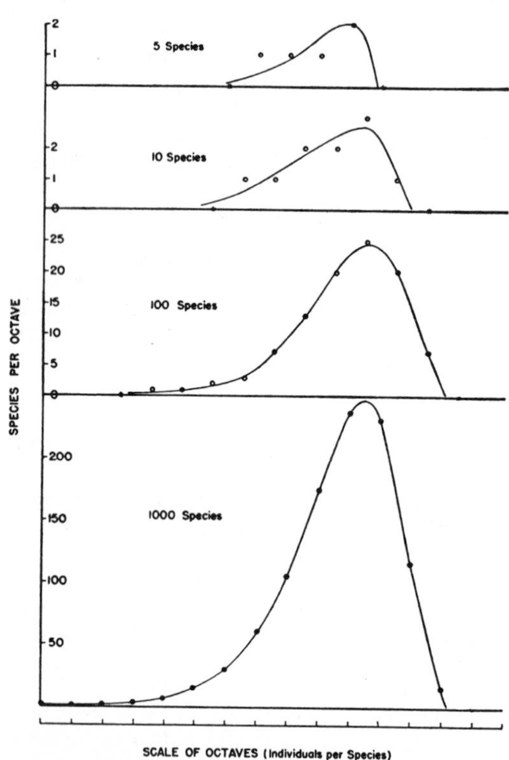

FIG. 22. The MacArthur distribution for 5, 10, 100, and 1000 species.

It lies at an immense distance below the points for gregarious or colonially-nesting birds and for the "contagious" plant communities. Since the effects of contagion may be expected to smooth out, and die out, with very large areas or very large ensembles of species, in the overall picture, while the MacArthur distribution, even for 1000 species (see Fig. 22) gives a σ-value not much higher than for 100 species, the distribution would not apply to such aggregrations of species, as MacArthur himself has said.

I therefore suggest somewhat tentatively that the MacArthur distribution is not the norm, and that the frequent agreement of Kohn's diagrams therewith is due to the cone-shells being somewhat over-regularized in their spacial distribution. We may note here Kohn's own comment that in a number of his diagrams "common species are too common and rare ones too rare" to agree with the MacArthur prediction. This statement amounts to saying that the logarithmic standard deviation of abundances is too high, which means that his points (not those I have plotted in Fig. 17) are often above the MacArthur line, perhaps approaching the canonical line.

It was, I think, MacArthur's original view that a genuine "community" might correspond approximately with his predictions, while fortuitous aggregations would more likely come close to the lognormal. This presents us with the problem of defining a genuine community. Superficially one might imagine that a heronry or a colony of gulls and terns and cormorants like Belknap's is a "community" since in some sense the birds "attract" one another in much the same way as human communities are brought together. But we see from Fig. 17 and from the data in Tables XI and XII, that it is precisely such communities that depart most from the MacArthur prediction. We should have to define a community as a group of persons, animals, or plants that insist on holding their fellows at arm's length, or that are repelled by one another.

Margaret Perner's breeding birds at Cleveland, Ohio

Since the MacArthur formula seems more appropriate to territorial species than others, I thought we might examine one instance in more detail, and for that purpose use Perner's (1955) data on 25 species of nesting birds. This time we plot simply the nests per species as ordinate against ordinal rank (of increasing commonness) as abscissa (Figure 23), and the MacArthur prediction is plotted on the same graph. The departures of the observations from the prediction are not random, but systematic, so that this aggregation of species, though negatively contagious, does not agree well with prediction.

Other pecularities of the Species-Area curve for samples

Samples, especially small ones, have a number of peculiarities, partly because they are samples and therefore likely to be incomplete or truncated distributions, and partly because they are small, especially in numbers of species, and therefore likely to be affected by contagion, positive or negative. Whereas the Species-Area curves for isolates can be understood comparatively easily, those for samples may present greater difficulties. Even in the matter of graphing the plots there are problems, and in interpreting the results there are worse uncertainties.

Methods and problems of plotting the Species-Area curves

This subject is almost a monopoly of plant ecologists, who set out "quadrats" of various sizes and count the species in the quadrats. They

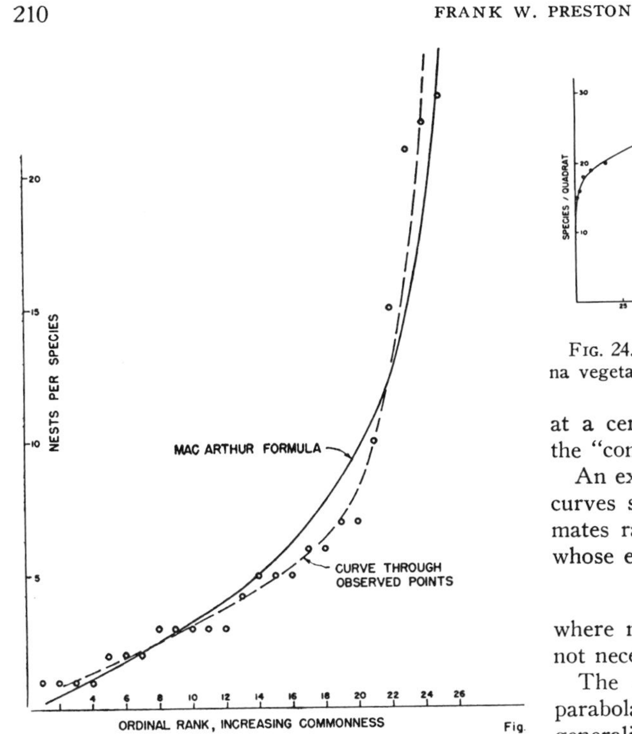

FIG. 23. Perner. Breeding Birds of a Cleveland, Ohio, park.

usually use quadrats that increase in size in a more or less geometric progression, for instance in the area ratios of 1, 2, 4, 8, 16, ... This is an advisable procedure, because the species count increases only slowly with increasing area. The geometric progression implies that psychologically or subconsciously, the operator is working with the logarithm of the area and not with the area itself, yet many investigators have plotted the results on an "arithmetical" basis with number of species as ordinate against area as abscissa (e.g. Cain 1938, Hopkins 1957). This results in most of the points being crowded into a narrow, nearly vertical, line near the origin and being sparsely distributed farther to the right.

The kink in Cain's curve

Numerous plottings by the above writers and others are available to any inspector, so I have chosen to use one that Cain (1959, p. 110) did not plot, though he gave the data from which it may be plotted (Figure 24). Braun-Blanquet and others (see Hopkins 1957, pp. 441-443) as well as Cain at one time, believed there was a definite "break" in this curve, the initial part being essentially vertical and the later part being essentially horizontal, as if the curve "saturated"

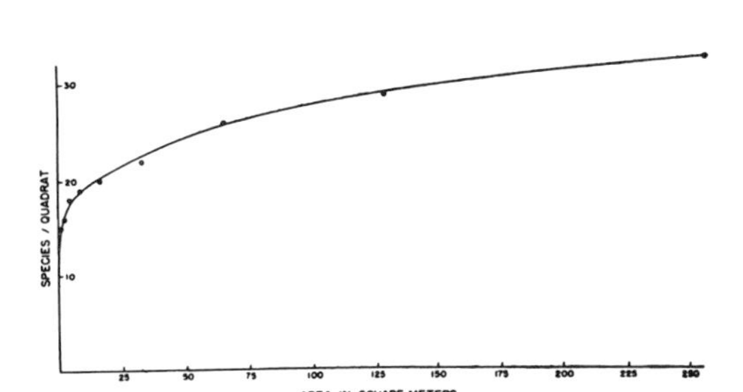

FIG. 24. Cain & Castro. Species-Area curve for savanna vegetation in Para, Brazil.

at a certain level that gave the total species in the "community" or "stand."

An examination of this and the many published curves suggests strongly that the curve approximates rather closely to a "generalized parabola," whose equation is

$$y^n = ax \tag{29}$$

where n is an exponent greater than unity, and not necessarily an integer.

The ordinary "conic-section" or "quadratic" parabola has $n = 2$ in the above equation. The generalized parabola has other values, the most interesting ones being higher than 2, often much higher. The quartic parabola for instance has $n = 4$, and the curve we have illustrated strongly suggests a quartic. Since y, or N, the number of species, is necessarily positive, and so is x, or A, the area of the various quadrats, the index n may take all positive values and the curve will remain real. Whatever the value of n, when limited to positive values the curve will pass through the origin and, provided n is greater than unity, it will there be tangential to the y-axis. The higher the value of n the closer it will hug the axis and the sharper will be the bend when it breaks away from it. Thus in a sense the break is real. The curve never becomes parallel to the x-axis, but continues indefinitely to climb, though always at a decreasing rate with increasing size of area.

In order to help visualize the "higher" parabolas, I have graphed several of them in Figure 25 and by setting $a = 1$ in equation (29), I have caused all curves to pass through the points 0,0 and 1,1. This leaves n as the only variable parameter, and it will be seen that as n increases the curve is steeper near the origin and flatter at the larger coordinates.

It is a property of the ordinary quadratic parabola that its curvature is greatest at the vertex, or origin in this case, but this is not true of the higher parabolas. Nonetheless all of them have a single

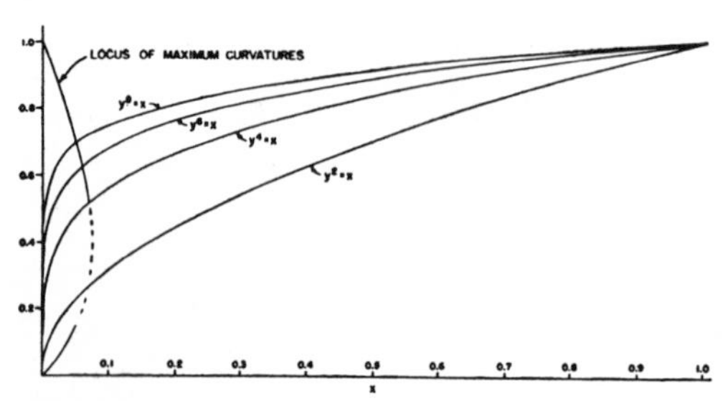

Fig. 25. A family of generalized parabolas, and the locus of points of maximum curvature.

point of sharpest curvature. In this sense there is a theoretical "break" in the curve, though it is not a discontinuity of any sort.

The curvature of any curve at any point is given by

$$\frac{1}{R} = \frac{d^2y}{dx^2} \Big/ \left\{1 + \left(\frac{dy}{dx}\right)^2\right\}^{3/2} \tag{30}$$

and if we differentiate this once more and set

$$\frac{d(1/R)}{dx} = 0 \text{ or } \frac{dR}{dx} = 0 \tag{31}$$

and solve for x, we can define the point of maximum curvature in terms of the abscissa, or by solving for y we can define it in terms of the ordinate, which is usually (in our problem) more satisfactory.

In this case, we find that the point is given by

$$y^{2n-2} = \frac{a^2}{n^2}\left(\frac{n-2}{2n-1}\right) \tag{32}$$

On Fig. 25 I have plotted the locus of the maximum curvature. It is probably the break-point that Cain was seeking. In practice, however, it is not easy to determine such a point by graphical methods. It is much easier, even with a curve perfectly free from experimental or statistical errors, to find the point by equation (32), and for that purpose we have first to find the value of n. Clearly, if the curve really is a parabola, this is most easily done by taking logarithms of both ordinate and abscissa, whereon the equation becomes

$$n \log y = \log a + \log x \tag{33}$$

which is a linear relation between log (Species) and log (Area): i.e. the curve ought to be a straight line. This log-log plotting I have elsewhere (Preston 1960) called an Arrhenius plot-

[2] I am indebted to Dr. R. E. Mould of Preston Laboratories, Inc., now American Glass Research, Inc., for this result.

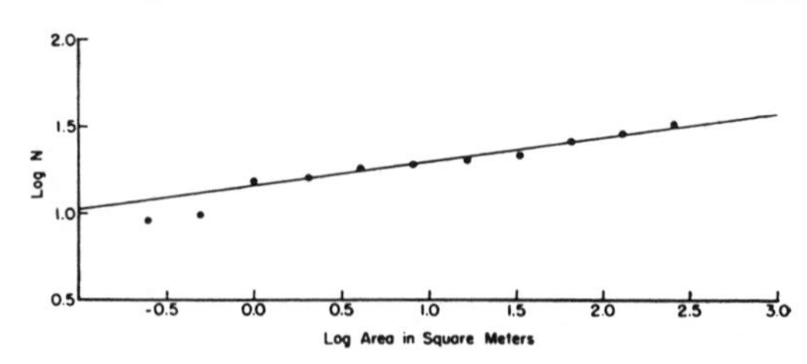

Fig. 26. The Cain & Castro Species-Area curve on a log-log (Arrhenius) basis.

ting. Using it on the data of G.A. Black and S. A. Cain as reported in Cain and Castro (1959), for savanna vegetation in Pará, Brazil, we get Figure 26, which, except for the first 2 points, is a pretty fair straight line. The slope is given by $k = 1/n = 0.14$, or $n = 7$ approximately, and the curve, over 2½ orders of magnitude, is a pretty good generalized parabola.

The Gleason plotting

Plant ecologists seem to have made less use of the Arrhenius "log-log" plotting than of the Gleason "semi-log" plotting. It is necessary, in order to space the points fairly uniformly across the graphs, to take the logarithm of the area, but it is not essential to take logarithms of the number of species. Furthermore, because species increase so slowly with increasing area, the logarithm of the number of individuals is a linear function of the number of species over rather wide intervals. Thus if the Arrhenius plot gives a straight line, so will the Gleason plot, over an interval of an order of magnitude or more in the size of quadrat or time of observation.

In Figure 27 we illustrate this point by means of Thomas' data on the 10-years of breeding bird counts on the 65 acres of Neotoma in south-central Ohio. The same data are plotted by the Arrhenius and by the Gleason methods. In each case the abscissa is a logarithmic scale of years-of-observation. The ordinate for the lower curve is

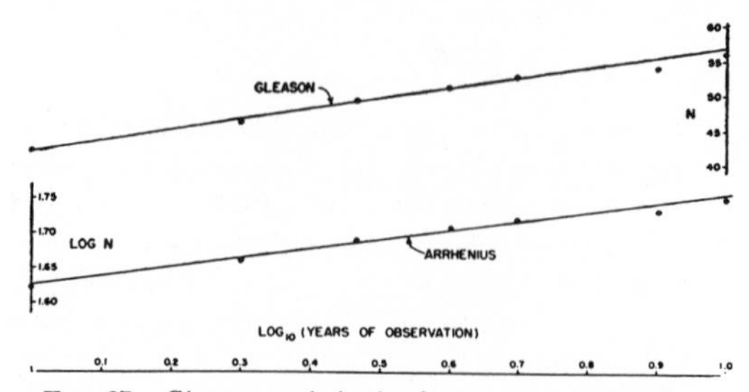

Fig. 27. Gleason and Arrhenius curves compared over a range of one order of magnitude (3.3 doublings or octaves).

logarithmic and is indicated at the left; for the upper curve it is arithmetical and is indicated at the right. Both curves are satisfactorily straight. On a log-log basis the index $k = 1/n = 0.132$.

If the range of abscissae is large enough, both curves cannot continue to be straight, unless k is so small that both curves are substantially horizontal. For if the log-log curve is straight, the Gleason curve will be concave upwards, and if the Gleason curve is straight, the Arrhenius curve will be concave downwards.

Brian Hopkins' curves (1955)

In order to examine whether the Gleason or the Arrhenius plot is more satisfactory in practice, on a purely empirical basis, it is necessary to have data over a range of areas of far more than one or 2 orders of magnitude. Hopkins (1955, 1957) gives a tabulation of a dozen communities or stands of plants over a nominal range of 8 orders of magnitude or a little more. In his 1955 paper he uses the Gleason semi-log plotting, and the curves tend to be concave upwards and of steadily increasing slope. This suggests that our Arrhenius log-log plot might be better, or at least more instructive, but in neither paper does Hopkins mention it.

With any form of plotting or computing, we run into trouble with the smallest areas, of 0.01 cm^2 and 0.07 cm^2 respectively. Here only 1 or 2 species, or even less than one species, are involved. These points I have had to ignore. The situation is better, but still statistically not too happy, with the other areas below one m^2, for we usually have less than 10 species, sometimes only 4 or 5. These, however, may conceivably make a "community" or at any rate a "stand."

The use of the log-log plot, though logical, is itself a source of some suspicion, to the extent that a liberal use of logarithms tends to reduce any monotonic function to a straight line, and this is especially the case when the dependent variable (the number of species) increases slowly with the independent variable (area) as it does in vegetation stands. The curve is going to approximate not only a straight line, but a horizontal line. Fortunately, Hopkins covered a very wide range of areas, and in most cases took 50 samples at each size of area in order to strike an average for his "point." Thus comparatively small departures from the graduating line will usually be meaningful, especially if the departures are systematic, to one side of the line for several points in succession.

We may summarise the outcome thus:

Hopkins' stand #2. Grassland at Wrynose Pass, Lake District, England: 55 species (bottom, Fig-

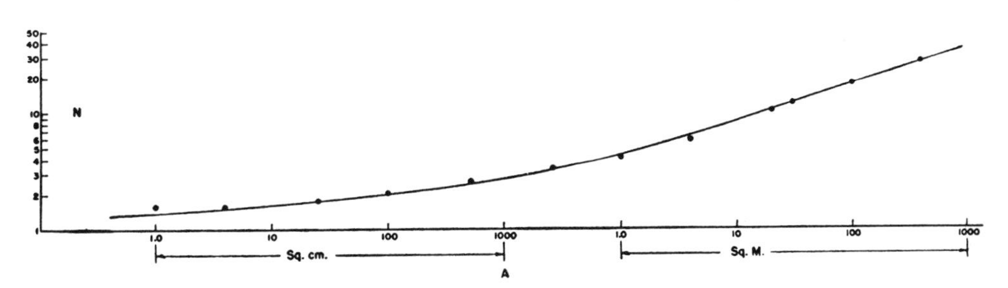

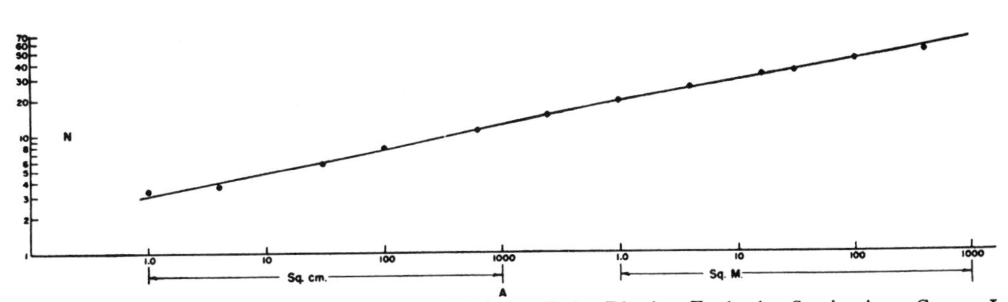

FIG. 28. Lower, Hopkins, Stand #2. Grassland in the Lake District, England. Species-Area Curve. In the following Figures, the ordinate (N) is the number of species on the various observed areas, plotted logarithmically. These are all log-log plottings (i.e. "Arrhenius plottings"). Upper, Hopkins, Stand #3, Beech Wood, Chiltern Hills.

ure 28). This is a good straight line over its whole length of 6½ orders of magnitude. The slope, k, is 0.193.

Hopkins' stand #3. Beechwood in the Chiltern Hills: 28 species. The curve seems to be concave upwards, steepening towards the right, but the last 5 or 6 points lie quite well on a straight line with a slope, $k = 0.327$. This is an unexpectedly high value, more appropriate to a set of isolates than to a sample.

Hopkins' stand #4. Blanket Bog in Mayo: 47 species. From 1 cm^2 to 4 m^2 the curve is close to a straight line, with a slope of $k = 0.245$. Then it changes abruptly to another much flatter slope, with $k = 0.097$ (Figure 29).

Hopkins' stand #5. Bog at Rannoch, Perthshire: 48 species. This somewhat resembles #4, the initial slope is $k = 0.203$, changing abruptly at 1/4 m^2 to a flatter slope, $k = 0.135$. In both #4 and #5 there is evidence of systematic departure on this flatter slope, as if it is slightly concave upwards (Figure 30).

Hopkins' stand #6. Pine woods at Rannoch, Perthshire. This set of points is successfully graduated with 2 straight lines, but here the upper slope is the steeper: $k = 0.159$ changing to $k = 0.229$. This shows that the curve can be "concave" upwards or downwards.

Hopkins' stand # 11. Blanket bog in the Pennine uplands, England: 46 species. Like the other bogs, this is initially a good straight line (from 1 cm^2 to 25 m^2) changing abruptly then to a flatter slope, defined only by 3 points. The initial slope, valid over 5½ orders of magnitude gives $k = 0.218$.

Thus our k values come out as follows:

	N_{max}	k
Grassland (#2)	55	0.193
Woodland (#3, Beech)	28	0.327, less at low areas
Woodland (#6, Pine)	42	0.229, less at low areas, where k = 0.159
Bog (#4)	47	0.245, less at greater areas, where k = 0.097
Bog (#5)	48	0.203, less at greater areas, where k = 0.135
Bog (#11)	46	0.218, less at greater areas

The average slope of the more trustworthy-looking sections of the curves is $k = 0.24$, a little above Williams (1943b) and not far below the theoretical value for a series of isolates, but well below it if we take account of the flatter slopes frequently present.

If we can interpret these curves at all, it seems as though they tend on the average to come close to being straight lines. Some are quite straight, the 2 woodland areas are concave upwards and the 3 bogs are concave downwards. The average, in fact, of all 12 of Hopkins' stands is very close to

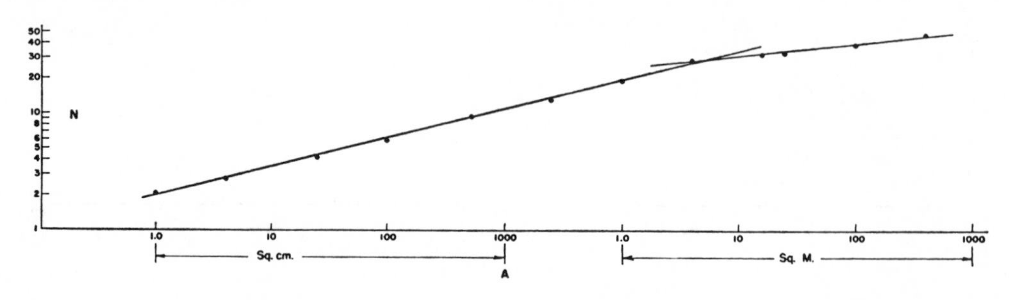

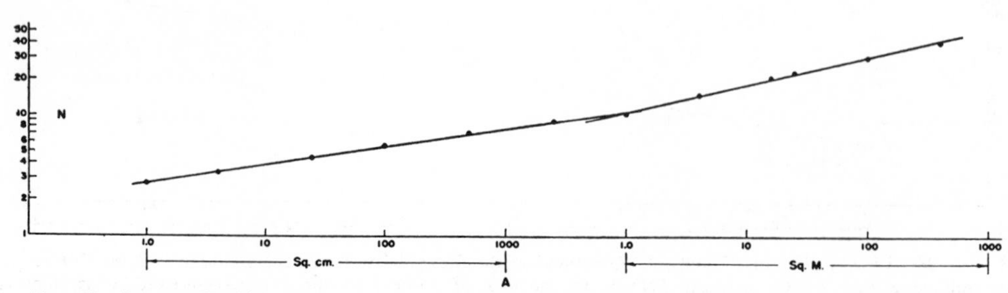

FIG. 29. Lower, Hopkins, Stand #6, Pine Woods. Perthshire. Species-Area Curve. Upper, Hopkins, Stand #4, Blanket Bog, County Mayo.

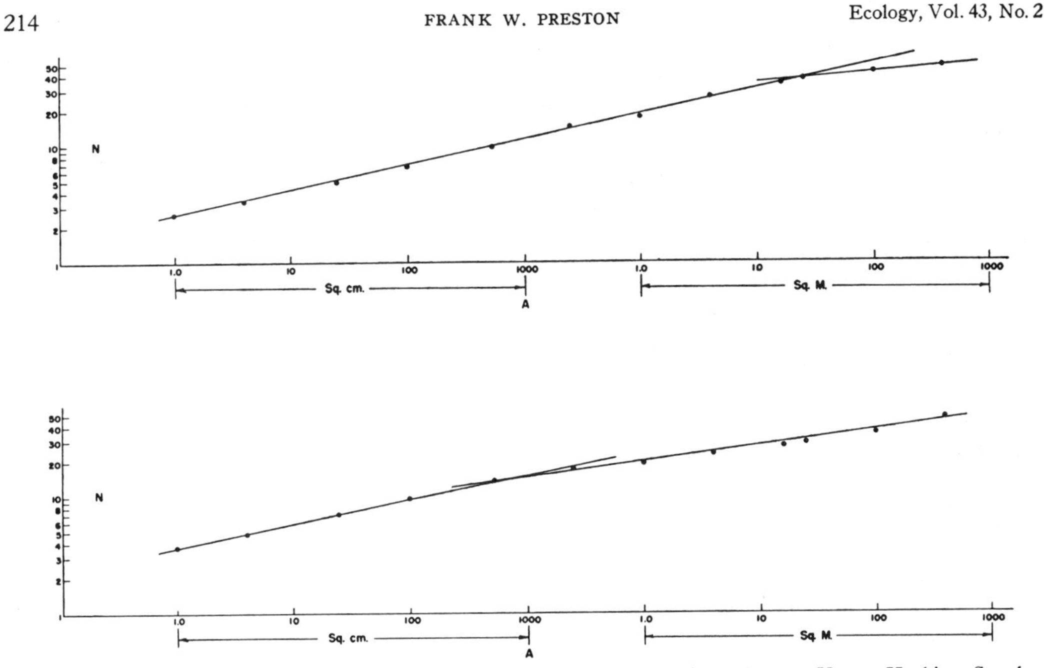

FIG. 30. Lower, Hopkins, Stand #5, Bog in Perthshire. Species-Area Curve. Upper, Hopkins, Stand #11, Blanket Bog, Pennine Uplands.

a straight line over nearly 6 orders of magnitude, with an exponent of about $k = 0.16$.

It would seem possible that bogs in Britain may have a limited flora, and that by the time we reach a few square meters in area we are beginning to exhaust the flora. On the other hand, woodlands there seem to get off to a slow start, and begin to show what they can do about the time we decide, at a hundred square meters or so, to call off our survey. The grassland seems to strike a very happy balance between the two.

It should be noticed that all samples are "small" in number of species and, therefore, positive or negative contagion can markedly affect the slope which can assume a new value, higher or lower (depending on the sign of the contagion), when the contagion begins to smooth out. However that is not the only thing that can bring about a change in slope since, as we have seen earlier, a constant slope for a long distance should imply, in a lognormal ensemble, that the universe is expanding as fast as the sample. To expect it to keep exact pace with the sample for any great distance seems unreasonable. Thus the curve may steepen and flatten from time to time, and one of Hopkins' stands, his #8, actually appears to do this. The rest seem to exhibit a single change, usually somewhat abrupt, but in one case gradual.

The interpretation I tentatively put upon these results, and upon the fact that most of the examples have an index or exponent of k ($= 1/n$) between about 0.15 and 0.24, is that from each area we have a sample of 60% to 80% of this species in the "universe" we are instantaneously sampling.

Preston. Birds of the nearctic and neotropical regions

Lest the botanists think they have a monopoly on problems dealing with species-area curves, we may note that the birds of 2 major zoogeographic regions produce similar results. Over a very wide range of areas, we found (Preston 1960) that the index k for the nearctic was around 0.12 and for the neotropical was about 0.16, and again we interpreted this to mean that after we had "collected" some 70% or 80% of the universe we were sampling, the universe started to expand pari-passu with our further observations.

Vestal's sigmoid

In 1949 Vestal in the U.S.A. and Archibald in Britain (quoted by Hopkins 1955) reached the conclusion that in the case of vegetation stands there was a tendency for the species-log area (Gleason) curve to be sigmoid; it began at a low slope, steepened considerably, and then became less steep. Hopkins (1955) is very dubious about the reality of this. I see no reason why the effect should not sometimes exist; indeed if the dominant factor in controlling the slope is the phenomenon of the "expanding universe," then failure of the universe and sample to keep exact step must produce steepenings and flattenings of the curve which can sometimes take the rather simple form of sigmoids, or S-shaped curves.

The evidence as presented by Vestal and by Archibald (or Hopkins 1955, Fig. 13) is on a Gleason-plot basis, but when the range of area is modest, the same phenomenon would appear in a log-log (Arrhenius) plot. Archibald's curve however covers a wide range. The possibility that such curves may exist can hardly be disputed on theoretical grounds; how often they occur in practice is a matter for observation.

We may note, however, that when the curve gets off to a slow start, an almost horizontal line at or near one species per quadrat, suggests that the "universe" is nearly a pure stand. We must expect that sooner or later, as we expand our quadrats, there will be a marked steepening of slope showing that the stand is not a pure one. This happened with Hopkins' example #3, our Figure 28, a woodland, and when the curve steepened it acquired so high a slope that it is doubtful if it could keep it up for long. If the quadrats could have been extended a few more orders of magnitude (which may not have been physically possible in recent centuries) it is almost certain that the slope would have flattened again to some important extent, and then we should have had a Vestal sigmoid.

All the species-area curves here replotted from Hopkins' data are plotted to the same scale, and if we superpose them we get Figure 31. It is difficult to resist the speculation that if we had data beyond the experimental limit of 400 m^2, say up to 10^6 m^2, we should find all the curves following closely a single line with a slope or index of 0.169, and indicating that at 1 km^2 the flora of the British Isles tends to amount to 250 species or thereabouts.

If the same index continued valid up to 2000 km^2 (= 800 square miles), the area of such a county as Leicestershire, the number of species should be about 880. An earlier count gave 890, but Horwood and Gainsborough (1933) report a total of about 1400, including, however, the "aliens," which are numerous.

The curves are actually closer together at the upper limit of observation (400 square meters) than they are at 1 cm^2, and much closer than at 1 m^2. It is not till we get above 100 m^2 that all the curves represent 20 species or more, and this suggests that the divergences below this area are largely the properties of samples that are statistically too small. This seems to be an almost universal property, or shortcoming, of botanical quadrats.

Summary

We see then that "samples" have some properties in common with isolates, and in some respects they differ radically from them. One of the most important of these differences is the slope of the Arrhenius plot. For isolates it is theoretically about 0.26 to 0.28 depending on the number of species involved, and we found that in practice it is often somewhere near this figure. For samples it is much less, sometimes no more than half this figure. This leads to an important zoogeographical conclusion to be discussed in the next section.

—To Be Continued—

Literature Cited and Acknowledgments will be found at the end of Part II of this paper.

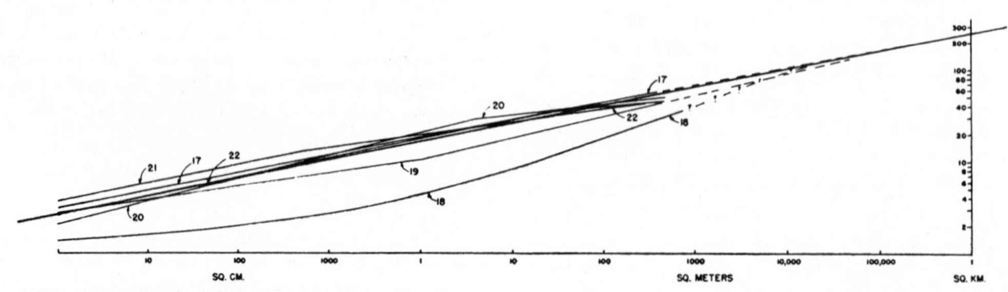

FIG. 31. Hopkins' 6 curves superposed. The heavier line that projects beyond both ends may be a sort of average to which all 6 ultimately trend.

bution of individuals on its own soil, and yet the isolation is sufficiently imperfect that the island is in equilibrium with other nearby islands, then z approximates to 0.27. This last value holds for the (families of) birds of the major Sclater-Wallace Zoogeographical regions, including the Australasian region. The value for mammal families is higher, around 0.4, indicating that mammals are more easily isolated than birds, and reaches a value around 0.9 for the Australian mammal fauna.

9. The extirpation and extinction of species can be understood as a purely mathematical matter, as a property of the fluctuating size of habitable areas. Subspeciation and speciation can similarly be understood, since there is a definite relation between number of species and number of individuals, or between species and area.

10. The mechanism, or possibly the biology, by which the adjustment of total species to total area is brought about is considered briefly in terms of the interpenetration of species and the adjustment of "niches."

11. A suspicion exists that a lognormal, canonical, distribution of individuals among species may in some fashion be analogous to Pareto's law of the natural distribution of wealth among the individuals of a human community. This argument will not appeal to everyone, and is at present somewhat nebulous.

As we emphasized at the outset, this paper is not so much an attempt to account for the facts of nature as an attempt to trace the consequences of a hypothesis. These consequences we can then compare with the facts of nature, and in many cases we seem to find good, and often interesting, agreement. Thus the original hypothesis may have some practical utility. It seems to me practically certain that there is a better way of formulating the hypothesis in purely algebraic language, and this might lead to other interesting conclusions and perhaps to a coherent theory embracing many ecological phenomena.

References

Archibald, E. E. A. 1949. The specific character of plant communities. J. Ecol. **37**: 260-274.

de Beaufort, L. F. 1951. Zoogeography of the land and inlet waters. London, Sedgwick and Jackson.

Beebe, W. 1924. Galapagos, world's end. New York, Putnam.

Belknap, J. B. 1951. Census of breeding birds on an island in an open lake. Audubon Field Notes **5**: 327-328.

Benson, C. W. 1960. Birds of the Comoro Islands. Ibis: 103b.

Bond, J. 1936. Birds of the West Indies Baltimore, Waverly Press.

———. 1956. Check-list of the birds of the West Indies. Academy of Natural Sciences of Philadelphia.

Buell, M. F. and J. Cantlon. 1951. A study of two forest stands in Minnesota with an interpretation of the prairie-forest margin. Ecology **32**: 294-316.

Burt, W. A. 1958. History and affirmities of recent land mammals of Western North America. Zoogeography, Am. Assoc. for Advancement of Science. Pub. 51.

Cain, S. 1938. The species area curve. Am. Midland Naturalist **19**: 573-581.

———**, and G. M. Castro.** 1959. Manual of vegetation analysis. New York, Harper.

Cole, L. C. 1946. A theory for analysing contagiously distributed populations. Ecology **27**: 329-341.

Darlington, P. J. Jr. 1957. Zoogeography. New York, John Wiley Sons.

Fisher, J. 1952. Bird numbers. South Eastern Naturalist and Antiquary **57**: 1-10.

Gibbs, J. W. 1931. Collected works. New York, London, and Toronto: Longmans, Green and Co.

Hairston, N. G. 1959. Species abundance and community organization. Ecology **40**: 404-416.

Hald, A. 1952. Statistical theory with engineering applications. New York. John Wiley & Sons.

Hatt, R. T., J. Van Tyne, L. C. Stuart, C. H. Pope, A. B. Grobman. 1948. Island life in Lake Michigan. Bull. 27, Cranbrook Institute of Science.

Henderson, N. 1956. A blackbird roost in Oberlin. Bull. Cleveland Audubon Soc. **2**: 13-14.

Hicks, L. E. 1935. A ten year study of a breeding bird population in central Ohio. Am. Midland Naturalist **16**: 177-186.

Hopkins, B. 1955. The Species-area relations of plant communities. J. Ecology **43**: 409-426.

———. 1957. The concept of minimal Area. J. Ecology **45**: 441-449.

Horwood, A. R. and C. W. F. Noel. 1933. The flora of Leicestershire and Rutland. Oxford University Press.

Kendeigh, S. C. 1946. Breeding birds of the beech-maple-hemlock community. Ecology **27**: 226-245.

Kerr, J. G. 1950. A naturalist in the Gran Chaeo. Cambridge University Press.

Kohn, A. J. 1959. The ecology of *Conus* in Hawaii. Ecol. Monographs **29**: 47-90.

Kroeber, A. L. 1916. Floral relations among the Galapagos Islands. Berkeley, University of California press.

Lowan, A. A. 1942. Tables of probability functions. Vol. II. National Bureau of Standards.

MacArthur, R. H. 1957. On the relative abundance of bird species. Proc. Nat. Acad. Sci. **43**: 293-295.

Martin, P. 1958. Pleistocene ecology and biogeography of North America. Zoogeography, AAAS.

Mayr, E. 1944. The birds of Timor and Sumba. Bull. Amer. Museum Nat. Hist. **83**: 123-194.

Merikallio, E. 1958. Finnish birds, their distribution and numbers. Societas pro Fauna et Flora Fennica, Helsinki.

Oosting, H. 1942. An ecological analysis of the plant communities of Piedmont, North Carolina, Amer. Midland Naturalist **28**: 1-26.

Ferner, M. 1955. A breeding bird population in North Chagrin Metropolitan Park, Cleveland, Ohio. Bull. Cleveland Audubon Soc. Vol. 2, No. 3.

Preston, F. W. 1948. The commonness, and rarity, of species. Ecology **29**: 254-283.

———. 1950. Gas laws and wealth laws. Scientific Monthly **71**: 309-311.

———. 1957. Analysis of Maryland statewide bird counts, Maryland Birdlife (Bull. Maryland Ornithological Soc.) **13**: 63-65.

———. 1958. Analysis of the Audubon Christmas bird counts in terms of the lognormal curve. Ecology **39**: 620-624.

———. 1960. Time and space and the variation of species. Ecology **41**: 785-790.

Rand, A. L. 1936. The distribution and habits of Madagascar birds. Bull. Amer. Museum Nat. Hist. **72**: 143-499.

Rensch, B. 1959. Evolution above the species level. London, Methuen.

Roosevelt, T. 1910. African game trails. New York, Charles Scribner's Sons.

Saunders, A. A. 1936. Ecology of the birds of Quaker Run Valley, Allegany State Park, New York.

Simpson, G. G. 1940. Mammals and land bridges. J. Washington Acad. Sci. p. 158.

———. 1957. The principles of classification and a clasification of mammals. Bull. Amer. Museum Nat. Hist. 85.

———. 1960. Notes on the measurement of faunal resemblance. Am. J. Science. Bradley Volume **258A**: 300-311.

Vestal, A. G. 1949. Minimum areas for different vegetations. Illinois Biological Monographs XX.

Walkinshaw, L. H. 1947. Breeding birds of brushy field, woodlots and pond (10 year summary). Audubon Field Notes **1**: 214-216.

Wallace, A. R. 1876. The geographical distribution of animals. London, MacMillan.

Wetmore, A. 1949. Ornithology. Encyclopedia Britannica.

Williams, A. B. 1947. Breeding birds of a climax beech-maple forest, 15 year summary. Audubon Feld Notes **1**: 205-210.

Williams, C. B. 1943b. Area and number of species. Nature **152**: 264-267.

———. 1953. The Abundance of species in a world population. J. Animal Ecology **22**: 14-31.

Yule, G. U. 1925. A mathematical theory of evolution based on the conclusion of J. C. Willis. Phil. Trans. Royal Soc. London **B. 213**: 21-87.

ACKNOWLEDGMENTS

I am indebted to many friends for help with this paper. In some cases their names appear in the text at appropriate places, but in others the help is diffused through a large part of the text. My thanks to Prof. Henry Oosting, Prof. Murray Buell, Prof. Stanley A. Cain, and Dr. O. E. Jennings for help with botanical points; to Mr. Harold Mayfield, Dr. E. S. Thomas and Dr. Milton B. Trautman, Prof. Robert MacArthur and Dr. Alan J. Kohn, Prof. Ernst Mayr, Mr. W. E. Clyde Todd, and Dr. Kenneth C. Parkes for information on zoological matters; to Dr. C. B. Williams and Prof. LaMont C. Cole for help on ecological points. I am indebted to Prof. J. L. Glathart of Albion, Michigan, and Mr. Russell D. Southwick of American Glass Research, Inc. for reading the text, to Prof. W. Burt and Prof. G. G. Simpson for reading parts of it, to Mr. R. R. Lehnerd for help with the figures, and to my secretaries Mrs. John D. Povlick, Jr. and Miss Marjorie L. Swallop for many computations and much typing. And not least I am indebted to a couple of anonymous reviewers for valuable suggestions in respect both of this paper and its companion, "Time and space and the variation of species."

Concluding Remarks

G. EVELYN HUTCHINSON

Yale University, New Haven, Connecticut

This concluding survey[1] of the problems considered in the Symposium naturally falls into three sections. In the first brief section certain of the areas in which there is considerable difference in outlook are discussed with a view to ascertaining the nature of the differences in the points of view of workers in different parts of the field; no aspect of the Symposium has been more important than the reduction of areas of dispute. In the second section a rather detailed analysis of one particular problem is given, partly because the question, namely, the nature of the ecological niche and the validity of the principle of niche specificity has raised and continues to raise difficulties, and partly because discussion of this problem gives an opportunity to refer to new work of potential importance not otherwise considered in the Symposium. The third section deals with possible directions for future research.

THE DEMOGRAPHIC SYMPOSIUM AS A HETEROGENEOUS UNSTABLE POPULATION

In the majority of cases the time taken to establish the general form of the curve of growth of a population from initial small numbers to a period of stability or of decline is equivalent to a number of generations. If, as in the case of man, the demographer is himself a member of one such generation, his attitude regarding the nature of the growth is certain to be different from that of an investigator studying, for instance, bacteria, where the whole process may unfold in a few days, or insects, where a few months are required for several cycles of growth and decline. This difference is apparent when Hajnal's remarks about the uselessness of the logistic are compared with the almost universal practice of animal demographers to start thinking by making some suitable, if almost unconscious, modification of this much abused function.

[1] I wish to thank all the participants for their kindness in sending in advance manuscripts or information relative to their contributions. All this material has been of great value in preparing the following remarks, though not all authors are mentioned individually. Where a contributor's name is given without a date, the reference is to the contribution printed earlier in this volume. I am also very much indebted to the members (Dr. Jane Brower, Dr. Lincoln Brower, Dr. J. C. Foothills, Mr. Joseph Frankel, Dr. Alan Kohn, Dr. Peter Klopfer, Dr. Robert MacArthur, Dr. Gordon A. Riley, Mr. Peter Wangersky, and Miss Sally Wheatland) of the Seminar in Advanced Ecology, held in this department during the past year. Anything that is new in the present paper emerged from this seminar and is not to be regarded specifically as an original contribution of the writer.

The human demographer by virtue of his position as a slow breeding participant observer, and also because he is usually called on to predict for practical purposes what will happen in the immediate future, is inevitably interested in what may be called the microdemography of man. The significant quantities are mainly second and third derivatives, rates of change of natality and mortality and the rates of change of such rates. These latter to the animal demographer might appear as random fluctuations which he can hardly hope to analyse in his experiments. What the animal demographer is mainly concerned with is the macrodemographic problem of the integral curve and its first derivative. He is accustomed to dealing with innumerable cases where the latter is negative, a situation that is so rare in human populations that it seems to be definitely pathological to the human demographer. Only when anthropology and archaeology enter the field of human demography does something comparable to animal demography, with its broad, if sometimes insufficiently supported generalisations and its fascinating problems of purely intellectual interest, emerge. From this point of view the papers of Birdsell and Braidwood are likely to appeal most strongly to the zoologist, who may want to compare the rate of spread of man with that considered by Kurtén (1957) for the hyena.

It is quite likely that the difference that has just been pointed out is by no means trivial. The environmental variables that affect fast growing and slow growing populations are likely to be much the same, but their effect is qualitatively different. Famine and pestilence may reduce human populations greatly but they rarely decimate them in the strict sense of the word. Variations, due to climatic factors, of insect populations are no doubt often proportionately vastly greater. A long life and a long generation period confer a certain homeostatic property on the organisms that possess them, though they prove disadvantageous when a new and powerful predator appears. The elephant and the rhinoceros no longer provide models of human populations, but in the early Pleistocene both may have done so. The rapid evolution of all three groups in the face of a long generation time is at least suggestive.

It is evident that a difference in interest may underlie some of the arguments which have enlivened, or at times disgraced, discussions of this subject. Some of the most significant modern work has arisen from an interest in extending the concepts of the struggle for existence put forward as an evolutionary mechanism by Darwin practically a century ago. Such work, of which Lack's recent contributions provide a distinguished example, tends to concentrate on relatively stable interacting populations in as undisturbed communition as possible. Another fertile field of research has been provided by the sudden increases in numbers of destructive animals, often after introduction or disturbance of natural environments. Here more than one point of view has been apparent. Where emphasis has been on biological control, that is, a conscious rebuilding of a complex biological association, a view point not unlike that of the evolutionist has emerged—where emphasis has been placed on the actual events leading to a very striking increase or decrease in abundance, given the immediate ecological conditions, the latter have appeared to be the most significant variables. Laboratory workers have moreover tended to keep all but a few factors constant, and to vary these few systematically. Field workers have tended to emphasize the ever changing nature of the environment. It is abundantly clear that all these points of view are necessary to obtain a complete picture. It is also very likely that the differences in initial point of view are often responsible for the differences in the interpretation of the data.

The initial differences of point of view are not the only difficulty. In the following section an analysis of a rather formal kind of one of the concepts frequently used in animal ecology, namely that of the *niche*, is attempted. This analysis will appear to some as compounded of equal parts of the obvious and the obscure. Some people however may find when they have worked through it, provided that it is correct, that some removal of irrelevant difficulties has been achieved. It is not necessary in any empirical science to keep an elaborate logicomathematical system always apparent, any more than it is necessary to keep a vacuum cleaner conspicuously in the middle of a room at all times. When a lot of irrelevant litter has accumulated the machine must be brought out, used, and then put away. It might be useful for those who argue that the word environment should refer to the environment of a population, and those who consider it should been the environment of an organism, to use the word both ways for a couple of months, writing "environment" when a single individual is involved, "Environment" when reference is to a population. In what follows the term will as far as possible not be used, except in the non-committal adjectival form environmental, meaning any property outside the organisms under consideration.

The Formalisation of the Niche and the Volterra-Gause Principle

Niche space and biotop space

Consider two independent environmental variables x_1 and x_2 which can be measured along ordinary rectangular coordinates. Let the limiting values permitting a species S_1 to survive and reproduce be respectively x'_1, x''_1 for x_1 and x'_2, x''_2 for x_2. An area is thus defined, each point of which corresponds to a possible environmental state permitting the species to exist indefinitely. If the variables are independent in their action on the species we may regard this area as the rectangle ($x_1 = x'_1$, $x_1 = x''_1$, $x_2 = x'_2$, $x_2 = x''_2$), but failing such independence the area will exist whatever the shape of its sides.

We may now introduce another variable x_3 and obtain a volume, and then further variables $x_4 \ldots x_n$ until all of the ecological factors relative to S_1 have been considered. In this way an n-dimensional hypervolume is defined, every point in which corresponds to a state of the environment which would permit the species S_1 to exist indefinitely. For any species S_1, this hypervolume $\mathbf{N}_1$ will be called the *fundamental niche*[2] of S_1. Similarly for a second species S_2 the fundamental niche will be a similarly defined hypervolume $\mathbf{N}_2$.

It will be apparent that if this procedure could be carried out, all X_n variables, both physical and biological, being considered, the fundamental niche of any species will completely define its ecological properties. The fundamental niche defined in this way is merely an abstract formalisation of what is usually meant by an ecological niche.

As so defined the fundamental niche may be regarded as a set of points in an abstract n-dimensional $\mathbf{N}$ space. If the ordinary physical space $\mathbf{B}$ of a given biotop be considered, it will be apparent that any point $p(\mathbf{N})$ in $\mathbf{N}$ can correspond to a number of points $p_i(\mathbf{B})$ in $\mathbf{B}$, at each one of which the conditions specified by $p(\mathbf{N})$ are realised in $\mathbf{B}$. Since the values of the environmental variables $x_1\ x_2 \ldots x_n$ are likely to vary continuously, any subset of points in a small elementary volume $\Delta\mathbf{N}$ is likely to correspond to a number of small elementary volumes scattered about in $\mathbf{B}$. Any volume $\mathbf{B}'$ of the order of the dimensions of the mean free paths of any animals under consideration is likely to contain points corresponding to points in various fundamental niches in $\mathbf{N}$.

Since $\mathbf{B}$ is a limited volume of physical space comprising the biotope of a definite collection of species S_1, $S_2 \cdots S_n$, there is no reason why a given point in $\mathbf{N}$ should correspond to any points in $\mathbf{B}$. If, for any species S_1, there are no points in

[2] This term is due to MacArthur. The general concept here developed was first put forward very briefly in a footnote (Hutchinson, 1944).

B corresponding to any of the points in $\mathbf{N}_1$, then **B** will be said to be *incomplete* relative to S_1. If some of the points in $\mathbf{N}_1$ are represented in **B** then the latter is *partially incomplete* relative to S_1, if all the points in $\mathbf{N}_1$ are represented in **B** the latter is *complete* relative to S_1.

Limitations of the set-theoretic mode of expression. The following restrictions are imposed by this mode of description of the niche.

1. It is supposed that all points in each fundamental niche imply equal probability of persistance of the species, all points outside each niche, zero probability of survival of the relevant species. Ordinarily there will however be an optimal part of the niche with markedly suboptimal conditions near the boundaries.

2. It is assumed that all environmental variables can be linearly ordered. In the present state of knowledge this is obviously not possible. The difficulty presented by linear ordering is analogous to the difficulty presented by the ordering of degrees of belief in non-frequency theories of probability.

3. The model refers to a single instant of time. A nocturnal and a diurnal species will appear in quite separate niches, even if they feed on the same food, have the same temperature ranges etc. Similarly, motile species moving from one part of the biotop to another in performance of different functions may appear to compete, for example, for food, while their overall fundamental niches are separated by strikingly different reproductive requirements. In such cases the niche of a species may perhaps consist of two or more discrete hypervolumes in **N**. MacArthur proposed to consider a more restricted niche describing only variables in relation to which competition actually occurs. This however does not abolish the difficulty. A formal method of avoiding the difficulty might be derived, involving projection onto a hyperspace of less than n-dimensions. For the purposes for which the model is devised, namely a clarification of niche-specificity, this objection is less serious than might at first be supposed.

4. Only a few species are to be considered at once, so that abstraction of these makes little difference to the whole community. Interaction of any of the considered species is regarded as competitive in sense 2 of Birch (1957), negative competition being permissible, though not considered here. All species other than those under consideration are regarded as part of the coordinate system.

Terminology of subsets. If $\mathbf{N}_1$ and $\mathbf{N}_2$ be two fundamental niches they may either have no points in common in which case they are said to be *separate*, or they have points in common and are said to *intersect.*

In the latter case:

$(\mathbf{N}_1 - \mathbf{N}_2)$ is the subset of $\mathbf{N}_1$ of points not in $\mathbf{N}_2$

$(\mathbf{N}_2 - \mathbf{N}_1)$ is the subset of $\mathbf{N}_2$ of points not in $\mathbf{N}_1$

$\mathbf{N}_1 \cdot \mathbf{N}_2$ is the subset of points common to $\mathbf{N}_1$ and $\mathbf{N}_2$, and is also referred to as the *intersection subset.*

Definition of niche specificity. Volterra (1926, see also Lotka 1932) demonstrated by elementary analytic methods that under constant conditions two species utilizing, and limited by, a common resource cannot coexist in a limited system.[3] Winsor (1934) by a simple but elegant formulation showed that such a conclusion is independent of any kind of finite variations in the limiting resource. Gause (1934, 1935) confirmed this general conclusion experimentally in the sense that if the two species are forced to compete in an undiversified environment one inevitably becomes extinct. If there is a diversification in the system so that some parts favor one species, other parts the other, the two species can coexist. These findings have been extended and generalised to the conclusion that two species, when they co-occur, must in some sense be occupying different niches. The present writer believes that properly stated as an empirical generalisation, which is true except in cases where there are good reasons not to expect it to be true,[4] the principle is of fundamental importance and may be properly called the Volterra-Gause Principle. Some of the confusion surrounding the principle has arisen from the concept of two species not being able to co-occur when they occupy identical niches. According to the formulation given above, identity of fundamental niche would imply $\mathbf{N}_1 = \mathbf{N}_2$, that is, every point of $\mathbf{N}_1$ is a member of $\mathbf{N}_2$ and every point of $\mathbf{N}_2$ a member of $\mathbf{N}_1$. If the two species S_1 and S_2 are indeed valid species distinguishable by a systematist and not freely interbreeding, this is so unlikely that the case is of no empirical interest. In terms of the set-theoretic presentation, what the Volterra-Gause principle meaningfully states is that for any small element of the intersection subset $\mathbf{N}_1 \cdot \mathbf{N}_2$, there do not exist in **B** corresponding small parts, some inhabited by S_1, others by S_2.

Omitting the quasi-tautotogical case of $\mathbf{N}_1 = \mathbf{N}_2$, the following cases can be distinguished.

(1) $\mathbf{N}_2$ is a proper subset of $\mathbf{N}_1$ ($\mathbf{N}_2$ is "inside" $\mathbf{N}_1$)
 (a) competition proceeds in favor of S_1 in all the elements of **B** corresponding to $\mathbf{N}_1 \cdot \mathbf{N}_2$; given adequate time only S_1 survives.
 (b) competition proceeds in favor of S_2 in all elements of **B** corresponding to some part of the intersection subset and both species survive.

(2) $\mathbf{N}_1 \cdot \mathbf{N}_2$ is a proper subset of both $\mathbf{N}_1$ and $\mathbf{N}_2$; S_1 survives in the parts of **B** space

[3] I regret that I am unable to appreciate Brian's contention (1956) that the Volterra model refers only to interference, and the Winsor model to exploitation.

[4] *cf.* Schrödinger's famous restatement of Newton's First Law of Motion, that a body perseveres at rest or in uniform motion in a right line, except when it doesn't.

corresponding to ($\mathbf{N}_1 = \mathbf{N}_2$), S_2 in the parts corresponding to ($\mathbf{N}_2 = \mathbf{N}_1$), the events in $\mathbf{N}_1 = \mathbf{N}_2$ being as under I, with the proviso that no point in $\mathbf{N}_1 \cdot \mathbf{N}_2$ can correspond to the survival of both species.

In this case the two difference subsets ($\mathbf{N}_1 - \mathbf{N}_2$) and ($\mathbf{N}_2 - \mathbf{N}_1$) are, in Gause's terminology, refuges for S_1 and S_2 respectively.

If we define the realised niche $\mathbf{N}'_1$ of S_1 in the presence of S_2 as ($\mathbf{N}_1 - \mathbf{N}_2$), if it exists, plus that part of $\mathbf{N}_1 \cdot \mathbf{N}_2$ as implies survival of S_1, and similarly the realised niche $\mathbf{N}'_2$ of S_2 as ($\mathbf{N}_2 - \mathbf{N}_1$), if it exists, plus that part of $\mathbf{N}_1 \cdot \mathbf{N}_2$ corresponding to survival of S_2, then the Volterra-Gause principle is a statement of an empirical generalisation, which may be verified or falsified, that realised niches do not intersect. If the generalisation proved to be universally false, the falsification would presumably imply that in nature resources are never limiting.

Validity of the Gause-Volterra Principle. The set-theoretic approach outlined above permits certain refinements which, however obvious they may seem, apparently require to be stated formally in an unambiguous way to prevent further confusion. This approach however tells us nothing about the validity of the principle, but merely where we should look for its verification or falsification.

Two major ways of approaching the problem have been used, one experimental, the other observational. In the experimental approach, the method (*e.g.* Gause, 1934, 1935; Crombie, 1945, 1946, 1947) has been essentially to use animal populations as elements in analogue computers to solve competition equations. As analogue computers, competing populations leave much to be desired when compared with the more conventional electronic machines used for instance by Wangersky and Cunningham. At best the results of laboratory population experiments are qualitatively in line with theory when all the environmental variables are well controlled. In general such experiments indicate that where animals are forced by the partial incompleteness of the **B** space to live in competition under conditions corresponding to a small part of the intersection subset, only one species survives. They also demonstrate that the identity of the survivor is dependent on the environmental conditions, or in other words on which part of the intersection subset is considered, and that when deliberate niche diversification is brought about so that at least one non-intersection subset is represented in **B**, two species may co-occur indefinitely. It would of course be most disturbing if confirmatory models could not be made from actual populations when considerable trouble is taken to conform to the postulates of the deductive theory.

The second way in which confirmation has been sought, namely by field studies of communities consisting of a number of allied species also lead to a confirmation of the theory, but one which may need some degree of qualification. Most work has dealt with pairs of species, but the detailed studies on *Drosophila* of Cooper and Dobzhansky (1956) and of Da Cunha, El-Tabey Shekata and de Olivera (1957), to name only two groups of investigators, the investigation of about 18 species of *Conus* on Hawaiian reef and littoral benches (Kohn, in press) and the detailed studies of the food of six co-occurring species of *Parus* (Betts, 1955) indicate remarkable cases among many co-occurring species of insects, mollusks and birds respectively. However much data is accumulated there will almost always be unresolved questions relating to particular species, though the presumption from this sort of work is that, in any large group of sympatric species belonging to a single genus or subfamily, careful work will always reveal ecological differences. The sceptic may reply in two ways, firstly pointing out that the quasi-tautological case of $\mathbf{N}_1 = \mathbf{N}_2$ has already been dismissed as too improbable to be of interest, and that when a great deal of work has to be done to establish the difference, we are getting as near to niche identity as is likely in a probabilistic world. Occasionally it may be possible to use indirect arguments to show that the differences are at least evolutionarily significant. Lack (1947b) for instance points out that in the Galapagos Islands, among the heavy billed species of *Geospiza*, where both *G. fortis* Gould, and *G. fuliginosa* Gould co-occur on an island, there is a significant separation in bill size, but where either species exists alone, as on Crossman Island and Daphne Island the bills are intermediate and presumably adapted to eating modal sized food. This is hard to explain unless the small average difference in food size believed to exist between sympatric *G. fortis* and *G. fuliginosa* is actually of profound ecological significance. The case is particularly interesting as most earlier authors have dismissed the significance of the small alleged differences in the size of food taken by the species. Few cases of specific ecological difference encountered outside *Geospiza* would appear at first sight so tenuous as this.

A more important objection to the Volterra-Gause principle may be derived from the extreme difficulty of identifying competition as a process actually occurring in nature. Large numbers of cases can of course be given in which there is very strong indirect evidence of competitive relationships between species actually determining their distribution. A few examples may be mentioned. In the British Isles (Hynes, 1954, 1955) the two most widespread species of *Gammarus* in freshwater are *Gammarus deubeni* Lillj: and *G. pulex* (L.). The latter is the common species in England and most of the mainland of Scotland, the former is found exclusively in Ireland, the Shetlands, Orkneys and most of the other Scottish Islands and in Cornwall. On northern mainland Scotland only

G. lacustris Sars is found. Both *deubeni* and *pulex* occur on the Isle of Man and in western Cornwall. Only in the Isle of Man have the two species been taken together. It is extremely probable that *pulex* is a recent introduction to that island. *G. deubeni* is well known in brackish water around the whole of northern Europe. It is reasonable to suppose that the fundamental niches of the two species overlap, but that within the overlap *pulex* is successful, while *deubeni* with a greater tolerance of salinity has a refuge in brackish water. Hynes moreover shows that *G. pulex* has a biotic (reproductive) potential two or three times that of *deubeni* so that in a limited system inhabitable by both species, under constant conditions *deubeni* is bound to be replaced by *pulex*. This case is as clear as one could want except that Hynes is unable to explain the absence of *G. deubeni* from various uninhabited favorable localities in the Isle of Man and elsewhere. Hynes also notes that Steusloff (1943) had similar experiences with the absence of *Gammarus pulex* in various apparently favorable German localities. Ueno (1934) moreover pointed out that *Gammarus pulex* (*sens. lat.*) occurs abundantly in Kashmir up to 1600 meters, and is an important element in the aquatic fauna of the Tibetan highlands to the east above 3800 miles, but is quite absent in the most favorable localities at intermediate altitudes. These disconcerting empty spaces in the distribution of *Gammarus* may raise doubts as to the completeness of the picture presented in Hynes' excellent investigations.

Another very well analysed case (Dumas, 1956) has been recently given for two sympatric species of *Plethodon*, *P. dunni* Bishop, and *P. vehiculum* (Cooper), in the Coastal Ranges of Oregon. Here experiments and field observations both indicate that *P. dunni* is slightly less tolerant of low humidity and high temperature than is *P. vehiculum*, but when both co-occur *dunni* can exclude *vehiculum* from the best sites. However under ordinary conditions in nature the number of unoccupied sites which appear entirely suitable is considerable, so that competition can not be limiting except in abnormally dry years.

In both these cases, which are two of the best analysed in the literature, the extreme proponent of the Volterra-Gause principle could argue that if the investigator was equipped with the sensory apparatus of *Gammarus* or *Plethodon* he would know that the supposedly suitable unoccupied sites were really quite unsuitable for any self respecting member of the genus in question. This however is pure supposition.

Even in the rather conspicuous case of the introduction of *Sciurus caroliniensis* Gmelin and its spread in Britain, the popular view that the bad bold invader has displaced the charming native *S. vulgaris leucourus* Kerr, is apparently mythological. Both species are persecuted by man; *S. caroliniensis* seems to stand this persecution better than does the native red squirrel and therefore tends to spread into unoccupied area from which *S. vulgaris leucourus* has earlier retreated (Shorten, 1953, 1954).

Andrewartha (see also Andrewartha and Birch, 1954) has stressed the apparent fact that while most proponents of the competitive organisation of communities have emphasised competition for food, there is in fact normally more than enough food present. This appears, incidentally, most strikingly in some of Kohn's unpublished data on the genus *Conus*.

The only conclusion that one can draw at present from the observations is that although animal communities appear qualitatively to be constructed as if competition were regulating their structure, even in the best studied cases there are nearly always difficulties and unexplored possibilities. These difficulties suggest that if competition is determinative it either acts intermittently, as in abnormally dry seasons for *Plethodon*, or it is a more subtle process than has been supposed. Thus Lincoln Brower (*in press*) investigating a group of species of North American *Papilio* in which one eastern polyphagous species is replaced by three western oligophagous species, has been impressed by the lack of field evidence for any inadequacy in food resources. He points out however, that specific separation of food might lower the probability of local high density on a given plant, and so the risk of predation by a bird that only stopped to feed when food was abundant (*cf.* de Ruiter, 1952).

Unfortunately there is no end to the possible erection of hypothesis fitted to particular cases that will bring them within the rubric of increasingly subtle forms of competition. Some other method of investigation would clearly be desirable. Before drawing attention to one such possible method, the expected limitations of the Volterra-Gause principle must be examined.

Cases where the Volterra-Gause principle is unlikely to apply. (a) Skellam (1951; see also Brian, 1956b) has considered a model in which two species occur one of (S_1) much lower reproductive potential than the other (S_2). It is assumed that if S_1 and S_2 both arrive in an element of the biotops S_1 always displaces S_2, but that excess elements are always available at the time of breeding and dispersal so that some are never occupied S_1. In view of the higher reproductive potential, S_2 will reach some of these and survive. The model is primarily applicable to annual plants with a definite breeding season, random dispersal of seeds and complete seasonal mortality so all sites are cleared before the new generation starts growing, S_2 is in fact a limiting case of what Hutchinson (1951, 1953) called a fugitive species which could only be established in randomly vacated elements of a biotop. Skellam's model requires clearing of sites by high death rate, Hutchinson's qualitative statement a formation of transient sites by random small catastrophes in the biotop. Otherwise the two concepts developed independently are identical.

(b) When competition for resources becomes a contest rather than a scramble in Nicholson's admirable terminology, there is a theoretical possibility that the principle might not apply. If the breeding population be limited by the number of territories that can be set up in an area, and if a number of unmated individuals without breeding territory are present, food being in excess of the overall requirements, it is possible that territories could be set up by any species entirely independent of the other species, the territorial contests being completely intraspecific. Here a resource, namely area, is limiting but since it does not matter to one species if another is using the area, no interspecific competition need result. No case appears yet to be known, though less extreme modifications of the idea just put forward have apparently been held by several naturalists. Dr. Robert MacArthur has been studying a number of sympatric species of American warblers of the genus *Dendroica* which might be expected to be as likely as any organism to show the phenomenon. He finds however very striking niche specificity among species inhabiting the same trees.

(c) The various cases where circumstances change in the biotop reversing the direction of competition before the latter has run its course. Ideally we may consider two extreme cases with regard to the effect of changing weather and season on competition. In natural populations living for a time under conditions simulating those obtaining in laboratory cultures in a thermostat, if the competition time, that is, the time needed to permit replacement of one species by another, is very short compared with the periods of the significant environmental variables, then complete replacement will occur. This can only happen in very rapidly breeding organisms. Proctor (1957) has found that various green algae always replace *Haematococcus* is small bodies of water which never dry up, though if desiccation and refilling occur frequently enough the *Haematococcus* which is more drought resistant than its competitors will persist indefinitely. If on the contrary the competition time is long compared with the environmental periods, then the relevant environmental determinants of competition will tend to be mean climatic parameters, showing but secular trends in most cases, and competition will inevitably proceed to its end unless some quite exceptional event intervenes.[5]

[5] If there were really three species of giant tortoise (Rothschild, 1915) on Rodriguez, and even more on Mauritius, and if these were sympatric and due to multiple invasion (unlike the races on Albemarle in the Galapagos Islands) it is just conceivable that the population growth was so slow that mixed populations persisted for centuries and that the completion of competition had not occurred before man exterminated all the species involved.

Between the two extreme cases it is reasonable to suppose that there will exist numerous cases in which the direction of competition is never constant enough to allow elimination of one competitor. This seems likely to be the case in the autotrophic plankton of lakes, which inhabits a region in which the supply of nutrients is almost always markedly suboptimal, is subject to continual small changes in temperature and light intensity and in which a large number of species may (Hutchinson, 1941, 1944) coexist.

There is interesting evidence derived from the important work of Brian (1956a) on ants that the completion of competitive exclusion is less likely to occur in seral than in climax stages, which may provide comparable evidence of the effect of environmental changes in competition. Moreover whenever we find the type of situation described so persuasively by Andrewartha and Birch (1954) in which the major limitation on numbers is the length of time that meteorological and other conditions are operating favorably on a species, it is reasonable to suppose that interspecific competition is no more important than intraspecific competition. Much of the apparent extreme difference between the outlook of, for instance, these investigators, or for that matter Milne on the one hand, and a writer such as Lack (1954) on the other, is clearly due to the relationship of generation time to seasonal cycle which differs in the insects and in the birds. The future of animal ecology rests in a realisation not only that different animals have different autecologies, but also that different major groups tend to have fundamental similarities and differences particularly in their broad temporal relationships. The existence of the resemblances moreover may be quite unsuspected and must be determined empirically. In another place (Hutchinson, 1951) I have assembled such evidence as exists on the freshwater copepoda, which seem to be reminiscent of birds rather than of phytoplankton or of terrestrial insects in their competitive relationships.

It is also important to realize, as Cole has indicated in the introductory contribution to this Symposium, that the mere fact that the same species are usually common or rare over long periods of time and that where changes have been observed in well studied faunas such as the British birds or butterflies they can usually be attributed to definite environmental causes in itself indicates that the random action of weather on generation is almost never the whole story. Skellam's demonstration that such action must lead to final extinction must be born in mind. It is quite possible that the change in the phytoplankton of some of the least culturally influenced of the English Lakes, such as the disappearance of *Rhizosolenia* from Wastwater (Pearsall, 1932), may provide a case of random extinction under continually reversing competition. The general evidence of considerable stability under most conditions would suggest that competitive action of some sort is nearly always of significance.

Rarity and commonness of species and the non-intersection of realised niches. Several ways of approaching the problem of the rarity and commonness of species have been suggested (Fisher, Corbet and Williams, 1943; Preston, 1948; Brian, 1953; Shinozaki and Urata, 1953). In all these approaches relatively simple statistical distributions have been fitted to the data, without any attempt being made to elucidate the biological meaning of such distribution. Recently however MacArthur (1957) has advanced the subject by deducing the consequences of certain alternative hypotheses which can be developed in terms of a formal theory of niches.

It has been pointed out in a previous paragraph that the Volterra-Gause principle is equivalent to a statement that the realised niches of co-occurring species are non-intersecting. Consider a **B** space containing an equilibrium community of n species $S_1 S_2 \cdots S_n$, represented by numbers of individuals $N_1 N_2 \cdots N_n$. For any species S_K it will be possible to identify in **B** a number of elements, each of which corresponds to a whole or part of $\mathbf{N}'_K$ and to no other part of **N**. Suppose that at any given moment each of these elements is occupied by a single individual of S_K, the total volume of B which may be regarded as the specific biotop of S_K will be $N_K \Delta\mathbf{B}\ (S_K)$, $\Delta\mathbf{B}(S_K)$ being the mean volume of **B** occupied by one individual of S_K. Since the biotop is in equilibrium with respect to the n species present, all possible spaces will be filled so that

$$\mathbf{B} = \sum_{K=1}^{n} N_K \Delta\mathbf{B}\ (S_K)$$

We do not know anything *a priori* about the distribution of $N_1 \Delta\mathbf{B}(S_1)$, $N_2 \Delta\mathbf{B}(S_2) \cdots N_n \Delta\mathbf{B}(S_n)$, except that these different specific biotops are taken as volumes proportional to $N_1 N_2 \cdots N_n$, which is a justifiable first approximation if the species are of comparable size and physiology. In general some of the species will be rare and some common. The simplest hypothesis consistent with this, is that a random division of **B** between the species has taken place.

Consider a line of finite length. This may be broken at random into n parts by throwing $(n - 1)$ random points upon it. It would also be possible to divide the line successively by throwing n random pairs of points upon it. In the first case the division is into non-overlapping sections, in the second the sections overlap. MacArthur, whose paper may be consulted for references to the mathematical procedures involved, has given the expected distributions for the division of a line by these alternative methods (Fig. 1). He has moreover shown that with certain restrictions the distribution (I) which corresponds to non-intersecting specific biotops and so to non-intersecting realised niches, fits certain multispecific biological associations extremely well. The form of this distribution is independent of the number of dimensions in **B**. The alternative distribution with overlapping specific biotops predicts fewer species of intermediate rarity and more of great rarity than is actually found; proceding from the linear case (II), to division of an area or a volume, accentuates this discrepancy. Two very striking cases in which distribution I fits biological multispecific populations are given in Figures 2 and 3 from MacArthur and in Figure 4 from the recent studies of Dr. Alan Kohn (in press).

The limitation which is imposed by the theory is that in all large subdivisions of **B** the ratio of total number of individuals $(m = \sum_{i=1}^{n} N_i)$ to

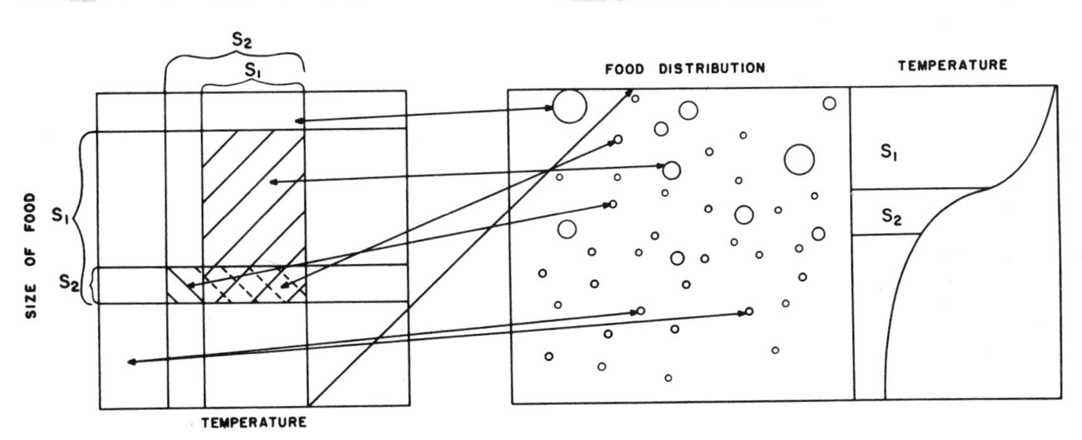

FIGURE 1. Two fundamental niches defined by a pair of variables in a two-dimensional niche space. Only one species is supposed to be able to persist in the intersection subset region. The lines joining equivalent points in the niche space and biotop space indicate the relationship of the two spaces. The distribution of the two species involved is shown on the right hand panel with a temperature depth curve of the kind usual in a lake in summer.

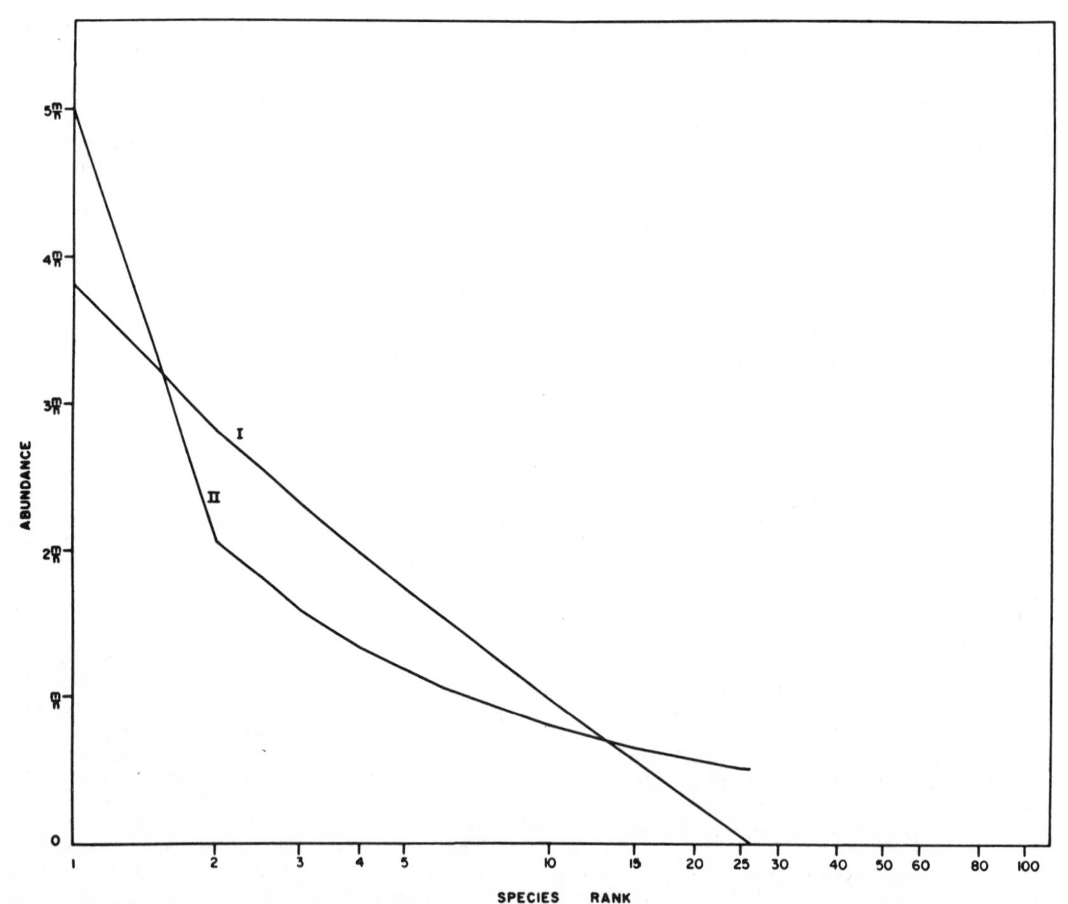

FIGURE 2. Rank order of species arranged per number of individuals according to the distributions I and II considered by MacArthur.

total number of species (n) must remain constant. This is likely to be the case in any biotop which is what may be termed *homogeneously diverse*, that is, in which the elements of the environmental mosaic (trees, stones, bushes, dead logs, etc.) are small compared with the mean free paths of the organisms under consideration. When a heterogeneously diverse area, comprising for instance stands of woodland separated by areas of pasture, is considered it is very unlikely that the ratio of total numbers of individuals to number of species will be identical in both woodland and pasture (if it occasionally were, the fact that both censuses could be added would not be of any biological interest). MacArthur finds that at least some bodies of published data which do not fit distribution I as a whole, can be broken down according to the type of environment into subcensuses which do fit the distribution. Data from moth traps and from populations of diatoms on slides submerged in rivers would not be expected to fit the distribution and in fact do not do so.[6] Such collection methods certainly sample very heterogeneously diverse areas.

The great merit of MacArthur's study is that it attempts to deduce operationally distinct differences between the results of two rival hypotheses, one of which corresponds essentially to the extreme density dependent view of interspecific interaction, the other to the opposite view. Although certain simplifying assumptions must be made in the theoretical treatment, the initial results suggest that in stable homogeneously diverse biotops the abundances of different species are arranged as if the realised niches were non-overlapping; this does not mean that populations may not exist under other conditions which would depart very widely from MacArthur's findings.

The problem of the saturation of the biotop. An important but quite inadequately studied aspect

[6] I am indebted to Dr. Ruth Patrick for the opportunity to test some of her diatometer censuses.

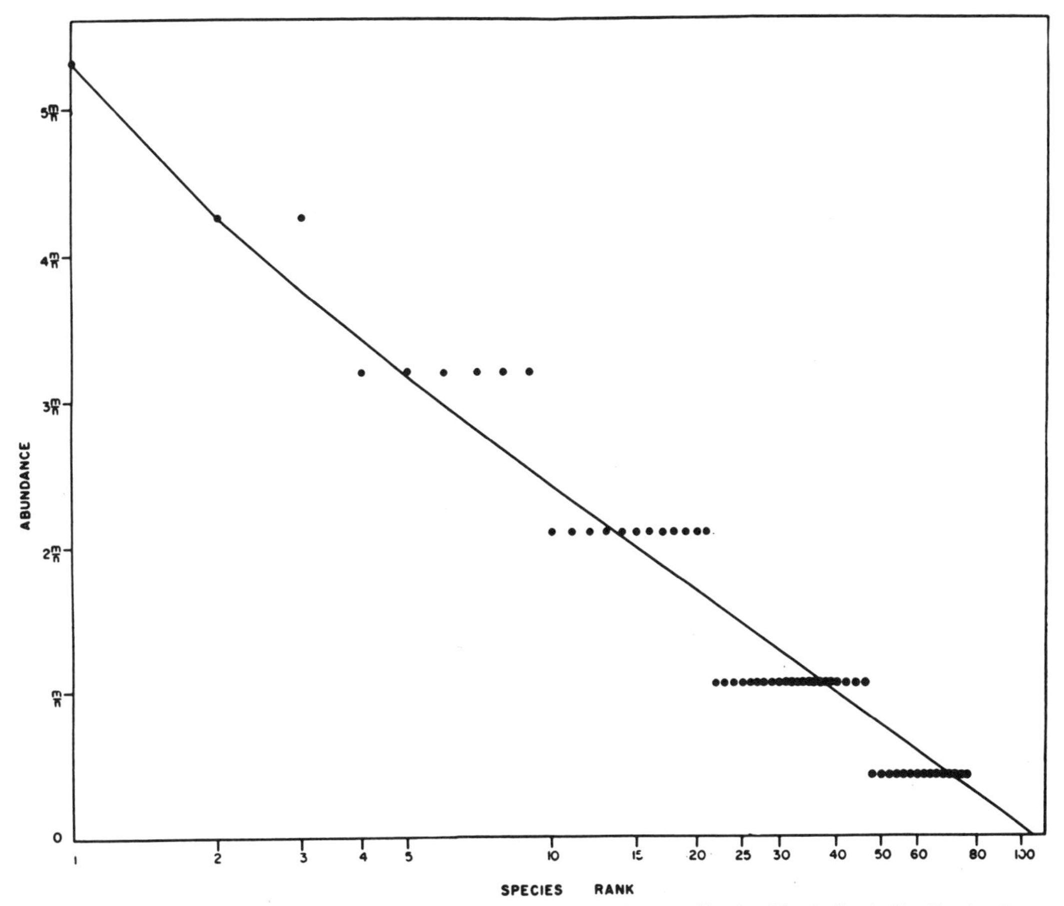

FIGURE 3. Rank order of species of birds in a tropical forest, closely following MacArthur's distribution I.

of niche specificity is that of the number of species that a given biotop can support. The nature of this problem can be best made clear by means of an example.

The aquatic bugs of the family *Corixidae* are of practically world wide distribution. Omitting a purely Australasian subfamily, they may be divided into the *Micronectinae* which are nearly always small, under 5 mm long and the *Corixinae* of which the great majority of species are over 5 mm long. Both subfamilies probably feed largely on organic detritus, though a few of the more primitive members of the *Corixinae* are definite predators. Some at least suck out the contents of algal cells, but unlike the other Heteroptera they can take particulate matter of some size unto their alimentary tracts. There is abundant evidence that the organic content of the bottom deposits of the shallow water in which these insects live is a major ecological factor regulating their occurrence. No *Micronectinae* occur in temperate North America and in the Old World this subfamily is far more abundant in the tropics while the *Corixinae* are far more abundant in the temperate regions (Lundblad, 1934; Jaczewski, 1937). Thus in Britain there are 30 species of *Corixinae* and three of *Micronectinae* (Macan, 1956), in peninsular Italy 20 or 21 species of *Corixinae* and five of *Micronectinae* (Stickel, 1955), in non-Palaeartic India about a dozen species of *Corixinae* and at least ten species of *Micronectinae* (Hutchinson, 1940) and in Indonesia (Lundblad, 1934) only three *Corixinae* and 14 *Micronectinae*. A reasonable explanation of this variation in the relative proportions of the two subfamilies is suggested by the findings of Macan (1938) and the more casual observations of other investigators that *Micronecta* prefers a low organic subtratum; in tropical localities the high rate of decomposition would reduce the organic content.

In certain isolated tropical areas at high altitudes, notably Ethiopia and the Nilghiri Hills of southern India the decline in the numbers of *Micronectinae* with increasing altitudes, and so

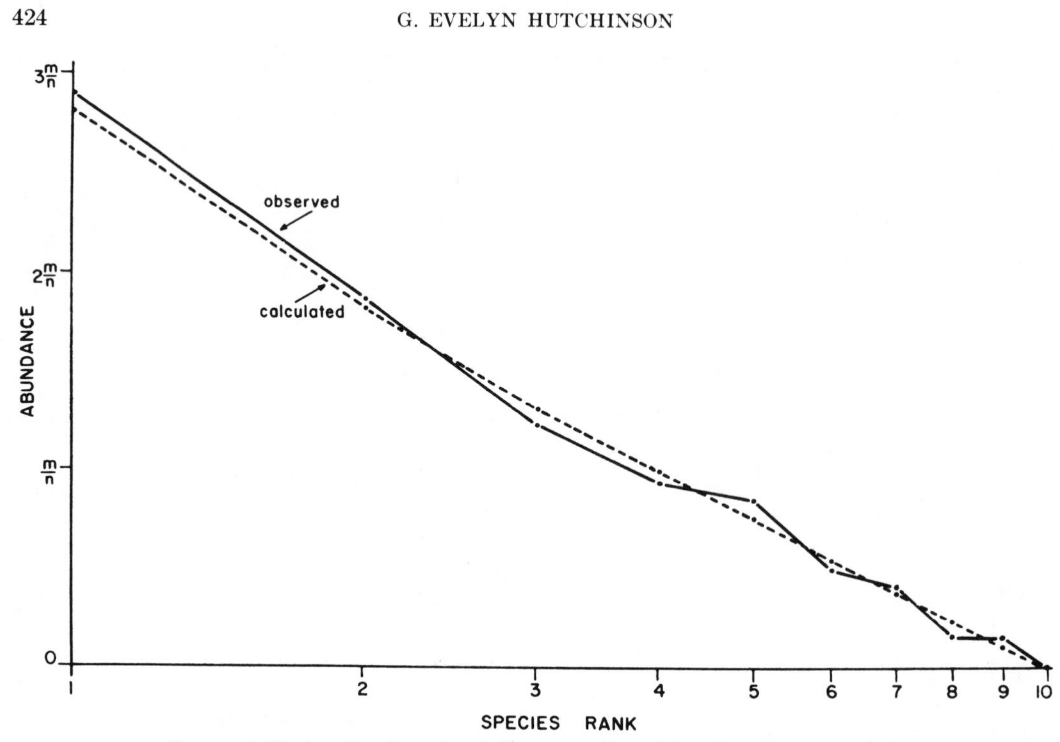

FIGURE 4. Rank order of species of *Conus* on a littoral bench in Hawaii (Kohn).

lower average water temperatures, is most noticable, but there is no increase in the number of *Corixinae*, presumably because the surrounding fauna is not rich enough to have permitted frequent invasion and speciation. Thus in the Nilghiri Hills between 2100 and 2300 m, intense collecting yielded three *Corixinae* of which two appear to be endemic, and one non-endemic species of *Micronecta*. Very casual collecting below 1000 m in south India has produced two species of *Corixinae* and five species of *Micronectinae*. The question raised by cases like this is whether the three Nilghiri *Corixinae* fill all the available niches which in Europe might support perhaps 15 or 20 species, or whether there are really empty niches. Intuitively one would suppose both alternatives might be partly true, but there is no information on which to form a real judgment. The rapid spread of introduced species often gives evidence of empty niches, but such rapid spread in many instances has taken place in disturbed areas. The problem clearly needs far more systematic study than it has been given. The addition and the replacement of species of fishes proceeding down a river, and the competitive situations involved, may provide some of the best material for this sort of study, but though much data exists, few attempts at systematic comparative interpretation have been made (*cf.* Hutchinson, 1939).

THE FUTURE OF COMPARATIVE DEMOGRAPHIC STUDIES

Perhaps the most interesting general aspect of the present Symposium is the strong emphasis placed on the changing nature of the populations with which almost all investigators deal. In certain cases, notably in the parthenogenetic crustacean *Daphnia* (Slobodkin, 1954), it is possible to work with clones that must be almost uniform genetically, but all the work on bisexual organisms is done under conditions in which evolution may take place. The emergence in Nicholson's experiments of strains of *Lucilia* in which adult females no longer need a protein meal before egg laying provides a dramatic example of evolution in the laboratory; the work reported by Dobzhansky, by Lewontin, and by Wallace, in discussion, shows how experimental evolution, for which subject the Carnegie Laboratory at Cold Spring Harbor was founded, has at last come into its own.

So far little attention has been paid to the problem of changes in the properties of populations of the greatest demographic interest in such experiments. A more systematic study of evolutionary change in fecundity, mean life span, age and duration of reproductive activity and length of post reproductive life is clearly needed. The most interesting models that might be devised would be those in which selection operated in favor of low fecundity, long pre-reproductive life and on any aspect of post-reproductive life.

There is in many groups, notably *Daphnia*, dependence of natality on food supply (Slobodkin, 1954) though the adjustment can never be instantaneous and so can lead to oscillations. In the case of birds the work of the Oxford school (Moreau, 1944; Lack, 1947a, 1954 and many papers quoted in the last named) indicates that in many birds natality is regulated by natural selection to correspond to the maximum number of young that can be reared in a clutch. In some circumstances the absolute survival of young is greater when the fecundity is low than when it is high. The peculiar nature of the subpopulations formed by groups of nestlings in nests makes this reasonable. Slobodkin (1953) has pointed out that in certain cases in which migration into numerous limited areas is possible, a high reproductive rate might have a lower selective advantage than a low rate. Actually in a very broad sense the bird's nest is a device to formalise the numerous limited areas, the existence of which permits such a type of selection. It should be possible with some insects to set up population cages in which access to a large number of very small amounts of larval food is fairly difficult for a fertile female. If the individual masses of larval food were such that there was an appreciable chance that many larvae on a single mass would die of starvation while a few larvae would survive, it is possible that selection for low fecundity might occur. This experiment would certainly imitate many situations in nature.

The evolutionary aspects of the problem raised by those cases where there is a delay of reproductive activity after adult morphology has been achieved is much harder to understand. Some birds though they attain full body size within a year (or in the case of most passerines in the nest) are apparently not able to breed until their third or later year. It is difficult to see why this should be so. In any given species there may be good endocrinological reasons for the delay, but they can hardly be evolutionarily inevitable. The situation has an abvious *prima facie* disadvantage, since most birds have a strikingly diagonal survivorship curve after the first year of life and this in itself indicates little capacity for learning to live. One would have supposed that in the birds, mainly but not exclusively large sea birds, which show the delay, any genetic change favoring early reproduction would have a great selective advantage. Any experimental model imitating this situation would be of great interest.

The problem of possible social effects of long post-reproductive life, which can hardly be subject to direct selection, provides another case in which any hints from changes in demographic parameters in experiments would be most helpful. The experimental study of the evolutionary aspects of demography is certain to yield surprises. While we have Nicholson's work, in which the amplitude of the oscillation in *Lucilia* populations appear to be increased or at least not decreased as a result of the evolution he has observed, though the minima are less low and the variation less regular, we do not know if this sort of effect is likely to be general. Utida's elegant work on bean weevils appears to be consistent with some evolutionary damping of oscillations which would be theoretically a likely result.

The most curious case of a genetic change playing a regular part in a demographic process is certainly that in rodents described by Chitty. In view of the large number of simple ways which are now available to explain regular oscillations in a population, it is extremely important to heed Chitty's warning that the obvious explanation is not necessarily the true one. To the writer, this seems to be a particular danger in human demography, though the mysteries of variation of the human sex ratio, so clearly expounded by Colombo, should be a warning against over-simple hypotheses, for here no reasonable hypotheses have been suggested. Human demography relies too much on what psychologists call intervening variable theory. The reproducing organisms are taken for granted; when their properties change, either as the result of evolution or of changes in learned behaviour, the results are apt to be upsetting. The present "baby boom" is such an upset, and here a tendency to over-simplified thinking is also apparent. If, as appears clear at least for parts of North America, the present birth rate is positively correlated with economic position, it is easy to suppose that couples now have as many children as they can afford, just as most small birds appear to do. There is, however, a difference. If at any economic level a four child family was desired, but occasionally owing to the imperfections of birth control a five child family was actually achieved, we should not expect the fifth child to have a negligible expectation of life at birth, so that the total contribution to the population per family would be the same from a four and a five child family. Yet this is exactly what Lack and Arn (1947) found for the broods of the Alpine swift *Apus melba*. In man the criterion is never purely economic; it is not how large a brood can be reared, but how large a brood the parents think they can rear without undue economic sacrifice. Such a method of setting limits to natality is obviously extremely complicated. It involves an equilibrium between a series of desires, partly conscious, partly unconscious, and a series of estimates of present and future resources. There is absolutely no reason to suppose that the mean desired family size determined in such a way is a simple function of economics, uninfluenced by a vast number of other cultural factors. The assumption that a large family is *per se* a good thing is obviously involved; this may be accepted individually by most parents even though it is at present a very dubious assumption on general grounds of social well being. Part of the acceptance of such an assumption is certain to be due to unconscious factors. Susannah Coolidge in a remarkable, as yet unpublished, essay,[7] "Population *versus* People," suggests that for many women a new pregnancy is an occasion for a temporary shifting of some of the responsibility for the older children away from the mother, and so is welcomed. She also suspects that it may be an unconscious expression of disappointment over, or repudiation of, the older children and so be essentially a repeated neurotic symptom. Moreover, the present highly conspicuous fashion for maternity, certainly a healthy reaction from the seclusion of upper-class pregnant women a couple of generations ago, is also quite likely fostered by those business interests which seem to believe that an indefinitely expanding economy is possible on a non-expanding planet.

An adequate science of human demography must take into account mechanism of these kinds, just as animal demography has taken into account all the available information on the physiological ecology and behaviour of blow flies, *Daphnia* and bean weevils. Unhappily, human beings are far harder to investigate than are these admirable laboratory animals; unhappily also, the need becomes more urgent daily.

References

ANDREWARTHA, H. G., and BIRCH, L. C., 1954, The Distribution and Abundance of Animals. Chicago, University of Chicago Press. xv, 782 pp.

BETTS, M. M., 1955, The food of titmice in oak woodland. J. Anim. Ecol. *24:* 282–323.

BIRCH, L. C., 1957, The meanings of competition. Amer. Nat. *91:* 5–18.

BRIAN, M. V., 1953, Species frequencies from random samples in animal populations. J. Anim. Ecol. *22:* 57–64.

1956a, Segregation of species of the ant genus *Myrmica*. J. Anim. Ecol. *25:* 319–337.

1956b, Exploitation and interference in interspecies competition. J. Anim. Ecol. *25:* 339–347.

COOPER, D. M., and DOBZHANSKY, TH., 1956, Studies on the ecology of *Drosophila* in the Yosemite region of California. I. The occurrence of species of *Drosophila* in different life zones and at different seasons. Ecology *37:* 526–533.

CROMBIE, A. C., 1945, On competition between different species of graminivorous insects. Proc. Roy. Soc. Lond. *132*B*:* 362–395.

1946, Further experiments on insect competition. Proc. Roy. Soc. Lond. *133*B*:* 76–109.

1947, Interspecific competition. J. Anim. Ecol. *16:* 44–73.

DA CUNHA, A. B., EL-TABEY SHEKATA, A. M., and DE OLIVIERA, W., 1957, A study of the diet and nutritional preferences of tropical of *Drosophila*. Ecology *38:* 98–106.

DUMAS, P. C., 1956, The ecological relations of sympatry in *Plethodon dunni* and *Plethodon vehiculum*. Ecology *37:* 484–495.

FISHER, R. A., CORBET, A. S., and WILLIAMS, C. B., 1943, The relation between the number of species

[7] I am greatly indebted to the author of this work for permission to refer to some of her conclusions.

and the number of individuals in a ransom sample of an animal population. J. Anim. Ecol. *12:* 42–58.

GAUSE, G. F., 1934, The struggle for existence. Baltimore, Williams & Wilkins. 163 pp.

1935, Vérifications expérimentales de la théorie mathématique de la lutte pour la vie. Actualités scientifiques *277*. Paris. 63 pp.

HUTCHINSON, G. E., 1939, Ecological observations on the fishes of Kashmir and Indian Tibet. Ecol. Monogr. *9:* 145–182.

1940, A revision of the Corixidae of India and adjacent regions. Trans. Conn. Acad. Arts Sci. *33:* 339–476.

1941, Ecological aspects of succession in natural populations. Amer. Nat. *75:* 406–418.

1944, Limnological studies in Connecticut. VII. A critical examination of the supposed relationship between phytoplankton periodicity and chemical changes in lake waters. Ecology *25:* 3–26.

1951, Copepodology for the ornithologist. Ecology *32:* 571–577.

1953, The concept of pattern in ecology. Proc. Acad. Nat. Sci. Phila. *105:* 1–12.

HYNES, H. B. N., 1954, The ecology of *Gammarus deubeni* Lilljeborg and its occurrence in fresh water in western Britain. J. Anim. Ecol. *23:* 38–84.

1955, The reproductive cycle of some British freshwater Gammaridae. J. Anim. Ecol. *24:* 352–387.

JACZEWSKI, S., 1937, Allgemeine Zügeder geographischen Verbreitung der Wasserhemiptera. Arch. Hydrobiol. *31:* 565–591.

KOHN, A. J., The ecology of *Conus* in Hawaii. (Yale Dissertation 1057, in press.)

KURTÉN, B., 1957, Mammal migrations, cenozoic stratigraphy, and the age of Peking man and the australopithecines. J. Paleontol. *31:* 215–227.

LACK, D., 1947a, The significance of clutch-size. Ibis *89:* 302–352.

1947b, Darwin's finches. Cambridge, England. x, 208 pp.

1954, The natural regulation of animal numbers. Oxford, The Clarendon Press, viii, 343.

LACK, D., and ARN, H., 1947, Die Bedeutung der Gelegegrösse beim Alpensegler. Ornith. Beobact. *44:* 188–210.

LOTKA, A. J., The growth of mixed populations, two species competing for a common food supply. J. Wash. Acad. Sci. *22:* 461–469.

LUNDBLAD, O., 1933, Zur Kenntnis der aquatilen und semi-aquatilen Hemipteren von Sumatra, Java, und Bali. Arch. Hydrobiol. Suppl. *12:* 1–195, 263–489.

MACAN, T. T., 1938, Evolution of aquatic habitats with special reference to the distribution of Corixidae. J. Anim. Ecol. *7:* 1–19.

1956, A revised key to the British water bugs (Hemiptera, Heteroptera) Freshwater Biol. Assoc. Sci. Publ. *16:* 73 pp.

MACARTHUR, R. H., 1957, On the relative abundance of bird species. Proc. Nat. Acad. Sci. Wash. *43:* 293–295.

MOREAU, R. E., 1944, Clutch-size: a comparative study, with special reference to African birds. Ibis *86:* 286–347.

PEARSALL, W. H., 1932, The phytoplankton in the English lakes II. The composition of the phytoplankton in relation to dissolved substances. J. Ecol. *20:* 241–262.

PRESTON, F. W., 1948, The commonness, and rarity, of species. Ecology *29:* 254–283.

PROCTOR, V. W., 1957, Some factors controlling the distribution of *Haematococcus pluvialis*. Ecology, *in press*.

ROTHSCHILD, LORD, 1915, On the gigantic land tortoises of the Seychelles and Aldabra-Madagascar group with some notes on certain forms of the Mascarene group. Novitat. Zool. *22:* 418–442.

RUITER, L. DE, 1952, Some experiments on the camouflage of stick caterpillars. Behaviour *4:* 222–233.

Shinozaki, K., and Urata, N., 1953, Researches on population ecology II. Kyoto Univ. (not seen; ref. MacArthur, 1957).

Shorten, M., 1953, Notes on the distribution of the grey squirrel (*Sciurus carolinensis*) and the red squirrel (*Sciurus vulgaris leucourus*) in England and Wales from 1945 to 1952. J. Anim. Ecol. *22:* 134–140.

1954, Squirrels. (New Naturalist Monograph 12.) London 212 pp.

Skellam, J. G., 1951, Random dispersal in theoretical populations. Biometrika *38:* 196–218.

Slobodkin, L. B., 1953, An algebra of population growth. Ecology *34:* 513–517.

1954, Population dynamics in *Daphnia obtusa* Kurz. Ecol. Monogr. *24:* 69–88.

Steusloff, V., 1943, Ein Beitrag zur Kenntniss der Verbreitung und der Lebensräume von *Gammarus*-Arten in Nordwest-Deutschland. Arch. Hydrobiol. *40:* 79–97.

Stickel, W., 1955, Illustrierte Bestimmungstabellen der Wanzen. II. Europa. Hf. 2, 3. pp. 40–80 (Berlin, apparently published by author).

Ueno, M., 1934, Yale North India Expedition. Report on the amphipod genus *Gammarus*. Mem. Conn. Acad. Arts Sci. *10:* 63–75.

Volterra, V., 1926, Vartazioni e fluttuazioni del numero d'individui in specie animali conviventi. Mem. R. Accad. Lincei ser. 6, *2:* 1–36.

Winsor, C. P., 1934, Mathematical analysis of the growth of mixed populations. Cold Spr. Harb. Symp. Quant. Biol. *2:* 181–187.

PAPER 10

THE POPULATION CONSEQUENCES OF LIFE HISTORY PHENOMENA

By LAMONT C. COLE

Department of Zoology, Cornell University

PREFACE

FEW branches of biology have attracted more analytical mathematical treatment than has the study of populations. Despite this, one may read in the most complete treatise of ecology yet published (Allee, et al., 1949, p. 271) that "theoretical population ecology has not advanced to a great degree in terms of its impact on ecological thinking." This unfortunate gap between the biologists and the mathematicians has elicited comments which need not be repeated in detail here (Allee, 1934; Gause, 1934; Allee et al., 1949, p. 386). The neglect of the analytical methods by biologists may be attributed in part to the tendency of writers in this field to concentrate on the analysis of human populations and in part to skepticism about the mathematical methods of analysis. Early analyses of population growth (Verhulst, 1838, 1845; Pearl and Reed, 1920) employed human populations as examples, although it is clear from other publications (e.g., Pearl and Miner, 1935; Pearl, 1937) that comparative and general population studies were the principal interest of some of these students. Similarly, the pioneer works of Lotka (1907b, 1910, 1925) were very general in conception but made their greatest impact in the field of demography (Dublin and Lotka, 1925; Dublin, Lotka, and Spiegelman, 1949). The skepticism expressed by biologists toward theoretical studies has ranged from antagonism (Salt, 1936) to approval given with the warning that "... for the sake of brevity and to avoid cumbersome expressions, variables are omitted and assumptions made in the mathematical analyses which are not justified by the biological data" (Allee, 1934). It may be unfortunate that warnings about mathematical oversimplification are especially pertinent in connection with the study of interactions between species (Ross, 1911; Lotka, 1920, 1925; Volterra, 1927, 1931; Nicholson and Bailey, 1935; Thompson, 1939), which is just that portion of the subject which has remained most closely associated with general ecology. Hence we have a situation in which the analytical theories which are recognized by ecologists deal with complex phenomena and are susceptible to cogent criticisms (e.g., Smith, 1952) while the simpler analysis of the ways in which differences between the life histories of species may result in different characteristics of their populations has remained relatively unexplored. It is the purpose of the present paper to consider some parts of this neglected branch of ecology which has been called "biodemography" by Hutchinson (1948).

It is possible, but often impracticable, to compute exactly the characteristics of the hypothetical future population obtained by assuming an unvarying pattern of the pertinent life history features which govern natality and mortality. It is often more practicable to employ approximate methods of computation of the type which have aroused skepticism among biologists. It will be shown that the two approaches can be reconciled and that for many cases of ecological interest they lead to identical conclusions. Some of these conclusions reached by the writer have appeared surprising when first encountered, and they seem to give a new perspective to life history studies. They also suggest that pertinent bits of information are frequently ignored in life history studies simply because their importance is not generally recognized.

The total life history pattern of a species has meaning in terms of its ability to survive, and ecologists should attempt to interpret these meanings. The following sections are intended primarily to indicate some of the possibilities in this direction. The writer wishes to express his gratitude to Professor Howard B. Adelmann for a critical reading of the manuscript of this paper, for suggesting numerous ways of clarifying the text and improving terminology, and for translating from the Latin parts of the text from Fibonacci (1202). Thanks are also due to Professors Robert J. Walker and Mark Kac who have been consulted about technical mathematical questions raised by the writer while considering various phases of this subject.

INTRODUCTION

If it is to survive, every species must possess reproductive capacities sufficient to replace the existing species population by the time this population has disappeared. It is obvious that the ability of the ancestors of existing species to replace themselves has been sufficient to overcome all environmental exigencies which have been encountered and, therefore, that the physiological, morphological, and behavioral adaptations that enable offspring to be produced and to survive in sufficient numbers to insure the persistence of a species are of fundamental ecological interest.

On the other hand, it is conceivable that reproductive capacity might become so great as to be detrimental to a species. The many deleterious effects of overcrowding are well known. It also seems obvious that a species which diverts too large a proportion of its available energies into unnecessary, and therefore wasteful, reproduction would be at a disadvantage in competition with other species.

In this paper it will be regarded as axiomatic that the reproductive potentials of existing species are related to their requirements for survival; that any life history features affecting reproductive potential are subject to natural selection; and that such features observed in existing species should be considered adaptations, just as purely morphological or behavioral patterns are commonly so considered.

Some of the more striking life history phenomena have long been recognized as adaptations to special requirements. The great fecundity rather generally found in parasites and in many marine organisms is commonly regarded as an adaptation insuring the maintenance of a population under conditions where the probability is low that any particular individual will establish itself and reproduce successfully. Again, parthenogenesis obviously favors the rapid growth of a population because every member of a population reproducing in this fashion can be a reproductive female. In turning seasonally to parthenogenesis, organisms like cladocerans and aphids are responding in a highly adaptive way during a limited period of time when the environmental resources are sufficient to support a large population. Parthenogenesis, hermaphroditism, and purely asexual reproduction may clearly offer some advantages under conditions that restrict the probability of contacts between the sexes. Protandry, as exhibited, for example, by some marine molluscs, and various related phenomena where population density affects the sex ratio (Allee et al., 1949, p. 409) may be considered as compromise devices providing the advantages of biparental inheritance while maintaining an unbalanced sex ratio which makes most of the environmental re-resources available to reproductive females.

Reproductive potentialities may be related to the success of a species in still other ways. It was an essential part of Darwin's thesis that the production of excess offspring provided a field of heritable variations upon which environmental conditions could operate to select the most favorable combinations. A high degree of fecundity may also aid the dispersal of species. An extreme example of this is afforded by the ground pine, *Lycopodium* (Humphreys, 1929), whose light wind-borne spores may be scattered literally over the whole face of the earth and so make it likely that all favorable habitats will come to be occupied. Another adaptational interpretation of the overproduction of offspring postulates that the excessive production of young fish which are frequently cannibalistic is a form of maternal provisioning, the majority of the young serving merely as food for the few that ultimately mature.

Many additional examples of life history phenomena that have been regarded as adaptive could be cited. Here, however, we wish rather to call attention to the striking fact that in modern ecological literature there have been relatively few attempts to evaluate quantitatively the importance of specific features of life histories. The apparent mathematical complexity of the general problem is undoubtedly partly responsible for this. When the biologist attempts to compute from observed life history data the numbers of organisms of a particular type that can be produced in a given interval of time he may find it necessary to make assumptions which biologists in general would hesitate to accept. And even with these simplifying assumptions the computations may become so tedious as to make the labor involved seem unjustifiable in view of the seemingly academic interest of the result. In particular, such computations involve biological parameters which are not necessarily fixed characteristics of the species and which are not ordinarily expressible in convenient mathematical form. It is necessary to know the way in which the chance of dying (or of surviving) and the reproductive activities vary during the life span of an individual. These quantities are nicely summed up by the familiar life-table function, survivorship (l_x), which is defined as the probability of surviving from birth to some age x, and by the age-specific birth rate (b_x), which is defined as the mean number of offspring produced during the interval of age from age x to age $x + 1$. The biologist immediately recognizes that these quantities vary with environmental conditions and that he cannot expect to obtain a realistic result if he must assume, for example, that the probability of surviving a day, a week, or a month, is the same for individuals born in the autumn as for those born in the spring. He also recognizes that the population consists of discrete units and that offspring are produced in batches (here called litters whether in plants or animals) rather than continuously; hence he necessarily regards with suspicion any formulation of the problem in terms of differential equations where these considerations are apparently ignored.

Actually a tremendous variability is observed in life history phenomena which could affect the growth of populations. Some organisms are semelparous, that is to say, they reproduce only once in a lifetime and in these semelparous forms reproduction may occur at the age of only 20 minutes in certain bacteria (Molisch, 1938), of a few hours in many protozoa, or of a few weeks or months in many insects. Many semelparous plants and animals are annuals; in other semelparous organisms reproduction may occur only after a number of years of maturation, for example, two or more years in dobson flies and Pacific salmon, and many years in "century plants" (*Agave*) and the periodic cicada or "17-year locust" (*Magicicada septendecim*). The number of potential offspring produced by semelparous individuals varies from two in the case of binary fission to the literally trillions (2×10^{13}) of spores produced by a large puffball (*Calvatia gigantea*).

In iteroparous forms, that is to say, those which reproduce more than once in a lifetime, the period of maturation preceding the first production of prospective offspring may vary from as little as a few days in small crustaceans to over a century in the giant sequoia (U. S. Forest Service, 1948), and practically any intermediate value may be encountered. After the first reproduction has occurred in iteroparous organisms it may be repeated at various intervals—for example, daily (as in some tapeworms), semiannually, annually, biennially, or irregularly (as in man). As in semelparous organisms, the litter size of iteroparous forms may also vary greatly; here it may vary from one (as is usual, for example, in man, whales, bovines, and horses) to many thousands (as in various fishes, tapeworms, or trees). The litter size may be constant in a species, vary about some average, or change systematically with the age of the parent, in which case it may increase to some maximum (as in tapeworms) or climb to a maximum and then decline as in some cladocerans (Banta et al., 1939; Frank, 1952). Furthermore, individuals may live on after their reproduction has ceased completely, and this post-reproductive period may amount to more than one-half of the normal life span (Allee et al., 1949, p. 285).

There is similar variability in the potential longevity of individual organisms. Man, various turtles, and trees may survive more than a century, while, on the other hand, the life span of many other species is concluded in hours or days. Innumerable intermediate values of course occur.

Additional sources of variation (such as biased sex ratios and the occurrence of asexual reproduction in developmental stages so as to result in the production of many offspring from one egg or spore) force the conclusion that the number of theoretical combinations of observed life history phenomena must greatly exceed the number of known species of organisms. And if all these phenomena have potential adaptive importance the interpretation of the possible merits of the particular combination of features exhibited by a species presents a problem of apparent great complexity.

The usual mathematical approach to the problem of potential population growth is straightforward. It is assumed that the growth of a population at any instant of time is proportional to the size of the population at that instant. If r is the factor of proportionality and P_x represents the population size at any x time this leads to the differential equation

$$\frac{dP}{dx} = rP \qquad (1)$$

which upon integration gives:

$$P_x = Ae^{rx} \qquad (1')$$

where A is a constant. This is an equation of continuous compound interest at the rate r or of a geometric progression where the ratio between the sizes of the populations in two consecutive time intervals, say years, is e^r.

While formulas (1) and (1′) represent only the usual starting point for mathematical discussions of population growth, they already exhibit points about which there has been, and still is, a great deal of controversy. Explicit statements to the effect that human populations potentially increase by geometric progression can be traced back at least to Capt. John Graunt (1662), who estimated that a human population tends to double itself every 64 years (which would correspond to $r = .0108$ in formula 1). This belief in geometric progression as the form of potential population increase was endorsed by numerous students prior to the great controversy initiated by Malthus in 1798 (see review by Stangeland, 1904). Among these early writers we may here note only Linnaeus (1743), who considered the problem of geometric increase in the progeny of an annual plant, and Benjamin Franklin (1751), who estimated that the population of "America" could double at least every 20 years (corresponding to $r = .035$), and who clearly regarded the geometric nature of potential population increase as a general organic phenomenon.

The great controversy over growth in human populations which was initiated by the publication in 1798 of Malthus' *Essay on Population* engendered numerous arguments regarding geometric progression as the potential form of population growth. This controversy is still alive and in much its original form, with the "Neo-Malthusian" position maintaining that potential population growth is indeed in the form of a geometric progression, whereas the capacity of the environment to absorb population is necessarily limited, and with their opponents denying both the geometric progression and the finite capacity of the environment. Essentially the modern arguments against the Malthusian thesis, although not presented in modern concise form, are to be found in the treatise by Sadler (1830) which, whatever its shortcomings from the modern point of view, contains in places (especially in the appendix to Book IV) a very remarkable pre-Darwinian statement of such ecological phenomena as food chains, species interactions, and the balance of numbers between predators and prey.

The entire problem of potential population growth and its relationship to the resources of the environment is clearly one of the fundamental problems of ecology, but one which has never been adequately summarized in a way to reconcile the mathematical approaches, such as those of Lotka (1925), Volterra (1927), Kuczynski (1932, 1935), Kostitzin (1939), and Rhodes (1940), and the purely biological approaches which have concentrated on life history features such as longevity, fecundity, fertility, and sex ratios. In the present paper we will consider the mathematical form of potential population growth and certain subsidiary phenomena and the way in which these are related to particular life history phenomena. It is hoped that this will bring to attention some of the possible adaptive values of observed life history phenomena and will lead ecologists to a greater consideration of population problems which are essentially ecological. Life history features do in fact control potential population growth, as Sadler recognized, but the quantitative relationships have still been so insufficiently elucidated that even today ecologists generally do not attempt to answer queries such as the following, written by Sadler in 1830 (Vol. 2, p. 318):

"For instance, how would those who have the folly to suppose that population in this country advances too fast by one per cent., so operate, had they even their wish, as to diminish the number of marriages by one in one hundred, or otherwise contract the fecundity of the existing number by about one twenty-fifth part of a birth each, or calculate, upon their own erroneous suppositions, the term of that postponement of marriage on which they insist so much, so as to produce this exact effect? The very idea is, in each instance, absurd to the last degree."

FUNDAMENTAL CONSIDERATIONS

Sadler (1830) makes clear in numerous places his belief that "... the geometrical ratio of human increase is, nevertheless, in itself, an impossibility ..." (Vol. 2, p. 68). However, when one examines his argument it is apparent that he is not actually opposing the principle that with fixed life history features populations would grow at compound interest, but rather is proposing the thesis that life history features change with population density, e.g., his fundamental thesis: "The prolificness of human beings, otherwise similarly circumstanced, varies inversely as their numbers" (Vol. 2, p. 252). Some of Sadler's computations assuming fixed ages at marriage and fecundity rates, in fact, lead to geometric progressions.

The modern conception of population growth regards the *potential* rate of increase as a more or less fixed species characteristic (cf. Chapman, 1935) governed by life history features; but it considers that this potential rate is ordinarily only partially realized, the "partial potential" characteristic of a particular situation being dependent on environmental conditions. Ecologists commonly associate this concept of "biotic potential" with the name of Chapman (1928, 1935), but actually the concept of populations as systems balanced between a potential ability to grow and an "environmental resistance" dates back at least to the Belgian statistician Quetelet (1835), who considered (p. 277) that potential population growth is a geometric progression, while the resistance to population growth (by analogy with a body falling through a viscous medium) varies as the square of the rate of growth. Only three years later Quetelet's student and colleague Verhulst (1838) set forth the thoroughly modern concept that potential population growth is a geometric progression corresponding to our formula (1′), and that the environmental resistance varies inversely with the unexploited opportunities for growth. By this conception, if K represents the capacity of the environment or the ultimate size which the population can attain, the resistance to population growth increases as $K - P$, the amount of space remaining to be occupied, decreases. As the simplest case Verhulst considered that the resistance is related in a linear manner to the remaining opportunities for growth and thus derived the familiar logistic function as a representation of population growth (for discussion see Allee et al., 1949).

The modern mathematical formulation of population growth, as given, for example, by Rhodes (1940), proceeds by expressing the environmental resistance as some function of population size, $f(P)$, and writing a differential equation of the type

$$\frac{dP}{dx} = rPf(P). \tag{2}$$

By employing different functions for $f(P)$, any number of population growth laws may be derived and the mathematical connection between P and x determined, providing equation (2) can be integrated. Rhodes gives several examples of the procedure.

Formula (1′), the equation of the geometric progression representing population growth in an unlimited environment, represents the special case of formula (2) where the factor $f(P)$ is replaced by a constant, most conveniently by the constant value unity. By the foregoing interpretation it is clear that the constant r must be regarded as a quantity of fundamental ecological significance. It is to be interpreted as the rate of true compound interest at which a population would grow if nothing impeded its growth and if the age-specific birth and death rates were to remain constant.

Quite recently a number of ecologists have recognized the importance of a knowledge of the value of r for non-human populations and have computed its value for various species by employing empirical values of age-specific birth rates and survivorship (Leslie and Ranson, 1940; Birch, 1948; Leslie and Park, 1949; Mendes, 1949; Evans and Smith, 1952). While Chapman's term "biotic potential" would seem to have ecological merit as the name for this parameter r it has been variously called by Lotka the "true," the "incipient," the "inherent," and the "intrinsic" rate of increase, and by Fisher (1930) the "Malthusian parameter" of population increase. Probably for the sake of stabilizing nomenclature it is advisable to follow the majority of recent writers and refer to r as "the intrinsic rate of natural increase."

In the works of Dublin and Lotka (1925),

Kuczynski (1932), and Rhodes (1940) on human populations and in the papers mentioned above dealing with other species, the value of r has typically been determined by some application of three fundamental equations developed by Lotka (1907a, b; Sharpe and Lotka, 1911). He showed that if the age-specific fecundity (b_x) and survivorship (l_x) remained constant, the population would in time assume a fixed or "stable" age distribution such that in any interval of age from x to $x + dx$ there would be a fixed proportion (c_x) of the population. Once this stable age distribution is established the population would grow exponentially according to our formula (1′) and with a birth rate per head, β. Then the following equations relate these quantities:

$$\int_0^{\infty} e^{-rx} l_x b_x \, dx = 1 \tag{3}$$

$$\int_0^{\infty} e^{-rx} l_x \, dx = 1/\beta \tag{4}$$

and

$$\beta e^{-rx} l_x = c_x. \tag{5}$$

While the use of formulas (3), (4), and (5) to compute the value of r often presents practical difficulties owing to the difficulty of approximating the functions l_x and b_x by a mathematical function, and also because the equations usually must be solved by iterative methods, it may fairly be stated that Lotka's pioneer work establishing these relationships provided the methods for interpreting the relationships between life history features and their population consequences.

However, the exceedingly important ecological questions of what potential advantages might be realized if a species were to alter its life history features have remained largely unexplored. Doubtless, as already noted, this is largely to be explained by a certain suspicion felt by biologists toward analyses such as those of Lotka, which seem to involve assumptions very remote from the realities of life histories as observed in the field and laboratories. A particularly pertinent statement of this point of view is that of Thompson (1931), who recognized the great practical need for methods of computing the rate of increase of natural populations of insects adhering to particular life history patterns but who insisted that the reproductive process must be dealt with as a discontinuous phenomenon rather than as a compound interest phenomenon such as that of formula (1′). His methods of computation were designed to give the exact number of individuals living in any particular time period and, while he recognized that the population growth can be expressed in an exponential form such as (1′), he rejected its use on these grounds:

"In the first place, the constant (r) cannot be determined until the growth of the population under certain definite conditions has been studied during a considerable period; in the second place, no intelligible significance can be attached to the constant after its value has been determined; in the third place, the growth of the population is considered in this formula to be at every moment proportional to the size of the population, which is not true except with large numbers and over long periods and cannot be safely taken as a basis for the examination of experimental data."

In the following sections of the present paper an effort will be made to reconcile these two divergent points of view and to show under what conditions Thompson's "discontinuous" approach and the continuous methods lead to identical results. Practical methods of computation can be founded on either scheme, and there are circumstances where one or the other offers distinct advantages. It is hoped that a theoretical approach to population phenomena proceeding from exact computational methods will clarify the meaning of some of the approximations made in deriving equations such as (3), (4), and (5) by continuous methods, and will stimulate students of ecology to a greater interest in the population consequences of life history phenomena.

Before proceeding to a discussion of potential population growth, one point which has sometimes caused confusion should be mentioned. This concerns the sex ratio and the relative proportions of different age classes in the growing population. Once stated, it is obvious that if a population is always growing, as are the populations in the models used for determining potential population growth, then each age and sex class must ultimately come to grow at exactly the same rate as every other class. If this were not the case the disproportion between any two classes would come to exceed all bounds; the fastest growing class would continue indefinitely to make up a larger and larger proportion of the total population. It is thus intuitively recognizable that with fixed life history features there must ultimately be a fixed sex ratio and a stable age distribution. In discussing potential population growth it is often convenient to confine our attention to females or even to a restricted age class, such as the annual births, while recognizing that the ultimate growth rate for such a restricted population segment must be identical to the rate for the entire population.

SIMPLEST CASES OF POPULATION GROWTH

Non-overlapping generations

The simplest possible cases of population growth from the mathematical point of view are those in which reproduction takes place once in a lifetime and the parent organisms disappear by the time the new generation comes on the scene, so that there is no overlapping of generations. This situation occurs in the many plants and animals which are annuals, in those bacteria, unicellular algae, and protozoa where reproduction takes place by fission of one individual to form two or more daughter individuals, and in certain other forms. Thus in the century plants (*Agave*) the plant dies upon producing seeds at an age of four years or more, the Pacific salmon (*Oncorhynchus*) dies after spawning, which occurs at an age of two to eight years (two years in the pink salmon *O. gorbuscha*), and cicadas breed at the end of a long developmental period which lasts from two years (*Tibicen*) to 17 years in *Magicicada*. For many other insects with prolonged developmental stages such as neuropterans and mayflies potential population growth may be considered on the assumption that generations do not overlap.

In these cases, perhaps most typically illustrated in the case of annuals, the population living in any year or other time interval is simply the number of births which occurred at the beginning of that interval. Starting with one individual which is replaced by b offspring each of which repeats the life history pattern of the parent, the population will grow in successive time intervals according to the series: $1, b, b^2, b^3, b^4, \cdots b^x$. Hence the number of "births," say B_x, at the beginning of any time interval, T_x, is simply b^x which is identical with the population, P_x, in that interval of time. If the population starts from an initial number P_0 we have:

$$P_x = P_0 b^x \tag{6}$$

which is obviously identical with the exponential formula (1′), $P_x = Ae^{rx}$, where the constant A is precisely P_0, the initial population size, and $r = \ln b$; the intrinsic rate of increase is equal to the natural logarithm of the litter size.

If litter size varies among the reproductive individuals, with each litter size being characteristic of a fixed proportion of each generation, it is precisely correct to use the average litter size, say $\bar{b}$, in the computations, so that we have $r = \ln \bar{b}$. Furthermore, if not all of the offspring are viable, but only some proportion, say l_1, survive to reproduce, we shall have exactly $r = \ln \bar{b}^{l_1}$. Thus, mortality and variations in litter size do not complicate the interpretation of population growth in cases where the generations do not overlap. On the other hand, even in species which reproduce only once, if the generation length is not the same for all individuals, this will lead to overlapping generations, and the simple considerations which led to formula (6) will no longer apply. In other words, we can use an average figure for the litter size b but not for the generation length x. It will be shown in the next section, however, that the more general formula (1′) is still applicable.

In these simplest cases the assumption of a geometric progression as the potential form of population growth is obviously correct, and numerous authors have computed the fantastic numbers of offspring which could potentially result from such reproduction. For example, according to Thompson (1942), Linnaeus (1740?) pointed out that if only two seeds of an annual plant grew to maturity per year, a single individual could give rise to a million offspring in 20 years. (In all editions available to the present writer this interesting essay of Linnaeus' is dated 1743, and the number of offspring at the end of twenty years is stated by the curious and erroneous figure 91,296.) That is, $P_{20} = 2^{20} = e^{20 \ln 2} = 1{,}048{,}576$. Additional examples are given by Chapman (1935, p. 148).

Formulas (1′) or (6) may, of course, also be used in an inverse manner to obtain the rate of multiplication when the rate of population growth is known. For the example given by Molisch (1938, p. 25), referring to diatoms reproducing by binary fission where the average population was observed to increase by a factor of 1.2 per day, we have $1.2 = e^{x \ln 2}$, where x is the number of generations per day. Solving for $1/x$, the length of a generation, we obtain $1/x = \dfrac{\ln 2}{\ln 1.2} = \dfrac{.69315}{.18232} = 3.8$ days.

Overlapping generations

Interest in computing the number of offspring which would be produced by a species adhering to a constant reproductive schedule dates back at least to Leonardo Pisano (=Fibonacci) who, in the year 1202, attempted to reintroduce into Europe the study of algebra, which had been neglected since the fall of Rome. One of the problems in his *Liber Abbaci* (pp. 283–84 of the 1857 edition) concerns a man who placed a pair of rabbits in an enclosure in order to discover how many pairs of rabbits would be produced in a year. Assuming that each pair of rabbits produces another pair in the first month and then reproduces once more, giving rise to a second pair of offspring in the second month, and assuming no mortality, Fibonacci showed that the number of pairs in each month would correspond to the series

1, 2, 3, 5, 8, 13, 21, 34, 55, etc.,

where each number is the sum of the two preceding numbers. These "Fibonacci numbers" have a rather celebrated history in mathematics, biology, and art (Archibold, 1918; Thompson, 1942; Pierce, 1951) but our present concern with them is merely as a very early attempt to compute potential population growth.

Fibonacci derived his series simply by following through in words all of the population changes occurring from month to month. One with sufficient patience could, of course, apply the same procedure to more complicated cases and could introduce additional variables such as deductions for mortality. In fact, Sadler (1830, Book III) did make such computations for human populations. He was interested in discovering at what ages persons would have to marry and how often they would have to reproduce to give some of the rates of population doubling which had been postulated by Malthus (1798). To accomplish this, Sadler apparently employed the amazingly tedious procedure of constructing numerous tables corresponding to different assumptions until he found one which approximated the desired rate of doubling.

Although we must admire Sadler's diligence, anyone who undertakes such computations will find that it is not difficult to devise various ways of systematizing the procedure which will greatly reduce the labor of computation. By far the best of these methods known to the present writer is that of Thompson (1931), which was originally suggested to him by H. E. Soper.

In the Soper-Thompson approach a "generation law" (G) is written embodying the fixed life history features which it is desired to consider. The symbol T^x stands for the xth interval of time, and a generation law such as $G = 2T^1 + 2T^2$ would be read as "two offspring produced in the first time interval and two offspring produced in the second time interval." This particular generation law might, for example, be roughly applicable to some bird such as a cliff swallow, where a female produces about four eggs per year. Concentrating our attention on the female part of the population, we might wish to compute the rate of population growth which would result if each female had two female offspring upon attaining the age of one year and had two more female offspring at the age of two years. The fundamental feature of the Thompson method is the fact that the expression:

$$\frac{1}{1 - G} \tag{7}$$

is a generating function which gives the series of births occurring in successive time intervals. In the algebraic division the indices of the terms T^1, T^2, etc., are treated as ordinary exponents and the number of births occurring in any time interval T^x is simply the coefficient of T^x in the expansion of expression (7). Thus, for our example where $G = 2T^1 + 2T^2$ we obtain:

$$\frac{1}{1 - 2T^1 - 2T^2} = 1 + 2T^1 + 6T^2 + 16T^3 + 44T^4 + 120T^5 + 328T^6 + \cdots,$$

showing that one original female birth gives rise to 328 female offspring in the sixth year. The series could be continued indefinitely to obtain the number of births any number of years hence. However, in practice it is not necessary to continue the division. In the above series the coefficient of each term is simply twice the sum of the coefficients of the two preceding terms; hence the generation law gives us the rule for extending the series. $G = 2T^1 + 2T^2$ instructs us to obtain each new term of the series by taking twice the preceding term plus twice the second term back. In the case of the Fibonacci numbers we would have $G = T^1 + T^2$, telling us at once that each new term is the sum of the two preceding it.

From the birth series we can easily obtain the series enumerating the total population. If each individual lives for λ years, the total population in T^x will be the sum of λ consecutive terms in the expansion of the generating function. Multiplying formula (7) by the length of life expressed in the form $1 + T^1 + T^2 + T^3 + \cdots + T^{\lambda-1}$ will give the population series. In our above example if we assume that each individual lives for three years, although, as before, it only reproduces in the first two, we obtain for the population

$$\frac{1 + T^1 + T^2}{1 - 2T^1 - 2T^2} = 1 + 3T^1 + 9T^2 + 24T^3 + 66T^4 + 180T^5 + 492T^6 + \cdots,$$

a series which still obeys the rule $G = 2T^1 + 2T^2$.

Thompson's method for obtaining the exact number of births and members of the population in successive time intervals is very general. As in the case of non-overlapping generations, the coefficients in the generation law may refer to average values for the age-specific fecundity. Also the length of the time intervals upon which the computations are based can be made arbitrarily short, so that it is easy to take into account variations in the age at which reproduction occurs. For the above example; time could have been measured in six-month periods rather than years so that the generation law would become $G = 2T^2 + 2T^4$, with the same results already obtained.

Furthermore, the factor of mortality can easily be included in the computations. For example, suppose that we wish to determine the rate of population growth for a species where the females have two female offspring when they reach the age of one, two more when they reach the age of two, and two more when they reach the age of three. Neglecting mortality, this would give us the generation law $G = 2T^1 + 2T^2 + 2T^3$. If we were further interested in the case where not all of the offspring survive for three years, the coefficients in the generation law need only be multiplied by the corresponding survivorship values. For example, if one-half of the individuals die between the ages of one and two, and one half of the remainder die before reaching the age of three we would have $l_1 = 1$, $l_2 = 1/2$, $l_3 = 1/4$, and the above generation law would be revised to $G = 2T^1 + T^2 + 1/2T^3$. The future births per original individual would then be

$$\frac{1}{1 - G} = 1 + 2T^1 + 5T^2 + 25/2T^3 + 31T^4 + 151/2T^5 + \cdots.$$

Very generally, if the first reproduction for a species occurs at some age α and the last reproduction occurs at some age ω, and letting b_x and l_x represent respectively the age-specific fecundity and survivorship, we may write the generation law as:

$$G = l_\alpha b_\alpha T^\alpha + l_{\alpha+1} b_{\alpha+1} T^{\alpha+1} + \cdots + l_\omega b_\omega T^\omega = \sum_{x=\alpha}^{\omega} l_x b_x T^x. \quad (8)$$

Therefore, in the Thompson method we have a compact system of computation for obtaining the exact number of births and the exact population size at any future time, assuming that the significant life history features (α, ω, l_x, and b_x) do not change.

Not all of the possible applications of Thompson's method have been indicated above. For example, formula (7) may be used in an inverse manner so that it is theoretically possible to work back from a tabulation of births or population counts made in successive time intervals and discover the underlying generation law. Formulas (7) and (8), together with the procedure of multiplying the birth series by the length of life expressed as a sum of T^x values, provide the nucleus of the system and offer the possibility of analyzing the potential population consequences of essentially any life-history phenomena. The system has the merit of treating the biological units and events as discontinuous variates, which, in fact, they almost always are. The members of populations are typically discrete units, and an event such as reproduction typically occurs at a point in time with no spreading out or overlapping between successive litters. While survivorship, l_x, as a population quantity, is most realistically regarded as continuously changing in time, the product $l_x b_x$ which enters our computations by way of formula (8) is typically discontinuous because of the discontinuous nature of b_x.

It is quite obvious that equations of continuous variation such as (1′) are often much more convenient for purposes of computation than the series of values obtained by expanding (7). This is especially true in dealing with the life histories of species which have long reproductive lives. In writing a generation law for man by (8) we should have to take α at least as small as 15 years and ω at least as great as 40 years, since for the population as a whole reproduction occurs well outside of these extremes and it would certainly be unrealistic to regard b_x as negligibly small anywhere between these limits. Thus there would be at least 25 terms in our generation law, and the computations would be extremely tedious. By selecting special cases

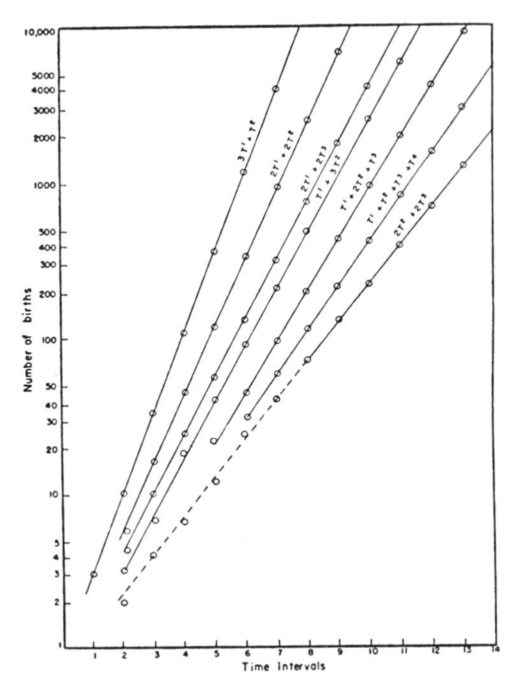

FIG. 1. EXACT VALUES OF POPULATION GROWTH IN TERMS OF BIRTHS PER UNIT TIME UNDER SEVERAL GENERATION LAWS, WHEN EACH FEMALE HAS A TOTAL OF FOUR FEMALE OFFSPRING

In each case it is assumed that a single female exists at time zero and produces her four progeny on or before her fourth birthday. The plotted points represent exact values as determined by Thompson's method. To the extent that the points for any generation law fall on a straight line in this logarithmic plot, they can be represented by the exponential growth formula (1′), and the slope of each line is a measure of the intrinsic rate of natural increase (r).

for study it is sometimes possible greatly to simplify the procedures. For example, if one is interested in the case where there is no mortality during the reproductive span of life and where the litter size is a constant, say b, the expression for the generation law (8) can be simplified to:

$$G = bT^{\alpha} + bT^{\alpha+1} + \cdots bT^{\omega} = \frac{bT^{\alpha} - bT^{\omega+1}}{1 - T}.$$

Since one can also write the length of life as

$$1 + T^1 + T^2 + \cdots + T^{\lambda-1} = \frac{1 - T^{\lambda}}{1 - T},$$

the generating function for the total population simplifies to

$$\frac{1 - T^{\lambda}}{1 - T - bT^{\alpha} + bT^{\omega+1}}.$$

This last formula is much more convenient for computations than one containing 25 terms or so in the denominator, but it applies only to a very special case and is much less convenient than formula (1′). Consequently, great interest attaches to these questions: can (1′) be used as a substitute for (7)? (i.e., does Thompson's method lead to a geometric progression?) and, if it is so used, can the constants, particularly r, be interpreted in terms of life-history features?

THE GENERALIZATION OF THOMPSON'S METHOD

Fig. 1 shows the exact values, as determined by Thompson's method, of the birth series arising from several generation laws (life-history patterns) which have in common the feature that in each case every female produces a total of four female offspring in her lifetime and completes her reproductive life by the age of four "years." The number of births is plotted on a logarithmic scale, hence if it can be represented by formula (1′), $P = Ae^{rx}$ or, logarithmically, $\ln P = \ln A + rx$, the points should fall on a straight line with slope proportional to r. It is apparent from Fig. 1 that after the first few time intervals the points in each case are well represented by a straight line. Therefore, except in the very early stages, formula (1′) does give a good representation of potential population growth. The question remains, however, as to whether we can meet Thompson's objection to (1′) and attach any intelligible significance to the constants of the formula. From Fig. 1 it is obvious that the lines do not, if projected back to time 0, indicate exactly the single individual with which we started. Thus, in these cases the constant A cannot be precisely P_0 as was the case with non-overlapping generations.

Before proceeding to interpret the constants of formula (1′) for the case of overlapping generations it will be well to notice one feature of Fig. 1 which is of biological interest. In the literature of natural history one frequently encounters references to the number of offspring which a female can produce per lifetime, with the implication that this is a significant life-history feature. The same implication is common in the literature dealing with various aspects of human biology, where great emphasis is placed on the analysis of total family size. From Fig. 1 it will be seen that this datum may be less significant from the standpoint of contributions to future population than is the age schedule upon which these offspring are produced. Each life history shown in Fig. 1 represents a total production of four offspring within four years of birth, but the resulting rates of potential population growth are very different for the different schedules. It is clear that the cases of most rapid population growth are associated with a greater concentration of reproduction into the early life of the mother. This is intuitively reasonable because we are here dealing with a compound interest phenomenon and should expect greater yield in cases where "interest" begins to accumulate early. However, the writer feels that this phenomenon is too frequently overlooked in biological studies, possibly because of the difficulty of interpreting the phenomenon quantitatively.

In seeking to reconcile the continuous and discontinuous approaches to potential population growth, let us first note that Thompson's discontinuous method corresponds to an equation of finite differences. We have seen above that the generation law gives us a rule for indefinitely extending the series representing the population size or the number of births in successive time intervals by adding together some of the preceding terms multiplied by appropriate constants. If we let $f_{(x)}$ represent the coefficient of T^x in the expansion of the generating function (7) and, for brevity, write in (8) $V_x = l_x b_x$, then our population series obeys the rule:

$$f_{(x)} = V_\alpha f_{(x-\alpha)} + V_{\alpha+1} f_{(x-\alpha-1)} + \cdots + V_\omega f_{(x-\omega)}, \quad (9)$$

which may be written in the alternative form,

$$f_{(x+\omega)} - V_\alpha f_{(x+\omega-\alpha)} - V_{\alpha+1} f_{(x+\omega-\alpha-1)} - \cdots - V_\omega f_{(x)}, = 0. \quad (10)$$

Thus for our "cliff swallow" example, where we had $G = 2T^1 + 2T^2$ we have

$$f_{(x)} = 2f_{(x-1)} + 2f_{(-2)} \text{ or,}$$
$$f_{(x+2)} - 2f_{(x+1)} - 2f_{(x)} = 0.$$

Formula (10) represents the simplest and best understood type of difference equation, a homogeneous linear difference equation with constant coefficients. It is outside the scope of the present paper to discuss the theory of such equations, which has been given, for example, by Jordan (1950). By the nature of our problem as summarized in formula (9), all of our V_x values are either equal to zero or are positive real numbers and all of the signs of the coefficients in (9) are positive: features which considerably simplify generalizations. By virtue of these facts it can be shown that there is always a "characteristic" algebraic equation corresponding to (10). This is obtained by writing ρ^x for $f_{(x)}$ and dividing through by the ρ value of smallest index. This gives

$$\rho^\omega - V_\alpha \rho^{\omega-\alpha} - V_{\alpha+1}\rho^{\omega-\alpha-1} \cdots - V_\omega = 0 \quad (11)$$

an algebraic equation which has the roots ρ_1, ρ_2, etc.

The general solution of the corresponding difference equation (10) is

$$f_{(x)} = C_1\rho_1^x + C_2\rho_2^x + \cdots + C_n\rho_n^x \quad (12)$$

where the C's are constants to be determined by the initial conditions of the problem. Formula (12) is precisely equivalent to Thompson's method and is a general expression for the number of births or the population size in any future time interval.

As an example we may consider the case where $G = 2T^1 + 2T^2$. The difference equation, as already noted, is $f_{(x+2)} - 2f_{(x+1)} - 2f_{(x)} = 0$ and the characteristic algebraic equation is $\rho^2 - 2\rho - 2 = 0$ which is a quadratic equation with the roots $\rho_1 = 1 + \sqrt{3}$, and $\rho_2 = 1 - \sqrt{3}$. Hence the general solution is $f_{(x)} = C_1(1+\sqrt{3})^x + C_2(1-\sqrt{3})^x$. To determine the constants C_1 and C_2 we look at the beginning a of the seriesnd note that we have $f_{(0)} = 1$ and $f_{(1)} = 2$. Substituting these values in the general solution we obtain $C_1 = \frac{\sqrt{3}+1}{2\sqrt{3}}$ and $C_2 = \frac{\sqrt{3}-1}{2\sqrt{3}}$. Therefore, the general expression for the number of births in time interval T^x is

$$f_{(x)} = \frac{\sqrt{3}+1}{2\sqrt{3}}(1+\sqrt{3})^x + \frac{\sqrt{3}-1}{2\sqrt{3}}(1-\sqrt{3})^x$$

which can be simplified to $f_{(x)} = \frac{\rho_1^{x+1} - \rho_2^{x+1}}{\sqrt{3}} = \rho_1^x + \rho_1^{x-1}\rho_2 + \cdots + \rho_2^x$.

In order to have the difference equation (12) correspond to the equation of exponential growth (1′), the ratio between populations in successive time intervals must assume a constant value giving

$$\frac{f_{(x+1)}}{f_{(x)}} = e^r. \quad (13)$$

By the nature of our problem, as already noted, the potential population growth is always positive, so that any limit approached by the ratio $\frac{f_{(x+1)}}{f_{(x)}}$ must be a positive real number.

It is beyond the scope of the present paper to discuss the conditions, for difference equations in general, under which this ratio does approach as a limit the largest real root of the characteristic algebraic equation. (See, for example, Milne-Thompson, 1933, chap. 17). Dunkel (1925) refers to the homogeneous equation with real constant coefficients corresponding to our formulas (10) and (11). The algebraic equation (11) has a single positive root which cannot be exceeded in absolute value by any other root, real or complex. Using (12) to express the ratio between successive terms, we have

$$\frac{f_{(x+1)}}{f_{(x)}} = \frac{C_1\rho_1^{x+1} + C_2\rho_2^{x+1} + \cdots + C_n\rho_n^{x+1}}{C_1\rho_1^{x} + C_2\rho_2^{x} + \cdots + C_n\rho_n^{x}}. \quad (14)$$

If we let ρ_1 represent the root of (11) of greatest absolute value and divide both numerator and denominator of (14) by $C_1\rho_1^x$ we obtain

$$\frac{f_{(x+1)}}{f_{(x)}} = \rho_1\left[\frac{1+\frac{C_2}{C_1}\left(\frac{\rho_2}{\rho_1}\right)^{x+1}+\frac{C_3}{C_1}\left(\frac{\rho_3}{\rho_1}\right)^{x+1}+\cdots+\frac{C_n}{C_1}\left(\frac{\rho_n}{\rho_1}\right)^{x+1}}{1+\frac{C_2}{C_1}\left(\frac{\rho_2}{\rho_1}\right)^{x}+\frac{C_3}{C_1}\left(\frac{\rho_3}{\rho_1}\right)^{x}+\cdots+\frac{C_n}{C_1}\left(\frac{\rho_n}{\rho_1}\right)^{x}}\right]. \quad (15)$$

The expressions in parentheses are all less than unity, on the assumption that ρ_1 is the largest root, and the entire expression in brackets approaches unity as x increases. Consequently we have, for x large

$$\frac{f_{(x+1)}}{f_{(x)}} \sim \rho_1 \sim e^r. \quad (16)$$

This then explains the shape of the potential birth and population series as illustrated in Fig. 1. In the very early stages population growth is irregular, because the expressions in (12) and (15) involving the negative and complex roots of (11) are still large enough to exert an appreciable influence. As x increases, the influence of these other roots becomes negligible and the population grows exponentially, conforming to (16). In considering potential population growth we are concerned with the ultimate influence of life-history features, and the equation of geometric progression or compound interest does actually represent the form of potential population growth. We are interested only in the single positive root of (11) for the purpose of determining the constant r, and this can readily be computed with any desired degree of precision by elementary algebraic methods.

Having established the relationship of formula (13) or (16), it is easy to reconcile Thompson's discontinuous approach to population growth with Lotka's continuous approach, as exemplified by formulas (3), (4), and (5).

Employing formula (9) we may write the ratio between populations in successive time intervals as

$$\frac{(f_{x+1})}{f_{(x)}} = V_\alpha\frac{f_{(x-\alpha+1)}}{f_{(x)}} + V_{\alpha+1}\frac{f_{(x-\alpha)}}{f_{(x)}} + \cdots + V_\omega\frac{f_{(x-\omega+1)}}{f_{(x)}}.$$

Substituting the relationship given by (13), this becomes

$$e^r = V_\alpha e^{-r(\alpha-1)} + V_{\alpha+1}e^{-r\alpha} + \cdots + V_\omega e^{-r(\omega-1)}, \text{ or}$$

$$1 = V_\alpha e^{-r\alpha} + V_{\alpha+1}e^{-r(\alpha+1)} + \cdots + V_\omega e^{-r\omega}.$$

Replacing V_x by its equivalent, l_xb_x, this is

$$1 = \sum_{x=\alpha}^{\omega} e^{-rx}\, l_x b_x. \quad (17)$$

Formula (17) is the precise equivalent in terms of finite time intervals of Lotka's equation (3) for infinitesimal time intervals. In Lotka's equation, as in (17), the limits of integration in practice are α and ω since b_x is zero outside of these limits. Formula (17) was in fact employed by Birch (1948) as an approximation to (3) in his method of determining r for an insect population. The only approximation involved in our derivation of (17) is the excellent one expressed by formula (13); otherwise the formula corresponds to Thompson's exact computational methods. It is hoped that recognition of this fact will make some of the approaches of population mathematics appear more realistic from the biological point of view.

Formulas (4) and (5), originally due to Lotka, are also immediately derivable from the relationship (13). In any time interval, T_x, we may say that the population members aged 0 to 1 are simply the births in that interval, say B_x. The population members aged 1 to 2 are the survivors of the births in the previous interval, that is $l_1 B_{x-1}$, or employing (13), $l_1 B_x e^{-r}$. Quite generally, the population members aged between z and $z + 1$ are the survivors from the birth z intervals previous, or $l_z B_x e^{-rz}$. If λ is the extreme length of life for any population members ($l_{\lambda+1} = 0$) we have for the total population

$$P_x = B_x(1 + l_1 e^{-r} + l_2 e^{-2r} + \cdots + l_\lambda e^{-r\lambda} = B_x \sum_{x=0}^{\lambda} e^{-rx} l_x.$$

The birth rate per individual, β, is B_x/P_x, therefore,

$$1/\beta = \sum_{0}^{\lambda} e^{-rx} l_x \qquad (18)$$

which is the equivalent in finite time intervals of Lotka's equation (4). Also the proportion, c_z, of the population in the age range z to $z + 1$ is $\dfrac{l_z B_x e^{-rz}}{P_x}$ which is simply,

$$c_z = \beta e^{-rz} l_z. \qquad (5)$$

COMPUTATIONAL METHODS

In the following sections we will examine some of the population effects which are the consequences of particular life history patterns. Probably the most significant comparisons are those involving the effects of life-history features on the intrinsic rate of natural increase, r. Of course, any change in r is accompanied by other effects, such as those on the age-structure and on the population birth-rate. However, the intrinsic rate of increase is a parameter of fundamental ecological importance. If a species is exposed to conditions which would favor the ability to outbreed competitors or where exceptional hazards limit the probability that an individual will become established, we might expect to find life-history adjustments tending to increase the value of r. Conversely, if a species has evolved life-history features of a type tending to hold down the intrinsic rate of increase, a fertile field of inquiry may be opened regarding the selective factors to which such a species is subject.

It is probably fairly obvious to anyone that in general a species might increase its biotic potential by increasing the number of offspring produced at a time (litter size), by reducing mortality at least until the end of active reproductive life, by reproducing oftener, by beginning reproduction at an earlier age, or by minimizing any wastage of environmental resources on sterile members of the population. Any biologist will at once recognize, however, that a great deal of evolution (an extreme case is the evolution of sterility in the social insects) has proceeded in precisely the wrong direction to increase biotic potential by some of these devices. Presumably, this can only mean that the optimum biotic potential is not always, or even commonly, the maximum that could conceivably be achieved by selecting for this ability alone. Comparative life-history studies appear to the writer to be fully as meaningful in evolutionary terms as are studies of comparative morphology or comparative physiology.

Although a great many empirical data on life histories have been accumulated, attempts to interpret these data comparatively have lagged far behind the corresponding efforts in morphology and physiology. The methods exhibited in the preceding parts of the present paper are adaptable for the quantitative interpretation of life history features and, while the number of conceivable life-history patterns is infinite, we propose to examine some of the cases which appear to possess particular ecological interest.

The life-history features with which we are concerned are the age at which reproduction begins (α), the litter size and frequency of reproduction (both summarized by a knowledge of the function b_x, which can also be computed so as to take account of the sex ratio), the maximum age at which reproduction occurs (ω), survivorship (l_x), and maximum longevity (λ). Corresponding to any given set of values for these quantities there is a definite value for the intrinsic rate of natural increase (r) and a definite stable age distribution of the population (c_x). In general, these population features will be altered by any alteration of the life-history features and we wish to examine some of these possible changes quantitatively.

The most efficient way of making the desired computations will vary from problem to problem. Thompson's method (formulas (7) and (8)), could be used to obtain exact population values arising from any life history, but the computations would in many cases be exceedingly laborious and would actually uield no more information about the ultimate course of population growth than would be obtained by solving (11) for the positive root.

In either case it will usually be most efficient to measure time in terms of the shortest interval between the pertinent life-history events with which we are concerned.

Except in very special cases, it is necessary to use iterative methods for obtaining the value of r corresponding to particular life-history patterns. In most cases the solutions are quite rapidly obtained by employing a calculating machine and detailed tables of natural logarithms (e.g., Lowan, 1941) or of the exponential function (e.g., Newman, 1883). In the majority of the cases considered by the writer, the most efficient procedure has been to rewrite formula (17) in the form:

$$e^{r\alpha} = V_\alpha + V_{\alpha+1}e^{-r} + V_{\alpha+2}e^{-2r} + \cdots + V_\omega e^{-r(\omega-\alpha)} \qquad (19)$$

and then to obtain the sum of the series on the right-hand side of (19) for different patterns of variation in the function $l_x b_x = V_x$. This method corresponds exactly to the discontinuous approach, granting only that potential population growth is a geometric progression, and it leads to relatively simple equations in a number of the cases of greatest ecological interest.

A more general approach from the standpoint of formal mathematics can be obtained by rewriting (3) in the form of a Stieltjes integral (Widder, 1940). We may define a maternity function $M(x)$ representing the average number of offspring which an individual will have produced by the time it has attained any age x, and such that its derivative with respect to time is $\left(V_x \frac{d}{dx}M(x) = V_x\right)$. We then have

$$\int_0^\infty e^{-rx}\,dM(x) = 1 \qquad (20)$$

which can represent cases where $V_{(x)}$ is either continuous or discontinuous because the integral vanishes for values where V_x is discontinuous. When V_x can be expressed as a function of time (x), formula (20) is identical with (3) and the use of the Laplace transformation, a procedure of considerable importance in engineering and physical mathematics, makes it possible to avoid the numerical integration and express V_x as a function of r. If V_x is considered as a series of single impulses regularly spaced from α to ω, equation (20) assumes the form (17). Laplace transformations for a number of functions are tabulated by Churchill (1944) and Widder (1947) and, no doubt, there are cases where this procedure would lead to simpler iterative solutions than those obtained from equation (19). For the cases considered in the present paper, however, the solution of equation (19) generally leads to somewhat simpler results.

In dealing with any particular life-history pattern the computational method of choice may depend upon the types of features to be investigated. The pure numbers α, ω, and λ typically offer no particular computational problems, as they are assigned different values, but this is not always the case with the functions b_x and l_x.

In the cases considered by the writer the intervals between successive periods of reproduction have been considered to be equal. There is no particular difficulty in altering this assumption so as to consider cases where the frequency of reproduction varies with age, but regular spacing seems to be so much more usual in nature as well as representing a limiting case that it seems to merit first consideration. Litter size often does vary with the age of the parent organism, and this fact may introduce complexities into the behavior of the function V_x. In this case also, it appears that the ecologically most interesting cases are those in which the average litter size is a constant. Furthermore, as will become apparent in later sections, the first few litters produced by an organism so dominate its contribution to future population growth that later changes in litter size would have only very minor population consequences. In dealing with empirical data on human populations attempts have been made to express analytically the changes in b_x with age [cf. "Tait's law" that fertility declines in a linear manner (Yule, 1906; Lotka, 1927)] but for the present we shall consider that b_x assumes only the values zero and some constant, b.

The shape of the survivorship (l_x) curve is more difficult to deal with in a realistic manner. Pearl and Miner (1935) originated the classification of survivorship curves which is most employed for ecological purposes (cf. Deevey, 1947; Allee et al., 1949). The "physiological" survivorship curve is the limiting type in which each individual lives to some limit characteristic of the species and the age at death (λ) is regarded as a constant. In this case $l_x = 1$ when $x \leqslant \lambda$ and $l_x = 0$ when $x > \lambda$. This is the simplest case for computations, and actual cases are known which approach this type. Furthermore, there are other types of survivorship curves of ecological interest which may be treated in the same manner. In what Deevey (1947) calls Type III there is an extremely heavy early mortality with the few survivors tending to live out a "normal life span." For the computation of r we are only concerned (cf. formula 19) with survivorship during the reproductive span of life, and it appears likely that "Type III" curves can be treated as constant throughout this age range without serious error. Another interesting type of survivorship curve which appears to be consistent with empirical data at least on some wild populations (cf. Jackson, 1939; Deevey, 1947; Ricker, 1948) is that in which a constant proportion of the population dies in each interval of age. This, of course, implies that life expectancy is independent of age, an assumption which cannot in general be considered realistic but which might apply to catastrophic causes of mortality. When this type of l_x curve applies, the V_x values will be in geometric progression and the right side of formula (19) can be summed as easily as in the case where V_x is constant. This case is, therefore, easily dealt with.

The type of survivorship curve usually observed in actual cases is a reverse sigmoid curve, interpreted by Deevey as intermediate between the "physiological" type and the geometric progression. This can be interpreted in various ways as a "wearing-out" curve. Gompertz (1825) attempted to find an analytical form on the assumption that the ability of individuals to "resist destruction" decreases as a geometric progression with age. Elston (1923) has reviewed formulas proposed to represent human mortality; none of these has proved generally applicable, despite great complexity in some cases. Another approach is to assume that some sort of a "vital momentum" (Pearl, 1946) or ability to survive is distributed among the members of the population in the form of a bell-shaped or "normal" frequency distribution. This point of view is a familiar and controversial one in the recent literature on bio-assay problems (Finney, 1947, 1949; Berkson, 1944, 1951) and, at least to the extent that a bell-shaped curve can represent the empirical distribution of ages at death, a probit function or a logit function (Berkson, 1944) can be used to represent l_x.

In the present paper we are concerned primarily with the limiting cases or the *potential* meaning of life-history phenomena. Consequently the writer has chosen to deal with survivorship curves of the physiological type and thus to investigate the ultimate effects of life-history phenomena for a species which is able to reduce mortality during the reproductive part of the life span to a negligible value. Our general conclusions will not be seriously altered even by rather startling drastic alterations of this assumption, and, in any case, our results will indicate the maximum gain which a species might realize by altering its life-history features.

Perhaps the most fundamental type of life-history pattern to be investigated in terms of population consequences is that in which the individuals are assumed to produce their first offspring at the age of α "years" with the mean litter size being a constant, b. A second litter is produced at age $\alpha + 1$ and an additional litter in each subsequent interval of age out to, and including, age ω. The total number of litters produced per individual is then $n = \omega - \alpha + 1$.

We then have, from (19),

$$e^{r\alpha} = b(1 + e^{-r} + e^{-2r} + \cdots + e^{-r(\omega-\alpha)}).$$

The expression in parentheses is a geometric progression the sum of which is $\frac{1 - e^{-rn}}{1 - e^{-r}}$. Consequently, the general implicit equation for r under these conditions may be written

$$1 = e^{-r} + be^{-r\alpha} - be^{-r(n+\alpha)} \qquad (21)$$

which may be solved by trial and error by employing a table of the descending exponential function.

Alternative formulas corresponding to (21) may be obtained by the use of the Laplace transformation. In the case where reproduction is considered to occur as a series of regularly spaced impulses, this approach leads to formula (21). Another approach is to consider that $V_x = 0$ when $x < \alpha$, $V_x = b$ when $\alpha \leqq x \leqq \omega$, and $V_x = 0$ when $x > \omega$. The Laplace transformation of a step-function is then employed, leading to the formula

$$\frac{be^{-r\alpha}}{r} - \frac{be^{-r(\omega+1)}}{r} = 1. \qquad (22)$$

Formula (22) and formula (21) would be identical under the condition that $r + e^{-r} = 1$, which is approximately true when r is small. If one desires more nearly to reconcile the continuous and discontinuous approaches in this case, he may note that in formula (21) he is finding the area under a "staircase-shaped" curve with the first vertical step located at $x = \alpha$, whereas in formula (22) he is finding the area under a straight line paralleling the slope of the staircase. It is apparent that the two areas will be more nearly identical if the straight line is started about one-half unit of time earlier. If we substitute in (22) $\alpha - \frac{1}{2}$ for α and $\omega - \frac{1}{2}$ for ω we obtain a formula which gives results for practical purposes identical with those obtained from (21). The formulas are about equally laborious to solve, and the writer has employed (21) for the following computations because of its more obvious relationship to the exact computational methods.

POSSIBLE VALUES OF REPEATED REPRODUCTION (ITEROPARITY)

One of the most significant of the possible classifications of life histories rests on the distinction between species which reproduce only once in a lifetime and those in which the individuals reproduce repeatedly. This being the case, it is very surprising that there seem to be no general terms to describe these two conditions. The writer proposes to employ the term semelparity to describe the condition of multiplying only once in a lifetime, whether such multiplication involves fission, sporulation, or the production of eggs, seeds, or live young. Thus nearly all annual plants and animals, as well as many protozoa, bacteria, insects, and some perennial forms such as century plants and the Pacific salmon, are semelparous species. The contrasting condition will be referred to as iteroparity. Iteroparous species include some, such as small rodents, where only two or three litters of young are produced in a lifetime, and also various trees and tapeworms where a single individual may produce thousands of litters. The distinction between annual and perennial plants is doubtless the most familiar dichotomy separating semelparous and iteroparous species, but general consideration of the possible importance of these two distinct reproductive habits illustrates some points of ecological and evolutionary interest. For purposes of illustration we shall first consider cases where the time interval between reproductive efforts is fixed at one year.

Many plants and animals are annuals. This is true, for example, of many of the higher fungi and seed plants, of insects, and even of a few vertebrates. One feels intuitively that natural selection should favor the perennial reproductive habit because an individual producing seeds or young annually over a period of several years obviously has the potential ability to produce many more offspring then is the case when reproduction occurs but once. It is, therefore, a matter of some interest to examine the effect of iteroparity on the intrinsic rate of natural increase in order to see if we can find an explanation for the fact that repeated reproduction is not more general.

Let us consider first the case of an annual plant (or animal) maturing in a single summer and dying in the fall at the time of reproduction. We have seen earlier (formula 16, seq.) that if b is the number of offspring produced by such an annual the intrinsic rate of increase would be the natural logarithm of b. We wish to determine by how much this would be increased if the individual were to survive for some additional years, producing b offspring each year. Obviously, an annual species with a litter size of one (or an average of one female per litter in sexual species) would merely be replacing current population and no growth would be possible ($\ln 1 = 0$); therefore, when the litter size is one the species must necessarily be iteroparous.

The most extreme case of iteroparity, and the one exhibiting the absolute maximum gain which could be achieved by this means, would be the biologically unattainable case of a species with each individual producing b offspring each year for all eternity and with no mortality. In this case we have $\alpha = 1$ "year" and, since ω is indefinitely large, the final term $be^{-r(\omega+1)}$ in equation (21) becomes zero. Thus we have

$$r = \ln(b + 1) \tag{23}$$

which is to be contrasted with $r = \ln(b)$ for the case of an annual. *For an annual species, the absolute gain in intrinsic population growth which could be achieved by changing to the perennial reproductive habit would be exactly equivalent to adding one individual to the average litter size.* Of course, this gain might be appreciable for a species unable to increase its average litter size. The extreme gain from iteroparity for a species with a litter size of two would be ($\ln 3/\ln 2$) or an increase of about 58 per cent, for a species with a litter size of four the increase would be about 16 per cent, but for one producing 30 offspring in a single reproductive period the extreme gain would amount to less than one per cent. It seems probable that a change in life history which would add one to the litter size would be more likely to occur than a change permitting repeated reproduction, which in many cases would necessitate adjustments to survive several seasons of dormancy. It appears that for the usual annual plants and insects with their relatively high fecundity any selective pressure for perennial reproduction as a means of increasing biotic potential must be negligible.

The above conclusion, which appears surprising when first encountered, arouses curiosity as to why iteroparity exists at all. Perhaps some species are physiologically unable to increase their fecundity. This must, however, be unusual and we are led to investigate whether the situation would be different for a species with a prolonged period of development preceding reproduction. One thinks immediately of the giant Sequoias which require a century to mature and begin reproduction but which, once started, produce large numbers of seeds biennially for centuries.

In order to investigate this question we may again compare the intrinsic rate of increase for a single reproduction with that corresponding to an infinite number of reproductions. This procedure will, of course, tend to overestimate the possible gain from iteroparity although it will set an upper limit, and the first few reproductive periods so dominate the situation that even for very modest litter sizes there is a negligible difference between the results of a very limited number of reproductive periods and an infinite number.

For α not necessarily equal to one, formula (21) gives

$$b = e^{r\alpha} - e^{r(\alpha-1)}, \qquad (24)$$

an implicit equation for r which must be solved by iterative means.

Fig. 2 was constructed from formula (24) to show the relationship between the age at which reproduction begins (α) and the litter size (b) in terms of the possible gain in intrinsic rate of increase which could be achieved by iteroparity. The ordinates represent the proportionate increase in the value of r which could be achieved by changing from a single reproductive effort at age α to an infinite number at ages α, $\alpha + 1$, $\alpha + 2$, etc. The curves all slope upward, indicating that species with long pre-reproductive periods could gain more

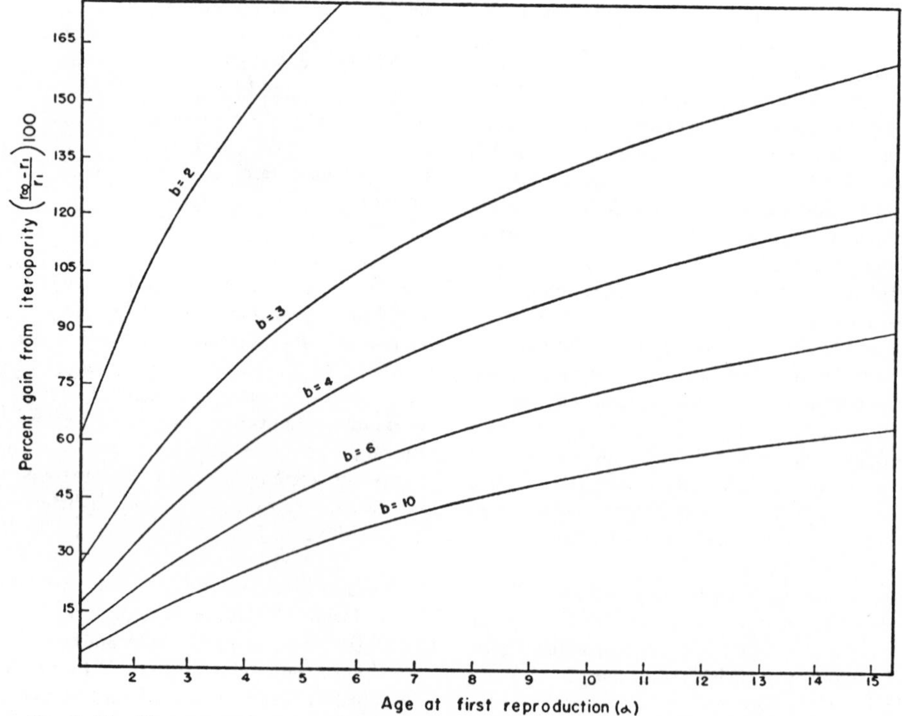

FIG. 2. THE EFFECTS OF LITTER SIZE (b) AND AGE AT MATURITY (α) ON THE GAINS ATTAINABLE BY REPEATED REPRODUCTION

The litter size, b, is the number of female offspring per litter in the case of sexual species.

from iteroparity than forms which mature more rapidly. The tendency of the curves to flatten out with large values of α, however, indicates that the advantages of repeated reproduction increase somewhat less rapidly as the pre-reproductive period is prolonged.

The relationship of iteroparity to litter size is clearly illustrated by Fig. 2. When the litter size is small, as shown by the curve for $b = 2$ (which would correspond to a litter of four individuals when the sex ratio is 1:1), iteroparity can yield important gains in biotic potential, and the possible gains are greater the longer maturity is delayed. The possible advantages diminish quite rapidly as litter size is increased, although it is clear that iteroparity as contrasted with a single reproductive effort would always add something to biotic potential.

Fig. 2 suggests that for semelparous species with large litters there would be very slight selective pressure in favor of adopting the iteroparous habit, and that for iteroparous species with large litters there would be little selection against loss of the iteroparous habit, especially in forms which mature rapidly. On the other hand, in a species which is established as iteroparous there would be slight selection for increasing fecundity or if litter size is relatively large, even against loss of fecundity. This perhaps explains the notoriously low level of viability among the seeds of many trees.

From these considerations it is obvious that when a species could benefit by an increase in the intrinsic rate of natural increase, this advantage might be achieved either by increasing fecundity in a single reproductive period or by adopting the iteroparous habit. A selective advantage would accrue to a mutation altering the life history in either of these directions, and it is an interesting field for speculation as to which type of mutation might be most likely to occur. In this connection it may be interesting to determine the amount of increase in litter size which, for a semelparous species, would be equivalent to retaining the initial litter size but becoming iteroparous.

From (6) we have seen that the intrinsic rate of increase for a semelparous species is defined by $e^{r\alpha} = b$. We wish to find an equivalence factor (E) which will indicate by how much b must be increased to make the value of r for a semelparous species equal to that in formula (21) referring to an iteroparous species. By neglecting the last term in (21) so as to consider the most extreme case of iteroparity and substituting $Eb = e^{r\alpha}$, we obtain

$$E = \frac{1}{1 - e^{-r}} = \frac{e^{r\alpha}}{b} \tag{25}$$

where the value of r must be obtained by solving equation (21). When E is plotted against α for various values of b, as shown in Fig. 3, the resulting curves are essentially straight lines.

Fig. 3 illustrates some interesting points bearing on the life histories of organisms, such as tapeworms and many trees, which are iteroparous in addition to producing large litters. From the arrangement of Fig. 2 one might suspect that the iteroparous habit would provide very little advantage to a species that could produce a thousand or so offspring in a single litter, but Fig. 3 indicates that the selective value of iteroparity may be greatly increased when the pre-reproductive part of the life span is prolonged.

A mature tapeworm may produce daily a number of eggs on the order of 100,000 and may continue this for years (Allee et al., 1949, p. 272; Hyman, 1951). With so large a litter size one wonders if iteroparity in this case may not represent something other than an adaptation for increasing biotic potential. Perhaps the probability that a tapeworm egg (or a *Sequoia* seed) will become established may be increased by distributing the eggs more widely in time and space, and this could conceivably be the reason for the iteroparous habit. No definite answer to this problem is possible at present, but Fig. 3 indicates that a knowledge of the length of the life cycle from egg to egg is an essential datum for considering the question. In at least some tapeworms a larva may grow into a mature worm and reproduce at an age of 30 days (Wardle and McLeod, 1952). If this represented the length of the entire life cycle, then Fig. 3 indicates, assuming $b = 100{,}000$, that a threefold increase in litter size would be the equivalent of indefinite iteroparity. However, with the larval stage in a separate host, the average life cycle must be much longer. If the total cycle requires as much as 100 days, Fig. 3 shows that it would require almost an eight-fold increase in litter size (a single reproductive effort producing 790,167 offspring) to yield the same biotic potential as iteroparity with a litter size of 100,000. Obviously, it is possible, when the life cycle is sufficiently prolonged, to reach a point where any attainable increase in litter size would be less advantageous for potential population growth than a change to the iteroparous

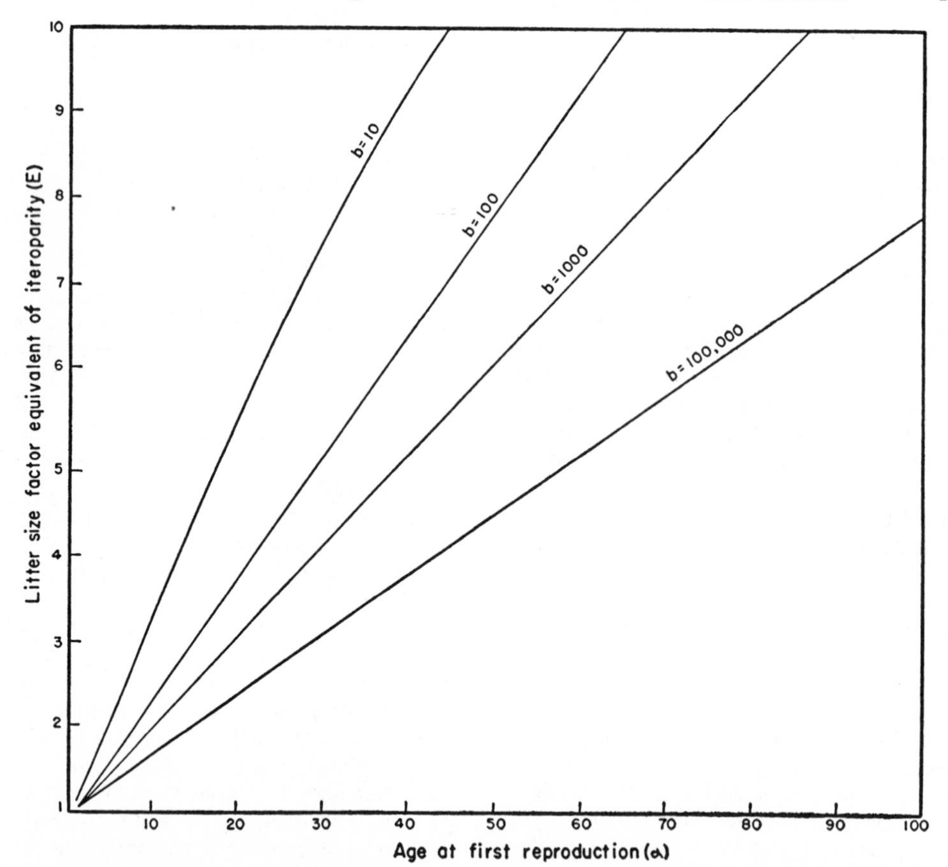

FIG. 3. THE CHANGES IN LITTER SIZE WHICH WOULD BE REQUIRED TO ACHIEVE IN A SINGLE REPRODUCTION THE SAME INTRINSIC RATE OF INCREASE THAT WOULD RESULT FROM INDEFINITE ITEROPARITY

b represents the litter size for an iteroparous species and the ordinate scale (E) represents the factor by which b would have to be multiplied to attain the same intrinsic rate of increase when each female produces only one litter in her lifetime.

habit. Hence a selective pressure can operate in favor of iteroparity even when the litter size is large. It is clear from Fig. 3, noting the greater slope of the lines representing smaller litter sizes, that in these cases the point will be reached more quickly at which the potential gains from iteroparity outweigh those attainable by increasing the litter size.

Man has a life cycle which is rather unusual in that it combines a long pre-reproductive period with a very small litter size; the very conditions under which iteroparity should be most advantageous. Everyone is, of course, aware that multiple births occur in man but with such a low frequency in the population that they are of negligible importance in population phenomena. It is also rather generally accepted that there is a hereditary basis for the production of multiple births. The question arises as to why increased litter size should not become more common simply as a result of increased contributions to subsequent population resulting from the increase in biotic potential associated with large litters. It should be of interest, therefore, to determine how large a litter would have to be produced in a single reproductive effort to provide an intrinsic rate of increase equal to that resulting from three or more single births.

In the case of man we may rather confidently accept the value $b = \frac{1}{2}$ to signify that the average number of female offspring produced per human birth, and which will ultimately mature, is one-half. Accepting this value means that a mother must on the average produce two "litters" merely to replace herself (to give $r = 0$), so we shall examine the intrinsic rate of increase only for cases where n, the total number of births, is greater than two. To examine the maximum gain attainable by iteroparity we assume that successive births are spaced one year apart and obtain the value of r from formula (21), employing different values of n and α. It is easily seen that the necessary litter size, say b', to give the same value of r by means of a single reproductive effort at age α, would be precisely $e^{r\alpha}$.

The value of r from formula (21) corresponding to three annual births beginning at age 12 is .0312. At the other extreme, if the first of the three births occurs at age 30 we obtain $r = .0131$. The corresponding values of $e^{r\alpha} = b'$ are successively 1.41 and 1.48. Under these conditions *it would require essentially a three-fold increase in litter size to achieve in one reproductive effort the same biotic potential as that obtained from three successive births.* The same conclusion is obtained when we consider larger numbers of births. In the case of man very little could be gained by increasing the litter size by any reasonable amount and it is probable that the biological risk involved in producing multiple births is more than sufficient to outweigh the very slight gain in biotic potential which could be obtained by this means. This would not be the case if the pre-reproductive period was drastically shortened, so we see that even in the case of man there is an interaction of life-history phenomena such that the importance of any conceivable change can only be evaluated through consideration of the total life-history pattern.

THE EFFECT OF TOTAL PROGENY NUMBER

In the preceding section we compared the two possible means by which an increase in total progeny number might lead to an increase in biotic potential. Our general conclusion was that the relative importance of changes in litter size and changes in the number of litters produced depends upon the rate of maturation. For species which mature early a modest change in litter size might be the equivalent of drastic changes in litter number but the possible value of iteroparity increases as the pre-reproductive part of the life span

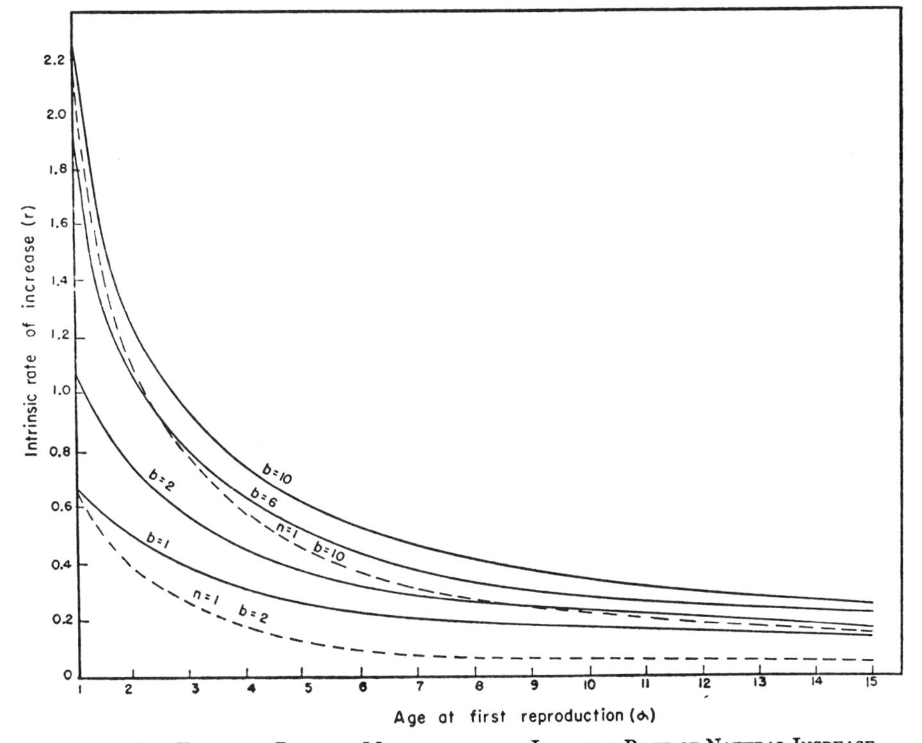

FIG. 4. THE EFFECT OF DELAYED MATURITY ON THE INTRINSIC RATE OF NATURAL INCREASE

The two broken lines represent semelparous species. The solid lines represent indefinitely iteroparous species where each female, after producing her first litter of size b, produces another similar litter in every succeeding time interval.

is lengthened. The importance of discovering the age at which reproduction begins has commonly been overlooked by students of natural history, hence it appears worthwhile to explore the matter further by examining the actual values of the intrinsic rate of natural increase corresponding to specified patterns of reproduction.

Fig. 4 was constructed from formula (21) to show, for several litter sizes, how the intrinsic rate of increase, r, is affected by lengthening the pre-reproductive period, α. Both semelparous ($n = 1$) and indefinitely iteroparous ($n = \infty$) species are illustrated. The striking feature of Fig. 4 is the way in which the lines representing different litter sizes converge as α increases. This occurs whether there is a single reproduction per lifetime or an infinite number, hence it is a general phenomenon. This supplements our earlier conclusions by suggesting that in species where reproductive maturity is delayed there should be relatively slight selection pressure for increased litter size. Here we are referring, of course, to the effective litter size or number of offspring which are capable of maturing. In cases where early mortality is very high, as is known to be the case with many fishes, it might require a tremendous increase in fecundity to produce a very small increase in effective litter size, and such increases might not be very important from the population standpoint. For example, a semelparous species reproducing at age 20 and with an effective litter size of 10 would have, $r = 0.120$. A ten-fold increase in litter size, to $b = 100$, would give $r = 0.231$ or an increase in biotic potential of 92 per cent. Another ten-fold increase to $b = 1000$ would give $r = 0.345$, or a gain of 50 per cent. The diminishing returns attainable by increasing litter size are obvious. For an iteroparous species reproducing first at age 20 and thereafter in each subsequent time interval, the increase in effective litter size from 10 to 100 would give only a 50 per cent increase in biotic potential and a further ten-fold increase in litter size would increase r by only another 35 per cent. In late-maturing species the litter size must be great enough to make it highly probable that *some* of the progeny will mature, but any further increases in fecundity will yield rapidly diminishing returns.

It is also clear from Fig. 4 that for any fixed litter size the biotic potential could be increased by shortening the period of maturation. Any specified amount of decrease in the pre-reproductive period will, however, be most effective for species where this part of the life span is already short.

Fig. 5 illustrates the way in which the two factors of length of the pre-reproductive part of the life span (α) and the number of offspring produced interact to determine the intrinsic rate of natural increase. These values were also computed from formula (21), in this case considering the litter size, b, as a constant with the value one-half. The figure then applies to species which, like man, produce one offspring at a time and where one-half of these offspring are females. Under these conditions it obviously requires two births just to replace the parents, but the population consequences of producing more than two offspring per lifetime vary tremendously with the age at which reproduction begins.

The female of the extinct passenger pigeon presumably produced her first brood consisting of a single egg at the age of one year. From the steep slope of the line representing $\alpha = 1$ in Fig. 5 it is clear that, beyond the minimum of two eggs per average female, several additional eggs produced in successive years would each add very appreciably to the value of r. Accordingly, a relatively slight reduction of the life expectancy for such a species might greatly reduce the biotic potential. The flattening out of the curves in Fig. 5 again illustrates the fact that each litter contributes less to potential population growth than the one preceding it. However, in a case such as that of the passenger pigeon even the seventh and eighth annual "litters" would add appreciable increments to the value of r.

Fig. 5 also shows that as the age at maturity increases the possible gains in biotic potential attainable by producing many offspring rapidly diminish. When $\alpha = 3$, as in the economically important fur-seal, each pup contributes much less to biotic potential than was the case for the eggs of the passenger pigeon. Nevertheless, it is apparent from the figure that if the life expectancy for females should be reduced to seven or eight years (corresponding to the fifth or sixth pup), or less, the species would be in a vulnerable position. The curve is steep in this portion of the graph and relatively slight changes in average longevity could produce disproportionately large population effects.

The lowest curves in Fig. 5 represent ages at maturity falling within the possible range for man. The curves come close together as α increases, so that in this range a change of a year or two in the age at which reproduction begins is less significant than in the case of a species that matures more

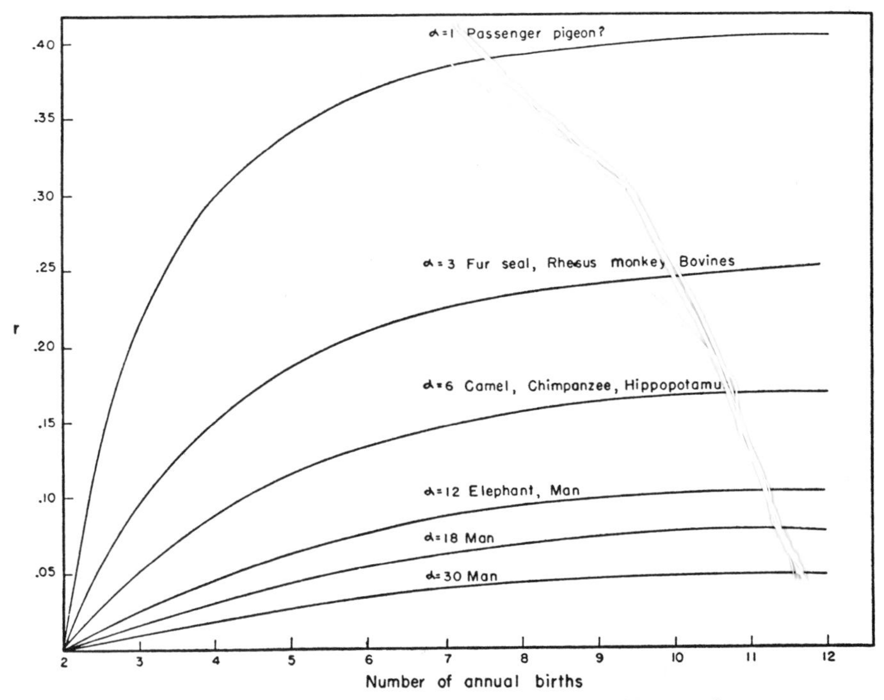

FIG. 5. THE EFFECTS OF PROGENY NUMBER ON THE INTRINSIC RATE OF NATURAL INCREASE WHEN THE LITTER SIZE IS ONE ($b = \frac{1}{2}$)

The ordinate scale shows the intrinsic rate of increase for species which produce an average of one-half female offspring per litter. For any given total progeny number, the intrinsic rate (r) is seen to be greatly affected by the age (α) at which the first offspring is produced.

rapidly. Furthermore, the curves flatten out rapidly so that families which are very large by ordinary standards actually contribute little more to potential population growth or to future population than do families of quite moderate size. As an explicit illustration consider the intrinsic rate of increase which would result if, on the average, human females produced their first offspring at the age of 20 and had a total of five children spaced at one-year intervals. In this case we would have $r = 0.042$. If, on the other hand, we assume that, instead of producing only five children, the females could live forever producing a child each year we would obtain $r = 0.0887$. Under these conditions we conclude that in terms of biotic potential five children are almost one-half (actually 47 per cent) the equivalent of an infinite number. With larger values of α the effect of very large families would be even further reduced. From these considerations the writer feels that human biologists, as well as other natural historians, often overemphasize the importance of total number of progeny while underestimating the significance of the age at which reproduction begins. It is impossible to conclude that one segment of the population is contributing more to future population than is some other segment without examining the total life-history pattern. Age at marriage could, in studying human populations, be a more significant datum than total family size.

The foregoing discussion suggests that a species such as man which is characterized by a long period of maturation and a small litter size can exhibit considerable variability in the details of its life history without greatly affecting the intrinsic rate of natural increase. The population consequences to be anticipated if the average age at which reproduction begins were to be altered by a few years or if the average number of progeny per female were slightly altered are much less striking than is the case for many other species. This implies that the intrinsic rate of increase should be relatively constant over the range of possible variations in the life-history features for man.

Such a conclusion would seem to be of both practical and theoretical interest and to merit closer examination.

The intrinsic rate of increase for man would be the rate of compound interest at which a human population, unrestrained by environmental resistance, would grow. We have already noted that Franklin (1751) estimated that a human population could double in 20 years and that this would correspond to $r = .035$. Malthus (1798) estimated that an unrestrained human population such as that of the United States at that time could double in 25 years. Malthus' estimate corresponds to $r = .0277$ which is remarkably close to the value of $r = .0287$ obtained by Lotka (1927) using much more refined methods for estimating the rate of increase prevailing in 1790 in the United States. Pearl and Reed (1920) fitted a logistic curve to the population figures for the United States, and their equation gives the value $r = .03134$. Additional examples of estimates based on empirical data could be given, but these are sufficient to suggest that the value of the intrinsic rate of increase for man is not far from 0.03.

Fig. 6 was constructed from formula (21) in order to examine the question whether or not the life-history features of man would actually lead us to anticipate the approximate value, $r = 0.03$. Fig. 6 suggests rather definitely that the value $r = 0.02$ is too low, since it falls well below the obvious reproductive capabilities of humans. If females, on the average, had their first child at the age of 12 years (which is possible, cf. Pearl, 1930, p. 223) it would require an average of 2.6 surviving annual births per female to correspond to the rate $r = .02$. This curve is quite flat, so that if the first birth was delayed until the age of 20 years, which seems to be roughly the beginning of the semi-decade of maximum human fertility (Pearl, 1939), three annual births would still be adequate to give $r = .02$. Even if the first birth is delayed until the age of 28 years, this intrinsic rate of increase calls

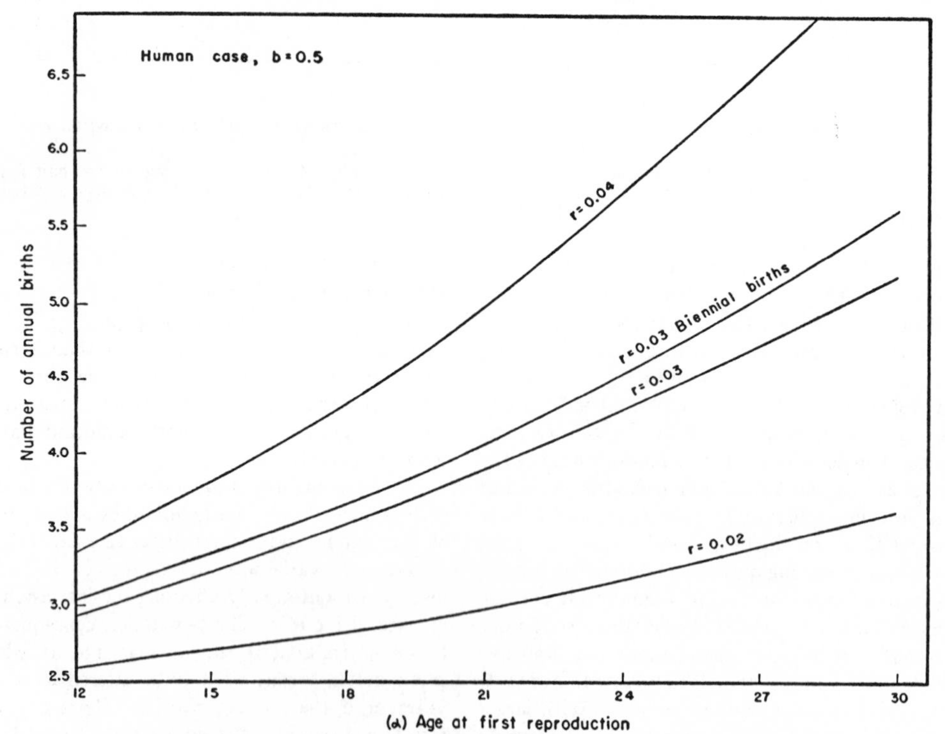

FIG. 6. AVERAGE REPRODUCTIVE PERFORMANCES REQUIRED TO GIVE SPECIFIED VALUES OF r, THE INTRINSIC RATE OF NATURAL INCREASE, IN HUMAN POPULATIONS

Assuming the average number of female offspring per human birth to be one-half ($b = 0.5$), this graph shows the extent to which total progeny number would have to be altered to maintain a specified intrinsic rate of increase while shifting the age at which reproduction begins. The figure also makes it possible to estimate the intrinsic rate of increase for a population when the average reproductive performance per female is known.

for an average progeny number of only 3.5. Clearly, man can easily exceed this average level of performance. On the other hand, the value $r = .04$ seems to require reproductive performance which would be astonishingly high as an average condition. If the first birth occurred when the mother was aged 13 years, an average of 3.5 children per female would suffice to give $r = .04$. However, the curve turns upward and if the first birth was delayed until the age of 20 it would require an average of 4.8 children to obtain this value. A delay of one more year, to the age of 21 for the first child, would increase the necessary mean progeny number to 5.0, while six children would be necessary if the first birth came when the mother was 25 years old. This intrinsic rate of increase then seems to call for exceptional rather than average reproductive performance, and the writer believes that Fig. 6 would lead us to expect that the intrinsic rate of increase for man lies between the limits .02 and .04 and might be estimated at about .03. Of course, the figure makes no allowance for mortality or for spacing the births at intervals of more than one year, so in actual cases the reproductive performance would have to be somewhat greater than indicated. The interval between births, however, is less critical than might be anticipated. The reproductive performances necessary to give $r = .03$ are shown both for one-year spacing and for two-year spacing between births and, to the writer, at least, either of these curves appears to represent a more reasonable picture of average human reproduction than do the cases representing the higher and lower intrinsic rates of increase.

THE POPULATION BIRTH-RATE AS A CONSEQUENCE OF LIFE-HISTORY PHENOMENA

We have noted earlier formulas (4) and (5) which were originally derived by Lotka (1907a, b; Sharpe and Lotka, 1911) and which show that when life-history features remain constant from generation to generation the population will ultimately settle down to a "fixed" or "stable" age distribution and will exhibit a fixed birth rate. We have also noted that this conclusion could be expected intuitively and can be obtained (formula 18) from discontinuous computational methods, once it is established that the potential form of population growth is a geometric progression. These potential consequences appear to provide the best justification for studying the life histories of various species, yet when such studies are conducted it is common practice not to attempt any interpretation in terms of population phenomena.

It is evident from formulas (3), (4), and (5) that the birth rate and the stable age distribution are tied together with the intrinsic rate of increase and that any extensive discussion of the way in which changes in life-history features would affect these population features might repeat many of the points already covered. Consequently, we shall here note very briefly the relationships between life-history features and the resultant phenomena of birth rates and age structure.

If we consider a "closed" population, which changes in size only through the processes of birth and death, it is apparent that the intrinsic rate of increase, r, in formula (1), $\frac{dP}{dx} = rP$, must represent the difference between the instantaneous birth rate and the instantaneous death rate. In practice, however, we are more interested in a finite rate of population change. If we employ formula (1′) to express the rate of population growth and consider that the changes result entirely from the birth rate (BR) and the death rate (DR), we obtain:

$$BR - DR = \frac{P_{x+1} - P_x}{P_x} = e^r - 1. \tag{26}$$

A birth rate, β, appropriate to this approach has already been defined by formula (18). However, because we are dealing with finite time intervals the B_x births regarded as occurring at the beginning of some time interval, T_x, should properly be credited to the P_{x-1} individuals living in the previous time interval. The birth rate would, therefore, be:

$$BR = e^r\beta. \tag{27}$$

On the other hand, the death rate should properly be the ratio of the D_x deaths in interval T_x to the total population exposed to the risk of death; that is D_x/P_x. Hence the simple relationship of formulas (5) and (26) can be misleading, especially when there is a rapid population turnover. For example, the birth rate β in Lotka's formula (5) is by definition identical with c_0, the fraction of the population aged between zero and one. Consequently β can never exceed unity no matter how many offspring are on the average produced per individual during a time interval.

In practical population problems the crude birth rate is often observed and employed as a criterion of the state of the population. This practice can be misleading, especially in comparing species which differ widely in their life-history features. It would be redundant to undertake a detailed analysis of the way in which life-history phenomena affect the crude birth rate because, as is evident from formulas (5), (26), and (27), any change which affects the value of r also affects the population birth-rate. However, the birth rate is subject to additional influences and these may be briefly examined.

The methods used in computing the value of r (formulas 17, 21, and 22) have involved survivorship only out to the age ω at which reproduction ceases. In species having a post-reproductive part of the life span, the crude birth rate corresponding to a given value of r will be reduced simply because post-reproductive individuals accumulate and are counted as part of the population on which the computations are based. It is clear, therefore, that any increase in longevity (λ) which is not accompanied by an increase in ω will tend to lower the observed birth rate; conversely, the birth rate must affect the age structure of the population.

If the maximum longevity for a species is λ, a continued birth rate of $1/\lambda$ would just suffice to leave a replacement behind at the time each individual dies. Hence $1/\lambda$ represents an absolute minimum for a steady birth rate which is capable of maintaining the population. For example, if it were possible to keep every human female alive for 100 years a birth rate as low as 0.01 or 10 births per thousand population per annum (assuming ½ of the offspring to be females), could theoretically suffice for population maintenance. However, for human females the value of ω is not much beyond 40 years; in general, if a female has not produced a replacement by the age of 40 she is not going to do so. The latter consideration might lead one to

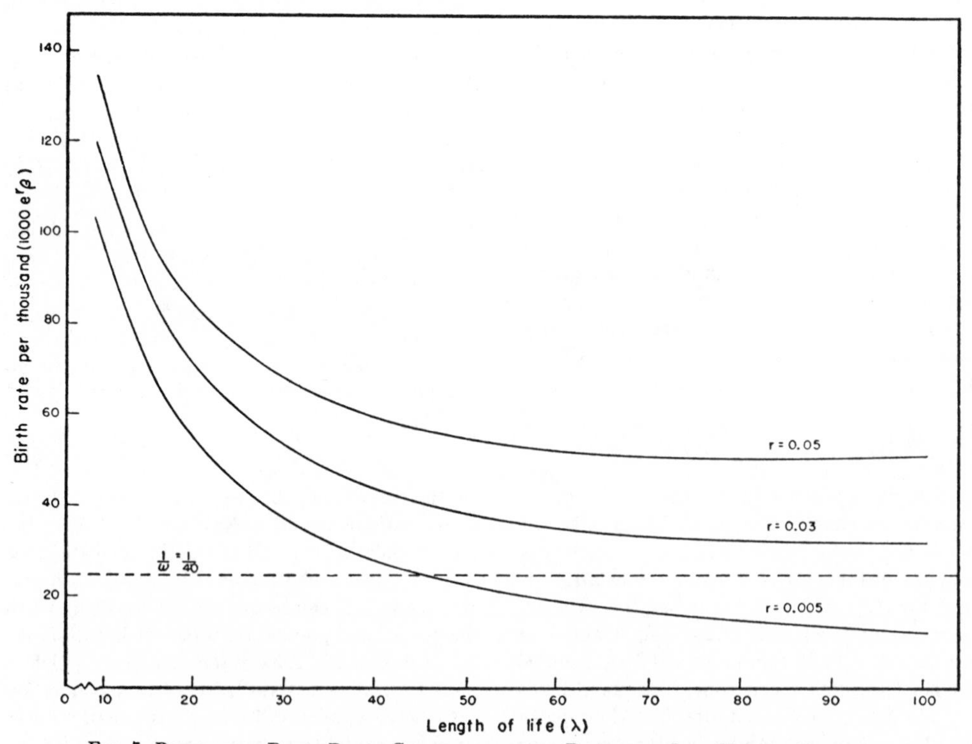

FIG. 7. POPULATION BIRTH RATES CORRESPONDING TO DIFFERENT LIFE HISTORY PATTERNS

The figure shows that the intrinsic rate of natural increase (r), the birth rate, and mean longevity (λ) are all interdependent. The broken line represents the minimum birth rate which would maintain a population if females did not survive beyond the age of 40 "years," which is here used as an estimate of the normal age (ω) at which reproduction is concluded. As longevity is increased, slow population growth becomes possible even when birth rates fall below $1/\omega$. Such low birth rates, however, have sometimes caused unwarranted concern about excessive population increase when abnormal conditions have temporarily reduced the death rate so that there is a large excess of births over deaths.

designate as a minimum maintenance birth rate the value $1/\omega$, or 0.025 in the case of man. But it is clear from the foregoing sections that offspring produced prior to the time that a female reaches the age of 40 will already have begun to "accumulate interest" before the mother can reach the age of 100. The compound interest nature of potential population growth complicates the relationship between birth rate and life-history phenomena and makes it conceivable that populations of species where $\lambda > \omega$ could even continue to grow while exhibiting birth rates lower than $1/\omega$.

No end of interesting combinations of birth rates, death rates, and life-history phenomena might merit consideration, but a single simple example will be selected here to illustrate the general relationship between changing longevity and population birth-rates. If we assume, as before, a physiological type of survivorship such that each individual that is born lives to attain its λth birthday but dies before reaching the age of $\lambda + 1$, we can sum the right-hand side of formula (19) as a geometric progression and combine this with formula (27) to obtain the following expression for the population birth rate (BR):

$$BR = e^r\beta = \frac{e^r - 1}{1 - e^{-r\lambda}}. \quad (28)$$

Formula (28) shows, as would be expected, that as longevity (λ) is increased, with the other life-history features remaining unchanged, the birth rate will fall and approach as a limit the value $e^r - 1$, which is in accord with (26) when the death rate is set equal to zero.

Fig. 7 was constructed from formula (28) to illustrate the interrelations between BR, r, and λ within a range of values of life history features roughly applicable to man. The condition for a stationary population would, of course, correspond to $r = 0$, while all positive values of r correspond to growing populations. r must become negative when the birth rate falls below $1/\lambda$, and it will be noted that the birth rates below about 20 per thousand which are sometimes observed in human populations (see tabulation in Allee et al., 1949, p. 288) must, unless they represent abnormal temporary phenomena, correspond to populations with very low potential growth rates. The curve for $r = .005$, in fact, does not differ greatly from $1/\lambda$. The curves in Fig. 7 flatten out rapidly for large values of λ, so that drastic and generally unattainable increases in longevity would be required to make such low birth rates compatible with appreciable population growth.

Looking at these relationships from a different point of view, Fig. 7 shows that a reduction in longevity, such as might result from reducing the life expectancy of game animals, can be expected to result in an increased birth rate even if the intrinsic rate of increase is unchanged. It does not seem worthwhile at present to attempt a quantitative estimate of these relationships because the assumption of a physiological type of survivorship curve is probably not even approximately true for game animals. When more realistic estimates of survivorship are available, however, the type of relationship illustrated in Fig. 7 may assume practical importance.

THE STABLE AGE DISTRIBUTION

The age structure of a population often is a matter of considerable practical concern. In economically valuable species such as timber, game animals, and commercial fishes certain age classes are more valuable than others, and it would be desirable to increase the proportion of the most valuable age classes in a population. Similarly, certain age classes of noxious organisms may be more destructive than others and the relative numbers of these destructive individuals will be governed by life-history phenomena which may conceivably be subject to alteration by control measures. In human populations, also, it is sometimes a matter of concern that the proportion of the population falling within the age limits most suitable for physical labor and military service seems to be below optimum. An article in the New York *Times* for September 24, 1950, headed "population shift in France traced—Study finds too many aged and very young in relation to total of workers" illustrates the potential importance of a knowledge of the age structure of populations.

The mathematical basis for relating the age structure of a population to life-history features was established in Lotka's first paper on population analysis (1907); and in the same year Sundbärg (1907) reached the conclusion that a human population reveals its condition (tendency to grow or decline) through its age structure. These important conclusions have not been sufficiently noted by ecologists. When the mortality factors affecting a population are altered either through natural environmental changes or through human exploitation or attempts at control there will in general result a change in the age structure of the population, and this may be observable even before changes in population size or in birth rates provide evidence of the consequences of the changed mortality factors. The subject of age structure is a large and difficult one because the various combinations of life-history features, birth rates, death rates, and age structure are analogous to a multidimensional figure where a change imposed in any one feature induces changes in all of the others. The subject has been considered most in connection with human populations, and some empirical generalizations have been obtained which may profitably be examined by means of the computational methods we have been employing. For the purpose of illustrating the general character of the relationships involved, one species will serve as well as another.

For illustrative purposes we may proceed as in the preceding sections and consider the stable age distribution for cases where survivorship is of the physiological type. Letting c_x represent the fraction of the population aged between x and $x + 1$, we may employ formulas (5) and (28) directly to obtain, for $x < \lambda$:

$$c_x = \frac{e^{-rx}(1 - e^{-r})}{1 - e^{-r\lambda}}. \qquad (29)$$

From formula (29) it is apparent that if the extreme longevity for a population is altered without changing the intrinsic rate of increase, the effect on the age structure will be of a very simple type. An increase in longevity from λ_1 to λ_2 will simply reduce the proportion of the population in each age category below λ_1 by the constant proportion $\frac{1 - e^{-r\lambda_1}}{1 - e^{-r\lambda_2}}$. Consequently, the effect will be most noticeable on the youngest age classes, because these are the largest classes.

Changes in the value of r affect the stable age distribution in a more complex way than do changes in λ, although the general result of increasing the value of r will be to increase the proportion of young in the population, with a corresponding decrease in the proportion of older individuals. Fig. 8 illustrates this effect for three values of r,

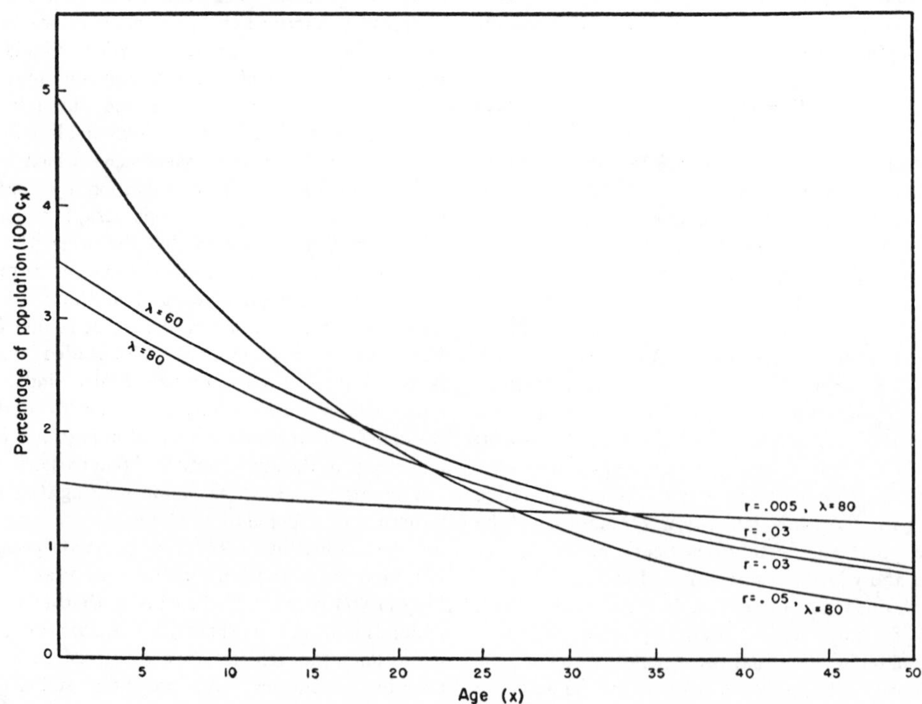

FIG. 8. THE STABLE AGE DISTRIBUTION, OR PROPORTION OF THE POPULATION FALLING WITHIN EACH INTERVAL OF AGE, IS SHOWN HERE AS A FUNCTION OF THE INTRINSIC RATE OF INCREASE (r) AND THE LENGTH OF LIFE (λ)

These relationships have important ecological correlaries. See text for a discussion relating these to the "optimum yield" problem.

assuming that λ in formula (29) remains constant at 80 "years."

The way in which population age-structure is affected by changes in the value of r and λ, as illustrated by formula (29) and Fig. 8, permits some qualitative conclusions which are of interest in connection with the "optimum yield problem" (for discussion, see Allee et al., 1949, p. 377). If man (or some other species) begins exploitation of a previously unexploited population, the age structure will be affected in a definite manner. An obvious result of increased predation will be to decrease the average longevity, corresponding to a decrease in λ. We have seen earlier (Fig. 7) that this will ordinarily have the effect of increasing the population birth rate, and Fig. 8 and formula (29) show that another effect will be to increase the proportion of young in the population, the very youngest age classes being most affected. However, the increased mortality may also affect the value of r. If the population is initially in equilibrium with the capacity of its environment, its total life history pattern will be adjusted to the effective value, $r = 0$. If exploitation is not too intense, it has the effect of making additional environmental resources available to the surviving members of the population and thus of stimulating population increase; the population "compensates" (see Errington, 1946) for the increased mortality by increasing the value of r. Fig. 8 shows that this will have the effect of increasing the proportion of young in the population, thus supplementing the effect of exploitation on λ. On the other hand, the increase in r will have the effect of reducing the proportion of aged individuals in opposition to the effect of reduced λ which is to increase all of the age classes which still persist. It seems clear that the most obvious population consequence of such exploitation will be to increase the proportion of young members of the population. If predation or exploitation becomes still more intense, as in the case of "overfishing" (Russell, 1942), it will reduce the effective value of r, and, of course, still further reduce average longevity. The decrease in r will tend to decrease the proportion of young individuals, but the decrease in λ will tend to increase this proportion. However, both changes will tend to raise the proportion of older individuals in the population, and this combined effect can then be expected to be the most obvious corollary of overfishing. These conclusions, of course, greatly oversimplify a complex phenomenon. In order to make quantitative estimates of these effects it would be necessary to have detailed information about the life-history features, especially survivorship under the conditions of increased predation. Nevertheless, these qualitative conclusions show the type of effect to be expected when populations are subjected to increased predation, and they suggest that observations of the changes in the age structure of populations may provide valuable evidence of over-exploitation, or, from the opposite point of view, of the effectiveness of control measures. Bodenheimer (1938) comments on the fact that ecologists have neglected this important subject.

From Fig. 8 it will be noticed that changes in life-history features produce their greatest effects on the extreme age classes, whereas the curves representing different patterns are close together in the "middle" age range. The same phenomenon is evident, for example, in a graph (Fig. 27) reproduced by Dublin, Lotka, and Spiegelman (1949) from Lotka (1931) to show the age structure of human populations corresponding to stationary, increasing, and decreasing populations. This suggests that the proportion of a population falling in the middle portion of the life span may be relatively independent of factors which produce drastic shifts in the ratio of very old to very young. This was first postulated by Sundbärg (1907), who concluded that it was "normal" for about 50 per cent of a human population to fall in the age range between 15 years and 50 years. Sundbärg distinguishes three primary population types based on the age distribution of the remaining 50 per cent of the population. In the "progressive" type of population there is a strong tendency for increase and the ratio of young (aged under 15) to old (aged over 50) is, by Sundbärg's criteria, about 40 to 10. In the "regressive" type, exhibiting a tendency toward population decrease, the corresponding ratio of young to old is about 20 to 30, while a "stationary" type with the ratio about 33 to 17 shows no particular tendency either to grow or to decrease. When first encountered Sundbärg's conclusion appears surprising, but in actual human populations differing as radically, for example, as those of Sweden and India, the proportion of the population aged between 15 and 50 is remarkably close to 50 per cent (see tabulation in Pearl, 1946, p. 78). It appears to the writer that this conclusion should be of great interest to students of human populations. The age class between 15 and 50 years includes the bulk of the workers and persons of military age, and Sundbärg's conclusion implies that the size of this class relative to the remainder of the population must be determined by life-history features which cannot readily be deliberately controlled.

For other species also, the life span may be meaningfully divided into three primary age classes, pre-reproductive (aged 0 to α), reproductive (aged α to ω), and post-reproductive (aged ω to λ), which differ considerably in their biological significance. If we continue with our assumption of physiological survivorship we can obtain the relative sizes of these three age classes directly from formula (29), and Sundbärg's generalization offers an interesting empirical pattern with which to compare our results. For the relative sizes of the three fundamental age classes formula (29) gives:

$$\left.\begin{aligned} \text{Pre-reproductive} &= \frac{1 - e^{-r\alpha}}{1 - e^{-r\lambda}} \\ \text{Reproductive} &= \frac{e^{-r\alpha} - e^{-r\omega}}{1 - e^{-r\lambda}} \\ \text{Post-reproductive} &= \frac{e^{-r\omega} - e^{-r\lambda}}{1 - e^{-r\lambda}} \end{aligned}\right\} \quad (30)$$

By putting $\alpha = 15$ and $\omega = 50$, we may examine the relationship between r and λ for Sundbärg's primary population types.

The generalization that about 50 per cent of a human population normally falls in the age range from 15 to 50 is in accord with formulas (30). When $r = .03$ the value of λ to give a stable age distribution with just 50 per cent in the middle age range would be 59 years, but an increase in longevity to 85 years would only reduce this class to 45 per cent of the population. The same general conclusion applies when r is small. The values $r = .005$ and $\lambda = 63$ years correspond to 55 per cent aged 15 to 50 years, and λ would have to be increased to 81 years to reduce this to 45 per cent. It appears that over the usual range of values of human longevity and potential population growth Sundbärg's generalization is very good. This is shown graphically in Fig. 9. It is noteworthy that when the length of life is about 70 years, Sundbärg's generalization holds over a wide range of values of r. In other words, this ratio of "middle-aged" to total population is quite insensitive to changes in other life history features.

Figure 9 illustrates Sundbärg's population criteria for values of $r > 0$. It is clear that the

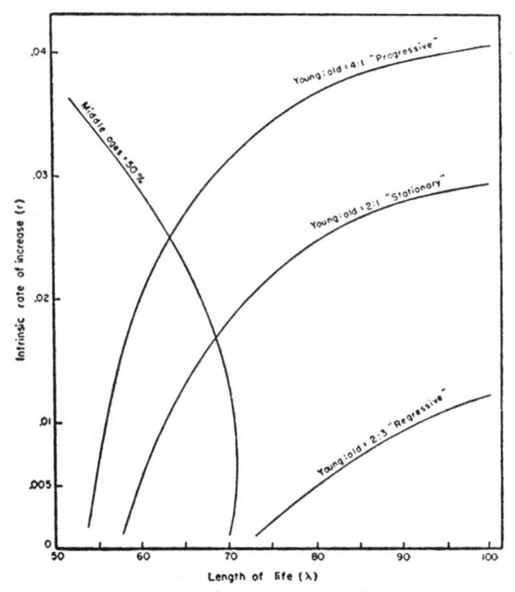

FIG. 9. SUNDBÄRG'S POPULATION CRITERIA IN RELATION TO LIFE-HISTORY FEATURES

The "middle age" range here consists of individuals aged 15 years to 50 years.

"regressive" type of population structure is not consistent with populations possessing a strong tendency to increase and, in fact, for usual longevity figures, will ordinarily correspond to decreasing populations ($r < 0$). On the other hand, the "progressive" type of structure does correspond to large values of r even when λ is not particularly large. The "stationary" type of population structure is less well defined. Any population for which the life-history features correspond to $r > 0$ will tend to grow, and the ratio of two "old" to one "young" can correspond to a large value of r when λ is large. The value of r is, of course, independent of the length of the post-reproductive part of the life span, whereas the age structure is not. For any given value of r, which is determined by the life-history features of individuals aged below ω, the effect of an increase in longevity will be to decrease the proportion of young and, to a lesser extent, the proportion of reproductive members of the population.

At first glance, the type of interactions shown in Fig. 9 might suggest that if the age structure of a population were artificially shifted in the "regressive" direction, for example, through migration or improvements in public health, there would result a reduction in the value of r. This, however, is not the case, since r is completely determined by life-history events occurring before age ω. Sharpe and Lotka (1911) showed that a characteristic of the "stable" age distribution is the fact that it will become reestablished after temporary displacements. However, practical problems arise in analyzing actual populations because phenomena such as birth rates and death rates may give a very misleading picture of population trends when the age structure is displaced from the stable type and has not yet had time to reestablish a stable age distribution. From Fig. 7 we have noted that in human populations birth rates on the order of 20 per thousand per annum would in general be little more than sufficient to maintain a population. But if the age distribution is displaced from the stable type in the direction of increased frequency of young individuals, the death rate may be temporarily reduced, so that there is a great excess of births over deaths and the population can grow temporarily even if the age-specific birth rates are too low to maintain the population permanently at a constant size. Under these conditions the population is "aging," and in human populations this phenomenon has a number of interesting and important corollaries (see Dublin, Lotka, and Spiegelman, 1949). Fisher (1930) has considered the problem in a more general sense, noting that the apparent value of r given by formula (3) will be incorrect when the age structure is displaced. Fisher suggests the possibility of measuring population size not in terms of the number of individuals present but in terms of "reproductive value," where the "value" of an individual represents his remaining potentialities for contributing to the ancestry of future generations. Some such approach has great potential value for ecological studies of natural populations, but its possibilities in this direction seem not yet to have been explored.

EMPIRICAL APPLICATIONS

In the preceding sections we have been concerned with the influence of specific life-history patterns on the characteristics of populations. In order to examine these effects we have made simplifying assumptions by regarding some of the life-history features as fixed in a certain way while we examined the results of varying other features. While this procedure oversimplifies the biological situation as it exists in actual populations, the writer regards it as a sound way of investigating the meaning of life-history features. The same general attitude may be traced back to Robert Wallace (1753) in his book which profoundly influenced Malthus. Wallace pointed out that "mankind do not actually propagate according to the rules in our tables, or any other constant rule . . ." but he emphasized that tables of potential population growth are still valuable because they permit us to evaluate the influences restraining population growth.

Wallace, then, was a pioneer in appreciating the potential value of comparisons between empirical and theoretical population phenomena. In modern actuarial practice, population data are subjected to involved mathematical treatments which are sometimes considered to represent biological laws and at other times to be merely empirical equations, but which, in any case, are known to yield results of practical value. In the words of Elston (1923, p. 68):

"... it seems to me that even though there be governing causes of mortality that may result in a true law of mortality, any group of lives studied is so heterogeneous, due to differences in occupation, climate, sanitary conditions, race, physical characteristics, etc., that any formula must in practice be considered to be merely a generalization of what is actually happening."

The number of different combinations of life-history features of the type we have been discussing is essentially infinite, and it is out of the question to make detailed examinations of any great proportion of these from the theoretical standpoint. However, we have seen that certain population features, such as the prevailing age distribution and the intrinsic rate of increase, summarize a great deal of information about the potentialities of the population and its relationship to its particular and immediate environment. As mentioned earlier, the recent ecological literature demonstrates that ecologists are becoming interested in determining such features as the intrinsic rate of increase for non-human populations. These computations may have practical value in dealing with valuable or noxious species, and they possess great theoretical interest for ecologists. For example, the logistic equation has been widely employed to represent population growth in a variety of organisms (Allee et al., 1949; Pearl, 1927); and it has also been attacked (Yule, 1925; and succeeding discussion, Gray, 1929; Hogben, 1931; Smith, 1952), on the grounds that it is too versatile and can be made to fit empirical data that might arise from entirely different "laws" of population growth. This criticism is, of course, directed at the fact that the curve-fitter has three arbitrary constants at his disposal in seeking to obtain a good fit. One of these constants is r, the intrinsic rate of increase. The grounds for accepting or rejecting the logistic equation as a law of population growth (or for seeking some other law) would be greatly strengthened if the value of r was computed directly from observed life-history features and independently of the data on population size.

The computational methods employed in the preceding sections suggest several possible ways of computing the value of r from empirical life-history data. The usual procedure for such computations has been a tedious one based on formula (3) (see Lotka's appendix to Dublin and Lotka, 1925), although Birch (1948) employed formula (17) as an approximation to (3) for his computations. The methods discussed in the preceding sections suggest that a logical procedure for obtaining an empirical value of r would be to observe age-specific birth rates and survivorship under the environmental conditions of interest, and from these to write a "generation law" so that formula (11) can be employed. The single positive root of (11) is e^r and this can be estimated to any desired degree of accuracy without great difficulty even for species where the reproductive life is prolonged. When one is actually solving equation (11) it is strikingly brought to one's attention that the final terms representing reproduction in later life are relatively unimportant in influencing the value of r. This once again reinforces our conclusion that reproduction in early life is of overwhelming importance from the population standpoint, and should be much more carefully observed in field and laboratory studies than has usually been the case.

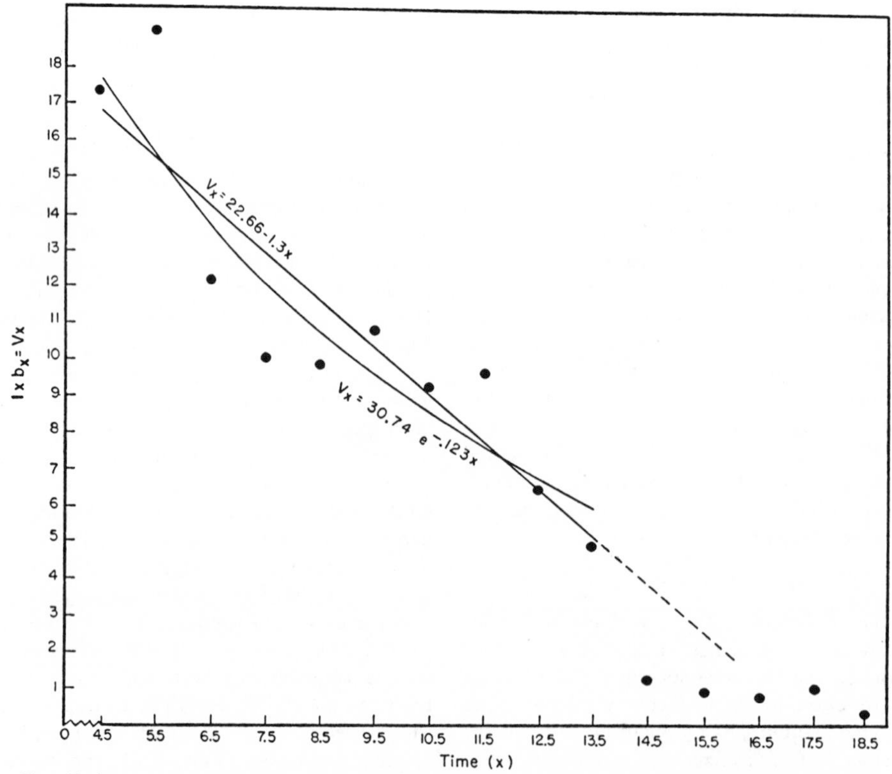

FIG. 10. A STRAIGHT LINE AND AN EXPONENTIAL FUNCTION FITTED TO THE DATA OF BIRCH (1948)

These data are suitable only for illustrating computational methods. Birch observed the age-specific fecundity rates (b_x) for the rice weevil *Calandra oryzae* living under constant conditions of temperature and moisture. He did not observe survivorship (l_x) but carried over the shape of the survivorship curve from the flour beetle *Tribolium confusum*. The black dots represent the product $l_x b_x$.

An alternative approach for obtaining empirical values of r consists of fitting the V_x ($= l_x b_x$) values with mathematical curves of types which make it possible either to obtain the sum of the series on the right-hand side of equation (19) or to employ Laplace transformations to solve equations (3) or (20) by simple iterative methods. Different formulas could be fitted to different sections of the V_x curve, and, when good fits can be obtained with combinations of straight lines, step-functions, exponential functions, and other simple formulas, this procedure will often lead to easy ways of solving for r. As a very simple example of this procedure we may consider the data of Birch (1948), for which he employed formula (17) and, by computations which are given in detail in his paper, obtained the value $r = 0.762$.

The V_x data employed by Birch are shown graphically in Fig. 10 by means of the black dots. The two curves, a straight line and a simple exponential function, were fitted by the method of least squares to the values up to $x = 13.5$. Neither of the functions gives an extremely good fit to the observed data, but, on the other hand, one may question whether the irregularities in the empirical data are not partly artifacts which should be smoothed out when one is attempting to estimate the value of r for a species. In any case, it appears worthwhile to compare Birch's value of $r = .762$ with that obtained by use of the empirically fitted curves.

Using first the exponential curve, if we set $V_x = Ke^{ax}$, formula (19) may be written in the form:

$$Ke^{\alpha(a-r)} = \frac{1 - e^{a-r}}{1 - e^{n(a-r)}}. \qquad (31)$$

When reproduction occurs several times, so that n is fairly large, the denominator in equation (31) becomes for practical purposes unity and may be ignored. In this case we have $\alpha = 4.5$, $K = 30.74$, $a = -.123$, and $n = 9$, so we will ignore the denominator. The equation is easily solved by iterative means, using a table of the exponential function, and the value obtained is $r = .758$, which differs from Birch's value by about one-half of one per cent.

The right-hand side of equation (19) can also be summed for the linear case where $V_x = a + bx$, but in this case the Laplace transformation employed with formula (3) yields a slightly simpler equation:

$$\frac{ar + b}{r^2} = e^{r\alpha}. \qquad (32)$$

Putting $a = 22.66$, $\alpha = 4.5$, and $b = -1.3$ in equation (32), and solving by iterative means we obtain $r = .742$, which differs from Birch's value by nearly three per cent, although this seems to be good agreement in view of the crude approximations employed.

In many practical applications dealing with natural and experimental populations some approximation to the value of r such as those presented above may be all that can be justified by the accuracy of the data. The estimate of r could undoubtedly be improved by fitting different portions of the V_x curve with different functions. Another refinement suggested by observations presented earlier in this paper would be to fit the empirical curves by a method which would give greater weight to the earlier points which have more influence on the value of r than do the later points. Any detailed discussion of these empirical applications would be out of place in the context of the present paper. The writer, however, anticipates that ecologists will in the future devote more attention to the interrelations of life-history features and population phenomena, and it is to be hoped that some of the approaches which have been indicated will accelerate trends in this direction.

SUMMARY

Living species exhibit a great diversity of patterns of such life-history features as total fecundity, maximum longevity, and statistical age schedules of reproduction and death. Corresponding to every possible such pattern of life-history phenomena there is a definitely determined set of population consequences which would ultimately result from adherence to the specified life history. The birth rate, the death rate, and the age composition of the population, as well as its ability to grow, are consequences of the life-history features of the individual organisms. These population phenomena may be related in numerous ways to the ability of the species to survive in a changed physical environment or in competition with other species. Hence it is to be expected that natural selection will be influential in shaping life-history patterns to correspond to efficient populations.

Viewed in this way, comparative studies of life histories appear to be fully as meaningful as studies of comparative morphology, comparative psychology, or comparative physiology. The former type of study has, however, been neglected from the evolutionary point of view, apparently because the adaptive values of life-history differences are almost entirely quantitative. The recent ecological literature does show a trend toward the increasing application of demographic analysis to non-human populations, but the opposite approach of deducing demographic consequences from life-history features has been relatively neglected. The present paper is presented with the hope that this situation can be changed. In other fields of comparative biology it is usual to examine individual characteristics and to regard these as possible adaptations, and the writer believes that life-history characteristics may also be profitably examined in this way.

It is possible by more or less laborious methods to compute the exact size and composition of the population which at any future time would be produced by any given initial population when the life-history pattern of the individual organisms is regarded as fixed. Thus it is possible to make an exact evaluation of the results of changing any life-history feature, and the value of this type of analysis may be apparent to those biologists who distrust the usual demographic procedures.

Starting with exact computational methods, it has been shown that early population growth may exhibit irregularities or cyclic components which are identifiable with negative or complex roots of an algebraic equation, but that these components vanish in time so that potential population growth is ultimately a geometric progression. Having established this fact, it is shown that the exact computational methods and the more convenient approximate methods lead to identical conclusions when considered over the long time scale which is of interest in adaptational and evolutionary considerations.

Some life-history patterns of ecological interest are examined and compared by means of relatively simple formulas derived from a consideration of the form of potential population growth. The results have bearing on the possible adaptive value of genetically induced changes of life-history features. It is suggested that this type of approach may add to the value of life-history studies and that an awareness of the possible meanings of empirical life-history data may aid in planning such studies by insuring that all pertinent information will be recorded. One of the most striking points revealed by this study is the fact that the age at which reproduction begins is one of the most significant characteristics of a species, although it is a datum which is all too frequently not recorded in the literature of natural history.

The number of conceivable life-history patterns is essentially infinite, if we judge by the possible combinations of the individual features that have been observed. Every existing pattern may be presumed to have survival value under certain environmental conditions, and the writer concludes that the study of these adaptive values represents one of the most neglected aspects of biology.

LIST OF LITERATURE

Allee, W. C. 1934. Recent studies in mass physiology. *Biol. Rev.*, 9: 1–48.

——, A. E. Emerson, O. Park, T. Park, and K. P. Schmidt. 1949. *Principles of Animal Ecology.* Saunders, Philadelphia.

Archibold, R. C. 1918. A Fibonacci series. *Amer. math. Monthly*, 25: 235–238.

Banta, A. M., T. R. Wood, L. A. Brown, and L. Ingle. 1939. Studies on the physiology, genetics, and evolution of some Cladocera. *Publ. Carneg. Inst.*, 39.

Berkson, J. 1944. Application of the logistic function to bio-assay. *J. Amer. statist. Ass.*, 39: 357–365.

——. 1951. Why I prefer logits to probits. *Biometrics*, 7: 327–329.

Birch, L. C. 1948. The intrinsic rate of natural increase of an insect population. *J. Anim. Ecol.*, 17: 15–26.

Bodenheimer, F. S. 1938. *Problems of Animal Ecology.* Oxford Univ. Press, London.

Chapman, R. N. 1928. The quantitative analysis of environmental factors. *Ecology*, 9: 111–122.

——. 1938. *Animal Ecology.* McGraw-Hill, New York.

Churchill, R. V. 1944. *Modern Operational Mathematics in Engineering.* McGraw-Hill, New York.

Deevey, E. S., Jr. 1947. Life tables for natural populations of animals. *Quart. Rev. Biol.*, 22: 283–314.

Dublin, L. I., and A. J. Lotka. 1925. On the true rate of natural increase as exemplified by the population of the United States, 1920. *J. Amer. statist. Ass.*, 20: 305–339.

——, ——, and M. Spiegelman. 1949. *Length of Life. A Study of the Life Table*, rev. ed. Ronald Press, New York.

Dunkel, O. 1925. Solutions of a probability difference equation. *Amer. math. Monthly*, 32: 354–370.

ELSTON, J. S. 1923. Survey of mathematical formulas that have been used to express a law of mortality. *Rec. Amer. Inst. Actuar.*, 12: 66–95.

ERRINGTON, P. L. 1946. Predation and vertebrate populations. *Quart. Rev. Biol.*, 21: 144–177, 221–245.

EVANS, F. C., and F. E. SMITH. 1952. The intrinsic rate of natural increase for the human louse, *Pediculus humanus* L. *Amer. Nat.*, 86: 299–310.

FIBONACCI, L. (1857). *Liber Abbaci di Leonardo Pisano publicati da Baldasarre Boncompagni.* Tipografia delle Scienze mathematiche e fisiche, Roma.

FINNEY, D. J. 1947. The principles of biological assay. *J. R. statist. Soc.*, 109: 46–91.

——. 1949. The choice of a response metameter in bio-assay. *Biometrics*, 5: 261–272.

FISHER, R. A. 1930. *The Genetical Theory of Natural Selection.* Oxford Univ. Press, London.

FRANK, P. W. 1952. A laboratory study of intraspecies and interspecies competition in *Daphnia pulicaria* (Forbes) and *Simocephalus vetulus* O. F. Müller. *Physiol. Zool.*, 25: 178–204.

FRANKLIN, B. 1751. Observations concerning the increase of mankind and the peopling of countries. In *The Works of Benjamin Franklin*, ed. by J. Sparks, 1836, Vol. 2: 311–321. Hilliard, Grey, Boston.

GAUSE, G. F. 1934. *The Struggle for Existence.* Williams & Wilkins, Baltimore.

GOMPERTZ, B. 1825. On the nature of the function expressive of the law of human mortality. *Phil. Trans.*, 36: 513–585.

GRAUNT, J. 1662. *Natural and Political Observations Mentioned in a Following Index and Made upon the Bills of Mortality. . . .* Printed by T. Roycroft, London.

GRAY, J. 1929. The kinetics of growth. *Brit. J. exp. Biol.*, 6: 248–274.

HOGBEN, L. 1931. Some biological aspects of the population problem. *Biol. Rev.*, 6: 163–180.

HUMPHREYS, W. J. 1929. *Physics of the Air*, 2nd ed. McGraw-Hill, New York.

HUTCHINSON, G. E. 1948. Circular causal systems in ecology. *Ann. N. Y. Acad. Sci.*, 50: 221–248.

HYMAN, L. H. 1951. *The Invertebrates. Vol. 2. Platyhelminthes and Rhynchocoela. The Acoelomate Bilateria.* McGraw-Hill, New York.

JACKSON, C. H. N. 1939. The analysis of an animal population. *J. Anim. Ecol.*, 8: 238–246.

JORDAN, C. 1950. *Calculus of Finite Differences*, 2nd ed. Chelsea, New York.

KOSTITZIN, V. A. 1939. *Mathematical Biology.* Harrap, London.

KUCZYNSKI, R. R. 1932. *Fertility and Reproduction.* Falcon Press, New York.

——. 1935. *The Measurement of Population Growth.* Sidgwick & Jackson, London.

LESLIE, P. H., and R. M. RANSON. 1940. The mortality, fertility, and rate of natural increase of the vole (*Microtus agrestis*) as observed in the laboratory. *J. Anim. Ecol.*, 9: 27–52.

——, and T. PARK. 1949. The intrinsic rate of natural increase of *Tribolium castaneum* Herbst. *Ecology*, 30: 469–477.

LINNAEUS, C. 1743. Oratio de telluris habitabilis. In *Amoenitates Academicae seu Dissertationes Variae. . . .* Editio tertia curante J. C. D. Schrebero, 1787, Vol. 2: 430–457. J. J. Palm, Erlangae.

LOTKA, A. J. 1907a. Relation between birth rates and death rates. *Science*, 26: 21–22.

——. 1907b. Studies on the mode of growth of material aggregates. *Amer. J. Sci.*, 24: 119–216.

——. 1910. Contributions to the theory of periodic reactions. *J. phys. Chem.*, 14: 271–274.

——. 1925. *Elements of Physical Biology.* Williams & Wilkins, Baltimore.

——. 1927. The size of American families in the eighteenth century and the significance of the empirical constants in the Pearl-Reed law of population growth. *J. Amer. statist. Ass.*, 22: 154–170.

——. 1931. The structure of a growing population. *Human Biol.*, 3: 459–493.

——. 1934. *Théorie Analytique des Associations Biologiques.* Hermann, Paris.

LOWAN, A. N., Technical director. 1941. *Table of Natural Logarithms.* 4 vols. Federal Works Agency, Works Projects Administration for the City of New York. Rept. of Official Project No. 15-2-97-33.

MALTHUS, T. R. 1798. *An Essay on the Principle of Population as it Affects the Future Improvement of Society, with Remarks on the Speculations of Mr. Godwin, M. Condorcet, and Other Writers.* Printed for J. Johnson in St. Paul's Churchyard, London.

MENDES, L. O. 1949. Determinação do potencial biotico da "broca do café"—*Hypothenemus hampei* (Ferr.)—e consideracões sôbre o crescimento de sua população. *Ann. Acad. bras. Sci.*, 21: 275–290.

MILNE-THOMPSON, L. M. 1933. *The Calculus of Finite Differences.* Macmillan, London.

MOLISCH, H. 1938. *The Longevity of Plants.* E. H. Fulling, Lancaster Press, Lancaster, Pa.

NEWMAN, F. W. 1883. Table of the descending exponential function to twelve or fourteen places of decimals. *Trans. Camb. phil. Soc.*, 13: 145–241.

NICHOLSON, A. J., and V. A. BAILEY. 1935. The balance of animal populations. Part I. *Proc. zool. Soc. Lond.*, 1935: 551–598.

PEARL, R. 1925. *The Biology of Population Growth.* Knopf, New York.

——. 1930. *Introduction to Medical Biometry and Statistics*, 2nd ed. Saunders, Philadelphia.

——. 1937. On biological principles affecting populations: human and other. *Amer. Nat.*, 71: 50–68.

——. 1939. *The Natural History of Population.* Oxford Univ. Press, London.

——. 1946. *Man the Animal.* Principia Press, Bloomington, Ind.

——, and L. J. REED. 1920. On the rate of growth of the population of the United States since 1790 and its mathematical representation. *Proc. natl. Acad. Sci. Wash.*, 6: 275–288.

——, and J. R. MINER. 1935. Experimental studies on the duration of life, XIV. The comparative mortality of certain lower organisms. *Quart. Rev. Biol.*, 10: 60–79.

PIERCE, J. C. 1951. The Fibonacci series. *Sci. Monthly*, 73: 224–228.

QUETELET, A. 1835. *Sur l'Homme et le Dèveloppment de ses Facultès ou Essai de Physique Sociale.* Bachelier, Imprimeur-Libraire, Paris.

RHODES, E. C. 1940. Population mathematics. *J. R. statist. Soc.*, 103: 61–89, 218–245, 362–387.

RICKER, W. E. 1948. *Methods of Estimating Vital Statistics of Fish Populations.* Indiana Univ. Pubs. Sci. Ser., 15, Bloomington.

ROSS, R. 1911. *The Prevention of Malaria*, 2nd ed. Dutton, New York.

RUSSELL, E. S. 1942. *The Overfishing Problem.* Cambridge Univ. Press, Cambridge.

SADLER, M. T. 1830. *The Law of Population: A Treatise, in Six Books; in Disproof of the Superfecundity of Human Beings, and Developing the Real Principle of their Increase.* Murray, London.

SALT, G. 1936. Experimental studies in insect parasitism. IV. The effect of superparasitism on populations of *Trichogramma evanescens*. *Brit. J. exp. Biol.*, 13: 363–375.

SHARPE, F. R., and A. J. LOTKA. 1911. A problem in age distribution. *Phil. Mag.*, 21: 435–438.

SMITH, F. E. 1952. Experimental methods in population dynamics, a critique. *Ecology*, 33: 441–450.

STANGELAND, C. E. 1904. *Pre-Malthusian Doctrines of Population; a Study in the History of Economic Theory.* Columbia Univ. Stud. Hist. Econ. pub. Law, Vol. 21, no. 3.

SUNDBÄRG, A. G. 1907. *Bevölkerungsstatistik Schwedens 1750–1900.* P. A. Norstedt and Söner, Stockholm.

THOMPSON, D'ARCY W. 1942. *On Growth and Form*, new ed. Cambridge Univ. Press, Cambridge.

THOMPSON, W. R. 1931. On the reproduction of organisms with overlapping generations. *Bull. ent. Res.*, 22: 147–172.

——. 1939. Biological control and the theories of the interactions of populations. *Parasitology*, 31: 299–388.

U. S. Forest Service. 1948. The woody-plant seed manual. *Misc. Publ. U. S. Dep. Agric.*, No. 654.

VERHULST, P. F. 1838. Notice sur la loi que la population suit dans son accroisissement. *Corresp. math. phys., A. Quetelet*, 10: 113–121.

——. 1845. Recherches mathématiques sur la loi d'accroissement de la population. *Mém. Acad. R. Belg.*, 18: 1–38.

VOLTERRA, V. 1927. Variazioni e fluttuazioni del numero d'individui in specie animali conviventi. *Mem. Accad. Lincei*, 324 (ser. sesta). Vol. 2.

——. 1931. Variation and fluctuations of the number of individuals in animal species living together. In *Animal Ecology* (Chapman, R. N.). McGraw-Hill, New York.

WALLACE, R. 1753. *A Dissertation on the Numbers of Mankind in Antient and Modern Times: in which the Superior Populousness of Antiquity is Maintained.* Printed for G. Hamilton and J. Balfour, Edinburgh.

WARDLE, R. A., and J. A. MCLEOD. 1952. *The Zoology of Tapeworms.* Univ. of Minnesota Press, Minneapolis.

WIDDER, D. V. 1940. *The Laplace Transform.* Princeton Univ. Press, Princeton.

——. 1947. *Advanced Calculus.* Prentice-Hall, New York.

YULE, G. U. 1906. On the changes in the marriage- and birth-rates in England and Wales during the past half century; with an inquiry as to their probable causes. *J. R. statist. Soc.*, 69: 88–132.

——. 1925. The growth of population and the factors which control it. *J. Roy. statist. Soc.*, 88: 1–58 (related discussions, pp. 58–90).

Biological Populations with Nonoverlapping Generations: Stable Points, Stable Cycles, and Chaos

Abstract. *Some of the simplest nonlinear difference equations describing the growth of biological populations with nonoverlapping generations can exhibit a remarkable spectrum of dynamical behavior, from stable equilibrium points, to stable cyclic oscillations between 2 population points, to stable cycles with 4, 8, 16, . . . points, through to a chaotic regime in which (depending on the initial population value) cycles of any period, or even totally aperiodic but bounded population fluctuations, can occur. This rich dynamical structure is overlooked in conventional linearized analyses; its existence in such fully deterministic nonlinear difference equations is a fact of considerable mathematical and ecological interest.*

In some biological populations (for example, man), growth is a continuous process and generations overlap; the appropriate mathematical description involves nonlinear differential equations. In other biological situations (for example, in 13-year periodical cicadas), population growth takes place at discrete intervals of time and generations are completely nonoverlapping; the appropriate mathematical description is in terms of nonlinear difference equations. For a single species, the simplest such differential equations, with no time delays, lead to very simple dynamics: a familiar example is the logistic, $dN/dt = rN(1 - N/K)$, with a globally stable equilibrium point at $N = K$ for all $r > 0$.

It is the purpose of this report to point out that many of the corresponding difference equations of population biology have been discussed inadequately, as having either a stable equilibrium point or being unstable, with growing oscillations (*1*, *2*). In fact, some of the very simplest nonlinear difference equations even for single species exhibit a spectrum of dynamical behavior which, as the intrinsic growth rate r increases, goes from a stable equilibrium point, to stable cyclic oscillations between 2 population points, to stable cycles with 4 points, then 8 points, and so on, through to a regime which can only be described as chaotic (a term coined by J. A. Yorke). For any given value of r in this chaotic regime there are cycles of period 2, 3, 4, 5, . . . , n, . . . , where n is any positive integer, along with an uncountable number of initial points for which the system does not eventually settle into any finite cycle; whether the system converges on a cycle, and, if so, which cycle, depends on the initial population point (and of course some of the cycles may be attained only from infinitely unlikely initial points). Figure 1 aims to illustrate this range of behavior.

Specifically, consider the simple nonlinear equation

$$N_{t+1} = N_t \exp[r(1 - N_t/K)] \qquad (1)$$

This is considered by some people (*2*, *3*) to be the difference equation analog of the logistic differential equation, with r and K the usual growth rate and carrying capacity, respectively. The stability character of this equation, as a function of increasing r, is set out in Table 1 and illustrated by Fig. 1.

Another example is

$$N_{t+1} = N_t[1 + r(1 - N_t/K)] \qquad (2)$$

This quadratic form is probably the simplest nonlinear equation one could write. Although discussed by various people (*4*, *5*) as the analog of the logistic differential equation, Eq. 2 is less satisfactory than Eq. 1 by virtue of its unbiological feature that the population can become negative if at any point N_t exceeds $K(1+r)/r$. Thus, stability properties here refer to stability within some specific neighborhood, whereas in Eq. 1, for example, the stable equilibrium point at $N = K$ is globally stable (for all $N > 0$) for $2 > r > 0$. With this proviso, the stability behavior of Eq. 2 is strikingly similar to that of Eq. 1; see Table 1.

That such single species difference equations should describe populations going from stable equilibrium points to stable cycles as r increases is not surprising, in view of the general engineering precept that excessively long time delays in otherwise stabilizing feedback ~~mechanisms~~ can lead to "instability" or, more precisely, to stable limit cycles (*5*, chapter 4; *6*). What is remarkable, and disturbing, is that the simplest, purely deterministic, single species models give essentially arbitrary dynamical behavior once r is big enough ($r > 2.692$ for Eq. 1, $r > 2.570$ for Eq. 2). Such behavior has previously been noted in a meteorological context (*7*), and doubtless has other applications elsewhere. For population biology in general, and for temperate zone insects in particular, the implication is that even if the natural world were 100 percent predictable, the dynamics of populations with "density dependent" regulation could nonetheless in some circumstances be indistinguishable from chaos, if the intrinsic growth rate r were large enough.

The detailed analysis substantiating these remarks, and deriving Table 1, will be set out in the technical literature. A very brief outline is as follows: (i) For the general nonlinear difference equation

$$N_{t+1} = f(N_t) \qquad (3)$$

the locally stable equilibrium point or points can be found by the conventional techniques of linearized stability analysis. For Eq. 1, a fully nonlinear analysis can be given by observing that $V_t = (N_t - K)^2$ is a Lyapunov function, with the properties $V_t \geqslant 0$ and $\Delta V_t \equiv V_{t+1} - V_t \leqslant 0$ for all $N_t > 0$, for $2 > r > 0$: this ensures that the equilibrium point is globally stable. (ii) Next, the possible occurrence of cycles with period 2 may be studied for the equation

$$N_{t+2} = f[f(N_t)] \qquad (4)$$

For Eqs. 1 and 2 this has a unique nontrivial equilibrium solution, $N^* = K$, for $r < 2$, corresponding to the above stable point; as r increases above 2 this solution of Eq. 4 becomes unstable, and (as

Table 1. Dynamics of a population described by the difference equations 1 or 2.

Dynamical behavior	Value of the growth rate, r: Equation 1	Value of the growth rate, r: Equation 2
Stable equilibrium point	$2 > r > 0$*	$2 > r > 0$
Stable cycles of period 2^n		
2-point cycle	$2.526 > r > 2.000$†	$2.449 > r > 2.000$
4-point cycle	$2.656 > r > 2.526$‡	$2.544 > r > 2.449$
8-point cycle	$2.685 > r > 2.656$	$2.564 > r > 2.544$
16, 32, 64, . . .	$2.692 > r > 2.685$	$2.570 > r > 2.564$
Chaotic behavior. (Cycles of arbitrary period, or aperiodic behavior, depending on initial condition.)	$r > 2.692$§	$r > 2.570$

* See Fig. 1a. † See Fig. 1b. ‡ See Fig. 1c. § See Fig. 1, d, e, and f.

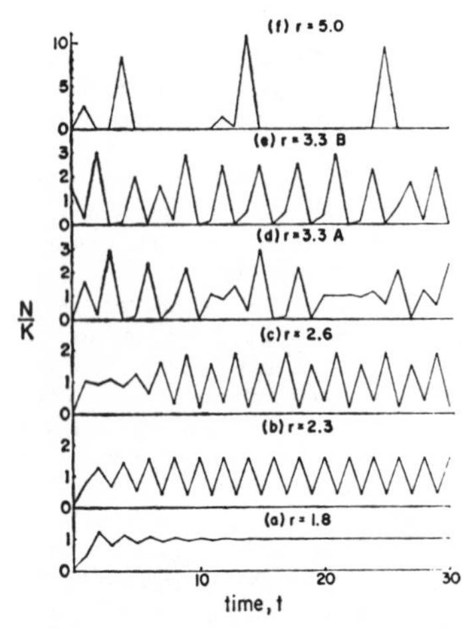

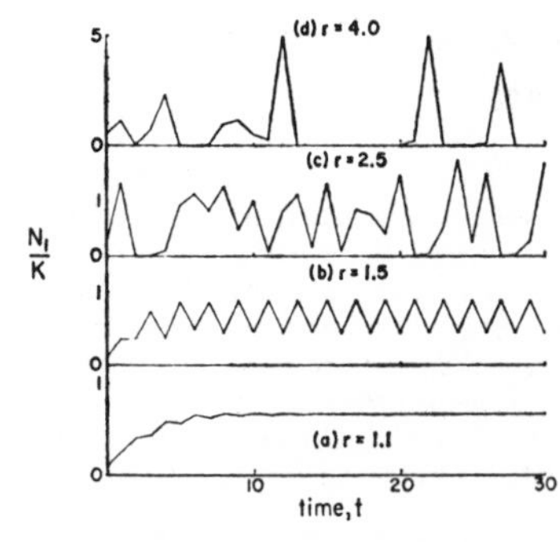

Fig. 1 (left). (a to f) Spectrum of dynamical behavior of the population density, N_t/K, as a function of time, t, as described by the difference Eq. 1 for various values of r. Specifically: (a) $r = 1.8$, stable equilibrium point; (b) $r = 2.3$, stable 2-point cycle; (c) $r = 2.6$, stable 4-point cycle; (d to f) in the chaotic regime, where the detailed character of the solution depends on the initial population value, with (d) $r = 3.3$ ($N_0/K = 0.075$), (e) $r = 3.3$ ($N_0/K = 1.5$), (f) $r = 5.0$ ($N_0/K = 0.02$). Fig. 2 (right). Stability character of the difference equation model of two-species competition, Eq. 5. Specifically, the figure is for $r_1 = r$, $r_2 = 2r$, $K_1 = K_2 = K$, $\alpha_{11} = \alpha_{22} = 1$, $\alpha_{12} = \alpha_{21} = \alpha$: under these conditions the criterion for a stable point, Eq. 7, reduces to the requirements $\alpha < 1$ (as for the analogous Lotka-Volterra differential equation), together with $[3 - (1 + 8\alpha^2)^{\frac{1}{2}}]/[2(1 - \alpha)] > r > 0$. The first population, expressed as N_1/K, is shown as a function of time for $\alpha = 0.5$ and several values of r: (a) $r = 1.1$; (b) $r = 1.5$; (c) $r = 2.5$; (d) $r = 4.0$.

illustrated by Fig. 1b) bifurcates into a pair of points, between which the population alternates in a 2-point cycle which is stable provided $2 < r < 2.526$ for Eq. 1, and $2 < r < 2.449$ for Eq. 2. Beyond this, the 2-point cycle in turn becomes unstable and each of the points bifurcates into 2 further points, giving a stable 4-point cycle (for example, Fig. 1c), and so on. (iii) As r continues to increase, there is a limit to this process whereby cycles of period 2^n become unstable and bifurcate into stable cycles of period 2^{n+1}. This limiting value of r, r_c say, may be calculated [either by brute force, or by analytic methods developed in (8)], and is as set out in the final line in Table 1. For $r > r_c$, there ensues a regime of chaos, in which there exist an uncountable number of initial points N_0 for which the system does not eventually settle into any cycle (that is, is not "asymptotically periodic"). (iv) In particular, at yet larger values of r ($r > 3.102$ for Eq. 1, and $r > 2.828$ for Eq. 2), Eqs. 1 and 2 may be shown to have cycles with period 3; that is, solutions such that $N_{t+3} = N_t \neq N_{t+1} \neq N_{t+2}$. But Li and Yorke (9) have recently proved an elegant and abstract mathematical theorem, which states that if the general difference equation, Eq. 3, has a 3-point cycle, it necessarily follows that for the same parameter values there are cycles of period n, where n is any positive integer, and furthermore there exist an uncountable number of initial points for which the system is not even asymptotically periodic. Li and Yorke's general theorem for cycles of period 3 may be extended (8) to show that equations of the generic form of 1 and 2 will enter a regime of chaos, with an uncountable number of cycles of integral period along with an uncountable number of aperiodic solutions, beyond the limiting value r_c defined above.

The dynamical behavior of Eqs. 1 and 2 in this chaotic regime, $r > r_c$, is illustrated in Fig. 1, d, e, and f. Figure 1, d and e, are for the same value of r, and differ only in their initial population value. Note that either of these figures, if looked at only over particular short time intervals, could convey the impression of being locked into a 3-point cycle; around this value of r there is a tendency to be "captured" into almost-periodic 3-point cycles, in between episodes of apparently chaotic behavior. A detailed understanding of these properties remains an interesting mathematical problem, related to that of determining what fraction of the totality of initial points converge to a 3-point cycle, what fraction to a 5-point cycle, and so on, ending with a determination of the fraction of initial points which lead to aperiodic behavior. For relatively large values of r beyond r_c (for example, Fig. 1f) the population variations become more severe, although the mean population value may be shown to remain around K; as r becomes larger, this mean value is increasingly constituted of a few fairly large population values, together with long sequences of very low population values.

The above discussion is restricted to single species systems obeying difference equations. However, similar considerations are likely to apply, a fortiori, to multispecies situations.

As one among many possible examples, consider a simple difference equation model for competition between two species

$$N_1(t+1) = N_1(t)\exp\{r_1[K_1 - \alpha_{11}N_1(t) - \alpha_{12}N_2(t)]/K_1\} \quad (5a)$$

$$N_2(t+1) = N_2(t)\exp\{r_2[K_2 - \alpha_{21}N_1(t) - \alpha_{22}N_2(t)]/K_2\} \quad (5b)$$

Just as Eq. 1 may be regarded as a difference equation analog of the logistic, Eq. 5 may be regarded as an analog of the familiar Lotka-Volterra differential equation model for two-species competition. As usual, r_i are the growth rates, K_i the carrying capacities, and α_{ij} the competition coefficients. The dynamical properties of such Lotka-Volterra differential equations are straightforward: the two species coexist, with a globally stable equilibrium point, if and only if

$$D > 0 \quad (6)$$

where D is defined as $D = \alpha_{11}\alpha_{22} - \alpha_{12}\alpha_{21}$. Failing this, one or the other species is extinguished. But for the system of difference equations, Eq. 5, the criterion for the existence of a stable two-species equilibrium point is more restrictive, namely

$$A > D > 0 \quad [\text{if } A < 2] \quad (7a)$$

$$A > D > 2A - 4 \quad [\text{if } A > 2] \quad (7b)$$

Here D is as above, and A is defined as $A = (\alpha_{11}K_2/r_2N_2^*) + (\alpha_{22}K_1/r_1N_1^*)$, with N_1^* and N_2^* the equilibrium solutions of Eq. 5. [The methodology for stability analysis of such two-species difference equations is indicated elsewhere (6)]. If the right-hand side of Eq. 7a is violated, one of the species is eliminated, as in the differential equation model, Eq. 6. If any of the other inequalities in Eq. 7 is transgressed, the two species continue to coexist, but there is no longer a stable equilibrium point. Numerical studies reveal a regime of stable cycles, giving way to one of apparent chaos, as for the single species systems discussed in detail above. The behavior of the system of Eq. 5 in these various regimes is illustrated in Fig. 2.

Equations 1 and 2 are two of the simplest nonlinear (density dependent) difference equations that can be written down. Their rich dynamical structure, and in particular the regime of apparent chaos wherein cycles of essentially arbitrary period are possible, is a fact of considerable mathematical and ecological interest, which deserves to be more widely appreciated. Without an understanding of the range of behavior latent in such deterministic difference equations, one could be hard put to make sense of computer simulations or time-series analyses in these models.

ROBERT M. MAY
Biology Department, Princeton University, Princeton, New Jersey 08540

References and Notes

1. Previous work in this general area of population biology consists largely of remarks on the relation between differential equation models and difference equation models [H. R. Van der Vaart, *Bull. Math. Biophys.* **35**, 195 (1973); R. M. May, *Am. Nat.* **107**, 46 (1972); J. M. Smith, *Models in Ecology* (Cambridge Univ. Press, Cambridge, 1974)] and linearized analyses of Eqs. 1 and 2 showing the equilibrium point to be locally stable only if $2 > r > 0$ [L. M. Cooke (2) for Eq. 1; J. M. Smith, *Models in Ecology*, for Eq. 2].
2. L. M. Cooke, *Nature (Lond.)* **207**, 316 (1965).
3. A. Macfadyen, *Animal Ecology: Aims and Methods* (Pitman, London, ed. 2, 1963).
4. J. M. Smith, *Mathematical Ideas in Biology* (Cambridge Univ. Press, Cambridge, 1968); J. R. Krebs, *Ecology: The Experimental Analysis of Distribution and Abundance* (Harper & Row, New York, 1972).
5. R. M. May, *Stability and Complexity in Model Ecosystems* (Princeton Univ. Press, Princeton, N.J., 1973).
6. ———, M. P. Hassell, G. R. Conway, T. R. E. Southwood, *J. Anim. Ecol.*, in press.
7. E. N. Lorenz, *Tellus* **16**, 1 (1964); *J. Atmos. Sci.* **20**, 448 (1963).
8. R. M. May and G. F. Oster, in preparation.
9. T.-Y. Li and J. A. Yorke, *SIAM (Soc. Ind. Appl. Math.) J. Math. Anal.*, in press.
10. I am indebted to many people, and particularly to J. A. Yorke, for stimulating discussions.

3 June 1974

Hycanthone Analogs: Dissociation of Mutagenic Effects from Antischistosomal Effects

Abstract. N-*Oxidation at the diethylamino group of hycanthone, of lucanthone, and of two chlorobenzothiopyranoindazoles resulted in a marked reduction in mutagenic activity, while antischistosomal activity was retained or even enhanced. Introduction of chlorine into the 8-position of benzothiopyranoindazoles reduced acute toxicity but had no effect on chemotherapeutic potency. These dissociations of biological activities indicate that safer antischistosomal compounds of this class can be developed.*

The geographical distribution and the nature of human schistosomiasis require special care in the selection of chemotherapeutic agents for the treatment of this infection. More than 200 million human subjects are infected with schistosomes and the incidence is on the increase. Even a low frequency of delayed serious complications, produced by mutagenic, teratogenic, and carcinogenic actions of a drug, can involve a large absolute number of individuals. Populations infected with schistosomes are not protected by national drug laws or regulatory agencies. Moreover, in an undetermined number, and possibly the majority, of subjects infected with *Schistosoma hematobium*, overt clinical and pathological manifestations disappear in adulthood (*1*). This must be taken into account when considering a drug for the mass treatment of children whose life expectancies are longer and whose reproductive potentials are greater than those of adults. As was stated by Rubidge *et al.* (*2*), "urinary tract bilharziasis is a relatively mild disease in South Africa and serious sequelae are rare. Hence, therapy must be safe."

It is estimated that during the past 6 years, in Brazil, Africa, and the Middle East, at least 700,000 human subjects infected with *S. hematobium* and *S. mansoni* have been treated with the antischistosomal thioxanthenone derivative hycanthone (the drug is ineffective in infections produced by *S. japonicum* prevalent in mainland China and the Philippines) (*3*). Reports from a variety of laboratories have indicated that hycanthone is mutagenic (*4*) and teratogenic (*5*), and that it induces prophage (*6*), mitotic crossing-over (*7*), cytogenic changes (*8*), and malignant transformations (*9*); hycanthone is carcinogenic in mice infected with *S. mansoni* (*10*). As pointed out by Firminger (*11*), a report (*12*) which seemingly did not support the last observation was based on such a small number of animals that no significant negative results could have been obtained. Since a number of compounds chemically related to hycanthone exhibit antischistosomal activity, the question arose whether structural alterations can bring about a dissociation of undesirable toxicological properties from chemotherapeutic activity. Data summarized below indicate that this is the case.

Fig. 1. Mutagenic activity: none detectable (less than 0.1 percent as active as hycanthone). Antischistosomal activity: intramuscular, 0.4; oral, 0.3.

A chloroindazole analog (IA-4, structure in Table 1) of hycanthone has the same antischistosomal activity in mice as hycanthone (*13*), while its acute toxicity and its hepatoxicity are lower (*13, 14*). Compound IA-4 failed to induce demonstrable malignant transformations in cells infected with Rauscher virus (*9*). Its mutagenic activity was found to be lower in *Salmonella* (*15*), bacteriophage T4 (*15*), and mouse lymphoblasts (*16*); no mutagenic effects were detected in yeast (*17*); no cytogenetic effects were detected in rat bone marrow cells (*18*). Furthermore, in contrast to hycanthone and to a number of chemical carcinogens, IA-4 failed to induce breaks in rat liver DNA (*19*). Another indazole analog (IA-3) had lower antischistosomal activity; but since there is decreased acute toxicity, the chemotherapeutic index of IA-3 approximately equals that of IA-4 (*13*).

We found that chloro substitution in position 8 produced a marked decrease in the acute toxicity of the indazole analogs for mice. For example, the median intramuscular lethal dose (LD_{50}) of IA-3 and of IA-4 was more than seven times higher than that of the corresponding deschloro derivatives.

In further studies of the effect of structural modifications on antischistosomal activity and on mutagenicity, *N*-oxides of active thioxanthenones and benzothiopyranoindazoles were prepared. The parent bases were oxidized with *m*-chloroperbenzoic acid in dichloromethane solution, and after chromatography (Al_2O_3) the *N*-oxides so obtained were converted to their water-soluble methanesulfonate salts. *N*-Oxidation at the diethylaminoethyl group consistently resulted in a marked reduction in mutagenicity for *Salmonella*

Vol. 100, No. 916 The American Naturalist November-December, 1966

ON OPTIMAL USE OF A PATCHY ENVIRONMENT

ROBERT H. MACARTHUR AND ERIC R. PIANKA

Department of Biology, Princeton University, Princeton, New Jersey

There is a close parallel between the development of theories in economics and population biology. In biology, however, the geometry of the organisms and their environment plays a greater role. Different phenotypes have different abilities at harvesting resources, and the resources are distributed in a patchwork in three dimensions of the environment. In this paper we undertake to determine in which patches a species would feed and which items would form its diet if the species acted in the most economical fashion. Hopefully, natural selection will often have achieved such optimal allocation of time and energy expenditures, but such "optimum theories" are hypotheses for testing rather than anything certain. Some aspects of dietary and patch utilization have been treated in rather different ways by Hutchinson and MacArthur (1959) and by MacArthur and Levins (1964). The best empirical support for the model to be presented is that given by McNab (1963).

The basic procedure for determining optimal utilization of time or energy budgets is very simple: an activity should be enlarged as long as the resulting gain in time spent per unit food exceeds the loss. When any further enlargement would entail a greater loss than gain no such enlargement should take place. The problem is to find which components of a time or energy budget increase and which decrease as certain activities are enlarged.

Consider, first, the optimal number of kinds of items (such as prey species) in the diet. We assume here that the environment is "fine-grained," that is, that during search for food the prey species are located in the proportion in which they occur. In a later paragraph we deal with patchy environments where this is not true. We divide the time spent, per item eaten, into two components: time for search, and time for pursuit capture and eating. (The difference is that the animal searches a fine grained environment for all kinds of items simultaneously but pursues captures and eats them one at a time.) Suppose that the predator already includes N kinds of prey in its diet. Then we may subdivide its time, T_N, per item eaten, into a search time T_N^S and a pursuit (plus capture and eating) time, T_N^P. We can do the same for the predator if he were to enlarge his diet to include $N + 1$ kinds of prey. Writing both down in symbols

$$\left.\begin{aligned} T_{N+1} &= T_{N+1}^S + T_{N+1}^P \\ T_N &= T_N^S + T_N^P \\ \hline \Delta T_N &= \Delta T_N^S + \Delta T_N^P \end{aligned}\right\} \quad \text{(all times are per item of food).}$$

We can subtract and find the change, ΔT_N, in total time which accompanies enlarging the diet from N to $N + 1$ items. If, in some way, the items can be ranked from most profitable to least profitable, then the optimal diet can be calculated by proceeding through the ranked list of items until ΔT_N first becomes negative. At this point no further enlargement should be contemplated. This gives the clue to the method of ranking: it should proceed from items of highest harvest per unit time to those of lowest. More specifically, we notice that ΔT^S is always negative, for the larger the variety of acceptable items, the less the search time, per unit of food. The pursuit time may increase, however, as new, hard-to-catch items are added to the diet. Hence ΔT^P may be positive. In Fig. 1a, 1b, we plot samples of $\Delta T_N{}^P$ and $-\Delta T_N{}^S$. For comparison purposes we actually plot the *reduction* in search time $\Delta S = -\Delta T_N{}^S$ against the increase in pursuit time $\Delta P = \Delta T_N{}^P$, since where these intersect there will be no further benefit from enlarging the diet. Notice that in both Fig. 1a and 1b, the items are ranked so that the reduction in pursuit time exceeds the gain in pursuit time by the greatest amount, i.e., the vertical distance between ΔS and ΔP curves is decreasing. In summary the optimal diet is the first value of N such that ΔT_N is negative, which is the first value of N to the right of the intersection point in the Fig. 1a, 1b.

The ΔS of Fig. 1 is calculated from assuming enlarging a diet from N to $N + 1$ equally abundant species reduces mean search time from $\frac{1}{N}$ to $\frac{1}{N+1}$, and so $\Delta S = \frac{1}{N} - \frac{1}{N+1}$. The ΔP curve measures the adaptations of the species for the items and must be empirically determined. The arrow indicates the optimal diet; when the species eats four kinds of resources an enlargement of the diet would, for the first time, cause a greater increase (in pursuit time) than decrease (in search time). The $\Delta S'$ curve is double the height of the ΔS curve indicating the effect of halving the density of each resource species. The optimal diet should be expanded to five species of prey. It would also be possible to indicate a more specialized predator by a steeper ΔP curve. This specialized predator should be less sensitive to changes in food density.

The exact shape of the curves is usually unknown and certainly varies from situation to situation. Hence no general prediction of the exact diet is worth attempting. However some interesting comparative predictions can be made. When the search time is multiplied by a constant factor, its decrease is also multiplied by that factor; if T^S in eq. 1 is multiplied by k, so is ΔT^S. Hence, in a productive environment where search time is uniformly reduced, its decrease is reduced; although the pursuit time, which is a function of the abilities of predator and prey, is unaltered and, according to the figure, the optimal diet becomes more restricted. Thus organisms which have low search/pursuit ratios should be more restricted in diet, whether the reason be high food density or high mobility of the prey. Recher has some evidence for this from herons (personal communication).

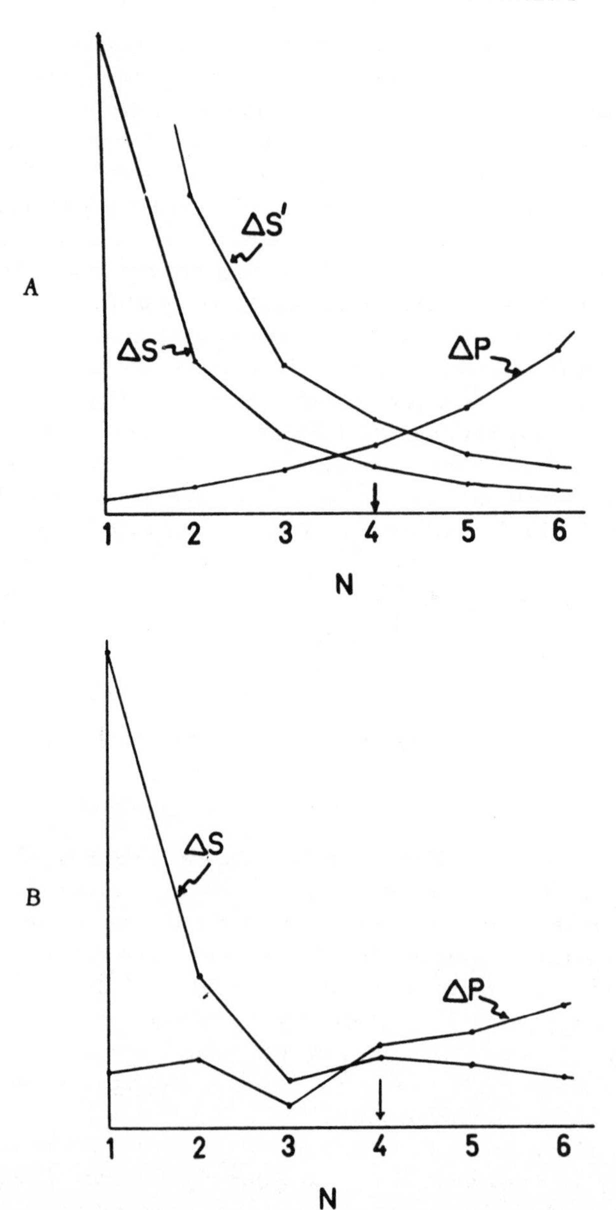

FIG. 1A. Equinumerous resource species. The decrease, ΔS, in mean search time and the increase, ΔP, in mean pursuit time which would accompany enlarging the diet from N to $N + 1$ species of prey are plotted for a hypothetical situation.

FIG. 1B. Resource species not equally numerous. The symbols are the same as in Fig. 1A, but the curves are no longer monotonic. The same qualitative conclusions hold.

The optimal use of patches of habitat is in many ways parallel to that of items in the diet within a patch. Now the time, per item caught, spent within suitable patches is an increasing function of the number of kinds of patches on the species' itinerary (for as the itinerary is enlarged to include less suitable patches the hunting time clearly increases). We call this hunting time H, and denote by ΔH the *increase* in hunting time per item which accompanies enlarging the itinerary to include the next most satisfactory patch type. The time spent travelling between suitable patches (or to and from the nest if no hunting is done on the trips), is divided by the harvest to give the travelling time, T, per item caught. This is clearly a decreasing function of the number of patch types in the itinerary, and we denote by ΔT the *decrease* which accompanies enlarging the itinerary by one more patch type. The patch types are ranked from most productive (i.e., most prey calories caught per unit time) to least; as before, this is equivalent to a ranking which orders $\Delta H - \Delta T$ from largest to smallest. When the patches are about equally common we get something like Fig. 2; if they are unequally common, Fig. 2 will be modified in the way Fig. 1b was obtained from 1a, with no change in the qualitative predictions.

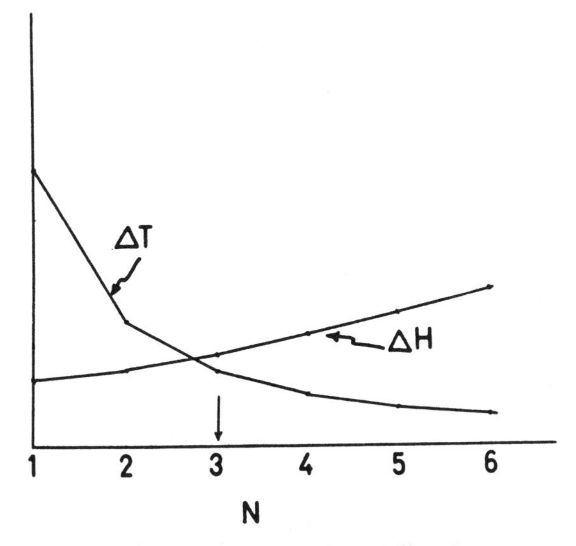

FIG. 2. The decrease, ΔT, associated with adding the next patch type, in the mean travelling time T (per prey item) across unsuitable patches and the increase, ΔH, in mean hunting time within suitable patches (per prey item) due to adding the next patch type, are plotted against the number of types of environmental patches in the species' itinerary. ΔT and ΔH measure the changes which accompany enlarging from N to $N+1$. As before, the arrow indicates the optimal utilization.

The qualitative predictions from Fig. 2 are not quite parallel to those from Fig. 1. The effect of increased productivity (or more precisely increased food density) is no longer unambiguous, for both ΔH and ΔT are lowered. However, species with large pursuit/search ratios will have their ΔH curves lowered by a reduced amount, for only the search time decreases with increased food density. Hence pursuers more than searchers, should show restricted patch utilization where food is dense.

A second interesting prediction involves the use of different sized patches. Two environments which differ not in the proportion or quality of their patches but only in the sizes of the patches (e.g., a checkerboard with one acre squares in one, and 1/4 acres squares in the other) will not have different H or ΔH, curves, for these are calculated per item caught and we postulated that the quality of the patches is unchanged. The ΔT curves will differ, though, for travelling distance between patches varies linearly with the linear dimension of the patch, while hunting area within a patch varies as its square. Hence, larger patches offer smaller travel time per unit hunting time, and thus have lower ΔT curves. Hence, by drawing an imaginary ΔT curve lower on Fig. 2, we see that larger patches are used in a more specialized way than are smaller ones, everything else being equal. This patch size effect should be greatly reduced in territorial species, since the travel time to and from the nest is independent of patch size; and only the travelling time from patch to patch across unsuitable ones drops in importance as patches are made larger. Hence, while feeding young in the nest, parents should exhibit nearly the same choice of patches whether they be large or small, but after the nesting, individuals in a large-patch environment should restrict their patch utilization. [Hutchinson (1959, 1965) has commented on other aspects of the relationship between an animal's size and the "grain-size" or texture of the environment.]

The effect of competitors is to reduce the density of some kinds of prey species in some patches. Curiously enough, an optimal predator faced with competition, should respond by shrinking its patch utilization but not (conspicuously) its diet! To see this, we consider the diet first. If a dietary item becomes rare, due to a competitor, its inclusion in the diet will have only a very slight effect on mean search and mean pursuit times. Hence ΔS and ΔP will be reduced toward zero for the item preceding the rare one and then will rise again (i.e., enlargement to include the rare one will cause little change). However, the reduction in the ΔS and ΔP will be roughly proportional to the reduction in abundance of the prey, so that if ΔS exceeded ΔP before the competitor entered, it will afterwards, also. In other words, any dietary item worth eating in the absence of competition is still worth eating afterwards. Patches, on the other hand, are a different story. For if food within one kind of patch becomes scarce, due to competition or any other cause, then to increase the itinerary to include these patches of scarce food means to increase the mean hunting time sharply. ΔH will then show a sharp peak of increase corresponding to the impoverished patch type. ΔT will be independent of the quality of the new patch and thus will not change with competition. Hence the ΔH curve may jump above the ΔT curve at an earlier point, causing a reduction in the optimal patch itinerary.

Next we ask whether the patch structure of the environment imposes any limit on the similarity of coexisting competitors. The answer is yes. Briefly, when the gain to a jack-of-all-trades in reduced travelling time makes up for his lower hunting efficiency compared to the patch specialists, then the jack-of-all-trades will outcompete both specialists. To be more precise, the specialists' harvest of food, per day, is kDH where k is the hunting rate, D the food density, and H the hunting time, per day. The jack-of-all-trades harvests $k'DH'$ where $k' < k$ (a jack-of-all-trades is a master of none) and $H' > H$ (the jack-of-all-trades, by feeding in a wider variety of patches, travels less between suitable ones and has more time left for hunting). Thus, if $k'H' > kH$ $\left(\text{i.e., } \frac{k'}{k} > \frac{H}{H'}\right)$ then the jack-of-all-trades can reduce the food density to a lower value of D and still harvest enough to maintain itself. At this value of D the specialists cannot harvest fast enough and so they are eliminated. Or looked at in another way, the specialists can only be expected to coexist and resist invasion by the more generalized foragers if $\frac{H}{H'} > \frac{k'}{k}$; and, if the specialists become too similar to each other, their hunting rates (k) become closer to the hunting rate of the jack-of-all-trades (k'), with the result that they become more susceptible to invasion and competitive replacement.

One further justification of the whole scheme is worth adding here: the proof that the optimal use by a species is essentially independent of the subtlety of the recognized differences between patches or diet items. In other words, the results are not artifacts of the classification of patch or diet types. Suppose, for instance, that Fig. 2 is calculated from an environment which one biologist considers to be a checkerboard of one acre squares of types A, B, C, D, . . . in decreasing order of preference. A second biologist classifies the same environment into quarter acre squares, those formerly labelled B now being of types b_1, b_2, b_3, b_4 and so on. Assuming all a's are still preferable to all b's and these to all c's, then Fig. 2 remains unchanged except that the abscissa should be four times as finely subdivided. Point b_4 will coincide with B, c_4 with C and so on, and the optimal strategy will be essentially independent of the fineness of the subdivision. If there were inverted rankings (say $b_1 < b_2 < c_1 < b_3 \ldots.$) then there will be changes in the optimal strategy, but these will reflect real differences in food concentration and are not simply artifacts of the naming of patches.

Our conclusions may be summarized by Table 1.

SUMMARY

A graphical method is discussed which allows a specification of the optimal diet of a predator in terms of the net amount of energy gained from a capture of prey as compared to the energy expended in searching for the prey.

The method allows several predictions about changes in the degree of specialization of the diet as the numbers of different prey organisms change. For example, a more productive environment should lead to more restricted diet in numbers of different species eaten. In a patchy environment, however, this will not apply to predators that spend most of their time searching. Moreover, larger patches are used in a more specialized way than smaller patches.

TABLE 1

Factors favoring increased specilization

Of diet	Of patches
Leave ΔP constant, but lower ΔS curve	Leave ΔH constant, but lower ΔT curve
1. Greater food density	1. Greater food density (pursuing species only*)
2. Increased mobility of animal, or decreased environmental resistance to movement, etc.	2. Increased mobility of animal, or decreased environmental resistance to movement, more contiguous patch structure, etc.
	3. Increased patch size relative to organism's size (less pronounced in territorial forms)
Leave ΔS constant, but raise ΔP curve	Leave ΔT constant, but raise ΔH curve
1. Increased differences between prey types, or increased specialization of pursuing behavior	1. Increased differences between patch types (or sizes), or more restricted hunting technique
2. Increased mobility of prey, or greater difficulty in pursuit	2. Increased mobility of prey, or greater difficulty of capturing it
	3. Reduction of food density in some patches by competition

*The hunting time is only independent of food density if it is all pursuit time and none search. The extent to which this is approximated determines our confidence in this effect.

ACKNOWLEDGMENTS

These ideas have been discussed with Drs. R. C. Lewontin and R. Levins, and will be developed further in a paper on the evolution of the niche by Levins and MacArthur. One author has been supported by a grant from the National Science Foundation, the other by a grant from the National Institutes of Health. A referee made helpful comments.

LITERATURE CITED

Hutchinson, G. E. 1959. Homage to Santa Rosalia, or why are there so many kinds of animals? Amer. Natur. 93:145-159.

______. 1965. The ecological theater and the evolutionary play. Yale Univ. Press, New Haven and London.

Hutchinson, G. E., and R. H. MacArthur. 1959. A theoretical ecological model of size distributions among species of animals. Amer. Natur. 93:117-125.

MacArthur, R. H., and R. Levins. 1964. Competition, habitat selection, and character displacement in a patchy environment. Proc. Nat. Acad. Sci. 51:1207-1210.

McNab, B. K. 1963. Bioenergetics and the determination of home range size. Amer. Natur. 97:133-140.

institution was re-modelled and co-ordinated in 1893, when the first principal was appointed, but it was not until 1901 that the educational work was organised for development on the lines of a university college with the provision of a curriculum for the external examinations of the University of London.

The College was placed upon the list of university institutions in receipt of grant from H.M. Treasury as from August 1, 1922, when it was incorporated under its present name as a company limited by guarantee, and the college buildings and halls of residence were transferred to it by the Exeter City Council. From that time its progress has been rapid, the number of degree students in the four years ending 1924–25 having been 96, 139, 187, and 211 respectively. The total number of full-time students in 1924–25 was 332, of whom 221 were in the teachers' training department. Residential halls provide accommodation for 134 women and 110 men students. Part-time students numbered 38 and occasional students 40. There are departments of biology, chemistry, classics, education, English, geography, history and economics, law, pure and applied mathematics, modern languages, music, philosophy, physics, and extra-mural studies. The total income was 29,067*l.*, including parliamentary grants 12,317*l.*, grants from local education authorities 10,384*l.*, tuition and examination fees 5040*l.*, income from endowment 674*l.*, and from other sources 1112*l.* The book value of its land, buildings, and permanent equipment is 81,433*l.*, and its endowment investments 11,603*l.*

These figures do not suggest that the College is likely to qualify soon for full university status, but it might conceivably join with the Technical Schools, Plymouth, the Seale-Hayne Agricultural College, Newton Abbot, and the Camborne School of Mines, to form a federal university. An important scheme for co-operation with the Technical Schools, Plymouth, has been worked out providing for degree and diploma courses in civil, electrical, marine, and mechanical engineering and in commerce at Plymouth, and the extension of the law teaching and extra-mural work already carried on there by the College. An appeal was launched in October 1925 for 100,000*l.* for the equipment and endowment.

UNIVERSITY COLLEGE, HULL.

The plans for the proposed University College for Hull, for which the Right Hon. T. R. Ferens gave 250,000*l.*, provide for an organisation somewhat similar in scope to that of University College, Southampton, with the addition of a department of agriculture and, eventually, departments of shipbuilding and applied chemistry of the oil, colour, gas, and spirit industries.

Lest the account already given of the policy of the University Grants Committee in regard to proposals for establishing new universities should be misunderstood, it must be added that the Committee is careful to point out that its "view of what is prudent at one particular stage of our history betokens no lack of sympathy with the general desire for a wider avenue to university education, or with the ambitions of certain large and populous cities to rival the more fortunate communities which already possess universities of their own." The Committee hopes that "as returning prosperity enables the schemes of local education authorities under the Education Act to be carried into effect, the local colleges will play an increasingly distinguished part in the higher education of the people, and will steadily raise the level of national knowledge and culture. It may well be that some of these will, in course of time, establish a claim to university rank and receive charters as independent universities."

Fluctuations in the Abundance of a Species considered Mathematically.[1]

By Prof. VITO VOLTERRA, For. Mem. R.S., President of the R. Accademia dei Lincei.

A CONSIDERATION of biological associations, or of the mutual interactions between two or more species associated together, has led me to certain mathematical results which may be set forth as follows.

The first case I have considered is that of two associated species, of which one, finding sufficient food in its environment, would multiply indefinitely when left to itself, while the other would perish for lack of nourishment if left alone; but the second feeds upon the first, and so the two species can co-exist together.

The proportional rate of increase of the eaten species diminishes as the number of individuals of the eating species increases, while the augmentation of the eating species increases with the increase of the number of individuals of the eaten species. Having determined the laws of this increase and diminution, it is possible to establish two differential equations of the first order, non-linear, which can be integrated. The integrals reveal the fact that the numbers of individuals of the two species are periodic functions of the time, with equal periods but with different phases, so that each species goes through a cycle relative to the other during a period, a process which may be called the 'fluctuation of the two species.' Figs. 1 and 2 give representations of different possible cycles, corresponding to different initial values of the number of individuals of the two species: ordinates representing the eating, and abscissæ the eaten species.

The co-ordinates of a point on a cycle are the concurrent values of the numbers of individuals of the two species, those of the central point Ω being the mean values; and the following laws have been deduced from integration of the differential equations which represent the fluctuation:

I. The fluctuation of the two species is periodic, the period depending only on the coefficients of increase and of destruction of the two species, and on the initial numbers of the individuals of the two species.

II. The average numbers of the two species tend to constant values, whatever the initial numbers may have been, so long as the coefficients of increase or of destruction of the two species and also the coefficients of protection and attack remain constant. (Laws I. and II. are illustrated in Fig. 2.)

III. If we try to destroy individuals of both species uniformly and proportionately to their number, the average number of individuals of the *eaten* species grows and the average number of the eating species diminishes (see Fig. 1). But increased protection of

[1] V. Volterra.—Variazioni e fluttuazioni del numero di individui in specie animali conviventi.—*Memorie della R. Accademia dei Lincei* (Cl. di Sci. Fis. etc.), ser. 6, vol. ii. fasc. 3, 85 pp., 1926.

the eaten species increases the average numbers of both.

In the case of small fluctuations, we have the following approximate laws :

(1) Small fluctuations are isochronous, *i.e.* their period is not sensibly affected either by the initial number of individuals, or by the conditions of protection and offence.

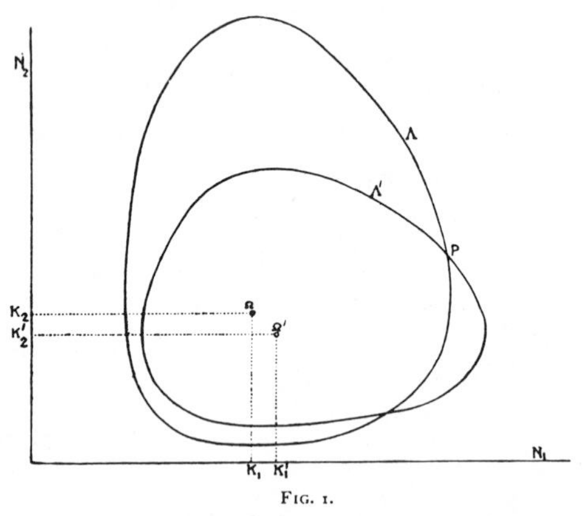

FIG. 1.

(2) The period of fluctuation is proportional to the product of the square roots of the time required for the first species to double itself, and for the second species to reduce itself to half. If the first species doubles itself in the time t_1 and the second species is reduced to half in the time t_2, the period is $T = \frac{2\pi}{\log_e 2}\sqrt{t_1 t_2} = 9{\cdot}06\sqrt{t_1 t_2}$.

(3) The steady destruction of individuals of the eating species accelerates the fluctuation, and the destruction of individuals of the eaten species retards it.

With the contemporaneous and uniform destruction of individuals of the two species, the ratio between the amplitude of the fluctuation of the eaten species and that of the eating species tends to increase.

In Fig. 1 are represented two cycles, the second of which corresponds to a perturbation produced in the first by a constant and proportionate destruction of the individuals of the two species. The centre Ω' of the perturbed curve is displaced, in respect to the centre Ω of the primitive curve, downwards and to the right ; this reveals an augmentation of the average number of individuals of the first species, and a diminution of the average number of the second.

Law III. is undoubtedly the most interesting of all, because it affords the best actual verification so far found of the theory. For Dr. U. d'Ancona, comparing fishery statistics in the Adriatic Sea before the War, during the War (when fishing almost ceased), and after fishing was resumed at the end of the War, has ascertained that the voracious species (selachians), which feed on other fishes, had increased during the War as compared with the preceding and following periods, while the contrary had been the case for the number of individuals of the eaten species.[2] In other words, a complete closure of the fishery was a form of 'protection' under which the voracious fishes were much the better and prospered accordingly, but the ordinary food-fishes, on which these are accustomed to prey, were worse off than before. This is in agreement with Fig. 1, and with Law III. My theoretical researches, which I was induced to undertake by the statistical studies begun by Dr. d'Ancona, correspond accordingly with his results.

Charles Darwin had an intuition of these phenomena in relation to the struggle for existence when in Chap. iii. of his "Origin of Species" he wrote : "The amount of food for each species of course gives the extreme limit to which each can increase ; but very frequently it is not the obtaining food, but the serving as prey to other animals, which determines the average number of a species. Thus there seems to be little doubt that the stock of partridges, grouse, and hares on any very large estate depends chiefly on the destruction of vermin. If not one head of game were shot during the next twenty years in England, and at the same time if no vermin were destroyed, there would in all probability be less game than at present, although hundreds of thousands of game animals are now annually shot."

Law III. is, however, true only up to a certain limit. It is evident that if the destruction of both species continue, their exhaustion will ensue. It is therefore necessary to ascertain up to just what point it is profitable to destroy both species in order to obtain the greatest augmentation in the average number of the eaten species. We arrive in this manner at a curious example of a mathematical ***upper limit*** without the existence of a *maximum*. There is in fact a limit of destruction beyond which both species are exhausted. If we remain below it, the average number of the eaten species grows as this limit is approached ; but once the limit is reached, the eating species tends to exhaustion and the fluctuation ceases, while the number of individuals of the eaten species tends asymptotically towards a value which is less than the average formerly reached.

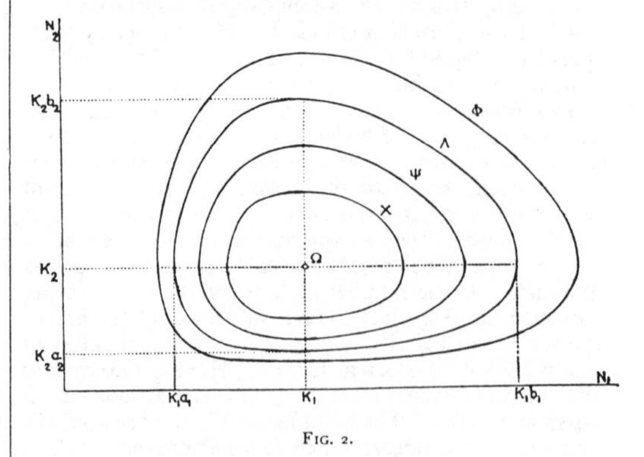

FIG. 2.

Besides the case dealt with above, a study of variations in the number of individuals of two associated species can also be made in all cases in which the species interact either favourably or injuriously, in all possible degrees or combinations. All such cases can be classified in distinct types, and in each of these it is possible to

[2] U. d' Ancona.—Dell' influenza della stasi peschereccia del periodo 1914-1918 sul patrimonio ittico dell' Alto Adriatico.—*Memorie del R. Comitato Talassografico Italiano* (in course of publication).

follow the numerical variations of the two species by the help of formulæ, or of diagrams to correspond. It is easy to see from these diagrams *which species is winning in the struggle for existence, and which of them is in process of extinction.*

Again, it is certain that many facts of medical interest may be classed among phenomena resulting from concurrent and reciprocal action between different species—between the human species and pathogenic germs, between parasitic species and those on which they are parasites. The periodicity of epidemics may be connected with the same phenomena.

A second part of my investigation is devoted to a general study of biological association, where any number of species may be living together. I have studied two types of association, and have called them *conservative* and *dissipative associations.*

For the first or conservative association, equivalent values may be assigned to the different species so that the destruction of a certain number of individuals belonging to one species by another species, to its own benefit, corresponds to an increase in numbers of the latter species, in the precise ratio which the said equivalents express. Moreover, in a conservative association, the number of individuals of every species has no influence on its own augmentation. The case of two species, already dealt with: the case of n species, where individuals of the first eat those of the second, and the latter those of the third, and so on to the nth; the case of four species so connected that the first eats the second and the second is also eaten by the third, which in its turn is eaten by the fourth; all these cases are examples of biological conservative associations, if we neglect actions between individuals of the same species. The variation in numbers of the conservative associations depends on a system of differential quadratic equations associated with a skew-symmetric determinant.

Owing to the peculiar properties of these skew-symmetric determinants, and to the differences between those of odd and those of even order, we must treat in different ways the cases where odd and where even numbers of biological species are associated together. When the number is odd, then, if the coefficients allow a 'stationary state,' we shall always have fluctuations such as to maintain the number of individuals of each species between positive limits. These limits are dependent on initial conditions, which may be so assumed as to restrict these limits to any extent we please.

The average numerical values of the different species tend, in periods of time of infinite duration, towards the corresponding values for the stationary state, and are therefore independent of the initial values.

The case of an odd number of species does not correspond to a condition of stability for a strictly conservative system.

In the case of associations which I call *dissipative,* if a stationary state exists, the variations will be fluctuations which slowly extinguish themselves, or are asymptotical. From an analytical point of view, the dissipative association is characterised by a definite positive quadratic form.

The analogous form is null in the case of a conservative system. The dissipative actions work in a way analogous to friction in a mechanical system.

Therefore the terms conservative and dissipative may also be applied to the fluctuations, which in the first case continue to exist, and in the second are dissipated. It will be observed that to prove the existence of the fluctuations, we follow an analysis different from that used in elastic or electric vibrations, the equations here employed being not linear but quadratic.

Applying this theory to a particular case, suppose three different species living together in a limited area, such as an island. Of these three species the first eats the second, which in turn eats the third. We may take for example a carnivorous species, feeding upon a herbivorous animal, which in turn feeds on a certain plant species—assuming that for the last the same method may be used which we apply to animals. The same method may also be employed in the case of insects parasitic upon plants, and of parasites of such parasites.

If we suppose that system to be conservative, that is, if we neglect the actions between individuals of the same species, we may have two different cases which are distinguished by the values of the coefficients occurring in the equations:

(1) The food which reaches the carnivorous species through the herbivorous is not sufficient to maintain the carnivorous species; and so the latter is exhausted, while the herbivorous animal and the plant tend to a periodical fluctuation.

(2) The vegetable species grows indefinitely. This case is, however, incompatible with the limitation of the island, which does not allow the indefinite multiplication of the plant. It is therefore necessary in this case to suppose the system to be dissipative, admitting that the coefficient of increase of the vegetable species is dependent on the number of existing plants, and then we have three cases: Either (1) both animal species are exhausted; or (2) only the carnivorous species is exhausted, while the herbivorous and the plant tend to a fluctuation of gradually diminishing amplitude or to an asymptotic variation; or (3) all three species live together without exhausting themselves, but vary asymptotically in a common fluctuation of gradually diminishing amplitude, the characteristic elements of which can be determined.

Side by side with the general theory, we may make various special inquiries. Thus, for example, we may suppose the coefficient of increase of the species to have an annual period, a supposition tending to establish a law of forced fluctuations superposed on the free fluctuations of the biological association considered.

We may also study how exhaustion takes place of a species in a biological conservative association of an uneven number of species; or in general how exhaustion takes place of species in dissipative associations; or what perturbation is produced when a new species is introduced into an association in equilibrium.

Seeing that a great number of biological phenomena are characteristic of *associations* of species, it is to be hoped that this theory may receive further verification and may be of some use to biologists.

PAPER **14**

RANDOM DISPERSAL IN THEORETICAL POPULATIONS

By J. G. SKELLAM

The Nature Conservancy, London

SYNOPSIS

The random-walk problem is adopted as a starting point for the analytical study of dispersal in living organisms. The solution is used as a basis for the study of the expansion of a growing population, and illustrative examples are given. The law of diffusion is deduced and applied to the understanding of the spatial distribution of population density in both linear and two-dimensional habitats on various assumptions as to the mode of population growth or decline. For the numerical solution of certain cases an iterative process is described and a short table of a new function is given. The equilibrium states of the various analytical models are considered in relation to the size of the habitat, and questions of stability are investigated. A mode of population growth resulting from the random scattering of the reproductive units in a population discrete in time, is deduced and used as a basis for a study on interspecific competition. The extent to which the present analytical formulation is applicable to biological situations, and some of the more important biological implications are briefly considered.

1. INTRODUCTION

1·1. It is now fifty years since the publication of *The Origin of the British Flora* by Clement Reid (1899). In it is suggested an interesting numerical problem on the rate of dispersal of plants. Reid states: 'Though the post-glacial period counts its thousands of years, it was not indefinitely long, and few plants that merely scatter their seed could advance more than a yard in a year, for though the seed might be thrown further, it would be several seasons before an oak for instance, would be sufficiently grown to form a fresh starting point. The oak, to gain its present most northerly position in North Britain after being driven out by the cold, probably had to travel fully six hundred miles, and this without external aid would take something like a million years.'

1·2. At the end of the last century, biologists, unlike physicists, rarely formulated such problems in terms of simplified abstract models, due no doubt to the comparatively greater complexity of biological systems. A beginning might have been made on the subject of dispersal, for much of the necessary mathematical technique had been developed already, and, in fact, had been utilized by Maxwell (1860) in developing a kinetic theory of gases based on the behaviour of an infinity of perfectly elastic spheres moving at random. The present century has witnessed the great success of the analytical method to quote only the work of Fisher (1930), Haldane (1932) and Wright (1931) in evolutionary genetics, and of Volterra (1931), Lotka (1925, 1939) and Kostitzin (1939) in ecology. Nevertheless, biologists as a whole have been reluctant to develop the analytical as distinct from the purely statistical approach, and apart from the pioneer work of Karl Pearson (1906) and of Brownlee (1911), the mathematical aspects of the problem of dispersal have not received the attention they deserve.

2. RANDOM DISPERSAL AND DIFFUSION

2·1. *Random displacement.* The mathematical methods employed in the present treatment are largely the natural outcome of the solution of the well-known random-walk problem associated with Pearson (1906), Rayleigh (1919) and Kluyver (1905), and reviewed more recently by Chandrasekhar (1943). In the simplest form of this problem, a particle on a line jumps one place to the right or one to the left, both being equally likely. It is easily seen that the probability distribution is:

Position	$\bar{3}$	$\bar{2}$	$\bar{1}$	0	1	2	3
After 1 jump	.	.	$\frac{1}{2}$	.	$\frac{1}{2}$	.	.
After 2 jumps	.	$\frac{1}{4}$	.	$\frac{2}{4}$	.	$\frac{1}{4}$	.
After 3 jumps	$\frac{1}{8}$	.	$\frac{3}{8}$	.	$\frac{3}{8}$	.	$\frac{1}{8}$

and so on, with the general result that after n jumps the distribution is binomial with variance equal to n. As n becomes large the distribution 'tends to' normality.

A natural extension of the problem to N dimensions, in which the particle moves on a rectangular lattice, yields 'in the limit' an N-variate normal distribution which is the product of N independent distributions.

An alternative model for the two- or three-dimensional analogue consists of a randomly kinked chain with rotatable links. If one end is fixed, the probability distribution of the position of the free end tends to normality when the number of links is large (see Uspensky, 1937). In the two-dimensional case with only seven links it has been shown by actual calculation by Pearson (1906) that the normal distribution provides a remarkably good approximation.

The model considered by Pearson is, however, somewhat artificial when applied to migration in that the links (or flights) are taken as being equal in size. In practice they will vary. The important special case where the probability distribution of the end-point of a flight is bivariate-normal (with the starting point of the flight as the centre) has been considered by Brownlee (1911). A solution of the more general case is outlined below.

2·2. *A spreading population.* Consider a surface of unlimited extent approximating to a Euclidean plane with co-ordinate axes OX and OY, and in the immediate neighbourhood of the origin let there be to start with a single self-reproducing particle of a species not otherwise represented. The probability distribution of the position of a particular descendant after n generations may be deduced once the characteristic function of the probability distribution of offspring about their parent is given. For a parent located at (ξ, η), let this function be

$$\phi(t, u \mid \xi, \eta) = \exp\{i\xi t + i\eta u + \Sigma(\kappa_{rs} t^r u^s i^{r+s}/r!\,s!)\}.*$$

The κ_{rs} are assumed independent of ξ and η, for it is reasonable to suppose that the location of the parent does not affect the *form* of the distribution of the offspring. Let the cumulative distribution function for the position of a particle in the nth generation be $F_n(\xi, \eta)$ with $\phi_n(t, u)$ as the corresponding characteristic function. Assuming that the generations do not overlap in time, we then have

$$\phi_{n+1}(t, u) = \iint \phi(t, u \mid \xi, \eta)\, dF_n(\xi, \eta) = \phi(t, u \mid 0, 0) \iint e^{i\xi t + i\eta u}\, dF_n(\xi, \eta) = \phi(t, u \mid 0, 0)\, \phi_n(t, u).$$

* Here and below in this paper the customary parentheses surrounding terms in the denominator, to the right of the solidus, have been omitted. Thus $\ldots /r!\,s!\}$ should be printed $\ldots /(r!\,s!)\}$ in accordance with the common convention. It is considered, however, that no confusion will result here from such omissions.

Clearly $\phi_n = [\phi(t, u \mid 0, 0)]^n$. It is apparent that all cumulants increase n-fold (as in the summation of independent variables). Standardization of the scale is effected by changing $n\kappa_{rs}$ to $n\kappa_{rs}/n^{\frac{1}{2}(r+s)}$. The approach to normality is obviously very rapid when the distribution of offspring about their parent is roughly normal to begin with. In the present treatment we assume that $\kappa_{01} = \kappa_{10} = 0$ so that there is no drift. Nevertheless, it should be noted that a slight systematic drift, no matter how small, is *ultimately* the most important cause of displacement, that is, when n is chosen sufficiently large.

The distribution of the position of a particle of the nth generation will henceforth be taken as

$$dF_n(x, y) = \pi^{-1}a^{-2}n^{-1}\exp\{-[x^2+y^2]/na^2\}\,dx\,dy. \tag{1}$$

Since we shall be concerned mainly with absolute distances from the origin, it is convenient to make the polar transformation $x = r\cos\theta$, $y = r\sin\theta$. From (1) we then obtain

$$dF_n = \pi^{-1}a^{-2}n^{-1}\exp\{-r^2/na^2\}\,r\,dr\,d\theta \quad (0 \leqslant \theta < 2\pi, 0 \leqslant r < \infty). \tag{2}$$

If r_0 is the distance from the origin of an individual value chosen at random, then the probability that $r-\frac{1}{2}dr \leqslant r_0 < r+\frac{1}{2}dr$ is obtained by integrating over θ, giving the radial probability density

$$f(r \mid n, a^2) = \exp\{-r^2/na^2\}\,2r/na^2 \quad (0 \leqslant r < \infty). \tag{3}$$

The parameter a^2 is to be interpreted as the mean-square dispersion per generation, analogous with the mean-square velocity in Maxwell's distribution. The maximum likelihood estimate of a^2 based on ν observed values of r is simply $\widehat{a^2} = \Sigma r^2/n\nu$.

Of the population spread out after n generations, that proportion lying outside a circle of radius R is

$$p = \int_R^\infty \exp[-r^2/na^2]\,2r\,dr/na^2 = \exp\{-R^2/na^2\}. \tag{4}$$

In a subsequent argument it is essential to make p very small indeed. By choosing $p = 1/N_n$, where N_n is the final population after n generations, it follows that only one particle can be expected outside a circle of radius $R_n = (na^2\log N_n)^{\frac{1}{2}}$. If during this period the population has increased geometrically in accordance with the relation $N = e^{n\gamma^2}$ (where $c = \gamma^2$ may be termed the Malthusian parameter, as in Fisher (1930)), we obtain immediately the relation

$$R_n/n = a\gamma. \tag{5}$$

If a circular contour is drawn in the XY plane through all points having an arbitrarily chosen constant areal population density, then as time passes, the contour expands outwards. From the condition, $\gamma^2 n - \log n - r^2/na^2 =$ constant, it is easily seen that

$$r_{n+1} - r_n \sim a\gamma. \tag{5a}$$

The ultimate velocity of propagation of this 'wave-front' (D. G. Kendall, 1948) is therefore the same as the constant velocity of expansion of the circle within which all but one of the population can be 'expected' to lie.

2·3. *A numerical illustration.* Let us consider the application of equation (4) to Reid's problem. We can clearly establish a rigorous conclusion in the form of an inequality provided that we can fix appropriate bounds to the various parameters.

The oak does not produce acorns until it is sixty or seventy years old, and even then it is not mature. It then produces acorns over a period of several hundred years. The rate of population expansion is obviously very much less than it would be if all the immediate offspring of a single individual could come into existence in the sixtieth year of its life. We are certainly safe in taking the generation time as not less than sixty years. Now according to De Geer (1910) the final recession of the ice started at about 18,000 B.C., and yet in Roman Britain (Tansley, 1939, p. 171) oak forests were apparently well established throughout the country. It appears safe to take $n \leqslant 300$.

The number of acorns produced by an oak in the course of its life must be prodigious. Nevertheless, under typical present-day conditions almost every acorn which falls to the ground in autumn is consumed by birds and mammals (Watt, 1919). Of those remaining, some decay and many fail to germinate. Mortality among the young seedlings is very high, and even when not overshadowed it seems that only about 1 % of the seedlings are likely to survive the next three years. Though there has been much speculation, little is really known about the biological conditions following the last glaciation. Even so, it seems perfectly safe to assume that the average number of mature daughter oaks produced by a single parent oak did not exceed 9 million. To most biologists such a figure must appear unnecessarily high, but we must remember that it is the rate of population growth at the periphery of the advancing 'wave' which matters most, and here there will be least competition.

The fact that the British Isles are of finite extent with irregular coastline does not detract from the power of the argument, for it is obvious that such a shape in no way encourages the advancing 'wave-front'.

We then have $R/a < 300\sqrt{(\log 9{,}000{,}000)} = 1200$.

In the original form of the problem as stated by Reid, R is given as 600 miles. It then follows that a (the root mean square distance of daughter oaks about their parent) $> \frac{1}{2}$ mile. On these premises the conclusion which Reid reached appears inescapable—namely, that animals such as rooks must have played a major role as agents of dispersal.

If, however, we regard the upper bounds of a, n, γ (given above) as collectively excessive, we are driven to accept the view that the last glaciation was not as extensive as was previously believed (W. B. Wright, 1937), and to suppose that the oak population regenerated from scattered pockets which survived in favourable valleys.

At this point it is interesting to consider the actual succession of forest trees since glacial times as revealed by pollen analysis (Godwin, 1934). In late sub-arctic and pre-boreal times, birches with their small, winged fruits, willows with their plumed seeds, and pine with its winged seeds, were able to spread quickly and were plentiful. Hazel, however, did not reach its maximum until Boreal times, and oak with still larger fruits not until much later. It is quite possible that the relatively different rates of dispersal were not unimportant in determining the order of succession.

2·4. *The dispersal of small animals.* The results already deduced can be applied equally well to the dispersal of small animals such as earthworms and snails. For example, if the mean square 'dispersion' per minute of a wingless ground beetle wandering at random is 1 yard^2, then after a season of 6 months, say, without rest, the R.M.S.D. of the resulting probability distribution is only 500 yards. The probability that after 6 months the beetle wanders more than a mile from the starting point is less than 8 in a million. Without external aid a period of time equivalent to 1,000,000 seasons would be required to raise the R.M.S.D. to 290 miles. The figures, however, are slightly deceptive, for the final position arrived at is in general a great deal nearer the origin than the farthermost position previously reached. Nevertheless, calculations of this kind clearly indicate in these cases the importance in the long run of rare accidental displacements due to external agencies such as high gales, floating plant material, the muddy feet of birds and mammals.

2·5. *Empirical confirmation.* In practice there is rarely sufficient information to construct the contours of population density with accuracy. One contour, however, can sometimes be drawn—that for the low 'threshold' density (depending on the thoroughness of the survey) at which the population begins to escape notice altogether.

Equation (4), derived initially on theoretical grounds, is well illustrated by the spread of the muskrat, *Ondatra zibethica* L., in central Europe since its introduction in 1905. Fig. 1, based on Ulbrich (1930), shows the apparent boundaries for certain years. If we are prepared to accept such a boundary as being representative of a theoretical contour, then we must regard the area enclosed by that boundary as an estimate of πr^2. The relation between the time and $\sqrt{\text{area}}$ is shown graphically in Fig. 2.

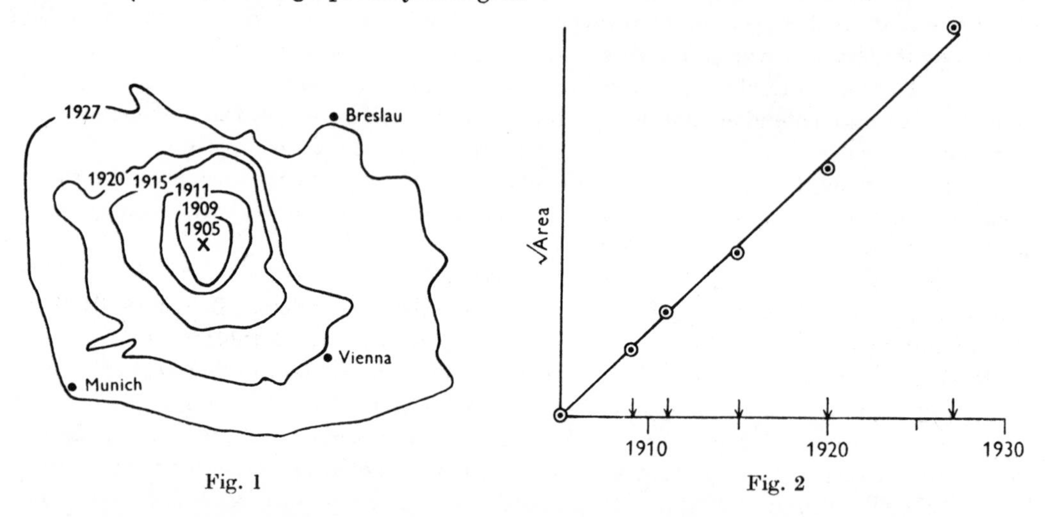

Fig. 1

Fig. 2

2·6. *The analogy with diffusion.* If a random particle suffers a displacement ϵ in any direction at regular intervals of time $(t, t+\omega, t+2\omega, \ldots)$, and if the probability density is denoted by ψ, it is clear that $\psi(x, y, t+\omega)$ is the mean value of $\psi(\xi, \eta, t)$ for all (ξ, η) on a circle of radius ϵ around (x, y). That is,

$$\psi(x, y, t+\omega) = \frac{1}{2\pi}\int_0^{2\pi} \psi(x+\epsilon\cos\theta, y+\epsilon\sin\theta, t)\, d\theta.$$

Expanding by Taylor's theorem and noting that

$$\int_0^{2\pi} \cos\theta\, d\theta = \int_0^{2\pi} \sin\theta\, d\theta = \int_0^{2\pi} \sin\theta\cos\theta\, d\theta = 0,$$

$$\int_0^{2\pi} \cos^2\theta\, d\theta = \int_0^{2\pi} \sin^2\theta\, d\theta = \pi,$$

we obtain for infinitesimally small ϵ and ω the relation

$$\frac{\partial\psi}{\partial t} = \frac{1}{4}\frac{\epsilon^2}{\omega}\nabla^2\psi, \tag{6}$$

where $$\nabla^2 = \frac{\partial^2}{\partial x^2} + \frac{\partial^2}{\partial y^2}.$$

The partial differential equation (6) is of course well known in connexion with the conduction of heat and related problems in mathematical physics (Jeans, 1921). The one-dimensional form has been applied by R. A. Fisher (1922) to the problem of the distribution of gene differences, and the three-dimensional form by Rashevsky (1948) to problems arising from the diffusion of materials from and into living cells.

It should be noted that the fundamental differential equation (6) is satisfied by the distribution

$$\psi = \pi^{-1}a^{-2}t^{-1}\exp\{-(x^2+y^2)/ta^2\},$$

already considered (see equation (1)), with the unit of time as one generation and $a^2 = \epsilon^2/\omega$.

It is apparent that many ecological problems have a physical analogue, and that the solution of these problems will require treatments and the use of functions with which we are already very familiar. Unlike most of the particles considered by physicists, however, living organisms reproduce, and members of the same and of different species interact. As a result the equations of mathematical ecology are often of a new and unusual kind (see, for example, Volterra (1931)), and require special treatment.

3. Dispersal and population density

3·1. *Modes of population growth.* Many problems on dispersal cannot be formulated unless some law of population growth (in the absence of dispersal) is assumed. As long as the population is small or shows a natural tendency to decrease, the Malthusian law $dN/dt = cN$ is usually satisfactory. If the population is not small the Pearl-Verhulst logistic law is more appropriate.

This law may be written in the form

$$\frac{dN}{dt} = cN - lN^2, \tag{7}$$

where, following Kostitzin (1939), c may be termed the coefficient of increase and l the limiting coefficient. Since a stable population level exists at $M \equiv c/l$, the law is often written in the form

$$\frac{dN}{dt} = cN(M-N)/M. \tag{7a}$$

For our present purpose it is convenient to use the concept of population density denoted by Ψ, a function of time and place. In general, the vital coefficients will also vary. The logistic law may then be written

$$\frac{\partial}{\partial t}\Psi(x,y,t) = c_1(x,y)\,\Psi - c_2(x,y)\,\Psi^2. \tag{8}$$

In the simplest cases where the vital coefficients remain constant throughout the habitat under consideration, the unit of population density may be chosen for convenience as that prevailing when $t \to \infty$, and the law then takes the very simple form

$$\frac{d\Psi}{dt} = c\Psi(1-\Psi). \tag{9}$$

When conditions become extremely unfavourable and ultimate survival not possible, the logistic law (7) may still be applied with c negative $(= -g^2)$. The population declines rapidly unless maintained by immigration. Such a population may be termed negative-logistic.

A convenient change of scale then gives

$$\frac{d\Psi}{dt} = -g^2\Psi(1+\Psi).$$

Two further laws of population growth applicable to populations which are discrete in time are deduced in §§4·1 and 4·3.

3·2. *Centres of multiplication and extinction.* We now consider the distribution in space of the density of a population which by reason of random dispersal, tends to flow from regions in which conditions are favourable to regions in which they are unfavourable. Whether a steady state can exist or not depends on the ability of the population in the favourable regions to make good the decline in the unfavourable ones.

Mathematically it is simpler to consider a linear habitat such as a coastal zone (Fisher, 1937) or the bank of a river. In these cases we are justified in assuming that an adjustment has been made in the vital coefficients to allow for diffusion inland, for this danger may be regarded as one of the many hazards associated with the habitat. Moreover, as will be apparent later, the mathematical form of the Malthusian law and of the logistic law (7) persists almost unchanged when the rate of diffusion inland is deducted from the rate of population growth.

The two-dimensional problems which most readily lend themselves to mathematical treatment are those with radial symmetry. Rough approximations to circles are sometimes afforded by islands, hill tops, marshes, patches of woodland, and approximations to annuli are provided at the margins of lakes, by the zones on hills of a conical or hemispherical shape, or even by the zonation created by rabbits around their individual burrows or around such patches of woodland as they infest (see, for example, Tansley, 1939, pp. 587, 505, 138, 141).

3·3. *Malthusian populations in linear habitats.* The one-dimensional form of (6) expressing the effect of random dispersal is

$$\frac{\partial\psi}{\partial t} = \frac{1}{2}\frac{\epsilon^2}{\omega}\frac{\partial^2\psi}{\partial x^2}. \tag{10}$$

In terms of population density, with $a^2 = \epsilon^2/\omega$, we then have as the combined effect of dispersal and population growth

$$\frac{\partial\psi}{\partial t} = \tfrac{1}{2}a^2\frac{\partial^2\Psi}{\partial x^2} + c\Psi. \tag{11}$$

The condition for a steady state, assuming it to exist, is

$$\frac{d^2\Psi}{dx^2} + \frac{2c}{a^2}\Psi = 0. \tag{12}$$

Case (*i*). Bordering on a suitable habitat is an unfavourable zone extending indefinitely in one direction:

$$c = -g^2 \quad \text{in} \quad x_b \leqslant x < \infty, \quad \Psi(x_b) = B, \quad \Psi(\infty) = 0.$$

Solution: $\Psi = B\,e^{-m(x-x_b)}$, where $m = g\sqrt{2}/a$.

Case (*ii*). An unfavourable interval is flanked at both ends by a suitable habitat:

$$c = -g^2 \quad \text{in} \quad -x_b \leqslant x \leqslant x_b, \quad \Psi(-x_b) = \Psi(x_b) = B.$$

Solution: $\Psi = B\cosh mx/\cosh mx_b.$

Case (iii). An apparently suitable habitat is flanked on either side by conditions which are *extremely* unfavourable. (It is assumed, for simplicity, that the organisms show no marked negative tactic response to the changed conditions, so that there is no appreciable reflexion at the boundary):

$$c = \gamma^2 \quad \text{in} \quad -x_b \leqslant x \leqslant x_b, \quad \Psi(-x_b) = \Psi(x_b) = 0.$$

Solutions that are non-negative but not zero everywhere in the interval are possible only when $x_b = \frac{1}{2}\pi a/\gamma\sqrt{2}$, and we then have

$$\Psi = C\cos(x\gamma\sqrt{2}/a). \tag{13}$$

When x_b is greater than the critical value given above, the population increases indefinitely, and when x_b is less than this value the population declines to extinction. For consider the partial differential equation (11) with $c = \gamma^2$ subject to the boundary conditions

$$\Psi(x_b, t) = \Psi(-x_b, t) = 0 \quad (0 \leqslant t < \infty),$$

and an arbitrary initial state developed as a Fourier half-range sine series

$$\Psi(x, 0) = \sum_{s=1}^{\infty} A_s \sin(\tfrac{1}{2}s\pi[x + x_b]/x_b).$$

The solution is

$$\Psi(x, t) = \sum_s A_s e^{k_s t} \sin(\tfrac{1}{2}s\pi[x + x_b]/x_b),$$

where $k_s = \gamma^2 - \frac{1}{2}a^2(\frac{1}{2}s\pi/x_b)^2$. The first of the k_s is obviously the greatest, so that in the course of time the solution is dominated by its first term, which tends to ∞, (13), or 0 as $x_b \gtreqless \frac{1}{2}\pi a/\gamma\sqrt{2}$.

3·4. *Logistic populations in linear habitats.* From (10) and (9) we obtain the fundamental equation

$$\frac{\partial\Psi}{\partial t} = \tfrac{1}{2}a^2\frac{\partial^2\Psi}{\partial x^2} + \gamma^2\Psi(1-\Psi). \tag{14}$$

Mathematically this is the same as an equation of R. A. Fisher (1937) given as

$$\frac{\partial p}{\partial t} = k\frac{\partial^2 p}{\partial x^2} + mpq$$

in connexion with the problem of gene flow in a linear population. Gene frequency p corresponds to population density and the term in $pq \equiv p(1-p)$ expressing the effect of natural selection corresponds to the term in $\Psi(1-\Psi)$ expressing logistic growth. The results deduced here in connexion with ecological problems automatically apply to the genetic analogues where they exist. Whereas, however, Ψ may exceed unity, p is restricted to $0 \leqslant p \leqslant 1$. Negative solutions are inadmissible in both types of problem.

The condition for a steady state is

$$\frac{d^2\Psi}{dz^2} + \Psi(1-\Psi) = 0, \quad \text{where} \quad z = x\gamma\sqrt{2}/a, \tag{15}$$

or in the case of a negative-logistic population (see (3·1))

$$\frac{d^2\Psi}{dz^2} - \Psi(1+\Psi) = 0, \quad \text{where} \quad z = xg\sqrt{2}/a. \tag{16}$$

The substitution $u = (\Psi_0 - \Psi)/\Psi_0(1-\Psi_0)$, where $\Psi_0 \equiv \Psi(x)|_{x=0}$,

converts (15) into

$$u'' = 1 + \lambda u - \tfrac{1}{4}(1-\lambda^2)u^2, \quad \text{where} \quad \lambda = 2\Psi_0 - 1. \tag{17}$$

Case (i). $c = \gamma^2, \quad 0 < \Psi_0 < 1, \quad \left.\frac{d\Psi}{dx}\right|_{x=0} = 0.$

Since $\frac{d^2u}{dz^2} = \frac{du}{dz}\frac{d}{du}\left(\frac{du}{dz}\right)$ we find on integrating (17), subject to the condition $du/dz = 0$ when $u = 0$, that

$$\frac{1}{2}\left(\frac{du}{dz}\right)^2 = u + \tfrac{1}{2}\lambda u^2 - \tfrac{1}{12}(1-\lambda^2)\, u^3.$$

Hence
$$|z|\sqrt{2} = \int_0^u \{\tau(1-\alpha^2\tau)(1+\beta^2\tau)\}^{-\frac{1}{2}}\, d\tau,$$

where
$$4\alpha^2 = \left(\frac{4-\lambda^2}{3}\right)^{\frac{1}{2}} - \lambda \quad \text{and} \quad 4\beta^2 = \left(\frac{4-\lambda^2}{3}\right)^{\frac{1}{2}} + \lambda.$$

In terms of Legendre's elliptic integrals, of which there are tables (1825), we then have the solution

$$z\rho 2^{-\frac{1}{2}} = F(k, \tfrac{1}{2}\pi) - F(k, \operatorname{arc}\cos \alpha\sqrt{u}),$$

where
$$\rho^2 = \alpha^2 + \beta^2 \quad \text{and} \quad k^2 = 2\beta^2 \bigg/ \left(\frac{4-\lambda^2}{3}\right)^{\frac{1}{2}}.$$

The inversion of this result leads to a simpler expression in terms of the Jacobian elliptic function sd (y) denoting $\sin \operatorname{am} y/\Delta \operatorname{am} y$. We thereby write

$$\sqrt{u} = \operatorname{sd}(2^{-\frac{1}{2}}\rho z, k)/\rho. \tag{18}$$

In practice, however, it is no more troublesome to compute a particular solution by iteration. For under the condition $\left.\frac{d\Psi}{dz}\right|_{z=0} = 0$ we may write equation (15) in the form

$$\Psi(z) = \Psi(0) - \int_0^z d\tau \int_0^\tau \Psi(\zeta)\,[1-\Psi(\zeta)]\, d\zeta.$$

Approximate values of Ψ are substituted on the right and the integrations performed numerically as in §3·9.

If we make Ψ_0 vanishingly small and proceed to determine z_b such that $\Psi(z_b) = 0$, we find that $u(z_b) \to 1$ and $\rho \to 2^{-\frac{1}{2}}$. Equation (18) passes formally into $2^{-\frac{1}{2}} = \sin \frac{1}{2}z_b$, confirming the result $x_b = \frac{1}{2}\pi a/\gamma\sqrt{2}$ given in §3·3 (iii).

Case (ii). $c = \gamma^2, \quad 1 < \Psi_0 < \frac{3}{2}, \quad \left.\frac{d\Psi}{dx}\right|_{x=0} = 0.$

Following the same procedure as before we find

$$|z|\sqrt{2} = \int_0^u \{\tau(1+\alpha^2\tau)(1+\beta^2\tau)\}^{-\frac{1}{2}}\, d\tau,$$

where
$$4\alpha^2 = \lambda - \left(\frac{4-\lambda^2}{3}\right)^{\frac{1}{2}} \quad \text{and} \quad 4\beta^2 = \lambda + \left(\frac{4-\lambda^2}{3}\right)^{\frac{1}{2}}.$$

Hence
$$2^{-\frac{1}{2}}\beta z = F(k, \operatorname{arc}\tan \beta\sqrt{u}), \quad \text{where} \quad k = \left(\frac{4-\lambda^2}{3}\right)^{\frac{1}{4}} \bigg/ \beta\sqrt{2}. \tag{19}$$

Alternatively
$$\sqrt{u} = \operatorname{tn}(2^{-\frac{1}{2}}\beta z, k)/\beta.$$

Since
$$F(k,\theta) \equiv \int_0^{\sin\theta} \{(1-t^2)(1-k^2t)\}^{-\frac{1}{2}}\,dt,$$
$$F(1,\theta) = \arg\tanh(\sin\theta) = gd^{-1}\theta.$$

In the important case when Ψ_0 is close to 1, so that β^2 is near $\frac{1}{2}$ and k near 1, we find that (19) passes into
$$\tfrac{1}{2}z = gd^{-1}(\arctan\sqrt{(\tfrac{1}{2}u)}) \quad \text{or} \quad \sinh\tfrac{1}{2}z = \sqrt{(\tfrac{1}{2}u)},$$
with the result that
$$\cosh z = 1+u = (\Psi-1)/(\Psi_0-1),$$
a conclusion which may be conjectured directly from the differential equation itself.

Other cases. Solutions in terms of elliptic functions are obtained in a similar manner after making appropriate substitutions. No special treatment is required in the case of a negative logistic population maintained by immigration in a region of extinction. We merely write $\Psi+1$ instead of Ψ in the corresponding solution for the positive logistic case.

3·5. *Malthusian populations in two-dimensional habitats.* The fundamental equation expressing the combined effect of diffusion and geometric increase in a habitat throughout which conditions are constant is
$$\frac{\partial\Psi}{\partial t} = \tfrac{1}{4}a^2\nabla^2\Psi + c\Psi(x,y,t). \tag{20}$$

In radially symmetrical cases we may write
$$\frac{\partial\Psi}{\partial t} = \frac{1}{4}a^2\left(\frac{\partial^2\Psi}{\partial r^2}+\frac{1}{r}\frac{\partial\Psi}{\partial r}\right) + c\Psi(r,t), \tag{21}$$
where $r^2 = x^2+y^2$.

Case (i). A circular region of multiplication enclosed in a zone of *absolute* extinction:
$$c = \gamma^2 \quad \text{in} \quad 0 \leqslant r < r_b, \quad 0 \leqslant t < \infty; \quad \Psi(r_b,t) = 0.$$
To ensure that the solution is finite at the origin we impose the further condition $\left.\frac{\partial}{\partial r}\Psi\right|_{r=0} = 0.$
The arbitrary initial state may be expressed as a Bessel series
$$\Psi(r,0) = \sum_s A_s J_0(rj_s/r_b),$$
where j_s is the sth zero of J_0 and A_s is determined in the usual Fourier manner by reason of the orthogonality relation
$$\int_0^1 yJ_0(yj_s)\,J_0(yj_\sigma)\,dy = 0 \quad (s \neq \sigma).$$
The solution is
$$\Psi(r,t) = \sum_s A_s e^{k_s t} J_0(rj_s/r_b),$$
where
$$k_s = \gamma^2 - \tfrac{1}{4}a^2 j_s^2/r_b.$$
Clearly $\Psi \sim A_1 e^{k_1 t} J_0(rj_1/r_b)$, tending to 0 or ∞ as $r_b \lesseqgtr j_1 a/2\gamma$. It appears again that habitats below a certain critical size, depending on the conditions, are insufficient to maintain a species undergoing unrestricted random dispersal.

Case (ii). Let (ξ,η) be the co-ordinates of the place of origin of an individual of the parental generation for which the density function in space is $\Psi(\xi,\eta)$. Let the probability that a particular one of its offspring originates in the interval $(x \pm \frac{1}{2}dx;\ y \pm \frac{1}{2}dy)$ be
$$\alpha(x-\xi, y-\eta \mid a^2)\,dx\,dy.$$

Let the probability that such an offspring reaches maturity be $P(x,y)$. For most simple, small populations P may be taken independent of Ψ. Then if on the average the number of original offspring per parent is ν we may define a viability function $V(x,y) \equiv \nu P(x,y)$. The condition for a steady state from generation to generation is the integral equation

$$\Psi(x,y) = V(x,y) \int_{-\infty}^{\infty} \int_{-\infty}^{\infty} \Psi(\xi,\eta)\,\alpha(x-\xi, y-\eta \mid a^2)\,d\xi\,d\eta. \tag{22}$$

For the purpose of the present simplified model we take α to be the function

$$\pi^{-1}a^{-2}\exp\{-[(x-\xi)^2+(y-\eta)^2]/a^2\},$$

a choice justified by previous arguments. The function $V(x,y)$ needs to be positive for all values of the variable, and preferably continuous. It will then be seen that the choice $V(x,y) = \exp\{(b^2-x^2-y^2)/v\}$ not only gives a reasonable spatial distribution of viability, but also renders the integral equation tractable. Viability is greatest at the origin, $V_0 = \exp\{b^2/v\}$. The circle $x^2+y^2 = b^2$ divides the plane into two parts, and in the absence of dispersal, the population inside the circle would increase and the population outside would decline.

The integral equation (22) then has a solution of the form

$$\Psi(\xi,\eta) = A\pi^{-1}\theta^{-1}\exp\{-(\xi^2+\eta^2)/\theta\}. \tag{23}$$

Since the double integral is the product of independent integrals and since

$$\int_{-\infty}^{\infty} \alpha(\xi \mid \theta_1)\,\alpha(x-\xi \mid \theta_2)\,d\xi = \alpha(x \mid \theta_1+\theta_2),$$

where α is a normal function, we find on substituting (23) into (22) and equating coefficients that

$$\frac{1}{\theta} = \frac{1}{v}+\frac{1}{\theta+a^2} \quad\text{and}\quad \frac{1}{\theta} = \frac{V_0}{\theta+a^2}, \quad\text{where}\quad \log V_0 = b^2/v.$$

From these it follows that $\qquad b^2 = a^2V_0\log V_0/(V_0-1)^2.$

In the sense indicated earlier, b is the radius of the region of multiplication. Unless this value is attained the population must decline to extinction.

Case (iii). An infinite zone of extinction around a single circular area of multiplication:

$$c = -g^2 \quad\text{in}\quad r_b \leqslant r < \infty, \quad \Psi(r_b) = B, \quad \Psi(\infty) = 0.$$

The steady state is given by

$$\frac{d^2\Psi}{dr^2}+\frac{1}{r}\frac{d\Psi}{dr}-m^2\Psi = 0 \quad\text{where}\quad m = 2g/a. \tag{24}$$

The solution is $\qquad \Psi(r) = A\int_0^{\infty} e^{-rm\cosh u}\,du = AK_0(rm),$

where K_0 is the modified Bessel function of the second kind of order zero, and $A = B/K_0(r_b m)$. It may be noted that $K_0(z) \sim e^{-z}(\pi/2z)^{\frac{1}{2}}$ for large z. When r_b is very large the problem reduces to a linear one (§3·3 (i)), the radial component of the R.M.S.D./generation being $a/\sqrt{2}$.

As a numerical illustration suppose that in the outer zone the population shows a natural tendency to decline at a rate of very nearly 4 % per generation ($g = 0{\cdot}2$). If the radius of the reservoir is 1000 yards and the R.M.S.D./generation 100 yards, then the population density at a point 1000 yards outside the boundary compared with that at the boundary will be approximately $e^{-2000m}\,2000^{-\frac{1}{2}}/(e^{-1000m}\,1000^{-\frac{1}{2}}) = 0{\cdot}01295$, in close agreement with the more accurate value 0·01312 based on the tables of K_0 given in Watson (1944).

Case (iv). A circular region of extinction:

$$c = -g^2 \quad \text{in} \quad 0 \leqslant r \leqslant r_b, \quad \Psi'(0) = 0, \quad \Psi(r_b) = B.$$

Solution: $$\Psi(r) = BI_0(rm)/I_0(r_b m),$$

where I_0 is the modified Bessel function of the first kind of order zero.

Case (v). An annulus of multiplication in a region of *absolute* extinction:

$$c = \gamma^2 \quad \text{in} \quad r_a \leqslant r \leqslant r_b, \quad \Psi(r_a) = \Psi(r_b) = 0.$$

A stationary non-negative solution exists for every fixed value r_a provided that the value of r_b is made to depend on r_a. If μ denotes $2\gamma/a$ the required solution is

$$\Psi(r) = A[J_0(\mu r)\,Y_0(\mu r_a) - J_0(\mu r_a)\,Y_0(\mu r)],$$

where Y_0 is the Bessel function of the second kind (of Weber's type) of order zero.

If $h = r_b/r_a$, it may be seen that μr_a is the first root (x_0) of the equation

$$J_0(hx)\,Y_0(x) - J_0(x)\,Y_0(hx) = 0. \tag{25}$$

The relation between r_b/r_a and $\mu r_b - \mu r_a$ is that between h and $(h-1)\,x_0$, the form in which the zeros of (25) are tabulated in Jahnke & Emde (1945). As r_a increases from zero, $\mu(r_b - r_a)$ moves from j_{01} to π, and the problem rapidly degenerates to the linear case.

3·6. *Logistic populations in two-dimensional habitats.* The combined effect of diffusion and logistic growth in radially symmetrical habitats may be expressed by

$$\frac{\partial \Psi}{\partial t} = \gamma^2 \left(\frac{\partial^2 \Psi}{\partial z^2} + \frac{1}{z}\frac{\partial \Psi}{\partial z} + \Psi(1-\Psi) \right), \tag{26}$$

where $z = 2r\gamma/a$. In order to describe stationary states we shall require solutions of the non-linear differential equation

$$\frac{d^2\Psi(z)}{dz^2} + \frac{1}{z}\frac{d\Psi}{dz} + \Psi(1-\Psi) = 0, \tag{27}$$

especially those that are finite at the origin, in which case

$$\Psi'(0) \equiv \frac{d}{dz}\Psi \bigg|_{z=0} = 0. \tag{28}$$

The equation, not being reducible to a Painlevé transcendent, is not free from movable critical points (Valiron, 1945, pp. 293–4). The general march of solutions finite at the origin is shown in Fig. 3. Since $\Psi(1-\Psi)$ is positive if and only if $0 < \Psi < 1$, it follows from the differential equation that the turning points of all solutions in the strip $0 < \Psi < 1$ are maxima. Similarly, the only turning points for which $\Psi > 1$ or $\Psi < 0$ are minima.

The spatial distribution of population density is illustrated by *AB* in the case of a region of extinction, by *CD* in a circular region of multiplication, and by *EF* in the case of an annular region of multiplication.

Particular solutions of (27) subject to stated boundary conditions may be computed without much labour by the iterative method outlined in §3·9.

3·7. *The function* $Q(z; q)$ is defined as the solution of equation (27) subject to condition (28) and the relation $Q(0; q) = q$. We are concerned here with the real variable only.

If $U(z) \equiv (q - Q(z; q))/q(1-q)$, the differential equation satisfied by U is

$$zU'' + U' = z[1 + \lambda U - \tfrac{1}{4}(1-\lambda^2)\,U^2], \quad \text{where} \quad \lambda = 2q-1. \tag{29}$$

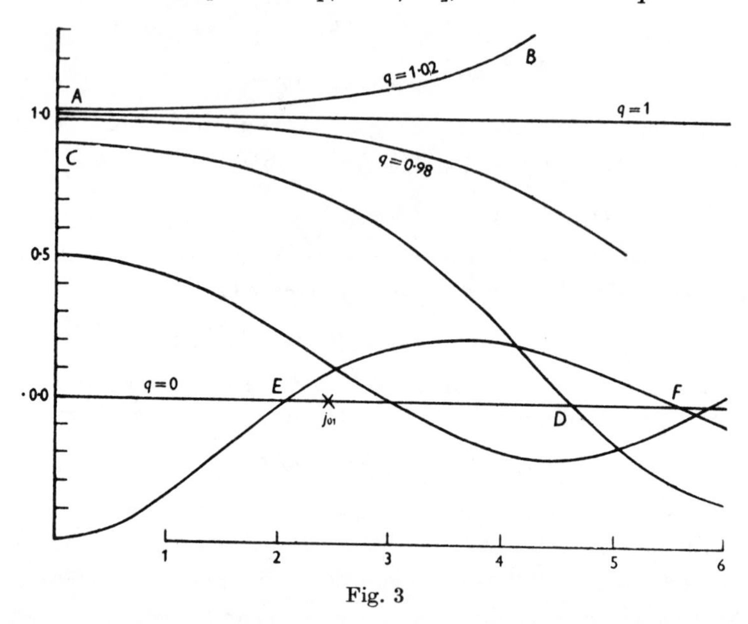

Fig. 3

Hence U tends to $1 - J_0(z)$ as $q \to 0$ and to $I_0(z) - 1$ as $q \to 1$. For computational purposes we may therefore seek an approximation to U in the immediate neighbourhood of $z = 0$ in the form

$$U = S(z) - 4q(1-q)\,\mathscr{C}(z),$$

where

$$S(z) = \sum_1^\infty \lambda^{n-1} z^{2n}/2^2 4^2 \dots (2n)^2,$$

may be regarded as a rough approximation

$$= [I_0(z\sqrt{\lambda}) - 1]/\lambda \quad \text{when} \quad \lambda > 0$$

$$= [1 - J_0(z\sqrt{|\lambda|})]/|\lambda| \quad \text{when} \quad \lambda < 0$$

$$= \tfrac{1}{4}z^2 \quad \text{when} \quad \lambda = 0,$$

and where

$$\mathscr{C}(z) = \sum_n C_{2n} z^{2n}/2^2 4^2 \dots (2n)^2$$

is a series of correction terms whose coefficients C_{2n} are to be determined. Only even powers are required, as may be seen either by differentiating (29) an even number of times or by considering the function $\mathscr{U}(z) \equiv U(-z)$ which is a solution of (29) with the same initial conditions. By differentiating $2n+1$ times by Leibnitz's theorem, making $z = 0$, and noting that $U_n \equiv \left.\dfrac{d^n}{dz^n} U(z)\right|_{z=0} = 0$ if n is odd, we obtain

$$(2n+2)\,U_{2n+2} = (2n+1)\left[\lambda U_{2n} - \tfrac{1}{4}(1-\lambda^2)\sum_{j=1}^{n-1}\binom{2n}{2j} U_{2j} U_{2n-2j}\right].$$

By substituting

$$A_{2j}/2^2 \dots (2j)^2 \quad \text{for} \quad U_{2j}/(2j)!, \quad \text{where} \quad A_{2j} = \{\lambda^{j-1} - (1-\lambda^2)\, C_{2j}\},$$

we find

$$C_{2n+2} = \lambda C_{2n} + \frac{1}{4} \sum_{j=1}^{n-1} \binom{n}{j}^2 A_{2j} A_{2n-2j}.$$

Thereby we obtain in succession:

$$C_2 = C_4 = 0, \quad C_6 = 1, \quad C_8 = 11\lambda/2, \quad C_{10} = (61\lambda^2 - 16)/2,$$

$$C_{12} = \lambda(847\lambda^2 - 507)/4, \quad C_{14} = (3820\lambda^4 - 3677\lambda^2 + 488)/2,$$

$$C_{16} = \lambda(176{,}245\lambda^4 - 234{,}718\lambda^2 + 67{,}857)/8,$$

$$C_{18} = (2{,}536{,}967\lambda^6 - 4{,}317{,}594\lambda^4 + 1{,}978{,}563\lambda^2 - 162{,}816)/8.$$

Computed values of U/z^2 are given in Table 1 for equidistant values of q and of z^2. With this arrangement interpolation can be carried out reasonably safely.

Table 1. $[q - Q(z;\, q)]/z^2 q(1-q)$

z^2 \ q	0·0	0·1	0·2	0·3	0·4	0·5	0·6	0·7	0·8	0·9	1·0
0	0·2500	0·2500	0·2500	0·2500	0·2500	0·2500	0·2500	0·2500	0·2500	0·2500	0·2500
1	0·2348	0·2376	0·2405	0·2435	0·2466	0·2496	0·2528	0·2559	0·2592	0·2626	0·2661
2	0·2204	0·2255	0·2309	0·2364	0·2423	0·2483	0·2546	0·2613	0·2682	0·2754	0·2831
3	0·2068	0·2138	0·2211	0·2289	0·2373	0·2461	0·2556	0·2658	0·2767	0·2884	0·3010
4	0·1940	0·2023	0·2113	0·2211	0·2317	0·2432	0·2558	0·2696	0·2847	0·3014	0·3199
5	0·1819	0·1913	0·2016	0·2130	0·2255	0·2395	0·2551	0·2725	0·2922	0·3145	0·3399
6	0·1705	0·1806	0·1919	0·2046	0·2189	0·2351	0·2535	0·2746	0·2990	0·3275	0·3609
7	0·1597	0·1704	0·1824	0·1961	0·2118	0·2300	0·2511	0·2758	0·3052	0·3404	0·3831
8	0·1496	0·1605	0·1731	0·1876	0·2044	0·2243	0·2479	0·2761	0·3105	0·3530	0·4065
9	0·1400	0·1511	0·1639	0·1790	0·1969	0·2181	0·2439	0·2759	0·3150	0·3654	0·4312

3·8. *The stability of stationary states.* In the case of Malthusian populations with positive parameter (§§ 3·3 (iii), 3·5 (i)), it was apparent that the stationary state was an unstable one. Though we were able to draw valid conclusions with regard to the critical size of a habitat for a population on the point of extinction, the model is inadequate in that it does not provide a stable stationary state for a population which is safely established. The assumption of logistic growth is free from this objection.

For let Ω be a solution of (27) subject to

$$\Omega(z_a) = A, \quad \Omega(z_b) = B \quad (0 < z_a < z_b),$$

and let $\Psi(z, t)$ satisfy (26) subject to

$$\Psi(z_a, t) = A, \quad \Psi(z_b, t) = B.$$

(The proof for the special case where the conditions at $z_a = 0$ are $\Omega' = \frac{\partial}{\partial z}\Psi = 0$ differs only in trivial details and is omitted.)

Then from our knowledge of the physical situation we should expect $\Psi \to \Omega$ as $t \to \infty$. Provided that the necessary biological condition holds, that $\Omega(z) \geqslant 0$ in $z_a \leqslant z \leqslant z_b$ (though not zero everywhere), it can at least be shown that Ω describes a stable equilibrium.

The function $\Delta(z,t) \equiv \Psi(z,t) - \Omega(z)$ satisfies

$$\frac{1}{\gamma^2}\frac{\partial \Delta}{\partial t} = \frac{\partial^2 \Delta}{\partial z^2} + \frac{1}{z}\frac{\partial \Delta}{\partial z} + \Delta(1 - 2\Omega - \Delta). \tag{30}$$

In dealing with questions of stability we need only consider such $\Delta(z, 0)$ as are infinitesimal variations of Ω. So long as Δ remains small, equation (30) remains effectively

$$\frac{1}{\gamma^2}\frac{\partial \Delta}{\partial t} = \frac{\partial^2 \Delta}{\partial z^2} + \frac{1}{z}\frac{\partial \Delta}{\partial z} + \Delta(1 - 2\Omega), \tag{31}$$

with the conditions $\Delta(z_a, t) = \Delta(z_b, t) = 0 \quad (0 \leqslant t < \infty)$.

We are therefore led to consider solutions of

$$\frac{d^2 f}{dz^2} + \frac{1}{z}\frac{df}{dz} + (\Lambda + 1 - 2\Omega) f = 0 \tag{32}$$

which satisfy $f(z_a) = f(z_b) = 0$, with $\Omega(z)$ as given. The Sturm-Liouville theorems are immediately applicable. Reduction to the standard form adopted by Titchmarsh (1946) may be brought about by the substitution $v = f\sqrt{z}$, though in this case there are advantages in following Ince (1927) and expressing (32) as

$$\frac{d}{dz}(zf') + z(\Lambda + 1 - 2\Omega) f = 0. \tag{33}$$

The values of Λ for which solutions are possible will be denoted by Λ_i ($i = 0, 1, 2, \ldots$) and the corresponding solutions by $f_i(z)$, the suffix being the number of zeros between z_a and z_b. We now prove that Λ_0 is positive.

By multiplying (33) by Ω, substituting f_0 for f, and integrating we find

$$[z\Omega f_0']_{z_a}^{z_b} - \int_{z_a}^{z_b} z\Omega' f_0' dz + (\Lambda_0 + 1)\int_{z_a}^{z_b} z f_0 \Omega\, dz - 2\int_{z_a}^{z_b} z f_0 \Omega^2 dz = 0.$$

By subtracting from this a similar relationship obtained in the corresponding manner from (27) with Ω as the dependent variable, we then have

$$\Lambda_0 \int_{z_a}^{z_b} z f_0 \Omega\, dz = \int_{z_a}^{z_b} z f_0 \Omega^2 dz - [\Omega z f_0']_{z_a}^{z_b}. \tag{34}$$

The sign of $f_0'(z_a)$ = minus sign of $f_0'(z_b)$ = sign of both integrands, provided that Ω is positive throughout (z_a, z_b). Hence

$$\Lambda_0 \geqslant \int_{z_a}^{z_b} z f_0 \Omega^2 dz \bigg/ \int_{z_a}^{z_b} z f_0 \Omega\, dz$$

and is positive. By Sturm's oscillation theorems, $\Lambda_0 < \Lambda_1 < \Lambda_2 \ldots$, so that all characteristic numbers are positive.

From (33) we have $$f_j \frac{d}{dz}(zf_i') + z(\Lambda_i + 1 - 2\Omega) f_j f_i = 0$$

and a similar relation with i and j interchanged. On integrating (the first term by parts) and subtracting, we find

$$(\Lambda_i - \Lambda_j)\int_{z_a}^{z_b} z f_i f_j dz = 0. \tag{35}$$

The fundamental set of solutions constitute an orthogonal system, and systems of the above kind are known to be closed (Ince, 1927). Any arbitrary function such as $\Delta(z, 0)$ has the formal development $\Delta = \sum_i A_i f_i(z)$, where the coefficients are given in the usual Fourier manner by

$$A_i = \int_{z_a}^{z_b} z\Delta(z) f_i(z)\, dz \Big/ \int_{z_a}^{z_b} z f_i^2(z)\, dz.$$

If $\Delta(z)$ is integrable over (z_a, z_b) and of bounded variation in the neighbourhood of $z = \zeta$, it may be shown that the Sturm-Liouville development given above converges to

$$\tfrac{1}{2}[\Delta(\zeta+0)+\Delta(\zeta-0)] \quad \text{at} \quad z = \zeta.$$

In problems of the type being discussed at present it is, however, permissible to assume the less general condition that $\Delta(z)$ has a continuous second derivative, and in this case the series is absolutely and uniformly convergent in (z_a, z_b) (Ince, 1927).

It now follows that the solution of (31) is

$$\Delta(z,t) = \sum_{i=0}^{\infty} A_i e^{-\Lambda_i \gamma^2 t} f_i(z).$$

As $t \to \infty, \Delta \to 0$, since all $\Lambda_i > 0$.

3·9. *An iterative method.* Consider, for example, the equation

$$\frac{d}{dz}(z\Psi') + zc_1(z)\Psi - zc_2(z)\Psi^2 = 0.$$

This may be written

$$\Psi(Z) = \Psi(b) + b\Psi'(b)\log(Z/b) - \int_b^Z \frac{dX}{X}\int_b^X (zc_1\Psi - zc_2\Psi^2)\, dz. \tag{36}$$

The limiting form of this equation as $b \to 0$ provided that $\Psi'(b)$ remains finite is

$$\Psi(Z) = \Psi(0) - \int_0^Z \frac{dX}{X}\int_0^X (zc_1\Psi - zc_2\Psi^2)\, dz. \tag{37}$$

If approximate values of $\Psi(z)$ are substituted on the right-hand side of (36) or (37) and the integration performed numerically, we obtain greatly improved values on the left. It is an easy matter to continue the solution step by step obtaining rough values as required by simple graphical extrapolation. Because of the remarkable rapidity of convergence it is soon found unnecessary to repeat the calculations for the earlier values (which automatically reproduce themselves).

For numerical illustration let $b = 0$, $\Psi'(0) = 0$, $\Psi(0) = 0{\cdot}5$, $c_1(z) = c_2(z) = 1$, so that the equation is

$$0{\cdot}5 - \Psi(Z) = \int_0^Z \frac{dX}{X}\int_0^X z\Psi(1-\Psi)\, dz.$$

The initial rough values are arranged in column (ii) of Table 2. Using intervals of z no smaller than $\frac{1}{2}$, and only the elementary integration formulae:

$$\int_0^h f(x)\, dx = \tfrac{1}{12}h[5f(0) + 8f(h) - f(2h)],$$

$$\int_{a-h}^{a+h} f(x)\, dx = \tfrac{1}{3}h[f(a-h) + 4f(a) + f(a+h)],$$

we find that a single application of the process reduces the sum of the squares of the errors to less than 1/800 its original value.

The method can be applied even if the c's are variable with finite discontinuities at a finite number of points in (b_1, b_2) provided that the interval of argument employed is not unreasonably large. Given $\Psi(b_1)$ and $\Psi'(b_1)$ new values of $\Psi(z)$ can be established progressively along (b_1, b_2). Given $\Psi(b_1)$ and $\Psi(b_2)$, however, it appears necessary to plot several integral curves radiating from the point $(b_1, \Psi(b_1))$ each with an assumed value of $\Psi'(b_1)$. After each trial this value is systematically adjusted until the condition at $z = b_2$ is satisfied.

Table 2

(i) z	(ii) approx. Ψ	(iii) $z\Psi(1-\Psi)$	(iv) $\int_0^X$ col. (iii)	(v) $\frac{1}{X}$ col. (iv)	(vi) $\int_0^Z$ col. (v)	(vii) $0{\cdot}5-$(vi)	(viii) Ψ by series
0·0	0·50	0·00000	0·000000	0·000000	0·000000	0·50000	0·50000
0·5	0·48	0·12480	0·031333	0·062666	0·015711	0·48429	0·48438
1·0	0·44	0·24640	0·124267	0·124267	0·062489	0·43751	0·43761
1·5	0·36	0·34560	0·274000	0·182667	0·139445	0·36056	0·36060
2·0	0·26	0·38480	0·459867	0·229935	0·243301	0·25670	0·25681
2·5	0·13	0·28275	0·635258	0·254103	0·365530	0·13447	0·13462
3·0	0·01	0·02970	0·717450	0·239150	0·490884	0·00912	0·00924

3·10. *A more general problem.* When the vital coefficients vary from place to place in an entirely arbitrary manner the equation we have to consider is

$$\frac{\partial \Psi}{\partial t} = \tfrac{1}{4}a^2\nabla^2\Psi + c_1(x,y)\,\Psi - c_2(x,y)\,\Psi^2. \tag{38}$$

Orthodox analytical methods appear inadequate, even in one-dimensional or radially symmetrical cases. In these cases an easy extension of the argument of §3·8 shows that a sufficient but not necessary condition for the stability of solutions is that $c_2 > 0$ in the habitat considered. Because of this, simple methods or numerical solution akin to 'relaxation' were tried out for such cases, but, because of the ultimate slowness of convergence, were abandoned in favour of the method of §3·9.

3·11. *Conditions at a common boundary.* In the present treatment Ψ is regarded as continuous, and the c's as continuous except at certain points. But this is only a convenient abstraction. In nature there are discontinuities almost everywhere (see Denjoy, 1937, p. 8), and it might be thought theoretically more desirable to construct our analytical model with functions of a more general nature. The scientist, however, is not so much concerned with the behaviour of Ψ and c at a particular point as in their mean value in a small neighbourhood of that point.

Consider the situation at the point $x = b$ in a linear habitat where the 'diffusivity' is uniform throughout but where the vital coefficients in $b' \leqslant x < b$ differ from those in $b < x \leqslant b''$. (The case where one subinterval is a region of absolute extinction is excluded, and this is best treated as a limiting case with c_1 tending to $-\infty$.) An appeal to our fundamental assumptions will show that unless Ψ is the same at $b-0$ and $b+0$, the rate of population flow across the boundary would be infinitely great, proceeding in such a direction as to restore the continuity of Ψ. Similarly, unless $\partial\Psi/\partial x$ is the same at $b-0$ as at $b+0$ the rate of change of density in an infinitesimally small neighbourhood of b would be infinitely great, proceeding in such a way as to equalize the derivatives at $b-0$ and $b+0$. A similar argument holds in a radially symmetrical habitat or the more general two-dimensional one.

When employing the iterative method of §3·9 it is not necessary to define the vital coefficients at $x = b$ unless b coincides with a tabular value of the argument. In such a case it is desirable to define $c(b) = \frac{1}{2}[c(b+0)+c(b-0)]$. The continuity of c is tacitly implied in the use of formulae of numerical integration.

4. REPRODUCTIVE CAPACITY AND POPULATION DENSITY

4·1. *A law of population growth.* Strictly the logistic law is applicable only to populations which are continuous in time. It will be shown that where competition is most marked among seedlings distributed at random, a somewhat different law holds for annual plants with the analogue of the logistic law as a limiting case.

The following symbols are used:

H is a coefficient denoting the suitability of the habitat as a whole, and is resolvable into separate components. For example, $H = Wsp/\Xi$, where

W is the number of spots of ground suitable in nature and size to support one plant of the species concerned to maturity. Such a spot will be termed a cell. Cells may be isolated or adjacent in groups of two or more. In the present simplified treatment, all cells are regarded as of the same surface area, namely, one unit.

Ξ is an arbitrary area constructed to include all the cells. It is assumed that this area is insulated in the sense that it receives no seeds from without.

p denotes the proportion of seeds which actually fall in Ξ. It is assumed that the probability distribution of the seeds is even throughout Ξ.

s is the probability that a 'seeded cell' rears a seedling to maturity.

Γ denotes reproductive capacity, the number of fertile seeds produced on the average per plant.

χ denotes relative density, defined here as the ratio of actual population density to the hypothetical maximum that the habitat could support, were every cell seeded. That is, $\chi = N/Ws$, where N is the actual population number.

In the interests of completeness it might be thought desirable to introduce more ecological factors than are considered here. In most cases it will be found that the mathematical form of the final result is unchanged, particularly when the appropriate modifications are incorporated into the definitions.

The probability distribution of the number of seeds per cell will be Poissonian with parameter $N\Gamma p/\Xi = N\Gamma H/Ws = \Gamma H\chi$. The proportion of cells with at least one seed is then $1-e^{-\Gamma H\chi}$ (neglecting minor irregularities due to sampling). By reason of the definition of χ this expression is the relative density of the resulting population. Using suffixes to distinguish between the values of χ in successive generations, we then have the law

$$\chi_{n+1} = 1-e^{-\Gamma H\chi_n}. \tag{39}$$

The ultimate stable value χ_∞ satisfies the equation

$$\log(1-\chi)+\Gamma H\chi = 0. \tag{40}$$

Typical growth 'curves' are illustrated in Fig. 4. From (40) it will be seen that as $\chi \to 1$ from lower values $\Gamma H \sim \log\frac{1}{1-\chi} \to \infty$, so that when χ is large a very considerable increase in reproductive capacity is required to bring about a perceptible increase in the population.

4·2. *Further applications.* The relation between the area 'covered' by a randomly moving insect and the distance traversed by that insect (Nicholson & Bailey, 1935) is closely analogous with the relation between area 'seeded' and the number of seeds produced by an annual plant. In many species of insect the population one year depends on the number of places suitable for oviposition encountered by the adult population of the previous year. In the simplest cases of this type, it is to be expected that relation (39) will be applicable.

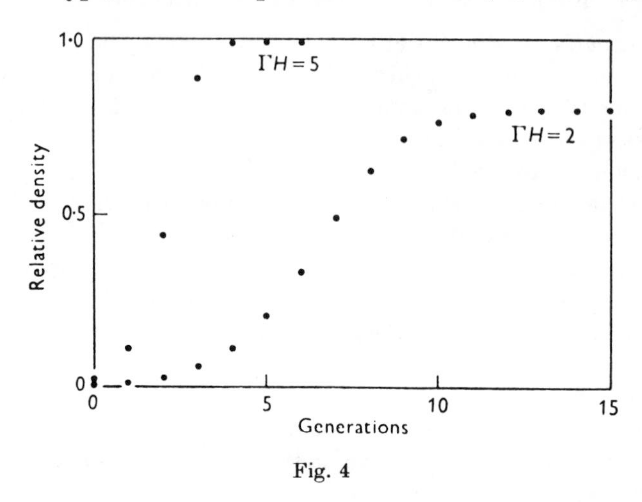

Fig. 4

Somewhat similar but more elaborate relationships have been deduced (see Jordan & Burrows, 1945; Zinsser & Wilson, 1932) in connexion with the dissemination of infectious diseases.

4·3. *Discrete logistic growth.* The recurrence relationship (39) also arises in the solution of a problem in evolutionary genetics (Skellam, 1948). Provided that $\Gamma H\chi$ is small we have as a good approximation

$$\chi_{n+1} = 1 - \frac{1-\frac{1}{2}\Gamma H\chi_n}{1+\frac{1}{2}\Gamma H\chi_n}. \tag{41}$$

Since $\chi_\infty = 2(\Gamma H - 1)/\Gamma H$ we may express this approximation as

$$\chi_{n+1} - \chi_n = \tfrac{1}{2}\chi_{n+1}(\chi_\infty - \chi_n),$$

corresponding to the differential equation (7*a*). The finite difference equation is of the Ricatti type (Milne-Thomson, 1933) and has the solution

$$\chi_n = \frac{\chi_0\chi_\infty}{\chi_0 + (\chi_\infty - \chi_0)(\Gamma H)^{-n}}. \tag{42}$$

This expression is in fact the logistic law for a discrete variable. For writing $N_n/N_\infty = \chi_n/\chi_\infty$, $c = \log \Gamma H$, and denoting the independent variable by t with its origin at the centre of symmetry, we obtain the standard form $N(t) = N(\infty)/(1+e^{-ct})$.

A simple guide to the values of ΓH for which (41) is a fair approximation to (39) is afforded by the comparison of the upper asymptotes in the two cases for the same fixed values of ΓH, as shown in Table 3.

It will be seen that χ_∞ is an increasing function of H and therefore of W, the number of suitable spots in Ξ. It therefore appears, other things being equal, that in the richer habitats a greater proportion of the available cells are actually utilized. The inference may well be extended to perennial populations.

Table 3

$\Gamma H = 1{\cdot}0258$		1·0536	1·1157	1·277	2·0
By (40) By (41)	0·0500 0·0504	0·1000 0·1017	0·2000 0·2074	0·400 0·433	0·797 1·000

4·4. *Two species in competition*. The classical case of two closely similar logistic species competing for the same food has been studied by Gause (1935) and Volterra (1931). The case we shall consider here is that of two closely related species of annual plant, S and S', competing in the same habitat. In order to investigate the extent to which a disadvantage in direct competition may be offset by a superiority in reproductive capacity, we shall assume the extreme condition that individual S' seedlings always fail to establish themselves in immediate competition with S seedlings (for example, the S seed may germinate earlier). All other ecological factors will be regarded as the same for both populations.

Under these conditions the S population is not adversely affected by the presence of S', and we may write

$$\Gamma H\chi + \log(1-\chi) = 0. \tag{43}$$

The number of cells unoccupied by S in the seedling stage is then $W(1-\chi)$, and these are the only ones available to S'.

It follows as in (4·2) that for an equilibrium

$$N' = sW(1-\chi)\left[1-e^{-N'\Gamma' p/\Xi}\right]. \tag{44}$$

In order that the unit in terms of which we express the relative population density of S' is independent of χ, we shall denote N'/sW by χ_0'. Equation (44) then becomes

$$\chi_0' = (1-\chi)\left[1-e^{-\Gamma'\chi_0' H}\right]. \tag{45}$$

Hence, from (43) and (45),

$$\Gamma'/\Gamma = \chi \log\left(1-\frac{\chi_0'}{1-\chi}\right) \Big/ \chi_0' \log(1-\chi). \tag{46}$$

If now we allow $\chi_0' \to 0$ we find that $\Gamma'/\Gamma \to -\chi/(1-\chi)\log(1-\chi)$. This will be termed the critical value of Γ'/Γ. Unless it is exceeded S' cannot coexist with S. When S is not dense it

Table 4*a*

χ	0·0+	0·2	0·4	0·6	0·8	0·9	0·99
Critical Γ'/Γ	1·0	1·12	1·31	1·64	2·49	3·91	21·5

Table 4*b*

χ \ χ_0'	0+	0·1	0·2	0·3	0·4	0·5	0·6
0·1	1·055	1·118	1·193	1·283	1·395	1·539	1·738
0·5	1·443	1·610	1·842	2·20	2·90	∞	—

is possible for S' to survive (despite its extreme inferiority in direct competition) with only a slightly superior reproductive capacity, but when S is very dense, it is impossible for S' to survive unless its reproductive capacity is many times greater than that of S. In the case of very small populations the 'safety level' of Γ'/Γ will be somewhat greater than the critical value, because of the danger of random extinction. Table 4*a* gives the critical values of Γ'/Γ for various values of χ. Table 4*b* gives the values of Γ'/Γ required to maintain the S' population at various levels in the two cases $\chi = 0{\cdot}1$ and $0{\cdot}5$.

4·5. *Two species and two habitats.* As a simple illustration of the application of these relations consider two habitats with coefficients $H_1 = 0{\cdot}001$ and $H_2 = 0{\cdot}01$, the same for both species, and let $\Gamma = 400$ and $\Gamma' = 2000$. The seeds of S', being more numerous, might well have smaller food reserves. Whatever the cause it is assumed as before that S' seedlings do not establish themselves in *immediate* competition with S seedlings.

Since $\Gamma H_1 = 0{\cdot}4 < 1$, S cannot survive in the first habitat but does so in the second with $\chi = 0{\cdot}9803$. Hence $-\chi/(1-\chi)\log(1-\chi) = 12{\cdot}7 > \Gamma'/\Gamma = 5$. As a result S' cannot compete successfully in the second habitat though in the first it appears safely established with $\chi' = 0{\cdot}797$.

In ecological terms we could say that though the requirements of both species are the same, the species with the greater reproductive capacity is better suited to the poorer habitat, and the species with the greater ability to establish itself against competition at the seedling stage is better adapted to life in the richer habitat.

5. Biological discussion

1. The analytical model developed here assumes that dispersal is effectively at random. This is at least approximately true for large numbers of terrestrial plants and animals. The behaviour patterns of certain animals may be such, however, as to tend to lead them to more favourable conditions. To deal adequately with these cases it will be necessary to introduce a further complication into the theoretical model somewhat analogous to gravitational attraction. Nevertheless, in most instances the range of an animal's perception is small compared with its powers of dispersal, and even the more intelligent may not discriminate between two parts of a habitat differing considerably in their effect on survival. Local irregularities in the character of the environment act as stimuli initiating repeated tactic displacements, the ultimate cumulative effect of which is scarcely distinguishable from a blind randomness.

2. If a population is initially located at a centre, spreads as with a Brownian motion, and undergoes unrestricted growth, the rate of radial expansion of its contours is approximately constant (see equations 5, 5*a*). A similar law applies to the flow of a gene, but here it should be noted that γ^2 for a new advantageous mutation will almost always be very small. The oft-repeated statement that a new advantageous gene spreads through the population replacing the former wild-type allelomorph is probably true (if no more than a reasonable period of time is allotted) only in those cases where the powers of dispersal are considerable in relation to the area occupied. In assessing the roles played by various processes in bringing about speciation, it might be borne in mind that fragmentation of the distribution area is by no means a necessary condition for different accumulations of genetic diversity to arise in the same species in different parts of a comparatively uniform habitat, and ecological adaptation can always occur when the rates of gene flow are insufficient to dissipate the genetic advances being made.

3. Just as the area/volume ratio is an important concept in connexion with continuance of metabolic processes in small organisms, so is the perimeter/area concept (or some equivalent relationship) important in connexion with the survival of a community of mobile individuals. Though little is known from the study of field data concerning the laws which connect the distribution in space of the density of an annual population with its powers of dispersal, rates of growth and the habitat conditions, it is possible to conjecture the nature of this relationship in simple cases. The treatment of §3 shows that if an isolated terrestrial habitat is less than a certain critical size the population cannot survive. If the habitat is slightly greater than this the surface which expresses the density at all points is roughly dome-shaped, and for very large habitats this surface has the form of a plateau.

4. The logistic law of population growth is often compared with that of autocatalysis (D'Arcy Thompson, 1942) and the inference made that population size is limited by the available food supply. In the present paper it is shown from spatial considerations alone that many populations which are discrete in time can be expected to satisfy the logistic law—at least as a close approximation.

5. The problem of the relationship between the reproductive capacity of plants and their habitat conditions has been studied very thoroughly by Salisbury (1942), but it is by no means certain that the thesis he adopts (pp. 159, 160, 231) is entirely tenable. The analytical approach of §§4·3–4·5, though no doubt over-simplified, provides us immediately with a possible explanation of the well-known paradox that certain species flourish only in unpromising situations, and might be developed and applied with advantage to throw light on related problems in this confused and bewildering subject.

REFERENCES

Brownlee, J. (1911). *Proc. R. Soc. Edinb.* **31**, 262.

Chandrasekhar, S. (1943). *Rev. Mod. Phys.* **15**, 1–89.

De Geer, G. (1910). *Int. Geol. Congr.* Stockholm.

Denjoy, A. (1937). *Théorie des Fonctions de Variables Réelles*, **1**. Paris: Hermann.

Fisher, R. A. (1922). *Proc. R. Soc. Edinb.* **42**, 327.

Fisher, R. A. (1930). *The Genetical Theory of Natural Selection.* Oxford: Clarendon Press.

Fisher, R. A. (1937). *Ann. Eugen., Lond.*, **7**, 355–69.

Gause, G. F. (1935). *Vérifications experimentales de la théorie mathématique de la lutte pour la vie.* Paris: Hermann.

Godwin, H. (1934). *New Phytol.* **33**.

Haldane, J. B. S. (1932). *The Causes of Evolution.* Appendix. London: Longmans, Green.

Ince, E. L. (1927). *Ordinary Differential Equations.* London: Longmans, Green.

Jahnke, E. & Emde, F. (1945). *Tables of Functions*, 4th ed., p. 205. New York: Dover Publications.

Jeans, J. (1921). *The Dynamical Theory of Gases.* Cambridge University Press.

Jordan, E. O. & Burrows, W. (1945). *Textbook of Bacteriology*, 14th ed. Philadelphia: Saunders.

Kendall, D. G. (1948). *Proc. Camb. Phil. Soc.* **44**, 591–4.

Kluyver, J. C. (1905). *Proc. K. Akad. Vet. Amst.* 25 Oct., pp. 341–50.

Kostitzin, V. A. (1939). *Mathematical Biology.* London: Harrap.

Legendre, A. M. (1825). *Traité des Fonctions Elliptiques*, **2**. Paris. Tables reissued with an Introduction by K. Pearson (1934). Cambridge University Press.

Lotka, A. J. (1925). *Elements of Physical Biology.* Baltimore.

Lotka, A. J. (1939). *Théorie Analytique des Associations Biologiques*, Part II. Paris: Hermann.

Maxwell, J. C. (1860). *Phil. Mag.* [4], **19**, 31.

Milne-Thomson, L. M. (1933). *The Calculus of Finite Differences.* London: Macmillan.

Nicholson, A. J. & Bailey, V. A. (1935). *Proc. Zool. Soc. Lond.*, pp. 551–98.

Pearson, K. (1906). Mathematical Contributions to the Theory of Evolution, XV. *Drapers' Company Research Memoirs.*

Rashevsky, N. (1948). *Mathematical Biophysics.* University of Chicago Press.

Rayleigh, Lord (1919). *Phil. Mag.* [6], **37**, 321.
Reid, C. (1899). *The Origin of the British Flora.* London: Dulau.
Salisbury, E. J. (1942). *The Reproductive Capacity of Plants.* London: Bell.
Skellam, J. G. (1948). *Proc. Camb. Phil. Soc.* **45**, 364.
Tansley, A. G. (1939). *The British Islands and their Vegetation.* Cambridge University Press.
Thompson, D'Arcy (1942). *Growth and Form.* Cambridge University Press.
Titchmarsh, E. C. (1946). *Eigenfunction Expansions associated with Second-Order Differential Equations.* Oxford: Clarendon Press.
Ulbrich, J. (1930). *Die Bisamratte.* Dresden: Heinrich.
Uspensky, J. V. (1937). *Introduction to Mathematical Probability.* New York: McGraw-Hill.
Valiron, G. (1945). *Equations Fonctionnelles.* Paris: Masson.
Volterra, V. (1931). *Leçons sur la Théorie Mathématique de la Lutte pour la Vie.* Paris: Gauthier-Villars.
Watson, G. N. (1944). *Treatise on Bessel Functions.* Cambridge University Press.
Watt, A. S. (1919). On the Causes of Failure of Natural Vegetation in British Oakwoods. *J. Ecol.* **7** (nos. 3 and 4), 173.
Wright, S. (1931). *Genetics*, **16**, 97–159.
Wright, W. B. (1937). *The Quaternary Ice Age.* London: Macmillan.
Zinsser, H. & Wilson, E. B. (1932). *J. Prev. Med.* **6**, 497–514.

PART 3
Theses, Antitheses, and Syntheses

Conversational Biology and Ecological Debate

Joel G. Kingsolver and Robert T. Paine

Scientific literature represents a written record of the ongoing dialogue among scientists. Like any science, ecology is a social as well as an intellectual enterprise, and the lively exchange of ideas is as integral a part of our science as laboratory experiments or field observations. Casual observation of any scientific meeting hall or seminar room reveals that armchair, and bar stool, ecology continues to be alive and well, despite its bad press.

This dialogue among scientists is not mere polite conversation. As Tansley (1935, p. 285) puts it, "bluntness makes for conciseness and has other advantages"; and the concise and clear statement of an idea is often the best way to make, or refute, a point. (Tansley qualifies his statement—"always provided that [bluntness] is not malicious and does not overstep the line which separates it from rudeness"—but this suggestion is not always heeded.) A hypothesis is suggested to explain some set of observations or phenomena, with specific and perhaps testable predictions. An alternative is proposed that provides an equal or better explanation. A debate ensues, motivating observations and experiments that can distinguish between the alternative hypotheses. Sometimes the answer is clear and unambiguous, and one camp in the debate is declared winner; more often a reformulation of the question occurs, and a synthesis of the alternatives is possible. With luck, this cycle of thesis, antithesis, synthesis leads to scientific progress in a sort of biased, random walk toward understanding.

The papers in this section provide some classic examples of this dialectical process. Some, like Harper (1967), succinctly synthesize ideas that were "in the air" at the time, and have provided a clear research program for generations of graduate students. Others, like Hairston, Smith, and Slobodkin (1960), present a novel interpretation of an old controversy, sparking productive debate that continues today. Yet others, like Tansley (1935), deal a death blow (at least temporarily) to some cherished ecological truth, or open a new field of inquiry. All of them defined or redefined the direction of ecological thought and research in their time. It is therefore worth considering briefly each of their contributions in the context of its period.

Any attempt to provide a limited list of "classic" ecological papers is certain to provoke disagreement. In a multivolume work many if not all of the papers cited in our discussion would legitimately be considered classics of the genre; the six papers reprinted here serve as selected examples from this larger set. Other students may find different insights, implications, and historical contexts in these writings than we have: These differences, of course, are debatable.

Tansley's (1935) contribution, "The use and abuse of vegetational concepts and terms," is a discussion and tribute to the power of words in scientific debates. He begins by declaring Clements "the greatest individual creator of the modern science of vegetation" (p. 285), then proceeds to dismantle systematically Clements's (1916) most central idea: the ecological community as a complex organism. Tansley maintains that the analogy between community succession and the progressive development of an individual organism does not hold for many plant communities. Instead, he argues that the succession in a community represents a trajectory of a dynamic system that may have multiple equilibrium points (see Lotka 1925; Gleason 1926 [part 1 in this volume]).

Extending this idea of a dynamic system, Tansley goes on to coin and define the term "ecosystem": "the whole *system* (in the sense of physics), including not only the organism-complex, but also the whole complex of physical factors forming what we call the environment of the biome . . . Though the organisms may claim our primary interest, when we are trying to think fundamentally we cannot separate them from their special environment, with which they form one physical system" (1935, p. 299). Tansley thus presents the view of communities and ecosystems as dynamic systems that has dominated ever since.

Tansley's ecosystem concept, in which biotic and abiotic factors interact dynamically as components of "one physical system," led directly to considerations of the flux of energy through ecosystems (see for example the introduction in Lindeman 1942 [part 1 in this volume]). In the 1950s and 1960s, more quantitative and systematic studies were conducted on the energy fluxes of both biotic and abiotic components of particular ecosystems (for example, Odum 1957; Teal 1962 [below, part 5]). Bormann and Likens (1967) demonstrated that this "systems" approach can be extended to the study of nutrient (biogeochemical) cycles by documenting the crucial interactions between hydrological and nutrient cycling in small watersheds. Bormann and Likens (1967) also argued that the *dynamics* of energy and nutrient flux are most clearly revealed by "experiments at the ecosystem level . . . Experiments can be designed to determine whether logging, burning, or use of pesticides or herbicides have an appreciable effect on net nutrient losses from the system" (p. 427). Perturbation experiments of this sort (for example, Likens et al. 1970 [below, part 6]) have added considerably to our understanding of ecosystems as dynamic systems, just as Tansley had envisioned.

The most general legacy of Tansley's work, however, lies in the methodological approach he advocates. Tansley identifies the key question: "What researches have been stimulated or assisted by the concept of 'the complex organism' *as such*" (p. 305). Tansley demonstrates how the proliferation of vegetational terms from polyclimax to plagioseres is the result of taking the "complex organism" metaphor too literally. When holistic views of ecosystems later regained importance (see for example, Odum 1964), these ideas were no longer based on the complex organism metaphor but on the quantitative methods of systems analysis and now emphasized the homeostatic features of dynamic systems (Odum 1969). This debate about "holism" and "reductionism" continues to the present in various forms in ecology. Tansley's comments on holism as a methodological tool versus a philosophical position could be read with profit by modern students of communities and ecosystems.

David Lack's (1954) book *The Natural Regulation of Animal Numbers* represents both a beginning and a return to the past. Lack starts with an essential observation from Darwin: the constancy of animal numbers, relative to what their intrinsic rates of increase will allow. Lack suggests that this constancy is the result of two processes: "first, that the reproductive rate of each species is a result of natural selection . . . ; and secondly, that the critical mortality factors are density-dependent" (p. 5). Lack emphasizes that the ecological features of organisms can and have evolved; and with appropriate acknowledgment of Darwin's *Origin of Species,* he advocates an evolutionary approach to the study of population regulation.

Lack's emphasis on what we now call evolutionary ecology mirrored an emerging view about the connections among natural selection, competition, speciation, and diversity. Competition for resources between closely related species could lead to an evolutionary divergence, or displacement, of the species for phenotypic characters relevant to resource use (Lack 1944; Mayr 1949). Brown and Wilson's (1956) review of biogeographic data, comparing populations in allopatry with those in sympatry, suggested that such character displacement may be widespread. (Experimental demonstrations of character displacement have proven more elusive, for example, Waage 1979.) Conversely, these workers argued that character displacement provides indirect, historical evidence that interspecific competition has occurred in the past, and that the outcome of evolution is the reduction of such competition, facilitating ecological coexistence of closely related species. Hutchinson's (1957 [above, part 2]) redefinition of the ecological niche as a geometric hypervolume with both biotic and abiotic dimensions helped to clarify these relationships between ecological displacement, coexistence, and the evolution of niches within communities.

The synthesis of these ideas into a general theory of organismic diversity is eloquently and forcefully forwarded in Hutchinson's (1959) "Homage to Santa Rosalia," reprinted in this section. Beginning with some natural history observations of aquatic bugs in a small pond, Hutchinson moves quickly to the general scientific issue: ecologists need to develop a theory that predicts why there are 10^6 animal species, rather than 10^4 or 10^8. (Current estimates suggest that 10^7 or so is closer to the truth.) As a first guess, Hutchinson outlines some important processes that generate and limit animal diversity.

Hutchinson's synthesis emphasizes the central place of niche diversification in generating and maintaining diversity: "The process of natural selection, coupled with isolation and later mutual invasions of ranges leads to the evolution of sympatric species, which at equilibrium occupy distinct niches" (1959, p. 146). Using available data on character displacement, Hutchinson suggests that a ratio of 1:3 between relevant metric characters is generally sufficient to allow two species to co-occur. For a given volume and dimensionality of niche space, this ratio sets a limit to the number of animal species in a community. This is the beginning of a quantitative theory of species diversity based on the biological attributes of individual organisms.

Another important part of Hutchinson's answer relies on the view that community stability increases with the number of links in its food web (MacArthur 1957). Thus, Hutchinson concludes that "the reason why there are so many species of animals is at least partly because a complex trophic organization of a community is more stable than a simple one" (1959, p. 155). It would take another classic (May 1973) to show that the relationship between community complexity and stability is not so straightforward. But Hutchinson's "Homage" helped establish a framework for a generation of ecologists in which studies of niche partitioning, species packing, and character displacement were integral components in the understanding of animal diversity (cf. Schoener 1974).

Lack's *Natural Regulation* also reflected changing views of how animal populations are regulated. Previous suggestions about the "natural" regulation of populations had been implicitly group selectionist, of the "for the good of the species" variety. Lack's important insight was that an animal's life history, including clutch size, was the result of individual selection, and that this life history could contribute to population regulation. The analysis of Cole (1954 [above, part 2]) and Smith (1954) provided a mathematical basis for this view, establishing the interplay between evolution by natural selection and population regulation via density-dependent mechanisms.

These studies fed directly into a continuing debate about the relative importance of density-dependence and density-independence in population regulation (see for example, Nicholson and Bailey 1935 [above, part 1]; Davidson and Andrewartha 1948 [below, part 5]). Hutchinson's "Homage" provided an elegant conceptual model in which population regulation and community structure was largely determined by the limiting similarity of competing species. But Andrewartha and Birch (1954) argued just as forcefully that abiotic factors determine distribution and abundance. The pitch and volume of this debate was heard most clearly at the 1957 symposium at Cold Spring Harbor; and discussions at this symposium helped provoke a short but brilliant note from Hairston, Smith, and Slobodkin (1960) (hereafter HSS).

As an elegant example of synthesis, HSS's note is hard to beat both for brevity and impact. It proposed an ingenious resolution to a long-standing debate by reconciling opposing viewpoints; like most such resolutions, it also launched a whole new set of debates. And it achieved all this in less than 1,500 words.

HSS's key insight was that the principal factor limiting population growth depends on the trophic level in question. They use a few general observations to identify this main factor limiting each trophic level as a whole. They then suggest that many disagreements about the importance of resource limitation and of competition in natural populations are the results of comparing species at different trophic levels.

HSS clearly did not eliminate debate about the importance of density-dependent regulation. In fact, their note provoked rebuttals several times its length (Murdoch 1966; Ehrlich and Birch 1967). However, their argument does emphasize how different processes dominate at different trophic levels (see also Paine 1966 [below, part 6]; Janzen 1970). Their observation that "the world is green" provided the challenge that helped stimulate important work on the interactions of herbivores and plants (see Ehrlich and Raven 1964 [this section]).

The most fundamental legacy of Lack's *Natural Regulation* and *Darwin's Finches* (1944), as the natural extension of the neodarwinian synthesis of evolution in the 1930s and 1940s, has been the renewed emphasis on individual selection in shaping the ecological features of animals. Harper's (1967) presidential address to the British Ecological Society was a sustained argument for a plant population ecology founded on the study of individuals. Harper's call for a Darwinian approach to plant ecology represented in part a straightforward application of well developed concepts from animal evolutionary ecology and demography as an alternative to vegetational studies of plant communities (see Watt 1947; Bray and Curtis 1956 [below, part 4]).

Harper's important insight, however, was to emphasize precisely those properties of plants —plasticity and vegetative reproduction—that had hindered the development of plant demography. After reviewing a growing literature which showed that plant demography could be usefully studied in terms of ramets, Harper focuses on the ecological and evolutionary consequences of plasticity of growth and forms. He highlights the importance of self-thinning for both survivorship curves and the population distribution of fitnesses. He illustrates how the growth and form of plants make them ideal for studies of resource allocation and life history strategies. And he argues convincingly that intra- and interspecific competition and population dynamics of plants can be analyzed quantitatively in terms of de Wit (1960) and ratio diagrams.

Harper's conclusion makes his agenda obvious: "My aim has been to try to show that much of what is exciting to me in the science of plant ecology in the late sixties has a highly respectable origin in the ecological thinking of Darwin . . . *The Origin of Species* states precise questions which plant ecologists can usefully spend the next hundred years answering" (1967, p. 268–69). The successful expansion and maturation of plant population biology over the next 20 years was clearly a vindication of this view.

Lack, Harper, and Hutchinson all focused on evolution for single species or for sets of competing species. The evolutionary consequences of ecological interactions across trophic levels was first systematically explored in Ehrlich and Raven's (1964) classic study of butterflies and plants. The idea that plant secondary compounds evolved as defenses against insect herbivores had been suggested by Fraenkel (1956, 1959); and much of Ehrlich and Raven's paper is a comprehensive survey of patterns of host plant use by different taxa of butterflies. To explain these patterns, they define the process of coevolution: the reciprocal interaction in which evolution of new chemical defenses in a plant taxon is followed by evolution of new means of circumventing those defenses in specialist butterfly herbivores. Noting that related butterfly herbivores are consistently associated with certain plant taxa, Ehrlich and Raven conclude that repeated coevolutionary interactions of this sort have been critical to the diversification of both butterflies and flowering plants. This coevolutionary "arms race" represents a much more dynamical view of the evolution of community diversity than that of Hutchinson (1959) and MacArthur (1958 [below, part 5]), in which the number of coexisting competitors is strictly limited by the volume of available niche space.

Ehrlich and Raven's hypothesis of reciprocal evolutionary interactions as a prime factor in evolutionary diversification had great impact on studies of insect-plant dynamics, chemical ecology, and predator-prey (host-parasite) evolution in general. It also provided an alternative interpretation of HSS's observations that the "world is green": the world is green not because insects aren't hungry, but because plants aren't edible (for example, Ehrlich and Birch 1967). Together with the more ecological contributions of Paine (1966 [below, part 6]), Brooks and Dodson (1965 [ibid.]), and Janzen (1970), Ehrlich and Raven helped reaffirm the power of predators in ecological and evolutionary communities.

Lack's approach to population regulation explicitly assumed that the ecological traits that have evolved are those which contribute to increased fitness. Cole's (1954 [above, part 2]) mathematical analysis of life history traits emphasized how such traits may evolve to maximize the intrinsic rate of increase of individuals. The notion that evolution will lead to certain "optimal" traits that maximize fitness formed the basis for the use of strategy reasoning in evolutionary ecology. Because competition among individuals for limited food resources was a primary focus of Lack, Hutchinson, MacArthur, and other ecologists of the time, the application of strategic thinking and optimality to animal feeding seemed a natural extension.

Schoener's (1971) influential review and synthesis of the theory of feeding strategies reflected the emergence of two other important trends in the 1960s and 1970s. The first was the development of more quantitative descriptions of the processes of feeding and foraging by individuals pioneered by Hollings (1959 [below, part 4]). The second was the identification of appropriate ecological "currencies," such as energy (Porter and Gates 1969 [ibid.]; Heinrich and Raven 1972) and carbon (Mooney 1972), that may be proportional to fitness in certain circumstances. These conceptual advances were essential for the development of general theories of optimal feeding and foraging.

Although models of optimal foraging had been developed (for example, Emlen 1966; MacArthur and Pianka 1966 [below, part 6]), Schoener's contribution provided both a model of feeding strategies and a framework for developing and testing such models. His framework consists of three basic steps—choosing a currency, choosing the cost-benefit functions, and solving for the optimum. He demonstrates how this framework can be applied to specific situations. After developing the strategy model, he considers the important limiting cases of energy maximizers and time minimizers and then uses these to make model predictions about optimal diet, foraging space, and foraging group size.

Schoener's synthesis had a considerable and immediate impact because it provided both a theoretical framework and a set of specific predictions directly relevant to empirical studies in behavioral ecology. Although substantive criticisms of optimality approaches have been raised (for example, Gould and Lewontin 1978), the study of foraging strategies remains an important part of behavioral ecology today and has also been integrated into studies of predator-prey interactions.

Although not reprinted here, a brief discussion of Schoener's (1974) review of resource partitioning is instructive, because it illustrates in miniature the process and progress of scientific debate. Hutchinson's (1957) redefinition of the ecological niche emphasized the relation of character differences, resource utilization, and competition. He also proposed that limits on the similarity among competitors regulated species diversity. Schoener asks, "What evidence demonstrates that the pattern of resource utilization among species results from competition?" (1974, p. 27); he addresses this question along three lines. First, he suggests more rigorous criteria for identifying resource partitioning in terms of the distribution of niche differences. Second, he reviews the many observational studies documenting ecological differences among related species, most of them done after Hutchinson's 1957 "Concluding Remarks." Third, Schoener reviews and extends the mathematical models that had been developed to address the relations of competition, limiting similarity, and species coexistence. These models blend the Lotka-Volterra equations, competition coefficients that incorporate the resource utilization curves with respect to individuals or phenotypes, and genetic parameters that allow explicit consideration of evolutionary change (cf. MacArthur and Levins 1964, 1967; Roughgarden 1972, 1974). Schoener's synthesis of these empirical and theoretical results clearly shows the progress made in addressing the basic issues outlined by Hutchinson.

The impact of Schoener's review on ecology today is equally instructive. The study of resource partitioning and the observational approach that Schoener emphasized play a less dominant role in ecological research today than they did fifteen years ago. In part, this reflects an increasing diversification of approaches and study systems in ecology; it also is the result of both the successes and failures of Hutchinson's "Homage" and "Remarks" as a research agenda for ecology. But Schoener's most general conclusion—"we require a holistic theory that draws upon models at the individual and population levels" (1974, p. 37)—still finds widespread and enthusiastic support among ecologists. The historical legacy of even a classic paper is often unpredictable, and always debatable.

If the legacy of truly classic papers is difficult to predict, what can be said for crystal-ball ecology? Is it possible to pick the classics from the immediate past and the near future, to distinguish them from *r*-selected notions which burst onto the scene, elicit a flurry of interest, and then disappear? What ongoing debates and research agendas will continue to be fruitful into the twenty-first century? Choosing specific papers is presumptuous, but we can identify the basic ingredients from the many concepts and controversies that interest and provoke ecologists today. We suggest here four areas in which classic antitheses and syntheses are already written, or are likely to be written in the near future.

The notion of a "null" hypothesis has a rich tradition in philosophy and statistics. MacArthur's (1957) paper on species abundance provided an early example in which a statistically random distribution was generated to which data were compared. Although both MacArthur's models and conclusions were extensively criticized, the general approach of comparing data to a derived null distribution remained generally unappreciated by ecologists. More than two decades later, ironically triggered by the single-minded emphasis on the importance of competition by MacArthur himself, null models were born again in the papers of Simberloff (1980) and Strong (1980). The ensuing debate about the role of competition in structuring communities was certainly blunter than Tansley (1935) might have wished. But it has left some enduring messages for ecologists: experimental tests should be better designed (Hurlbert 1984), and greater consideration should be given to alternative interpretations of observational data (Price et al. 1984).

Are ecological assemblages equilibrial at the spatial and temporal scales accessible to most ecological research? The basic issue is perhaps best illustrated by Caswell's (1978) contrast of closed, equilibrial versus open, nonequilibrial models. The notion of ecological communities as dynamic systems far from equilibirum has found a receptive audience and generated important research (though the idea is essentially an extension and modernization of Andrewartha and Birch 1954). Thus, disturbance has been rediscovered as an important biological force in many environments (Pickett and White 1985). Disturbance generates a spatially-dynamic pattern in which ecological processes occur (Paine and Levin 1981); it can play a major role in shaping life histories (Werner and Caswell 1977); and its intensity may influence community diversity (Connell 1978). However, the general role of disturbance and patchiness in stabilizing or destabilizing ecological systems remains uncertain (Kareiva 1986). At the moment, the potential for debate is greater than it is for synthesis.

Descriptive studies (such as food webs) show that multispecies interactions are commonplace if not universal, but most experimental and theoretical work until recently has focused on two-species' systems and pair-wise interactions. What new insights will be revealed once this traditional pair-wise approach is extended? While the answer is still unknown, there is new enthusiasm among ecologists for tackling the dynamics of truly complex communities. Mutualism has finally been added to the list of ecologically important interactions for communities (Boucher et al. 1982). Indirect effects are increasingly recognized as significant to understanding how natural communities are organized (Carpenter 1988). However, it is not clear how ecologists will build on the early attempts of Gause (1934) and later of Vandermeer (1969), Wilbur (1972) and Neill (1974) to unravel complex interactions. At the least, this will require a synthesis of experimental and theoretical approaches to the problem.

Finally, we cannot ignore the role of individual variation. While evolutionary and natural history studies continue to document ubiquitous variation among individuals in ecologically important traits, population and community ecology deal only with "average" individuals. The importance of growth form and size variation in plants for demography and competition has recently received increasing attention (Turner and Rabinowitz 1983). Several recent models have incorporated behavioral variation among individuals into population dynamics and interspecific competition (Hassell and May 1985; Lomnicki 1987). But the basic question remains: is individual variation merely "noise," or does it qualitatively alter the outcome of ecological interactions? Here the points of debate have not yet been clearly articulated, much less synthesized.

What papers of the past fifteen years will be generally accepted as classics at the one-hundredth anniversary of The Ecological Society of America? Hindsight may be more accurate than foresight, but the former is burdened by acquired prejudices while the latter is less fettered by history and knowledge. Because a general synthesis is not currently attainable at any ecological level, our discipline will remain enjoyably enmeshed in the polar forces of hypothesis and antithesis.

Literature Cited

Andrewartha, H. G., and L. C. Birch. 1954. *The distribution and abundance of animals.* Chicago: University of Chicago.

Bormann, F. H., and G. E. Likens. 1967. Nutrient cycling. *Science* 155:424–29.

Boucher, D. H., S. James, and K. H. Keeler. 1982. The ecology of mutualism. *Ann. Rev. Ecol. Syst.* 13: 315–47.

Brown, W. L., and E. O. Wilson. 1956. Character displacement. *Syst. Zool.* 5:49–64.

Carpenter, S. R., ed. 1988. *Complex interactions in lake communities.* New York: Springer-Verlag.

Caswell, H. 1978. Predator-mediated coexistence: A nonequilibrium model. *Amer. Nat.* 112:127–54.

Clements, F. F. 1916. Plant succession: An analysis of the development of vegetation. Publication no. 242. Washington, D.C.: Carnegie Institution.

Connell, J. H. 1978. Diversity in tropical rain forests and coral reefs. *Science* 199:1302–1310.

Darwin, C. 1859. *On the origin of species by means of natural selection.* Facsimile of first edition, Harvard University Press, 1964.

Ehrlich, P. R., and L. C. Birch. 1967. The "balance of nature" and "population control." *Amer. Nat.* 101:97–107.

Emlen, J. M. 1966. The role of time and energy in food preference. *Amer. Nat.* 100:611–17.

Fraenkel, G. 1956. Insects and plant biochemistry: The specificity of food plants for insects. *Proc. 4th Int'l. Congr. Zool.*, pp. 383–87.

——— 1959. The raison d'etre of secondary plants substances. *Science* 129:1466–1470.

Gause, G. F. 1934. *The struggle for existence.* Reprint 1964, New York: Hafner Press.

Gould, S. J., and R. C. Lewontin. 1978. The spandrels of San Marco and the Panglossian paradigm: A critique of the adaptationist programme. *Proc. Roy. Soc. Lond.* B205:581–98.

Hassell, M. P. and R. M. May. 1985. From individual behavior to population dynamics. In R. M. Sibley and R. H. Smith, eds. *Behavioral ecology: Ecological consequences of adaptive behavior.* Oxford: Blackwell.

Heinrich, B., and P. H. Raven. 1972. Energetics and pollination ecology. *Science* 176:597–602.

Hurlbert, S. H. 1984. Pseudoreplication and the design of ecological field experiments. *Ecol. Monogr.* 54:187–211.

Janzen, D. 1970. Herbivores and the number of tree species in tropical forests. *Amer. Nat.* 104:501–28.

Kareiva, P. 1986. Patchiness, dispersal, and species interactions: Consequences for communities of herbivorous insects. In J. Diamond and T. J. Case. *Community ecology.* New York: Harper & Row.

Lack, D. 1944. *Darwin's finches.* Oxford: Oxford University Press.

Lack, D. 1944. Ecological aspects of species formation in passerine birds. *Ibis* 86:260–86.

Lomnicki, A. 1987. *Population ecology of individuals.* Princeton: Princeton University Press.

Lotka, A. J. 1925. *Elements of physical biology.* Reprinted 1956. London: Dover.

MacArthur, R. H. 1957. On the relative abundance of bird species. *Proc. Nat. Acad. Sci.* 43:293–95.

MacArthur R. H., and R. Levins. 1964. Competition, habitat selection, and character displacement in a patchy environment. *Proc. Nat. Acad. Sci.* 51:1207–10.

MacArthur, R. H., and R. Levins. 1967. The limiting similarity, convergence, and divergence of coexisting species. *Amer. Nat.* 101:377–85.

May, R. M. 1973. *Stability and complexity in model ecosystems.* Princeton: Princeton University Press.

Mayr, E. 1949. Speciation and selection. *Proc. Amer. Phil. Soc.* 93:514–19.

Mooney, H. A. 1972. The carbon balance of plants. *Ann. Rev. Ecol. Syst.* 3:315–46.

Murdoch, W. W. 1966. Community structure, population control, and competition: A critique. *Amer. Nat.* 100:219–26.

Neill, W. E. 1974. The community matrix and interdependence of the competition coefficients. *Amer. Nat.* 108:399–408.

Odum, E. P. 1964. The new ecology. *Bioscience* 14:14–16.

——— 1969. The strategy of ecosystem development. *Science* 164:262–70.

Odum, H. T. 1957. Trophic structure and productivity of Silver Springs, Florida. *Ecol. Monogr.* 27:55–112.

Paine, R. T., and S. A. Levin. 1981. Intertidal landscapes and the dynamics of pattern. *Ecol. Monogr.* 51:145–78.

Pickett, S. T. A., and P. S. White. 1985. *The ecology of natural disturbances and patch dynamics.* New York: Academic Press.

Price, P. W., W. S. Gaud, and C. N. Slobodchikoff. 1984. Introduction: Is there a new ecology? Pp. 1–11 in P. W. Price, W. S. Gaud, and C. N. Slobodchikoff, eds., *A new ecology.* New York: Wiley-Interscience.

Roughgarden, J. 1972. Evolution of niche width. *Amer. Nat.* 106:683–718.

——— 1974. Species packing and the competition function with illustrations from coral reef fish. *Theor. Pop. Biol.* 5:163–86.

Schoener, T. W. 1974. Resource partitioning in ecological communities. *Science* 185:27–39.

Simberloff, D. 1980. A succession of paradigms in ecology: Essentialism to materialism and probabilism. *Synthese* 43:3–39.

Smith, F. E. 1954. Quantitative aspects of population growth. Pp. 277–94 in E. J. Boell, ed., *Dynamics of growth processes.* Princeton: Princeton University Press.

Strong, D. R. 1980. Null hypotheses in ecology. *Synthese* 43:271–85.

Turner, M. D., and D. Rabinowitz. 1983. Factors affecting frequency distributions of plant mass: The absence of dominance and suppression in competing monocultures of *Festuca paradoxa*. *Ecology* 64:469–75.

Vandermeer, J. 1969. The competitive structure of communities: An experimental approach with protozoa. *Ecology* 50:362–71.

Waage, J. K. 1979. Reproductive character displacement in *Calopteryx* (Odonata: Calopterygidae). *Evolution* 33:104–16.

Werner, P. A. and H. Caswell. 1977. Population growth rates and age- versus stage-distribution models for teasel (*Dipsacus sylvestris*). *Ecology* 58: 1103–1111.

Wilbur, H. 1972. Competition, predation, and the structure of the *Ambystoma-Rana sylvatica* community. *Ecology* 53:3–21.

Wit, C. T. de. 1960. On competition. *Versl. landbouwk. Onderz. Ned.* 66:8–26.

PAPER 15

THE USE AND ABUSE OF VEGETATIONAL CONCEPTS AND TERMS

A. G. Tansley

Oxford University, England

CONTENTS

It is now generally admitted by plant ecologists, not only that vegetation is constantly undergoing various kinds of change, but that the increasing habit of concentrating attention on these changes instead of studying plant communities as if they were static entities is leading to a far deeper insight into the nature of vegetation and the parts it plays in the world. A great part of vegetational change is generally known as *succession*, which has become a recognised technical term in ecology, though there still seems to be some difference of opinion as to the proper limits of its connotation; and it is the study of succession in the widest sense which has contributed and is contributing more than any other single line of investigation to the deeper knowledge alluded to.

It is to Henry Chandler Cowles that we owe, not indeed the first recognition or even the first study of succession, but certainly the first thorough working out of a strikingly complete and beautiful successional series (1899), which together with later more comprehensive studies ('01, '11) brought before the minds of ecologists the reality and the universality of the process in so vivid a manner as to stimulate everywhere—at least in the English-speaking world—that interest and enthusiasm for the subject which has led and is leading to such great results. During the first decade of this century indeed Cowles did far more than any one else to create and to increase our knowledge of succession and to deduce its general laws. By acute and thorough observation and by lucid exposition he became the great pioneer in the subject. It is therefore natural and fitting that my contribution to a volume intended to express the honour and affection in which Cowles is held by his fellow botanists should deal with this subject.

In 1920 and in 1926 I wrote general articles ('20, '29)[1] on this and some related topics. My return to the subject to-day is immediately stimulated by the appearance of Professor John Phillips' three articles in the *Journal of Ecology* ('34, '35) which seem to me to call rather urgently for comment and criticism. At the same time I shall take the opportunity of trying to clarify some of the logical foundations of modern vegetational theory.

If some of my comments are blunt and provocative I am sure my old friend Dr. Clements and my younger friend Professor Phillips will forgive me. Bluntness makes for conciseness and has other advantages, always provided that it is not malicious and does not overstep the line which separates it from rudeness. And at the outset let me express my conviction that Dr. Clements has given us a theory of vegetation which has formed an indispensable foundation for the most fruitful modern work. With some parts of that theory and of its expression, however, I have never agreed, and when it is pushed to its logical limit and perhaps beyond, as by Professor Phillips, the revolt becomes irrepressible. But I am sure nevertheless that Clements is by far the greatest individual creator of the modern science of vegetation and that history will say so. For Phillips' work too, and particularly for his intellectual energy and single-mindedness, I have a great admiration.

Phillips' articles remind one irresistibly of the exposition of a creed—of a closed system of religious or philosophical dogma. Clements appears as the major prophet and Phillips as the chief apostle, with the true apostolic fervour in abundant measure. Happily the *odium theologicum* is entirely absent: indeed the views of opponents are set out most fully and fairly, and the heresiarchs, and even the infidels, are treated with perfect courtesy. But while the survey is very complete and almost every conceivable shade of opinion which is or might be held is considered, there is a remarkable lack of any sustained criticism of opponents' arguments. Only here and there, as for instance in dealing with Gillman's and Michelmore's specific contentions, and in a few other places, does the author present scientific *arguments*. He is occupied for the most part in giving us the pure milk of the Clementsian word, in expounding and elaborating the organismal theory of vegetation.

[1] The latter was not published till 1929 owing to the long delay in the appearance of the Proceedings of the International Congress of Plant Sciences at Ithaca, N. Y. It was unfortunate too that certain misprints appeared in the paper because the proof corrections were not incorporated in the published text. Since some of these misprints destroy the sense intended it may be useful to call attention to them here.

P. 677, third line from bottom: Insert "these" after "All".
P. 684, line 2: delete second comma.
P. 685, line 2: for "criticism" read "criterion".
line 13: for "cause" read "causes".
line 14, third word from end; for "of" read "on".

This exposition, with its very full citations and references, is a useful piece of work, but it invites attack at almost every point.

The three articles are respectively devoted to " Succession," " Development and the Climax" and " the Complex Organism." The greater part of the third article is mainly concerned with the relation of this last concept to the theory of " holism " as expounded by General Smuts and others, and is really a confession of the holistic faith. As to the repercussions of this faith on biology I shall have something to say in the sequel. But first let me deal with " Succession " and " Development and the Climax."

Succession

My own views on succession are given fairly fully in my two papers already mentioned. In the first place I consider that the concept of succession can be given useful scientific signifiance only if we can trace in the sequences of vegetation " certain uniformities which we can make the subject of investigation, comparison, and the formulation of laws " ('29). In a paper also read at the Ithaca Congress, Cooper ('26) takes the view that since succession is the universal process of vegetational change " all vegetational changes must of necessity be successional." But I think the concept of succession involves not *merely* change, but the recognition of a *sequence of phases* (admittedly continuous from one phase to another) subject to ascertainable laws: otherwise why do we employ the term succession instead of change? And also I cannot admit that catastrophic changes due to external factors form parts of succession. Suppose an area of forest (*A*) to be suddenly invaded and devastated but not completely destroyed by a herd of elephants which then departs to other feeding grounds. Suppose that after partial regrowth (*B*) the vegetation of the same area is completely destroyed by a volcanic eruption and that on the volcanic ash which has buried *B* a new vegetation (*C*) appears. Can *A, B* and *C* be usefully regarded as parts of *any* succession? Cooper calls the catastrophes " landmarks." I should say they were clearly *interruptions,* each initiating a new succession (sere). I think Cooper is somewhat obsessed by his image of universal vegetational change as a " braided stream," just as Clements and Phillips are obsessed by their " complex organism." A stream is continuous, *therefore* all vegetational change must also be continuous. Succession (according to my definition) *is* continuous, but it may be interrupted by catastrophes unrelated to successional processes, which last are subject to ascertainable laws. The stream analogy has its points, particularly the separation and re-uniting of currents, but it breaks down as applied to the entire history of vegetation on the earth, just because of the catastrophes; nor do I find it constructively very helpful in considering the processes of succession itself.

In 1926 (p. 680) I proposed to distinguish between *autogenic succession,* in which the successive changes are brought about by the action of the plants themselves on the habitat, and *allogenic succession* in which the changes are brought about by external factors. " It is true of course (I wrote) and must never be forgotten, that actual successions commonly show a mixture of these two classes of factors—the external and the internal " (p. 678). I think now that I should have gone farther than this and applied my suggested new terms in the first place to the factors rather than to the successions. It is the fact, I think, that autogenic and allogenic factors are present in all successions; but there is often a clear preponderance of one or the other, and where this is so we may fairly apply the terms, with any necessary qualifications, to the successions themselves. I went on to contend, as indeed I had already done in 1920 (pp. 136–9) though without using the terms, that only to autogenic succession can we apply the concept of development of what I called a " quasi-organism " (= climax vegetation), but that this developmental (or autogenic) succession is the normal typical process in the gradual production of climax vegetation.

Phillips, following Clements, contends, on the other hand, that " succession is due to biotic reactions only, and is always progressive . . . succession being developmental in nature, the process must and can be progressive only " ('34, p. 562); and again, " succession is the expression of development " ('35, II, p. 214).

Now here we are concerned first of all with the use of words. If we choose to confine the use of the term succession to the series of phases of vegetation which lead up to a climatic climax, for example the various " priseres " from bare rock or water to forest, then it naturally follows that the process is " progressive only." If in addition we conceive of vegetation as an organism, of which the climax is the adult and the earlier phases of the prisere are successive larval forms, then also succession is clearly " developmental in nature," is " the expression of development." But if, on the other hand, we apply the term, as I do, and as I think most ecologists naturally do, to *any* series of vegetational phases following one another in one area, repeating themselves everywhere under similar conditions, and clearly due in each case to the same or a similar set of causes, then to say that " succession must and can be progressive only," or that it is always and everywhere developmental, is clearly contrary to the fact.

Most of the controversy about the possibility of " retrogressive succession " depends simply on this difference in the use of the word. It is true that Clements ('16, pp. 146–63) successfully showed that the phenomena represented by some of the looser uses of " retrogression " were more properly described as destruction of (for example) the climax phase, or of the dominants of the climax phase, a destruction which would normally initiate a subsere leading again to the climax if the vegetation were then let alone. But if on the other hand there is what Phillips would call a " continuative cause " at work which gradually leads to the degradation of vegetation to a lower type it seems to me that the phenomenon is properly called retrogressive succession. Here I should include the continuous effect of grazing animals which may gradually reduce forest to grassland, the gradual leaching and concomitant raw humus formation which may ultimately reduce forest to heath, gradual increase of drainage leading to the replacement of a more luxuriant and mesophytic by a poorer and more xerophytic vegetation, or a gradual waterlogging which also leads to a change of type and usually the replacement of a "higher" by a "lower" one. All these are perfectly well-established vegetational processes. To me they are clear examples of allogenic retrogressive successions, and I cannot see how their title can be denied except by an arbitrary and unnatural limitation of the meaning of the word succession. All the processes mentioned certainly involve destruction, but they also involve the invasion, ecesis and growth of new species. "Destruction" by itself is not a criterion: does not all *progressive* succession, as Cooper ('26, p. 402) has pointed out, involve constant destruction of the plants of the earlier phases?

In the discussion referred to Clements ('16, pp. 155–9) questions the reality of the retrogressive changes posited by European ecologists in the conversion of forest into heath, in the absence of violent destruction or of change of climate. Along with his insistence on the prime importance of the water-relations in succession goes a refusal to accept the possibility of a gradual change in the soil factors as a result of progressive leaching without change of climate. We may agree with Clements that strict proof of the reality of a retrogression caused in this way must be lacking unless and until we have the results of long-continued observation and properly controlled experiment with the appropriate quantitative data; and we may also agree that "biotic factors" have not always been satisfactorily excluded from the demonstration of examples supposed to be primarily due to leaching. But we can say from numerous observations in the oceanic and sub-oceanic regions of Europe that retrogression due to leaching and concomitant soil and vegetational changes is extremely probable—at least as probable as many successions which have been inferred rather than demonstrated. And to these examples I should add the retrogression of life form involved in the gradual conversion of forest to heath or grassland and of heath to grassland due to persistent grazing.

I agree with Clements that the invasion and destruction of forest (or heath) by Sphagnum bog is not properly considered as retrogression. I should call it the conquest and suppression of a "higher" type of community by a "lower" one, owing to the peculiar nature of the latter. That the power to effect this invasion and conquest is largely due to the power of Sphagnum to hold water and to carry water with it as it invades, is certainly true, and also that Sphagnum thereby establishes a new hydrophytic habitat, which may become the starting point of a new hydrarch "prisere." But such events cannot quite be *equated,* as Clements would equate them, with the formation of new " bare " (water) areas. Sphagnum is after all a plant, and the dominant of very extensive and important communities. Under certain conditions, which are due partly to climate and partly to topography, it may retain its dominance indefinitely. I myself should not hesitate to describe it as the primary dominant of a distinct plant formation, but then I am a heretic (or should I say a schismatic?) ('20, pp. 139–145). The weakness of this discussion of Clements, which is both able and ingenious, seems to me to reside partly in his too exclusive insistence on the water factor (which we all admit to be of prime importance), partly on his rather undiscriminating use of " destruction," but very largely on the assumption which governs the whole argument, and, as it seems to me, is quite illegitimate, that vegetation *is* an organism and therefore *must* obey the laws of development of what we commonly know as organisms.

Catastrophic destruction, whether by " natural " agencies or by man, does, I think, remove the phenomena from the field of the proper connotation of succession, because catastrophes are unrelated to the causes of the vegetational changes involved in the actual process of succession. They are only initiating causes, as Clements rightly insists: they clear the field, so to speak, for a new succession. That is why I have insisted on gradualness as a character of succession. Gradualness in effect is the mark of the action of " continuative " causes.

Development and the Quasi-Organism

The word development may be used in a very wide sense: thus we speak of the development of a theme or of the development of a situation, though always, I think, with the implication of becoming more complex or more explicit. Always, too, it is some kind of *entity* which develops, and in biology it is particularly to the growth and differentiation of that peculiarly well defined entity the individual organism that we apply the term. Hence we can perfectly well speak in a general way of the development of any piece of vegetation that has the character of an entity, such as marsh or forest, and in common language we actually do so; but we should use the term as part of the theory of vegetation, of a body of well-established and generally acceptable concepts and laws, only if we can recognise in vegetation a number of sufficiently well-defined entities whose development we can trace, and the laws of whose development we can formulate.

In 1920 I enquired whether we could recognise such entities in vegetation, and I analysed the whole topic in considerable detail and with considerable care. To the best of my knowledge that analysis has not been seriously criticised or impugned, and I may be permitted to think it holds the field, though various divergent opinions unsupported by arguments have since been expressed. Briefly my conclusion was that mature well-integrated plant communities (which I identified with plant associations) had enough of the characters of organisms to be considered as *quasi-organisms,* in the same way that human societies are habitually so considered. Though plant communities are not and cannot be so highly integrated as human societies and still less than certain animal communities such as those of termites, ants and social bees, the comparison with an organism is not merely a loose analogy but is firmly based, at least in the case of the more complex and highly integrated communities, on the close inter-relations of the parts of their structure, on their behaviour as wholes, and on a whole series of other characters which Clements ('16) was the first to point out. In 1926 (p. 679) I called attention to another important similarity which, it seems to me, greatly strengthens the comparison between plant community and organism—the remarkable correspondence between the species of a plant community and the genes of an organism, both aggregates owing their "phenotypic" expression to development in the presence of all the other members of the aggregate and within a certain range of environmental conditions.

But this position is far from satisfying Clements and Phillips. For them the plant community (or nowadays the "biotic community") *is* an organism, and he who does not believe it departs from the true faith.

Here we are back again at the question of the meanings of words. Professor Phillips writes as if he believed words to have perfectly precise and invariable meanings, and that a given verbal proposition *must* either be true or not true, whereas in fact a proposition obviously has different meanings according to the exact connotation of the words employed. The word organism can be applied very widely indeed. Thus we have Professor Whitehead's "Philosophy of Organism" and a whole school of "organicist" philosophers: many have not hesitated to call the universe an organism. Indeed it would seem from the quotations given in the Oxford "New English Dictionary" that the application of the term primarily to individual animals and plants did not begin till less than a century ago. Professor Phillips undoubtedly has some such wide conception in the back of his mind, and indeed his confession in Part III ('35) of the holistic faith and his citations of organicist philosophers make it certain that he has. But he should remember that he is writing primarily for ecologists, who are biologists, and that the modern biologist *means* by an organism an individual animal or plant, and would usually refuse to apply the term to anything else. At the most we may be able to get the average biologist to admit that plant (or biotic) communities have *some* of the characters of organisms, and that it may be permissible to apply to them some such term as quasi-organism. That I think would be a useful gain because I believe (with Clements and Phillips) the idea to be of great service.

There is no need to weary the reader with a list of the points in which the biotic community does *not* resemble the single animal or plant. They are so obvious and so numerous that the dissent expressed and even the ridicule poured on the proposition that vegetation *is* an organism are easily understood. Of course Clements and Phillips reply that no one asserts that the plant community is an *individual* organism. In the more recent phrase it is a " complex organism "—a thoroughly bad term, as it seems to me, for it is firmly associated in the minds of biologists with the " higher " animals and plants—the mammals and spermaphytes. In any case it is, in my judgment, impossible to get the proposition generally accepted. Whether it is true or untrue depends entirely on the connotation of " organism," and as to that the present generation of biologists have a firmly established use from which they will not depart—and I think they are right. We need a word for the peculiarly definite, sharply limited and unique type of organisation embodied in the individual animal or plant, and " organism " is the accepted term.

It may be said, as I imagine Cooper would say, that even such a term as " quasi-organism " is quite unnecessary if we keep the concept of " climax," which is very widely accepted. I do not agree, because climax does not suggest *organisation,* and the organisation of a mature complex plant association is a very real thing. The relatively stable climax community is a complex whole with more or less definite structure, *i.e.,* inter-relation of parts adjusted to exist in the given habitat and to co-exist with one another. It has come into being through a series of stages which have approximated more and more to dynamic equilibrium in these relations. This surely *is* " organisation," and organisation of the same type as, though by no means identical with, that of the single animal or plant. The organising factors are on the one hand the total net action of the effective environmental factors, on the other the combined actions of the individual organisms themselves. Phillips aptly quotes Karzinkin (1927) working on the " biocenoses " of animals living on water plants. Karzinkin found that changes in the external biota or in the constituents of the biocenosis disturb its equilibrium; but while the disturbance may be long-continued and complicated, equilibrium is ultimately again attained. It is possible therefore to speak of a " biocenosis " only when it reacts as a whole on the changes of the external and also of the internal factors. Cooper, who says ('26, p. 402) that progress in vegetational change is developmental " not because the vegetation unit is an organism but because it is made up of organisms undergoing development," adds that the progress of the whole is " subject to modifications due to mass action." It is precisely this " mass action," together with the actions due to the close and often delicate interlocking of the functions of the constituent organisms, which gives coherence to the aggregation, forces us to call it a " unit," justifies us in considering it as an organic entity, and makes it reasonable to speak of the development *of* that entity.

That this " development " is something very different from the ontogeny of a plant or animal (though even here there are also striking similarities) goes without saying. The adult quasi-organism can develop from beginnings which are totally opposed—a phenomenon completely alien from the ontogeny of a plant or animal—it can be hydrarch or xerarch; and the constituents of the " developmental stages " are quite different from the constituents of the

" adult." Starting from the type of the individual organism we have here something so different that it is no wonder there is refusal to call it by the same name, but at the same time something like enough to justify a related name.

I can only conclude that the term " quasi-organism " is justified in its application to vegetation, but that the terms " organism " or " complex organism " are not.

Climaxes

Professor Phillips' treatment of the concept of climax is open to nearly the same criticism as his treatment of succession. Just as he will only have one kind of succession, which is always progressive, and entirely caused by the " biotic reactions " of the community, so he will have only one kind of climax, the climatic climax, of which there is only one in each climatic region. He rather ingenuously suggests that the adjective " climatic " had better be dropped: it is misleading to the uninitiated. Since there is only one kind of climax why qualify the word? The suggestion would be unanswerable if we all agreed with him!

First there are some ecologists who believe there may be more than one climax in a climatic region, each with distinct dominants. This is the so-called " polyclimax theory," opposed to the " monoclimax " doctrine of Clements and Phillips, which supposes that there is only one " true " climax in each " climatic region," and that this should therefore be called *the* climax.

Now the so-called " polyclimax theory " takes what appear to be permanent types of vegetation under given conditions and calls them climaxes, because they are culminations of successions. The usual view is that under the " typical " climatic conditions of the region and on the most favourable soils the climatic climax is reached by the succession; but that on less favourable soils of special character different kinds of stable vegetation are developed and remain in possession of the ground, to all appearance as permanently as the climatic climax. These are called *edaphic climaxes,* because the differentiating factor is a special soil type. Similarly special local climates determined by topography (*i.e.,* land relief) determine *physiographic climaxes.* But we may go farther than this and say that the incidence and maintenance of a decisive " biotic factor " such as the continuous grazing of animals may determine a *biotic climax.* And again we may speak of a *fire climax* when a region swept by constantly recurrent fires shows a vegetation consisting only of species able to survive under these trying conditions of life; or of a *mowing climax* established as a result of the regular periodic cutting of grasses or sedges. In each case the vegetation appears to be in equilibrium with *all* the effective factors present, including of course the climatic factors, and the climax is named from the special factor differentiating the vegetation from the climatic climax. The edaphic climaxes correspond in general with Schimper's edaphic formations.

I should not myself call the usage embodied in this terminology a "theory" of any kind. It is simply an empirical terminology applied to what seem rather obvious facts of vegetational distribution. The word climax is used in its simple and natural signification of a culmination of development—a permanent or apparently permanent condition reached when the vegetation is in equilibrium with all the incident factors.

Clements realised from the first ('16) that vegetation existed which was neither climatic climax nor part of a sere actually moving towards it, but might be in a permanent or quasi-permanent condition in some sense "short of" the climax, and all such vegetation he called *sub-climax*. He used this term in two senses, for an actual seral stage which would normally lead to the climatic climax, and for a type of climax "subordinate to" the climatic climax. It was pointed out that this double use was undesirable, and that if we confined the term subclimax to the former case, terms were wanted for permanent or quasi-permanent vegetation which did not closely represent a particular phase of a sere leading to the climatic climax, but were dominated by species that did not enter into any of the "normal" seres. For such climaxes Clements has now ('34, p. 45) proposed the word *proclimax, i.e.,* vegetation which appears *instead of* the climatic climax, or as he would say, instead of *the* climax. This I think is an unobjectionable term, but it does not specify the factors which have differentiated the different types of this sort of climax.

Godwin ('29) has insisted that the factors which prevent a sere from reaching the climatic climax not only *arrest* the sere, but also *deflect* it from its normal course, which may be re-entered when these factors are removed. He is sceptical of the existence of subclimaxes in the strict sense, and prefers to speak of "deflected succession." We might call such successions, which undoubtedly exist, *plagioseres, i.e.,* "bent" or "twisted" seres, and if the vegetation really does come into equilibrium with the deflecting factor, of a *plagioclimax,* if such terms are considered useful.

As expounded by Phillips the "monoclimax theory" explains away the existence of what some of us are accustomed to call edaphic and physiographic climaxes within a climatic region in two ways. Either these supposed climaxes are not climaxes at all but stages in a sere leading to *the* climax, whose movement has been *retarded,* perhaps for a long time, by the edaphic or physiographic factors, or they are mere variations of "the formation" (the climatic climax). It is not to be supposed and is not in fact the case, it is argued, that either climate or soil will be absolutely uniform within a great climatic region, which often extends for many hundreds of miles. The climatic formation (*the* formation according to the "monoclimax theory") is often "a veritable mosaic" of vegetation (Clements). This of course is quite true: the only question is, *how great differences* are we to admit as mere variations within the formation? The difficulty disappears of course if we *define* a formation—a climatic climax—as *all* permanent vegetation within the climatic region and are therefore willing to swallow such differences, however great. But is this sound empirical method? It is not rather a case of making the facts fit the theory? Is it not sounder scientific method *first* to recognise, describe and study all the relationships of actually existing vegetation, and *then* to see how far they fit or do not fit any general hypothesis we may have provisionally adopted?

Most of the kinds of vegetation which some of Phillips' colleagues in Africa consider as separate formations Phillips declares to be seral stages—examples of retarded succession, and if they are not that then they are variations of the climax. It is impossible for one who has not studied this vegetation at first hand to decide which is right—Phillips or his critics. My general impression after reading the discussion, so far as it has gone, is that not enough is known of the behaviour of the vegetation in question to enable one to be at all sure which view interprets the facts more naturally. It is possible that Phillips is right in his particular interpretations, for some of which he seems to make a good case. His general view seems to be that the so-called " edaphic climaxes " or " edaphic formations " are *never* permanent, but *always* seral stages, in which the succession may be delayed for a longer or shorter time, but which will always ultimately progress to the climatic climax. If this were true they would be excluded from Clements' category of " proclimaxes," which is intended to be applied ('34, p. 45) to climaxes produced by such allogenic factors as fire or grazing. If on the other hand edaphic factors are really capable of holding vegetation in a permanent or quasi-permanent equilibrium—and I am far from being convinced that they are not—then, as it seems to me, such vegetation is quite reasonably included in the general concept of the " proclimax," though it is clear that specific edaphic factors stand in a relationship to vegetation different from that of fire or grazing, both because they form part of the " original " environment and because they themselves usually undergo continuous change.

Here we encounter a complication which has not hitherto, so far as I know, received any adequate consideration in the literature—I mean the influence of the modern theory of soil development on the theory and classification of vegetation. It is a simple and attractive idea that development of the soil profile runs *pari passu* with development of the vegetation it bears, and that consequently the mature climatic soil type corresponds and co-exists with the climatic climax community. It is however quite premature and probably untrue to make any such general assertion. It may very well be that in particular cases such a correspondence actually exists. But on the other hand, even when profile development under the influence of climate is perfectly normal and regular, the climatic climax community may establish itself long before the soil is mature, and may not be substantially altered by the later stages of profile maturation. Again a climatic climax may establish itself on a soil which is *kept immature* by geological and physiographic causes, as on a steep slope. And finally it is now generally agreed by pedologists that some rocks, owing to the simplicity of their composition, produce soils which can *never* form the normal climatic mature profile, and these may or may not bear the typical climatic climax vegetation. Whether any deviating communities which they may bear should be included as *parts* of the climatic climax should depend, as it seems to me, on the *extent* of that deviation. If it is wide, involving for example the dominance of different life forms, to assert that such vegetation *must* be part of the climatic climax *because* it appears in the same climatic region is surely to force the facts into a bed of Procrustes, to classify vegetation arbitrarily and unnaturally in the interests of a pre-conceived theory. Exactly the same is true of vegetation determined by any other edaphic factor, *e.g.*, permanent waterlogging for part at least of the year, or high soil acidity due to the poverty of the subsoil in basic ions or to the high rate of leaching in a highly permeable soil—which checks the maturation of the soil or diverts its course and thus prevents the appearance of climatic climax communities. There is no evidence that such kinds of vegetation represent stages of seres which will lead to climatic climax, nor can they be naturally regarded as parts of that climax.

On the other hand Bourne ('34) would have us regard every distinct variation of the climatic formation as a separate climax, *e.g.*, the spruce forests of the Vosges and of the Jura. No doubt they differ, as he says, quite markedly in certain respects which may be very important to a forester and for detailed ecological studies; and they may perhaps be suitably distinguished as separate *climax sociations*. But his general view reminds one of the taxonomists who will attend to nothing but " microspecies," losing sight of the higher grades of the taxonomic hierarchy.

I have even heard the argument that immature topography, for example, the slope of a hill, bears immature vegetation, and that since the slope will eventually disappear because it will ultimately be worn down to the base level of erosion, its vegetation must be regarded as seral. But this is surely to assert that tectonic and vegetational development must always run *pari passu*, whereas their time factors are usually widely different. They are very far from always keeping step, and immature topography is actually often clothed with climax vegetation, though Cowles ('01) has cited some striking cases of correlated development between physiography and vegetation.

I plead for empirical method and terminology in all work on vegetation, and avoidance of generalised interpretation based on a theory of what *must* happen because " vegetation is an organism."

" The Complex Organism "

Professor Phillips' third article ('35, III) is devoted to a discussion of the " complex organism," otherwise known as " the biotic community " (or " biome " of Clements) in the light of the doctrines of emergent evolution and of holism. On the biotic community he had already written ('31) and so also have Shelford ('31) and others.

I have already expressed a certain amount of scepticism of the soundness of the conception of the biotic community ('29, p. 680), without giving my reasons at all fully. It seems necessary now to state the grounds of my scepticism, and at the same time to make clear that I am not by any means wholly opposed to the ideas involved, though I think that these are more naturally expressed in another way.

On linguistic grounds I dislike the term biotic *community*. A "community," I think it will be generally agreed, implies *members,* and it seems to me that to lump animals and plants together as *members* of a community is to put on an equal footing things which in their whole nature and behaviour are too different. Animals and plants are not common members of anything except the organic world (in the biological, not the "organicist" sense). One would not speak of the potato plants and ornamental trees and flowers in the gardens of a human community as *members* of that community, although they certainly enter into its constitution—it would be different without them. There must be some sort of *similarity,* though not of course *identity,* of nature and status between the members of a community if the term is not to be divorced too completely from its common meaning. It may of course be argued by advocates of the term that the disparity of nature and behaviour between autotrophic plants and parasites—fungal or phanerogamic—is nearly as great as between animals and plants. But it may be rejoined that "human parasites" are well known in the societies of men, and that though it may well be held that a human society would get on better without them, yet they are in some sense members of the community. Though fungi are so different from autotrophic plants that they have even been regarded as forming a third "kingdom," distinct from both animals and plants, they are at least a good deal closer to green plants than they are to animals; and parasitic phanerogams undoubtedly form a link in nature and behaviour between parasitic fungi and autophytes, while saprophytic fungi are brought within the conceptual framework as "members" of a complex community such as a forest without any violence at all. Between all these organisms and the members of the animal kingdom there is however a very big gap in every respect.

Animal ecologists in their field work constantly find it neceessary to speak of *different* animal communities living in or on a given plant community, and this is a much more natural conception, formed in the proper empirical manner as a direct description of experience, than the "biotic community." Some of the animals belonging to these various animal communities have very restricted habitats, others much wider ones, while others again such as the larger and more active predaceous birds and mammals range freely not only through an entire plant community but far outside its limits. For these reasons also, the practical necessity in field work of separating and independently studying the animals communities of a "biome," and for some purposes the necessity of regarding them as external factors acting on the plant community—I cannot accept the concept of the *biotic* community.

This refusal is however far from meaning that I do not realise that various " biomes," the whole webs of life adjusted to particular complexes of environmental factors, are real " wholes," often highly integrated wholes, which are the living nuclei of *systems* in the sense of the physicist. Only I do not think they are properly described as " organisms " (except in the "organicist " sense). I prefer to regard them, together with the whole of the effective physical factors involved, simply as *" systems."*

I have already criticised the term " organism " as applied to communities of plants or animals, or to " communities " of plants *and* animals, on the ground that while these aggregations have *some* of the qualities of organisms (in the biological sense) they are too different from these to receive the same unqualified appellation. And I have criticised the term " complex organism " on the ground that it is already commonly applied to the species or individuals of the higher animals and plants. Professor Phillips' third article ('35, III) is largely devoted to an exposition and defence of the concept of " the complex organism." According to the organicist philosophy, which he seems to espouse, though he does not specifically say so, he is perfectly justified in calling the whole formed by an integrated aggregate of animals and plants (the " biocenosis," to use the continental term) an " organism," provided that he includes the physical factors of the habitat in his conception. But then he must also call the universe an organism, and the solar system, and the sugar molecule and the ion or free atom. They are all organised " wholes." The nature of what biologists call living organisms is wholly irrelevant to this concept. They are merely a special kind of " organism."

With the philosophical aspects of Phillips' discussion I cannot possibly deal adequately here. They involve, as indeed he recognises, some of the most difficult and elusive problems of philosophy. The doctrine of " emergent evolution," stated in a particular way, I hold to be perfectly sound, and some, though not all, of the ideas contained in Smuts' holism I think are acceptable and useful. But on the scientific, as distinct from the philosophical plane, I do think a good deal of fuss is being made about very little. For example—" newness springing from the interaction, interrelation, integration and organisation of qualities . . . could not be predicted from the sum of the particular qualities or kinds of qualities concerned: integration of the qualities thus results in the development of a whole different from, unpredictable from, their mere summation." Can one in fact form any clear conception of what " mere summation " can mean, as contrasted with the actual relations and interactions observed between the components of an integrated system? Has "mere summation " any meaning at all in this connexion? What we *observe* is juxtaposition and interaction, with the resulting emergence of what we call (and I agree *must* call) a " new " entity. And who will be so bold as to say that this new entity, for example the molecule of water and its qualities, would be unpredictable, if we really understood *all* the properties of hydrogen and oxygen atoms and the forces brought into play by their union? Unpredictable by us with our present knowledge, yes; but *theoretically* unpredictable, surely not. When an inventor makes a new machine, he is just as certainly making a new entity, but he can predict with accuracy what it will be and what it will do, because within the limits of his purpose he *does* understand the whole of the relevant properties of his materials and knows what their interactions will be, given a particular set of spatial relations which he arranges.

In discussing General Smuts' doctrine of "holism" Phillips lays stress on the whole as a *cause*, "holism" is called the fundamental factor operative towards the creation of wholes in the universe." It is an "operative cause" and an "inherent, dynamic characteristic" in communities. All but those who take "a static view of the structure, composition and life of communities—cannot fail to be impressed with the fundamental nature of the *factor of holism* innate in the very being of community, a factor of *cause*" (italics in the original).

How is this view justified? "At different levels the whole reacts upon habitat, changing (ameliorating) this for higher level wholes: the reaction of a whole, taken into account with its particular habitat and with the interrelations existing among its constituent organisms, shows as emergent changes in the habitat that are different from the sum of the changes that the constituent organisms would undergo were these not in communal association"[2] ('35, III, p. 498).

In this statement, we may note, it is not the mysterious "factor" called "holism" but the *particular* "whole" which is supposed to act as cause. Perhaps the "factor of holism" is intended as an abstraction from the effects of all the particular observed wholes. There is here again the artificial antithesis of an abstraction, "the sum of the changes that the constituent organisms would undergo" if they were not "in communal association," with what actually takes place in the community. Such a "sum" is quite unreal, there can be no meaning in considering the total activities *under unspecified conditions* of a particular lot of organisms taken together unless they *are* "in communal association." And if they are, they act upon one another, modify one another's actions, and produce new actions which are jointly dependent on two or more components. And it is precisely the sum of these modified and new actions which constitutes what we call, and rightly call, the activity of the community as a whole, because they depend upon the existence of that particular association of organisms with that particular habitat.

Is the community then the "cause" of its own activities? Here we touch

[2] Phillips however seems to think his statement is open to logical objection, but adds that "the accumulation of ecological evidence is becoming so impressive that I am not seriously perturbed by the strictures of pure logic." Surely it is his business either to show that the logic referred to is bad logic, or else to *be* "seriously perturbed" by it.

the very difficult philosophical question of the meaning of causation, which I cannot possibly attempt to discuss here. In a certain sense however, the community as a whole may be said to be the " cause " of its own activities, because it represents the aggregation of components the sum (or more properly the synthesis) of whose actions we call the activities of the community—actions which would not be what they are unless the components were associated in the way in which they are associated. So far we may concede Phillips' contention. But it is important to remember that these activities of the community are *in analysis* nothing but the synthesised actions of the components in association. We have simply shifted our point of view and are contemplating a new entity, so that we now, quite properly, regard the totality of actions as the activity of a higher unit.[3]

It is difficult to resist the impression that Professor Phillips' enthusiastic advocacy of holism is not wholly derived from an objective contemplation of the facts of nature, but is at least partly motived by an imagined future " whole " to be realised in an ideal human society whose reflected glamour falls on less exalted wholes, illuminating with a false light the image of the " complex organism."

The Ecosystem

I have already given my reasons for rejecting the terms " complex organism " and " biotic community." Clements' earlier term " biome " for the whole complex of organisms inhabiting a given region is unobjectionable, and for some purposes convenient. But the more fundamental conception is, as it seems to me, the whole *system* (in the sense of physics), including not only the organism-complex, but also the whole complex of physical factors forming what we call the environment of the biome—the habitat factors in the widest sense. Though the organisms may claim our primary interest, when we are trying to think fundamentally we cannot separate them from their special environment, with which they form one physical system.

It is the systems so formed which, from the point of view of the ecologist, are the basic units of nature on the face of the earth. Our natural human prejudices force us to consider the organisms (in the sense of the biologist) as the most important parts of these systems, but certainly the inorganic " factors " are also parts—there could be no systems without them, and there is constant interchange of the most various kinds within each system, not only between the organisms but between the organic and the inorganic. These *ecosystems,* as we may call them, are of the most various kinds and sizes. They form one category of the multitudinous physical systems of the universe, which range from the universe as a whole down to the atom. The whole method of science, as H. Levy ('32) has most convincingly pointed

[3] If this statement is applied to the individual organism, it of course involves the repudiation of belief in any form of vitalism. But I do not understand Professor Phillips to endow the " complex organism " with a " vital principle."

out, is to isolate systems mentally for the purposes of study, so that the series of *isolates* we make become the actual objects of our study, whether the isolate be a solar system, a planet, a climatic region, a plant or animal community, an individual organism, an organic molecule or an atom. Actually the systems we isolate mentally are not only included as parts of larger ones, but they also overlap, interlock and interact with one another. The isolation is partly artificial, but is the only possible way in which we can proceed.[4]

Some of the systems are more isolated in nature, more autonomous, than others. They all show organisation, which is the inevitable result of the interactions and consequent mutual adjustment of their components. If organisation of the possible elements of a system does not result, no system forms or an incipient system breaks up. There is in fact a kind of natural selection of incipient systems, and those which can attain the most stable equilibrium survive the longest. It is in this way that the dynamic equilibrium, of which Professor Phillips writes, is attained. The universal tendency to the evolution of dynamic equilibria has long been recognised. A corresponding idea was fully worked out by Hume and even stated by Lucretius. The more relatively separate and autonomous the system, the more highly integrated it is, and the greater the stability of its dynamic equilibrium.

Some systems develop gradually, steadily becoming more highly integrated and more delicately adjusted in equilibrium. The ecosystems are of this kind, and the normal autogenic succession is a progress towards greater integration and stability. The " climax " represents the highest stage of integration and the nearest approach to perfect dynamic equilibrium that can be attained in a system developed under the given conditions and with the available components.

The great regional climatic complexes of the world are important determinants of the primary terrestrial ecosystems, and they contribute *parts* (components) to the systems, just as do the soils and the organisms. In any fundamental consideration of the ecosystem it is arbitrary and misleading to abstract the climatic factors, though for purposes of separation and classification of systems it is a legitimate procedure. In fact the climatic complex has more effect on the organisms and on the soil of an ecosystem than these have on the climatic complex, but the reciprocal action is not wholly absent. Climate acts on the ecosystem rather like an acid or an alkaline " buffer " on a chemical soil complex.

Next comes the soil complex which is created and developed partly by the subjacent rock, partly by climate, and partly by the biome. Relative maturity of the soil complex, conditioned alike by climate, by subsoil, by physiography and by the vegetation, may be reached at a different time from that at which the vegetation attains its climax. Owing to the much greater local variation of subsoil and physiography than of climate, and to the fact that some of the

[4] The mental isolates we make are by no means all coincident with physical systems, though many of them are, and the ecosystems among them.

existing variants prevent the climatic factors from playing the full part of which they are capable, the developing soil complex, jointly with climate, may determine variants of the biome. Phillips' contention that soil never does this is too flatly contrary to the experience of too many ecologists to be admitted. Hence we must recognise ecosystems differentiated by soil complexes, subordinate to those primarily determined by climate, but none the less real.

Finally comes the organism-complex or biome, in which the vegetation is of primary importance, except in certain cases, for example many marine ecosystems. The primary importance of vegetation is what we should expect when we consider the complete dependence, direct or indirect, of animals upon plants. This fact cannot be altered or gainsaid, however loud the trumpets of the " biotic community " are blown. This is not to say that animals may not have important effects on the vegetation and thus on the whole organism-complex. They may even alter the primary structure of the climax vegetation, but usually they certainly do not. By all means let animal and plant ecologists study the composition, structure, and behaviour of the biome together. Until they have done so we shall not be in possession of the facts which alone will enable us to get a true and complete picture of the life of the biome, for both animals and plants are components. But is it really necessary to formulate the unnatural conception of biotic *community* to get such co-operative work carried out? I think not. What we have to deal with is a *system,* of which plants and animals are components, though not the only components. The biome is determined by climate and soil and in its turn reacts, sometimes and to some extent on climate, always on soil.

Clements' " prisere " ('16) is the gradual development of an ecosystem as we may see it taking place before us to-day. The gradual attainment of more complete dynamic equilibrium (which Phillips quite rightly stresses) is the fundamental characteristic of this development. It is a particular case of the universal process of the evolution of systems in dynamic equilibrium. The equilibrium attained is however never quite perfect: its degree of perfection is measured by its stability. The atoms of the chemical elements of low atomic number are examples of exceptionally stable systems—they have existed for many millions of millennia: those of the radio-active elements are decidedly less stable. But the order of stability of all the chemical elements is of course immensely higher than that of an ecosystem, which consists of components that are themselves more or less unstable—climate, soil and organisms. Relatively to the more stable systems the ecosystems are extremely vulnerable, both on account of their own unstable components and because they are very liable to invasion by the components of other systems. Nevertheless some of the fully developed systems—the " climaxes "—have actually maintained themselves for thousands of years. In others there are elements whose slow change will ultimately bring about the disintegration of the system.

This relative instability of the ecosystem, due to the imperfections of its equilibrium, is of all degrees of magnitude, and our means of appreciating and measuring it are still very rudimentary. Many systems (represented by vegetation climaxes) which appear to be stable during the period for which they have been under accurate observation may in reality have been slowly changing all the time, because the changes effected have been too slight to be noted by observers. Many ecologists hold that *all* vegetation is *always* changing. It may be so: we do not know enough either to affirm or to deny so sweeping a statement. But there may clearly be minor changes within a system which do not bring about the destruction of the system as such.

Owing to the position of the climate-complexes as primary determinants of the major ecosystems, a marked change of climate must bring about destruction of the ecosystem of any given geographical region, and its replacement by another. This is the *clisere* of Clements ('16). If a continental ice-sheet slowly and continuously advances or recedes over a considerable period of time all the zoned climaxes which are subjected to the decreasing or increasing temperature will, according to Clements' conception, move across the continent " as if they were strung on a string," much as the plant communities zoned round a lake will move towards its centre as the lake fills up. If on the other hand a whole continent desiccates or freezes many of the ecosystems which formerly occupied it will be destroyed altogether. Thus whereas the prisere is the development of a single ecosystem *in situ,* the clisere involves their destruction or bodily shifting.

When we consider long periods of geological time we must naturally also take into account the progressive evolution and rise to dominance of new types of organism and the decline and disappearance of older types. From the earlier Palaeozoic, where we get the first glimpses of the constitution of the organic world, through the later Palaeozoic where we can form some fairly comprehensive picture of what it was like, through the Mesozoic where we witness the decline and dying out of the dominant Palaeozoic groups and the rise to prominence of others, the Tertiary with its overwhelming dominance of Angiosperms, and finally the Pleistocene ice-age with its disastrous results for much of the life of the northern hemisphere, the shifting panorama of the organic world presents us with an infinitely complex history of the formation and destruction of ecosystems, conditioned not only by radical changes of land surface and climate but by the supply of constantly fresh organic components. We can never hope to achieve more than a fragmentary view of this history, though doubtless our knowledge will be very greatly extended in the future, as it has been already notably extended during the last 30 years. In detail the initiation and development of the ecosystems in past times must have been governed by the same principles that we can recognise to-day. But we gain nothing by trying to envisage in the same concepts such very different processes as are involved in the shifting or destruction of ecosystems on the one hand and the development of individual systems on the other. It is true, as Cooper insists ('26), that the changes of vegetation on the earth's surface form a continuous story: they form in fact only a part of the story of the changes of the surface of this planet. But to analyse them effectively we must split up the story and try to focus its phases according to the various kinds of process involved.

Biotic Factors

Professor Phillips makes a point of separating the effect of grazing herbivorous animals *naturally* belonging to the "biotic community," e.g., the bison of the North American prairie or the antelopes, etc., of the South African veld, from the effect of grazing animals introduced by man. The former are said to have co-operated in the production of the short grass vegetation of the Great Plains, which has even been called the *Bison-Bouteloa* climax, and to have kept back the forest from invading the edges of the grassland formation. The latter are supposed to be merely destructive in their effects, and to play no part in any successional or developmental process. This is perhaps legitimate as a description of the ecosystems of the world before the advent of man, or rather with the activities of man deliberately ignored. It is obvious that modern civilised man upsets the "natural" ecosystems or "biotic communities" on a very large scale. But it would be difficult, not to say impossible, to draw a natural line between the activities of the human tribes which presumably fitted into and formed parts of "biotic communities" and the destructive human activities of the modern world. Is man part of "nature" or not? Can his existence be harmonised with the conception of the "complex organism"? Regarded as an exceptionally powerful biotic factor which increasingly upsets the equilibrium of preexisting ecosystems and eventually destroys them, at the same time forming new ones of very different nature, human activity finds its proper place in ecology.

As an ecological factor acting on vegetation the effect of grazing heavy enough to prevent the development of woody plants is essentially the same effect wherever it occurs. If such grazing exists the grazing animals are an important factor in the biome actually present whether they came by themselves or were introduced by man. The dynamic equilibrium maintained is primarily an equilibrium between the grazing animals and the grasses and other hemicryptophytes which can exist and flourish although they are continually eaten back.

Forest may be converted into grassland by grazing animals. The substitution of the one type of vegetation for the other involves destruction of course, but not merely destruction: it also involves the appearance and gradual establishment of new vegetation. It is a successional process culminating in a climax under the influence of the actual combination of factors present and since this climax is a well-defined entity it is also the development of that entity. It is true of course that when man introduces sheep and cattle he protects them by destroying carnivores and thus artificially maintains the ecosystem whose essential feature is the equilibrium between the grassland and the grazing animals. He may also alter the position of equilibrium by feeding his animals not only on the pasture but also partly away from it, so that their dung represents food for the grassland brought from outside, and the floristic composition of the grassland is thereby altered. In such ways *anthropogenic ecosystems* differ from those developed independently of man. But the essential formative processes of the vegetation are the same, however the factors initiating them are directed.

We must have a system of ecological concepts which will allow of the inclusion of *all* forms of vegetational expression and activity. We cannot confine ourselves to the so-called " natural " entities and ignore the processes and expressions of vegetation now so abundantly provided us by the activities of man. Such a course is not scientifically sound, because scientific analysis must penetrate beneath the forms of the " natural " entities, and it is not practically useful because ecology must be applied to conditions brought about by human activity. The " natural " entities and the anthropogenic derivates alike must be analysed in terms of the most appropriate concepts we can find. Plant community, succession, development, climax, used in their wider and not in specialised senses, represent such concepts. They certainly involve an abstraction of the vegetation as such from the whole complex of components of the ecosystem, the remaining components being regarded as factors. This abstraction is a convenient isolate which has served and is continuing to serve us well. It has in fact many, though by no means all, of the qualities of an organism. The biome is a less convenient isolate for most purposes, though it has some uses, and it is not in the least improved by being called a " biotic community " or a " complex organism," terms which are illegitimately derived and which introduce misleading implications.

Methodological Value of the Concepts Relating to Successional Change

There can be no doubt that the firm establishment of the concept of succession has led directly to the creation of what is now often called dynamic ecology and that this in its turn has greatly increased our insight into the nature and behaviour of vegetation. The simplest possible scheme involves a succession of vegetational stages (the prisere of Clements) on an initially " bare " area, culminating in a stage (the climax) beyond which no further advance is possible under the given conditions of habitat (in the widest sense) and in the presence of the available colonising species. If we recognise that the climax with its whole environment represents a system in relatively stable dynamic equilibrium while the preceding stages are not, we have already the *essential framework* into which we can fit our detailed investigations of particular successions. Unless we use this framework, unless we recognise the universal tendency of the system in which vegetation is the most conspicuous component to attain dynamic equilibrium by the most complete adjustment possible of all the complexes involved we have no key to correct interpretation of the observed phenomena, which are open to every kind of misinterpretation. From the results of detailed investigations of successions, which incidentally throw a great deal of new light on existing vegetation whose nature and status were previously obscure, we may deduce certain general laws and formulate a number of useful subsidiary concepts. So far the concept of succession has proved itself of prime methodological value.

The same can scarcely be said of the concept of the climax as an organism and all that flows from its strict interpretation. On the contrary this leads to the dogmatic theses that development of the "complex organism" can *never* be retrogressive, because retrogression in development is supposed to be contrary to the nature of an organism, and that edaphic or biotic factors can *never* determine a climax, because this would cut across the conception of the climatic climax as *the* "complex organism."

Phillips says ('35, II, p. 242) that "the utility of the climax in Clements' sense would be greatly impaired were we to attempt to isolate from it the concept of the community as a complex organism. Its natural dynamic utility for orientation of research in succession, development and classification would be distinctly diminished." And again ('35, III, p. 503), "The biotic community is an organism, a highly complex one: this concept is fundamental to a natural setting and classification of the profoundly important processes of succession, development and attaining of dynamic equilibrium."

What is the justification for such statements? What researches have been stimulated or assisted by the concept of "the complex organism" *as such?* Professor Phillips seems to have in mind co-operative work in which plant and animal ecologists take part. But nobody denies the necessity for investigation of *all* the components of the ecosystem and of the ways in which they interact to bring about approximation to dynamic equilibrium. That is the prime task of the ecology of the future.

We cannot escape the conclusion that the supposed methodological value of the concept of the "complex organism," contrasted with the value of succession, development, climax and ecosystem, is a false value, and can only mislead. And it is false because it is based either on illegitimate extension of the biological concept of organism [5] (Clements) or on a confusion between the biological and "organicist" uses of the word (Phillips).

[5] Clements is quoted as saying that biologists present at the evolution of multicellular from unicellular organisms would have denied that they *were* organisms, because they were *different*. Perhaps; but from our superior vantage point we can assert with perfect confidence that the so-called "complex organism" is vastly *more* different from either multicellular or unicellular organisms than they are from one another.

Conclusions

Succession is a continuous process of change in vegetation which can be separated into a series of phases. When the dominating factors of change depend directly on the activities of the plants themselves (autogenic factors) the succession is *autogenic:* when the dominating factors are external to the plants (allogenic factors) it is *allogenic*. The successions (priseres) which lead from bare substrata to the highest types of vegetation actually present in a climatic region (progressive) are primarily autogenic. Those which lead away from these higher forms of vegetation (retrogressive) are largely allogenic, though both types of factor enter into all successions.

A *climax* is a relatively stable phase reached by successional change. Change may still be proceeding within a climax, but if it is too slow to appreciate or too small to affect the general nature of the vegetation, the apparently stable phase must still be called a climax. The highest types of vegetation characteristic of a climatic region and limited only by climate form the *climatic climax*. Other climaxes may be determined by other factors such as certain soil types, grazing animals, fire and the like.

The term *development* may be applied, as in ordinary speech, to the appearance of any well-defined vegetational entity; but the term is more strictly applied to the autogenic successions leading to climaxes, which have several features in common with the development of organisms. Such climaxes may be considered as *quasi-organisms*.

The concept of the "biotic community" is unnatural because animals and plants are too different in nature to be considered as members of the same community. The whole complex of organisms present in an ecological unit may be called the *biome*.

The concept of the "complex organism" as applied to the biome is objectionable both because the term is already in common use for an individual higher animal or plant, and because the biome is not an organism except in the sense in which inorganic systems are organisms.

The fundamental concept appropriate to the biome considered together with all the effective inorganic factors of its environment is the *ecosystem*, which is a particular category among the physical systems that make up the universe. In an ecosystem the organisms and the inorganic factors alike are *components* which are in relatively stable dynamic equilibrium. Succession and development are instances of the universal processes tending towards the creation of such equilibrated systems.

From the standpoint of vegetation biotic factors, in the sense of decisive influences of animal action, are a legitimate and useful conception. Of these biotic factors heavy and continuous grazing which changes and stabilises the vegetation is an outstanding example.

The supposed methodological value of the ideas of the biotic community and the complex organism is illusory, unlike the values of plant community, succession, development, climax and ecosystem, the concepts of which form the essential framework into which detailed studies of successional processes must be fitted.

Literature Cited

Bourne, R. 1934. Some ecological conceptions. *Empire Forestry Journal* **13**: 15–30.

Clements, F. E. 1916. Plant Succession. *Carnegie Inst. Wash. Publ. 242.*

——. 1934. The relict method in dynamic ecology. *Jour. Ecol.* **22**: 39–68.

Cooper, W. S. 1926. Fundamentals of vegetational change. *Ecology* **7**: 391–413.

Cowles, H. C. 1899. The ecological relations of the vegetation on the sand dunes of Lake Michigan. *Bot. Gaz.* **27**: 95–391.

——. 1901. The physiographic ecology of Chicago and vicinity. *Bot. Gaz.* **31**: 73–108.

——. 1911. The causes of vegetative cycles. *Bot. Gaz.* **51**: 161–183.

Godwin, H. 1929. The subclimax and deflected succession. *Jour. Ecol.* **17**: 144–147.

Levy, H. 1932. The Universe of Science. London.

Phillips, John. 1931. The biotic community. *Jour. Ecol.* **19**: 1–24.

——. 1934. Succession, development, the climax and the complex organism: an analysis of concepts. I. *Jour. Ecol.* **22**: 554–571.

——. 1935. Idem II, III. Ibid. **23**: 210–246, 488–508.

Shelford, V. E. 1931. Some concepts of bioecology. *Ecology* **12**: 455–467.

Tansley, A. G. 1920. The classification of vegetation and the concept of development. *Jour. Ecol.* **8**: 118–144.

——. 1929. Succession: the concept and its values. *Proc. Intern. Congress Plant Sciences, 1926.* Pp. 677–686.

PAPER **16**

THE
AMERICAN NATURALIST

Vol. XCIII | May-June, 1959 | No. 870

HOMAGE TO SANTA ROSALIA
or
WHY ARE THERE SO MANY KINDS OF ANIMALS?*

G. E. HUTCHINSON

Department of Zoology, Yale University, New Haven, Connecticut

When you did me the honor of asking me to fill your presidential chair, I accepted perhaps without duly considering the duties of the president of a society, founded largely to further the study of evolution, at the close of the year that marks the centenary of Darwin and Wallace's initial presentation of the theory of natural selection. It seemed to me that most of the significant aspects of modern evolutionary theory have come either from geneticists, or from those heroic museum workers who suffering through years of neglect, were able to establish about 20 years ago what has come to be called the "new systematics." You had, however, chosen an ecologist as your president and one of that school at times supposed to study the environment without any relation to the organism.

A few months later I happened to be in Sicily. An early interest in zoogeography and in aquatic insects led me to attempt to collect near Palermo, certain species of water-bugs, of the genus *Corixa*, described a century ago by Fieber and supposed to occur in the region, but never fully reinvestigated. It is hard to find suitable localities in so highly cultivated a landscape as the Concha d'Oro. Fortunately, I was driven up Monte Pellegrino, the hill that rises to the west of the city, to admire the view. A little below the summit, a church with a simple baroque facade stands in front of a cave in the limestone of the hill. Here in the 16th century a stalactite encrusted skeleton associated with a cross and twelve beads was discovered. Of this skeleton nothing is certainly known save that it is that of Santa Rosalia, a saint of whom little is reliably reported save that she seems to have lived in the 12th century, that her skeleton was found in this cave, and that she has been the chief patroness of Palermo ever since. Other limestone caverns on Monte Pellegrino had yielded bones of extinct pleistocene *Equus*, and on the walls of one of the rock shelters at the bottom of the hill there are beautiful Gravettian engravings. Moreover, a small relic of the saint that I saw in the treasury of the Cathedral of Monreale has a venerable and

*Address of the President, American Society of Naturalists, delivered at the annual meeting, Washington, D. C., December 30, 1958.

petrified appearance, as might be expected. Nothing in her history being known to the contrary, perhaps for the moment we may take Santa Rosalia as the patroness of evolutionary studies, for just below the sanctuary, fed no doubt by the water that percolates through the limestone cracks of the mountain, and which formed the sacred cave, lies a small artificial pond, and when I could get to the pond a few weeks later, I got from it a hint of what I was looking for.

Vast numbers of Corixidae were living in the water. At first I was rather disappointed because every specimen of the larger of the two species present was a female, and so lacking in most critical diagnostic features, while both sexes of the second slightly smaller species were present in about equal number. Examination of the material at leisure, and of the relevant literature, has convinced me that the two species are the common European *C. punctata* and *C. affinis*, and that the peculiar Mediterranean species are illusionary. The larger *C. punctata* was clearly at the end of its breeding season, the smaller *C. affinis* was probably just beginning to breed. This is the sort of observation that any naturalist can and does make all the time. It was not until I asked myself why the larger species should breed first, and then the more general question as to why there should be two and not 20 or 200 species of the genus in the pond, that ideas suitable to present to you began to emerge. These ideas finally prompted the very general question as to why there are such an enormous number of animal species.

There are at the present time supposed to be (Muller and Campbell, 1954; Hyman, 1955) about one million described species of animals. Of these about three-quarters are insects, of which a quite disproportionately large number are members of a single order, the Coleoptera.[1] The marine fauna although it has at its disposal a much greater area than has the terrestrial, lacks this astonishing diversity (Thorson, 1958). If the insects are excluded, it would seem to be more diverse. The proper answer to my initial question would be to develop a theory at least predicting an order of magnitude for the number of species of 10^6 rather than 10^8 or 10^4. This I certainly cannot do. At most it is merely possible to point out some of the factors which would have to be considered if such a theory was ever to be constructed.

Before developing my ideas I should like to say that I subscribe to the view that the process of natural selection, coupled with isolation and later mutual invasion of ranges leads to the evolution of sympatric species, which at equilibrium occupy distinct niches, according to the Volterra-Gause principle. The empirical reasons for adopting this view and the correlative view that the boundaries of realized niches are set by competition are mainly indirect. So far as niches may be defined in terms of food, the subject has been carefully considered by Lack (1954). In general all the indirect evi-

[1]There is a story, possibly apocryphal, of the distinguished British biologist, J. B. S. Haldane, who found himself in the company of a group of theologians. On being asked what one could conclude as to the nature of the Creator from a study of his creation, Haldane is said to have answered, "An inordinate fondness for beetles."

dence is in accord with the view, which has the advantage of confirming theoretical expectation. Most of the opinions that have been held to the contrary appear to be due to misunderstandings and to loose formulation of the problem (Hutchinson, 1958).

In any study of evolutionary ecology, food relations appear as one of the most important aspects of the system of animate nature. There is quite obviously much more to living communities than the raw dictum "eat or be eaten," but in order to understand the higher intricacies of any ecological system, it is most easy to start from this crudely simple point of view.

FOOD CHAINS

Animal ecologists frequently think in terms of food chains, of the form *individuals of species S_1 are eaten by those of S_2, of S_2 by S_3, of S_3 by S_4*, etc. In such a food chain S_1 will ordinarily be some holophylic organism or material derived from such organisms. The simplest case is that in which we have a true *predator chain* in Odum's (1953) convenient terminology, in which the lowest link is a green plant, the next a herbivorous animal, the next a primary carnivore, the next a secondary carnivore, etc. A specially important type of predator chain may be designated Eltonian, because in recent years C. S. Elton (1927) has emphasized its widespread significance, in which the predator at each level is larger and rarer than its prey. This phenomenon was recognized much earlier, notably by A. R. Wallace in his contribution to the 1858 communication to the Linnean Society of London.

In such a system we can make a theoretical guess of the order of magnitude of the diversity that a single food chain can introduce into a community. If we assume that in general 20 per cent of the energy passing through one link can enter the next link in the chain, which is overgenerous (cf. Lindeman, 1942; Slobodkin in an unpublished study finds 13 per cent as a reasonable upper limit) and if we suppose that each predator has twice the mass, (or 1.26 the linear dimensions) of its prey, which is a very low estimate of the size difference between links, the fifth animal link will have a population of one ten thousandth (10^{-4}) of the first, and the fiftieth animal link, if there was one, a population of 10^{-49} the size of the first. Five animal links are certainly possible, a few fairly clear cut cases having been in fact recorded. If, however, we wanted 50 links, starting with a protozoan or rotifer feeding on algae with a density of 10^6 cells per ml, we should need a volume of 10^{26} cubic kilometers to accommodate on an average one specimen of the ultimate predator, and this is vastly greater than the volume of the world ocean. Clearly the Eltonian food-chain of itself cannot give any great diversity, and the same is almost certainly true of the other types of food chain, based on detritus feeding or on parasitism.

Natural selection

Before proceeding to a further consideration of diversity, it is, however, desirable to consider the kinds of selective force that may operate on a food chain, for this may limit the possible diversity.

It is reasonably certain that natural selection will tend to maintain the efficiency of transfer from one level to another at a maximum. Any increase in the predatory efficiency of the nth link of a simple food chain will however always increase the possibility of the extermination of the $(n-1)^{th}$ link. If this occurs either the species constituting the nth link must adapt itself to eating the $(n-2)^{th}$ link or itself become extinct. This process will in fact tend to shortening of food chains. A lengthening can presumably occur most simply by the development of a new terminal carnivore link, as its niche is by definition previously empty. In most cases this is not likely to be easy. The evolution of the whale-bone whales, which at least in the case of *Balaenoptera borealis*, can feed largely on copepods and so rank on occasions as primary carnivores (Bigelow, 1926), presumably constitutes the most dramatic example of the shortening of a food chain. Mechanical considerations would have prevented the evolution of a larger rarer predator, until man developed essentially non-Eltonian methods of hunting whales.

Effect of size

A second important limitation of the length of a food chain is due to the fact that ordinarily animals change their size during free life. If the terminal member of a chain were a fish that grew from say one cm to 150 cms in the course of an ordinary life, this size change would set a limit by competition to the possible number of otherwise conceivable links in the 1-150 cm range. At least in fishes this type of process (metaphoetesis) may involve the smaller specimens belonging to links below the larger and the chain length is thus lengthened, though under strong limitations, by cannibalism.

We may next enquire into what determines the number of food chains in a community. In part the answer is clear, though if we cease to be zoologists and become biologists, the answer begs the question. Within certain limits, the number of kinds of primary producers is certainly involved, because many herbivorous animals are somewhat eclectic in their tastes and many more limited by their size or by such structural adaptations for feeding that they have been able to develop.

Effects of terrestrial plants

The extraordinary diversity of the terrestrial fauna, which is much greater than that of the marine fauna, is clearly due largely to the diversity provided by terrestrial plants. This diversity is actually two-fold. Firstly, since terrestrial plants compete for light, they have tended to evolve into structures growing into a gaseous medium of negligible buoyancy. This has led to the formation of specialized supporting, photosynthetic, and reproductive structures which inevitably differ in chemical and physical properties. The ancient Danes and Irish are supposed to have eaten elm-bark, and sometimes sawdust, in periods of stress, has been hydrolyzed to produce edible carbohydrate; but usually man, the most omnivorous of all animals, has avoided almost all parts of trees except fruits as sources of food, though various individual species of animals can deal with practically every tissue of many arboreal species. A major source of terrestrial diversity was thus introduced by the evolution of almost 200,000 species of flowering plants, and the three quarters of a million insects supposedly known today are in part a product of that diversity. But of itself merely providing five or ten kinds of food of different consistencies and compositions does not get us much further than the five or ten links of an Eltonian pyramid. On the whole the problem still remains, but in the new form: why are there so many kinds of plants? As a zoologist I do not want to attack that question directly, I want to stick with animals, but also to get the answer. Since, however, the plants are part of the general system of communities, any sufficiently abstract properties of such communities are likely to be relevant to plants as well as to herbivores and carnivores. It is, therefore, by being somewhat abstract, though with concrete zoological details as examples, that I intend to proceed.

INTERRELATIONS OF FOOD CHAINS

Biological communities do not consist of independent food chains, but of food webs, of such a kind that an individual at any level (corresponding to a link in a single chain) can use some but not all of the food provided by species in the levels below it.

It has long been realized that the presence of two species at any level, either of which can be eaten by a predator at a level above, but which may differ in palatability, ease of capture or seasonal and local abundance, may provide alternative foods for the predator. The predator, therefore, will neither become extinct itself nor exterminate its usual prey, when for any reason, not dependent on prey-predator relationships, the usual prey happens to be abnormally scarce. This aspect of complicated food webs has been stressed by many ecologists, of whom the Chicago school as represented by Allee, Emerson, Park, Park and Schmidt (1949), Odum (1953) and Elton (1958), may in particular be mentioned. Recently MacArthur (1955) using an ingenious but simple application of information theory has generalized the points of view of earlier workers by providing a formal proof of the increase in stability of a community as the number of links in its food web increases.

MacArthur concludes that in the evolution of a natural community two partly antagonistic processes are occurring. More efficient species will replace less efficient species, but more stable communities will outlast less stable communities. In the process of community formation, the entry of a new species may involve one of three possibilities. It may completely displace an old species. This of itself does not necessarily change the stability, though it may do so if the new species inherently has a more stable population (cf. Slobodkin, 1956) than the old. Secondly, it may occupy an unfilled niche, which may, by providing new partially independent links, increase stability. Thirdly, it may partition a niche with a pre-existing species. Elton (1958) in a fascinating work largely devoted to the fate of species accidentally or purposefully introduced by man, concludes that in very diverse communities such introductions are difficult. Early in the history of a community we may suppose many niches will be empty and invasion will proceed easily; as the community becomes more diversified, the process will be progressively more difficult. Sometimes an extremely successful invader may oust a species but add little or nothing to stability, at other times the invader by some specialization will be able to compete successfully for the marginal parts of a niche. In all cases it is probable that invasion is most likely when one or more species happen to be fluctuating and are underrepresented at a given moment. As the communities build up, these opportunities will get progressively rarer. In this way a complex community containing some highly specialized species is constructed asymptotically.

Modern ecological theory therefore appears to answer our initial question at least partially by saying that there is a great diversity of organisms because communities of many diversified organisms are better able to persist than are communities of fewer less diversified organisms. Even though the entry of an invader which takes over part of a niche will lead to the reduction in the *average* population of the species originally present, it will also lead to an increase in stability reducing the risk of the original population being at times underrepresented to a dangerous degree. In this way loss of some niche space may be compensated by reduction in the amplitude of fluctuations in a way that can be advantageous to both species. The process however appears likely to be asymptotic and we have now to consider what sets the asymptote, or in simpler words why are there not more different kinds of animals?

LIMITATION OF DIVERSITY

It is first obvious that the processes of evolution of communities must be under various sorts of external control, and that in some cases such control limits the possible diversity. Several investigators, notably Odum (1953) and MacArthur (1955), have pointed out that the more or less cyclical oscillations observed in arctic and boreal fauna may be due in part to the communities not being sufficiently complex to damp out oscillations. It is certain that the fauna of any such region is qualitatively poorer than that of warm temperate and tropical areas of comparable effective precipitation. It is probably considered to be intuitively obvious that this should be so, but on analysis the obviousness tends to disappear. If we can have one or two species of a large family adapted to the rigors of Arctic existence, why can we not have more? It is reasonable to suppose that the total biomass may be involved. If the fundamental productivity of an area is limited by a short growing season to such a degree that the total biomass is less than under more favorable conditions, then the rarer species in a community may be so rare that they do not exist. It is also probable that certain absolute limitations on growth-forms of plants, such as those that make the development of forest impossible above a certain latitude, may in so acting, severely limit the number of niches. Dr. Robert MacArthur points out that the development of high tropical rain forest increases the bird fauna more than that of mammals, and Thorson (1957) likewise has shown that the so-called infauna show no increase of species toward the tropics while the marine epifauna becomes more diversified. The importance of this aspect of the plant or animal substratum, which depends largely on the length of the growing season and other aspects of productivity is related to that of the environmental mosaic discussed later.

We may also inquire, but at present cannot obtain any likely answer, whether the arctic fauna is not itself too young to have achieved its maximum diversity. Finally, the continual occurrence of catastrophes, as Wynne-Edwards (1952) has emphasized, may keep the arctic terrestrial community in a state of perennial though stunted youth.

Closely related to the problems of environmental rigor and stability, is the question of the absolute size of the habitat that can be colonized. Over much of western Europe there are three common species of small voles, namely *Microtus arvalis*, *M. agrestis* and *Clethrionomys glareolus*. These are sympatric but with somewhat different ecological preferences.

In the smaller islands off Britain and in the English channel, there is only one case of two species co-occurring on an island, namely *M. agrestis* and Clethrionomys on the island of Mull in the Inner Hebrides (Barrett-Hamilton and Hinton, 1911–1921). On the Orkneys the single species is *M. orcadensis*, which in morphology and cytology is a well-differentiated ally of *M. arvalis*; a comparable animal (*M. sarnius*) occurs on Guernsey. On most of the Scottish Islands only subspecies of *M. agrestis* occur, but on Mull and Raasay, on the Welsh island of Skomer, as well as on Jersey, races of Clethrionomys of somewhat uncertain status are found. No voles have reached Ireland, presumably for paleogeographic reasons, but they are also absent from a number of small islands, notably Alderney and Sark. The last named island must have been as well placed as Guernsey to receive *Microtus arvalis*. Still stranger is the fact that although it could not have got to the Orkneys without entering the mainland of Britain, no vole of the *arvalis* type now occurs in the latter country. Cases of this sort may be perhaps explained by the lack of favorable refuges in randomly distributed very unfavorable seasons or under special kinds of competition. This explanation is very reasonable as an explanation of the lack of Microtus on Sark, where it may have had difficulty in competing with *Rattus rattus* in a small area. It would be stretching one's credulity to suppose that the area of Great Britain is too small to permit the existence of two sympatric species of Microtus, but no other explanation seems to have been proposed.

It is a matter of considerable interest that Lack (1942) studying the populations of birds on some of these small British islands concluded that such populations are often unstable, and that the few species present often occupied larger niches than on the mainland in the presence of competitors. Such faunas provide examples of communities held at an early stage in development because there is not enough space for the evolution of a fuller and more stable community.

NICHE REQUIREMENTS

The various evolutionary tendencies, notably metaphoetesis, which operate on single food chains must operate equally on the food-web, but we also have a new, if comparable, problem as to how much difference between two species at the same level is needed to prevent them from occupying the same niche. Where metric characters are involved we can gain some insight into this extremely important problem by the study of what Brown and Wilson (1956) have called *character displacement* or the divergence shown when two partly allopatric species of comparable niche requirements become sympatric in part of their range.

I have collected together a number of cases of mammals and birds which appear to exhibit the phenomenon (table 1). These cases involve metric characters related to the trophic apparatus, the length of the culmen in birds and of the skull in mammals appearing to provide appropriate measures. Where the species co-occur, the ratio of the larger to the small form varies from 1.1 to 1.4, the mean ratio being 1.28 or roughly 1.3. This latter figure may tentatively be used as an indication of the kind of difference necessary to permit two species to co-occur in different niches but at the same level of a food-web. In the case of the aquatic insects with which I began my address, we have over most of Europe three very closely allied species of Corixa, the largest *punctata*, being about 116 per cent longer than the middle sized species *macrocephala*, and 146 per cent longer than the small species *affinis*. In northwestern Europe there is a fourth species, *C. dentipes*, as large as *C. punctata* and very similar in appearance. A single observation (Brown, 1948) suggests that this is what I have elsewhere (Hutchinson, 1951) termed a fugitive species, maintaining itself in the face of competition mainly on account of greater mobility. According to Macan (1954) while both *affinis* and *macrocephala* may occur with *punctata* they never are found with each other, so that all three species never occur together. In the eastern part of the range, *macrocephala* drops out, and *punctata* appears to have a discontinuous distribution, being recorded as far east as Simla, but not in southern Persia or Kashmir, where *affinis* occurs. In these eastern localities, where it occurs by itself, *affinis* is larger and darker than in the west, and superficially looks like *macrocephala* (Hutchinson, 1940).

This case is very interesting because it looks as though character displacement is occurring, but that the size differences between the three species are just not great enough to allow them all to co-occur. Other characters than size are in fact clearly involved in the separation, *macrocephala* preferring deeper water than *affinis* and the latter being more tolerant of brackish conditions. It is also interesting because it calls attention to a marked difference that must occur between hemimetabolous insects with annual life cycles involving relatively long growth periods, and birds or mammals in which the period of growth in length is short and of a very special nature compared with the total life span. In the latter, niche separation may be possible merely through genetic size differences, while in a pair of ani-

TABLE 1

Mean character displacement in measurable trophic structures in mammals (skull) and birds (culmen); data for Mustela from Miller (1912); Apodemus from Cranbrook (1957); Sitta from Brown and Wilson (1956) after Vaurie; Galapagos finches from Lack (1947)

	Locality and measurement when sympatric	Locality and measurement when allopatric	Ratio when sympatric
Mustela nivalis	Britain; skull ♂ 39.3 ♀ 33.6 mm.	(*boccamela*) S. France, Italy ♂ 42.9 ♀ 34.7 mm.; (*iberica*) Spain, Portugal ♂ 40.4 ♀ 36.0	♂ 100:128
M. erminea	Britain; " ♂ 50.4 ♀ 45.0	(*hibernica*) Ireland ♂ 46.0 ♀ 41.9	♀ 100:134
Apodemus sylvaticus	Britain; " 24.8	unnamed races on Channel Islands 25.6–26.7	100:109
A. flavicollis	Britain; " 27.0		
Sitta tephronota	Iran; culmen 29.0	races east of overlap 25.5	100:124
S. neumayer	Iran; " 23.5	races west of overlap 26.0	
Geospiza fortis	Indefatigable Isl.; culmen 12.0	Daphne Isl. 10.5	100:143
G. fuliginosa	Indefatigable Isl.; " 8.4	Crossman Isl. 9.3	
Camarhynchus parvulus	James Isl.; " 7.0	N. Albemarle Isl. 7.0	James 100:140:180
	Indefatigable Isl.; " 7.5	Chatham Isl. 8.0	100:129
C. psittacula	S. Albemarle Isl.; " 7.3	Abington Isl. 10.1	Indefatigable 100:128:162
	James Isl.; " 9.8	Bindloe Isl. 10.5	100:127
	Indefatigable Isl.; " 9.6		
C. pallidus	S. Albemarle Isl.; " 8.5		
	James Isl.; " 12.6	N. Albemarle Isl. 11.7	S. Albemarle 100:116:153
	Indefatigable Isl.; " 12.1	Chatham Isl. 10.8	100:132
	S. Albemarle Isl.; " 11.2		
			Mean ratio 100:128

mals like *C. punctata* and *C. affinis* we need not only a size difference but a seasonal one in reproduction; this is likely to be a rather complicated matter. For the larger of two species always to be larger, it must never breed later than the smaller one. I do not doubt that this is what was happening in the pond on Monte Pellegrino, but have no idea how the difference is achieved.

I want to emphasize the complexity of the adaptation necessary on the part of two species inhabiting adjacent niches in a given biotope, as it probably underlies a phenomenon which to some has appeared rather puzzling. MacArthur (1957) has shown that in a sufficiently large bird fauna, in a uniform undisturbed habitat, areas occupied by the different species appear to correspond to the random non-overlapping fractionation of a plane or volume. Kohn (1959) has found the same thing for the cone-shells (Conus) on the Hawaiian reefs. This type of arrangement almost certainly implies such individual and unpredictable complexities in the determination of the niche boundaries, and so of the actual areas colonized, that in any overall view, the process would appear random. It is fairly obvious that in different types of community the divisibility of niches will differ and so the degree of diversity that can be achieved. The fine details of the process have not been adequately investigated, though many data must already exist that could be organized to throw light on the problem.

MOSAIC NATURE OF THE ENVIRONMENT

A final aspect of the limitation of possible diversity, and one that perhaps is of greatest importance, concerns what may be called the mosaic nature of the environment. Except perhaps in open water when only uniform quasi-horizontal surfaces are considered, every area colonized by organisms has some local diversity. The significance of such local diversity depends very largely on the size of the organisms under consideration. In another paper MacArthur and I (Hutchinson and MacArthur, 1959) have attempted a theoretical formulation of this property of living communities and have pointed out that even if we consider only the herbivorous level or only one of the carnivorous levels, there are likely, above a certain lower limit of size, to be more species of small or medium sized organisms than of large organisms. It is difficult to go much beyond crude qualitative impressions in testing this hypothesis, but we find that for mammal faunas, which contain such diverse organisms that they may well be regarded as models of whole faunas, there is a definite hint of the kind of theoretical distribution that we deduce. In qualitative terms the phenomenon can be exemplified by any of the larger species of ungulates which may require a number of different kinds of terrain within their home ranges, any one of which types of terrain might be the habitat of some small species. Most of the genera or even subfamilies of very large terrestrial animals contain only one or two sympatric species. In this connection I cannot refrain from pointing out the immense scientific importance of obtaining a really full insight into the ecology of the large mammals of Africa while they can still be studied under natural conditions. It is indeed quite possible that the results of studies on these wonderful animals would in long-range though purely practical terms pay for the establishment of greater reservations and National Parks than at present exist.

In the passerine birds the occurrence of five or six closely related sympatric species is a commonplace. In the mammal fauna of western Europe no genus appears to contain more than four strictly sympatric species. In Britain this number is not reached even by Mustela with three species, on the adjacent parts of the continent there may be three sympatric shrews of the genus Crocidura and in parts of Holland three of Microtus. In the same general region there are genera of insects containing hundreds of species, as in Athela in the Coleoptera and Dasyhelea in the Diptera Nematocera. The same phenomenon will be encountered whenever any well-studied fauna is considered. Irrespective of their position in a food chain, small size, by permitting animals to become specialized to the conditions offered by small diversified elements of the environmental mosaic, clearly makes possible a degree of diversity quite unknown among groups of larger organisms.

We may, therefore, conclude that the reason why there are so many species of animals is at least partly because a complex trophic organization of a community is more stable than a simple one, but that limits are set by the tendency of food chains to shorten or become blurred, by unfavorable physical factors, by space, by the fineness of possible subdivision of niches, and by those characters of the environmental mosaic which permit a greater diversity of small than of large allied species.

CONCLUDING DISCUSSION

In conclusion I should like to point out three very general aspects of the sort of process I have described. One speculative approach to evolutionary theory arises from some of these conclusions. Just as adaptative evolution by natural selection is less easy in a small population of a species than in a larger one, because the total pool of genetic variability is inevitably less, so it is probable that a group containing many diversified species will be able to seize new evolutionary opportunities more easily than an undiversified group. There will be some limits to this process. Where large size permits the development of a brain capable of much new learnt behavior, the greater plasticity acquired by the individual species will offset the disadvantage of the small number of allied species characteristic of groups of large animals. Early during evolution the main process from the standpoint of community structure was the filling of all the niche space potentially available for producer and decomposer organisms and for herbivorous animals. As the latter, and still more as carnivorous animals began to appear, the persistence of more stable communities would imply splitting of niches previously occupied by single species as the communities became more diverse. As this process continued one would expect the overall rate of evolution to have increased, as the increasing diversity increased the probability of the existence of species preadapted to new and unusual niches. It is reasonable to suppose that strong predation among macroscopic metazoa did not begin until the late Precambrian, and that the appearance of powerful predators led to the appearance of fossilizable skeletons. This seems the only reasonable hypothesis, of those so far advanced, to account for the relatively sudden appearance of several fossilizable groups in the Lower Cambrian. The process of diversification would, according to this argument, be somewhat autocatakinetic even without the increased stability that it would produce; with the increase in stability it would be still more a self inducing process, but one, as we have seen, with an upper limit. Part of this upper limit is set by the impossibility of having many sympatric allied species of large animals. These however are the animals that can pass from primarily innate to highly modifiable behavior. From an evolutionary point of view, once they have appeared, there is perhaps less need for diversity, though from other points of view, as Elton (1958) has stressed in dealing with human activities, the stability provided by diversity can be valuable even to the most adaptable of all large animals. We may perhaps therefore see in the process of evolution an increase in diversity at an increasing rate till the early Paleozoic, by which time the familiar types of community structure were established. There followed then a long period in which various large and finally large-brained species became dominant, and then a period in which man has been reducing diversity by a rapidly increasing tendency to cause extinction of supposedly unwanted species, often in an indiscriminate manner. Finally we may hope for a limited reversal of this process when man becomes aware of the value of diversity no less in an economic than in an esthetic and scientific sense.

A second and much more metaphysical general point is perhaps worth a moment's discussion. The evolution of biological communities, though each species appears to fend for itself alone, produces integrated aggregates which increase in stability. There is nothing mysterious about this; it follows from mathematical theory and appears to be confirmed to some extent empirically. It is however a phenomenon which also finds analogies in other fields in which a more complex type of behavior, that we intuitively regard as higher, emerges as the result of the interaction of less complex types of behavior, that we call lower. The emergence of love as an antidote to aggression, as Lorenz pictures the process, or the development of cooperation from various forms of more or less inevitable group behavior that Allee (1931) has stressed are examples of this from the more complex types of biological systems.

In the ordinary sense of explanation in science, such phenomena are explicable. The types of holistic philosophy which import *ad hoc* mysteries into science whenever such a situation is met are obviously unnecessary. Yet perhaps we may wonder whether the empirical fact that it is the nature of things for this type of explicable emergence to occur is not something that itself requires an explanation. Many objections can be raised to such a view; a friendly organization of biologists could not occur in a universe in which cooperative behavior was impossible and without your cooperation I could not raise the problem. The question may in fact appear to certain types of philosophers not to be a real one, though I suspect such philosophers in their desire to demonstrate how often people talk nonsense, may sometimes show less ingenuity than would be desirable in finding some sense in such questions. Even if the answer to such a question were positive, it might not get us very far; to an existentialist, life would have merely provided yet one more problem; students of Whitehead might be made happier, though on the whole the obscurities of that great writer do not seem to generate unhappiness; the religious philosophers would welcome a positive answer but note that it told them nothing that they did not know before; Marxists might merely say, "I told you so." In spite of this I suspect that the question is worth raising, and that it could be phrased so as to provide some sort of real dichotomy between alternatives; I therefore raise it knowing that I cannot, and suspecting that at present others cannot, provide an intellectually satisfying answer.

My third general point is less metaphysical, but not without interest. If I am right that it is easier to have a greater diversity of small than of large organisms, then the evolutionary process in small organisms will differ somewhat from that of large ones. Wherever we have a great array of allied sympatric species there must be an emphasis on very accurate interspecific mating barriers which is unnecessary where virtually no sympatric allies occur. We ourselves are large animals in this sense; it would seem very unlikely that the peculiar lability that seems to exist in man, in which even the direction of normal sexual behavior must be learnt, could have developed to quite the existing extent if species recognition, involving closely related sympatric congeners, had been necessary. Elsewhere (Hutchinson, 1959) I have attempted to show that the difficulties that *Homo sapiens* has to face in this regard may imply various unsuspected processes in human evolutionary selection. But perhaps Santa Rosalia would find at this point that we are speculating too freely, so for the moment, while under her patronage, I will say no more.

ACKNOWLEDGMENTS

Dr. A. Minganti of the University of Palermo enabled me to collect on Monte Pellegrino. Professor B. M. Knox of the Department of Classics of Yale University gave me a rare and elegant word from the Greek to express the blurring of a food chain. Dr. L. B. Slobodkin of the University of Michigan and Dr. R. H. MacArthur of the University of Pennsylvania provided me with their customary kinds of intellectual stimulation. To all these friends I am most grateful.

LITERATURE CITED

Allee, W. C., 1931, Animal aggregations: a study in general sociology. vii, 431 pp. University of Chicago Press, Chicago, Illinois.

Allee, W. C., A. E. Emerson, O. Park, T. Park and K. P. Schmidt, 1949, Principles of animal ecology. xii, 837 pp. W. B. Saunders Co., Philadelphia, Pennsylvania.

Barrett-Hamilton, G. E. H., and M. A. C. Hinton, 1911-1921, A history of British mammals. Vol. 2. 748 pp. Gurney and Jackson, London, England.

Bigelow, H. B., 1926, Plankton of the offshore waters of the Gulf of Maine. Bull. U. S. Bur. Fisheries 40: 1-509.

Brown, E. S., 1958, A contribution towards an ecological survey of the aquatic and semi-aquatic Hemiptera-Heteroptera (water-bugs) of the British Isles etc. Trans. Soc. British Entom. 9: 151-195.

Brown, W. L., and E. O. Wilson, 1956, Character displacement. Systematic Zoology 5: 49-64.

Cranbrook, Lord, 1957, Long-tailed field mice (Apodemus) from the Channel Islands. Proc. Zool. Soc. London 128: 597-600.

Elton, C. S., 1958, The ecology of invasions by animals and plants. 159 pp. Methuen Ltd., London, England.

Hutchinson, G. E., 1951, Copepodology for the ornithologist. Ecology 32: 571-577.

1958, Concluding remarks. Cold Spring Harbor Symp. Quant. Biol. 22: 415-427.

1959, A speculative consideration of certain possible forms of sexual selection in man. Amer. Nat. 93: 81-92.

Hutchinson, G. E., and R. MacArthur, 1959, A theoretical ecological model of size distributions among species of animals. Amer. Nat. 93: 117-126.

Hyman, L. H., 1955, How many species? Systematic Zoology 4: 142-143.

Kohn, A. J., 1959, The ecology of Conus in Hawaii. Ecol. Monogr. (in press).

Lack, D., 1942, Ecological features of the bird faunas of British small islands. J. Animal Ecol. London 11: 9-36.

1947, Darwin's Finches. x, 208 pp. Cambridge University Press, Cambridge, England.

1954, The natural regulation of animal numbers. viii, 347 pp. Clarendon Press, Oxford, England.

Lindeman, R. L., 1942, The trophic-dynamic aspect of ecology. Ecology 23: 399-408.

Macan, T. T., 1954, A contribution to the study of the ecology of Corixidae (Hemipt). J. Animal Ecol. 23: 115-141.

MacArthur, R. H., 1955, Fluctuations of animal populations and a measure of community stability. Ecology 35: 533-536.

1957, On the relative abundance of bird species. Proc. Nat. Acad. Sci. Wash. 43: 293-295.

Miller, G. S., Catalogue of the mammals of Western Europe. xv, 1019 pp. British Museum, London, England.

Muller, S. W., and A. Campbell, 1954, The relative number of living and fossil species of animals. Systematic Zoology 3: 168-170.

Odum, E. P., 1953, Fundamentals of ecology. xii, 387 pp. W. B. Saunders Co., Philadelphia, Pennsylvania, and London, England.

Slobodkin, L. B., 1955, Condition for population equilibrium. Ecology 35: 530-533.

Thorson, G., 1957, Bottom communities. Chap. 17 *in* Treatise on marine ecology and paleoecology. Vol. 1. Geol. Soc. Amer. Memoir 67: 461-534.

Wallace, A. R., 1858, On the tendency of varieties to depart indefinitely from the original type. *In* C. Darwin and A. R. Wallace, On the tendency of species to form varieties; and on the perpetuation of varieties and species by natural means of selection. J. Linn. Soc. (Zool.) 3: 45–62.

Wynne-Edwards, V. C., 1952, Zoology of the Baird Expedition (1950). I. The birds observed in central and southeast Baffin Island. Auk 69: 353–391.

COMMUNITY STRUCTURE, POPULATION CONTROL, AND COMPETITION

NELSON G. HAIRSTON, FREDERICK E. SMITH, AND LAWRENCE B. SLOBODKIN

Department of Zoology, The University of Michigan, Ann Arbor, Michigan

The methods whereby natural populations are limited in size have been debated with vigor during three decades, particularly during the last few years (see papers by Nicholson, Birch, Andrewartha, Milne, Reynoldson, and Hutchinson, and ensuing discussions in the Cold Spring Harbor Symposium, 1957). Few ecologists will deny the importance of the subject, since the method of regulation of populations must be known before we can understand nature and predict its behavior. Although discussion of the subject has usually been confined to single species populations, it is equally important in situations where two or more species are involved.

The purpose of this note is to demonstrate a pattern of population control in many communities which derives easily from a series of general, widely accepted observations. The logic used is not easily refuted. Furthermore, the pattern reconciles conflicting interpretations by showing that populations in different trophic levels are expected to differ in their methods of control.

Our first observation is that the accumulation of fossil fuels occurs at a rate that is negligible when compared with the rate of energy fixation through photosynthesis in the biosphere. Apparent exceptions to this observation, such as bogs and ponds, are successional stages, in which the failure of decomposition hastens the termination of the stage. The rate of accumulation when compared with that of photosynthesis has also been shown to be negligible over geologic time (Hutchinson, 1948).

If virtually all of the energy fixed in photosynthesis does indeed flow through the biosphere, it must follow that all organisms taken together are limited by the amount of energy fixed. In particular, the decomposers as a group must be food-limited, since by definition they comprise the trophic level which degrades organic debris. There is no a priori reason why predators, behavior, physiological changes induced by high densities, etc., could not limit decomposer populations. In fact, some decomposer populations may be limited in such ways. If so, however, others must consume the "left-over" food, so that the group as a whole remains food limited; otherwise fossil fuel would accumulate rapidly.

Any population which is not resource-limited must, of course, be limited to a level *below* that set by its resources.

Our next three observations are interrelated. They apply primarily to terrestrial communities. The first of these is that cases of obvious depletion of green plants by herbivores are exceptions to the general picture, in which the plants are abundant and largely intact. Moreover, cases of obvious mass destruction by meteorological catastrophes are exceptional in most areas. Taken together, these two observations mean that producers are neither herbivore-limited nor catastrophe-limited, and must therefore be limited by their own exhaustion of a resource. In many areas, the limiting resource is obviously light, but in arid regions water may be the critical factor, and there are spectacular cases of limitation through the exhaustion of a critical mineral. The final observation in this group is that there are temporary exceptions to the general lack of depletion of green plants by herbivores. This occurs when herbivores are protected either by man or natural events, and it indicates that the herbivores are able to deplete the vegetation whenever they become numerous enough, as in the cases of the Kaibab deer herd, rodent plagues, and many insect outbreaks. It therefore follows that the usual condition is for populations of herbivores *not* to be limited by their food supply.

The vagaries of weather have been suggested as an adequate method of control for herbivore populations. The best factual clues related to this argument are to be found in the analysis of the exceptional cases where terrestrial herbivores have become numerous enough to deplete the vegetation. This often occurs with introduced rather than native species. It is most difficult to suppose that a species had been unable to adapt so as to escape control by the weather to which it was exposed, and at the same time by sheer chance to be able to escape this control from weather to which it had not been previously exposed. This assumption is especially difficult when mutual invasions by different herbivores between two countries may in both cases result in pests. Even more difficult to accept, however, is the implication regarding the native herbivores. The assumption that the hundreds or thousands of species native to a forest have failed to escape from control by the weather despite long exposure and much selection, when an invader is able to defoliate without this past history, implies that "pre-adaptation" is more likely than ordinary adaptation. This we cannot accept.

The remaining general method of herbivore control is predation (in its broadest sense, including parasitism, etc.). It is important to note that this hypothesis is not denied by the presence of introduced pests, since it is necessary only to suppose that either their natural predators have been left behind, or that while the herbivore is able to exist in the new climate, its enemies are not. There are, furthermore, numerous examples of the direct effect of predator removal. The history of the Kaibab deer is the best known example, although deer across the northern portions of the country are in repeated danger of winter starvation as a result of protection and predator removal. Several rodent plagues have been attributed to the local destruction of predators. More recently, the extensive spraying of forests to kill caterpillars has resulted in outbreaks of scale insects. The latter are protected from the spray, while their beetle predators and other insect enemies are not.

Thus, although rigorous proof that herbivores are generally controlled by predation is lacking, supporting evidence is available, and the alternate hypothesis of control by weather leads to false or untenable implications.

The foregoing conclusion has an important implication in the mechanism of control of the predator populations. The predators and parasites, in controlling the populations of herbivores, must thereby limit their own resources, and as a group they must be food-limited. Although the populations of some carnivores are obviously limited by territoriality, this kind of internal check cannot operate for all carnivores taken together. If it did, the herbivores would normally expand to the point of depletion of the vegetation, as they do in the absence of their normal predators and parasites.

There thus exists either direct proof or a great preponderance of factual evidence that in terrestrial communities decomposers, producers, and predators, as whole trophic levels, are resource-limited in the classical density-dependent fashion. Each of these three can and does expand toward the limit of the appropriate resource. We may now examine the reasons why this is a frequent situation in nature.

Whatever the resource for which a set of terrestrial plant species compete, the competition ultimately expresses itself as competition for space. A community in which this space is frequently emptied through depletion by herbivores would run the continual risk of replacement by another assemblage of species in which the herbivores are held down in numbers by predation below the level at which they damage the vegetation. That space once held by a group of terrestrial plant species is not readily given up is shown by the cases where relict stands exist under climates no longer suitable for their return following deliberate or accidental destruction. Hence, the community in which herbivores are held down in numbers, and in which the producers are resource-limited will be the most persistent. The development of this pattern is less likely where high producer mortalities are inevitable. In lakes, for example, algal populations are prone to crash whether grazed or not. In the same environment, grazing depletion is much more common than in communities where the major producers are rooted plants.

A second general conclusion follows from the resource limitation of the species of three trophic levels. This conclusion is that if more than one species exists in one of these levels, they may avoid competition only if each species is limited by factors completely unutilized by any of the other species. It is a fact, of course, that many species occupy each level in most communities. It is also a fact that they are not sufficiently segregated in their needs to escape competition. Although isolated cases of non-overlap have been described, this has never been observed for an entire assemblage. Therefore, interspecific competition for resources exists among producers, among carnivores, and among decomposers.

It is satisfying to note the number of observations that fall into line with the foregoing deductions. Interspecific competition is a powerful selective force, and we should expect to find evidence of its operation. Moreover, the evidence should be most conclusive in trophic levels where it is necessarily present. Among decomposers we find the most obvious specific mechanisms for reducing populations of competitors. The abundance of antibiotic substances attests to the frequency with which these mechanisms have been developed in the trophic level in which interspecific competition is inevitable. The producer species are the next most likely to reveal evidence of competition, and here we find such phenomena as crowding, shading, and vegetational zonation.

Among the carnivores, however, obvious adaptations for interspecific competition are less common. Active competition in the form of mutual habitat-exclusion has been noted in the cases of flatworms (Beauchamp and Ullyott, 1932) and salamanders (Hairston, 1951). The commonest situation takes the form of niche diversification as the result of interspecific competition. This has been noted in birds (Lack, 1945; MacArthur, 1958), salamanders (Hairston, 1949), and other groups of carnivores. Quite likely, host specificity in parasites and parasitoid insects is at least partly due to the influence of interspecific competition.

Of equal significance is the frequent occurrence among herbivores of apparent exceptions to the influence of density-dependent factors. The grasshoppers described by Birch (1957) and the thrips described by Davidson and Andrewartha (1948) are well known examples. Moreover, it is among herbivores that we find cited examples of coexistence without evidence of competition for resources, such as the leafhoppers reported by Ross (1957), and the psocids described by Broadhead (1958). It should be pointed out that in these latter cases coexistence applies primarily to an identity of food and place, and other aspects of the niches of these organisms are not known to be identical.

SUMMARY

In summary, then, our general conclusions are: (1) Populations of producers, carnivores, and decomposers are limited by their respective resources in the classical density-dependent fashion. (2) Interspecific competition must necessarily exist among the members of each of these three trophic levels. (3) Herbivores are seldom food-limited, appear most often to be predator-limited, and therefore are not likely to compete for common resources.

LITERATURE CITED

Andrewartha, H. G., 1957, The use of conceptual models in population ecology. Cold Spring Harbor Symp. Quant. Biol. 22: 219–232.

Beauchamp, R. S. A., and P. Ullyott, 1932, Competitive relationships between certain species of fresh-water triclads. J. Ecology 20: 200–208.

Birch, L. C., 1957, The role of weather in determining the distribution and abundance of animals. Cold Spring Harbor Symp. Quant. Biol. 22: 217–263.

Broadhead, E., 1958, The psocid fauna of larch trees in northern England. J. Anim. Ecol. 27: 217–263.

Davidson, J., and H. G. Andrewartha, 1948, The influence of rainfall, evaporation and atmospheric temperature on fluctuations in the size of a natural population of *Thrips imaginis* (Thysanoptera). J. Anim. Ecol. 17: 200–222.

Hairston, N. G., 1949, The local distribution and ecology of the Plethodontid salamanders of the southern Appalachians. Ecol. Monog. 19: 47–73.

1951, Interspecies competition and its probable influence upon the vertical distribution of Appalachian salamanders of the genus Plethodon. Ecology 32: 266–274.

Hutchinson, G. E., 1948, Circular causal systems in ecology. Ann. N. Y. Acad. Sci. 50: 221–246.

1957, Concluding remarks. Cold Spring Harbor Symp. Quant. Biol. 22: 415–427.

Lack, D., 1945, The ecology of closely related species with special reference to cormorant (*Phalacrocorax carbo*) and shag (*P. aristotelis*). J. Anim. Ecol. 14: 12–16.

MacArthur, R. H., 1958, Population ecology of some warblers of northeastern coniferous forests. Ecology 39: 599–619.

Milne, A., 1957, Theories of natural control of insect populations. Cold Spring Harbor Symp. Quant. Biol. 22: 253–271.

Nicholson, A. J., 1957, The self-adjustment of populations to change. Cold Spring Harbor Symp. Quant. Biol. 22: 153–172.

Reynoldson, T. B., 1957, Population fluctuations in *Urceolaria mitra* (Peritricha) and *Enchytraeus albidus* (Oligochaeta) and their bearing on regulation. Cold Spring Harbor Symp. Quant. Biol. 22: 313–327.

Ross, H. H., 1957, Principles of natural coexistence indicated by leafhopper populations. Evolution 11: 113–129.

PAPER 18

BUTTERFLIES AND PLANTS: A STUDY IN COEVOLUTION[1]

PAUL R. EHRLICH AND PETER H. RAVEN
Department of Biological Sciences, Stanford University, Stanford, California

Accepted June 15, 1964

One of the least understood aspects of population biology is community evolution—the evolutionary interactions found among different kinds or organisms where exchange of genetic information among the kinds is assumed to be minimal or absent. Studies of community evolution have, in general, tended to be narrow in scope and to ignore the reciprocal aspects of these interactions. Indeed, one group of organisms is all too often viewed as a kind of physical constant. In an extreme example a parasitologist might not consider the evolutionary history and responses of hosts, while a specialist in vertebrates might assume species of vertebrate parasites to be invariate entities. This viewpoint is one factor in the general lack of progress toward the understanding of organic diversification.

One approach to what we would like to call coevolution is the examination of patterns of interaction between two major groups of organisms with a close and evident ecological relationship, such as plants and herbivores. The considerable amount of information available about butterflies and their food plants make them particularly suitable for these investigations. Further, recent detailed investigations have provided a relatively firm basis for statements about the phenetic relationships of the various higher groups of Papilionoidea (Ehrlich, 1958, and unpubl.). It should, however, be remembered that we are considering the butterflies as a model. They are only one of the many groups of herbivorous organisms coevolving with plants. In this paper, we shall investigate the relationship between butterflies and their food plants with the hope of answering the following general questions:

1. Without recourse to long-term experimentation on single systems, what can be learned about the coevolutionary responses of ecologically intimate organisms?

2. Are predictive generalities about community evolution attainable?

3. In the absence of a fossil record can the patterns discovered aid in separating the rate and time components of evolutionary change in either or both groups?

4. Do studies of coevolution provide a reasonable starting point for the understanding of community evolution in general?

FACTORS DETERMINING FOOD CHOICE

Before proceeding to a consideration of the relationships between butterfly groups and their food plants throughout the world, it is necessary briefly to consider some of the factors that determine the choice of food plants in this group and in phytophagous insects in general. Any group of phytophagous animals must draw its food supply from those plants that are available in its geographical and ecological range (Dethier, 1954). For instance, the butterflies are primarily a tropical group, and therefore there is a relatively greater utilization of primarily tropical than of temperate families of plants. The choice of oviposition site by the imago is also important. Many adult butterflies and moths lay their eggs on certain food plants with great precision as stressed by Merz (1959), but on the other hand, numerous "mistakes" have been recorded (e.g., Remington, 1952; Dethier, 1959). In such cases, larvae have either to find an appropriate plant or perish. There is an obvious selective advantage in oviposition on suitable

[1] This work has been supported in part by National Science Foundation Grants GB-123 (Ehrlich) and GB-141 (Raven).

plants, but inappropriate choices can be overcome by movement of the larvae. Furthermore, larvae feeding on herbs often consume the entire plant, and then must move even if the adult originally made an appropriate choice.

Larval choice therefore plays an important role in food plant relationships. An excellent review of a long series of experiments pertinent to this subject has recently been presented by Merz (1959); much of the following is based on his account. The condition of a given larva often has an effect on what foods it will or will not accept. In addition, many structural and mechanical characteristics of plants modify these relationships, mostly by limiting the acceptability of those plants in which they occur. For example, Merz (1959) found that larvae of *Lasiocampa quercus*, a moth that normally feeds along the edge of leaves, could not eat the sharply toothed leaves of holly (*Ilex*, Aquifoliaceae). When these same leaves were cut so that untoothed margins were presented, the larvae ate them voraciously. In other cases, larvae eat the young, soft leaves of plants but not the old, tough leaves of the same plants. Many Lycaenidae feed on flowers, and these butterflies may be unable to utilize the tough foliage of the same plants. Numerous similar examples could be given, but it must be borne in mind that chemical factors are operative in the same plants that present mechanical difficulties to larvae (Thorsteinson, 1960), and actually may be more important.

Chemical factors are of great general importance in determining larval food choice. In the first place, potential food sources are probably all nutritionally unbalanced to some extent (Gordon, 1961). The exploitation of a particular plant as a source of food thus involves metabolic adjustments on the part of an insect. These render the insect relatively inefficient in utilizing other sources of food and tend to restrict its choice of food plants. Secondly, many plants are characterized by the presence of secondary metabolic substances. These substances are repellent to most insects and may often be decisive in patterns of food plant selection (Thorsteinson, 1960). It has further been demonstrated that the chemical compounds that repel most animals can serve as trigger substances that induce the uptake of nutrients by members of certain oligophagous groups (Dethier, 1941, 1954; Thorsteinson, 1953, 1960). Presence of such repellent compounds may be correlated with the presence of the nutrients. Both odor and taste seem to be important.

The chemical composition of plants often changes with age, exposure to sunlight, or other environmental factors (Merz, 1959; Flück, 1963), and this may be critical for phytophagous insects (Dethier, 1954). For example, insects that feed on Umbelliferae prefer the old leaves, which appear to us less odorous than the young ones. Some insects that feed on alkaloid-rich species of *Papaver* (Papaveraceae) prefer the young leaves, which are relatively poor in alkaloids. Diurnal chemical cycles, influenced by exposure of the plant to sunlight, may be of prime importance in determining the habits of night-feeding groups, such as Argynnini.

Merz (1959, p. 159) has given a particularly interesting case of chemical repellents at the specific level. The larvae of the moth *Euchelia jacobaeae* feed on many species of *Senecio* (Compositae), but not on the densely glandular-hairy *S. viscosus*. When the glandular substance was dissolved in methyl alcohol, the larvae ate *S. viscosus*. When the same substance was painted on the leaves of other normally acceptable species of *Senecio*, these were refused. In an extensive study of the food plants of *Plebejus icarioides* (Lycaeninae), Downey (1961, 1962) showed that larvae would feed on any species of *Lupinus* (Leguminosae) in captivity, but populations in the field normally utilized only one or a few of the possible range of *Lupinus* species growing locally. This work suggests the subtle interaction of ecological, chemical, and mechanical factors that doubtless characterizes most natural situations. Relationships with predators (Brower, 1958), parasites (Downey, 1962), or, at least in the case of Lycaenidae, ants (Downey, 1962), may further modify patterns of food plant choice.

Despite all of these modifying factors, there is a general and long-recognized pattern running through the food plants of various groups of butterflies, and it is this pattern with which we shall be concerned. It certainly should not be inferred from anything that follows that all members of a family or genus of plants are equally acceptable to a given butterfly (for example, see Remington, 1952). We have placed our main emphasis on positive records, especially at the level of plant species and genera.

The Diversity of Butterflies

The butterflies comprise a single superfamily of Lepidoptera, the Papilionoidea. In comparison with many other superfamilies of insects they are uniform morphologically and behaviorally. Table 1 gives a rough idea of the taxonomic diversity of this superfamily.

Papilionoidea are divided into five families. Two of these, Nymphalidae and Lycaenidae, contain at least three-quarters of the genera and species; it is uncertain which family is the larger. Two smaller families, Pieridae and Papilionidae, include virtually all remaining butterflies. Pieridae, although containing many fewer genera and species than either Lycaenidae or Nymphalidae, form a prominent part of the butterfly fauna in many parts of the world, making up in number of individuals what they lack in number of kinds. Papilionidae are a group about half the size of Pieridae, but gain prominence through the large size of the included forms. The tiny family Libytheidae, closely related to Nymphalidae, is obscure to everyone except butterfly taxonomists.

The Papilionidae lead the butterflies in morphological diversity. Nymphalidae probably take second place, with Pieridae and Lycaenidae about tied for third. Difficult as this diversity is to estimate, it is clear that Papilionidae are a more heterogeneous group of organisms than any of the other families, whereas, considering the number of species and genera included, Lycaenidae are remarkably uniform. A rough idea of the phenetic relationships of the major groups of butterflies is given by Ehrlich (1958).

With food plant records from between 46 and 60% of all butterfly genera (table 1), it seems highly unlikely that future discoveries will necessitate extensive revisions of the conclusions drawn in this paper. The food plants of Riodininae are very poorly known, however, and it will be interesting to have more information about them and records for other outstanding "unknowns" such as *Styx* and *Pseudopontia*. It is, however, difficult to imagine any additional food plant record that would seriously distort the patterns outlined here.

Butterfly Food Plants

Sources of Information

The food plant information abstracted in this paper is derived principally from two sources. First, we have examined all the extensive and scattered literature that we could uncover with a library search and through the recommendations of various lepidopterists. Particularly helpful have been the volumes of Barrett and Burns (1951), Corbet and Pendlebury (1956), Costa Lima (1936), Ehrlich and Ehrlich (1961), Lee (1958), Seitz (1906–1927), van Son (1949, 1955), Wiltshire (1957), and Wynter-Blyth (1957), as well as the *Journal of the Entomological Society of South Africa*, the *Journal of the Lepidopterists' Society* (formerly *Lepidopterists' News*), and the *Journal of Research on the Lepidoptera*.

Our second major source of information has been provided by the following scientists, who have aided us in this ambitious undertaking not only by sending unpublished data and reprints of their works, but by helping to evaluate the validity of cer-

TABLE 1. *Summary of the taxonomic diversity of Papilionoidea*

Taxon	Approximate number of		Distribution
	Genera*	Species	
Papilionidae	24(22)	575–700	
Baroniinae	1(1)	1	Central Mexico
Parnassiinae	8(8)	45–55	Holarctic and Oriental; greatest diversity, Asia
Papilioninae	15(13)	480–640	Worldwide; mainly tropical. Greatest diversity, Old World tropics
Pieridae	58(40)	950–1,150	
Coliadinae	11(8)	225–250	Cosmopolitan; greatest diversity tropics outside of Africa
Pierinae	43(29)	650–750	Cosmopolitan; greatest diversity tropics
Dismorphiinae	3(3)	80–120	Primarily Neotropical; one small Palearctic genus
Pseudopontiinae	1(0)	1	West equatorial Africa
Nymphalidae	325–400 (ca. 202)	4,800–6,200	
Ithomiinae	30–40(10)	300–400	Neotropical; *Tellervo* Australian
Danainae	10–12(10)	140–200	Cosmopolitan; greatest diversity Old World tropics
Satyrinae	120–150 (ca. 70)	1,200–1,500	Cosmopolitan; greatest diversity extratropical
Morphinae	23–26(12)	180–250	Indomalayan and Neotropical
Charaxinae	8–10(8)	300–400	Tropicopolitan, few temperate
Calinaginae	1(1)	1	Oriental
Nymphalinae	125–150 (ca. 85)	2,500–3,000	Cosmopolitan
Acraeinae	8(6)	225–275	Tropical; greatest diversity, Africa
Libytheidae	1(1)	10	Cosmopolitan
Lycaenidae	325–425 (ca. 167)	5,800–7,200	
Riodininae	75–125(17)	800–1,200	Tropical, few Nearctic and Palearctic. Metropolis, Neotropical
Styginae	1(0)	1	Peruvian Andes
Lycaeninae	250–300 (ca. 150)	5,000–6,000	Cosmopolitan; greatest diversity, Old World tropics
Total	730–930 (ca. 432)	12,000–15,000	

* Number in parentheses indicates number of genera for which food plant records are available.

tain published records and commenting on other aspects of the work. The cooperation of these people has been truly extraordinary, and we are particularly indebted to them: Remauldo F. d'Almeida (Brazil), Peter Bellinger (USA), C. M. de Biezanko (Brazil), L. P. Brower (USA), C. A. Clarke (England), H. K. Clench (USA), J. A. Comstock (USA), C. G. C. Dickson (Africa), J. C. Downey (USA), Maria Etcheverry (Chile), K. J. Hayward (Argentina), T. G. Howarth (England), Taro Iwase (Japan), T. W. Langer (Denmark), C. D. MacNeill (USA), D. P. Murray (England), G. Sevastopulo (Africa), Takashi Shirozu (Japan), E. M. Shull (India), Henri Stempffer (France), V. G. L. van Someren (Africa), G. van Son (Africa).

John C. Downey (Southern Illinois University), Gordon H. Orians (University of Washington), T. A. Geissman (University of California, Los Angeles), and Robert F. Thorne (Rancho Santa Ana Botanic Garden) have been so kind as to read the entire manuscript. Their advice has been invaluable.

To our knowledge, the data assembled here represent the most extensive body of information ever assembled on the interactions between a major group of herbivorous animals and their food plants.

Evaluation of the Literature

Extreme care has been taken in associating insects with particular food plants, as the literature is replete with errors and unverified records. In evaluating records, preference has been given to those which are concerned with the entire life cycle of a particular insect on a wild plant. Laboratory experiments and records from cultivated plants demonstrate only potentialities, not necessarily natural associations. In the laboratory, larvae may be starved or plants abnormal. In the wild, larvae are often misidentified, especially if not reared to maturity (cf. Brower, 1958b). Even more serious is the lack of precise plant identifications, or their identification in the vernacular only, which almost inevitably leads to confusion (Jörgensen, 1932). Any serious student of phytophagous animals should preserve adequate herbarium specimens of the plants with which he is concerned (cf. Remington, 1952, p. 62); only by doing this can the records be verified. Despite the extremely erratic oviposition behavior often shown by butterflies (Dethier, 1959), oviposition records have all too frequently been accepted as being equivalent to food plant records. Finally, nomenclatural difficulties, including changes in name and careless misspellings (e.g., "Oleaceae" for Olacaceae), have given rise to serious errors. In the literature on butterfly food plants, errors have often been compounded when copied from one source to another, and they are difficult to trace back to their origins. All of these problems make quantitative comparisons unreasonable. We therefore have been exceedingly conservative about accepting records, and focused our attention primarily on broad, repeatedly verified patterns of relationship.

The Food Plants of Butterflies

In this section, we will first outline the main patterns of food plant choice for each family, and then discuss what bearing these patterns have on our interpretation of relationships within the various butterfly families. It is necessary to give the data in considerable detail, as no comprehensive survey on a world basis is available elsewhere.

Papilionidae.—There are three subfamilies. *Baronia brevicornis*, the only species of Baroniinae, occurs in Mexico and feeds on *Acacia* (Leguminosae; Vazquez and Perez, 1961). In Parnassiinae, all five genera of Zerynthiini (Munroe, 1960; Munroe and Ehrlich, 1960) feed on Aristolochiaceae, as does *Archon* (Parnassiini). *Hypermnestra* (Parnassiini) is recorded from *Zygophyllum* (Zygophyllaceae). *Parnassius* feeds on Crassulaceae and herbaceous Saxifragaceae, two closely related families, with one small group on Fumariaceae. In view of the discussion below, it is of interest that Zygophyllaceae are close relatives of Rutaceae, and Fumariaceae are rich in alkaloids similar to those of woody Ranales (Hegnauer, 1963).

The third and last subfamily, Papilioninae, is cosmopolitan but best developed in the Old World, and consists of three tribes: Troidini, Graphiini, and Papilionini. The eight genera of Troidini feed mostly on Aristolochiaceae, with individual species of *Parides* also recorded from Rutaceae, Menispermaceae, Nepenthaceae, and Piperaceae. *Parides* (*Atrophaneura*) *daemonius* is reported to feed on *Osteomeles* (Rosaceae), and *P.* (*A.*) *antenor*, a Malagasy butterfly that is the only representative of its tribe in the Ethiopian region, feeds on *Combretum* (Combretaceae). Both of these last-mentioned records need confirmation. At least two species of *Battus* have been recorded from Rutaceae in addition to the usual Aristolochiaceae. Records available for five of the seven genera of Graphiini

(*Eurytides, Graphium, Lamproptera, Protographium, Teinopalpus*) are mostly from Annonaceae, Hernandiaceae, Lauraceae, Magnoliaceae, and Winteraceae. This is clearly a closely allied group of plant families referable to the woody Ranales. In addition, some species of *Graphium* feed on Rutaceae, and others both on Apocynaceae (*Landolphia*) and Annonaceae (one of the latter also on *Sphedamnocarpus*, Malpighiaceae). Several species of *Eurytides* feed on *Vitex* (Verbenaceae) and one on *Jacobinia* (Acanthaceae). *Eurytides lysithous* feeds both on Annonaceae and on *Jacobinia*, and *E. helios* both on *Vitex* and Magnoliaceae. The bitypic Palearctic *Iphiclides* departs from the usual pattern for the group in feeding on a number of Rosaceae–Pomoideae.

The third and last tribe, Papilionini, consists only of the enormous cosmopolitan genus *Papilio*. Two of the five sections recognized by Munroe (1960; II, IV) are primarily on Rutaceae, with occasional records from Canellaceae, Lauraceae, and Piperaceae. Members of the circumboreal *Papilio machaon* group are not only on Rutaceae but also on Umbelliferae and *Artemisia* (Compositae). The African *P. demodocus*, in another group, is known to feed on Rutaceae and Umbelliferae, as well as *Pseudospondias* (Anacardiaceae), *Ptaeroxylon* (Meliaceae), and *Hippobromus* (Sapindaceae). Another African species, *P. dardanus*, is recorded from Rutaceae and also from *Xymalos* (Flacourtiaceae; Dickson, pers. comm.). The Asian and Australian *P. demoleus* is mostly on Rutaceae but also locally on *Salvia* (Labiatae) and *Psoralea* (Leguminosae). The other three sections (Munroe's I, III, V) are primarily associated with Annonaceae, Canellaceae, Hernandiaceae, Lauraceae, and Magnoliaceae, with a few records from Berberidaceae, Malvaceae (*Thespea*), and Rutaceae. The North American temperate *Papilio glaucus* group (sect. II) feeds not only on Lauraceae and Magnoliaceae like its more southern relatives but also on Aceraceae, Betulaceae, Oleaceae, Platanaceae, Rhamnaceae, Rosaceae, Rutaceae (*Ptelea*), and Salicaceae (Brower, 1958b).

In Papilionidae, Munroe and Ehrlich (1960) have argued that the red-tuberculate, Aristolochiaceae-feeding larvae of Papilioninae–Troidini and Parnassiinae–Zerynthiini, plus *Archon* (Parnassiinae–Parnassini) are so similar, and the likelihood of their converging on Aristolochiaceae so remote, that these probably represent the remnants of the stock from which the rest of Papilioninae and Parnassiinae were derived. Viewed in this context other food plants of these groups are secondary. The two remaining tribes of Papilioninae (Papilionini, Graphiini) are above all associated with the group of dicotyledons known as the woody Ranales. This is a diverse assemblage of plant families showing many unspecialized characteristics. Thorne (1963) has used the food plant relationships of Papilionidae as a whole to support his suggestion of affinity between Aristolochiaceae and Annonaceae (one of the woody Ranales). He appears to have established the existence of this similarity on morphological evidence. Likewise, similar alkaloids are shared by Aristolochiaceae and woody Ranales (Hegnauer, 1963; Alston and Turner, 1963, p. 170). Recently, Vazquez and Perez (1961) described the life cycle of *Baronia brevicornis*, the only member of the third subfamily of the group. *Baronia* feeds on *Acacia* (Leguminosae) and has tuberculate larvae like those of the forms that feed on Aristolochiaceae. Considering its morphological distinctness, and in accordance with the scheme of relationships presented by Munroe and Ehrlich (1960, p. 175), it appears likely that *Baronia* represents a phylogenetic line which diverged early from that leading to the rest of Papilionidae. It may thus be the only member of the family neither feeding on Aristolochiaceae nor descended from forms that did.

Following this reasoning, we suggest that the original transition to Aristolochiaceae opened a new adaptive zone for Papilionidae. Their further spread and the multiplication of species was accompanied by the exploitation of other presumably chemically similar plant groups, such as woody Ranales, in areas where Aristolochiaceae were poorly represented, like Africa today. The site of greatest diversity for both Aristolochiaceae and Papilionidae is Asia. It is likely that the major diversification of Papilionidae (involving differentiation into Parnassiinae and Papilioninae) took place after the evolution of Aristolochiaceae. When this might have been is entirely uncertain, despite the unfounded speculations of Forbes (1958).

Another interesting problem is posed by the many representatives of Papilionini and Graphiini feeding on Rutaceae, in addition to woody Ranales. Rutaceae are morphologically very different from woody Ranales, and have not been closely associated with them taxonomically. Recently, however, Hegnauer (1963) has pointed out that some Rutaceae possess the alkaloids widespread in woody Ranales, in addition to an unusually rich repertoire of other alkaloids. Earlier, Dethier (1941) showed the similarity between the attractant essential oils in Rutaceae and Umbelliferae. Some Rutaceae-feeding groups of *Papilio* seem to have shifted to Umbelliferae, especially outside of the tropics. Dethier also implicated some of the similar-scented species of *Artemisia* (Compositae), another plant group fed on by at least one species of *Papilio*. Although species of *Papilio* link these groups of plants, none is known to feed on Burseraceae, Cneoraceae, Simarubaceae, or Zygophyllaceae, families thought to be related to Rutaceae but not known to contain alkaloids or coumarins (Price, 1963). On the other hand, the record of *Papilio demodocus* on *Ptaeroxylon* (Meliaceae), in addition to numerous Rutaceae, would seem to indicate a promising plant to investigate for the alkaloids suspected (Price, 1963) in Meliaceae.

Pieridae.—Our discussion of Pieridae is based taxonomically on the generic review of Klots (1933) as modified by Ehrlich (1958). There are four subfamilies, but nothing is known of the biology of the monobasic West African Pseudopontiinae. Of the remaining three, Dismorphiinae, including the Neotropical *Dismorphia* and *Pseudopieris* and the Palearctic *Leptidia*, are recorded only on Leguminosae. Larval food plants are known for 7 of the 11 genera of Coliadinae. *Catopsilia, Phoebis, Anteos, Eurema,* and *Colias* are mostly associated with Leguminosae, but there are a few records from Sapindaceae, Guttiferae, Euphorbiaceae, Simarubaceae, Oxalidaceae, Salicaceae, Ericaceae, and Gentianaceae (the last three with northern and montane species of *Colias*). On the other hand, three genera are associated with non-leguminous plants: *Gonepteryx* with *Rhamnus* (Rhamnaceae), *Nathalis* with Compositae, and *Kricogonia* with *Guaiacum* (Zygophyllaceae). Nonetheless, Leguminosae are decidedly the most important food plants of Coliadinae.

Pierinae, the third subfamily, are divided into two tribes, Pierini (36 genera) and Euchloini (7). In Euchloini, the temperate *Anthocharis, Euchloë, Zegris,* and *Hesperocharis* (also from *Phrygilanthus*, Loranthaceae) feed on Cruciferae, the tropical *Pinacopteryx* and *Hebemoia* on Capparidaceae. For Pierini, 14 of the 23 genera for which we know something of the food plants (namely, *Appias, Ascia, Belenois, Ceporis, Colotis, Dixeia, Elodina, Eronia, Ixias, Leptosia, Pareronia, Pieris, Prioneris*, and *Tatochila*) are primarily on Capparidaceae in the tropics and subtropics and on Cruciferae in temperate regions. Some have occasionally been reported to feed on Resedaceae, Salvadoraceae, and Tropaeolaceae. There are also a very few scattered records from other plants, including one or two from Leguminosae. The basis for selecting Capparidaceae, Cruciferae, Resedaceae, Salvadoraceae, and Tropaeolaceae is relatively easy to comprehend, since all of these plants are known to contain mustard oil glucosides (thioglucosides) and the associated enzyme myrosinase which acts in the hydrolysis of glucosides to release mustard oils (Alston and Turner,

1963, p. 284–288). In an early series of food choice experiments, Verschaeffelt (1910) found that larvae of *Pieris rapae* and *P. brassicae* would feed on Capparidaceae, Cruciferae, Resedaceae, and Tropaeolaceae, as well as another family which contains mustard oils but upon which Pierinae are not known to feed in nature: Moringaceae. Verschaeffelt also found that these larvae would eat flour, starch, or even filter paper if it was smeared with juice expressed from *Bunias* (Cruciferae), and Thorsteinson (1953, 1960) showed that the larvae would eat other kinds of leaves treated with sinigrin or sinalbin (two common mustard oil glucosides) if the leaves were not too tough and did not contain other kinds of repellents. Very few butterflies outside Pierdinae feed on these plants, but there is one example in Lycaeninae. In addition, there are at least two records of *Phoebis* (Coliadinae) from Capparidaceae and Cruciferae. Numerous groups of insects other than butterflies are characteristically associated with this same series of plant families (Fraenkel, 1959).

It is not so easy to interpret scattered records of these pierine genera feeding on other plant families: *Belenois raffrayi* on *Rhus* (Anacardiaceae); *Nepheronia argyia* on Capparidaceae, but also *Cassipourea* (Rhizophoraceae) and *Hippocratea* (Hippocrateaceae), with *N. thalassina* reported only from *Hippocratea*; *Ascia monuste* on Rhamnaceae and *Cassia* (Leguminosae), as well as Capparidaceae; and *Tatochila autodice* on *Cestrum* (Solanaceae) and also *Medicago* (Leguminosae). Several species of *Appias* have been reported from different genera of Euphorbiaceae, whereas others feed both on Capparidaceae and Euphorbiaceae, but this probably can be explained somewhat more simply, since mustard oils have been reported in some genera of Euphorbiaceae (Alston and Turner, 1963, p. 285).

The remaining genera of Pierini fall mostly into what has been called the *Delias* group. Of these, *Catasticta* and *Archonias* have been recorded from *Phrygilanthus* (Loranthaceae) in South America, and *Delias*, a large Indo-Malaysian genus, from "*Loranthus*" (Loranthaceae) and *Exocarpus* of the closely related Santalaceae, with *D. aglaja* on *Nauclea* (Rubiaceae). *Aporia*, a large genus of temperate regions of the Old World, has several species on *Berberis* (Berberidaceae), and one on woody Rosaceae. *Pereute*, South American, feeds on *Ocotea* (Lauraceae; Jörgensen, 1932), Tiliaceae, and Polygonaceae. Two very peculiar genera of the *Delias* group are the monotypic Mexican *Eucheira*, which feeds on woody hard-leaved Ericaceae, and the bitypic western North American *Neophasia*, which feeds on various genera of Pinaceae. It is very interesting that *Cepora*, which falls into the *Delias* group morphologically, feeds on *Capparis* like many other Pierini.

Finally, the large, taxonomically isolated, Ethiopian *Mylothris* feeds on Loranthaceae and Santalaceae (*Osyris*), with *M. bernice rubricosta* on *Polygonum* (Polygonaceae).

It is difficult to understand the reasons for large groups of Pierinae being associated both with plants that possess mustard oils and with Loranthaceae–Santalaceae; neither morphological nor biochemical evidence has been adduced to link these two groups of plants. Perhaps the Loranthaceae-feeders represent an old offshoot of Pierinae; in any case it would appear that the main diversification of this group occurred after it became associated with Capparidaceae–Cruciferae.

Nymphalidae.—This enormous family is divided into eight subfamilies which will be discussed one by one in the succeeding paragraphs.

Ithomiinae are primarily American, and there feed only on Solanaceae (many genera). The Indo-Malaysian *Tellervo*, only Old World representative of the group, which is segregated as a distinct tribe Tellervini, has been recorded from *Aristolochia* (Aristolochiaceae). The identity of the plant was inferred from the fact that papilionid larvae normally associated with *Aristolochia* were found on it with *Tellervo*. Solanaceae are rich in alkaloids (as are

Aristolochiaceae), and are very poorly represented among the food plants of butterflies as a whole. The diversification of the ithomiines that feed on them has probably followed a pattern similar to that of Papilionidae on Aristolochiaceae and Pierinae on Capparidaceae–Cruciferae. Many other groups of insects feed primarily on Solanaceae (Fraenkel, 1959).

Danainae are a rather uniform cosmopolitan group, obviously related to Ithomiinae. The danaines feed primarily and apparently interchangeably on Apocynaceae and Asclepiadaceae. In addition, there are records of *Euploea, Ituna,* and *Lycorella* on Moraceae and of the last occasionally on *Carica* (Caricaceae). All of these plants have milky juice. There is also a single record of *Ituna ilione,* which normally feeds on *Ficus* (Moraceae), from *Myoporum* (Myoporaceae). Apocynaceae and Asclepiadaceae form a virtual continuum in their pattern of variation and can scarcely be maintained as distinct (Safwat, 1962). Both are noted for their abundant bitter glycosides and alkaloids (Alston and Turner, 1963, p. 258), and share at least some alkaloids (Price, 1963, p. 431) and pyridines with Moraceae. Thus it appears very likely that here too the acquisition of the ability to feed on Apocynaceae and Asclepiadaceae has constituted for Danainae the penetration of a new adaptive zone, in which they have radiated. Numerous distinctive segments of other insect orders and groups likewise feed on these two plant families.

Eleven genera of Morphinae are recorded from a variety of monocotyledons: Bromeliaceae, Gramineae (mostly bamboos), Marantaceae, Musaceae, Palmae, Pandanaceae, and Zingiberaceae. In contrast, most species of *Morpho* feed on dicotyledons, including Canellaceae, Erythroxylaceae, Lauraceae, Leguminosae, Menispermaceae, Myrtaceae, Rhamnaceae, and Sapindaceae, but *M. aega* feeds on bamboos (Gramineae) and *M. hercules* on Musaceae. Whether the progenitors of *Morpho* fed on dicotyledons or monocotyledons cannot be determined.

Closely related to Morphinae are the more temperate Satyrinae, an enormous group that feeds mostly on Gramineae (including bamboos and canes) and Cyperaceae, occasionally on Juncaceae. *Pseudonympha vigilans* feeds on *Restio* (Restionaceae), a family close to Gramineae, *Physcaneura pione* on Zingiberaceae, and *Elymnias* on Palmae. There are no records of this group from dicotyledons. Thus the phenetically similar Morphinae–Satyrinae assemblage is the outstanding example in butterflies of a group associated primarily with monocotyledons.

The distinctive tropicopolitan Charaxinae are often associated with woody Ranales (Annonaceae, Lauraceae, Monimiaceae, Piperaceae), but also with such diverse families as Anacardiaceae, Araliaceae (*Schefflera*), Bombacaceae, Celastraceae, Connaraceae, Convolvulaceae, Euphorbiaceae, Flacourtiaceae, Hippocrateaceae, Leguminosae, Linaceae, Malvaceae, Meliaceae, Melianthaceae, Myrtaceae, Proteaceae, Rhamnaceae, Rutaceae, Salvadoraceae (*Charaxes hansali*), Sapindaceae, Sterculiaceae, Tiliaceae, Ulmaceae, and Verbenaceae. Records (largely Sevastopulo and van Someren, pers. comm.) are available for about 50 African species of *Charaxes*, most of which are associated with dicotyledons. At least three feed on grasses (Gramineae), two of these on dicotyledons also.

The Oriental *Calinaga buddha*, the only species of Calinaginae, feeds on *Morus* (Moraceae).

Nymphalinae are a huge cosmopolitan group with relatively few "gaps" in their pattern of variation which would permit the recognition of meaningful subgroups (cf. Reuter, 1896; Chermock, 1950). The tribes do, however, display some significant patterns in their choice of food plants, with Heliconiini (Michener, 1942) and Argynnini feeding mostly on the Passifloraceae–Flacourtiaceae–Violaceae–Turneraceae complex of families, a closely related group of plants also important for Acraeinae. Acraeinae (see below), Heliconiini, and Argynnini are closely related phenetically, and their diversification may have taken place from a common ancestor associated with this particular assemblage of plants. No biochemical basis is known for the association of this series of four plant families, but we confidently predict that one eventually will be found (cf. also Gibbs, 1963, p. 63). Melitaeini are often associated with Acanthaceae, Scrophulariaceae and their wind-pollinated derivatives Plantaginaceae, and with Compositae and Verbenaceae. Nymphalini feed on plants of the same families as Melitaeini, but also very prominently on the Ulmaceae–Urticaceae–Moraceae group and the Convolvulaceae, Labiatae, Portulacaceae, and Verbenaceae. A single species in this group, however, *Nymphalis canace*, feeds on Liliaceae and Dioscoreaceae. Apaturini are associated chiefly with Ulmaceae, especially *Celtis*. Cyrestini (*Chersonesia, Cyrestis,* and *Marpesia*) and Gynaeciini (*Gynaecia* and *Historis,* but not *Callizona* and *Smyrna*) are often associated with Moraceae, and Hamadryini (*Ectima, Hamadryas*), Didonini (*Didonis*), Ergolini (*Byblia, Byblis, Ergolis, Eurytela, Mestra*), Eunicini (*Asterope, Catonephele, Eunica, Myscelia*), and Dynamini (*Dynamine*) mostly on the related Euphorbiaceae, which, like Moraceae, have milky sap. In addition to the Eunicini just mentioned, *Diaethria* (*Callicore, Catagramma*), *Epiphile, Haematera, Pyrrhogyra,* and *Temenis* feed almost exclusively on Sapindaceae. There is no obvious dominant theme for the last tribe, Limenitini, but it is interesting to note that two species of *Euphaedra* (*Najas*), a group that is mostly on Sapindaceae, are on *Cocos* and other Palmae. Additional families represented among the food plants of Nymphalinae are: Aceraceae, Amaranthaceae, Anacardiaceae, Annonaceae, Berberidaceae, Betulaceae, Bignoniaceae, Bombacaceae, Boraginaceae, Caprifoliaceae, Combretaceae, Corylaceae, Crassulaceae, Curcurbitaceae, Dilleniaceae, Dipterocarpaceae, Ebenaceae, Eleagnaceae, Ericaceae, Fagaceae, Gentianaceae, Geraniaceae, Guttiferae, Icacinaceae, Leguminosae (very uncommonly), Loranthaceae, Malvaceae, Melastomaceae, Melianthaceae, Menispermaceae, Myrtaceae, Oleaceae, Ranunculaceae (*Vanessa* on *Delphinium*), Rhamnaceae, Rosaceae, Rubiaceae, Sabiaceae, Salicaceae, Sapotaceae, Saxifragaceae, Sterculiaceae, Thymeleaceae, Tiliaceae, and Vitaceae.

Some of the butterflies in this group feed on a very wide range of plants, and most of the families mentioned in the above list are represented by one or at most a very few records. For example, *Euptoieta claudia* is known to feed on Berberidaceae (*Podophyllum*), Crassulaceae, Leguminosae, Linaceae, Menispermaceae, Nyctaginaceae, Passifloraceae, Portulacaceae, Violaceae, and even Asclepiadaceae (*Cyanchum*), and *Precis lavinia* is recorded from, among others, Acanthaceae, Bignoniaceae, Compositae, Crassulaceae, Onagraceae (*Ludwigia*), Plantaginaceae, Scrophulariaceae, and Verbenaceae.

Acraeinae, a rather small tropical group, are often associated with Passifloraceae–Flacourtiaceae–Violaceae–Turneraceae, as noted above, but also with Amaranthaceae, Compositae, Convolvulaceae, Leguminosae, Lythraceae, Moraceae, Polygonaceae, Rosaceae, Sterculiaceae, Urticaceae, and Vitaceae. In addition, *Acraea encedon* is reported from *Commelina* (Commelinaceae).

For the very diverse Nymphalidae as a whole, the following groups of plants are especially important: (1) Passifloraceae–Flacourtiaceae–Violaceae–Turneraceae; (2) Ulmaceae–Urticaceae–Moraceae, as well as the closely related (Thorne, pers. comm.) Euphorbiaceae; (3) Acanthaceae–Scrophulariaceae–Plantaginaceae. The second and third of these groups are represented among the food plants of other butterflies, such as Lycaeninae, but not abundantly. Conversely, as will be seen, the groups of food plants commonly represented in Lycaenidae—for example, Fagaceae, Leguminosae, Oleaceae, Rosaceae—are rare in Nymphalidae. Although each of these two families of butterflies is very wide in its choice of food plants, there is a distinctiveness to the two patterns which suggests a history of selection along different lines.

Libytheidae.—This small family consists of a single widespread genus, *Libythea,* which feeds almost exclusively on *Celtis* (Ulmaceae), but in southern Japan on *Prunus* (Rosaceae). *Libythea* is obviously closely related to Nymphalidae (Ehrlich, 1958), as has recently been confirmed by a quantitative study of adult internal anatomy (Ehrlich, unpubl.).

Lycaenidae.—An enormous group, the family Lycaenidae may be larger even than the Nymphalidae. Lycaenidae are in general poorly known from the standpoint of food plants (Downey, 1962). Our discussion is based largely on the classification of Clench (1955 and pers. comm.). Nothing is known of the life history of the Peruvian *Styx infernalis,* only member of Styginae. Of the two remaining subfamilies, Riodininae, divided into three tribes, will be discussed first. *Euselasia* (Euselasiini) has been recorded from *Mammea* (Guttiferae) and three genera of Myrtaceae. The Old World Hamearini consist of three genera, with *Dodona* and *Zemeros* on *Maesa* (Myrsinaceae) and *Hamearis* on *Primula* (Primulaceae). The two plant families are very closely related, with Myrsinaceae being primarily tropical and woody, Primulaceae primarily temperate and herbaceous. The third and largest tribe, Riodinini, is divided into four subtribes. *Abisara* (Abisariti) feeds, like the Hamearini, on Myrsinaceae; *Theope* (Theopiti) is on *Theobroma* (Sterculiaceae); and *Helicopus* (Helicopiti) is one of two members of the subfamily known to feed on a monocotyledon, in this case *Montrichardia* (Araceae). The remaining genera of Riodinini are in the exclusively New World Riodiniti, with very few records for a great many species. Plant families represented are Acanthaceae, Anacardiaceae, Aquifoliaceae, Chenopodiaceae, Compositae, Euphorbiaceae, Leguminosae, Moraceae, Myrtaceae, Polygonaceae, Ranunculaceae (*Clematis*), Rosaceae, Rutaceae, Sapindaceae, and Sapotaceae. Deserving special mention are the records of *Cariomathus* and *Rhetus* from Loranthaceae, *Napaea nepos* from *Oncidium* (Orchidaceae), and *Stalachtis* from *Oxpetalum* (Asclepiadaceae; cf. Jörgensen, 1932, p. 43, however, where it is suggested that associations of larvae of this group with ants may determine the food plant on which they are found). The scanty food plant records for this group are thus sufficiently diverse to suggest that further studies of food plants will be of considerable interest. The most salient feature is the occurrence of Hameaerini and Abisariti on Myrsinaceae and Primulaceae, two closely related families that are fed on by very few other butterflies.

Lycaeninae likewise consist of three tribes. Of these, Leptinini are African and feed on lichens, some of them (*Durbania, Durbaniopsis,* and *Durbaniella*) even on the low crustose lichens that grow on rocks. Liphyrini, almost entirely confined to the Old World tropics, are predaceous on aphids, coccids, ant larvae, membracids, and jassids. There are no reliable records of phytophagy in this group.

The largest of the three tribes, Lycaenini, presents a bewildering array of forms that can be separated only informally at present. Many of these larvae are closely associated with and tended by ants, and this association may modify their food plant relationships (Downey, 1962; Stempffer, pers. comm.). For the large *Plebejus* group (the "blues"), we have records of the food plants of 45 genera, and 33 of these are known to feed, at least in part, on Leguminosae. Records of special interest in this group include *Nacaduba* on several genera of Myrsinaceae and *Agriades* on Primulaceae; in this way they are like Hamaerini and Abisariti of Riodininae. *Chilades* and *Neopithecops* are recorded from Rutaceae. Four genera (*Philotes, Scolitantides, Talicada,* and *Tongeia*) are known to feed, at least in part, on Crassulaceae. *Catachrysops pandava* feeds not only on *Wagatea* and *Xylia* (Leguminosae) but also on *Cycas revoluta* (Cycadaceae), a cycad to which it does harm in gardens. *Hemiargus ceraunus* feeds on Marantaceae. Although most species of *Jamides* feed on Leguminosae, *J. alecto* feeds on Zingiberaceae (a monocotyledon).

In the Strymon group (Clench *in* Ehrlich and Ehrlich, 1961, plus *Strymonidia*), there is no obvious pattern, but there are several records of interest: *Dolymorpha* on *Solanum* (Solanaceae; Clench, unpubl.); *Eumaeus*, with *E. debora* on both *Dioon edule* (Cycadaceae) and *Amaryllis* (Liliaceae) and *E. atala* on both *Manihot* (Euphorbiaceae; Comstock, unpubl.) and on *Zamia integrifolia* (Cycadaceae); *Strymon melinus*, which feeds on a variety of dicotyledonous plants, but also on the flowers of *Nolina* (Liliaceae); and *Tmolus echion*, which feeds not only on *Lantana* (Verbenaceae), *Cordia* (Boraginaceae), *Datura* and *Solanum* (Solanaceae), *Hyptis* (Labiatae), and *Mangifera* (Anacardiaceae), but also on *Ananas* (Bromeliaceae). The impressive pattern of food plant radiation among the four subgenera of *Callophrys* deserves special mention, for subg. *Callophrys* and *Incisalia* feed mostly on angiosperms—Leguminosae, Polygonaceae, Rosaceae, and Ericaceae—but three species of *Incisalia* have switched to conifers, feeding on *Picea* and *Pinus* (Pinaceae). A third subgenus, *Mitoura*, feeds primarily on another group of conifers, Cupressaceae, with two species surprisingly on the pine mistletoes, *Arceuthobium* (Loranthaceae). Finally, *Callophrys* (*Sandia*) *macfarlandi*, the only species of its group, feeds on the flowers of *Nolina* (Liliaceae) in the southwestern United States.

Lycaena and *Heliophorus*, closely related, feed primarily on Polygonaceae throughout the nearly cosmopolitan but largely extratropical range of both groups.

The theclines, narrowly defined (Shirôzu and Yamamoto, 1956), have recently been treated by Shirôzu (1962), who has demonstrated that Fagaceae are the most important food plants, with a number of genera associated with Oleaceae. One genus (*Shirozua*) has become predaceous on aphids.

Among the remaining genera of Lycaeninae, a few points are especially noteworthy. The morphologically diverse South American group referred to "*Thecla*" is also extraordinarily diverse in its choice of food plants: Bromeliaceae, Celastraceae, Compositae, Euphorbiaceae, Leguminosae, Liliaceae, Malpighiaceae, Malvaceae, Sapotaceae, Solanaceae, and Ulmaceae. In addition to *Callophrys*, already mentioned, many distinctive and in some cases large genera feed primarily on Loranthaceae and the closely related Santalaceae: *Charana*, *Deudorix*, *Hypochrysops*, *Iolaus* s. str. (also often on *Ximenia*, Olacaceae), *Ogyris*, *Pretapa*, *Pseudodipsas*, *Rathinda*, and *Zesius*. It would appear that the epiphytic mistletoes and their relatives have constituted an important adaptive zone for a number of genera of Lycaeninae (as suggested by Clench, pers. comm.). Some species of *Iolaus* are on *Colocasia* (Araceae). Olacaceae, Loranthaceae, and Santalaceae are presumably closely related (Hutchinson, 1959), and interestingly share some acetylinic fatty acids (Sørensen, 1963) and lipids (Shorland, 1963). *Chliaria* feeds on the buds and flowers of a number of genera of Orchidaceae, and *Eooxylides*, *Loxura*, and *Yasoda* feed on *Smilax*, a hard-leaved member of Liliaceae, and the superficially similar *Dioscorea* (Dioscoreaceae). *Artipe* lives inside the fruits of *Punica* (Punicaceae), and *Bindahara* inside the fruits of *Salacia* (Celastraceae). Finally, *Aphnaeus* inhabits galleries hollowed out by ants in the twigs of *Acacia* (Leguminosae), where it feeds on fungi (van Son, pers. comm.)!

In summary, the plant families that are best represented among the food plants of Lycaenini are Ericaceae, Labiatae, Polygonaceae, Rhamnaceae, and Rosaceae. Other records from families not hitherto mentioned are: Aizoaceae, Amaranthaceae, Araliaceae, Betulaceae, Boraginaceae, Bruniaceae, Burseraceae, Caprifoliaceae, Caryophyllaceae, Chenopodiaceae, Cistaceae, Combretaceae, Convolvulaceae, Coriaraceae, Cornaceae, Diapensiaceae, Dipterocarpaceae, Ebenaceae, Eleagnaceae, Gentianaceae, Geraniaceae, Hamamelidaceae, Juglandaceae, Lauraceae, Lecithydaceae, Lythraceae, Meliaceae, Melianthaceae, Myricaceae, Oxalidaceae, Pittosporaceae, Plantaginaceae, Plumbaginaceae, Proteaceae, Rubiaceae, Saxifragaceae, Sterculiaceae, Styracaceae, Symplocaceae, Theaceae, Thymeleaceae, and Zygophyllaceae. As before it must be borne in mind that many of these listings represent single records only; for example, the widespread Holarctic *Celastrina argiolus* has been recorded from food plants belonging to at least 14 families of dicotyledons. Nonetheless, it should be evident that the pattern is very different from that of Nymphalinae, the only subfamily comparable to Lycaeninae in size.

Discussion

What generalities can be drawn from these observed patterns? We shall approach this question from the standpoint of the utilization of different plant groups by butterflies and see what light this throws on patterns of evolution in the two groups. Butterflies, of course, are only one of many phytophagous groups of organisms affecting plant evolution.

Within the appropriate ecological framework, our view of the immediate potentialities of studies of phytophagy has been stated clearly and succinctly by Bourgogne (1951, p. 330), who, speaking of the patterns of food plant choice in Lepidoptera, said: "Ces anomalies apparentes peuvent parfois démontrer l'existence, entre deux végétaux, d'une affinité d'ordre chimique, quelquefois même d'une parenté systématique. . . ." Thus, the choices exercised by phytophagous organisms may provide approximate but nevertheless useful indications of biochemical similarities among groups of plants. These do *not* necessarily indicate the plants' overall phenetic or phylogenetic relationships. The same can be said of the choice of arrow poisons by primitive human groups (Alston and Turner, 1963, p. 293) and of patterns of parasitism by fungi (Saville, 1954). In many of these cases, biochemists have not yet worked out the bases for the observed patterns, but as Merz (1959, p. 181) points out, we should nevertheless assume that they probably do have a chemical basis.

Now let us consider the groups of organisms utilized as food by butterfly larvae, starting with the most unusual diets. Two tribes of Lycaeninae have departed completely from the usual range of foods: Liptenini feed on lichens, Liphyrini are carnivorous. Many other Lycaeninae, however, are tended by ants and in some cases the larvae are brought into the ant nests. It would seem to be a relatively small step for such larvae to switch and feed on the ant grubs or fungi present in these nests. A number of species of the group exhibit well-developed cannibalism (Downey, 1962). Several Lycaenini, such as *Shirozua*, are carnivorous, and at least one species of *Aphnaeus* feeds on fungi in ant galleries. These transitional steps suggest the evolutionary pathways to the most divergent of butterfly larval feeding habits.

Among those groups of butterflies that feed on plants, none is known to feed on bryophytes or on Psilopsida, Lycopsida, or Sphenopsida, nor is any known from ferns. In fact, very few insects feed on ferns at all (cf. Docters van Leeuwen, 1958), a most surprising and as yet unexplained fact with no evident chemical or mechanical basis. At least one genus of moths, *Papaipema*, is known to feed on ferns, however (Forbes, 1958).

There are a few groups of butterflies that feed on gymnosperms. Two genera of Lycaenidae (*Catachrysops*; *Eumaeus*, two species) feed on Cycadaceae, but all three species involved also feed on angiosperms. *Neophasia* (Pierinae) and three species of *Callophrys* subg. *Incisalia* (Lycaeninae) feed on Pinaceae, while *Callophrys* subg. *Mitoura* feeds on Cupressaceae (and also on *Arceuthobium*, Loranthaceae, a mistletoe that grows on pines). It is well established that Cupressaceae and Pinaceae are chemically quite distinct (Erdtman, 1963, p. 120). Judging from the taxonomic distance between these butterfly groups, it can be assumed that butterflies feeding on gymnosperms had ancestors that fed on angiosperms.

An overwhelmingly greater number of butterfly larvae feed on dicotyledons than on monocotyledons. The only two groups primarily associated with monocotyledons are Satyrinae and Morphinae, closely related subfamilies of Nymphalidae. One genus of morphines (*Morpho*) is more often associated with dicotyledons, but we can think of no way to determine whether this represents a switch from previous monocotyledon feeding. No member of Papilionidae, Pieridae, or Libytheidae is known to feed on monocotyledons, but in Nymphalidae and Lycaenidae numerous genera do so in whole or in part. Among the Nymphalidae, several species of *Charaxes*, one of *Acraea*, one of *Nymphalis*, and two of *Euphaedra* (*Najas*) are known to feed on monocotyledons; all of these genera and some of the same species feed on dicotyledons also. In Lycaenidae, a number of very diverse groups, including at least two genera of Riodininae (*Helicopis* on Araceae, *Napaea* on Orchidaceae) and 11 genera of Lycaeninae (*Jamides* on Zingiberaceae; *Iolaus* on Araceae; *Tmolus* on Bromeliaceae; *Chliaria* on Orchidaceae; *Hemiargus* on Marantaceae; and *Callophrys, Eooxylides, Eumaeus, Strymon,* "*Thecla*," and *Yasoda* on Liliaceae) feed on monocotyledons. Representatives of many of these genera and in some cases the same species feed on dicotyledons also. This pattern strongly suggests that butterflies of two families have switched to monocotyledons from dicotyledons in a number of independent lines (probably at least 18).

A corollary to the observations presented above is that the diversity we see in modern butterflies has been elaborated against a dicotyledonous background. Indeed this is probably true for Lepidoptera as a whole (cf. Forbes, 1958). The dominant themes in this particular coevolutionary situation are therefore of considerable interest. We conclude from this relationship that the appearance of dicotyledons, as yet undated but surely pre-Cretaceous, must have antedated the evolutionary radiation that produced the modern lines of diversification in Lepidoptera and specifically in Papilionoidea. All utilization of foods other than dicotyledons by butterfly larvae (and probably by any Lepidoptera) is assumed to be the result of changes from an earlier pattern of feeding on dicotyledons.

In general, the patterns of utilization by butterflies of dicotyledonous food plants show a great many regularities. Certain relationships are very constant; the plants are usually fed upon by a single, phenetically coherent group of butterflies or several very closely related groups. As examples we have the Aristolochiaceae-feeding Papilionidae; Pierinae on Capparidaceae and Cruciferae; Ithomiinae on Solanaceae; Danainae on Apocynaceae and Asclepiadaceae; Acraeinae, Heliconiini, and Argynnini on Passifloraceae, Flacourtiaceae, Violaceae, and Turneraceae; and Riodininae–Hamearini and Abisariti on Myrsinaceae and Primulaceae. In many of these cases, the broad patterns observed probably support suggestions of overall phenetic similarity among the plants utilized and among the groups of butterflies concerned. Other clusterings on the basis of food plant choice like that of Ulmaceae, Urticaceae, Moraceae, and Euphorbiaceae by certain groups of Nymphalidae, probably also reflect phylogenetic relationship among the plants concerned. In several instances, the patterns of food plant choice of butterfly groups underscore the close relationship between certain sets of tropical woody and temperate herbaceous families elaborated by Bews (1927). Examples are Danainae, feeding interchangeably on Apocynaceae and Asclepiadaceae; Pierinae, on Capparidaceae and Cruciferae; and part of Riodininae, on Myrsinaceae and Primulaceae. In the first case, the families are generally thought to be closely related. On the other hand, Hutchinson (1959) widely separated the members of the second and third pairs of plant families in his system, but this disposition is considered inappropriate by almost all botanists since it is based upon his primary division of flowering plants into woody and herbaceous lines. In making such decisions based on larval food plants, it must be remembered that we are dealing only with an indirect measure of biochemical similarity. For example, Pierinae not only feed on Capparidaceae and Cruciferae, which most botanists would agree are closely related, but also on Salvadoraceae, which contain mustard oil glucosides but otherwise seem totally different from Capparidaceae and Cruciferae. More equivocal cases likewise occur. For example, Pierinae feed on Tropaeolaceae. Not only do Tropaeolaceae share mustard oil glucosides with Capparidaceae and Cruciferae, they likewise have in common the rare fatty acid, erucic acid. Can we, with Alston and Turner (1963, p. 287), dismiss this as coincidence, or do these groups of plants have more in common than is generally assumed? Finally, is there biochemical similarity between Loranthaceae–Santalaceae and Capparidaceae–Cruciferae, both common food plants of different groups of Pierinae (and of the genus *Hesperocharis*).

Whatever conclusions are drawn about the biochemical affinities of plants from the habits of phytophagous or parasitic organisms, little or no weight should be given to individual records. This is true not only because of the numerous sources of error enumerated earlier, but also because of the multiple explanations possible for such switches. For example, *Atella* (Nymphalinae) feeds on Flacourtiaceae and Salicaceae, among other plants. It is quite possible that these two families are fairly closely related, despite the greatly reduced anemophilous flowers of the latter (Thorne, pers. comm.). But to assume that the few records involved indicate biochemical similarity between the groups would be an unwarranted extension of the data; it would be far simpler and safer at that point to make comparative investigations of the biochemistry of the two plant families.

Patterns of food plant utilization provide evidence bearing on the relationship of Araliaceae and Umbelliferae. Some groups of Papilioninae, normally associated with Rutaceae, feed interchangeably on Umbelliferae, or in some cases have switched entirely to this family. As we have seen, these two plant families are chemically similar. But Araliaceae are close relatives of Umbelliferae (Rodríguez, 1957) despite their wide separation in the system of Hutchinson (1959), and are common in many regions where Papilioninae feed on Rutaceae. Despite this, there is not a single record of a papilionid butterfly (indeed very few butterflies of any kind) feeding on Araliaceae. An even more interesting relationship hinges on the suggestion that the three subfamilies of Umbelliferae—Apioideae, Hydrocotyloideae, and Saniculoideae—may represent three phylogenetic lines, derived independently from a group like the present-day Araliaceae. All records of Papilioninae from Umbelliferae are concerned with Apioideae, and indeed Dethier (1941) found that Umbelliferae-feeding papilionine larvae refused *Hydrocotyle* (Hydrocotyloideae). This very strongly suggests that biochemical analysis may go far in elucidating relationships within the Araliaceae–Umbelliferae complex, and that the chemical properties generally ascribed to Umbelliferae as a whole may be characteristic only of one subfamily, Apioideae. Araliaceae and Umbelliferae are known to share certain distinctive fatty acids (Alston and Turner, 1963, p. 121) and acetylinic compounds (Sørensen, 1963), and it might be very instructive to see how these were distributed in Umbelliferae outside of Apioideae.

In further evaluating the patterns of food plant choice in butterflies, it is important to consider those plant families, especially dicotyledons, which are absent or very poorly represented. One outstanding group is that partly characterized by Merz (1959, p. 169) as "Sphingidpflanzen" —plants fed on by moths of the family Sphingidae. These include, among others,

Onagraceae, Lythraceae, Balsaminaceae, Vitaceae, Rubiaceae, and Caprifoliaceae. The first two, and probably the third and fourth, are generally regarded as fairly closely related. Each one of the first five families (Rubiaceae only in part) is characterized by the abundant presence of raphides, bundles of needlelike crystals of calcium oxalate (see discussion in Gibbs, 1963). In a very interesting experiment, Merz (1959) offered mature leaves of *Vitis* (Vitaceae) to larvae of *Pterogon proserpina* (Sphingidae). Young larvae ate these leaves, their pointlike bites falling between the clusters of raphides. Older larvae, which make large slashing bites, could not avoid the raphides and did not eat the leaves. After the raphides were dissolved in very dilute hydrochloric acid, the leaves were accepted by larvae of all sizes. Although it cannot be proven that some other chemical repellent was not removed by this treatment, it is obvious that raphides offer considerable mechanical difficulty for phytophagous insects. A number of families of moths other than Sphingidae feed on this same series of plant families (Forbes, 1958).

Rubiaceae, one of the families mentioned above, is perhaps the most prominent family that is nearly absent from the records of butterfly food plants. Probably the third largest family of dicotyledons, with nearly 10,000 species, it is, like the butterflies themselves, mostly tropical. One can only speculate that some chemical factor, perhaps the rich representation of alkaloids, sharply restricts the ability of butterfly larvae to feed on plants of this family. In this respect, the similarities between the alkaloids of Apocynaceae (which however have milky juice) and Rubiaceae are of interest. Other dicotyledonous families that are very poorly represented or not represented at all among butterfly food plants include Begoniaceae, Bignoniaceae, Boraginaceae, Celastraceae, Cornaceae, Curcurbitaceae (with curcurbitacins, bitter-tasting terpenes), Gesneriaceae, Hydrophyllaceae, Loasaceae, Menispermaceae (rich in alkaloids), Myrtaceae, Polemoniaceae, Ranunculaceae (rich in alkaloids), and Theaceae. In addition, very few butterflies feed on Centrospermae, a group characterized both by its morphological and biochemical traits (summary in Alston and Turner, 1963, p. 141–143, 276–279). This group includes such large families as Amaranthaceae, Cactaceae, Caryophyllaceae, Chenopodiaceae, Nyctaginaceae, and Portulacaceae. Although no biochemical basis for this lack of utilization is known at present, one probably exists. For a family such as the enormous Compositae, poorly represented among the food plants of butterflies, the explanation may lie either in their chemical composition or largely extratropical distribution, or most likely a combination of these. One prominent family of monocotyledons that is practically unrepresented among butterfly food plants is Araceae (*Helicopis*, Riodininae, and species of *Iolaus*, Lycaeninae, are exceptions).

One can conclude only that at least some of the plant groups enumerated above have chemical or mechanical properties that render them unpalatable to butterfly larvae. Thus far the combination of circumstances permitting a shift into the adaptive zones represented by these groups has not occurred. The assumption that such a shift is theoretically possible is strengthened by the observation that nearly every one of these plant groups is fed upon by one or more families of moths.

Conclusions

A systematic evaluation of the kinds of plants fed upon by the larvae of certain subgroups of butterflies leads unambiguously to the conclusion that secondary plant substances play the leading role in determining patterns of utilization. This seems true not only for butterflies but for all phytophagous groups and also for those parasitic on plants. In this context, the irregular distribution in plants of such chemical compounds of unknown physiological function as alkaloids, quinones, essential oils (including terpenoids), glycosides (including cyanogenic substances and saponins), flavonoids, and even raphides (needlelike calcium oxalate crystals) is immediately explicable (Dethier, 1954; Fraenkel, 1956, 1959; Lipke and Fraenkel, 1956; Thorsteinson, 1960; Gordon, 1961).

Angiosperms have, through occasional mutations and recombination, produced a series of chemical compounds not directly related to their basic metabolic pathways but not inimical to normal growth and development. Some of these compounds, by chance, serve to reduce or destroy the palatability of the plant in which they are produced (Fraenkel, 1959). Such a plant, protected from the attacks of phytophagous animals, would in a sense have entered a new adaptive zone. Evolutionary radiation of the plants might follow, and eventually what began as a chance mutation or recombination might characterize an entire family or group of related families. Phytophagous insects, however, can evolve in response to physiological obstacles, as shown by man's recent experience with commercial insecticides. Indeed, response to secondary plant substances and extreme nutritional imbalances and the evolution of resistance to insecticides seem to be intimately connected (Gordon, 1961). If a recombinant or mutation appeared in a population of insects that enabled individuals to feed on some previously protected plant group, selection could carry the line into a new adaptive zone. Here it would be free to diversify largely in the absence of competition from other phytophagous animals. Thus the diversity of plants not only may tend to augment the diversity of phytophagous animals (Hutchinson, 1959), the converse may also be true.

Changes in food plant choice would be especially favored in situations where the supply of the "preferred" plant is sufficiently limited to be an important factor in the survival of the larvae. Such situations have been described by Dethier (1959), who showed that the density of *Aster umbellatus* (Compositae) plants was critical to the success of the larvae of *Chlosyne harrisii*. Similarly, in western Colorado, the density of the small plants of *Lomatium eastwoodiae* (Umbelliferae) is an important factor limiting population size in *Papilio indra* (T. and J. Emmel, pers. comm.). In these, and many similar situations, it is logical to assume that genetic variants able to utilize another food plant successfully would be relatively favored. This advantage would be much enhanced if genotypes arose that permitted switching to a new food plant sufficiently novel biochemically that it was not utilized, or little utilized, by herbivores in general.

The degree of physiological specialization acquired in genetic adjustment to feeding on a biochemically unusual group of plants would very likely also act to limit the choice of food available to the insect group in the general flora (Merz, 1959, p. 187; Gordon, 1961). As stressed by Brower (1958a), moreover, close relationships between insects and a narrow range of food plants may be promoted by the evolution of concealment from predators in relation to a single background. The food plant provides the substrate for the larvae, not just their food (Dethier, 1954, p. 38).

After the restriction of certain groups of insects to a narrow range of food plants, the formerly repellent substances of these plants might, for the insects in question, become chemical attractants. Particularly interesting is the work of Thorsteinson (1953), who found that certain mustard oil glucosides from Cruciferae would elicit feeding responses from larvae that fed on these plants if these glucosides were smeared on other, normally unacceptable, leaves. But if these glucosides were smeared on the alkaloid-rich leaves of *Lycopersicum* (Solanaceae), the larvae still refused them. Similarly, Sevastopulo (pers. comm.) was unable to induce the larvae of *Danais chrysippus* to eat anything but Asclepiadaceae even by smearing the leaves of other plants with the juice of *Calotropis* (Asclepiadaceae).

This illustrates clearly that the choice of

a particular food plant or of a spectrum of food plants may be governed by repellents present in other plants (Thorsteinson, 1960) as well as by attractants in the normal food plants; this fully accords with the model outlined above. It should not, however, be assumed without experimental verification that a particular secondary plant substance is an attractant or feeding stimulant for the insects feeding on plants that contain it. Indeed, for the beetle *Leptinotarsa*, the alkaloids of the Solanaceae on which it feeds serve as repellents (summary in Fraenkel, 1959, p. 1467–1468).

In view of these considerations, we propose a comparable pattern of adaptive radiation for each of the more or less strictly limited groups of butterflies enumerated above. It is likewise probable that the elaboration of biochemical defenses has played a critical role in the radiation of those groups of plants characterized by unusual accessory metabolic products.

Further, it can be pointed out that all groups of butterflies that are important in furnishing models in situations involving mimicry are narrowly restricted in food plant choice: Papilioninae–Troidini; Ithomiinae; Nymphalinae–Heliconiini; Acraeinae; and Danainae. This is in accordance with a long-standing supposition of naturalists and students of mimicry that the physiological shifts that enabled the butterfly groups to feed on these plants conferred a double advantage by making the butterflies in question unpalatable. These groups of butterflies have been selected for warning coloration, and once established, this conspicuousness would tend to put anything that would maintain their distastefulness at a selective premium.

Conversely, those groups of butterflies that furnish most of the Batesian mimics—Papilioninae–Papilionini and Graphiini; Satyrinae; Nymphalinae except Heliconiini; Pierinae; and Dismorphiinae—feed mostly on plant groups that are shared with other dissimilar groups of butterflies. It is somewhat surprising that Nymphalinae–Argynnini, which feed on the same plants as Heliconiini and Acraeinae, do not play any prominent role as models for mimetic forms. This may be in part because Argynnini are best developed in temperate regions, where butterflies and therefore mimetic complexes are less common. It may further be suggested that the mimicry supposed to exist between dark female forms of various species of *Speyeria* and the model *Battus philenor* may well be a case of Müllerian rather than Batesian mimicry. Indeed, the results of Brower (1958) with the model *Danais plexippus* and its mimic *Limenitis archippus* suggest that there is in fact no sharp line between Batesian and Müllerian mimicry. These should be thought of as the extremes of a continuum. Thus *Speyeria* females may be somewhat distasteful but can only acquire warning coloration when the selective balance is tipped by the presence of other distasteful forms. This follows logically from numerous experiments and observations on the behavior of predators with mimetic complexes, which rarely reveal an "either/or" type of response (cf. Swynnerton, 1919). Assuming a balance of this sort would resolve some of the difficulties of interpretation concerning the two kinds of mimicry (e.g., Sheppard, 1963, p. 145).

Numerous unusual feeding patterns scattered among butterfly families attest to the frequency of radiation into new groups of food plants. We can, however, only guess at the probability of future radiation in the new adaptive zone. For example, does *Stalachtis susanae* (Riodininae), which is known to feed on *Oxypetalum campestre* (Asclepiadaceae) in Argentina, represent the start of a new phylogenetic series of butterflies restricted to this group of plants? Probably not, since the examples of this sort of unusual feeding habit today far exceed the number of radiations observed in the past. Nevertheless, the close patterns of coadaptation we have discussed above must have started in a similar fashion. Comparable patterns can be found among plants with biochemical innovations, for example, *Senecio viscosus*, mentioned earlier, or the relationship of *Sedum acre* to other species of its genus (Merz, 1959, p. 160).

In viewing present-day patterns of food plant utilization, however, the historical aspect of the situation must not be neglected. A biochemical innovation might have had a considerable selective advantage for a group of plants in the Cretaceous. Such an advantage would, of course, have been in terms of the phytophagous animals and parasites present in the Cretaceous, and not necessarily those of the present day. The crossing of an adaptive threshold by a member of a living group of phytophagous animals would have an entirely different significance now than that which it would have had in the Cretaceous.

For example, even though a species of *Stalachtis* is able to feed on Asclepiadaceae, it shares the available supply of these plants with representatives of numerous other groups of phytophagous animals. These have, somewhere in the course of geological time, acquired the ability to feed on asclepiads. Further, if the phytophagous organisms switching early to milkweeds became protected from predators by their ingestion of distasteful plant juices, this initial advantage might have been overcome in the intervening years by corresponding changes in prospective predators. A species of bird long selected to like milkweed bugs might find milkweed-feeding *Stalachtis* a gourmet's delight.

As in the occupation of any adaptive zone, the first organisms to enter it have a tremendous advantage and are apt to have the opportunity to become exceedingly diverse before evolution in other organisms sharply restricts their initial advantage. In short, the nature of any adaptive zone is altered by the organisms that enter it. From our vantage point in time we view only the remnants, doubtless often disarranged if not completely shattered by subsequent events, of the great adaptive radiations of the past.

In view of these considerations, we cannot accept the theoretical picture of a generalized group of polyphagous insects from which specialized oligophagous forms were gradually derived. Just as there is no truly "panphagous" insect (cf. Fraenkel, 1959), so there is no universally acceptable food plant; and this doubtless has always been true. This statement is based on the chemical variation observed in plants and the physiological variation observed in insects. Leguminosae are important food plants for several groups of Lycaenidae and Pieridae, and woody Ranales are well represented among the food plants of Papilionidae and Nymphalidae; but this should not be taken to prove that these groups of plants are "inert" chemically or readily available to other phytophagous groups of insects. The initial radiation of butterfly taxa onto these groups may for a time have produced a pattern just as spectacular as, for example, the close association between Troidini and Aristolochiaceae seen today. We hold that plants and phytophagous insects have evolved in part in response to one another, and that the stages we have postulated have developed in a stepwise manner.

As suggested by Fraenkel (1956, 1959), secondary plant substances must have been formed early in the history of angiosperms. At the present day, many classes of organic compounds are nearly or quite restricted to this group of plants (for example, see Alston and Turner, 1963, p. 164; Harborne, 1963, p. 360; Paris, 1963, p. 357). We suggest that some of these compounds may have been present in early angiosperms and afforded them an unusual degree of protection from the phytophagous organisms of the time, relative to other contemporary plant groups. Behind such a biochemical shield the angiosperms may have developed and become structurally diverse. Such an assumption of the origin of angiosperms provides a cogent reason why one of many structurally modified groups of gymnosperms would have been able to give rise to the bewildering diversity of modern angiosperms, while most other lines became extinct. It seems at least as convincing to us as do theories based on the structural peculiarities of angiosperms. Although the chemical basis for the success of early angiosperms may no longer be discernible, it can be mentioned that woody Ranales, generally accepted as the most "primitive" assemblage of living angiosperms on other grounds, are as a group characterized by many alkaloids as well as by essential oils. Of course this might also be interpreted only to mean that the development of alkaloids has permitted this group to persist despite its many generalized features.

In turn, the fantastic diversification of modern insects has developed in large measure as the result of a stepwise pattern of coevolutionary stages superimposed on the changing pattern of angiosperm variation. With specific reference to the butterflies, one is tempted in terms of present-day patterns to place more emphasis on the full exploitation of diurnal feeding habits by the adults than on the penetration of any particular biochemical barrier by the larvae. On the other hand, phenetic relationships suggest that Papilionoidea (with Hesperioidea, the skippers) are representatives of a line that is amply distinct from all other living Lepidoptera (Ehrlich, 1958). Thus it is entirely possible that radiation onto a new food plant was decisive at the time Papilionoidea first diverged, even though the feeding habits of the order as a whole are now much wider than those of the butterflies alone. The impossibility of deciding objectively which groups of Papilionoidea are more primitive than others (cf. Ehrlich, 1958, p. 334–335) relegates the task of identifying the original group of food plants for butterflies to the realm of profitless speculation.

We would like to return now to the four general questions posed at the beginning of this paper. First, what have we learned of the reciprocal responses of butterflies and their food plants? The observed patterns clearly point to the critical importance of plant biochemistry in governing the relationships between the two groups. The degree of plasticity of chemoreceptive response and the potential for physiological adjustment to various plant secondary substances in butterfly populations must in large measure determine their potential for evolutionary radiation. Of secondary, but still possibly major importance, are mechanical plant defenses, and the butterflies' responses to them.

With respect to the second question on the generation of predictions the answer also seems clear. We cannot predict the results of any given interaction with precision—*Stalachtis* on Asclepiadaceae or *Neophasia* on pines may or may not form the basis for further patterns of radiation. On the other hand, the basis for a probabilistic statement of "Further radiation unlikely" seems to have been developed. A great many minor predictions can be made, such as the probable presence of alkaloids in *Ptaeroxylon* (Meliaceae), the solanaceous character of the food plants of unknown larvae of Ithomiinae, and so forth.

Although the data we have gathered permit us to make some reasonable sequence predictions about phylogenetic patterns (e.g., diversification of Apocynaceae and Solanaceae before Danainae and Ithomiinae, respectively), these predictions cannot be tested and the relationships cannot be specified further in the absence of a fossil record. The reconstruction of phylogenies on the basis of this sort of information would seem an unwarranted imposition on the data, since evolutionary rate and time are still inseparable.

In response to the fourth question, it seems to us that studies of coevolution provide an excellent starting point for understanding community evolution. Indeed the seeming ease with which our conclusions have been extended to include the complex interactions among plants, phytophagous organisms, mimics, models, and predators leads us to believe that population biologists should pursue similar studies of other systems. Many examples come to mind such as parasitoid–caterpillar, *Plasmodium*–

hemoglobin, tree–mycorrhizal fungus, in which stepwise reciprocal selective response is to be expected. Studying most of these systems experimentally tends to be difficult, and may be complicated by lack of repeatability in the results.

An approach to biology that is concerned with broad patterns quite possibly will lead to a better understanding of some other problems of community ecology. For example, biologists have long been interested in the reasons for the differences in species diversity between tropical and temperate areas. An important factor in maintaining these differences may be the sort of synergistic interactions between plants and herbivores we have been discussing. The selective advantage of living in a tropical climate is evident for insects, which are poikilothermal. Insects are much more abundant in the tropics than elsewhere and doubtless constitute the major class of herbivorous animals. The penetration of relatively cold environments or other environments requiring diapause is probably a rather recent occurrence in most insect groups. That these environments are not always readily entered is attested to by the repeated failure of insects such as the butterfly *Mestra amymone* to survive the winter in localities at the northern fringes of their ranges where summer colonies have been established (Ehrlich and Ehrlich, 1961).

The abundance of phytophagous insects in tropical regions would be expected to accentuate the pace of evolutionary interactions with plants. These interactions may have been the major factor in promoting the species diversity of both plants and animals observed in the tropics today. As this diversity was being produced, it became arrayed in richly varied mixtures of species with relatively great distances between individuals of any one plant species. As Grant (1963, p. 420–422) has suggested, this arrangement would have the additional advantage of providing a maximum degree of protection from epidemic outbreaks of plant diseases and plant pests. It must, however, also be mentioned that the relatively permissive tropical climate presumably allows a greater diversity of plant life forms and therefore secondarily of animals (Hutchinson, 1959, p. 150).

Probably our most important overall conclusion is that the importance of reciprocal selective responses between ecologically closely linked organisms has been vastly underrated in considerations of the origins of organic diversity. Indeed, the plant–herbivore "interface" may be the major zone of interaction responsible for generating terrestrial organic diversity.

Summary

The reciprocal evolutionary relationships of butterflies and their food plants have been examined on the basis of an extensive survey of patterns of plant utilization and information on factors affecting food plant choice. The evolution of secondary plant substances and the stepwise evolutionary responses to these by phytophagous organisms have clearly been the dominant factors in the evolution of butterflies and other phytophagous groups. Furthermore, these secondary plant substances have probably been critical in the evolution of angiosperm subgroups and perhaps of the angiosperms themselves. The examination of broad patterns of coevolution permits several levels of predictions and shows promise as a route to the understanding of community evolution. Little information useful for the reconstruction of phylogenies is supplied. It is apparent that reciprocal selective responses have been greatly underrated as a factor in the origination of organic diversity. The paramount importance of plant–herbivore interactions in generating terrestrial diversity is suggested. For instance, viewed in this framework the rich diversity of tropical communities may be traced in large part to the hospitality of warm climates toward poikilothermal phytophagous insects.

Literature Cited

Alston, R. E., and B. L. Turner. 1963. Biochemical systematics. Prentice-Hall, Englewood Cliffs, N. J.

Barrett, C., and A. N. Burns. 1951. Butterflies of Australia and New Guinea. N. H. Seward, Melbourne.
Bews, J. W. 1927. Studies in the ecological evolution of the angiosperms. New Phytol., **26**: 1–21, 65–84, 129–148, 209–248, 273–294.
Bourgogne, J. 1951. Ordre des Lépidoptères. Traité Zool., **10**: 174–448.
Brower, Jane Van Zandt. 1958. Experimental studies of mimicry in some North American butterflies. Part I. The monarch, *Danaus plexippus*, and viceroy, *Limenitis archippus archippus*. Evolution, **12**: 32–47.
Brower, L. P. 1958a. Bird predation and food plant specificity in closely related procryptic insects. Amer. Nat., **92**: 183–187.
——. 1958b. Larval food plant specificity in butterflies of the *Papilio glaucus* group. Lepidop. News, **12**: 103–114.
Chermock, R. L. 1950. A generic revision of the Limenitini of the world. Am. Midl. Nat., **43**: 513–569.
Clench, H. K. 1955. Revised classification of the butterfly family Lycaenidae and its allies. Ann. Carnegie Mus., **33**: 261–274.
Corbet, A. S., and H. M. Pendlebury. 1956. The butterflies of the Malay Peninsula. Ed. 2. Oliver and Boyd, London.
Costa Lima, A. M. da. 1936. Terceiro Catalogo dos Insectos que vivem nas plantas do Brasil. Directoria de Estatistica da Producção Secção de Publicade, Rio de Janeiro, p. 201–231.
Dethier, V. G. 1941. Chemical factors determining the choice of food plants by *Papilio* larvae. Amer. Nat., **75**: 61–73.
——. 1954. Evolution of feeding preferences in phytophagous insects. Evolution, **8**: 33–54.
——. 1959. Food-plant distribution and density and larval dispersal as factors affecting insect populations. Canad. Entom., **91**: 581–596.
Docters van Leeuwen, W. M. 1958. Zoocecidia. *In* Verdoorn, F., ed., Manual of pteridology. Martinus Nijhoff, The Hague, p. 192–195.
Downey, J. C. 1962. Host-plant relations as data for butterfly classification. Syst. Zool., **11**: 150–159.
—— and W. C. Fuller. 1961. Variation in *Plebejus icarioides* (Lycaenidae). I. Food plant specificity. J. Lepidop. Soc., **15**: 34–42.
Ehrlich, P. R. 1958. The comparative morphology, phylogeny and higher classification of the butterflies (Lepidoptera: Papilionoidea). Univ. Kansas Sci. Bull., **39**: 305–370.
—— and A. H. Ehrlich. 1961. How to know the butterflies. Wm. C. Brown, Dubuque.
Erdtman, H. 1963. Some aspects of chemotaxonomy. *In* Swain, T., ed., Chemical plant taxonomy. Academic Press, London, p. 89–125.
Flück, H. 1963. Intrinsic and extrinsic factors affecting the production of secondary plant products. *In* Swain, T., ed., Chemical plant taxonomy. Academic Press, London, p. 167–186.
Forbes, W. T. M. 1958. Caterpillars as botanists. Proc. Tenth Int. Congr. Ent., **1**: 313–317.
Fraenkel, G. 1956. Insects and plant biochemistry. The specificity of food plants for insects. Proc. 14th Int. Congr. Zool., p. 383–387.
——. 1959. The raison d'etre of secondary plant substances. Science, **129**: 1466–1470.
Gibbs, R. D. 1963. History of chemical taxonomy. *In* Swain, T., ed., Chemical plant taxonomy. Academic Press, London, p. 41–88.
Gordon, H. T. 1961. Nutritional factors in insect resistance to chemicals. Ann. Rev. Entom., **6**: 27–54.
Grant, V. 1963. The origin of adaptations. Columbia University Press, New York and London.
Harborne, J. B. 1963. Distribution of anthocyanins in higher plants. *In* Swain, T., ed., Chemical plant taxonomy. Academic Press, London, p. 359–388.
Hegnauer, R. 1963. The taxonomic significance of alkaloids. *In* Swain, T., ed., Chemical plant taxonomy. Academic Press, London, p. 389–427.
Hutchinson, G. E. 1959. Homage to Santa Rosalia *or* why are there so many kinds of animals. Amer. Nat., **93**: 145–159.
Hutchinson, J. 1959. The families of flowering plants. Ed. 2. 2 vols. Clarendon Press, Oxford.
Jörgensen, P. 1932. Lepidopterologisches aus Südamerika. Deutsch. Ent. Zeitschr. Iris, Dresden, **46**: 37–66.
Klots, A. B. 1933. A generic revision of the Pieridae (Lepidoptera). Entom. Amer. n.s., **12**: 139–242.
Lee, C. L. 1958. Butterflies. Academia Sinica (in Chinese).
Lipke, H., and G. Fraenkel. 1956. Insect nutrition. Ann. Rev. Ent., **1**: 17–44.
Merz, E. 1959. Pflanzen und Raupen. Über einige Prinzipien der Futterwahl bei Grossschmetterlingsraupen. Biol. Zentr., **78**: 152–188.
Michener, C. D. 1942. A generic revision of the Heliconiinae (Lepidoptera, Nymphalidae). Am. Mus. Novitates, **1197**: 1–8.
Munroe, E. 1960. The generic classification of the Papilionidae. Canad. Ent., Suppl., **17**: 1–51.
—— and P. R. Ehrlich. 1960. Harmonization of concepts of higher classification of the Papilionidae. J. Lepid. Soc., **14**: 169–175.
Paris, R. 1963. The distribution of plant glycosides. *In* Swain, T., ed., Chemical plant taxonomy. Academic Press, London, p. 337–358.

PINHEY, E. C. G. 1949. Butterflies of Rhodesia. Rhodesia Sci. Association, Salisbury.

PRICE, J. R. 1963. The distribution of alkaloids in the Rutaceae. *In* Swain, T., ed., Chemical plant taxonomy. Academic Press, London, p. 429–452.

REMINGTON, C. L. 1952. The biology of nearctic Lepidoptera I. Food plants and life-histories of Colorado Papilionoidea. Psyche, **59**: 61–70.

REUTER, E. 1896. Über die Palpen der Rhopaloceren. Acta Soc. Sci. Fennica, **22**: i–xvi, 1–577.

RODRÍGUEZ, R. L. 1957. Systematic anatomical studies on *Myrrhidendron* and other woody Umbellales. Univ. Calif. Publ. Bot., **29**: 145–318, pls. 36–47.

SAFWAT, FUAD M. 1962. The floral morphology of *Secamone* and the evolution of the pollinating apparatus in Asclepiadaceae. Ann. Missouri Bot. Gard., **49**: 95–129.

SAVILLE, D. B. O. 1954. The fungi as aids in the taxonomy of flowering plants. Science, **120**: 583–585.

SEITZ, A. 1906–1927. The macrolepidoptera of the world. Vols. 1, 5, 9, 13. Fritz Lehman Verlag and Alfred Kerner Verlag, Stuttgart.

SHEPPARD, P. M. 1963. Some genetic studies of Müllerian mimics in butterflies of the genus *Heliconius*. Zoologica, **48**: 145–154, pls. 1–2.

SHIRÔZU, T. 1962. Evolution of the food-habits of larvae of the thecline butterflies. Tyô to Ga (Trans. Lepidop. Soc. Japan), **12**: 144–162.

—— AND H. YAMAMOTO. 1956. A generic revision and the phylogeny of the tribe Theclini (Lepidoptera: Lycaenidae). Sieboldia, **1**: 329–421.

SHORLAND, F. B. 1963. The distribution of fatty acids in plant lipids. *In* Swain, T., ed., Chemical plant taxonomy. Academic Press, London, p. 253–303.

SØRENSEN, N. A. 1963. Chemical taxonomy of acetylinic compounds. *In* Swain, T., ed., Chemical plant taxonomy. Academic Press, London, p. 219–252.

SWYNNERTON, C. M. F. 1919. Experiments and observations bearing on the explanation of form and colouring, 1908–1913, Africa. J. Linn. Soc. Zool., **33**: 203–385.

THORNE, R. F. 1963. Some problems and guiding principles of angiosperm phylogeny. Amer. Nat., **97**: 287–306.

THORSTEINSON, A. J. 1953. The chemotactic responses that determine host specificity in an oligophagous insect (*Plutella maculipennis* (Curt.) Lepidoptera). Canad. J. Zool., **31**: 52–72.

——. 1960. Host selection in phytophagous insects. Ann. Rev. Ent., **5**: 193–218.

VAN SON, G. 1949. The butterflies of southern Africa. Part I, Papilionidae and Pieridae. Part II (1955), Danainae and Satyrinae.

VAZQUEZ G., LEONILA, AND PEREZ R., H. 1961. Observaciones sobre la biología de *Baronia brevicornis* Salv. (Lepidoptera: Papilionidae-Baroniinae). An. Inst. Biol. Mex., **22**: 295–311.

VERSCHAEFFELT, E. 1910. The cause determining the selection of food in some herbivorous insects. Proc. Acad. Sci., Amsterdam, **13**: 536–542.

WILTSHIRE, E. P. 1957. The Lepidoptera of Iraq. Rev. ed. Nicholas Kaye, London.

WYNTER-BLYTH, M. A. 1957. Butterflies of the Indian region. Bombay Nat. Hist. Soc., Bombay.

A DARWINIAN APPROACH TO PLANT ECOLOGY

By J. L. HARPER

Department of Agricultural Botany, University College of North Wales, Bangor
(*Being the Presidential Address to the British Ecological Society on* 5 *January* 1967)

The theory of evolution by natural selection is an ecological theory—founded on ecological observation by perhaps the greatest of all ecologists. It has been adopted by and brought up by the science of genetics, and ecologists, being modest people, are apt to forget their distinguished parenthood. Indeed, Darwinian plant ecology has been largely neglected and a changeling child nourished and brought to adulthood by Schimper and Warming who asked geographical questions about vegetation, and answered the questions by demonstrating correlations between climate and soils on the one hand and comparative physiology on the other.

By contrast with the 'vegetationalist' and his concern to describe and interpret areas of land, Darwin's ecological observations and the questions he asked were based on a consideration of individuals and populations—a preoccupation with numbers. 'Look at a plant in the midst of its range, why does it not double or quadruple its *numbers*?' '. . . if we wish in imagination to give the plant the power of increasing in *number*, we should have to give it some advantage over its competitors, or over the animals which prey on it'. 'Look at the most vigorous species; by as much as it swarms in *numbers*, by so much will it tend to increase still further' (Darwin, *Origin of Species*, Chapter III).*

These quotations, with their emphasis on numbers, pose problems of population biology—of a demography which has never gained a momentum in plant ecology, although it has played a vertebral role in animal ecology.

Two interlinked properties of higher plants have seriously hindered the development of plant demography—plasticity and vegetative reproduction. Darwin found a 26-year-old pine tree on heathland which 'had during many years tried to raise its head above the stems of the heath and had failed'. It is clearly not fair to count such a plant as a unit equal to a full grown tree in a population census. A mature plant of an annual weed such as *Chenopodium album* may produce four seeds or 100 000 seeds, depending on the nutrient and water status of the soil. It can therefore be argued that a statement about numbers of plants implies very little about the real nature of the population. Vegetative reproduction is a further obstacle to census making, because the vegetative offspring remain to some extent a part of the parent, often for a long period. When is such a ramet to be counted as an individual? Arbitrary decisions have to be made if plant populations are to become numerable and the arbitrariness of the decisions has often discouraged attempts to count plants.

These problems are, however, not peculiar to plants and have had to be faced in animal demography where they arise in only a slightly less acute form. Plasticity in individual size and reproductive capacity have to be taken into account in population studies of fish and even of *Drosophila*. Vegetative reproduction in *Hydra*, where the 'ramets' slowly develop independence, has not prevented its use in model population studies.

* All quotations in this paper are from Chapters III and IV of *The Origin of Species* (1859), the text being the Everyman edition of 1928. Italics are mine.

The very few population studies which have been made of plants in natural populations suggest that the difficulties are not overgreat in practice and the results can be very revealing. Tamm's study of perennial plants in forests and meadows is a classic (Tamm 1948, 1956) (Fig. 1a, b and c). Over a period of 13 years he counted 'individuals' of selected species in permanent quadrats. Many of the populations were remarkably constant in total numbers over this period and yet displayed a high turnover rate including losses both of seedlings and of mature (flowered) plants and gains from new recruits both as seedlings and from vegetative reproduction.

Tamm's study of a declining population of *Centaurea jacea* (Fig. 1c) is equally remarkable for the insight it gives to the process of disappearance of a species from a community. These data have been recalculated and presented as the change in plant numbers (as logarithms) with time (Fig. 2). The relationships are almost startlingly linear, indicating that the chance of an individual dying remains the same through the period of the study. It shows elimination as a steady, not jerky, process, with some individuals remaining vigorous enough to reproduce vegetatively to the bitter end. It suggests that the exodus of a species from a community is not due to the occasional occurrence of extreme conditions but to a slow process of elimination of which the causes act with surprisingly constant intensity. Such population changes may be characterized like the decay of a radioisotope, by a 'half-life'. The half-life of *C. jacea* calculated from Tamm's data is *c.* 1·9 years. Similar calculations made from more constant populations in Tamm's study show widely different half-life values. For example, the individuals of *Filipendula vulgaris* present at the start of Tamm's observations had a half-life of *c.* 18·4 years, and *Sanicula europaea* $>$ 50 years (Fig. 2). In these cases the population decay was at least matched by new recruitment from seedlings or vegetative reproduction.

Sagar (1959) made a shorter but more detailed study of populations of *Plantago lanceolata* in 'permanent' grassland near Oxford which had been under essentially constant management over the preceding 10 years. Using pantograph techniques for mapping plants and seedlings and following the fates of all the seedlings and rosettes in replicate sample plots A and B, he attempted to extract life table data (Table 1).

On both of the two replicate areas sampled, the number of plants of *P. lanceolata* increased but much more strikingly in Area B than in Area A. The behaviour of those plants which were present at the start of the observations was, however, very similar on the two areas. Of the plants recruited to the population as seedlings within the 2-year period, two-thirds died when less than 12 months old. The mature plants present at the start of the period of study had a half-life of *c.* 3·2 years. This is a measure of the dynamic character of such superficially stable vegetation.

Antonovics (1966) has studied the dynamics of population of *Anthoxanthum oderatum* in the open communities of metal mine spoils—in this case marking individuals present along line transects. This was again a short-term study and typical results are given in Fig. 3. As with many of Tamm's examples (and Sagar's plantains) the new recruitment from seedlings compensated for (might indeed be determined by) the rate of loss of the initial population. Such attempts to trace the fate of individuals within populations necessarily involve detailed and frequent mapping so that a seedling which appears where another has just died is recognized as a new individual, and that individuals do not both appear and die in the interval between observations. Probably the first serious observation of this sort was that of Darwin '. . . on a piece of ground three feet long and two wide, dug and cleared, and where there could be no choking from other plants, I marked all the seedlings of our native weeds as they came up, and out of 357 no less

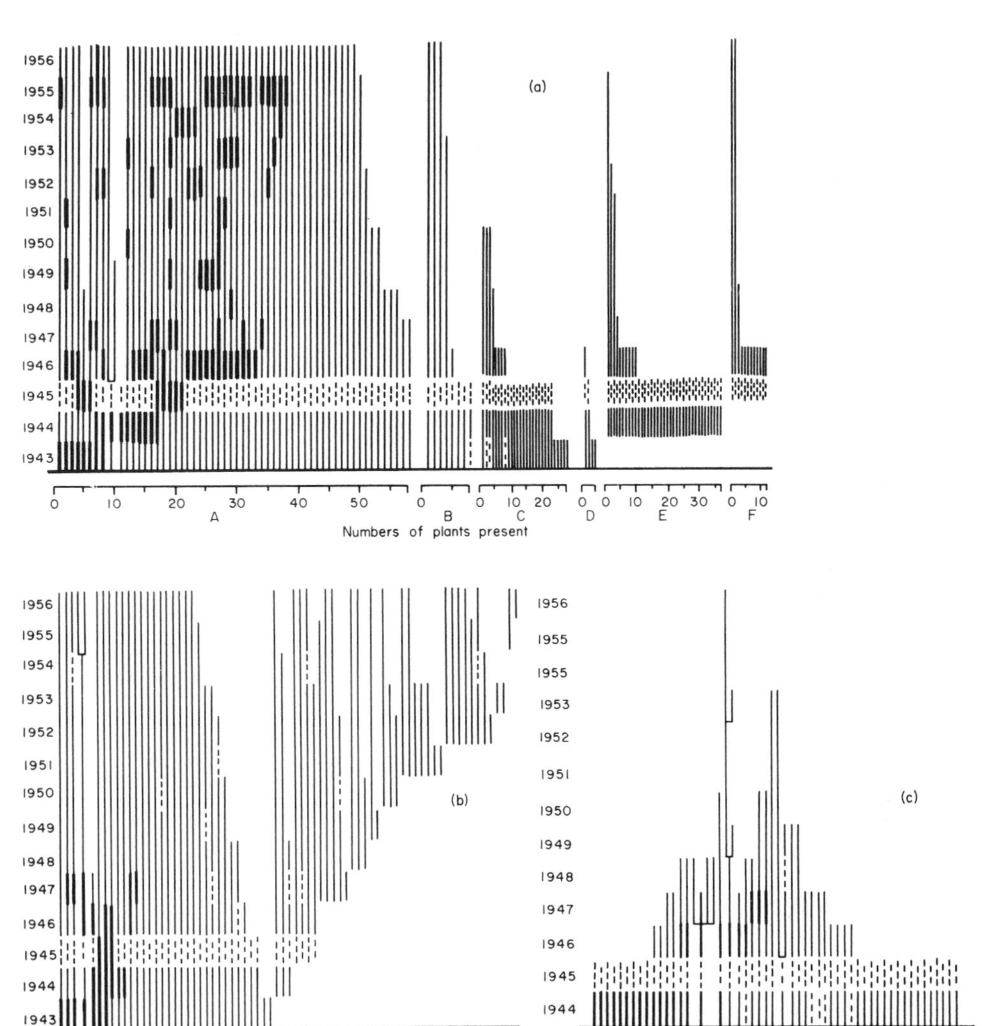

FIG. 1. The behaviour of populations of (a) *Sanicula europaea*, (b) *Filipendula vulgaris* and (c) *Centaurea jacea* within a $\frac{1}{4}$ m^2 quadrat in woodland from 1943 to 1956. Group A includes specimens that were large or intermediate in size at the first inspection, Group B rather small plants, Group C very small plants and Groups D–F different crops of seedlings. Heavy lines indicate that the plant flowered in that year. Branching of a line indicates vegetative reproduction. (Redrawn from Tamm 1956.)

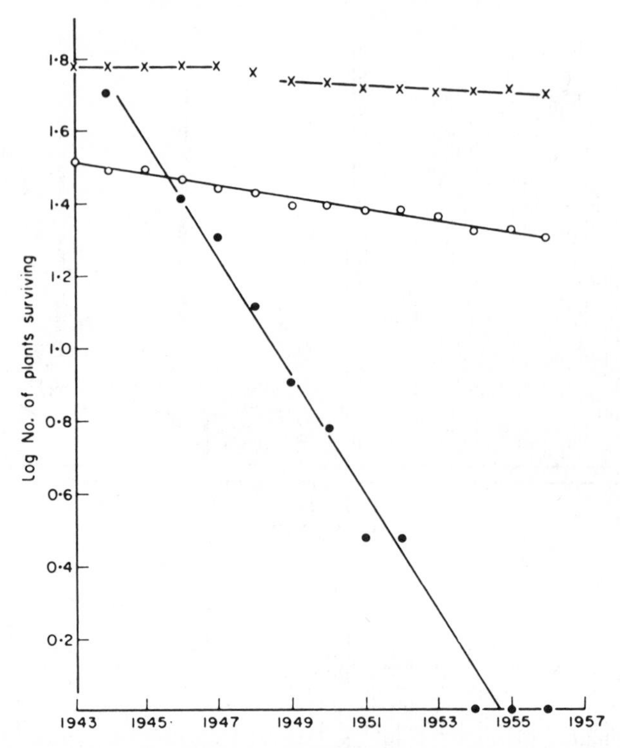

FIG. 2. The decay rate of populations of *Sanicula europaea* (×, half-life > 50 years), *Filipendula vulgaris* (○, half-life *c.* 18·4 years) and *Centaurea jacea* (●, half-life *c.* 19 years) calculated from data given by Tamm (1956).

Table 1. *The behaviour of a population of* Plantago lanceolata *in an alluvial meadow near Oxford (adapted from Sagar* 1959)

		Area A	Area B
(a)	Number of individuals per m^2 present, April 1957	238	149
(b)	Number of individuals per m^2 present, April 1959	249	211
(c)	Net gain	+11	+62
	% gain ($\frac{c}{a} \times 100$)	4·6	41·6
(d)	New individuals appearing between April 1957 and April 1959	576	487
(e)	Individuals lost from population between April 1957 and April 1959	565	425
(f)	Individuals present at April 1957 and surviving April 1959	83	52
(g)	Individuals present at April 1957 but lost by April 1959	155	97
(h)	Percentage survival of individuals from April 1957 to April 1959 ($\frac{f}{a} \times 100$)	34·9	34·9
(i)	Calculated half-life (years) of mature plants present at start of observation	3·2	3·2
(j)	Fate of plants appearing between April 1957 and April 1959		
	(i) % dying when less than 12 months old	68·2	66·2
	(ii) % dying when 12–24 months old	2·3	1·2
	(iii) % still surviving April 1959	29·5	32·6

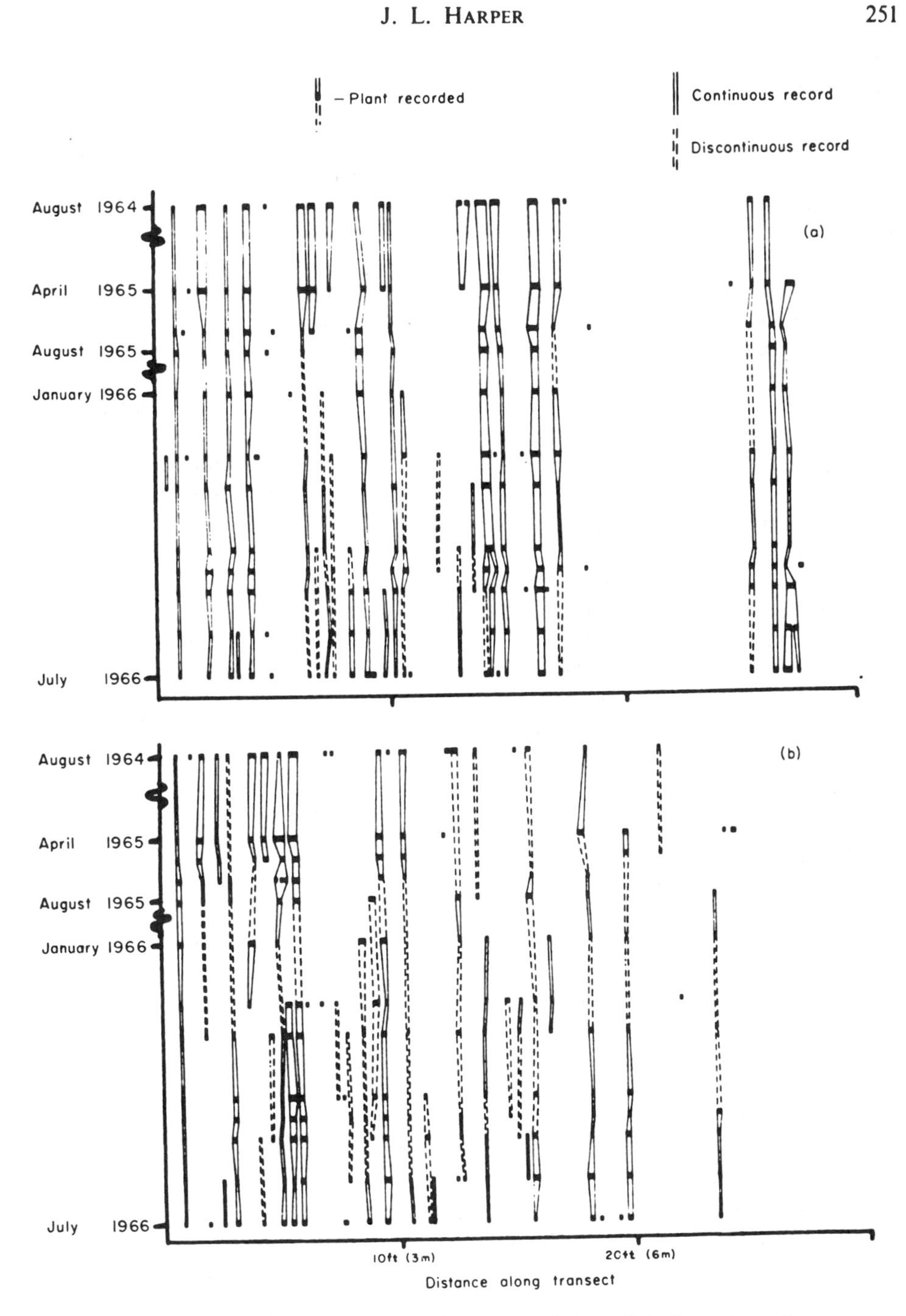

FIG. 3. (a) and (b) The pattern of change in two populations of *Anthoxanthum odoratum* over a period of 2 years in open vegetation on a copper mine site at Trelogan, Flintshire (from Antonovics 1966). Note abbreviation of time scale.

than 295 were destroyed, chiefly by slugs and insects.' Demographic study of this sort focuses attention on the life span of individuals, focuses attention on the causes as well as the time of death, and particularly on the enormous fungal and animal roles in plant mortality.

Measures of population turnover can only be obtained by the detailed observation of individuals—they are totally obscured by vegetational study and revealed by population studies only if plants are marked for repeated observations. They bring measurements of flux into ecological studies in terms which are meaningful to the selection geneticist and the evolutionist. Stable vegetation is then seen to be in a state of continuous flux in which the rates of turnover are critical characteristics of the stability.

The early seedling phases of a plant's life are usually considered the most risky, and the hazards are often strikingly exaggerated by increasing plant density; 'seedlings suffer most from germinating in ground already thickly stocked with other plants'. I have discussed the density dependent mortality of seedlings in other papers (Harper & McNaughton 1962; Harper 1960; Harper 1964a), and this self-regulating property of plant populations is becoming increasingly well documented although the ultimate causes of density-induced death are still very obscure.

One of the earliest attempts to study self-regulation of numbers in populations of plants was made by Sukatschev (1928), who sowed *Matricaria inodora* at two densities in fertilized and unfertilized soil. At the end of a season's growth the percentage loss from the population was greater at the higher density, and in the fertilized soil. (Sukatschev had also observed that the density of mature fir trunks in Leningrad forests declined steadily with *increasing* soil fertility.) Yoda *et al.* (1963) extended this type of observation on natural and artificial populations of plant species (see, for example, Fig. 4). They found that:

(a) The chance of a seed producing a mature plant declined with increasing density.

(b) That irrespective of the density of seeds sown there is a maximum population size of plants produced, and densities beyond this level cannot be realized no matter how many seeds are sown.

(c) The densities of overcrowded populations converge with the passage of time, irrespective of the differences in initial density. The converging densities are always lower on the more fertile soil.

(d) The converging (or asymptotic density) is closely correlated with plant size—so that plants having a certain average size always maintained a more or less similar level of surviving density regardless of the differences in stand age, initial density and fertilizer level.

A particular achievement of this group of Japanese workers was that they were able to formulate a hypothesis linking the numbers of plants and their weight in pure stands which could be expressed as a mathematical relationship—susceptible to experimental test—the major step in transforming a wordy descriptive science into a rigid discipline.

The empirically derived relationship of Yoda *et al.* is

$$w = Cp^{-3/2}$$

where w = mean weight per plant, and p = existing plant density or

$$y = wp = Cp^{-1/2}$$

where y = mean weight per unit area.

It remains to be seen whether this formal relationship holds good for a wider range of species than those studied by Yoda *et al.* which did, however, include such different forms as pure populations of *Betula* sp., *Pinus densiflora*, *Abies sachalinensis*, *Erigeron canadensis*, *Amaranthus retroflexus* and *Plantago asiatica*. It is of great interest to know how far this type of generalization, made for populations of a simple species, can be extended to include populations of several species. The studies of density-dependent mortality in *Papaver* spp. made by Harper & NcNaughton (1962) suggest that pure stands behave in essentially the same way as that suggested by Yoda *et al.*, but that in

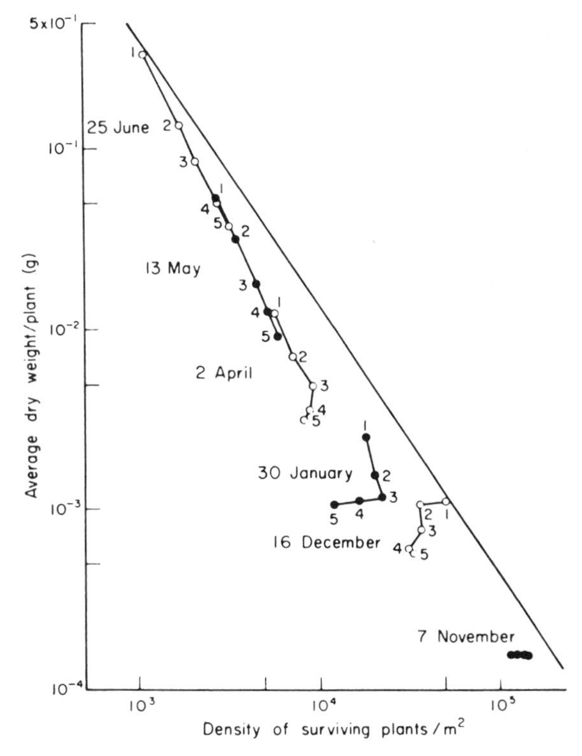

FIG. 4. Changes in numbers and individual plant weight of *Erigeron canadensis* with time. Observations on an abandoned field at Osaka, Japan. The field contained a steep fertility gradient and the plot numbers 1–5 represent an order of decreasing fertility which was exaggerated by the addition of N-P-K-Mg fertilizer in the ratio 5:4:3:2:1 on the plot numbers 1–5. Seed of *E. canadensis* was distributed evenly over the ground $1-2 \times 10^5$ seeds/m^2. (Redrawn from Yoda *et al.* 1963.)

mixed stands the processes of self-thinning and alien-thinning (intra- and interspecific effects) are subtly different. A very important aspect of the work of Yoda *et al.* is that it examines both the response of numbers and of individual plant size (i.e. both mortality and plasticity) to changing density. Most previous attempts to study density effects in plant populations have looked at changes in mean plant weight while density is artificially regulated, or have examined changes in plant numbers whilst ignoring plant weight. Because of the comment made earlier that the study of plant demography is hindered by plant plasticity, the attempt by Yoda *et al.* to take plasticity into account in a generalized theory is important. Their experiments suggest that mortality is a continuing risk through the life of the plant, continually adjusting the numbers in the population in relation to the increasing size of the plants. This to an extent contradicts Darwin's view that mortality is concentrated in the seedling stage, and much work is needed to clear up this point and obtain accurate life tables for a number of plant species.

'There is no exception to the rule that every organic being naturally increases at so high a rate that, if not destroyed, the earth would soon be covered by the progeny of a single pair.'

'Lighten any check, mitigate the destruction ever so little, and the number of the species will almost instantaneously increase to any amount.'

'A struggle for existence inevitably follows from the high rate at which all organic beings tend to increase.'

The manner in which potentially explosive plant populations are regulated or controlled in nature is obscure. The pattern of population growth is commonly taken to be logistic

$$\frac{dN}{dt} = rN\left(\frac{K - N}{K}\right)$$

Rate of growth of population = Intrinsic rate of natural increase ×
Degree of realization of the potential increase

In environments in which there is a recurrence of natural hazards, populations may spend most of their time recovering from the hazards—increasing their population size at a rate near to the intrinsic rate of natural increase. The size of such populations may frequently be a function of the magnitude of the last catastrophe and the time available for recovery. Regulation preventing an excessive population size may then be a relatively rare occurrence. This may well be true for many annual plant species of disturbed habitats such as arable weeds. More stable environments are likely to contain populations which spend most of their time near to the K value or saturation level of the population and then density-dependent regulating processes may control population growth continuously.

The intrinsic rate of natural increase of higher plants is a function of seed output and of vegetative reproduction. These two forms of increase represent different values of r, the one appropriate for increase over a broad geographical range and the other for immediate and local colonization. The higher (seed) value of r is associated not only with high risk but also with a spread of the risk over a large, sometimes very large, number of small capital investments (or bets!). The lower (vegetative) value of r is often associated with heavy and continuous capital investment, a 'cautious' policy of placing the investments and a low risk. I know of no attempts to compare the capital investment in seed and vegetative reproduction in any species that possesses both. Comparisons of the seed output of a range of species were made by Salisbury (1942) and these are invaluable starting material for a study of the reproductive strategy of higher plants. However, numbers and seed size are not all of the qualities needed to assess or compare reproductive strategy of different species and perhaps the most important is some measure of the proportion of the annual capital increment of a plant which is invested in reproduction. Cody (1966) and R. H. MacArthur (unpublished) have argued that the way in which the resources of an organism are proportioned may be of profound ecological importance, particularly in comparative biology. Cody was concerned with clutch size in birds and argued that available energy may be partitioned between three ecologically important ends (amongst others): (a) contributions to r, the intrinsic rate of natural increase (in plants this represents energy put into seed and ramet production); (b) contributions to competitive ability, adaptations in relation to the K value (in plants this presumably represents energy spent in putting leaves higher than neighbours—long petioles, tall stems, or in possessing roots which grow and search faster and further than neighbours); and (c) contributions to predator avoidance (which in plants presumably corresponds to energy spent producing unpalatable structures, defensive chemicals, spines, stinging hairs, etc.).

Few attempts have been made to compare the ways in which different species of plant allocate their limited resources. The procedural difficulties are immense—it is necessary to estimate roots (always an ecologist's nightmare) and to estimate tissues which are shed as the plant grows; there is much difficulty in determining seed output because so many plants shed their seeds over a long period instead of being neat and tidy like crops and holding all their seeds until harvest.

Two examples of attempts to partition a higher plant's activities are shown in Figs. 5 and 6. I suggest that these represent ways of describing the behaviour of a plant which will be of great ecological interest when sufficient examples have been studied for generalizations to be made. It should then be possible to answer such ecological questions as the following:

(1) Is the proportion of a plant's output that is devoted to reproduction higher in colonizing species than in those of mature habitats?

(2) Is the proportion of a plant's output that is devoted to reproduction fixed or plastic? Is it changed by inter- and intraspecific density stress?

(3) Does the proportion of a plant's output that is devoted to reproduction differ between plants of hazardous climatic conditions and those of more stable environments such as a tropical rain forest?

(4) Do plants adapted to competitive environments devote a greater proportion of energy to non-photosynthetic organs (such as support organs)?

(5) Do plants with vegetative reproduction sacrifice a proportion of the energy which would otherwise be expended on seed? Are the two processes competitive within the plant?

(6) What is the relative energy expended in producing a vegetative propagule and a seed? Can this expenditure be related to the relative risks of establishment by the two means and the relative ecological importance of local and long-distance spread?

(7) What is the expenditure on organs ancillary to reproduction—the economic cost to the plant of attractive flowers, a pappus or the massive woody cones of conifers? Is this expenditure a measure of the selective advantages due to possessing these organs?

A whole branch of plant ecology lies almost untouched in attempts to understand the significance of the strategy of reproduction (a very different matter from the tactics—the significance of specific dispersal mechanisms, etc.).

'The only difference between organisms which annually produce eggs or seeds by the thousand, and those which produce extremely few, is that slow breeders would require a few more years to people under unfavourable conditions, a whole district, be it ever so large.'

The strategy of the life cycle itself is an ecologically fascinating but neglected subject of study. Cole (1954) and, more recently, Lewontin (1965) have examined the consequences to population growth of changes in the life cycle of plants and animals. Their studies show that a high intrinsic rate of natural increase (r) can be obtained by producing a few offspring early in life—precocious reproduction is all important. For example, a population of which the individuals produce two offspring in the 1st year of life and then die, will have a potential rate of increase as high as if the individuals produce one offspring every year for ever.

It is instructive to examine the consequences of life cycle strategy in such a species as

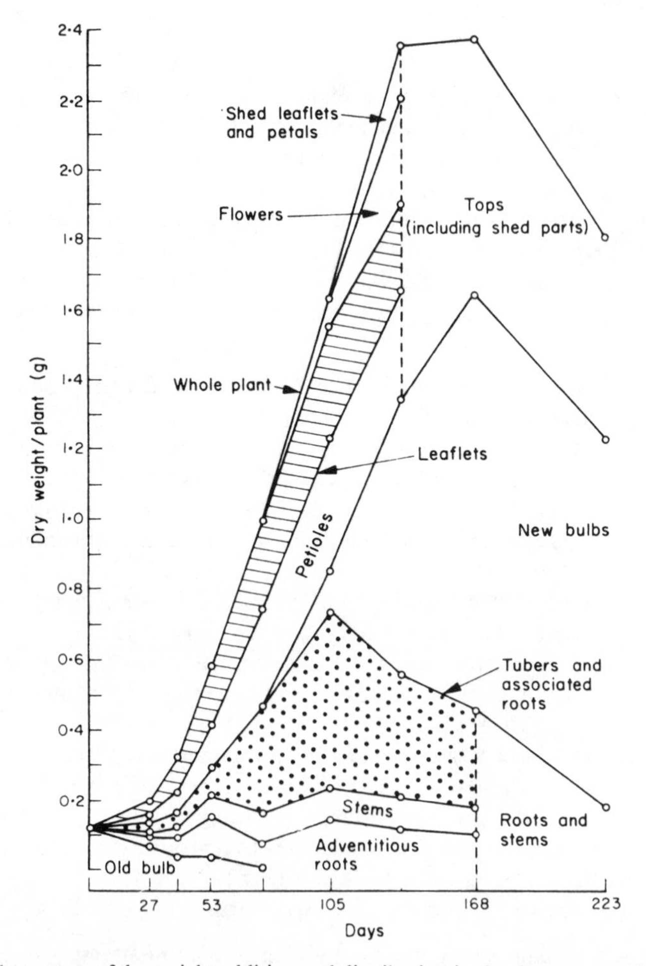

FIG. 5. The pattern of dry weight addition and distribution in *Oxalis pescaprae* L. from the time of planting of the bulb. (Redrawn from Michael 1965.)

the foxglove, which is commonly monocarpic, and under favourable conditions is biennial. In an unexploited environment an annual species producing S seeds per annum could theoretically show a population growth of 1, S, S^2, . . . , S^n in succeeding years. To achieve the same rate of population growth a biennial species would require to produce S^2 seeds at the end of each 2-year period. Thus if a biennial such as the foxglove (*Digitalis purpurea*) produces 100 000 seeds every 2 years, its annual counterpart would require to produce only 333 seeds to achieve the same population growth rate. However, if there is a significant mortality risk which is concentrated in the seed and seedling stage, the annual will experience this risk every year and the biennial only in its 1st year of growth. Thus if the 1st year mortality risk is high, a biennial with seed production $S^{<2}$ will maintain the same population growth rate as an annual with seed production S. Fig. 7 shows the calculated relationships between x, the probability of a seed producing a plant that survives through the first season, and p, the power by which the seed production S of an annual would need to be raised to permit equivalent biennial reproduction. This is shown for various values of S.

The following points emerge: (1) Where $x = 1$, and all seeds produce a mature plant, the biennial must bear the square of the number of the seeds of the annual. (2) As the chance of the plant surviving the 1st year decreases p decreases. (3) At $p = 1$ the biennial

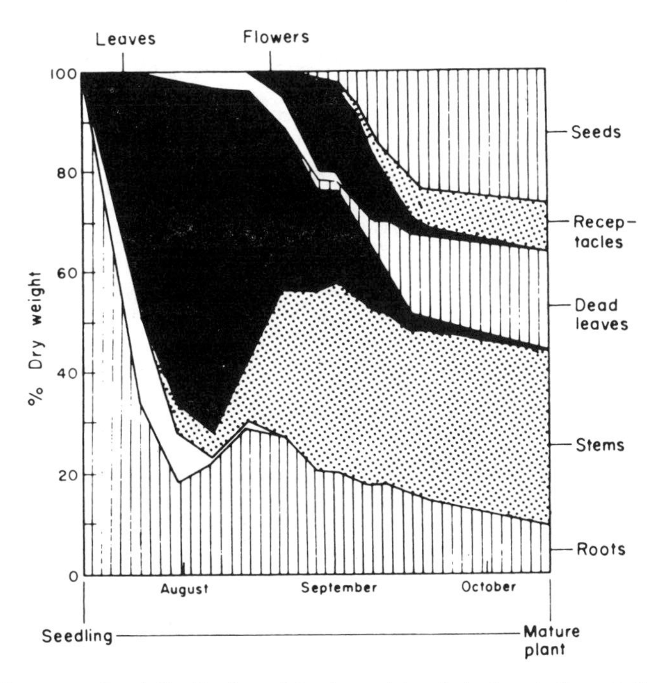

FIG. 6. The proportional distribution of the dry weight of plant parts through the life cycle of *Senecio vulgaris* (data of J. Ogden).

produces the same number of seeds as the annual—this corresponds to the state at which each plant, whether annual or biennial, leaves only one replacement. (4) When $p < 1$ the mortality rate exceeds the seed production and the populations decline. It is apparent that a population of annuals producing S seeds per plant will decline faster than one of biennials bearing S seeds per plant.

This sort of theoretical argument, even in the simple case argued above, emphasizes the rather subtle interplay between the strategy of the life cycle, the population parameters and the timing of the mortality risk.

Similar arguments can clearly be made for a wide variety of alternative strategies. For example: (a) If seed number is sacrificed in favour of seed size, what increase in survival value of a seed is required to maintain the same potential for population increase? (b) In the face of recurrent hazard conditions, of specified frequency and magnitude, what is the optimal fraction of seeds remaining dormant or undergoing dispersal (see Cohen 1966). (c) In what ways is the optimal strategy altered if overcrowding is a more common experience than unhindered population growth?

The great value of this type of theoretical approach lies in the extent to which it focuses attention on quantitative aspects of the life history and behaviour of species which are of acknowledged importance but have remained part of natural history rather than of science.

Even though a high intrinsic rate of natural increase is of great importance to a colonizing species—and such a high rate can be obtained by precocious reproduction

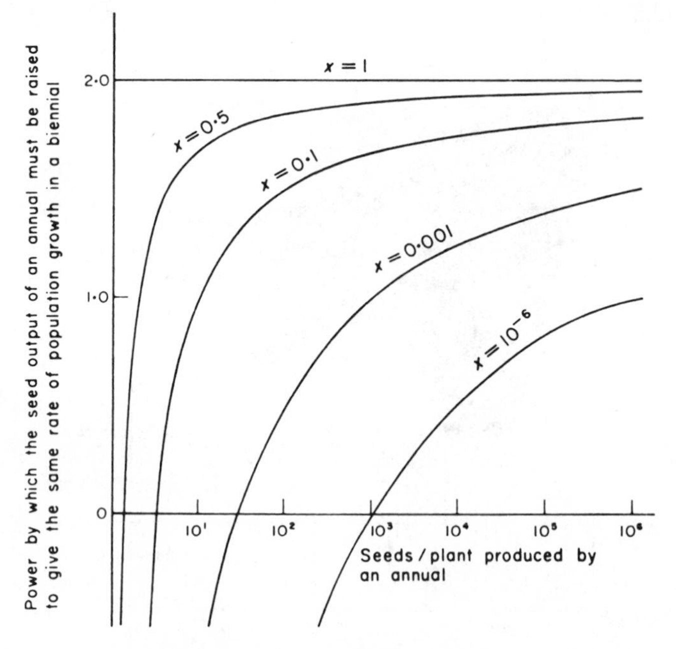

FIG. 7. The theoretical relationship between the seed output of an annual and a biennial for an equal intrinsic rate of natural increase at various values of density-independent mortality (x) where this risk is confined to the 1st year of development (data of R. Oxley).

—very few plants use all their reproductive output for immediate multiplication. Seed dormancy is widespread, and it is clear from an elementary consideration of life cycle strategy that a precociously produced seed loses the advantage of precocity if it is unable to start the new generation quickly. An annual colonizing species like *Avena fatua* loses much of its potential rate of population increase by the seed dormancy mechanism which ensures one to several years delay in germination. The adaptive gain from dormancy must be set against an adaptive loss in the intrinsic rate of natural increase. It is all the more interesting that *A. fatua* in the more predictable agriculture and climate of central California seems to lack seed dormancy (Harper 1965).

Much of Darwin's concept of the behaviour of organisms in nature centres around the concept of struggle, '. . . as more individuals are produced than can possibly survive, there must in every case be a struggle for existence, either one individual with another of the same species, or with the individuals of distinct species, or with the physical conditions of life.'

Intraspecific struggle, in which individual development is restricted because of interference from neighbours, has been studied as an agronomic problem—what is the optimal density of plants per unit area needed to achieve maximal dry matter production per area, or maximal economic yield per unit area? At low plant densities plants may not interfere with each other, but as density increases the growth of the population becomes limited by a shortage of environmental supply factors—such as light, water and nutrients—and the growth made by the population becomes a function of the availability of supplies rather than the number of individuals. In many agronomic experiments, self-thinning is absent, so that all density stress is absorbed in the plastic development of the individuals.

Various attempts have been made to fit mathematical relationships to the yield/density responses of crop plants. Amongst the most successful of these have been various forms of the reciprocal yield law

$$\frac{1}{w} = a + bx \text{ where } x = \text{density and } w = \text{mean plant weight.}$$

Various modifications of this basic law are found in the work of Kira, Ogawa & Sagazaki (1953), and other papers of this Japanese School, De Wit (1960) and Holliday (1960). This relationship, which has wide applicability to the study of the behaviour of plants in pure stands, assumes that: (1) the increase in plant dry weight is logistic, (2) the initial growth rate is independent of plant size, (3) the final yield per unit area is constant at high density, (4) time is measured from a common time of sowing, and (5) density-dependent mortality does not occur. If it does, a model similar to that of Yoda *et al.* (1963), discussed earlier, may be preferable.

The major disadvantage of this form of description of population behaviour is that it focuses attention on yield per unit area, or on mean plant behaviour, usually measured by dividing yield per unit area by the number of plants present. This obscures the existence of plant to plant variation. When populations of plants under density stress are examined by sampling and measuring individual plants, curious effects of density on the frequency distribution are revealed. Koyama & Kira (1956) showed that populations of plants which at low density showed normally distributed plant weight, progressively developed log-normal distributions with the passage of time and with increase in density. This phenomenon is shown in Fig. 8 for populations of a variety of fibre flax (Obeid 1965). Stern (1965) has shown the same phenomenon in populations of subterranean clover. It seems that under conditions of density stress not only is there a forced sharing of limited resources with a compensating plastic reduction in individual development, but that a hierarchy emerges amongst the individuals in the population. This hierarchy consists of a few large individuals and an excessive number of small. Such a hierarchy develops in natural habitats as well as in the model crop experiments. R. Oxley has recently shown strikingly log-normal distributions for capsule number per plant in *Digitalis purpurea* in natural habitats and R. Bunce for hazel shoot length in coppices. Apart from the significance of these observations to the interpretation of population samples (in a log-normal distribution the mean plant is *not* the most representative) they show that dominance and suppression may develop within a single species stand. The direct consequences of density stress on a plant population are therefore three-fold: (i) to elicit a plastic response from the individuals as they adjust to share limiting resources, (ii) to increase mortality, and (iii) to exaggerate differentials within the population and encourage a hierarchy of exploitation.

Just as in a population of a single species the stress of density intensifies the expression of small differences (genetic or environmental) between individuals, so in mixed populations density stress may exaggerate and exploit interspecific differences. The experimental models of De Wit (1960) are superbly designed to study the behaviour of two species in mixture and so to begin the exploration of natural diversity. In these models two species are sown together in varied proportions while the overall density of the sown or planted mixture is maintained constant. The behaviour of a species can then be compared in pure stand with its performance in variously proportioned mixtures, and the mutual aggressiveness of two forms may be measured.

Fig. 9 illustrates the use of such an experimental design for the study of the interaction of *Chenopodium album* with barley and with kale. The results of experiments designed

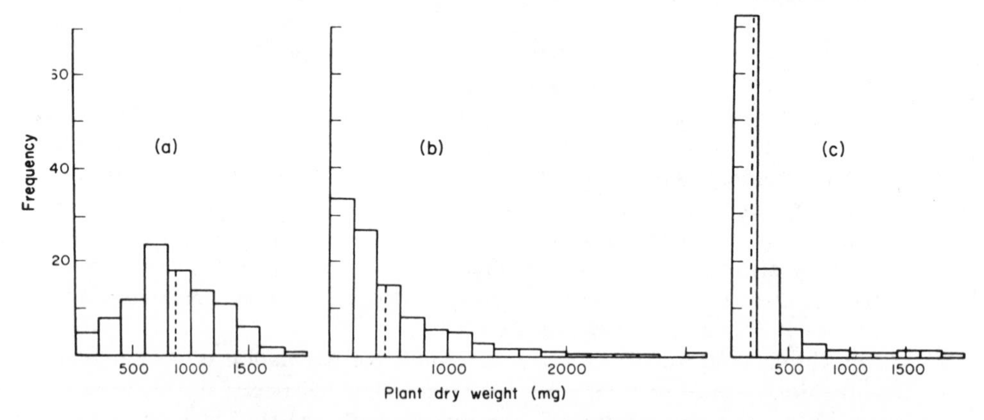

FIG. 8. The influence of density on the frequency distribution of individual plant dry weight in *Linum usitatissimum* sown at densities of (a) $60/m^2$, (b) $1440/m^2$ and (c) $3600/m^2$, and harvested at seed maturity (from Obeid 1965).

in this way may also be conveniently expressed as 'ratio diagrams' relating the sown proportions of two species to the harvested proportions. Such a diagram has considerable value in understanding both the progress of extinction of one species by another and the processes by which an equilibrium between species may be produced. Fig. 10 illustrates some idealized forms of ratio diagrams which are considered in more detail in De Wit (1960).

Darwin was clearly aware that in nature there is a frequent ousting of one form by another, either over long periods as witnessed in the palaeontological record, involving extinction, or over short periods, involving successional changes of vegetation and purely local elimination. At times he gives the impression that the struggle for existence between forms must lead to a winner and a loser. He sees this as particularly likely between closely related species or varieties. 'In the case of varieties of the same species, the struggle will generally be almost equally severe (*cf. members of the same species*). and we sometimes see the contest soon decided.' But he is also aware that in nature diversity is everywhere, and far from this involving perpetual struggles to extinction 'Battle within battle must be continually recurring with varying success; and yet in the

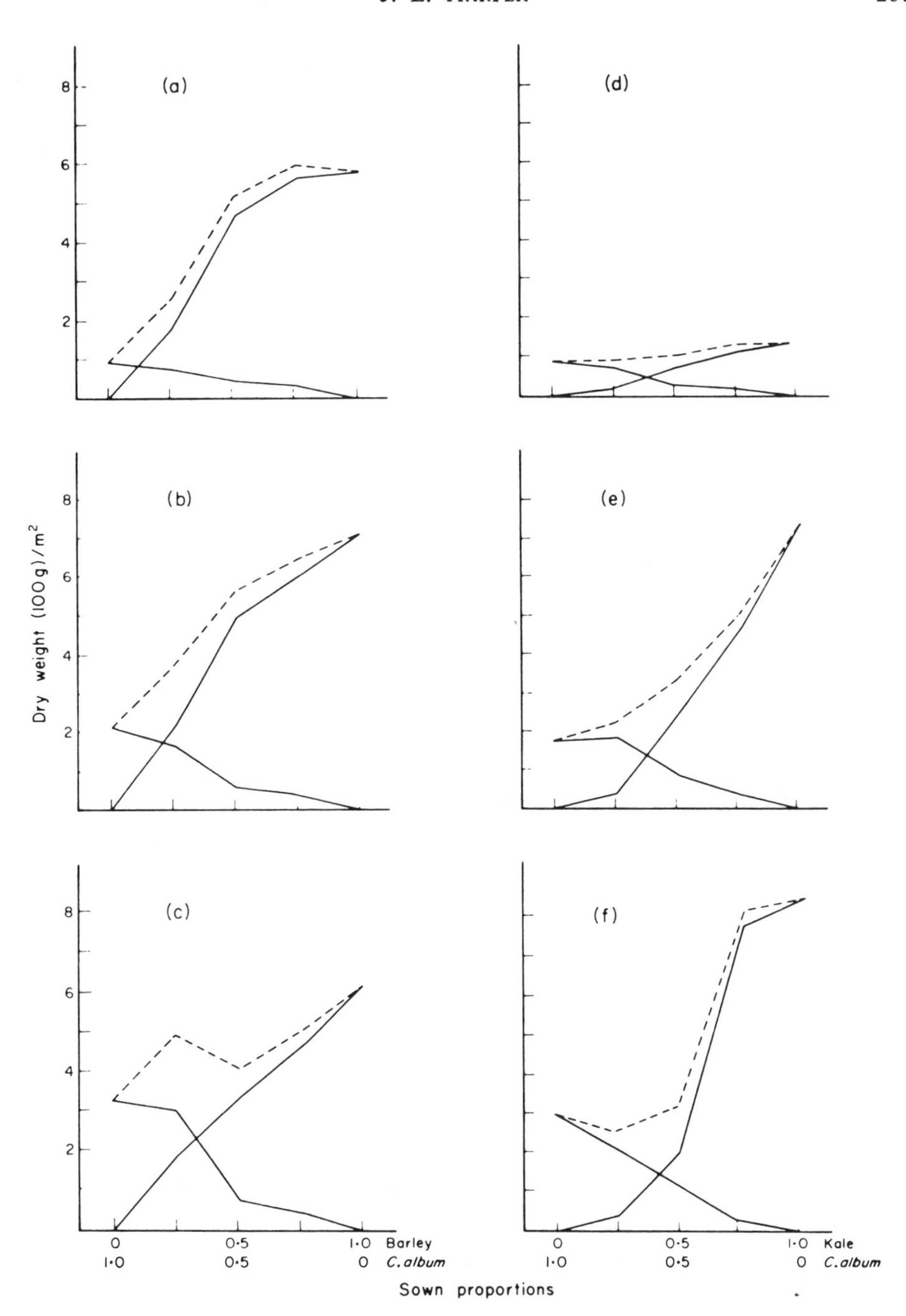

FIG. 9. The influence of proportion of two plant species in mixture at constant density on dry weight per unit area of each component and combined yield (– – – –). (a), (b) and (c) *Chenopodium album* and barley at three successive harvest dates. (d), (e) and (f) *C. album* and kale at three successive harvest dates (from Williams 1964).

long run the forces are so nicely balanced that the face of nature remains for long periods of time uniform, though assuredly the merest trifle would give the victory to one organic being over another.'

The existence of natural diversity implies that the struggle for existence is not regularly forced to decide between stronger and weaker brethren—and that the struggle between some forms living in the same area is either evaded or does not occur. This poses the question which has worried so many zoologists, most notably Gause (1934), Gause & Witt (1935) and Hutchinson (1966) and his various colleagues, what are the characteristics of species which permit them to co-occur in the diversity of nature? This has been

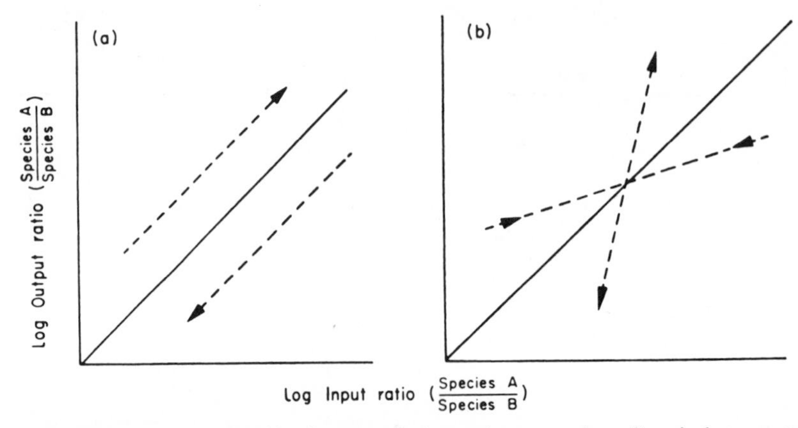

FIG. 10. Theoretical relationships between the input (sown or planted) ratio between two species and the output (seed harvested) ratio. (a) Frequency independence in which the empirical relationship has unit slope (– – →) lying above or below the continuous line which represents input ratio = output ratio. Such situations lead to the progressive extinction of one or other component of the mixture. (b) Frequency dependence in which the empirical relationships have a slope > or < unity (– – →) and intersect the line of input ratio = output ratio. A slope of > 1 implies that the population moves towards the extinction of the minority component. A slope of < 1 implies equilibration and stabilization of the mixed condition.

examined in great and fascinating detail in experiments and theoretical models and a conclusion is conveniently expressed by Hutchinson in the modified logistic equation of growth for two species living together:

$$\mathrm{d}N_1 = r_1 N_1\left(\frac{K_1 - N_1 - \alpha N_2}{K_1}\right)$$

$$\mathrm{d}N_2 = r_2 N_2\left(\frac{K_2 - N_2 - \beta N_1}{K_2}\right)$$

where N_1 and N_2 are the numbers of individuals of species S_1 and S_2.

The conditions under which both S_1 and S_2 survive is that $\alpha < K_1/K_2$ and $\beta < K_2/K_1$, or that the growth of each species population inhibits its further growth more than it inhibits that of the other species. This solution assumes a linear competition function, and there are likely to be further solutions where the function is not linear. This solution generalizes a host of differences between organisms which may permit them to evade a struggle to the death and so permit diversity. In animals it is easy to see that differences in food taken or in feeding habits, differences in habitat preference, differences in prey (or in predators) may prevent an exclusive struggle by focusing the intensive battles within rather than between the species. In plants it is less easy to see how a group of species which all require the same basic food requirements—light, water and mineral nutrients—may possess sufficiently diverse biologies to prevent a best species from excluding all others. There are, however, a number of experiments which show that mixtures of two plant species may form stable associations and indeed possess self-stabilizing properties.

Van den Bergh & De Wit (1960) grew *Phleum pratense* and *Anthoxanthum odoratum* together in field plots at a range of proportions and compared the ratio of tillers of the two species after the first winter with the ratio after the second winter (Fig. 11). In plots

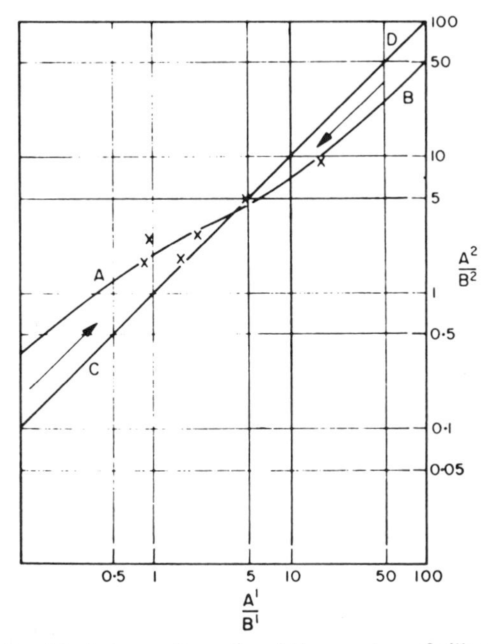

Fig. 11. The relationship between the ratio of the number of tillers of *Anthoxanthum odoratum* (A) and *Phleum pratense* (B) after the first winter (A^1/B^1) and after the second winter (A^2/B^2). (Redrawn from Van den Bergh & De Wit 1960).

in which *A. odoratum* had been in excess, the proportion of *Phleum pratense* increased. Where *P. pratense* was in excess, *Anthoxanthum odoratum* increased. Thus the mixture possessed self-stabilizing properties—the stable mixture under this set of environmental conditions being the point at which the experimentally obtained ratio line A–B on Fig. 11 intersects with the line of unit slope C–D (cf. Fig. 10). In a similar experiment performed in a controlled environment chamber, *A. odoratum* was at a slight advantage over *Phleum pratense* at all relative frequencies, and at all proportions and the population changed regularly towards increasing dominance by *Anthoxanthum*.

A stabilizing situation is also found in mixtures of *Lolium perenne* and *Trifolium repens*. Lieth (1960) has described how in pastures these two species form a moving mosaic in which patches of the mosaic dominated by grass tend to be invaded by clover, while patches dominated by clover are overrun by grass. Ennik (1960) grew *Lolium perenne* and *Trifolium repens* in mixtures of varying proportions in controlled environments at high and low light intensities. Ratio diagrams relating the proportion of the two species at the beginning and end of the treatment are shown in Fig. 12. At high light intensity the populations changed regularly towards extinction of the grass. At low light intensities the populations equilibrated, those with excess clover tending to increase their grass content, and those with excess grass tending to increase in clover. In this example the ecological differentiation between the species which permits this stabilization is almost certainly due to differences in their nitrogen nutrition.

A further example of stabilizing action in mixture has been demonstrated by P. D. Putwain for the sexes of the sexually dimorphic *Rumex acetosella*. In permanent grasslands this species only rarely reproduces by seed and the main means of spread is by

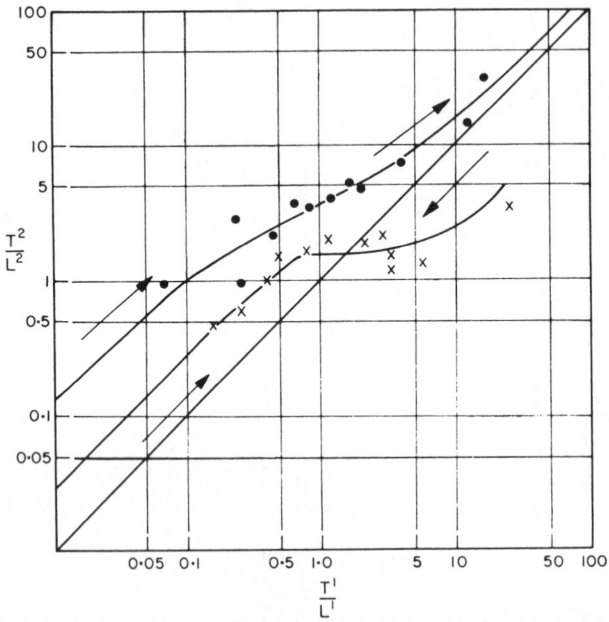

FIG. 12. The ratio of the length of stolons of *Trifolium repens* (T) and the number of tillers of *Lolium perenne* (L) in the first winter (T^1/L^1) plotted against the same ratio in the second winter (T^2/L^2). ●, At high light intensity; ×, at low light intensity. (Redrawn from Ennik 1960.) The same data are presented with calculated linear regressions in De Wit *et al.* (1960).

root buds. In dense stands the two sexes are therefore the equivalent of two species with different biologies, one having the whole task of seed production and the other the apparently light responsibility of producing pollen. It might be expected that two such forms would differ in aggression in the community and that in old pastures the 50 : 50 sex ratio of seedlings would become biased. In fact, natural populations have been shown by Löve (1944) and many others to have a sex ratio not departing significantly from equality.

An experiment was designed on the 'De Wit' model to study the population balance of the sexes in this species. Clonal populations of males and of females were obtained from root fragments and were grown at various proportions at low and high densities. The results of the high density treatment are shown in Fig. 13. Populations with an excess of males increased the proportion of females, and those with excess females increased the proportion of males. Experimental distortion of the sex ratio from equality led to readjustment towards equality as a result of the more vigorous reproduction by the minority sex.

In the three experiments described above, reproduction was clonal and the adjustment of balance between the species (or sexes) was by differential vegetative reproduction. All these examples illustrate the essential criterion for a balanced 'co-occurrence' of two species—that the minority component should always be favoured.

A similar model was shown by Harper & McNaughton (1962) for the survival of *Papaver* spp. sown in mixture. In this case each individual of a species in a mixture of two species had the greatest likelihood of becoming an adult plant when it was in the minority. In the discussion of the poppy experiments we contrasted 'self-thinning' (the

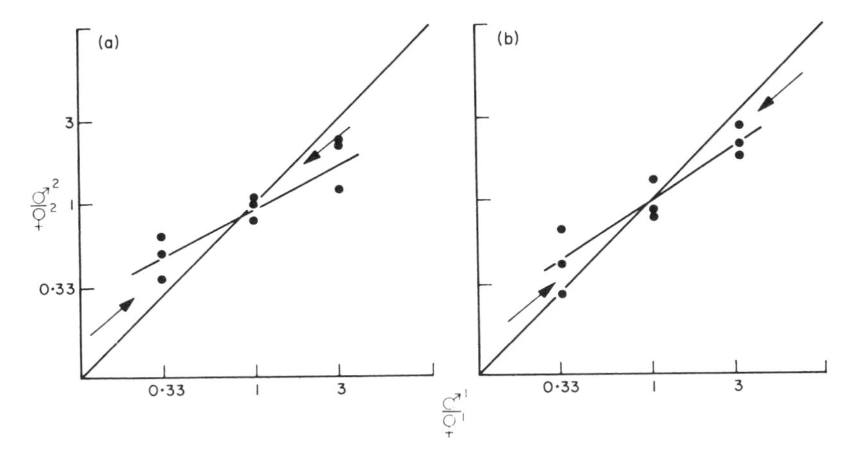

FIG. 13. The ratio of male to female plants of *Rumex acetosella* planted in the spring ($♂^1/♀^1$) plotted against the same ratio in the autumn ($♂^2/♀^2$). (a) Dry weight of above ground parts, (b) number of inflorescences (data from P. D. Putwain). Values for three replicates are shown individually.

reaction of a species to an increase in its own density) with 'alien-thinning' (the reaction of a species to the density of its associated species). These experiments suggested that each species in a mixture suffered more severely from intra- than from interspecific interference. It was argued that this was a necessary condition for stable diversity. It is interesting to find that a similar differentiation between self- and alien-thinning occurs in mixtures of *Medicago sativa* and *Trifolium pratense*. Black (1960) sowed these two species in variously proportioned mixtures, and recorded plant survival. From his data it is possible to calculate the chance of a seed forming a plant and this is shown for two ranges of density in Fig. 14. The nature of the differences between poppy species and between *Medicago sativa* and *Trifolium pratense* which alleviate interspecific interference are obscure. It is presumably an understanding of these critical differences between species that will provide the 'explanation' of stable diversity in nature.

In the various models of the behaviour of two species in mixture, the existence of a stable equilibrium depends on *frequency-dependent* 'competition'. This phenomenon has recently been recognized by Pavlovsky & Dobzhansky (1966) in *Drosophila* cultures, and, as they point out, it plays havoc with the Sewall Wright concept of adaptive values.

Selection genetics is therefore forced to take account of fundamental ecological phenomena and the convergence of population genetics and population ecology provides one of the most exciting fields of development in modern biology.

Recently there has been growing interest in the genetic consequences of a struggle for existence between two populations and the ecological significance of these changes. The experiments have all involved flies, and in a lecture on plant ecology I have assumed that *Drosophila* (by usage) and all flies (by analogy) are appropriate objects for botanical study!

Pimentel *et al.* (1965) allowed populations of blowflies and houseflies to multiply with a controlled food supply in a network of interconnected population cages. He

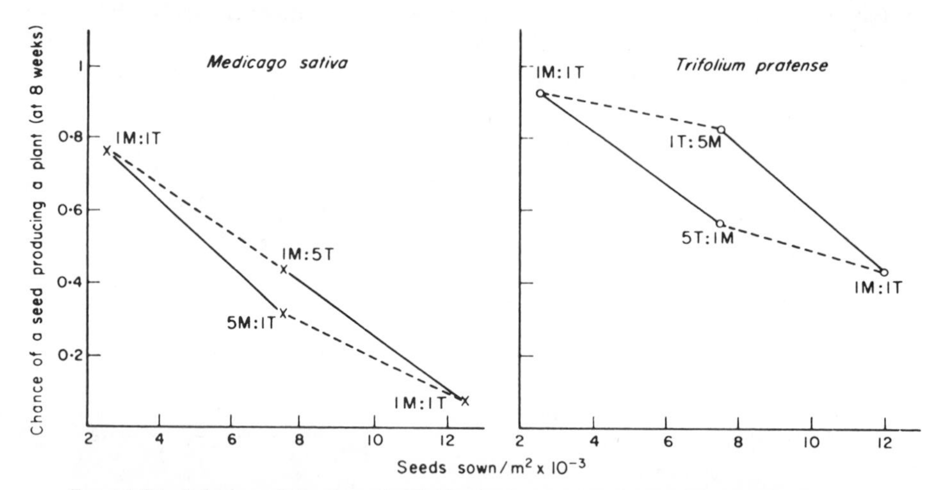

FIG. 14. The influence of density on 'self-thinning' and 'alien-thinning' in mixed populations of *Trifolium pratense* (T) and *Medicago sativa* (M). The extreme densities, 2500 and 12 500 seeds/m², were composed of equal numbers of the two species. The intermediate density, 7000 seeds/m², was either with a preponderance (5:1) of one species or the other.

The difference between the chance of a seed producing an 8-week-old plant at the two differently constituted intermediate densities reflects the different degrees of thinning in predominantly 'self' and predominantly 'alien' populations. In each diagram the broken line indicates 'alien-thinning' and the continuous line 'self-thinning' (i.e. inter- and intraspecific effects respectively). Calculated from data in Black (1960).

observed oscillation in the populations of the two species. By sampling and testing his two populations against each other at intervals he was able to show that there was an oscillation in the 'competitive ability' of the two species. He argued that in a mixture the minority species faces predominantly interspecific interference, and is therefore under selection to improve its performance relative to the majority species. The majority species, however, experiences predominantly intraspecific interference and hence selection within it is concentrated on qualities important in the intraspecific struggle. The minority species eventually gains the majority because of its newly selected qualities, and interspecific selection immediately concentrates on the new minority species. The two species thus oscillate in relative abundance. Pimentel shows that the amplitude of oscillations may become progressively damped and he suggests that the two species are attaining an equilibrium.

Seaton & Antonovics (1967) reared wild type and 'dumpy' *D. melanogaster* in milk bottle cultures. The two forms were reared in mixtures, but care was taken that mating did not occur between them (by removing virgin females and mating them with males of the same type from their own culture). Each succeeding generation was started with six wild type and six dumpy flies from the preceding experiment, and the food supply was limited. After four generations the two mutually selected populations were tested against each other and against stock populations using a 'De Wit' model for the experimental testing. The results are shown in Fig. 15.

The results of mutual selection were at first sight surprising. Instead of one type of fly acquiring increased success in a struggle for existence the selection process seems to

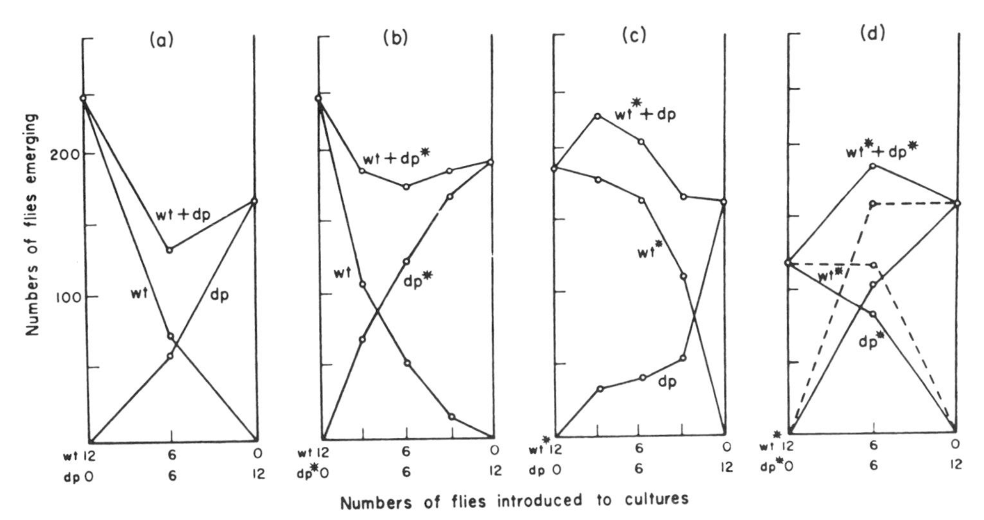

Fig. 15. The results of a replacement experiment (cf. Fig. 9) involving wild type and dumpy *Drosophila melanogaster* grown in pure cultures and mixtures. The two forms were prevented from mating with each other.

The success of each component in mixture and pure culture was assessed as the number of offspring hatched in the first generation of progeny. (a) Stock 'wild type' with stock 'dumpy'. (b) Stock 'wild type' with selected 'dumpy'. Selected 'dumpy' had experienced four repeated generations in mixed culture with 'wild type'. (c) Selected 'wild type' with stock 'dumpy'. Selected 'wild type' had experienced four repeated generations in mixed culture with 'dumpy'. (d) Selected 'wild type' with selected 'dumpy' (after one further generation of relaxed selection). – – –, Replicate experiment; wt, Stock wild type; wt*, selected wild type; dp, stock dumpy; dp*, selected dumpy. (From Seaton & Antonovics 1967.)

have resulted in an ability of the wild type and the 'dumpy' fly to avoid interference from each other. At the end of the experiment neither wild type nor dumpy had as depressive effect on the other, as did mutually unselected strains. Moreover, the mutually selected strains grown together achieved a greater output of offsprings than did either of the stock types grown in pure culture.

This would appear to be a case of direct selection for some difference in niche occupancy, the evolutionary process which Ludwig (1950) has called 'annidation'.

The experiments of Pimentel and of Seaton & Antonovics illustrate three different evolutionary solutions to the problem of intergroup struggle for limited resources: (i) extinction of one group, (ii) mutual oscillating inter-group selection leading to increased stability of the mixture (the Pimentel solution), and (iii) mutual divergence in behaviour leading to the avoidance of inter-group struggle (the Seaton–Antonovics solution).

Both (ii) and (iii) imply that in the process of natural selection the reaction to the other species may be a critical force '. . . . the structure of every organic being is related, in the most essential yet often hidden manner, to that of all the other organic beings, with which it comes into competition for food or residence, or from which it has to escape, or on which it preys.' Goodall (1966) has recently argued a similar point, though he despaired of the chance of demonstrating such coadaptive evolution in action.

The concept of natural selection operating between groups of organisms and adjusting their mutual strategies to permit increasing diversity and more efficient environmental exploitation strengthens the image of the community as an integrated whole rather than a Gleason-type assemblage of individuals.

The logical extension of the view that community stability and diversity depend on niche specialization, is that complex ecosystems are more efficient than simple in using environmental resources. Yet the rigid demonstration that mixtures of plant species outyield pure stands seems not to have been made. Darwin wrote 'It has been experimentally proved that if a plot of ground be sown with one species of grass, and a similar plot be sown with several distinct genera of grasses, a greater number of plants and a greater dry weight of herbage can be raised in the latter than in the former case.' More than 100 years later, the Director of the Grassland Research Institute writes 'No research, as far as I am aware, has yet shown that a mixture of two or more grass species, when sown together, will outyield the crop that can be obtained from a single bred variety' (Woodford 1966). Perhaps the answer to this contradiction lies in the degree to which the forms that are grown together have been mutually selected. It may be too much to hope that any pair of species grown together will already possess the precise biological differences needed to interniche or annidate effectively. Perhaps special breeding programmes are needed to develop 'ecological combining ability' (Harper 1964b) in agricultural crops.

With the exception of examples of mutual exploitation of the light environment shown by the studies of Salisbury (1916) in woodland—there are few demonstrations that stable mixed vegetation is compounded of ecologically complementary species. Recently, a number of experimental designs have been suggested which permit the 'ecological combining ability' of groups of species to be examined, in the same way that a plant breeder explores the 'genetic combining ability' of a range of genotypes (Sakai 1961; Williams 1962; McGilchrist 1965; Harper 1964b; Norrington Davies 1967). These are all based on diallell analysis, in which a number of species are grown in all possible combination of pairs and their yield as mixtures is contrasted with the yield of pure stands. In this way the ecological equivalents of the genetic concepts of dominance, recessiveness, overdominance and interaction can be detected and—more important—measured. These techniques seem likely to bring to experimental synecology a refinement and subtlety appropriate for a science which has outgrown its qualitative and descriptive youth.

It is impossible in the time of one lecture to do justice to the range of ecological thought and the stimulus to modern experimental ecology to be found in just two chapters of *The Origin of Species.* I have wholly omitted the fascinating matter of animal–plant interactions that play so large a part in Darwin's ecology. I am not amongst those very few ecologists since Darwin who are intellectually equipped to deal with the plant–animal interface. My aim has been to try to show that much of what is exciting to me in the science of plant ecology in the late sixties has a highly respectable origin in the ecological thinking of Darwin. 'It is never safe for a biologist to announce a discovery if he has not read and mastered *The Origin of Species*.' (Keith 1928). A presidential address is a rare opportunity for a personal expression of belief. My own is that *The Origin of Species* states precise questions which plant ecologists can usefully spend the next hundred years answering.

ACKNOWLEDGMENTS

I am grateful to J. Ogden, P. D. Putwain and R. Oxley for allowing me to use material which has not yet appeared in thesis form.

REFERENCES

Antonovics, J. (1966). *Evolution in adjacent populations.* Ph.D. thesis, University of Wales.

Black, J. N. (1960). An assessment of the role of planting density in competition between red clover (*Trifolium pratense* L.) and lucerne (*Medicago sativa* L.) in the early vegetative stage. *Oikos*, **11**, 26–42.

Bergh, J. P. van den & Wit, C. T. de (1960). Concurrentie tussen Timothee en Reukgras. *Meded. Inst. biol. scheik. Onderz. Landb Gewass.* **121**, 155–65.

Cody, M. L. (1966a). A general theory of clutch size. *Evolution, Lancaster, Pa.* **20**, 174–84.

Cohen, D. (1966). Optimising reproduction in a randomly varying environment. *J. theor. Biol.* **12**, 119–29.

Cole, L. C. (1954). The population consequences of life history phenomena. *Q. Rev. Biol.* **29**, 103–37.

Ennik, G. C. (1960). De concurrentie tussen witte klaver en Engels raaigras bij verschillen in lichtintensiteit en vochtvoorziening. *Meded. Inst. biol. scheik. Onderz. LandbGewass.* **109**, 37–50.

Gause, G. F. (1934). *The Struggle for Existence.* Baltimore.

Gause, G. F. & Witt, A. A. (1935). Behaviour of mixed populations and the problem of natural selection. *Am. Nat.* **69**, 596–609.

Goodall, D. W. (1966). The nature of the mixed community. *Proc. ecol. Soc. Aust.* **1**, 84–96.

Harper, J. L. (1960). Factors controlling numbers. *The Biology of Weeds* (Ed. by J. L. Harper), pp. 119–32. Oxford.

Harper, J. L. (1964a). The individual in the population. *J. Ecol.* **52**(Suppl.), 149–58.

Harper, J. L. (1964b). The nature and consequences of interference amongst plants. *Proc. XIth int. Conf. Genet.*, pp. 465–81.

Harper, J. L. (1965). Establishment, aggression and cohabitation in weedy species. *The Genetics of Colonising Species* (Ed. by H. G. Baker and G. L. Stebbins), pp. 243–268. New York.

Harper, J. L. & McNaughton, I. H. (1962). The comparative biology of closely related species living in the same area. VII. Interference between individuals in pure and mixed populations of *Papaver* species. *New Phytol.* **61**, 175–88.

Holliday, R. (1960). Plant population and crop yield. *Nature, Lond.* **186**, 22–4.

Hutchinson, G. E. (1965). *The Ecological Theater and the Evolutionary Play.* Newhaven and London.

Keith, (1928). Introduction to Everyman Edition of C. Darwin, *The Origin of Species.*

Kira, T., Ogawa, H. & Sagazaki, N. (1953). Intraspecific competition among higher plants. I. Competition–density–yield interrelationship in regularly dispersed populations. *J. Inst. Polytech. Osaka Cy Univ.* Ser. D, **4**, 1–16.

Koyama, H. & Kira, T. (1956). Intraspecific competition among higher plants. VIII. Frequency distribution of individual plant weight as affected by the interaction between plants. *J. Inst. Polytech. Osaka Cy Univ.* Ser. D, **7**, 73–94.

Lewontin, R. C. (1965). Selection for colonising ability. *The Genetics of Colonising Species* (Ed. by H. G. Baker and G. L. Stebbins), pp. 77–94. New York.

Lieth, H. (1960). Patterns of change within grassland communities. *The Biology of Weeds* (Ed. by J. L. Harper), pp. 27–39. Oxford.

Löve, A. (1944). Cytogenetic studies on *Rumex* sub-genus *acetosella*., *Hereditas*, **30**, 1–127.

Ludwig, W. (1950). Zur Theorie der Konkurrenz. Die Annidation (Einnischung) als fünfter Evolutionsfaktor. *Neue Ergebnisse in Probleme der Zoologie* (*Klattfestschrift*), 516–37.

McGilchrist, C. A. (1965). Analysis of competition experiments. *Biometrics*, **21**, 975–85.

Michael, P. W. (1965). Studies on *Oxalis pes-caprae* L. in Australia. I. Quantitative studies on oxalic acid during the life cycle of the pentaploid variety. *Weed Res.* **5**, 123–32.

Norrington Davies, J. (1967). The analysis of competition experiments. *J. appl. Ecol.* (In press).

Obeid, M. (1965). *Experimental models in the study of plant competition.* Ph.D. thesis, University of Wales.

Pavlovsky, O. & Dobzhansky, T. H. (1966). Genetics of natural populations. xxxvii. The coadapted system of chromosomal variants in a population of *Drosophila pseudoobscura. Genetics, Princeton,* **53,** 843–54.

Pimentel, D., Feinberg, E. G., Wood, P. W. & Hayes, J. T. (1965). Selection, spatial distribution, and the coexistence of competing fly species. *Am. Nat.* **94,** 97–109.

Sagar, G. R. (1959). *The biology of some sympatric species of grassland.* D.Phil. thesis, University of Oxford.

Sakai, K. (1961). Competitive ability in plants: its inheritance and some related problems. *Symp. Soc. exp. Biol.* **15,** 245–63.

Salisbury, E. J. (1916). The oak-hornbeam woods of Hertfordshire: a study in colonisation. *Trans. Herts nat. Hist. Soc. Fld Club,* **17.**

Salisbury, E. J. (1916). *The Reproductive Capacity of Plants.* London.

Seaton, A. J. P. & Antonovics, J. (1967). Population interrelationships. I. Evolution in mixtures of *Drosophila* mutants. *Heredity, Lond.* (In press).

Stern, W. R. (1965). The effect of density on the performance of individual plants in subterranean clover swards. *Aust. J. agric. Res.* **16,** 541–55.

Sukatschev, V. N. (1928). [*Plant communities*] (in Russian). Moscow.

Tamm, C. O. (1948). Observations on reproduction and survival of some perennial herbs. *Bot. Notiser,* **3,** 305–21.

Tamm, C. O. (1956). Further observations on the survival and flowering of some perennial herbs. *Oikos,* **7,** 273–92.

Williams, E. J. (1962). The analysis of competition experiments. *Aust. J. biol. Sci.* **15,** 509–25.

Williams, J. T. (1964). *Studies on the biology of weeds with special reference to the genus* Chenopodium *L.* Ph.D. thesis, University of Wales.

Wit, C. T. de (1960). On competition. *Versl. landbouwk. Onderz. Ned.* **66,** 8.

Wit, C. T. de, Ennik, G. C., Bergh, J. P. van den & Sonneveld, A. (1960). Competition and non-persistency as factors affecting the composition of mixed crops and swards. *Proc. 8th int. Grassld Congr.,* pp. 736–41.

Woodford, E. K. (1966). The need for a fresh approach to the place and purpose of the ley. *J. Br. Grassld. Soc.* **21,** 109–15.

Yoda, K., Kira T., Ogawa, H. & Hozumi, K. (1963). Self-thinning in overcowded pure stands under cultivated and natural conditions. *J. Biol. Osaka Cy Univ.* **14,** 107–29.

THEORY OF FEEDING STRATEGIES

Thomas W. Schoener
Biological Laboratories, Harvard University, Cambridge, Massachusetts

Natural history is replete with observations on feeding, yet only recently have investigators begun to treat feeding as a device whose performance—as measured in net energy yield/feeding time or some other units assumed commensurate with fitness—may be maximized by natural selection (44, 113, 135, 156, 181). The primary task of a theory of feeding strategies is to specify for a given animal that complex of behavior and morphology best suited to gather food energy in a particular environment. The task is one, therefore, of optimization, and like all optimization problems, it may be trisected: 1. *Choosing a currency*: What is to be maximized or minimized? 2. *Choosing the appropriate cost-benefit functions*: What is the mathematical form of the set of expressions with the currency as the dependent variable? 3. *Solving for the optimum*: What computational technique best finds extrema of the cost-benefit function? In this review, most of the following section is devoted to possible answers to the first problem. Then four key aspects of feeding strategies will be considered: (*a*) the optimal diet, (*b*) the optimal foraging space, (*c*) the optimal foraging period, and (*d*) the optimal foraging-group size. For each, possible cost-benefit formulations will be discussed and compared, and predictions derived from these will be matched with data from the literature on feeding. Because the third problem is an aspect of applied mathematics, it will be mostly ignored. Throughout, emphasis will be placed on strategic aspects of feeding rather than on what Holling (75) has called "tactics."

A GENERAL MODEL OF FEEDING STRATEGIES

Relation of Total Feeding Time to Total Net Energy Yield

Despite enormous variation in the ways animals gather food energy, one function always useful for description of feeding is that relating the total net energy yield E from feeding to T, the total time spent feeding during some long time period P (Figure 1a). This function should be monotone increasing at a monotone nonincreasing rate under several interpretations of optimal feeding:

1. *The expense in time and energy for feeding includes only that of pursuit, handling, and eating of prey, and not that of search.*—This kind of feeder [Type I (171)], while watching for food, simultaneously monitors

mates, territorial invaders and predators on the feeder itself. Thus it expends no more energy searching for food than it would have expended anyway in other activities and indeed may expend less; therefore, the cost-benefit functions exclude search. The most obvious biological counterpart of the mathematical formulation is the "sit-and-wait" predator (142), e.g. certain raptors (64) or lizards (95, 148); animals that discover food while engaged in more vigorous activities, e.g. search for mates, also might be classified here. If we know relative numbers of different kinds of prey to which the feeder is exposed[1] during P, we can approximate the optimal diet for a given energy requirement M by the following algorithm (171, 172). Arrange the kinds of items in decreasing order of net energy/feeding time, e_i/t_i (as computed below, Equation 1, p. 377). Select that kind of item of highest e_i/t_i and calculate how much energy the feeder will gain if it takes all or a constant fraction of such items to which it is exposed. If this is greater than M, the feeder should eat only that kind of item; if not, continue to add, in decreasing order of e_i/t_i kinds of items to the prospective diet and accumulate total net energy until the requirements are just satisfied. The resulting set of items is the optimal diet. The algorithm can be graphed on a plot of E versus T as a chain extending from the origin (Figure 1a): each segment, corresponding to a kind of item, has projections on the E axis equal to abundance times net energy per item, and on the T axis equal to abundance times feeding time per item. The slope of segment i is equal to e_i/t_i. Graphing the items in order of decreasing slope and then intersecting the chain by a line from the ordinate at M gives the optimal diet: all items between the intersection point and the origin should be eaten. We can also intersect the function with a line from the abscissa and find out how much energy is accumulated for a given time spent feeding.

Fluctuations in the total number of items N selected from during P that affect kinds of items in proportion to their abundances expand or contract the chain while preserving its slope. For such fluctuations, if Figure 1a graphs food at $N = 1$, to determine how much time must be spent feeding to satisfy M at some other abundance N', divide M by N', calculate T, then multiply T by N' (172).

2. *The expense in time and energy includes that for search.*—If areas were equal in richness and the feeder foraged such that areas were revisited only after complete resource renewal [Type II predators (171)], the relation of E to T would be linear. Its slope would monotonically decrease over part of the range of T if areas were visited in decreasing order of food abundance and if the feeder selected the same kinds of items from each,

[1] Since the number of items to which the feeder is exposed may depend on the amount of time spent pursuing, handling, and eating, a correction may have to be made if this time is large relative to the activity period or potential exposure time (171, p. 286).

because of increased search cost in the poorer patches. Such a plot, unlike that for the sit-and-wait predator, would also approximate the time course of feeding. In an uncertain environment, this foraging pattern should be optimal in the long run if the forager knew how to order feeding areas by their probable richness. Of course, an animal that knows exactly how many items to expect in each area during P could visit areas in any order; then the plot would not represent a time course. The slope would also monotonically decrease over part of the range of T if the feeder fed repeatedly over areas whose resources did not have time to completely renew.

$E(T)$ would not monotonically increase for animals that had to make some initial time and energy investment before any feeding could be done—for example, birds that must fly to feeding grounds or spiders that must build webs. Rather, $E(T)$ would bend downward before beginning to climb.

This model specifies how long an animal that is feeding optimally should take to gain a given energy requirement or how much energy it can gain in a given amount of feeding time. To fit this into a more inclusive model which maximizes fitness (e.g. the Malthusian parameter of Fisher 53), we must take into account additional factors. First, the degree to which net food energy increases fitness depends on when during the animal's lifetime it is acquired, as well as on the ability of the animal to satisfy its immediate requirements and allocate the remainder toward stored food, growth, or descendants. Second, the profit an animal acquires in feeding must be weighed against the loss of time, while feeding, to participate in activities such as search for mates or predator avoidance. We review briefly the major biological facts relevant to these two considerations.

Gain in Fitness from Net Food Energy

All organisms, before they can grow or reproduce, must acquire energy to meet their maintenance requirements. Typically, basal or standard metabolic rates are proportional to W^λ, where W is body weight and λs vary between 0.5 and 1 but cluster about 0.75 (summary in 171). If we assume for closely related organisms that (a) active metabolism equals some multiple (depending on intensity of activity) of standard or basal metabolism (see p. 378) and (b) these organisms partake in activities with the same distribution of intensities over some extended time, then we can approximate the long-term metabolic requirements as some constant C_m times W^λ. For captive and some wild animals, the approximation is supported to some extent by calculations of the caloric expenditure or weight of uniform food eaten (e.g. 68, 94, 140, 146, 150). Of course, C_m varies greatly between homeotherms and poikilotherms; this and size variation result in enormous differences in the weight of food, relative to their own weight, that animals must eat per day: fish may need only 2–4% (92), whereas shrews must consume close to their entire body weight (162).

The simple picture of a weight-dependent metabolic feeding requirement is complicated by several factors. Most important are effects of other forms of energy such as radiation and heat (7, 144). In a broad sense, part of the feeding strategy involves the degree to which an animal controls its metabolic expenditures to meet seasonal changes in thermal conditions and food availability. For example, pike (86), deer (178), and arctic beaver (2) may reduce food intake (and sometimes activity) during winter, and certain poikilotherms can markedly reduce metabolic rate during starvation (9, 60). Daily torpor, hibernation, and diapause are all devices animals use to escape high energy consumption during unfavorable periods. Another strategy found in some birds and mammals is to distribute the major energy-demanding activities evenly throughout the year (93, 181, 206). Further, in anticipation of unfavorable feeding conditions, animals may store food externally (58, 115, 181) or internally (68, 96, 132); the buffering effect of the latter should increase with size and hence often with age, since storage capacity should increase at a greater power of W than metabolic expense.

A host of studies relate increased growth or speed of development to increased intake of food (e.g. 24, 37, 40, 102, 108, 120, 121, 140, 146, 152, 153, 164, 195, 196, but see 165, 166). Some studies show a leveling off of growth as a critical food ration is approached (24, 37, 86, 108, 140, 146). If an increased growth rate implies earlier reproduction or reproduction at a larger size, fitness may be increased.

The increased energy requirement which reproduction entails—mostly premating activities for males, but energy devoted to eggs, gestating embryos (201), incubation (18), and lactation (164) for females, and possibly parental care for both sexes (102, 159)—suggests that there might be positive relationships between food intake and reproductive output. In fact, increased quality or quantity of food may: 1. *Increase the weight or nutritional reserves of reproducing individuals,* shown either experimentally [e.g. 37, 38, 40, 43, 196, 200 (except at low feeder densities)] or suggested by comparisons of regional or seasonal differences in productivity (e.g. 118, 120, 157). This increase may be directly related to offspring production in those animals (mostly poikilotherms) in which egg number is directly proportional to a power, usually 2.5–3.5, of body length. 2. *Increase the clutch or litter size,* which may correlate, but not always, with increased body size (5, 37, 38, 40, 42, 43, 70, 76, 102, 119, 120, 152, 153, 164, 175, 202, but see 95). 3. *Increase the size of newborn young* (38, 201, but see 70). 4. *Increase egg size* thereby providing the young with greater initial energy (5, 119, 195). However, egg size may often increase slightly or not at all and be inversely related to increase in egg number (102, 175). 5. *Increase the number of broods or litters per season* (77, 102, 133, 164, 202, but see 121) *or oviposition rate* (23).

Loss in Fitness from Lack of Time to Participate in Other Activities

Mosimann (128) proposes that search for mates may be random, whence number of contacts between mates increases directly with search time.

Where animals are not grouped into spatially localized pairs or harems or where population density is low, decrease in mating success, especially for males, might indeed be directly proportional to the time spent feeding over most of the range of feeding times; otherwise, it may everywhere increase at an increasing rate. At least for animals below a certain size, there is probably a positive relation between conspicuousness, such as usually occurs while feeding, and the chance of being eaten (166), especially by those predators that rely on prey movement either for entrapment, as do spiders (194), or for providing sensory cues (16, 66, 125, 176). Other competing activities, such as grooming, preening, and play, are probably less important. Thermoregulation is important, but the cost of suboptimal thermoregulation may be expressed as an energy deficit and added to feeding cost.

A General Model

Armed with the above facts and logic, we now outline a general model which shows the amount of time T_{op} during P an organism should spend feeding so as to maximize reproductive output. Choose P as some time period such that 1. if the animal does not feed at all during P, it has zero (or arbitrarily close to zero) expected future reproductive output, and 2. if the animal feeds sometime during P, it has nonzero (or above arbitrarily close to zero) expected future reproductive output. Then define three functions:

1. F, the function relating E, the total net energy yield, to T, the total time spent feeding during P. $E = F(T)$ (Figure 1a).

2. $G(E)$, the function relating the expected future reproductive output G (weighted, if appropriate, by its time distribution) to E, *if no time were spent acquiring that* E. G includes consideration of the ability of the animal to convert food into offspring, the value of growth, and long and short-term energy needs. $G > 0$ for $T > 0$.

3. $L(T)$, the function relating to T the difference L between G and the actual expected future reproductive output, given that the animal is in fact feeding T units of time to acquire E. Thus L measures the deficit in reproductive output that accrues because the animal spends time feeding rather than doing something else.

Then we wish to maximize $G - L$. A necessary condition for this is that

$$\frac{d}{dT}(G - L) = 0 \quad \text{or} \quad \frac{dG}{dF}\cdot\frac{dF}{dT} = \frac{dL}{dT}$$

Placing three restrictions on the form of these derivatives, we easily state sufficient conditions for a single relative maximum, though some absolute maximum will always exist. First, as justified above, dE/dT $(= dF/dT) >$ 0 and $d^2E/dT^2 \leq 0$. Second, dG/dE $(= dG/dF)$ is monotone nonincreasing and approaches (but may not reach) zero (Figure 1b). This is likely not to be strictly true for certain animals, but includes the decreasing or at least

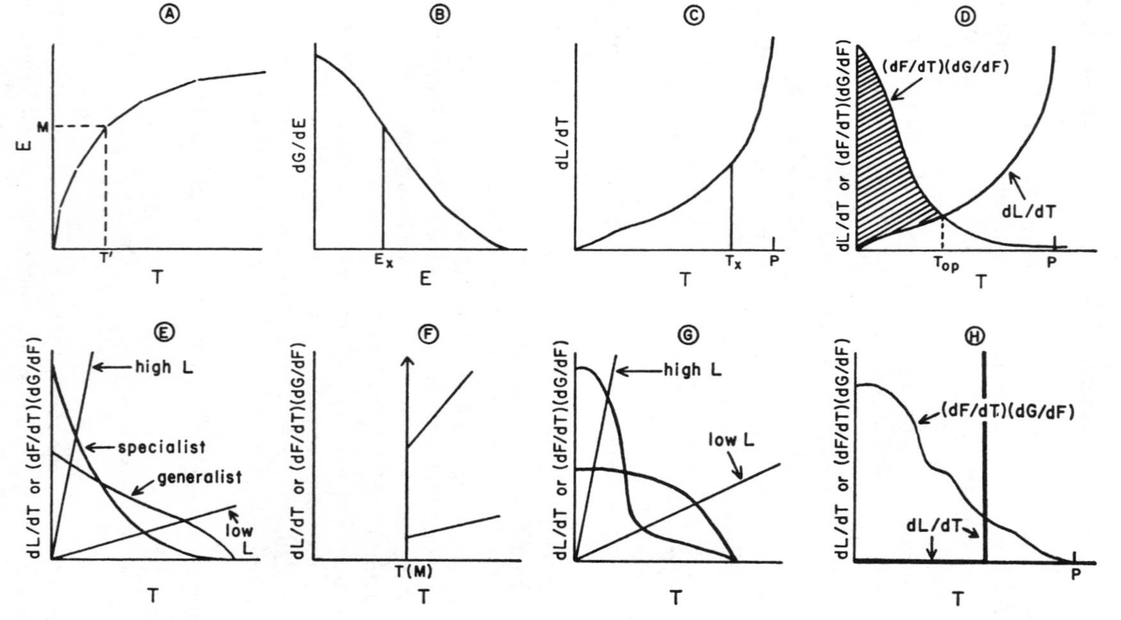

FIGURE 1. (*a*) $E = F(T)$. $F(T)$ gives the net energy yield to the feeder for a given time spent feeding or gives the time to acquire a given net energy. An optimal feeder will gather M energy requirements in T' time. (*b*) dG/dE as a function of E. $\int_0^{E_x} (dg/dE)\,dE =$ the future reproductive output from E_x if no time were spent acquiring that E_x. (*c*) dL/dT as a function of T. $\int_0^{T_x} (dL/dT)\,dT =$ the loss in future reproductive output from feeding T_x time units rather than doing something else. (*d*) Method for determining optimal time spent feeding when dL/dT and $(dG/dF)(dF/dT)$ are monotonic. (*e*) Whether the specialist or generalist should feed longer depends on the slope of dL/dT. If the two descending curves begin at the same ordinal intercept but otherwise keep the same relation to one another, the curve for the specialist could represent feeding during situations of high food density, and the curve for the generalist feeding during situations of low food density. (*f*) A time minimizer. (*g*) When $(dG/dF)(dF/dT)$ is very different for the two feeders, models of time minimizers are better approximations for gradual dL/dT (large difference on the T axis) than steep dL/dT. (*h*) An energy maximizer. Before dL/dT swings upward, the more time spent feeding, the better, unless $(dG/dF)(dF/dT)$ drops below zero first.

limited ability to convert energy into additional protoplasm or offspring as reviewed above. Third, dL/dT is monotone nondecreasing over its entire range (Figure 1c). This is the only assumption which may sometimes be grossly inappropriate, as would be the case, for example, if the probability of being eaten during some small time interval is proportional to the length of that interval and L is proportional only to the probability of being eaten by time T. Then $L = K(1 - e^{-\gamma T})$, $dL/dT = K\gamma\, e^{-\gamma T}$, where γ measures predation intensity. More likely, feeders will arrange schedules so that feeding is done during time of least predation, thus favoring the assumption. Loss in fitness due to lack of mating time is much more likely to favor the assumption: if number of mates contacted is proportional to $P - T$, dL/dT should be relatively constant; more likely, dL/dT will rise precipitously as $T \rightarrow P$. It is important to note that since L is defined as the difference between G and some non-negative quantity, dL/dT is to some extent dependent on the form of G. Figure 1d graphs an example where the three restrictions hold. The maximum $G - L$ is shaded and equals

$$\int_0^{T_{op}} \left[\frac{dG}{dF} \cdot \frac{dF}{dT} - \frac{dL}{dT} \right] dT$$

There are several ways we can think of the life history in terms of this scheme. The optimal animal, born with some amount of energy, proceeds through life gaining and expending energy according to some schedule that maximizes its total reproductive output. Though desirable, it is extremely difficult to make this continuous process discrete for models of feeding strategy. A weak interpretation of the model just presented is that it specifies the fittest phenotype from animals identical at the beginning of P but with different Ts during P, assuming the feeding time is scheduled identically for those animals after the conclusion of P. A stronger interpretation is to imagine the animal's life, beginning at birth, divided into a series of such Ps, each chosen as satisfying both the conditions mentioned above: At the beginning of each, the animal assesses its strategy in the face of alternative future conditions whose probabilities are known (i.e. genetically programed), but whose exact manifestation is unknown. This assessment might take the form of solving a dynamic programming problem, in which the animal on the basis of fairly exact information on its physiological state and the near future but with information on expected outcomes with great variance for the far future, "tries" all solutions for the remainder of its life history and comes up with the best feeding strategy for the P just beginning. At the commencement of each succeeding P, a reassessment is made on the basis of a different physiological state and a different (shorter) future reinterpreted on the basis of new cues. For example, if the animal is likely to lay a clutch of eggs shortly after the beginning of a particular P, much of the area under dG/dE in Figure 1b will be close to the origin and feeding will be at a premium. F, G, and L change through time, and G and

L should begin to decrease once reproduction has begun, though not necessarily monotonically. This model is only a very crude approximation, since animals must often change their assessment of G and L at intervals shorter than that P resulting in close-to-zero expected future reproductive output if no food energy is obtained. A different approach would be to choose P coincident with some natural cycle, e.g. a day; then, conditions 1 and 2 need not hold. Still another approach would be to discard discrete models altogether and try a continuous model—no such model has yet been devised, however.

It is obvious from the general model that an animal with F everywhere greater than that of a second is fitter if both have the same G and L. However, even if both had the same L and the capacity to produce the same total number of offspring, the one with the smaller F could be superior if its dG/dE were sufficiently greater than that of the first over low E, i.e., if it were more efficient there in conversion of intake into offspring. Indeed, it is likely that animals might specialize in one of feeding or conversion efficiency, thus representing equivalent strategies toward maximizing fitness. Other biologically agreeable results can be obtained by moving the curves around: a steeper dL/dT, which might reflect greater predation or difficulty in mating, causes feeding time to shorten; a greater dG/dE, which might reflect greater efficiency in producing offspring, has the opposite effect. A more bowed F, which might represent a more specialized feeder, causes feeding time to shorten if dL/dT is gradual and to lengthen if dL/dT is steep (Figure 1e).

While it is thus possible to make qualitative statements about fitness as related to F, G, and L, G and L incorporate difficult to measure and sometimes nonindependent variables so that their precise specification seems distant. A shortcut and considerable reduction in complexity can be obtained if we consider two limiting cases of the general model:

1. *Time minimizers* (Figure 1f).—These are animals whose fitness is maximized when time spent feeding to gather a given energy requirement M is minimized (171, 172). In terms of the general model, animals that have most of the area under dG/dE piled up over a very small range of E approach time minimizers. In the pure case, L can take any form, but no reproductive output is achieved until M energy units are gathered, and energy beyond M gathered during P does not further increase expected reproductive output; in this case, condition 2 for choice of P no longer holds. Animals with relatively fixed reproductive output per season (constant clutch size and brood number, for example) or often males as opposed to females may be approximated well by this case. So should animals for which dL/dT is relatively gradual, because then optimal feeding time varies greatly as a function of F and G (Figure 1g).

2. *Energy maximizers* (Figure 1h).—These are animals whose fitness is maximized when net energy is maximized for a given time spent feeding. In

terms of the general model, G can take any form, but dL/dT is assumed zero everywhere before or at a fixed T (which may be greater than P) and infinitely large past that T. Females and animals with variable clutch sizes may more suitably be represented by this asymptotic case.

OPTIMAL DIET

Components and Parameters

Although net energy/time has rarely been measured, animals have been observed which appeared to be feeding optimally, selecting food of greater biomass yield per unit feeding time (108), caloric value (160), ease of handling (21) or of catching (47, 79, 124, 139, 143, 161). All models of optimal diet require that kinds of items be able to be ranked according to some measure of profit to the feeder, usually net energy per unit feeding time. Because foragers often search for all items simultaneously, that function which ranks items often need not include search time and energy and may be written:

$$\frac{e_i}{t_i} = \frac{\text{Potential energy} - \text{pursuit costs} - \text{handling and eating costs}}{\text{Pursuit time} + \text{handling and eating time}} \qquad 1.$$

where digestion is assumed simultaneous with other feeding behavior. To include capture probability p, all terms for costs or benefits incurred after capture should be multiplied by p, whereas those preceding capture should be multiplied by p plus $(1-p)$ times the average fraction of the relevant cost expended on missed prey. Dividing expressions like 1 by the energy requirements of the feeder during P gives that fraction of the energy requirements contributed by item i per unit feeding time, a useful function for studies of optimal feeder size (158, 171).

Detailed discussion of expressions for search time and energy and the ingredients for Equation 1 have been given elsewhere (171) and can only be summarized and brought up to date here.

Search time.—Search time between two items of available food T_s can be most simply expressed as some constant over the density D of items. For random search, the value of the constant depends upon the speeds of predator and prey, the search area of the predator, and the diameter of the prey. Most commonly, expressions are given for the reciprocal of T_s, namely N_t the expected number of items encountered/time. Thus, for stationary prey, assuming no depletion of prey by the predator, Laing (104) gives for search over a two-dimensional surface: $N_t = 2V(\sigma_1 + \sigma_2)D$, where V is the speed of the predator, σ_1 is the radius of the perceptual field of the predator and σ_2 is the radius of the prey. This equation is, of course, identical to that for prey moving at speed V and predators motionless. The reaction field of a predator may be extremely large, especially where visual acuity is great: shrikes may react at 100 yards to a bumblebee or at over 200 yards to a mouse (29). For random search through a volume (185), $N_t = \pi V(\sigma_1 + \sigma_2)^2 D$. If prey are in random movement over a surface, V in Laing's equation can be replaced by the average relative speed: a very good approximation is $(V^2 + V_i^2)^{0.5}$, where V_i is prey speed (179). Holling (74) uses a slightly more complex version which adds to the "encounter path" a circle of radius $\sigma_1 + \sigma_2$; his "σ_1" is the average reaction distance of the predator. If prey sometimes detect predators before being themselves detected, the assumption of random movement of prey is violated and encounter is more complicated (166).

Factors reported to be associated with decreased search speed are low insolation (15), rough surface texture (166), small flock size (58, 126), high food density (132), time of day (132), and lower hunger (10).

Pursuit time.—This term, used here literally to mean the time from detection of prey to the time contact is made, may vanish for animals detecting prey by touch, but for others, such as wolves, pursuit ranges up to 5 miles (124). Obviously, for a motionless prey, pursuit time is just initial distance divided by some predator speed V_p; for a fleeing prey, the initial distance is multiplied by $V_p/(V_p - V_i)$, where V_i is prey escape speed. Appropriate predator speed must be chosen with care: certain predators, such as peregrine falcons in stoop, pursue prey at maximum speed (161); others, such as certain large cats, stalk prey over much of the distance (79, 101, 167). Additional complicating factors are acceleration and fatigue. Certain measures of speed are related to length in a power function with exponent 0.5–1 in fish and 0.33 in flying insects (13, 19, 80); however, relations for other taxa are nonexistent or unimodal (65, 67, summary in 171).

Search and pursuit energy.—The simplest assumption on expenditure of energy during activity is that it is proportional to W^λ, where λ is the same power as for basal or standard rate. This is supported by theoretical argument and data from some birds, mammals, and fish; however, deviations in λ in both directions occur (19, 51, 199, references in 57, 171, 190). While active metabolism can be over 20 $\times$ basal or standard rates (7), most ratios fall within 6–14, and food search may cost little more than rest (138). Dividing activity cost by speed produces expressions for cost per unit distance: minimal costs are calculated as proportional to $W^{0.77}$ in flying animals (193) and $W^{0.60}$ in mammals (190). Such costs are more likely to apply to search than rapid pursuit, which may even incur an oxygen debt (19). Exponential (20) or linear functions (190) relate speed to per gram energetic cost for similar animals.

Handling-and-eating time (THS) and cost per item.—THS may vary from milliseconds in wood storks (88) to major fractions of several-day periods in large carnivores (124, 167). As Dixon (37) showed for coccinellids, the predominant factor affecting THS is the relative size of feeder and food. One possible representation sets THS constant for food sizes i below a critical food size i_B determined by the size of the feeder, and above i_B it relates THS exponentially to $i - i_B$ (171). Cost per unit THS ought usually to fall below that for active pursuit, though it may exceed cost of ordinary locomotion (181). Another possible cost is removal of prey from the capture site to some location for consumption or storage—a short hop for the tube-dwelling blenny (71) or a long haul for the tiger (167). A final "handling" cost, ignored in extant models of feeding, is risk of mortality or injury from supposed prey—cougars have been kicked to death (56) and birds strangulated by large fish (163).

Pursuit and capture success.—Fairly similar animals show a good deal of variation in the percentage of attemps resulting in successful prey capture: e.g. on moose wolves average 7.8% (124) while wild dogs average 86% (50). The single most important factor determining success for raptors (161), coccinellid larvae (37), and probably many others is relative size of feeder and food, though mobility (61, 145), prey density (61), and predator experience or conspicuousness (37) have been shown or suggested to influence success. Much variation in reported success/attempt ratios must be merely a matter of definition, resulting in part from differences in the way animals select their prey. Often predators will "test" possible prey by chasing it, as do wolves (124), or nudging it for juices, as do sharks (184). Such testing behavior could be considered a part of food search, but regardless of nomenclature, if important it should be incorporated into the ranking functions.

Potential caloric content of the food.—This term may simply be $C_a i^\tau$, where C_a is some constant and i is food length. For dry weight of insects, τ is closer to 2 than 3, but if food is drawn from a uniformly proportioned series of animals such as might be found within a species, τ varies about 3 (many references). C_a should incorporate the number of calories that the animal can extract from the item per volume after having eaten it. Though not attempted so far, more complicated formulations might include distinguishing kinds of nutrients.

Relative abundance of kinds of food items.—Models of optimal diet must include relative abundances of food items. These may be set arbitrarily (113) or follow some empirical distribution in nature, such as the log normal for insects (171).

Models

Although this review deals mainly with strategic models, Holling (72) has analyzed optimal prey size in a way which illustrates well the principal alternative approach. From an analysis of forces in the mantid's forearm, he predicted that size of prey which it grasped best; the size can be considered a mechanical optimum. He then showed in behavioral experiments that this was in fact almost exactly the size eliciting the largest number of feeding responses. Thus the mantid is a well-integrated feeding machine, no matter what the forces shaping the overall design.

Following a quite different tack, MacArthur & Pianka (113) ask what range of dietary items natural selection should favor if animals fed optimally. This range can be predicted from a graph which plots two functions: ΔS, decrease in search time per kind of item eaten, and ΔP, increase in pursuit time per kind eaten (including handling and eating time), against kinds of items ranked along the abscissa in order of decreasing ($\Delta S - \Delta P$). In other words, they point out that search time per item eaten monotonically decreases as more kinds of items are added to the diet, while pursuit time per item eaten monotonically increases as harder-to-catch kinds are added. Optimally, items are added until the decrease just equals the increase, i.e., the two functions intersect, all kinds of items between the intersection point and the origin constituting the optimal diet. This model will always produce a maximum net energy yield/feeding time only if the ranking by ($\Delta S - \Delta P$) is the same as a ranking of kinds of items by net energy/pursuit time. The rankings are always equivalent if items are of equal net caloric value; if not, they may easily be different. In fact, the difference in pursuit time when the diet is enlarged to include less suitable items may be a gain and not a loss. For example, item A might have 10 net cal harvestable in 5 units time ($e_i/t_i = 2$), whereas item B might have 2 net cal harvestable in 3 units time ($e_i/t_i = 2/3$). Thus item A is superior even though it takes longer to eat and could be rejected first by the animal if its cost-benefit model only included time and not energy.

An energy-incorporating cost benefit function allowing the maximization of net energy/feeding time, where feeding time includes time for search, is

$$\frac{\sum_{i=N_1}^{N_2} p_i(U_i - C_i t_i) - C_s T_s}{\sum_{i=N_1}^{N_2} p_i(t_i) + T_s} \qquad 2.$$

where U_i is the utilizable energy in a single item of type i, C_i is the cost per unit time of feeding on item i, t_i is the time it takes to feed on item i, p_i is the encounter frequency of i, T_s is the search time between items of available food (whether eaten or not), and C_s is cost per unit T_s. The summations are taken over all i actually eaten, and maximization is achieved by suitable choice of the bounds N_1 and N_2. t_i may be broken down into the different kinds of activities associated with feeding on i, with appropriate Cs for each. This formulation assumes that the feeder has enough foraging area to satisfy its long-term energy requirements; otherwise, net energy per unit time would be less than specified by Equation 2.

All models discussed so far are in one sense deterministic: they assume that kinds of items can be characterized by single net energy and single feeding time, or if this is unrealistic, that the kinds can be subdivided more finely until the assumption is reasonable. A different, probabilistic approach has been followed by Emlen (44, see also 156). For each of two kinds of items i and j, he asks what fraction of the time animals should stop and eat rather than skip the item (i or j) before going on to a second item which is assumed to be invariably consumed. Assuming that optimally this fraction is equal to the probability that the first alternative provides a greater net energy/feeding time than the second, true always only if the feeder knows in advance of encounter the net energy and feeding time for the next item, it is then possible to state how changes in absolute and relative abundances should affect the proportion of i and j in the diet.

Levins & MacArthur (111) give another approach to food choice under environmental uncertainty for phytophagous insects. They ask what degree of selectivity of food types, unknown as to their suitability as food of offspring, will maximize expected production of offspring. Expected production of offspring is equal to the product of two quantities: first, the probability for the adult of finding a plant on which to lay eggs (which approaches one with increased time in an exponential function if the probability of finding a host is constant for small units of time), and second, the expected production from those eggs, which depends on the proportion of known and unknown suitable food and the proportion of unknown, unsuitable food. They find that polyphagy is more likely the greater the probability of failure to find a known suitable plant, the greater the proportion of unknown but suitable plants, and the lower the proportion of unknown, unsuitable plants.

Predictions and Data

Range of items eaten.—All models discussed in the last section predict that the lower the absolute abundance of food, the greater the range of items that should be taken. In MacArthur and Pianka's graph, lowered density increases ΔS and results in an intersection point on the dietary axis further from the origin. In Emlen's model, decreased density increases distance between items and causes the fraction of the time any item should be eaten to go to one. For Schoener's Type I predator (Figure 1a), lowering density is equivalent to raising the M intercept for purposes of selecting an optimal diet, which results in more kinds of items being taken. For Schoener's Type II predator (Equation 2), decreasing density increases mean search time between items and thereby makes it less profitable to skip any item, just as in Emlen's model. Levins & MacArthur's (111) model favors polyphagy with decreased overall density because of the lower probability per unit time of finding a known suitable host (see 110).

Evidence for decreased selectivity under conditions of low food abundance or availability has been found in fish (82), weaverbirds (204), blackbirds (136), swifts (103), and *Conus* molluscs (98).

What if kinds of items have abundances reduced differentially, as would be caused, for example, by the introduction of a competitor with different preferenda? MacArthur and Pianka give a graphical argument which shows that the range of items should either increase or remain the same, just as for a lowering of absolute abundance. This result does not depend on their assumption on energy: it is obviously true for Type I predators and easy to show from Equation 2 that if an item is worth taking in the absence of competition, it is worth taking in its presence. Let $e_i = U_i - C_i t_i$ and $T_s = K/D$, where D is food density and K is a constant incorporating rate of search and proportionality factors. Then item x should be worth taking if

$$\frac{p_x e_x + \Sigma p_i e_i - C_s(K/D)}{p_x t_x + \Sigma p_i t_i + (K/D)} > \frac{\Sigma p_i e_i - C_s(K/D)}{\Sigma p_i t_i + (K/D)}$$

Rearranging and cancelling terms, this is when

$$p_x e_x \Sigma p_i t_i + p_x e_x K/D > p_x t_x \Sigma p_i e_i - p_x t_x C_s K/D$$

Letting $N_i/N = p_i$ and $N/A = D$, where N_i is the absolute number of items i and N is the total number of items, we get

$$p_x e_x \Sigma N_i t_i + p_x e_x KA > p_x t_x \Sigma N_i e_i - p_x t_x C_s KA$$

Notice that the truth or falsity of this inequality does not depend on p_x; therefore, a decrease in p_x caused by competition has no effect on the inclusion of item x. This form of the inequality also shows that if p_x is lowered while some N_i are raised, as would be the case were absolute abundances to remain constant, the inequality no longer may hold, and item x may be dropped. Orians & Horn (137) have some indirect evidence for MacArthur and Pianka's argument. Note, however, that the animal whose diet widens after competition may no longer necessarily possess the fittest phenotype. Therefore, natural selection could cause a change in size or other characteristic allowing the new phenotype to shift and probably shrink its diet (171).

Differences between distributions of items in the diet and in nature have been explained by factors other than selection of items with highest e_i/t_i. Tinbergen (191) documents the existence of a "search image," in which predators learn by exposure to look for a certain kind of prey; the result is that items rare in nature are even rarer in the diet. Gibb (59) postulates "predation by expectation" to explain why birds sometimes leave habitats before preferred food is completely depleted; the result is that the commonest kinds of items are not as common in the diet as in nature. Other examples of environmentally induced selectivity exist for starfish (105), molluscs (131), sticklebacks (10), and cheetah (39); and Holling (73) has modeled some kinds of learning. The need for a balanced diet may also affect selectivity (59, 129, 191), sometimes causing items with greatest caloric yield to be underselected (121, 181). Models that allow for preference factors other than net energy and feeding times, but into which the latter can be incorporated, are currently being developed by A. P. Covich as a modification of the "indifference-curve" approach of economic analysis. Emlen (45) points out that item i with mean e/t less than that of item j may be taken in greater proportion if its variance in e/t is such as to cause some items of i to be better than any of j [a similar earlier result (44), resting on a miscancellation in a probability statement (his Equation 11), is incorrect].

Of the models discussed above, only that for a Type I predator directly allows for changes in diet with changes in the energy requirements (Figure 1a), though others can be expanded to include it. The prediction is that an increase in energy requirements should have the same effect as a decrease in food density on selectivity: in fact, birds select large prey from the habitat to feed young (155, 159), and the number of small prey brought to the nest may increase with brood size (159). Root (155) gives other examples of decreased selectivity when raising young, including increased diversity of foraging behavior and greater pursuit distances. A short term measure of energy requirements is hunger, which might be considered a device by which the organism monitors small temporal perturbations in food availability: it "tells" the animal what intake is needed by what future time to forestall a given degree of weakness or starvation. Thus the animal can more finely attune to fluctuating food supplies and feed on low-yield items only when necessary. Examples of hunger increasing the range of acceptable items, in terms of factors which should be correlated with net energy yield per unit time, are found for distance thresholds in mantids (74) and dragonfly larvae (145), for size thresholds in frogs (66), mantids (73), and dragonfly larvae (145), for effort per unit reward in weaverbirds (204), and for food type in rats (209), fish (10, 82), and otters (49).

Food size and distance from the predator.—An equation like 1, in which the exponential form discussed above (p. 379) is used for handling and swallowing time and cost per unit distance increases at a decreasing rate with predator size, can predict the best food sizes for a given-sized predator and how these sizes vary with pursuit distance (171):

1. *Food sizes should decrease with decreasing predator size, but asymptotically.* Many studies, too numerous to be cited here, support the first part of this prediction, both within and between species—though there are some exceptions. Evidence for the asymptote occurs in lizards (168, 170, 174) and hawks (187), though not in seabirds (4).

2. *A predator at a given distance should take prey both larger and smaller than any at a greater distance, but the large size limit should decline less with increasing distance than the small size limit should increase* (see also 156). Preliminary evidence for this effect exists in *Anolis* lizards. This energetic consideration could explain why certain birds bring their nestlings larger food than they themselves eat (159, 203). Brawn (17) found that with increasing distances trout struck mostly at larger items; though she explains this by perceptual limitation, energetics could also play a role. As a

corollary, species which pursue their prey over relatively great distances should take larger prey relative to their own size than those which do not, a result borne out for raptorial birds (169). A related fact is that oilbirds, which travel great distances for fruit, choose especially high-energy items (183).

3. Distributions of prey sizes eaten by a predator from a uniform size-abundance distribution of available prey should be more negatively skewed (i.e. have a long left tail) *if the predator pursues its prey over greater distances or is relatively large than if the predator pursues its prey less or is relatively small* (171). Ivlev's experiments (82) support this prediction, and Rosenzweig (156) showed that "active" pursuers (weasels) are more specialized in prey size than certain confamilial searchers (skunks).

Optimal kinds of feeders.—

1. Specialists versus generalists: These terms, invested with many meanings, need to be redefined with each use (127). Most commonly, an animal is said to be more generalized for prey types if it eats (*a*) a greater range of food types, (*b*) a greater variance of types, or (*c*) a greater "breadth" of types, as measured by information-theoretic *H*s or some other way (110). A second definition of *generalist,* often nonindependent of the first, is an animal, such as the mallard (134), with a great repertoire of feeding behavior. A third definition can be made on the basis of relative ability in extracting energy from food, as plotted on an *E(T)* plot (Figure 1a) : a species is more generalized than another if its *E* approaches a greater asymptote, but initially at a slower rate.

When is the dietary specialist or generalist optimal? One way in which an animal can be said to be more generalized than a second is if it is omnivorous, eating both animal and plant food, rather than one or the other. While the degree of omnivority in nature varies with relative availability of the two kinds of food, as might be produced by seasonal or habitat effects, it also relates to characteristics of the feeder. For example, young, largely herbivorous animals which feed for themselves often take more animal matter than adults (31, 32, 54, 174); that this is also sometimes true of altricial birds (169, 204, but see 133) suggests a higher protein need in growing animals as the explanation. However, it is also possible that larger animals are forced toward herbivory because not enough energy can be gathered from animal food to meet their energy requirements (31, 123, 169). Unless food is very abundant, larger animals should also usually eat a greater range of food sizes than smaller ones (171), which they seem to do (84, 95, 107, 156, 168, 169, 170, 174, 194, 197). Prey species diversity may also be greater for larger animals (47, 54, 90, 133, 184), unless small prey are sufficiently more diverse than large (8, 149). Differences in energy requirements between poikilotherms and homeotherms may produce corresponding differences in diversity of prey eaten: thus temperate snakes (28) are often highly stenophagous compared to temperate birds of the same weight. Of course, differences in the abundance and clumping (82) of food may also cause differences in the taxonomic diversity of prey; this is perhaps why predatory arthropods (e.g. 194) are often more catholic than herbivorous arthropods. Fluctuations in food abundance are also commonly related to the optimal degree of specialization. Especially where food species are affected differently and unpredictably, taxonomic generalists should be favored, perhaps explaining the greater specialization in tropical as compared to temperate latitudes among animals such as fish (91). Fluctuations in food density uniform over all food types may favor generalists or specialists (in the sense of meaning 3 above), depending on whether animals approach time minimizers or energy maximizers (172).

2. *Large versus small animals:* Independently of the range of items in the diet, we can ask when large or small animals should be optimal. Rosenzweig & Sterner (158) calculate from measurements of husking times that small seed-eating heteromyid rodents should harvest their energy requirements in less time than large ones. In contrast, Pearson (141) observes that large sea birds spend less time fishing per day and per chick than do smaller ones. In a theoretical analysis of time minimizers, Schoener (171) hypothesizes for certain predators that (*a*) larger animals should often satisfy their energy requirements faster than most smaller ones when food is abundant and slower when food is rare, and (*b*) if competitors reduce food abundance by a relatively constant percentage over all food types, convergence in size may be favored, whereas more differential reduction may favor divergence. Conditions favoring sexual dimorphism and other intraspecific size polymorphisms are also hypothesized (171). However, results are qualitative, and it is possible that when the real parameters are used, certain differences between kinds of predators will be minimal or absent. For instance, the energy expended in feeding on an item may be trivial compared to the return. Martinsen (117) found that a shrew needed to expend only 0.35 kcal in handling a 30 g vole to gain its 40 kcal of energy. However, that some animals have lost weight or starved on suboptimal natural food, sometimes even while eating continuously, suggests that the return may often not be so disproportionately great (8, 21, 22).

Feeding rates.—Strategic models of feeding have had little to say on feeding rates, though there exist other kinds of models which specify rates for animals feeding at or near mechanical capacity. For example, Leong & O'Connell (109) empirically derive and support models which predict rates proportional to a power of length that varies with type of feeding. The best-known models of feeding rate are those relating it to food density in the "functional response." As reviewed by Holling (73), early explanations of the functional response were on the basis of available time: if food is discovered in proportion to density times the total available search time, and if search time decreases by an amount equal to pursuit, handling, and eating time times number of prey attacked, then a convex functional response is generated. This asymptotes at the ratio, total time spent feeding: time to feed on one item, or the number of items that could be eaten if search time between items were zero. Holling (74) has given much more complicated models for the functional response, piecing together experimentally determined expressions for the components of predation and using a measure of hunger to predict predatory responses. Alternatively, it is possible, using the general model for maximizing fitness presented above, to solve for the number of items eaten as a function of food density. Rates and asymptotes will then depend on the predator's ability to convert food into net energy and energy into offspring, the probability of being eaten while feeding, and other factors affecting F, G, and L.

OPTIMAL FORAGING SPACE

Components and Parameters

In this section we ask how the animal can maximize fitness as a function of its home range size, patch selection, foraging perch or path, and exclusiveness of feeding area. To answer this question, we would like to specify the temporo-spatial abundance of food resources—what the resource gradients look like, where patches occur, how fast resources depleted for a given area renew, how predictable the appearance of abundant food is in space and time, and what the mappings of efficient pathways are for search and pursuit. Mathematical representation of such components, however, are rare: McLay (122) recently proposed an exponential distribution to describe the abundance of drifting food along the length of a flowing stream. In addition to those for food, we may wish described, among others, gradients for climate, predation, shelter from predation, mates, and competitors.

Models

A first-order approach to the problem of home range size, adopted by McNab (123) and Schoener (169), is to imagine that available food is proportional to area traversed and that the feeder occupies an area just large enough to supply its energy requirements. Hence, knowledge of food density, food selectivity, and metabolic rate of the feeder is sufficient to specify home range size. For uniformly dispersed, temporally constant food items, it is realistic to use average density in the computation of home range size. However, if food is variable in space and time, it may have the appropriate long-term average density, but fluctuate so that it is sometimes insufficient to satisfy the short-term energy demands of the feeder. In such cases, home range size should be enlarged over that predicted for the simplest case (34). In fact, the possibility of patchiness underscores the primary weakness of the home-range concept: an animal may have a large home range yet use intensely only a small part (e.g. 12, 48, 99, 116). Further, "range" is laden with drawbacks such as estimation difficulties. Statistically more appealing is "home variance," calculated on distance from the center of gravity of points occupied by the animal. This concept is biologically attractive, too, provided utilization contours aren't too misshapen; otherwise, evenness of occupancy could be calculated over arbitrarily sized subdivisions.

A second approach to modeling foraging space is specification of the maximum distance from a vantage point an animal should go to eat an item (171). As discussed above for Type I predators, items are indexed by size and distance, and once the optimal set of items is known, so is the maximum distance. For some animals this may be equivalent to home range radius; for others, it may be proportional to it.

A third approach treats foraging patches similarly to items of food. Just as in their model of optimal diet, MacArthur & Pianka (113) plot ΔT, loss in traveling time between patches per item, and ΔH, gain in hunting time per item, both as functions of range of patches utilized. As poorer patches are added, ΔT decreases and ΔH increases; the intersection of the two denotes the optimal patch range. The model makes the same assumption about the ordering of patches as for food items: patches are ordered according to decreasing ($\Delta T - \Delta H$). Since much of the energy lost while foraging in poorer patches is due to increased search time, this ranking should closely correspond to a ranking in net energy/feeding time, unless kinds of patches vary too much in average net energy of items they contain.

MacArthur (112) has introduced in other ways the idea that an animal should minimize traveling time between feeding areas, most recently in the concept of "connectedness." For a sit-and-wait predator, a feeding area could be defined as more connected the smaller the average distance that the predator would have to travel from the vantage point (or every point) to other points in the area or volume, divided by the average geometric distance from the perch to the same points—thus lizards inhabiting the branches of a tree live in a less connected space than lizards inhabiting the ground. Points may, if appropriate, be weighted by resource abundance. Connectedness might also be defined for searchers according to the degree to which they can continuously harvest food while moving along their feeding path.

Some animals show searching behavior that appears to maximize food gain: for example, when food is discovered, coccinellid beetles (6) and sticklebacks (10) increase turning rate, the better to exploit concentrations; and during the dry season when food is patchier, flocks of *Quelea* change from random to directed movement, whereby birds at the rear constantly fly ahead of the rest, producing a "gigantic roller moving across the plains" (204). Little formal theory now exists for search movement, but A. R. Kiester and M. L. Slatkin are presently modeling movement along gradients, and Rohlf & Davenport (154) have approached the problem from a stimulus-response point of view. Movements over long distances, especially in response to changes in food abundance, are also part of the overall foraging strategy (127): for example, some birds must "weigh" a possible increased food supply per bird against the increased energy needed in colder climates (33, 205).

Models of the optimal foraging space may need to incorporate yet another factor: depletion of the foraging area by competing animals. This can be reduced by the feeder at a cost, that of territorial defense, and the economics of defendability have been reviewed by Brown (25) and Brown & Orians (26). A mathematical approach is as follows:

Suppose that the time spent by an animal searching through an area for invaders is T_{SD}, invaders are encountered randomly per unit time in proportion to J (the area per unit time of search, analogous to that for feeding given above) and N/A (the density of invaders), a fraction f of the invaders that are encountered are chased from the territory, P is the total period during which invaders enter the area of defense A, $qA^{0.5}$ is the rate at which invaders enter the territory if only the defender is present (thus proportional to the circumference of A), and H is the maximum density of invaders in A. Then, assuming a linear decline in the invasion rate with increasing density of invaders, the density of invaders at equilibrium,

$$N_e/A = \frac{qA^{0.5}HP}{HJfT_{SD} + PqA^{0.5}}$$

(Figure 2). Suppose that R_D is the energy requirement of the defender and R_I the average energy requirement of an invader during P, D_P is the food available during P per unit area, and defender and invaders gather food over A, then the number of items available to defender and invaders is D_pA. If F (Figure 1a) is the exponential function $X(1 - e^{-\alpha T})$, then the combined time for invaders and defender required to eat those items necessary to satisfy the requirements of invaders and defender is

$$\frac{-D_PA}{\alpha} \ln\left[1 - \frac{R_D + R_IN_e}{XD_PA}\right]$$

Assuming invaders and defender eat identical distributions of items, time spent by the defender alone in eating is

$$T_E = \frac{-D_PA}{\alpha} \ln\left[1 - \frac{R_D + R_IN_e}{XD_PA}\right]\left(\frac{R_D}{R_D + R_IN_e}\right)$$

Using the first three terms of Taylor's series for expansion of

$$\ln\left[1 - \frac{R_D + R_IN_e}{XD_PA}\right]$$

about 1, valid if animals are not too close to eating all their available food,

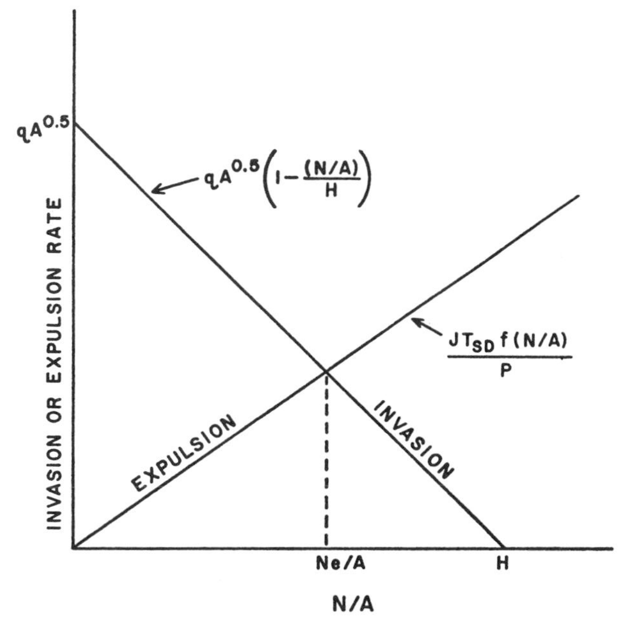

FIGURE 2. Territorial equilibrium: rates of invasion and expulsion are plotted against invader density; N_e/A is the equilibrium density of invaders.

$$T_E^* = \frac{R_D}{X\alpha}\left(1 + \frac{R_I N_e + R_D}{2XD_P A}\right).$$

Suppose that C_D is the cost to the defender of expelling a single invader, C_{SD} is the cost of searching for invaders per unit time, and M is the energy requirement of the defender during P excluding that spent in defense or feeding, then $R_D = M + T_{SD}C_{SD} + JfT_{SD}C_D(N_e/A)$. If T_D is the time to expel a single invader, we wish to minimize

$$T_{FD}(f, A, T_{SD}) = \frac{R_D}{X\alpha}\left(1 + \frac{R_D + R_I N_e}{2XAD_P}\right) + T_{SD} + (N_e/A)T_{SD}JfT_D \quad 3.$$

This is a very complex expression, even though it contains oversimplifications such as an invasion rate linear and independent of D_P.

PREDICTIONS AND DATA

Home range size.—Analyses of home range size for broad groups of animals support to some degree simple interpretations based on energy requirements, selectivity, and food density. First, mammals (123), birds (3, 169), and lizards (148, 198) all show increased home range or territory size with increased body weight and therefore with average energy requirements; additionally, Pearson (141) found that, except for the largest species studied, larger sea-bird species traveled greater distances from the nest for food. Second, for mammals of the same size, grazers and browsers have smaller home ranges than hunting herbivores (123), and herbivores, smaller than carnivores (169); for birds of the same size, predators have larger home ranges than omnivores or herbivores (169). All are expected on the basis of surmised relative abundance of acceptable food. Third, McNab (123) found that herbivorous mammals increase home range size at a rate proportional to $W^{0.63}$, where W is body weight, nearly the same factor as for metabolic rates. Schoener (169) found a small sample of herbivorous birds consistent with this effect, but both predatory birds and mammals have much steeper rates; Turner et al (198), after adjusting for sample-size effects, found a greater slope for lizards as well, most of which are predatory. The result is expected if density of preferred food shows no relation to W in herbivores but decreases with W for predators, as would be the case, for example, if prey sizes were distributed log-normally. Differences in foraging or home range sizes within species have been shown or inferred to be related to differences in food productivity among birds (29, 77, 89, 132, 186, references in 26, 169) and mammals (1, 11, 48, 189). In accordance with predictions on patchiness of food, animals feeding on fruiting trees tend often to have large home ranges (34, 183).

If home range size is proportional to maximum distance traveled for food for a vantage area, the ranking functions discussed above (p. 377) predict that relatively efficient pursuers should have larger home ranges than others of the same size: this may be why *Accipiter* hawks have larger ranges than similarly sized *Buteo* hawks, or why mammals have smaller home ranges than similarly sized birds (169). Furthermore, variations in abundance, as might be encountered in moving over the area occupied by a population of such birds, should cause large variation in home range size, relative to that in prey size range, for more as opposed to less efficient pursuers—a prediction also borne out by hawks (171).

Patch utilization.—From their model, MacArthur & Pianka (113) show that a decrease in overall food density should not affect range of patches as much as range of items in the diet, because both mean hunting time per item and traveling time between patches per item should increase, the former because search time increases per item, the latter if fewer items are taken per patch or if patches are reduced in size or number so that it takes longer to get from one to the next. Further, animals spending most of their feeding time searching should be affected more than those spending it pursuing, handling, and eating. The contrasting ways that density affects diet and patches result in the "compression hypothesis" (113, 114): when competitors differentially reduce food density, the range of patches utilized should shrink because some patches are then worse than others, but range of food types within patches should not decrease and in fact may expand, as shown above (p. 380). This hypothesis is meant to apply to the first stage of species contact, before natural selection has acted on phenotypes to alleviate the overlap—it is supported by some *Anolis* lizards recently introduced onto Bermuda (173).

Smith (182) has studied the order in which squirrels utilize cones from various species of trees, which could be considered "patches" of habitat. He finds that trees whose cones contain the most energy are harvested species by species, without discriminating high- and low-yield trees within the same species. He argues that the cost of sampling such trees to determine the best intraspecific ranking would not be outweighed by the benefit gained. Trees of species of low average energy per cone should be more often discriminated, however, and in at least one habitat this has proved to be the case.

Home range overlap and defense.—How food is defended may influence the optimal degree of overlap between sexes. Smith (181) calculates that a pair of squirrels definding one territory should spend

$$\int_0^{\sqrt{2}r} 4\pi R^2 dR \quad \text{or} \quad \sqrt{2} \text{ times that amount} \int_0^{r} 4\pi R^2 dR$$

of time and energy bringing food to a central point that two squirrels on separate territories of radius r and the same combined area should spend; the squirrels he studied in fact show the latter system.

Optimal extent and area of defense can be calculated from Equation 3. Brown (25) has pointed out that territories should be economical and physically defendable. By inspection of Equation 3, it is obvious that at sufficiently high food density D_P, the degree of defense f should equal zero. At low food density, f may or may not equal zero depending on R_I in relation to T_D and to terms of R_D. At $f = 0$, A should be indefinitely large. For a given optimal positive value of f, if J, the area searched per unit time, decreases while all other parameters and variables stay constant, f should increase correspondingly, unless of course $f = 1$. If invasion rate is proportional only to the density of invaders, then inspection of Equation 3 shows that if it is economical to defend a territory at a given food density, it is also economical to defend at all lower food densities. If, however, invasion declines to an intercept on the abscissa which is directly proportional to food density (i.e. substitute D_PH for H in 3)—implying that food productivity as well as number of invaders determines the chance of invasion—then under some circumstances a lowering of food density will first favor switching from lack of defense to defense, while a further lowering will result in switching back to no defense (under this assumption on invasion, defense may still be profitable at very high food density). It has been argued that birds which need to range over huge areas should not find it worthwhile to defend territories. For this model, for A sufficiently large, optimal f should equal zero, as can be seen by passing to the limit as A goes to infinity in Equation 3. Other properties of this and similar models will be examined more carefully elsewhere.

OPTIMAL FEEDING PERIOD

No formal theory yet covers optimal placement of feeding periods over the activity cycle. Obviously basic components are metabolic costs of activity under different climatic conditions and time distributions of climatic factors, food, and predator abundance. A host of observations indicate that climate will be overwhelmingly important in models of optimal feeding period. For example, many animals have bimodal feeding periods during the warmer season of the year and unimodal ones during the colder. Another important consideration, especially for animals which require food at frequent intervals, is time from prior feeding periods.

It is clear by comparison with models of optimal diet for time minimizers that animals should often expand their feeding periods toward less profitable times when 1. food is sparse, 2. they have relatively great energy requirements, and 3. they can best convert food into descendants. Williams (207) gives examples of restricted feeding or other activity during favorable overall conditions for arthropods and cattle; and Murton et al (132) have calculated that during winter wood pigeons on pasture spend 95% of daylight feeding, while during most of the summer they spend less than 10%.

OPTIMAL FORAGING-GROUP SIZE

The possible relation of group size to feeding has received much recent attention. Three components stand out in these analyses, namely, those relating group size to 1. some measure of foraging efficiency, 2. probability of predation, and 3. the defendable area per unit cost of defense, all computed per individual.

Hypotheses and Data

Hypotheses on the selective value of group foraging are of three kinds: 1. The group may hinder feeding, but there is some overriding advantage to gregariousness. 2. Animals aggregate in proportion to concentration of food, and group size per se may be of neutral selective effect. 3. Groups allow animals to feed more efficiently. We consider each in turn.

Foraging efficiency may decline.—According to Goss-Custard (61), certain wading birds may be hindered in their feeding by other individuals for two major reasons: prey are more readily driven to inaccessible places and resources are depleted too quickly; those species which detect prey by touch are less affected overall and so occur in denser aggregations. Crook, Aldrich-Blake, and co-workers (1, 35, 36) reason that in certain taxa of monkeys the largest groups are found on savannas because there chance of predation is high, smaller groups are allowed in forests by reduced predation and possibly favored there by a more uniform temporo-spatial dispersion of food, and small groups are necessary in extremely arid areas where food is very sparse. In gelada baboons, troop size parallels food abundance especially closely. Hobson and others have shown that various species of fish disperse when feeding (71, 109) or after they have been deprived of food for a time (81) and may reassemble into schools when attacked. Omnivorous and herbivorous birds (169) and omnivorous monkeys (35) have been argued as suffering least from a grouping situation. In all cases, grouping seems to hinder feeding, and the amount of grouping is thus inversely proportional to its feeding cost. The compensatory advantages hypothesized for grouping include enhancement of navigational ability (106), social stimulation of breeding (never successfully confirmed, 78), and most especially, reduction of predation (e.g. 1, 35, 61, 71, 78, 102, 130).

There may be no change in foraging efficiency.—Certain animals, such as peccaries (41), various birds (147), sea lions (52), and chimpanzees (83, 151, 189), congregate on patches of abundant food—thus their dispersion is proportional to that of their food. Horn (78) has stated the conditions under which birds should nest at the same place, viz, in a colony, regardless of any advantage of grouping qua grouping: During each of one or more intervals of time "critical" for feeding, food is clumped at some point in the foraging area (the point possibly varying with time) such that at least two pairs can feed more efficiently there than at any other point. Horn has computed the optimal location for nests in this area by minimizing a second moment for food distribution at points in the area, weighted by the fraction of time each point is better for foraging than any other point. Hamilton et al (63, also Vol. 1 of this series) argue that for animals such as starlings, which disperse daily from a central roost to feed, those traveling the greater distances are compensated for their increased cost of dispersal by the gain in feeding efficiency once on the feeding grounds, due to a food supply less depleted by other birds—this hypothesis applies, of course, to groups whose foraging areas do not overlap too much.

Grouping increases foraging efficiency.—Animals may accidentally or purposely increase effective availability of prey. Rand (147) and Moynihan (130) suggest that group foraging may flush more insects per bird than solitary foraging. Piscivorous birds sometimes engage in "group drives," whereby prey are forced into shallow water so as to be easily eaten (46).

A second possible benefit of group size is the increased size of prey that can be captured—hence it is a consequence of adaptive radiation in communities. In the Canidae, species which hunt singly or in pairs nearly always prey on smaller animals than do those hunting in large packs (97). Moreover, an ability within species for groups to feed on large prey more successfully than can solitary individuals is known among certain birds (147) and hyenas (100). However, relations between group size and prey size for multimember groups, except for lions, have not been found (39, 101). Most investigators on wolves (87, 124) surmise from observation that, although larger packs may be able to overcome prey, optimal pack size is often lower than that which actually occurs. Here, possibly "kin selection" (62) favors a pack size suboptimal for feeding. Group size in wolves and lions may be limited by starvation of the most subordinate individuals during times of food scarcity (87, 208).

The elaborate communication devices facilitating cooperative group foraging in social insects are well known (188), and some have been given mathematical representation (14). Sometimes these groups of genetically similar animals, such as acacia or leaf-cutting ants, practice "management" of their food resources (30, 85); Slobodkin (180) suggests a similar possibility for less closely related predators.

Another advantage of grouping may be an increased ability to defend the feeding area. The circumstances for which this should result in a lower feeding and defending time T_{FD} per individual can be investigated with Equation 3 for time minimizers (for a different approach, see Hamilton & Watt in Vol. 1 of this series). If no invaders were on the area ($N_e = 0$), then if area increased in proportion to group size, eating time (T_E) per group member would be constant; however, if $N_e > 0$, T_E can vary with group size. Changes in T_{FD} with changing group size depend on whether defense is by the group acting in concert, as for many monkeys, or defense is performed individually, as often for acorn woodpeckers (115). Suppose animals act in concert such that the number of invaders contacted per unit time is the same function of invader density for all defender group sizes but the time T_{SD} and cost C_{SD} of search and defense for each of certain (perhaps all) individuals is constant regardless of group size. Then for those individuals, the total time spent eating and defending must increase if area is directly proportional to group size and f is constant, because the first (T_E) and third terms in Equation 3 increase while the second stays the same. In contrast, suppose that members act individually such that the sum of their separate search times is constant regardless of group size. Then the second and third terms in Equation 3 decrease with group size but the first may increase or decrease. The first term is affected by several factors but is most sensitive to R_I the food requirements of an invader—if R_I is too high, then group defense is still uneconomical. Suppose, finally, that area increases proportionately to group size but each individual spends the same time searching for invaders regardless of group size. Then the second term is constant, but the first and third decrease with increasing group size, thereby implying the bigger the group the better. Of course, it will rarely be optimal in the sense of minimizing Equation 3 for area to increase in direct proportion to group size while T_{SD} and f remain constant, as assumed for some of these results.

Another advantage of communal foraging—the prevention of overlap in foraging areas—is generally ascribed to flocks of birds (126, 127, 177). The advantage may work by reducing overlap between individuals of different or the same species. As Morse (126, 127) points out, in a mixed-species flock individuals of each species know what other species are in the flock and so can opportunistically extend their foraging niches without the costly trial and error (including increased aggressive encounters) incurred by solitary foragers. For species with but one member per mixed-species flock, intraspecific overlap would not be less than for solitary individuals if flocks moved randomly, unless the different flock compositions resulted for the given species in differences in feeding zone from flock to flock. For species with several individuals per flock (including monospecific flocks), group foraging may be advantageous because the larger the group, the greater the degree to which the entire area is known as to its suitability. We can show this for two flocks that are trying to forage in such a way as to search all nonutilized space in a given area before returning to places already searched. Suppose A is the total area available for foraging, S_L is the area of search of a large flock per unit time, S_S is the same area for a small flock, and $S_L > S_S$, either because the large flock's area is larger or because it moves faster (e.g. 126). Then the probability P_L at any time t (before the entire area has been searched exactly once) for the large group of encountering for the first time an area depleted prior to t by the small group is equal to the fraction of the area used by the small group divided by the fraction of the area unknown to the large group; P_S is defined analogously:

$$P_L = \frac{tS_S/A}{(A - S_L t)/A} \quad \text{and} \quad P_S = \frac{tS_L/A}{(A - S_S t)/A}$$

Then $P_S > P_L$ when $S_L(A/t - S_L) > S_S(A/t - S_S)$, where $(A/t - S_L)$ and $(A/t - S_S)$ are positive, and $(S_L + S_S) < A/t$ (note that before the entire area has been searched exactly once, $A > S_L t + S_S t$). Because the factors on the left add to the same constant as do those on the right (A/t), the inequality must hold, since $S_L > S_S$. The same argument can be extended to areas with more than two groups. This calculation assumes that the geometry of exploitation is such that smaller flocks don't exploit the area in such a pattern as to reduce to zero the probability of a large flock finding a contiguous nondepleted area large enough to support the foraging area per unit time of its flock—otherwise, certain individuals in the large flock would be foraging over depleted areas and the advantage may shift to smaller flocks.

A final advantage for communal feeding—that of increasing the area monitored for food—has been proposed for those animals whose food appears in large patches at erratic time intervals. For example, perch feeding in schools may better locate food, and they appear to increase search area in winter when food is patchy and rarer by expanding the volume of the school (69); chimpanzee and spider monkey groups may split and reassemble at abundant food to better locate irregularly seasonal fruit (35, 151); and various birds may use flocking to exploit temporary food (34, 88, 147, 204). For the well-documented case of the weaver *Quelea quelea,* Ward (204) argues that communal roosting extends the area of food-monitoring to hundreds of square kilometers; thus birds which have run out of food on a particular day are able to follow directed flocks the next day to a new food source. The roost gains in numbers as the dry season progresses and nomadism may eventually develop as food becomes scarcer. For birds which search for food from a central point, it is obvious that above a group size of two the long-term amount of food found per bird, if food is proportional to area searched, must decline, and at a rate increasing with the width of an individual's search path w and decreasing with the maximum outward distance l that can be flown per day; in fact, the radius of the maximum total area that can be monitored by all the birds is

$$\sqrt{\left(\frac{\omega}{2}\right)^2 + l^2}$$

Hence, as Ward (204) points out, for foraging efficiency to be maximized, patches which are discovered and used by a group must be, at least sometimes, larger than those smaller groups would have time to utilize completely. Probably group size is actually *above* that for optimal feeding, since there should be a group size at which the addition of a new bird is of great advantage to that bird but only slightly disadvantageous for a bird in the flock; if the flock is large and the number of invaders sufficiently small, it is unlikely that any individual would derive long-term benefit, and thus be altruistic in Trivers' (192) sense, by defending the roost against a bird that should be quite determined to stay, even though this defense may increase the gain per bird in the group.

ACKNOWLEDGMENTS

I thank J. E. Cohen, M. Gadgil, D. H. Morse, J. Roughgarden, and R. L. Trivers for reading a previous draft and supplying many useful ideas. I also thank A. Covich, C. S. Holling, A. L. Turnbull, and G. W. Salt for bibliographical information, and J. L. Brown, D. H. Morse, G. H. Orians, and C. C. Smith for allowing me to see unpublished manuscripts. This work was supported by NSF Grant B 019801X and the Harvard University Society of Fellows.

LITERATURE CITED

1. Aldrich-Blake, F. P. G. 1970. Problems of social structure in forest monkeys. In *Social Behaviour in Birds and Mammals,* ed. J. H. Crook, 79–101. London & New York: Academic
2. Aleksiuk, M., Cowan, I. M. 1969. The winter metabolic depression in arctic beavers (*Castor canadensis* Kuhl) with comparisons to California beavers. *Can. J. Zool.* 47:965–79
3. Armstrong, J. T. 1965. Breeding home range in the nighthawk and other birds: its evolutionary and ecological significance. *Ecology* 46:619–29
4. Ashmole, N. P. 1968. Body size, prey size, and ecological segregation in five sympatric tropical terns (Aves:Laridae). *Syst. Zool.* 17: 292–304
5. Atwood, A. D. 1960. Nutrition and reproductive capacity in fish. *Proc. Nutr. Soc.* 19:23–28
6. Banks, C. J. 1957. The behavior of individual coccinellid larvae on plants. *J. Anim. Behav.* 5:12–24
7. Bartholomew, G. A. 1968. Body temperature and energy metabolism. In *Animal Function: Principles and Adaptations,* ed. M. S. Gordon, 290–354. London: MacMillan
8. Bedard, J. 1969. Feeding of the least, crested, and parakeet auklets around St. Lawrence Island, Alaska. *Can. J. Zool.* 47:1025–50
9. Belkin, D. A. 1965. Reduction of metabolic rate in response to starvation in the turtle *Sternothaerus minor. Copeia* 1965(3): 367–68
10. Beukema, J. J. 1968. Predation by the three-spined stickleback (*Gasterosteus aculeatus* L,): the influence of hunger and experience. *Behavior* 31:1–126
11. Bider, J. R. 1961. An ecological study of the hare *Lepus americanus. Can. J. Zool.* 39:81–103
12. Birkenholz, D. C. 1963. A study of the life history and ecology of the round-tailed muskrat (*Neofiber alleni* True) in north central Florida. *Ecol. Monogr.* 33:187–213
13. Bonner, J. T. 1965. *Size and Cycle.* Princeton: Princeton Univ. Press. 219 pp.
14. Bossert, W. H., Wilson, E. O. 1963. The analysis of olfactory communication among animals. *J. Theor. Biol.* 5:443–69
15. Bostic, D. L. 1966. Food and feeding behavior of the teiid lizard, *Cnemidophorus hyperthrus beldingi. Herpetologica* 22:23–31
16. Bragg, A. N. 1957. Some factors in the feeding of toads. *Herpetologica* 13:189–91
17. Brawn, V. M. 1969. Feeding behavior of cod (*Gadus morhua*). *J. Fish. Res. Bd. Can.* 26:583–96
18. Brenner, F. J. 1967. Seasonal correlations of reserve energy of the redwinged blackbird. *Bird-Banding* 38:195–211
19. Brett, J. R. 1965. The relation of size to rate of oxygen consumption and sustained swimming speed of sockeye salmon (*Oncorhynchus nerka*). *J. Fish. Res. Bd. Can.* 22:1491–1501
20. Brett, J. R., Sutherland, D. B. 1965. Respiratory metabolism of pumpkinseed (*Lepomis gibbosus*) in relation to swimming speed. *J. Fish. Res. Bd. Can.* 22:405–409
21. Brink, C. H., Dean, F. C. 1966. Spruce seed as a food of red squirrels and flying squirrels in interior Alaska. *J. Wildl. Manage.* 30:503–12
22. Broadhead, E. 1958. The psocid fauna of larch trees in northern England—an ecological study of mixed species populations exploiting a common resource. *J. Anim. Ecol.* 27:217–63
23. Broadhead, E., Wapshere, A. J. 1966. *Mesopsocus* populations on larch in England—the distributions and dynamics of two closely-related coexisting species of Psocoptera sharing the same food resource. *Ecol. Monogr.* 36:327–88
24. Brocksen, R. W., Davis, G. E., Warren, C. E. 1968. Competition, food consumption, and production of sculpins and trout in laboratory stream communities. *J. Wildl. Manage.* 32:51–75
25. Brown, J. L. 1964. The evolution of diversity in avian territorial systems. *Wilson Bull.* 6:160–69

26. Brown, J. L., Orians, G. H. 1970. Spacing patterns in mobile animals. *Ann. Rev. Ecol. Syst.* 1: 239–62
27. Brown, L., Amadon, D. 1968. *Eagles, Hawks and Falcons of the World,* Vol. 1. New York: McGraw Hill. 411 pp.
28. Burghardt, G. M. 1969. Comparative prey-attack studies in newborn snakes of the genus *Thamnophis. Behavior* 33:77–114
29. Cade, T. J. 1967. Ecological and behavioral aspects of predation by the northern shrike. *Living Bird* 6:43–86
30. Cherrett, J. M. 1968. The foraging behavior of *Atta cephalotes* L. (Hymenoptera, Formicidae). I. Foraging pattern and plant species attacked in tropical rain forest. *J. Anim. Ecol.* 37:387–403
31. Clark, D. B., Gibbons, J. W. 1969. Dietary shift in the turtle *Pseudemys scripta* (Schoepff) from youth to maturity. *Copeia* 1969(4): 704–6
32. Collias, N. E., Collias, E. C. 1963. Selective feeding by wild ducklings of different species. *Wilson Bull.* 75:6–14
33. Cox, G. W. 1961. The relation of energy requirements of tropical finches to distribution and migration. *Ecology* 42:253–66
34. Crook, J. H. 1965. The adaptive significance of avian social organizations. *Symp. Zool. Soc. London* 14:181–218
35. Crook, J. H. 1970. The socio-ecology of primates. In *Social Behavior in Birds and Mammals,* ed. J. H. Crook, 103–66. London & New York: Academic
36. Crook, J. H., Aldrich-Blake, P. 1968. Ecological and behavioural contrasts between sympatric ground dwelling primates in Ethiopia. *Folia Primat.* 8:192–227
37. Dixon, A. F. G. 1959. An experimental study of the searching behavior of the predatory coccinellid beetle *Adalia decempunctata* (L.). *J. Anim. Ecol.* 28:259–81
38. Dixon, A. F. G. 1970. Quality and availability of food for a sycamore aphid population. *Brit. Ecol. Soc. Symp.* 10:271–87
39. Eaton, R. L. 1970. Hunting behavior of the cheetah. *J. Wildl. Manage.* 34:56–67
40. Ebert, T. A. 1968. Growth rates of the sea urchin *Strongylocentrotus purpuratus* related to food availability and spine abrasion. *Ecology* 49:1075–91
41. Eddy, T. A. 1961. Foods and feeding patterns of the collared peccary in southern Arizona. *J. Wildl. Manage.* 25:248–57
42. Edmondson, W. T. 1965. Reproductive rate of planktonic rotifers as related to food and temperature in nature. *Ecol. Monogr.* 35:61–111
43. Eisenberg, R. M. 1970. The role of food in the regulation of the pond snail *Lymnaea elodes. Ecology* 51:680–84
44. Emlen, J. M. 1966. The role of time and energy in food preference. *Am. Natur.* 100:611–17
45. Emlen, J. M. 1968. Optimal choice in animals. *Am. Natur.* 102:385–90
46. Emlen, S. T., Ambrose, H. W. III. 1970. Feeding interactions of Snowy Egrets and Red-breasted Mergansers. *Auk* 87:164–65
47. Erlinge, S. 1967. Food habits of the fish-otter *Lutra lutra* L. in Swedish habitats. *Viltrevy Swed. Wildl.* 4:371–443
48. Erlinge, S. 1967. Home range of the otter *Lutra lutra* L. in southern Sweden. *Oikos* 18:186–209
49. Erlinge, S. 1968. Food studies on captive otters *Lutra lutra* L. *Oikos* 19:259–70
50. Estes, R. D., Goddard, J. 1967. Prey selection and hunting behavior of the African wild dog. *J. Wildl. Manage.* 31:52–70
51. Farmer, G. J., Beamish, F. W. H. 1969. Oxygen consumption of *Tilapia nilotica* in relation to swimming speed and salinity. *J. Fish. Res. Bd. Can.* 26:2807–21
52. Fiscus, C. H., Baines, G. A. 1966. Food and feeding behavior of Steller and California sea lions. *J. Mammal.* 47:195–200
53. Fisher, R. A. 1958. *The Genetical Theory of Natural Selection.* New York: Dover. 291 pp.
54. Gangwere, S. K. 1961. A monograph on food selection in Orthoptera. *Trans. Am. Entomol. Soc.* 87: 67–230
55. Gangwere, S. K. 1965. Food selection in the Oedipodine grasshopper *Arphia sulphurea* (Fabricius). *Am. Midl. Natur.* 74:67–75

56. Gashwiler, J. S., Robinette, W. L. 1957. Accidental fatalities of the Utah cougar. *J. Mammal.* 38: 123–26
57. Gerking, S. D. 1967. *The Biological Basis of Freshwater Fish Production.* Oxford: Blackwell. 495 pp.
58. Gibb, J. A. 1960. Populations of tits and goldcrests and their food supply in pine plantations. *Ibis* 102:163–208
59. Gibb, J. A. 1962. L. Tinbergen's hypothesis of the role of specific search images. *Ibis* 104:106–11
60. Glass, N. R. 1968. The effect of time of food deprivation on the routine oxygen consumption of largemouth black bass *(Micropterus salmoides). Ecology* 49: 340–43
61. Goss-Custard, J. D. 1970. Feeding dispersion in some overwintering wading birds. In *Social Behaviour in Birds and Mammals,* ed. J. H. Crook, 1–35. New York: Academic
62. Hamilton, W. D. 1964. The genetical evolution of social behavior. *J. Theor. Biol.* 7:1–52
63. Hamilton, W. J. III, Gilbert, W. M., Heppner, F. H., Planck, R. J. 1967. Starling roost dispersal and a hypothetical mechanism regulating rhythmical animal movement to and from dispersal centers. *Ecology* 48: 825–33
64. Haverschmidt, F. 1962. Notes on the feeding habits and food of some hawks of Surinam. *Condor* 64: 154–58
65. Hayward, B., Davis, R. 1964. Flight speeds in western bats. *J. Mammal.* 45:236–42
66. Heatwole, H., Heatwole, A. 1968. Motivational aspects of feeding behavior in toads. *Copeia* 1968(4): 692–98
67. Heckrotte, C. 1967. Relations of body temperature, size, and crawling speed of the common garter snake *Thamnophis s. sirtalis. Copeia* 1967(4):759–63
68. Helms, C. W. 1968. Food, fat, and feathers. *Am. Zool.* 8:151–67
69. Hergenrader, G. L., Hasler, A. D. 1968. Influence of changing seasons on schooling behavior of yellow perch. *J. Fish. Res. Bd. Can.* 25:711–16
70. Hester, F. J. 1964. Effects of food supply on fecundity in the female guppy *Lebistes reticulatus* (Peters). *J. Fish. Res. Bd. Can.* 21: 757–64
71. Hobson, E. S. 1968. *Predatory behavior of some shore fishes in the Gulf of California. US Dep. Interior Fish Wildl. Serv. Bur. Sport Fish. Wildl. Res. Rep.* 73: 1–92
72. Holling, C. S. 1964. The analysis of complex population processes. *Can. Entomol.* 96:335–47
73. Holling, C. S. 1965. The functional response of predators to prey density and its role in mimicry and population regulation. *Mem. Entomol. Soc. Can.* 45:1–60
74. Holling, C. S. 1966. The functional response of invertebrate predators to prey density. *Mem. Entomol. Soc. Can.* 48:1–86
75. Holling, C. S. 1968. The tactics of a predator. *Symp. Roy. Entomol. Soc. London* 4:47–58
76. Holmes, R. T. 1966. Feeding ecology of the red-backed sandpiper (*Calidris alpina*) in Arctic Alaska. *Ecology* 47:32–45
77. Honer, M. R. 1963. Observations on the barn owl (*Tyto alba guttata*) in the Netherlands in relation to its ecology and population fluctuations. *Ardea* 51:158–95
78. Horn, H. S. 1968. The adaptive significance of colonial nesting in the Brewer's blackbird (*Euphagus cyanocephalus*). *Ecology* 49: 682–94
79. Hornocker, M. G. 1970. An analysis of mountain lion predation upon mule deer and elk in the Idaho primitive area. *Wildl. Monogr.* 1970(21):5–39
80. Houde, E. D. 1969. Sustained swimming ability of larvae of walleye (*Stizostedion vitreum vitreum*) and yellow perch (*Perca flavescens*). *J. Fish. Res. Bd. Can.* 26: 1647–59
81. Hunter, J. R. 1966. Procedure for analysis of schooling behavior. *J. Fish. Res. Bd. Can.* 23:547–62
82. Ivlev, V. S. 1961. *Experimental Ecology of the Feeding of Fishes.* New Haven: Yale Univ. Press. 302 pp.
83. Izawa, K. 1970. Unit groups of chimpanzees and their nomadism in the savanna woodland. *Primates* 11:1–46

84. Jackson, P. B. N. 1961. The impact of predation, especially by the tiger fish (*Hydocyon vittatus* Cast.) on African freshwater fishes. *Proc. Zool. Soc. London* 136:603–22

85. Janzen, D. H. 1967. Interaction of the bull's-horn acacia (*Acacia cornigera* L.) with an ant inhabitant (*Pseudomyrmex ferruginea* F. Smith) in eastern Mexico. *Univ. Kan. Sci. Bull.* 47(6):315–558

86. Johnson, L. 1966. Experimental determination of food consumption of pike, *Esox lucius*, for growth and maintenance. *J. Fish. Res. Bd. Can.* 23:1495–1505

87. Jordan, P. A., Shelton, P. C., Allen, D. L. 1967. Numbers, turnover, and social structure of the Isle Royale wolf population. *Am. Zool.* 7:233–52

88. Kahl, M. P. Jr. 1964. Food ecology of the wood stork (*Mycteria americana*) in Florida. *Ecol. Monogr.* 34:97–117

89. Kale, H. W. 1965. Ecology and bioenergetics of the long-billed marsh wren in Georgia salt marshes. *Publ. Nuttall Ornithol. Club* 5:1–142

90. Keast, A. 1970. Food specializations and bioenergetic interrelations in the fish faunas of some small Ontario waterways. In *Marine Food Chains*, ed. J. H. Steele, 377–411. Edinburgh: Oliver & Boyd

91. Keast, A., Webb, D. 1966. Mouth and body form relative to feeding ecology in the fish fauna of a small lake, Lape Opinicon, Ontario. *J. Fish. Res. Bd. Can.* 23: 1845–74

92. Keast, A., Welsh, L. 1968. Daily feeding periodicities, food uptake rates, and dietary changes with hour of day in some lake fishes. *J. Fish. Res. Bd. Can.* 25:1133–44

93. Kendeigh, S. C. 1949. Effect of temperature and season on energy resources of the English Sparrow. *Auk* 66:113–27

94. Kendeigh, S. C. 1970. Energy requirements for existence in relation to size of bird. *Condor* 72: 60–65

95. Kennedy, J. P. 1956. Food habits of the rusty lizard *Sceloporus olivaceus* Smith. *Tex. J. Sci.* 8:328–49

96. King, J. R., Farner, D. S. 1966. The adaptive role of winter fattening in the white-crowned sparrow with comments on its regulation. *Am. Natur.* 100:403–18

97. Kleiman, D. G. 1966. The comparative social behavior of the Canidae. *Am. Zool.* 6:182

98. Kohn, A. J. 1968. Microhabitats, abundance and food of *Conus* on atoll reefs in the Maldive and Chagos Islands. *Ecology* 49: 1046–61

99. Kolenosky, G. B., Johnston, D. H. 1967. Radio-tracking timber wolves in Ontario. *Am. Zool.* 7:289–303

100. Kruuk, H. 1966. Clan-system and feeding habits of Spotted Hyenas (*Crocuta crocuta* Erxleben). *Nature* 209:1257–58

101. Kruuk, H., Turner, M. 1967. Comparative notes on predation by lion, leopard, cheetah and wild dog in the Serengeti area, East Africa. *Mammalia* 31:1–27

102. Lack, D. 1968. *Ecological Adaptations for Breeding in Birds*. London: Methuen. 409 pp.

103. Lack, D., Owen, D. F. 1955. The food of the swift. *J. Anim. Ecol.* 24:120–36

104. Laing, J. 1938. Host-finding by insect parasites. II. The chance of *Trichogramma evanescens* finding its hosts. *J. Exp. Biol.* 51: 281–302

105. Landenberger, D. E. 1968. Studies on selective feeding in the Pacific starfish *Pisaster* in southern California. *Ecology* 49:1062–75

106. Larkin, P. A., Walton, A. 1969. Fish school size and migration. *J. Fish. Res. Bd. Can.* 26:1372–74

107. Latta, W. C., Sharkey, R. F. 1966. Feeding behavior of the American merganser in captivity. *J. Wildl. Manage.* 30:17–23

108. LeBrasseur, R. J. 1969. Growth of juvenile chum salmon (*Oncorhynchus keta*) under different feeding regimes. *J. Fish. Res. Bd. Can.* 26: 1631–45

109. Leong, R. J. H., O'Connell, C. P. 1969. A laboratory study of particulate and filter feeding of the northern anchovy (*Engraulis mordax*). *J. Fish. Res. Bd. Can.* 26:557–82

110. Levins, R. 1968. *Evolution in Changing Environments*. Princeton: Princeton Univ. Press. 120 pp.

111. Levins, R., MacArthur, R. 1969. An hypothesis to explain the incidence of monophagy. *Ecology* 50: 910–11
112. MacArthur, R. H. 1968. The theory of the niche. In *Population Biology and Evolution,* ed. R. C. Lewontin, 159–76. Syracuse: Syracuse Univ. Press
113. MacArthur, R. H. Pianka, E. R. 1966. On optimal use of a patchy environment. *Am. Natur.* 100: 603–9
114. MacArthur, R. H., Wilson, E. O. 1967. *The Theory of Island Biogeography.* Princeton: Princeton Univ. Press. 203 pp.
115. MacRoberts, M. H. 1970. Notes on the food habits and food defense of the acorn woodpecker. *Condor* 72:196–204
116. Martinsen, D. L. 1968. Temporal patterns in the home ranges of chipmunks (*Eutamias*). *J. Mammal.* 49:83–91
117. Martinsen, D. L. 1969. Energetics and activity patterns of short-tailed shrews (*Blarina*) on restricted diets. *Ecology* 50:505–10
118. Martin, N. V. 1966. The significance of food habits in the biology, exploitation and management of Algonquin Park, Ontario, lake trout. *Trans. Am. Fish. Soc.* 95: 415–22
119. Martin, N. V. 1970. Long-term effects of diet on the biology of the lake trout and the fishery in Lake Opeongo, Ontario. *J. Fish. Res. Bd. Can.* 27:125–46
120. McFadden, J. T., Cooper, E. L., Andersen, J. K. 1965. Some effects of environment on egg production in brown trout (*Salmo trutta*). *Limnol. Oceanogr.* 10: 88–95
121. McLaren, I. A. 1969. Population and production ecology of zooplankton in Ogac Lake, a landlocked fiord on Baffin Island. *J. Fish. Res. Bd. Can.* 26:1485–559
122. McLay, C. 1970. A theory concerning the distance traveled by animals entering the drift of a stream. *J. Fish. Res. Bd. Can.* 27: 359–70
123. McNab, B. K. 1963. Bioenergetics and the determination of home range size. *Am. Natur.* 97:133–40
124. Mech, L. D. 1966. The wolves of Isle Royale. *Fauna of National Parks of US Fauna Ser.* 7:1–210
125. Metzgar, L. H. 1967. An experimental comparison of screech owl predation on resident and transient white-footed mice (*Peromyscus leucopus*). *J. Mammal.* 48:387–91
126. Morse, D. H. 1970. Ecological aspects of some mixed-species foraging flocks of birds. *Ecol. Monogr.* 40:119–68
127. Morse, D. H. 1971. The insectivorous bird as an adaptive strategy. *Ann. Rev. Ecol. Syst.* 2:177–200
128. Mosimann, J. E. 1958. The evolutionary significance of rare matings in animal populations. *Evolution* 12:246–61
129. Moss, R. 1968. Food selection and nutrition in ptarmigan (*Lagopus mutus*). *Symp. Zool. Soc. London* 21:207–16
130. Moynihan, M. 1962. The organization and probable evolution of some mixed species flocks of neotropical birds. *Smithson. Misc. Collect.* 143(7):1–140
131. Murdoch, W. W. 1969. Switching in general predators: experiments on predator specificity and stability of prey populations. *Ecol. Monogr.* 39:335–54
132. Murton, R. K., Isaacson. A. J., Westwood, J. V. 1963. The feeding ecology of the wood pigeon. *Brit. Birds* 56:345–75
133. Newton, I. 1967. The adaptive radiation and feeding ecology of some British finches. *Ibis* 109: 33–98
134. Olney, P. J. S. 1963. The autumn and winter feeding biology of certain sympatic ducks. *Trans. 6th Congr. Int. Union Game Biol.* 309–20
135. Orians, G. H. 1961. The ecology of blackbird (*Agelaius*) social systems. *Ecol. Monogr.* 31:285–312
136. Orians, G. H. 1966. Food of nestling yellow-headed blackbirds, Cariboo Parklands, British Columbia. *Condor* 68:321–37
137. Orians, G. H., Horn, H. S. 1969. Overlap in foods and foraging of four species of blackbirds in the potholes of central Washington. *Ecology* 50:930–38
138. Owen, R. B. Jr. 1970. The bioenergetics of captive blue-winged teal under controlled and outdoor conditions. *Condor* 72:153–63

139. Ozoga, J. J., Harger, E. M. 1966. Winter activities and feeding habits of northern Michigan coyotes. *J. Wildl. Manage.* 30:809–18
140. Paloheimo, J. E., Dickie, L. M. 1966. Food and growth of fishes. II & III. *J. Fish. Res. Bd. Can.* 23:869–908, 1209–48
141. Pearson, T. H. 1968. The feeding biology of sea-bird species breeding on the Farne Islands, Northumberland. *J. Anim. Ecol.* 37: 521–52
142. Pianka, E. R. 1966. Convexity, desert lizards and spatial heterogeneity. *Ecology* 47:1055–59
143. Pimlott, D. H. 1967. Wolf predation and ungulate populations. *Am. Zool.* 7:267–78
144. Porter, W. P., Gates, D. M. 1969. Thermodynamic equilibria of animals with environment. *Ecol. Monogr.* 39:245–70
145. Pritchard, G. 1965. Prey capture by dragonfly larvae (Odonata:Anisoptera). *Can. J. Zool.* 43:271–89
146. Rafail, S. Z. 1968. A statistical analysis of ration and growth relationship of plaice (*Pleuronectes platessa*). *J. Fish. Res. Bd. Can.* 25:717–32
147. Rand, A. L. 1954. Social feeding behavior of birds. *Fieldiana Zool.* 36:1–71
148. Rand, A. S. 1967. Ecology and social organization in the iguanid lizard *Anolis lineatopus. US Nat. Mus. Proc.* 122:1–79
149. Recher, H. F. 1966. Some aspects of the ecology of migrant shorebirds. *Ecology* 47:393–407
150. Reichle, D. E. 1968. Relation of body size to food intake, oxygen consumption, and trace element metabolism in forest floor arthropods. *Ecology* 49:538–42
151. Reynolds, V., Reynolds, F. 1965. Chimpanzees of the Budongo Forest. In *Primate Behavior,* ed. I. DeVore, 368–424. New York: Holt, Rinehart & Winston
152. Reynoldson, T. B. 1960. A quantitative study of the population biology of *Polycelis tenuis* (Ijima) (Turbellaria, Tricladida). *Oikos* 11:125–41
153. Rivard, I. 1962. Some effects of prey density on survival, speed of development, and fecundity of the predaceous mite *Melichares dentriticus* (Berl.) (Acarina: Aceosejidae). *Can. J. Zool.* 40: 1233–36
154. Rohlf, F. J., Davenport, D. 1969. Simulation of simple models of animal behavior with a digital computer. *J. Theor. Biol.* 23: 400–24
155. Root, R. B. 1967. The niche exploitation pattern of the blue-gray gnatcatcher. *Ecol. Monogr.* 37: 317–50
156. Rosenzweig, M. L. 1966. Community structure in sympatric carnivora. *J. Mammal.* 47:602–12
157. Rosenzweig, M. L. 1968. The strategy of body size in mammalian carnivores. *Am. Midl. Natur.* 80: 299–315
158. Rosenzweig, M. L., Sterner, P. W. 1970. Population ecology of desert rodent communities: body size and seed-husking as bases for heteromyid coexistence. *Ecology* 51:217–24
159. Royama, T. 1966. Factors governing feeding rate, food requirement and brood size of nestling great tits *Parus major. Ibis* 108:313–47
160. Rozin, P., Mayer, J. 1961. Regulation of food intake in the goldfish. *Am. J. Physiol.* 201: 968–74
161. Rudebeck, G. 1950. The choice of prey and modes of hunting of predatory birds with special reference to their selective effect. *Oikos* 2:65–88. 3:200–31
162. Rudge, M. R. 1968. The food of the common shew *Sorex araneus* L. (Insectivora: Soricidae) in Britain. *J. Anim. Ecol.* 37:565–81
163. Ryder, R. A. 1950. Great Blue Heron killed by a carp. *Condor* 52:40–41
164. Sadleir, R. M. F. S. 1969. *The Ecology of Reproduction in Wild and Domestic Mammals.* London: Methuen. 321 pp.
165. Salt, G. W. 1968. The feeding activity of *Amoeba proteus* on *Paramecium aurelia. J. Protozool.* 15: 275–80
166. Salt, G. W. 1967. Predation in an experimental protozoan population (*Woodruffia-Paramecium*). *Ecol. Monogr.* 37:113-44
167. Schaller, G. B. 1967. *The Deer and the Tiger.* Chicago: Univ. Chicago Press. 370 pp.
168. Schoener, T. W. 1967. The ecologi-

cal significance of sexual dimorphism in size in the lizard *Anolis conspersus*. *Science* 155:474–77

169. Schoener, T. W. 1968. Sizes of feeding territories among birds. *Ecology* 49:123–41
170. Schoener, T. W. 1968. The *Anolis* lizards of Bimini: resource partitioning in a complex fauna. *Ecology* 49:704–26
171. Schoener, T. W. 1969. Models of optimal size for solitary predators. *Am. Natur.* 103:277–313
172. Schoener, T. W. 1969. Optimal size and specialization in constant and fluctuating environments: an energy-time approach. *Brookhaven Symp. Biol.* 22:103–14
173. Schoener, T. W. 1970. Nonsynchronous spatial overlap of lizards in patchy habitats. *Ecology* 51:408–18
174. Schoener, T. W., Gorman, G. C. 1968. Some niche differences among three species of Lesser Antillean anoles. *Ecology* 49:819–30
175. Scott, D. P. 1962. Effect of food quantity on fecundity of rainbow trout *Salmo gairdneri*. *J. Fish. Res. Bd. Can.* 19:715–31
176. Sexton, O. J. 1960. Experimental studies of artificial Batesian mimics. *Behavior* 15:244–52
177. Short, L. I. 1961. Interspecies flocking of birds of montane forest in Oaxaca, Mexico. *Wilson Bull.* 73:341–47
178. Silver, H., Colovos, N. F., Holter, J. B., Hayes, H. H. 1969. Fasting metabolism of white-tailed deer. *J. Wildl. Manage.* 33:490–98
179. Skellam, J. G. 1958. The mathematical foundations underlying the use of line transects in animal ecology. *Biometrics* 14:385–400
180. Slobodkin, L. B. 1968. How to be a predator. *Am. Zool.* 8:43–51
181. Smith, C. C. 1968. The adaptive nature of social organization in the genus of tree squirrels *Tamiasciurus*. *Ecol. Monogr.* 38:31–63
182. Smith, C. C. 1970. The coevolution of pine squirrels (*Tamiasciurus*) and conifers. *Ecol. Monogr.* 40:349–71
183. Snow, D. W. 1962. The natural history of the oilbird, *Steatornis caripensis*, in Trinidad, W. I. Part 2. Population, breeding ecology and food. *Zoologica* 47:199–221
184. Springer, S. 1960. Natural history of the sandbar shark *Eulamia milberti*. *US Fish Wildl. Serv. Fish. Bull. 178* 61:1–38
185. Stanley, J. 1932. A mathematical theory of the growth of populations of the flour beetle *Tribolium confusum* Duv. *Can. J. Res.* 6:632–71
186. Stenger, J. 1958. Food habits and available food of ovenbirds in relation to territory size. *Auk* 75:335–46
187. Storer, R. W. 1966. Sexual dimorphism and food habits in three North American accipiters. *Auk* 83:423–36
188. Sudd, J. H. 1967. *An introduction to the behavior of ants*, Chap. 5, 77–113. London: Edward Arnold
189. Suzuki, A. 1969. An ecological study of chimpanzees in a savanna woodland. *Primates* 10:103–48
190. Taylor, C. R., Schmidt-Nielsen, K., Raab, J. L. 1970. Scaling of energetic cost of running to body size in mammals. *Am. J. Physiol.* 219:1104–7
191. Tinbergen, L. 1960. The natural control of insects in pinewoods. I. Factors influencing the intensity of predation by songbirds. *Arch. Neer. Zool.* 13:265–343
192. Trivers, R. L. 1971. The evolution of reciprocal altruism. *Quart. Rev. Biol.* In press
193. Tucker, V. A. 1970. Energetic cost of locomotion in animals. *Comp. Biochem. Physiol.* 34:841–46
194. Turnbull, A. L. 1960. The prey of the spider *Linyphia triangularis* (Clerck) (Araneae, Linyphiidae). *Can. J. Zool.* 38:859–73
195. Turnbull, A. L. 1962. Quantitative studies of the food of *Linyphia triangularis* Clerck (Araneae: Linyphiidae). *Can. Entomol.* 94:1233–49
196. Turnbull, A. L. 1965. Effects of prey abundance on the development of the spider *Agelenopsis potteri* (Blackwell) (Araneae: Agelenidae). *Can. Entomol.* 97:141–47
197. Turner, F. B. 1959. An analysis of the feeding habits of *Rana p. pretiosa* in Yellowstone Park, Wyoming. *Am. Midl. Natur.* 61:403–13
198. Turner, F. B., Jennrich, R. I., Weintraub, J. D. 1969. Home ranges and body size of lizards. *Ecology* 50:1076–81

199. Utter, J. M., LeFebvre, E. A. 1970. Energy expenditure for free flight by the purple martin (*Progne subis*). *Comp. Biochem. Physiol.* 35:713–19
200. Valiela, I. 1969. An experimental study of the mortality factors of larval *Musca autumnalis* DeGeer. *Ecol. Monogr.* 39:199–225
201. Verme, L. J. 1963. Effect of nutrition on growth of white-tailed deer fawns. *Trans. 28th N. Am. Wildl. Nat. Res. Conf.* 1963:431–43
202. Verme, L. J. 1969. Reproductive patterns of white-tailed deer related to nutritional plane. *J. Wildl. Manage.* 33:881–87
203. Verner, J. 1965. Time budget of the male long-billed marsh wren during the breeding season. *Condor* 67:125–39
204. Ward, P. 1965. Feeding ecology of the black-faced dioch *Quelea quelea* in Nigeria. *Ibis.* 107:173–214, 326–349
205. West, G. C. 1960. Seasonal variation in the energy balance of the tree sparrow in relation to migration. *Auk* 77:306–29
206. West, G. C. 1968. Bioenergetics of captive willow ptarmigan under natural conditions. *Ecology* 49: 1035–45
207. Williams, G. 1962. Seasonal and diurnal activity of harvestmen (Phalangida) and spiders (Araneida) in contrasted habitats. *J. Anim. Ecol.* 31:23–42
208. Wright, B. S. 1960. Predation on big game in East Africa. *J. Wildl. Manage.* 24:1–15
209. Young, P. T. 1945. Studies of food preference, appetite and dietary habit V. *Comp. Psychol. Monogr.* 19:1–58

PART 4 Methodological Advances

New Approaches and Methods in Ecology

James H. Brown

Ecologists have the reputation—and many practitioners perpetuate it—of looking upon methodology with some disdain. We need only toothpicks, pipecleaners, and a field notebook—and perhaps now a microcomputer—to do our research. Creative brainpower and dedicated fieldwork substitute for technique. This image is incorrect and inimical to progress, especially in the present age of technological science.

The following papers show that as a discipline ecology has never really disdained methodology. On the contrary, in ecology, just as in other fields of science, the development and application of new techniques have led to major theoretical and empirical advances. Methodological advances have given us new mathematical techniques with which to build and test models, new ways to measure organisms and their present and past environments, and new insights into the structure and function of ecological systems. The papers that follow are unrepresentative in that they do not indicate the enormous influence of modern technology on ecology within the last decade or so. But they do show how new techniques can open whole avenues for investigation, by making it possible either to ask new questions or to obtain different answers to old ones.

Reconstructing Ecological History

The first paper by the Swedish naturalist Lennart von Post showed that almost perfectly preserved pollen grains can be recovered from the anaerobic sediments of peat bogs. Furthermore, when von Post identified these pollen grains, he found that they represented plant species that did not presently occur in the vicinity of the bogs. In fact, the pollen in the layered sediments recorded a sequence of change in the dominant plant genera from the end of the last glacial period to the present.

Von Post's paper is far more than a description of how to extract and identify fossil pollen. Although his analyses are crucially dependent on these methods, von Post's paper pays little attention to technique. Most of it is concerned with describing the changes in composition of plant assemblages during postglacial times and with drawing inferences about the shifting climatic regimes that caused the vegetation to change.

Von Post's methods of collecting, identifying, depicting (in diagrams showing the changing proportions of the dominant genera in the pollen record as a function of time), and interpreting the pollen in anaerobic sediments still serve as the basis for the modern discipline of palynology. His work also laid the foundation for contemporary paleoecology, the discipline that attempts to provide a long-term historical perspective in ecology by characterizing the composition of past communities, determining the kinds of environments where they occurred, and documenting the rate and causes of changes in the abundance and distribution of species. Of course biologists had known for several centuries that petrified fossils recorded the remains of ancient organisms. Many of these fossils were of distinctive plants and animals that no longer occur anywhere on earth, but a few were of forms closely related to contemporary organisms. The paleontologists who studied these fossilized organisms had always been interested in reconstructing the kinds of environments where they lived. But these paleontological investigations, unlike the study of fossil pollen and subsequent, conceptually similar work on plant macrofossils in packrat middens (for example, Wells and Berger 1967; Wells 1983; Van Devender 1986), and on planktonic organisms in anaerobic lake and ocean sediments (for example, Stockner and Benson 1967; Stoermer et al. 1985), have had little impact on ecology.

There are at least two reasons for this. First, paleontology has traditionally been allied with geology. It has focused on questions of taxonomy, stratigraphy, and evolution—what occurred when, and what were the antecedents and descendents—rather than on ecological relationships. There were good reasons for this. Often fossils were found in isolation or in small associations of the same kinds of organisms, so that reconstructions of past communities and environments were difficult. The fact that the majority of fossils were millions of years old and obviously had no living close relatives compounded this problem. In contrast, the fossil pollen deposits preserved large samples of past communities, were usually only a few thousand years old, and contained remains indistinguishable from contemporary forms. This made the composition and environmental context of these past assemblages much more interpretable.

The second reason for the limited impact of paleontology on ecology is that the fossil record preserved in rocks provides only limited insight into the history of contemporary organisms and their ecological relationships. The plants and animals that lived tens and hundreds of millions of years ago were so different from contemporary organisms that it is difficult to make inferences about their physiology, behavior, and ecology, either from their fossilized remains or from characteristics of their distant living relatives. Even when reconstructions are attempted, they are of limited value in understanding the ecology of the contemporary biota which is comprised of different kinds of organisms. In contrast, the fact that pollen and other fossils preserved in anaerobic sediments were only a few thousand years old and indistinguishable from still-living species meant that they offered powerful insights into the history of contemporary ecological systems. If we assume that the physiology, behavior, and ecology of the fossilized organisms were as identical to living species as the structures that were preserved, then the ecological relationships of contemporary species and communities can provide powerful insights into past ecological relationships. Perhaps more importantly, these past associations give invaluable information on the histories of contemporary species and communities.

The legacy of von Post's study has been the recognition that the pollen preserved in lakes and bogs, the plant parts in woodrat middens, and the skeletons of planktonic organisms in lake and ocean sediments provides a superb record of the last 20,000 years. This period includes the dramatic climatic changes from Pleistocene to the present and the increasing local to global impacts of *Homo sapiens*. The fossils record the dramatic changes in species distributions and associations as present communities were being assembled. These changes are currently under study by a growing number of paleo- and historical ecologists. Some of their most exciting discoveries are reviewed in recent papers by Huntly and Birks (1983), Wells (1983), Davis (1986), and Van Devender (1986).

Age- and Stage-structured Populations

By the 1930s, Pearl (1925), Lotka (1925), Volterra (1926), and others had laid the foundations of a mathematical theory for ecology. But then, as now, there was great debate about the adequacy of necessarily simplified mathematical models to describe the complexities of nature. A continuing problem has been how to measure empirically the parameters of the models so as to test the theory. The r and K of Pearl's logistic equation and the alphas that Lotka and Volterra used to parameterize the strength of interaction between competitors or between predator and prey made for tractable mathematics and appealed to biological intuition. It was far from clear, however, how these quantities could be measured in the simplified conditions of the laboratory, not to mention in the realistically complicated situations of the field. This remains a major problem. These quantities still bedevil attempts to develop a realistic, testable ecological theory (for example, Levins 1966; Levin 1981).

The earliest mathematical models dealt with population dynamics: the growth, regulation, and interactions of one or two populations. The simple models, such as Pearl's logistic and Lotka and Volterra's equations for two-species interactions, ignored age structure and other complications that affect population growth. But it soon became obvious that any serious attempt to characterize the growth and regulation of real populations would have to deal with these problems. Thus the science of demography was born. Pearl, Lotka, and others made early conceptual advances, showing how variation in fecundity and mortality with age affect the rate of population growth. It remained for Patrick Leslie, a young mathematician working in Elton's Bureau of Animal Populations in Oxford, to make the seminal methodological contribution.

Techniques of matrix algebra had been developed by mathematicians in the 1920s and 1930s and Lotka had applied some of them to problems of population dynamics. In 1945 Leslie developed the mathematical treatment of life tables that is still widely used today. He demonstrated the robust power of a simple matrix containing only diagonal elements of probabilities of surviving from age x to $x + 1$ and one row of elements containing the average number of offspring born to individuals of age x. Leslie showed that if these survival probabilities and fecundities remained constant, the matrix could be used to calculate the proportional changes in each age class, the stable age distribution, the time required to approach this distribution, and the rate of population increase, r, for any initial age distribution. In a later paper, Leslie (1948) applied similar matrix techniques to the more realistic but more complicated case of density-dependent regulation, in which fecundity or survival varies with population density.

The impact of Leslie's contribution stems from two features of his paper. First, it shows how one elegantly simple mathematical technique can be used to make all the important demographic projections. Second, it shows how this technique can be applied to basic life table data of the sort that a laboratory or field ecologist can realistically expect to obtain. Only two years after the publication of Leslie's seminal paper, Deevey (1947) published an influential review of animal demography, showing how different kinds of data collected by field biologists could be used to construct life tables for natural populations. Deevey's paper helped to cement the relationship between demographic theory and data by calling attention to the careful field studies of Lack (summarized a few years later in an influential book on population regulation, Lack 1954), Hatton (1938), and others, and by characterizing the different types of survivorship curves that are depicted in every ecology textbook.

The enduring power of the "Leslie Matrix" is illustrated not only by its continuing use in population ecology but also by its use by the multibillion dollar life insurance industry to set age-specific premium schedules that will earn a profit. So fundamental was Leslie's contribution that subsequent conceptual advances in demography have been modest. One important insight was the recognition that, in most plants and some animals, the demographic contribution of an individual may depend more on its body size, physical condition, or social status than on its chronological age (Leftkovitch 1965; Werner and Caswell 1977; Sauer and Slade 1987). It is fairly straightforward to incorporate these effects by modifying the Leslie Matrix to incorporate stage-specific rather than age-specific mortality and fecundity terms. When this is done, entries in the matrix are no longer necessarily confined to the diagonal, because it is possible to change several stages in one time interval and to regress as well as advance in stage.

Measuring Population Parameters

Central to all considerations of population growth and regulation is the intrinsic rate of natural increase, r. Malthus (1798) recognized that any population has the intrinsic capacity to increase multiplicatively (exponentially) in an unlimited environment. When growing exponentially, the rate of increase is equal to the population size times some rate constant, r, that depends on characteristics of both the organism and the environment. As the theory of population dynamics and demography developed in the early-twentieth century, it became important to measure r empirically for populations in the laboratory and in the field. Such measurements were necessary not only to test the theories, but also to assess the variation in patterns of population dynamics between different kinds of organisms and between similar organisms living in different environments.

In 1948 L. Charles Birch was a young ecologist who had just returned to Australia after a fellowship at the Bureau of Animal Population in Oxford. Birch had already begun his lifelong investigations of insect populations with H. G. Andrewartha, and during his stay in Oxford he was strongly influenced by his interactions with Leslie and by the writings of Lotka. With this background, he set out to measure the intrinsic rate of increase and other demographic parameters for the rice weevil, *Calandra oryzae*, under controlled conditions in the laboratory.

Birch's paper is notable for its clarity and completeness. Others had previously acquired different kinds of demographic data for populations in both the laboratory and the field (reviewed by Deevey 1947). But Birch's was the first comprehensive analysis of the demography of a population growing exponentially under carefully controlled conditions. Birch assembled a life table of age-specific mortality and fecundity values and then proceeded to calculate the net reproductive rate, average generation time, stable age distribution, and instantaneous birth and death rates, in addition to the intrinsic rate of increase. He used the summation approximation of Euler's equation (which he attributed to Lotka) to calculate r from the life table data.

Birch's paper made two other important contributions. First, it measured the effect of temperature on the intrinsic rate of increase. This showed that even when food and other resources were in unlimited supply, environmental conditions such as temperature can greatly influence the rate of population growth. This presaged his later work on the effect of the abiotic environment on the dynamics of insect populations. Second, in the discussion section of his paper, Birch considered the effect of variation in life history traits on the growth of the population. He demonstrated that reducing the duration of the immature stages (decreasing the age to first reproduction) could greatly increase the intrinsic rate of increase. Thus, in this empirical treatment he anticipated one of Cole's important theoretical results (Cole 1954, reprinted in part 2 of this volume).

The methodological advances of Leslie, Deevey, Birch, and their contemporaries showed how to apply the theory of demography to empirical studies of population dynamics. This body of work made the mathematical models of Pearl, Lotka, Volterra, Nicholson, and the other early mathematical ecologists accessible to evaluation by laboratory and field ecologists. By the middle of this century a firm theoretical and methodological foundation had been laid for the study of populations. The empirical studies of population regulation, the great debate about density dependence and density independence, and theoretical models of life history tactics that occupied so many ecologists during the 1950s and 1960s were built on this foundation.

The Behavior and Impact of Predators

By the late 1960s, debates about population regulation were raging. Many ecologists were searching for a single general mechanism that would regulate the density of all populations. Andrewartha and Birch (1954) were championing the role of the fluctuating abiotic environment, Lack (1954) and Milne (1957) advocated the importance of food supply, and intra- and interspecific competition, and Errington (1946) and Varley (1947) were arguing the role of predation. It is in this background that a young Canadian ecologist, C. S. (Buzz) Holling, set out to do a comprehensive laboratory and field study of predation on an irruptive insect pest. He analyzed predation by small mammals on pupae of pine sawflies.

Many features of Holling's paper make it important and innovative. Some of these features are its elegant combination of general theory and specific observations and experiments, its choice of deliberately simplified laboratory and field (a pine plantation) systems to use as empirical models, its inspired use of laboratory measurements under controlled conditions for accurate quantification and of field studies to supply a realistic context, and the community-level approach that took account of multiple predators and alternative prey. But the paper's greatest contribution is the breakdown of predation into behavioral and population components, which can be studied in isolation in controlled situations and then reassembled to understand their effects on the dynamics of the predator-prey system as a whole. In this one paper Holling developed several themes that were to become major research programs during the next two decades.

Holling distinguished two levels of predation: (1) the behavioral acts performed by an individual predator searching for and consuming prey; and (2) the population processes of mortality and fecundity influencing the growth of predator and prey populations. He quantified both the behavioral and population components in terms of the number of individual prey killed as a function of prey density; he termed these functional response and numerical response, respectively. Holling used graphical models to describe the range of dynamics that these behavioral and population responses might exhibit under different conditions. These, especially the functional response curves, remain a central element of predation theory today.

Holling's treatment of the functional response anticipated the theoretical and empirical studies of optimal foraging that were to follow in the next two decades (for example, MacArthur and Pianka 1966 [above, part 2]; see also Emlen 1966, Pyke et al. 1978, Stephens and Krebs 1986). In his laboratory and field studies, Holling considered the effects of several prey characteristics on the functional response of predators. In evaluating the interacting effects of prey density and availability of alternative prey on the number of prey killed, Holling called attention to two variables that are central to most subsequent theoretical and empirical studies of foraging.

Holling's attempt to reduce predation to its behavioral and population components for analysis and then to reassemble these components to understand whole-system dynamics typifies many reductionist approaches to ecology. One can argue about the value of this kind of reductionism. I suggest that Holling was much more successful in reducing predation to its behavioral components than in reassembling these elements to explain and predict population dynamics. This probably reflects a general limit to the usefulness of reductionism in studies of complex ecological systems, but it does not diminish the magnitude of Holling's contribution.

A Biophysical Approach to Physiological Ecology

Physiological animal ecology has been criticized, with some justification, for "showing that animals can live where they do." As a discipline, physiological ecology has historically been rather isolated from the mainstream of population, community, and ecosystem ecology. Much of its early work was really comparative physiology. It was more concerned with mechanisms of adaptation to physical environmental stresses than with accounting for the role of the abiotic environment in influencing abundance and distribution. Andrewartha and Birch (1954) and others had emphasized the role of abiotic conditions in limiting insect populations, but rarely was any attempt made to reduce the response at the level of the population to the components of performance at the level of individual organisms (as Holling did in the case of predation).

Although it is easy to follow the development of ideas about the ecological consequences of energy relations of individual animals (for example, Cowles and Bogert 1944; Scholander et al. 1950*a*, *b*; Gates 1962; Porter et al. 1973), it is difficult to isolate a single seminal publication. The early work of David Gates (for example, 1962) emphasized the physics of energy exchange between organisms and their environment. This research received added emphasis and ecological relevance when Gates collaborated with young Warren Porter, who had been trained in comparative physiology and had displayed an aptitude for mathematics. We have chosen to reprint the 1969 monograph by Porter and Gates because it illustrates three important contributions: the use of mathematical models to identify the variables that affect energy exchange; the use of physical models to measure the interacting influences of key variables under laboratory and field conditions; and the development of the concept of climate space to characterize the range of physical conditions where an individual can survive and reproduce. All three of these approaches are still widely used today.

The mathematical models revealed the sensitivity of thermodynamic relationships of organisms to such environmental factors as solar radiation and wind speed. They led to increased insights into the important roles of postural changes and microhabitat selection in allowing individuals to maintain favorable energy relations in otherwise unfavorable, even uninhabitable, environments.

The physical models, cast in metal from molds made from real animals, permitted detailed measurements of physical variables to be made under both highly controlled conditions in the laboratory and much more complex but also more realistic conditions in the field. They provided insights into the roles of biological processes, missing in the inanimate models, that regulate energy exchange. These roles include metabolism, pulmonary and cutaneous evaporation, changes in peripheral circulation and in insulating fur and feathers, and postural adjustments.

Porter and Gates developed the concept of climate space to synthesize the information from their mathematical models, physical models, and experiments with living organisms. The climate space quantifies the physical dimensions of "fundamental niche" (sensu Hutchinson 1959 [above, part 3]). It predicts the range of physical conditions under which a living individual can maintain thermal equilibrium and hence potentially survive and reproduce (assuming that the requirements of all stages of the life history are included). Like Holling's functional response, the concept of climate space illustrates the valuable insights that can come from a reductionist approach which quantifies relationships between individual organisms and important components of their environment.

Porter and Gates' paper was an early sign of the tendency of recent physiological animal ecology to become more ecological. Increasing emphasis continues to be placed on the performance of individual organisms under realistic field conditions, as well as on physical features of the environment that affect survival and reproduction. This new tradition is represented not only by research on energy exchange (for example, Porter et al. 1973; Huey and Slatkin 1976; Tracy 1976), but also on behavioral thermoregulation and microclimate selection (for example, Tracy et al. 1979; Kingsolver 1983), biochemical and physiological bases of foraging and predator escape behavior (for example, Christian and Tracy 1981; Huey and Bennett 1986), energetic costs of reproduction (for example, Hails and Bryant 1979; Ricklefs and Williams 1984), and abiotic factors that limit geographic distributions (for example, Porter and Tracy 1983; Root 1988).

Like the comparable study of animals, early research on the physiological ecology of plants was largely comparative physiology. It documented variation in the basic processes of photosynthesis, respiration, and transpiration, and offered adaptive explanations for correlations between physiology and environment. The realization that a plant's light, carbon, water, and nutrient economy are closely linked and that they influence all components of individual fitness (for example, Mooney 1972), however, eventually stimulated research on more ecological questions; what are the adaptive tradeoffs of different photosynthetic/metabolic pathways, such as C_3, C_4, and CAM; what are the costs and benefits of secondary compounds; why do leaves and roots leak substantial quantities of organic compounds; what are the economic tradeoffs in wind versus animal pollination; how much is invested in seed dispersal structures; what are the costs and benefits of maintaining microbial symbionts? Recently Tilman (1982, 1988; see also Grime 1979) has explored the relationships between physiological mechanisms and plant community organization. His theoretical and empirical studies have focused on the interacting effects of nutrient and water availability below ground and light availability above ground.

Multivariate Analyses of Plant Community Structure

Plant community ecology is another discipline that developed for a while in relative isolation from the rest of ecology. Although Shreve (1914), Clements (1916), Gleason (1917, 1926), and Whittaker (1956, 1967) used plant associations to address some of the most fundamental questions about the abundance, distribution, and diversity of species, a large influential group of plant ecologists seemingly became preoccupied with describing quantitatively the composition of plant communities.

The monograph by Bray and Curtis typifies the so-called phytosociological or ordination approach to plant community ecology. The paper developed and applied a new method to classify quantitatively plant assemblages on the basis of some measure of the abundance or importance of the component species. The technique developed here, in contrast to earlier methods, was multivariate; it arrayed the assemblages along multiple, independent, orthogonal axes of decreasing explanatory power. Although based on a coefficient of similarity rather than on relationships among the abundances of the species, the method had much in common with principal components analysis, factor analysis, discriminant analysis, detrended correspondence analysis, and other multivariate techniques that subsequently have been widely used for this and other purposes.

The Bray and Curtis monograph illustrates both the strengths and weaknesses of the phytosociological approach. The method orders the communities in *n*-dimensional space on the basis of species composition. The first two or three axes are correlated with gradients of disturbance and physical conditions, and individual species usually exhibit unimodal peaks of abundance when plotted on these axes. But as it is used here, the technique is very descriptive. While results of the ordination may raise interesting questions about the environmental determinants of the axes, the position of individual species, and the coexistence of certain combinations of species, they do not answer these questions or provide definitive tests of most hypotheses.

A cynic might ask whether three decades of plant ordination studies have had any real impact on the rest of ecology. We offer two answers. First, phytosociological studies have indeed had less direct influence on other areas of ecology than other contemporaneous research in plant ecology such as gradient studies (for example, Whittaker 1956, 1967; Werner and Platt 1976), studies of plant animal interactions (for example, Ehrlich and Raven 1964 [above, part 3]; Heinrich and Raven 1972; Janzen 1970), and studies of plant strategies in relation to spatial distribution and successional dynamics (for example, Harper 1967 [above, part 3], and 1977; Horn 1971; Grime 1979; and Tilman 1982, 1988). Second, perhaps the main contribution of ordination studies was to introduce other ecologists to a battery of powerful multivariate analytical and statistical techniques. These have now found wide application in many areas of ecology.

New Perspectives on Succession and Human Ecology

Eugene Odum is perhaps rivaled only by Robert MacArthur as the most influential ecologist of the last few decades. Odum's greatest influence has been felt through his textbook *Fundamentals of Ecology*, which first appeared in 1953 and is now in its third edition. But Odum's influence hardly ends with this famous book; it also includes training an illustrious cadre of students and making important conceptual and methodological contributions. In fact, the emergence of ecosystem science as a major subdiscipline of ecology in the second half of this century can be attributed in large part to Odum's energetic leadership. Many of the qualities of Odum and his ecosystem approach to ecology are epitomized in his paper "The strategy of ecosystem development," which appeared in the prestigious journal *Science* in 1969 [this section].

Few papers in ecology have been as influential and as controversial as Odum's. By the time this article was written, American ecology had become divided into two competing schools—ecosystem and evolutionary ecology—under the leadership of its two most eminent ecologists, Odum and MacArthur. Odum's "The strategy of ecosystem development" can be seen as an attempt to bridge the ideological gap between these two subdisciplines. Odum used the phenomenon of ecological succession to speculate about general principles that govern the structure and dynamics of ecological systems. He tried to develop relationships between, on the one hand, the processes of energy flow and nutrient cycling as they traditionally had been studied by ecosystem scientists, and, on the other hand, the phenomena of population growth, community organization, and adaptation as they had been studied by evolutionary ecologists.

Although "The strategy of ecosystem development" tried to present an integrated conceptual model of the interrelationships among the various components of the successional process, this paper did not succeed in achieving synthesis and rapprochement between ecosystem and evolutionary ecology. On the one hand, Odum's attempt to place these ecosystem-level phenomena in a strategic and evolutionary framework was ridiculed by evolutionary ecologists as naive because it seemed to require that a process analogous to natural selection operate at the community or ecosystem level. The paper reinforced the prejudice of many population and community ecologists that most ecosystem ecologists had little appreciation and understanding of evolution. In recent years, however, some evolutionary ecologists and other biologists have begun to challenge the orthodox view that natural selection acts only at the level of individual organisms within populations. D. S. Wilson (for example, Wilson and Sober 1989) and others have revived the Clementsian concept of the superorganism and suggested that at least some kinds of ecological communities may exhibit the kinds of evolutionary strategies envisioned by Odum.

On the other hand, most ecosystem ecologists welcomed Odum's speculations, not only because they presented a new, synthetic perspective on the old topic of succession, but also because they made a number of explicit, testable predictions about basic energetic and biogeochemical dynamics. While several of these specific hypotheses have been disproven, Odum's general approach has been fruitful. Odum's paper, together with the nearly contemporaneous Hubbard Brook study (Likens et al. 1970 [below, part 6]) and the International Biological Program (IBP), played a major role in stimulating ecosystem ecology to become an experimental, hypothesis-testing science.

In addition to advancing new perspectives on successional processes, Odum's paper applied some of these concepts to human ecology. Traditionally, ecologists have emphasized the study of natural systems—"natural" defined as wild and as little influenced by human activities as possible. Odum eloquently argued for an ecology that would include humans and all their activities as integral parts of ecosystems. He pointed out that humans cannot exist apart from nature, and that apparent conflicts between human beings and nature must be resolved or both parties would suffer.

Odum used a number of examples to illustrate the "relevance of ecosystem development theory to human ecology." He prophetically warned about the dangers of managing ecosystems for stability, and he accurately predicted the problems that have resulted from preventing the annual flooding of the Everglades. He pointed out the economic and environmental benefits of reducing the dependency of agriculture on chemical pesticides and fertilizers. And, most ambitiously, he argued for a form of land use zoning that would limit the nature and distribution of human activities on the landscape in an effort to bring the human population and economy into balance with the earth's life support systems. It is a testimony to Odum's visionary concept of human ecology that most of the problems that he addressed are even more relevant today than they were when his paper was written more than two decades ago.

Other Methodological Advances

How many important methodological contributions are not represented by the papers reprinted in this volume and not mentioned above? Many. Some of them have been omitted because their overall impact has been limited (for example, replacement series competition experiments, Wit 1960), because their origins are obscure, or at least difficult to associate with a single publication (for example, mark-recapture and other methods of population estimation; computer simulation), or because their real application to ecology has been seen only since 1975 (for example, applications of graph theory in food web studies, Cohen 1978; use of doubly-labeled water to measure metabolitic rates of free-living animals, Nagy 1987). Other methodological advances are described in the same publications that make major conceptual contributions. Many of these are exemplified in other papers reprinted in this volume: optimization models (MacArthur and Pianka 1966 [part 2], nonlinear mathematical techniques (May 1974 [part 2]), laboratory experiments (Park 1948, Huffaker 1958 [both in part 6]; see also Gause 1934), and field experiments (Clausen, Keck, and Heisey 1948; Connell 1961 [part 6]). Clearly, any perception that progress in ecological science did not depend on new methods is incorrect.

But to a large extent the perception of ecology as a low-tech discipline was accurate up to 1975. But this view is no longer true. Like virtually all flourishing modern sciences, contemporary ecology is highly dependent on new technological developments and applications. We can give only a few examples here: computers which analyze large data sets, make difficult calculations, and perform complex simulations; particle counters which measure the abundance of small organisms; stable isotopes to follow chemical compounds of different origins through food chains and between habitats; remote sensing to map the distribution of soils and vegetation; automated gas chromatography to identify the defensive chemicals of plants; lasers used in combination with fourier analyses to measure atmospheric gas concentrations over heterogeneous landscapes; and geographic information systems to analyze spatial arrays of data.

The contrast between these current methodologies and the historically important ones highlighted in this section is evident. It is interesting to speculate about the content of a future classics volume written fifteen years from now, containing seminal contributions up to 1990. One thing seems certain about such a future volume: methodology would receive greater emphasis, and many of the methodological advances would be technologically-oriented.

Literature Cited

Andrewartha, H. G., and L. C. Birch. 1954. The distribution and abundance of animals. Chicago: University of Chicago Press.

Christian, K. A., and C. R. Tracy. 1981. The effect of the thermal environment on the ability of hatchling Galapagos land iguanas to avoid predation during dispersal. *Oecologia* 49:218–23.

Cements, F. E. 1916. Plant succession: An analysis of the development of vegetation. Publication no. 242. Washington, D.C.: Carnegie Institution.

Cohen, J. E. 1978. *Food webs and niche space*. Princeton: Princeton University Press.

Cowles R. B., and C. M. Bogert. 1944. A preliminary study of the thermal requirements of desert reptiles. *Bull. Amer. Mus. Nat. Hist.* 83:261–96.

Davis, M. B. 1986. Climatic instability, time lags, and community disequilibrium. Pp. 269–84 *in* J. Diamond and T. J. Case, ed., *Community ecology*. New York: Harper and Row.

Deevey, E. S. 1947. Life tables for natural populations of animals. *Quart. Rev. Biol.* 22:283–314.

Emlen, J. M. 1966. The role of time and energy in food preferences. *Amer. Nat.* 100:611–17.

Errington, P. L. 1946. Predation and vertebrate populations. *Quart. Rev. Biol.* 21:144 77.

Gates, D. M. 1962. *Energy exchange in the biosphere*. New York: Harper and Row.

Gause, G. J. 1934. *The struggle for existence*. Baltimore: Williams and Wilkins.

Gleason, H. A. 1917. The structure and development of the plant association. *Bull. Torrey Bot. Club* 44: 463–81.

——— 1926. The individualistic concept of plant association. *Bull. Torrey Bot. Club* 53:7–26.

Grime, J. P. 1979. Plant strategies and vegetation processes. Chichester: John Wiley and Sons.

Hails, C. J., and D. M. Bryant. 1979. Reproductive energetics of a free-living bird. *J. Anim. Ecol.* 48: 471–82.

Harper, J. L. 1977. *Population biology of plants*. London: Academic Press.

Hatton, H. 1938. Essais de biomomie explicative sur quelques especes intercotidales d'alques at d'animaux. *Annal. Inst. Ocean.* 17:241–348.

Heinrich, B., and P. H. Raven. 1972. Energetics and pollination ecology. *Science* 176:597–602.

Horn, H. S. 1971. *The adaptive geometry of trees*. Princeton: Princeton University Press.

Huey, R. B., and A. F. Bennett. 1986. A comparative approach to field and laboratory studies in biology. Pp. 82–98 in M. E. Feder and G. V. Lauder, eds., *Predator-prey relationships: Perspectives and approaches from the study of lower vertebrates*. Chicago: University of Chicago Press.

Huey, R. B., and M. Slatkin. 1976. Costs and benefits of lizard thermoregulation. *Quart. Rev. Biol.* 51: 363–84.

Huntley, B., and H. J. B. Birks. 1983. An atlas of past and present pollen maps for Europe: 0–13,000 years ago. Cambridge: Cambridge University Press.

Janzen, D. H. 1970. Herbovores and the number of tree species in tropical forests. *Amer. Nat.* 104: 501–28.

Kingsolver, J. G. 1983. Thermoregulation and flight in *Colias* butterflies: Elevational patterns and mechanistic limitations. *Ecology* 64:534–45.

Lack, D. 1954. *The natural regulation of animal numbers*. London: Oxford University Press.

Leftkovitch, L. P. 1965. The study of population growth in organisms grouped by stages. *Biometrics* 21:1–18.

Leslie, R. A. 1948. Some further notes on the use of matrices in population mathematics. *Biometrika* 35:213–35.

Levin, S. A. 1981. The role of theoretical ecology in the description and understanding of populations in heterogeneous environments. *Amer. Zool.* 21:865–75.

Levins, R. 1966. Strategy of model building in population biology. *Amer. Sci.* 54:421–31.

Lotka, A. J. 1925. *Elements of physical biology*. Baltimore: Williams and Wilkins.

Malthus, T. R. 1798. *An essay on the principle of popula-*

tion. London: J. Johnson.

Milne, A. 1957. The natural control of insect populations. *Can. Ent.* 89:193–213.

Mooney, H. A. 1972. The carbon balance of plants. *Ann. Rev. Ecol. Syst.* 3:315–46.

Nagy, K. A. 1987. Field metabolic rate and food requirement scaling in mammals and birds. *Ecol. Monogr.* 57:111–28.

Pearl, R. 1925. *The biology of population growth.* New York: Knopf.

Porter, W. P., J. W. Mitchell, W. A. Beckman, and C. B. DeWitt. 1973. Behavioral implications of mechanistic ecology: Thermal and behavioral modeling of desert ectotherms and their environment. *Oecologia* 13:1–54.

Porter, W. P., and C. R. Tracy. 1983. Biophysical analyses of energetics, time-space utilization, and distributional limits. Pp. 55–83 in R. B. Huey, E. R. Pianka, and T. W. Schoener, eds., *Lizard ecology.* Cambridge, Mass.: Harvard University Press.

Pyke, G. H., H. R. Pulliam, and E. L. Charnov. 1978. Optimal foraging: A selective review of theory and tests. *Quart. Rev. Biol.* 52:137–54.

Ricklefs, R. E., and J. B. Williams. 1984. Daily energy expenditure and water-turnover rate of adult European starlings (*Sturnus vulgaris*) during the nesting cycle. *Auk* 101:707–16.

Root, T. 1988. Energetic constraints on avian distributions and abundances. *Ecology* 69:330–39.

Sauer, J. R., and N. A. Slade. 1987. Uinta ground squirrel demography: Is body mass a better categorical variable than age? *Ecology* 68:642–50.

Scholander, P. F., R. Hock, V. Walters, F. Johnson, and L. Irving. 1950*a*. Heat regulation in some arctic and tropical mammals and birds. *Biol. Bull.* 99:237–58.

Scholander, P. F., V. Walters, R. Hoch, and L. Irving. 1950*b*. Body insulation of some arctic and tropical mammals and birds. *Biol. Bull.* 99:225–36.

Shreve, F. 1914. A montane rainforest: A contribution to the physiological plant geography of Jamaica. Carnegie Institute, Publication 199.

Stephens, D. W., and J. R. Krebs. 1986. *Foraging theory.* Princeton: Princeton University Press.

Stockner, J. G., and W. W. Benson. 1967. The succession of diatom assemblages in recent sediments of Lake Washington. *Limn. and Ocean.* 12:513–32.

Stoermer, E. F., J. A. Wolin, C. L. Scheleske, and D. J. Conley. 1985. An assessment of ecological changes during the recent history of Lake Ontario based on siliceous algal microfossils preserved in the sediments. *J. Phycology* 21:257–76.

Tilman, D. 1982. *Resource competition and community structure.* Princeton: Princeton University Press.

——— 1988. Plant strategies and the dynamics and structure of plant communities. Princeton: Princeton University Press.

Tracy, C. R. 1976. A model of the dynamic exchanges of water and energy between a terrestrial amphibian and its environment. *Ecol. Monogr.* 46: 293–326.

Tracy, C. R., B. J. Tracy, and D. S. Dobkin. 1979. The role of posturing in behavioral thermoregulation by black dragons (*Hagenius brevistylus* Selys; Odonata). *Physio. Zool.* 52:565–71.

Van Devender, T. R. 1986. Climatic cadences and the composition of Chihuahuan Desert communities: The Late Pleistocene packrat midden record. Pp. 285–99 in J. Diamond and T. J. Case, eds., *Community ecology.* New York: Harper and Row.

Varley, G. C. 1947. Ecological aspects of population regulation. *Trans. IXth Int. Cong. Ent.* 2:210–14.

Volterra, V. 1926. Variations and fluctuations of the numbers of individuals of animal species living together. Reprinted 1931 in R. N. Chapman, *Animal ecology.* New York: McGraw-Hill.

Wells, P. V. 1983. Paleobiogeography of montane islands in the Great Basin: Observations on the last 20,000 years. *Ecol. Monogr.* 53:341–82.

Wells, P. V., and R. Berger. 1967. Late Pleistocene history of coniferous woodland in the Mojave Desert. *Science* 155:1640–1647.

Werner, P. A., and W. J. Platt. 1976. Ecological relationships of co-occurring goldenrods (*Solidago:* Compositae). *Amer. Nat.* 110:959–71.

Werner, P. P., and H. Caswell. 1977. Population growth rates and age *versus* stage-distribution models for teasel (*Dipsacus sylvestris* Huds.). *Ecology* 58:1103–1111.

Whittaker, R. H. 1956. The vegetation of the Great Smoky Mountains. *Ecol. Monogr.* 26:1–80.

——— 1967. Gradient analysis of vegetation. *Biol. Rev.* 42:207–64. Cambridge: Cambridge University Press.

Wilson, D. S., and E. Sober. 1989. Reviving the superorganism concept. *J. Theor. Biol.* 136:337–56.

Wit, C. T. de 1960. On competition. *Versl. landbouwk. Onderz. Ned.* 660:1–82.

PAPER 21

FOREST TREE POLLEN IN SOUTH SWEDISH PEAT BOG DEPOSITS,

BY

Lennart von POST.

Translation by Margaret Bryan DAVIS and Knut FÆGRI, introduction by Knut FÆGRI and Johs. IVERSEN.

Introduction.

The semi-centennial of the meeting at which *Lennart von Post* first presented pollen analysis to the scientific community is a fitting occasion to present to an international public that lecture which became the basis of a new field of rich scientific activity. Truly, palynology of to-day is much more than *von Post's* « pollen statistics », but one may safely say that without the notable success of his technique in Quaternary vegetational history research the other branches of palynology had hardly developed like they did. And although there is abundant information that fossil pollen had been observed long before the 1916 lecture, and even that percentage calculations had been carried out, anybody who reads *von Post's* presentation will realize that it represents a decisive, a fundamental step forward, that pollen analysis before and after were two different things.

Unfortunately for the development of pollen analysis, the lecture was never published in an international language, and even in Scandinavia it is not easily accessible, being, as it is, tucked away

POLLEN ET SPORES, VOL. IX, N° 3.

in the proceedings of a scientific meeting which, to make things even worse, took place during World War I when communications were disrupted and many scientists had other things to look after. Consequently, while the lecture formed the basis of a strong and vigorous school of research, it became known more by its implications than by its own wording - like so many classics.

There was also in *von Post's* mentality an element of what might perhaps, lacking a better term, be called aristocratic arrogance : to propagate for his ideas was beyond him. He knew they were good, and he presumed that the rest of the scientific world would recognize that too, published in Swedish or not.

It was also part of the same picture that he never attracted many pupils. As a matter of fact, there is hardly a dozen persons active in palynology to-day who can claim *von Post* as their teacher. Nor was he an easy teacher. His clarity of expression—which everyone can see in his publications—reflected a corresponding clarity of mind and thought, and he never accepted muddleheadedness in others. Opposition, yes, but dullness, NO ! He was a man of strong, almost passionate feelings, and he had no time for fools or persons he considered fools.

The only time *von Post* had the international audience one feels he should always have had, was in 1933 when — most characteristically — others had arranged a course in which took part 9 students from other countries in addition to a few of his Swedish pupils. Perhaps that was the happiest time of his scientific life. He was still at the top of his creative activity, and he found within this international group a background on which to expose his ideas and discuss their consequences.

Although he used to consider himself a pupil of *Rutger Sernander's* (they had published together as early as in 1910), and even if he was in possession of the admirable classical Swedish botanical tradition (which protected him from making botanical mistakes), *Lennart von Post* was primarily a Quaternary geologist. To him, pollen analysis was a method of investigating the problem of climatic change during the Quaternary period. When he realized that his plan to make a world-wide vegetational history survey could not be carried out during his life-time, he turned to a geological approach, namely that of the size of eustatic sealevel changes, hoping to reach his goal by means of a short-cut. Through a terribly complicated system of intense investigations he hoped to be able to unravel the intricacies of iso- and eustatic movements. Unfortunately, his death stopped the work, and it will forever remain a torso.

In addition to his critical acumen and the brilliancy of his ideas, *von Post's* investigations owe their penetrating depth to the fact that they all focused on the same great problem. Although limited by being a geologist, he fully saw that the future development of pollen analysis had to be thought in the refined morphological and, even more so, the ecological problems of which he himself did not possess a sufficent command. The evaluation of the new, botanical, trends in pollen analysis, his last contribution to a scientific journal, is in reality more of a monument to his own breadth of vision than an evaluation of the book he was reviewing.

Mentality and physique contributed to make *Lennart von Post* a legend even during his lifetime. On the other hand, his agility and swiftness, even grace of movements were completely unexpected in a person whose stature could only be characterised as obese, a physical characteristic that was counteracted by the strong lines of his head and face.

In accordance with his whole mentality he was not very interested in international meetings, nor had he travelled much outside of Sweden. Apart from an invited journey to Britain the two most notable exceptions were the Vienna INQUA-meeting in 1936 and the 6th International Botanical Congress in Amsterdam in 1935. At the latter he gave a highly appreciated lecture, and hardly any of those present will forget the chairman's desperate attempt by means of the red lights provided for the purpose to indicate that his time was up. Nobody could stop *Lennart von Post* from developing an idea once he got started.

Lennart von Post was a very human man, with human fallibilities and weaknesses, but above and beyond that he was one of thel intrinsically great men, and we hope that our translation of his classical lecture will convey also to others something of the sense of greatness we felt who were fortunate enough to know him.

Lecture to the 16th convention of Scandinavian naturalists in Kristiana (Oslo) 1916.

One of the most important goals of all geological research is the determination of the age of the strata under investigation. To this end, investigations of peat bogs have followed two, fundamentally different lines of approach, especially during the last 25 years. One group of workers, using *Japetus Steenstrup* as a model, has focused on isolating and identifying plant remains from peat. On the basis of the floral lists thus obtained, they have established a series of zones within peat sequences that purport to show the main features of the immigration of the vegetation. This line of research can be termed paleofloristic because the history of the development of the flora is the central theme of investigation.

In contrast to the paleofloristic study of peat bogs, the members of the paleophysiognomic school, led by *Rutger Sernander* in Sweden and *C. A. Weber* in Germany, attempt to elucidate the ontogeny of each bog through careful investigation of its stratigraphy. For the paleophysiognomist the central problems are distinguishing genetically different peat types, and identifying the parent plant formations (from which they originated). Each peat profile is thus translated into a series of plant communities that succeed each other, and through which the plant physiognomic development at the sampling site manifests itselfs. Dissecting various parts of a peat bog in this manner, a more or less complete picture of the history of formation of the bog is obtained. From this record one can later deduce possible changes in water level and other phenomena. If similar changes are observed regularly in several bogs, they can be used as evidence for general geographical changes, changes of sea level, etc. In the paleophysiognomist's view, studies of this type must precede any discussion of the implication of fossils contained in peat bogs, or conclusions regarding the general developmental history of the vegetation.

To what extent is it now possible to give geological dates to peat-bog sequences on the basis of results obtained by the two schools ?

It must be admitted that so far results are rather limited. It can be said that at least in some parts of nothern Europe, particularly south Sweden, paleophysiognomists have succeded in establisheing certain developmental features of bogs that are conditioned by general causes. To a large extent, they have even succeeded in placing these changes in relation to the general geological development. However of the four climatic epochs registered in peat-bogs, found by *Sernander* and the « Uppsala school », as it is called, and termed the boreal, atlantic, subboreal and subatlantic by *Axel Blytt,* only the latter two manifest themselves regularly as physiognomic changes of bog plant formations. In some cases, as for example in spring-fed bogs, the boreal and atlantic epochs can indeed be recognized stratigraphically. But in the majority of bogs, especially those resulting from filling-in of former lakes, detailed investigation of the total developmental history is necessary to determine where the boreal-atlantic and the atlantic-subboreal contacts are situated. Even then the determination is only approximate.

Only one true stratigraphic marker, occurring regularly, even in isolated profiles, has thus far been found in our peat bogs. This is the subboreal-subatlantic contact, or the horizon that corresponds to what *Sernander* has called the postglacial climatic deterioration. In most cases this is developed as a pronounced stratigraphic contact, in which a regression of one type or another is recorded in the plant-physiognomic development of the bog. Sometimes a terrestrial peat — for example a forest peat or a short-sedge peat — underlies a telmatic or limnic deposit, such as a tall-sedge peat, a lake peat or a gyttja. Sometimes a deep water gyttja or an alluvial clay covers a telmatic fen-peat or a gyttja from shallow water. Or else there is the « Grenzhorizon » described by *Weber* from the North Germany high moors with slightly humified young Sphagnum peat resting on strongly decomposed older Sphagnum peat or even raw-humus from a heath.

However this horizon — the Grenzhorizon in the raised bogs and its equivalent in the fen deposits — is the only one that can be used at all regularly as a starting point for dating peat using the stratigraphic-paleophysiognomic method. *Sernander* has shown that it corresponds more or less to the transition between the Bronze and the Iron Ages. Except for a small number of cases we lack real stratigraphic horizons for all of the earlier part of Postglacial time. Frequently the paleophysiognomist has to resort to less exact age estimates, based on a complete study of the ontogenetic history of the site.

Paleofloristic peat bog investigations have been even less successful in establishing a general rationale for dating peat deposits. *Sernander* has repeatedly emphasized that the well-known zonal division established by *Steenstrup* cannot be generalized to all north European bogs. Even in *Steenstrup's* own country, where just a few years ago his immigration sequence was considered beyond doubt, voices have recently been raised questioning its validity. In the following I hope also to demonstrate that the *Steenstrup* scheme, particularly in its more precise form as presented by *Gunnar Andersson,* is a construction that only to a small degree corresponds to the actual facts.

It should also become obvious from my presentation that one simply cannot expect generalizable results when these are based on macroscopic plant remains alone. This is what has customarily been donne, particularly in *Gunnar Andersson's* paleobotanical investigations.

The fossil content of a peat comes primarily from the parent formation. Thus the first information one can get from a list of fossils sieved out of peat — and in my opinion the most important information — is identification of the plant-physiognomic composition of the parent formation. But in addition to the sedentary, autochtonous fossils, there are also to a variable degree imbedded with them allochtonous constituents transported by animals or wind or water. It is this assemblage which more than any other should yield information about the composition of the regional vegetation.

The information from fossil drift material will of necessity be rather incomplete and will always represent information gained more or less by chance. The primary constituent of the drift will be material carried by water, if one can neglect dispersal by wind and by animals, both of which play a comparatively subordinate role. Two considerations are of importance. First, in addition to remains of open-water and shore vegetation, water-born drift contains appreciable quantities only of those upland plants whose remains are dispersed by wind. This fact was demonstrated by *Sernander*. Second, water-born drift can only be incorporated into those sediments formed below highwater line of a basin, i.e. in limnic and telmatic. Obviously only a part of the total drift transported by water throughout the year can be incorporated into telmatic sediments, which are only inundated during certain periods. And in terrestrial and ombrogenous peat types — forest peat and the true high-moor formations — water drift is never represented, or is represented only under very exceptional conditions.

It must be obvious therefore that macroscopic remains, for example of forest trees, preserved in peat bogs, are not easily interpreted as a reliable record of the history of these plants. Yet they form the primary basis for the *Steenstrup* zonal divisons. Even more obvious is the fact that these fossils do not give a secure basis for dating peat sequences, nor — even less — for dating individual peat samples. The absence of a certain tree species in a deposit in no way proves that the same tree was not a constituent of the vegetation of the area at the time the deposit was formed. The fossil remains of beech, hornbeam and yew occur so rarely in peat bog deposits that *Gunnar Andersson* has not found it possible to discuss the history of immigration of these trees on a paleobotanical basis. The discussion to follow will show that macroscopic fossil material has also given unsatisfactory information about pine, alder, elm, lime, oak, hazel and spruce as well. Yet these trees and forest elements are extremely important for the elucidation of the history of vegetation.

Thus the question arises : is it possible to use fossil material preserved in peat deposits for tracing the postglacial history of a terrestrial plant at all ? Yes, it is, but only if a plant is represented by fossil remains evenly, and in great quantity, uninfluenced by the genesis of the enclosing peat type. Only under these conditions could the absence of a fossil have true indicator value, and only under these conditions might possibility exist for reliable evaluation of any changes that may have occurred in the plant's frequency.

Every observer of nature is well aware that pollen of certain anemophilous plants, especially trees, is scattered over the terrain in enormous quantities during the flowering period. In the forests and in their vicinity this pollen can form deposits several millimetres thick, covering the ground evenly, and in lakes and rivers the pollen rain even causes a sort of water bloom.

Fossil pollen should consequently meet the requirements for an exact elucidation of the history of the vegetation. A few peat-bog investigators have realized this, among them *Früh* in Switzerland, *Weber* and *Stoller* in Germany and *Harald Lindberg* in Finland. And since the end of the 1890's Professor *Gustav Lagerheim* in Stockholm has given extensive help to Swedish and Danish peat bog investigators by identifying fossil pollen. He is also the first to analyze fossil tree pollen quantitatively, although he has done this only a few cases. Actually the method outlined briefly by *Holst* (in Postglaciala Tidsbestämningar) who used *Lagerheim's* analyses of sample series he collected from south Scanian peat bogs, is precisely the one I have applied systematically in the investigations I shall now present. I am very grateful personally to Professor *Lagerheim*, not only for the initial information concerning pollen morphology and fossil pollen in general, but also for much advice and instruction that he has given subsequently.

The fossil pollen flora of peat bogs includes well preserved and fully identifiable pollen of the majority of our genera of forest trees, viz. *Pinus, Picea, Salix, Betula, Alnus, Corylus, Carpinus, Quercus, Fagus, Ulmus, Tilia, Acer*, and *Fraxinus*. Thus *Populus* is the only one of our more important forest trees that is completely missing, although among those I enumerated I have found only single grains of *Acer* and *Fraxinus*. Although the pollen types enumerated can be identified generically, the species cannot be distinguished. Thus *Betula* species, *Alnus* species, *Salices* etc. can unfortunately only be treated collectively.

Both the morphological character of the fossil pollen and the technique of pollen investigations will be passed over here. With regard to the latter I shall only mention that the preparations are made from untreated peat by rapid boiling in potassium hydroxide and that even dry samples can be used without any difficulties worth mentioning. In investigating the pollen flora I have generally found that a magnification of about 200 × is most convenient. If one is studying only the coniferous pollen, 50-60 × is sufficient. For investigating finer details, sculpturing, perforations and the like a greater magnification should be available, for instance 400-500 ×.

The percentage composition of the pollen flora is calculated on the basis of the total number of pollen grains in a series of preparations. The preparations are moved by the mechanical stage and the counting is carried out line by line.

Pollen is differently preserved in different peats. In certain peats, especially those that are aerated during their genesis, such as certain types of forest peat, small sedge peat, etc., and also some *Carex* root peat, pollen is very scarce and what is found is more or less crumpled and corroded. In such cases it is to be suposed that pollen has been destroyed to a greater or less extent, and no conclusions can be drawn in regard to the original composition of the pollen flora, at least not before the differential resistance to decay of various pollen types has been investigated. In other types of deposits, e.g. all limnic deposits, especially gyttja, and in *Sphagnum* peat, pollen shows no traces of destruction. The pollen grains are clear and demonstrate well-preserved form and sculpturing and also occur in great abundance. In certain gyttjas, for example, I have calculated the frequency of tree pollen alone at 2000-4000 grains per mm^3. In his investigation of peat-bogs around Hornborgasjön, *Sandegren* showed that the pollen content of *Sphagnum* peat is definitely proportional to the degree of destruction of the peat, as if the pollen were concentrated by increasing destruction of the peat. I can corroborate this observation, and add that the state of preservation of pollen even in the most humified *Sphagnum* peat is unchanged. For example this is demonstrated by samples of regenerative *Sphag-*

num peat taken out of both strongly humified heath layers and adjacent pool layers for purposes of comparison ; the pollen flora registers no change in composition with the degree of humification of the peat.

There is no need to work for long investigating fossil pollen to discover that both the composition of the pollen flora, and pollen abundance are completely constant and characteristic for each peat sample. To investigate the extent to which this really is the case I controlled for variation in most of the 300 samples I analysed in the course of this investigation. This was done by making comparisons at various stages in each analysis, for instances after 1/8, 1/4, 1/2, 1, 2 preparations, etc., thus doubling the quantity for each new calculation. It proved the rule with very few exceptions that, first, the total pollen content in a given sample is astonishingly constant in one preparation after the other, and that second, the percentage composition of the pollen flora is determined with sufficient exactness even when few more than 100 grains have been counted. By continued counting in many cases I was able to show that the percentage figures after this point are certain within limits of a few units.

In those deposits where the state of preservation of the pollen indicates that no destruction has occurred, the composition of the pollen flora reflects the average composition of the pollen rain at the time the deposit was laid down. And the pollen rain in turn must reflect the general character of the forest. This idea assumes, of course, that the observation points are chosen to avoid purely local influences that might disturb the general picture. Variations from year to year, undoubtedly very important, resulting from the varying intensity of flowering of the trees, are integrated of course because each sample presumably corresponds to a much longer period of time than one year, especially if small comprehensive samples have been collected by design.

If the fossil pollen flora gives an expression of the composition of the forest vegetation, then by following the changes of the pollen flora step by step through a sediment profile, it should be possible to observe the secular changes to which the average composition of the forest has been subject through the course of time. And to the extent that these changes follow general rules when great numbers of profiles are investigated from different areas, it should in fact be possible to pin down the postglacial history of the forest. Conversely it should also be possible to work out a reliable paleofloristic basis for the geological dating of sediments with forest tree pollen in them.

With these considerations in mind, I started at the beginning of the year 1915 to make a systematic investigation of the pollen flora in a series of South Swedish peat-bogs. The material was taken from the archives of peat-bog profiles at the Swedish Geological Survey, from samples that had been collected over the course of time, many of them in connection with experimental work related to a planned statistical inventory of the quantity of available peat in South Sweden.

I selected a series of bogs lying along a line through the middle of South Sweden from Scania up to southwestern Närke. I also got a valuable complement to this main transect partly from some analyses from Hornborgasjön which Dr. *Sandegren* carried out according to my instructions, and partly from two peat bogs that Dr. *Sandegren* analyzed subsequent to geological mapping in northeastern Västergötland and southeastern Värmland. In addition, a sample series from Femsölyng in northeastern Zealand (Denmark) was included for comparison, as this is the bog where *Emil Chr. Hansen* in his day made his famous discoveries of fossil remains of beech.

In selecting peat bogs for the main transect I established the following principles :

1. The bogs had to be large enough so that a sample series from their center was at least 200-300 meters from the nearest point of dry ground.

2. The majority had to lie above the maximum limit of the marine transgression, or at least above the limit of the *Ancylus* lake, in order that their profiles could comprise the whole or at least the greater part of postglacial time. Where possible they should be former lakes, with lateglacial lake sediments as their basal layer.

3. The general features of the sediments and of the developmental history of the bogs had to be known from at least one carefully measured profile.

4. The subboreal — subatlantic contact was clearly developed, if possible as a Grenzhorizon sensu *Weber*.

5. The profiles to be analyzed could not include any type of peat in which one might suspect that destruction of pollen had occured. As far as possible the profiles consisted of gyttjas and Sphagnum-peats, sediments in which according to my earlier experience pollen is extremely well preserved. Furthermore, there had to have been as little forest on the bogs themselves as possible, at least not over a long period, nor in the immediate vicinity of the sampling points. It has not always been possible to avoid this.

The pollen flora in samples taken from the central part of these bogs has now been analyzed partly with the help of Dr. *Sandegren* and Dr. *Leo von zur Mühlen*. The samples were usually taken at intervals of 20-25 cm. From the results of the analyses, curve-diagrams were constructed, with the sediment profile along the Y-axis and the pollen percentages of the various kinds of trees along the X-axis.

The percentage figures were calculated on the basis of the sum of forest tree pollen. As I wanted to be able to read the relative displacements among the competing forest types — not only among individual tree species — I have treated the mixed oak-forest pollen, that is, the sum of pollen percentages of oak, lime and elm (ash and maple were not included for the time being as they are too sporadic) as a unit on a par with pine, birch, spruce, beech etc. But obviously I noted down the elements of this mixed oak-forest individually, and introduced their individual curves into the diagrams, using a different signature from that of the main curves.

For the same reason I have not included hazel pollen in the sum, but instead have only given its frequency as percent of the sum. This is because hazel occurs mostly as a shrub layer in mixed oak-forest, and forms only exceptionally a separate community competitive with other forest types. This is the rule, and it is noticeable how the hazel curve conforms in most cases with that of the mixed oak-forest. But in some bogs, especially in early postglacial time, the curves show that hazel occurred locally without correlation with the mixed oak-forest, as a kind of edaphic precursor. Nevertheless it has proved most correct to treat hazel as a complement to the oak-forest, and not to consider it on a part with it or with other forest types.

Obviouly it would have been desirable from various points of view to use figures giving the absolute frequencies of the various pollen types instead of the relative measures that are indicated by percentage figures. But the total quantity of pollen in different samples varies a great deal, obviously because the peat was deposited at different rates and therefore different samples have different time values. Absolute figures are thus not directly comparable. And although certain types of peat are on the average richer in pollen, others poorer, the amplitudes of variation within different peat types are on the whole too great to allow the establishment at present of reduction coefficients. Thus I have found pollen contents in Sphagnum peat from 10-20 to about 900 pollen grains per preparation (size : 18 × 32 mm), in gyttjas from about 100 to about 4000, and in carr peat with well-preserved pollen from 50-

100 to about 6000. For this reason it has not been possible in my investigation to tackle the question of the area coverage of the forest as a whole at different times. On the basis of my material I venture to say that important general shifts hardly could have taken place in the area of forested land, at least during the older parts of the postglacial period within the area I have investigated. No regular changes of total pollen content have manifested themselves that are related to the position of the samples in the profile. It is especially noteworthy that the series from the basal layers that represent the very oldest postglacial period do not show an abnormally low pollen content relative to the upper parts of the profiles. Forest growth obviously had achieved its maximal rate at the beginning of the postglacial interval.

If in accordance with my working hypothesis, the pollen curves do register the relative displacements between the forest types, then diagrams from closely neighbouring profiles should correspond even in detail. I have investigated this in several cases, and I have found that this is really so. Obviously one cannot ask that diagrams constructed from samples collected at large vertical intervals should show detailed correspondence. If samples are half a meter apart in younger Sphagnum-peat, for example, which formed at a rate now estimated at 5-10 cm per century, they represent rather widely separated forest generations. There can only be a remote possibility in a case like this that the same generations of forest are represented in parallel series, and therefore one might expect that the details of the curves will vary.

Conditions are different, however, if samples are taken so densely that there is likelihood that most forest generations are represented, for example at 5-10 cm intervals in younger Sphagnum-peat. I shall restrict myself to the presentation of a single example, taken from Rönneholms bog in Scania. Two profiles from this bog have been analyzed in the way indicated. The profiles are situated about 250 m from each other. Both consist of 90 cm of younger Sphagnum-peat, overlying older Sphagnum-peat represented in its lower levels by a forest bog peat with pine-stumps. In the younger Sphagnum-peat, the curves for beech, mixed oak-forest, pine and birch show practically identical courses. The pollen frequencies of beech, and even of hornbeam, however, are slightly higher in the profile that is closer to dry ground. The alder curves are rather similar, while the curves of hazel correspond somewhat less, although even in that case there is a certain resemblance.

In the older Sphagnum-peat, which formed more slowly, the samples investigated correspond to forest generations more distantly separated in time than in the younger peat. In addition, pine grew on the bog at and near the profile points. In spite of these two factors, the diagrams also show a rather good correspondance within the older Sphagnum-peat. It seems remarkable that the local pine forest, although influencing the course of the pine curve in a noticeable way, does not disturb the diagram completely as might have been expected.

This is a phenomenon I have observed frequently in my diagrams, which indicates that it really is the general composition of the ancient forest within the total region that determines the composition of the pollen rain. The local pollen in its more restricted meaning is quantitatively subordinate. It will be shown later that this is true to an even greater degree in regard to pollen that appears to have been transported over great distances.

Having given this survey of the basic premises and methods of my investigations, I am now prepared to present a short discussion of the results. Limitations of time prevent me from demonstrating all the diagrams in detail (1).

To demonstrate general features of the curves within the various diagrams, which are on the whole astonishingly uniform, I have constructed instead a generalized diagram, based on the averages of the individual curves. To construct this diagram I first calculated the average thicknesses of the portions of the profiles above, and beneath, the subboreal-subatlantic contact. These average thicknesses were used as the ordinate axis on the generalized diagram. Subsequently the pollen percentages for the levels corresponding to 0.1, 0.5, 1.0, 1.5 meters, etc. above and beneath the subboreal-subatlantic contact were read from the original curves by interpolation. The averages calculated from each group of numbers obtained in this way were used to build up the generalized diagram in the usual manner, without any type of smoothing. We see that these curves are so regular, that one is entitled to consider the average diagram as at least an approximate demonstration of the general features in the development of the pollen flora:that is to

(1) At the lecture a graphic demonstration of the primary diagrams was given, in which each separate pollen curve was shown from the main transect through South Sweden. Percentages from each bog were plotted along the horizontal axes with the bogs placed in accordance to their correct distances from each other, while the curves for individual pollen types from each profile were placed above each other vertically. Each pollen type thus formed a horizontal column within which the individual curves from the different sites were oriented according to the stratigraphically determined, subboreal-subatlantic contact. This demonstration, giving all the details, cannot be included here. — (The actual chart used to hang in *von Post's* laboratory for many years, but has apparently now been lost. — K. F.).

say the development within the series that I have treated, after smoothing out local variations. This is the case particularly as these local variations are quite insignificant, at least when it comes to the general course of the curves. To avoid too much variation within the groups of primary diagrams from which the averages were calculated, I have only used those profiles that correspond with certainty to the whole of the period the average diagram is supposed to represent, that is to the two epochs from the beginning of the post-glacial through the subboreal, and from beginning of the subatlantic until the present. Furthermore only one diagram was used in instances where there were several from the same tract. The basis for the younger part of the average diagram is 12 primary diagrams, for the older 9.

Among the 8 main curves in the average diagram (the *Salix* curve was dropped for lack of space as it is on the whole rather low and does not say very much) three groups can be separated that are distinct in the general course of their curves. One group is formed by the birch and pine curves, the second by the curves for mixed oak-forest, alder and hazel, and the third by the curves of beech, spruce and hornbeam.

Birch and pine pollen exhibit high frequencies at the beginning of postglacial time, with a subsequent decline. They do not increase again until subatlantic time, and even then birch occurs in high frequency only in the very first part of the period.

Curves for mixed oak-forest, alder and hazel start with low values at the base of the diagram and then increase rapidly as birch and pine pollen frequencies decrease. They reach their maximum at or a little below the subboreal-subatlantic contact ; above that, especially after the post-glacial climatic deterioration, their frequency diminishes.

Beech, hornbeam and spruce are indeed represented, although sporadically, even within the older part of the diagram. But with the onset of subatlantic time these species and pine, and to begin with birch as well, begin to increase. The increase occurs at the expense of mixed oak-forest and alder, hazel, and later even birch. In some of the original diagrams the subatlantic levels contain pollen of *Salix* in frequencies of 5-6 %, once even 8 %. This pollen type only occasionally reaches frequencies as high as 1-2 % in other parts of the profiles, except for the basal layers.

Within the mixed oak-forest elm reaches its highest pollen frequency in the lower parts of the older part of the diagram, lime at or a little above this, and oak just below the subboreal-subatlantic contact. Oak is regularly absent from the basal layers of the profiles. In the subatlantic part of the diagram elm and lime pollen play a subordinate role in comparison to oak. In diagrams from the high parts of Småland and from Middle Sweden, lime and elm pollen is almost completely absent above the subboreal-subatlantic contact.

The development of the fossil pollen flora as I have indicated it is valid without exception in all the sample series I have investigated so far. On the other hand, there are obviously important differences of degree in regard to the magnitude of relative displacements between pollen types, and in the general habitus of the pollen flora within various areas. For example pollen of mixed oak-forest, alder, and hazel is quantitatively most important within the older parts of the Scanodanish profiles, whereas the pine-birch group tends to dominate the pollen flora within the high country in Småland. The bogs in Middle Sweden occupy an intermediate position in this respect. I have already pointed out that in the southernmost profiles hazel sometimes appears to substitute for mixed oak-forest in the beginning of postglacial time. Within the subatlantic to recent layers, beech plays the dominant role farthest to the south. It diminishes successively towards the north to be replaced in northern Småland by spruce, which next to pine is the most characteristic element in the pollen flora of the youngest horizons in Middle Sweden. Both beech and hornbeam, the latter in Scania alone reaching a pollen frequency of 10 %, are represented by pollen even in the older subatlantic layers in the northernmost profiles investigated (even in southwestern Närke, northernmost Västergötland and southeastern Värmland, i.e. north of the present limits of the trees in question). In the same way spruce pollen, although more locally and in significant frequencies only in Femsölyng (9 %), has been encountered in individual samples within the subatlantic layers of the southernmost bogs.

It must be admitted that the average diagram constitutes an artificial and schematic picture. But nevertheless, because of the way in which it has been prepared, it presents a more or less correct expression of the general laws governing the changes in pollen flora within the whole of the investigated area. On the other hand the proportions between frequencies of different pollen types which appear in the average diagram have no exact counterparts within any special region. It is my intention in the future, as the pollen-analytic material grows, to establish similar average diagrams for smaller areas. These areas should obviously be limited in such a way that each of the « area-diagrams » is based on primary material that is fully homogenous in regard to composition of pollen flora.

When it comes to a decision as to which conclusions can validly be drawn from the pollen diagrams, one should keep in mind first and foremost that a diagram can only give trends of frequency changes between forest types. As long as we have no indices to express the relative pollen productivity of the various trees, nor to express the different degrees to which their pollen is dispersed, we have no right to seek in the percentage figures an adequate expression of the composition of the forest communities.

And a further point : to interpret the pollen curves correctly one should clarify to one's own satisfaction the extent to which pollen dispersal over very great distances influences the composition of the pollen rain. On several occasions various investigators have maintained that fossil pollen could hardly be used to prove anything about vegetation history. This argument was made because it was considered established that conifer pollen, for example, could be transported hundreds of kilometers. The observations on which this view was based are hardly fully convincing, however, although one might easily expect that wind should be capable of transporting pollen grains, which are light and in some cases equipped with devices for floating in the air over considerable distances. This may be the case. But even on the basis of the material I have presently at my disposal, it would be possible to find sufficient proof for the statement, that pollen transported great distances does not play any important role in the pollen rain of a reasonably forested area.

Thus the spruce pollen frequency in the youngest layer of Köpingemyr, situated close to the present southern limit for spontaneous occurrences of spruce, is only 2 %. About 25 km south of this, in Rönneholms bog, only individual spruce pollen grains were found in the topmost samples. Similar observations have been made near the southern limit of spruce on Gotland. In subboreal peat in Femsölyng the pollen frequency of pine in three successive samples was 1 %, although pine stumps prove that pine occurred in the immediate neighbourhood of the profile point, and although pine at the time in question grew amply on the peat-bogs of southern Scania. In subrecent samples 5 cm below the present surface in Frinnaryds mosse and Dagsmosse, pollen of oak, alder and hazel have either not been found, or only in frequencies of one or two percent, even though these trees are at present no rarities in the surrounding area. And even in Scania and on Sjælland the mixed oak-forest in the topmost samples reaches 10 % in only one case (Femsölyng). Beech pollen, which frequently reaches percentages of 50-70 % in the upper parts of the Scanodanian profiles, amounts to about 5 % or less in middle Småland where beech occurs quite frequently. In the profile from Dagsmosse, 4 km from the not insignificant occurrences of beech at Alvastra, the subrecent sample contains less than 1 % beech pollen. More examples could be added to the ones cited here. However, even these show that pollen transported great distances and pollen from relatively subordinate forest elements play such a small quantitative part in the pollen rain, that there is a greater risk that a sparsely-occurring species is not represented in a pollen analysis, than that the species is over-represented because of pollen dispersal from occurrences at a distance. As long as the analysis is not conducted in greater detail than I have done in my investigation so far, pollen transported great distances apparently falls entirely within the margin of error of the frequency figures. One would also have predicted that the quantity of pollen dependent upon transport from afar would be extremely small, and insignificant in relation to the quantity of pollen from within the area.

At present one cannot possibly say what level of detail of analysis will be sufficient to cause pollen transported from afar to appear more or less regularly. But the examples quoted above show that one is obviously on the safe side if one supposes that a regular occurrence of one or two per cent of a certain pollen type corresponds to a more or less sporadic occurrence of the tree in question within the area. I presume that the total pollen content of the sample is not smaller than normal for the area, and also that the state of preservation excludes secondary changes of the pollen flora by partial destruction.

Under these conditions I find it justifiable to draw conclusions from the diagram that first and foremost confirm the idea *Holst* and also I myself stated earlier : that the warmth-demanding trees alder, elm, lime and hazel immigrated to Sweden during the very earliest part of the postglacial epoch, and that oak came only slightly later. The diagrams very distinctly contradict the supposition that real birch or pine periods without southern forms existed after the end of the *Dryas* period. This of course does not preclude the possiblity that birch and, in the southernmost part of the area even pine, were the physiognomically dominant trees at the beginning. Actually, I agree with *Holst* that pine was apparently living in South Sweden during late glacial time. The rather rich occurrence of pine pollen in layers containing *Dryas,* and also the high absolute pine pollen figures in the lowermost postglacial layers are hardly to be explained without this supposition.

On the other hand one may argue against this refusal to accept the birch and pine periods of *Steenstrup* by suggesting that these periods might be represented in my profiles by very slowly formed layers or even by stratigraphic gaps. The sequences themselves speak against such a supposition, as the transition to the late-glacial layers in most cases takes place without any sharp delimitation. There is no suggestion of a stratigraphic hiatus or discordance. In addition I have found exceptionally high *Salix* pollen contents in the lowermost samples of many sequences. I would interpret this to mean that the subarctic willows were dying out at the time the hazel-mixed oak forest and alder appeared for the first time.

I am in a position, however, to give additional, and according to my opinion more decisive proof, that the southern forest element really came in at the same time as, or very shortly after, the disappearance of the late-glacial flora.

In order to bring the pollen curves as far as possible into chronologic relation to the main stages of changes of sea-level, I investigated by means of pollen analysis a number of samples, collected chiefly by *Munthe* and *Holst,* from peat and gyttja sequences that are covered by shore bars of the Ancylus lake and Litorina sea. At all six localities that I have investigated so far for peat under the Ancylus bar, in the area of the Lake Vättern, at Kalmarsund, and on Gotland, I found in these layers in addition to ample and dominating pine and birch, pollen of southern trees, sometimes alder, sometimes elm, sometimes lime or some of these together, always hazel, but never oak. The frequency is low (maximum 6-7 %, usually less), but in the cases when series of samples have been analyzed it has proved absolutely constant, and to correspond also to that of the lowermost part of the diagrams. Thus even before the time of the Ancylus maximum, the tree types referred to grew in Southern Sweden in a frequency which seems to correspond more or less to the present frequencies within the northern parts of the high country in Småland, for example. This is not a new result. In many places under the Ancylus shore bars, macroscopic remains had already been found of such southern forms as *Carex pseudocyperus, Cladium mariscus, Iris pseudacorus* and *Lycopus europaeus.* On the other hand, *Dryas* has been found under the Ancylus shore bar at Fröjel in Gotland, at or some few meters above the Ancylus level at Laxå in Närke, and at Rangiltorp i Ostergötland. These occurrences show that the lateglacial flora, at least in certain areas of southern Sweden, persisted until near the time of the Ancylus maximum.

I have thus succeeded in showing that the maximum of the Ancylus transgression occurred after alder, elm, lime and hazel immigrated to southern Sweden, but apparently before the arrival of oak. But on the other hand I have not succeeded so far in fixing the position of the Litorina maximum relative to the pollen diagrams. The only fact that I have been able to establish so far is that the southern forest element in Scania reached very high frequency before this time. In peat several metres under the present sea-level, and in peat covered by deposits of the Litorina transgression, I have found pollen frequencies for mixed oak forest up to 50 %, for alder up to 74 %, and for hazel up to 55 %. In the submarine peat or peat covered by older Litorina deposits, elm and lime pollen dominates the mixed oak forest sum. In peat under the maximum Litorina wall, more than half of this sum is contributed by oak pollen.

In my attempt to establish the position of the Litorina maximum in the pollen diagrams, however, I investigated a number of samples from the Gotland-Kalmarsund area in addition to the Scanian samples. This work showed that there is a definite difference between « pollen spectra » for these samples and for the Scanian ones. In the former the pollen frequencies for mixed oak forest, alder and hazel do not exceed 7 %, 15 %, and 14 %, respectively, and on an average (from 6 stations) they are 4 %, 8 %, and 9 %. For the corresponding Scanian samples (also 6 stations) the averages are for the mixed oak forest 31 %, for alder 38 % and for hazel 26 %. As there cannot be any temporal difference between these two groups of samples the difference must be due to the geographical conditions during the time immediately preceding the Litorina maximum. To be more specific it must be due to the fact that the open Baltic within the Gotland-Kalmersund area had a very strong cooling effect on its coastal area. With one single exception, viz. Dagsmosse, which also lies within the coastal area of the receding Ancylus lake, the diagrams of the main transect belong clearly to the Scanian group of samples with high frequencies for the elements of the southern forest types. The reason for the depressing effect of the sea on the southern elements might be that its temperature was depressed by cold water coming for a long period off the melting remains of glacial ice in Norrland. Immediately before the maximum of the Litorina transgression, southern trees started becoming more important in Gotland, as is shown for example in a pollen diagram from Mästermyr. The differences between the areas in question and Gotland, at least were apparently partly equalized by that time.

The condition mentioned here is of importance because it suggests an explanation for the discrepancy which otherwise exists in regard to the temperature conditions of the older postglacial period, between the ocean fauna in western Scandinavia, and the evidence from vegetation.

It is well known that the former demonstrates a rather slow immigration of exigent species, while the latter suggests a very rapid rise of temperature in the interior. If continued investigations should show that the vegetation of coastal areas maintained a more northern character for a comparatively long period during postglacial time, this difference is explained.

Admittedly I have not yet proved with absolute certainty that the first appearance of southern forest elements was simultaneous within the area I have investigated. But both the similarity of the diagrams, and the relationship I showed between the Ancylus maximum and the time of the first appearance of these trees in various parts of the country, speak in favor of this idea. The rapid appearance of southern plants simultaneous with, or almost immediately following the disappearance of the *Dryas* flora, must be considered a phytogeographical anomaly unless it is explained as resulting from a very fast improvement of climate. This seems to me as good as proof for the contemporaneity of the first appearances of southern elements. Through the investigations of *G. de Geer* on the recessions of the ice margin in Middle Sweden, we know that a rapid rise of temperature took place some time after the edge of the ice had left its stagnation stage at the Middle Swedish terminal moraines. I correlated this previously (1909) with the early appearance of southern trees in Närke, and proposed to call this « the late-glacial (or better « finiglacial ») climatic improvement ». There is thus an analogy to the postglacial climatic deterioration (Sernander), which also seems to have taken place very fast, geologically speaking.

If this view should prove to be true, we shall have in the alder, elm, lime, and hazel pollen limit an easily identifiable and reliable paleofloristic marker-horizon. This horizon lies within the older parts of plant-bearing deposits of somewhat older geological age than the maximum extension of the Ancylus lake. Another similar, although somewhat less distinct horizon is represented by the oak pollen limit, corresponding to the time shortly after the Ancylus maximum.

Another horizon which is, in my opinion already established, is the beech-spruce pollen limit. However, one must distinguish between an empirical and a rational pollen limit. The latter is indicated by that point in the profile at which the spruce and beech pollen frequencies start to increase steadily toward the surface. It regularly coincides with the subboreal-subatlantic contact, i.e. it corresponds to the postglacial climatic deterioration.

On the other hand by the empirical spruce and beech pollen limits, we mean the levels at which the pollen types in question start to occur regularly with a frequency of at least 1 %. These limits should be of the same age within large areas ; at least the empirical pollen limit of spruce should belong in most cases to the late subboreal. It is the empirical spruce pollen limit that I have used since 1909 in order to date and correlate middle Swedish peat-bog sequences. Within Närke I have been able to determine its age as late neolithic time, i.e. coinciding with a late part of subboreal time. However, as I shall show presently, it is not unlikely that the empirical spruce pollen limit in other parts of southern Sweden will prove to be of greater geological age.

Because of the fact that spruce and beech pollen occur as individual pollen grains or in low percentage frequencies even within the older parts of the profile I have been obliged in this case to introduce the distinction between empirical and rational pollen limits.

In pollen diagrams from Scania and Småland, beech pollen has been found down to levels equivalent to one metre below the sub-boreal-subatlantic limit in the average diagram. Usually these occurrences are only of individual pollen grains. Sometime there are higher frequencies at individual levels, for instance 5 % in Bengtsboda bog. In my view these scattered occurrences must have come from scattered stands of beech or individual beech trees, that occurred here and there before the mass-dispersal of the tree started. We also know from finds of macroscopic beech remains that beech lived in southern Scandinavia long before the postglacial climatic deterioration. Beech nuts have been found in a north Scanian grave from older Bronze Age (*Montelius*), and beech stumps in sub-boreal layers in Löberödsmosse in Scania (*Sernander*), and at Ransbaeck in Vendsyssel, *Axel Jessen* found beech remains in a « marine ooze » which he thinks corresponds to 50 % of the Litorina maximum. This is only to mention a few examples.

It seems possible, although so far it can in no way be proved or disproved by geological facts, that the first appearance of beech in southern Scandinavia occurred during the Atlantic period, as *Sernander* thought.

Concerning spruce, its pollen attains low frequencies much deeper in the profiles than pollen of beech (mostly individual pollen grains, reaching 1-2 %, exceptionally more, only in the late subboreal layers). In the average diagram these individual spruce pollen finds form a continuous curve down to three metres below the subboreal-subatlantic limit. And even in submarine peat (Ystads hamn) and in peat under the Litorina shore bar both in Scania and in the Gotland-Kalmarsund area, I have observed individual spruce pollen grains.

In order to try to find out if this spruce pollen occurring in very low frequences should be considered due to transport from afar, or should be considered indication of sporadic appearance of spruce in southern Sweden even before this tree started to form forests, I have made per mille calculations of the content of spruce pollen in the older parts of some of the profiles in the main transect. Under my supervision these calculations have been carried out by Dr. *Sandegren* and Dr. *von zur Mühlen* in the following way : After the samples were analyzed in the usual way and the percentage figures of pollen were found, they counted pine and spruce pollen only under low magnification in a series of preparations, until the pine pollen number constituted more than 10 times the percentage figure of pine. Usually they counted enough pine pollen to make the figure correspond to 1200 or 1500 tree pollen grains in total. Up to 46 preparations from individual samples were scanned. From these counts the per mille figures for spruce were calculated and curves constructed in the usual way (1). The curves show interruptions and gradual waxing and waning up to 10 ‰. In my opinion the consistently individual nature of the curves points more to separate spruce occurrences which appeared and disappeared through time than to even pollen transport from a spruce area at some distance. As far as I am concerned I am inclined to believe that spruce had a history similar to beech. Beech pollen occurs below the rational beech pollen limit in a completely analogous way and beech immigration is demonstrated, as I mentioned, by macroscopic finds. I believe that spruce also came in from the south comparatively early in postglacial time, as Holst maintained at one time, and that during the greater part of postglacial time it lived at scattered stations, even in parts of south Scandinavia where it no longer occurs, since sporadic spruce pollen was found in Southern Scania and Femsölyng. The 0/00-curve of spruce in Femsölyng is really the one that speaks most strongly for long distance dispersal. Here spruce pollen occurs only in two parts of the profile, the lower part of the younger Sphagnum-peat, and at the base of the older Sphagnum-peat from a level one meter

(1) ‰-curves were demonstrated with the lecture, but have been left out here.

under the Grenz horizon. The per mille curve conforms in detail to the percentage curve for pine.

In regard to spruce, however, there are also macroscopic finds that prove beyond doubt that spruce lived in various parts of southern Sweden before the postglacial climatic deterioration. In subboreal stump layers spruce has been found at a considerable number of stations, for instance in the lakes in northern Småland investigated by Gavelin, in Dagsmosse, and in several peat-bogs in Närke and Uppland. At the Stone Age site at Åloppe in Uppland, *S. Lindquist* refers to floor covers of spruce branches with needles and cones. These were found in huts from the older passage grave period, consequently early subboreal, corresponding to about 50 % of the Litorina maximum. From the older parts of the postglacial period, however, there is so far only the occurrence of a rope of bast found by *Holst* in Kallsjömossen in Scania, which was probably made from spruce, although this cannot be said with complete certainty, according to *Bengt Jonsson* (and also to Dr. *Nils Sylvén*. who reinvestigated the find at my request).

The opinion that I have presented here, that spruce immigrated early and at least in part from the south, is in complete opposition to the present theory. The theory that the tree immigrated to Scandinavia from the east late in geological time has been elevated almost to a dogma. Actually all the facts that support this theory are easily combined with my viewpoint.

The facts are :

1. The absence of macroscopic finds of spruce in older plant-bearing deposits.

2. The presence of a southern and western limit for the distribution of spruce in Scandinavia and its absence as a spontaneous tree in Denmark and northwestern Germany.

3. The fact that spruce is spreading at its present southwest limit, as shown by *Hesselman* and *Schotte* with support of historical documents.

None of these facts are contradicted if one constructs the history of spruce and beech in southern Sweden in the following way with the support of the pollen diagrams.

Both trees (like *Carpinus*, incidentally) started to occur as subordinate elements of the forests at a rather early time in the postglacial epoch. They were subordinated by competition from forest elements which were then favoured by the climate, especially the mixed oak-forest. During subboreal time when a great number of peat bogs dried out, new habitats opened up for both trees, particularly spruce, and a first general increase of their frequencies took place. The comparatively cold and moist climate of the subatlantic period depressed the competitive power of the mixed oak-forest, but favoured spruce and beech. Together with hornbeam and pine these species started spreading at the cost of the mixed oak-forest, beeach spreading up to and north of its present northern limit, spruce continuously as far south as Middle Småland, and occasionally even further, as to Sjælland. Due most probably to climatic conditions, beech got the upper hand in the south, and spruce in the north. Beech and hornbeam were then by spruce displaced from the northern parts of the area into which they had started to spread at the time of the climatic deterioration ; hornbeam was displaced completely, and beech to a great extent. For its part, spruce left scattered occurrences within the true beech area. In this way a rather indefinite border came into being between the great beech and spruce areas. In later (subrecent) times spruce has slowly penetrated to the south. Meanwhile because of human activity, beech has further diminished in abundance within the northern parts of its present distribution area in South Sweden, as shown by *Wibeck* for the Ostbo and Västbo districts in Småland.

The supposition that beech and spruce were rather generally distributed as subordinate forest elements is necessary to explain the fact, shown by the pollen diagrams, that both of these trees at the time of the postglacial deterioration increased in frequency, with great rapidity, throughout the whole of South Sweden, simultaneously each within its own special area. When and from which direction they originally came in seems from this point of view to be of lesser importance.

The main points of the preceding account can now be summarized :

First it can be stated that nothing remains from the immigration scheme of *Steenstrup* and *Gunnar Andersson*, and the paleofloristic zonal division based upon it, except for the order of arrival of the different forest trees.

The following basis has been established for paleofloristic dating of postglacial sequences :

Postglacial time (perhaps postarctic time is a more suitable expression as I include in this the latter part of *G. De Geer's* finiglacial melting period) can be divided into two paleofloristically distinct sub-epochs. On the basis of the forest elements that characterize them phytogeographically and climatologically, these subepochs can be called respectively the time of mixed oak-forests, and the time of beech-spruce forests. The limit between them is the rational beech-spruce pollen limit, which has been shown to correspond to the time of the postglacial climatic deterioration. The beech-spruce forest time corresponds to *Sernander's* subatlantic period ; the time of mixed oak-forest comprises the boreal, atlantic and subboreal periods, i.e. the postglacial (or rather postarctic) warm period (*Sernander*). It extends from a time just preceding the maximum of the Ancylus lake, to a time more or less corresponding to the transition between the Bronze and Iron Ages.

In southern Sweden deposits of the various epochs can be characterized paleofloristically as follows :

The very oldest deposits, from the time before the Ancylus maximum, are characterized by a pollen flora with dominating pine and birch, and low frequencies of alder, elm, lime or hazel.

At a later time oak pollen is present in the pollen flora, at first as a subordinate element occurring with first elm and later lime pollen, each of which occurs relatively copiously. Towards the end of the time of mixed oak-forests, oak dominates within the mixed oak-forest sum, and in late subboreal sequences spruce or beech pollen often occurs in low frequencies. There are no definite horizons resulting from pollen analyses within the time of mixed oak-forest. Even if it proves possible in the future, as implied by my diagrams, to divide this epoch into elm time, lime time and oak time, the limits between these sub-epochs remain diffuse and probably of somewhat different age from one area to another.

The time of beech-spruce forests can be distinguished from the preceding by its higher frequencies of beech and spruce pollen. Especially characteristic in the youngest deposits are the low frequencies, or absence — depending upon the area — of pollen of mixed oak-forest, alder and hazel. To a certain extent the late subatlantic and subrecent formations resemble the very oldest postartic deposits ; they are distinct from them, however, in containing spruce or beech pollen, and also by the complete dominance of oak within the mixed oak-forest sum wherever it occurs. Frequently elm and lime pollen is completely absent.

Obviously there are a great number of questions of various kinds that are tempting to discuss. But I will restrict myself to pointing out a few potential ways in which the working methods of stratigraphic paleophysiognomic peat bog investigation might be made more penetrating. These improvements would lead to a more profound knowledge of phytogeographic history and the development of climate during the postarctic period.

It should be possible, particularly when the necessary experience has been obtained regarding the relationship between composition of pollen rain and real frequency of various forest elements, to use pollen evidence for discussions of the character of the climate during various parts of postarctic time. This will enable us to determine at least some of the main features of the climate with greater certainty than has been possible so far. Even in the present material some intimations in this regard suggest themselves. The northern limits for *Alnus glutinosa, Tilia parvifolia,* and *Ulmus montana* in European Russia, as indicated by *Köppen,* are more or less coincident, inasmuch as first one and then the other limit is the northernmost. On the whole these limits correspond to the August isotherm for + 14 to + 15°C, just as the limit for oak corresponds with the + 16°C August isotherm. It is tempting, at least as a working hypothesis, to correlate these data with the first occurrence of the trees mentioned in the diagrams. The resulting conclusion would be that the August temperature in southern Sweden at the time shortly before the Ancylus maximum reached + 14 to + 15°C, and that shortly after it rose to + 16°C.

The succession from elm, to lime, to oak that was demonstrated within the mixed oak-forest levels also invites comparison with the present distribution of these trees in Russia. If one compares Köppen's data on their occurrence at the border of the South Russian steppe-area, one finds that the range of *Quercus pedunculata* extends farthest south towards the real steppe : *Tilia parvifolia* stops somewhat farther north, and *Ulmus montana* still further north. It seems possible that the marked retreat of elm and lime throughout subboreal time can be explained by the fact that the climate of this time, as *Sernander* has maintained, was continental, and to a certain degree comparable with the climate that is now found in the transitional belt between the forest and steppe regions in Russia.

These suggestions must only be considered as tentative working hypotheses, however. Their confirmation or disproof must be left to future, more comprehensive investigations. But the characteristic general course of the pollen curves, and the distinct influence the postglacial climatic deterioration proves to have had on the development of the forests, both raise the hope that with this method the character of the postarctic climate and its development will be understod in more detail and with considerably greater precision. This effort will be aided by the fact that it will be possible to discuss each region separately.

When sufficient material has been worked through we should be able to produce comparative forest maps for at least some parts of the postarctic period. The analyses from the peats under the

Ancylus and Litorina shore bars already represent initial data for the beginning of the postarctic period and for the older Litorina period. And for the time immediately before and immediately after the postglacial deterioration we can produce fragmentary maps even now.

I have shown in the above discussion that pollen diagrams from different profiles can be correlated without reference to the peat stratigraphy by means of the variations in the pollen flora. In the examples I gave from Rönneholm's bog, the correlation was made over a distance of only a few hundred metres, and it is improbable that really detailed correlations will be possible at much greater distance. But certain persistent local changes in forest composition might conceivably be followed throughout areas of somewhat larger extent. In this way one might get local marker horizons in pollen diagrams, by means of which dates from archeologic finds or favourable stratigraphic conditions could be transferred to sequences near by. I expect pollen analysis may thus become an important role in helping with detailed investigation of the development of peat-bogs and bog areas.

This closes the report on the pollen analytic investigations I have carried out thus far. The investigations have only been done in a preliminary kind of way. Certainly the working methods can be much improved ; in this respect it will be especially desirable to investigate the feasibility of producing comparable absolute frequency figures. And additional, more comprehensive studies of the present day pollen rain must be carried out before the pollen diagrams can be interpreted in detail.

However, I feel sufficiently confident of the method to close with a request to my Scandinavian colleagues to collect as many samples as possible for investigation of the fossil pollen flora in connection with their work in various areas. I believe that this category of fossils, which until this time has been quite overlooked, can if treated systematically give a much expanded knowledge of our late Quaternary sedimentary deposits and their history.

ON THE USE OF MATRICES IN CERTAIN POPULATION MATHEMATICS

By P. H. LESLIE, *Bureau of Animal Population, Oxford University*

CONTENTS

1. Introduction

If we are given the age distribution of a population on a certain date, we may require to know the age distribution of the survivors and descendants of the original population at successive intervals of time, supposing that these individuals are subject to some given age-specific rates of fertility and mortality. In order to simplify the problem as much as possible, it will be assumed that the age-specific rates remain constant over a period of time, and the female population alone will be considered. The initial age distribution may be entirely arbitrary; thus, for instance, it might consist of a group of females confined to only one of the age classes.

The method of computing the female population in one unit's time, given any arbitrary age distribution at time t, may be expressed in the form of $m+1$ linear equations, where m to $m+1$ is the last age group considered in the complete life table distribution, and when the same unit of age is adopted as that of time. If

n_{xt} = the number of females alive in the age group x to $x+1$ at time t,

P_x = the probability that a female aged x to $x+1$ at time t will be alive in the age group $x+1$ to $x+2$ at time $t+1$,

F_x = the number of daughters born in the interval t to $t+1$ per female alive aged x to $x+1$ at time t, who will be alive in the age group 0–1 at time $t+1$,

then, working from an origin of time, the age distribution at the end of one unit's interval will be given by

$$\begin{aligned}
\sum_{x=0}^{m} F_x n_{x0} &= n_{01} \\
P_0 n_{00} &= n_{11} \\
P_1 n_{10} &= n_{21} \\
P_2 n_{20} &= n_{31} \\
\vdots \quad & \quad \vdots \\
P_{m-1} n_{m-1,0} &= n_{m1}
\end{aligned}$$

or, employing matrix notation, $Mn_0 = n_1$, where n_0 and n_1 are column vectors giving the age distribution at $t = 0$ and 1 respectively, and the matrix

$$M = \begin{bmatrix} F_0 & F_1 & F_2 & \cdots & . & F_k & F_{k+1} & \cdots & F_{m-1} & F_m \\ P_0 & . & . & \cdots & . & . & . & \cdots & . & . \\ . & P_1 & . & \cdots & . & . & . & \cdots & . & . \\ . & . & P_2 & \cdots & . & . & . & \cdots & . & . \\ \cdots & \cdots & \cdots & \cdots & \cdots & \cdots & \cdots & \cdots & \cdots & \cdots \\ . & . & . & \cdots & P_{k-1} & . & . & \cdots & . & . \\ . & . & . & \cdots & . & P_k & . & \cdots & . & . \\ \cdots & \cdots & \cdots & \cdots & \cdots & \cdots & \cdots & \cdots & \cdots & \cdots \\ . & . & . & \cdots & . & . & . & \cdots & P_{m-1} & . \end{bmatrix} \quad 0 < P_x < 1;\ F_x \geqq 0.$$

This matrix is square and consists of $m+1$ rows and $m+1$ columns. All the elements are zero, except those in the first row and in the subdiagonal immediately below the principal diagonal. The P_x figures all lie between 0 and 1, while the F_x figures are by definition necessarily positive quantities. Some of the latter, however, may be zero, their number and position depending on the reproductive biology of the species we happen to be considering in any particular case, and on the relative span of the pre- and post-reproductive ages. If $F_m = 0$, the matrix M is singular, since the determinant $| M | = 0$.

Since $Mn_0 = n_1$, and $Mn_1 = M^2_{n_0} = n_2$, etc., the age distribution at time t may be found by pre-multiplying the column vector $\{n_{00}\, n_{10}\, n_{20} \ldots n_{m0}\}$, i.e. the age distribution at $t = 0$, by the matrix M^t. Moreover, it will be seen that with the help of the jth column of M^t the age distribution and number of the survivors and descendants of the $n_{j-1,0}$ individuals, who were alive at $t = 0$, can readily be calculated. Thus, $n_{j-1,0}$ times the sum of the elements in the jth column of M^t gives the number of living individuals contributed to the total population at time t by this particular age group.

2. Derivation of the matrix elements

The basic data, from which the numerical elements of this matrix may be derived, are given usually in the form of a life table and a table of age specific fertility rates. To take the P_x figures first; if at $t = 0$ there are n_{x0} females alive in the age group x to $x+1$, the survivors of these will form the $x+1$ to $x+2$ age group in one unit's time, and thus $P_x n_{x0} = n_{x+1,1}$. Then it is usually assumed (e.g. Charles, 1938, p. 79; Glass, 1940, p. 464) that

$$P_x = \frac{L_{x+1}}{L_x},$$

where

$$L_x = \int_x^{x+1} l_x \, dx,$$

or the number alive in the age group x to $x+1$ in the stationary or life table age distribution. This method of computing the survivors in one unit's time would be exact if the distribution of those alive within a particular age group was the same as in the life-table distribution.

The F_x figures are more troublesome, and in the numerical example which will be given later they were obtained from the basic maternal frequency figures (m_x = the number of live daughters born per unit of time to a female aged x to $x+1$) by an argument which ran as follows. Consider the n_{x0} females alive at $t = 0$ in the age group x to $x+1$, and let us suppose that they are concentrated at the midpoint of the group, $x+\frac{1}{2}$. During the interval of time 0–1 some of these individuals are dying off, and at $t = 1$ the $n_{x+1,1}$ survivors can be regarded as concentrated at the age $x+1\frac{1}{2}$. Although these deaths are taking place continuously, we may assume them all to occur around $t = \frac{1}{2}$, so that at this latter time the number of females alive in the age group we are considering changes abruptly from n_{x0} to $n_{x+1,1} = P_x n_{x0}$. Then during the time interval 0–$\frac{1}{2}$ these n_{x0} females will have been exposed to the risk of bearing daughters, and the number of the latter they will have given birth to per female alive will be given by the maternal frequency figure for the ages $x+\frac{1}{2}$ to $x+1$. This figure may be obtained by interpolating in the integral curve of the m_x values, and thus expressing the latter in $\frac{1}{2}$ units of age throughout the reproductive span instead of in single units. The daughters born during the interval of time 0–$\frac{1}{2}$ will be aged $\frac{1}{2}$–1 at $t = 1$, the number of them surviving at this time being determined approximately by multiplying the appropriate $m_{x+\frac{1}{2}}$ figure by the factor $2\int_{\frac{1}{2}}^{1} l_x dx$ according to the given life table. Similarly, each of the $P_x n_{x0}$ females during the interval of time $\frac{1}{2}$–1 give birth to $m_{x+1-x+1\frac{1}{2}}$ daughters, the survivors of which form part of the 0–$\frac{1}{2}$ age group at $t = 1$. The survivorship factor is in this case taken to be $2\int_{0}^{\frac{1}{2}} l_x dx$.

Combining these two steps together we obtain a series of F_x figures, which may be defined as the number of daughters alive in the age group 0–1 at $t = 1$ per female alive in the age group x to $x+1$ at $t = 0$. Putting

$$k_1 = 2\int_{0}^{\frac{1}{2}} l_x dx, \quad k_2 = 2\int_{\frac{1}{2}}^{1} l_x dx,$$

then

$$F_x = (k_2 m_{x+\frac{1}{2}-x+1} + k_1 P_x m_{x+1-x+1\frac{1}{2}}),$$

and

$$\sum_{x=0}^{m} F_x n_{x0} = n_{01},$$

the total number of daughters alive aged 0–1 at $t = 1$.

3. Numerical example

In order to see whether the P_x and F_x figures obtained in this way from the basic data give a reasonably accurate estimate of the population in one unit's time, a numerical example was worked out for an imaginary rodent population, the species chosen being the brown rat, *Rattus norvegicus*. Full details of the basic life table and fertility table which were used are given in an appendix, together with a short account of the genesis of these tables and the methods employed to estimate the rate of natural increase (r) and the stable age distribution. Compared with man, the fertility of this imaginary rat population was relatively very great; thus, the gross reproduction rate was 31·21 daughters and the net rate (R_0) 25·66, the life table used being a reasonably good one. The inherent rate of natural increase was estimated to be 0·44565 per head per month of 30 days, and the stable age distribution was so overladen with young that the proportion of females in the post-reproductive age groups was negligible. Some 74·45 % of the females were younger than 3 months, at which age breeding was assumed to commence.

By definition the Malthusian age distribution is stable; that is to say, once a population subject to the given rates of fertility and mortality achieves this form of distribution, it continues to increase e^r times every unit of time and the proportions of the population alive in each group remain constant. Thus, in the present example, given 100,000 females distributed as to age in the stable form at $t = 0$, the number alive in each age group in 1 month's time can be immediately calculated by multiplying each element in the original distribution by 1·561505. This 'true' age distribution at $t = 1$ is compared in Table 1 with that obtained by operating on the original distribution with the P_x and F_x figures, which are given in Table 5 of the Appendix.

The agreement between the true and estimated age distributions is remarkably close. It might be expected that the principal errors would occur in the early age groups, since the

Table 1

(1) (Units of 30 days) Age group	(2) Population at $t=0$ Stable age distribution	(3) Expected population at $t=1$ Col. 2 × 1·561505	(4) Population at $t=1$ Estimated by operating on col. 2 with the matrix M
0–	37,440	58,463	58,374
1–	22,595	35,282	35,455
2–	14,417	22,512	22,519
3–	9,227	14,408	14,406
4–	5,903	9,218	9,218
5–	3,775	5,895	5,895
6–	2,413	3,768	3,768
7–	1,542	2,408	2,407
8–	984	1,537	1,537
9–	627	979	980
10–	399	623	623
11–	254	397	396
12–	161	251	251
13–	101	158	159
14–	64	100	99
15–	40	62	62
16–	25	39	39
17–	15	23	24
18–	9	14	14
19–	6	9	8
20–	3	5	5
Total	100,000	156,151	156,239

The span of the reproductive ages is from 3 to 21 months.

P_x figures are based on the stationary age distribution which is clearly very different from the stable form. However, as will be seen from Table 1, the biggest error from this cause is due to the first P_0 which overestimates the number alive in the 1–2 age group at $t = 1$ by some 0·5 %. The F_x figures underestimate the number alive in the 0–1 group by 0·2 %, and the total population is overestimated by 0·06 %. On the whole these results are satisfactory and, judging from this example, it would seem that the matrix M operating on a given age distribution should give a reasonable estimate of the population in one unit's time, provided that the unit of time and age chosen be not too coarse as compared with the life span of the species. The degree of cumulative error which is introduced by continued operation with the matrix will be considered later.

4. PROPERTIES OF THE BASIC MATRIX

The matrix M is square and of order $m+1$; it is not necessary, however, in what follows to consider this matrix as a whole. For, if $x = k$ is the last age group within which reproduction occurs, F_k is the last F_x figure which is not equal to zero. Then, if the matrix be partitioned symmetrically at this point,

$$M = \begin{bmatrix} A & \cdot \\ B & C \end{bmatrix}.$$

The submatrix A is square; B is of order $(m-k) \times (k+1)$; C again is square consisting of $m-k$ rows and columns, the only numerical elements being in the subdiagonal immediately below the principal diagonal. The remaining submatrix is of order $(k+1) \times (m-k)$ and consists only of zero elements. Then in forming the series of matrices M^2, M^3, M^4, etc.,

$$M^t = \begin{bmatrix} A^t & \cdot \\ f(ABC) & C^t \end{bmatrix}.$$

The submatrix C is, however, of such a type that $C^{m-k} = 0$, so that M^t, $t \geqq m-k$, will have all its last $m-k$ columns consisting of zero elements. This is merely an expression of the obvious fact that individuals alive in the post-reproductive ages contribute nothing to the population after they themselves are dead. It is the submatrix A which is principally of interest, and in the mathematical discussion which follows, attention is focused almost entirely on it and on age distributions confined to the prereproductive and reproductive age groups.

The matrix A is of order $(k+1) \times (k+1)$, where $x = k$ is the last age group in which reproduction occurs, and written in full,

$$A = \begin{bmatrix} F_0 & F_1 & F_2 & F_3 & \cdots & F_{k-1} & F_k \\ P_0 & \cdot & \cdot & \cdot & \cdots & \cdot & \cdot \\ \cdot & P_1 & \cdot & \cdot & \cdots & \cdot & \cdot \\ \cdot & \cdot & P_2 & \cdot & \cdots & \cdot & \cdot \\ \cdots & \cdots & \cdots & \cdots & \cdots & \cdots & \cdots \\ \cdot & \cdot & \cdot & \cdot & \cdots & P_{k-1} & \cdot \end{bmatrix}.$$

This matrix is non-singular, since the determinant $|A| = (-1)^{k+2}(P_0 P_1 P_2 \ldots P_{k-1} F_k)$. There exists, therefore, a reciprocal matrix of the form

$$A^{-1} = \begin{bmatrix} \cdot & P_0^{-1} & \cdot & \cdot & \cdots & \cdot \\ \cdot & \cdot & P_1^{-1} & \cdot & \cdots & \cdot \\ \cdot & \cdot & \cdot & P_2^{-1} & \cdots & \cdot \\ \cdots & \cdots & \cdots & \cdots & \cdots & \cdots \\ \cdot & \cdot & \cdot & \cdot & \cdots & P_{k-1}^{-1} \\ F_k^{-1} & -(P_0 F_k)^{-1} F_0 & -(P_1 F_k)^{-1} F_1 & -(P_2 F_k)^{-1} F_2 & \cdots & -(P_{k-1} F_k)^{-1} F_{k-1} \end{bmatrix}.$$

Thus, given an initial age distribution n_{x0} $(x = 0, 1, 2, 3, \ldots, k)$ at $t = 0$, in addition to the forward series of operations An_0, $A^2 n_0$, $A^3 n_0$, ..., etc., there is also a backward series $A^{-1} n_0$, $A^{-2} n_0$, $A^{-3} n_0$, ..., etc. There is, however, a fundamental difference between these; for, whereas the forward series can be carried on for as long as we like, given any initial age distribution, the backward series can only be performed so long as n_{xt} remains $\geqq 0$, since a negative number of individuals in an age group is meaningless. Apart from this limitation, it is possible to foresee that the reciprocal matrix might be of some use in the solution of certain types of problem.

5. Transformation of the co-ordinate system

Hitherto an age distribution n_{xt} has been regarded as a matrix consisting of a single column of elements. For simplicity in notation, this column vector will now be termed the vector ξ and different ξ's will be distinguished by different subscripts (ξ_a, ξ_x, etc.). We may picture an age distribution as a vector having a certain magnitude and related to a definite direction in a vector space, the space of the ξ's. The different age distributions which may arise in the case of any particular population will be assumed to be ξ's all radiating from a common origin. The numerical elements of a ξ vector are thus taken to be the co-ordinates of a point in multi-dimensional space referred to a general Cartesian co-ordinate system, in which the reference axes may make any angles with one another. At this point in the argument another type of vector will be introduced, which in matrix notation will be written as a row vector, and which will be termed the vector η. There is an intimate relationship between this new type and the old, for, associated with each vector ξ_a, there is a uniquely determined vector η_a, and vice versa. The inner or scalar product, $\eta_a \xi_a$, is the square of the length of the vector ξ_a. Either we may picture each of these vectors as associated with a different kind of vector space, the space of the ξ's and the dual space of the η's, which are not entirely disconnected but related in a special way; or, alternatively, we may regard them as two different kinds of vector associated with the same vector space. The relationship between η and ξ is precisely the same as that between covariant and contravariant vectors in differential geometry.

If we pass from our original co-ordinate system to a new frame of reference, and the variables η and ξ undergo the non-singular linear transformations,

$$\eta = \phi H, \quad \xi = H^{-1}\psi, \quad |H| \neq 0,$$

it can be seen that since the variables are contragredient, $\eta\xi = \phi\psi$, so that the square of the length of a vector remains invariant. Moreover, since the result of operating on a vector ξ_a with the matrix A is, in general, another vector ξ_b, where ξ_a and ξ_b are both referred to the original co-ordinate system, it follows that in the new frame of reference which is defined by the linear transformations given above, the relationship

$$A\xi_a = \xi_b$$

becomes

$$HAH^{-1}\psi_a = \psi_b,$$

or

$$B\psi_a = \psi_b.$$

Thus, in the new frame of reference the matrix $B = HAH^{-1}$ operating on the vector ψ_a is equivalent to the matrix A operating on the vector ξ_a in the original frame.

It is convenient, for the purposes of studying the matrix A and of performing any numerical computations with it, to transform the variables η and ξ in the above way, choosing the matrix H so as to make $B = HAH^{-1}$ as simple as possible. For $B^t = (HAH^{-1})^t = HA^tH^{-1}$, and since A is non-singular, by the reversal law, $(HAH^{-1})^{-1} = HA^{-1}H^{-1}$. Thus, if $f(A)$ is a rational integral function of A, $f(B) = f(HAH^{-1}) = Hf(A)H^{-1}$; and the properties of matrix functions $f(A)$ can be studied by means of the simpler forms $f(B)$. Moreover, the matrices A and B have the same characteristic equation and, therefore, the same latent roots. For $B - \lambda I = H(A-\lambda I)H^{-1}$ and, forming the determinants of both sides,

$$|B-\lambda I| = |H||A-\lambda I||H|^{-1},$$

so that the characteristic equation is

$$|A-\lambda I| = |B-\lambda I| = 0.$$

If, in the present case, the transforming matrix is taken to be

$$H = \begin{bmatrix} (P_0P_1P_2\dots P_{k-1}) & . & . & \dots & . & . & . \\ . & (P_1P_2P_3\dots P_{k-1}) & . & \dots & . & . & . \\ . & . & (P_2P_3\dots P_{k-1}) & \dots & . & . & . \\ \dots & \dots & \dots & \dots & \dots & \dots & \dots \\ . & . & . & \dots & (P_{k-2}P_{k-1}) & . & . \\ . & . & . & \dots & . & P_{k-1} & . \\ . & . & . & \dots & . & . & 1 \end{bmatrix}$$

in which, it is to be noted, the only numerical elements lie in the principal diagonal and are derived entirely from the life table, then

$$B = HAH^{-1} = \begin{bmatrix} F_0 & P_0F_1 & P_0P_1F_2 & P_0P_1P_2F_3 & \dots & (P_0P_1P_2\dots P_{k-1})F_k \\ 1 & . & . & . & \dots & . \\ . & 1 & . & . & \dots & . \\ . & . & 1 & . & \dots & . \\ . & . & . & 1 & \dots & . \\ \dots & \dots & \dots & \dots & \dots & \dots \\ . & . & . & . & \dots\ 1 & . \end{bmatrix}.$$

Comparing this matrix B with the original form A, it can be seen that the latter has been simplified to the extent that the original P_x figures in the principal subdiagonal are now replaced by a series of units, and the matrix A has been reduced to the rational canonical form $B = HAH^{-1}$ (see Turnbull & Aitken, 1932, chap. v). In this way any computations with the matrix A are made easier, and we may work henceforward in terms of ϕ and ψ vectors together with the matrix B, instead of with the original η and ξ vectors, and the matrix A. Any results obtained in this new system of co-ordinates may be transformed back again to the original system whenever necessary. It is evident that by suitably enlarging H the original matrix M may be transformed in a similar way.

This linear transformation of the original co-ordinate system is equivalent biologically to the transformation of the original population we were considering into a new and completely imaginary type which, although intimately connected with the old, has certain quite different properties. Thus, it can be seen from the transformed matrix B that the individuals in this new population, instead of dying off according to age as the original ones did, live until the whole span of life is completed, when they all die simultaneously. This is indicated by the P_x figures being now all equal to unity; an individual alive in the age group x to $x+1$ at $t = 0$ is certain of being alive at $t = 1$, excepting in the last age group of all where none of the individuals will be alive in one unit's time. Accompanying this somewhat radical change in the life table, there is a compensatory adjustment made in the rates of fertility so that the new population has the same inherent power of natural increase (r) as that of the old. This follows from the fact that the latent roots of the matrices A and B are the same, and, as will be shown later, the dominant latent root is closely related to the value of r obtained by the usual methods of computation. Insomuch as the transformation is reversible and $A = H^{-1}BH$, it can be seen that by changing H we could transform the canonical form B, if we wished, into another matrix in which the P_x subdiagonal might be a specified set of figures derived from some other form of life table. But, for our present purposes, the canonical form B, in which all the P_x figures are units, offers advantages over any other matrix of a similar type owing to the greater ease with which it can be handled.

6. Relation between the canonical form B and the $L_x m_x$ column

The actual computation of the matrix B by way of the steps indicated in the theoretical development is by no means difficult, although it is a somewhat tedious process, particularly if the matrix is of a large order. The numerical elements in the first row of B for the brown rat are given in Table 5 of the Appendix. These values were obtained from the F_x and P_x figures which have already been used in the numerical example in § 3 and which will be found in the same table. Further reflection suggested, however, that instead of first of all obtaining A and then transforming to B, a short cut could be taken which would save labour and which also would tend to eliminate some of the small cumulative errors arising in the longer method.

The series of values P_0, P_0P_1, $P_0P_1P_2$, ..., $(P_0P_1P_2 \dots P_{k-1})$ by which the individual F_x figures are multiplied in order to obtain the first row of B, is essentially a stationary age distribution. For, since by definition,

$$P_x = \frac{L_{x+1}}{L_x},$$

$$(P_0P_1P_2 \dots P_x) = \frac{L_{x+1}}{L_0},$$

where $L_0 = \int_0^1 l_x dx$. Hence the required series of multipliers is given by a stationary age distribution in which only one individual is alive in the age group 0–1. Now, the F_x figures, as defined in § 2, already contain within them some allowance not only for the probability of survival during the first unit of life, but also for the fact that some adult individuals in each age group are dying off during the interval of time 0–1. The process of multiplying F_x by $(P_0P_1P_2 \dots P_{x-1})$ is thus analogous to the formation of the $L_x m_x$ column, by means of which the net reproduction rate is estimated. The chief difference between the first row of B and the $L_x m_x$ distribution is that in the former the maternal frequency is expressed as between the ages of $x+\frac{1}{2}$ to $x+1\frac{1}{2}$, instead of between x to $x+1$ as in the latter. If each element $(P_0P_1P_2 \dots P_{x-1}F_x)$ of the first row of B is regarded as centred at the age of $x+1$, the sum, mean and seminvariants of this 'distribution' may be estimated and compared with the values which are obtained from the $L_x m_x$ column in the process of calculating r by the usual methods. In the present numerical example the results of this comparison were as follows:

Parameter	$L_x m_x$ column	First row of B
Sum (R_0)	25·65786	25·6603
Mean	9·60604	9·5948
m_2	14·14397	14·1839
m_3	22·15696	21·9358
$m_4-3m_2^2$	−117·6480	−117·920

After allowing for the small cumulative errors which might be expected to occur in the calculation of the matrix elements, there is a substantial agreement between the respective estimates. This agreement strongly suggests that if we had wished to pass immediately to the matrix B without going through the laborious process of calculating the F_x and P_x figures, the elements of the first row could have been obtained by forming a new $L_x m_x$ column in which the age group limits were shifted a half unit later in life. This could readily be done by interpolating in the integral curve of the $L_x m_x$ values for the ages $x+\frac{1}{2}$. This method of forming the first row of B has been adopted in other instances, when the matrix A was not of any immediate interest. It proved to be relatively quick and certainly less laborious than the method of first establishing A and then transforming to B which was the one used in the present numerical example.

7. The stable age distribution

The result of operating on an age distributon ψ_x with the matrix B is, in general, a different distribution ψ_y. But, in the special case when the relation between the two distributions is such that

$$B\psi_a = \lambda\psi_a,$$

where λ is an algebraic number, then ψ_a may be said to be a stable age distribution appropriate to the matrix B. For the sake of brevity it will be referred to as a stable ψ. Similarly for initial row vectors, if

$$\phi_a B = \lambda\phi_a,$$

then ϕ_a is said to be a stable ϕ.

The matrix equation defining a stable ψ may be written as $k+1$ linear equations, of which the ith is

$$\sum_{j=1}^{k+1} b_{ij} n_j - \lambda n_i = 0,$$

where n_i $(i = 1, 2, \ldots, k+1)$ are the co-ordinates of the stable ψ, and b_{ij} the element in the ith row and jth column of B. Eliminating the n_i from this system of equations, we obtain the characteristic equation of B, namely,

$$| B - \lambda I | = 0;$$

and, expanding this determinant in powers of λ, we have in the present case,

$$\lambda^{k+1} - F_0\lambda^k - P_0F_1\lambda^{k-1} - P_0P_1F_2\lambda^{k-2} - \ldots - (P_0P_1 \ldots P_{k-2})F_{k-1}\lambda - (P_0P_1 \ldots P_{k-1})F_k = 0.$$

The $k+1$ roots λ_a of this equation are the latent roots of B, and corresponding to each distinct λ_a there is a pair of stable vectors, ϕ_a and ψ_a, determined except for an arbitrary scalar factor.

Once a latent root λ_a has been determined, it is a comparatively simple matter to find the appropriate stable ψ_a and ϕ_a vectors. Thus, it is easily shown that the stable ψ_a is the column vector $\{\lambda_a^k \lambda_a^{k-1} \lambda_a^{k-2} \ldots \lambda_a 1\}$. A short method of estimating ϕ_a is the following. Suppose, to take a simple case, that

$$B = \begin{bmatrix} a & b & c & d \\ 1 & . & . & . \\ . & 1 & . & . \\ . & . & 1 & . \end{bmatrix}$$

and let y_x $(x = 1, 2, 3, 4)$ be the elements of the stable ϕ_a appropriate to the root λ_a. Then

$$\begin{aligned} \phi_a B &= [ay_1 + y_2 \quad by_1 + y_3 \quad cy_1 + y_4 \quad dy_1] \\ &= [\lambda_a y_1 \quad \lambda_a y_2 \quad \lambda_a y_3 \quad \lambda_a y_4]. \end{aligned}$$

By equating similar elements and putting $y_1 = 1$, $y_4 = d/\lambda_a$, $y_3 = \frac{c+y_4}{\lambda_a}$, etc., it is easy to see how the required row vector can be built up. Having in this way obtained the stable ψ and ϕ vectors for the matrix B, they may be transformed to the appropriate stable ξ and η for the matrix A by means of the relations

$$\eta = \phi H, \quad \xi = H^{-1}\psi.$$

The characteristic equation of the matrix B, when expanded, is of degree $k+1$ in λ, and once B has been obtained this equation can immediately be written down, since the numerical coefficients of λ^k, λ^{k-1}, λ^{k-2}, etc., are merely the elements of the first row taken with a negative sign. Since there is only one change of sign in this equation, only one of the latent roots will be real and positive. Excluding the rather special case when the first row of B has only a single non-zero element, and taking the more usual type of matrix which will be met with, namely, that for a species breeding continuously over a large proportion of its total life span, it will be found that the modulus of this root (λ_1) is greater than any of the others,

$$|\lambda_1| > |\lambda_2| > |\lambda_3| > \ldots > |\lambda_{k+1}|,$$

the remaining roots being either negative or complex.

This dominant latent root λ_1, which will be $\gtreqless 1$ according as to whether the sum of the elements in the first row of B is $\gtreqless 1$, is the one which is principally of interest. Since it is real and positive, it is the only root which will give rise to a stable ψ or ξ vector consisting of real and positive elements. It is this stable ξ_1 associated with the dominant root λ_1 which is ordinarily referred to as the stable age distribution appropriate to the given age specific rates of fertility and mortality. Since

$$A^t\xi_1 = \lambda_1^t \xi_1,$$

it can be seen that the latent root λ_1 of the matrix A and the value of r obtained in the usual way from

$$\int_0^\infty e^{-rx} l_x m_x dx = 1,$$

are related by

$$\log_e \lambda_1 = r.$$

From the mathematical point of view, however, the negative and complex roots of the characteristic equation are of importance in the further theoretical development. Moreover, as will be shown later, the stable vectors associated with them are not entirely without interest. Two main cases then arise: when the remaining roots are all distinct, and when there are repeated roots. For the present it will be assumed that the latent roots of the matrix are all distinct.

8. Properties of the stable vectors

Before proceeding further it is necessary to mention briefly the reasons why the methods given above for the computation of the stable ψ and ϕ vectors were adopted, apart from their simplicity in practice. If the $k+1$ distinct roots of the characteristic equation are known, we may form a set of $k+1$ matrices $f(\lambda_a)$ by inserting in turn the numerical value of each root in the matrix $[B-\lambda_a I]$. The adjoint of $f(\lambda_a)$ is

$$F(\lambda_a) = \prod_{b \neq a} [B-\lambda_b I] \quad \text{and} \quad f(\lambda_a)\,F(\lambda_a) = 0.$$

It may be shown that the stable ψ_a appropriate to the root λ_a can be taken proportional to any column, and the stable ϕ_a proportional to any row of the matrix $F(\lambda_a)$ (see e.g. Frazer, Duncan & Collar, 1938, chap. III). Moreover, $F(\lambda_a)$ is a matrix product of the type $\psi\phi$, where the ψ vector is given by the first column and the ϕ vector by the last row of $F(\lambda_a)$, each divided by the square root of the element in the bottom left-hand corner; and the trace of the matrix is equal to the scalar product $\phi\psi$. Now $[B-\lambda_a I]$ is a square matrix of order $k+1$ with only zero elements below and to the left of the principal subdiagonal, which itself consists of units. The product of k such matrices, which gives $F(\lambda_a)$, will have therefore a unit in the bottom left-hand corner. Since the stable ϕ_a and ψ_a vectors obtained by the methods suggested in § 7 have respectively their first and last elements $=1$, it follows that

$$\psi_a\phi_a = F(\lambda_a), \quad \phi_a\psi_a = \text{trace } F(\lambda_a).$$

The stable vectors may now be normalized. If the scalar product, $\phi_a\psi_a = z^2$, say, then

$$\frac{\phi_a}{|z|}\frac{\psi_a}{|z|} = 1.$$

From now on it will be assumed that the stable vectors appropriate to each of the latent roots have been normalized in this way.

These vectors have the following important properties:

(1) The $k+1$ stable ψ are linearly independent. There is thus no such relationship, with non-zero coefficients c, as

$$c_1\psi_1 + c_2\psi_2 + c_3\psi_3 + \ldots + c_{k+1}\psi_{k+1} = 0.$$

(2) The scalar product of a stable ψ, ψ_a with the associated vector of another stable ψ, ψ_b is zero, i.e.

$$\phi_b\psi_a = 0 \quad (a \neq b).$$

The normalized stable ψ thus form a set of $k+1$ independent and mutually orthogonal vectors of unit length.

(3) Any arbitrary ψ—ψ_x say—can be expanded in terms of the stable ψ, thus

$$\psi_x = c_1\psi_1 + c_2\psi_2 + c_3\psi_3 + \ldots + c_{k+1}\psi_{k+1},$$

where the coefficients c may be either real or complex. Similarly an arbitrary vector ϕ_x can be expanded in terms of the stable ϕ.

9. THE SPECTRAL SET OF OPERATORS

The matrix product $\psi_a\phi_a$ of the normalized stable vectors associated with the latent root λ_a will be termed the matrix S_a. From the relationships which have already been given, it can be seen that S_a is merely the adjoint matrix $F(\lambda_a)$ of the previous section after each element in the latter has been divided by the sum of the elements in the principal diagonal; in other words it is the normalized $F(\lambda_a)$. In the case of all the latent roots being distinct, there are thus $k+1$ matrices S_a, and these S_a form a spectral set of operators with the following properties:

$$S_a^2 = S_a, \quad S_aS_b = 0 \quad (a \neq b), \quad \sum_{a=1}^{k+1} S_a = I.$$

Moreover, if $f(B)$ is a polynomial of the matrix B, we have by Sylvester's theorem (Turnbull & Aitken, 1932, chap. VI, § 8)

$$f(B) = \sum_{a=1}^{k+1} f(\lambda_a)\,S_a,$$

so that the matrix

$$B = \lambda_1S_1 + \lambda_2S_2 + \ldots + \lambda_{k+1}S_{k+1}, \quad \text{and} \quad B^t = \lambda_1^tS_1 + \lambda_2^tS_2 + \ldots + \lambda_{k+1}S_{k+1}.$$

If the latent roots in the expansion of B are raised to a high power, the term associated with the positive real root predominates over all the others, so that when t is large, we have approximately

$$B^t = \lambda_1^t S_1.$$

In any particular case the power to which B will have to be raised in order that this equation should be approximately true, will depend both on the order of the matrix and on the relative magnitude of the dominant root as compared with that of the remaining roots.

At this point it is possible to attach some biological meaning to one of the ϕ or η row vectors, which in the first place were introduced into the theory for reasons of symmetry, and which were defined solely in terms of their mathematical properties. If at a given moment a transformed population has an arbitrary age distribution ψ_x, and the sequence $B\psi_x, B^2\psi_x, \ldots, B^t\psi_x$ is formed, it can be seen that when t is large and ψ_x is expanded in terms of the stable ψ, we have approximately

$$B^t\psi_x = c_1\lambda_1^t\psi_1.$$

Thus, a population with any arbitrary age distribution tends ultimately to approach the stable form appropriate to the given rates of fertility and mortality, provided that these age-specific rates remain constant. This theorem is, of course, well known; and it is clear that the achievement of the stable form of age distribution associated with the dominant latent root is very unlikely to occur in practice, except in the case when the initial distribution is already of that form or exhibits only small departures from it. Now, it has already been shown that the sums of the columns of a matrix B^t provide a measure of the contributions made to the population at time t per individual alive in the respective age groups at $t = 0$. When t is large, the matrix B^t is equivalent to the matrix S_1 multiplied by a scalar factor. From the way in which this latter matrix was constructed by the outer multiplication of ψ_1 and ϕ_1, it is evident that the sums of the columns of S_1 are proportional to the vector ϕ_1. Thus, transforming back again to the original co-ordinate system, the stable η_1 associated with the dominant latent root provides a measure of the relative contributions per head made to the stable population by the individual age groups.

10. Reduction of B to classical canonical form

From the $k+1$ stable ψ a matrix Q can be constructed, whose columns are the stable ψ arranged, reading from left to right, in descending order of the moduli of the roots with which they are associated. Corresponding to every pair of complex roots, $u \pm iv$, there will be in this matrix a pair of columns consisting of complex elements, the one column being the conjugate complex of the other. Some of the columns associated with the negative roots may be purely imaginary owing to the normalization of the corresponding ψ and ϕ vectors. In a similar way a matrix U may be formed, whose rows, reading from above down, are the stable ϕ arranged in the same order. Since the stable ϕ and ψ are normalized, and $\phi_a\psi_b = 0$ for $a \neq b$,

$$UQ = I,$$

and, therefore, U and Q are reciprocal matrices. By premultiplying and postmultiplying respectively with U and Q, the matrix B may be reduced to the classical canonical form C, in which the only elements lie in the principal diagonal and consist of the latent roots arranged in the order prescribed above. This reduction of B to a purely diagonal form by means of the collineatory transformation $UBQ = C$ is, however, only possible in the type of matrix we are considering, when the latent roots are all distinct.

The expansion of an arbitrary ψ_x in terms of the stable ψ,

$$\psi_x = c_1\psi_1 + c_2\psi_2 + \ldots + c_{k+1}\psi_{k+1},$$

may be written in matrix notation as $\psi = Qc,$

where c is the column vector $\{c_1c_2c_3 \ldots c_{k+1}\}$. Similarly, the expansion of the vector ϕ_x associated with ψ_x may be written $\phi = dU,$

where d is the row vector $[d_1d_2d_3 \ldots d_{k+1}]$. This is again a transformation to another co-ordinate system, but this time the reference axes are at right angles to one another. Since the variables transform contragrediently, $dc = \phi\psi = \eta\xi$. At this point it is necessary to make some assumption as to the relationship between the elements of the vectors d and c. Since these elements may be either real or complex, it will be assumed that

$$d = \bar{c}',$$

where the row vector $\bar{c}'$ is the transposed conjugate complex of the column vector c. Hence, the square of the length of a vector referred to this orthogonal co-ordinate system is given by $\bar{c}'c$, a number which is essentially real and non-negative. (The assumption that $d = c'$ will be found, in the particular case studied here, to lead to values of $c'c$ which, although real, may be negative.)

11. The relation between ϕ and ψ vectors

Since $\psi_x = Qc_x$, and the associated $\phi_x = \bar{c}'_x U$, it may be seen from the relations given in the two previous sections that $\bar{\phi}'_x = \bar{U}'U\psi_x = G\psi_x.$

The matrix $G = \bar{U}'U$ is symmetrical and all its elements are real numbers, those in the principal diagonal being necessarily positive in sign. It therefore remains unaltered after transposition. Since the elements of the vector ψ_x, which is by definition an age distribution transformed by the matrix H, are also necessarily real, we may write

$$\phi_x = \psi'_x G.$$

The role of the matrix G is therefore the same as that of the double covariant metric tensor g_{mn} in the tensor calculus. It transforms any ψ vector into its associated ϕ vector. This process is reversible, the reciprocal matrix being given by $G^{-1} = Q\bar{Q}'$.

The magnitude of a vector ψ_x is defined by the equation

$$x = (\psi'_x G\psi_x)^{\frac{1}{2}},$$

where the square root is taken with a positive sign. If we have two vectors ψ_x and ψ_y, both radiating from the common origin, the angle between them is given by

$$\cos\theta = \frac{\psi'_x G\psi_y}{xy},$$

from which it follows that when $\psi'_x G\psi_y = \phi_x\psi_y = 0$, the two vectors are at right angles to one another, and when $\cos\theta = 1$ their directions are the same. If, in the last equation, we take ψ_y to be the stable vector ψ_1 associated with the dominant latent root λ_1, then knowing the magnitude of a vector $B^t\psi_x$ and the angle which it makes with the ψ_1 axis, we can obtain a graphical representation of the way in which a particular age distribution approaches the stable form.

196 *On the use of matrices in certain population mathematics*

The matrix G also defines the angles between the reference axes of the co-ordinate system. If we introduce into the vector space of the ψ's a system of reference axes defined by the unit column vectors

$$\begin{aligned} e_1 &= \{1 \quad 0 \quad 0 \quad \dots \quad 0\} \\ e_2 &= \{0 \quad 1 \quad 0 \quad \dots \quad 0\} \\ e_3 &= \{0 \quad 0 \quad 1 \quad \dots \quad 0\} \\ &\vdots \\ e_{k+1} &= \{0 \quad 0 \quad 0 \quad \dots \quad 1\} \end{aligned}$$

then the distance from the origin of the unit point e_j is

$$Oe_j = \sqrt{g_{jj}},$$

and the angle between any two of the co-ordinate axes is given by

$$\cos \theta_{ij} = \frac{g_{ij}}{\sqrt{(g_{ii} g_{jj})}},$$

where, in both cases, g_{ij} is the element in the ith row and jth column of G.

By transforming back to the original co-ordinate system, the metric matrix associated with the vector space of the ξ's will be found to be

$$G_\xi = HG_\psi H.$$

Hitherto we have been chiefly concerned with an operator B which, acting in the vector space of the ψ's, has the power of transforming a vector ψ_x into what is in general a new vector ψ_y. We may now have reason to inquire how the associated vectors ϕ_x and ϕ_y are related in the vector space of the ϕ's, whenever

$$B\psi_x = \psi_y.$$

Since
$$\psi_x = G^{-1}\phi'_x \quad \text{and} \quad \psi_y = G^{-1}\phi'_y,$$

we have
$$GBG^{-1}\phi'_x = \phi'_y,$$

and hence, by transposing,
$$\phi_x G^{-1} B' G = \phi_y.$$

Thus, the matrix which transforms ϕ_x into ϕ_y is not the same as that which transforms ψ_x into ψ_y. In order to distinguish these two operators, they will be referred to as B_ϕ and B_ψ respectively. In the few numerical examples which have been worked out, the matrix B_ϕ differed greatly from the rational canonical form B_ψ, and consisted of $(k+1)^2$ real elements, some of which were negative. In addition to the relationship

$$B_\phi = G^{-1} B'_\psi G,$$

it was also found that in the case of distinct latent roots

$$B_\phi = \bar{\lambda}_1 S_1 + \bar{\lambda}_2 S_2 + \bar{\lambda}_3 S_3 + \dots + \bar{\lambda}_{k+1} S_{k+1},$$

where the S_a matrices are the spectral set of operators defined in § 9 and $\bar{\lambda}_a$ is the conjugate complex of the latent root λ_a. It may be seen from this expansion of B_ϕ that the necessary condition for $B_\phi = B_\psi$ is that all the latent roots of B should be real, the one positive and the remainder negative. (It is to be noted that we are dealing here with the case of distinct latent roots; it would appear that even if all the λ were real, $B_\phi \neq B_\psi$ in the case of repeated roots.) Unless, however, the matrix B is of a small order, it is unlikely that this condition would be fulfilled, since in the more usual-sized matrix we shall be dealing with in the case of human or other mammalian populations, some of the roots will almost certainly be complex.*

It seems unlikely that the equations given in this section will have very much practical application at the moment; they have been included merely to fill in the picture of the relationship between the two types of vector. For all ordinary purposes no one would choose to work in terms of ϕ vectors and the operator B_ϕ, instead of the more obvious ψ vectors and the more simple matrix form B_ψ. Nevertheless, since it has been necessary to assume that there are such vectors as η or ϕ associated with every ξ or ψ, and since these vectors play such an important part in the mathematical theory, the question naturally arises as to what significance must be attached to them from the biological point of view. Have they in fact any real meaning at all? Or must they be regarded purely as mathematical abstractions? At the end of § 9 it has been suggested that the row vector associated with the stable ξ_1 appropriate to the dominant latent root is a measure of the contributions made to the stable population per individual female alive in the respective age groups of the initial distribution; but this is a special case and the interpretation offered here is not applicable, even in a wider form, to η vectors in general. It may well be, of course, that the latter as a class have no concrete meaning: and that in seeking to define them in terms of some property or characteristic of an age distribution one is merely attempting the impossible. But the fact of one η vector having been defined in non-mathematical terms, even though on further consideration some revision may be needed of the actual definition given here, suggests that impossible may perhaps be too final a word to use in this connexion.

12. Case of repeated latent roots

When any of the latent roots other than the real positive dominant root are repeated, a number of the relations given in the previous sections no longer hold good and certain equations must therefore be modified. Suppose a root λ_a has a multiplicity s, and consider the matrix $f(\lambda_a)$ such as, to take a simple example,

$$f(\lambda_a) = \begin{bmatrix} a-\lambda_a & b & c & d \\ 1 & -\lambda_a & 0 & 0 \\ 0 & 1 & -\lambda_a & 0 \\ 0 & 0 & 1 & -\lambda_a \end{bmatrix}.$$

Then, since the determinant $|f(\lambda_a)| = 0$ and at least one of the first minors of order 3 is not equal to zero, the above matrix has rank 3 and, therefore, nullity 1. Hence it can be seen that $f(\lambda_a)$ of whatever order it may be has nullity 1. Since $f(\lambda_a)$ is thus simply degenerate, there is only one stable ψ appropriate to the s equal roots λ_a (see e.g. Frazer, Duncan & Collar, 1938, chap. III).

Certain consequences immediately follow. Since the matrices Q and U cannot be constructed in the way given in § 10, the reduction of B to a purely diagonal matrix by means of the collineatory transformation UBQ can no longer be carried out. Neither is the expansion of B in terms of the spectral set of S_a matrices, nor the expansion of an arbitrary ψ_x

* The interesting theoretical case of the matrix A or B having a number of its latent roots real and positive, with the remainder real and negative, is outside the scope of the present study. The necessary conditions for this to be true would involve a number of the F_x figures becoming negative, a case not considered here, but which biologically might be held to correspond with the destruction of eggs, or the very young, by certain age groups, e.g. as observed by Chapman (1933) in experimental populations of the flour beetle, *Tribolium confusum*.

in terms of the stable ψ, possible in the actual forms given in §§ 9 and 8. We may, however, proceed as follows.

The matrix Q is essentially an alternant which has been postmultiplied by a diagonal matrix N, the elements in the latter being given by the reciprocals of the scalar factors $|z|$ by which the stable vectors were divided in the process of normalization $\left(\frac{\phi_a}{|z|}\frac{\psi_a}{|z|} = 1\right)$, so that $Q = XN$, where

$$X = \begin{bmatrix} \lambda_1^k & \lambda_2^k & \dots & \lambda_{k+1}^k \\ \lambda_1^{k-1} & \lambda_2^{k-1} & \dots & \lambda_{k+1}^{k-1} \\ \vdots & \vdots & & \vdots \\ \lambda_1 & \lambda_2 & \dots & \lambda_{k+1} \\ 1 & 1 & \dots & 1 \end{bmatrix}.$$

When a root λ_a is repeated s times, s of the columns in X become the same, and therefore the matrix becomes singular. In place of this alternant matrix we have the confluent alternant form (see Turnbull & Aitken, 1932, chap. VI) in which the s columns ($s = 0, 1, 2, 3, \dots, s-1$) corresponding to the repeated root λ_a are written

$$\begin{matrix} \lambda_a^k & k\lambda_a^{k-1} & \frac{k(k-1)\lambda_a^{k-2}}{2!} & \dots \\ \lambda_a^{k-1} & (k-1)\lambda_a^{k-2} & \frac{(k-1)(k-2)}{2!}\lambda_a^{k-3} & \dots \\ \vdots & \vdots & \vdots & \dots \\ \vdots & \vdots & 3\lambda_a & \dots \\ \vdots & 2\lambda_a & 1 & \dots \\ \lambda_a & 1 & 0 & \dots \\ 1 & 0 & 0 & \dots \end{matrix}$$

the column s being obtained from column 0 (the non-normalized stable ψ_a) by the operation $\left(\frac{d}{d\lambda_a}\right)^s \Big/ s!$. This confluent alternant form of X is non-singular and therefore a reciprocal matrix can be determined (see Aitken, 1939, § 50). The general classical canonical form of B obtained by the collineatory transformation, $X^{-1}BX = C$, has corresponding to the repeated root λ_a a diagonal submatrix:

$$\begin{bmatrix} \lambda_a & 1 & \cdot & \cdot & \cdot & \cdot \\ \cdot & \lambda_a & 1 & \cdot & \cdot & \cdot \\ \cdot & \cdot & \lambda_a & 1 & \cdot & \cdot \\ \dots & \dots & \dots & \dots & \dots & \dots \\ \cdot & \cdot & \cdot & \cdot & \lambda_a & 1 \\ \cdot & \cdot & \cdot & \cdot & \cdot & \lambda_a \end{bmatrix}.$$

The matrix product of a column of X with the appropriate row of the reciprocal X^{-1} gives as before S_a and the $k+1$ S_a form a spectral set with the same properties as those defined in § 9, except that

$$B \neq \sum_{a=1}^{k+1} \lambda_a S_a.$$

In place of this expansion of B in terms of the S_a matrices, we have in the case of repeated roots the confluent form of Sylvester's theorem, for details of which reference may be made to Frazer, Duncan & Collar, 1938, chap. III. Apart from this modification, however, we may obtain by inspection of the S_a the factors by which the respective columns of X must be divided in order to express this matrix in a form comparable to that of Q in § 10. Similarly, when the respective rows of X^{-1} are multiplied by the appropriate factors, the matrix U is found and hence $G = \overline{U}'U$ can be constructed. (It is to be noted that $(\overline{X^{-1}})' X^{-1}$ is neither equal to, nor directly proportional to $\overline{U}'U$.) An arbitrary ψ_x can be expanded in terms of the column vectors of Q, though in the case of only one column associated with the repeated root λ_a does the relationship $B\psi_a = \lambda_a \psi_a$ hold.

13. The approach to the stable age distribution

A stable age distribution appropriate to the matrix B has been defined mathematically by the equation

$$B\psi = \lambda\psi,$$

and it has already been shown that since only one latent root of B is real and positive, only one of the stable ψ will consist of real and positive elements. But, in addition to this Malthusian age distribution, it is also of some interest to inquire whether any significance can be attached to the remaining stable ψ associated with the negative and complex roots of the characteristic equation.

Any age distribution ψ_x, the elements of which are necessarily $\geqq 0$, may be expressed as a vector of deviates from the stable ψ_1 associated with the dominant latent root, and we may therefore write the expansion of ψ_x in terms of the stable ψ as

$$(\psi_x - c_1\psi_1) = c_2\psi_2 + c_3\psi_3 + \ldots + c_{k+1}\psi_{k+1} = \psi_d,$$

where the coefficients c are given by the vector $c = U\psi_x$. Thus, the way in which a particular type of age distribution will approach the stable form may be studied by means of the vector ψ_d.

Among the terms occurring in the right-hand side of this expression there will be, corresponding to each negative root, a single term $c_a\psi_a$ which will consist of real elements alternately positive and negative in sign. (Even if the normalized ψ_a is imaginary this term will consist of real numbers, since in this case c_a will also become imaginary.) Moreover, corresponding to every pair of complex roots there will be a pair of terms $(c_m\psi_m + c_n\psi_n)$ which taken together will also give a single vector with real elements. This follows from the fact that c_m is the conjugate complex of c_n owing to the way in which the matrix U is constructed. Then, apart from the scalar c_1 which must necessarily be > 0, some of the remaining coefficients $c_2, c_3, \ldots, c_{k+1}$ in the expansion of ψ_d may be zero. The first and most obvious case is when they are all zero, and the age distribution ψ_x is therefore already of the stable form. But, if either

$$\psi_d = c_a\psi_a,$$

where ψ_a corresponds to a negative latent root, or

$$\psi_d = c_m\psi_m + c_n\psi_n,$$

where ψ_m and ψ_n are associated with a conjugate pair of complex roots, then it follows that the age distribution ψ_x will, as time goes on, approach the stable form in a particular way defined by either

$$B^t\psi_d = c_a\lambda^t\psi_a \quad \text{or} \quad B^t\psi_d = c_m\lambda^t\psi_m + c_n\overline{\lambda}^t\psi_n,$$

in which λ^t for a pair of complex roots $u+iv$ with modulus r may be written in the form of $r^t(\cos\theta t \pm i\sin\theta t)$. Thus, the negative and complex latent roots of B serve to determine a number of age distributions which are of some interest owing to the fact that they will approach the Malthusian form in what may be termed a stable fashion.

Since $|\lambda_1|>|\lambda_2|>|\lambda_3|>\ldots>|\lambda_{k+1}|$, the vector of deviates ψ_d will tend towards zero as $t\to\infty$ whenever $\lambda_1 \lesseqgtr 1$. Thus, in the case of a stationary population, any ψ_x will converge to the stable form of age distribution. But if $\lambda_1>1$, there is a possibility of one or more of the remaining roots having a modulus $\geqq 1$, e.g. $|\lambda_2|\geqq 1$. In the latter case there may be certain age distributions with $c_2 \neq 0$ for which the amplitude of the deviations from the stable form tend either to increase ($|\lambda_2|>1$), or to remain constant ($|\lambda_2|=1$). From the practical point of view, however, we may still say that a population with such an age distribution approaches or becomes approximately equal to the stable population, since λ_1^t is much greater than λ_2^t when t is large.

14. Special case of the matrix with only a single non-zero F_x element

The interesting case of the matrix A having only a single non-zero element in the first row has been illustrated in a numerical example by Bernardelli (1941).* This author has also used a matrix notation in the mathematical appendix to his paper, and the form of his basic matrix is the same as that referred to here as M or A. It is not clear, however, from the definitions which he gives whether he regards the elements in the first row of his matrix as being constituted by the maternal frequency figures (m_x) themselves, or by a series of values similar to those defined here as the F_x figures. He refers to them merely as the specific fertility rates for female births.

In discussing the causes of population waves, Bernardelli describes a hypothetical species, such as a beetle, which lives for only three years and which propagates in the third year of life. He assumes, for the sake of argument, that—to employ the terminology used here—$P_0=\frac{1}{2}$ and $P_1=\frac{1}{3}$, and that 'each female in the age 2–3 produces, on the average, 6 new living females'. Assuming for the moment that he is here defining a F_x figure, we may write this system of mortality and fertility rates as

$$A=\begin{bmatrix}0&0&6\\ \frac{1}{2}&0&0\\ 0&\frac{1}{3}&0\end{bmatrix},\quad B=HAH^{-1}=\begin{bmatrix}0&0&1\\1&0&0\\0&1&0\end{bmatrix}.$$

The characteristic equation expanded in terms of λ is $\lambda^3-1=0$; and the latent roots are therefore $1,\ -\frac{1}{2}\pm\frac{\sqrt{3}}{2}i$, all three being of equal modulus. The matrix A has the interesting properties

$$A^2=A^{-1},\quad A^3=I,$$

so that any initial age distribution repeats itself regularly every three years. Thus, as Bernardelli shows, a population of 3000 females distributed equally among the three age

* At this point I should like to acknowledge the gift of a reprint of this paper, which was received by the Bureau of Animal Population at a time when I was in the middle of this work, and when I was just beginning to appreciate the interesting results which could be obtained from the use of matrices and vectors: also a personal communication from Dr Bernardelli, received early in 1942, at a time when it was difficult to reply owing to the developments of the war situation in Burma. Although the problems we were immediately interested in differed somewhat, this paper did much to stimulate the theoretical development given here, and it is with great pleasure that I acknowledge the debt which I owe to him.

groups becomes a total population of 6833 at $t = 1$; of 5166 at $t = 2$; and again 3000 distributed equally among the age groups at $t = 3$. Unless a population has already an initial age distribution in the ratio of $\{6:3:1\}$, no approach will be made to the stable form associated with the real latent root, and the vector of deviates ξ_d will continue to oscillate with a stable amplitude, which will in part depend on the form of the initial distribution. Although this numerical example refers specifically to a stationary population, it is evident that a similar type of argument may be developed in the case when $|\lambda| > 1$ and $A^3 = \lambda^3 I$.

We have assumed here that his definition of the fertility rate refers to a F_x figure. But, if we were to interpret the words quoted above as referring to a maternal frequency figure, namely that every female alive between the ages 2–3 produces on the average 6 daughters per annum, then the results become entirely different. For, deriving the appropriate F_x figures by the method described in § 2, the matrix is now

$$A = \begin{bmatrix} 0 & 1 & 3 \\ \frac{1}{2} & 0 & 0 \\ 0 & \frac{1}{3} & 0 \end{bmatrix}, \quad B = HAH^{-1} = \begin{bmatrix} 0 & \frac{1}{2} & \frac{1}{2} \\ 1 & 0 & 0 \\ 0 & 1 & 0 \end{bmatrix},$$

and the latent roots are $1, -\frac{1}{2} \pm \frac{1}{2}i$. The modulus of the pair of complex roots is $1/\sqrt{2}$, which is < 1, so that every age distribution will now converge to the stable form associated with the real root. Thus, to take the same example as before, 3000 females distributed equally among the age groups will tend towards a total population of 4000 distributed in the ratio of $\{6:3:1\}$, and it was found that this age distribution would be achieved at approximately $t = 23$. During the approach to this stable form periodic waves are apparent both in the age distribution and in the total number of individuals, but these oscillations are now damped, in contrast with the results obtained with the first type of matrix.

This simple illustration serves to emphasize the importance which must be attached to the way in which the basic data are defined and to the marked difference which exists between what are termed here the m_x and F_x figures. Nevertheless, apart from the question of the precise way in which the definition of the fertility rates is to be interpreted in this example, the first type of matrix with only a single element in the first row does correspond to the reproductive biology of certain species. Thus, in the case of many insect types the individuals pass the major portion of their life span in various immature phases and end their lives in a short and highly concentrated spell of breeding. The properties of this matrix suggest that any stability of age structure will be exceptional in a population of this type, and that even if the matrix remains constant we should expect quite violent oscillations to occur in the total number of individuals.

15. NUMERICAL COMPARISON WITH THE USUAL METHODS OF COMPUTATION

From the practical point of view it will not always be necessary to estimate the actual values of all the stable vectors and of the associated matrices which are based on them. Naturally, much will depend on the type of information which is required in any particular case. In order to compute, for instance, the matrices U, Q and G, it is necessary first of all to determine all the latent roots of the basic matrix. The ease with which these may be found depends very greatly upon the order of the matrix. Thus, in the numerical example for the brown rat used previously in § 3, the unit of age and time is one month and the resulting square matrix A is of order 21. To determine all the 21 roots of the characteristic equation would be a formidable undertaking. It might be sufficient in this case to estimate the positive real root and the stable vector associated with it. On the other hand, it is possible to reduce the size of the matrix by taking a larger unit of age, and in some types of problem, where extreme accuracy is not essential, a unit say three times as great might be equally satisfactory, which would reduce the matrix for the rat population to the order of 7 × 7. It is not too difficult to find all the roots of a seventh degree equation by means of the root-squaring method (Whittaker & Robinson, 1932, p. 106). But the reduction of the matrix in this way will generally lead to a value of the positive real root which is not the same as that obtained from the larger matrix, and it is therefore necessary to see by how much these values may differ owing to the adoption of a larger unit of time.

Another important point which must be considered is the following. By expressing the age specific fertility and mortality rates in the form of a matrix and regarding an age distribution as a vector, an element of discontinuity is introduced into what is ordinarily taken to be a continuous system. Instead of the differential and integral calculus, matrix algebra

Table 2

Age group (units of 30 days)	'True' stable age distribution	Matrix stable age distribution	Age group (units of 30 days)	'True' stable age distribution	Matrix stable age distribution
0–	37,440	37,362	12–	161	160
1–	22,595	22,644	13–	101	101
2–	14,417	14,444	14–	64	63
3–	9,227	9,238	15–	40	40
4–	5,903	5,906	16–	25	24
5–	3,775	3,775	17–	15	15
6–	2,413	2,412	18–	9	9
7–	1,542	1,540	19–	6	5
8–	984	982	20–	3	3
9–	627	626			
10–	399	398			
11–	254	253	Total	100,000	100,000

is used, a step which leads to a great economy in the use of symbols and consequently to equations which are more easily handled. Moreover, many quite complicated arithmetical problems can be solved with great ease by manipulating the matrix which represents the given system of age specific rates. But the question then arises whether these advantages may not be offset by a greater degree of inaccuracy in the results as compared with those obtained from the previous methods of computation. It is not easy, however, to settle this point satisfactorily. In the way the usual equations of population mathematics are solved, a similar element of discontinuity is introduced by the use of age grouping. Thus, in the case of a human population, if we were estimating the inherent rate of increase in the ordinary way, we should not expect to obtain the same value of r from the data grouped in five year intervals of age as that from the data grouped in one year intervals. The estimates of the seminvariants would not be precisely the same in both cases. Nevertheless, the estimate from the data grouped in five year intervals is usually considered to be sufficiently accurate for all ordinary purposes, and there is little doubt that if we merely require the inherent rate of increase and the stable age distribution, these methods of computation are perfectly satisfactory when applied to human data. But, in the case of rodents, and probably also other species with high gross and net reproduction rates, it will be found that even a 4th degree equation in r with the coefficients based on the seminvariants of the $L_x m_x$ distribution is, in many examples, not sufficient to give an accurate estimate of the rate of increase, and it is necessary to arrive at a better value of r in a somewhat roundabout way. Here, the determination of the positive real root of the characteristic equation for the matrix, once the latter has been established, may be even quicker than finding a solution from the $L_x m_x$ column by a method such as that described in the appendix.

In order to compare the values of r obtained from the characteristic equation of the matrix with those obtained from the $L_x m_x$ column, both methods were used in the numerical example for the brown rat, and a comparison was also made between the values when the data were grouped in 1 month and in 3 month age intervals. In addition, the stable age distribution appropriate to the positive real root of the matrix was also calculated in both cases. The results were as follows.

(*a*) *One month unit of grouping; matrix of order* 21×21. Using the method of computation indicated in the appendix, the value of r was estimated to be 0·44565 per month of 30 days. The positive real root of the characteristic equation was $\lambda_1 = 1{\cdot}56246$, whence $r = 0{\cdot}44626$, a value which differs from the former only in the fourth decimal place. The appropriate age distributions, expressed per 100,000, are given in Table 2, the 'true' stable being obtained from

$$n_x = 100{,}000 b \int_x^{x+1} e^{-rx} l_x \, dx.$$

The agreement between these distributions is very good, although the one derived from the matrix shows certain small rhythmical departures from the 'true' distribution particularly in the earlier age groups. The maximum difference between them in this region, however, is not greater than 2·2 per thousand. Since the matrix stable distribution is proportional to the columns of the matrix $H^{-1} S_1 H$, which in turn is proportional to A^t when t is large, this agreement between the two distributions also indicates that the cumulative errors which might be expected in forming the series A^2, A^3, A^4, etc., owing to the P_x figures being based on the life table age distribution, are not very serious. Judging by this example it seems that satisfactory estimates of the inherent rate of increase and of the stable age distribution may be made from a large order matrix.

(*b*) *Three months unit of grouping; matrix of order* 7×7. Clearly there are several ways in which a large order matrix may be condensed into one of a smaller order. The method which was used in the present instance was to construct the first row of the condensed canonical form B by interpolating in the integral curve of the original $L_x m_x$ column (1 month units of grouping) for the ages 4·5, 7·5, 10·5, etc., and taking the first differences of the seven values thus obtained. Since interpolation was not very satisfactory in the earlier part of the integral curve—the differences converged rather slowly in this region—the elements were expressed to only three places of decimals. (Some preliminary transformation of the integral $L_x m_x$ figures might have been better in this case.) The characteristic equation, expanded in terms of λ, was found to be

$$\lambda^7 - 1{\cdot}756\lambda^6 - 6{\cdot}899\lambda^5 - 7{\cdot}203\lambda^4 - 5{\cdot}344\lambda^3 - 3{\cdot}244\lambda^2 - 1{\cdot}110\lambda - 0{\cdot}102 = 0.$$

It will be seen that the sum of the coefficients, $R_0 = 25{\cdot}658$, which is necessarily the same as the original net reproduction rate owing to the way in which the coefficients were derived. For interest, the seven roots of this equation were then determined by the root-squaring method, using 4-place tables of logarithms and Barlow's tables of squares (Whittaker & Robinson, 1932, p. 110). The approximate values of the roots, arranged in descending order of their moduli, are

$$\lambda_1 = 4{\cdot}016,$$
$$\lambda_2 = -1{\cdot}032,$$
$$\lambda_3\lambda_4 = -0{\cdot}0215 \pm 0{\cdot}6762i \quad (\text{mod.} = 0{\cdot}6765),$$
$$\lambda_5\lambda_6 = -0{\cdot}5245 \pm 0{\cdot}3486i \quad (\text{mod.} = 0{\cdot}6298),$$
$$\lambda_7 = -0{\cdot}135.$$

Thus, apart from the positive real root, there are two negative and two pairs of complex roots. It is interesting to note in passing that in this example the modulus of the second latent root is >1 (vide § 13). From the value of the dominant root we find $r = 0{\cdot}4634$ per head per month of 30 days, an estimate of the rate of increase which is 1·71 % per month higher than that from the large order matrix.

Table 3. *Stable age distributions*

Age group (in months)	$100{,}000b\int_x^{x+3} e^{-rx}l_x dx$	From condensed matrix	Summation of 'true' distribution
0–	75,960	75,762	74,452
3–	18,055	18,267	18,905
6–	4,514	4,519	4,939
9–	1,120	1,111	1,280
12–	273	267	326
15–	64	61	80
18–	14	13	18
Total	100,000	100,000	100,000

The net reproduction rate given by the new $L_x m_x$ column which was obtained by working in units of three months, was 25·6162, a figure somewhat lower than the original one of 25·6579. The rate of increase, estimated in a similar way to the former example for one month age units, was $r = 0{\cdot}46034$, again a higher figure, though one of much the same order as that obtained from the condensed matrix.

The appropriate age distributions are given in Table 3, together with the 'true' distribution of Table 2 summed in three month age groups.

Compared with the last column, both of the stable distributions for the data grouped in three month age intervals are tilted towards the younger age classes, so that the number of immature females (<3 months of age) is overestimated, while the remaining age groups are underestimated. The distributions derived from the integral and from the matrix are again of much the same order, and the differences between them and the last column, although not very great, are quite marked.

The four estimates of the inherent rate of increase which have obtained from these numerical data may be compared in the following table.

In both cases the estimates from the $L_x m_x$ column and from the matrix agree very well: for a given unit of grouping both methods would seem to give comparable results. The differences between the estimates made by the same method are much greater, and the effect of increasing the unit of grouping, and in this way shortening the labour, is to increase the value of r quite appreciably. Whether, or not, we should regard these estimates for the three months age grouping as satisfactory would depend on the degree of accuracy required in any particular calculation. It must be remembered, however, that the basic numerical data are of rather an extreme type in this example. It is doubtful whether any naturally living rat population would have so good a life table and so high a degree of fertility as that assumed for this imaginary population. In fact it was for these very reasons that these data were

	$L_x m_x$ column	Matrix	Difference
1 month age groups	0·44565	0·44626	0·00061
3 ,,	0·46034	0·4634	0·0031
Difference	0·01469	0·0171	

chosen as the basis of the numerical calculations in this work. For, if it could be shown that the two methods of computation gave comparable results in this case, it was felt that an even better agreement should be obtained in less extreme examples and more particularly with data relating to populations whose rate of increase is nearer to the stationary state. Although in this example the larger unit of grouping leads to rather unsatisfactory estimates of the rate of increase and the stable age distribution, it seems probable that, for the reasons just given, the differences would be less for instance in the case of human data. Hence, the question of the unit to be adopted is likely to become of less importance in the type of data more commonly met with, though it would be necessary to work out an example for such a population in order to check this point.

There is, however, one way to avoid this difficulty of the working unit for populations with a high relative rate of increase. For example, returning to the numerical data used here, supposing that it was necessary in some particular problem to have a fairly high degree of accuracy in the results, but that the work involved in manipulating the large order matrix of 21×21 was too excessive. It might be sufficient in the case we are imagining to know the age distribution of the population in three month age groups at some particular time in the future, which we will take to be a multiple of three. Then, once the real latent root (λ_1) of the large matrix and its associated stable vector have been determined, it is possible to construct a small order matrix of 7×7 which has λ_1^3 as its dominant root and therefore the same real stable age distribution as the larger matrix, only expressed in three month instead of in one month age units. It is convenient to carry out the calculation in terms of the canonical form B and of ψ vectors. Having determined the dominant latent root and the stable vector for the larger matrix, the elements of the first three rows of B^3 are then written down and summed in columns. This can be done very quickly in the present example, where reproduction does not start until the age of 3 months, for the third row of B^3 is the same as the first row of B; the second row is merely the first row of B shifted one age group to the left; and similarly again for the first row. The sums of the columns are then weighted with the number alive in the appropriate age group in the stable population (ψ_1 vector), and by summing the weighted column totals in groups of three and taking the weighted mean, we obtain the elements of the first row of a 7×7 matrix which has λ_1^3 as its dominant latent root. Thus, the characteristic equation of the original matrix condensed in this way was

$$\lambda^7 - 1{\cdot}5056\lambda^6 - 6{\cdot}4694\lambda^5 - 7{\cdot}2047\lambda^4 - 5{\cdot}5371\lambda^3 - 3{\cdot}4537\lambda^2 - 1{\cdot}3451\lambda - 0{\cdot}1447 = 0,$$

and, out of idle curiosity, all the seven roots were extracted in order to compare them with those of the previous example of a condensed matrix. The estimation in this case was carried out to a higher degree of accuracy. The results were:

$$\begin{aligned} \lambda_1 &= 3{\cdot}81452, \\ \lambda_2 &= -1{\cdot}02526, \\ \lambda_3\lambda_4 &= -0{\cdot}5905 \pm 0{\cdot}3782i \quad (\text{mod.} = 0{\cdot}70125), \\ \lambda_5\lambda_6 &= 0{\cdot}0280 \pm 0{\cdot}6879i \quad (\text{mod.} = 0{\cdot}68847), \\ \lambda_7 &= -0{\cdot}15876. \end{aligned}$$

The dominant root of the original matrix was 1·56246 and the cube root of λ_1 is 1·56248. The remaining roots may be compared with those given for the previous example. The two negative roots are very similar and, in the second case, the two pairs of complex roots appear to have changed places, the real part of one pair becoming positive instead of negative. Although the cube root of λ_1 is equal to the dominant root of the original matrix, it is unfortunately not true that a similar relationship holds for the remaining six values of λ. There is, for instance, no negative latent root > 1 for the larger form in this actual example.

This point, however, raises an extremely interesting question. For a given series of data a finite matrix of a relatively small order may be constructed, as in the first example given here of a condensed matrix. Supposing that the order of this matrix is increased step by step and that in each case the latent roots are found. Then, in this approach to an infinite matrix, how do the latent roots behave and what relation does the array of roots in each case bear to those of the preceding steps? For the purposes of comparison it will be necessary to express the roots in terms of some suitable unit of time, e.g. per month or per year. So far as the real positive root is concerned, it seems likely that the series of individual roots will approach nearer and nearer to a limiting value. For the root λ_1 is the ratio N_{t+h}/N_t, or the number of times the stable population has increased at the end of the interval of time h. Then, expressing λ_1 in the chosen unit of time, we have $\Lambda_1 = (\lambda_1)^{1/h}$, and taking logarithms;

$$\log_e \Lambda_1 = \frac{\log_e N_{t+h} - \log_e N_t}{h},$$

so that, when the interval of time becomes very small, corresponding to a matrix of a very large order, and $h \to 0$, the right-hand side of this equation approaches the limit $\frac{1}{N}\frac{dN}{dt} = \rho$, the true instantaneous relative rate of increase of the stable population. This argument is put forward with a certain amount of diffidence; it is only too easy for the biologist to overlook some flaw which will be immediately obvious to the trained mathematician. But, even if it were a valid argument for the behaviour of the dominant root, it can hardly be extended in this form to the case of the remaining roots; and thus the main question is left unanswered. From the point of view of the biologist, it would be interesting to know whether with an increase in the size of the matrix the array of secondary roots tends to coalesce round certain values of $\lambda^{1/h}$.

16. FURTHER PRACTICAL APPLICATIONS

If we wish merely to estimate the inherent rate of increase and the stable age distribution appropriate to some system of age specific fertility and mortality rates, there is evidently little to choose between the matrix and the ordinary methods of computation. The advantages of expressing the basic rates in the form of a matrix are more clearly seen in considering the type of problem such as the following. Let us suppose that a species of mammal at a certain season of the year invades a fresh environment where there is an ample food supply, a freedom from predators, and plenty of space to accommodate any rapid increase in numbers which might take place. Under these conditions it might be assumed for theoretical purposes that some age specific rates of fertility and mortality would remain approximately constant over a period of time. The age distribution of these immigrants then becomes of some importance owing to the effect which it must necessarily have on the future course of events. For this initial distribution must clearly be very different from that which would ultimately be established in the case of a species, such as a rodent, with possibly a very rapid rate of increase, since nestlings will not be represented in it and young individuals may be present in only relatively small numbers. Supposing then that we have a number of such populations subject to the same age schedules of fertility and mortality, but differing in the age distribution of the original immigrants, we may have reason to enquire how far the development of these populations is affected over a limited period of time by the varying form of this initial distribution, assuming for simplicity that no further waves of immigration occur.

If an estimate of the number and age distribution of the female population at successive intervals of time is alone required, the answer for any form of initial distribution is readily obtained once the series of matrices M, M^2, M^3, ..., M^t have been constructed. But, in addition, we may require to know the changes which might be expected to occur in the birth rate and death rate, and also, for example, in some such rate as the percentage of adult females pregnant, a figure which is one of the simplest measures we have of the degree of fertility among wild populations. Again, in a species like the wild rat we never know the exact age of individuals caught in the field, and thus the only measure of the form of the female age distribution is the percentage of immature females, provided, of course, we are sampling the complete population. Some method is therefore required for calculating such rates at successive intervals of time.

Once the age distribution of the female population at time t is known, an estimate of the expected number of female births per unit of time may be obtained by operating on the age distribution with the maternal frequency figures. Thus, in matrix notation we may write, the number of female births equals $m_x M^t \xi_0$, where ξ_0 is the initial age distribution and the m_x figures are treated as a row vector. Similarly the estimated number of deaths per unit of time may be obtained with the help of the age specific death rates (D_x). The relative rate of increase calculated in this way is not necessarily exact, but it may be sufficiently accurate for our present purposes. As an example of the degree of error involved in this method, we may compare the values of the stable birth rate and death rate, as given in the appendix, with those derived from the matrix stable distribution in Table 2 by operating with the m_x and D_x figures. The latter were in this case computed from the stationary age distribution and the d_x column of the life table ($D_x = d_x/L_x$). The results were as follows:

	'True' values from appendix	By operating with m_x and D_x on matrix stable distribution
Birth-rate (b)	0·51265	0·51257
Death-rate (d)	0·06700	0·06154
$b-d=r$	0·44565	0·45103

The rate of increase is overestimated by about 5·4 per thousand per month, the principal error being in the death rate. This discrepancy is due to the fact that the number of deaths under 2 months of age is underestimated by applying age specific death rates, which are based on the stationary age distribution, to the stable population grouped in one month intervals at these ages. The difference between these distributions happens to be quite marked in this example. The degree of error, however, is not very great; and in the type of problem we are considering, when the age distribution of a population may take any form, this seems to be the only practical method of estimating the rate of increase.*

Supposing, then that in the case of the rat population used here as a numerical illustration, we wished to estimate the number of females, the birth rate and death rate, and the percentage of immature females at monthly intervals up to—say—7 months from the origin of the time scale, when the initial immigration is assumed to take place. Since the jth column of the matrix gives the age distribution of the survivors and the surviving descendants per individual female alive in the age group $j-1$ to j at $t=0$, the sum of the elements in this column gives the number of times the original population in this age group has increased, or decreased, at time t. The percentage of immature females may be obtained once the sum of the first three elements in the column is known (reproduction begins at the age of 3 months in this example); and the number of births and deaths per unit of time may be found by operating on the column with the m_x and D_x figures. Each of these totals, of course, will have to be multiplied in the end by the number of females alive in this age group at $t=0$. Since the initial age distribution may be of any form under the conditions of the problem, it will be necessary first of all to calculate these four totals for every column of each of the seven matrices M^t. Now, to add up the elements forming each column of a matrix is equivalent to premultiplying the matrix by a row vector of units; the sum of the first three elements may be obtained by premultiplying with a row vector of which the first three elements are units and the remainder zeros; and similarly the numbers of births and deaths are found with the help of the row vectors m_x and D_x. Thus, the operations which it is necessary to perform on

* Another similar method, which avoids the actual calculation of the number of deaths by means of the age specific death rates, is suggested by the following relationship. If the transformed age distribution $\psi = H\xi$, where H is the matrix defined in § 5, is operated on with a row vector which consists of the $L_x m_x$ figures (Appendix, Table 4), and the resulting scalar is divided by the sum of the elements of ψ, an estimate is obtained of the relative rate of increase of a population with an age distribution ξ in the original co-ordinate system. This follows from the properties of the transformed population discussed at the end of § 5 and from the relationship between the first row of the canonical form $B = HAH^{-1}$ and the $L_x m_x$ column (§ 6). In the transformed population the death rate $=0$, and the maternal frequency is given by $L_x m_x$. Thus, by transforming back again to the original co-ordinate system

$$r = \frac{[L_x m_x]\, H\xi}{[1]\, H\xi},$$

where [1] defines a row vector of units. Taking ξ as the matrix stable distribution of Table 2, and calculating the row vectors $[L_x m_x]\, H$ and $[1]\, H$, the value of r was estimated by this method to be 0·44468.

each of the columns of M^t may be written as the matrix R, which will consist of $m+1$ columns and n rows, the number of the latter depending on the number of operations. Then the required totals for the matrix M^t will be given by

$$Z^t = RM^t = RMMM \dots M,$$

and it is easy to see how the Z matrices may be built up in succession without calculating the actual matrices M^2, M^3, M^4, etc. Once the series of Z matrices have been constructed, we can obtain from $Z^t\xi_0$ the necessary figures from which the required rates at time t for a population with an initial age distribution ξ_0 may be calculated. Moreover, if we wish, the contributions made, for instance, to the total number of births or deaths by any particular age group in the initial distribution can also be determined.

The computations in this illustration have been greatly simplified by the assumption that the system of age specific fertility and mortality rates remains constant. In the case when the basic matrix M is changing with time and the age distribution at time t is given by $M_t \dots M_3 M_2 M_1 \xi_0$, some of the rows of R will also be varying. Hence the series of Z matrices could not be built up without first computing $M_2 M_1$, $M_3 M_2 M_1$, etc. The latter, however, may often be of interest in themselves. For, if each column of M^t—or $M_t \dots M_3 M_2 M_1$ in the case of a variable matrix—is multiplied by the number of females alive in the appropriate age group at $t = 0$, the complete age structure of the population at time t is represented in the form of a two-dimensional array. Since the sum of the elements in each row is the total number alive in the age group x to $x+1$ at time t, the number contributed to this total by each age group at $t = 0$ is given by the individual entries.

APPENDIX

(1) *The tables of mortality and fertility*

The basic life table and fertility table which have been used in the numerical part of this study are given in Table 4. The adult l_x figures from the age of 2 months onwards are based on the mortality observed among the females of a domesticated brown rat stock housed at the Wistar Institute, Philadelphia. According to the data for 26 generations of this laboratory stock published by King (1939) it appears that out of 1384 females alive at the age of 2 months (60 days), 1337 were alive at 12 months, and 984 at 20 months. This information gave three points on the l_x curve, supposing that these survivors could be regarded as ordinates at these exact ages. In order to interpolate for other ages, a logistic type of curve was fitted to the data, the values of the constants being chosen so that the curve passed through these three points. The l_x values in Table 4 are given by

$$l_x = \frac{0{\cdot}85156355}{1+0{\cdot}00101065e^{0{\cdot}30016x}}, \quad \text{for} \quad x \geqq 2.$$

Although the original data did not extend beyond the age of 20 months, by which time the vast majority of the females had ceased breeding (King, 1939), this l_x curve was extrapolated to later ages, whenever necessary, simply for the purposes of this theoretical study.

The degree of infant mortality assumed here, namely 15 % between birth and the age of 2 months, is entirely arbitrary; it represents a moderate degree of loss at these early ages. Some care, however, was taken to weld the infant mortality smoothly on to the remainder of the l_x curve, and it was assumed that the number of deaths according to age (d_x) decreased geometrically between birth and the age of 2 months. The actual calculations for these age groups were carried out in units of 1/8 of a month and the resulting l_x curve was integrated by means of Simpson's rule. The same method of numerical integration was used also for the adult part of the life table in order to obtain the L_x figures.

The fertility table is partly artificial and was constructed in the following way. The gross reproduction rate of these domesticated brown rats was estimated from the data published by King (1939) to have been just under 10 litters for the later generations, when the stock was thoroughly adapted to life in the laboratory. The frequency of litter production according to the age of the mother has been found by the author (unpublished observations) to be represented closely by a Pearsonian type I curve in the case of certain litter fertility tables, for example in the cross-albino rat, the vole and some human populations with a high degree of fertility; and, moreover, the values of β_1 and β_2 were very similar for all three species. The actual equation for the curve used here may be written

$$y = y'x^{1\cdot5-1}(a-x)^{2\cdot5-1},$$

and the range was assumed to be from 3 to 21 months, which for grouping purposes represents the span of the reproductive ages observed by King in this Wistar strain of brown rats. The ordinates of the integral curve of the above equation were taken from the tables of the incomplete Beta function and, in this way, a column was formed which gave a gross reproduction rate of 10 litters. The individual entries were then multiplied by the mean number of daughters per litter according to the age of the mother, which was recorded by King, and thus the m_x figures in Table 4 were obtained. The gross reproduction rate is 31·21 daughters and the net rate 25·66. These tables of fertility and mortality were originally constructed in order to determine the relative rate of increase and the type of stable age distribution which might be expected in a brown rat population living under more or less optimum conditions.

(2) *Calculation of the rate of increase*

Some difficulty was experienced in obtaining a satisfactory estimate of the rate of increase (r) from the usual solution (Dublin & Lotka, 1925) of the equation:

$$\int_0^\infty e^{-rx} l_x m_x dx = 1.$$

The 4th degree equation in r with the numerical coefficients based on the seminvariants of the $L_x m_x$ distribution, the estimates of which are given in § 6, was

$$4\cdot90199r^4 + 3\cdot69283r^3 - 7\cdot07198r^2 + 9\cdot60604r - 3\cdot2448498 = 0,$$

and the real root was found to be 0·42447. This value of r was, however, clearly too low and a better estimate had to be obtained in a rather roundabout way, since it was thought that the use of higher moments than the fourth would be unsatisfactory in the present example.

If the force of mortality represented by the original life table is increased by a constant factor (r) which is independent of age, the new life table is

$$l'_x = e^{-rx} l_x,$$

and the net reproduction rate will be given by $R_r = \Sigma L'_x m_x$, where L'_x are the integrals of the new l'_x curve. Clearly, the greater r is taken to be, the smaller R_r becomes. Then, suppose that the relation between R_r and r is given by

$$\log_e R_r = a + br + cr^2 + dr^3 + \ldots,$$

Table 4

Age (x) Units of 30 days	Life table		Fertility table	
	l_x	L_x	m_x	$L_x m_x$
0	1·00000	0·46544	—	—
0·5	0·88706	0·43489	—	—
1	0·85882	0·42725	—	—
1·5	0·85176	0·42534	—	—
2	0·85000	0·84973	—	—
3	0·84945	0·84910	1·1342	0·96305
4	0·84871	0·84824	2·0797	1·76408
5	0·84772	0·84708	2·6596	2·25289
6	0·84638	0·84553	2·8690	2·42582
7	0·84458	0·84344	2·9692	2·50434
8	0·84217	0·84063	2·9535	2·48280
9	0·83893	0·83687	2·8143	2·35520
10	0·83459	0·83184	2·6114	2·17227
11	0·82881	0·82515	2·2455	1·85287
12	0·82113	0·81629	2·0533	1·67609
13	0·81098	0·80463	1·7971	1·44600
14	0·79768	0·78940	1·5561	1·22839
15	0·78039	0·76975	1·2175	0·93717
16	0·75821	0·74472	0·9548	0·71106
17	0·73018	0·71342	0·6610	0·47157
18	0·69548	0·67512	0·4043	0·27295
19	0·65354	0·62953	0·1846	0·11621
20	0·60434	0·57696	0·0435	0·02510
21	0·54859	—	—	—
Total	—	—	31·2086	25·65786

where $a = \log_e R_0$, and the constants b, c, d, etc. are to be determined. It was assumed in the present instance that a 4th degree polynomial in r would be sufficient, and four new life tables were constructed taking r to be in turn 0·1, 0·2, 0·4 and 0·5. The L'_x integrals were obtained by Simpson's rule for the reproductive ages and the four values of R_r calculated. The equations for finding the values of the constants were:

$$0{\cdot}0001e + 0{\cdot}001d + 0{\cdot}01c + 0{\cdot}1b = -0{\cdot}8934974,$$
$$0{\cdot}0016e + 0{\cdot}008d + 0{\cdot}04c + 0{\cdot}2b = -1{\cdot}6708610,$$
$$0{\cdot}0256e + 0{\cdot}064d + 0{\cdot}16c + 0{\cdot}4b = -2{\cdot}9792984,$$
$$0{\cdot}0625e + 0{\cdot}125d + 0{\cdot}25c + 0{\cdot}5b = -3{\cdot}5490542,$$

whence

$$b = -9{\cdot}617235,$$
$$c = 7{\cdot}371816,$$
$$d = -5{\cdot}698291,$$
$$e = 2{\cdot}062332.$$

Inserting these values of the constants in the equation for $\log_e R_r$, the value of r for which $\log_e R_r = 0$ was found to be 0·44565.

The stable birth rate was estimated in the usual way from

$$\frac{1}{b} = \int_0^\infty e^{-rx} l_x \, dx.$$

212 *On the use of matrices in certain population mathematics*

The integrals were computed by means of Simpson's rule, treating the age groups 0–2 separately from the remainder of the life table, the units adopted being 1/4 month for the early ages compared with 1 month for the later. It was found that

$$\int_0^{21} e^{-0\cdot44565x} l_x\,dx = 1\cdot95064,$$

$$b = 0\cdot51265,$$

and hence the death rate,

$$d = 0\cdot06700.$$

The value of r is so high in this case that the error in the estimate of b due to neglecting the ages from 21 onwards would only be in the last figure. The stable age distribution is given in Tables 1 and 2 of the text. Owing to the way in which the value of r was determined, it will be found that the birth rate of the stable population obtained by operating on this age distribution with the maternal frequency figures is precisely the same as that given by the above integral.

(3) *Numerical values of the matrix elements*

The numerical elements of the matrices A and B to which reference has been made in §§ 3 and 6 of the text are given in Table 5.

Table 5

Age group x	Matrix A		Matrix B
	P_x	F_x	Elements of first row
0–	0·94697	0	0
1–	0·99665	0	0
2–	0·99926	0·3964	0·3741
3–	0·99899	1·4939	1·4089
4–	0·99863	2·1777	2·0517
5–	0·99817	2·5250	2·3756
6–	0·99753	2·6282	2·4682
7–	0·99667	2·6749	2·5059
8–	0·99553	2·6018	2·4293
9–	0·99399	2·4419	2·2698
10–	0·99196	2·1865	2·0202
11–	0·98926	1·9044	1·7454
12–	0·98572	1·7259	1·5648
13–	0·98107	1·4918	1·3332
14–	0·97511	1·2415	1·0885
15–	0·96748	0·9522	0·8141
16–	0·95797	0·7141	0·5907
17–	0·94631	0·4618	0·3659
18–	0·93247	0·2518	0·1888
19–	0·91649	0·0901	0·0630
20–		0·0035	0·0022

I wish to thank Professor M. Greenwood, F.R.S. and Dr J. O. Irwin for their kindness in reading the manuscript, and also Dr H. Motz for many fruitful discussions. The study arose out of some research work which is being carried out by the Bureau of Animal Population with the aid of a grant from the Agricultural Research Council.

REFERENCES

AITKEN, A. C. (1939). *Determinants and Matrices*, 1st ed. Edinburgh: Oliver and Boyd.

BERNARDELLI, H. (1941). Population waves. *J. Burma Res. Soc.* **31**, no. 1, 1–18.

CHAPMAN, R. N. (1933). The causes of fluctuations of populations of insects. *Proc. Hawaii Ent. Soc.* **8**, no. 2, 279–92.

CHARLES, ENID (1938). The effect of present trends in fertility and mortality upon the future population of Great Britain and upon its age composition. Chap. II in *Political Arithmetic*, ed. Lancelot Hogben. London: Allen and Unwin.

DUBLIN, L. I. & LOTKA, A. J. (1925). On the true rate of natural increase as exemplified by the population of the United States, 1920. *J. Amer. Statist. Ass.* **20**, 305–39.

FRAZER, R. A., DUNCAN, W. J. & COLLAR, A. R. (1938). *Elementary Matrices*. Camb. Univ. Press.

GLASS, D. V. (1940). *Population Policies and Movements in Europe*. Oxford: Clarendon Press.

KING, HELEN D. (1939). Life processes in grey Norway rats during fourteen years in captivity. *Amer. Anat. Mem.* no. 17, 1–72.

TURNBULL, H. W. & AITKEN, A. C. (1932). *An Introduction to the Theory of Canonical Matrices*. London and Glasgow: Blackie and Son.

WHITTAKER, E. T. & ROBINSON, G. (1932). *The Calculus of Observations*, 2nd ed. London and Glasgow: Blackie and Son.

THE INTRINSIC RATE OF NATURAL INCREASE OF AN INSECT POPULATION

By L. C. BIRCH*, *Zoology Department, University of Sydney*

CONTENTS

1. INTRODUCTION

The intrinsic rate of increase is a basic parameter which an ecologist may wish to establish for an insect population. We define it as the rate of increase per head under specified physical conditions, in an unlimited environment where the effects of increasing density do not need to be considered. The growth of such a population is by definition exponential. Many authors, including Malthus and Darwin, have been concerned with this and related concepts, but there has been no general agreement in recent times on definitions. Chapman (1931) referred to it as 'biotic potential', and although he does state in one place that biotic potential should in some way combine fecundity rate, sex ratio and survival rate, he never precisely defined this expression. Stanley (1946) discussed a somewhat similar concept which he called the 'environmental index'. This gives a measure of the relative suitability of different environments, but it does not give the actual rate of increase of the insect under these different conditions. An index for the possible rate of increase under different physical conditions would at the same time provide a measure of the relative suitability of different environments. Birch (1945*c*) attempted to provide this in an index combining the total number of eggs laid, the survival rate of immature stages, the rate of development and the sex ratio. This was done when the author was unaware of the relevance of cognate studies in human demography. A sounder approach to insect populations based on demographic procedures is now suggested in this paper. The development of this branch of population mathematics is principally due to A. J. Lotka. From the point of view of the biologist, convenient summaries of his fundamental contributions to this subject will be found in Lotka (1925, Chapter 9; 1939 and 1945). A numerical example of the application of Lotka's methods in the case of a human population will be found in Dublin & Lotka (1925). The parameter which Lotka has developed for human populations, and which he has variously called the 'true' or 'inherent' or 'intrinsic' rate of natural increase, has obvious application to populations of animals besides the human species. The first determination of the intrinsic rate of increase of an animal other than man was made by Leslie & Ranson (1940). They calculated the 'true rate of natural increase' of the vole, *Microtus agrestis*, from age-specific rates of fecundity and mortality determined under laboratory conditions. With the use of matrices Leslie has extended these methods and, as an example, calculated the true rate of natural increase of the brown rat, *Rattus norvegicus* (Leslie, 1945). The author is much indebted to Mr Leslie for having drawn his attention to the possible application of actuarial procedures to insect populations. He has been completely dependent upon him for the methods of calculation used in this paper.

Before proceeding to discuss the reasons for the particular terminology adopted in this paper, it is necessary first to consider the true nature of the parameter with which we are concerned.

* This investigation was carried out at the Bureau of Animal Population, Oxford University, during the tenure of an overseas senior research scholarship from the Australian Science & Industry Endowment Fund.

16 *The intrinsic rate of natural increase of an insect population*

2. BIOLOGICAL SIGNIFICANCE OF THE INTRINSIC RATE OF NATURAL INCREASE

The intrinsic rate of increase is best defined as the constant '*r*' in the differential equation for population increase in an unlimited environment,

$$dN/dt = rN,$$

or in the integrated form $N_t = N_0 e^{rt}$,

where N_0 = number of animals at time zero,
N_t = number of animals at time t,
r = infinitesimal rate of increase.

The exponent r is the difference between the birth-rate (b) and the death-rate (d) in the population ($r = b - d$). In some circumstances it may be more useful to know the finite rate of increase, i.e. the number of times the population multiplies in a unit of time. Thus, in a population which is increasing exponentially, if there are N_t individuals at time t then in one unit of time later the ratio

$$\frac{N_{t+1}}{N_t} = e^r$$
$$= \text{antilog}_e\, r = \lambda.$$

Hence the finite rate of increase (λ) is the natural antilogarithm of the intrinsic (infinitesimal) rate of increase.

Any statement about the rate of increase of a population is incomplete without reference to the age distribution of that population, unless every female in it happens to be producing offspring at the same rate at all ages, and at the same time is exposed to a chance of dying which is the same at all ages. In such an inconceivable population the age of the individuals obviously has no significance. In practice, a population has a certain age schedule both of fecundity and of mortality. Now a population with *constant* age schedules of fecundity and mortality, which is multiplying in an unlimited environment, will gradually assume a fixed age distribution known as the stable age distribution' (Lotka, 1925, p. 110). When this age distribution is established the population will increase at a rate $dN/dt = rN$. Thus the parameter r refers to a population with a stable age distribution. The consideration of rates of increase in terms of the stable age distribution was one of the most important advances in vital statistics. In any other sort of population the rate of increase varies with time until a stable age distribution is assumed. There is, for example, no simple answer to the question: what is the rate of increase of x newly emerged adult insects in an unlimited environment? The rate will vary with time as immature stages are produced until the population has a stable age distribution. The rate of increase in the first generation might be given, but that is a figure of limited value. On the other hand, the maximum rate that it can ever maintain over an indefinite period of time is given by the rate of increase in a population of stable age distribution. That rate is therefore the true intrinsic capacity of the organism to increase. Thompson (1931) rejected the use of the exponential formula in the study of insect populations in preference for a method of dealing with the rate of increase as a 'discontinuous phenomenon'. His paper should be consulted for the reasons why he considers a single index unsatisfactory in relation to the particular problems with which he was concerned.

If the 'biotic potential' of Chapman is to be given quantitative expression in a single index, the parameter r would seem to be the best measure to adopt, since it gives the intrinsic capacity of the animal to increase in an unlimited environment.* But neither 'biotic potential' nor 'true rate of natural increase' can be regarded as satisfactory descriptive titles. The word 'potential' has physical connotations which are not particularly appropriate when applied to organisms. There is a sense in which it might be better used with reference to the environment rather than the organism. Contrary to what it seems to imply, the 'true rate of natural increase' does not describe the actual rate of increase of a population at a particular point in time, unless the age distribution of that population happens to be stable. But it does define the intrinsic capacity of that population, with its given regime of fecundity and mortality, to increase. This point is clearly made by Dublin & Lotka (1925). More recently, Lotka (1945) has dropped the use of 'true rate of natural increase' for the more precise 'intrinsic rate of natural increase'. It would seem desirable that students of populations should adopt the same terminology, irrespective of the animals concerned, and as 'intrinsic rate of natural increase' is more truly descriptive of the parameter r than other alternatives, its use is adopted in this paper.

The intrinsic rate of increase of a population may be calculated from the age-specific fecundity† and survival rates observed under defined environmental conditions. For poikilothermic animals these rates vary with physical factors of the environment such as temperature and humidity. Furthermore, within any

* For a discussion of the relative merits of this and other parameters in human demography reference should be made to Lotka (1945).

† Fecundity rate is used to denote the rate at which eggs are laid by a female. Some eggs laid are infertile and so do not hatch. The percentage 'infertility' is included in the mortality rate of the egg stage. It is usual amongst entomologists to denote the percentage of fertile eggs as the 'fertility rate'. Demographers, on the other hand, use 'fertility rate' to denote the rate of live births. Since 'fertility rate' has this other usage in entomology the term 'fecundity rate' is used throughout this paper as synonymous with the 'fertility rate' of the demographers.

given combination of physical factors, the fecundity and survival rates will vary with the density of the animals. Hence it is possible to calculate an array of values of r at different densities. But particular significance attaches to the value of r when the fecundity and survival rates are maximal, i.e. when density is optimal, for this gives the maximum possible rate of increase *within* the defined physical conditions. *Between* the whole array of physical conditions in which the animal can survive there is a zone where fecundity and survival rates are greatest and where, therefore, the intrinsic rate of increase will be greatest too. The zone within which the intrinsic rate of increase is a maximum may be referred to as the optimum zone. This is an arbitrary use of the word optimum and it does not imply that it is always to the advantage of the animal to increase at the maximum possible rate. The maximum intrinsic rate of increase under given physical conditions has importance from two points of view. It has a theoretical value, since it is the parameter which necessarily enters many equations in population mathematics (cf. Lotka, 1925; Volterra, 1931; Gause, 1934; Crombie, 1945). It also has practical significance. The range of temperature and moisture within which the insect can multiply is defined most precisely by that range within which the parameter exceeds zero. This will define the maximum possible range. In nature the range of physical conditions within which the species may be found to multiply may be less, since it is possible that effects of density and interspecific competition may reduce this range, and also the range of the optimum zone. These considerations are, however, beyond the scope of this paper; some discussion of them will be found in a review paper by Crombie (1947).

There are some important differences in the orientation with which the demographer and the student of insect populations face their problems. In human populations the parameter r varies in different civilizations and at different times in one civilization, depending upon customs, sanitation and other factors which alter mortality and fecundity rates. The maximum possible value of r does not enter into most demographic studies. In a population which is growing logistically the initial rate of increase is theoretically the maximum intrinsic rate of increase, and this latter value can be determined indirectly by calculating the appropriate logistic curve. Lotka (1927) has done this for a human population and so arrived at an estimate of a physiological maximum for man. This has theoretical interest only. In insect populations, on the other hand, the maximum value for the intrinsic rate of increase does assume considerable theoretical and practical significance, as has already been pointed out. The entomologist can readily determine the maximum values and this is his obvious starting-point. But the determination of r at different stages in the population history of an insect, whether in an experimental population or in the field, offers many practical difficulties which have not yet been surmounted for any single species. The values which the entomologist has difficulty in determining are those which are most readily obtained for human populations. The crude birth-rates and crude death-rates of the population at specific stages in its history are precisely those indices with which the demographer works. His census data provides him with the actual age distribution which is something not known empirically for a single insect species. He can have a knowledge of age distribution even at intercensal periods, and under civilized conditions he can also determine the age-specific rates of fecundity and mortality which were in operation during any particular year. In insect populations this is at present impossible; one can only keep a number of individuals under specified conditions and determine their age-specific rates of fecundity and survival, and from these data r can be calculated.

The fact that populations in nature may not realize the maximum value of their intrinsic rate of natural increase, does not negate the utility of this parameter either from a theoretical or a practical point of view. Having determined this parameter, the next logical step is to find out the extent to which this rate of increase is realized in nature. It is conceivable that some species, such as those which infest stored wheat or flour, may increase exponentially when liberated in vast quantities of these foodstuffs. This would imply that the insects could move out of the area in which they were multiplying with sufficient speed to escape density effects and that they had no gregarious tendencies. An exponential rate of increase may also occur in temperate climates in some plant-feeding species which only multiply in a short period of the year in the spring. In seasons with abundant plant growth the insect population may be far from approaching any limitation in the resources of the environment before the onset of summer retards the rate of increase. The population counts of *Thrips imaginis* in some favourable seasons in South Australia suggest such a picture (Davidson & Andrewartha, 1948).

3. CALCULATION OF THE INTRINSIC RATE OF NATURAL INCREASE

(a) *Experimental data required*

The calculation of r is based on the female population; the primary data required being as follows:

(1) The female life table giving the probability at birth of being alive at age x. This is usually designated l_x ($l_0 = 1$).

(2) The age-specific fecundity table giving the

mean number of female offspring produced in a unit of time by a female aged x. This is designated m_x.

In the calculation of the stable age distribution the age-specific survival rates (l_x) of both the immature stages and the reproductive stages are required. For the calculation of r the life table of the adult and only the total survival of the immature stages (irrespective of age) are needed. In practice, the age-specific fecundity rates m_x will be established for some convenient interval of age, such as a week. If N eggs are laid per female alive between the ages x to $x+1$ in the unit of time chosen, then m_x simply equals $\frac{1}{2}N$ when sex ratio is unity. It is assumed that this value occurs at the mid-point of the age group.

A numerical example is worked out for the rice weevil *Calandra* (*Sitophilus*) *oryzae* (L.) living under optimum conditions (29° C. in wheat of 14% moisture content). Data for the rates of development and survival of the immature stages, and the age-specific fecundity rates were obtained from Birch (1945*a*, *b*). The life table of adult females has not been determined experimentally, only the mean length of adult life being known. However, an estimate was obtained for purposes of these calculations by adapting the known life table of *Tribolium confusum* Duval (Pearl, Park & Miner, 1941) to *Calandra oryzae*, making the necessary reduction in the time scale. Since the mean length of life of *Tribolium confusum* in this life table was 198 days and the mean length of life of *Calandra oryzae* at 29° was 84 days, one '*Calandra* day' has been taken as equivalent to 2·35 '*Tribolium* days'. To this extent the example worked out is artificial, but, for reasons which will become evident later in the paper, it is unlikely that the error so introduced in the estimate of r is of much significance.

Before proceeding to outline direct methods of estimating r two other parameters must first be mentioned: the net reproduction rate and the mean length of a generation.

(*b*) *The net reproduction rate*

This is the rate of multiplication in one generation (Lotka, 1945) and is best expressed as the ratio of total female births in two successive generations. This we shall call R_0 and so follow the symbolism of the demographers. R_0 is determined from age-specific fecundity and survival rates and is defined as

$$R_0 \doteq \int_0^\infty l_x m_x dx,$$

where l_x and m_x are as already defined.

The method of calculating R_0 is set out in Table 1. The values of l_x are taken at the mid-point of each age group and age is given from the time the egg is laid. Since the survival rate of the immature stages was 0·90 the life table of adults reckoned from 'birth', i.e. oviposition, was the product: l_x for adults $\times$ 0·90. Development from the egg to emergence of the adult from the grain lasts 28 days and 4·5 weeks is the mid-point of the first week of egg laying. The product $l_x m_x$ is obtained for each age group and the sum of these products $\Sigma l_x m_x$ is the value R_0. In this particular example $R_0 = 113·6$. Thus a population of *Calandra oryzae* at 29° will multiply 113·6 times *in each generation.*

Table 1. *Showing the life table (for oviposition span) age-specific fecundity rates and the method of calculating the net reproduction rate* (R_0) *for* Calandra oryzae *at* 29° *in wheat of* 14% *moisture content. Sex ratio is equal*

Pivotal age in weeks (x)	(l_x)	(m_x)	($l_x m_x$)
4·5	0·87	20·0	17·400
5·5	0·83	23·0	19·090
6·5	0·81	15·0	12·150
7·5	0·80	12·5	10·000
8·5	0·79	12·5	9·875
9·5	0·77	14·0	10·780
10·5	0·74	12·5	9·250
11·5	0·66	14·5	9·570
12·5	0·59	11·0	6·490
13·5	0·52	9·5	4·940
14·5	0·45	2·5	1·125
15·5	0·36	2·5	0·900
16·5	0·29	2·5	0·800
17·5	0·25	4·0	1·000
18·5	0·19	1·0	0·190
			$R_0 = 113·560$

The comparison of two or more populations by means of their net reproduction rates may be quite misleading unless the mean lengths of the generations are the same. Two or more populations may have the same net reproduction rate but their intrinsic rates of increase may be quite different because of different lengths of their generations. Consider, for example, the effect of moving the $l_x m_x$ column in Table 1 up or down by a unit of age, R_0 remains the same but it is obvious that the generation times are now very different. For these reasons the parameter R_0 has limited value and it must always be considered in relation to the length of the generation (T).

(*c*) *The mean length of a generation*

The relation between numbers and time in a population growing exponentially is given by

$$N_T = N_0 e^{rT}.$$

When T = the mean length of a generation, then from the definition of net reproduction rate $N_T/N_0 = R_0$, hence

$$R_0 = e^{rT},$$

and

$$T = \frac{\log_e R_0}{r}.$$

It follows that an accurate estimate of the mean length of a generation cannot be obtained until the value of r is known. For many purposes, however, an approximate estimate of T which can be calculated independently of r may be of use. Thus, although oviposition by the female is extended over a period of time, it may be considered as concentrated for each generation at one point of time, successive generations being spaced T units apart (Dublin & Lotka, 1925). For approximate purposes therefore it may be defined as

$$T=\frac{\Sigma x l_x m_x}{\Sigma l_x m_x}.$$

We may thus consider the figures for the product $l_x m_x$ given in the last column of Table 1 as a frequency distribution of which the individual items are each concentrated at the mid-point of each age group. The mean of this distribution is the approximate value of T. In this particular example

$$T=\frac{943\cdot09}{113\cdot56}=8\cdot3 \text{ weeks.}$$

If this were an accurate estimate of T we could proceed to calculate the value of r since, from the above equation relating R_0, r and T, we have

$$r=\frac{\log_e R_0}{T}$$

$$=\frac{\log_e 113\cdot56}{8\cdot30}=0\cdot57 \text{ per head per week.}$$

It will become evident in what follows that this is an underestimate of r owing to the approximate estimate of T. The procedure does, however, serve to illustrate the nature of the parameter, and in some cases where r is small it may be a sufficiently accurate means of calculation (cf. for example, Dublin & Lotka, 1925). We shall proceed in the next section to an accurate method for the calculation of r.

(d) *The calculation of 'r'*

A population with constant age schedules of fecundity and mortality will gradually approach a fixed form of age distribution known as the stable age distribution (p. 16). Once this is established the population increases at a rate $dN/dt=rN$ and the value of r may be calculated from the equation

$$\int_0^\infty e^{-rx} l_x m_x dx=1.$$

For the derivation of this formula reference must be made to Lotka (1925) and the bibliography therein. The usual methods of calculation may be found in Dublin & Lotka (1925, Appendix) or Lotka (1939, p. 68 *et seq.*). For high values of r, these methods may not be particularly satisfactory (Leslie & Ranson, 1940; Leslie, 1945, Appendix), and the computations, moreover, become very tedious. Some approximations to the rigorous procedures are justified in so far as the determination of the primary data which enter the above formula is of course subject to considerable error, arising from the normal variation in the organisms and conditions to which they are subjected in the experiments. It was considered that an estimate of r, calculated to the second decimal place, was sufficient in these circumstances. The following approximate method was therefore adopted. It has the merit of being both simple and fast.

As an approximation we may write

$$\Sigma e^{-rx} l_x m_x=1.$$

Here x is taken to be the mid-point of each age group and the summation is carried out over all age groups for which $m_x>0$. A number of trial values are now substituted in this equation, in each case calculating a series of values e^{-rx} and multiplying them by the appropriate $l_x m_x$ values for each age group. By graphing these trial values of r against the corresponding summation values of the left-hand side of the above expression, we may find the value of r which will make

$$\Sigma e^{-rx} l_x m_x \rightarrow 1.$$

The whole procedure is greatly simplified by the use of 4-figure tables for powers of e (e.g. Milne-Thomson & Comrie 1944, Table 9). Since these tables only give the values of $e^{\pm x}$ at intervals of 0·01 in the argument x up to $e^{\pm 6}$, it may be convenient to multiply both sides of the equation by a factor e^k in order to work with powers of e which lie in the more detailed parts of the table. Thus, in the present example, k was taken as 7:

$$e^7 \Sigma e^{-rx} l_x m_x = e^7$$

$$\Sigma e^{7-rx} l_x m_x = 1097.$$

A value of r was now sought which would make the left-hand side of this expression equal to 1097. The actual process of carrying out this simple computation is exemplified in Table 2. The summation of the expression is not carried beyond the age group centred at 13·5 because of the negligible contribution of the older age groups. It has already been mentioned that r is an infinitesimal rate of increase not to be confused with a finite rate of increase λ which equals antilog$_e$ r. In this particular example $r=0\cdot76$ and λ therefore has a value 2·14. In other words the population will multiply 2·14 times *per week*.

By reference to Table 2 it is clear that the relative weights with which the different age groups contribute to the value of r are given by the values $l_x m_x e^{7-rx}$ at each age group. It is of particular interest to observe the relation between values at successive age intervals (Table 3). The value of r is 56% accounted for by the first week of adult life. The first 2 weeks combined contribute 85% towards the final value and the first 3 weeks combined total 94%. The 13·5th week, on the other hand, contributes 0·02%. It should not be inferred that adults 13·5 weeks old are of no importance since their eggs will eventually give rise to adults in the productive age categories. The biological significance of Table 3 is that the intrinsic rate of increase is determined to a much greater extent by the rate of oviposition in the first couple of weeks of adult life than by the total number of eggs laid in the life span of the adult, even although only 27 % of the total number of eggs are laid in the first 2 weeks. With

Table 2. *Showing the method of calculating r for* Calandra oryzae *at 29° by trial and error substitutions in the expression* $\Sigma e^{7-rx} l_x m_x = 1097$

Pivotal age group (x)	$l_x m_x$	$r=0{\cdot}76$		$r=0{\cdot}77$	
		$7-rx$	e^{7-rx}	$7-rx$	e^{7-rx}
4·5	17·400	3·58	35·87	3·53	34·12
5·5	19·090	2·82	16·78	2·76	15·80
6·5	12·150	2·06	7·846	1·99	7·316
7·5	10·000	1·30	3·669	1·22	3·387
8·5	9·875	0·54	1·716	0·45	1·5683
9·5	10·780	−0·22	0·8025	−0·32	0·7261
10·5	9·250	−0·98	0·3753	−1·09	0·3362
11·5	9·570	−1·74	0·1755	−1·86	0·1557
12·5	6·490	−2·50	0·0821	−2·62	0·0728
13·5	4·940	−3·26	0·0384	−3·39	0·0337
$\sum_{4\cdot5}^{13\cdot5} e^{7-rx} l_x m_x =$			1108		1047

r lies between 0·76 and 0·77 and by graphical interpretation = 0·762.

Table 3. *The contribution of each age group to the value of r when* $r=0{\cdot}76$

Pivotal age group (x)	$l_x m_x e^{7-rx}$	Percentage contribution of each age group
4·5	624·1	56·33
5·5	320·3	28·91
6·5	95·3	8·60
7·5	36·7	3·31
8·5	17·0	1·53
9·5	8·7	0·78
10·5	3·5	0·32
11·5	1·7	0·15
12·5	0·5	0·05
13·5	0·2	0·02
	1108·0	100·00

each successive week, eggs laid make a lessened contribution to the value of r. In this particular case this can be expressed by stating that for each egg laid in the first week of adult life it would require 2·1 times as many in the second week to make the same contribution to the value of r, $(2{\cdot}1)^2$ in the third week and $(2{\cdot}1)^{n-1}$ in the nth week. The ratio 2·1 : 1 is the ratio between successive weighting values e^{7-rx} (per egg) in Table 2. The importance of the first few weeks is further intensified by the fact that egg laying is at a maximum then. From these considerations it follows that in determining oviposition rates experimentally, the rates in early adult life should be found with the greatest accuracy. Of corresponding importance is the accurate determination of the pivotal age for the first age category in which eggs are laid. In the example being cited an error of half a week causes an error of 8 % in the estimate of r.

The calculations were repeated ignoring the adult life table. The value of r was then 0·77. Since the imposition of an adult life table only makes a difference of 1 % in the value of r it is evident that the life table is of little importance in this example. This is due to the fact already noted that the major contribution to the value of r is made by adults in early life, and during early adult life survival rate is at a maximum. The life table may assume quite a different importance in a species with a different type of age schedule of fecundity or when the value of r is lower.

4. THE STABLE AGE DISTRIBUTION

With a knowledge of the intrinsic rate of increase and the life table it is possible to calculate the stable age distribution and the stable female birth-rate of the population. Thus if c_x is the proportion of the stable population aged between x and $x+dx$, and b is the instantaneous birth-rate

$$c_x = be^{-rx} l_x,$$

and

$$1/b = \int_0^\infty e^{-rx} l_x dx.$$

For the usual methods of computation reference should be made again to Dublin & Lotka (1925). Mr Leslie has, however, pointed out to me another method of calculation which saves much of the numerical integration involved in the more usual methods. At the same time it is sufficiently accurate for our present purpose. If at time t we consider a stable population consisting of N_t individuals, and if during the interval of time t to $t+1$ there are B_t female births, we may define a birth-rate

$$\beta = B_t/N_t.$$

Then if we define for the given life table (l_x) the series of values L_x by the relationship $L_x = \int_x^{x+1} l_x dx$ (the stationary or 'life table' age distribution of the actuary),* the proportion (p_x) of individuals aged between x and $x+1$ in the stable population is given by

$$p_x = \beta L_x e^{-r(x+1)},$$

$$1/\beta = \sum_{x=0}^{m} L_x e^{-r(x+1)},$$

where $x=m$ to $m+1$ is the last age group considered in the complete life table age distribution. It will be noticed that the life table (l_x values) for the complete

* For a discussion of L_x see Dublin & Lotka (1936).

age span of the species are required for the computation of p_x and β. But where r is high it will be found that for the older age groups the terms $L_x e^{-r(x+1)}$ are so small and contribute so little to the value of β that they can be neglected.

The calculations involved are quite simple and are illustrated in the following example for *Calandra oryzae* at 29° (Table 4). Actually, in the present example, instead of calculating the values of L_x, the values of l_x were taken at the mid-points of each age group. This was considered sufficiently accurate in the present instance. It should also be pointed out that whereas only the total mortality of immature stages was required in the calculation of r, the age specific mortality of the immature stages is needed

Table 4. *Calculation of the stable age distribution of* Calandra oryzae *at 29° when $r = 0{\cdot}76$*

Age group (x)	L_x	$e^{-r(x+1)}$	$L_x e^{-r(x+1)}$	Percentage distribution $100\beta L_x e^{-r(x+1)}$	
0–	0·95	0·4677	0·4443150	54·740	95·5 % total immature stages
1–	0·90	0·2187	0·1968300	24·249	
2–	0·90	0·10228	0·0920520	11·341	
3–	0·90	0·04783	0·0430470	5·304	
4–	0·87	0·02237	0·0194619	2·398	4·5 % total adults
5–	0·83	0·01046	0·0086818	1·070	
6–	0·81	0·00489	0·0039609	0·488	
7–	0·80	0·002243	0·0017944	0·221	
8–	0·79	0·001070	0·0008453	0·104	
9–	0·77	0·000500	0·0003850	0·047	
10–	0·74	0·000239	0·0001769	0·022	
11–	0·66	0·000110	0·0000726	0·009	
12–	0·59	0·000051	0·0000301	0·004	
13–	0·52	0·000024	0·0000125	0·002	
14–	0·45	0·000011	0·0000050	0·001	
			$1/\beta = 0{\cdot}8116704$	100·000	

for the calculation of the stable age distribution. In this example the total mortality of the immature stages was 10 %—and 98 % of this mortality occurred in the first week of larval life (Birch, 1945*d*). Hence the approximate value of L_x for the mid-point of the first week will be 0·95 and thereafter 0·90 for successive weeks of the larval and pupal period (column 2, Table 4). The stable age distribution is shown in the fifth column of Table 4. This column simply expresses the fourth column of figures as percentages. It is of particular interest to note the high proportion of immature stages (95·5 %) in this theoretical population. This is associated with the high value of the intrinsic rate of natural increase. It emphasizes a point of practical importance in estimating the abundance of insects such as *C. oryzae* and other pests of stored products. The number of adults found in a sample of wheat may be quite a misleading representation of the true size of the whole insect population. Methods of sampling are required which will take account of the immature stages hidden inside the grains, such, for example, as the 'carbon dioxide index' developed by Howe & Oxley (1944). The nature of this stable age distribution has a bearing on another practical problem. It provides further evidence to that developed from a practical approach (Birch, 1946) as to how it is possible for *C. oryzae* to cause heating in vast bulks of wheat, when only a small density of adult insects is observed. It is not an unreasonable supposition that the initial rate of increase of insects in bulks of wheat may approach the maximum intrinsic rate of increase and therefore that the age distribution may approach the stable form. Nothing, however, is known about the actual age distribution in nature at this stage of an infestation.

5. THE INSTANTANEOUS BIRTH-RATE AND DEATH-RATE

We have already defined a birth-rate β by the expression

$$1/\beta = \sum_{x=0}^{m} L_x e^{-r(x+1)}.$$

This is not, however, the same as the instantaneous birth-rate (b) where $r = b - d$. In personal communications Mr Leslie has provided me with the following relationship between these two birth-rates.

$$b = \frac{r\beta}{e^r - 1}.$$

Thus, in the example for *C. oryzae*, we have $1/\beta = 0{\cdot}81167$ (Table 4), $r = 0{\cdot}76$ and thus $b = 0{\cdot}82$ and the difference between r and b is the instantaneous death-rate $(d) = 0{\cdot}06$.

The instantaneous birth-rate and death-rate are widely used by students of human populations. The insect ecologist is more likely to find greater use for the finite rate of increase λ (natural antilog$_e$).

6. THE EFFECT OF TEMPERATURE ON 'r'

As an illustration of the way in which the value of r varies with temperature and the corresponding changes in rate of development, survival and fecundity, an estimate of r for *C. oryzae* has been made for two temperatures (23° and 33·5° C.) on either side of the optimum (29°). The span of adult life at 23° is about the same as at the optimum 29° and so the same life table has been applied. Even although the egg laying is more evenly distributed throughout adult life the life table makes little difference to the value of r. Furthermore, the first 2 weeks of adult life carry a weight of 59 % of the total weight of all age groups in the determination of the value of r. For every egg laid in the first week of adult life it would require 1·5 times as many eggs in the second week to make the same contribution to the value of r, and 2·3 times as many eggs in the third week to have the same effect. The relative weight of each week decreases less with successive weeks at 23° than at 29°. This is associated with the lower oviposition rates and the longer duration of the immature stages at 23°.

At 33·5° egg laying ceases after the fourth week of adult life, the mortality of adults during these 4 weeks is not high and so the estimate of r obtained without a life table may not be very different from the true value.

Table 5. *Showing the values of l_x, m_x and the estimate of r for* Calandra oryzae *at* 23° *and* 33·5°

23°			33·5°		
Pivotal age in weeks (x)	l_x	m_x	Pivotal age in weeks (x)	l_x	m_x
0·5 ... 6·5 Immature stages	0·90	—	0·5 ... 8·5 Immature stages	0·25	—
7·5	0·87	9·0	9·5	0·25	6·0
8·5	0·83	11·0	10·5	0·25	3·5
9·5	0·81	11·5	11·5	0·25	3·0
10·5	0·80	12·0	12·5	0·25	1·0
11·5	0·79	11·5			
12·5	0·77	13·0	r=0·12 per head per week		
13·5	0·74	11·5			
14·5	0·66	11·0			
15·5	0·60	10·0			
16·5	0·52	11·0			
17·5	0·45	12·5			
18·5	0·36	10·5			
19·5	0·29	11·5			
20·5	0·25	4·0			
21·5	0·19	2·0			

With adult life table r=0·43 per head per week.
Without adult life table r=0·44 " "

7. DISCUSSION

In order for a species to survive in a particular environment it may need to have evolved a certain minimum value for its intrinsic rate of natural increase. If its rate of increase is less than this it may succumb in the struggle for existence. It does not necessarily follow that the higher the intrinsic rate of increase the more successful will the species be. Evolution may operate to select species with an intrinsic rate of increase which is both large enough to enable them to compete successfully with other species and small enough to prevent a rate of multiplication which would exhaust the food supply in the environment. Whatever is the minimum necessary value of 'r' it could be attained along more than one route, since r has a number of component variables; the length of development of the immature stages, the survival rate of the immature stages, the adult life table and the age-specific fecundity schedule. These components enter into the value of r with various weights, and it is suggested in the discussion which follows that a knowledge of their relative contributions may provide a clue to the significance of the life patterns characteristic of different species. There is clearly a pattern in the seasonal environment too, which must be considered at the same time. A hot dry period, for example, may necessitate a prolonged egg stage. In an environment which has relatively uniform physical conditions all the year round, these complicating factors are at a minimum, e.g. a tropical forest or the micro-environment of a stack of wheat.

(1) Consider first the length of the immature stages (non-reproductive period) in relation to the span of egg laying and the age schedule of fecundity. The earlier an egg is laid in the life of the insect the greater is the contribution of that particular egg to the value of r. In illustration we may consider the age schedule of fecundity for *C. oryzae* at 29°. Since over 95% of the value of r is determined by the eggs laid in the first 4 weeks of adult life (Table 3) we can, for purposes of illustration, ignore the remaining period.

At 29° the immature stages of *C. oryzae* last 4 weeks and the maximum rate of egg laying is 46 eggs per week (Table 1). Now the same value of r (0·76) is given in a number of imaginary life cycles by reducing the length of the immature stages along with a reduction in the rate of egg laying and alternatively by increasing both the length of the immature stages and the number of eggs laid (ordinary figures, Table 6). In Table 6 the age schedule of fecundity is kept proportionate in each case. In the extreme examples if the immature stages could develop in a week, an oviposition maximum of 5 eggs per week would give the same rate of increase as the imaginary insect which took 6 weeks to develop and had an oviposition maximum of 204 eggs per week. The imaginary life cycles have been calculated from the ratio (2:1:1) for successive weighting values (e^{7-rx}) in Table 2.

The question might now be asked, what determines the particular combination which the species happens to possess? In the specific example in question, if the larva took 6 weeks to develop the adult would need to lay 200 eggs per week. But it now becomes necessary to consider the behaviour pattern, for *C. oryzae* bores a hole in the grain of wheat for every egg which is laid. The whole process of boring and egg laying occupies about 1 hr. per egg. So that with this particular mode of behaviour 24 eggs per day would be an absolute maximum. There must of course also be some physiological limit to egg production. For a larger insect the physiological limit might be less restrictive provided that the size of the egg does not increase proportionately with the size of the insect. In considering this possibility, ecological considerations are important, for *C. oryzae* is adapted to complete its development within a grain of wheat and a size limit is set by the length of the grain. There is, in fact, a strain which is found in maize kernels in Australia and this is considerably larger than the so-called 'small strain' (Birch, 1944). Furthermore, the larger insect would probably require a longer time to complete development (which is actually the relationship observed between the small and large strains) and this would operate to reduce the value of r. In considering the possibilities in the opposite direction, there is obviously a limit below which the length of development could not be reduced any further. A species of smaller size could doubtless develop in a shorter time and on this merit might be a more successful mutation. But the question then arises whether a smaller species could command muscles and mandibles of sufficient strength to chew whole grain. Grain-feeding species of beetles which are smaller than *C. oryzae* are in fact scavengers rather than feeders on sound grain. Thus it would seem that a balance is struck somewhere between the minimum time necessary for development and the maximum possible rate of egg laying, and this is conditioned by the behaviour pattern of the insect and the particular ecology of its environment.

For a maximum value of r the optimum age schedule of fecundity is one which has an early maximum. In an imaginary schedule for *C. oryzae* a concentration of 71 eggs in the first week of egg laying would give the same value of r (0·76) as 141 eggs distributed over 4 weeks.

Age schedule of fecundity of Calandra oryzae *at 29°*

Weeks	1	2	3	4	Total
Actual	40	46	30	25	141
Imaginary	71	0	0	0	71

The relative advantage of this type of fecundity schedule is less, the smaller the value of r. At 23° the value of r for *C. oryzae* is 0·43 and the eggs laid in the first week of adult life are worth $(1{\cdot}5)^{n-1}$ eggs in the nth week (compare this with the value of $(2{\cdot}1)^{n-1}$ when $r=0{\cdot}76$). The actual oviposition time curve at 23° has no distinct peak as at 29° (Tables 1 and 5).

There is a wide variation in the nature of the age schedule of fecundity amongst different species of insects with perhaps the tsetse fly and the lucerne flea illustrating contrasting extremes. Whereas tsetse flies (*Glossina*) deposit single larvae spaced at intervals of time, the 'lucerne flea' (*Smynthurus viridis*) deposits its eggs in one or two batches of as many as 120 at a time (Maclagan, 1932). The particular advantage of this mode of oviposition must be tremendous, and is probably responsible in part for the great abundance of this collembolan and possibly other members of the same order, which, as a whole, are among the most abundant insects in nature. This is of course speculative and much more information is required before any generalizations can be made. Another interesting category are the social insects, since only one female of the population (in termites and the hive-bee) or a few (in social wasps) are reproductives. A theoretical consideration of the relative merits of one queen and many queens might throw more light on the evolution of these systems, especially as they relate to differences in behaviour.

The relation between the length of the pre-reproductive stages and the nature of the age-fecundity schedule is in part dependent upon the nature of the seasonal changes in the environment. Life histories may be timed so that the reproductive and feeding stages coincide with the least hostile season of the year. Diapause, aestivation and hibernation are some of the adaptations which ensure this. They have particular significance too in determining the age distribution of the initial population in the reproductive

Table 6. *Showing the actual relation between the length of the immature stages and the age schedule of fecundity for* Calandra oryzae *at 29° (black figures) and some theoretical possibilities which would give the same intrinsic rate of increase ($r=0{\cdot}76$). The length of the immature stages is shown in the left of the table; figures in the body of the table are number of eggs per week*

	Pivotal age in weeks									
0·5	1·5	2·5	3·5	4·5	5·5	6·5	7·5	8·5	9·5	Total
1 week	4	5	3	3	—	—	—	—	—	15
2 weeks		9	10	7	6	—	—	—	—	32
3 weeks			19	22	14	12	—	—	—	67
4 weeks				**40**	**46**	**30**	**25**	—	—	**141**
5 weeks					84	97	63	53	—	297
6 weeks						117	204	132	111	564

season. Consideration of this is left to a later section of the discussion.

(2) For the calculation of the maximum intrinsic rate of increase the life table of the species from deposition of the egg (or larva) to the end of egg-laying life in the adult must be known. The starting-point of this life table is thus the stage which corresponds to the point of 'birth'. Deevey (1947) has noted that the point of universal biological equivalence for animals is doubtless fertilization of the ovum. But, from the point of view of the number of animals in the population and for purposes of calculating r, a knowledge of pre-birth mortality is not required. For the calculation of r the life table beyond the end of reproductive life has no significance, but a knowledge of age-specific post-reproductive survival until the point of death is needed, on the other hand, for the calculation of the stable age distribution and the instantaneous birth-rate. This is evident from a consideration of the method of calculation shown in Table 4. The post-reproductive life assumes negligible significance in this particular example, but its importance in such calculations increases as r approaches a value of zero.

The relative importance of the survival pattern (i.e. the shape of the l_x curve) in determining the value of r is itself a function of r. When r is small its value may be dependent to a significant extent on the oviposition in late adult life, when it is large it is mostly determined by the oviposition rates of adults in early adult life. When the intrinsic rate of increase is high and the life table of the adult follows the typical diagonal pattern (e.g. Pearl *et al.* 1941) with no high mortality in early adult life, consideration of the adult life table is of little importance in calculating r. This is because survival rate is high in the ages which contribute most to the value of r. In species which have a low intrinsic rate of increase the life table may assume more significance in determining the value of that rate of increase. More data are required before the importance of this point can be established. The pattern of survival which gives a maximum value of r has its maxima in the pre-reproductive and early reproductive stages. A knowledge of total survival of the immature stages is of course essential in all cases. More attention might well be given by entomologists to securing life table data than has been given in the past. Without it no true picture of the intrinsic rate of increase can be obtained.

(3) There remains to be considered the age distribution of the population in relation to its capacity to increase in numbers. In a population in an unlimited environment the stable age distribution is the only one which gives an unvarying value of r. For this reason the stable age distribution is the only sound basis on which to make comparisons between different values for rates of increase (whether between different species or one species under different physical conditions). The actual age distribution of a population in nature may be quite different and its consideration is of importance in determining the initial advantage one form of distribution has over another. In an unlimited environment these initial differences in age composition are eventually ironed out. A population which initiates from a number of adults at the peak age of egg laying clearly has a higher initial rate of increase than one which starts from the same number in all different stages of development.

These considerations may be of most importance in temperate climates where there is a definite seasonal occurrence of active stages. The stage in which the insect overwinters or oversummers will determine the age distribution of the population which initiates the seasonal increase in the spring or in the autumn (whichever the case may be). The pea weevil, *Bruchus pisorum*, in California hibernates as an adult. With the first warm days in the spring the adults leave their overwintering quarters under bark and fly into the pea fields (Brindley, Chamberlin & Hinman, 1946). Following a meal of pollen they commence oviposition on the pea crops. This mode of initiating the spring population would be far more effective than one which started with the same number of insects in the egg stage. The overwintering adults begin their reproductive life at a much later age than the adults in the next generation. It would be of interest to know whether the age schedule of fecundity (taking the commencement of egg laying as zero age) is the same for both generations. This is a point which does not appear to have been investigated for insects which hibernate as adults. It is clearly of much importance in determining the intrinsic rate of increase of successive generations.

Overwintering as pupae must theoretically rank as the second most effective age distribution for initiating spring increase. Many species which overwinter as pupae would have a higher mortality if they overwintered as adults. The corn-ear worm, *Heliothis armigera*, for example, can hardly be conceived as overwintering as an adult moth in the North American corn belt. In the northern part of this belt even the pupae which are protected in the soil are unable to survive the winter. Recolonization evidently takes place each year from the warmer south (Haseman, 1931).

Overwintering in the egg stage (in hibernation or in diapause) is common in insects. Here again it is difficult to imagine the other stages of these orders as successfully hibernating. The grasshopper, *Austroicetes cruciata* (Andrewartha, 1944), and the majority of aphids are examples of this. A minority of aphid species are, however, able to overwinter as apterae by finding protection in leaf axils and similar niches (Theobald, 1926), some others are enabled to survive as adults by virtue of their symbiosis with ants (Cutright, 1925). The relatively vulnerable aphids find protection in the nests of ants. The ants not only carry them to their nests, but feed them during the winter months and with the return of spring plant them out on trees again!

Overwintering as nymphs or larvae is a rarer phenomenon except with species which can feed and grow at low temperatures and so are not hibernators. The active stages of the lucerne flea, *Smynthurus viridis*, for example, can be found in the winter in Australia (Davidson, 1934). The seasonal cycle of the reproducing population commences with the first rains in autumn; this population being initiated with oversummering eggs. The eggs are the only stage which are resistant to the dryness and high temperatures of the summer months. Species of insects with hardy adult stages like the weevil, *Otiorrhynchus cribricollis* (Andrewartha, 1933), aestivate as adults. An interesting case of a butterfly, *Melitaea phaeton*, aestivating as a quarter-grown larva at the base of its food plant is described by Hovanitz (1941).

The preceding examples illustrate how the age distributions of initiating populations vary in seasonal species. This depends on the nature of the overwintering or oversummering stage. The particular stage may have been selected in nature not only by virtue of its resistance to unfavourable physical conditions but also in relation to its merits in initiating rapid establishment of a population in the spring and autumn. The calculation of the initial and subsequent rates of increase of populations with these different types of age distribution is considerably more complicated than the calculation of intrinsic rates of increase for populations with stable age distributions. This problem is not dealt with in this paper and the reader is referred to Leslie (1945, p. 207 *et seq.*) for an outline of the principles involved in such calculations.

The length of the developmental stages, the age schedule of fecundity, the life table of the species and the age distribution of initiating populations present a pattern which has adaptive significance for the species. The analytic study of the intrinsic rate of increase of a species (as exemplified by *Calandra oryzae*) may throw light on the evolutionary significance of the life pattern of different species. Such a study must necessarily be related to the behaviour pattern of the insect and the type of environment it lives in. Nor can the importance of effects of density and competition be overlooked. These are, of course, studies in themselves beyond the scope of this paper.

8. SUMMARY

The parameter known as the intrinsic rate of natural increase, which was developed for demographic analyses by A. J. Lotka, is introduced as a useful concept for the study of insect populations. It is suggested that for the sake of uniformity of terminology in population biology and for precision of definition, that the term 'intrinsic rate of natural increase' might be considered more appropriate than an alternative term 'biotic-potential' which is more frequently used in relation to insect populations. The intrinsic rate of natural increase is defined as the exponent 'r' in the exponential equation for population increase in an unlimited environment. The rate of increase of such a population is given by $dN/dt = rN$. The parameter r refers to the rate of increase of a population with a certain fixed age distribution known as the stable age distribution. Both the intrinsic rate of natural increase and the stable age distribution may be calculated from the age-specific survival rates (life table) and age-specific fecundity rates. The methods of calculation are exemplified with data for the rice weevil, *Calandra oryzae* (L.), and some adapted from the flour beetle, *Tribolium confusum* Duval. It is shown in this example that the intrinsic rate of natural increase is determined to a much greater extent by the rate of oviposition in the first 2 weeks of adult life than by the total number of eggs laid in the entire life time. The oviposition rates in the first 2 weeks account for 85 % of the value of r whereas only 27 % of the total number of eggs are laid in that time. With each successive week in the life of the adult, eggs laid make a lessened contribution to the value of r. The methods of calculation of r provide a means of determining the extent to which the various components—the life table, the fecundity table and the length of the pre-reproductive stages—enter into the value of r. It is suggested that analyses of this sort may provide a clue to the life patterns characteristic of different species.

The importance of the age distribution of populations which initiate seasonal increase in the autumn and spring is discussed. These age distributions depend on the nature of the overwintering or oversummering stage. It is suggested that this particular stage, whether it be adult, larva, pupa or egg, has been selected by virtue not only of its resistance to the unfavourable season, but also in relation to its merits in initiating rapid establishment of a population in the succeeding season.

It is shown how the value of r for *Calandra oryzae* varies with temperature. Four other parameters are also defined: the net reproduction rate, the mean length of a generation, the infinitesimal birth-rate and the infinitesimal death-rate. The methods of calculation of these parameters are also exemplified with data for *C. oryzae*.

9. ACKNOWLEDGEMENTS

Grateful acknowledgement is made to the Director of the Bureau of Animal Population, Oxford University, Mr C. S. Elton, for the facilities of the Bureau

26 *The intrinsic rate of natural increase of an insect population*

which were placed at the author's disposal during his term there as a visiting worker. Mr Elton provided much encouragement during the investigation. It is a pleasure to acknowledge too the inspiration and help of Mr P. H. Leslie of the Bureau of Animal Population. His direction was indispensable in all mathematical and actuarial aspects of the paper and his critical examination of the manuscript was much to its advantage.

REFERENCES

Andrewartha, H. G. (1933). 'The bionomics of *Otiorrhynchus cribricollis* Gyll.' Bull. Ent. Res. 24: 373–84.

Andrewartha, H. G. (1944). 'The distribution of plagues of *Austroicetes cruciata* Sauss. (Acrididae) in relation to climate, vegetation and soil.' Trans. Roy. Soc. S. Aust. 68: 315–26.

Birch, L. C. (1944). 'Two strains of *Calandra oryzae* L. (Coleoptera).' Aust. J. Exp. Biol. Med. Sci. 22: 271–5.

Birch, L. C. (1945*a*). 'The influence of temperature on the development of the different stages of *Calandra oryzae* L. and *Rhizopertha dominica* Fab. (Coleoptera).' Aust. J. Exp. Biol. Med. Sci. 23: 29–35.

Birch, L. C. (1945*b*). 'The influence of temperature, humidity and density on the oviposition of the small strain of *Calandra oryzae* L. and *Rhizopertha dominica* Fab.' Aust. J. Exp. Biol. Med. Sci. 23: 197–203.

Birch, L. C. (1945*c*). 'The biotic potential of the small strain of *Calandra oryzae* and *Rhizopertha dominica*.' J. Anim. Ecol. 14: 125–7.

Birch, L. C. (1945*d*). 'The mortality of the immature stages of *Calandra oryzae* L. (small strain) and *Rhizopertha dominica* Fab. in wheat of different moisture contents.' Aust. J. Exp. Biol. Med. Sci. 23: 141–5.

Birch, L. C. (1946). 'The heating of wheat stored in bulk in Australia.' J. Aust. Inst. Agric. Sci. 12: 27–31.

Brindley, T. A., Chamberlin, J. C. & Hinman, F. G. (1946). 'The pea weevil and methods for its control.' U.S. Dept. Agric. Farmers' Bull. 1971: 1–24.

Chapman, R. N. (1931). 'Animal ecology with especial reference to insects.' New York.

Crombie, A. C. (1945). 'On competition between different species of graminivorous insects.' Proc. Roy. Soc. B, 132: 362–95.

Crombie, A. C. (1947). 'Interspecific competition.' J. Anim. Ecol. 16: 44–73.

Cutright, C. R. (1925). 'Subterranean aphids of Ohio.' Ohio Agric. Exp. Sta. Bull. 387: 175–238.

Davidson, J. (1934). 'The "lucerne flea" *Smynthurus viridis* L. (Collembola) in Australia.' Bull. Coun. Sci. Industr. Res. Aust. 79: 1–66.

Davidson, J. & Andrewartha, H. G. (1948). 'Annual trends in a natural population of *Thrips imaginis* Bagnall (Thysanoptera).' (In the Press.)

Deevey, E. S. (1947). 'Life tables for natural populations of animals.' Biometrics, 3: 59–60.

Dublin, L. I. & Lotka, A. J. (1925). 'On the true rate of natural increase as exemplified by the population of the United States, 1920.' J. Amer. Statist. Ass. 20: 305–39.

Dublin, L. I. & Lotka, A. J. (1936). 'Length of life.' New York.

Gause, G. F. (1934). 'The struggle for existence.' Baltimore.

Haseman, L. (1931). 'Outbreak of corn earworm in Missouri.' J. Econ. Ent. 24: 649–50.

Hovanitz, W. (1941). 'The selective value of aestivation and hibernation in a Californian butterfly.' Bull. Brooklyn Ent. Soc. 36: 133–6.

Howe, R. W. & Oxley, T. A. (1944). 'The use of carbon dioxide production as a measure of infestation of grain by insects.' Bull. Ent. Res. 35: 11–22.

Leslie, P. H. (1945). 'On the use of matrices in certain population mathematics.' Biometrika, 33: 183–212.

Leslie, P. H. & Ranson, R. M. (1940). 'The mortality, fertility and rate of natural increase of the vole (*Microtus agrestis*) as observed in the laboratory.' J. Anim. Ecol. 9: 27–52.

Lotka, A. J. (1925). 'Elements of physical biology.' Baltimore.

Lotka, A. J. (1927). 'The size of American families in the eighteenth century and the significance of the empirical constants in the Pearl-Reed law of population growth.' J. Amer. Statist. Ass. 22: 154–70.

Lotka, A. J. (1939). 'Théorie analytique des associations biologiques. Deuxième Partie. Analyse démographique avec application particulière à l'espèce humaine.' Actualités Sci. Industr. 780: 1–149.

Lotka, A. J. (1945). 'Population analysis as a chapter in the mathematical theory of evolution.' In LeGros Clark, W. E. & Medawar, P. B., 'Essays on Growth and Form', 355–85. Oxford.

Maclagan, D. S. (1932). 'An ecological study of the "lucerne flea" (*Smynthurus viridis*, Linn.)–I.' Bull. Ent. Res. 23: 101–90.

Pearl, R., Park, T. & Miner, J. R. (1941). 'Experimental studies on the duration of life. XVI. Life tables for the flour beetle *Tribolium confusum* Duval.' Amer. Nat. 75: 5–19.

Stanley, J. (1946). 'The environmental index, a new parameter as applied to *Tribolium*.' Ecology, 27: 303–14.

Theobald, F. V. (1926). 'The plant lice or the Aphididae of Great Britain.' Vol. 1. Ashford.

Thompson, W. R. (1931). 'On the reproduction of organisms with overlapping generations.' Bull. Ent. Res. 22: 147–72.

Thomson, L. M. Milne- & Comrie, L. J. (1944). 'Standard four-figure mathematical tables.' London.

Volterra, V. (1931). 'Leçons sur la théorie mathématique de la lutte pour la vie.' Paris.

The Components of Predation as Revealed by a Study of Small-Mammal Predation of the European Pine Sawfly[1]

By C. S. Holling

Forest Insect Laboratory, Sault Ste. Marie, Ont.

INTRODUCTION

The fluctuation of an animal's numbers between restricted limits is determined by a balance between that animal's capacity to increase and the environmental checks to this increase. Many authors have indulged in the whimsy of calculating the progressive increase of a population when no checks were operating. Thus Huxley calculated that the progeny of a single *Aphis* in the course of 10 generations, supposing all survived, would "contain more ponderable substance than five hundred millions of stout men; that is, more than the whole population of China", (in Thompson, 1929). Checks, however, do occur and it has been the subject of much controversy to determine how these checks operate. Certain general principles—the density-dependence concept of Smith (1955), the competition theory of Nicholson (1933)—have been proposed both verbally and mathematically, but because they have been based in part upon untested and restrictive assumptions they have been severely criticized (e.g. Andrewartha and Birch 1954). These problems could be considerably clarified if we knew the mode of operation of each process that affects numbers, if we knew its basic and subsidiary components. Predation, one such process, forms the subject of the present paper.

Many of the published studies of predation concentrate on discrete parts rather than the whole process. Thus some entomologists are particularly interested in the effect of selection of different kinds of prey by predators upon the evolution of colour patterns and mimicry; wildlife biologists are similarly interested in selection but emphasize the role predators play in improving the condition of the prey populations by removing weakened animals. While such specific problems should find a place in any scheme of predation, the main aim of the present study is to elucidate the components of predation in such a way that more meaning can be applied to considerations of population dynamics. This requires a broad study of the whole process and in particular its function in affecting the numbers of animals.

Such broad studies have generally been concerned with end results measured by the changes in the numbers of predator and prey. These studies are particularly useful when predators are experimentally excluded from the environment of their prey, in the manner adopted by DeBach and his colleagues in their investigations of the pests of orchard trees in California. This work, summarized recently (DeBach, 1958) in response to criticism by Milne (1957), clearly shows that in certain cases the sudden removal of predators results in a rapid increase of prey numbers from persistently low densities to the limits of the food supply. Inasmuch as these studies have shown that other factors have little regulatory function, the predators appear to be the principal ones responsible for regulation. Until the components of predation are revealed by an analysis of the processes leading to these end results, however, we will never know whether the conclusions from such studies apply to situations other than the specific predator—prey relationship investigated.

Errington's investigations of vertebrate predator—prey situations (1934, 1943, 1945 and 1956) suggest, in part, how some types of predation operate. He has

[1]Contribution from the Dept. of Zoology, University of British Columbia and No. 547, Forest Biology Division, Research Branch, Department of Agriculture, Ottawa, Canada. Delivered in part at the Tenth International Congress of Entomology, Montreal, 1956.

postulated that each habitat can support only a given number of animals and that predation becomes important only when the numbers of prey exceed this "carrying capacity". Hence predators merely remove surplus animals, ones that would succumb even in the absence of natural enemies. Errington exempts certain predator-prey relations from this scheme, however, and quotes the predation of wolves on deer as an example where predation probably is not related to the carrying capacity of the habitat. However logical these postulates are, they are only indirectly supported by the facts, and they do not explain the processes responsible.

In order to clarify these problems a comprehensive theory of predation is required that on the one hand is not so restrictive that it can only apply in certain cases and on the other not so broad that it becomes meaningless. Such a comprehensive answer requires a comprehensive approach, not necessarily in terms of the number of situations examined but certainly in terms of the variables involved, for it is the different reactions of predators to these variables that produce the many diverse predator-prey relations. Such a comprehensive approach is faced with a number of practical difficulties. It is apparent from the published studies of predation of vertebrate prey by vertebrate predators that not only is it difficult to obtain estimates of the density of predator, prey, and destroyed prey, but also that the presence of many interacting variables confuses interpretation.

The present study of predation of the European pine sawfly, *Neodiprion sertifer* (Geoff.) by small mammals was particularly suited for a general comprehensive analysis of predation. The practical difficulties concerning population measurement and interpretation of results were relatively unimportant, principally because of the unique properties of the environment and of the prey. The field work was conducted in the sand-plain area of southwestern Ontario where Scots and jack pine have been planted in blocks of up to 200 acres. The flat topography and the practice of planting trees of the same age and species at standard six-foot spacings has produced a remarkably uniform environment. In addition, since the work was concentrated in plantations 15 to 20 years of age, the closure of the crowns reduced ground vegetation to a trace, leaving only an even layer of pine needles covering the soil. The extreme simplicity and uniformity of this environment greatly facilitated the population sampling and eliminated complications resulting from changes in the quantity and kind of alternate foods of the predators.

The investigations were further simplified by the characteristics of the prey. Like most insects, the European pine sawfly offers a number of distinct life-history stages that might be susceptible to predation. The eggs, laid in pine needles the previous fall, hatch in early spring and the larvae emerge and feed upon the foliage. During the first two weeks of June the larvae drop from the trees and spin cocoons within the duff on the forest floor. These cocooned sawflies remain in the ground until the latter part of September, when most emerge as adults. A certain proportion, however, overwinter in cocoons, to emerge the following autumn. Observations in the field and laboratory showed that only one of these life-history stages, the cocoon, was attacked by the small-mammal predators, and that the remaining stages were inacessible and/or unpalatable and hence completely escaped attack. These data will form part of a later paper dealing specifically with the impact of small mammal predation upon the European pine sawfly.

Cocooned sawflies, as prey, have some very useful attributes for an investigation of this kind. Their concentration in the two-dimensional environment of the duff-soil interface and their lack of movement and reaction to predators considerably simplify sampling and interpretation. Moreover, the small mammals' habit of making a characteristically marked opening in the cocoon to permit removal of the insect leaves a relatively permanent record in the ground of the number of cocooned sawflies destroyed. Thus, the density of the destroyed prey can be measured at the same time as the density of the prey.

Attention was concentrated upon the three most numerous predators—the masked shrew, *Sorex cinereus cinereus* Kerr, the short-tail shrew, *Blarina brevicauda talpoides* Gapper, and deer mouse, *Peromyscus maniculatus bairdii* Hoy and Kennicott. It soon became apparent that these species were the only significant predators of the sawfly, for the remaining nine species trapped or observed in the plantations were either extremely rare or were completely herbivorous.

Here, then, was a simple predator-prey situation where three species of small mammals were preying on a simple prey—sawfly cocoons. The complicating variables present in most other situations were either constant or absent because of the simple characteristics of the environment and of the prey. The absence or constancy of these complicating variables facilitated analysis but at the expense of a complete and generally applicable scheme of predation. Fortunately, however, the small-mammal predators and the cocoons could easily be manipulated in laboratory experiments so that the effect of those variables absent in the field situation could be assessed. At the same time the laboratory experiments supported the field results. This blend of field and laboratory data provides a comprehensive scheme of predation which will be shown to modify present theories of population dynamics and to considerably clarify the role predators play in population regulation.

I wish to acknowledge the considerable assistance rendered by a number of people, through discussion and criticism of the manuscript: Dr. I. McT. Cowan, Dr. K. Graham and Dr. P. A. Larkin at the University of British Columbia and Dr. R. M. Belyea, Mr. A. W. Ghent and Dr. P. J. Pointing, at the Forest Biology Laboratory, Sault Ste. Marie, Ontario.

FIELD TECHNIQUES

A study of the interaction of predator and prey should be based upon accurate population measurements, and in order to avoid superficial interpretations, populations should be expressed as numbers per unit area. Three populations must be measured—those of the predators, prey, and destroyed prey. Thus the aim of the field methods was to measure accurately each of the three populations in terms of their numbers per acre.

Small-Mammal Populations

Since a complete description and evaluation of the methods used to estimate the density of the small-mammal predators forms the basis of another paper in preparation, a summary of the techniques will suffice for the present study.

Estimates of the number of small mammals per acre were obtained using standard live-trapping techniques adapted from Burt (1940) and Blair (1941). The data obtained by marking, releasing and subsequently recapturing animals were analysed using either the Lincoln index (Lincoln, 1930) or Hayne's method for estimating populations in removal trapping procedures (Hayne, 1949). The resulting estimates of the number of animals exposed to traps were converted to per acre figures by calculating, on the basis of measurements of the home range of the animals (Stickel, 1954), the actual area sampled by traps.

The accuracy of these estimates was evaluated by examining the assumptions underlying the proper use of the Lincoln index and Hayne's technique and by comparing the efficiency of different traps and trap arrangements. This analysis showed that an accurate estimate of the numbers of *Sorex* and *Blarina* could be obtained using Hayne's method of treating the data obtained from trapping with bucket traps. These estimates, however, were accurate only when the populations had not been disturbed by previous trapping. For *Peromyscus*, Lincoln-index estimates obtained from the results of trapping with Sherman traps provided an ideal way of estimating numbers that was both accurate and unaffected by previous trapping.

N. sertifer Populations

Since small-mammal predation of *N. sertifer* was restricted to the cocoon stage, prey populations could be measured adequately by estimating the number of cocoons containing living insects present immediately after larval drop in June. This estimate was obtained using a method outlined and tested by Prebble (1943) for cocoon populations of the European spruce sawfly, *Gilpinia hercyniae* (Htg.), an insect with habits similar to those of *N. sertifer*. Accurate estimates were obtained when cocoons were collected from sub-samples of litter and duff distributed within the restricted universe beneath the crowns of host trees. This method was specially designed to provide an index of population rather than an estimate of numbers per acre. But it is obvious from this work that any cocoon-sampling technique designed to yield a *direct* estimate of the number of cocoons per acre would require an unpractically large number of sample units. It proved feasible in the present study, however, to convert such estimates from a square-foot to an acre basis, by stratifying the forest floor into three strata, one comprising circles with two-foot radii around tree trunks, one comprising intermediate rings with inner radii two feet and outer radii three feet, and one comprising the remaining area (three to five feet from the tree trunks).

At least 75 trees were selected and marked throughout each plantation, and one or usually two numbered wooden stakes were placed directly beneath the crown of each tree, on opposite sides of the trunk. Stakes were never placed under overlapping tree crowns. The four sides of each stake were lettered from A to D and the stake was placed so that the numbered sides bore no relation to the position of the trunk. Samples were taken each year, by collecting cocoons from the area delimited by one-square-foot frames placed at one corner of each stake. In the first year's sample the frames were placed at the AB corner, in the second year's at the BC corner, etc. Different-sized screens were used to separate the cocoons from the litter and duff.

Cocoons were collected in early September before adult sawflies emerged and those from each quadrat were placed in separate containers for later analysis. These cocoons were analysed by first segregating them into "new" and "old" categories. Cocoons of the former category were a bright golden colour and were assumed to have been spun in the year of sampling, while those of the latter were dull brown in colour and supposedly had been spun before the sampling year. These assumptions proved partly incorrect, however, for some of the cocoons retained their new colour for over one year. Hence the "new" category contained enough cocoons that had been spun before the sampling year to prevent its use, without correction, as an estimate of the number of cocoons spun in the year of sampling. A correction was devised, however, which reduced the error to negligible proportions.

This method provided the best available estimate of the number of healthy cocoons per acre present in any one year. The population figures obtained ranged from 39,000 (Plot 1, 1954) to 1,080,000 (Plot 2, 1952) cocoons per acre.

Predation

Small-mammal predation has a direct and indirect effect on *N. sertifer* populations. The direct effect of predation is studied in detail in this paper. The indirect effect, resulting from the mutual interaction of various control factors (parasites, disease, and predators) has been discussed in previous papers (Holling, 1955, 1958b).

The direct effect of predation was measured in a variety of ways. General information was obtained from studies of the consumption of insects by caged animals and from the analysis of stomach contents obtained from animals trapped in sawfly-infested plantations. More particular information was obtained from the analysis of cocoons collected in the regular quadrat samples and from laboratory experiments which studied the effect of cocoon density upon predation.

The actual numbers of *N. sertifer* cocoons destroyed were estimated from cocoons collected in the regular quadrat samples described previously. As shown in an earlier paper (Holling, 1955), cocoons opened by small mammals were easily recognized and moreover could be classified as to species of predator. These estimates of the number of new and old cocoons per square foot opened by each species of predator were corrected, as before, to provide an estimate of the number opened from the time larvae dropped to the time when cocoon samples were taken in early September.

It has proved difficult to obtain a predation and cocoon-population estimate of the desired precision and accuracy. The corrections and calculations that had to be applied to the raw sampling data cast some doubt upon the results and conclusions based upon them. It subsequently developed, however, that a considerable margin of error could be tolerated without changing the results and the conclusions that could be derived from them. In any case, all conclusions based upon cocoon-population estimates were supported and substantiated by results from controlled laboratory experiments.

LABORATORY TECHNIQUES

Several experiments were conducted with caged animals in order to support and expand results obtained in the field. The most important of these measured the number of cocoons consumed by *Peromyscus* at different cocoon densities. These experiments were conducted at room temperature (ca. 20°C) in a screen-topped cage, 10′ x 4′ x 6″. At the beginning of an experiment, cocoons were first buried in sand where the lines of a removable grid intersected, the grid was then removed, the sand was pressed flat, and a metal-edged levelling jig was finally scraped across the sand so that an even 12 mm. covered the cocoons. A single deer mouse was then placed in the cage together with nesting material, water, and an alternate food—dog biscuits. In each experiment the amount of this alternate food was kept approximately the same (i.e. 13 to 17 gms. dry weight). After the animal had been left undisturbed for 24 hours, the removable grid was replaced, and the number of holes dug over cocoons, the number of cocoons opened and the dry weight of dog biscuits eaten were recorded. Consumption by every animal was measured at either four or five different densities ranging from 2.25 to 36.00 cocoons per sq. ft. The specific densities were provided at random until all were used, the consumption at each density being measured for three to six consecutive days. Ideally the size of the cage should remain constant at all densities but since this would have required over 1,400 cocoons at the highest density, practical considerations necessitated a compromise whereby the cage was shortened at the higher densities. In these experiments the total number of cocoons provided ranged from 88 at the lowest density to 504 at the highest. At all densities, however, these numbers represented a surplus and no more than 40 per cent were ever consumed in a single experiment. Hence consumption was not limited by shortage of cocoons, even though the size of the cage changed.

The sources and characteristics of the cocoons and *Peromyscus* used in these experiments require some comment. Supplies of the prey were obtained by collecting cocoons in sawfly-infested plantations or by collecting late-instar larvae and allowing them to spin cocoons in boxes provided with foliage and litter. Sound cocoons from either source were then segregated into those containing healthy, parasitized, and diseased prepupae using a method of X-ray analysis (Holling, 1958a). The small male cocoons were separated from the larger female cocoons by size, since this criterion had previously proved adequate (Holling, 1958b). To simplify the experiments, only male and female cocoons containing healthy, living prepupae were used and in each experiment equal numbers of cocoons of each sex were provided, alternately, in the grid pattern already described.

Three mature non-breeding male deer mice were used in the experiments. Each animal had been born and raised in small rearing cages 12 x 8 x 6 in. and had been isolated from cocoons since birth. They therefore required a period to become familiar with the experimental cage and with cocoons. This experience was acquired during a preliminary three-week period. For the first two weeks the animal was placed in the experimental cage together with nesting material, water, dog biscuits and sand, and each day was disturbed just as it would be if an experiment were in progress. For the final week cocoons were buried in the sand at the first density chosen so that the animal could learn to find and consume the cocoon contents. It has been shown (Holling, 1955, 1958b) that a seven-day period is more than ample to permit complete learning.

THE COMPONENTS OF PREDATION

A large number of variables could conceivably affect the mortality of a given species of prey as a result of predation by a given species of predator. These can conveniently be classified, as was done by Leopold (1933), into five groups:

(1) density of the prey population.
(2) density of the predator population.
(3) characteristics of the prey, e.g., reactions to predators, stimulus detected by predator, and other characteristics.
(4) density and quality of alternate foods available for the predator.
(5) characteristics of the predator, e.g., food preferences, efficiency of attack, and other characteristics.

Each of these variables may exert a considerable influence and the effect of any one may depend upon changes in another. For example, Errington (1946) has shown that the characteristics of many vertebrate prey species change when their density exceeds the number that the available cover can support. This change causes a sudden increase in predation. When such complex interactions are involved, it is difficult to understand clearly the principles involved in predation; to do so we must find a simplified situation where some of the variables are constant or are not operating. The problem studied here presents such a situation. First, the characteristics of cocoons do not change as the other factors vary and there are no reactions by the cocooned sawflies to the predators. We therefore can ignore, temporarily, the effect of the third factor, prey characteristics. Secondly, since the work was conducted in plantations noted for their uniformity as to species, age, and distribution of trees, there was a constant and small variety of possible alternate foods. In such a simple and somewhat sterile environment, the fourth factor, the density and quality of alternate foods, can therefore be initially ignored, as can the fifth factor, characteristics of the predator, which is really only another way of expressing factors three and four. There are thus only two basic variables affecting predation in this instance, i.e., prey density and predator density. Furthermore, these are the only essential ones, for the remainder, while possibly important in affecting the amount of predation, are not essential to describe its fundamental characteristics.

The Basic Components

It is from the two essential variables that the basic components of predation will be derived. The first of these variables, prey density, might affect a number of processes and consumption of prey by individual predators might well be one of them.

The data which demonstrate the effect of changes of prey density upon consumption of cocooned sawflies by *Peromyscus* were obtained from the yearly cocoon quadrat samples in Plots 1 and 2. In 1951, Dr. F. T. Bird, Laboratory of Insect Pathology, Sault Ste. Marie, Ont., had sprayed each of these plots with a low concentration of a virus disease that attacked *N. sertifer* larvae, (Bird 1953). As a result, populations declined from 248,000 and 1,080,000 cocoons per acre, respectively, in 1952, to 39,000 and 256,000 in 1954. Thus predation values at six different cocoon densities were obtained. An additional sample in a neighbouring plantation in 1953 provided another value.

Predation values for *Sorex* and *Blarina* were obtained from one plantation, Plot 3, in one year, 1952. In the spring of that year, virus, sprayed from an aircraft flying along parallel lines 300 feet apart, was applied in three concentrations, with the lowest at one end of the plantation and the highest at the other. An area at one end, not sprayed, served as a control. When cocoon populations were sampled in the autumn, a line of 302 trees was selected at right angles to the lines of spray and the duff under each was sampled with one one-square-foot quadrat. The line, approximately 27 chains long, ran the complete length of the plantation. When the number of new cocoons per square foot was plotted against distance, discrete areas could be selected which had fairly constant populations that ranged from 44,000 to 571,000 cocoons per acre. The areas of low population corresponded to the areas sprayed with the highest concentration of virus. In effect, the plantation could be divided into rectangular strips, each with a particular density of cocoons. The width of these strips varied from 126 to 300 feet with an average of 193 feet. In addition to the 302 quadrats examined, the cocoons from another 100 quadrats were collected from the areas of lowest cocoon densities. Thus, in this one plantation in 1952, there was a sufficient number of different cocoon densities to show the response of consumption by *Sorex* and *Blarina* to changes of prey density.

The methods used to estimate predator densities in each study plot require some further comment. In Plots 1 and 2 this was done with grids of Sherman traps run throughout the summer. In Plot 3 both a grid of Sherman traps and a line of snap traps were used. This grid, measuring 18 chains by 4 chains, was placed so that approximately the same area sampled for cocoons was sampled for small mammals. The populations determined from these trapping procedures were plotted against time, and the number of "mammal-days" per acre, from the start of larval drop (June 14) to the time cocoon samples were made (Aug. 20-30), was determined for each plot each year. This could be done with *Peromyscus* and *Blarina* since the trapping technique was shown to provide an accurate estimate of their populations. But this was not true for *Sorex*. Instead, the number of *Sorex*-days per acre was approximated by dividing the number of cocoons opened at the highest density by the known number consumed by caged *Sorex* per day, i.e. 101. Since the number of cocoons opened at the highest cocoon density was

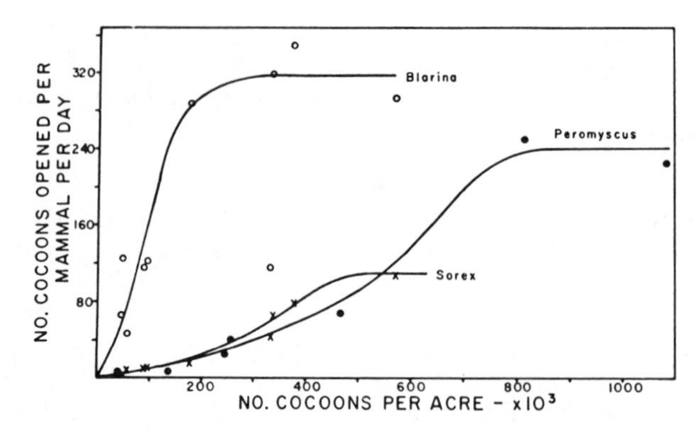

Fig. 1. Functional responses of *Blarina, Sorex* and *Peromyscus* in plots 1, 2, and 3.

151,000 per acre, then the number of *Sorex*-days per acre should be 151,000/101 = 1,490. This is approximately 10 times the estimate that was obtained from trapping with Sherman traps. When the various trapping methods were compared, estimates from Sherman trapping were shown to underestimate the numbers of *Sorex* by about the same amount, i.e. one-tenth.

With estimates of the numbers of predators, prey and destroyed prey available, the daily number of prey consumed per predator at different cocoon densities can be calculated. As seen in Fig. 1, the number of cocoons opened by each species increased with increasing cocoon density until a maximum daily consumption was reached that corresponded approximately to the maximum number that could be consumed in any one day by caged animals. For *Sorex* this of course follows from the method of calculation. The rates at which these curves rise differ for the different species, being greatest for *Blarina* and least for *Peromyscus.* Even if the plateaus are equated by multiplying points on each curve by a constant, the rates still decrease in the same order, reflecting a real difference in species behaviour.

The existence of such a response to cocoon density may also be demonstrated by data from the analysis of stomach contents. The per cent occurrence and per cent volume of the various food items in stomachs of *Peromyscus* captured immediately after larval drop and two months later is shown in Table I. When cocoon densities were high, immediately after larval drop, the per cent occurrence and per cent volume of *N. sertifer* material was high. Two months later when various cocoon mortality factors had taken their toll, cocoon densities were lower and

Table I

Stomach contents of *Peromyscus* trapped immediately before larval drop and two months later

Time trapped	Approx. no. cocoons per acre	No. of stomachs	Analysis	Plant	*N. sertifer*	Other insects	All insects
June 16–21	600,000	19	% occurrence	37%	95%	53%	100%
Aug. 17–19	300,000	14		79%	50%	64%	86%
June 16–21	600,000	19	% volume	5%	71%	24%	95%
Aug. 17–19	300,000	14		47%	19%	34%	53%

TABLE II

Occurrence of food items in stomachs of *Microtus* trapped before and after larval drop

Time trapped	Plant		*N. sertifer*		All insects	
	No. of stomachs	% occurrence	No. of stomachs	% occurrence	No. of stomachs	% occurrence
before larval drop	25	100%	2	8%	2	8%
after larval drop	29	100%	8	28%	11	38%

N. sertifer was a less important food item. The decrease in consumption of *N. sertifer* was accompanied by a considerable increase in the consumption of plant material and a slight increase in the consumption of other insect material. Plants and other insects acted as buffer or alternate foods. *Microtus*, even though they ate few non-plant foods in nature, also showed an increase in the per cent occurrence of *N. sertifer* material in stomachs as cocoon density increased (Table II). Before larval drop, when cocoon densities were low, the incidence of *N. sertifer* in *Microtus* stomachs was low. After larval drop, when cocoon densities were higher, the incidence increased by 3.5 times. Even at the higher cocoon densities, however, *N. sertifer* comprised less than one per cent of the volume of stomach contents so that this response to changes in prey density by *Microtus* is extremely low.

The graphs presented in Fig. I and the results of the analyses of stomach contents leave little doubt that the consumption of cocooned sawflies by animals in the field increases with increase in cocoon density. Similar responses have been demonstrated in laboratory experiments with three *Peromyscus*. As shown in Fig. 2, the number of cocoons consumed daily by each animal increased with increase in cocoon density, again reaching a plateau as did the previous curves. Whenever the number of prepupae consumed did not meet the caloric requirements, these were met by consumption of the dog biscuits, the alternate food provided. Only one of the animals (A) at the highest density fulfilled its caloric requirements by consuming prepupae; the remaining animals (B and C) consumed

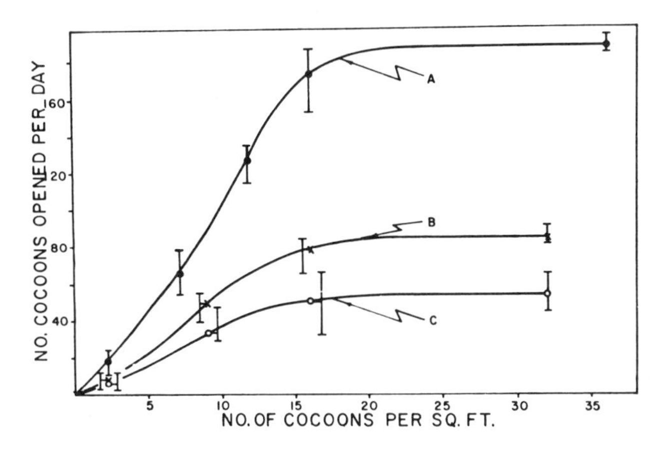

Fig. 2. Functional responses of three caged *Peromyscus* (means and ranges shown).

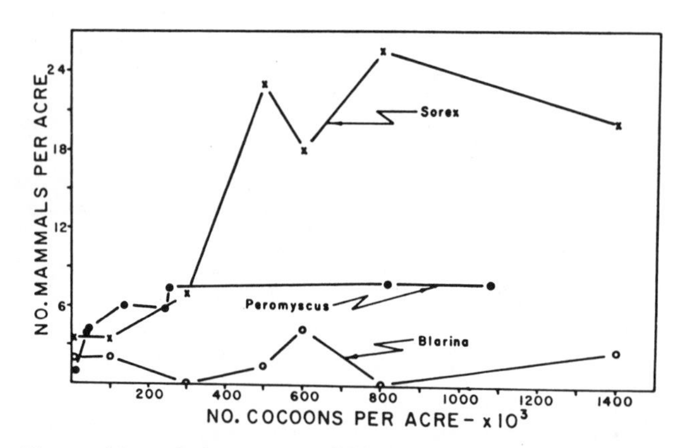

Fig. 3. Numerical responses of *Blarina, Sorex* and *Peromyscus.*

less than one-half the number of sawflies they would consume if no alternate foods were present. The cocoons used in experiments involving animals B and C, however, had been spun 12 months earlier than those involving animal A. When the characteristics of the functional response are examined in another paper, it will be shown that the strength of stimulus from older cocoons is less than that from younger cocoons, and that these differences are sufficient to explain the low consumption by animals B and C. The shape of the curves and the density at which they level is very similar for all animals, so similar that multiplying points along any one curve by the proper constant will equate all three. These curves are very similar to the ones based upon field data. All show the same form, the essential feature of which is an S-shaped rise to a plateau.

The effect of changes of prey density need not be restricted exclusively to consumption of prey by individual predators. The density of predators may also be affected and this can be shown by relating the number of predators per acre to the number of cocoons per acre. Conclusions can be derived from these relations but they are tentative. The data were collected over a relatively short period of time (four summers) and thus any relationship between predator numbers and prey density may have been fortuitous. Only those data obtained in plantations over 12 years old are included since small mammal populations were most stable in these areas. The data for the three most important species of predators are shown in the curves of Fig. 3, where each point represents the highest summer population observed either in different plantations or in the same plantation in different years.

The densities of *Blarina* were lowest while those of *Sorex* were highest. In this situation, *Blarina* populations apparently did not respond to prey density, for its numbers did not noticeably increase with increase in cocoon density. Some agent or agents other than food must limit their numbers. Populations of *Peromyscus* and *Sorex,* on the other hand, apparently did initially increase with increase in cocoon density, ultimately ceasing to increase as some agents other than food became limiting. The response of *Sorex* was most marked.

Thus two responses to changes of prey density have been demonstrated. The first is a change in the number of prey consumed per predator and the second is a change in the density of predators. Although few authors appear to recognize the existence and importance of *both* these responses to changes of prey density, they have been postulated and, in the case of the change of predator density, demonstrated. Thus Solomon (1949) acknowledged the two-fold nature of the response to changes of prey density, and applied the term *functional response* to the change in the number of prey consumed by individual predators, and the term *numerical response* to the change in the density of predators. These are apt terms and, although they have been largely ignored in the literature, they will be adopted in this paper. The data available to Solomon for review did not permit him to anticipate the form the functional response of predators might take, so that he could not assess its importance in population regulation. It will be shown, however, that the functional response is as important as the numerical.

It remains now to consider the effect of predator density, the variable that, together with prey density, is essential for an adequate description of predation. Predator density might well affect the number of prey consumed per predator. Laboratory experiments were designed to measure the number of cocoons opened by one, two, four, and eight animals in a large cage provided with cocoons at a density of 15 per square foot and a surplus of dog biscuits and water. The average number of cocoons opened per mouse in eight replicates was 159, 137, 141 and 159 respectively. In this experiment, therefore, predator density apparently did not greatly affect the consumption of prey by individual animals. This conclusion is again suggested when field and laboratory data are compared, for the functional response of *Peromyscus* obtained in the field, where its density varied, was very similar to the response of single animals obtained in the laboratory.

In such a simple situation, where predator density does not greatly affect the consumption by individuals, the total predation can be expressed by a simple, additive combination of the two responses. For example, if at a particular prey density the functional response is such that 100 cocoons are opened by a single predator in one day, and the numerical response is such that the predator density is 10, then the total daily consumption will be simply 100 x 10. In other situations, however, an increase in the density of predators might result in so much competition that the consumption of prey by individual predators might drop significantly. This effect can still be incorporated in the present scheme by adopting a more complex method of combining the functional and numerical responses.

This section was introduced with a list of the possible variables that could affect predation. Of these, only the two operating in the present study – prey and predator density – are essential variables, so that the basic features of predation can be ascribed to the effects of these two. It has been shown that there are two responses to prey density. The increase in the number of prey consumed per predator, as prey density rises, is termed the functional response, while the change in the density of predators is termed the numerical response. The total amount of predation occurring at any one density results from a combination of the two responses, and the method of combination will be determined by the way predator density affects consumption. This scheme, therefore, describes the effects of the basic variables, uncomplicated by the effects of subsidiary ones. Hence the two responses, the functional and numerical, can be considered the basic components of predation.

The total amount of predation caused by small mammals is shown in Fig. 4, where the functional and numerical responses are combined by multiplying the number of cocoons opened per predator at each density by the number of effective mammal-days observed. These figures were then expressed as percentages opened. This demonstrates the relation between per cent predation and prey density during the 100-day period between cocoon formation and adult emergence. Since the data obtained for the numerical responses are tentative, some reservations must be applied to the more particular conclusions derived

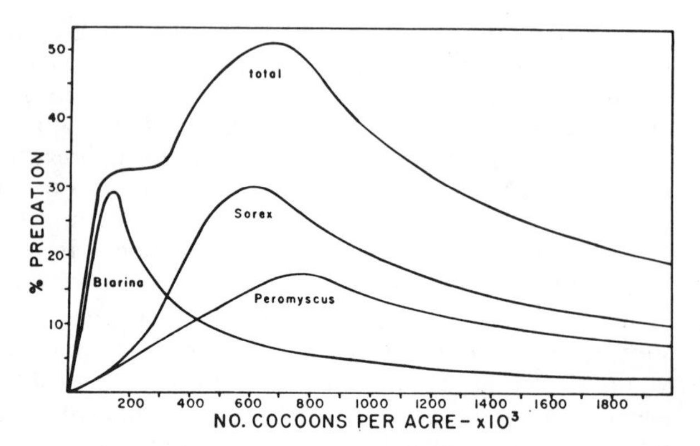

Fig. 4. Functional and numerical responses combined to show the relation between per cent predation and cocoon density.

from this figure. The general conclusion, that per cent predation by each species shows an initial rise and subsequent decline as cocoon density increases holds, however. For this conclusion to be invalid, the numerical responses would have to decrease in order to mask the initial rise in per cent predation caused by the S-shaped form of the functional responses. Thus from zero to some finite cocoon density, predation by small mammals shows a direct density-dependent action and thereafter shows an inverse density-dependent action. The initial rise in the proportion of prey destroyed can be attributed to both the functional and numerical responses. The functional response has a roughly sigmoid shape and hence the proportion of prey destroyed by an individual predator will increase with increase in cocoon density up to and beyond the point of inflection. Unfortunately the data for any one functional response curve are not complete enough to establish a sigmoid relation, but the six curves presented thus far and the several curves to be presented in the following section all suggest a point of inflection. The positive numerical responses shown by *Sorex* and *Peromyscus* also promote a direct density-dependent action up to the point at which predator densities remain constant. Thereafter, with individual consumption also constant, the per cent predation will decline as cocoon density increases. The late Dr. L. Tinbergen apparently postulated the same type of dome-shaped curves for the proportion of insects destroyed by birds. His data were only partly published (1949, 1955) before his death, but Klomp (1956) and Voûte (1958) have commented upon the existence of these "optimal curves". This term, however, is unsatisfactory and anthropocentric. From the viewpoint of the forest entomologist, the highest proportion of noxious insects destroyed may certainly be the optimum, but the term is meaningless for an animal that consumes individuals and not percentages. Progress can best be made by considering predation first as a behaviour before translating this behaviour in terms of the proportion of prey destroyed. The term "peaked curve" is perhaps more accurate.

Returning to Fig. 4, we see that the form of the peaked curve for *Blarina* is determined solely by the functional response since this species exhibited no numerical response. The abrupt peak occurs because the maximum consumption of prepupae was reached at a very low prey density before the predation was "diluted" by large numbers of cocoons. With *Sorex* both the numerical and functional responses are important. Predation by *Sorex* is greatest principally because of the marked numerical response. The two responses again determine the form of the peaked curve for *Peromyscus*, but the numerical response, unlike that of *Sorex*, was not marked, and the maximum consumption of cocoons was reached only at a relatively high density; the result is a low per cent predation with a peak occurring at a high cocoon density.

Predation by all species destroyed a considerable number of cocooned sawflies over a wide range of cocoon densities. The presence of more than one species of predator not only increased predation but also extended the range of prey densities over which predation was high. This latter effect is particularly important, for if the predation by several species of predators peaked at the same prey density the range of densities over which predation was high would be slight and if the prey had a sufficiently high reproductive capacity its density might jump this vulnerable range and hence escape a large measure of the potential control that could be exerted by predators. Before we can proceed further in the discussion of the effect of predation upon prey numbers, the additional components that make up the behaviour of predation must be considered.

The Subsidiary Components

Additional factors such as prey characteristics, the density and quality of alternate foods, and predator characteristics have a considerable effect upon predation. It is necessary now to demonstrate the effect of these factors and how they operate.

There are four classes of prey characteristics: those that influence the caloric value of the prey; those that change the length of time prey are exposed; those that affect the "attractiveness" of the prey to the predator (e.g. palatability, defence mechanisms); and those that affect the strength of stimulus used by predators in locating prey (e.g. size, habits, and colours). Only those characteristics that affect the strength of stimulus were studied experimentally. Since small mammals detect cocoons by the odour emanating from them (Holling, 1958b), the strength of this odour perceived by a mammal can be easily changed in laboratory experiments by varying the depth of sand covering the cocoons.

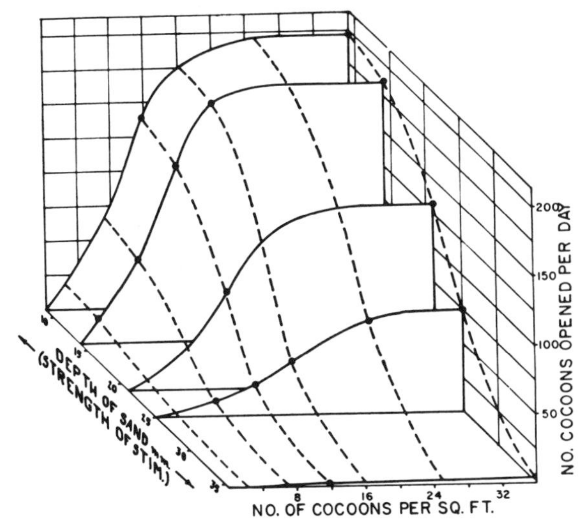

Fig. 5. Effect of strength of stimulus from cocoons upon the functional response of one caged *Peromyscus*. Each point represents the average of three to six replicates.

One *Peromyscus* was used in these experiments and its daily consumption of cocoons was measured at different cocoon densities and different depths of sand. These data are plotted in Fig. 5. Since the relation between depth of sand and strength of stimulus must be an inverse one, the depths of sand are reversed on the axis so that values of the strength of stimulus increase away from the origin. Each point represents the mean of three to six separate measurements. Decreasing the strength of the perceived stimulus by increasing the depth of sand causes a marked decrease in the functional response. A 27 mm. increase in depth (from nine to 36 mm.), for example, causes the peak consumption to drop from 196 to four cocoons per day. The daily number of calories consumed in all these experiments remained relatively constant since dog biscuits were always present as alternate food. The density at which each functional-response curve levels appear to increase somewhat as the strength of stimulus perceived by the animal decreases. We might expect that the increase in consumption is directly related to the increase in the proportion of cocoons in the amount of food available, at least up to the point where the caloric requirements are met solely by sawflies. The ascending portions of the curves, however, are S-shaped and the level portions are below the maximum consumption, approximately 220 cocoons for this animal. Therefore, the functional response cannot be explained by random searching for cocoons. For the moment, however, the important conclusion is that changes in prey characteristics can have a marked effect on predation but this effect is exerted through the functional response.

In the plantations studied, cocoons were not covered by sand but by a loose litter and duff formed from pine needles. Variations in the depth of this material apparently did not affect the strength of the perceived odour, for as many cocoons were opened in quadrats with shallow litter as with deep. This material must be so loose as to scarcely impede the passage of odour from cocoons.

The remaining subsidiary factors, the density and quality of alternate foods and predator characteristics, can also affect predation. The effect of alternate foods could not be studied in the undisturbed plantations because the amount of these "buffers" was constant and very low. The effect of quality of alternate foods on the functional response, however, was demonstrated experimentally using one *Peromyscus*. The experiments were identical to those already described except that at one series of densities an alternate food of low palatability (dog biscuits) was provided, and at the second series one of high palatability (sunflower seeds) was provided. When both foods are available, deer mice select sunflower seeds over dog biscuits. In every experiment a constant amount of alternate food was available: 13 to 17 gms. dry weight of dog biscuits, or 200 sunflower seeds.

Fig. 6 shows the changes in the number of cocoons opened per day and in the amount of alternate foods consumed. The functional response decreased with an increase in the palatability of the alternate food (Fig. 6A). Again the functional response curves showed an initial, roughly sigmoid rise to a constant level.

As cocoon consumption rose, the consumption of alternate foods decreased (Fig. 6B) at a rate related to the palatability of the alternate food. Each line indicating the change in the consumption of alternate food was drawn as a mirror image of the respective functional response and these lines closely follow the mean of the observed points. The variability in the consumption of sunflower seeds at any one cocoon density was considerable, probably as a result of the extreme variability in the size of seeds.

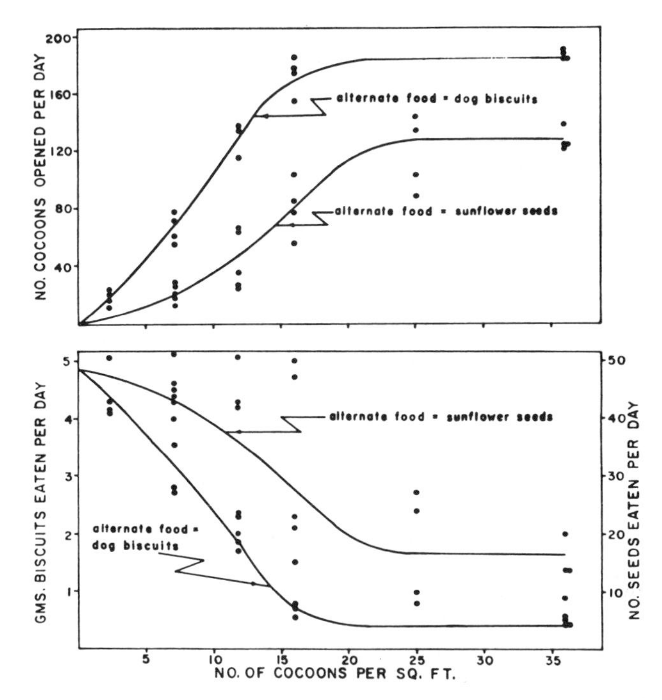

Fig. 6. Effect of different alternate foods upon the functional response of one *Peromyscus*. *A* (upper) shows the functional responses when either a low (dog biscuits) or a high (sunflower seeds) palatability alternate food was present in excess. *B* (lower) shows the amount of these alternate foods consumed.

Again we see that there is not a simple relation between the number of cocoons consumed and the proportion of cocoons in the total amount of food available. This is most obvious when the functional response curves level, for further increase in density is not followed by an increase in the consumption of sawflies. The plateaus persist because the animal continued consuming a certain fixed quantity of alternate foods. L. Tinbergen (1949) observed a similar phenomenon in a study of predation of pine-eating larvae by tits in Holland. He presented data for the consumption of larvae of the pine beauty moth, *Panolis griseovariegata*, and of the web-spinning sawfly *Acantholyda pinivora*, each at two different densities. In each case more larvae were eaten per nestling tit per day at the higher prey density. This, then, was part of a functional response, but it was that part above the point of inflection, since the proportion of prey eaten dropped at the higher density. It is not sufficient to explain these results as well as the ones presented in this paper by claiming, with Tinbergen, that the predators "have the tendency to make their menu as varied as possible and therefore guard against one particular species being strongly dominant in it". This is less an explanation than an anthropocentric description. The occurrence of this phenomenon depends upon the strength of stimulus from the prey, and the amount and quality of the alternate foods. Its proper explanation must await the collection of further data.

We now know that the palatability of alternate foods affects the functional response. Since the number of different kinds of alternate food could also have

TABLE III

The effect of alternate foods upon the number of cocoons consumed per day by one *Peromyscus*

Alternate food	No. of exp'ts	No. of cocoons opened	
		$\bar{X}$	$S.E._{\bar{x}}$
none	7	165.9	11.4
dog biscuits	5	143.0	8.3
sunflower seeds	8	60.0	6.2
sunflower seeds and dog biscuits	8	21.5	4.2

an important effect, the consumption of cocoons by a caged *Peromyscus* was measured when no alternate foods, or one or two alternate foods, were present. Only female cocoons were used and these were provided at a density of 75 per sq. ft. to ensure that the level portion of the functional response would be measured. As in the previous experiments, the animal was familiarized with the experimental conditions and with cocoons for a preliminary two-week period. The average numbers of cocoons consumed each day with different numbers and kinds of alternate foods present are shown in Table III. This table again shows that fewer cocoons were consumed when sunflower seeds (high palatability) were present than when dog biscuits (low palatability) were present. In both cases, however, the consumption was lower than when no alternate foods were available. When two alternate foods were available, i.e., both sunflower seeds and dog biscuits, the consumption dropped even further. Thus, increase in both the palatability and in the number of different kinds of alternate foods decreases the functional response.

DISCUSSION

General

It has been argued that three of the variables affecting predation—characteristics of the prey, density and quality of alternate foods and characteristics of the predators – are subsidiary components of predation. The laboratory experiments showed that the functional response was lowered when the strength of stimulus, one prey characteristic, detected from cocoons was decreased or when the number of kinds and palatability of alternate foods was increased. Hence the effect of these subsidiary components is exerted through the functional response. Now the numerical response is closely related to the functional, since such an increase in predator density depends upon the amount of food consumed. It follows, therefore, that the subsidiary components will also affect the numerical response. Thus when the functional response is lowered by a decrease in the strength of stimulus detected from prey, the numerical response similarly must be decreased and predation will be less as a result of decrease of the two basic responses.

The density and quality of alternate foods could also affect the numerical response. Returning to the numerical responses shown in Fig. 3, if increase in the density or quality of alternate foods involved solely increase in food "per se", then the number of mammals would reach a maximum at a lower cocoon density, but the maximum itself would not change. If increase in alternate foods also involved changes in the agents limiting the numerical responses (e.g. increased cover and depth of humus), then the maximum density the small mammals could attain would increase. Thus increase in the amount of alternate foods could increase the density of predators.

Increase in alternate foods *decreases* predation by dilution of the functional response, but *increases* predation by promoting a favourable numerical response. The relative importance of each of these effects will depend upon the particular problem. Voûte (1946) has remarked that insect populations in cultivated woods show violent fluctuations, whereas in virgin forests or mixed woods, where the number of alternate foods is great, the populations are more stable. This stability might result from alternate foods promoting such a favourable numerical response that the decrease in the functional response is not great enough to lower predation.

The importance of alternate foods will be affected by that part of the third subsidiary component – characteristics of the predators – that concerns food preferences. Thus an increase in plants or animals other than the prey will most likely affect the responses of those predators, like the omnivore *Peromyscus*, that are not extreme food specialists. Predation by the more stenophagous shrews, would only be affected by some alternate, animal food.

Food preferences, however, are only one of the characteristics of predators. Others involve their ability to detect, capture, and kill prey. But again the effect of these predator characteristics will be exerted through the two basic responses, the functional and numerical. The differences observed between the functional responses of the three species shown earlier in Fig. 1 undoubtedly reflect differences in their abilities to detect, capture, and kill. The amount of predation will similarly be affected by the kind of sensory receptor, whether visual, olfactory, auditory, or tactile, that the predator uses in locating prey. An efficient nose, for example, is probably a less precise organ than an efficient eye. The source of an undisturbed olfactory stimulus can only be located by investigating a gradient in space, whereas a visual stimulus can be localized by an efficient eye from a single point in space – the telotaxis of Fraenkel and Gunn (1940). As N. Tinbergen (1951) remarked, localization of direction is developed to the highest degree in the eye. Thus the functional response of a predator which locates prey by sight will probably reach a maximum at a much lower prey density than the response of one that locates its prey by odour. In the data presented by Tothill (1922) and L. Tinbergen (1949), the per cent predation of insects by birds was highest at very low prey densities, suggesting that the functional responses of these "visual predators" did indeed reach a maximum at a low density.

The Effect of Predation on Prey Populations

One of the most important characteristics of mortality factors is their ability to regulate the numbers of an animal – to promote a "steady density" (Nicholson, 1933; Nicholson and Bailey, 1935) such that a continued increase or decrease of numbers from this steady state becames progressively unlikely the greater the departure from it. Regulation in this sense therefore requires that the mortality factor change with change in the density of the animal attacked, i.e. it requires a direct density-dependent mortality (Smith, 1935, 1939). Density-independent factors can affect the numbers of an animal but alone they cannot *regulate* the numbers. There is abundant evidence that changes in climate, some aspects of which are presumed to have a density-independent action, can lower or raise the numbers of an animal. But this need not be regulation. Regulation will only result from an interaction with a density-dependent factor, an interaction

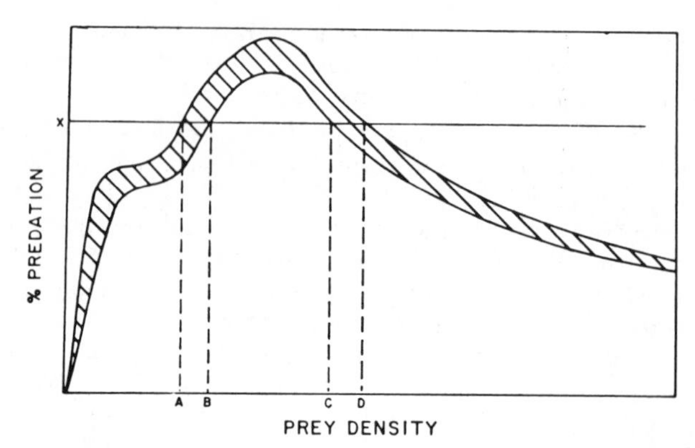

Fig. 7. Theoretical model showing regulation of prey by predators. (see text for explanation).

that might be the simplest, i.e. merely additive. Recently, the density-dependent concept has been severely criticized by Andrewartha and Birch (1954). They call it a dogma, but such a comment is only a criticism of an author's use of the concept. Its misuse as a dogma does not militate against its value as a hypothesis.

We have seen from this study that predation by small mammals does change with changes in prey density. As a result of the functional and numerical responses the proportion of prey destroyed increases from zero to some finite prey density and thereafter decreases. Thus predation over some ranges of prey density shows a direct density-dependent action. This is all that is required for a factor to regulate.

The way in which predation of the type shown in this study can regulate the numbers of a prey species can best be shown by a hypothetical example. To simplify this example we will assume that the prey has a constant rate of reproduction over all ranges of its density, and that only predators are affecting its numbers. Such a situation is, of course, unrealistic. The rate of reproduction of real animals probably is low at low densities when there is slight possibility for contact between individuals (e.g. between male and female). It would rise as contacts became more frequent and would decline again at higher densities when the environment became contaminated, when intraspecific stress symptoms appeared, or when cannibalism became common. Such changes in the rate of reproduction have been shown for experimental populations of *Tribolium confusum* (MacLagan, 1932) and *Drosophila* (Robertson and Sang, 1944). Introducing more complex assumptions, however, confuses interpretations without greatly changing the conclusions.

This hypothetical model is shown in Fig. 7. The curve that describes the changes in predation with changes in prey density is taken from the actual data shown earlier in Fig. 4. It is assumed that the birth-rate of the prey at any density can be balanced by a fixed per cent predation, and that the variation in the environment causes a variation in the predation at any one density. The per cent predation necessary to balance the birth-rate is represented by the horizontal line, x%, in the diagram and variation in predation is represented by the thickness of the mortality curve. The death-rate will equal the birth-rate at two density ranges, between A and B and between C and D. When the densities of the prey are below A, the mortality will be lower than that necessary to balance reproduction and the population will increase. When the densities of the animal are between B and C, death-rate will exceed birth-rate and the populations will decrease. Thus, the density of the prey will tend to fluctuate between densities A and B. If the density happens to exceed D, death-rate will be lower than birth-rate and the prey will increase in numbers, having "escaped" the control exerted by predators. This would occur when the prey had such a high rate of reproduction that its density could jump, in one generation, from a density lower than A to a density higher than D. If densities A and D were far apart, there would be less chance of this occurring. This spread is in part determined by the number of different species of predators that are present. Predation by each species peaks at a different density (see Fig. 4), so that increase in the number of species of predator will increase the spread of the total predation. This will produce a more stable situation where the prey will have less chance to escape control by predators.

Predation of the type shown will regulate the numbers of an animal whenever the predation rises high enough to equal the effective birth-rate. When the prey is an insect and predators are small mammals, as in this case, the reproductive rate of the prey will be too high for predation *alone* to regulate. But if other mortality occurs, density-independent or density-dependent, the total mortality could rise to the point where small mammals were contributing, completely or partially, to the regulation of the insect.

Predation of the type shown will produce stability if there are large numbers of different species of predators operating. Large numbers of such species would most likely occur in a varied environment, such as mixed woods. Perhaps this explains, in part, Voûte's (1946) observation that insect populations in mixed woods are less liable to show violent fluctuations.

I cannot agree with Voûte (1956 and 1958) that factors causing a peaked mortality curve are not sufficient for regulation. He states (1956) that "this is due to the fact that mortality only at low densities increases with the increase of the population. At higher densities, mortality decreases again. The growth of the population is at the utmost slowed down, never stopped". All that is necessary for regulation, however, is a rise in per cent predation over some range of prey densities and an *effective* birth-rate that can be matched at some density by mortality from predators.

Neither can I agree with Thompson (1930) when he ascribes a minor role to vertebrate predators of insects and states that "the number of individuals of any given species (i.e. of vertebrate predators) is . . . relatively small in comparison with those of insects and there is no reason to suppose that it varies primarily in function of the supply of insect food, which fluctuates so rapidly that it is impossible for vertebrates to profit by a temporary abundance of it excepting to a very limited extent". We know that they do respond by an increase in numbers and even if this is not great in comparison with the numerical response of parasitic flies, the number of prey killed per predator is so great and the increase in number killed with increase in prey density is so marked as to result in a heavy proportion of prey destroyed; a proportion that, furthermore, increases initially with increase of prey density. Thompson depreciates the importance of the numerical response of predators and ignores the functional response.

In entomological literature there are two contrasting mathematical theories of regulation. Each theory is based on different assumptions and the predicted results are quite different. Both theories were developed to predict the interaction between parasitic flies and their insect hosts but they can be applied equally well to predator-prey relations. Thompson (1939) assumes that a predator has a limited appetite and that it has no difficulty in finding its prey. Nicholson (1933) assumes that predators have insatiable appetites and that they have a specific capacity to find their prey. This searching capacity is assumed to remain constant at all prey densities and it is also assumed that the searching is random.

The validity of these mathematical models depends upon how closely their assumptions fit natural conditions. We have seen that the appetites of small mammal predators in this study are not insatiable. This fits one of Thompson's assumptions but not Nicholson's. When the functional response was described, it was obvious that predators did have difficulty in finding their prey and that their searching ability did not remain constant at all prey densities. Searching by small mammals was not random. Hence in the present study of predator-prey relations, the remaining assumptions of both Thompson and Nicholson do not hold.

Klomp (1956) considers the damping of oscillations of animal numbers to be as important as regulation. If the oscillations of the numbers of an animal affected by a delayed density-dependent factor (Varley, 1947) like a parasite, do increase in amplitude, as Nicholson's theory predicts (Nicholson and Bailey, 1935), then damping is certainly important. It is not at all certain, however, that this prediction is true. We have already seen that the assumptions underlying Nicholson's theory do not hold in at least some cases. In particular he ignores the important possibility of an S-shaped functional response of the type shown by small mammal predators. If the parasites did show an S-shaped functional response, there would be an *immediate* increase in per cent predation when host density increased, an increase that would modify the effects of the delayed numerical response of parasites emphasized by Nicholson and Varley. Under these conditions the amplitude of the oscillations would not increase as rapidly, and might well not increase at all. An S-shaped functional response therefore acts as an intrinsic damping mechanism in population fluctuations.

Oscillations undoubtedly do occur, however, and whether they increase in amplitude or not, any extrinsic damping is important. The factor that damps oscillations most effectively will be a concurrent density-dependent factor that reacts immediately to changes in the numbers of an animal. Predation by small mammals fulfils these requirements when the density of prey is low. The consumption of prey by individual predators responds immediately to increase in prey density (functional response). Similarly, the numerical response is not greatly delayed, probably because of the high reproductive capacity of small mammals. Thus if the density of a prey is low, chance increases in its numbers will immediately increase the per cent mortality caused by small mammal predation. When the numbers of the prey decrease, the effect of predation will be immediately relaxed. Thus, incipient oscillations can be damped by small-mammal predation.

We have seen that small mammals theoretically can regulate the numbers of prey and can damp their oscillations under certain conditions. Insufficient information was obtained to assess precisely the role of small mammals as predators of *N. sertifer* in the pine plantations of southwestern Ontario, however. Before the general introduction of a virus disease in 1952 (Bird, 1952, 1953), the sawfly was exhausting its food supplies and 70 to 100% defoliation of Scots, jack and red pines was observed in this area. Predators were obviously not regulating the numbers of the sawfly. After the virus was introduced, however, sawfly populations declined rapidly. In Plot 1, for example, their numbers declined from 248,000 cocoons per acre in 1952 to 39,000 per acre in 1954. The area was revisited in 1955 and larval and cocoon population had obviously increased in this plot, before the virus disease could cause much mortality. It happened, however, that *Peromyscus* was the only species of small mammal residing in Plot 1 and it is interesting that similar increases were not observed in other plantations where sawfly numbers had either not decreased so greatly, or where shrews, the most efficient predators, were present. These observations suggest that predation by shrews was effectively damping the oscillations resulting from the interaction of the virus disease with its host.

Types of Predation

Many types of predation have been reported in the literature. Ricker (1954) believed that there were three major types of predator-prey relations, Leopold (1933) four, and Errington (1946, 1956) two. Many of these types are merely minor deviations, but the two types of predation Errington discusses are quite different from each other. He distinguishes between "compensatory" and "noncompensatory" predation. In the former type, predators take a heavy toll of individuals of the prey species when the density of prey exceeds a certain threshold. This "threshold of security" is determined largely by the number of secure habitable niches in the environment. When prey densities become too high some individuals are forced into exposed areas where they are readily captured by predators. In this type of predation, predators merely remove surplus animals, ones that would succumb even in the absence of enemies. Errington feels, however, that some predator-prey relations depart from this scheme, so that predation occurs not only *above* a specific threshold density of prey. These departures are ascribed largely to behaviour characteristics of the predators. For example, he does not believe that predation of ungulates by canids is compensatory and feels that this results from intelligent, selective searching by the predators.

If the scheme of predation presented here is to fulfill its purpose it must be able to explain these different types of predation. Non-compensatory predation is easily described by the normal functional and numerical responses, for predation of *N. sertifer* by small mammals is of this type. Compensatory predation can also be described using the basic responses and subsidiary factors previously demonstrated. The main characteristic of this predation is the "threshold of security". Prey are more vulnerable above and less vulnerable below this threshold. That is, the strength of stimulus perceived from prey increases markedly when the prey density exceeds the threshold. We have seen from the present study that an increase in the strength of stimulus from prey increases both the functional and numerical responses. Therefore, below the "threshold of security" the functional responses of predators will be very low and as a result there will probably be no numerical response. Above the threshold, the functional response will become marked and a positive numerical response could easily occur. The net effect will result from a combination of these functional and numerical responses so that per cent predation will remain low so long as there is sufficient cover and food available for the prey. As soon as these supply factors are approaching exhaustion the per cent predation will suddenly increase.

Compensatory predation will occur (1) when the prey has a specific density level near which it normally operates, and (2) when the strength of stimulus perceived by predators is so low below this level and so high above it that there is a marked change in the functional response. Most insect populations tolerate considerable crowding and the only threshold would be set by food limitations. In addition, their strength of stimulus is often high at all densities. For *N. sertifer* at least, the strength of stimulus from cocoons is great and the threshold occurs at such high densities that the functional responses of small mammals are at their maximum. Compensatory predation upon insects is probably uncommon.

Entomologists studying the biological control of insects have largely concentrated their attention on a special type of predator – parasitic insects. Although certain features of a true predator do differ from those of a parasite, both predation and parasitism are similar in that one animal is seeking out another. If insect parasitism can in fact be treated as a type of predation, the two basic responses to prey (or host) density and the subsidiary factors affecting these responses should describe parasitism. The functional response of a true predator is measured by the number of prey it destroys; of a parasite by the number of hosts in which eggs are laid. The differences observed between the functional responses of predators and parasites will depend upon the differences between the behaviour of eating and the behaviour of egg laying. The securing of food by an individual predator serves to maintain that individual's existence. The laying of eggs by a parasite serves to maintain its progenies' existence. It seems likely that the more a behaviour concerns the maintenance of an individual, the more demanding it is. Thus the restraints on egg laying could exert a greater and more prolonged effect than the restraints on eating. This must produce differences between the functional responses of predators and parasites. But the functional responses of both are similar in that there is an upper limit marked by the point at which the predator becomes satiated and the parasite has laid all its eggs. This maximum is reached at some finite prey or host density above zero. The form of the rising phase of the functional response would depend upon the characteristics of the individual parasite and we might expect some of the same forms that will be postulated for predators at the end of this section. To summarize, I do not wish to imply that the characteristics of the functional response of a parasite are identical with those of a predator. I merely wish to indicate that a parasite has a response to prey density – the laying of eggs – that can be identified as a functional response, the precise characteristics of which are unspecified.

The effects of host density upon the number of hosts parasitized have been studied experimentally by a number of workers (e.g., Ullyett, 1949a and b; Burnett, 1951 and 1954; De Bach and Smith, 1941). In each case the number of hosts attacked per parasite increased rapidly with initial increase in host density but tended to level with further increase. Hence these functional response curves showed a continually decreasing slope as host density increased and gave no indication of the S-shaped response shown by small mammals. Further information is necessary, however, before these differences can be ascribed solely to the difference between parasitism and predation. It might well reflect, for example, a difference between an instinctive response of an insect and a learned response of a mammal or between the absence of an alternate host and the presence of an alternate food.

The numerical response of both predators and parasites is measured by the way in which the number of adults increases with increase in prey or host density. At first thought, the numerical response of a parasite would seem to be so intimately connected with its functional response that they could not be separated. But the two responses of a predator are just as intimately connected.

The predator must consume food in order to produce progeny just as the parasite must lay eggs in order to produce progeny.

The agents limiting the numerical response of parasites will be similar to those limiting the response of predators. There is, however, some difference. During at least one stage of the parasites' life, the requirements for both food and niche are met by the same object. Thus increase in the amount of food means increase in the number of niches as well, so that niches are never limited unless food is. This should increase the chances for parasites to show pronounced numerical responses. The characteristics of the numerical responses of both predators and parasites, however, will be similar and will range from those in which there is no increase with increase in the density of hosts, to those in which there is a marked and prolonged increase.

A similar scheme has been mentioned by Ullyett (1949b) to describe parasitism. He believed that "the problem of parasite efficiency would appear to be divided into two main phases, viz.: (a) the efficiency of the parasite as a mortality factor in the host population, (b) its efficiency as related to the maintenance of its own population level within the given area". His first phase resembles the functional response and the second the numerical response. Both phases or responses will be affected, of course, by subsidiary components similar to those proposed for predation—characteristics of the hosts, density and quality of alternate hosts, and characteristics of the parasite. The combination of the two responses will determine the changes in per cent parasitism as the result of changes in host density. Since both the functional and numerical responses presumably level at some point, per cent parasitism curves might easily be peaked, as were the predation curves. If these responses levelled at a host density that would never occur in nature, however, the decline of per cent parasitism might never be observed.

The scheme of predation revealed in this study may well explain all types of predation as well as insect parasitism. The knowledge of the basic components and subsidiary factors underlying the behaviour permits us to imagine innumerable possible variations. In a hypothetical situation, for example, we could introduce and remove alternate food at a specific time in relation to the appearance of a prey, and predict the type of predation. But such variations are only minor deviations of a basic pattern. The major types of predation will result from major differences in the form of the functional and numerical responses.

If the functional responses of some predators are partly determined by their behaviour, we could expect a variety of responses differing in form, rate of rise, and final level reached. All functional responses, however, will ultimately level, for it is difficult to imagine an individual predator whose consumption rises indefinitely. Subsistence requirements will fix the ultimate level for most predators, but even those whose consumption is less rigidly determined by subsistence requirements (e.g., fish, Ricker 1941) must have an upper limit, even if it is only determined by the time required to kill.

The functional responses could conceivably have three basic forms. The mathematically simplest would be shown by a predator whose pattern of searching was random and whose rate of searching remained constant at all prey densities. The number of prey killed per predator would be directly proportional to prey density, so that the rising phase would be a straight line. Ricker (1941) postulated this type of response for certain fish preying on sockeye salmon, and De Bach and Smith (1941) observed that the parasitic fly, *Muscidifurax raptor*, parasitized puparia of *Musca domestica*, provided at different densities, in a similar fashion. So few prey were provided in the latter experiment, however, that the initial linear rise in the number of prey attacked with increase in prey density may have been an artifact of the method.

A more complex form of functional response has been demonstrated in laboratory experiments by De Bach and Smith (1941), Ullyett (1949a) and Burnett (1951, 1956) for a number of insect parasites. In each case the number of prey attacked per predator increased very rapidly with initial increase in prey density, and thereafter increased more slowly approaching a certain fixed level. The rates of searching therefore became progressively less as prey density increased.

The third and final form of functional response has been demonstrated for small mammals in this study. These functional responses are S-shaped so that the rates of searching at first increase with increase of prey density, and then decrease.

Numerical responses will also differ, depending upon the species of predator and the area in which it lives. Two types have been demonstrated in this study. *Peromyscus* and *Sorex* populations, for example, increased with increase of prey density to the point where some agent or agents other than food limited their numbers. These can be termed direct numerical responses. There are some cases, however, where predator numbers are not affected by changes of prey density and in the plantations studied *Blarina* presents such an example of no numerical response. A final response, in addition to ones shown here, might also occur. Morris *et al.* (1958) have pointed out that certain predators might decrease in numbers as prey density increases through competition with other predators. As an example of such inverse numerical responses, he shows that during a recent outbreak of spruce budworm in New Brunswick the magnolia, myrtle, and black-throated green warblers decreased in numbers. Thus we have three possible numerical responses – a direct response, no response, and an inverse response.

The different characteristics of these types of functional and numerical responses produce different types of predation. There are four major types conceivable; these are shown diagramatically in Fig. 8. Each type includes the three possible numerical responses – a direct response (a), no response (b), and an inverse response (c), and the types differ because of basic differences in the functional response. In type 1 the number of prey consumed per predator is assumed to be directly proportional to prey density, so that the rising phase of the functional response is a straight line. In type 2, the functional response is presumed to rise at a continually decreasing rate. In type 3, the form of the functional response is the same as that observed in this study. These three types of predation may be considered as the basic ones, for changes in the subsidiary components are not involved. Subsidiary components can, however, vary in response to changes of prey density and in such cases the basic types of predation are modified. The commonest modification seems to be Errington's compensatory predation which is presented as Type 4 in Fig. 8. In this figure the vertical dotted line represents the "threshold of security" below which the strength of stimulus from prey is low and above which it is high. The functional response curves at these two strengths of stimulus are given the form of the functional responses observed in this study. The forms of the responses shown in Types 1 and 2 could also be used, of course.

The combination of the two responses gives the total response shown in the

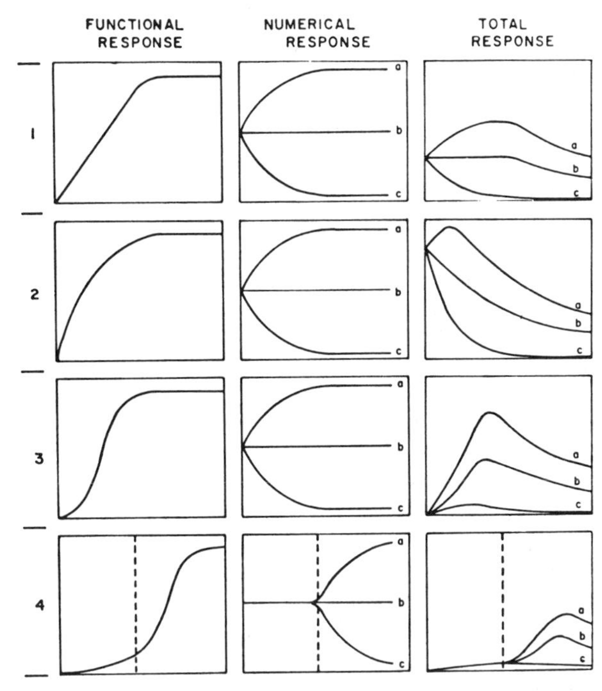

Fig. 8. Major types of predation.

final column of graphs of Fig. 8. Both peaked (curves 1a; 2a; 3a, b, c; 4a, b, c) and declining (1b, c; 2b, c) types of predation can occur, but in the absence of any other density-dependent factor, regulation is possible only in the former type.

This method of presenting the major types of predation is an over-simplification since predator density is portrayed as being directly related to prey density. Animal populations, however, cannot respond *immediately* to changes in prey density, so that there must be a delay of the numerical response. Varley (1953) pointed this out when he contrasted "delayed density dependence" and "density dependence". The degree of delay, however, will vary widely depending upon the rate of reproduction, immigration, and mortality. Small mammals, with their high reproductive rate, responded so quickly to increased food that the delay was not apparent. In such cases the numerical response graphs of Fig. 8 are sufficiently accurate, for the density of predators in any year is directly related to the density of prey in the same year. The numerical response of other natural enemies can be considerably delayed, however. Thus the density of those insect parasites that have one generation a year and a low rate of immigration results from the density of hosts in the preceding generation.

In these extreme cases of delay the total response obtained while prey or hosts are steadily increasing will be different than when they are steadily decreasing. The amount of difference will depend upon the magnitude and amount of delay of the numerical response, for the functional response has no element of delay.

SUMMARY AND CONCLUSIONS

The simplest and most basic type of predation is affected by only two variables – prey and predator density. Predation of cocooned *N. sertifer* by small mammals is such a type, for prey characteristics, the number and variety of alternate foods, and predilections of the predators do not vary in the plantations where *N. sertifer* occurs. In this simple example of predation, the basic components of predation are responses to changes in prey density. The increase in the number of prey consumed per predator, as prey density rises, is termed the functional response. The change in the density of predators, as a result of increase in prey density, is termed the numerical response.

The three important species of small mammal predators (*Blarina*, *Sorex*, and *Peromyscus*) each showed a functional response, and each curve, whether it was derived from field or laboratory data, showed an initial S-shaped rise up to a constant, maximum consumption. The rate of increase of consumption decreased from *Blarina* to *Sorex* to *Peromyscus*, while the upper, constant level of consumption decreased from *Blarina* to *Peromyscus* to *Sorex*. The characteristics of these functional responses could not be explained by a simple relation between consumption and the proportion of prey in the total food available. The form of the functional response curves is such that the proportion of prey consumed per predator increases to a peak and then decreases.

This peaked curve was further emphasized by the direct numerical response of *Sorex* and *Peromyscus*, since their populations rose initially with increase in prey density up to a maximum that was maintained with further increase in cocoon density. *Blarina* did not show a numerical response. The increase in density of predators resulted from increased breeding, and because the reproductive rate of small mammals is so high, there was an almost immediate increase in density with increase in food.

The two basic components of predation — the functional and numerical responses — can be affected by a number of subsidiary components: prey characteristics, the density and quality of alternate foods, and characteristics of the predators. It was shown experimentally that these components affected the amount of predation by lowering or raising the functional and numerical responses. Decrease of the strength of stimulus from prey, one prey characteristic, lowered both the functional and numerical responses. On the other hand, the quality of alternate foods affected the two responses differently. Increase in the palatability or in the number of kinds of alternate foods lowered the functional response but promoted a more pronounced numerical response.

The peaked type of predation shown by small mammals can theoretically regulate the numbers of its prey if predation is high enough to match the effective reproduction by prey at some prey density. Even if this condition does not hold, however, oscillations of prey numbers are damped. Since the functional and numerical responses undoubtedly differ for different species of predator, predation by each is likely to peak at a different prey density. Hence, when a large number of different species of predators are present the declining phase of predation is displaced to a higher prey density, so that the prey have less chance to "escape" the regulation exerted by predators.

The scheme of predation presented here is sufficient to explain all types of predation as well as insect parasitism. It permits us to postulate four major types of predation differing in the characteristics of their basic and subsidiary components.

REFERENCES

Andrewartha, H. G. and L. C. Birch. 1954. The distribution and abundance of animals. *The Univ. of Chicago Press*, Chicago.

Bird, F. T. 1952. On the artificial dissemination of the virus disease of the European sawfly, *Neodiprion sertifer* (Geoff.). *Can. Dept. Agric., For. Biol. Div., Bi-Mon. Progr. Rept.* 8(3): 1-2

Bird, F. T. 1953. The use of a virus disease in the biological control of the European pine sawfly, *Neodiprion sertifer* (Geoff.). *Can. Ent.* 85: 437-446.

Blair, W. F. 1941. Techniques for the study of mammal populations. *J. Mamm.* 22: 148-157.

Buckner, C. H. 1957. Population studies on small mammals of southeastern Manitoba. *J. Mamm.* 38: 87-97.

Burnett, T. 1951. Effects of temperature and host density on the rate of increase of an insect parasite. *Amer. Nat.* 85: 337-352.

Burnett, T. 1954. Influences of natural temperatures and controlled host densities on oviposition of an insect parasite. *Physiol. Ecol.* 27: 239-248.

Burt, W. H. 1940. Territorial behaviour and populations of some small mammals in southern Michigan. *Misc. Publ. Univ. Mich. Mus. Zool.* no. 45: 1-52.

De Bach, P. 1958. The role of weather and entomophagous species in the natural control of insect populations. *J. Econ. Ent.* 51: 474-484.

De Bach, P., and H. S. Smith. 1941. The effect of host density on the rate of reproduction of entomophagous parasites. *J. Econ. Ent.* 34: 741-745.

De Bach, P., and H. S. Smith. 1947. Effects of parasite population density on rate of change of host and parasite populations. *Ecology* 28: 290-298.

Errington, P. L. 1934. Vulnerability of bob-white populations to predation. *Ecology* 15: 110-127.

Errington, P. L. 1943. An analysis of mink predation upon muskrats in North-Central United States. *Agric. Exp. Sta. Iowa State Coll. Res. Bull.* 320: 797-924.

Errington, P. L. 1945. Some contributions of a fifteen-year local study of the northern bob-white to a knowledge of population phenomena. *Ecol. Monog.* 15: 1-34.

Errington, P. L. 1946. Predation and vertebrate populations. *Quart. Rev. Biol.* 21: 144-177, 221-245.

Fraenkel, G., and D. L. Gunn. 1940. The orientation of animals. Oxford.

Hayne, D. W. 1949. Two methods for estimating population from trapping records. *J. Mamm.* 30: 339-411.

Holling, C. S. 1955. The selection by certain small mammals of dead, parasitized, and healthy prepupae of the European pine sawfly, *Neodiprion sertifer* (Goeff.). *Can. J. Zool.* 33: 404-419.

Holling, C. S. 1958a. A radiographic technique to identify healthy, parasitized, and diseased sawfly prepupae within cocoons. *Can. Ent.* 90: 59-61.

Holling, C. S. 1958b. Sensory stimuli involved in the location and selection of sawfly cocoons by small mammals. *Can. J. Zool.* 36: 633-653.

Klomp, H. 1956. On the theories on host-parasite interaction. *Int. Union of For. Res. Organizations, 12th Congress*, Oxford, 1956.

Leopold, A. 1933. Game management. Charles Scribner's Sons.

Lincoln, F. C. 1930. Calculating waterfowl abundance on the basis of banding returns. *U.S. Dept. Agric.* Circular 118.

MacLagan, D. S. 1932. The effect of population density upon rate of reproduction, with special reference to insects. *Proc. Roy. Soc. Lond.* 111: 437-454.

Milne, A. 1957. The natural control of insect populations. *Can. Ent.* 89: 193-213.

Morris, R. F., W. F. Chesire, C. A. Miller, and D. G. Mott. 1958. Numerical response of avian and mammalian predators during a gradation of the spruce budworm. *Ecology* 39(3): 487-494.

Nicholson, A. J. 1933. The balance of animal populations. *J. Anim. Ecol.* 2: 132-178.

Nicholson, A. J., and V. A. Bailey. 1935. The balance of animal populations. Part 1, *Proc. Zool. Soc. Lond.* 1935, p. 551-598.

Prebble, M. L. 1943. Sampling methods in population studies of the European spruce sawfly, *Gilpinia hercyniae* (Hartig.) in eastern Canada. *Trans. Roy. Soc. Can.*, Third Series, Sect. V. 37: 93-126.

Ricker, W. E. 1941. The consumption of young sockeye salmon by predaceous fish. *J. Fish. Res. Bd. Can.* 5: 293-313.

Ricker, W. E. 1954. Stock and recruitment. *J. Fish. Res. Bd. Can.* 11: 559-623.

Robertson, F. W., and J. H. Sang. 1944. The ecological determinants of population growth in a *Drosophila* culture. I. Fecundity of adult flies. *Proc. Roy. Soc. Lond.*, B., 132: 258-277.

Solomon, M. E. 1949. The natural control of animal populations. *J. Anim. Ecology* 18: 1-35.

Stickel, L. F. 1954. A comparison of certain methods of measuring ranges of small mammals. *J. Mamm.* 35: 1-15.

Thompson, W. R. 1929. On natural control. *Parasitology* 21: 269-281.

Thompson, W. R. 1930. The principles of biological control. *Ann. Appl. Biol.* 17: 306-338.

Thompson, W. R. 1939. Biological control and the theories of the interactions of populations. *Parasitology* 31: 299-388.

Tinbergen, L. 1949. Bosvogels en insecten. *Nederl. Boschbouwe. Tijdschr.* 21: 91-105.

Tinbergen, L. 1955. The effect of predators on the numbers of their hosts. *Vakblad voor Biologen* 28: 217-228.

Tinbergen, N. 1951. The study of instinct. Oxford.

Tothill, J. D. 1922. The natural control of the fall webworm (*Hyphantria cunea* Drury) in Canada. *Can. Dept. Agr. Bull.* 3, new series (Ent. Bull. 19): 1-107.

Ullyett, G. C. 1949a. Distribution of progeny by *Cryptus inornatus* Pratt. (Hym. Ichneumonidae). *Can. Ent.* 81: 285-299, 82: 1-11.

Ullyett, G. C. 1949b. Distribution of progeny by *Chelonus texanus* Cress. (Hym. Braconidae). *Can. Ent.* 81: 25-44.

Varley, G. C. 1947. The natural control of population balance in the knapweed gall-fly (*Urophora jaceana*). *J. Anim. Ecol.* 16: 139-187.

Varley, G. C. 1953. Ecological aspects of population regulation. *Trans. IXth Int. Congr. Ent.* 2: 210-214.

Voûte, A. D. 1946. Regulation of the density of the insect populations in virgin forests and cultivated woods. *Archives Neerlandaises de Zoologie* 7: 435-470.

Voûte, A. D. 1956. Forest entomology and population dynamics. *Int. Union For. Res. Organizations*, Twelfth Congress, Oxford.

Voûte, A. D. 1958. On the regulation of insect populations. *Proc. Tenth Int. Congr. of Ent.* Montreal, 1956.

(Received April 16, 1959)

THERMODYNAMIC EQUILIBRIA OF ANIMALS WITH ENVIRONMENT[1]

WARREN P. PORTER[2] AND DAVID M. GATES
Missouri Botanical Garden
2315 Tower Grove Avenue, St. Louis, Missouri 63110
and
Washington University, St. Louis, Missouri 63130

TABLE OF CONTENTS

INTRODUCTION

All processes of life, all physiological events, do work and expend energy. A continuing supply of energy is necessary for an animal to live. One of the primary means by which the environment influences animals is through the exchange of energy. If the animal takes in more energy than it gives out it will get warmer, overheat and perish. If the animal loses more energy than it gains it will cool and not survive. An animal may warm or cool for a limited period of time, but on the average, over an extended period of time, an animal must be in energy balance with its environment. The mobility of most animals permits them to seek the environment most compatible with their physiological requirements for energy. The factors of the environment which are primarily responsible for energy flow to an animal are several, e.g. radiation (sunlight, skylight, and radiant heat), air temperature, wind, and humidity. There are other environmental factors, such as the gravitational, electrical, or magnetic fields, which might determine energy flow between animals and environment but may be ignored as generally not significant to the energy budget of an animal. The microclimate around an animal is thought of as a four dimensional space in which the four independent variables—radiation, wind, air temperature, and humidity—are acting simultaneously and are each time dependent.

The exchange of energy between an animal and its environment affects the body temperature of the aimal. The body temperature is considered here as a dependent variable. The rate of moisture loss by the animal is in part determined by environmental conditions and in part by the animal's physiology. The metabolic rate, although primarily a physiological property of the animal involving chemical energy, may also depend somewhat on environmental conditions. Hence, in working with relations between an animal and its environment, one is confronted with two or more dependent variables of the animal and four or more independent variables of the environment. Generally, the problem of interaction between an animal and its environment can be no simpler than this six or seven dimensional space. These variables of body temperature, rate of moisture loss, metabolic heat, radiation, air temperature, wind, and humidity must be written in a self-consistent framework or else there is no basis for comparison. This self-consistent framework is energy. There is no other relation than energy which connects the environmental factors with the animal response. All energy flow quantities used in this article are expressed in calories /cm^2 of animal surface/minute.

The outermost surface of skin, fur, or feathers of an animal is the transducing surface across which the environment interacts with the internal physiology. The streams of energy flowing to and from this surface are shown in Fig. 1. There may be direct sunlight, S, scattered skylight, s, and reflected light, r(S + s). There always is infrared thermal radiation emitted by the soil, rock, and vegetation surfaces of the environment, R_g, and from the atmosphere, R_a. The animal will lose energy by radiation from its surface according to the radiation law, $\varepsilon\sigma T_r^4$, where ε is the emissivity of the skin, fur, or feather surface, σ is the Stefan-Boltzmann constant, and T_r is the surface temperature of the radiating outermost surface. The actual physical surface from which this radiation occurs is often very irregular and complex, varying up and down with the structure of the fur,

[1] Manuscript first submitted December 2, 1968. Accepted for publication March 12, 1969.
[2] Present address: Department of Zoology, Zoology Research Building, 1117 W. Johnson Street, Madison, Wisconsin 53706.

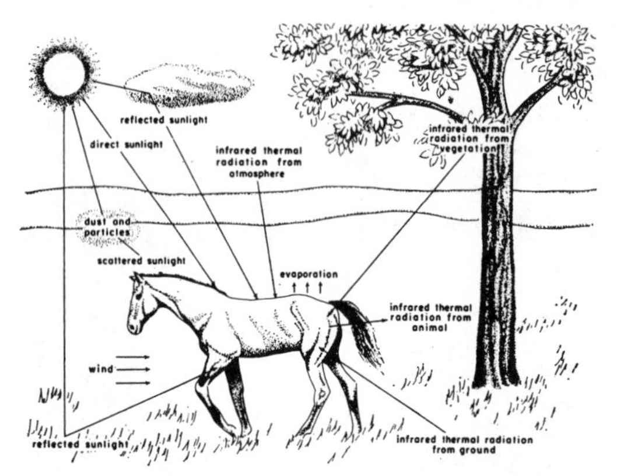

FIG. 1. Streams of energy between an animal and the environment.

feathers, or skin. The wind may blow with a speed, V, expressed in cm sec^{-1}. If the air temperature, T_a, is cooler than the surface temperature, T_r, of the animal, then energy is lost from the animal to the air by convection. The term convection includes conduction in the boundary layer of air next to the animal surface and mass movement of air beyond the adhering boundary layer. If the air is warmer than the animal surface temperature, then convection will add heat to the animal. When there is no wind movement, convection still occurs in the form of free convection set up by the temperature differential between the animal surface and the air. In addition to convection heat may be exchanged by direct conduction if the animal is in contact with a substrate at a different temperature.

Energy is lost vaporizing water within the respiratory tract and expelling it from the animal. The amount of water lost in grams per unit time when multiplied by the latent heat of vaporization of water at the appropriate temperature (580 calories per gram at 30°C), gives the rate of water loss in energy units of cal sec^{-1} or cal min^{-1}. The respiratory moisture loss is E_{ex} and moisture lost by sweating is E_{sw}. All quantities are expressed finally in terms of unit surface area as cal cm^{-2} min^{-1}. In addition to the streams of energy described above, energy is lost by defecation and urination. This is usually a small part of the total energy flow.

The model formulated here is simple. The animal is treated for conceptual purposes as a series of concentric cylinders. We know that an animal is not, in fact, simply concentric cylinders, but has appendages and irregular shapes. Yet a great deal is understood concerning the energy budget of an animal from a simple model to which complications can be added later.

The analysis given here is for steady state situations. However, much of the time an animal is in a transient energy state as it moves about in the environment. Yet while in transient states the energy budget for the animal must average within the environmental limits permitted by the steady state requirements for survival. An animal may search for food in an environment which is intolerable to it as a steady state situation, providing it is in this environment for a small period of time compared to the body time constant of the animal.

The purpose of this study is to put on a quantitative basis the interaction between animals and environment. From this our intention was four-fold; to establish the relative importance of each environmental factor and animal property to the animal energy budget; to assess the allowable errors in measurement of these factors and properties; to predict with reasonable accuracy the surface and core temperatures for any animal in any specified environment; and, knowing the physiological requirements and physical properties of the animal, to predict the limits of the climatic habitat it must occupy to survive.

The difficulty of dealing with and describing the effects of the environment on organisms has been realized for a long time (Gagge 1936; Winslow, Herrington, & Gagge 1937; and Riemerschmid 1943). In some instances careful experiments have been performed to establish some of the relations of the environment to the heat load on organisms (Digby 1955; Church 1960 a, b; Birkebak 1966; Bartlett & Gates 1967; and Norris 1967) but it is impractical to examine experimentally all the possible combinations of radiation, air temperature, wind velocity, animal size or dimension, feather or fur thickness, metabolic rate, and evaporative water loss. Moreover, one would have little insight into the relative importance of each of the environmental and animal parameters in a given situation until the full experiment was completed. There are considerable experimental difficulties in trying to simulate accurately many natural environments in which animals exist. In addition, the allowable errors in the measurements of environmental quantities and animal properties have not been evaluated in the context of all the variables acting simultaneously. For example, how precisely must we know the conductivity of fat or flesh? Does the same thickness of fur confer equal advantages to animals of different dimensions? Because of the need to try to understand why and how the environments of this planet interact with its animals, we have devised an elementary mathematical model of environment-animal energy exchange based on the physics of heat transfer and the known physical and physiological properties of animals. The predictive power of such a model is enormous, as is demonstrated here.

THE ENERGY BUDGET

The total surface of an animal in a steady state situation must have a net energy flow of zero. The sources of energy flow between an animal and the environment are metabolism, moisture loss or gain, radiation, convection, and conduction. The energy budget of an animal is written in the following form:

$$\text{Energy In} = \text{Energy Out}$$
$$M + Q_{abs} = \varepsilon\sigma T_r^4 + h_c(T_r - T_a) + E_{ex} + E_{sw} \pm C \pm W \quad (1)$$

where M is the metabolic rate, Q_{abs} is the amount of radiation absorbed by the animal surface, h_c is the convection coefficient, C is heat conducted to the substrate, W is work done, and the other quantities have been identified in the text. Most of these terms are very complex in themselves and are described in more detail below. All terms in the equation are expressed in cal cm^{-2} min^{-1}. For each new set of given conditions each of the terms in Eqn. (1) has a particular value and the surface temperature T_r of the animal adjusts so as to balance the energy budget. If anyone term changes in value, the surface temperature will readjust to a new equilibrium value. All quantities vary with time and may change simultaneously.

Metabolic rates for specific animals were taken from the literature. In most cases the values published are in cc $O_2(g\text{-}hr)^{-1}$. To convert them to cal min^{-1}, an assumption was made about the type of food each specific animal was consuming; e.g. shrews were assumed to be metabolizing protein only. The caloric energy released when one cc of O_2 is consumed depends upon whether protein, fat, or carbohydrate is being oxidized. The metabolic heat released is known (Prosser & Brown, 1961) for protein, 4.5 cal (cc of $O_2)^{-1}$; fat, 4.8 cal (cc of $O_2)^{-1}$; and carbohydrate, 5.05 cal (cc of $O_2)^{-1}$. In order to convert metabolic rate from cc $O_2(g\text{-}hr)^{-1}$ to cal min^{-1}, the literature value was multiplied by the weight of the animal, then by the caloric equivalent of the food typically consumed, then by a unit conversion to convert hours to minutes.

Surface area was computed by procedures described by Birkebak (1966). For animals smaller than 10 cm in diameter possessing small appendages, their appendage surface area was neglected. Appendage surface areas for larger limbs were determined using a mean diameter of each limb segment and its length. Total surface area was divided into the metabolic rate in cal min^{-1} to get heat production per unit area.

The metabolic rate of an animal is governed by genetic limitations to its physiology and by particular environmental conditions. The metabolic rate may be nearly constant or it may vary over a considerable range of values. All animals will lose water by breathing and some animals will sweat or salivate. The rates of water loss by various means and under various conditions must be known to understand the energy budget of an animal.

An animal receives in the natural habitat several streams of radiation including direct sunlight, skylight, reflected sunlight and thermal radiation. These radiations, R_i, are of various intensities and spectral compositions and come from many sources and directions. The animal surface presents various areas, A_i, to the different incident fluxes of radiation. The animal surface absorbs the various incident fluxes with mean absorptivities, $\bar{a}_i$. The amount of radiation absorbed, Q_{abs}, by an animal whose total surface area is A, is given by:

$$AQ_{abs} = \sum_{i=1}^{n} a_i A_i R_i = \bar{a}_1 A_1 S + \bar{a}_2 A_2 s + \bar{a}_3 A_3 r(S + s) + \bar{a}_4 A_4 R_g + \bar{a}_5 A_5 R_a \quad (2)$$

where r is the reflectivity of the underlying surface to sunlight and skylight. It is, of course, possible that several of the a_i's are equal and perhaps some of the A_i's equal one another, but in general they are different. The mean absorptivities to the various streams of incident radiation represent the coupling between the animal surface and the radiation field. If the animal is black to all wavelengths of incident radiation, the coupling of the animal surface temperature and the animal energy budget to the incident radiation is strong. If the animal is white to all wavelengths of radiation, then its surface temperature is decoupled from the incident radiation. Values of the absorptivities for different animals are given later in this paper.

Because the radiation from each source has a unique spectral quality at any given moment in time, the mean absorptivity, $\bar{a}_i$, of the animal surface to each incident flux of radiation is different. The spectral quality of the radiation from various sources is shown in Fig. 2 and the spectral absorptivities for various selected animals are given in Fig. 3. The mean value of the absorptivity of an animal surface to incident monochromatic radiation, ${}_{\lambda}R_i$, at each wavelength, λ, is given by:

$$\bar{a}_i = \frac{\int_{\lambda} a_{\lambda} R_i d\lambda}{R_i d_{\lambda}} \quad (3)$$

where ${}_{\lambda}a$ is the monochromatic absorptivity of the animal surface.

To estimate carefully the amount of radiation absorbed by an animal's total surface is generally very difficult and involves all of the complexities expressed in Eqns. (2) and (3). Yet such estimates are made reasonably well if the intensities of the streams of radiation are known and if the mean absorptivities are known. These quantities are readily measurable and when not measured are estimated fairly well.

The transfer of energy by convection is the result of a temperature difference between the animal surface temperature and the air temperature nearby. All surfaces have adhering to them a boundary layer of air which is the transition zone for the temperature gradient, for the moisture gradient, and for the dif-

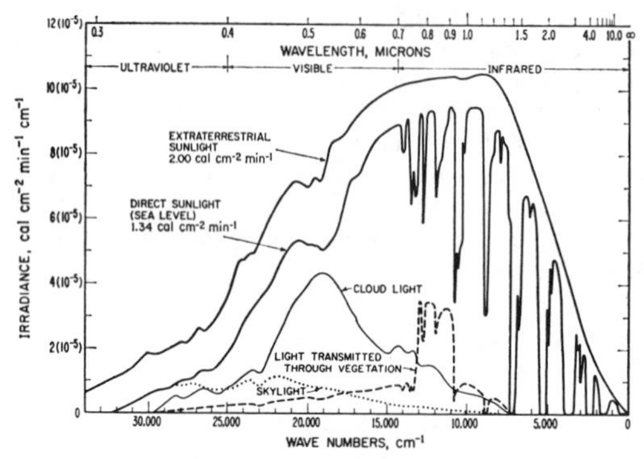

FIG. 2. The spectral distribution of extraterrestrial sunlight, direct sunlight at the Earth's surface, light from an overcast, skylight, and sunlight transmitted through the vegetation canopy.

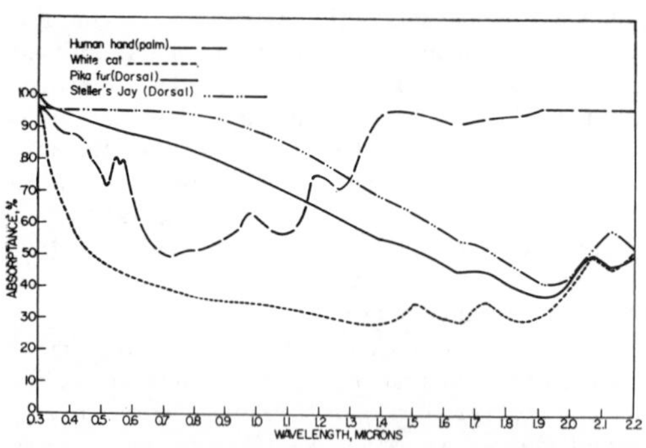

FIG. 3. Spectral absorptance of human hand, white cat, Pika (dorsal), and Steller's Jay (dorsal) as a function of wavelength.

fusion of gases between the animal surface and the free air. The thickness of the boundary layer is approximately proportional to the two-thirds root of the dimension (diameter) of the animal and depends upon the roughness, shape, and orientation of the surface. Experiments have shown the thickness of the boundary layer of cylinders to vary inversely with the one-third root of the wind speed. Heat is conducted from the animal surface across the boundary layer until free air movement beyond the layer carries the heat away by convection. Since the thicker the boundary layer the less the convection, the convection coefficient is inversely proportional to the boundary layer thickness. The convection coefficient is given by:

$$h_c = k \frac{V^{1/3}}{D^{2/3}} \tag{4}$$

where V is the wind speed in cm sec^{-1}, D is the diameter of the animal's body in cm and k is a constant which has units such that h_c is in cal cm^{-2} min^{-1} $°C^{-1}$. For a smooth cylinder with axis perpendicular to the direction of the wind flow, $k = 6.17 \times 10^{-3}$ (Gates, 1962). Strictly speaking, k should be evaluated experimentally for each animal, but throughout this work the cylinder value is used.

HEAT TRANSFER WITHIN THE ANIMAL

Heat is conducted from the surface of the animal inward to the body core if the surface temperature, T_r, is greater than the body temperature, T_b. Heat is conducted outward to the surface if the core temperature is greater than the surface temperature. A schematic diagram of the concentric cylinder model used for the analysis presented here is shown in Fig. 4. There is an inner core where the metabolic energy, M, is generated at a temperature, T_b, and from which respiratory moisture loss expels energy, E_{ex}, from the body cavity. This central core is surrounded by a fatty layer of thickness, d_b, and conductivity, K_b. The fatty layer terminates at the skin at temperature, T_s. The skin may or may not be surrounded by fur or feathers. If it is, then the fur or feathers are represented by another concentric cylinder of thickness, d_f, and conductivity, K_f. During steady state conditions, the energy crossing each concentric cylindrical surface must be the same.

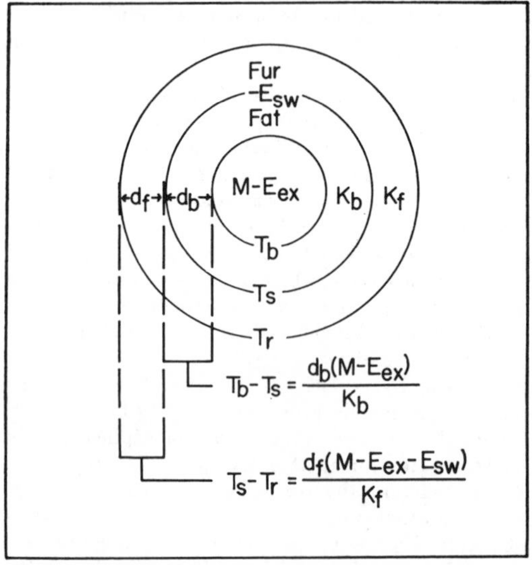

FIG. 4. Concentric cylinder model of animal for heat transfer analysis. M = metabolism, E_{ex} = respiratory moisture loss, E_{sw} = moisture loss by sweating, T_b = body temperature, T_s = skin temperature, T_r = radiant surface temperature, K_b = conductivity of fat, K_f = conductivity of fur or feathers, d_b = thickness of fat, and d_f = thickness of fur or feathers.

The net amount of energy generated within the body is $(M - E_{ex})$. This amount of energy must be conducted through the fatty layer, across which there is a temperature difference $(T_b - T_s)$. Hence,

$$M - E_{ex} = \frac{K_b}{d_b}(T_b - T_s) \tag{5}$$

At the skin surface an amount of energy, E_{sw}, is lost to sweating or evaporation of moisture. The amount of energy conducted through the layer of fur or feathers, across which there is a temperature differential $(T_s - T_r)$, must be the same energy conducted across the fat, $(M - E_{ex})$, less the energy lost by sweating:

$$M - E_{ex} - E_{sw} = \frac{K_f}{d_f}(T_s - T_r) \tag{6}$$

The total temperature difference between the internal body temperature, T_b, and the external radiating surface temperature, T_r, is:

$$T_b - T_r = (T_b - T_s) + (T_s - T_r)$$
$$= \frac{d_b}{K_b}(M - E_{ex}) + \frac{d_f}{K_f}(M - E_x - E_{sw}) \tag{7}$$

Eqn. (7) defines the ability of any animal to maintain a given temperature differential between inside and outside in terms of the following basic properties: fat, fur, or feather thicknesses and conductivities; metabolic rate; and moisture loss by breathing and sweating. The conductivity of fat is generally five to eight times greater than the conductivity for fur or feathers. Hence, for fur bearing animals and for birds, the fatty tissue adds little in the way of insulation. But for animals without fur or feathers, the fat layer must provide all of the insulation.

For animals without fur or feathers, Eqn. (7) reduces to the following when the radiating surface becomes the skin surface at temperature T_s:

$$T_b - T_s = \frac{d_b}{K_b}(M - E_{ex}) \tag{8}$$

The energy budget of the skin surface for an animal without fur or feathers is the same as represented in Eqn. (1) except with T_s replacing T_r.

The ability of an animal to support a given temperature differential is represented in Figs. 5 and 6. Here, all combinations of values known for M, E_{sw}, E_{ex}, d_b, and d_f are considered to determine the maximum and minimum values for $(T_b - T_s)$ and $(T_s - T_r)$. The conductivity of fat with total (hypothetical) vasoconstriction is not variable and the conductivity of fur or feathers can change with compactness and with moisture content. However, the conductivity of fat and fur was assumed to be constant. The thicknesses d_b and d_f were varied when necessary to take account of heat transport by the circulatory system. For the results presented here it was our desire to calculate the outer limits for an animal's tolerance.

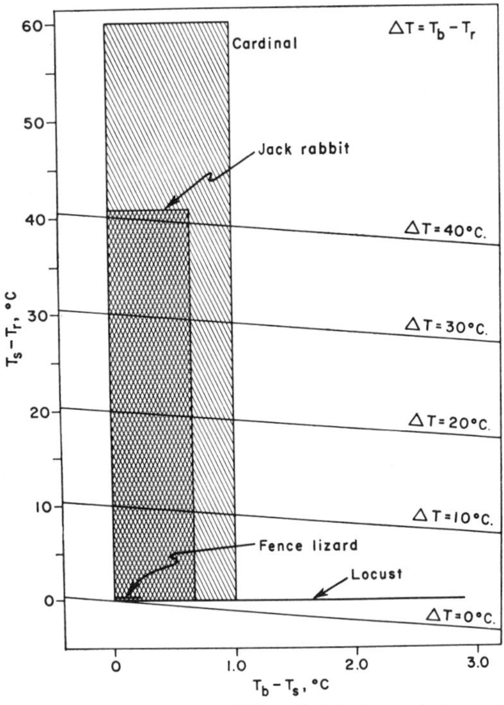

FIG. 5. Temperature differential between body and radiant surface temperature for animals. Abscissa is temperature difference across the body fat. Ordinate is temperature difference across the fur or feather layer.

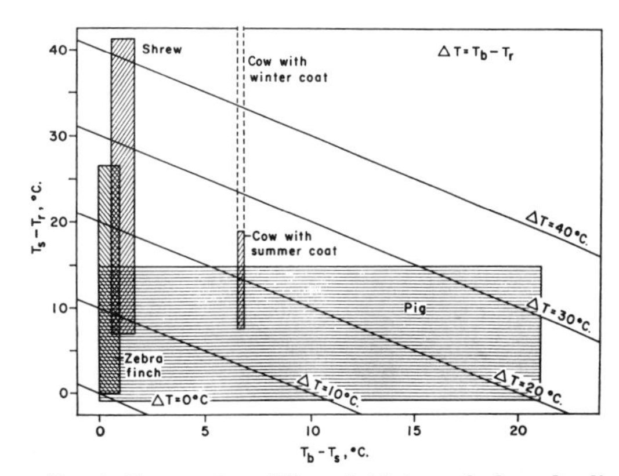

FIG. 6. Temperature differential between body and radiant surface temperature for animals. Abscissa is temperature difference across the body fat. Ordinate is temperature difference across the fur or feather layer.

To do this we would use the maximum thickness of fat or assumed conductivity of fur or feathers to be the conductivity of air at the maximum thickness of fur or feathers. The conductivity of fat used was 0.0294 cal cm^{-1} min^{-1} $°C^{-1}$ (Hatfield & Pugh, 1951) and the conductivity of air was 0.0036 cal cm^{-1} min^{-1}

$°C^{-1}$ (Gates, 1962). The error allowable for d_b or d_f depends on the particular animal. The affect of an error on a particular calculation can be determined only with reference to the climate space diagram for each animal.

The vertical axis of Figs. 5 and 6 represents the temperature differential which the animal can maintain between skin and free air across fur or feathers and the adhering boundary layer of air. The horizontal axis represents the temperature differential which the animal can maintain between the body cavity and the skin across the layer of fat. The fine lines shown in Figs. 5 and 6 represent the sum of the values of the axes or the total temperature difference $(T_b - T_r)$. These two figures show the enormous range of values of $(T_b - T_r)$ or $(T_b - T_s)$ for various animals. The values of the parameters used for the calculations are given in Table 1.

ENERGY EXCHANGE FOR ANIMAL SURFACE

The energy exchange for an animal surface is given by Eqn. (1). Substituting into Eqn. (1) the convection coefficient expressed by Eqn. (4), one gets the following:

$$M + Q_{abs} = \varepsilon\sigma T_r^4 + k(V^{1/3}/D^{2/3})\ (T_r - T_a) + E_{ex} + E_{sw} \pm C \pm W \quad (9)$$

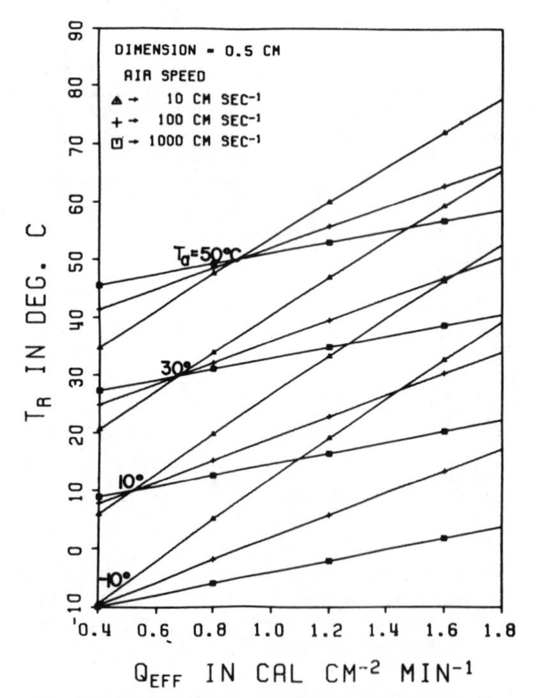

FIG. 7. Relations between radiant surface temperature of a small animal of diameter 0.5 cm as a function of $Q_{eff} = M + Q_{abs} - E_{ex} - E_{sw}$ where M = metabolic rate, Q_{abs} = radiation absorbed by animal surface, E_{ex} = respiratory moisture loss rate, and E_{sw} = sweating moisture loss rate. T_a = air temperature.

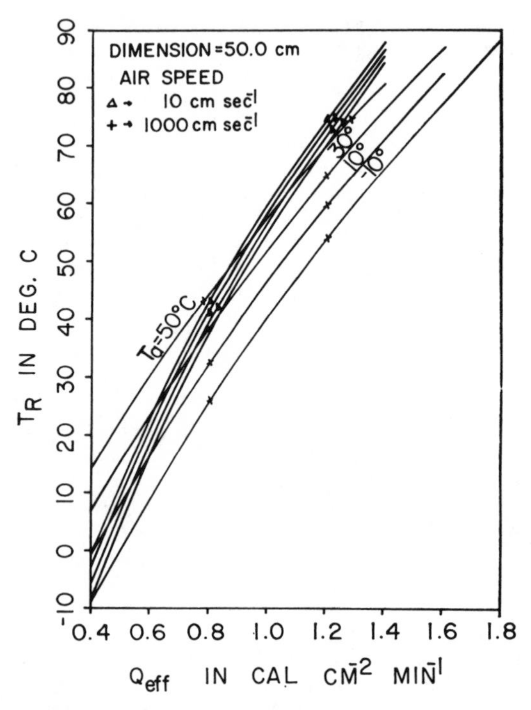

FIG. 8. Relations between radiant surface temperature of a large animal of diameter 50 cm as a function of $Q_{eff} = M + Q_{abs} - E_{ex} - E_{sw}$ where M = metabolic rate, Q_{abs} = radiation absorbed by animal surface, E_{ex} = respiratory moisture loss rate, and E_{sw} = sweating moisture loss rate. T_a = air temperature.

The work done by the animal is considered zero for the calculations given here. The influence of the environmental variables, air temperature, radiation, and wind speed on the radiant surface temperature of the animal is seen from Eqn. (9). It is desired to visualize the relation between the radiant surface temperature, T_r, and the independent variables of T_a, M, Q_{abs}, E_{ex}, E_{sw}, C, and D. Obviously some simplification must be made. For this purpose Q_{eff}, the net effective energy available for reradiation and convection, is defined as:

$$Q_{eff} = M + Q_{abs} - E_{ex} - E_{sw} \quad (10)$$

Hence:

$$Q_{eff} = \varepsilon\sigma T_r^4 + k(V^{1/3}/D^{2/3})(T_r - T_a) \quad (11)$$

where the conductance term is considered negligible. It is always easy to put the conductance term in when needed. It is now necessary to understand the dependence of T_r on Q_{eff}, T_a, V, and D.

The relation between T_r and Q_{eff} is shown in Figs. 7 and 8 for an animal of diameter 0.5 cm and 50.0 cm respectively at air temperatures of −10, 10, 30, and 50°C. In Fig. 7, wind speeds of 10, 100, and 1,000 cm sec^{-1} are given, while in Fig. 8, only 10 and 10,000 cm sec^{-1} are shown. The lines at different wind speeds must intersect at the point where $T_r = T_a$.

The relation between T_r and Q_{eff} is always steeper at low wind speeds than at high wind speeds and steeper for a large dimension than for a small dimension. The reason for this is that T_r is more tightly coupled to the air temperature by convection when the wind speed is higher or when the dimension is smaller, or both.

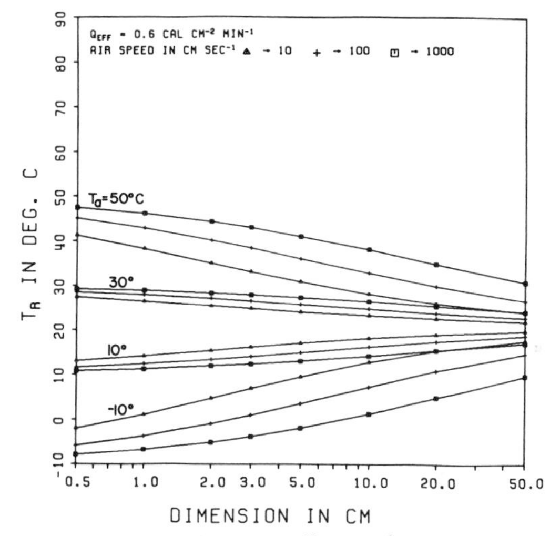

FIG. 9. Relations between radiant surface temperatures of animals as a function of animal diameter for given air temperature, T_a, and $Q_{eff} = M + Q_{abs} - E_{ex} - E_{sw} = 0.6$ cal cm^{-2} min^{-1} where M = metabolic rate, Q_{abs} — radiation absorbed by animal surface, E_{ex} = respiratory moisture loss rate, and E_{sw} = sweating moisture loss rate.

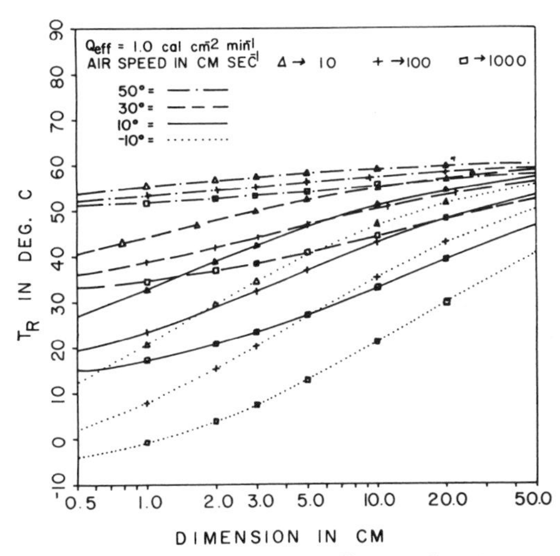

FIG. 10. Relations between radiant surface temperatures of animals as a function of animal diameter for indicated air temperatures and $Q_{eff} = M + Q_{abs} - E_{ex} - E_{sw} = 1.0$ cal cm^{-2} min^{-1} where M = metabolic rate, Q_{abs} = radiation absorbed by animal surface, E_{ex} = respiratory moisture loss rate, and E_{sw} = sweating moisture loss rate.

The relation between T_r and the diameter of the animal is shown in Figs. 9 and 10 at a Q_{eff} of 0.6 and 1.0 cal cm^{-2} min^{-1} respectively for wind speeds of 10,100, and 1,000 cm sec^{-1} and air temperatures of −10, 10, 30, and 50°C. The curves for the three wind speeds converge at each air temperature for small dimension because of convection, but for large dimension, where convection has little influence, all curves for all wind speeds and air temperatures converge at the value of T_r, corresponding to the blackbody temperature of the Q_{eff}. Hence, when Q_{eff} is 0.6 cal cm^{-2} min^{-1}, the curves converge to 20°C and when Q_{eff} is 1.0 cal cm^{-2} min^{-1}, they converge to 60°C at large dimensions. These numbers simply mean that the animal has net energy supplied by radiation absorbed plus metabolic heat less energy consumed in evaporative water loss equivalent to the energy supplied by a blackbody at these temperatures. At a fixed air temperature of 30°C and wind speed of 100 cm sec^{-1}, the relation between T_r, dimension, and Q_{eff} is shown in Fig. 11. Similar plots can be made at any other air temperature.

RADIATION ABSORBED

The average radiation absorbed as a function of the air temperature was calculated for a cylinder with specific absorptivity to sunlight. The irradiation of a cylinder can represent reasonably well the average irradiation of an animal. The purpose here is to get generalized approximations to the average

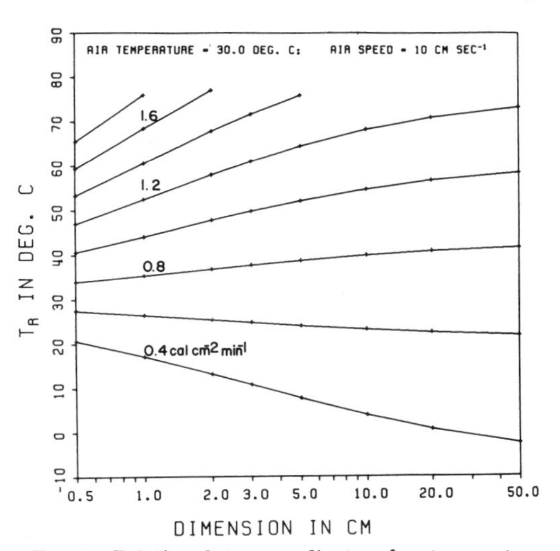

FIG. 11. Relations between radiant surface temperatures of animals as a function of animal diameters for various values of $Q_{eff} = M + Q_{abs} - E_{ex} - E_{sw}$ where M = metabolic rate, Q_{abs} = radiation absorbed by animal surface, E_{ex} = respiratory moisture loss rate, and E_{sw} = sweating moisture loss rate.

amount of radiation absorbed under various circumstances. The sun is a point source of radiation, the skylight is scattered light from the upper hemisphere, reflected light is from the lower hemisphere, and thermal radiation emitted by the ground surface is from the lower hemisphere, and thermal radiation emitted by the sky comes from the upper hemisphere. To get the average radiation from the sun on the upward facing half of the cylinder, one must make the following calculation:

$$\bar{S} = \frac{2S\int_0^{\pi/2} \cos\vartheta \, d\vartheta}{\pi} = \frac{2S}{\pi} \tag{12}$$

where S is the incident stream of direct sunlight and ϑ is the angle between the radial direction of an element on the surface of the cylinder and the direction of the sun. The integral is over a 90° quadrant of the cylinder and the two in the numerator gives the radiation on two quadrants or upper half cylinder. The extended sources of radiation, such as the skylight and the thermal radiation from ground and atmosphere, irradiate the cylinder fairly uniformly, The lower half of the cylinder may receive some reflected light, $r(S + s)$, which averages somewhere between what a diffuse reflecting surface would give and what a specular reflecting surface would produce. The absorptivity to sunlight and skylight of various animals may vary from as low as 0.2 to nearly 1.0, but the absorptivity to infrared thermal radiation is almost always between 0.95 and 1.0. For the purpose of these calculations, which necessarily must be approximate, the absorptivities to sunlight, skylight, and reflected light are assumed to be the same and equal to a and the absorptivities to thermal radiation equal to 1.0. On this basis, the following calculations were made to give the average radiation absorbed by the cylinder.

$$\bar{Q}_{abs} = \frac{a(2/\pi)S + as + ar(S + s) + R_a + R_g}{2} \tag{13}$$

The amount of direct sunlight changes with the air temperature for the following reasons. Warm air temperatures generally imply summer conditions or tropical conditions; but either of these is true when the sun is high in the sky at small zenith angle during midday. For these conditions, the amount of sunlight reaching the surface is generally high. When very low air temperatures exist, it generally implies winter conditions and the sun is low in the sky at midday. The amount of sunlight incident at the earth's surface is low for these low temperature conditions. Exceptions to this situation include the low temperatures of high mountain regions of low and middle latitudes in the summer when the direct sunlight is intense. These exceptional cases can be easily worked out separately. On this basis, an estimate was made of $\bar{Q}_{abs}$ as generally related to air

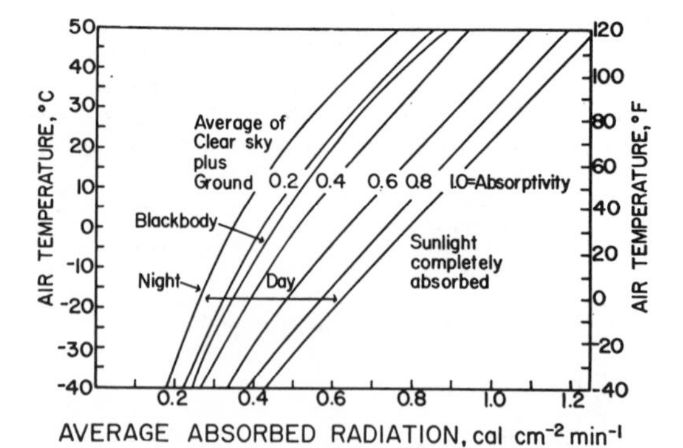

FIG. 12. Relations between the air temperature and the amount of radiation absorbed by a cylinder of various absorptivities to direct, scattered, and reflected sunlight. The absorptivity to the infrared thermal radiation from atmosphere and ground is unity. The intensity of radiation from a blackbody at the air temperature is shown. At night all radiation incident is longwave infrared. The ground radiates as a blackbody near to air temperature, but the clear sky radiates as a graybody at a much lower temperature than the air. Hence, the average radiation on a cylinder at night receiving radiation from ground and sky is less than the blackbody radiation at the air temperature.

temperature for values of absorptivity from 0.2 to 1.0. The results of this calculation are given in Fig. 12. If the absorptivity to sunlight is 0 or there is no sunlight, e.g. night-time conditions, then only thermal radiation from the ground and sky is incident on the cylinder and the average radiation absorbed is midway in value between the two fluxes. This situation is shown by the line in Fig. 12 marked "clear sky plus ground." The flux of atmospheric radiation as a function of the air temperature was estimated from Swinbank's (1963) observations and empirical formula. The thermal radiation from the ground was estimated by assuming the ground surface temperature to be near to the air temperature. This is considered a reasonable approximation, although the ground surface may be warmer or cooler than the air, depending upon various circumstances. Also shown in Fig. 12 is the line representing radiation from a blackbody at the air temperature.

CLIMATE SPACE

One of the ultimate goals for the energy budget analysis of animals is to predict the climate space which any given animal must occupy in order to survive. An animal may actually occupy a more restrictive climate space than predicted here for other reasons such as territorial limits, food supply, etc.; but an animal cannot occupy a larger climate space than that permitted by his properties. The appropriate climate space is identified by means of the analysis presented in the previous sections. The climate space is the four dimensional space which the animal can endure, formed of radiation, air temperature, wind and humidity, each acting simultaneously. It is the combination of these variables which will make the energy budget of the animal compatible with its body temperature requirements. The relations between T_r and the external environmental factors have been demonstrated as well as the relations between the internal physiological characteristics of the animal, including the animal's body temperature, and the radiant surface temperature, T_r. If one knows the body temperature and the extreme values it can have for survival, the metabolic rate, the water loss, the absorptivity, the diameter, and the thickness of fat and of fur or feathers, one can determine the simultaneous set of values of climate factors for survival of the animal. First, Eqn. (7) is solved for T_r and then Eqn. (9) is solved for the various combinations of Q_{abs}, T_a, and V for a given dimension. The humidity is not included in the calculations presented here. Allowance is made for the fact that an animal can vary its metabolic rate, water loss rate, fur or fat thickness and insulation, body temperature and sometimes absorptivity to radiation by means of color change. One must know these properties for an animal and what variation to expect for extreme conditions. Usually this information is difficult to obtain and often one or more values are missing from the information available for a particular animal.

The relations which must exist among the climate variables to give a certain body temperature or range of body temperatures for a specific animal with specific properties are now described. The relations could extend over all temperature, radiation, and wind values but are limited in the figures to the extremes of air temperature, radiation, and wind normally encountered by the animal.

Desert Iguana

The relatively large, white desert iguana, *Dipsosaurus dorsalis,* has been extensively studied and values of its properties are readily available. This animal will tolerate a maximum body temperature of 46°C (Norris, 1953). For the calculations presented here, a maximum body temperature of 45°C is used. The minimum body temperature which the animal will tolerate, according to experience by the senior author, is 3°C. A *Dipsosaurus* of 50.8 grams had a surface area of 174.8 cm^2 (Norris, 1967). For our calculations, we used an animal weighing 67 grams with a surface area of 180 cm^2. The metabolic rates used here are from Dawson & Bartholomew (1958) and the evaporative water loss from Templeton (1960). The fat thickness used is taken to be skin thickness and was measured by the authors. The diameter of the desert iguana is typically about 1.5 cm. Spectral absorptivity measurements of *Dipsosaurus* were made by Porter (1967) and by Norris (1967). The average absorptivity of the surface of a dark colored *Dipsosaurus* is about 0.8 and of a light colored lizard, 0.6.

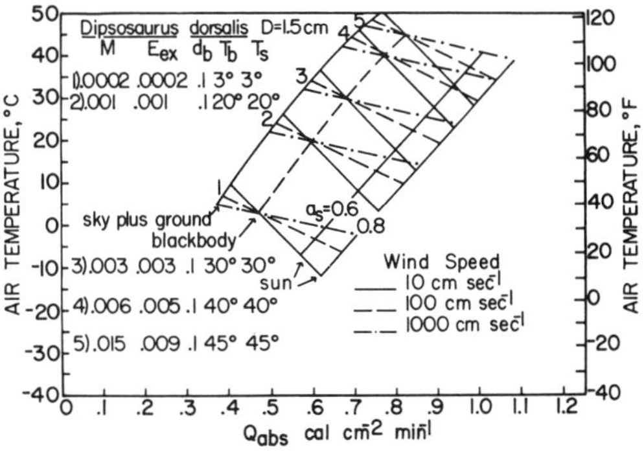

FIG. 13. Climate diagram for *Dipsosaurus dorsalis* showing relations between air temperature, radiation absorbed and wind speed for constant body and radiant surface temperatures at actual values of metabolic and water loss rates.

The climate space diagram for *Dipsosaurus dorsalis* is shown in Fig. 13. The data used for the various parameters are listed in the figure. A small poikilotherm like a lizard has relatively little physiological control of body temperature. The minimum tolerated body temperature of 3°C is given for minimum metabolic rate and minimum water loss according to the lines marked (1). These lines would indicate that in still air this lizard could not survive a night-time air temperature of less than 10°C if it were exposed to the clear night sky and not less than 5°C if in a wind of 1,000 cm sec^{-1}. A blackbody cavity, e.g. a burrow, would have to be at 3°C or greater for the lizard to survive. It is likely that this is lower than he could withstand for extended periods of time. In full sun in still air, the lizard could not survive air temperatures less than —12°C if dark, nor —7°C if light colored. In a wind of 1,000 cm sec^{-1} in sunshine, it could not withstand less than —2°C if dark, nor 0°C if light. The maximum lethal limits are between 28°C and 33°C air temperatures in full sunshine in still air for dark and light colored animals respectively, are about 40°C in wind of 1,000 cm sec^{-1}. Normally a lizard near the ground surface will not encounter a wind of this magnitude because of the great wind sheer near the surface. At night, the *Dipsosaurus,* when exposed to clear sky, could sustain air temperatures between 47 and 51°C. However, air temperatures of this magnitude do not occur at night. *Dipsosaurus* in the shade is an environment approximating a blackbody cavity. The maximum shade temperature is equal to the maximum body temperature of 45°C.

The smaller the animal the more nearly horizontal are the lines in the climate space diagram since the convection coefficient is large and the animal surface temperature is coupled tightly to the air temperature. From the climate space diagram one can predict the environmental conditions which force the desert iguana to move to a more favorable situation. If the desert iguana is in still air in full sun and the air temperature goes above 30°C, he must go into partial shade in order to endure the conditions. If the air temperature at the ground surface in the vicinity of the desert iguana goes above 45°C in the shade, he will need to seek a burrow at a cooler temperature. *Dipsosaurus* may prefer a behavioral limit, such as the set of lines No. (4), rather than the maximum lethal limit as set by lines No. (5) (Fig. 12). In this instance, the desert iguana will seek cooler habitats when the air temperature exceeds 40°C in the shade. The behavioral limits may be well within the lethal limits set by the climate space diagram.

Masked Shrew

The smallest of mammals are the shrews, which are sometimes characterized by high metabolic rates and the necessity to search for food incessantly. *Sorex cinereus,* the masked shrew, is distributed throughout the northern United States, Canada, and Alaska. The metabolic rates, fur thickness, and body temperatures are taken from the work of Morrison, Ryser, & Dawe (1959). Minimum body temperature tolerated is about 37.5°C and maximum body temperature 41°C. The surface area of an extended shrew, for which the calculations here were made, was 20.6 cm^2, and of a curled up shrew, 13.6 cm^2. Surface areas were based on the work of Morrison and Teitz (1957). The absorptivity for the shrew was measured by us.to be 0.8. The climate space diagram for *Sorex cinereus* is shown in Fig. 14. The most striking feature of the climate space diagram is its demonstration of the low maximum temperature limit for the shrew. Since the shrew lives on the ground surface amidst leaf litter and in burrows, it is rarely subjected to any significant amount of wind. Water loss from the shrew has not been measured and values used here are assumed. The set of curves marked (2) and (3)

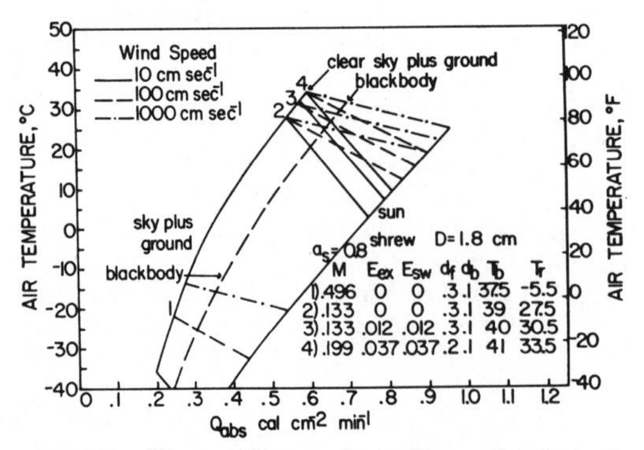

FIG. 14. Climate diagram for a Shrew showing relations between air temperature, radiation absorbed, and wind speed for constant body and radiant surface temperatures at actual values of metabolic and water loss rates.

are for metabolic rate and moisture losses near the thermal neutral zone. In still air, curve (3) would show that in blackbody conditions the shrew could not withstand an air temperature greater than 24°C. If the shrew is exposed to the cold night sky, it can withstand an air temperature of up to 31°C in still air. In full sunlight with still air, the shrew cannot endure an air temperature greater than 6°C for extended periods of time. Morrison, Ryser, & Dawe (1959) did not measure moisture loss, but measured metabolic rates, body temperatures, and air temperatures. We assumed $T_r = T_a$ in curves (3) and (4) to compute water loss in order to make our calculated body temperatures agree with those of Morrison, Ryser, and Dawe (1959) for the metabolic rates and chamber temperatures they used. If the set of lines, No. (4), are valid, then a shrew in still air could sustain blackbody temperatures not to exceed 27°C. It is very doubtful that *Sorex cinereus* can withstand this temperature for any sustained length of time, since this requires dumping a substantial amount of water. It is more likely that the real temperature maximum for this shrew is between 20 and 24.5°C, as given by curves (2) and (3) for still air.

The low temperature limit shown for *Sorex cinereus* is based on its maximum metabolic rate of 0.496 cal cm^{-2} min^{-1}, maximum fur thickness, no moisture loss (which is unreal, although it could be very low), minimum body temperature 37.5°C, and minimum surface area of a curled animal. This would indicate a blackbody temperature lower limit of —41°C. In sunshine, the shrew could go to a slightly lower temperature. It is very doubtful that a shrew can actually maintain itself under these conditions very long and more likely that the minimum air temperature is perhaps —10 to —20°C. The shrew spends the winter under snow and when snow cover is not available would need to spend very cold periods beneath leaf litter or in the soil.

Zebra Finch

The Zebra finch is a small seed-eating bird from Australia which has been imported to other countries of the world as pets. The Zebra finch is an easy experimental bird for which good physiological data exist. The majority of the data are from Cade (1964), who gives body size, weight, metabolic rates, and water loss. Further metabolic rates are given by Hamilton & Heppner (1967). Body temperatures, feather thickness, and fat thickness as well as weight (11.5 grams), surface area (31.4 cm^2), diameter (2.5 cm), and absorptivities were obtained by us. The birds are naturally nearly white and have an absorptivity between 0.21 and 0.27. When dyed black, their absorptivity is 0.59.

The climate space diagram for the Zebra finch is shown in Fig. 15. The bird is small and does not have very thick feather covering. The result is that the white Zebra finch cannot survive very low air temperatures in still air where the limit in the sun

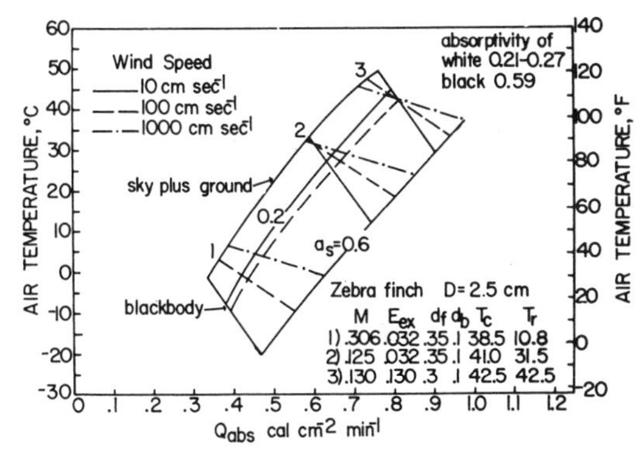

FIG. 15. Climate diagram for a Zebra Finch showing relations between air temperature, radiation absorbed, and wind speed for constant body and radiant surface temperatures at actual values of metabolic and water loss rates.

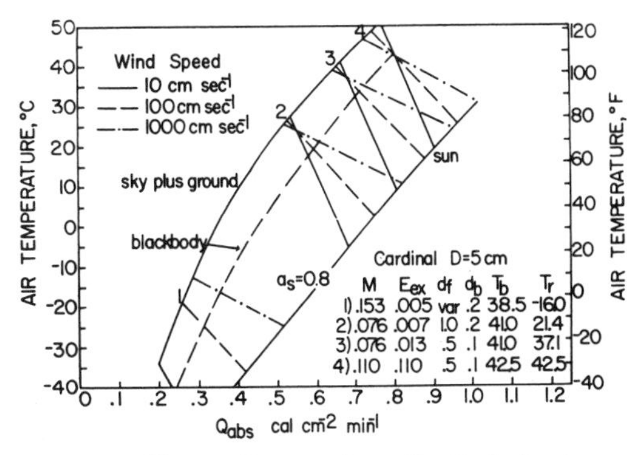

FIG. 16. Climate diagram for a Cardinal showing relations between air temperature, radiation absorbed, and wind speed for constant body and radiant surface temperatures at actual values of metabolic and water loss rates.

is about —10°C, but in the shade about —7.5°C. The lower limit for the black Zebra finch in the full sunlight is —20°C for still air. Obviously, the bird could not fly at these air temperatures and survive. When flying or in wind, these birds would not withstand air temperatures less than about 0 to 3°C in full sunlight, nor less than about 7°C at night. The white Zebra finch can withstand relatively high air temperatures in full sunlight, up to 42.5°C, but the black-dyed Zebra finch could only withstand 30°C in still air.

Kentucky Cardinal

The Kentucky cardinal, *Richmondena cardinalis,* a brilliantly colored red bird with a top notch, has in recent years extended its range westward and northeastward from its original range in the southern United States and Mexico. All metabolic and water loss rates, body temperatures, and weights for the cardinal are from Dawson (1958). The dimensions ($D = 5$ cm) and surface area (110 cm^2) are from Birkebak (1966) and from our own work. The thickness of feathers and fat are from Birkebak (1966) and from our own measurements. The absorptivity of 0.8 was measured by us with the spectrophotometer.

The climate space diagram for the cardinal is shown in Fig. 16. Many interesting things can be understood by studying this diagram carefully. At the high temperature limit the cardinal would not withstand full sunlight in still air for extended periods of time at an air temperature exceeding 20°C, but in wind of 100 cm sec^{-1} could withstand up to 26°C and at 1,000 cm sec^{-1}, up to 31°C. A cardinal will perch in a tree top singing where there is generally some air movement when air temperatures are moderate, but must fly into the shade when air temperatures become high. The cardinal can withstand very high air temperatures (up to 50°C in still air) at night when exposed to clear sky. With wind of 1,000 cm sec^{-1}, the warmest night-time air temperature for the cardinal is about 46.5°C. The evaporative water loss given for the lines No. (4) is not from Dawson, but was calculated by us to be the water loss which must occur for the bird to withstand an air temperature of 42.5°C.

At the low temperature limit, the cardinal can increase its metabolic rate, decrease its water loss, fluff out its feathers to increase insulation, and perhaps tuck its bill under its wing, which would further reduce surface area and minimize water loss. According to our calculations based on the best data available, the cardinal could withstand very low temperatures in still air. Since the lowest temperatures for most habitats occur about 0600 on winter days, it is apparent that the cardinal will not have sunshine available at this time and will be in a blackbody habitat or one of lower radiation flux. The lower limit for a cardinal with blackbody conditions would be about —40°C in still air and —16°C in wind of 1,000 cm sec^{-1}. As is true with the other animals, we do not feel very certain about the precision of the lower limits for the climate space of the cardinal, but it may be a reasonable estimate. The uncertainty has to do with the amount and quality of the insulation. The lower limits indicated here seem to be compatible with the known climate conditions within the winter range of the cardinal.

The behavior of the cardinal for the thermal neutral range is shown by the set of lines No. (2) and (3). Other examples of thermal neutral conditions of metabolism, water loss rate, and thermal insulation (thickness of fat or feathers) could be given but they will be between those shown here. It is seen from Fig. 16 that the cardinal in sunshine during thermal neutral body conditions generally must have cool or moderate temperatures. If the cardinal is not in full sun but is outside on an overcast day, he can withstand moderate to warm temperatures without becoming stressed thermally. When exposed to the full sun, the cardinal can endure 15°C warmer temperatures with a wind of 1,000 cm sec^{-1} than when in relatively still air (10 cm sec^{-1}) and remain in a thermal neutral condition. However, when exposed to an environment which is approximately a blackbody at the air temperature, a wind of 1,000 cm sec^{-1} permits the cardinal to be in only slightly warmer air than when in still air for the same metabolic rate and water loss rate.

Sheep

The sheep, *Ovis aries,* is an animal with unusual thermal insulation given by its thick covering of fleece. Sheep at the St. Louis stockyards were measured to give an average diameter of 25 cm unshorn, a surface area of 1.06 m^2 and a mean weight of 69.6 Kg. Metabolic rates were obtained from Joyce & Blaxter (1964) and from Brockway, McDonald & Pullar (1965). It was known from the work of Slee (1966) that a shorn sheep in a cold room could withstand a fairly low temperature. A metabolic rate was computed for the known values of T_b, T_s, and d_f for the coldest temperatures encountered by the sheep in Slee's work. The metabolic rate obtained was compared favorably with values given by Joyce & Blaxter (1964) and by Brockway, McDonald & Pullar (1965). Water loss rates are from Brockway, McDonald & Pullar (1965). The thickness of fur was taken from Slee (1966) for the intermediate case and reasonable estimates of the thickness were made for the extreme cases. The fat thickness used was based on the average of measurements made on slaughtered sheep at the St. Louis stockyards. The body temperature for the upper limit was 41.7°C, as given by Lee & Robinson (1941) and for the minimum and thermal neutral conditions given as 39.5°C by Joyce & Blaxter (1964). The absorptivity of fresh fleece from the stockyards as measured with our spectrophotometer was 0.7.

The climate space diagram for the sheep is given

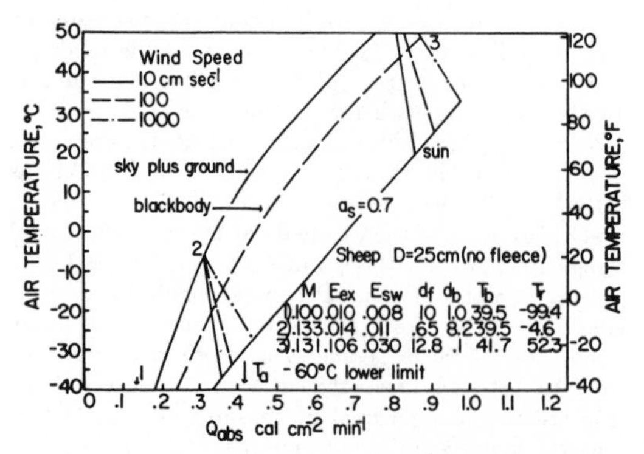

FIG. 17. Climate diagram for a Sheep showing relations betwen air temperature, radiation absorbed, and wind speed for constant body and radiant surface temperatures at actual values of metabolic and water loss rates.

in Fig. 17. Because of the larger body size of the sheep, the amount of heat transferred by convection is small, the sheep temperature is decoupled from the air temperature, and the lines in the climate space diagram become steeper than for smaller animals. Consider for a moment the upper limit for a sheep. In still air a sheep in full sun cannot endure for long an air temperature above about 20°C or with a small amount of air movement (100 cm sec^{-1}) above about 24°C. If temperatures are substantially greater than these values, sheep are to be found in partial shade out of the full solar irradiation. A sheep may endure blackbody conditions to air temperatures as high as 44 to 49°C; and if the animals can stand in the shade of a tree or a cliff and radiate to a clear cold sky, they can endure air temperatures of 55°C or greater. The sheep tended by the Navajo Indians in the arid southwest no doubt take advantage of the latter situation and should be found to behave in precisely this way. Sheep which are being grazed high in the mountains may be in the full sun during midday, but usually there is some wind and the air temperature is not much above 25°C. Because of the steepness of the limiting lines, a small reduction in the flux of incident radiation will make an enormous difference in the maximum air temperature which can be endured. The lower limit which can be tolerated by sheep is very low indeed and indicates that sheep may endure almost any degree of cold and wind, providing they have a thick coat of fleece and suffiicent food to maintain their metabolic rate.

Pig

The domesticated pig, *Sus scrofa,* is a medium sized animal with considerable body fat and relatively little fur covering. The body dimensions for a pig were obtained by measurements of adult pigs at the St. Louis stockyards. The mean diameter is taken as 36 cm, surface area 1.36 m^2, and body weight 102.8 Kg. The thicknesses of fat and hair covering were averages of measurements made at the stockyards. The metabolic rates, water loss rates, and body temperatures were obtained from Mount's chapter in Hafez (1968). Lee & Robinson (1941) give a maximum body temperature of 41°C. The absorptivity of the pig was measured with the spectrophotometer to be 0.59 for a white pig and 0.79 for a black pig. Kelly, Bond, & Heitman (1954) measured the spectral reflectance of white, red, and black pigs. They found the white pig to absorb 49% of incident solar radiation, the red about 75%, and the black about 93%. Their measurements did not cover the full spectrum of solar radiation, but only from 0.4 to 1.0μ. High reflectance of near infrared wavelengths beyond their range of measurements would reduce the average absorptivity to solar radiation and bring their measurements more in line with ours.

The climate space diagram is shown in Fig. 18.

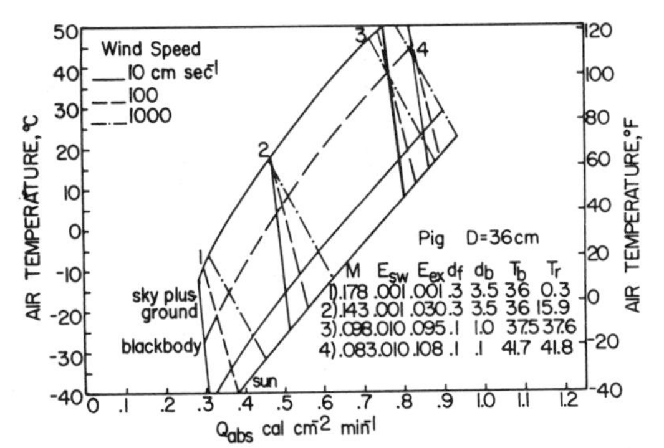

FIG. 18. Climate diagram for a Pig showing relations between air temperature, radiation absorbed, and wind speed for constant body and radiant surface temperatures at actual values of metabolic and water loss rates.

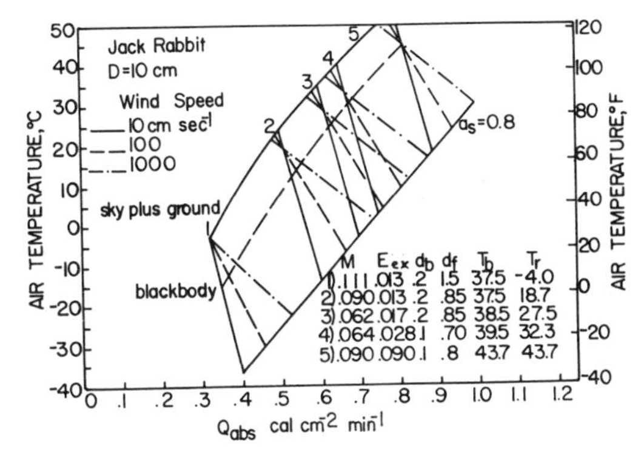

FIG. 19. Climate diagram for a Jack Rabbit showing relations between air temperature, radiation absorbed, and wind speed for constant body and radiant surface temperatures at actual values of metabolic and water loss rates.

The lines are steep because of the relatively large body size. The diagram would indicate that a pig should not be in full sun when the air temperature is above about 23°C with wind of 100 cm sec^{-1}. A pig could endure partial shade at air temperatures of 35°C, full shade up to 40°C, and air temperatures as high as 45° to 50°C only if he is fully shaded from the sun but radiating to the cold sky as well. Kelly, Bond, & Heitman (1954) clearly recognized the advantage of constructing animal shelters which would shade the animals from the sun but let them also exchange radiation with the cold north sky. These investigators state that "beef cattle of European breeds cannot usually live without shade in the summers of the Imperial Valley of California." They found that swine at an "environmental" temperature of 38°C were losing energy, primarily through evaporation of water. This is consistent with our calculations.

Differences in fur color are expected to make notable differences in the limiting climate conditions for the animals when exposed to full sun. The climate space diagram would indicate the black pig would find it difficult to endure full sun when the air temperature was above about 10°C, the red pig above about 15°C, and the white pig above about 25°C when there is little air movement. Pigs enjoy wallows and the opportunity they provide for losing heat by conduction and evaporation.

Pigs lose heat rapidly during severe winter conditions because of their lack of fur insulation. Thus at extremely low air temperatures and during severe winter conditions pigs must have shelter. In still air with blackbody conditions, pigs can withstand an air temperature of —27°C and in wind of 1,000 cm sec^{-1}, only down to —15°C.

Jack Rabbit

The jack rabbit, *Lepus californicus,* is an animal of western America which lives in extreme environments. Most of the metabolic data, water loss information, surface area of 1740 cm^2, weight of 2.3 Kg, and body temperatures are derived from Schmidt-Nielsen, et al. (1965). The body diameter of 10 cm was estimated fom general information available concerning the jack rabbit. Schmidt-Nielsen, et al. (1965) give the maximum body temperature as 43.7°C and the minimum as 37.5°C. The absorptance of the fur was not measured, but assumed to be approximately 0.8.

The climate space diagram for the jack rabbit is shown in Fig. 19. In still air the jack rabbit will not be in full sun at air temperatures above 20°C except during transient conditions. If there is strong air movement, 1,000 cm sec^{-1}, the jack rabbit may not be in full sun in steady state at air temperatures above 30°C. The jack rabbit, which is predominantly nocturnal, could endure night temperaturs as high as 45 to 50°C. During very hot days, e.g. days with $T_a = 40°C$, the jack rabbit must be in deep shade and may rest in a position whereby it could cool radiatively to the cold north sky. In any event, the jack rabbit will not be found in full sun during hot days for extended periods of time.

The jack rabbit may endure down to —15°C with considerable cover, which protects it from radiative loss to the sky. In sunshine in the winter this rabbit may sustain temperatures as low as —35°C, if there is very little wind. In strong wind they are limited to air temperatures above —20°C.

Hypothetical Animals

It is of great interest to understand the climatic limitations associated with an animal with any specified properties. The analytical method permits us to explore the entire continuum of properties and climate limitations for poikilotherms and homeotherms.

The first example to be considered here is a lizard, a poikilotherm, whose metabolic rate and evaporative water loss are considered so small as to be essentially

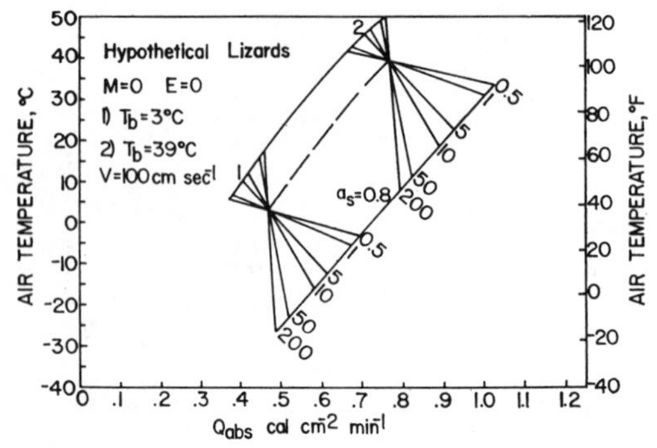

FIG. 20. Climate diagram for hypothetical lizards of varying body size from 0.5 to 200 cm showing relations between air temperature and radiation absorbed at constant wind speed of 100 cm sec^{-1}.

zero. When this is true, thermal insulation does not affect the body temperature for steady state conditions and Heath's (1964) beer can analogy is essentially correct. Fig. 20 shows the climate space for air flow of 100 cm sec^{-1} for lizards of various dimensions, assuming a maximum body temperature of 39°C and a minimum body temperature of 3°C. This diagram would indicate that a small lizard can withstand a higher air temperature limit than can a large lizard when in full sunlight. This is the result of the small lizard's losing heat more rapidly by convection than the large lizard. The large lizard will endure a lower air temperature in full sunlight than will a small lizard for a given body temperature. When considering actual habitats, one must use the air temperature at the level of the lizard body in the immediate vicinity of the body. If the lizard crawls into a burrow or under a fallen tree, this habitat approximates a blackbody at the temperature of the burrow or space around the lizard. Our discussion of steady state climate space does not take into account transient states which are often governing animal behavior. We can only say that the lizard in transient states must average out within its steady state requirements.

The lizard may spend the night in a habitat which approximates blackbody conditions for which the absolute minimum environmental temperature is 3°C and the maximum environmental temperature is 39°C. If the lizard is exposed to the celar night sky, then the air temperature must be warmer than 3°C in order that the lizard body temperature not drop below 3°C. On the other hand, for conditions of maximum body temperature the air temperature may exceed 39°C if the lizard is exposed to the clear night sky. Normally a lizard would not encounter such a condition.

The large lizard, because of his greater heat capacity, has a body temperature which responds more slowly to changes in environmental conditions than does the small lizard. If he is already warm, the large lizard can spend more time in cold air moving around seeking a position in the sun in order to bask.

One can conclude from Fig. 20 that a large lizard can sustain a large change of air temperature but a relatively narrow change of radiation absorbed when maintaining a constant body temperature. On the other hand, a small lizard can maintain its body temperature constant with a large change of radiation absorbed if the air temperature changes by a small amount only. For either a small or a large lizard, an increase in radiation absorbed requires a decrease of air temperature for a constant body temperature. For a given air temperature, a large lizard in sunshine will always have a warmer body temperature than a small lizard.

The results of calculations leading to the above Fig. 20 apply directly to behavioral observations by the senior author of small lizards in the deserts of southern California. Fairly early in the morning when the ground is still cold and there is little wind, many small lizards, e.g. *Uta stansburiana,* the side-blotched lizard, climb onto rocks to warm up. Why don't they stay on the ground to absorb the sun's rays? From Fig. 20 it is now quantitatively evident why they should climb onto rocks. They are tightly coupled convectively to the air temperature and radiation does not have as strong an effect on them as it does on large animals. By climbing up on a rock and turning their back or side to the sun, they not only maximize the amount of radiant energy absorbed but perhaps more importantly they get above the coldest layer of air at the soil surface (Gates, 1962). The rock has its own boundary layer, too, which on the sunny side of the rock will be near the surface temperature of the rock and will in part insulate the lizard from colder air provided there is little wind.

As the desert warms up and the wind begins to blow, the rock and the air layer around it now become cooler than the substrate with its boundary layer of hot air. In the heat of the day the senior author has frequently observed lizards, e.g. *Callisaurus draconoides,* the zebra-tailed lizard, on top of rocks on the side opposite the sun with their bodies pointing at the sun. Besides minimizing the radiation intercepted from the sun, they have placed themselves in the best possible convective environment. Not only are the animals up in cooler air, but in a higher wind speed regime (Gates, 1962). Fig. 20 shows us that a small drop in air temperature is exceedingly important to a lizard with a small characteristic dimension.

The potential climate relations for homeotherms is explored by means of several diagrams, since a homeotherm may or may not have fur or feather insulation and of course may be of various sizes and have varying metabolic rates. Fig. 21 displays the climate space for a small homeotherm (D = 5 cm) without fur or feathers whose body temperature is fixed

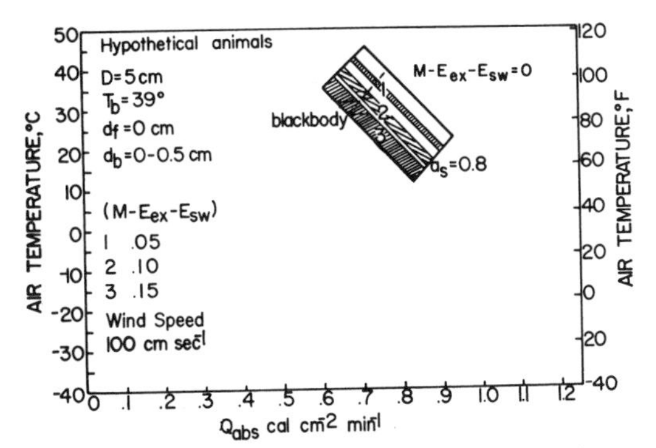

FIG. 21. Climate diagram for hypothetical small homeotherm with no fur or fat insulation showing relations between air temperature and radiation absorbed for a constant wind speed of 100 cm sec^{-1}. Three values of $M - E_{ex} - E_{sw}$ are given in cal cm^{-2} min^{-1}. Each crosshatched area is bounded by lines for minimum and maximum amount of fat.

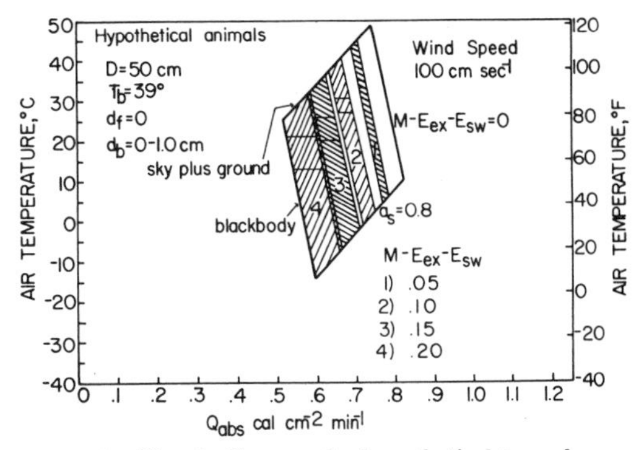

FIG. 22. Climate diagram for hypothetical large homeotherm with no fur or fat insulation showing relations between air temperature and radiation absorbed for a constant wind speed of 100 cm sec—^{1}. Three values of $M - E_{ex} - E_{sw}$ are given in cal cm^{-2} min^{-1}. Each cross-hatched area is bounded by lines for minimum and maximum amount of fat.

at 39°C and has variable amounts of fat and variable quantities of metabolic rate minus moisture loss rate. The diagram is calclulated for a gentle breeze of 100 cm sec^{-1}. The uppermost boundary for Fig. 21 and the next three figures is for a condition $M - E_{ex} - E_{sw} = 0$. Each of the shaded areas represents climate relations necessary to maintain body temperature at 39°C for varying amounts of $(M - E_{ex} - E_{sw})$. It is notable that a small homeotherm with no fur or feather insulation has enormous difficulty maintaining its body temperature nearly constant. Put another way, the climate to which a small homeotherm is restricted is very narrow for survival. The hypothetical homeotherm illustrated in Fig. 21 would not survive a night-time environmental or blackbody temperature less than about 28°C. The small homeotherm without insulation must have a very high metabolic rate in order to survive temperate or cool conditions.

The allowable error in the values of d_b can be indicated from Fig. 21. If the maximum net metabolic rate $(M - E_{ex} - E_{sw})$ is 0.05 cal cm^{-2} min^{-1} an error in measurement of d_b of 0.5 cm is not a serious error compared to the same error in d_b for an animal with a net metabolic rate of 0.15 cal cm^{-2} min^{-1} where the affect is 3 times as serious.

The climate space for a large homeotherm without fur or feather insulation and body temperature of 39°C is shown in Fig. 22. The lines of constant surface temperature for constant body temperature are now very much steeper than for the small animal. The large homeotherm without fur or feather insulation can withstand a blackbody temperature as low as 13°C and maintain a fixed body temperature of 39°C when $M - E_{ex} - E_{sw} = 0.20$ cal cm^{-2} min^{-1}. It is easy for the reader to visualize what the climate space will do if a fixed body temperature of 32°C is required rather than 39°C by noting that the intersection of the dashed blackbody line with the upper boundary line marked $(M - E_{ex} - E_{sw} = 0)$ is the body temperature of the animal. If one cuts out an overlay of the climate space parallelogram and slides it downward to the left along the blackbody line until the intersection is at the new body temperature, say 32°C, one can see the new climate requirements of air temperature and radiation absorbed. Hence, if 32°C is the body temperature required rather than 39°C, the suggested procedure will show that the lowest blackbody temperature compatible for survival, if $M - E_{ex} - E_{sw} = 0.20$ cal cm^{-2} min^{-1}, is now 6°C. The decrease in minimum air temperature for survival is the same amount as the decrease in required body temperature. Fig. 22 illustrates once again how very difficult it is for a homeotherm without fur or feathers to survive large variations of climate. The fine horizontal lines seen clearly in Fig. 22 and less clearly in Fig. 21 are limits of the air temperature as prescribed by compatible day and night maximum air temperatures. Rarely will a night-time air temperature maximum exceed a daytime temperature maximum, although we know it occasionally does happen for strong advective conditions. In other words, if the daytime temperatures with much sunlight available are very cold, the night-time temperature is not likely to be very warm. The animal may go from cold daytime air temperatures to warm night-time conditions by moving into a burrow or under a log. The parallelogram between two fine horizontal lines within a given crosshatched zone is the region of compatible climate conditions encountered in the air above ground day and night.

The climate space for a small homeotherm with fur or feather insulation and a body temperature of 39°C is shown in Fig. 23. By comparing this figure

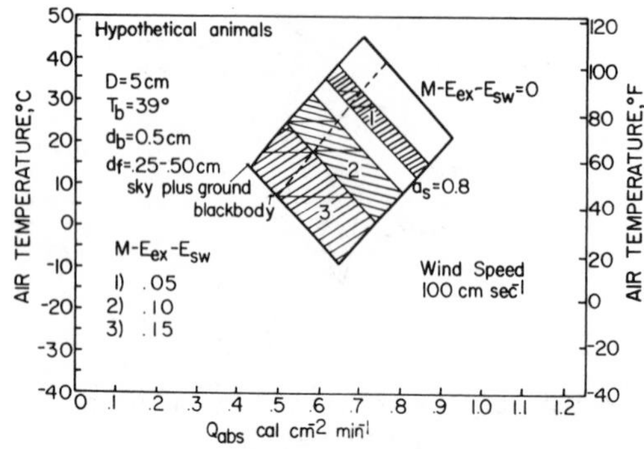

FIG. 23. Climate diagram for hypothetical small homeotherm with fur or feather insulation showing relations between air temperature and radiation absorbed for a constant wind speed of 100 cm sec^{-1}. Three values of $M - E_{ex} - E_{sw}$ are given in cal cm^{-2} min^{-1}. Each cross-hatched area is bounded by lines for minimum and maximum amount of fur or feathers.

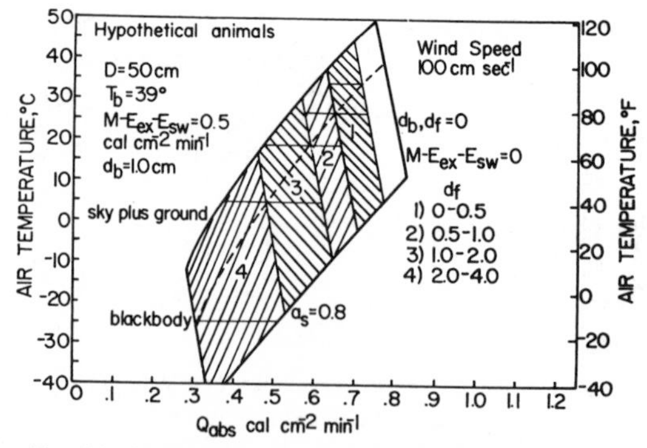

FIG. 24. Climate diagram for hypothetical large homeotherm with fur or feather insulation showing relations between air temperature and radiation absorbed for a constant wind speed of 100 cm sec^{-1}. Three values of $M - E_{ex} - E_{sw}$ are given in cal cm^{-2} min^{-1}. Each crosshatched area is bounded by lines for minimum and maximum amount of fur or feathers.

with Fig. 21, the significant importance of fur and feather insulation for a homeotherm is seen immediately. The well insulated homeotherm can withstand much lower temperatures for the same values of $(M - E_{ex} - E_{sw})$ than can the homeotherm without good insulation.

The climate space for a large homeotherm with fur or feather insulation and a body temperature of 39°C is shown in Fig. 24. Here the use of a good thickness of fur or feather insulation permits the large homeotherm to maintain its body temperature at 39°C when the air temperature becomes quite low. Comparing Figs. 21 and 22, and also 23 and 24, one notices how much steeper are the constant surface temperature lines for constant body temperature for the large homeotherm than for the small homeotherm. The large animal cannot withstand as high an air temperature in full sunshine at the temperature maximum as can the small animal and the small animal requires more radiation or a higher metabolic rate at low air temperatures for survival than does the large animal.

SUMMARY

Any organism must be in thermodynamic equilibrium when averaged over a reasonable length of time in order to survive. Environmental factors affecting the exchange of energy between an organism and the environment are sunlight, skylight, reflected light, thermal radiation, air temperature, air movement, and water vapor pressure of the air. Animal properties which affect the exchange of energy are metabolic rate, moisture loss rate, conductance of fat, fur, or feathers, absorptivity to radiation, and body size, shape, and orientation. The mechanisms affecting energy exchange are radiation, convection, conduction, and evaporation.

A mathematical model for the energy budgets of animals is derived based on simple concentric cylinder approximations to body geometry. From the physiological requirements for a certain body temperature or temperature range, the simultaneous values of radiation, air temperature, and wind speed are derived in order that the animal is in energy balance. The three dimensional space of radiation absorbed, wind speed, and air temperature is referred to as the climate space which the animal must occupy in order to survive. Climate spaces for the following animals are derived: desert iguana, masked shrew, zebra finch, cardinal, jack rabbit, pig, and sheep. In addition, the climate spaces for families of hypothetical poikilotherms and homeotherms are derived.

ACKNOWLEDGMENTS

We thank LaVerne Papian for invaluable assistance in computer programming. W. Rees Davis made many reflectance measurements and reduced the spectral data for computer processing. We thank Jim Hauser, of the St. Louis Museum of Science and Natural History, for providing stuffed animals used in the wind tunnel experiments. We thank Dr. Joel Wallach, of the St. Louis Zoological Park, for samples of fresh skin for reflectance measurements.

We also thank C. L. Burns and John Guffey, of Swift and Company, for the use of their facilities to obtain detailed measurements of surface areas and fur and fat thicknesses from freshly killed animals and for the use of fresh hides for spectral measurements. We thank Howard White, of St. Louis Order Buyers, for fresh hog skins and for the opportunity to make field measurements of live animals to spot check some of our predictions. We thank Julie Gates for illustrations. The senior author would like to thank Ron Alderfer and S. Elwynn Taylor for

TABLE 1. Values of parameters

Parameters used for calculations. M, E_{ex}, and E_{sw} in cal cm^{-2} min^{-1}, d_b, d_f in cm, and ΔT_f and ΔT_b in °C.

	M		E_{ex}		E_{sw}		d_b		d_f		ΔT_f		ΔT_b	
	Max	Min	Max	Min	Max	Min	Max	Min	Max	Min	Max	Min	Max	Min
Shrew	0.496	0.199	0	0.037	0	0.037	0.1	0.1	0.3	0.2	41.3	6.9	1.7	0.6
Cow														
Summer	0.149	0.149	0.013	0.006	0.083	0.008	1.4	1.4	0.5	0.5	18.8	7.4	6.8	6.5
Winter	0.149	0.149	0.013	0.006	0.083	0.008	1.4	1.4	2.7	2.7	101.0	39.8	6.8	6.5
Pig	0.178	0.083	0.001	0.108	0.001	0.010	3.5	0.1	0.3	0.1	14.7	-1.0	21.0	-0.1
Zebra Finch	0.306	0.130	0.032	0.130	0	0	0.1	0.1	0.35	0.30	26.6	0	0.9	0
Locust	0.86	0.0001	0.020	0	0	0	0.1	—	—	—	—	—	2.9	0
Cardinal	0.153	0.110	0.005	0.110	0	0	0.2	0.1	1.5	0.5	61.7	0	1.0	0
Jack Rabbit	0.111	0.090	0.013	0.090	0	0	0.2	0.1	1.5	0.8	40.8	0	0.7	0
Fence Lizard	0.01	—	0.003	—	—	—	0.1	—	—	—	—	—	0.5	—

TABLE 2. Wind tunnel experiments.

	Characteristic Dimension (cm)	Air Temperature (°C)	Wind Speed (cm/sec)	Q_{abs} (cal/cm^2-min)	Mean Absorptivity	Zenith Angle (deg.)	T_r Calculated (°C)	T_r Measured (°C)
Siberian Snow Leopard (*Panthera uncia*)	23.0	27.5	259	1.45	0.78	90	61.7	58
Gray Fox (*Urocyon cinereo argenteus*)	13.5	27.8	269	1.10	0.66	90	53.9	53
Glaucous-Winged Gull (*Larus glaucescens*)	3.0	28.0	268	0.81	0.48	20	33.4	34
Starling (*Sturnus vulgaris*)	4.6	27.8	268	1.29	0.85	90	54.0	56
Cardinal (*Richmondena cardinalis*)	4.6	27.5	268	0.97	0.76	32	40.9	42
Eastern Meadowlark (*Sturnella magna*)	5.7	27.2	268	1.07	0.78	45	46.3	50
Bobwhite (*Colinus virginianus*)	6.4	27.2	268	1.12	0.78	55	49.4	48

helpful discussions and advice during the course of this investigation.

This research was supported by the Center for the Biology of Natural Systems, Washington University, under PHS Grant No. ES 00139-03 and the computer facilties of Washington University under NSF Grant No. G 22296.

LITERATURE CITED

BARTLETT, P. N. & D. M. GATES. 1967. The energy budget of a lizard on a tree trunk. Ecology **48**: 315-322.

BIRKEBAK, R. C. 1966. Heat transfer in biological systems. Int. Rev. General and Exptl. Zool. **2**: 269-344.

BROCKWAY, J. M., J. D. MCDONALD, & J. D. PULLAR. 1965. Evaporative heat loss mechanisms in sheep. J. Physiol. **179**: 554-568.

CADE, T. J. 1964. Water and salt balances in granivorous birds. In Proc. 1st Int. Symp. Thirst in regulation of body water. M. J. Wagner, ed. Macmillan, New York. pp. 237-256.

CHURCH, N. S. 1960a. Heat loss and body temperatures of flying insects. I. Heat loss by evaporation of water from the body. J. Exp. Biol. **37**: 171-185.

———. 1960b. Heat loss and the body temperatures of flying insects. II. Heat conduction within the body and its loss by radiation and convection. J. Exp. Biol. **37**: 186-212.

DAWSON, W. R. 1958. Relation of oxygen consumption and evaporative water loss to temperature in the cardinal. Physiol. Zool. **31**: 37-48.

DAWSON, W. R. & G. A. BARTHOLOMEW. 1958. Metabolic and cardiac responses to temperature in the lizard *Dipsosaurus dorsalis*. Physiol. Zool. **31**: 100-111.

DIGBY, P. S. B. 1955. Factors affecting the temperature excess of insects in sunshine. J. Exp. Biol. **32**: 279-298.

GAGGE, A. P. 1936. The linearity criterion as applied to partitional calorimetry. Amer. J. Physiol. **116**: 656-668.

GATES, D. M. 1962. Energy exchange in the biosphere. Harper and Row, N. Y. 151 pp.

———. 1968. Sensing biological environments with a portable radiation thermometer. Applied Optics. **7**: 1803-1809.

HAFEZ, E. S. E., ed. 1968. Adaptation of domestic animals. Lea and Febiger. Philadelphia. 415 pp.

HAMILTON, W. J. III & F. HEPPNER. 1967. Radiant solar energy and the function of black homeotherm pigmentation: an hypothesis. Science **155**: 196-197.

HATFIELD, H. S. & L. G. C. PUGH. 1951. Thermal conductivity of human fat and muscle. Nature **168**: 918-919.

HEATH, J. E. 1964. Reptilian Thermoregulation: Evaluation of field studies. Science **146**: 784-785.

JOYCE, J. P. & K. L. BLAXTER. 1964. Respiration in sheep in cold environments. Res. Vet. Sci., **5**: 506-516.

KELLY, C. F., T. E. BOND, & H. HEITMAN. 1954. The role of thermal radiation in animal ecology. Ecology **35**: 562-569.

LEE, D. H. K. & K. ROBINSON. 1941. Reactions of the sheep to hot atmospheres. Proc. Roy. Soc. Queensland **8**: 189-200.

MORRISON, P. R. & F. TEITZ. 1957. Cooling and thermal conductivity in three small Alaskan mammals. J. Mammal **38**: 78-86.

———, F. A. RYSER & A. R. DAWE. 1959. Studies on the physiology of the masked shrew, *Sorex cinereus*. Physiol. Zool. **32**: 256-271.

MOUNT, L. E. 1968. Adaptation of Swine. Adaptation of domestic animals; E. S. E. Hafez, ed. pp. 277-291. Lea and Febiger, Philadelphia.

NORRIS, K. S. 1953. The ecology of the desert iguana, *Dipsosaurus dorsalis*. Ecology **34**: 265-287.

———. 1967. Color adaptation in desert reptiles and its thermal relationships. Symposium on Lizard Ecology. W. Milstead, ed. 162-229.

PORTER, W. P. 1967. Solar radiation through the living body wall of vertebrates with emphasis on desert reptiles. Ecol. Monogr. **37**: 273-296.

PROSSER, C. L., & F. A. BROWN, JR. 1961. Comparative Animal Physiology. W. B. Saunders, Philadelphia. 688 pp.

RIEMERSCHMID, G. 1943. Some aspects of solar radiation in its relation to cattle in South Africa and Europe. Onderstepoort J. Vet. Sci. and Animal Industry **18**: 327-353.

ROBINSON, K. & D. H. K. LEE. 1941. Reactions of the pig to hot atmospheres. Proc. Roy. Soc. Queensland **8**: 145-158.

SCHMIDT-NIELSEN, K., T. J. DAWSON, H. T. HAMMEL, D. HINDS, & D. C. JACKSON. 1965. The jack rabbit—a study in its desert survival. Hualradets Skrifter **48**: 125-142.

SLEE, J. 1966. Variation in the response of shorn sheep to cold exposures. Anim. Prod., **8**: 425-434.

SWINBANK, W. C. 1963. Longwave radiation from clear skies. Quart. J. Roy. Met. Soc. **89**: 339-348.

TEMPLETON, J. R. 1960. Respiration and water loss at higher temperatures in the desert iguana, *Dipsosaurus dorsalis*. Physiol. Zool. **33**: 136-145.

WINSLOW, C. E. A., L. P. HERRINGTON, & A. P. GAGGE. 1937. Physiological reactions of the human body to varying environmental temperatures. Am. J. Physiol. **120**: 1-22.

APPENDIX I—*WIND TUNNEL EXPERIMENTS*

The convection coefficient, h_c, used in the energy budget equation was the engineering value for smooth cylinders (Gates, 1962). Most animals, however, are furred or feathered. To evaluate the error inherent in using a convection coefficient for smooth cylinders and applying it to feathered and furred cylinders, we compared measured and predicted surface temperatures of stuffed animals in a wind tunnel. Surface temperatures were measured with a Barnes portable radiometer (Gates, 1968). To predict surface temperature, the Q_{abs} had to be determined. Spectral absorption as a function of wavelength of the animal's surface was determined from reflectance measurements made by a Beckman DK-2A spectroreflectometer. Energy as a function of wavelength from overhead floodlamps was determined from manufacturers' specifications. Beer's Law (Porter, 1967) was used to determine the energy available to the animal after it passed through the plexiglas. Absorption coefficients of plexiglas were determined using plexiglas transmission data from the DK-2A spectroreflectometer (Porter, 1967). The monochromatic energy available at the animal's surface times the monochromatic absorption of the surface was integrated over the whole spectrum by a high speed digital computer. The energy absorbed (by a plane normal to the incident light) times the cosine of the zenith angle, where the zenith angle is the angle from the vertical, gives the actual amount of radiation absorbed. The temperature of the plexiglas sheet between the lights and the animals was measured with the portable radiometer pointed up at the plexiglas from the position of the animal's back. By using the radiometer to view temperatures in the hemisphere viewed by the back of the animal, the infrared energy striking the back of the animal can be determined. Wind speed was measured with a Hastings wind meter at the level of the back and about 3 cm upwind of the back. Air temperature was measured with a small thermocouple in the same location. The characteristic dimension was taken across the body at the point of measurement. The fur or feathers were not depressed except to insure that good contact was occurring with the calipers. A ruler was used for the very large animals.

The energy budget equation was solved for T_r, assuming a convection coefficient for a smooth cylinder at right angles to the wind. The actual surface temperature of the animals was measured with the radiometer. Table 2 lists the animals studied and the data which were used in the energy budget predictions. The animals were oriented normal to the wind. Agreement between calculated and measured surface temperatures was generally good.

The predicted surface temperatures of a stuffed whooping crane chick and a stuffed ostrich chick were very different from measured values. These differences were due to convection coefficients which differ from the values for a smooth cylinder. To reduce the error in Q_{abs} and to solve the energy budget equation for approximate convection coefficients for the oddly shaped bodies of the chicks, a large radiant tunnel was installed at the end of the wind tunnel. Q_{abs} in the tunnel consisted only of longwave infrared radiation which is nearly perfectly absorbed. The h_c's for the whooping crane chick and the ostrich chick were determined using a non-linear regression program on a high speed digital computer. The data for the program came from ten experiments on each of the birds in the radiant tunnel.

AN ORDINATION OF THE UPLAND FOREST COMMUNITIES OF SOUTHERN WISCONSIN*

J. ROGER BRAY† AND J. T. CURTIS
Department of Botany, University of Minnesota, Minneapolis, Minnesota
Department of Botany, University of Wisconsin, Madison, Wisconsin

TABLE OF CONTENTS

INTRODUCTION

A renewed interest in objective and quantitative approaches to the classification of plant communities has led, within the past decade, to an extensive examination of systematic theory and technique. This examination, including the work of Sörenson (1948), Motyka *et al.* (1950), Curtis & McIntosh (1951), Brown & Curtis (1952), Ramensky (1952), Whittaker (1954, 1956), Goodall (1953a, 1954b), deVries (1953), Guinochet (1954, 1955), Webb (1954), Hughes (1954) and Poore (1956) has accompanied theoretic studies in taxonomy [Fisher (1936), Womble (1951), Clifford & Binet (1954), Gregg (1954)] and in statistics (Isaacson 1954). It is a conclusion of many of these studies that nature of unit variation is a major problem in systematics, and that whether this variation is discrete, continuous, or in some other form, there is a need for application of quantitative and statistical methods. In ecologic classification, an increased use of ordinate systems, which has been stimulated by the development of more efficient sampling techniques and the collection of stand data on a large scale, has prompted the proposal of the term "ordination" (Goodall 1953b). Goodall (1954a) has defined ordination as "an arrangement of units in a uni- or multi-dimensional order" as synonymous with "Ordnung," (Ramensky 1930), and as opposed to "a classification in which units are arranged in discrete classes." The present study is an attempt to examine the upland forests of southern Wisconsin in relation to a suspected multidimensional community structure by the use of ordination method and in so doing, to review the theoretic position necessary to such an examination.

* This work was supported, in part, by the Research Committee of the Graduate School of the University of Wisconsin, from funds supplied by the Wisconsin Alumni Research Foundation.

The Junior F. Hayden Fund of the University of Minnesota provided partial support of publication costs.

† Present address: Department of Botany, University of Toronto, Toronto, Canada.

LITERATURE REVIEW

The application of quantitative techniques to community classification has been based, in part, upon the assumption that quantitative community composition, as determined by suitable sampling methods, can be a primary basis for the building of ordination systems. This assumption was emphasized by Gleason (1910) in an examination of the relationship between biotic and physical factors. Gleason quoted Spalding (1909), "The establishment of a plant in the place which it occupies is conditioned quite as much by the influence of other plants as by that of the physical environment," and concluded from his own observations, ". . . the differentiation of definite associations is mainly due to the interrelation of the component plants; and the physical environment is as often the result as the cause of vegetation." Further emphasis upon vegetation in itself was made by Clements & Goldsmith (1924), and more recently, by Mitsudera (1954), who regard the community as an instrument which, if properly examined and manipulated, might be a key to the relation of biotic and physical phenomena. Cain (1944), Goodall (1954a, 1954b), Whittaker (1954) and Williams (1954) have questioned the relevance of considering single physical environmental factors apart from an environmental complex. Insistence upon the study of vegetation on its own level has been reiterated in many studies with recent examples in Shreve (1942), Curtis & McIntosh (1951), Brown & Curtis (1952), and Ramensky (Pogrebnjak 1955).

Included in information on community structure are two basic relationships: that of individual species to each other, and that of stands or plots as a whole to each other. These relationships have determined the development of two complementary but separate approaches to the problem of classification.

The species approach stresses the degree of mutual occurrence of a species with other species. This approach received an early quantitative background in the work of Forbes (1907a&b, 1925) who proposed a coefficient of associate occurrence which he applied to the study of bird and fish communities. Further development of the concept of associate occurrence led to the elaboration of indexes of interspecific association (Dice 1948; Cole 1949; Nash 1950) and to the construction of indexes of relative species occurrence (deVries, 1953, 1954; Bray 1956a). Indexes of interspecific association have been applied, in classification, (1) to correlate species with host specificity or with environment (Agrell 1945; Hale 1955), (2) to determine community groupings by the selection of groups of species with high interspecific correlations (Stewart & Keller 1936; Tuomikoski 1942; Sörenson 1948; Goodall 1953a; Hosokawa 1955-1956) and (3) to determine degree of amplitudinal overlap of species as an indication of kind of community variation (Gilbert & Curtis 1953). Indexes of relative species occurrence have been used as the basis for a spatial ordination of the species (deVries 1953, 1954) or for an objective assignment of species adaptation values (Bray 1956a).

The second basic approach, that of correlation among stands as a whole, can be roughly divided into three methods. The first uses information other that that derived from the vegetation in order to establish a primary series of gradients or regimes along which a subsequent vegetation alignment is undertaken (Wiedemann 1929; Vorobyov & Pogrebnjak 1929; Pogrebnjak 1930; Ramensky 1930; Hansen 1930; Whittaker 1956). These regimes do not usually represent direct physical environmental factors, but more often express environmental complexes of interrelated factors, such as soil moisture, or they follow environmental controls, such as elevation, which determine a complex of factors. Ramensky, for example, used previously established soil moisture and soil nutrient regimes to construct a primary stand ordination from which "Functional Averages" were extracted by a series of eliminations of aberrations in the compositional stand data. These averages were considered the median conditions of the biocoenosis and served as bases for final stand orientation. The work of Ramensky included some of the first intergrading bell-shaped species distributions to be demonstrated along vegetational gradients.

The second stand method is the use of objective techniques to show relationships among stands which have previously been classified into discrete units, usually within the Braun-Blanquet system. This use was given an early formulation in the work of Lorenz (1858) who was apparently the first to apply quantitative methods in community classification when he compared various kinds of moors on the basis of "per cent of species similarity." Later techniques, including those of Kulczyński (1929), Motyka *et al.* (1950), Raabe (1952) and Hanson (1955) use Jaccard's Coefficient of Community or one of its quantitative modifications to show the compositional similarity of units on various hierarchical classification levels.

The third method attempts, from a direct analysis of quantative vegetational data, to demonstrate degree of relationship by the construction of compositional gradients which are independent of environmental or other considerations. The use of the various techniques of factor analysis (Goodall 1954b), of stand weighting devices based on the assignment of species adaptation values (Curtis & McIntosh 1951; Brown & Curtis 1952; Parmalee 1953; Kucera & McDermott 1955; Horikawa & Okutomi 1955), and of attempts to utilize directly indexes of quantitative coefficients of community (Whittaker 1952; Bray 1956a) or indexes of occurrence probability (Kato *et al.* 1955) are examples of the above approach.

Previous Treatment of the Upland Forest of Wisconsin

A linear ordination of the stands of the upland forest of southern Wisconsin was presented by Curtis & McIntosh (1951). Subsequent studies were made in which soil fungi (Tresner *et al.* 1954) were arranged along this ordination and in which corticolous cryptogams were related in part to the ordination and in part to host specificity (Hale 1955). Other studies were made of the forest herbs (Gilbert 1952), of autecological characteristics of herbs (Randall 1951), and of the savanna transition into prairie (Bray 1955).

Limitations to a linear presentation became apparent from continued Wisconsin field work. One was the observation of ecological substitution in which two separate species alternated in sharing what appeared to be identical ranges of environmental tolerance (McIntosh 1957). Further reason to suspect the existence of a possible multidimensional structure came from a growing realization of the importance of past history in determining the composition of any stand.

Source of Data

All of the 59 stands of this study were sampled by the same methods. The trees were measured by the random pairs technique (Cottam & Curtis 1949) using 40 points and 80 trees per stand. The characters here used are absolute number of trees per acre and total basal area per acre, both on a species basis. The shrubs and herbaceous plants were sampled by 20 quadrats, each 1 m. sq., laid at every other point. The character used is simple frequency.

The stands employed were selected from the large number available by a stratified random procedure, so devised as to give an equal number of stands from each major geographic portion of the southwestern one-half of Wisconsin. All stands were at least 15A in size, were on upland sites upon which rain water did not accumulate, and were in reasonably undisturbed condition. As actually applied, this last criterion meant that the stands were ungrazed, had not been subject to fire within the recent past, and had never been logged to such an extent as to create large openings in the canopy. In most cases, a few trees had been removed at various intervals in the past, as witnessed by an occasional stump. The limited logging probably created serious errors in the measured amounts of *Juglans nigra,* since this high-value species was deliberately searched for on an intensive scale during World War I. It is believed that the population densities of the remaining species did not vary greatly from those which would normally be produced by natural death and windthrow. A very few stands were totally undisturbed for at least the past 50 yrs.

The sampling methods employed for both the trees and the understory were not, as could be expected, completely free from error. Estimates of sampling error were made by repeated sampling of the same stand, using different investigators in both the same and in different years. Two extensive series of such tests, in a maple woods and an oak woods, showed a standard error of 10.8% for the individual tree species and 7.1% for the understory plants. On this basis a conservative estimate of over-all error in the individual stand measures of about 10% seems reasonable. Obviously, this error would be much less for the most common species and greater for the rare species (Cottam *et al.* 1953).

We are indebted to Dr. Orlin Anderson, Dr. R. T. Brown, Dr. Margaret L. Gilbert, Dr. George H. Ware, Dr. Richard T. Ward, and especially to Dr. R. P. McIntosh for their aid in the collection of the original data. Professor Grant Cottam of the University of Wisconsin and Professor J. W. Tukey of Princeton University were very generous with their time and advice on various problems.

Taxonomic nomenclature in the present paper is after Gleason (1952).

TREATMENT OF THE DATA

NATURE OF THE APPROACH

The ordination approach was selected for the present study in order to provide statements (1) which depict, with a sufficient degree of quantitative exactness, the compositional structure of a community and (2) which might be able to give some initial indication of the over-all patterns of interaction between biologic and physical phenomena. The possibility of using ordination statements to suggest causal reactions is dependent on the concept of physical and biotic factors interacting in a relationship in which each factor is, to some degree, mutually determined by the others. There is, therefore, as is often noted in ecologic writing, no simple cause and effect relationship between physical phenomena (as primarily causal) and biotic phenomena (as primarily effectual), especially in the more complex environments. There is, rather, instead of a domain which is determined by a small number of independent factors (that is, a system of mechanist causality), a field of interrelated units and events (configurational causality). The ordination of this field is, then, a plotting of the changes in some biotic and/or physical features from area to area within the system, or, in another sense, a mapping of its complexity, Such a mapping indicates, by the relative proximity of different features and their varying spatial patterns, the degree to which the features may participate in a mutually determined complex of factors. With the completion of this mapping, it may then be possible to apply statistical tools which indicate more fully the causal interactions in any one part of the ordination.

Of the two major approaches to ordination study, that of stand or of species orientation, it was decided to use a technique which gave theoretic spatial relations of stands as a first result. With such a framework, the distribution patterns of individual species can be easily studied by directly plotting some measure of their behavior in each stand. A similar plotting can be made for measures of environmental or historic factors in each stand, or for general descriptive features. If the ordination is originally based upon species rather than stand relationships, then the location of relative stand position (and as a consequence, correlation with environmental features) becomes more difficult.

USE OF SCORE SHEETS

If the degree of similarity of stands, one to another, is to be assessed, then some decision must be made as to what criteria are to be used in judging this similarity. There is wide agreement amongst temperate-zone ecologists that community comparisons must be made on a floristic basis and that environmental or other features are not valid for primary comparisons. Unfortunately, this agreement does not extend much beyond the general idea of floristics. Clements and the Anglo-American school generally recommend the use of the dominants as the main criterion of community or stand relationship, while Braun-Blanquet and his adherents use characteristic species of high fidelity, even though these may be small, rare or otherwise dynamically insignificant members of the assemblage. Lippmaa (1939), Daubenmire (1954) and others use synusia, either singly or in combination, for their characterization of community resemblance. In no case has the total flora of a given stand been used, since the determination of all of the bacteria, soil fungi, soil algae, liverworts, mosses, lichens and vascular plants is usually beyond the facilities at the command of ecologists. The principle is well recognized, therefore, that the entire species complement is not needed for meaningful statements about community composition. The disagreement lies in the question of where to draw the line short of the total flora.

In the similarly complex problem of soil classification, it is possible that the degree of resemblance between a series of soil samples might be determined by applying a single test to each sample. Such tests might measure the texture of the soil, its percentage content of sand, or some other simple character. It would then be possible to arrange the group in descending order, as from very sandy through sandy-loam to non-sandy. The sandy loams in the center would have certain features in common and would differ greatly from the two extremes, but they might well include soils which differed widely among themselves. The application of a second test to the samples, such as fertility level, would serve to differentiate soils which were of similar texture, but, even so, the resulting groups might not be homogeneous—they might differ in organic matter content, pH, color, or other characters. A similar problem was recognized by Pirie (1937), "In an earlier part of this essay the transition from living to non-living was compared to the transition from green to yellow or from acid to alkaline. If this comparison were valid, it would be possible to lay down a precise but arbitrary dividing line. But as it has been shown that "life" cannot be defined in terms of one variable as colours can, the comparison is not strictly valid and any arbitrary division would have to be made on the basis of the sum of a number of variables any one of which might be zero." The best approach might therefore be to apply a series of tests, each examining some pertinent or important aspect of soil makeup. A study of the results of the series of tests would give a firm basis for comparison of the original group and would easily pick out the soils most nearly related to each other and least closely related to others. It might be desirable to weight the test results, in order of their importance. The calcium content is more important than the sodium content in most temperate forest soils and its results might be appropriately weighted to show this importance. Statistically, the degree of relationship could be shown by a suitable measure of the correlation between the sets of test values for any two soil samples.

The same procedure can be employed in the floristic analysis of plant communities. A standard series of tests can be applied to each stand and the results for pairs of stands correlated with each other. If this is done for all stands in the series, then similar stands should have high mutual correlation values, but dissimilar stands should show low correlation. If an appropriate measure of correlation is used, then the resulting index should give a linear measure of the difference between any two stands.

One test of the similarity of two stands could be the quantity of *Acer saccharum* that each contains, analogous to a determination of the calcium content of soil samples. A series of such tests, using other dominants, would increase the precision of the comparison. Similarly, a series of herbs and shrubs, chosen to include some which were restricted to the formation under study and others known to be sensitive to the varying conditions present in different stands of the formation, could be added to the list of tests.

In the current studies, measurements of 26 species were used as the tests. Twelve of these species were dominant trees, while the remainder were herbs and shrubs. The trees included all of the species with a presence value above 33% in the 59 stands studied. The herbs and shrubs were chosen at random in blocks along the original continuum gradient from among those species which were neither overly common nor rare and which had shown clear cut distribution patterns in previous studies. Weighting of the 12 dominant tree species was accomplished by using two separate measurements of each species (absolute density per acre, and absolute dominance per acre) as independent tests.

There were, thus, 38 tests employed for each stand: frequency (in 1 m quadrats) of 14 species of herbs and shrubs, density of 12 species of trees, and dominance (in square inches of basal area at breast height) of the same 12 trees. The results of the tests for each stand were recorded in a separate score sheet for that stand. (Of course, many of the test scores on a given sheet were zero, when one or more test species was missing from that stand.) The test scores were in different units, since the original measurements were made in three different classes. This discrepancy was rectified by expressing each score as a percentage of the maximum value attained by that test on any of the sheets. These corrected scores thus indicated, in comparable units, the behavior of each test species in relation to its optimum behavior in the entire series. Since the number of test scores and the sum of these scores varied from stand to stand, the scores for each stand were adjusted to a relative basis. They finally indicated, therefore, the relative amount contributed by each test organism to the combined score for the stand. In one stand, for example, species A may contribute 17.1% of the total score while species B may represent only 1.9%. The adjusted scores on a relative basis appear to offer the best basis for making comparisons between stands (Whittaker 1952).

INDEX OF SIMILARITY

The choice of a suitable index is largely dependent on the choice of an ordination technique. There are, however, several characteristics of available indexes which should be examined. The standard correlation coefficient, "r," incorporates a square transformation which leads to the weighting of the importance of entries with high values. Thus, if a pair of stands have one or two species in common which have high score values, the stands will have a high correlation coefficient regardless of the relative similarity or dissimilarity of their lesser species. It can be shown that high values of "r" cover a wide range in interstand variation and are relatively insensitive in the medium to high areas of stand similarity. There is a high sensitivity in the lower range of coefficient values of "r," but this sensitivity lies in an area where the ecologic differences between stands are not very significant due to the residue of widely plastic species which inhabit many of the stands of any geographic area. Of the available indexes of similarity employed in phytosociology, both Gleason's quantitative modification of Jaccard's Coefficient of Community (Gleason 1920) and Kulczyński's index (1927) can be shown to have a greater ability to differentiate stands within the area of medium to high similarity than has the correlation coefficient. When the sum of score values is relative and equals 100, both Gleason's and Kulczyński's coefficients can be expressed in the terms later used by Motyka *et al.* (1950) as $C = \frac{2w}{a+b}$ where a is the sum of the quantitative measures of the plants in one stand, b is the similar sum for a second stand, and w is the sum of the lesser value for only those species which are in common between the two stands (Oosting 1956). Thus, if two stands by chance had exactly the same scores for exactly the same species, the index would be 1.00, since (a) and (b) would be equal and both would equal (w). If there were no species in common, then the index would be zero. The range from no resemblance to complete identity is appropriately covered by the range from 0 to 1. This index appears to be the best approximation yet available to a linear measure of relationship.

As used in the present study, the index reduces simply to (w), or the sum of the lesser scores for those species which have a score above zero in both stands. This is due to the use of relative scores, such that (a) plus (b) is always 2.00 in every pair of stands and $\frac{2w}{2.00} = w$. In practice, the score sheets were so arranged that the final adjusted scores were recorded in the last column on the extreme edge of the sheet. One sheet could then be superimposed, in turn, on every other sheet in a slightly offset position, and the lesser values added on a machine for all tests where a positive value was present on both sheets.

When a large number of stands are studied, the calculation of the w index becomes burdensome, since there are $n \times \frac{n-1}{2}$ comparisons to be made. Thus, for 10 stands, 45 comparisons are needed, while for 100 stands, 4950 are required. In such cases, recourse should be had to electronic calculators, using punch cards as score sheets. In the present case, a complete comparison was made by hand for 59 stands, resulting in 1711 values of the w index. These values were arranged by stand number in a matrix. We would be happy to correspond with anyone who is interested in obtaining a copy of the matrix for further work.

THE ORDINATION METHOD

The use of stand data and of a summation of a series of tests of these stands has been outlined as the quantitative basis for the present study. Of the ordination techniques which were reviewed, many depend upon the use of a previous knowledge and sometimes classification of either the vegetation or physical environment. Although this use is not necessarily undesirable in ordination studies, it is apparent that a technique which can extract an ordination directly from the available data would be best suited to the present study. One quantitative and completely objective technique which makes this extraction is factor analysis.

Factor analysis seeks to draw "functional unities" (factors) from an oriented table (i.e. matrix) of correlation coefficients. These coefficients can be calculated, as in Goodall (1954b), from a correlation among species, which are correspondent to tests in factor analysis; or, if the relationship among stands is needed, the use of direct interstand correlation is permissible (Tucker 1956). In either case, the standard techniques of factor analysis are applicable. The extraction of a functional unity is followed, in most factor techniques, by the computation of a new matrix, called the residual matrix, from which the next unity can be obtained. This series of extractions results in a number of linear vectors, called the factor matrix, and an attempt is then made to identify each vector with an underlying cause. Factor analysis is used in areas where no hypotheses are available about the causal nature of the domain, and is based upon the assumption, according to Thurstone (1947), that ". . . a variety of phenomena within a domain are related and that they are determined, at least in part, by a relatively small number of functional unities or factors." When applied to ordination study, however, these functional unities are not, as is emphasized by Goodall (1954b), direct environmental forces but are rather, sociologic factors. A sociologic factor is defined as "an element in the description of the composition of the vegetation," and it may or may not be related to environmental factors.

The application of factor analysis to the present study was carefully considered, but was rejected for the following reasons: (1) the heavy computational load involved in handling 59 stands, (2) the disadvantages, which have been discussed above, in applying "r" (the correlation coefficient) to stand data, and (3) a hesitancy in interpreting factor anaylsis when applied to stand data especially in regard to the difficulties noted by Goodall (1954b), "though the factors may be statistically orthogonal, they are not biologically independent; the interpretation thus becomes more complicated." The construction of a preliminary empiric method with the following criteria therefore seemed desirable: (1) vegetation structure is regarded as a possible key to the nature of the interaction of factors, and, as such, must be studied on its own level, (2) an extraction of an ordination directly from objectively derived data without previous classification would be desirable.

The basis of a technique which might satisfy the above criteria is the same as that for ordination systems in general: the degree of phytosociologic relationship between stands can be used to indicate the distance by which they should be separated within a spatial ordination. The degree of relationship of vegetation units has usually been measured by some estimate of the similarity of stand composition, with a high degree of similarity signifying a close spatial proximity. The technique which will be outlined attempts, therefore, to extract from a matrix of measurements of interstand similarity, a spatial pattern in which the distance between stands is related to their degree of similarity.

Given a matrix of values of distance between points in Euclidean space, it is possible, without a prior knowledge of their location, to reconstruct their spatial placement (Torgerson 1952). This reconstruction depends upon simple techniques in which, in two dimensional space, for example, 3 points not in the same location and not on the same line, are used to locate the other points by their relationship to the 3 reference points. If, for example, there were in a matrix, 4 points, of which A, B, and C were each separated by a distance of 40 units and point D was separated from A and B by 20 units and from C by 34.64 units, then the position of these points could be established, with A, B, and C forming the apicies of an equilateral triangle and D occurring midway on line AB.

When coefficients of community are used, however, as indicators of spatial distance, then exact interstand distances are not available, since the position of a stand in relation to another stand occurs within an area of uncertainty originating in the sampling error made in surveying the stands. Furthermore, it is likely that stands occupy proximate instead of exact theoretic positions in relation to each other. The occurrence of stands within an area of uncertainty results in a matrix of estimated proximate distances. If such proximate distances are available, then the location of stand D in the previous example might be in a different position relative to reference stands A, B, and C as compared to three other reference stands (if exact interstand distances were available, the positions would be the same regardless of the choice of reference stands). The technique to be developed, therefore, is a preliminary attempt to derive an ordination from estimates of proximate interstand distance which gives a single spatial configuration most closely approximating the matrix distances.

This technique depends upon the selection of a pair of reference stands for the determination of stand positions on any one axis. Given proximate interstand distances, the choice of reference stands is of crucial importance. In making this choice, it is evident that reference stands are comparable, in part, to sighting points as used in plane-table surveying and that those stands which are furthest apart will be more accurate for judging interstand distance than those which are in close proximity. This accuracy is especially desirable because of the area of uncertainty in which each stand fluctuates relative to the positions of neighboring stands. If these fluctuations are greater than the actual distance apart of the reference stands, then the resulting ordination will reveal only these fluctuations. It is necessary for any ordination that the sphere of fluctuation for any stand be small in relation to the space occupied by the ordination as a whole. The choice of reference stands should be, therefore, of those stands which are furthest apart and as a consequence, have the greatest sensitivity to over-all compositional change.

AXIS CONSTRUCTION

The ordinate location of points in space by the use of reference stands is illustrated by the ordination of five points whose hypothetical interstand distances are shown in the lower-left of Table 1. The distances, although hypothetical, represent exact spatial distances. They were determined by inverting the estimates of stand similarity which appear in the upper-right of the table so that a high degree of similarity was represented by a low degree of spatial separation. The inversions were accomplished by subtracting each index of similarity from a maximum similarity value of 100.

To locate stands between a pair of selected reference stands, a line connecting the reference stands is drawn to scale on a piece of blank paper, and the position of every other stand is projected onto this line. The projection is accomplished by rotating two arcs representing the distance of the projected stand from each of the reference stands, and then projecting the point of arc intersection perpendicularly onto the axis. Applying the criterion of the greatest degree of spatial separation as determining the choice of reference stands, Table 1 shows stands number 1 and 2 to have a maximum separation of 99.9 units. These stands were selected, therefore, as the x axis reference stands and are placed in Fig. 1 at a distance of 99.9 units. Stand 3 in Table 1 is 70 units from reference stand 1 and the same distance from reference stand 2. Stand 3 is located in Fig. 1, therefore, at the intersection of arcs with radii of 70 units and bases at points 1 and 2. Two such intersections are possible in a two dimensional ordination, and the points of intersection are projected perpendicularly onto the x axis, as shown in Fig. 1, to give an x axis location of 50 units. Stand 4 can also be located along the x axis by the arc intersection and projection technique; it occurs at 62 units along the x axis. Stand 5 is similarly located, after intersection and projection, at 62 x axis units. It can be proven geometrically that a constellation of points in n space can be projected perpendicularly onto the line connecting the two reference points which are furthest distant in the constellation by using the above technique.

TABLE 1. Matrix of hypothetical exact interpoint distances. The upper-right portion of the table shows hypothetical data on point similarity for an exact spatial system. The lower-left portion shows data on similarity which were inverted to show interpoint distance.

Stand No.	1	2	3	4	5
1	....	0.1	30	30	30
2	99.9	....	30	50	50
3	70	70	....	17.8	79.6
4	70	50	82.2	....	35.2
5	70	50	20.4	64.8	....

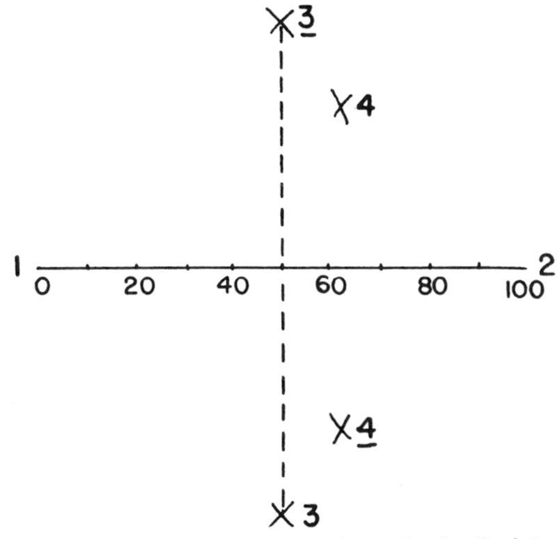

FIG. 1. Demonstration of stand location by the intersection and projection technique. Stand 3 is located at the intersection of arcs with radii of 70 units, and bases at stands 1 and 2. Its projected position on the x axis is midway between stands 1 and 2.

A second axis can be constructed by the same method using a line on the paper erected at a right angle to the x axis. Two new reference stands are selected which are in close proximity on the x axis, but which are nevertheless separated by a great interstand distance. In the matrix in Table 1, stands 3 and 4 fit such criteria showing an x axis separation of 12 units and an interstand distance of 82.2 units. If stand 3 is assigned a location at its upper arc intersection point in Fig. 1, then the interstand distance of 82.2 units of stands 3 and 4 indicates that the proper location of stand 4 is at its lower arc intersection point (intersection of underlined number 4, Fig. 1.). The location of stands 3 and 4 is shown, in relation to the x axis reference stands, in Fig. 2.

Stands 3 and 4 are, therefore, separated by a projected distance of 12 units on the x axis in Fig. 1, but are, nevertheless, separated as shown in Fig. 2 by an interstand distance of 82.2 units. The x axis proximity of stands 3 and 4 and their high spatial separation indicate that they might be used

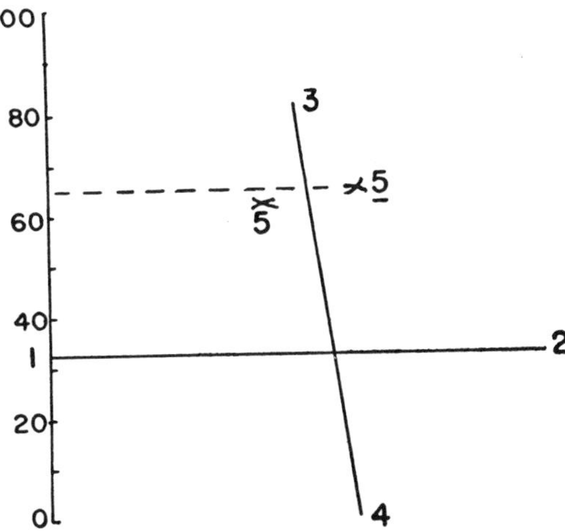

FIG. 2. Demonstration of y axis construction and stand location. Stand 5 is located at one of the two intersections of arcs 20.4 units from reference stand 3 and 64.8 units from reference stand 4. The distance of 50 units from stand 2 to stand 5 indicates the point of intersection to the right is the correct point for projection onto the y axis.

as reference stands for y axis construction. In Fig. 2, a y axis location of stand 5, which is 20.4 units from reference stand 3 and 64.8 units from reference stand 4, is illustrated. The arc intersections of stand 5 with reference stands 3 and 4 shows, after projection, two separate y axis locations, one of which, because the line connecting stands 3 and 4 is not perpendicular to the x axis, is incorrect. To determine the correct position of stand 5, the distance of stand 5 from the reference stands of the x axis should be consulted. The distance of stand 5 from reference stand 2 is 50 units which indicates that the arc intersection to the right (intersection in Fig. 2 with underlined no. 5) which is 50 units from stand 2, is the correct point for projection onto the y axis. Stand 5 is located, therefore, after perpendicular projection at a distance of 65 units on the y axis in Fig. 2.

By following the outlined technique, the information on interpoint distances in Table 1 has been used to indicate the spatial location of the points in a two-dimensional ordinate system. If such distance information necessitated the use of more than 2 dimensions, then further dimensions could be constructed, using the same technique.

APPLICATION OF THE METHOD

X AXIS CONSTRUCTION

The ordination of the upland forests depends upon selecting stands from the complete matrix which can be used as reference points for the location of the other stands. Since the available interstand distances in the complete matrix represent only proximate distances, it is obvious that any resulting stand ordination will only approximate interstand relationships. It is assumed, however, that there is a certain degree of order within the matrix and that the interstand distances are not a series of random numbers. Only a limited number of stand positions for any one stand should, therefore, be possible. As noted before, these positions should fluctuate within a narrow area if an ordination is to present a meaningful approximation of the matrix measurements.

Before the reference stands for the first axis can be selected, coefficient of community values, as found in the complete matrix, must be inverted so that a low coefficient of community value represents a relatively greater spatial separation and a high value represents a close proximity. Although the coefficient has a range from 0 to 100, the error involved in sampling a stand makes it unlikely that any of a series of replicate samples from the same stand will show the maximum value. Two stands were sampled, each 7 times, using the field methods described earlier, and coefficients of community were calculated among these samplings. The mean coefficient of community within each of the 7 replications was 82 which indicated a mean error of index reproducibility of around 20 index units. A value of 80 was, therefore, considered to represent the maximum coefficient value, that is, the value for two indentical stands; the highest value actually found in the complete matrix for 59 stands was 79.

To convert coefficients of community to interpoint distance values, an inversion was accomplished by subtracting each coefficient value from the fixed maximum of 80. Stands which had a coefficient of 0 were separated, therefore, by a maximum distance of 80 units. All subsequent mention of stand interpoint distances refers to the inverted coefficients of community.

Using the criterion of the greatest spatial separation for the choice of reference stands, an examination of the inverted values showed 3 pairs of stands, numbers 35 and 136, 00 and 137, and 33 and 138, to be separated by the maximum distance of 80 units. Since these 3 stand pairs each showed coefficients of community of zero indicating no similarity, their relationship to the other stands in the study was examined to determine whether they were completely unrelated to the other stands. It is suggested that a stand pair with a 0 coefficient of community be used as a reference pair only if each member of the pair shows a value greater than 0 with all stands which are not members of reference pairs. By establishing this criterion, the choice of a stand pair member which shows no relationship to other stands, and which, therefore, contributes nothing to a knowledge of their relative spatial location, will be avoided. This criterion can be met in relation to stand pairs 35-136, 00-137, and 33-138 each member of which shows a relation greater than 0 to every other stand in the study.

The stand pairs selected above appear to represent 2 sets of related stands, numbers 136, 137 and 138 and numbers 00, 33, and 35. The index values between each of the members of each set were highly significantly correlated, as tested with r. Each of the 3 pairs of reference stands were used, therefore, to ordinate stands along the x axis on the supposition that the use of several sets of stand data might include more information from the matrix than if a separate set were used, and might reduce the fluctuations from non-exact distance measurements.

Since each of the 3 pairs of reference stands were separated by the maximum distance of 80 units, a line of 80 units was drawn connecting each of the reference pairs. The ordination was accomplished by the technique outlined above of arc rotation and of the projection of the point of arc intersection onto the x axis. Stand 89, for example, showed a coefficient of community of 39 and 17 with stands 00 and 137. The inverse of these values, representing spatial separation, is 41 and 63, and stand 89 was located, therefore, at the intersections of arcs with radii of 41 and 63, respectively. These intersections were projected perpendicularly onto the x axis at a position of 25.5 units from reference stand 00 and of 54.5 units from reference stand 137.

The final x axis position of a stand was determined as the median position for the stand in relation to the three pairs of reference stands. There was a close similarity in the x axis positions of the stand in relation to the three pairs of reference stands. Stand 89, for example, had x axis positions of 54.5, 48.5, and 46.5, and was assigned an x axis location of 48.5. The final median values presenting x axis location are shown in column 1 of Table 2.

Table 2. Stand locations in three dimensions

Stand No.	Ordination Axis			Stand No.	Ordination Axis		
	X	Y	Z		X	Y	Z
00	79.0	35.5	37.0	93	53.0	54.0	52.0
01	55.0	58.5	41.0	95	4.0	45.75	40.5
03	21.0	55.0	32.0	96	37.5	51.75	33.0
04	32.5	61.25	43.0	100	65.0	30.25	48.0
05	63.0	38.0	50.5	101	62.0	21.0	52.0
06	46.5	33.5	41.5	102	49.5	40.75	45.5
09	45.0	38.75	44.0	103	61.0	33.5	48.0
15	67.5	32.75	58.0	104	68.0	41.0	66.0
16	14.0	39.0	19.0	105	28.0	41.0	23.0
17	21.0	56.5	40.0	106	16.5	44.5	14.0
18	20.5	40.75	28.0	107	47.0	55.5	19.5
19	66.0	27.75	57.5	108	43.0	57.5	28.0
20	63.0	39.5	55.0	109	60.0	60.25	21.0
21	17.0	38.25	13.5	110	18.5	57.75	48.5
23	71.0	53.5	50.5	111	47.0	61.75	33.0
24	56.5	59.0	54.0	112	37.0	42.0	50.0
25	68.5	25.0	59.0	114	31.5	70.75	41.0
26	46.5	61.75	30.5	117	27.0	61.0	45.5
31	76.0	35.25	44.5	118	13.5	66.5	53.0
33	74.0	46.25	39.0	119	18.0	72.5	39.0
35	77.0	38.5	32.0	120	19.5	62.0	28.5
41	11.0	39.0	39.0	121	10.5	64.25	41.0
73	14.5	53.75	31.5	127	55.0	45.75	55.5
85	16.5	62.5	47.5	128	61.0	43.75	32.5
86	71.0	43.5	30.0	136	8.5	47.0	30.0
87	42.0	50.5	47.0	137	7.5	36.75	52.0
88	73.5	37.0	57.0	138	7.5	39.0	34.0
89	48.5	49.0	60.5	151	30.5	47.25	43.0
91	70.0	54.25	39.5	185	47.0	17.5	40.0
92	62.5	25.0	61.5	...	....	.....	....

Y AXIS CONSTRUCTION

The choice of reference stands for the second axis is based upon criteria which are, in part, similar to those used in the choice of the first dimension reference stands: stands separated by the greatest interpoint distance and by the least projected x axis distance can be expected, if chosen as reference stands, to give the greatest spatial separation to the other stands. Stand pairs which most closely fit the above criteria are likely to be central in axis location since by the mechanics of the arc intersection and projection technique, the more nearly stands are found toward the center of any axis, the greater is the probability that they will have spatial separation in the new dimension of a relatively great distance. Conversely, stands located towards one of the ends of the axis are less likely to be distantly related, since by sharing a relatively high relationship with the reference stands towards which they are found, they are, therefore, more likely to be related to each other.

A test is suggested for the selection of y axis reference stands in which the value of stand separation on previous axes is subtracted from the index value of interpoint distance, with the highest value considered to be the most suitable. Such a test weights a low separation on previous axes as of equal importance with a high degree of interpoint distance. The importance of choosing reference stands in close proximity on previous axes is illustrated in subsequent z axis construction in which the condition of non-exact interpoint distances makes it impossible to correct for non-perpendicular axes. This test was applied to the 59 stands in the study and stand pair 111 and 185, which are separated by a projected x axis distance of 0 units (Table 2) and by an interpoint distance value of 47 (as inverted from a coefficient of community of 33), gave a maximum value of 47. Another stand pair, 26 and 185, also gave a high value (46.5) by the above test, with separations of 0.5 and 47 respectively. Stands 26 and 111 were found to be highly significantly correlated in their relationship to the other stands, and the use of several reference sets again appeared feasible. Stands 26 and 185, and 111 and 185 were, therefore, selected as y axis reference stands. Since these two sets of stands were separated by only 0.5 and 0.0 units respectively on the x axis, projections after arc rotation were made directly onto the line connecting each reference pair, and this line was considered the y axis. Final y axis location for each stand was determined by taking the mean position for each stand on the two constructed axes. These axis positions are shown in column 2 of Table 2.

Z AXIS CONSTRUCTION

With the completion of the y axis, a search was made for stands which had relatively similar x and y axis positions, but which were, nevertheless, separated by relatively great interstand distances. The same test of maximum axis separation was made as in the choice of the y axis reference stands, and a pair of stands, numbers 89 and 107 were found to give the highest value. Stands 89 and 107 were separated by an interstand distance of 41 units and by projected distances of 0.5 on the x axis and of 8.5 on the y axis for a test value of 32.5. The line connecting reference stands 89 and 107, although not perpendicular to the y axis and, therefore, not exactly parallel to the z axis was, nevertheless, used as a base line onto which to project the arc intersections. This was necessary since the sampling error involved in the area of uncertainty surrounding each stand made it impossible to apply formulae which would correct the effect of a non-perpendicular axis. This error also prohibited the consultation of distances to the x or y axis reference stands, since some stands to be projected onto the z axis were found to be equidistant from the x and y reference stands. No choice could, therefore, be made (as is illustrated in the location of stand 5 in Fig. 2) between upper and lower intersection points. Arc intersections were, therefore, projected directly onto the line connecting reference stands 89 and 107. The projections onto this line were considered z axis stand locations, and are shown in column 3 of Table 2.

RESULTS

Using the values in Table 2, each of the 59 stands studied was located on a two dimensional graph by the intersection of its values on any two of the three axes. In Fig. 3, for example, each point on the graph

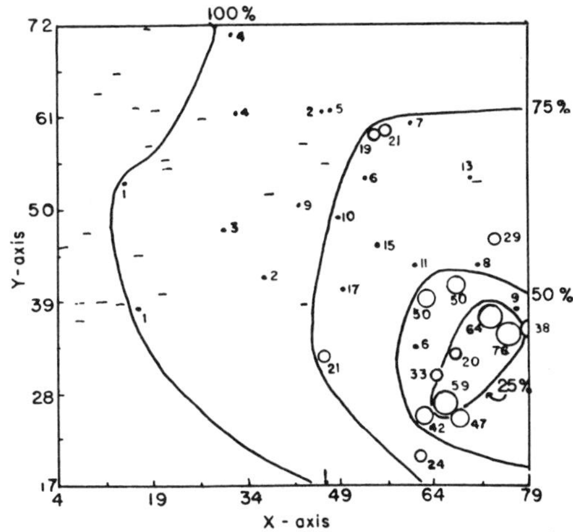

FIG. 3. Demonstration of dominance distribution of *Tilia americana* within the x-y ordination. Each circle or dash represents a stand location. Actual dominance figures in basal area per 100 sq. in. per acre at breast height are given beside each stand location. Values in the upper 25% are represented by the largest circle, values in the 50 to 26% quartile by medium sized circles, values in the 75-51% quartile by small circles, and values in the 100 to 76% quartile by dots. Contour lines are drawn around the 4 quartile lower limits in such a manner as to include all examples of the indicated size class whether or not lesser size class values are present.

represents a stand with its locus determined by its values in Table 2 for the x and y axes. Similar plottings were made for the stand locations on the x and z axes and the y and z axes. These 3 graphs can be thought of as 3 views (front, top, and side) of a three-dimensional cube, within which the stations are located at the intersections of lines projected from each axis. The actual construction of three-dimensional models is very time consuming (Fig. 7).

Once the stands are located in a two- or three-dimensional configuration, it becomes easy to study the behavior of individual species within the stands. In Fig. 3, for example, the actual basal area per acre for *Tilia americana* is plotted on the x-y

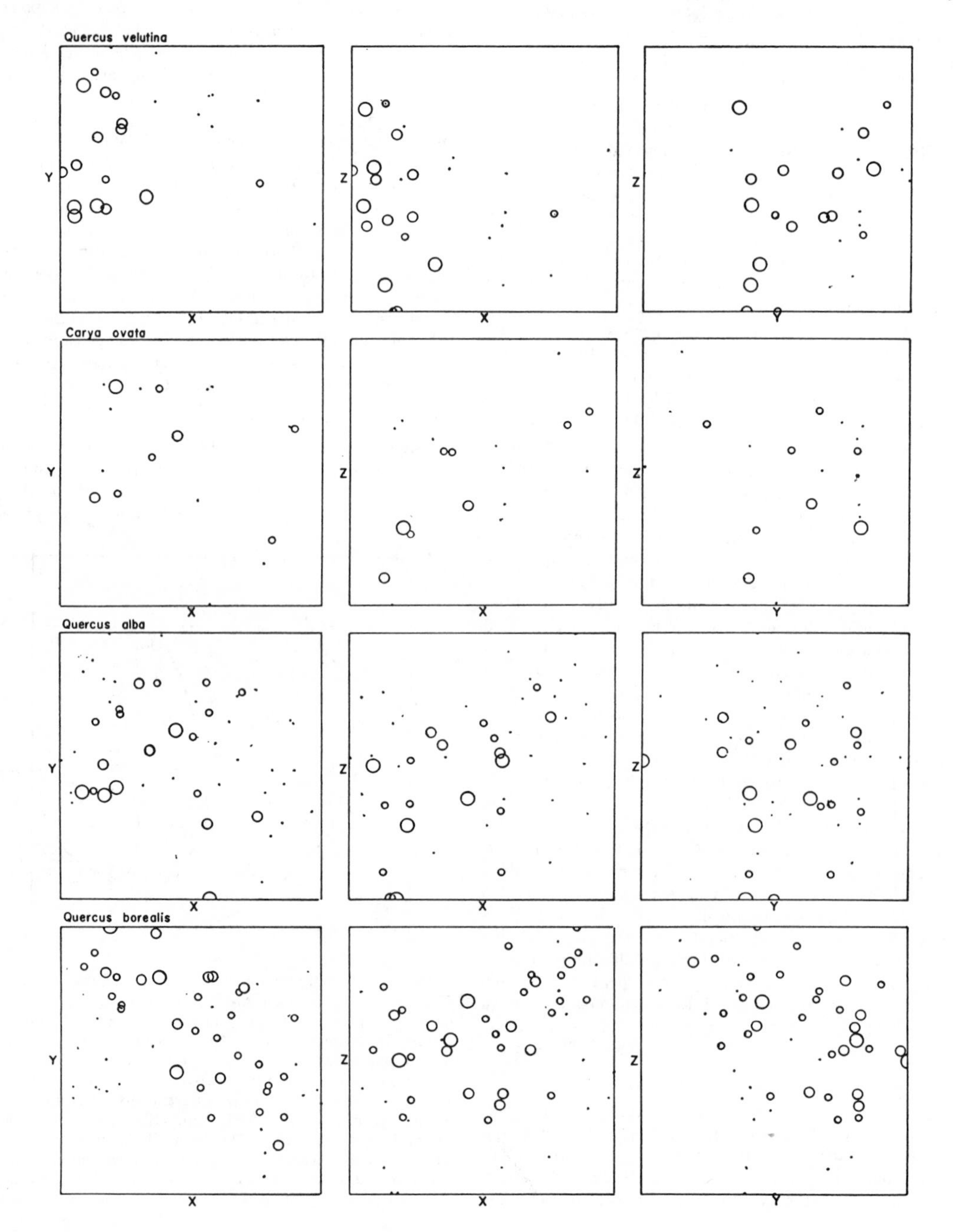

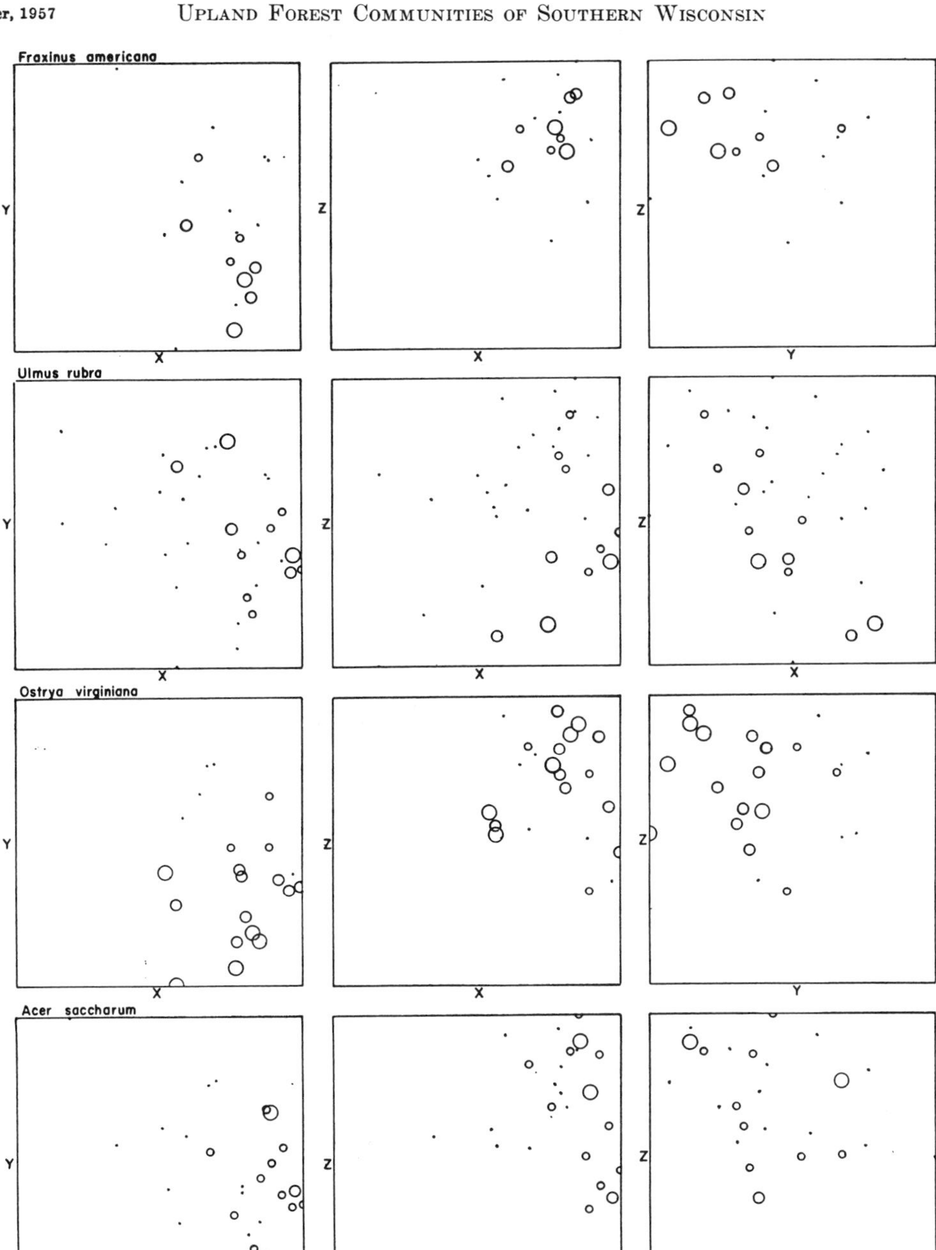

FIG. 4. Dominance behavior of the 8 most important tree species (other than *Tilia americana*) within each of the 3 views of the ordination. Size of circle corresponds to the quartile size class distribution illustrated in FIG. 3. Dominance per acre at the 50% level in sq. in. is as follows: *Acer saccharum* 8,000; *Carya ovata* 1,000; *Fraxinus americana* 2,200; *Ostrya virginiana* 400; *Quercus alba* 5,700; *Quercus borealis* 7,600; *Quercus velutina* 4,000; *Ulmus rubra* 3,200.

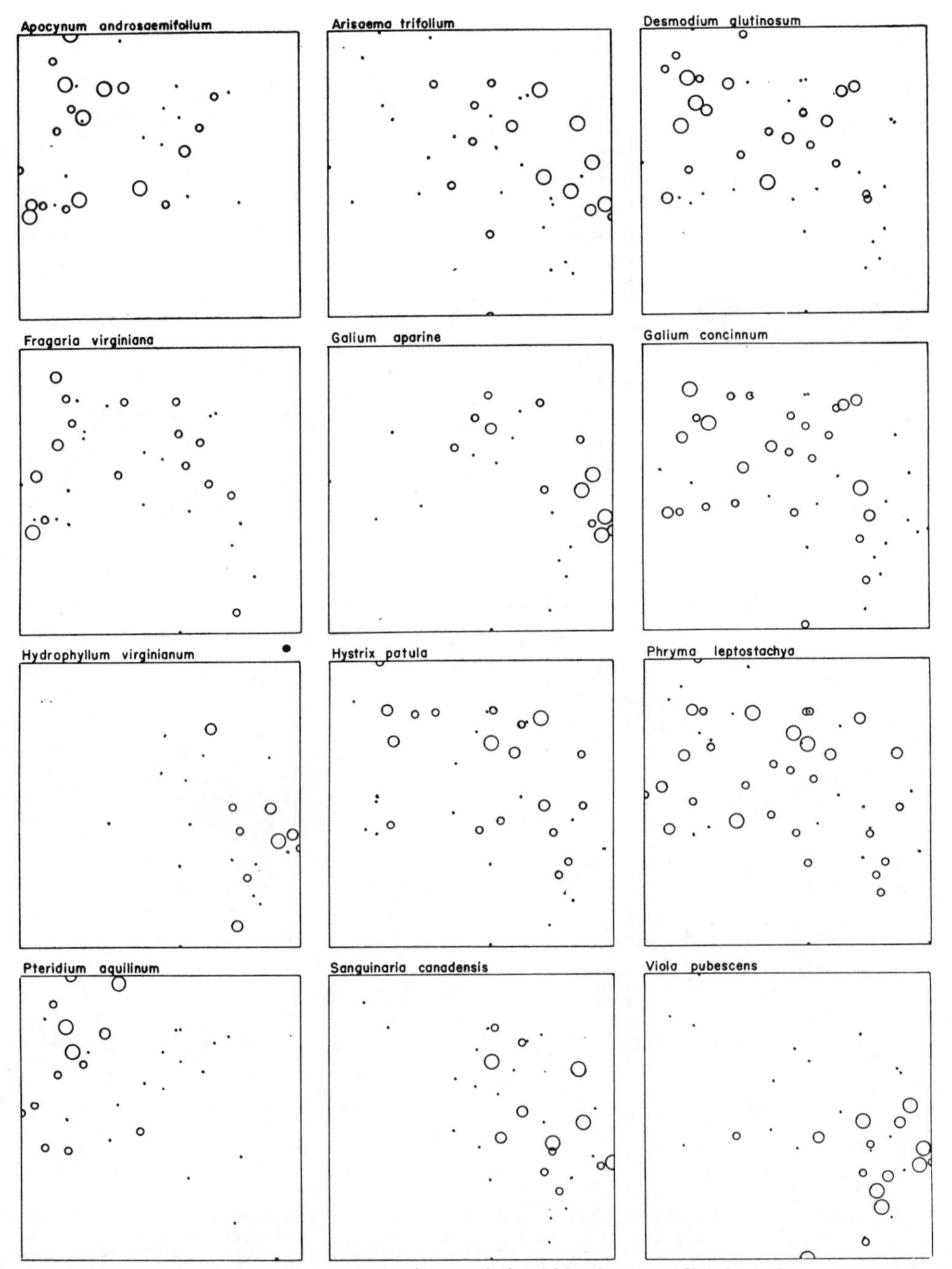

FIG. 5. Frequency behavior of score herbs (12 species) within the x-y ordination. Size of circle corresponds to the quartile size class distribution illustrated in FIG. 3. Frequency at the 50% level is as follows: *Apocynum androsaemifolium* 10; *Arisaema trifolium* 30; *Desmodium glutinosum* 40; *Fragaria virginiana* 25; *Galium aparine* 50; *Galium concinnum* 47; *Hydrophyllum virginianum* 25; *Hystrix patula* 10; *Phryma leptostachya* 30; *Pteridium aquilinum* 27; *Sanguinaria canadensis* 35; *Viola pubescens* 37.

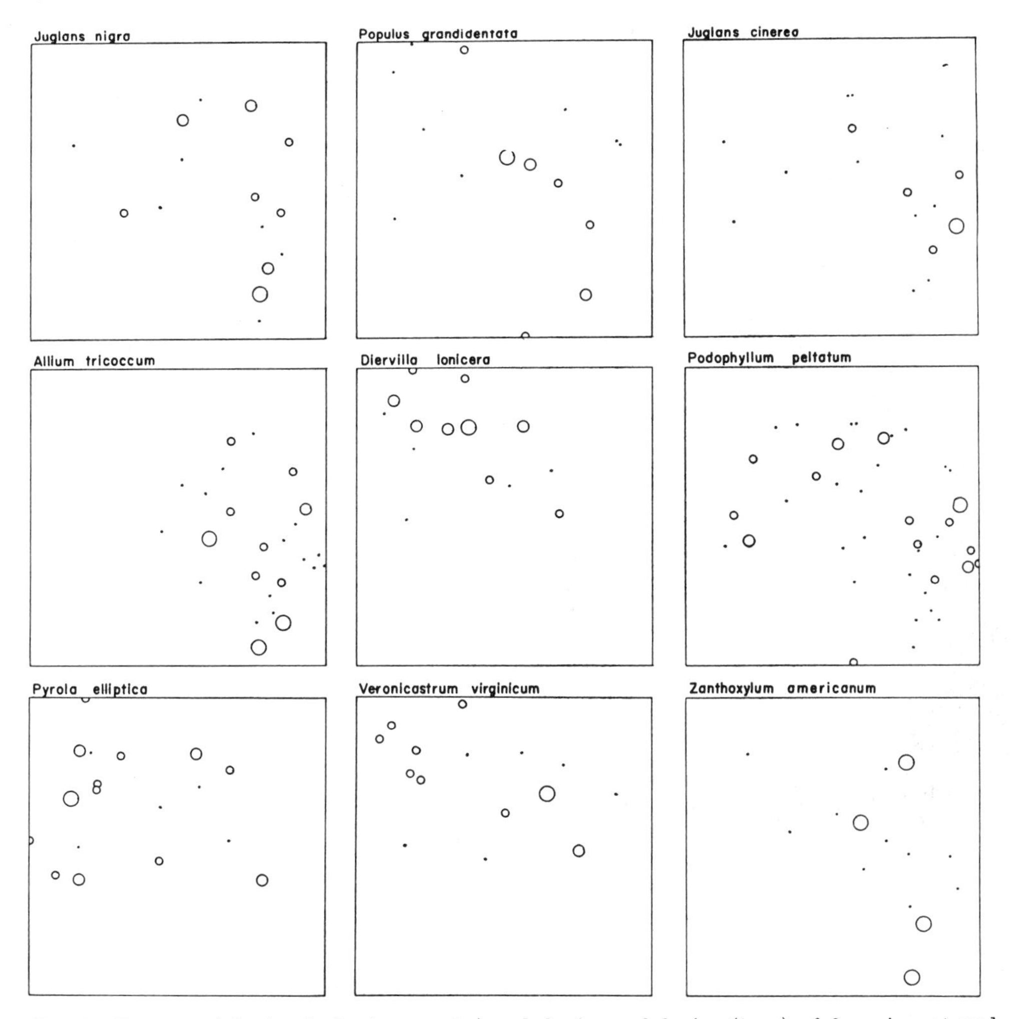

FIG. 6. Frequency behavior (understory species) and dominance behavior (trees) of 9 species not used in ordination construction. Size of circles corresponds to the quartile size class distribution illustrated in FIG. 3. Dominance per acre at the 50% level in sq. in. is as follows: *Juglans cinerea* 2,100; *Juglans nigra* 800; *Populus grandidentata* 1,200. Frequency at the 50% level is as follows: *Allium tricoccum* 10; *Diervilla lonicera* 10; *Podophyllum peltatum* 30; *Pyrola elliptica* 15; *Veronicastrum virginicum* 10; *Zanthoxylum americanum* 20.

graph. The values at each point are the measured basal areas for *Tilia* as taken from the field data for each stand. Those stands in which *Tilia* was absent are indicated by a dash. The basal areas have been put into size classes, as indicated by the circles of different size. In this and all similar figures in the paper, the largest circle includes the top 25% of all of the values; the next size, the third 25%; the smallest circle, the second 25%; and the solid points, the first or lowest 25% of the values. At a glance, therefore, it is apparent that *Tilia* reaches its highest importance in a very small portion of the possible area and that the stands with lesser dominance of this species are spread out from it in a pattern of decreasing occurrence. The contour lines on Fig. 3 have been drawn in such a way as to include all examples of the indicated size class, regardless of whether lesser size classes are also present. They, therefore, indicate the area within which the species may reach the indicated level of domination.

Similar graphs were made for all axis combinations for all species used on the test sheet and for a large number of species not used in the ordination. Fig. 4 illustrates the 3 views for the 8 most important tree species, while Fig. 5 shows x-y views only for the 12 herbs used on the score sheet. Nine species not used in constructing the ordination are shown in Fig. 6 in x-y views.

A three-dimensional representation of the behavior of *Quercus borealis* is given in Fig. 7. The 3 sizes of spheres indicate the top 3 quartiles of dominance per acre. No differentiation is made between the lowest quartile and the stands not containing the species; both are indicated by holes which appear as dots in the figure. A comparison of Fig. 7 with the appropriate views of *Q. borealis* in Fig. 4 will show how these separate presentations may be used to gain a visual image of the ordinations in three-dimensional space.

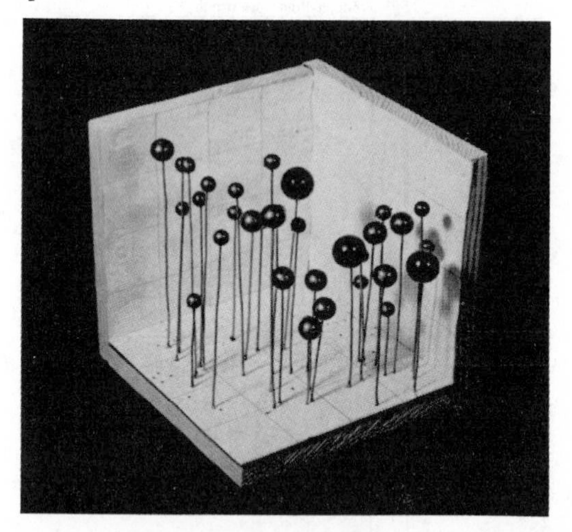

Fig. 7. Three-dimensional model of the dominance behavior of *Quercus borealis* within the ordination. The 3 sizes of spheres indicate the top 3 quartiles of dominance per acre. Stands of the lowest quartile and without the species are represented with holes which appear as dots in the figure. The x axis is on base of model at front from left to right; y axis on base from front to rear; z axis in vertical plane from below to above.

Certain measures of the environment, including soil analyses made upon a pooled sample from 3 random collections of A_1 layer in each stand are given in Table 6 for the stands used in the study. The soil nutrient analyses were made by the State Soils Laboratory, Madison, Wisconsin. Water retaining capacity was determined by the method outlined by Partch (1949).

DISCUSSION

The Mechanical Validity of the Ordination

The biologic and environmental results of the ordination in Table 2 should be assessed by criteria which are consistent with the assumptions upon which the ordination technique is based. The first of these assumptions is that the degree of compositional similarity between stands can be used as comparative distances to indicate spatial locations for these stands. The complexity of stand relationship and of the forces which influence community structure is of such magnitude that a matrix of comparative distance cannot, as previously noted, be oriented in a single exact configuration. The complexity of stand relationship is not chaotic, however, and it is assumed that each stand fluctuates within a fairly limited area in its compositional (and spatial) relationships, although this area is enlarged by the sampling error of the techniques used in field survey. The ordination is further based on the assumption that by establishing a set of exact criteria for stand selection, the number of interdependent causal complexes acting within the community can be limited. This limitation increases the probability that reference stands can be selected which are oriented along the lines of major changes in community structure, but it does not necessarily prevent a certain loss of matrix information by the selection of reference stands which reflect, in part, independent causal happenings which are unrelated to the major complexes. It is evident, therefore, that the validity of the ordination should be tested on its ability to approximate (but not exactly reproduce) the estimates of stand similarity, and perhaps to show stand alignments which correlate with the available estimates of physical environmental features.

Table 3. Interstand distances of the reference stands from matrix and ordination. The first column lists all pair combinations of reference stands; second column shows coefficients of community inverted to represent interpoint distance; third column shows distances between the stands in the ordination.

Stand Pairs	Matrix Distance	Ordination Distance	Stand Pairs	Matrix Distance	Ordination Distance
00- 35	16	06	89-107	41	41
00- 33	25	12	89-111	35	30
00- 89	41	40	89-185	46	38
00-107	37	42	89- 26	38	33
00-111	48	42	89-136	64	50
00-185	49	37	89-138	64	49
00- 26	58	42	89-137	63	44
00-136	79	72	107-111	25	15
00-138	78	72	107-185	46	43
00-137	80	73	107- 26	25	13
35- 33	21	11	107-136	57	41
35- 89	55	42	107-138	58	45
35-107	48	37	107-137	70	54
35-111	58	38	111-185	47	45
35-185	55	37	111- 26	21	03
35- 26	55	38	111-136	53	41
35-136	80	69	111-138	65	46
35-138	79	69	111-137	70	50
35-137	79	72	185- 26	48	45
33- 89	52	33	185-136	64	49
33-107	51	35	185-138	61	45
33-111	56	32	185-137	69	46
33-185	63	39	26-136	64	41
33- 26	61	33	26-138	59	45
33-136	77	66	26-137	72	51
33-138	80	67	136-138	42	09
33-137	78	68	136-137	34	24
			138-137	30	18

To assess the approximation of ordination distances to coefficient distances, 58 stand pairs were selected at random and their interpoint distances in the ordination compared with their coefficient of community values in the matrix. Interstand distance between two points (with locations x_1, y_1, and z_1, and x_2, y_2, and z_2) was determined by the formula $\sqrt{(x_1 - x_2)^2 + (y_1 - y_2)^2 + (z_1 - z_2)^2}$. The comparison of the 58 stand pairs showed a correlation value of −.35 which is significant at the 1% level. The correlation is negative since ordination distance between stands, which shows low values for a high similarity in composition, was compared with coefficient of community values which show high values for a high similarity. The highly significant correlation demonstrates the tendency for the ordination to approximate the stand relationships in the matrix. The check of ordination distance compared to coefficient distance was also applied to the 11 reference stands. Coefficient of community values were first inverted by the method previously discussed to represent degree of spatial separation and are presented in the first column of Table 3. The second column of the table shows interpoint distance in the ordination as calculated by the above formula. The correlation coefficient of these two distances for 55 stand pairs is +.73 which is significant at the 0.1% level.

The test of the ordination as a whole is that it approximates the interstand distance relationships in the matrix of coefficients of community. For any individual axis, however, additional assurance is necessary that it has contributed new and meaningful separations of the stands. A correlation test was, therefore, applied among the stand locations of the 3 axes. Stand locations along the x and y axis and along the y and z axis were found to be uncorrelated. Stand locations on the x and z axis were, however, correlated at the 5% level, though not at the 1% level. It was, therefore, possible that the z axis repeated, in part, information previously revealed in the x axis. It was decided that this repetition was not sufficiently great to justify discarding the z axis for the following reasons: (1) As will be demonstrated, the z axis showed meaningful separations of species midpoint locations which were not available on previous axes. (2) The x-z species distribution patterns in Figs. 4 & 9 showed little tendency toward a linear arrangement which would result if the axes were perfectly correlated. (3) Of 10 environmental measurements which were tested with each axis, 7 were correlated with the x axis, but, of these 7, only 2 were also correlated with the z axis. One of these two correlations was of a ratio which had a different basis on the z axis than on the x axis. There was also an environmental feature which correlated with the z axis, but not with the x axis.

The third test of ordination validity is whether the x, y, and z axes lead to a randomization of stand location. Such a randomization would obscure any differences in species or environmental behavior. It is apparent that if this were the case, then the species midpoints on each of these axes, as shown in Table 4, would have been in the same location, which they clearly are not. A random unordered stand orientation would probably also result in few or no environmental correlations, but as seen from Table 7, this does not happen. Every environmental feature is correlated with at least one of the axes.

TABLE 4. Location of species midpoints on ordination axes. Midpoints are mean axis locations of dominace values and represent point at which species reaches its optimum importance with respect to size.

Species	X	Y	Z
Quercus macrocarpa	16.3	46.4	38.3
Quercus velutina	17.4	49.3	41.0
Carya ovata	30.1	44.8	29.5
Prunus serotina	31.5	46.5	30.1
Quercus alba	35.5	46.7	32.9
Quercus borealis	40.9	51.8	42.7
Ulmus americana	44.8	33.5	41.9
Populus grandidentata	47.6	45.4	49.4
Juglans nigra	56.6	41.0	43.4
Ostrya virginiana	57.7	34.9	50.4
Fraxinus americana	62.2	31.8	45.5
Juglans cinerea	63.1	40.4	43.7
Carya cordiformis	63.3	43.6	42.7
Tilia americana	64.9	37.5	49.9
Ulmus rubra	67.8	42.9	38.3
Acer saccharum	68.4	40.5	48.8

The determination of species midpoints referred to above was made by finding the mean quantitative behavior (in this case, absolute basal area per acre) in each of 10 equal gradient sections, weighting the mean value by axis position, summing these weighted values, and dividing by the sum of the quantitative behaviors. These midpoint values for dominance per acre are shown in Table 4. They indicate the point at which each species reaches its optimum importance, at least with respect to size.

The differing relationships of species with each other along the 3 gradients can be seen from Table 4. Along the x axis, for example, *Acer saccharum* and *Ulmus rubra* occupy almost identical positions, and both are separated from *Ostrya virginiana* by over 10 units, yet, on the z axis, *Acer* and *Ulmus* are separated by over 10 units, while *Acer* is less than 2 units distant from *Ostrya*. Similarly, *Ulmus americana* and *Quercus borealis* which are less than 4 units distant on the x axis, are separated by over 18 units on the y axis, while *Juglans nigra* and *Ostrya virginiana* which are 1.1 units apart on the x axis, are separated by 6.1 and 7.0 units on the y and z axis, respectively.

The distances between the basal area per acre midpoints of the species is shown, for 3 dimensions, in the upper-right of Table 5. This table can be used as a basis for a spatial ordination of the species which is comparable to the patterns presented in deVries (1953). A drawing of such an ordination with midpoint locations the same as on the 3 axes of Table 4

Table 5. Species midpoints—interpoint distances in three dimensions. The upper-right of the table shows the distances between the dominance midpoints of species in three dimensions. The lower-left indicates whether there was greatest separation of midpoints in the first, second, or third dimension.

	Q.m.	Q.v.	C.o.	P.s.	Q.a.	Q.b.	U.a.	P.g.	J.n.	O v.	F.a.	J.c.	C.c.	T.a.	U.r.	A.s.
Quercus macrocarpa	..	4.1	16.4	17.3	19.9	25.6	31.4	33.2	41.0	44.7	48.7	47.5	47.3	50.7	51.6	53.5
Quercus velutina	2	...	17.7	18.0	20.0	23.7	31.7	31.6	40.2	43.8	48.3	46.6	46.3	49.6	50.8	52.3
Carya ovata	1	1	....	2.3	6.7	18.4	22.3	26.5	30.1	36.0	38.1	36.2	35.7	41.0	38.8	43.1
Prunus serotina	1	1	2	....	4.8	16.6	22.0	25.1	28.9	35.2	37.4	34.9	34.3	39.8	37.4	41.8
Quercus alba	1	1	1	1	....	12.3	18.5	20.5	24.2	30.7	33.1	30.3	29.8	32.5	33.0	37.1
Quercus borealis	1	1	3	3	3	....	18.7	11.4	19.1	25.0	29.3	24.9	23.8	28.8	28.7	30.3
Ulmus americana	1	1	1	1	2	2	....	14.3	14.1	15.5	17.9	19.6	21.1	22.0	25.1	25.5
Populus grandidentata	1	1	3	3	3	3	2	....	11.7	14.6	20.3	17.3	17.2	19.0	23.2	21.4
Juglans nigra	1	1	1	1	1	1	1	1	....	9.3	10.9	6.5	7.2	11.1	12.4	13.0
Ostrya virginiana	1	1	1	1	1	2	2	2	3	....	7.4	10.2	12.9	7.5	17.7	12.2
Fraxinus americana	1	1	1	1	1	1	1	1	2	3	....	8.8	12.2	7.6	14.3	11.2
Juglans cinerea	1	1	1	1	1	1	1	1	1	3	2	....	3.3	6.3	7.6	7.3
Carya cordiformis	1	1	1	1	1	1	1	1	1	2	2	2	....	9.6	6.3	8.5
Tilia americana	1	1	1	1	1	1	1	1	1	1	2	3	3	....	13.1	4.7
Ulmus rubra	1	1	1	1	1	1	1	1	1	3	2	3	1	3	....	10.8
Acer saccharum	1	1	1	1	1	1	1	1	1	1	2	1	3	1	3	....

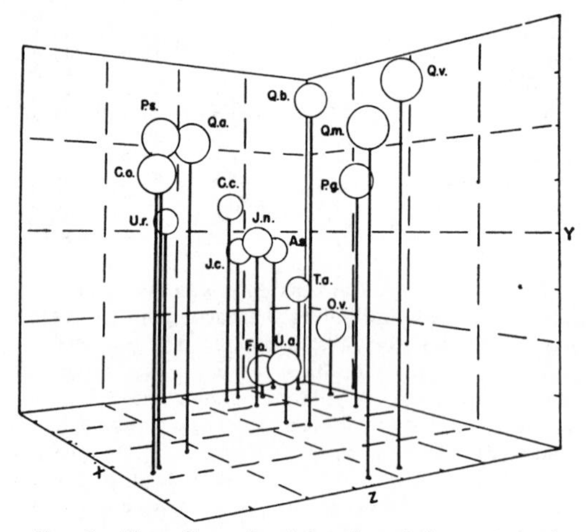

Fig. 8. Three-dimensional drawing of the center point locations of dominance behavior of tree species. Determination of center point locations explained in text. Point on frame closest to reader is lowest value for all 3 axes. The traditional position of the axes was changed to prevent the hiding of some species in the drawing.

is presented in Fig. 8. Both the upper-right of Table 5 and Fig. 8 show community relationships of species by suggesting an estimate of the relative correspondence of their stand locations; the less distant is the degree of midpoint separation, the more likely are the species to occur in the same stands.

The lower left of Table 5 lists the axis for each stand pair on which there is the greatest spatial separation in their dominance midpoints. In spite of the greater importance of the x axis (with a maximum distance between midpoints of 52.1 as compared to 20.0 and 20.9 for the y and z axes), there are 33 of a total of 120 stand pairs which give a greater separation on the y or z axis than on the x axis. These separations, in many cases, complemented the results of field observations, and gave indication that a biologic interpretation of the meaning of the axes might be possible.

The Biologic Validity of the Ordination

The ordination, therefore, by the use of a technique which is open to modification, gave one (but certainly not the only possible) approximation to the information on stand similarity in the matrix of coefficients of community. This approximation was made in 3 dimensions which were demonstrated to give species locations and environmental correlations which were non-random and non-repetitive. The ultimate test of the value of the ordination is, however, a biologic one, and succeeding discussion will attempt to utilize the ordination in examining the nature of the community and of its factorial relationships.

The views of species distribution in Figs. 4, 5, and 6 can be used in creating a mental image of the species as they appear in three dimensions, an image which will reveal the same effect as that of the 3 dimensional pattern in Fig. 7. From these visualizations and from Fig. 7, it is evident that each species shows all or part of an atmospheric distribution, that is, one in which there is an increasing concentration (number of points) and importance (size of points) of the species as the center of the distribution is reached. Away from the center, the decrease in numbers and sizes of the points is not always uniform in all directions but the species distributions, nevertheless, suggest that an idealized distribution would show a concentration and size of points diminishing outward in all directions from a dense center to an area beyond a sparse periphery where the points no longer occur.

An atmospheric distribution is a form which can be expected from an extension of the frequently expressed concept of ecologic amplitude into more than 2 dimensions. This concept postulates a mini-

TABLE 6. Environmental measurements by stands. Average canopy estimates from a number of stations within each stand. Soil analyses from pooled samples of 3 random collections of the A_1 layer in each stand. Nutrient values in pounds per acre at a soil depth of 7 in.; W.R.C. is water retaining capacity; OM is organic matter.

Stand No.	Percent Canopy	Depth A_1 (inch)	pH	WRC%	Ca (lbs.)	K (lbs.)	P (lbs.)	NH_4 (lbs.)	OM%	1/10 Ca / K
00	96	4.25	7.0	53	......	280	50	..	....	...
01	99	4.0	7.3	85	6,000	185	45	20	....	3.2
03	65	4.25	5.8	93	4,000	255	60	30	....	1.6
04	70	2.5	6.0	75	50	160	60	45	....	0.3
05	88	5.25	7.0	87	8,000	230	60	20	....	3.5
06	98	3.0	7.0	96	8,000	250	95	30	....	3.2
09	90	2.5	7.4	77	12,000	200	105	20	7.0	6.0
15	..	3.5	6.0	58	......	180	40	..	6.0	...
16	..	3.0	5.6	36	2,400	100	50	..	5.0	2.4
17	78	1.7	5.7	100	6,000	130	40	40	5.2	4.6
18	..	3.0	7.0	64	......	220	70	..	....	...
19	..	7.0	6.7	85	9,400	140	50	..	....	6.7
20	97	8.0	7.2	54	6,000	230	135	20	....	2.6
21	..	3.0	5.4	48	1,100	150	60	..	....	0.7
23	..	2.5	6.8	107	10,000	130	60	..	9.0	7.7
24	..	4.0	5.4	67	4,000	180	40	..	4.0	2.2
25	97	3.5	6.0	71	600	155	130	40	8.5	0.4
26	88	2.25	6.2	93	5,000	200	75	20	....	2.5
31	87	3.25	7.3	75	10,000	430	200	10	....	2.3
33	97	6.0	7.4	66	10,000	370	80	30	....	2.7
35	88	3.5	6.1	75	5,000	190	70	10	....	2.6
41	82	1.5	4.9	79	3,000	260	50	25	....	1.2
73	..	....	...	...	......	...	...	..	....	...
85	..	....	...	...	......	...	...	..	....	...
86	98	5.0	6.9	57	5,500	185	60	15	....	3.0
87	85	....	6.2	53	5,000	190	70	30	....	2.6
88	98	....	7.7	83	10,000	130	30	10	....	7.7
89	..	6.0	8.0	42	11,000	250	...	..	....	4.4
91	..	1.0	...	107	......	240	...	..	9.0	...
92	..	....	...	38	......	...	...	..	....	...
93	..	1.0	6.0	125	15,000	260	...	..	13.5	5.7
95	70	1.5	7.0	68	3,400	240	50	..	4.0	1.4
96	..	0.5	5.8	79	2,300	160	...	..	6.5	1.4
100	..	3.0	7.0	94	15,000	180	...	..	11.5	8.3
101	..	1.5	7.0	94	7,000	160	...	..	7.5	4.3
102	..	4.0	6.3	72	......	60	...	..	....	...
103	95	2.5	7.0	67	6,000	205	10	20	5.5	2.9
104	94	3.0	6.9	61	6,000	190	60	10	....	3.2
105	72	1.0	7.1	66	4,500	185	60	25	5.5	2.4
106	92	3.0	7.0	68	5,000	190	80	25	5.5	2.6
107	..	2.0	5.5	56	1,000	150	...	..	2.5	0.7
108	..	5.0	5.7	56	6,200	265	75	60	....	2.3
109	80	3.5	7.4	84	8,000	340	80	35	....	2.4
110	85	3.0	6.5	60	8,000	200	60	25	....	4.0
111	..	0.5	5.5	67	1,300	140	...	..	2.0	0.9
112	85	1.25	6.0	85	1,500	180	60	50	....	0.8
114	..	....	6.0	123	4,200	320	...	..	12.5	1.3
117	80	3.5	7.1	134	10,000	270	80	25	....	3.8
118	83	3.0	6.4	113	7,000	205	60	20	15.0	3.4
119	65	1.5	6.9	139	6,000	250	30	20	....	2.4
120	..	....	...	68	......	...	...	..	....	...
121	..	1.0	6.5	74	5,000	170	...	..	6.5	2.9
127	..	4.0	...	107	......	...	...	..	....	...
128	..	1.0	...	162	......	...	...	..	....	...
136	..	5.0	7.1	89	8,000	185	20	15	....	4.3
137	71	1.5	5.5	54	2,000	180	30	15	....	1.1
138	..	1.0	5.6	81	6,000	180	40	30	....	3.3
151	..	1.5	5.3	133	6,000	180	30	35	....	3.3
185	..	0.25	5.8	88	5,000	180	40	15	....	2.8

mum, optimum, and maximum behavior for each species in relation to the dynamics of community structure, and has often been demonstrated in 2 dimensions by contour-shaped ("solid normal") patterns and in one dimension by bell-shaped ("normal") patterns. It can be shown that a compression of an atmospheric distribution into 2 dimensions will give contour-shaped patterns as shown in Fig. 3 and in Wiedemann (1929), Ramensky (1930), Pogrebńjak (1955), and Whittaker (1956). Further compression into one dimension will yield the bell-shaped patterns demonstrated in the original linear treatment of the upland forest (Curtis and McIntosh 1951).

Fig. 6, which shows species patterns similar in form to the two preceding figures, demonstrates the relevance of the ordination to the entire plant population of the upland hardwoods. The 9 species in Fig. 6 include all of the minor tree species for which adequate data were available and an unbiased selection of shrub and herb species. Although none of the species in Fig. 6 contributed to the placement of stands in the ordination, they, nevertheless, show the same atmospheric distributions outlined above. The ability of the ordination to give meaningful patterns to species not used in its construction is considered as both a basic test of the usefulness of the ordination and as a demonstration of the feasibility of gradient construction by the consideration of less than the total species complement.

Figs. 4 through 6 show each species to have an individual pattern, different in size and location, although fairly similar in shape to those of other species. The distribution and the relationship of the patterns within the ordination is clearly one of continuous variation, as was previously demonstrated in the linear continuum of Curtis and McIntosh. The species used in the ordination, as well as those in Fig. 6 which were examined after the ordination was completed, can be described, therefore, as having patterned, non-random distributions within the prescribed geographic, environmental, and physiognomic limits of the study. Each of these distributions moves outward from central areas of high density to peripheral areas of sparse density, and this movement reveals along 1, 2, or 3 dimensions corresponding bell-shaped, contour-shaped, or atmospheric distributions. Each species has a separate area of location, and within this area its distribution is interspersed to varying degrees with other species distributions so that there is a continuous change in stand composition from any part of the ordination to any other part.

The above description supports, to a large degree, conclusions from work completed in a diversity of geographic regions and vegetations, including the studies of Gleason (1926), Vorobyov & Pogrebńjak (1929), Ramensky (1930), Sörenson (1948), Sjörs (1950), Curtis & McIntosh (1951), Whittaker (1951), deVries (1953), Goodall (1954a), Guinochet (1954), Webb (1954), Churchill (1955), Horikawa & Okutomi (1955), Poore (1956), and Hewetson (1956). It is suggested that evidence for the individualist theory of species distribution and for the continuum nature of community structure is now sufficiently compelling to require studies concerned with community structure to examine the relative continuity or discontinuity of their material and to use quantitative methods which will permit this examination. At a minimum, the examination would include a sampling of at least one analytic character in a sufficient number of stands to allow comparisons of quantitative composition. If an apparent grouping of stands into a discrete unit (i.e. association, etc.) is suspected, then this unit should be tested against samples of related vegetations to determine whether there are separate groups of stands with a certain range of variation within each group, or whether this variation is great enough to obscure the boundaries between the groups. If the latter is true, the application of ordination methods is necessary. A reasonable approach to the treatment of phytosociologic material about which little is known might be to check carefully the homogeneity of each stand, and then to apply an ordination technique. If clumps of stands are shown along the resultant gradients, it would then be possible to regard these clumps as castes (associations, etc.) and determine the suitable parameters necessary for the future classification of each caste.

One reason for reexamining the upland hardwoods of Wisconsin was the diversity of interpretations which various readers gave the original paper. Thus, some correspondents questioned if the linear continuum was not a statistically advanced restatement of succession, while others (Horikawa & Okutomi 1955) regarded it as a demonstration of relationships which were independent of succession. The continuous compositional variation which was demonstrated was at times assumed to apply to spatial transition as seen in the field (Churchill 1955), as contrasted to theoretic variation in the structure of communities regardless of their microgeographic relationships. In clarification of the above interpretations, it should be evident from the present study that the dimensions of the ordination are purely compositional and cannot necessarily be directly related to factors or to complexes of factors. Successional change is only one among many causal forces which have shaped the species distribution patterns of Figs. 4 through 6. The dimensions represent an approximation of the changes in compositional structure which are present within the community and are not spatial transitions as they exist in the field, although, as noted by Gleason (1926) and as is evident to many field workers, there are often natural areas where the salient features of a continuum can be observed.

The Nature of the Gradients

It is likely that the degree to which there are separations in more than one dimension was in part dependent upon the qualifications used to differentiate the community initially. Had sufficient knowledge been available before the study was begun, it might have been possible to eliminate those stands from the study which had relatively poor subsoil drainage and one of the additional dimensions might, thereby, have been eliminated. On the other hand, had there been an expansion of the data to include poorly drained forests and those subject to inundation, then a more complete picture of the southern forest as a whole could have been given. The view of Ashby (1948), that the most important aspect of community study is the original delineation of the study area, is very applicable to the present study.

As previously noted, the 3 dimensions of the ordination represent compositional gradients which are not likely to be related to any single causal agency. The probability that every factor is a constant influence on the structure of a stand, and the interrelated nature of biotic and physical factors, prohibits the identification of single causal mechanisms. In spite of these limitations, however, certain over-all patterns are evident in the 3 axes, patterns which relate to broad bio-physical complexes and to history. It is noteworthy that some of these patterns of species and of environmental features had not been suspected before the application of multidimensional technique.

Thus, in general, the x axis duplicates the original linear continuum of Curtis & McIntosh and shows a complex of conditions which include gradients from higher to lower light intensity and evaporation, gradients from lower to higher soil moisture and relative humidity, and gradients from more to less widely fluctuating soil and air temperature. The order of species along the x axis is basically determined by an over-all linear direction in the many paths of community development within the upland forest. These paths follow a network of successional patterns which are mainly related neither to primary nor secondary succession but to recovery from past disturbance. This disturbance, in the form of fire, reduced the forest in many places to a savanna or barrens condition in which there were scattered oaks and/or oak brush and roots (Cottam 1949; Bray 1955). During this reduction, the more terminal species which were also the more fire susceptible, were replaced by less terminal and by initial species which were the least fire susceptible. The coincidence that *Quercus macrocarpa* is both the most initial and fire resistant species, and that *Acer saccharum* is the most terminal and fire susceptible species, suggests that the ultimate explanation for the composition of a forest stand in upland Wisconsin is largely an historic one. The longer a stand has been free of fire (and other disturbance forces) and the more favorable the habitat in which it occurs, the more likely it can develop to a maple-basswood forest. The x axis shows, therefore, mainly the relationship of the community to major past disturbance factors and to its own developmental recovery from these factors.

To examine correlations with physical measurements of the environment, a check was made, using r, the correlation coefficient, of all available environmental data with each of the three dimensions. The results are shown in Table 7. Since both environmental data and stand locations showed approximately normal distributions, the significance of correlation was checked with the t test by examining the hypothesis that correlation (ρ) = 0. The values in Table 7 are the probability that $\rho = 0$, and that there is, therefore, no correlation. Thus, a probability of <.01 is a basis for rejecting the hypothesis and is considered a highly significant correlation. A value of <.05 is considered a significant correlation, while values of >.05 are considered to represent no significant correlation. By the above interpretation, for example, percent canopy is positively correlated with the x axis at a highly significant level, while it is not correlated with either the y or z axis.

The values in Table 7 show a highly significant correlation between percent canopy, depth of A_1, organic matter, pH, Ca, P, $\frac{.1Ca}{K}$ and the x axis. Water retaining capacity, K and NH_4 are not, however, correlated with the x axis. An examination of the values in Table 6 indicated the majority of the features related to the x axis, were positively correlated with each other. It is highly probable that these correlations represent the measured aspects of an increasingly mesic environment which accompany the successional recovery of the forest from past disturbance. While only a general interpretation can be made of this recovery at present, it is clear that the dynamics of factorial interrelationships along the x axis offer a broad area for future research.

The y axis seems to be correlated, in part, with the influence of surface and sub-surface drainage, and, consequently, with a soil moisture complex, with internal soil air space and aeration probably also

TABLE 7. Environmental correlations with three ordination axes. A_1 is depth of A_1 in inches; W.R.C. is water retaining capacity; O.M. is organic matter; Values in body of table are the probability that there is no correlation.

	Percent Canopy	A_1	pH	W.R.C.	Ca	K	P	NH_4	O.M.	$\frac{.1\ Ca}{K}$
x..........	+ <.001	+ <.01	+ <.01	>.05	+ <.001	>.05	+ <.001	>.05	+ <.01	+ <.001
y..........	>.05	>.05	>.05	+ <.001	>.05	>.05	>.05	+ <.01	>.05	>.05
z..........	>.05	>.05	>.05	>.05	>.05	− <.05	>.05	>.05	+ <.05	+ <.01

influential. The 3 tree species with midpoints towards the lower end of the y axis in Table 4 are *Fraxinus americana, Ulmus americana,* and *Ostrya virginiana* and are, of the species occurring in the upland forests, the most tolerant of poor drainage and aeration. Species which are found toward the upper end of the y axis are predominantly oaks, such as *Quercus borealis, Q. alba, Q. macrocarpa,* and *Q. velutina,* which are intolerant of inundation and poor aeration. Species with midpoints towards the center of the axis, such as *Acer saccharum, Juglans nigra, Ulmus rubra,* and *Carya cordiformis* are mesic species with an intermediate tolerance of poor drainage and aeration.

Two environmental features, W.R.C. and NH_4 show a highly significant positive correlation with the y axis. Water retaining capacity is represented with its highest values occurring towards the upper end of the axis assumed above to have the better internal drainage. The anomaly of high W.R.C. associated with good drainage is perhaps explained by the differences in past history of the stands along the y axis. It was found that W.R.C. has a significant negative correlation with depth of the A_1 layer. This correlation is apparently related to the circumstance that stands with high water-retaining capacities have a sharply demarcated boundary between the A_1 and A_2 layers, while the opposite stands tend to have a diffuse boundary, with the organic matter gradually decreasing in amount. These differences represent trends towards mor humus and typical podzols on the one hand and mull humus and gray-brown podzolic or brown forest soils on the other. They probably reflect past history to the extent that stands at the upper end of the y axis (mor humus, high W.R.C.) have been occupied by mixed conifer-hardwood forests in more recent postglancial times than stands at the other end, which may have developed on savannas or grasslands with a deep layer of incorporated humus. Substantiating evidence for this was seen in the distribution patterns of *Diervilla lonicera, Goodyera pubescens, Maianthemum canadense, Pteridium aquilinum* and *Pyrola elliptica* which closely matched the distribution of the higher values of water retaining capacity. All of these species currently reach their optimum in the conifer forests of northern Wisconsin and are only incidental members of the southern forest under discussion. The mean W.R.C. in stands in which 3 or more of the above species occurred in the quadrat samples is 103 as compared to a mean of 78 for the remaining stands.

The strong positive correlation of NH_4 with the y axis is not related to pH, since these two features show no correlation with each other. There is, however, a likelihood that the NH_4 correlation is, in part, controlled by the soil moisture and aeration complex discussed above in relation to tree distribution. The assumed poorly aerated soils towards the lower part of the y axis have, perhaps, a relatively higher proportion of denitrifying bacteria. With an increase in internal drainage and in soil oxygen along the y axis, there is a proportional increase in nitrifying bacteria and, therefore, an indirect increase in NH_4. The y axis, therefore, appears to represent a complex of internal soil drainage and aeration factors which is coincident with a cliseral historic factor reflecting post-glacial vegetational changes.

The z axis apparently represents, in part, the influence of recent disturbance, with species which benefit by disturbance being separated from the species with which they are proximate on other axes. Both *Carya ovata* and *Prunus serotina* at the lower end of the z axis are gap phase species (Watt 1947, Bray 1956b) which take advantage of oak wilt gaps and of grazing in initial forests. *Populus grandidentata* is another gap phase species and is found toward the upper end of the z axis. Its ability to enter areas opened by fire has long been noted (Chamberlin 1877) and observed in the field. *Ulmus rubra,* and to a lesser extent *Carya cordiformis,* apparently do well in some disturbed intermediate and early terminal forests and, as a result, are pulled away from their proximity to *Acer saccharum* to which they are closely adjacent on the previous axes. *Quercus alba* and *Quercus borealis* are separated on the z axis to a greater extent than on the x or y axes, a separation which might be related to the ability of red oak to increase in importance in maple-basswood forests which have been burned, forests from which white oak had been eliminated earlier by successional developments.

The values in Table 7 show 3 environmental features, K, O.M., and $\frac{.1Ca}{K}$ to be correlated with the z axis. Potassium shows a significant negative correlation which might be related to amount and content of loess, in that K is characteristically high in loessial soils in the Middle West and stands towards the upper end of the z axis might have an increasingly shallow or absent loessial layer. The highly significant positive correlation between the ratio $\frac{.1Ca}{K}$ and the z axis is illustrated by values which more than double from one end of the axis to the other. This correlation is the result of a Ca content which is not correlated along the axis coupled with the decreasing K content noted above. There is, therefore, a consequent increase in the ratio. Along the x axis there is a similar $\frac{.1Ca}{K}$ correlation, but it is produced by an uncorrelated K content which is accompanied by an increasing Ca content. No connection is directly apparent between this ratio and the growth of plants although it may reflect amount of available K. Thus, availability of K is considerably decreased with increased Ca, if K is tending toward low levels of fertility (Lutz & Chandler 1946). Such low fertility levels are possible in the decreasing K concentration along the z axis which is accompanied by a uniformly high Ca level. A similar non-availability

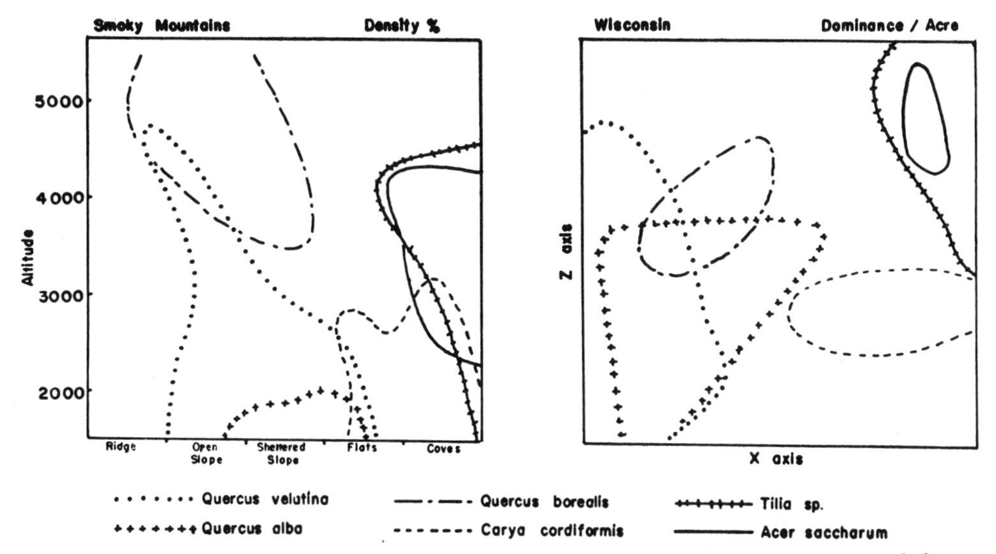

FIG. 9. View of contour drawing of trees within the x-z ordination compared with nomograms of the same or similar species in the Smoky Mountains (Whittaker 1956). Contour lines in the Wisconsin ordinations are made at the 25% (upper) size class level. Dominance values at this level in sq. in. are as follows: *Acer saccharum* 12,000; *Carya cordiformis* 400; *Quercus alba* 8,500; *Quercus borealis* 11,400; *Quercus velutina* 6,-000; *Tilia americana* 5,700.

of K correlated with the x axis is not as likely, since the non-correlated K concentration along this axis indicates there is no area where low K values are consistently found with high Ca values.

The reasons for the significant positive z axis correlation with organic matter are not readily apparent. The overwhelming probability that all the measured environmental factors plus numerous non-measured factors are interacting with each other and with biotic forces, and that plants are responding to the interactions rather than to single factors indicates that attempts to pinpoint the axes too closely are doomed to failure.

The difficulty of assigning definite factorial meanings to the ordination axes is further emphasized by a consideration of Fig. 9. This represents a comparison of the two-dimensional behavior patterns of a series of tree species which occurred in the Wisconsin study and also in Whittaker's 1956 analysis of the Great Smoky Mountains vegetation. The species of Tilia are slightly different in the two regions. The patterns on the Great Smoky graph are taken from Whittaker's nomograms. They portray either the 10% contour of species density or the highest value attained by the species if it was less than 10% and thus represent favorable or optimum conditions for growth. The patterns on the Wisconsin graph are the contours of the top quartile of dominances per acre. The absolute levels in square inches per acre for this 75% level are indicated in the legend of the figure. This may seem an unfair comparison since only selected contours are used, but a study of the full set of curves in Whittaker's original graphs in relation to the full diagrams for the Wisconsin data only strengthens the comparison.

With the exception of *Quercus alba,* the patterns in the two graphs are remarkably similar, especially if the axes of the Wisconsin graph are rotated slightly to the right. The exact meaning of this coincidence is not clear. A relationship between the x axis and Whittaker's moisture gradient seems reasonable enough, but there is no immediately evident connection between the factors usually associated with elevation and any possible interpretation of the z axis. All of the stands in the Wisconsin study were within a few hundred feet of each other in elevation and did not differ significantly in average temperature as measured by nearby weather stations.

The similarity in species distribution patterns between Wisconsin and Tennessee suggests that the biotic forces within a community have a high degree of homeostatic stability in their interactions with other causal agencies. Such stability is a feature of the dynamics of open systems in which, although a wide variety of factorial configurations may exist, there are apparently only a limited number of system structures which are possible. In biotic communities a limitation in structure means each species is confined to a definite community position in relation to other species with the unlikehood that all combinations of species can occur. It is the limited number of stand compositions which permits the application of quantitative methods to community classification.

FURTHER USES OF THE ORDINATION

Although the main purpose of the study was to investigate the phytosociology of community structure, the ordination results can be used as a basis for further research and for investigations in related fields. The value of the original upland hardwoods continuum in animal ecology (Bond 1955) and in plant ecology (Tresner, Backus & Curtis 1954; Gilbert & Curtis 1953; Hale 1955) has been demonstrated. The construction of additional dimensions should facilitate any further work which is done by providing a more complex pattern of community structure. Thus, studies of life histories of individual species can be given an exact community background against which to correlate those changes in the morphology, vitality, manner and means of dissemination, and perhaps taxonomy and phenology of the species which are determined by differences in community composition. Other important phytosociologic features which can be directly related to differences in community structure are (1) the distribution patterns of species, both as to variation in population means and in manner of dispersion, as demonstrated in Whitford (1949) and (2) the variation in interspecific association relationships.

The nature of the causal role of factors shaping a community is an ultimate goal in ecology. The determination of community structure, as was suggested earlier, is perhaps a key to these causal relationships. Goodall (1954b) notes, "It is the high correlation between different environmental factors that often suggests a deceptively simple relationship between plant distribution and the environment. There is much to be said for the view that the complexes of environmental factors determining plant distribution can be indicated and measured better indirectly, through the plants themselves, than by direct physical measurements; this is, of course, the idea behind the use of "phytometers" in agricultural meteorology, and the attention devoted to phenology." By separating unlike stands along several different compositional gradients, possible extremes in causal reactions are determined and the location of crucial junctures of change in community structure is possible. Elaborate and expensive tests can then be made at these junctures which would have been impossible to apply to the community as a whole. A further interpretation of the southern upland hardwoods would then, perhaps, be able to demonstrate the interactive nature of a series of factors as they change the structure of the community and, in turn, are changed in their own expression.

SUMMARY

The relationships of 59 stands of upland hardwood forest in southern Wisconsin were studied by means of a new ordination technique. Quantitative measurements of 26 different species were used as the basis of the ordination. These 26 species included the 12 most important trees and 14 herbs and shrubs. The quantitative characters used were absolute density per acre and absolute basal area per acre for the trees and simple frequency for the herbs and shrubs. These measurements were arranged on a "score" sheet for each stand after having been put on a comparable and relative basis. The degree of phytosociologic similarity of any stand to each of the other stands was assessed by Gleason's coefficient of community. Such coefficients were calculated for each stand with each other stand (1711 values).

It was considered that the degree of similarity of two stands as shown by their coefficient could be translated into a spatial pattern in which the inverse of the coefficient was equated with linear distance. Using groups of stands which were least similar as end points of an axis, the relative positions of all other stands were located along this axis by a geometric technique of arc projection, with the radii of the arcs given by the inverted coefficients of similarity. When this was done, it was seen that some stands were equidistant from the two end groups but nevertheless were themselves unlike. Using the most distantly related of these, a second axis was erected by the same method as the first. The position of all stands on this new axis was then determined. Further inspection revealed that a few stands were about equally spaced between the ends of both the new axis and the first axis but were still dissimilar. These were used to produce a third axis. The three axes were at right angles to each other, so the positions of any given stand on each axis could be projected to give a three-dimensional locus for that stand. When all stands were thus located in three-dimensional space, a highly significant correlation was found between their actual, measured spacings and the original coefficient of community values between them.

Given the framework of the spatial distribution of the stands, the behavior of individual species was readily examined by plotting a measured phytosociologic value of a species at the loci of the stands of its occurrence. It was found that all species (including an unbiased selection of those not used in the original 26 on which the ordination was based), formed atmospheric distributions with high or optimum values in a restricted portion of the array, surrounded by decreasing values in all directions. No two species showed the same location, but each was interspersed to varying degree with other species, in a continuously changing pattern.

The three axes of the ordination are compositional gradients, and their structure is interpreted as not the result of a causal determination by the physical environment but of an interaction between organisms and environment. A preliminary attempt was made to describe patterns of this interaction although an exact description remains for future study. Thus, the major (x) axis appeared to represent, in part, patterns of developmental recovery from major past disturbances. This recovery was correlated with features of an increasingly shaded and mesic environment including canopy cover and the following soil features: depth of A_1 layer, organic matter, pH,

Ca, P, and $\frac{1 \text{ Ca.}}{\text{K}}$. The second axis was related, in part, with surface and subsurface drainage and with soil aeration and it showed correlation with soil water retaining capacity and NH_4. The third axis was apparently related to recent disturbance and the influence of gap phase replacement, and it showed correlation with the soil feautres of K, $\frac{.1 \text{ Ca,}}{\text{K}}$ and organic matter. The possibility that the axes may have different interaction patterns than suggested above is evident in the similarity between the two-dimensional distributions of a number of tree species in the Wisconsin stands and in Whittaker's Smoky Mountain stands which were arranged along topographic and altitudinal gradients.

LITERATURE CITED

Agrell, I. 1945. The Collemboles in nests of warm-blooded animals, with a method for sociological analysis. Kungl. Fysiograf. Sallisk, Handl. **56**: 1-19.

Ashby, E. 1948. Statistical ecology. II—A reassessment. Bot. Rev. **14**: 222-234.

Bond, R. R. 1955. Ecological distribution of breeding birds in the upland forests of southern Wisconsin. Ph.D. Thesis. Univ. of Wisc.

Bray, J. R. 1955. The savanna vegetation of Wisconsin and an application of the concepts order and complexity to the field of ecology. Ph.D. Thesis. Univ. of Wisc.

———. 1956a. A study of mutual occurrence of plant species. Ecology **37**: 21-28.

———. 1956b. Gap phase replacement in a maple-basswood forest. Ecology **37**: 598-600.

Brown, R. T. & J. T. Curtis. 1952. The upland conifer-hardwood forests of northern Wisconsin. Ecol. Monog. **22**: 217-234.

Cain, S. A. 1944. Foundations of plant geography. New York.

Chamberlin, T. C. 1877. Native vegetation of eastern Wisconsin. *In* Chamberlin's Geology of Wisconsin. **2**: 176-187.

Churchill, E. D. 1955. Phytosociological and environmental characteristics of some plant communities in the Umiat region of Alaska. Ecology **36**: 606-627.

Clements, F. E. & G. W. Goldsmith. 1924. The phytometer method in ecology; the plant and community as instruments. Carnegie Inst. Wash. Publ. **356**.

Clifford, H. T. & F. E. Binet. 1954. A quantitative study of a presumed hybrid swarm between *Eucalyptus elaeophora* and *E. goniocalyx*. Austral. Jour. Bot. **2**: 325-337.

Cole, L. C. 1949. The measurement of interspecific association. Ecology **30**: 411-424.

Cottam, G. 1949. The phytosociology of an oak woods in southwestern Wisconsin. Ecology **30**: 271-287.

Cottam, G. & J. T. Curtis. 1949. A method for making rapid surveys of woodlands by means of pairs of randomly selected trees. Ecology **30**: 101-104.

Cottam, G., J. T. Curtis, & B. W. Hale. 1953. Some sampling characteristics of a population of randomly dispersed individuals. Ecology **34**: 741-757.

Curtis, J. T. & R. P. McIntosh. 1951. An upland forest continuum in the prairie-forest border region of Wisconsin. Ecology **32**: 476-496.

Daubenmire, R. 1954. Vegetation classification. Veröffentl. des Geobot. Inst. Rübel. **29**: 29-34.

deVries, D. M. 1953. Objective combinations of species. Acta Bot. Neerlandica **1**: 497-499.

———. 1954. Constellation of frequent herbage plants, based on their correlation in occurrence. Vegetatio **5 & 6**: 105-111.

Dice, L. R. 1948. Relationship between frequency index and population density. Ecology **29**: 389-391.

Fisher, R. A. 1936. The use of multiple measurements in taxonomic problems. Ann. Eugen., Lond. **7**: 179-188.

Forbes, S. A. 1907a. An ornithological cross-section of Illinois in autumn. Ill. State Lab. Nat. Hist. Bull. **7**: 305-335.

———. 1907b. On the local distribution of certain Illinois fishes: An essay in statistical ecology. Ill. State Lab. Nat. Hist. Bull. **7**: 273-303.

———. 1925. Method of determining and measuring the associative relations of species. Science **61**: 524.

Gilbert, M. L. 1952. The phytosociology of the understory vegetation of the upland forests of Wisconsin. Ph.D. Thesis. Univ. of Wisc.

Gilbert, M. L. & J. T. Curtis. 1953. Relation of the understory to the upland forest in the prairie-forest border region of Wisconsin. Wisc. Acad. Sci. Arts Lett. Trans. **42**: 183-195.

Gleason, H. A. 1910. The vegetation of the inland sand deposits of Illinois. Ill. State Lab. Nat. Hist. Bull. **9**: 23-174.

———. 1920. Some applications of the quadrat method. Torrey Bot. Club Bull. **47**: 21-33.

———. 1926. The individualistic concept of the plant association. Torrey Bot. Club Bull. **53**: 7-26.

———. 1952. The new Britton and Brown illustrated flora of the northeastern United States and adjacent Canada. Lancaster, Penna.: Lancaster Press.

Goodall, D. W. 1953a. Objective methods for the classification of vegetation. I. The use of positive interspecific correlation. Austral. Jour. Bot. **1**: 39-63.

———. 1953b. Factor analysis in plant sociology. Paper read to Third Internat. Biometric Conf. in Bellagio. Sept. 4, 1953.

———. 1954a. Vegetational classification and vegetational continua. Angew. Pflanzensoziologie, Wien. Festschrift Aichinger **1**: 168-182.

———. 1954b. Objective methods for the classification of vegetation. III. An essay in the use of factor analysis. Austral. Jour. Bot. **2**: 304-324.

Gregg, J. R. 1954. The language of taxonomy. New York.

Guinochet, M. 1954 Sur les fondements statistiques de la phytosociologie et quelques unes de leurs consequences. Veröffentl. des Geobot. Inst. Rübel. **29**: 41-67.

———. 1955. Logique et dynamique du peuplement végétal. Paris: Masson et Cie.

Hale, M. E., Jr. 1955. Phytosociology of corticolous cryptogams in the upland forest of southern Wisconsin. Ecology **36**: 45-63.

Hansen, H. M. 1930. Studies on the vegetation of Iceland. In The Botany of Iceland (ed. Rosenvinge & Warming). **3**(10):1-186. Copenhagen.

Hanson, H. C. 1955. Characteristics of the *Stipa comata-Bouteloua gracilis-Bouteloua curtipendula* association in northern Colorado. Ecology **36:** 269-280.

Hewetson, C. E. 1956. A discussion on the "climax" concept in relation to the tropical rain and deciduous forest. Empire Forestry Rev. **35:** 274-291.

Horikawa, Y. & K. Okutomi. 1955. The continuum of the vegetation on the slopes of Mt. Shiroyama, Iwakuni City, Prov. Suwo. The Seibutsugakkaishi **6:** 8-17. (Japanese, English summary.)

Hosokawa, T. 1955-1956. An introduction of 2 x 2 table methods into the studies of the structure of plant communities (on the structure of the beech forests, Mt. Hiko of S. W. Japan). Jap. Jour. Ecol. **5:** 58-62; 93-100; 150-153. (Japanese, English summary.)

Hughes, R. E. 1954. The application of multivariate analysis to (a) problems of classification in ecology; (b) the study of the interrelationships of the plant community and environment. Congr. Int'l. de Bot. Rap. et Comm. **8:** (sect. 7/8): 16-18.

Isaacson, S. L. 1954. Problems in classifying populations. *In* Statistics and Mathematics in Biology. (O. Kempthorne, et al., eds.). Ames, Iowa.

Kato, M., M. Toriumi & T. Matsuda. 1955. Mosquito larvae at Mt. Kago-bo near Wakuya, Miyagi Prefecture, with special reference to the habitat segregation. Ecol. Rev. **14:** 35-39. (Japanese, English summary.)

Kucera, C. L. & R. E. McDermott. 1955. Sugar maple-basswood studies in the forest-prairie transition of central Missouri. Amer. Midland Nat. **54:** 495-503.

Kulczyński, S. 1927. Zespoły roślin w Pieninach (Die Pflanzenassoziation der Pieninen). Internat. Acad. Polon. Sci., Lettr. Bull., Classe Sci. Math. et Nat., ser. B. Sci. Nat. Suppl. **2:** 1927: 57-203.

Lippmaa, T. 1939. The unistratal concept of plant communities. Amer. Midland Nat. **21:** 111-145.

Lorenz, J. R. 1858. Allgemeine Resultate aus der pflanzengeographischen und genetischen Untersuchung der Moore in präalpinen Hügellande Salzburg's. Flora **41:** 209 et seq.

Lutz, H. J. & R. F. Chandler. 1946. Forest soils. New York.

McIntosh, R. P. 1957. The York Woods, a case history of forest succession in southern Wisconsin. Ecology **38:** 29-37.

Mitsudera, M. 1924. Studies on phytometer II-Phytometer method in communities. Jap. Jour. Ecol. **4:** 113-120. (Japanese, English summary.)

Motyka, J., B. Dobrzański & S. Zawadzki. 1950. Wstępne badania nad łagami południowowschodniej Lubelszczyzny. (Preliminary studies on meadows in the southeast of the province Lublin.) Univ. Mariae Curie-Skłodowska Ann. Sect. E. **5** (13): 367-447.

Nash, C. B. 1950. Associations between fish species in tributaries and shore waters of western Lake Erie. Ecology **31:** 561-566.

Oosting, H. J. 1956. The study of plant communities. 2d ed. San Francisco: Freeman.

Parmalee, G. W. 1953. The oak upland community in southern Michigan. Ecol. Soc. Amer. Bull. **34:** 84. (Abst.).

Partch, M. L. 1949. Habitat studies of soil moisture in relation to plants and plant communities. Ph.D. Thesis. Univ. of Wisc.

Pirie, N. W. 1937. The meaninglessness of the terms 'life' and 'living'. *In* Perspectives in Biochemistry, (eds. J. Needham & D. E. Green). Cambridge.

Pogrebnjak, P. S. 1930. Über die Methodik der Standortsuntersuchungen in Verbindung mit den Waldtypen. Internatl. Congr. Forestry Exp. Stations, Stockholm Proc. **1929:** 455-471.

———. 1955. Osnovi lesnoj tipologie (Foundations of forest typology). Publ. Acad. of Sci. of the Ukrainian S. S. R., Kiev.

Poore, M. E. D. 1956. The use of phytosociological methods in ecological investigations. IV. General discussion of phytosociological problems. Jour. Ecol. **44:** 28-50.

Raabe, E. W. 1952. Über den "Affinitätswert" in der Pflanzensoziologie. Vegetatio **4:** 53-68.

Ramensky, L. G. 1930. Zur Methodik der vergleichenden Bearbeitung und Ordnung von Pflanzenlisten und anderen Objekten, die durch mehrere, verschiedenartig wirkende Faktoren bestimmt werden. Beitr. z. Biol. der Pflanz. **18:** 269-304.

———. 1952. On some principal aspects of present day geobotany. Bot. Zhur. **37:** 181-201. (Russian.)

Randall, W. E. 1951. Interrelations of autecological characteristics of woodland herbs. Ecol. Soc. Amer. Bull. **32:** 57. (Abst.)

Shreve, F. 1942. The desert vegetation of North America. Bot. Rev. **8:** 195-246.

Sjörs, H. 1950. Regional studies in North Swedish mire vegetation. Botaniska Notiser. **1950** (2): 173-222.

Sörenson, Th. 1948. A method for establishing groups of equal magnitude in plant sociology based on similarity of species content. Acta K. Danske Vidensk. Selsk. Biol. Skr. J. **5:** 1-34.

Spalding, V. M. 1909. Problems of local distribution in arid regions. Amer. Nat. **43:** 472-486.

Stewart, G. & W. Keller. 1936. A correlative method for ecology as exemplified by studies of native desert vegetation. Ecology **17:** 500-514.

Thurstone, L. L. 1947. Multiple-factor analysis. Chicago: The Univ. of Chicago Press.

Torgerson, W. S. 1952. Multidimensional scaling: I. Theory and method. Psychometrika **17**(4): 401-419.

Tresner, H. D., M. P. Backus & J. T. Curtis. 1954. Soil microfungi in relation to the hardwood forest continuum in southern Wisconsin. Mycologia **46:** 314-333.

Tucker, L. G. 1956. Personal communication.

Tuomikoski, R. 1942. Untersuchunger über die Untervegetation der Bruckmoore in Ostfinnland. I. Zur Methodik der pflanzensoziologischen Systematik. Soc. Zool.-Bot. Fenn. Vanamo. Ann. Bot. **17:** 1-203.

Vorobyov, D. V. & P. S. Pogrebnjak. 1929. Lisovyi tipologitchnyi visnatchnik ukrainskogo polissya. (A key to Ukrainian forest types.) Trudi z lisovoi dosvidoni spravy na Ukraini **11.**

Watt, A. S. 1947. Pattern and process in the plant community. Jour. Ecol. **35:** 1-22.

Webb, D. A. 1954. Is the classification of plant communities either possible or desirable? Bot. Tidsskr. **51:** 362-370.

Whitford, P. B. 1949. Distribution of woodland plants in relation to succession and clonal growth. Ecology **30:** 199-208.

Whittaker, R. H. 1951. A criticism of the plant association and climatic climax concepts. Northw. Sci. **25:** 17-31.

———. 1952. A study of summer foliage insect communities in the Great Smoky Mountains. Ecol. Monog. **23:** 1-44.

———. 1954. Plant populations and the basis of plant indication. Angewandte Pflanzensoziologie, Wien. Festschrift Aichinger **1:** 183-206.

———. 1956. Vegetation of the Great Smoky Mountains. Ecol. Monog. **26:** 1-80.

Wiedemann, E. 1929. Die ertragskundliche und waldbauliche Brauchbarkeit der Waldtypen nach Cajander im sachsischer Erzgebirge. Allgemeine Forst und Jagd-Zeitung. **105:** 247-254.

Williams, C. B. 1954. The statistical outlook in relation to ecology. Jour. Ecol. **42:** 1-13.

Womble, W. H. 1951. Differential systematics. Science **114:** 315-322.

PAPER 27

6. E. L. Thorndike, *Animal Intelligence* (Macmillan, New York, 1911).
7. E. D. Adrian, *The Physical Background of Perception* (Oxford Univ. Press, London, 1947).
8. D. H. Raab and H. W. Ades, *Amer. J. Psychol.* **59**, 59 (1946); M. R. Rosenzweig, *ibid.*, p. 127; R. A. Butler, I. T. Diamond, W. D. Neff, *J. Neurophysiol.* **20**, 108 (1957); R. F. Thompson, *ibid.* **23**, 321 (1960).
9. W. D. Neff and I. T. Diamond, in *Biological and Biochemical Bases of Behavior*, H. F. Harlow and C. N. Woolsey, Eds. (Univ. of Wisconsin Press, Madison, 1958); R. B. Masterton and I. T. Diamond, *J. Neurophysiol.* **27**, 15 (1964).
10. J. M. Warren, *Ann. Rev. Psychol.* **16**, 95 (1965).
11. K. S. Lashley, in *Symposia of the Society for Experimental Biology* (Academic Press, New York, 1950), vol. 4.
12. K. Z. Lorenz, *ibid.*, vol. 4.
13. G. Elliot Smith, *The Evolution of Man* (Oxford Univ. Press, London, 1924); C. J. Herrick, *Brain of the Tiger Salamander* (Univ. of Chicago Press, Chicago, 1948).
14. G. G. Simpson, *Bull. Amer. Museum Nat. Hist.* **85**, 1307 (1945); A. S. Romer, *Vertebrate Paleontology* (Univ. of Chicago Press, Chicago, 1945); J. Z. Young, *The Life of Vertebrates* (Oxford Univ. Press, London, 1962).
15. W. C. Hall and I. T. Diamond, *Brain, Behavior Evolution* **1**, 181 (1968).
16. J. Kaas, W. C. Hall, I. T. Diamond, "Dual representation of the retina in the cortex of the hedgehog in relation to architectonic boundaries and thalamic projections," in preparation.
17. R. A. Lende and K. M. Sadler, *Brain Res.* **5**, 390 (1967).
18. P. Abplanalp and W. Nauta, personal communication.
19. J. Altman and M. B. Carpenter, *J. Comp. Neurol.* **116**, 157 (1961); D. K. Morest, *Anat. Record* **151**, 390 (1965); E. C. Tarlov and R. Y. Moore, *J. Comp. Neurol.* **126**, 403 (1966).
20. H. S. Gasser and J. Erlanger, *Amer. J. Physiol.* **88**, 581 (1929).
21. G. H. Bishop, *J. Nervous Mental Disease* **128**, 89 (1959).
22. C. J. Herrick, *Brains of Rats and Men* (Univ. of Chicago Press, Chicago, 1926).
23. R. P. Erickson, W. C. Hall, J. A. Jane, M. Snyder, I. T. Diamond, *J. Comp. Neurol.* **131**, 103 (1967).
24. J. E. Rose and C. N. Woolsey, *Electroencephalog. Clin. Neurophysiol.* **1**, 391 (1949).
25. Unpublished experiments in our laboratory.
26. M. Snyder, W. C. Hall, I. T. Diamond, *Psychonomic Sci.* **6**, 243 (1966); M. Snyder and I. T. Diamond, *Brain Behavior Evolution* **1**, 244 (1968).
27. H. Killackey, I. T. Diamond, W. C. Hall, G. Hudgins, *Federation Proc.* **27**, 517 (1968).
28. J. E. Rose and C. N. Woolsey, *J. Comp. Neurol.* **91**, 441 (1949); ——, in *Biological and Biochemical Bases of Behavior*, H. F. Harlow and C. N. Woolsey, Eds. (Univ. of Wisconsin Press, Madison, 1958); I. T. Diamond, K. L. Chow, W. D. Neff, *J. Comp. Neurol.* **109**, 349 (1958).
29. L. J. Garey and T. P. S. Powell, *Proc. Roy. Soc. London Ser. B* **169**, 107 (1967); M. Glickstein, R. A. King, J. Miller, M. Berkley, *J. Comp. Neurol.* **130**, 55 (1967).
30. The following abbreviations are used in the figures and figure legends: *A*, auditory area of cortex; *CM* + *Pf*, centromedian + parafascicular nuclei; *GL*, lateral geniculate nucleus; *GM*, medial geniculate nucleus; *Hab*, habenula; *Hip*, hippocampus; *L*, lateral group of nuclei; *LP*, lateroposterior nucleus; *MD*, mediodorsal nucleus; *PC*, posterior commissure; *Po*, posterior nucleus; *Pul*, pulvinar nucleus; *R*, reticular nucleus; *S I*, somatic area I; *S II*, somatic area II; *SC*, superior colliculus; *T*, temporal area of cortex; *TO*, optic tract; *V*, ventral group; *V I*, visual area I; *V II*, visual area II; *VGL*, ventral lateral geniculate; *VL*, ventrolateral nucleus; *VM*, ventromedial nucleus; *VP*, ventroposterior nucleus; *ZI*, zona incerta.

31. The research described was supported by grant MH-04849 from the National Institute of Mental Health. For their helpful comments we thank Doctors G. Bishop, J. Kaas, G. Kimble, N. Guttman, J. Jane, R. B. Masterton, and M. Warren.

The Strategy of Ecosystem Development

An understanding of ecological succession provides a basis for resolving man's conflict with nature.

Eugene P. Odum

The principles of ecological succession bear importantly on the relationships between man and nature. The framework of successional theory needs to be examined as a basis for resolving man's present environmental crisis. Most ideas pertaining to the development of ecological systems are based on descriptive data obtained by observing changes in biotic communities over long periods, or on highly theoretical assumptions; very few of the generally accepted hypotheses have been tested experimentally. Some of the confusion, vagueness, and lack of experimental work in this area stems from the tendency of ecologists to regard "succession" as a single straightforward idea; in actual fact, it entails an interacting complex of processes, some of which counteract one another.

As viewed here, ecological succession involves the development of ecosystems; it has many parallels in the developmental biology of organisms, and also in the development of human society. The ecosystem, or ecological system, is considered to be a unit of biological organization made up of all of the organisms in a given area (that is, "community") interacting with the physical environment so that a flow of energy leads to characteristic trophic structure and material cycles within the system. It is the purpose of this article to summarize, in the form of a tabular model, components and stages of development at the ecosystem level as a means of emphasizing those aspects of ecological succession that can be accepted on the basis of present knowledge, those that require more study, and those that have special relevance to human ecology.

Definition of Succession

Ecological succession may be defined in terms of the following three parameters (*1*). (i) It is an orderly process of community development that is reasonably directional and, therefore, predictable. (ii) It results from modification of the physical environment by the community; that is, succession is community-controlled even though the physical environment determines the pattern, the rate of change, and often sets limits as to how far development can go. (iii) It culminates in a stabilized ecosystem in which maximum biomass (or high information content) and symbiotic function between organisms are maintained per unit of available energy flow. In a word, the "strategy" of succession as a short-term process is basically the same as the "strategy" of long-term evolutionary development of the biosphere—namely, increased control of, or homeostasis with, the physical environment in the sense of achieving maximum protection from its perturbations. As I illustrate below, the strategy of "maximum protection" (that is, trying to achieve maximum support of complex biomass structure) often conflicts with man's goal of "maximum

The author is director of the Institute of Ecology, and Alumni Foundation Professor, at the University of Georgia, Athens. This article is based on a presidential address presented before the annual meeting of the Ecological Society of America at the University of Maryland, August 1966.

production" (trying to obtain the highest possible yield). Recognition of the ecological basis for this conflict is, I believe, a first step in establishing rational land-use policies.

The earlier descriptive studies of succession on sand dunes, grasslands, forests, marine shores, or other sites, and more recent functional considerations, have led to the basic theory contained in the definition given above. H. T. Odum and Pinkerton (*2*), building on Lotka's (*3*) "law of maximum energy in biological systems," were the first to point out that succession involves a fundamental shift in energy flows as increasing energy is relegated to maintenance. Margalef (*4*) has recently documented this bioenergetic basis for succession and has extended the concept.

Changes that occur in major structural and functional characteristics of a developing ecosystem are listed in Table 1. Twenty-four attributes of ecological systems are grouped, for convenience of discussion, under six headings. Trends are emphasized by contrasting the situation in early and late development. The degree of absolute change, the rate of change, and the time required to reach a steady state may vary not only with different climatic and physiographic situations but also with different ecosystem attributes in the same physical environment. Where good data are available, rate-of-change curves are usually convex, with changes occurring most rapidly at the beginning, but bimodal or cyclic patterns may also occur.

Bioenergetics of Ecosystem Development

Attributes 1 through 5 in Table 1 represent the bioenergetics of the ecosystem. In the early stages of ecological succession, or in "young nature," so to speak, the rate of primary production or total (gross) photosynthesis (P) exceeds the rate of community respiration (R), so that the P/R ratio is greater than 1. In the special case of organic pollution, the P/R ratio is typically less than 1. In both cases, however, the theory is that P/R approaches 1 as succession occurs. In other words, energy fixed tends to be balanced by the energy cost of maintenance (that is, total community respiration) in the mature or "climax" ecosystem. The P/R ratio, therefore, should be an excellent functional index of the relative maturity of the system.

So long as P exceeds R, organic matter and biomass (B) will accumulate in the system (Table 1, item 6), with the result that ratio P/B will tend to decrease or, conversely, the B/P, B/R, or B/E ratios (where $E = P + R$) will increase (Table 1, items 2 and 3). Theoretically, then, the amount of standing-crop biomass supported by the available energy flow (E) increases to a maximum in the mature or climax stages (Table 1, item 3). As a consequence, the net community production, or yield, in an annual cycle is large in young nature and small or zero in mature nature (Table 1, item 4).

Comparison of Succession in a Laboratory Microcosm and a Forest

One can readily observe bioenergetic changes by initiating succession in experimental laboratory microecosystems. Aquatic microecosystems, derived from various types of outdoor systems, such as ponds, have been cultured by Beyers (*5*), and certain of these mixed cultures are easily replicated and maintain themselves in the climax state indefinitely on defined media in a flask with only light input (*6*). If samples from the climax system are inoculated into fresh media, succession occurs, the mature system developing in less than 100 days. In Fig. 1 the general pattern of a 100-day autotrophic succession in a microcosm based on data of Cooke (*7*) is compared with a hypothetical model of a 100-year forest succession as presented by Kira and Shidei (*8*).

During the first 40 to 60 days in a typical microcosm experiment, daytime net production (P) exceeds nighttime respiration (R), so that biomass (B) accumulates in the system (*9*). After an early "bloom" at about 30 days, both rates decline, and they become approximately equal at 60 to 80 days. the B/P ratio, in terms of grams of carbon supported per gram of daily carbon production, increases from less than 20 to more than 100 as the steady state is reached. Not only are autotrophic and heterotrophic metabolism balanced in the climax, but a large organic structure is supported by small daily production and respiratory rates.

While direct projection from the small laboratory microecosystem to open nature may not be entirely valid, there is evidence that the same basic trends that are seen in the laboratory are characteristic of succession on land and in large bodies of water. Seasonal successions also often follow the same pattern, an early seasonal bloom characterized by rapid growth of a few dominant species being followed by the development later in the season of high B/P ratios, increased diversity, and a relatively steady, if temporary, state in terms of P and R (*4*). Open systems may not experience a decline, at maturity, in total or gross productivity, as the space-limited microcosms do, but the general pattern of bioenergetic change in the latter seems to mimic nature quite well.

These trends are not, as might at first seem to be the case, contrary to the classical limnological teaching which describes lakes as progressing in time from the less productive (oligotrophic) to the more productive (eutrophic) state. Table 1, as already emphasized, refers to changes which are brought about by biological processes *within* the ecosystem in question. Eutrophication, whether natural or cultural, results when nutrients are imported into the lake from *outside* the lake—that is, from the watershed. This is equivalent to adding nutrients to the laboratory microecosystem or fertilizing a field; the system is pushed back, in successional terms, to a younger or "bloom" state. Recent studies on lake sediments (*10*), as well as theoretical considerations (*11*), have indicated that lakes can and do progress to a more oligotrophic condition when the nutrient input from the watershed slows or ceases. Thus, there is hope that the troublesome cultural eutrophication of our waters can be reversed if the inflow of nutrients from the watershed can be greatly reduced. Most of all, however, this situation emphasizes that it is the entire drainage or catchment basin, not just the lake or stream, that must be considered the ecosystem unit if we are to deal successfully with our water pollution problems. Ecosystematic study of entire landscape catchment units is a major goal of the American plan for the proposed International Biological Program. Despite the obvious logic of such a proposal, it is proving surprisingly difficult to get tradition-bound scientists and granting agencies to look beyond their specialties toward the support of functional studies of large units of the landscape.

Food Chains and Food Webs

As the ecosystem develops, subtle changes in the network pattern of food chains may be expected. The manner in which organisms are linked together through food tends to be relatively sim-

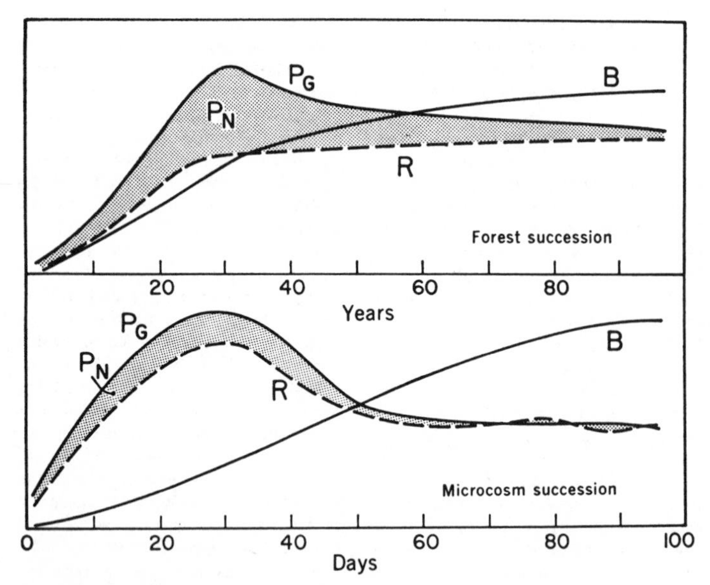

Fig. 1. Comparison of the energetics of succession in a forest and a laboratory microcosm. P_G, gross production; P_N, net production; R, total community respiration; B, total biomass.

ple and linear in the very early stages of succession, as a consequence of low diversity. Furthermore, heterotrophic utilization of net production occurs predominantly by way of grazing food chains—that is, plant-herbivore-carnivore sequences. In contrast, food chains become complex webs in mature stages, with the bulk of biological energy flow following detritus pathways (Table 1, item 5). In a mature forest, for example, less than 10 percent of annual net production is consumed (that is, grazed) in the living state (*12*); most is utilized as dead matter (detritus) through delayed and complex pathways involving as yet little understood animal-microorganism interactions. The time involved in an uninterrupted succession allows for increasingly intimate associations and reciprocal adaptations between plants and animals, which lead to the development of many mechanisms that reduce grazing—such as the development of indigestible supporting tissues (cellulose, lignin, and so on), feedback control between plants and herbivores (*13*), and increasing predatory pressure on herbivores (*14*). Such mechanisms enable the biological community to maintain the large and complex organic structure that mitigates perturbations of the physical environment. Severe stress or rapid changes brought about by outside forces can, of course, rob the system of these protective mechanisms and allow irruptive, cancerous growths of certain species to occur, as man too often finds to his sorrow. An example of a stress-induced pest irruption occurred at Brookhaven National Laboratory, where oaks became vulnerable to aphids when translocation of sugars and amino acids was impaired by continuing gamma irradiation (*15*).

Radionuclide tracers are providing a means of charting food chains in the intact outdoor ecosystem to a degree that will permit analysis within the concepts of network or matrix algebra. For example, we have recently been able to map, by use of a radiophosphorus tracer, the open, relatively linear food linkage between plants and insects in an early old-field successional stage (*16*).

Diversity and Succession

Perhaps the most controversial of the successional trends pertain to the complex and much discussed subject of diversity (*17*). It is important to distinguish between different kinds of diversity indices, since they may not follow parallel trends in the same gradient or developmental series. Four components of diversity are listed in Table 1, items 8 through 11.

The variety of species, expressed as a species-number ratio or a species-area ratio, tends to increase during the early stages of community development. A second component of species diversity is what has been called equitability, or evenness (*18*), in the apportionment of individuals among the species. For example, two systems each containing 10 species and 100 individuals have the same diversity in terms of species-number ratio but could have widely different equitabilities depending on the apportionment of the 100 individuals among the 10 species—for example, 91-1-1-1-1-1-1-1-1-1 at one extreme or 10 individuals per species at the other. The Shannon formula,

$$-\Sigma \frac{ni}{N} \log_2 \frac{ni}{N}$$

where *ni* is the number of individuals in each species and *N* is the total number of individuals, is widely used as a diversity index because it combines the variety and equitability components in one approximation. But, like all such lumping parameters, Shannon's formula may obscure the behavior of these two rather different aspects of diversity. For example, in our most recent field experiments, an acute stress from insecticide reduced the number of species of insects relative to the number of individuals but increased the evenness in the relative abundances of the surviving species (*19*). Thus, in this case the "variety" and "evenness" components would tend to cancel each other in Shannon's formula.

While an increase in the variety of species together with reduced dominance by any one species or small group of species (that is, increased evenness) can be accepted as a general probability during succession (*20*), there are other community changes that may work against these trends. An increase in the size of organisms, an increase in the length and complexity of life histories, and an increase in interspecific competition that may result in competitive exclusion of species (Table 1, items 12–14) are trends that may reduce the number of species that can live in a given area. In the bloom stage of succession organisms tend to be small and to have simple life histories and rapid rates of reproduction. Changes in size appear to be a consequence of, or an adaptation to, a shift in nutrients from inorganic to organic (Table 1, item 7). In a mineral nutrient-rich environment, small size is of selective advantage, especially to autotrophs, because of the greater surface-to-volume ratio. As the ecosystem develops, however, inorganic nutrients tend to become more and more tied up in the biomass (that is, to become intrabiotic), so that the selective advantage shifts to larger organisms (either larger individuals of the same species or larger species, or both) which have greater storage capacities and more complex life histories, thus are adapted to exploiting seasonal or periodic releases of nutrients or other resources. The question of whether the seemingly direct relationship between organism size and stability is the result of positive feedback or is merely fortuitous remains unanswered (*21*).

Thus, whether or not species diversity continues to increase during succession will depend on whether the increase in potential niches resulting from increased biomass, stratification (Table 1, item 9), and other consequences of biological organization exceeds the countereffects of increasing size and competition. No one has yet been able to catalogue all the species in any sizable area, much less follow total species diversity in a successional series. Data are so far available only for segments of the community (trees, birds, and so on). Margalef (*4*) postulates that diversity will tend to peak during the early or middle stages of succession and then decline in the climax. In a study of bird populations along a successional gradient we found a bimodal pattern (*22*); the number of species increased during the early stages of old-field succession, declined during the early forest stages, and then increased again in the mature forest.

Species variety, equitability, and stratification are only three aspects of diversity which change during succession. Perhaps an even more important trend is an increase in the diversity of organic compounds, not only of those within the biomass but also of those excreted and secreted into the media (air, soil, water) as by-products of the increasing community metabolism. An increase in such "biochemical diversity" (Table 1, item 10) is illustrated by the increase in the variety of plant pigments along a successional gradient in aquatic situations, as described by Margalef (*4, 23*). Biochemical diversity within populations, or within systems as a whole, has not yet been systematically studied to the degree the subject of species diversity has been. Consequently, few generalizations can be made, except that it seems safe to say that, as succession progresses, organic extrametabolites probably serve increasingly important functions as regulators which stabilize the growth and composition of the ecosystem. Such metabolites may, in fact, be extremely important in preventing populations from overshooting the equilibrial density, thus in reducing oscillations as the system develops stability.

The cause-and-effect relationship between diversity and stability is not clear and needs to be investigated from many angles. If it can be shown that biotic diversity does indeed enhance physical stability in the ecosystem, or is the result of it, then we would have an important guide for conservation practice. Preservation of hedgerows, woodlots, noneconomic species, noneutrophicated waters, and other biotic variety in man's landscape could then be justified on scientific as well as esthestic grounds, even though such preservation often must result in some reduction in the production of food or other immediate consumer needs. In other words, is variety only the spice of life, or is it a necessity for the long life of the total ecosystem comprising man and nature?

Nutrient Cycling

An important trend in successional development is the closing or "tightening" of the biogeochemical cycling of major nutrients, such as nitrogen, phosphorus, and calcium (Table 1, items 15–17). Mature systems, as compared to developing ones, have a greater capacity to entrap and hold nutrients for cycling within the system. For example, Bormann and Likens (*24*) have estimated that only 8 kilograms per hectare out of a total pool of exchangeable calcium of 365 kilograms per hectare is lost per year in stream outflow from a North Temperate watershed covered with a mature forest. Of this, about 3 kilograms per hectare is replaced by rainfall, leaving only 5 kilograms to be obtained from weathering of the underlying rocks in order for the system to maintain mineral balance. Reducing the volume of the vegetation, or otherwise setting the succession back to a younger state, results in increased water yield by way of stream outflow (*25*), but this

Table 1. A tabular model of ecological succession: trends to be expected in the development of ecosystems.

Ecosystem attributes	Developmental stages	Mature stages
Community energetics		
1. Gross production/community respiration (P/R ratio)	Greater or less than 1	Approaches 1
2. Gross production/standing crop biomass (P/B ratio)	High	Low
3. Biomass supported/unit energy flow (B/E ratio)	Low	High
4. Net community production (yield)	High	Low
5. Food chains	Linear, predominantly grazing	Weblike, predominantly detritus
Community structure		
6. Total organic matter	Small	Large
7. Inorganic nutrients	Extrabiotic	Intrabiotic
8. Species diversity—variety component	Low	High
9. Species diversity—equitability component	Low	High
10. Biochemical diversity	Low	High
11. Stratification and spatial heterogeneity (pattern diversity)	Poorly organized	Well-organized
Life history		
12. Niche specialization	Broad	Narrow
13. Size of organism	Small	Large
14. Life cycles	Short, simple	Long, complex
Nutrient cycling		
15. Mineral cycles	Open	Closed
16. Nutrient exchange rate, between organisms and environment	Rapid	Slow
17. Role of detritus in nutrient regeneration	Unimportant	Important
Selection pressure		
18. Growth form	For rapid growth ("*r*-selection")	For feedback control ("*K*-selection")
19. Production	Quantity	Quality
Overall homeostasis		
20. Internal symbiosis	Undeveloped	Developed
21. Nutrient conservation	Poor	Good
22. Stability (resistance to external perturbations)	Poor	Good
23. Entropy	High	Low
24. Information	Low	High

greater outflow is accompanied by greater losses of nutrients, which may also produce downstream eutrophication. Unless there is a compensating increase in the rate of weathering, the exchangeable pool of nutrients suffers gradual depletion (not to mention possible effects on soil structure resulting from erosion). High fertility in "young systems" which have open nutrient cycles cannot be maintained without compensating inputs of new nutrients; examples of such practice are the continuous-flow culture of algae, or intensive agriculture where large amounts of fertilizer are imported into the system each year.

Because rates of leaching increase in a latitudinal gradient from the poles to the equator, the role of the biotic community in nutrient retention is especially important in the high-rainfall areas of the subtropical and tropical latitudes, including not only land areas but also estuaries. Theoretically, as one goes equatorward, a larger percentage of the available nutrient pool is tied up in the biomass and a correspondingly lower percentage is in the soil or sediment. This theory, however, needs testing, since data to show such a geographical trend are incomplete. It is perhaps significant that conventional North Temperate row-type agriculture, which represents a very youthful type of ecosystem, is successful in the humid tropics only if carried out in a system of "shifting agriculture" in which the crops alternate with periods of natural vegetative redevelopment. Tree culture and the semiaquatic culture of rice provide much better nutrient retention and consequently have a longer life expectancy on a given site in these warmer latitudes.

Selection Pressure: Quantity versus Quality

MacArthur and Wilson (*26*) have reviewed stages of colonization of islands which provide direct parallels with stages in ecological succession on continents. Species with high rates of reproduction and growth, they find, are more likely to survive in the early uncrowded stages of island colonization. In contrast, selection pressure favors species with lower growth potential but better capabilities for competitive survival under the equilibrium density of late stages. Using the terminology of growth equations, where r is the intrinsic rate of increase and K is the upper asymptote or equilibrium population size, we may say that "r selection" predominates in early colonization, with "K selection" prevailing as more and more species and individuals attempt to colonize (Table 1, item 18). The same sort of thing is even seen within the species in certain "cyclic" northern insects in which "active" genetic strains found at low densities are replaced at high densities by "sluggish" strains that are adapted to crowding (*27*).

Genetic changes involving the whole biota may be presumed to accompany the successional gradient, since, as described above, quantity production characterizes the young ecosystem while quality production and feedback control are the trademarks of the mature system (Table 1, item 19). Selection at the ecosystem level may be primarily interspecific, since species replacement is a characteristic of successional series or seres. However, in most well-studied seres there seem to be a few early successional species that are able to persist through to late stages. Whether genetic changes contribute to adaptation in such species has not been determined, so far as I know, but studies on population genetics of *Drosophila* suggest that changes in genetic composition could be important in population regulation (*28*). Certainly, the human population, if it survives beyond its present rapid growth stage, is destined to be more and more affected by such selection pressures as adaptation to crowding becomes essential.

Overall Homeostasis

This brief review of ecosystem development emphasizes the complex nature of processes that interact. While one may well question whether all the trends described are characteristic of all types of ecosystems, there can be little doubt that the net result of community actions is symbiosis, nutrient conservation, stability, a decrease in entropy, and an increase in information (Table 1, items 20–24). The overall strategy is, as I stated at the beginning of this article, directed toward achieving as large and diverse an organic structure as is possible within the limits set by the available energy input and the prevailing physical conditions of existence (soil, water, climate, and so on). As studies of biotic communities become more functional and sophisticated, one is impressed with the importance of mutualism, parasitism, predation, commensalism, and other forms of symbiosis. Partnership between unrelated species is often noteworthy (for example, that between coral coelenterates and algae, or between mycorrhizae and trees). In many cases, at least, biotic control of grazing, population density, and nutrient cycling provide the chief positive-feedback mechanisms that contribute to stability in the mature system by preventing overshoots and destructive oscillations. The intriguing question is, Do mature ecosystems age, as organisms do? In other words, after a long period of relative stability or "adulthood," do ecosystems again develop unbalanced metabolism and become more vulnerable to diseases and other perturbations?

Relevance of Ecosystem Development Theory to Human Ecology

Figure 1 depicts a basic conflict between the strategies of man and of nature. The "bloom-type" relationships, as exhibited by the 30-day microcosm or the 30-year forest, illustrate man's present idea of how nature should be directed. For example, the goal of agriculture or intensive forestry, as now generally practiced, is to achieve high rates of production of readily harvestable products with little standing crop left to accumulate on the landscape—in other words, a high P/B efficiency. Nature's strategy, on the other hand, as seen in the outcome of the successional process, is directed toward the reverse efficiency—a high B/P ratio, as is depicted by the relationship at the right in Fig. 1. Man has generally been preoccupied with obtaining as much "production" from the landscape as possible, by developing and maintaining early successional types of ecosystems, usually monocultures. But, of course, man does not live by food and fiber alone; he also needs a balanced CO_2–O_2 atmosphere, the climatic buffer provided by oceans and masses of vegetation, and clean (that is, unproductive) water for cultural and industrial uses. Many essential life-cycle resources, not to mention recreational and esthetic needs, are best provided man by the less "productive" landscapes. In other words, the landscape is not just a supply depot but is also the *oikos*—the home—in which we must live. Until recently mankind has more or less taken for granted the gas-exchange, water-purification, nutrient-cycling, and other protective functions of self-maintaining ecosystems, chiefly because neither his numbers nor his environmental manipulations have been great enough to affect regional and global balances. Now, of course, it is painfully evident that such balances are being affected, often detrimentally. The "one problem, one solution approach" is no longer adequate and must be replaced by some form of ecosystem analysis that considers man as a part of, not apart from, the environment.

The most pleasant and certainly the safest landscape to live in is one containing a variety of crops, forests, lakes, streams, roadsides, marshes, seashores, and "waste places"—in other words, a mixture of communities of different ecological ages. As individuals we more or less instinctively surround our houses with protective, nonedible cover (trees, shrubs, grass) at the same time that we strive to coax extra bushels from our cornfield. We all consider the cornfield a "good thing," of course, but most of us would not want to live there, and it would certainly be suicidal to cover the whole land area of the biosphere with cornfields, since the boom and bust oscillation in such a situation would be severe.

The basic problem facing organized society today boils down to determining in some objective manner when we are getting "too much of a good thing." This is a completely new challenge to mankind because, up until now, he has had to be concerned largely with too little rather than too much. Thus, concrete is a "good thing," but not if half the world is covered with it. Insecticides are "good things," but not when used, as they now are, in an indiscriminate and wholesale manner. Likewise, water impoundments have proved to be very useful man-made additions to the landscape, but obviously we don't want the whole country inundated! Vast man-made lakes solve some problems, at least temporarily, but yield comparative little food or fiber, and, because of high evaporative losses, they may not even be the best device for storing water; it might better be stored in the watershed, or underground in aquafers. Also, the cost of building large dams is a drain on already overtaxed revenues. Although as individuals we readily recognize that we can have too many dams or other large-scale environmental changes, governments are so fragmented and lacking in systems-analysis capabilities that there is no effective mechanism whereby negative feedback signals can be received and acted on before there has been a serious overshoot. Thus, today there are governmental agencies, spurred on by popular and political enthusiasm for dams, that are putting on the drawing boards plans for damming every river and stream in North America!

Table 2. Contrasting characteristics of young and mature-type ecosystems.

Young	Mature
Production	Protection
Growth	Stability
Quantity	Quality

Society needs, and must find as quickly as possible, a way to deal with the landscape as a whole, so that manipulative skills (that is, technology) will not run too far ahead of our understanding of the impact of change. Recently a national ecological center outside of government and a coalition of governmental agencies have been proposed as two possible steps in the establishment of a political control mechanism for dealing with major environmental questions. The soil conservation movement in America is an excellent example of a program dedicated to the consideration of the whole farm or the whole watershed as an ecological unit. Soil conservation is well understood and supported by the public. However, soil conservation organizations have remained too exclusively farm-oriented, and have not yet risen to the challenge of the urban-rural landscape, where lie today's most serious problems. We do, then, have potential mechanisms in American society that could speak for the ecosystem as a whole, but none of them are really operational (*29*).

The general relevance of ecosystem development theory to landscape planning can, perhaps, be emphasized by the "mini-model" of Table 2, which contrasts the characteristics of young and mature-type ecosystems in more general terms than those provided by Table 1. It is mathematically impossible to obtain a maximum for more than one thing at a time, so one cannot have both extremes at the same time and place. Since all six characteristics listed in Table 2 are desirable in the aggregate, two possible solutions to the dilemma immediately suggest themselves. We can compromise so as to provide moderate quality and moderate yield on all the landscape, or we can deliberately plan to compartmentalize the landscape so as to simultaneously maintain highly productive and predominantly protective types as separate units subject to different management strategies (strategies ranging, for example, from intensive cropping on the one hand to wilderness management on the other). If ecosystem development theory is valid and applicable to planning, then the so-called multiple-use strategy, about which we hear so much, will work only through one or both of these approaches, because, in most cases, the projected multiple uses conflict with one another. It is appropriate, then, to examine some examples of the compromise and the compartmental strategies.

Pulse Stability

A more or less regular but acute physical perturbation imposed from without can maintain an ecosystem at some intermediate point in the developmental sequence, resulting in, so to speak, a compromise between youth and maturity. What I would term "fluctuating water level ecosystems" are good examples. Estuaries, and intertidal zones in general, are maintained in an early, relatively fertile stage by the tides, which provide the energy for rapid nutrient cycling. Likewise, freshwater marshes, such as the Florida Everglades, are held at an early successional stage by the seasonal fluctuations in water levels. The dry-season drawdown speeds up aerobic decomposition of accumulated organic matter, releasing nutrients that, on reflooding, support a wet-season bloom in productivity. The life histories of many organisms are intimately coupled to this periodicity. The wood stork, for example, breeds when the water levels are falling and the small fish on which it feeds become concentrated and easy to catch in the drying pools. If the water level remains high during the usual dry season or fails to rise in the wet season, the stork will not nest (*30*). Stabilizing water levels in the Everglades by means of dikes, locks, and impoundments, as is now advocated by some, would, in my opinion, destroy rather than preserve the Everglades as we now know them just as surely as complete drainage would. Without periodic drawdowns and fires, the shallow basins would fill up with organic matter and succession would proceed from the present pond-and-prairie condition toward a scrub or swamp forest.

It is strange that man does not readily recognize the importance of recurrent changes in water level in a natural situation such as the Everglades when similar pulses are the basis for some of his most enduring food culture systems (*31*). Alternate filling and draining of ponds has been a standard procedure in fish culture for centuries in Europe and the Orient. The flooding, draining, and soil-aeration procedure in rice culture is another example. The rice paddy is thus the cultivated analogue of the natural marsh or the intertidal ecosystem.

Fire is another physical factor whose periodicity has been of vital importance to man and nature over the centuries. Whole biotas, such as those of the African grasslands and the California chaparral, have become adapted to periodic fires producing what ecologists often call "fire climaxes" (*32*). Man uses fire deliberately to maintain such climaxes or to set back succession to some desired point. In the southeastern coastal plain, for example, light fires of moderate frequency can maintain a pine forest against the encroachment of older successional stages which, at the present time at least, are considered economically less desirable. The fire-controlled forest yields less wood than a tree farm does (that is, young trees, all of about the same age, planted in rows and harvested on a short rotation schedule), but it provides a greater protective cover for the landscape, wood of higher quality, and a home for game birds (quail, wild turkey, and so on) which could not survive in a tree farm. The fire climax, then, is an example of a compromise between production simplicity and protection diversity.

It should be emphasized that pulse stability works only if there is a complete community (including not only plants but animals and microorganisms) adapted to the particular intensity and frequency of the perturbation. Adaptation—operation of the selection process—requires times measurable on the evolutionary scale. Most physical stresses introduced by man are too sudden, too violent, or too arrhythmic for adaptation to occur at the ecosystem level, so severe oscillation rather than stability results. In many cases, at least, modification of naturally adapted ecosystems for cultural purposes would seem preferable to complete redesign.

Prospects for a Detritus Agriculture

As indicated above, heterotrophic utilization of primary production in mature ecosystems involves largely a delayed consumption of detritus. There is no reason why man cannot make greater use of detritus and thus obtain food or other products from the more protective type of ecosystem. Again, this would represent a compromise, since the short-term yield could not be as great as the yield obtained by direct exploitation of the grazing food chain. A detritus agriculture, however, would have some compensating advantages. Present agricultural strategy is based on selection for rapid growth and edibility in food plants, which, of course, make them vulnerable to attack by insects and disease. Consequently, the more we select for succulence and growth, the more effort we must invest in the chemical control of pests; this effort, in turn, increases the likelihood of our poisoning useful organisms, not to mention ourselves. Why not also practice the reverse strategy—that is, select plants which are essentially unpalatable, or which produce their own systemic insecticides while they are growing, and then convert the net production into edible products by microbial and chemical enrichment in food factories? We could then devote our biochemical genius to the enrichment process instead of fouling up our living space with chemical poisons! The production of silage by fermentation of low-grade fodder is an example of such a procedure already in widespread use. The cultivation of detritus-eating fishes in the Orient is another example.

By tapping the detritus food chain man can also obtain an appreciable harvest from many natural systems without greatly modifying them or destroying their protective and esthetic value. Oyster culture in estuaries is a good example. In Japan, raft and long-line culture of oysters has proved to be a very practical way to harvest the natural microbial products of estuaries and shallow bays. Furukawa (*33*) reports that the yield of cultured oysters in the Hiroshima Prefecture has increased tenfold since 1950, and that the yield of oysters (some 240,000 tons of meat) from this one district alone in 1965 was ten times the yield of natural oysters from the entire country. Such oyster culture is feasible along the entire Atlantic and Gulf coasts of the United States. A large investment in the culture of oysters and other seafoods would also provide the best possible deterrent against pollution, since the first threat of damage to the pollution-sensitive oyster industry would be immediately translated into political action!

The Compartment Model

Successful though they often are, compromise systems are not suitable nor desirable for the whole landscape. More emphasis needs to be placed on compartmentalization, so that growth-type, steady-state, and intermediate-type ecosystems can be linked with urban and industrial areas for mutual benefit. Knowing the transfer coefficients that define the flow of energy and the movement of materials and organisms (including man) between compartments, it should be possible to determine, through analog-computer manipulation, rational limits for the size and capacity of each compartment. We might start, for example, with a simplified model, shown in Fig. 2, consisting of four compartments of equal area, partitioned according to the basic biotic-function criterion—that is, according to whether the area is (i) productive, (ii) protective, (iii) a compromise between (i) and (ii) or (iv), urban-industrial. By continually refining the transfer coefficients on the basis of real world situations, and by increasing and decreasing the size and capacity of each compartment through computer simulation, it would be possible to determine objectively the limits that must eventually be imposed on each compartment in order to maintain regional and global balances in the exchange of vital energy and of materials. A systems-analysis procedure provides at least one approach to the solution of the basic dilemma posed by the question "How do we determine when we are getting too much of a good thing?" Also it provides a means of evaluating the energy drains imposed on ecosystems by pollution, radiation, harvest, and other stresses (*34*).

Implementing any kind of compartmentalization plan, of course, would require procedures for zoning the landscape and restricting the use of some land and water areas. While the principle of zoning in cities is universally accepted, the procedures now followed do not work very well because zoning restrictions are too easily overturned by short-term economic and population pressures. Zoning the landscape would require a whole new order of thinking. Greater use of legal measures providing for tax relief, restrictions on use, scenic easements, and public ownership will be required if appreciable land and water areas are to be held in the "protective" categories. Several states (for example, New Jersey and California), where pollution and population pressure are beginning to hurt, have made a start in this direction by enacting "open space" legislation designed to get as much unoccupied land as possible into a "protective" status so that future uses can be planned on a rational and scientific basis. The United States as a whole is fortunate in that large areas of the country are in national forests, parks, wildlife refuges, and so on. The fact that such areas, as well as the bordering oceans, are not quickly exploitable gives us time for the accelerated ecological study and programming needed to determine what proportions of different types of landscape provide a safe balance between man and nature. The open oceans, for example, should forever be allowed to remain protective rather than productive territory, if Alfred Redfield's (*35*) assumptions are correct. Redfield views the oceans, the major part of the hydrosphere, as the biosphere's governor, which slows down and controls the rate of decomposition and nutrient regeneration, thereby creating and maintaining the highly aerobic terrestrial environment to which the higher forms of life, such as man, are adapted. Eutrophication of the ocean in a last-ditch effort to feed the populations of the land could well have an adverse effect on the oxygen reservoir in the atmosphere.

Until we can determine more precisely how far we may safely go in expanding intensive agriculture and urban sprawl at the expense of the protective landscape, it will be good insurance to hold inviolate as much of the latter as possible. Thus, the preservation of natural areas is not a peripheral luxury for society but a capital investment from which we expect to draw interest. Also, it may well be that restrictions in the use of land and water are our only practical means of avoiding overpopulation or too great an exploitation of resources, or both. Interestingly enough, restriction of land use is the analogue of a natural behavioral control mechanism known as "territoriality" by which

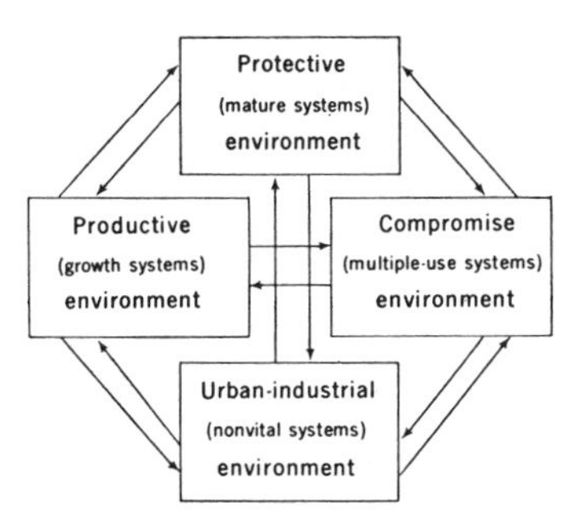

Fig. 2. Compartment model of the basic kinds of environment required by man, partitioned according to ecosystem development and life-cycle resource criteria.

many species of animals avoid crowding and social stress (*36*).

Since the legal and economic problems pertaining to zoning and compartmentalization are likely to be thorny, I urge law schools to establish departments, or institutes, of "landscape law" and to start training "landscape lawyers" who will be capable not only of clarifying existing procedures but also of drawing up new enabling legislation for consideration by state and national governing bodies. At present, society is concerned—and rightly so—with human rights, but environmental rights are equally vital. The "one man one vote" idea is important, but so also is a "one man one hectare" proposition.

Education, as always, must play a role in increasing man's awareness of his dependence on the natural environment. Perhaps we need to start teaching the principles of ecosystem in the third grade. A grammar school primer on man and his environment could logically consist of four chapters, one for each of the four essential kinds of environment, shown diagrammatically in Fig. 2.

Of the many books and articles that are being written these days about man's environmental crisis, I would like to cite two that go beyond "crying out in alarm" to suggestions for bringing about a reorientation of the goals of society. Garrett Hardin, in a recent article in *Science* (*37*), points out that, since the optimum population density is less than the maximum, there is no strictly technical solution to the problem of pollution caused by overpopulation; a solution, he suggests, can only be achieved through moral and legal means of "mutual coercion, mutually agreed upon by the majority of people." Earl F. Murphy, in a book entitled *Governing Nature* (*38*), emphasizes that the regulatory approach alone is not enough to protect life-cycle resources, such as air and water, that cannot be allowed to deteriorate. He discusses permit systems, effluent charges, receptor levies, assessment, and cost-internalizing procedures as economic incentives for achieving Hardin's "mutually agreed upon coercion."

It goes without saying that the tabular model for ecosystem development which I have presented here has many parallels in the development of human society itself. In the pioneer society, as in the pioneer ecosystem, high birth rates, rapid growth, high economic profits, and exploitation of accessible and unused resources are advantageous, but, as the saturation level is approached, these drives must be shifted to considerations of symbiosis (that is, "civil rights," "law and order," "education," and "culture"), birth control, and the recycling of resources. A balance between youth and maturity in the socio-environmental system is, therefore, the really basic goal that must be achieved if man as a species is to successfully pass through the present rapid-growth stage, to which he is clearly well adapted, to the ultimate equilibrium-density stage, of which he as yet shows little understanding and to which he now shows little tendency to adapt.

References and Notes

1. E. P. Odum, *Ecology* (Holt, Rinehart & Winston, New York, 1963), chap. 6.
2. H. T. Odum and R. C. Pinkerton, *Amer. Scientist* **43**, 331 (1955).
3. A. J. Lotka, *Elements of Physical Biology* (Williams and Wilkins, Baltimore, 1925).
4. R. Margalef, *Advan. Frontiers Plant Sci.* **2**, 137 (1963); *Amer. Naturalist* **97**, 357 (1963).
5. R. J. Beyers, *Ecol. Monographs* **33**, 281 (1963).
6. The systems so far used to test ecological principles have been derived from sewage and farm ponds and are cultured in half-strength No. 36 Taub and Dollar medium [*Limnol. Oceanog.* **9**, 61 (1964)]. They are closed to organic imput or output but are open to the atmosphere through the cotton plug in the neck of the flask. Typically, liter-sized microecosystems contain two or three species of nonflagellated algae and one to three species each of flagellated protozoans, ciliated protozoans, rotifers, nematodes, and ostracods; a system derived from a sewage pond contained at least three species of fungi and 13 bacterial isolates [R. Gordon, thesis, University of Georgia (1967)]. These cultures are thus a kind of minimum ecosystem containing those small species originally found in the ancestral pond that are able to function together as a self-contained unit under the restricted conditions of the laboratory flask and the controlled environment of a growth chamber [temperature, 65° to 75°F (18° to 24°C); photoperiod, 12 hours; illumination, 100 to 1000 footcandles].
7. G. D. Cooke, *BioScience* **17**, 717 (1967).
8. T. Kira and T. Shidei, *Japan. J. Ecol.* **17**, 70 (1967).
9. The metabolism of the microcosms was

monitored by measuring diurnal *p*H changes, and the biomass (in terms of total organic matter and total carbon) was determined by periodic harvesting of replicate systems.
10. F. J. H. Mackereth, *Proc. Roy. Soc. London Ser. B* **161**, 295 (1965); U. M. Cowgill and G. E. Hutchinson, *Proc. Intern. Limnol. Ass.* **15**, 644 (1964); A. D. Harrison, *Trans. Roy. Soc. S. Africa* **36**, 213 (1962).
11. R. Margalef, *Proc. Intern. Limnol. Ass.* **15**, 169 (1964).
12. J. R. Bray, *Oikos* **12**, 70 (1961).
13. D. Pimentel, *Amer. Naturalist* **95**, 65 (1961).
14. R. T. Paine, *ibid.* **100**, 65 (1966).
15. G. M. Woodwell, *Brookhaven Nat. Lab. Pub. 924(T-381)* (1965), pp. 1–15.
16. R. G. Wiegert, E. P. Odum, J. H. Schnell, *Ecology* **48**, 75 (1967).
17. For selected general discussions of patterns of species diversity, see E. H. Simpson, *Nature* **163**, 688 (1949): C. B. Williams, *J. Animal Ecol.* **22**, 14 (1953); G. E. Hutchinson, *Amer. Naturalist* **93**, 145 (1959); R. Margalef, *Gen. Systems* **3**, 36 (1958); R. MacArthur and J. MacArthur, *Ecology* **42**, 594 (1961); N. G. Hairston, *ibid.* **40**, 404 (1959); B. C. Patten, *J. Marine Res. (Sears Found. Marine Res.)* **20**, 57 (1960); E. G. Leigh, *Proc. Nat. Acad. Sci. U.S.* **55**, 777 (1965); E. R. Pianka, *Amer. Naturalist* **100**, 33 (1966); E. C. Pielou, *J. Theoret. Biol.* **10**, 370 (1966).
18. M. Lloyd and R. J. Ghelardi, *J. Animal Ecol.* **33**, 217 (1964); E. C. Pielou, *J. Theoret. Biol.* **13**, 131 (1966).
19. G. W. Barrett, *Ecology* **49**, 1019 (1969).
20. In our studies of natural succession following grain culture, both the species-to-numbers and the equitability indices increased for all trophic levels but especially for predators and parasites. Only 44 percent of the species in the natural ecosystem were phytophagous, as compared to 77 percent in the grain field.
21. J. T. Bonner, *Size and Cycle* (Princeton Univ. Press, Princeton, N.J., 1963); P. Frank, *Ecology* **49**, 355 (1968).
22. D. W. Johnston and E. P. Odum, *Ecology* **37**, 50 (1956).
23. R. Margalef, *Oceanog. Marine Biol. Annu. Rev.* **5**, 257 (1967).
24. F. H. Bormann and G. E. Likens, *Science* **155**, 424 (1967).
25. Increased water yield following reduction of vegetative cover has been frequently demonstrated in experimental watersheds throughout the world [see A. R. Hibbert, in *International Symposium on Forest Hydrology* (Pergamon Press, New York, 1967), pp. 527–543]. Data on the long-term hydrologic budget (rainfall input relative to stream outflow) are available at many of these sites, but mineral budgets have yet to be systematically studied. Again, this is a prime objective in the "ecosystem analysis" phase of the International Biological Program.
26. R. H. MacArthur and E. O. Wilson, *Theory of Island Biogeography* (Princeton Univ. Press, Princeton, N.J., 1967).
27. Examples are the tent caterpillar [see W. G. Wellington, *Can. J. Zool.* **35**, 293 (1957)] and the larch budworm [see W. Baltensweiler, *Can. Entomologist* **96**, 792 (1964)].
28. F. J. Ayala, *Science* **162**, 1453 (1968).
29. Ira Rubinoff, in discussing the proposed sea-level canal joining the Atlantic and Pacific oceans [*Science* **161**, 857 (1968)], calls for a "control commission for environmental manipulation" with "broad powers of approving, disapproving, or modifying all major alterations of the marine or terrestrial environments. . . ."
30. See M. P. Kahl, *Ecol. Monographs* **34**, 97 (1964).
31. The late Aldo Leopold remarked long ago [*Symposium on Hydrobiology* (Univ. of Wisconsin Press, Madison, 1941), p. 17] that man does not perceive organic behavior in systems unless he has built them himself. Let us hope it will not be necessary to rebuild the entire biosphere before we recognize the worth of natural systems!
32. See C. F. Cooper, *Sci. Amer.* **204**, 150 (April 1961).
33. See "Proceedings Oyster Culture Workshop, Marine Fisheries Division, Georgia Game and Fish Commission, Brunswick" (1968), pp. 49–61.
34. See H. T. Odum, in *Symposium on Primary Productivity and Mineral Cycling in Natural Ecosystems*, H. E. Young, Ed. (Univ. of Maine Press, Orono, 1967), p. 81; ———, in *Pollution and Marine Ecology* (Wiley, New York, 1967), p. 99; K. E. F. Watt, *Ecology and Resource Management* (McGraw-Hill, New York, 1968).
35. A. C. Redfield, *Amer. Scientist* **46**, 205 (1958).
36. R. Ardrey, *The Territorial Imperative* (Atheneum, New York, 1967).
37. G. Hardin, *Science* **162**, 1243 (1968).
38. E. F. Murphy, *Governing Nature* (Quadrangle Books, Chicago, 1967).

Visual Receptors and Retinal Interaction

Haldan Keffer Hartline

The neuron is the functional as well as the structural unit of the nervous system. Neurophysiology received an impetus of far-reaching effect in the 1920's, when Adrian and his colleagues developed and exploited methods for recording the activity of single neurons and sensory receptors. Adrian and Bronk were the first to analyze motor function by recording the activity of single fibers dissected from a nerve trunk and Adrian and Zotterman the first to elucidate properties of single sensory receptors (*1*). These studies laid the foundations for the unitary analysis of nervous function.

My early interest in vision was spurred by another contribution from Adrian's laboratory: his study, with R. Matthews, of the massed discharge of nerve impulses in the eel's optic nerve (*2*). I aspired to the obvious extension of this study: application of unitary analysis to the receptors and neurons of the visual system.

Oscillograms of the action potentials in a single nerve fiber are now commonplace. The three shown in Fig. 1 are from an optic nerve fiber whose retinal receptor was stimulated by light, the relative values of which are given at the left of each record. One of the earliest results of unitary analysis was to show that higher intensities are signaled by higher frequencies of discharge of uniform nerve impulses.

In 1931, when C. H. Graham and I sought to apply to an optic nerve the technique developed by Adrian and Bronk for isolating a single fiber, we made a fortunate choice of experimental animal (*3*). The xiphosuran arachnoid, *Limulus polyphemus*, commonly called "Horseshoe crab," abounds on the eastern coast of North America (*4*). These "living fossils" have lateral compound eyes that are coarsely faceted and connected to the brain by long optic nerves. The optic nerve in the adults can be frayed into thin bundles which are easy to split until just one active fiber remains. The records in Fig. 1 were obtained from such a preparation.

The sensory structures in the eye of *Limulus* from which the optic nerve fibers arise are clusters of receptor cells, arranged radially around the dendritic process of a bipolar neuron (eccentric cell) (*5*). Each cluster lies behind its corneal facet and crystalline cone, which give it its own, small visual field (Fig. 2). Each such ommatidium, though not as simple as I once thought, seems to act as a functional receptor unit. Restriction of the stimulating light to one facet elicits discharge in one fiber—the axon of the bipolar neuron

The author is professor of biophysics at the Rockefeller University in New York City. This article is the lecture he delivered in Stockholm, Sweden, 12 December 1967, when, with Regnar Arthur Granit and George Wald, he received the Nobel Prize in medicine or physiology. It is published here with the permission of the Nobel Foundation and also is included in the complete volume of *Les Prix Nobel en 1967*, as well as in the series Nobel Lectures (in English), published by the Elsevier Publishing Company, Amsterdam and New York.

PART

5

Case Studies in Natural Systems

Lessons from Nature

Robert K. Peet

Observations of nature are key to all scientific progress. Science cannot be deduced from first principles; without facts, no progress can be made and no theories tested. However, the various sciences differ in the interplay between observation and theory. In the physical sciences, broad generalizations often derive from a few simple observations, though creation of the conditions necessary for the required observations may be far from trivial as in the case of experiments with particle accelerators. Biological systems are generally far more complex than physical systems, owing in part to their dependence on information stored in the genetic code. Consequently, molecular and cellular biologists often resort to the use of model systems where a single species such as *Drosophila melanogaster, Escherichia coli,* or *Saccharomyces cerevisiae* is studied in detail and assumed to represent the general features of numerous other more-or-less related species.

Ecologists must deal with still more complex systems. Not only does the ecological system depend on the genetic information of the numerous interacting species, but it also depends on the spatial arrangement and physical attributes of all individuals, in addition to attributes of the physical environment that can contribute in important ways to ecological processes. In short, the intrinsic complexity of ecosystems forces ecologists to search for generalizations that are contingent on numerous and often quite specific initial conditions (MacArthur 1972; May 1986). Consequently, progress in ecology has depended greatly on numerous case studies, in which investigators have studied a natural system or process in detail. Where a study of a single species will often suffice for general understanding in molecular biology, numerous systems must be studied and then compared before ecological generalizations can be accepted. The papers presented in this section are exemplary case studies of natural systems or processes, studies of the sort that ecology as a field needs to accumulate so that more general patterns can be sought.

Papers become classics for many reasons. Although the six papers reprinted in this section are perhaps most striking in their diversity of content, they all share at least two attributes. First, in each case the study has a strong descriptive component made possible by the willingness of the author to grapple with the complexities of nature in the field. Second, in each case the author has combined synthesis with vision, anticipating or even consciously shaping the future development of ecology.

Many case studies in ecology have had a significant and long-lasting impact. Because of the descriptive character of most of these studies, they tend to be long and often spread over several publications, making it difficult to reprint many of the better-known examples. Although space limitations have kept the number of reprinted papers in this section to six, I have tried to provide a more comprehensive overview by discussing a few other papers equally deserving of inclusion here.

Two different types of research can be recognized among the papers discussed in this part. The first consists of in-depth studies of single sites. While there is certainly a conceptual focus for each of these studies, the papers gain classic status in large part because the depth of understanding that the author(s) achieved about his or her system led to significant and unanticipated new insights. The second group contains papers where the focus was not on a particular system, but rather on a particular process as observed at several sites. In these studies, classic status is merited principally because the results or approaches were shown to be broadly applicable. These two categories provide a logical basis for organizing comments on the papers.

Case studies of specific systems

Case studies of natural systems have been important across the breadth of the discipline of ecology. To illustrate this universal reliance on case studies, I have chosen to discuss examples from three divergent areas: regulation of animal populations, energy flow in ecosystems, and the structure of plant communities.

The great body of theory on the population biology of animals, built up during the first three decades of the twentieth century through the work of such pioneering ecologists as Pearl, Gause, Lotka, Volterra, and Nicholson and Bailey, was largely based on the idea that animal numbers are regulated in a density-dependent fashion. When in the 1940s and early 1950s this body of theory first encountered data generated by extensive case studies of natural populations, the results were not altogether pleasing. The theory in many cases fell short of explaining the results, and the ecologists of the day were not always cool and rational in their efforts to reconcile the discrepancies.

Two Australian entomologists, H. G. Andrewartha and L. C. Birch, played a particularly important role in forcing theorists to examine cold, hard facts. Through a series of case studies conducted during the 1940s, they tried to develop statistical models to predict population sizes of economically important insect species. The first of their two most famous studies focused on grasshopper plagues in southern Australia (Birch and Andrewartha 1941). In this study, population eruptions of the grasshopper *Austroicetes cruciata* were correlated with weather conditions. The second classic study (Davidson and Andrewartha 1948*a,b*, the second of which is reprinted in this section), focused on the apple-blossum thrip (*Thrips imaginis*). Both studies were interpreted as demonstrating density-independent population control.

Thrips were initially of interest to workers at the Waite Institute of the University of Adelaide because their populations were known to vary dramatically from year to year, and when abundant they could cause considerable damage to the apple crop. Roses planted around the institute were observed to serve as thrip traps since they attracted adults, yet did not present satisfactory conditions for growth of immature stages. Almost every day for seven years workers at the institute collected 20 rose flowers and counted the number of thrips each contained. During this period they tallied nearly 1.4 million thrips. However, not yet satisfied, knowing that nearly 35 years of records were needed to accurately measure average monthly rainfall for the Adelaide area with its irregular climatic regime, they continued for another seven years, counting thrips from ten roses daily during the spring and summer.

As expected, the peak in the number of thrips was remarkably variable from year to year. Armed with the data from the Waite Institute study, Davidson set out to determine the association between the thrip's population size and weather. The amount of work necessary can hardly be imagined by ecologists trained in an age of electronic data processing. Davidson died before completing the task, and the work was taken up by H. G. Andrewartha. What Andrewartha eventually found was that a combination of four easily measured environmental variables could account for 78% of the variance in the logarithm of the geometric mean number of thrips per rose for the 30 days preceding the population peak. In short, he found no evidence that the thrip populations were controlled by density, but instead they appeared controlled by climate.

This study and others Andrewartha and Birch conducted on Australian insects led them to become highly skeptical of the importance that most practicing ecologists placed on competition as the factor regulating animal population size. Ultimately, this skepticism led them to write a highly influential book that appeared in 1954, *The distribution and abundance of animals.* The success of their book and the visibility it gave their earlier case studies undoubtedly played a major role in bringing the thrips example to the attention of the ecological community and ultimately into textbooks as a classic case study.

Andrewartha and Birch steadfastly held that the thrips example showed no evidence of density-dependent control. They may, in fact, have been wrong. The summer, fall, and early winter population levels were relatively consistent from year to year, and the minimum levels may well have been controlled by density (Varley et al. 1973). Furthermore, they had no way to assess the number of flowers available in a given year, and thus no way to assess the resource size to which the thrips may have been responding. Alternatively, perhaps only in extreme years when populations reached sufficiently high spring peaks did competition become important, an idea advocated effectively for many types of populations by Wiens (for example, 1977). Regardless, the case studies cited by Andrewartha and Birch (1954), of which the thrips example is perhaps the best known, forced a rethinking of the primacy of biological interactions, especially competition, as a fundamental assumption of population ecology.

The history of research on energy flows in natural systems can easily be documented in terms of classic case studies. Raymond Lindeman (1942, reprinted above in part 1) was the first to try to quantify all the major energy flows between trophic levels in a natural system and to address the issue of trophic efficiency. However, as Lindeman admitted, his landmark paper was more a theoretical essay than an empirical case study (Cook 1977). Fifteen years later H. T. Odum (1957) published a study of Silver Spring, Florida (unfortunately too long to be reprinted here) which was the first relatively complete documentation of energy flows in a natural system. Odum's study demonstrated the feasibility of studying the functioning of whole ecosystems and set the stage for the development of the modern field of ecosystem analysis.

H. T. Odum received his doctorate from Yale University, where he worked with G. E. Hutchinson. Hutchinson was among the first to recognize the importance of energy and material flows in ecosystems. When Lindeman's now-classic paper on the trophic-dynamic concept was initially rejected by *Ecology* (Cook 1977), Hutchinson argued persuasively for its eventual acceptance. Hutchinson was also the first to use radioactive tracers in nutrient cycling studies (Hutchinson and Bowen 1950). Odum absorbed the enthusiasm for understanding ecosystem processes that characterized his mentor and chose to work on strontium cycling for his dissertation. Shortly after leaving Yale, he teamed up with his brother Eugene to study the trophic structure and productivity of a coral reef community. The resulting monograph (Odum and Odum 1955) is a classic in its own right, but because of the complexity of coral reef systems, the study did not provide a complete picture of energy flows.

Silver Spring provided the advantages of being a discrete, well-bordered system and of having a relatively simple trophic structure. Odum decomposed the system into functionally similar groups of organisms and proceeded to obtain reasonably accurate measurements of all the major energy transfers. There were relatively few conceptual advances involved. The data could not be compared to those from other studies because there were no other studies. Instead, this paper provided the baseline against which to compare all future studies. If there was a conceptual advance, it was the relatively simple one of viewing the ecosystem as a series of boxes connected by arrows, each defined as a transfer function. If there was a surprise in the results, it was the importance of decomposer organisms in energy flow.

Shortly after Odum published his Silver Spring study, John Teal published a similar study of a cold temperate zone spring in Massachusetts (1957). In this paper he was able to compare his results with those from Silver Spring as well as partial results from Lake Mendota, Wisconsin, and Cedar Bog Lake, which Lindeman had studied in Minnesota. This marked the start of comparative studies of ecosystem dynamics. When Teal moved to the recently established biological laboratory at Sapelo Island, Georgia, it was only natural for him to apply the energy flow approach to the extensive salt marsh systems adjoining the island. The results of his work on energy flows in the Sapelo salt marsh were presented in his 1962 paper in *Ecology,* reprinted here.

The salt marsh system proved somewhat less tractable than the cold spring system Teal had earlier studied, in part because it was an open system with the tides repeatedly carrying materials in and out. Teal also recognized that ecologically very different species could make up a single trophic level. Accordingly, he went beyond what had been done in the earlier studies of springs and quantified energy flows between guilds composed of functionally similar organisms rather than just recognizing simple trophic levels.

Teal's analysis was necessarily crude. In many cases his values for specific energy fluxes were little more than educated guesses. Nonetheless, the study presented a first approximation of the energy flows in the marsh, and the results provided a number of surprises. The loss of energy through respiration by the dominant plant, *Spartina,* was over 70%. A large portion (47%) of the net primary production was consumed and respired by bacteria. And finally, 45% of the net production could not be accounted for and was assumed lost from the system by export with the tides. These are the sorts of unexpected but potentially highly significant results that can emerge from a carefully conducted case study. Today Teal's paper is a classic largely because of these unexpected results and the subsequent research they inspired (summarized in Pomeroy and Wiegert 1981).

As Jonathan Roughgarden (1989) has observed, "The central question of community ecology was posed decades ago: Do the populations at a site consist of all those that happen to arrive there, or of only a special subset—those with properties allowing their coexistence?" This question has been particularly important in plant community ecology. One of the best known case studies, that of Robert Whittaker on the vegetation of the Great Smoky Mountains (unfortunately too long to reprint here), specifically addressed this issue.

Whittaker was a graduate student in the zoology department of the University of Illinois in the years immediately following World War II, a time when the old Clementsian ideas of highly integrated and consistent community organization (supported at Illinois by Victor Shelford and Charles Kendeigh) were being challenged by Gleason's individualistic hypothesis (championed at Illinois by Arthur Vestal). When Kendeigh offered Whittaker the opportunity to participate in a series of studies on the animal communities of the Great Smoky Mountains, Whittaker saw an opportunity to rigorously test these two opposing views of community organization (Westman and Peet 1982). He later claimed to have entered with the bias that neither view was strictly true, but rather that "species should tend to evolve toward coadapted groupings, each representing a favorable, balanced pattern of species interactions adapted to some range of environments" (Whittaker 1972).

To test these theories, Whittaker set out to document patterns of community structure among foliage-eating insects, searching for zones of stable composition separated by zones of "community-level hybridization" (Whittaker 1972, see also 1952). However, he quickly came to realize that he would not be able to make much sense of his insect data if he did not first determine the underlying pattern of plant species distributions. He then proceeded to collect some 300 vegetation samples distributed along transects that traversed gradients of elevation and topographic position. During the summer of 1947 he completed the necessary fieldwork for essentially two different dissertations (Westman and Peet 1982).

When, back in Illinois, Whittaker examined the plant species distributions in detail, he found to his surprise that all the evidence supported the Gleasonian view that species varied independently. Whittaker described this work in a doctoral dissertation in zoology which may be unique in that it contains no mention of any animal. The work was eventually published in 1957 in *Ecological Monographs.*

The implications of Whittaker's work were numerous and profound. Almost a half century of American literature in plant ecology had been built around the Clementsian framework, and now the framework had been found to be seriously flawed. Gleason's views were first proposed in 1917 and were more fully elaborated in 1926, but they had never been tested. I have chosen to discuss Whittaker's Smoky Mountain vegetation paper in part because of the central role it played in the eventual death of the Clementsian viewpoint and the transformation of plant community ecology into a population-oriented field. But perhaps more important for this section is the central role case studies of vegetation played in Whittaker's brilliant career. He wrote numerous synthetic review papers, but almost always they were based on insights he gained from a few intensive field studies, in particular his work in the Smoky Mountains of Tennessee, the Siskiyou Mountains of Oregon (Whittaker 1960), and the Santa Catalina Mountains of Arizona (for example, Whittaker and Niering 1965).

To be fair, it was not Whittaker alone who was responsible for the eventual acceptance of Gleason's individualistic hypothesis. At the same time Whittaker was studying plant and animal communities in the Smokies, John Curtis and his students at the University of Wisconsin, particularly Robert McIntosh, were conducting a now classic case study of the vegetation of southern Wisconsin. They published earlier than Whittaker (Curtis and McIntosh 1950, 1951) and probably had greater influence in converting plant ecology to the individualistic perspective. They fell short of Whittaker only by developing less fully the implications of their ideas in synthetic reviews.

While in some cases the authors of classic case studies have anticipated the significance of their work from the outset, often their lasting contributions have arisen from unanticipated observations. One example, which builds on the contributions of Whittaker's work described above, is found in the palynological studies of Margaret Davis (1969).

Prior to Davis's study of the pollen record in Rogers Lake, Connecticut, most palynological studies examined the percentages of pollen attributable to each species, from which their authors extrapolated directly to the vegetation of the landscape surrounding the basin at the time of deposition. Davis realized that a more direct, quantitative approach was needed if pollen data were to give accurate information on the quantitative composition of the vegetation which generated that pollen. In particular, her classic 1969 study of Rogers Lake was designed to demonstrate how quantification of pollen influx rates could provide a much more accurate interpretation of the abundance of species in the region than could be obtained from simple percentages.

Davis's Rogers Lake study remains a classic in palynology, but probably not for the reasons she would have anticipated when she wrote the paper. Subsequent studies of pollen influx rates have shown these to be closely tied to peculiarities of sedimentation within a lake basin and often to exhibit much greater noise than simple percentage data (see Prentice 1988). Strikingly, Davis was among the first to urge caution in the use of influx data (Davis, Brubaker, and Webb 1973).

The lasting contributions of Davis's paper derive from two sets of ideas it introduced. First, contemporary pollen influx data were used specifically to establish quantitative relationships between pollen data and species abundance in neighboring forests. This early work led directly to the efforts of Tom Webb (who had been a post-doc in Davis's lab) and others to adapt multivariate statistical techniques for predicting vegetation from pollen data (summarized in Prentice 1988).

Secondly, Davis attempted to interpret the results of the study by looking for vegetation that produced analogous pollen assemblages in other regions, but she concluded that some assemblages did not have modern analogs. Furthermore, she observed that all the dominant species of an assemblage did not arrive simultaneously, but rather their migration rates were strongly individualistic. These observations led her to state in the concluding sentences of the paper that "Only during the last 2000 years have the pollen assemblages in southern Connecticut been sufficiently similar to modern assemblages to permit comparison. This suggests that the forest communities there, as they are presently recognized, may be of very recent origin." While these sentences were initially overlooked by most readers of her paper, the idea they expressed lead Davis and a few others to realize that a broad geographical perspective was needed to clarify the history of modern assemblages. The next logical step was obvious to Davis; she proceeded to compile North American pollen records and synthesize them into maps showing the different rates and patterns of postglacial migrations among the various tree species of eastern North America (Davis 1976, 1981). The maps clearly illustrate the individualistic behavior of eastern North American tree species and forced a major rethinking of coevolution and the importance of equilibrium conditions in plant communities. In essence, this work provides the strongest evidence we have for the Gleasonian view of the plant community advocated by Whittaker in his Smoky Mountain vegetation study.

Process Studies

In 1947, A. S. Watt presented a presidential address to The British Ecological Society which he titled "Pattern and process in the plant community" (reprinted in this section). Watt did not present original data in his address but instead summarized and synthesized a series of case studies which he had carried out during his long and productive career. In his address Watt suggested that plant communities could be described from two points of view: "for diagnosis and classification, and as a working mechanism." His primary concern was with the latter, which he viewed as underrepresented in the literature of the time. Despite a half century of research on community dynamics, very little had been written on the internal dynamics and regeneration of plant communities.

In each of the seven studies summarized in his address, Watt observed processes of small-scale cyclic succession occurring within stable climax communities. The examples were carefully selected to represent a broad range of vegetation types including bogs, grasslands, tundra, and forest. Today we are accustomed to thinking of the plant community as a dynamic entity varying in time and space and composed, as Watt suggested, of numerous patches in various stages of development (see Pickett and White 1985). However, in Watt's time it was more common to think of the climax as stable and homogeneous.

Watt was not the first to point out the dynamic and patchy nature of mature natural communities. Aubreville (1938) had made this point quite forcefully for tropical forests a decade earlier, and Jones (1945) had extended this view to old-growth temperate forests. Watt's contributions pointed out how universal such cyclical processes are and emphasized the need to study spatial pattern and regeneration processes in seemingly stable, equilibrium systems. He was able to do this so effectively because he could draw on a host of case studies he had conducted earlier in his career.

Many of Watt's examples have been reexamined in the years since he proposed cyclical processes as the norm in mature plant communities. Many of his interpretations have been shown, upon careful scrutiny, to be flawed (for example, Miles 1981; Marrs 1986; Prentice et al. 1987; Gimmingham 1988). However, as Gimmingham (1988) states in his reexamination of cycles in *Calluna* heath, "This . . . in no way detracts from his achievement. . . . The idea has been of utmost value in many ways. It has provided a strong stimulus to the investigation of pattern in plant communities, by offering an explanation to be considered where habitat heterogeneity was insufficient to account for the mosaic. It provided a framework on which to base detailed research on the dynamics of even-aged stands. . . ."

Watt taught us to think of the death of an individual plant (or sessile organism) and its subsequent replacement as part of the internal dynamics of the community. Of course, disturbances come in all sizes, ranging from the death of a single plant to the synchronous death of a large number of individuals resulting from some form of catastrophic disturbance such as fire. Consequently, just as the death of a single tree can be viewed as part of the natural dynamics of a forest, so can forest fires and other large, seemingly catastrophic disturbances be viewed as recurrent events in the natural disturbance regime of a landscape. The characterization of the frequency and size distributions of disturbance events has become a major focus of contemporary ecology as it provides a meaningful way of comparing the dynamics of structurally or taxonomically dissimilar ecosystems (Peet 1991). Among the best-studied disturbance regimes are those generated by wildfires (see Heinselman 1981) and treefalls (see Runkle 1985).

Research on gap processes started slowly following Watt's paper, with just a few early contributions (for example, Bray 1956). The field then entered a period of exponential growth with the result that a large 1985 book on "patch-dynamics" hardly started to cover the field (Pickett and White 1985). The exceptional present-day vitality of research aimed at understanding spatial processes in communities is perhaps best understood as resulting from the merging of the vigorous line of research derived from Watt's seminal paper with the equally vigorous line derived from the theoretical contributions of Skellam (1951) (discussed in part 2 of this volume) and represented by papers such as those of Horn and MacArthur (1972) and Levin and Paine (1974). The vigor of the field serves to illustrate the mutually reinforcing roles of theory and case studies in the development of ecology.

Numerous other conceptual advances can be traced to Watt's patch-dynamics model, of which there is space to mention only a few examples. The highly successful forest simulation models of Botkin et al. (1972) and Shugart (for example, 1984) are based on the patch as the functional unit. Bormann and Likens (1979) expanded on Watt's work to build a conceptual model of forest development for summarizing changes in nutrient dynamics during succession. It is probably safe to say that virtually all recent ecological studies which invoke in some way spatial pattern or heterogeneity owe at least a small debt to the intellectual environment that was born of Watt's classic paper.

Ecology in 1954 was altogether different as a science from what it is today. The books by Lack and by Andrewartha and Birch were typical of the discipline in that each presented a series of examples to support a particular point of view. In fact, Watt's paper on pattern and process represents one of the more effective ecological publications of this age of narrative scholarship, a style necessarily still dominant in some academic disciplines, such as history. With this approach, the side perceiving itself to have the better examples generally claims to have won the debate. Only rarely is theory developed to incorporate the seemingly divergent examples, and even more rarely is theory actually tested. It was during these times that Robert MacArthur entered graduate school at Yale University.

G. Evelyn Hutchinson, MacArthur's advisor, was at that time particularly concerned with the question of how communities are structured and what conditions permit species to coexist (Hutchinson 1951, 1957, 1959). With Hutchinson's encouragement, MacArthur undertook for his doctoral dissertation a case study of how the seemingly very similar warbler species that inhabit the spruce forests of Maine manage to coexist. The fact that the species do coexist and are recognized as different would lead almost any ecologist to expect that MacArthur would find differences between them. He did, concluding that "there are differences of feeding position, behavior, and nesting date which reduce competition," and that these together ". . . with slight differences in habitat preference and perhaps a tendency for territoriality to have a stronger regulating effect upon the same species than upon others, permit the coexistence of the species." This paper marked the start of a major shift in research style and approach that characterized ecology in the late 1950s and early 1960s.

MacArthur, more than any other ecologist of his time, introduced to ecology the power of the hypothetical-deductive approach. He developed a theory and used it to formulate hypotheses which he then proceeded to test. During the period preceding MacArthur's dissertation (1950–56), only some 5% of the papers in *Ecology* tested predictions, whereas 20 years later 50% of the papers in that journal explicitly tested predictions (Fretwell 1975). While MacArthur did not singlehandedly bring about this transformation, he was clearly the most influential and visible person involved, and his dissertation provided an impressive example that served as a model for many future studies.

In MacArthur's study of warblers one can identify many approaches and themes that were to recur during his career and which led to important advances in the discipline. In the introduction to his 1957 paper, MacArthur developed a number of theoretical ideas on species coexistence. Many of these ideas were further developed in mathematical form in his later writings (for example, 1969, 1970; May and MacArthur 1972), papers which in turn have generated a huge theoretical literature (see Abrams 1983). His discussion of the specific differences important in allowing warblers to coexist led to a major research thrust on how species partition resources (reviewed in Schoener 1974, 1986).

A final example further illustrates how MacArthur's paper may have helped shape the future development of ecology. MacArthur wished to demonstrate the role of density-dependent events in regulating warbler populations. He observed that if a population tended to return to an equilibrium value, there would be fewer sequential increases or decreases in it than one would expect if increases and decreases were distributed randomly. While the meaning of this test is debatable, the combination of theory and hypothesis testing is typical of MacArthur's methodology.

It is striking that in the decade following MacArthur's premature death, two camps developed in animal community ecology. One camp advocated his approach of developing analytical caricatures of nature (exemplified by some of the papers in Cody and Diamond 1975), while a second group called for a more rigorous formulation of null hypotheses and statistical tests (exemplified by some of the papers in Strong et al. 1984). The important components of both approaches are clearly presented in MacArthur's classic study of coexistence in warblers.

The final paper reprinted in this section, written by J. L. Brooks and S. I. Dodson, was originally published in *Science* in 1965. Like most papers published in *Science*, it was short and reported only one primary observation; yet, like most of the papers discussed in this part, it had a major impact on the future development of a large branch of ecology, in this case, limnology.

Brooks and Dodson observed that among lakes in eastern Connecticut, those containing the introduced marine fish *Alosa pseudoharengus*, or alewife, lacked the larger species of zooplankton characteristic of the region, whereas these larger zooplankters were present in lakes lacking *Alosa* and were known to have previously been present in lakes now having *Alosa*. They speculated that the larger zooplankters were more visible to the planktivorous *Alosa* and were thus selectively removed from the lakes, a pattern now widely accepted as true for many lakes. They further hypothesized that in the absence of fish, the larger zooplankters have a competitive advantage over the smaller ones because they can consume a larger range of food sizes, and because they have a lower metabolic demand per unit mass, a proposition that remains of considerable interest but which has not been universally accepted (see Stein et al. 1988).

Studies of trophic interactions in lakes have not been the same since Brooks and Dodson published their paper. Suddenly the importance that one trophic level might have on the composition and structure of another could not be ignored. This led to discussions of what are today called top-down and bottom-up theories of community structure, and ultimately to a developing theory of trophic cascades and complex trophic interactions (summarized in Carpenter 1988). In effect, Brooks and Dodson's paper was the "keystone predator" paper of limnology and has played much the same role that Paine's 1966 experimental study of starfish removal (discussed and reprinted in the next section) played for students of sessile organisms. Both papers demonstrated how a predator could control the structure of the trophic level upon which it feeds. In fact, these two seminal papers were published almost simultaneously and each probably reciprocally reinforced the impact of the other.

Concluding Remarks

Aside from representing case studies of natural systems, the papers discussed in this section are as diverse as the field of ecology itself. Some are remembered for synthesizing diverse results, others for simple but provocative observations or insights, and still others because they have served as models for subsequent studies. Each has served to inspire a considerable body of new research.

That the classic studies we have chosen for this section have little in common is a necessary consequence of the innovative approaches and unanticipated results that made them classics. In some cases the selected studies were conducted because the author expected the results to bear on conceptual or theoretical issues of great interest. In others, the author had simply observed an area about which little was known but which had the potential to yield new insights. Certainly all of the authors have demonstrated their intuitive ability to identify critical processes or phenomena for study, as well as ideal natural systems for their observations. These abilities derive in no small part from extensive knowledge of natural systems and the natural history of the component organisms, linked with a desire to occasionally step back from the details in an effort to see larger-scale patterns and mechanisms.

Theory, methods, and experiments are all critical for a healthy science. However, ecologists in particular, being students of a science of diversity requiring contingent generalizations, must exercise great care not to get so caught up in the glamour of abstract theory or experimental manipulation as to lose sight of the roots of their science in the observation of natural systems and processes. A deep understanding of many natural systems and the processes occurring in them, which can only be obtained through an interplay of observation and experimentation, will be necessary for ecology to acquire the foundation it needs to become a mature, predictive science.

Literature Cited

Abrams, P. 1983. The theory of limiting similarity. *Ann. Rev. Ecol. Syst.* 14:359–76.

Andrewartha, H. G., and L. C. Birch. 1954. *The distribution and abundance of animals.* Chicago: University of Chicago Press.

Aubréville, A. M. 1938. La forêt coloniale: les forêts de l'Afrique occidentale française. *Ann. Acad. Sci. Colon.*, Paris 9:1–245.

Birch, L. C., and H. G. Andrewartha. 1941. The influence of weather on grasshopper plagues in South Australia. *J. Dep. Agr. S. Aust.* 45:95–100.

Bormann, F. H., and G. E. Likens. 1979. *Pattern and process in a forested ecosystem.* New York: Springer-Verlag.

Botkin, D. B., J. F. Janak, and J. R. Wallis. 1972. Some ecological consequences of a computer model of forest growth. *J. Ecol.* 60:849–72.

Bray, J. R. 1956. Gap phase replacement in a maple-basswood forest. *Ecology* 37:598–600.

Brooks. J. L., and S. I. Dodson. 1965. Predation, body size, and composition of plankton. *Science* 150:28–35.

Carpenter, S. R., ed. 1988. *Complex interactions in lake communities.* New York: Springer-Verlag.

Cody, M. L., and J. M. Diamond, eds. 1975. *Ecology and the evolution of communities.* Cambridge, Mass.: Belknap Press.

Cook, R. E. 1977. Raymond Lindeman and the trophic-dynamic concept in ecology. *Science* 198: 22–26.

Curtis, J. T., and R. P. McIntosh. 1950. The interrelations of certain analytic and synthetic phytosociological characters. *Ecology* 31:434–55.

——— 1951. An upland forest continuum in the prairie-forest border region of Wisconsin. *Ecology* 32:476–96.

Davidson, J., and H. G. Andrewartha. 1948*a*. Annual trends in a natural population of *Thrips imaginis* (Thysanoptera). *J. Anim. Ecol.* 17:193–99.

——— 1948*b*. The influence of rainfall, evaporation and atmospheric temperature on fluctuations in the size of a natural population of *Thrips imaginis* (Thysanoptera). *J. Anim. Ecol.* 17:200–22.

Davis, M. B. 1969. Climatic changes in southern Connecticut recorded by pollen deposition at Rogers Lake. *Ecology* 50:409–22.

——— 1976. Pleistocene biogeography of temperate deciduous forests. *Geoscience and Man* 13:13–26.

——— 1981. Quaternary history and the stability of forest communities. Pp. 132–53 in D. C. West, H. H. Shugart, and D. B. Botkin, eds., *Forest succession: Concepts and application.* New York: Springer-Verlag.

Davis, M. B., L. B. Brubaker, and T. Webb III. 1973. Calibration of absolute pollen influx. Pp. 9–25 in H. J. B. Birks and R. G. West, eds., *Quaternary plant ecology.* Oxford: Blackwell.

Fretwell, S. D. 1975. The impact of Robert MacArthur on ecology. *Ann. Rev. Ecol. Syst.* 6:1–14.

Gimmingham, C. H. 1988. A reappraisal of cyclical processes in *Calluna* heath. *Vegetatio* 77:61–64.

Gleason, H. A. 1917. The structure and development of the plant association. *Bull. Torrey Bot. Club* 44: 463–81.

——— 1926. The individualistic concept of the plant association. *Bull. Torrey Bot. Club* 53:7–26.

Heinselman, M. L. 1981. Fire intensity and frequency as factors in the distribution and structure of northern ecosystems. Pp. 7–57 in H. A. Mooney, T. M. Bonnicksen, N. L. Christensen, J. E. Lotan, and W. A. Reiners, eds., *Fire regimes and ecosystem properties.* U.S.D.A. Forest Service, Gen. Tech. Rep. WO-26.

Horn, H. S., and R. H. MacArthur. 1972. Competition among fugitive species in a harlequin environment. *Ecology* 53:749–52.

Hutchinson, G. E. 1951. Copepodology for the ornithologist. *Ecology* 32:571–77.

——— 1957. Concluding remarks. *Cold Spring Harbor Symp. Quant. Biol.* 22:415–27.

——— 1959. Homage to Santa Rosalia *or* Why are there so many kinds of animals? *Amer. Nat.* 93: 145–59.

Hutchinson, G. E., and V. T. Bowen. 1950. Limnological studies in Connecticut. IX. A quantitative radiochemical study of the phosphorus cycle in Linsley Pond. *Ecology* 31:194–203.

Jones, E. W. 1945. The structure and reproduction of the virgin forest of the North Temperate zone. *New Phytologist* 44:130–48.

Lack, D. 1954. *The natural regulation of animal numbers.* Oxford: Oxford University Press.

Levin, S. A., and R. T. Paine. 1974. Disturbance, patch formation, and community structure. *Proc. Nat. Acad. Sci. U.S.A.* 71:2744–2747.

Lindeman, R. L. 1942. The trophic-dynamic aspect of ecology. *Ecology* 23:399–418.

MacArthur, R. H. 1958. Population ecology of some warblers of northeastern coniferous forests. *Ecology* 39:599–619.

——— 1969. Species packing and what competition minimizes. *Proc. Nat. Acad. Sci. U.S.A.* 64: 1369–1371.

——— 1970. Species packing and competitive equilibrium for many species. *Theor. Pop. Biol.* 1:1–11.

——— 1972. *Geographical ecology: Patterns in the distribution of species.* New York: Harper and Row.

Marrs, R. H. 1986. The role of catastrophic death of *Calluna* in heathland dynamics. *Vegetatio* 66: 109–15.

May, R. M. 1986. The search for patterns in the balance of nature: Advances and retreats. *Ecology* 67:1115–1126.

May, R. M., and R. H. MacArthur. 1972. Niche overlap as a function of environmental variability. *Proc. Nat. Acad. Sci. U.S.A.* 69:1109–1113.

Miles, J. 1981. Problems in heathland and grassland dynamics. *Vegetatio* 46:61–74.

Odum, H. T. 1957. Trophic structure and productivity of Silver Springs, Florida. *Ecol. Monogr.* 27:55–112.

Odum, H. T., and E. P. Odum. 1955. Trophic structure and productivity of a windward coral reef community on Eniwetok Atoll. *Ecol. Monogr.* 25: 291–320.

Paine, R. T. 1966. Food web complexity and species diversity. *Amer. Nat.* 100:65–75.

Peet, R. K. 1991. Community structure and ecosystem function. In D. C. Glenn-Lewin, R. K. Peet and T. T. Veblen, eds., *Plant Succession: Theory and prediction.* London: Chapman and Hall (in press).

Pickett, S. T. A., and P. S. White, eds. 1985. The ecology of natural disturbance and patch dynamics. New York: Academic Press.

Pomeroy, L. R., and R. G. Wiegert, eds. 1981. The ecology of a salt marsh. Ecological Studies 38. New York: Springer-Verlag.

Prentice, I. C. 1988. Record of vegetation in time and space: The principles of pollen analysis. Pp. 17–42 in B. Huntley and R. Webb III, eds., *Vegetation history.* The Hage: Kluwer.

Prentice, I. C., O. van Tongeren, and J. T. de Smidt. 1987. Simulation of heathland vegetation dynamics. *J. Ecol.* 75:203–19.

Roughgarden, J. 1989. The structure and assembly of communities. Pp. 203–26 in J. Roughgarden, R. M. May, and S. A. Levin, eds., *Perspectives in ecological theory.* Princeton: Princeton University Press.

Runkle, J. R. 1985. Disturbance regimes in temperate forests. Pp. 17–33 in S. T. A. Pickett and P. S. White, eds., *The ecology of natural disturbance and patch dynamics.* New York: Academic Press.

Schoener, T. W. 1974. Resource partitioning in ecological communities. *Science* 185:27–39.

——— 1986. Overview: Kinds of ecological communities—ecology becomes pluralistic. Pp. 467–79 in J. Diamond and T. J. Case, eds., *Community ecology.* New York: Harper and Row.

Shugart, H. H. 1984. *A theory of forest dynamics: The ecological implications of forest succession models.* New York: Springer-Verlag.

Skellam, J. G. 1951. Random dispersal in theoretical populations. Biometrika 38:196–218.

Stein, R. A., T. Threlkeld, C. D. Sandgren, W. G. Sprules, L. Persson, E. E. Werner, W. E. Neill, and S. I. Dodson. 1988. Size structured interactions in lake communities. Pp. 161–79 in S. R. Carpenter, ed., *Complex interactions in lake communities.* New York: Springer-Verlag.

Strong, D. R. Jr., D. Simberloff, L. G. Abele, and A. B. Thistle, eds. 1984. *Ecological communities: Conceptual issues and the evidence.* Princeton: Princeton University Press.

Teal. J. M. 1957. Community metabolism in a temperate cold spring. *Ecol. Monogr.* 27:283–302.

——— 1962. Energy flow in the salt marsh ecosystem of Georgia. *Ecology* 43:614–24.

Varley, G. C., G. R. Gradwell, and M. P. Hassell. 1973. *Insect population ecology.* Oxford: Blackwell.

Watt, A. S. 1947. Pattern and process in the plant community. *J. Ecol.* 35:1–22.

Westman, W. E., and R. K. Peet. 1982. Robert H. Whittaker (1920–1980): The man and his work. *Vegetatio* 48:97–122.

Whittaker, R. H. 1952. A study of summer foliage insect communities in the Great Smoky Mountains. *Ecol. Monogr.* 22:1–44.

——— 1956. Vegetation of the Great Smoky Mountains. *Ecol. Monogr.* 26:1–80.

——— 1960. Vegetation of the Siskiyou Mountains, Oregon and California. *Ecol. Monogr.* 30:279–338.

——— 1972. A hypothesis rejected: The natural distribution of vegetation. Pp. 689–91 in W. A. Jensen and F. B. Salisbury, *Botany: An ecological approach.* Belmont, Calif.: Wadsworth.

Whittaker, R. H., and W. A. Niering. 1965. Vegetation of the Santa Catalina Mountains, Arizona: A gradient analysis of the south slope. *Ecology* 46:429–52.

Wiens, J. A. 1977. On competition and variable environments. *Amer. Sci.* 65:590–97.

THE INFLUENCE OF RAINFALL, EVAPORATION AND ATMOSPHERIC TEMPERATURE ON FLUCTUATIONS IN THE SIZE OF A NATURAL POPULATION OF *THRIPS IMAGINIS* (THYSANOPTERA)

By the late J. DAVIDSON and H. G. ANDREWARTHA*

From the Waite Agricultural Research Institute, University of Adelaide

(With 5 Figures in the Text)

1. INTRODUCTION

The apple-blossom thrips (*Thrips imaginis* Bagnall) is indigenous to southern Australia. Several times during the past 40 years mass outbreaks of the species have caused widespread damage to apple blossom in West Australia, South Australia, Victoria and New South Wales. During an outbreak large numbers of the insects leave the flowers of weeds and other plants where they have been breeding and enter the flowers and flower-buds of the apple; the movement may be more marked during warm dry days. Mass outbreaks usually occur simultaneously over a remarkably wide area. For example, in 1926 and again in 1931 the plague was widespread in all apple-growing areas of South Australia, Victoria, and also parts of New South Wales (Evans, 1932).

These areas are separated by almost a thousand miles, and it is difficult to imagine any set of causes other than long-term fluctuations in the weather which might operate simultaneously over such a wide area. The weather may be associated with the growth and decline of the thrips population directly by the influence of temperature and soil moisture on the rate of development and the survival rate of the insects, and indirectly by the influence of the same factors on the development of the plants which are hosts for *T. imaginis*.

There is a regular annual cycle in the growth and decline of the population, but the annual variation, both in the maximum size attained by the population and the date when the maximum is reached is considerable. For example, the average for November 1934 was 36 compared with 575 for November 1938 (Davidson & Andrewartha, 1948). Since it was considered that fluctuations in the physical environment of the insects may be important in causing fluctuations in the thrips population the present study was directed towards measuring the degree of association between the numbers of thrips present in the roses during spring and early summer and those components of the weather which were considered to be causally related to the size of the population. The method of partial regression has been used; in choosing the independent variates we have been guided by a knowledge of the biology of *T. imaginis* and of the climate of the Adelaide area.

The biology of *T. imaginis* was studied by a team of workers under the direction of the late Prof. J. Davidson and the results of their work were summarized by him (Davidson, 1936). The climate of the area has been extensively studied by a school of workers at the Waite Institute, and summaries of their published work have appeared from time to time in the Report of the Waite Institute.

2. MATERIAL

For 14 years during the period 1932–46 daily counts have been made of the numbers of adult *T. imaginis* in roses in the garden at the Waite Institute; during 1944 records were not kept owing to pressure of other work. During the 7 years 1932–8 a sample of 20 roses was taken daily throughout the year; from 1939 onwards a sample of 10 roses was taken daily during the spring and early summer (September–December); this is the season when the population attains its maximum size each year. Usually samples were not taken on Sundays nor on holidays; the number of sampling days for each of the months from September to December for the 14 years are shown in Table 1: altogether 1252 daily records were taken. The method of sampling was described and discussed elsewhere (Davidson & Andrewartha, 1948). The roses used for sampling were attractive to the adult thrips but were not favourable for the development of the immature stages. They served as a 'trap' which gave a satisfactory indication of the density of the population in the area.

* The late Prof. J. Davidson was the leader of a team of workers who spent several years studying the biology of *T. imaginis*. The present series of population counts was begun by J. W. Evans in April 1932: after his departure from South Australia they were continued by G. H. Andrewartha. Prof. Davidson was working on the data with the object of measuring the association between the density of the thrips population and the weather, but unfortunately the work was still unfinished when he died in August 1945. The work was taken over and brought forward to its present condition by H. G. Andrewartha.

The rainfall, evaporation and atmospheric temperature were measured at the meteorological station situated at the Waite Institute, a few chains away from the garden where the roses were growing. The records were made on standard instruments described in the Australian Meteorological Observers' Handbook.

Table 1. *The number of sampling days each month for 14 years*

Month	1932	1933	1934	1935	1936	1937	1938	1939	1940	1941	1942	1943	1945	1946	Total
Sept.	21	19	24	25	22	20	22	25	17	25	18	15	19	15	287
Oct.	25	25	24	26	24	26	26	24	27	25	27	25	26	26	356
Nov.	26	26	25	27	22	26	25	26	24	24	24	26	23	24	348
Dec.	21	20	18	26	14	21	20	13	13	24	17	21	15	18	261
Total	93	90	91	104	82	93	93	88	81	98	86	87	83	83	1252

3. METHOD

The total variability of the 1252 daily records may conveniently be ascribed to four components. Each year the numbers in the roses start from a low level at the end of the winter; they increase usually rapidly and continuously until they reach a maximum about the end of November and then decline again (Davidson & Andrewartha, 1948). The growth of the population with time within each year may be regarded as one component of the total variability of the daily counts. The numbers of thrips present at the end of the winter, the rate at which the numbers increase and the size of the maximum attained, may all vary from year to year; thus the second component of total variability may be specified as variation in the growth of the population from one year to another. The number of thrips recorded in the roses may be taken as one criterion of the numbers present in the garden: but this estimate of the population will be more variable than the true numbers of the thrips in the garden to the extent that the mean number of thrips per rose in the daily sample is not a constant proportion of the total number of thrips in the garden. A number of causes may operate to vary this proportion: for example, the thrips may be more active in seeking out the roses during a warm dry day than during a cold wet one. It is convenient to specify the third component of variability as the variation due to the variable 'activity' of the insects in seeking out the roses. Finally there is the residual variability which may be considered as being random with respect to the three components which have been selected for special study.

The method consisted essentially of studying the second and third components one at a time, after having first eliminated as much as was practicable of the variance associated with the other components. Before describing the method more particularly, it is appropriate to consider (without at this stage giving detailed consideration to the nature of the 'cause') the sort of relationship which may exist between the '*cause*' (various elements in the weather) and the '*effect*' (fluctuations in the numbers of thrips).

Experience indicates that the thrips numbers are not likely to change by a constant quantity for each increment in the causal agent but rather are they likely to change by a constant proportion. This may be expressed in symbols: the expected relationship would not be $Y=kx$, but would be of the form $Y=ka^x$; or, when several causes may be operating together, $Y=k\times a_1^{x_1}\times a_2^{x_2}\times a_3^{x_3}\ldots$. This expression assumes the form of the partial regression equation when logarithms are written in place of the original quantities:

$$\log Y=\log k+x_1 \log a_1+x_2 \log a_2 \ldots$$

In this expression $\log k$ is the logarithm of the geometric mean of y; and $\log a_1$, $\log a_2$ are the partial regression coefficients of y on x_1 and x_2 when the logarithm of y is used in the place of the original value of y.

Accordingly, all records of the daily counts of thrips per rose were converted to logarithms. It is fortunate that in this instance the logarithms of the thrips counts were distributed very nearly normally, whereas the distribution of the original counts was strongly skew (see Fig. 1). Thus the transformation to logarithms, which was made in the first instance for the practical purpose of satisfying the expectation that independent and dependent variates would be related in a particular way, also transforms the data to a form that is appropriate for statistical treatment.

The method of partial regression as described by Fisher (1941, p. 149) is admirably suited for the treatment of data of this sort. The use of the 'c' multipliers makes it practicable to include a number of independent variates and with little labour to discard successively the less important of them. Furthermore, very little extra labour is involved in using successively a number of different estimates of the dependent variate with the same series of independent variates. In this way the mutual relationships of the data may be explored with precision and with remarkably little extra labour.

There are two qualities of the partial regression coefficient which are worthy of emphasis in relation to the present study. The partial regression coefficient expresses the degree of association between its independent variate and the dependent variate when all other terms *included in the regression* are held at their means. For example, when the equation includes two terms which are not only closely associated in the same direction, with each other, but also with the dependent variate, the regression measures the degree of association which each has with the dependent variate independent of the influence of the other; but neither may have an appreciable effect independent of the influence of the other; and the total effect measured in the regression equation may not be large. The correct procedure is to discard one of them and leave the other to measure the influence of both; for the regression coefficient always includes in itself not

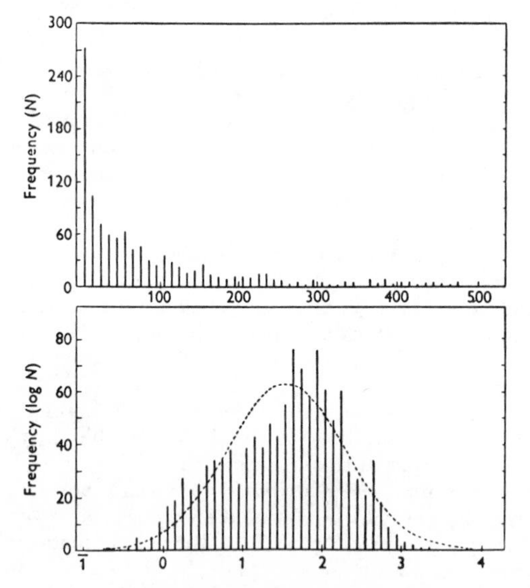

Fig. 1. The distribution of the daily counts of thrips per rose (above); and the distribution of the logarithms of the daily counts (below). The broken line represents the normal curve having the same mean and standard deviation as the observed data.

only a measure of the degree of association of the dependent variate with the particular independent variate but also of all terms *not included in the regression* which are associated with the independent variate.

The regression equation measures only association. Causality requires to be inferred from independent sources. In the present instance an adequate knowledge of the biology of *T. imaginis* has led to the selection of independent variates which have accounted for a substantial part of the total variance. In his pioneer work in this field Williams was dealing with groups of species, and the same degree of detail in postulating causal relationships was not always practicable with his material (Williams, 1940). Also with Williams's data the greater part of the total variance was due to variable activity of the insects in seeking out the trap: and relatively little of the total variance could be attributed to fluctuations in the abundance of the insects in the area. With the present paper the reverse is the case.

4. DAILY FLUCTUATIONS IN THE NUMBER OF THRIPS IN THE ROSES

In any one year during the period when the population is increasing rapidly in size the variability in the daily counts may be ascribed to: (*a*) natural increase with time in the number of thrips in the garden; (*b*) variable 'activity' of the thrips in seeking out the flowers; (*c*) residual or random variation.

(*a*) *The elimination of variability due to trend with time*

There are theoretical grounds for expecting the trend of the population with time to approximate to a logistic curve (Davidson, 1944; Davidson & Andrewartha, 1948). But the logistic formula is not easy to handle with all types of data, and has certain other disadvantages; so, for the present purpose, polynomial curves were used to eliminate trend with time. A curve of predetermined degree was calculated for the 60 days which preceded the day when the population attained its maximum. The period 60 days was chosen because this gave a satisfactorily long run of observations and for most years the population increased continuously and rapidly for a period of at least 60 days before it attained its maximum. The date of the 'peak' of the population was determined by the method described elsewhere (Davidson & Andrewartha, 1948): 15-day progressive means were calculated from the logarithms of the daily counts of thrips per rose and the maximum of the curve was considered to occur on the central day of the 15-day period having the highest mean value.

In order to avoid the labour of calculating polynomial coefficients from the full 60 daily records the data were grouped: 3-day intervals were chosen giving 20 groups. A curve of the third degree in x (days) was fitted to the means of the 20 groups, using Fisher's method of orthogonal polynomials. The trend of the data with time (i.e. the regression of log thrips on days) varied considerably from year to year. The figures in Table 2 indicate the extent and direction of this variation: the table sets out the regression coefficients for each of the 14 years. The table also shows that the residual variance was relatively high for some years. For example, a fourth term was fitted to the regression for 1932; E' was found to equal $0{\cdot}1257 \times 10^{-3}$, and was significant at the 1 % level of probability ($t = 4{\cdot}056$). Nevertheless, it was considered that the regressions for most years

Table 2. *Showing the regression coefficients for log thrips on time for the 60 days preceding the peak for each of 14 years. The residual variance (s^2) is also shown*

The data for $n=60$ were derived from the values for $n=20$ (see below)

	$n=20$					$n=60$			
Year	A'	$B'\times 10^3$	$C'\times 10^3$	$D'\times 10^3$	$s^2\times 10^3$	$B'\times 10^3$	$C'\times 10^3$	$D'\times 10^3$	No. of observations during 60-day period
1932	1·6010	47·6165	2·0119	−0·1810	43·595	15·8545	0·0741	−0·0066	50
1933	1·0605	49·1692	0·5611	−0·1838	62·341	16·3715	0·0207	−0·0067	51
1934	1·0095	57·7030	−3·5048	−0·2910	59·745	19·2130	−0·1291	−0·0106	47
1935	1·6195	42·8835	−5·2352	0·0639	25·410	14·2786	−0·1928	0·0023	51
1936	1·2630	35·9925	−1·3545	0·2039	52·779	11·9842	−0·0499	0·0074	46
1937	1·4620	50·6541	2·9358	−0·3345	38·297	16·8659	0·1081	−0·0122	51
1938	2·2750	26·0827	1·9640	−0·0605	39·052	8·6846	0·0723	−0·0022	50
1939	2·1495	53·0188	−3·2337	−0·0463	34·386	17·6533	−0·1191	−0·0017	50
1940	2·0250	8·0977	−0·9854	0·0419	27·826	2·6962	−0·0363	0·0015	49
1941	2·1370	28·1053	−2·4402	0·1045	51·115	9·3580	−0·0899	0·0038	49
1942	2·1645	38·0940	−2·2334	−0·2043	49·669	12·6839	−0·0823	−0·0074	49
1943	1·1695	56·9586	6·5943	−0·2475	28·152	18·9651	0·2429	0·0090	50
1945	1·6745	26·2444	−1·3426	0·0013	40·420	8·7384	−0·0494	0·0000	48
1946	2·1160	12·7970	0·4520	−0·1576	29·904	4·2609	0·0166	−0·0057	49
Mean	1·6948	38·1012	−0·4151	−0·0922	—	12·6863	−0·0153	−0·0034	—

would not be likely to be improved appreciably by the addition of further terms. Moreover, there are theoretical grounds for not carrying the regression too far. The trend of the population with time may be expected to approximate to a portion of a logistic curve: when the logarithms of the calculated values for the logistic curve, fitted to the present data and published elsewhere (Davidson & Andrewartha, 1948, Fig. 4A), were fitted to a polynomial curve of the third degree the regression accounted for all but a very small proportion of the variance (see Table 3). Accordingly, the regressions for eliminating the trend of the data with time were not carried beyond the cubic term.

Table 3. *Analysis of variance of regression of log of the calculated value for logistic curve on time*

Source of variance	Degrees of freedom	Sums of squares	Mean square
Linear term	1	2·9598	2·9598
Quadratic term	1	0·0005	0·0005
Cubic term	1	0·0029	0·0029
Total regression	3	2·9632	0·9877
Residual	16	0·0002	0·00001
Total	19	2·9634	—

At first it was intended to use the data from all the 14 years, but as it was proposed to calculate a 6-term partial regression the labour would have been heavy, and it was decided to use instead the first 8 years viz. 1932–9. The residual deviations from the trend lines provided the individual values for the dependent variate of the partial regression equation. Since the data were all expressed as logarithms, relative differences in the general level of thrips population from year to year had no influence on the results and it was appropriate to pool the data for the 8 years.

The calculations of the sums of squares and products were made by the short method using identities of the form

$$S(y-Y)^2 = Sy^2 - [nA^2 + B^2\, S\xi_1^2 + C^2 S\xi_2^2 + D^2 S\xi_3^2].$$

But there were 60 intervals along the x axis for the data for the partial regression whereas the polynomials had been calculated using 20 intervals for x. The polynomial coefficients appropriate to the value $n=60$ were estimated from those calculated for $n=20$ using identities of the form

$$3\, B_{20}^2 . S_{20}\xi_1^2 = B_{60}^2 . S_{60}\xi_1^2.$$

The conversion factors which were calculated in this way were:

$$A_{60} = A_{20}, \qquad C_{60} = 0{\cdot}036831\, C_{20},$$
$$B_{60} = 0{\cdot}332963\, B_{20}, \qquad D_{60} = 0{\cdot}036460\, D_{20}.$$

In order to have an indication of the accuracy of these estimates the 60 polynomial values for one variate (x_1 for 1936) were calculated out. The sum of the 60 estimated values agreed with the sum of the observed values to 6 significant figures, and it was considered likely that little accuracy had been lost by the labour-saving device of grouping the data before calculating the polynomial. The estimated values for the polynomial coefficients for $n=60$ are given in Table 2. The number of days on which records were taken during each 60-day period are also given in the last column of the table.

The data were non-orthogonal. Since the samples of roses were not taken on Sundays, and certain other days were also missed, there were always less than 60 observations in each period of 60 days. The missing records were allowed for by adding to the original estimates of sums of squares and products a quantity calculated from expressions of the form $S(2yY-Y^2)$ for the sums of squares, and $S(Yx+yX-XY)$ for the sums of products. These expressions were summed over all the missing records.

A similar procedure was followed with the independent variates. The 60 observations for each independent variate were divided into 20 groups and regression equations were obtained by calculating orthogonal polynomials to the third degree in x (days). Residual deviations from the trend lines provided the individual values of the independent variates of the partial regression equations.

temperature and rainfall. A preliminary 3-term partial regression was calculated using data from 4 years' records and only for the day immediately before the sampling day: daily maximum air temperature, daily total rainfall and changes in the barometric pressure between 9 a.m. and 3 p.m. were chosen as the three independent variates. The partial regression of thrips on changes in barometric pressure was small and somewhat less than half its standard error. It was therefore concluded that changes in barometric pressure were not exerting any appreciable influence on the numbers of thrips in the roses over and above that which could be related to temperature and rainfall. Changes in barometric pressure were excluded from further consideration.

Table 4. *Individual sums of squares and products contributed by each year 1932–9*

Variate	1932	1933	1934	1935	1936	1937	1938	1939	Total: 4 years	Total: 8 years
x_1	3547·23	4438·36	3325·82	3638·66	4244·16	3484·85	2814·25	3787·28	15109·10	29280·61
x_2	1581·08	1148·65	17771·50	13339·54	7073·56	3155·82	748·54	18339·02	19225·09	63157·71
x_3	3343·57	4407·76	3033·34	2766·13	3986·55	3201·53	2616·41	3966·77	13718·99	27322·06
x_5	3117·18	3945·19	2224·22	3552·16	2797·69	3346·98	2973·52	3905·24	13961·51	25862·18
x_1x_2	−302·32	−240·66	−479·57	−45·32	−1310·45	−140·99	142·72	302·45	−729·29	−2074·14
x_1x_3	1865·48	1656·57	1223·54	1531·95	2229·59	1583·56	603·48	2537·22	6637·56	13231·39
x_1x_5	295·16	−538·78	−280·99	−861·83	360·37	−62·95	−999·45	1323·79	−1168·40	−764·68
x_2x_3	473·10	499·00	1690·43	1659·32	−568·01	272·92	9399·94	2364·71	2904·34	15791·41
x_2x_5	−58·93	294·09	627·83	646·71	352·70	820·59	250·40	1448·54	1702·46	4381·93
x_3x_5	1698·76	1522·29	1299·64	1687·30	2075·90	1639·23	525·07	2670·97	6547·58	13119·16
x_1y	12·39	48·90	−5·32	37·24	23·09	44·25	24·40	51·18	142·78	236·13
x_2y	−35·82	−37·57	−6·41	−25·68	−7·00	−4·94	−3·15	−20·23	−104·01	−140·80
x_3y	11·87	40·31	−6·09	11·88	0·34	17·04	8·17	33·07	81·10	116·59
x_5y	31·80	29·28	−15·09	−12·08	−15·08	−2·99	1·01	18·82	46·01	35·67
y	3·125	2·601	3·186	2·014	2·794	1·635	2·188	1·926	9·375	19·469

(*b*) *Selection of independent variates*

The sample of roses was taken about 9 a.m. each day. It was unlikely that appreciable movement of the thrips had taken place earlier than this: the sample may therefore be regarded as an estimate of the number of thrips in the roses at the close of the previous day. The individual roses may have been attractive to the thrips during several days before the sample was taken; in selecting independent variates for the partial regression equation it is therefore necessary to consider causes which may have been operating several days before the sample was collected.

Experience with *T. imaginis* suggested that atmospheric temperature, rainfall and possibly changes in barometric pressure may influence the activity of the insects in seeking out the flowers. Changes in barometric pressure were known to be closely associated with both atmospheric temperature and rainfall and it was possible that changes in barometric pressure may not have an appreciable influence independent of

The daily maximum atmospheric temperature, recorded 1, 2 and 3 days before the sampling day and the total daily rainfall for the same 3 days were selected as independent variates for a 6-term partial regression equation. For ease of reference these will be referred to by symbols:

- x_1 Daily maximum atmospheric temperature in degrees Fahrenheit recorded one day before the sampling day.
- x_2 Total daily rainfall in units of 0·01 in. recorded 1 day before the sampling day.
- x_3 As for x_1 but recorded 2 days before the sampling day.
- x_4 As for x_2 but recorded 2 days before the sampling day.
- x_5 As for x_1 but recorded 3 days before the sampling day.
- x_6 As for x_2 but recorded 3 days before the sampling day.

It is not practicable to publish all the data from which the regression was calculated. Fig. 2 illustrates some of the data for 1935: the deviations of the observed values from the three trend lines represent respectively the individual values for the dependent variate (y) and the two independent variates (x_1) and (x_2) for 1935. Similarly, the individual values for the other independent variates (x_3, x_4, x_5 and x_6) were found by calculating the appropriate trend lines and taking the deviations of the observed values from them. The same process was repeated for each year under consideration. Table 4 lists the individual sums of squares and products contributed by each year and the totals from which the 'c' multipliers for the partial regression equations were calculated.

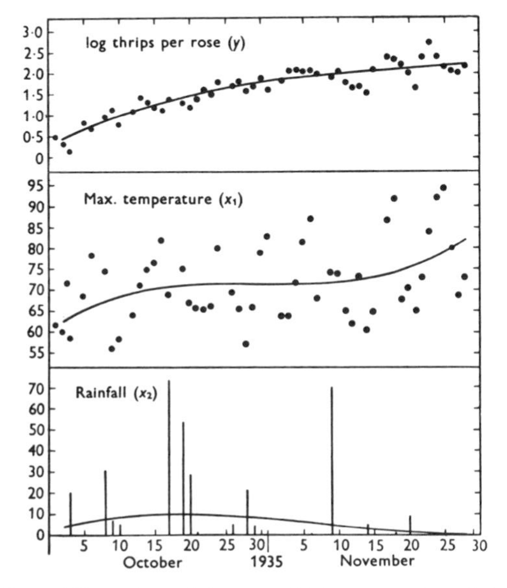

Fig. 2. The polynomial curves which were used to eliminate trend with time for y, x_1 and x_2 for 1935. They were calculated from the formulae:

$X_1 = 71{\cdot}805 + 0{\cdot}3359\ \xi_1 + 0{\cdot}008789\ \xi_2 + 0{\cdot}003627\ \xi_3$,

$X_2 = 5{\cdot}715 - 0{\cdot}2056\ \xi_1 - 0{\cdot}06132\ \xi_2 + 0{\cdot}002255\ \xi_3$,

$Y = 1{\cdot}6195 + 0{\cdot}04288\ \xi_1 - 0{\cdot}005235\ \xi_2 + 0{\cdot}00006393\ \xi_3$,

X_1 in degrees Fahrenheit; X_2 in units of 0·01 in.; Y in log thrip per rose.

Similar regressions were calculated for the other variates and for all the years 1932–9. The deviations from the curves provided the individual values for the variates in the partial regressions, calculated in § 4.

(c) *The partial regression of thrips numbers on maximum temperature and rainfall*

Because of the labour that would have been involved if the larger sample had been used, the 6-term partial regression was first calculated on the data for 4 years. The years 1932, 1933, 1935 and 1937 were chosen because they had the fewest missing records (see Table 2). The results are set out in Table 5. The partial regressions of y on x_4 and x_6 were small and insignificant, indicating that rain falling on the second and third day back from the sampling day was not having an appreciable influence on the numbers of thrips in the roses. These terms were excluded from the regression: the results of the new 4-term regression are set out in Table 6. The partial regressions of y on x_1 and x_2 were highly significant; the partial regression of y on x_3 was small and non-significant despite a relatively high total correlation of y on x_3 (see Table 4); the partial regression of y on x_5 though relatively small is significant at the 5 % level. These

Table 5. *Partial regressions of y on six independent variates. Calculated from data for 4 years*

Variate	Regression coefficients $b \times 10^2$	S.E. $\times 10^2$	t
x_1	0·9146	0·2023	4·520
x_2	−0·5541	0·1479	3·747
x_3	0·0180	0·2420	—
x_4	−0·0817	0·1363	—
x_5	0·4830	0·2092	2·309
x_6	−0·0854	0·1220	—

Analysis of variance

Source of variance	Degrees of freedom	Sums of squares	Mean square
Regression	6	2·1842	0·36403
Residual	181	7·1905	0·03973
Total	187*	9·3747	—

$Z = 1{\cdot}1077$; for $P = 0{\cdot}01$, $Z = 0{\cdot}5152$; for $P = 0{\cdot}05$, $t = 1{\cdot}960$.

* There were 203 sampling days during the 4 years: 4 degrees of freedom were lost with each polynomial calculated making 16 for the 4 years and leaving a total of 187 degrees of freedom for the partial regression.

Table 6. *Partial regressions of y on four independent variates. Calculated for data for 4 years*

Variate	Regression coefficients $b \times 10^2$	S.E. $\times 10^2$	t
x_1	0·9387	0·1992	4·713
x_2	−0·5513	0·1472	3·744
x_3	0·0346	0·2300	—
x_5	0·4591	0·2066	2·222

Analysis of variance

Source of variance	Degrees of freedom	Sums of squares	Mean square
Regression	4	2·1530	0·53824
Residual	183	7·2218	0·03946
Total	187	9·3747	—

$Z = 1{\cdot}3065$; for $P = 0{\cdot}01$, $Z = 0{\cdot}5999$; for $P = 0{\cdot}05$, $t = 1{\cdot}960$.

apparent anomalies may be explained by reference to Table 4. The data indicate a positive correlation of x_3 with y and also with x_1 and x_5, and the last two are also both positively correlated with y. In other words, x_3 accounts for no appreciable part of the variance of y in addition to that which is explicable in terms of x_1 and x_5, and of these two the dependence of x_3 on x_1 is the more important. The total regression of y on x_5 is small but positive, and the partial regression is enhanced by the negative correlation between x_1 and x_5. The figures in Table 4 also indicate that the small positive regression of y on x_5 was due largely to the influence of 2 of the 4 years, namely 1932 and 1933; and the negative correlation of x_1 with x_5 was not consistent for all 4 years.

The physical explanation for these statistical results depends upon the nature of the weather in the Adelaide district. The weather is dominated by a series of cyclonic and anti-cyclonic systems which drift across the continent with considerable irregularity but with an average periodicity of somewhere between 6 and 7 days. Consequently, the temperature of any one day is likely to resemble that of the day before or the day after and to differ from that recorded 3 days before or 3 days after. The difference may still be there though less marked for readings separated by 2 days. For example, this was the case with 1933 and 1935 but not with 1932 (Table 4).

The same 4-term partial regression was recalculated from a larger sample, i.e. using the data for all the 8 years 1932–9. The four independent variates were retained, because it was considered that the more precise estimates obtained from the larger sample might be appreciably different from those calculated from the smaller sample, particularly with respect to the partial regressions of y on x_3 and x_5. The results

Table 7. *Partial regression of y on four independent variates. Calculated from data for 8 years*

Variate	Regression coefficients: $b \times 10^2$	S.E. $\times 10^2$	t
x_1	0·7153	0·1626	4·400
x_2	−0·2498	0·1001	2·496
x_3	0·1693	0·2087	—
x_5	0·1154	0·1698	—

Analysis of variance

Source of variance	Degrees of freedom	Sums of squares	Mean square
Regression	4	2·2792	0·5698
Residual	360	17·1899	0·04775
Total	364*	19·4691	—

$Z = 1{\cdot}2397$; for $P = 0{\cdot}01$, $Z = 0{\cdot}5999$; for $P = 0{\cdot}01$, $t = 2{\cdot}576$.

* There were 396 sampling days altogether: since 8 polynomials were calculated 32 degrees of freedom were used up leaving a total of 364 for the partial regression.

are given in Table 7. With the larger sample the partial regression of y on x_5 has become smaller and non-significant; the partial regression of y on x_3 has become larger but is still non-significant. The data in Table 4 indicate that the change in the relative importance of x_5 may be partly due to a reduction in the size both of the negative correlation between x_1 and x_5 and the positive correlation between x_5 and y. In the light of these results, and of the discussion above, it seems likely that the results set out in Table 6 may have been unduly influenced by the inclusion of one or more 'unusual' years in the small sample.

Table 8. *Partial regression of y on three independent variates x_1, x_2 and x_3. Calculated from data for 8 years*

Variate	Regression coefficients: $b \times 10^2$	S.E. $\times 10^2$	t
x_1	0·6732	0·1502	4·481
x_2	−0·2642	0·0978	2·703
x_3	0·2534	0·1680	1·509

Analysis of variance

Source of variance	Degrees of freedom	Sums of squares	Mean square
Regression	3	2·2571	0·75236
Residual	361	17·2119	0·04768
Total	364	19·4690	—

$Z = 1{\cdot}3794$; for $P = 0{\cdot}01$, $Z = 0{\cdot}6651$; for $P = 0{\cdot}05$, $t = 1{\cdot}960$.

Table 9. *Partial regression of y on three independent variates x_1, x_2 and x_5. Calculated from data for 8 years*

Variate	Regression coefficients: $b \times 10^2$	S.E. $\times 10^2$	t
x_1	0·7967	0·1278	5·552
x_2	−0·2105	0·0875	2·405
x_5	0·1971	0·1367	1·443

Analysis of variance

Source of variance	Degrees of freedom	Sums of squares	Mean square
Regression	3	2·2478	0·7493
Residual	361	17·2212	0·04770
Total	364	19·4690	—

$Z = 1{\cdot}3770$; for $P = 0{\cdot}01$, $Z = 0{\cdot}6651$; for $P = 0{\cdot}05$, $t = 1{\cdot}960$.

Since the partial regressions of y on x_3 and x_5 account for no appreciable amount of the variance of y in addition to that which may be related to x_1 and x_2, it is best to eliminate x_3 and x_5 from the regression. But having regard to (*a*) the relatively high positive total correlation of x_3 with y, and (*b*) the significant value for the partial regression of y on x_5 given by the earlier calculation with the smaller sample, it is instructive to eliminate x_3 and x_5 separately and to consider the two 3-term regressions which are thereby obtained. The results are set out in Tables 8 and 9. The same conclusion is indicated by the two 3-term regressions as was reached from consideration of the 4-term regression, namely, that neither x_3 nor x_5 was influential independently of x_1 and x_2.

Nevertheless, it seems likely that (due to the irregular movement of the cyclonic and anti-cyclonic systems which dominate the weather in the Adelaide district) occasionally a period may occur during which the maximum temperature registered 3 days before the sampling day may influence the number of thrips in the sample. But with the present data the best expression of the variability of the number of thrips in roses in terms of maximum temperature and rainfall is given by a 2-term partial regression in which the independent variates are daily maximum atmospheric temperature and total daily rainfall occurring on the day before the sample was taken. The figures given in Table 10 indicate (reverting now from logarithms back to the original scale) that on the average the daily number of thrips per rose increased by about 25 % for each increase of 10° F. in the daily maximum atmospheric temperature, and decreased about 66 % for each increase of 1 in. in the total daily rainfall. But daily deviations of 10 or 20° F. from the trend line were quite usual, whereas daily deviations of 1 in. or even 0·5 in. were relatively uncommon; and the β coefficients given in column 5 of Table 10 indicate that on the average changes in maximum temperature were about three times as important as changes in rainfall in influencing the numbers of thrips in the roses.

Table 10. *Partial regression of log thrips per rose on maximum temperature and rainfall. Calculated on data for 8 years*

Variate	Regression coefficients			
	$b \times 10^2$	S.E. $\times 10^2$	t	β*
x_1	0·7925	0·1280	6·192	0·307
x_2	−0·1969	0·0871	2·260	−0·102

Change per 1° F. = 2·5 %.
Change per 0·01 in. rain = 0·66 %.

Analysis of variance

Source of variance	Degrees of freedom	Sum of squares	Mean square	Variance (%)
Regression	2	2·1486	1·0743	10·54
Residual	362	17·3204	0·0478	—
Total	364	19·4690	—	—

$Z = 1{\cdot}5557$; for $P = 0{\cdot}01$, $Z = 0{\cdot}7636$; for $P = 0{\cdot}05$, $t = 1{\cdot}960$.

* β is used in the sense defined by Snedecor (1937); i.e.

$$\beta = b \times \frac{\sqrt{[S(x-\bar{x})^2]}}{\sqrt{[S(y-\bar{y})^2]}}.$$

Table 10 shows that the partial regression accounted for about 10 % of the variance. This was a smaller proportion than was expected. It seems likely that the residual variance was not composed entirely of errors of random sampling but also included certain other components. For example, incomplete elimination of trend with time may result in a larger residual variance: gregariousness on the part of the insects, and fluctuations in the number of roses on the hedge are other possible sources of 'error' which may have been represented in the residual variance (Davidson & Andrewartha, 1948). But probably more important than all these must be counted the absence of precise information on the behaviour of *T. imaginis*. In the absence of precise experimental data on the behaviour of the insect there is no guide to suitable quantities which might be tried as additional independent variates. For the present the 2-term regression given in Table 10 must be taken as the best expression that can be got to relate the activity of *T. imaginis* to the daily weather.

Although the partial regressions were relatively small they were significant, and it is instructive to demonstrate the influence of temperature and rainfall on the activity of the thrips since this knowledge may lead to a better understanding of the behaviour of the thrips in seeking out apple blossom during an outbreak.

The remainder of this paper will be devoted to a discussion of the annual fluctuations in the numbers of thrips i.e. the second component of the variance discussed in § 3. The daily counts of thrips per rose provide an estimate of the number of thrips in the garden. This estimate includes an 'error' due to the variable activity of the insects in seeking out the flowers. For certain purposes it may be desirable to consider the data after this 'error' has been eliminated. The partial regression set out in Table 10 provides the means for allowing for part of this 'error'. This matter will be discussed in § 6; § 5 is concerned with a discussion of the second component of the variance discussed in § 3 without regard to the 'error' introduced by the variable activity of the insects in seeking out the flowers.

5. ANNUAL FLUCTUATIONS IN THE NUMBER OF THRIPS IN THE ROSES

The number of thrips in the roses usually reached a maximum about the end of November, but the date varied from 15 November in 1939 to 13 December in 1945. The size of maximum varied from 82 in 1934 to 791 in 1939: these figures refer to the means of the 15-day period during which the thrips were most numerous. There was also appreciable variation in the numbers present at the end of the winter and in the rate at which the population increased. It is not practicable to publish all the records in detail; Fig. 3 shows the mean curve for 14 years, in addition to the curves for 1934 and 1938 which may be taken as representing the two extremes. The curves were fitted to the logarithms of the daily counts for the 60 days preceding the 'peak' day. For 1934 the peak day was 5 December, for 1938 20 November and for the mean, 1 December. A further indication of the nature and extent of the variability of the data is given by the polynomial coefficients set out in Table 2. The mean

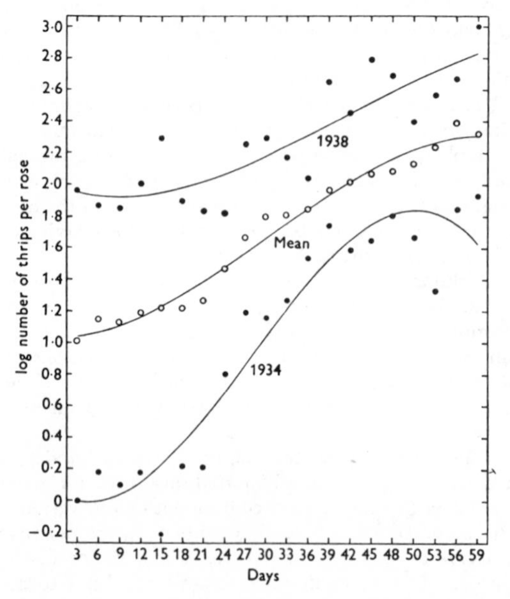

Fig. 3. The trend with time of the thrips populations for 1934, 1938, and the mean for 14 years. The observed values are the mean logarithms of the daily counts in groups of 3 days. The coefficients for the curves are given in Table 2. Thus the equation for the mean curve is

$$y = 1{\cdot}695 + 0{\cdot}0381\,\xi_1 - 0{\cdot}0042\,\xi_2 - 0{\cdot}0001\,\xi_3.$$

varied from 1·0095 to 2·2750; the linear terms were all positive but varied considerably in size; the quadratic and cubic terms were mostly negative but not consistently so: and they also varied considerably in size.

It is the purpose of this section to measure the association between the number of thrips present during the spring and early summer and certain components of the weather which had occurred during preceding months. The method of partial regression is appropriate. First it was necessary to proceed to the selection of suitable variates: as before we were guided by a knowledge of the biology of *T. imaginis* and of the weather of the Adelaide district.

(a) Selection of alternative dependent variates

It was pointed out at the beginning of § 3 that the variability in the daily records of thrips per rose may conveniently be ascribed to four components. The partial regressions which may be used to account for the variability due to annual fluctuations in the size of the thrips population will be the more satisfactory (i.e. they will account for a larger proportion of the variance) when the variance due to the other two components has been eliminated. Nevertheless, it may be necessary, for practical reasons, to ignore these components in certain instances. For example, we may require to predict the size of the thrips population at some specified time. For this purpose it was not practicable to include in the regression terms which could not be determined more than a few days in advance of the period to which the prediction applies. Hence the appropriate regressions were first calculated without 'correcting' the observed values for 'activity'.

Five quantities were selected as alternative dependent variates:

- y_1 The logarithm of the geometric mean of the daily counts of thrips per rose during the fifteen days about the peak day.*
- y_2 The logarithm of the geometric mean of the daily counts of thrips per rose for the 30 days preceding the peak.
- y_3 As for y_2, but using the data for the 60 days preceding the peak.
- y_4 The logarithm of the geometric mean of the daily counts for October.
- y_5 As for y_4 but using the data for November.

These quantities provided five alternative criteria for the number of thrips present during spring and early summer. The first three were designed to allow for the variable growth of the population within each year. There is a correlation of −0·82 between the height of the peak and the date on which it occurred (Davidson & Andrewartha, 1948). So a substantial part of the variance due to growth of the population within each year was eliminated by referring y_1, y_2 and y_3 to the date of the peak rather than to a fixed calendar date. This is illustrated by the smaller variances of y_1 and y_2 compared with those of y_4 and y_5 (Table 11). The variance associated with this source was further reduced by taking a suitably short section of the growth curve. The comparison between the variances of y_1 and y_2 on the one hand, and y_3 on the other, illustrates this point. These three quantities may be regarded as alternative criteria for the number of thrips present each year about the time that the

* The 'peak day' refers to the date when the curve obtained by taking 15-day progressive means of the daily counts reached its maximum.

population reached a maximum. Whatever their respective practical merits it is clear (anticipating the results which will be presented below) that y_3 was based on too long a period for the best results statistically.

There may be practical grounds for ignoring the date when the thrips population attained greatest density and studying instead the numbers of thrips present at some specified date in the calendar year. For example, the apple trees burst into blossom at about the same time each year and it may be desirable to predict the number of thrips likely to be present at apple blossom time. The variates y_4 and y_5 represent the number of thrips present at specified times of the year irrespective of the stage of development of the population at that time.

Table 11. *Showing the variates selected for the partial regressions to measure the association between the number of thrips present during spring and the weather during preceding months*

Year	Date of 'break' of season	Independent variates					Dependent variates				
		x_1	x_2	x_3	x_4	x_5	y_1	y_2	y_3	y_4	y_5
1932	27 Mar.	8·65	12·09	4·37	6·09	13·77	2·58	2·10	1·60	1·03	2·00
1933	11 Apr.	8·94	10·42	4·39	7·04	12·09	2·02	1·57	1·06	0·78	1·76
1934	3 June	0·00	6·33	6·55	7·07	10·42	1·82	1·67	1·01	0·14	1·50
1935	14 Mar.	7·54	13·92	5·48	7·15	6·33	2·32	2·05	1·62	1·15	2·04
1936	7 Apr.	4·24	11·54	3·94	6·48	13·92	1·92	1·50	1·26	0·86	1·45
1937	28 Mar.	6·77	12·37	4·37	7·16	11·54	2·52	2·06	1·46	1·16	2·33
1938	19 Feb.	8·99	18·74	2·03	7·29	12·37	2·84	2·53	2·28	2·11	2·71
1939	24 Feb.	7·92	18·34	2·71	6·82	18·74	2·87	2·67	2·15	2·27	2·66
1940	7 Apr.	4·82	10·89	2·38	7·85	18·34	2·26	2·06	2·03	1·93	2·06
1941	3 Apr.	2·54	12·96	6·48	6·81	10·89	2·65	2·42	2·14	1·46	2·22
1942	30 Mar.	8·27	12·62	6·05	7·28	12·96	2·69	2·59	2·15	1·76	2·59
1943	7 Apr.	4·67	9·63	4·36	6·96	12·62	2·33	1·77	1·19	0·55	1·77
1944	2 Apr.	—	—	—	—	—	—	—	—	—	—
1945	4 May	2·30	8·33	6·76	6·42	9·65	2·13	1·95	1·67	1·05	1·78
1946	18 Feb.	9·92	17·69	3·28	6·43	8·33	2·34	2·30	2·11	1·91	2·21
Variance	—	9·148	13·376	2·482	0·206	11·381	0·1092	0·1445	0·2020	0·3964	0·1633

y_1 Logarithm of geometric mean of the daily records of thrips per rose for 15 days about the peak.
y_2 Logarithm of geometric mean of the daily counts of thrips per rose for the 30 days before the peak.
y_3 As for y_2, but for 60 days before the peak.
y_4 Logarithm of geometric mean of daily counts of thrips per rose during October.
y_5 As for y_4 but for November.
x_1 Total rainfall between the break of the season and 31 May (in inches).
x_2 Total effective day-degrees (in units of 100 day-degrees) from the break of the season to 31 August.
x_3 Total rainfall (in inches) for September and October.
x_4 Total effective day-degrees (in units of 100 day-degrees) for September and October.
x_5 As for x_2 but for the year preceding the one when the samples of thrips were taken.

(b) Selection of independent variates

Unlike many other insect species which occur in the same climatic region *T. imaginis* has no resting or 'resistant' stage in its life cycle. The insects are active throughout the year; during winter development is slow due to low temperature; the mortality rate is high during summer due to dryness and lack of food; but during the spring and early summer the environment is more favourable. At this time of the year adequate temperature and soil moisture favour rapid development and the survival rate is high. Food is usually more plentiful (Davidson, 1936). Pollen is essential for growth and reproduction; practically no eggs were laid by females from whose diet pollen had been excluded (Andrewartha, 1935).

The flowers of a wide variety of plants serve as hosts for *T. imaginis*, but the most important belong to a group of annual weeds which are adapted to the climate of the Adelaide area (Evans, 1932). The seeds remain dormant during the long dry summer and germinate in the autumn when the dry season gives way to the rainy winter season. The 'break' of the season is usually quite distinct: rainfall which is adequate to germinate the over-summering seeds is usually followed by sufficient rainfall to maintain the seedlings and promote their growth. The plants make vegetative growth during the autumn and produce flowers in the spring; they set seeds and die as the dry summer season develops. On those occasions when the 'break' of the season occurs early in the autumn the plants make strong vegetative growth before the low temperature of winter retards development and usually commence flowering earlier in the spring, producing an abundance of flowers during spring and early summer. When the seasonal 'break' occurs late in the autumn the plants make little growth before the winter and the vegetative growth continues during the spring. On these occasions the plants commence flowering later and may produce a smaller crop of flowers. It may be expected that an abundance of flowers, particularly when they are present early in the spring, would favour the multiplication of *T. imaginis*.

Consequently, in selecting independent variates consideration was given to quantities which might represent the amount of growth made by the host plants of the thrips during autumn and winter. But first it was necessary to define the 'break' of the season with reference to meteorological records. The method used by Trumble (1937) to define the duration of the growing season for annual pasture species in southern Australia was adopted. Starting with the first substantial fall of rain in the autumn a progressive total of rainfall minus evaporation was calculated by adding daily rainfall and subtracting one-third of the daily evaporation from a free water surface.* Owing to the steep increase in the rate of daily rainfall and the steep decrease in the rate of daily evaporation at this time of the year there was usually a clearly defined date when the curve representing the progressive total of rainfall minus evaporation left the zero axis and remained consistently above it for the duration of the winter. This date was chosen to represent the 'break' of the season, i.e. it was considered that this was the date on which the seeds of the annual plants which are hosts for *T. imaginis* began to develop. For 10 of the 14 years the method indicated the 'break' of the season with no ambiguity. But on four occasions, namely 1937, 1938, 1941, and 1942, an unusual dry spell occurred after the apparent 'break' of the season, and the progressive total of rainfall minus evaporation remained at zero for a short period before definitely leaving the zero axis. Details of rainfall and evaporation for these periods in 1937, 1938, 1941 and 1942 are given in Table 12. From field notes kept by the farm manager at the Waite Institute, and from other considerations, it was clear that this short dry spell should be ignored and the growing season should be taken to begin with the dates indicated in the table. The estimated date of the 'break' of the season for the 14 years is given in Table 11.

It was considered unlikely that lack of rainfall during winter (June to August) would ever be critical for the growth of the host plants; but on some occasions the rainfall during autumn may have been less than optimal. Consequently, the first independent variate selected was the total rainfall recorded between the 'break' of the season and 31 May. The individual totals for the 14 years are given in inches of rainfall under the heading x_1 in Table 11.

Table 12. *Details of the rainfall and evaporation at the beginning of the growing season in* 1937, 1938, 1941 *and* 1942

Year	Date of dry spell	Duration of 'dry spell'* days	Total rain during first wet spell (in.)	Total rain during 'dry spell'* (in.)	One-third net evaporation during dry spell (in.)
1937	20–26 Apr.	7	0·99	0·04	0·25
1938	23 Mar.–15 Apr.	24	1·61	0·17	1·01
1941	14–16 Apr.	3	0·60	0·06	0·19
1942	22–26 Apr.	5	1·01	0·08	0·12

* Duration of 'dry spell' was arbitrarily defined as the period during which the progressive total rainfall minus evaporation remained at zero.

It was considered likely that the temperature experienced during the growing period would influence the amount of vegetative growth made by the host plants. The plants grow only slowly during the winter and there is probably a threshold below which no appreciable growth occurs. There was no experimental evidence to indicate where the threshold occurs with the plants under consideration and it was arbitrarily chosen at 48° F. (8·9° C.). Thus the second independent variate which was selected may be defined as the total number of 'thermal units' accumulated between the 'break' of the season and 31 August. The number of thermal units (or day-degrees) per day was calculated from the expression

$$\frac{\text{Maximum temperature } - 48}{2},$$

and the daily values were summed for all the days between the 'break' of the season and 31 August. In Table 11, under the heading x_2, the individual totals for the 14 years are given in units of 100 day-degrees.

The development of both the host plants and the thrips may be influenced by rainfall and temperature during the spring. In particular rainfall becomes less reliable as the dry season approaches and it was

* The standard Australian evaporimeter tank is used to measure evaporation at the Waite Institute. This is described in an Australian Meteorological Observers' Handbook.

considered that on occasion rainfall in the spring may have been less than optimal. Accordingly, two further quantities were selected as independent variates:

x_3 Total rainfall in inches during September and October.

x_4 Total accumulated 'thermal units' for September and October in units of 100 day-degrees.

Finally there was the possibility that the number of thrips present in the spring and early summer might have been influenced not only by the weather during the same year but also by events which had occurred the year before. For example, after a year which had favoured the strong development of host plants the seeds of these may be more numerous and more widely distributed; or there may be more thrips survive the summer and winter after a year when the insects had been more numerous. To allow for this possibility a fifth independent variate was selected:

x_5 The total 'thermal units' accumulated between the 'break' of the season and 31 August during the year preceding the sampling year.

It was not possible to select independent variates which would serve to discriminate between the influence of the weather on the host plants and its more direct influence on the insects themselves. For example, during the 'pupal' stage *T. imaginis* is susceptible to desiccation (Andrewartha, 1934): adequate soil moisture during the autumn while temperature is still adequate for development may reduce the mortality rate due to desiccation, and thus result in an increase in the size of the population at this time of the year. The records showed that there were relatively more thrips in the roses during the autumn for some years than for others (Davidson & Andrewartha, 1948). Thus the rainfall and temperature experienced in the autumn may influence the numbers of thrips present in the spring by causing an increase in the size of the autumn population with a corresponding larger initial population at the beginning of the next spring (Evans, 1934). In this case the direct influence of the weather on the insects would be operating in the same direction as the indirect influence on the host plants. During the spring when the number of niches potentially available to the thrips is increasing rapidly the multiplication of the insects may be favoured by warm dry weather which would increase the 'activity' of the thrips in seeking out new niches. But the continued development of the host plants may be favoured by moist cool weather. In this case the direct and indirect influences of the weather on the numbers of thrips may be operating in opposite directions. Considerations of this sort should be taken into account in considering the results of the partial regressions which are given below.

(*c*) *Results of partial regressions designed to account for annual fluctuations in the number of thrips in the roses*

The calculations for the partial regressions including all five independent variates were completed for only two of the dependent variates, namely y_1 and y_2. The results are set out in Tables 13 and 14. The regression with y_1 was barely significant; that for y_2 was significant at the 1% level. In both cases the

Table 13. *Partial regressions of* y_1 *on* x_1, x_2, x_3, x_4 *and* x_5

Variate	Regression coefficients *b*	S.E.	*t*
x_1	0·0199	0·0295	—
x_2	0·0903	0·0258	3·507
x_3	0·1514	0·0658	2·303
x_4	0·1041	0·1369	—
x_5	0·0503	0·0230	2·186

Analysis of variance

Source of variance	Degrees of freedom	Sum of squares	Mean square	Variance (%)
Regression	5	1·06000	0·2120	58·8
Residual	8	0·35600	0·0450	—
Total	13	1·420	—	—

$Z=0·7750$; for $P=0·05$, $Z=0·6525$; for $P=0·05$, $t=2·306$.

Table 14. *Partial regressions of* y_2 *on* x_1, x_2, x_3, x_4 *and* x_5

Variate	Regression coefficients *b*	S.E.	*t*
x_1	0·0027	0·0274	—
x_2	0·1267	0·0239	5·297
x_3	0·2091	0·0611	3·423
x_4	0·1918	0·1271	1·509
x_5	0·0527	0·0214	2·463

Analysis of variance

Source of variance	Degrees of freedom	Sum of squares	Mean square	Variance (%)
Regression	5	1·56837	0·3137	73·1
Residual	8	0·31063	0·0388	—
Total	13	1·879	—	—

$Z=1·0446$; for $P=0·01$, $Z=0·9459$; for $P=0·01$, $t=3·355$.

more important terms were x_2, x_3 and x_5. The coefficient b_1 was small and less than half its standard error. Apparently the amount of rain falling during autumn contributed no appreciable influence additional to that which could be attributed to the other four variates. So x_1 was eliminated giving a series of new partial regressions with four independent variates; these were calculated out for all five of the dependent variates. The results are given in Tables 15, 17–19 and 21.

When y_1 (the mean log of the number of thrips present during the 15 days about the maximum of the population curve) was taken as a criterion of the density of the population the partial regression containing the four terms x_2, x_3, x_4 and x_5 was significant at the 5 % level (Table 15). But it was clear that x_4 contributed nothing to the value of the regression. So x_4 was eliminated leaving x_2, x_3 and x_5.

Table 15. *Showing the partial regression of y_1 on x_2, x_3, x_4 and x_5*

Variate	Regression coefficients: b	S.E.	t
x_2	0·0988	0·0218	4·533
x_3	0·1376	0·0606	2·272
x_4	0·0971	0·1323	—
x_5	0·0470	0·0218	2·155

Analysis of variance

Source of variance	Degrees of freedom	Sum of squares	Mean square	Variance (%)
Regression	4	1·03953	0·2599	61·3
Residual	9	0·38047	0·0423	—
Total	13	1·420	—	—

$Z=0·9077$; for $P=0·01$, $Z=0·9299$, $t=3·250$; for $P=0·05$, $t=2·262$.

The three-term partial regression was now significant at the 1 % level and each term was individually significant at the 5 % level or better (Table 16). This three-term regression (which accounted for 63 % of the variance of y_1) was the best expression of the relationship between the numbers of thrips and their physical environment which could be got when y_1 was used as a criterion of density of the thrips population.

When y_2 (the mean log of the number of thrips present during the 30 days preceding the maximum of the population curve) was used the partial regression containing the four terms x_2, x_3, x_4 and x_5 was significant at the 1 % level; three of the four terms were individually significant at the 5 % level or better, and b_4 exceeded its standard error in the ratio 1·6; the regression accounted for 76 % of the total variance of y_2 (Table 17). When x_4 was excluded the resulting three-term regression was rather more significant ($Z=1·257$), but accounted for rather less of the total variance (72 %). Thus little, if anything, was to be gained by dispensing with x_4 and it was considered that the four-term regression given in Table 17 was the best expression of the relationship between the numbers of thrips and their physical environment that could be got when y_2 was used as a criterion of the density of the thrips population.

The partial regression of y_3 (mean log of the number of thrips present during the 60 days preceding the maximum of the population curve) on x_2, x_3, x_4 and x_5 is given in Table 18. The regression was significant at the 5 % level: but x_2 was the only term which was individually significant: and the regression accounted for only 56 % of the total variance of y_3. When x_4 was excluded the relationship was slightly improved ($Z=0·9268$ $P<0·05>0·01$), but x_2 was still the only term which was individually significant. Since y_3 was thus shown to be less satisfactory, on statistical grounds, than either y_1 or y_2, and as there was no biological reason for preferring y_3 to y_1 or y_2, y_3 was not considered any further.

Table 16. *Showing the partial regression of y_1 on x_2, x_3 and x_5*

Variate	Regression coefficients: b	S.E.	t	β*
x_2	0·0960	0·0210	4·579	1·063
x_3	0·1290	0·0580	2·223	0·615
x_5	0·0483	0·0212	2·277	0·493

Increase per unit increase in $x_2=31·9$ %†
Increase per unit increase in $x_3=42·8$ %
Increase per unit increase in $x_5=16·0$ %

Analysis of variance

Source of variance	Degrees of freedom	Sum of squares	Mean square	Variance (%)
Regression	3	1·01673	0·3389	63·1
Residual	10	0·40327	0·0403	—
Total	13	1·420	—	—

$Z=1·0645$; for $P=0·01$, $Z=0·9399$, $t=3·169$; for $P=0·05$, $t=2·228$.

$$* \quad \beta=b\times\frac{\text{S.D. of } x}{\text{S.D. of } y}.$$

$$\dagger \quad \text{Percentage increase}=\frac{100\,b}{0·3010}.$$

Table 17. *Showing the partial regression of y_2 on x_2, x_3, x_4 and x_5*

Variate	Regression coefficients: b	S.E.	t
x_2	0·1278	0·0197	6·490
x_3	0·2072	0·0548	3·784
x_4	0·1909	0·1196	1·596
x_5	0·0522	0·0197	2·652

Analysis of variance

Source of variance	Degrees of freedom	Sum of squares	Mean square	Variance (%)
Regression	4	1·56802	0·3920	76·1
Residual	9	0·31097	0·0346	—
Total	13	1·879	—	—

$Z=1·2146$; for $P=0·01$, $Z=0·9299$, $t=3·250$; for $P=0·05$, $t=2·262$.

The remaining two independent variates which were chosen as criteria of the number of thrips present ignored the date of the maximum of the population curve and thus included a substantial part of the variance due to natural increase of the population within each year. The regression of y_4 (mean log of the number of thrips during October) on x_2, x_3, x_4 and x_5 is given in Table 19. The regression was significant at the 1 % level and accounted for 75 % of the variance of y_4: but x_2 was the only term which was individually significant. With the progressive elimination of x_3, x_4 and x_5 the significance of the regression was increased. The total regression of y_4 on x_2 was significant at the 0·1 % level and the regression accounted for 66 % of the variance of y_4 (Table 20).

The partial regression of y_5 (mean log of the number of thrips during November) on x_2, x_3, x_4 and x_5 is given in Table 21. The regression was significant at the 1 % level: and the regression accounted for 77 % of the total variance. Three of the four terms were individually significant at the 5 % level or better, and the fourth (b_5) exceeded its standard error in the ratio of nearly 2. When x_5 was excluded the resulting three-term regression was less significant ($Z=1·197$): and the regression accounted for less of the total variance of y_5 (70 %). The four-term regression given in Table 21 was therefore the best expression of the relationship between the physical environment and the number of thrips present during November.

Table 18. *Partial regression of y_3 on x_2, x_3, x_4 and x_5*

Variate	Regression coefficients		
	b	S.E.	*t*
x_2	0·1313	0·0315	4·170
x_3	0·1690	0·0875	1·932
x_4	0·2152	0·1911	—
x_5	0·0477	0·0315	1·516

Analysis of variance

Source of variance	Degrees of freedom	Sum of squares	Mean square	Variance (%)
Regression	4	1·83189	0·4580	56·3
Residual	9	0·79411	0·0882	—
Total	13	2·626	—	—

$Z=0·8233$; for $P=0·01$, $Z=0·9299$, $t=3·250$; for $P=0·05$, $Z=0·6450$, $t=2·262$.

Table 19. *Partial regression of y_4 on x_2, x_3, x_4 and x_5*

Variate	Regression coefficients		
	b	S.E.	*t*
x_2	0·1668	0·0335	4·983
x_3	0·1140	0·0930	1·226
x_4	0·3100	0·2033	1·525
x_5	0·0680	0·0335	2·032

Analysis of variance

Source of variance	Degrees of freedom	Sum of squares	Mean squares	Variance (%)
Regression	4	4·25473	1·0637	74·8
Residual	9	0·89827	0·0998	—
Total	13	5·153	—	—

$Z=1·1833$; for $P=0·01$, $Z=0·9299$, $t=3·250$; for $P=0·05$, $t=2·262$.

Table 20. *Showing total regression of y_4 on x_2*

Variate	Regression coefficient		
	b	S.E.	*t*
x_2	0·1425	0·0279	5·105

Percentage increase per unit increase of $x_2=47·3$ %

Analysis of variance

Source of variance	Degrees of freedom	Sum of squares	Mean square	Variance (%)
Regression	1	3·52839	3·5284	65·8
Residual	12	1·62461	0·1354	—
Total	13	5·153	—	—

$Z=1·6302$; for $P=0·01$, $Z=1·1166$, $t=3·055$.

Table 21. *Partial regression of y_5 on x_2, x_3, x_4 and x_5*

Variate	Regression coefficients		
	b	S.E.	*t*
x_2	0·1247	0·0208	6·002
x_3	0·1530	0·0577	2·651
x_4	0·3156	0·1261	2·502
x_5	0·0405	0·0208	1·953

Analysis of variance

Source of variance	Degrees of freedom	Sum of squares	Mean square	Variance (%)
Regression	4	1·77712	0·4443	76·5
Residual	9	0·34588	0·0384	—
Total	13	2·123	—	—

$Z=1·2238$; for $P=0·01$, $Z=0·9299$, $t=3·250$; for $P=0·05$, $t=2·262$.

All the dependent variates (y_1 to y_5) used in this section included in themselves the variability associated with the variable 'activity' of the thrips in seeking out the roses. In the next section we shall show how some of the variability due to this cause may be eliminated and shall calculate two new quantities which may be regarded as improved estimates of the true density of the thrips population in the garden.

6. ANNUAL FLUCTUATIONS IN THE NUMBERS OF THRIPS AFTER THE ELIMINATION OF VARIANCE DUE TO 'ACTIVITY'

The methods employed in § 5 may provide the best information that can be got for the purpose of predicting the numbers of thrips likely to be present during the spring and early summer. But more precise information about the physical ecology of *T. imaginis* may be obtained by first eliminating the variance associated with the activity of the insects in seeking out the flowers. In other words, the estimate of the number of thrips present which was got by counting the number of thrips in the roses may be improved by subtracting a correction factor calculated from the regression equation obtained in § 4.

(*a*) *Calculation of the correction factor*

Each daily record of the number of thrips per rose may be considered to consist of two components: one component related to the 'real' number of thrips in the garden, the other to the activity of the thrips in seeking out the flowers. This may be written in symbols:

$$Y_{\text{observed}} = Y_{\text{real}} + Y_{\text{activity}}$$

or

$$Y_{\text{real}} = Y_{\text{observed}} - Y_{\text{activity}}.$$

The component $Y_{\text{act.}}$ was calculated by substitution in the regression equation obtained in § 4 (see Table 10):

$$Y_{\text{act.}} = 0{\cdot}007925\, x_1 - 0{\cdot}001969\, x_2. \quad \text{(i)}$$

In this equation all the quantities $Y_{\text{act.}}$, x_1 and x_2 are written about their means, and they are all deviations from the appropriate polynomial curves, calculated from the equations:

$$1000\, Y = 1694{\cdot}75 + 12{\cdot}6863\, \xi_1 - 0{\cdot}015287\, \xi_2 - 0{\cdot}0033618\, \xi_3, \quad \text{(ii)}$$

$$1000\, X_1 = 71401{\cdot}80 + 88{\cdot}6804\, \xi_1 + 8{\cdot}37664\, \xi_2 + 0{\cdot}012072\, \xi_3, \quad \text{(iii)}$$

$$1000\, X_2 = 5488{\cdot}60 - 25{\cdot}1385\, \xi_1 - 0{\cdot}88726\, \xi_2 + 0{\cdot}020080\, \xi_3. \quad \text{(iv)}$$

The coefficients for the equation for Y are means of the 14 values given in Table 2 under the heading '$n = 60$': those for X_1 and X_2 were derived in the same way from the data for x_1 and x_2. Since the means of the deviations from the curves represented by equations (ii), (iii) and (iv) were zero the quantities in equation (i) were simply the actual deviations from the appropriate mean polynomial curves: and the equation may be written

$$Y_{\text{act.}} = 0{\cdot}007925\,(x_1 - X_1) - 0{\cdot}001969\,(x_2 - X_2).$$

(*b*) *Selection of dependent variates*

The quantity $Y_{\text{act.}}$ was calculated for each day: there were 14 sets of 60 values to be calculated, one set of 60 for each year. The daily values of $Y_{\text{act.}}$ were then subtracted from the daily values of $Y_{\text{obs.}}$ to give a set of daily values for Y_{real}. The estimated values of Y_{real} were then substituted for the logarithms of the original daily counts and two new dependent variates were calculated, one related to the peak of the thrips population curve, the other to a fixed calendar date. These were:

y'_2 The logarithm of the geometric mean number of thrips per rose for the 30 days preceding the peak: using values for Y_{real} instead of $Y_{\text{obs.}}$.

y'_6 The logarithm of the geometric mean number of thrips per rose during the 30 days 17 October to 15 November: using values for Y_{real} in place of $Y_{\text{obs.}}$.

The individual values of y'_2 and y'_6 for the 14 years are shown in Table 22: the values for y_2 are included so that they may be compared with those for y'_2. The variate y'_2 is the counterpart of the variate y_2 which was used in § 5 except that the former has been 'corrected' for the influence of maximum temperature and rainfall on the activity of the thrips in seeking out the flowers. The regressions calculated with y'_2 will therefore be not only a more precise estimate of the relationship between the number of thrips and their environment, but by comparison with the regressions obtained with y_2 will also indicate the extent of the precision gained by the elimination of part of the variance associated with the activity of insects in seeking out the flowers. The variate y'_6 has no counterpart in § 5 for the very practical reason that it had to be chosen from a period which was wholly included in every one of the fourteen 60-day periods used for calculating the regression in Table 10. Since the earliest of these periods lasted from 17 September to 15 November (1939) and the latest from 15 October to 13 December (1946) it was not possible to include either the whole of October or the whole of November. The 30-day period 17 October to 15 November was chosen as a compromise.

Table 22. *Individual values for* y_2, y'_2 *and* y'_6

Variate	1932	1933	1934	1935	1936	1937	1938	1939	1940	1941	1942	1943	1945	1946
y_2	2·10	1·57	1·67	2·05	1·50	2·06	2·53	2·67	2·06	2·42	2·59	1·77	1·95	2·30
y'_2	2·08	1·57	1·65	2·03	1·54	2·04	2·51	2·72	2·10	2·40	2·57	1·79	1·93	2·28
y'_6	1·44	1·32	0·79	1·70	1·18	1·80	2·41	2·73	2·05	1·84	2·22	1·03	1·47	2·01

(c) Results of regressions calculated with corrected estimates of the density of the thrips population

The partial regression of y'_2 (corrected estimate of the number of thrips during the 30 days before the maximum of the population curve) on x_2, x_3, x_4 and x_5 is given in Table 23. The regression was significant at the 0·1 % level: two of the terms were individually

Table 23. *Partial regression of y'_2 on x_2, x_3, x_4 and x_5*

Variate	Regression coefficients			
	b	S.E.	t	β
x_2	0·1254	0·0185	6·79	1·224
x_3	0·2019	0·0514	3·93	0·848
x_4	0·1866	0·1122	1·66	0·226
x_5	0·0580	0·0185	3·08	0·511

Increase per unit increase in x_2 = 41·7 %
Increase per unit increase in x_3 = 67·1 %
Increase per unit increase in x_4 = 62·0 %
Increase per unit increase in x_5 = 18·9 %

Analysis of variance

Source of variance	Degrees of freedom	Sum of squares	Mean square	Variance (%)
Regression	4	1·5537	0·3884	78·4
Residual	9	0·2736	0·0304	—
Total	13	1·8273	—	—

$Z = 1·2739$, $P < 0·001$.

Table 24. *Partial regression of y'_6 on x_2, x_3, x_4 and x_5*

Variate	Regression coefficients			
	b	S.E.	t	β
x_2	0·1632	0·0229	7·13	1·091
x_3	0·1709	0·0636	2·69	0·492
x_4	0·3616	0·1389	2·60	0·300
x_5	0·0698	0·0229	3·05	0·430

Increase per unit increase in x_2 = 54·2 %
Increase per unit increase in x_3 = 56·8 %
Increase per unit increase in x_4 = 120·0 %
Increase per unit increase in x_5 = 23·2 %

Analysis of variance

Source of variance	Degrees of freedom	Sum of squares	Mean square	Variance (%)
Regression	4	3·4694	0·8674	84·4
Residual	9	0·4198	0·0466	—
Total	13	3·8892	—	—

$Z = 1·4621$, $P < 0·001$.

significant at the 1 % level, b_5 was significant at the 2 % level, and b_4 though not individually significant exceeded its standard error in the ratio 1·66. The regression accounted for 78 % of the variance of y'_2. The three-term regression from which x_4 had been excluded was slightly more significant ($Z = 1·3079$): and all the terms were individually significant at the 2 % level or better: but the regression accounted for rather less of the total variance (74 %). There was therefore little, if anything, to be gained by discarding x_4 and it was considered that the four-term regression given in Table 23 was the best expression of the relationship between the number of thrips and their environment that could be got when y'_2 was used as a criterion of the density of the thrips population.

The regression of y'_6 on x_2, x_3, x_4 and x_5 is given in Table 24. The regression was significant at the 0·1 % level: all the terms were individually significant at the 5 % level or better: and the regression accounted for 84 % of the total variance of y'_6. This was the most precise of all the relationships which have been derived in this paper judged not only by the significance of the regression as a whole ($Z = 1·4621$), and the significance of all the individual terms, but also by the proportion of the variance which the regression accounted for.

7. DISCUSSION

The influence of daily maximum temperature and daily rainfall on the activity of the thrips seeking out the flowers was discussed in § 4. The results presented in §§ 5 and 6 will be discussed under four headings.

(a) Alternative criteria for the density of the thrips population

Seven different quantities were used as alternative criteria for the density of the thrips population. Each is to be judged by a dual standard: it is relevant to ask of each how much information it gives of biological or practical matters, and how well it serves the purpose of statistical analysis. For example, if it is required to predict the number of thrips likely to be present on some future occasion, quantities like y'_2 or y'_6 will be of little interest since the correction factor for these cannot be calculated in advance. Similarly, quantities like y_1 or y_2 are of limited interest since the date of the maximum of the population curve cannot be known in advance. With *T. imaginis* the chief interest in prediction may centre around the time when the apple blossom is present; this is usually October. For other purposes it may be desirable to know the number of thrips likely to be present during November. Thus, for the practical purpose of prediction, the quantities y_4 and y_5 may prove most instructive.

But in seeking information about the ecology of *T. imaginis* it may be better to consider the density of the population about the time when it is at a maximum each year rather than during any arbitrary calendar period. The quantities which may serve this purpose are y_1, y_2 and y'_2. In this paper our chief purpose has been to explain the fluctuations in the growth of the population from year to year. The quantity y_1 may therefore not be the best criterion to choose since it is made up from an equal number of records taken before and after the maximum of the curve. The records taken after the maximum may reflect the influence of components in the environment directly associated with the decline of the population; and these may be quite different from those associated with its growth. The quantity y_2 may therefore be preferred to y_1 since it is made up entirely from records which were taken while the population curve was ascending, being in fact the mean of the 30 days preceding the maximum of the curve. The choice between y_2 and y'_2 needs to be made on statistical grounds since biologically the two quantities are exactly equivalent.

Of the several quantities related to the maximum of the population curve, y_3 was the least satisfactory on statistical grounds. The variance was greater than that of y_1 or y_2, doubtless because y_3 was calculated from the 60 days before the maximum of the curve and therefore included more of the variance due to variable growth within each year. The greater variance of y_3 was not offset by any stronger association with the independent variates and therefore the regression with y_3 was less precise than those with y_1 or y_2 (Tables 11, 17 and 18). The quantity y_1 was less satisfactory than y_2. The variance of y_1 was less than that of y_2 (Table 11), but this was more than offset by the weaker association of y_1 with the independent variates. (Compare the total sums of squares and residual sums of squares in Tables 15 and 17.) This was doubtless due to the inclusion in y_1 of records taken when the population was beginning to decline, whereas the independent variates were all chosen to explain the growth of the population. The regression calculated with y_2 was inferior to that with y'_2 doubtless because the latter was a better estimate of the number of thrips in the garden. The residual variance of y_2 was swollen by the retention of more of the variance due to 'activity' of the insects (compare residual mean squares in Tables 17 and 23). Thus of all the quantities related to the maximum of the population, y'_2 was the best on statistical grounds.

The regression given for y'_6 was statistically the most significant of all. This was largely due to the increased strength of the relationship between x_4 and y'_6 (compare the β coefficients, for example, in Tables 23 and 24). The partial regression of y'_6 on x_4 was larger than that of y'_2 on x_4, because both y'_6 and x_4 are related to calendar dates whereas y'_2 is related to the date of the maximum of the population curve which varied from 15 November in 1939 to 13 December in 1946. The least satisfactory (on statistical grounds) of all the alternative dependent variates was y_4. The number of thrips present during October was more variable than at any other time (see Table 11). It was not possible to demonstrate a significant association between y_4 and any of the individual components of the physical environment except x_2: and the regression of y_4 on x_2 accounted for no more than 66% of the total variance of x_4 (Table 20). Nevertheless, y_4 has been retained because it is the most instructive quantity that could be selected for the practical purpose of prediction.

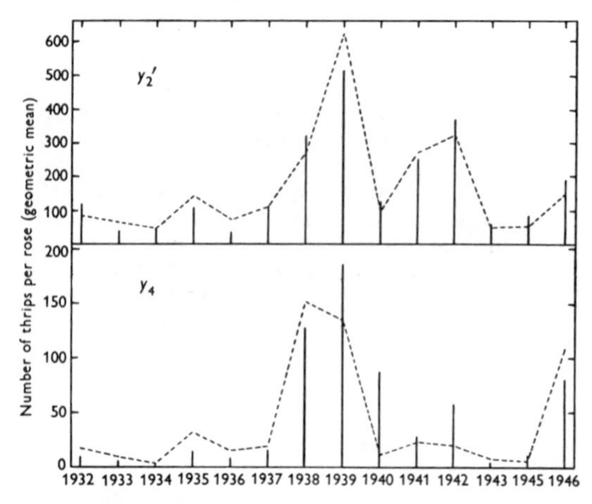

Fig. 4. The columns represent the geometric means of the daily counts of thrips per rose. The curves represent the estimated values for the same quantities calculated from the appropriate regression equations:

$$\log Y'_2 = -2{\cdot}390 + 0{\cdot}125\, x_2 + 0{\cdot}202\, x_3 + 0{\cdot}187\, x_4 + 0{\cdot}058\, x_5,$$

$$\log Y_4 = -0{\cdot}492 + 0{\cdot}142\, x_2.$$

In Fig. 4 the estimated and observed values for y'_2 and y_4 have been plotted. The agreement between the estimated and observed values for y'_2 is remarkable; for y_4 the agreement is not so close. The estimated values for y'_2 were got by substitution in the equation

$$\log Y'_2 = -2{\cdot}390 + 0{\cdot}125\, x_2 + 0{\cdot}202\, x_3 + 0{\cdot}187\, x_4 + 0{\cdot}058\, x_5,$$

which may be constructed from the information in Tables 11, 22 and 23. The estimated value for y_4 was got by substitution in the equation

$$\log Y_4 = -0{\cdot}492 + 0{\cdot}142\, x_2,$$

which may be constructed from the information in Tables 11 and 20.

The discussion which follows in §§ (*b*), (*c*) and (*d*) below has been based chiefly on the results of the regressions with y'_2, since this quantity is the most instructive on biological grounds and is to be preferred on statistical grounds to y_1, y_2 and y_3.

(b) *Relative importance of different components of the physical environment*

The density of the thrips population was influenced more by x_2 than by any other component of the physical environment. The β coefficients in Table 23 show that x_2 was about 1·5 times as influential as x_3, 5·4 times as influential as x_4, and 2·4 times as influential as x_5. The quantity x_2 was calculated by summing accumulated 'effective' temperature from the date of the 'break' of the season at the end of the autumn up to 31 August (see p. 210). Since temperature was considerably higher during autumn than during winter the size of x_2 was largely determined by the date of the 'break' of the season (see Table 11). When this occurred early x_2 was large. The biological significance of an early 'break' to the season was that it allowed the seeds of the annual plants which serve as hosts for *T. imaginis* to germinate early. The plants were then able to make strong growth before and during the winter and were ready to burst into flower as soon as the temperature began to rise in the spring. On the other hand, when germination was delayed by a late 'break' to the season the plants were still relatively immature at the end of winter and continued their vegetative growth as temperature increased during the spring. On these occasions flowering was delayed until late in the spring.

Since *T. imaginis* requires pollen for growth and reproduction, and since the mortality rate of the pupal stages may be high when soil moisture is inadequate, the favourable period for the multiplication of the insect was limited at the one end by the date in the spring when pollen became available in quantity and at the other end by the onset (about the end of November) of the dry summer season. It becomes clear that the close association of x_2 with the maximum density attained by the thrips population may be largely due to the value of x_2 as an indication of the date when pollen first became available in quantity in the spring, and thus as a measure of the duration of the favourable period for multiplication. This conception is illustrated in Fig. 5.

The broken lines in the figure were obtained by taking 15-day moving averages of the daily records of the number of thrips per rose, starting from the day when the population began to decline. The smooth curve is a hypothetical logistic curve.

It was shown that the lower portion of a logistic curve provided an adequate empirical fit to the data representing the mean growth of the population during the period September to December (Davidson & Andrewartha, 1948). The calculation was not carried beyond the point of inflexion of the curve because empirical data were not available beyond this point. Usually the onset of the summer drought calls a halt to the multiplication of the insects long before the population has had time to saturate its environment; empirical data on which to base a more complete curve are thus unobtainable.

The extrapolated curve which is given in Fig. 5 is therefore hypothetical. It is given because it is convenient to imagine the maximum density attained by the population each year as an ordinate on some (hypothetical) logistic curve. The height of the ordinate can then be seen to be determined by the value on the abscissa of the logistic curve which corresponds to the date when the population began to decline. In those years when this date corresponded to a high value on the abscissa the curve may be imagined as having had its origin earlier in the spring, and vice versa.

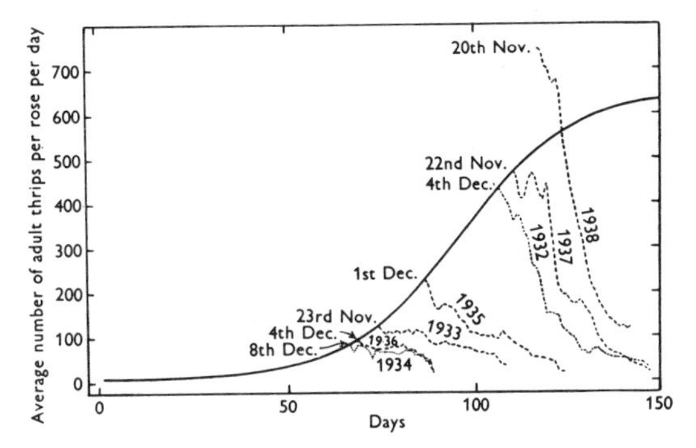

Fig. 5. Presenting a hypothetical explanation for the control of the population density by 'density-independent' components of the environment. The trend line is a hypothetical logistic curve. The broken lines are the descending parts of the observed population curves for each of the 7 years 1932–8. Each one has been moved along the abscissa so that its maximum will coincide with the hypothetical trend line. For further explanation see text.

This hypothesis is not capable of being demonstrated empirically, but it is consistent with two major facts established in § 6: (*a*) four 'density-independent' components of the physical environment accounted for 78 % of the variance of the maximum annual density of the population; (*b*) the most important individual component was x_2 which has been shown to serve as a measure of the earliness of appearance of blossoms on the plants which serve as hosts for *T. imaginis*.

The relatively large value for the maximum density attained in 1938, which exceeds the upper asymptote of the theoretical curve, may not be taken to indicate that in this year the insects may have saturated the environment. To argue that would be to attribute an empirical value to the diagram which it does not possess. Indeed it is likely that the potentialities of the environment may vary from year to year.* In 1938 the 'potential' maximum population may have been much higher than the observed maximum (Davidson & Andrewartha, 1948).

The influence of x_2 may not have been restricted to its relationship with the duration of the favourable period for multiplication. Two other possible influences are: (*a*) when the plants made vigorous growth early they may have produced more flowers in addition to producing them earlier; (*b*) usually the chief hazard for the insects during the summer is inadequate soil moisture. Rain falling early in the autumn may alleviate this hazard and allow a small increase in the population at this time of the year, with the result that there may be a larger initial population at the beginning of the next spring (Evans, 1934). The data did not permit of these several components being measured separately and the partial regression given for x_2 represents the influence of the sum of them all. But it is considered that the most influential component was the one associated with the growth of the host plants, and the tendency for the date on which they came into blossom to be advanced on those occasions when x_2 was large and retarded when x_2 was small.

Since the value for x_2 was largely determined by the date of the 'break' of the season we may conclude that the maximum density of the population in the spring was largely determined by the weather during the previous autumn; but the significance of the partial regressions for x_3 and x_4 show that the weather during the spring may modify the tendencies established during the autumn. This is illustrated in Fig. 4. Consider, for example, the data for 1942 and 1946. In 1942 the value for x_2 was small but the population attained a moderate density because x_3 and x_4 were both large; in 1946 the value for x_2 was large but this was offset by low values for x_3 and x_4 (see Table 11). With Y_4 the calculated values were obtained from x_2 alone, and the regression largely underestimated the numbers of thrips in 1942 and overestimated them in 1946 (Fig. 4, lower curve).

The β coefficients in Table 23 show that x_3 was the second most influential component. The value for x_3 was calculated by summing the total rainfall for September and October. At this time of the year temperature and evaporation normally increase and rainfall is required to maintain adequate soil moisture. With favourable soil moisture the mortality rate of the pupal stages of *T. imaginis* may be reduced and the development of the host plants may be promoted. The data did not permit the influence of these two components to be estimated separately. Both will contribute to a positive regression of y on x_3 and it is considered that the direct influence of soil moisture on the survival rate of the insects in the pupal stage may have been more important than its indirect influence through the host plants.

During the period September–October the number of niches available to *T. imaginis* increases enormously. The multiplication of the insect may be favoured by increased activity in seeking out new situations. It was shown in § 4 that rainfall reduces activity. Thus the regression of y on x_3 will contain a negative component associated with the activity of the insects in seeking out new situations. This means that the values given for b_3 underestimate the influence of soil moisture during September–October on the maximum density of the population.

The least important component (excluding x_1) was x_4, which measured the influence of temperature during September–October. Increasing temperature will reflect an increase in the rate of development of the insect (Andrewartha, 1936), an increase in its activity in seeking out new situations, and an increase in the rate of development of the host plants. The sum of the influence of all these phenomena will be represented by the regression of y on x_4; each of them will contribute a positive component to the regression, and experience does not suggest any influence which might contribute a negative component. Nevertheless, x_4 was the least influential of all the independent variates. Apparently temperature during September–October was usually adequate, and the fluctuations which occurred were not reflected in large changes in the density of the thrips population.

The significant regression of y on x_5 was interesting because it demonstrated a residual influence carried over from the year before. Again the data did not permit of the partitioning of x_5 into further components. Since x_5 was derived from the same quantities as x_2 but for one year earlier, and since x_2 was shown to be closely related to the density of the thrips population, the influence of x_5 might be attributed to its association with the thrips population of the year before. That is there might have been a large initial population of thrips at the beginning of the spring on those occasions when x_5 was large. This association was not apparent from the data. Considering the enormous mortality which occurred each summer it seems unlikely that the chief component of x_5 was its influence on the residual number of thrips. It is more likely that x_5 may have exercised its chief influence through the host plants. Since a large value for x_2 favoured the strong development of the host plants, it is likely that seeds would be more numerous and more widespread on those occasions when x_5 was large.

But the data neither confirm nor deny these speculations. They merely indicate that the sum of the influences represented by x_5 was the third most important of the four components of the physical environment which between them accounted for

* In this connexion consider also the discussion given below on the influence of x_5.

78 % of the total variance of the annual maximum density attained by the thrips population.

(c) Secular rhythm in the annual maximum density of the population

The observed values for y'_2 plotted in Fig. 4 indicate that the maximum density of the population tended to increase progressively for 3 or 4 years and then to decrease abruptly in the course of 1 year to a very low level from which it then gradually increased again. There was an outbreak of *T. imaginis* in 1931. So the population was known to be larger in 1931 than in either 1930 or 1932. Thus maxima occurred in 1931, 1935, 1939, 1942 and 1946. The only exception to the rule that the decline occupied only 1 year occurred after the outbreak of 1931. The average distance between maxima is 3·75 years. Do these data indicate that rhythm of the sort indicated in the sample is characteristic of the population of *T. imaginis* living in the Adelaide district?

The hypothesis which fits the observations most closely is that the density of the population was increasing progressively for 3 years and then declining abruptly every fourth year. If in such a hypothetical population we subtract the record for each year from the one ahead of it, positive and negative differences between the pairs of years would occur in the ratio of 3 positive to 1 negative. If, on the other hand, the quantities were occurring at random with respect to time there would be an equal number of positive and negative differences, and the distribution of groups of 4 differences would have the form: in every 120 trials there would be 1 group with 0 positive and 4 negative, 26 groups with 1 positive and 3 negative, 66 with 2 positive and 2 negative, 26 with 3 positive and 1 negative, and 1 group with 4 positive and 0 negative. This distribution has a mean of 2 positive (or negative) differences and a variance of 0·5. The standard error of a mean of 3·5 groups is thus $\sqrt{(0\cdot5/3\cdot5)}=0\cdot378$.

In the sample under consideration there were 9 positive and 5 negative differences (1944 being missing), giving a mean of 2·571 positive for a sample of 3·5 groups of 4. This differs from 2 by 0·571 which exceeds 0·378 in the ratio 1·51. A difference of this magnitude would have occurred in 13 % of random samples in which an excess of positive or negative differences was equally likely. But we have excluded (by hypothesis) the possibility of runs of negative differences and must therefore divide this probability by 2. A series of similar random samples of 15 years taken from a population in which the fluctuations were quite random with respect to time might have shown as much or more 'rhythm' as the present sample on 6 or 7 occasions in every 100. Thus the 'rhythm' shown in this sample is suggestive but does not provide conclusive evidence that oscillations of this sort are characteristic of the population of *T. imaginis* living in the Adelaide district.

Nevertheless, in this particular sample of 15 years, there was a rhythmic oscillation in the maximum density attained by the population. Periodic oscillations in a population have sometimes been related to the presence of predators in the environment (Nicholson, 1933; Smith, 1939). In the present instance no predators, parasites nor diseases of any importance were observed. Moreover, 78 % of the variance was related to four quantities which were derived entirely from meteorological records: and there is no evidence that any of the four terms in the regression concealed in themselves the influence of any competitor or predator of *T. imaginis*. A substantial part of the remaining 22 % is doubtless to be attributed to variance of sampling and departures from linear regression, leaving little that could be attributed to any other component of the environment whether physical or biotic.

Apart from periodic oscillations in the density of the population there is also the possibility that the maximum density of the population may have been changing slowly with time. To test this possibility orthogonal polynomials were calculated up to the 5th degree in x (years). The regression was not significant at any stage, nor were any of the terms significant. The quadratic term was negative and large enough to be suggestive, but even so accounted for a non-significant proportion of the variance. The sample of 14 years therefore did not provide any evidence of a long-term trend in the density of the population. It is therefore appropriate to conclude that the maximum annual density attained by the population had neither been increasing nor decreasing during the 14 years represented in this sample. Since four quantities derived entirely from meteorological records had accounted for 78 % of the variance this may only be an indirect way of showing that there had not been a trend with time in those particular components of the weather.

(d) The 'balance' of the population

The theory of the 'balance' of animal populations rests largely on experimental results got from populations growing in artificial environments in laboratories (Pearl, 1926; Chapman, 1928; Holdaway, 1932); or on hypothetical mathematical models (Nicholson, 1933; Nicholson & Bailey, 1935). The theory states that 'balance' may be maintained only by the operation of 'density-dependent' components of the environment, since these alone fulfil the requirements of 'competition'. Without 'competition' (brought about by the operation of components whose influence varies in proportion to the density of the population) an increasing population would ultimately become infinitely large and a decreasing population would ultimately become infinitesimally small and presumably become extinct. Since populations do not usually do either one or the other it is argued that all populations are in a state of 'balance' brought about by the operation of 'density-dependent' components in the environment.

An experimental population of host and predator living in an artificial environment which had been arranged to comply as nearly as possible with the requirements assumed for their mathematical models by Nicholson & Bailey, gave results during seven generations which closely approximated to those predicted by the theory (De Bach & Smith, 1941). The growth of the sheep population of South Australia followed a course which was consistent with the theory that 'density-dependent' components were of chief importance (Davidson, 1938). But the sheep also were living in an 'artificial' environment since it was to the farmer's advantage to maintain their population at the highest density consonant with a healthy development of the pastures. In the rather more 'natural' environments used by Gause and his associates to study the interaction of predator and prey, 'balance' was found to occur in two special cases where there was periodic immigration of hosts from a 'predator-free' reserve of hosts. In other cases the state of 'balance' did not develop: in some instances the predator died out while the host was still numerous, in others the predator first destroyed the host and then died out completely itself (Gause, Smaragdova & Witt, 1936). When we turn to the facts established by the present study of a natural population of *T. imaginis* we find that they are not altogether consistent with the theory as it stands at present.

Like many, if not most, insect populations in temperate climates the population of *T. imaginis* changed continuously and rhythmically with the changing seasons of the year (Davidson & Andrewartha, 1948). Each year about August it reached its lowest level: during the next 3 months the population grew rapidly reaching a maximum about the end of November, often at this time being several hundredfold more numerous than during August. During December the numbers declined abruptly and usually had reached a low level by January or February. Occasionally a slight temporary increase occurred during May–June, but in general the numbers continued to decline until August. This cycle was repeated with minor variations each year.

Since the population increased each year to a maximum and then declined, and since the size of the maximum attained each year showed no systematic trend with time it is relevant to ask: (*a*) what were the components of the environment which limited (or controlled) the growth of the population; and (*b*) was the population in a state of 'balance' with these components of the environment? If not there seems to be no sense in which the conception of balance may be applied to the observed fluctuations in the population at this time of the year.

This is indeed the central theme of the present paper. It has been shown that the maximum density attained by the population each year was related very closely to fluctuations in the weather during the preceding months. Four quantities derived entirely from meteorological records accounted for the major part of the annual variability in the thrips population. There was no evidence that the influence of these components of the environment varied with the density of the population, nor would one expect it to do so.

That four 'density-independent' components of the environment should have accounted for 78% of the variance of the population density ceases to be surprising when it is considered that the population did not normally attain anything like the density required to saturate the environment. On the contrary, during the favourable period for multiplication the number of flowers suitable for the breeding of *T. imaginis* increased enormously and usually very rapidly; and the density of the population relative to the available situations was so low that 'competition' was likely to be quite unimportant. The 'balance' (if that be a suitable term) is seen as a race against time; multiplication proceeded further and the population attained a greater density on those occasions when the favourable period was prolonged; multiplication was curtailed and the maximum density attained was lower when the favourable period was shorter. But on the average the physical environment remained favourable for far too short a period to permit the insects to multiply to a point where competition became important.

Nicholson & Bailey (1935) may have imagined such a situation when they wrote (p. 553): 'Moreover the competition which exists between searching animals is shown to produce certain effects such as balance and interspecific oscillation, which are independent of other environmental factors in their essential features and so must exist—unless the fundamental hypotheses given are wrong, or there is some factor capable of nullifying the effects of competition, which seems most improbable.' Experience suggests that the phenomenon which has been shown for *T. imaginis* to be 'capable of nullifying the effects of competition' may be quite commonplace, at least with species which live in temperate climates.

The theory of the balance of animal populations postulates the operation of 'density-dependent' components in the environment of a decreasing population. Otherwise the population would become infinitesimally small and die out. (Nicholson, 1933, p. 136). The population of *T. imaginis* was usually decreasing during the period December to August each year.

The mortality which occurred each year during the summer was tremendous. During the spring fields, gardens, headlands, roadsides, indeed the whole countryside may harbour flowering plants which may provide favourable situations for *T. imaginis*. During the summer these disappeared and the insects died out completely from the greater part of the terrain which they had previously occupied. The survivors were to be found in restricted local situations scattered thinly throughout the area of 'permanent' distribution of the species. The garden at the Waite Institute where our records were taken represented one such situation. Since the mortality each year was tremendous, yet never proceeded to the point where the species died out completely, it is relevant to ask: (*a*) what were the components of the environment which controlled the density of the population at this time of the year; and (*b*) was the population in a state of 'balance' with these components of the environment?

No attempt has been made to relate the minimum numbers observed to any selected components of the environment in the same way that the maximum density was studied. But observation indicates that predators, parasites and disease played a negligible part in controlling the population. Nor was there any evidence of intraspecific competition, since at this time of the year the flowers always contained far fewer thrips than the maximum they could support.

In the absence of precise statistical measurement the components of the environment which may be associated with the minimum density of the population may only be guessed at. But it would be a mistake to suppose that the components of the physical environment (e.g. summer drought) operate at an intensity which is independent of the density of the population. This becomes clear when the detailed geographic distribution of the insect is considered.

The geographic distribution of *T. imaginis* has not been studied in detail. But for the springtail, *Smynthurus viridis*, Davidson (1934) showed that there was a close relationship between the distribution of the insect and the distribution of certain climatic zones. He mapped four zones of decreasing degrees of favourableness. Outside the least favourable zone the insect was unable to survive the unfavourable season of the year even in the most favourable situations. Inside the most favourable zone no conceivable variation in the weather could destroy all the insects in the more favourable situations. A detailed study of the ecology of the grasshopper, *Austroicetes cruciata*, established the same principles. During the years 1936 and 1937 the insects colonized a large area in western Australia which was outside the zone where swarms normally occur. With the return of more 'normal' weather in 1938 the insects died out of this area but persisted in the zone of 'permanent' distribution (Andrewartha, 1944). In the 'grasshopper belt' of South Australia *A. cruciata* is able to increase in numbers during a number of years and then a drought may cause a 'crash'. The insect may die out completely from all but a few local situations. The complete annihilation of the grasshopper during a drought is prevented by the existence of scattered niches, be they ever so small, where some food remains (Andrewartha & Birch, 1948).

Similarly *Thrips imaginis* was reduced each year during the unfavourable season to local situations which were more favourable than the surrounding countryside. Some of these situations may be 'marginal' in that some thrips may survive in them some years. Outside the area of 'permanent' distribution of the species there may be some situations where some individuals may survive in some years but die out completely in other years. Inside the zone of permanent distribution there are some situations so favourable that no conceivable variation of weather will destroy all the insects in them. These local situations may be arranged in a graded series from the most favourable through the marginal down to those so unfavourable that all the insects will disappear from them during the unfavourable season every year.

At the end of spring most of the population occurred in situations which readily became unfavourable with the approach of summer. Thus, while the population was large, a small increase in the severity of the weather destroyed a large proportion of the population. As the numbers were reduced the insects were restricted to an increasing extent to the more favourable situations and relatively large increases in the severity of the weather destroyed relatively fewer of the total population. Indeed no conceivable variation in the weather is likely to destroy all the insects in these situations in the time available: by the time spring comes the population ceases to decline and begins to increase once again in response to the changing physical environment.

In this special sense the physical environment (as determined by the weather) becomes one of the 'density-dependent' components of the environment, and the facts observed with *T. imaginis* during the period of the year when the population was declining can be fitted into the general theory. But this interpretation requires a restatement of the theory with respect to detail since Nicholson (1933, pp. 135–6) clearly excludes climate from the list of possible 'density-dependent factors'.

8. SUMMARY

1. Daily counts of the numbers of *Thrips imaginis* in roses in the garden at the Waite Institute have been made continuously for 14 years. The roses used for sampling were highly attractive to the adult thrips

but were not favourable for the development of immature stages. They served as a 'trap' which gave a satisfactory indication of the density of the population in the area.

2. The method of partial regression was used to measure the degree of association between the numbers of thrips present during the spring and the weather experienced during the preceding months.

3. With the aid of precise knowledge of the biology of *T. imaginis* and the weather of the Adelaide area it was possible to select four components of the physical environment which were likely to be closely associated with the density of the thrips population in the spring. The analysis showed that 78 % of the variance of the population could be related to these four quantities.

4. The maximum density attained by the population in the spring is largely determined by the weather during the preceding autumn, but rainfall and temperature during the early spring may modify the tendencies established in the autumn.

5. 'Competition' plays little or no part in determining the maximum density attained in the spring. Rapid multiplication is possible only for a limited period during spring and early summer. During this period the insects increase rapidly but so does the number of situations available to them; and the favourable period normally ends long before the thrips have had time to saturate the environment. The onset of summer drought brings about a 'crash' and the numbers decline to the low level characteristic of late summer and winter. The 'balance' of the population is seen as a race against time with the increase in density being carried further in those years when the favourable period lasts longer, but never reaching the point where competition begins to be important. The annual fluctuations in the maximum density attained by the population are controlled almost entirely by 'density-independent' components of the environment.

6. During the period each year when the population is decreasing weather is still the most important influence at work to determine the density of the population. The population never decreases to zero, partly because during this phase of the 'population rhythm' weather operates as a 'density-dependent' component of the environment, and partly because the spring supervenes and the population begins increasing under the influence of the changing environment.

7. It was also shown in the earlier part of the paper that the movement of the thrips into the flowers is influenced by the daily weather. The movement was greater during warm dry days and less during cold wet ones.

9. ACKNOWLEDGEMENTS

It is a pleasure to acknowledge the help of Mr E. A. Cornish. With great patience and sympathy he guided us through the difficulties of statistical analysis throughout, but particularly in the earlier stages when it was necessary to sort and rearrange the masses of data which have gone into the making of this paper.

Mr L. C. Birch helped with criticism, particularly at the manuscript stage. The very heavy burden of routine calculation was carried largely by Miss Ruth Burns to whom we are deeply indebted. The diagrams were prepared by Miss Helen Brookes.

REFERENCES

Andrewartha, H. G. (1934). J. Coun. Sci. Industr. Res. Aust. 7: 239–44.
Andrewartha, H. G. (1935). J. Coun. Sci. Industr. Res. Aust. 8: 281–5.
Andrewartha, H. G. (1944). Trans. Roy. Soc. S. Aust. 68: 315–26.
Andrewartha, H. G. & Birch, L. C. (1948). Nature, Lond., 161: 447–8.
Andrewartha, H. V. (1936). J. Coun. Sci. Industr. Res. Aust. 9: 57–64.
Chapman, R. N. (1928). Ecology, 9: 111–22.
Davidson, J. (1934). Bull. Coun. Sci. Industr. Res. Aust. No. 79: 1–66.
Davidson, J. (1936). J. Agric. S. Aust. 39: 930–9.
Davidson, J. (1938). Trans. Roy. Soc. S. Aust. 62: 141–8.
Davidson, J. (1944). Aust. J. Exp. Biol. Med. Sci. 22: 95–103.
Davidson, J. & Andrewartha, H. G. (1948). J. Anim. Ecol. 17: 193–9.
De Bach, P. & Smith, H. S. (1941). Ecology, 22: 363–9.
Evans, J. W. (1932). Pamphl. Coun. Sci. Industr. Res. Aust. No. 30: 48 pp.
Evans, J. W. (1934). J. Coun. Sci. Industr. Res. Aust. 7: 61–9.
Fisher, R. A. (1941). 'Statistical methods for research workers.' Edinburgh, etc. (8th ed.).
Gause, G. F., Smaragdova, N. P. & Witt, A. A. (1936). J. Anim. Ecol. 5: 1–18.
Holdaway, F. G. (1932). Ecol. Monogr. 2: 262–304.
Nicholson, A. J. (1933). J. Anim. Ecol. 2: 132–78.
Nicholson, A. J. & Bailey, V. A. (1935). Proc. Zool. Soc. Lond. 551–98.
Pearl, R. (1926). 'The biology of population growth.' London.
Smith, H. S. (1939). Ecol. Monogr. 9: 311–20.
Snedecor, G. W. (1937). 'Statistical methods applied to experiments in agriculture and biology.' Ames, Iowa.
Trumble, H. C. (1937). Trans. Roy. Soc. S. Aust. 61: 41–62.
Williams, C. B. (1940). Trans. R. Ent. Soc. Lond. 90: 227–306.

and their drainage basins; hydrophysical approach to quantitative morphology. Bull. Geol. Soc. Amer. 56: 275-370.

Hubbs, C., R. A. Kuehne and J. C. Ball. 1953. The fishes of the upper Guadalupe River, Texas. Tex. Jour. Sci. 5: 216-244.

Klugh, A. B. 1923. A common system of classification in plant and animal ecology. Ecology 4: 366-377.

Kuehne, R. A. and R. M. Bailey. 1961. Stream capture and the distribution of the percid fish, *Etheostoma sagitta*, with geologic and taxonomic considerations. Copeia 1961: 1-8.

Margalef, Ramon. 1960. Ideas for a synthetic approach to the ecology of running waters. Internat. Revue ges. Hydrobiol. 45: 133-153.

Odum, E. P. 1959. Fundamentals of Ecology. Philadelphia, W. B. Saunders.

Pearse, A. S. 1939. Animal Ecology New York, McGraw-Hill.

Reid, G. K. 1961. Ecology of Inland Waters and Estuaries. New York, Reinhold.

Ruttner, F. 1953. Fundamentals of Limnology. Toronto, Univ. of Toronto Press.

Shelford, V. E. 1911. Ecological succession. I. Stream fishes and the method of physiographic analysis. Biol. Bull. 21: 9-35.

Stehr, W. C. and J. W. Branson. 1938. An ecological study of an intermittent stream. Ecology 19: 294-310.

Strahler, A. N. 1954. Quantitative geomorphology of erosional landscapes. C.-R. 19th Intern. Geol. Cong., sec. 13: 341-354.

Strahler, A. N. 1957. Quantitative analysis of watershed geomorphology. Trans. Amer. Geophys. Union 38: 913-920.

Thornbury, W. D. 1954. Principles of Geomorphology. New York, John Wiley & Sons.

Welch, P. S. 1952. Limnology. New York, McGraw-Hill.

ENERGY FLOW IN THE SALT MARSH ECOSYSTEM OF GEORGIA[1]

JOHN M. TEAL

Woods Hole Oceanographic Institution, Woods Hole, Massachusetts

INTRODUCTION

Along the coast of the United States from northern Florida to North Carolina runs a band of salt marsh bordered on the east by a series of sea islands and on the west by the mainland. The Marine Institute of the University of Georgia was established on one of these islands, Sapelo, and has tended to focus attention on the marsh. Several studies have provided data from which it is now possible to construct a picture of the energy flow through the organisms of this marsh.

Reasonably detailed studies of the energy flow, or trophic level production have been limited to a few natural ecosystems. These include Cedar Bog Lake, reported in the pioneer work of Lindeman (1942), and 2 fresh-water springs (Odum 1957, Teal 1958). There have been a number of studies of the energetics of laboratory populations (Richman 1958, Slobodkin 1959), and some theoretical comments upon energetics of populations and ecosystems (e.g. Patten 1959, Slobodkin 1960), but work on even the broad details of energy flow in natural ecosystems has lagged.

The present paper draws heavily upon the work of others. The authors are cited in the appropriate places but I wish here to express my appreciation for their cooperation.

[1] Contribution No. 38 from the University of Georgia Marine Institute, Sapelo Island, Georgia. This research was supported by funds from the Sapelo Island Research Foundation and by N.S.F. grant G-6156.

The physical and chemical features of the marsh have been described (Teal 1958, Teal & Kanwisher 1961) but I will briefly define 5 regions into which the marsh was divided in many of these studies (Figure 1).

Creek bank: muddy and/or sandy banks of tidal creeks between low water and the beginning of *Spartina* growth.

Streamside marsh: an area 1-3 m wide of closely spaced, tall *Spartina* located just above the bare creekbank.

Levee marsh: *Spartina* of intermediate height spacing atop the natural levees bordering the creeks.

Short-*Spartina* marsh: flat areas behind the levees with short, widely spaced *Spartina*.

Salicornia marsh: sandy areas near land where plants other than *Spartina* occur, among which *Salicornia* is conspicuous.

The relative areas of these various marsh types were measured on aerial photographs (Table V) to enable calculation of averages for the marsh as a whole.

The marsh fauna

Animals living in the marsh must be able to survive or avoid the great changes in salinity, temperature and exposure. Salinity of water flooding the marshes varies from 20 to 30 o/oo with values as low as 12 o/oo recorded in heads of creeks just after heavy rains. Salinity of water in the mud may be 5 o/oo in isolated areas where fresh water drains from the islands and 70 o/oo in isolated low areas during rainless summer periods. An average aquatic or soil animal must be able to withstand variations from 20 to 30 o/oo but probably escapes greater extremes by burrowing in the mud and/or migrating short distances.

The limited number of animals which have adapted to these extremes are relatively free from competing species and enemies. For example, mussels living in the marsh are bothered by neither snails nor echinoderms, which take great toll of the estuarine bivalves living only a few meters away. Ants and grasshoppers are each represented by only one common species, *Crematogaster clara* and *Orchelimum fidicinium* respectively, which is quite abundant in marsh areas optimal for it. Once adapted to the marsh, the lack of competition from similar animals has perhaps allowed them to occupy a broader niche and be more abundant than would otherwise be possible.

TABLE I. Known macro-fauna of a Georgia salt marsh listed by groups according to distribution and origin

(1a) Terrestrial species living in marsh
Orchelimum fidicinium Rehn & Hebard
Ischnodesmus sp.
Prokelisia marginata (Van Duzee)
Liburnia detecta Van Duzee
Tabanus spp.
Culicoides canithorax Hoffman
Dimicoenia spinosa (Loew)
Plagiopsis aneo-nigra (Loew)
Parydra vanduzeei (Creeson)
Chaetopsis aenea (Wiedermann)
Chaetopsis apicalis Johnson
Haplodictya setosa (Coquillett)
Mordellid sp.
Crematogaster clara Mayr
Camponotus pylartes fraxinicola M. R. Smith
Hyctia pikei Peckham
Seriolus sp.
Lycosa modesta (Keyserling)
Philodromus sp.
Grammonata sp.
Hyctia brina (Hentz)
Eustala sp.
Singa keyserlingi McCook
Tetragnatha vermiformis Emerton
Dictyna sp.
Rallus longirostris Boddaert
Termatodytes palustris (Wilson)
Ammospiza caudacuta (Gmelin)
A. maritima (Wilson)
Oryzomys palustris (Harlan)
Procyon lotor (Linne)
Mustela vison Schreber

(1b) Terrestrial or fresh-water species only on landward edge of marsh
Pachydiplax longipennis Burmeister
Pantala flavescens Fabrisius
Erythrodiplax verenice Drury
Anax junius Drury
Erythemis simplicicollis Say
Orphylella sp.
Platunus cincticollis (Say)
Kinesternum s. subrubrum (Lacepede)
Lutra canadensis (Schreber)

(2a) Estuarine species limited in marsh to low water level
Bouganvillia carolinensis (McCrady)
Campanularid sp.
Oerstedia dorsalis burger
Nolella stipata Gosse
Eteone alba Webster
Autolytus prolifer (O. F. M.)
Polydora ligni Webster
Heteromastis filiformis (Claparede)
Crassostrea virginica (Gmelin)
Mercenaria mercenaria (Linne)
Tagelus plebeius Solander
T. divisus Spengler
Mulinia lateralis Say
Epitomium rupicolum (Kurtz)
Balanus improvisus Darwin
Microprotopus maculatoides Shoemaker
Paracaprella sp.
Crangon heterochelis (Say)
Clibanarium vittatus (Bosc)
Molgula manhattensis (DeKay)

(2b) Estuarine species in streamside marsh
Nassarius obsoletus Say
Chthamalus fragilis Darwin
Neomysis americana (S. I. Smith)
Leptochelia rapax Harger
Cassidisca lumifrons (Richardson)
Gammarus chesapeakensis Bousfield
Melita nitida Smith
Callinectes sapidus Rathbun
Panopeus herbstii Milne-Edwards
Eurypanopeus depressus (Smith)
Malaclemys terrapin centrata (Latreille)

(2c) Estuarine species occurring well into marsh
Neanthes succinea (Frey & Leuckart)
Laeonereis culveri (Webster)
Streblospio benedicti Webster
Capitella capitata (Fab.)
Orchestia grillus (Bosc.)
O. Platensis Kroyer

(3a) Aquatic marsh species with planktonic larvae
Manayunkia aestuarina (Bourne)
Modiolus demissus Dillwyn
Polymesoda caroliniana Bosc
Littorina irrorata (Say)
Littoridina tenuipes (Couper)
Eurytium limosum (Say)
Sesarma reticulatum (Say)
S. cinereum (Bosc)
Uca pugilator (Bosc)
U. minax (LeConte)
U. pugnax (S. I. Smith)

(3b) Aquatic marsh species living entirely within marsh
Oligochaetes-3 spp.
Melampus bidentatus Say
Cyathura carinata (Kroyer)
Orchestia uhleri Shoemaker

Table I lists the marsh fauna divided into several groups: (1) typically terrestrial insects and arachnids subdivided into those occurring throughout the marsh and those confined to the landward edge, (2) the aquatic species with their center of abundance in the estuaries and 2a confined to regions near low water, 2b occurring in the streamside marsh, or 2c occurring throughout the marsh, (3) marsh species derived from aquatic ancestors with their centers of distribution within the marsh which are subdivided into those with planktonic larvae and those that spend their entire life cycle in the marsh.

The list shows that of the aquatic species, 33 or 60% are in groups 2a and 2b, estuarine forms that have managed to colonize the lowest portion of the salt marsh. They can survive only where periods of exposure at low tides are short. The individuals in the marsh are living at one edge of their species' distribution and their numbers are maintained by migrations from the surrounding waters. Those living above the mud are especially subject to damage by extremes of weather, and species that have penetrated farthest into the marsh are burrowers. The remaining aquatic species are either tolerant enough to inhabit the entire marsh although they are most common in the estuaries, group 2c, or are most common in the marshes themselves. But even among the latter, only 6 do not spend part of their life cycle in the estuaries. These are isopods, amphipods, oligochaetes and the pulmonate snail, *Melampus*, the last 2 of which are derived from fresh water or terrestrial rather than marine ancestors.

That part of the fauna derived from the land is at present the least well known, but the species so far encountered comprise nearly half of the marsh animals. They are, however, far less important in the energetics of the community than their aquatic counterparts, as will be seen below.

Most of the terrestrial species have made only slight adaptation to the marshes. They breath air and resist salinity changes and desiccation by means of their impervious exoskeleton. Most of the larger forms climb the marsh grass to escape rising tides, but can climb under water to seek refuge from birds. Some insects, such as the ant, *Crematogaster*, are easily drowned, but live within *Spartina* stems and can effectively plug the entrance to their nests.

The distribution of marsh species can be seen in Figure 1. The data are from the samples reported in this paper taken at the sites indicated as well as other samples and general collecting. Most of the insect and spider samples are from Smalley (unpublished). Estuarine species occur mostly near low water and in drainage channels and are reduced on the levees which may dry out between spring tides. The aquatic marsh species are distributed relatively evenly throughout the marsh. Terrestrial species are common throughout the grassy areas, with more species present on the higher ground and in the taller marsh-grass.

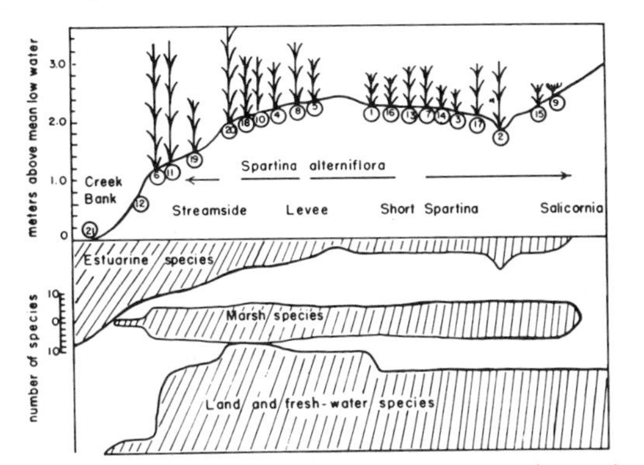

FIG. 1. Representative section of a Georgia salt marsh with horizontal scale distorted non-uniformly. Sample sites indicated by circled numbers. Site 2 represents the beginning of a drainage channel, not an isolated low spot. Symbols for grass are drawn to correct height for average maximum growth at those sites. The number of species of animals of 3 groups listed in Table I are plotted against sample sites. Names of marsh types used herein are also indicated.

The food web

The herbivorous faunas of many ecosystems can be divided into 2 groups, those which feed directly on living plants and those which feed on plants only after the plants have died and fallen to the ground (Odum and Smalley 1959). The marsh fauna may be grouped in a similar manner (Figure 2).

A group of insects lives and feeds directly upon the living *Spartina*: *Orchelium,* eating the tissues, and *Prokelisia,* sucking the plant juices. These and their less important associates support the spiders, wrens, and nesting sparrows. A different group lives at the level of the mud surface and feeds on the detritus formed by bacterial decomposition of *Spartina* and on algae. These mud dwelling groups function mostly as primary consumers, although the detritus also contains animal remains and numbers of the bacteria that help break the *Spartina* into small pieces. The carnivores preying on the algal and detritus group are principally mud crabs, raccoons, and rails.

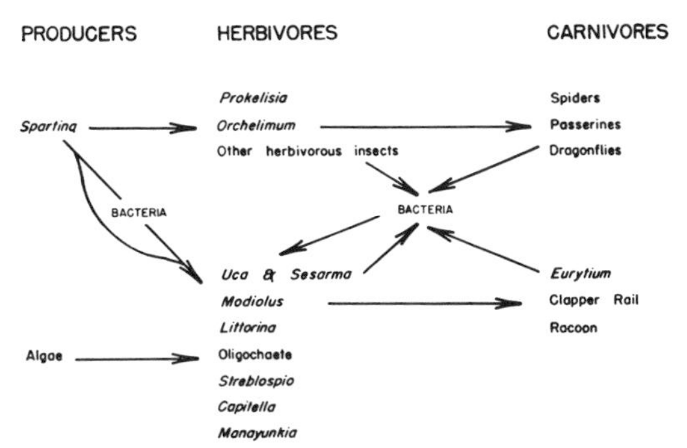

FIG. 2. Food web of a Georgia salt marsh with groups listed in their approximate order of importance.

The species of the detritus-algae feeding group that are important in the economy of the marsh are the fiddler crabs, oligochaetes, *Littorina,* and the nematodes among the deposit feeders, and *Modiolus* and *Manayunkia* among the suspension feeders. Thus, the community consists of 2 parts, one deriving its energy directly from the living *Spartina* and the other deriving its energy from detritus and algae.

Energy Flow by Trophic Groups

Methods

Some populations were sampled completely enough that production could be measured directly, e.g. *Spartina* and grasshoppers. In other cases production could not be determined from the sampling but was estimated either from turnover time, in which case production equals one maximum population per turnover period, or by assuming that the ratio between respiration and production for the group in question is 0.25 to 0.30 as has been found for other groups (Teal 1958, Slobodkin 1960). Respiration in air was the measure of energy degradation. Some of the mud dwelling forms live in completely anaerobic conditions (Teal & Kanwisher 1961), e.g. nematodes, and the energy degradation under anaerobic conditions is assumed to be equal to that in air. This is supported only by the observation of Wieser and Kanwisher (1961) that nematodes from anaerobic muds are as active under anaerobic conditions as under aerobic.

Primary production

The only higher plant of importance on the salt marsh is *Spartina alterniflora.* It grows over the entire marsh, is eaten by insects, then dies, decomposes and as detritus furnishes the food for much of the remaining fauna of the marsh. Smalley (1959) measured production of *Spartina* by harvesting and weighing plants at monthly intervals. Teal and Kanwisher (1961) measured respiration. Since the net production was determined by short-term harvesting, it is necessary to add the 305 kcal/m² yr consumed by the insects (see below) to arrive at the true net production. Table II shows that net production of *Spartina* comes to only 19% of gross production. There are indications (Odum 1961) that the production values are underestimated, but not the standing crops upon which respiration values are based. Furthermore, the *Spartina* used to measure respiration was collected in spring. If there is appreciable acclimation to temperature by this species, then summer values will be lower and winter values higher than indicated, but as summer contributes twice as much (4 vs. 2 months), an adjustment would lower the total figure for respiration. But assuming perfect summer acclimation would bring net production to only 24% of gross production.

TABLE II. Data for *Spartina* in Georgia salt marhes. Production figures from Smalley (1959), respiration rates from Teal and Kanwisher (1961)

Season		Short Spartina 42% of total area	Levee-Streamside 58% of total area
Winter 2 mo at 10°	Standing crop Respiration	300 fresh g/m² 235 kcal/m²	750 fresh g/m² 580 kcal/m²
Spring 3 mo at 17.5°	Standing crop Respiration	600 g/m² 1250 kcal/m²	1350 g/m² 2800 kcal/m²
Summer 4 mo at 26°	Standing crop Respiration	705 g/m² 6450 kcal/m²	3225 g/m² 29600 kcal/m²
Autumn 3 mo at 20°	Standing crop Respiration	900 g/m² 3240 kcal/m²	1800 g/m² 6480 kcal/m²
Production		2570 kcal/m² yr	8970 kcal/m² yr

Marsh average: net product on = 6580 kcal/m² yr
respiration = 28000 kcal/m² yr

gross production = 34580

In addition to *Spartina,* the algae living on the surface of the marsh mud contribute 1800 kcal/m² yr gross production and not less than 1620 kcal/m² yr net production (Pomeroy 1959).

Decomposition of *Spartina*

Before the *Spartina* is available to most of the marsh consumers it must be broken down by bacteria. Part of the *Spartina* crop decomposes in place on the marsh, especially that portion in the Short Spartina areas not subject to strong tidal currents. *Spartina* from the streamside marsh, however, is swept off by the water and carried back and forth until it is either decomposed in the water, stranded, or carried out to sea. Material stranded on the beach and representative of that carried out of the system consists mostly of stalks. The leaves have decomposed before leaving the marsh-estuarine system.

Burkholder and Bornside (1957) found that when marsh grass was confined in cages in a tidal creek over the winter, one-half of the dry weight was broken up and washed away after 6 months, by which time only the stems remained.

Pieces of dead standing *Spartina* were collected in mid-winter when decomposition was starting. Algae were not present, hence oxygen consumption was a measure of bacterial activity. From the initial rate of decomposition measured at 15° C, it was estimated that leaves submerged at every tide would be completely consumed in 2 months, stems in 3½ months. Leaves and stems that dry

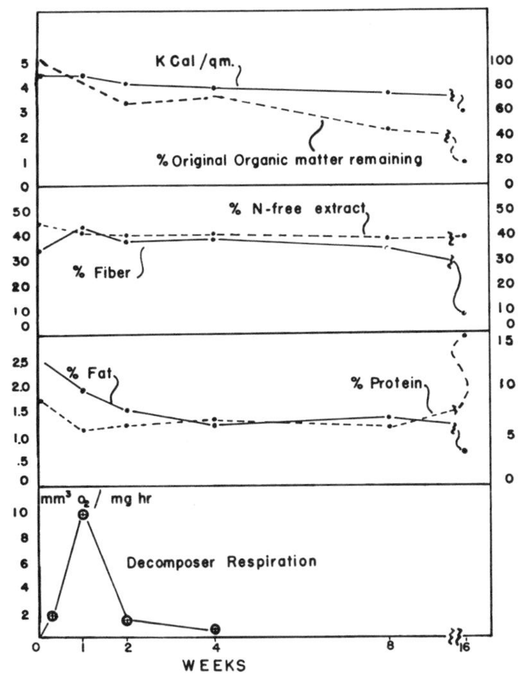

FIG. 3. Decomposition of marsh grass in sea water showing changes in oxygen consumption and composition of grass-bacteria mixture. Line between 8-week point and break shows correct slope.

out in periods between spring tides would last about twice as long.

To find out what changes in composition occur during the breakup of *Spartina,* 10 g of finely-dropped air-dried marsh grass were placed in 500 ml flasks with 200 ml sea water, inoculated with 1 ml of marsh mud and placed on a shaker in the dark at 20° C. The oxygen content of the flasks was not measured but in no case did the material go completely anaerobic. After periods of 0, 2, 7, 14, 28, 56 and 112 days samples were removed. Oxygen consumption of 2 samples was measured and the material from 4 flasks was lumped for analyses of moisture, fat, protein, crude fiber, nitrogen-free extract, ash and caloric content made by Law and Company, Atlanta, Georgia. The results are shown in Figure 3.

Respiration in the flasks rapidly reached a maximum as bacteria grew on the material liberated from crushed cells by the chopping. In 2 weeks this phase passed and bacterial action remained low for the remainder of the experiment. The initial phase is reflected in the decrease in fat and protein percentages but, while fat continued to decline slowly, protein concentration increased gradually until at 16 weeks it was twice as high as at the beginning. At the same time carbohydrates (N-free extracts) remained constant but fiber, principally cellulose, declined to less than ¼ of its initial value. The caloric content declined by 33%. During the period 82% of the organic matter was consumed.

If this is representative of what happens to the *Spartina* as it is changed from standing marsh grass to detritus then, although the total amount of material is decreased, the animal food value of what remains is increased. Bacteria attack the grass substances and convert a portion into bacterial protoplasm and in this process cellulose in the bacterial-detritus mixture decreases most swiftly and protein least swiftly.

The magnitude of the bacterial metabolism was calculated with figures for respiration of plankton in the estuarine waters and bacteria in the marsh sediments and on the standing *Spartina.* Ragotzkie (1958) found that plankton respiration averaged 1600 kcal/m^2 yr in estuarine waters. Since the turbidity of the water is high, there is very little phytoplankton and I assumed that all of this respiration represents bacterial action upon *Spartina* detritus. Since a planimetric survey of charts and aerial photos of the region showed that there is twice as much marsh as estuarine area in the system, this represents 800 kcal/m^2 yr of marsh. From Teal and Kanwisher (1961) we find that the bacteria in the marsh sediments degrade 2090 kcal/m^2 yr. The average respiration of bacteria on standing, dead *Spartina* is about 60 mm^3/gm hr, which, multiplied by the biomass of that *Spartina* (Smalley 1959) comes to 1000 kcal/m^2 yr. Thus the activities of bacteria account for 3890 kcal/m^2 yr averaged over the marsh area. This amounts to 59% of the available *Spartina.*

Besides bacteria, colorless blue-green algae are also active in the degradation of *Spartina.* Bits of partly decomposed *Spartina* were often found within the mud which was usually black just around them. Within these bits were often numerous filaments of what were apparently *Thioploca* and *Beggiatoa* or *Oscillatoria.* The algae were alive and active, living in a lightless, highly reduced environment as has been discussed by Pringsheim (1949).

Herbivorous insects

The salt marsh grasshopper and the plant hopper are the only important animals in this category. The grasshoppers respire 18.6 kcal and produce 10.8 kcal of tissue per m^2 per year which adds to an assimilation of 29.4 kcal/m^2 yr (Smalley 1960). The corresponding plant hopper figures are: respiration 205 kcal/m^2 yr, produc-

TABLE III. Summary of energy-flow for detritus-algae feeders in Georgia salt marsh

	Respiration	Production	Assimilation	Production / Assimilation
Crabs	171	35	206	17%
Annelids	26	9	35	25%*
Nematodes	64	21	85	25%*
Mussels	39	17	56	30%
Snails	72	8	80	10%
Totals	372	90	462	19.5%

* Assumed as means of calc. P.

tion 70 kcal/m^2 yr and assimilation 275 kcal/m^2 yr (Smalley 1959). Production of plant hoppers was not measured but calculated on the assumption that the ratio of production to assimilation equals 25%. The grasshoppers assimilate only about 30% of what they ingest which means they produce nearly 70 kcal/m^2 yr of fecal matter which can probably serve as food for some of the other marsh inhabitants.

Detritus-algae feeders

Table III gives a summary of the results for the various groups of animals that feed at the surface of the mud-eating Spartina-detritus, algae and to a lesser but unknown extent each other. The groups are considered in turn below.

Crabs

Uca pugilator, U. pugnax, and *Sesarma reticulatum* are the most conspicuous consumers in the marsh. One or more of these species is present in all parts of the marsh. They feed on the surface of the mud for the most part, picking up clawfuls of mud, sorting it with their mouthparts, spitting the rejected material into a claw and depositing it back on the mud and swallowing the remainder. If the spit is compared with the undisturbed mud surface, it is apparent that most of the sorting consists of rejecting larger particles. Spit from *U. pugilator* which live on sand consists of sand from which the smaller particles, the algae and detritus, are gone. In spit from *U. pugnax* feeding on mud rich in diatoms and nematodes most of the larger diatoms and nematodes were no longer distinguishable but the average particle size of the spit was larger than that of the mud and contained many bits of diatom shells. Apparently the large diatoms and nematodes were crushed and then the finer particles were swallowed and the larger ones rejected.

By comparing the amount of feces produced in a few hours by freshly collected animals with the normal rates of respiration and growth, the portion of the ingested food actually assimilated was estimated. Four measurements during the winter on groups of from 5 to 16 animals gave values from 23 to 31%. One measurement on 57 crabs in August gave a value of 75%. Algae and detritus are scarcer in summer than winter on the areas where these crabs were collected and apparently they assimilate a larger part of the digestible material when it is scarce. A parallel situation is found in copepods (Marshall and Orr 1955).

In areas of dense crab populations the entire surface of the marsh is worked over between successive high tides. The feces produced by the crabs feeding on muddy substrates contain about one-third more calories per gram than the mud; feces produced by crabs on sand about 10 times as many calories as the sand. Both by the working over of the marsh surface and the concentration of organic matter in their feces the crabs will have considerable influence upon other organisms, especially the nematodes, annelids and bacteria.

The crab populations were sampled by placing metal rings, 30 cm high, on the marsh while the tide was in and the animals were in or near their burrows, returning at low tide and removing

TABLE IV. Crab populations in g/m^2 in Georgia salt marshes followed by standard errors. Sizes: s=0-150 mg; m=150-500 mg; 1=>500 mg

Marsh type	Size	Winter	Spring	Summer	Autumn	Species
Creek bank	s	0	?	4.89± 1.04	0	*Uca pugilator*
	m	0	0	3.67±0.75	0	
	l	0	0	33.1 ±5.6	0	
Stream-side	s	0	0	4.8 ±2.8	0	*Sesarma reticulatum*
	m	0.7 ±0.4	3.65± 1.15	7.35± 1.15	*	
	l	10.2 ±1.5	*	*	*	
Stream-side	s	0	0	2.6 ±1.6	0	*Eurytium limosum*
	m	0	0	2.85±0.45	*	
	l	8.7 ±1.44	*	*	*	
Levee	s	0.5 ±0.3	?	0.8 ±0.8	0.41±0.14	*Sesarma reticulatum*
	m	0.5 ±0.5	0.2 ±0.2	2.34±0.45	1.4?	
	l	12.4 ±4.35	12.4 ±4.33	22.1 ±0.8	17.0?	
Levee	s	0.81±0.36	2.27±0.68	5.74±0.93	2.47±0.67	*Uca pugnax*
	m	3.0 ±1.5	9.8 ±2.1	7.14±0.65	5.0?	
	l	16.7 ±3.4	32.5 ±9.0	54.3 ±9.5	35.5?	
Levee	l	15.6 ±2.9	*	*	*	*Eurytium limosum*
Short Spartina	s	1.0 ±0.45	0	3.16±0.71	1.0 ±0.45	*Uca pugnax*
	m	2.30±0.60	1.95±0.65	5.00±0.85	3.65?	
	l	16.25±3.90	9.5 ±2.0	12.2 ±1.70	14.25?	
Salicornia-Distichlis	s	?	?	1.91±0.43	8.13±1.17	*Uca pugilator*
	m	3.6 ±0.5	*	*	*	
	l	114.9 ±20.4	*	*	*	

? indicates no samples were taken, numbers if entered are interpolations. When data for several seasons were pooled, the mean appears, followed by asterisks for other seasons involved in the average.

TABLE V. Respiration of Georgia salt marsh crabs by marsh type. Values are kcal/m²/season

Marsh type	Species	Winter	Spring	Summer	Autumn	Total	% by adults
Creek bank	U. pl.	0	0	69.5	0	69.5	66%
Streamside	S. r.	2.9	10.5	66.6	17.4	97.3	37%
Levee	S. r.	3.8	8.6	49.4	17.0	78.7	81%
	U. px.	5.9	50.0	139.6	51.2	246.6	67%
Short spartina	U. px.	5.5	11.6	48.0	23.3	88.4	56%
Salicornia	U. pl.	37.3	80.6	170	100	388	90%

everything within the ring, separating, counting and weighing the crabs. Rings of $\frac{1}{5}$ m² were used for adults which were picked out by hand but rings of only 0.018 m² were used for the young, which had to be separated by sieving. Crabs of more than 150 mg were ignored in the small samples and vice versa.

The sampling results are shown in Table IV. In general the biomass of crabs follows the same distribution as the numbers of species of marsh animals (Fig. 1). The *U. pugilator* on the creek bank are apparently completely killed during the autumn and replaced the following spring. It seems unlikely that they migrate as they are confined to sandy substrates (Teal 1958) and the Streamside, Levee and Short Spartina marshes are uniformly muddy (Teal and Kanwisher 1961). The values for the Salicornia marsh must be divided by 4 since only about ¼ of the area is occupied by plants and crab burrows, the rest being open sand flats not included in the samples where the crabs feed but do not live.

Table V lists values for respiration of the crabs by marsh types. Respiratory rates are from Teal (1959). The last column gives an idea of the relative importance of the large and small individuals to the population's energy degradation.

By assuming that the crab populations replace themselves annually the energy flow figures in Table VI were calculated.

TABLE VI. Energy flow of detritus-eating crabs in a Georgia salt marsh. Data in kcal/m²/yr

Marsh type	Relative area	Respiration	Production
Creek-bank	10%	70	28
Streamside	10%	97	19
Levee	35%	325	65
Short Spartina	40%	88	14
Salicornia	5%	97	20

Average respiration	171
Average production	35
Average assimilation	206
Production efficiency	17%

TABLE VII. Summary of annelid sampling in Georgia salt marsh

Marsh type	NOV-DEC SAMPLE		JUL-AUG SAMPLE	
	N	g/m²	N	g/m²
Creek bank	8	2.0 ±2.0	7	0.95 ±0.25
Streamside	15	2.9 ±0.6	10	1.0 ±0.53
Levee	35	1.8 ±0.28	38	2.2 ±0.32
Short Spartina	39	2.2 ±0.95	8	0.5 ±0.12
Salicornia	5	0.2 ±0.2	2	0.0
Marsh average		2.01 ±0.46		1.16 ±0.14

Annelids

The annelids in the marsh are mostly deposit feeders, feeding either from fixed burrows or working their way through the sediments like the oligochaetes, although *Manayunkia* is a filter-feeder. Although *Neanthes*, because of its jaws, might be thought to be predaceous, guts examined at different seasons revealed only diatoms, detritus, and mud and sand.

The annelids were sampled once in November-December and once in July-August. Five samples were taken with a plastic coring tube at spots chosen by taking a pair of random numbers, one indicating the distance north and the other, the distance east of a stake marking the southwest corner of a square meter plot selected at random from the marsh as a whole. The sites are indicated in Fig. 1 and may be seen on an aerial photograph of the marsh in Teal and Kanwisher (1961). Each core was divided into 3 parts at 2 cm intervals and the portion below 6 cm discarded. The annelids were removed by gentle washing in sea water in a sieve with 16 meshes/cm. All samples were examined within 24 hours of collection. Biomass was calculated by multiplying the average weight of each species by the number found in the sample. Average weights were determined for the more common species by weighing on a quartz helix as well as by measuring length and width and calculating the weight based on a specific density equal to sea water. The methods agreed within 15%. Insect larvae collected with the annelids are included in the figures (Table VII).

Table VIII shows the numbers of the most common annelids and insect larvae from selected representative sites. *Capitella*, the oligochaetes, *Streblospio* and *Manayunkia* made up most of the biomass, usually in that order. Two of these, *Capitella* and *Streblospio*, are characteristic not of the marshes but of the estuaries, indicating that the annelids have had to make relatively little adaptation to marsh life. In general they are most

Table VIII. Numbers of selected annelids and insect larvae/0.01 m² in representative marsh types in a Georgia salt marsh

Site	21	6	19	10	8	4	1	7	3
Winter Series									
Capitella capitata	0	10	30	70	50	30	0	10	0
Steblospio benedicti	50	150	80	40	50	0	20	30	0
Neanthes succinea	0	10	0	0	0	0	0	0	10
Manayunkia aestuarina	0	0	40	90	0	290	0	0	0
oligochaete	10	10	170	80	30	80	40	40	0
dipteran larvae	0	0	10	0	10	10	0	0	0
Summer Series									
Capitella capitata	30	26	14	56	136	38	12	30	0
Streblospio benedicti	14	38	40	6	4	0	4	0	0
Neanthes succinea	0	0	0	0	2	0	0	0	0
Manayunkia aestuarina	8	10	12	52	0	60	0	0	0
oligochaete	78	10	60	88	88	66	30	32	20
dipteran larvae	2	0	0	0	0	2	0	0	0

numerous in the most productive parts of the marsh as are other animals, except that they are somewhat scarcer on the highest and thus the driest parts of the levees.

Energy flow for the annelids was calculated on the basis of an average respiration rate of 400 mm³/gm hr (Zeuthen 1953) and a production equal to 25% of assimilation. The latter results in a turnover time of 1.6 months which is reasonable since the animals are between 20μ gm and 200μ gm in weight.

Nematodes

Using the relative areas of various marsh types from Table V, and the weights of nematodes from Teal and Wieser (1961) I calculate that there are about 2.76 g fresh weight/m². The samples were all taken in spring so it must be assumed that the nematode biomass does not change appreciably throughout the year. Wieser and Kanwisher (1961) found that there was only slightly more than a 2-fold variation in a marsh at Woods Hole, Mass. where the climate is considerably colder and more variable. Using the average respiratory rate of 540 mm³/fresh gm/hr (Teal and Wieser 1961) the nematodes would respire an average of 64 kcal/m² yr. Assuming their production to equal 25% of assimilation, production would be 21 kcal/m² yr which would amount to a turnover of the population every 1.6 months. This may be compared with turnovers of 1 year and 1 month quoted by Wieser and Kanwisher (1961) and Nielsen (1949) respectively.

Snails and Mussels

The mussels (Kuenzler 1961) respire on the average 39 kcal/m² yr and produce 17 kcal/m² yr. The small snails of the current year class had a production equal to 14% of their assimilation but Smalley's population data did not provide a production value for the adults which grew very slowly. Since there is some growth of adult snails as well as the formation of gametes, which amounts to ⅙ of the total production of the mussels (Kuenzler 1961), the snail production must be something less than 14%, and 10% is used here.

Secondary consumers

The population data for the mud crab, *Eurytium limosum*, were presented in Table III. An average respiration value of 21.9 kcal/m² yr was derived from data of Teal (1959) and a production of 5.3 kcal/m² yr was calculated in the same manner as for the other crabs. For the Clapper Rails, a production of 0.1 kcal/m² yr was calculated from the data of Oney (1954) for population and mortality, and an average respiration of 1.6 kcal/m² yr was calculated on the basis of the weight-metabolism curve of warm blooded animals in Hemmingsen (1950). So figured the total assimilation of the carnivores comes to 30.6 kcal/m² yr divided into 25.1 kcal/m² yr respiration and 5.5 kcal/m² yr production. The raccoons in the marsh have not been studied but on the basis of general observation during 4 years when I was in the marsh nearly every week, they are considered to have an assimilation equal to that of the rails.

As yet there is no study of the carnivorous birds and spiders that feed on the marsh insects. For purposes of this calculation they are assumed to take the same portion of their prey as do the predators feeding on the detritus-algae eaters.

Community Energy Flow

Figure 4 shows the energy flow for the marsh system calculated in the preceding sections. The value for light input is from Kimball (1929). The amount of light that is actually intercepted by the plants is unknown but it is obvious from colored aerial photographs that a considerable portion of the mud is exposed especially in the parts of the marsh not close to a creek. Nevertheless, gross production is 6.1% of the incident light energy. This may be compared with values of from 0.1 to 3.0% reported for various fresh-water and marine areas (Odum and Odum 1959, 1955). The salt marsh occupies a highly advantageous position where nutrients are plentiful and circulation is supplied by the tides.

However, as noted above, a large part of the gross production is metabolised by the plants themselves and net production over light is a little less than 1.4%. This is still high compared to other

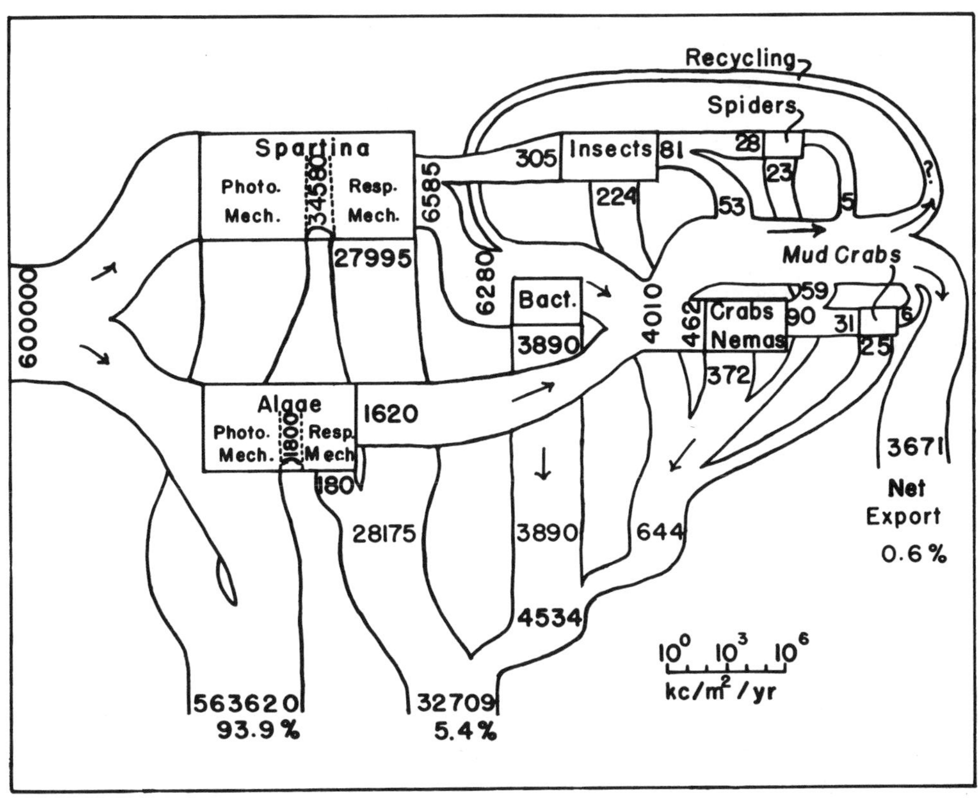

FIG. 4. Energy-flow diagram for a Georgia salt marsh.

systems although values as high as 6.0% have been reported for pine plantations (Ovington 1959).

The herbivorous insects assimilate 4.6% of their potential food, the net production of *Spartina.* They eat the plants directly and there is no significant time lag between production and consumption, since Georgia is far enough south that the *Spartina* grows to some extent throughout the year. The grasshoppers feed on it during the summer whereas the leaf hoppers are most abundant during the cooler seasons.

The relation between production and consumption for the algae-detritus feeders is more complex. Mud algae production is rather constant throughout the year (Pomeroy 1959) and algal turnover is much more rapid than *Spartina* turnover. When the algae-detritus feeders utilize algae there is little or no time lag between production and primary consumption, but when they feed on *Spartina* detritus and the associated bacteria, there is a definite time lag. Detritus is produced throughout the year as the older leaves die and are broken up, but most of the *Spartina* dies in autumn and winter after seed formation. Low winter temperatures retard formation of detritus and spread the supply out into the spring. At the beginning of summer, a new supply forms as the leaves of the spring growth die and decay. Whatever the actual time relationships may be, it is certain that there is considerable delay before detritus feeders can use the *Spartina* and the longer the delay, the less food remains.

These animals compensate for the variations in detrital supply by eating algae. The most conspicuous and abundant consumers of the marsh, the fiddler crabs, have perhaps the most omnivorous food habits. They can survive on detritus, algae, bacteria or animal remains (Teal 1958), i.e. they have a very unrestricted diet.

Odum and Odum (1959) and MacArthur (1955) have suggested that the fluctuations characteristic of certain communities may be correlated with the presence of few species as in arctic or desert areas. In the salt marsh we have a system with few species but one which seems to exhibit considerable stability. In the 5 years during which these studies were carried on there were no noticeable changes in population size in any of the important animals. (The microfauna of the soil is probably relatively little affected by the weather extremes which make life difficult for the larger animals and are not considered in this argument.)

Stability is a valuable asset for an ecosystem as it minimizes disturbance which might lead to partial or total extinction. Stability will therefore have selective value and ecosystems will tend to develop more stable configurations with time (Dunbar 1960). Salt marsh faunas and floras have had long periods in which to develop and although they have been greatly affected by the considerable changes in sea level which occurred during the Pleistocene they may not be considered as youthful as arctic areas. There has therefore been sufficient time for stability to develop and for species to adapt to the marsh conditions.

MacArthur (1955) shows that a community may achieve stability by having either many species with restricted diets or fewer species with broad diets. The former alternative will permit greater efficiency and, other things equal, will be the one selected. The salt marsh has, however, the 2nd alternative. Among the detritus-algae feeders there are only a few important species and they all have a very unrestricted diet. There are also only a few species among the carnivores that prey on the detrital feeders and they also have an unrestricted diet. Among the insects there are only 2 important species and though they feed only on *Spartina,* this is the only higher plant growing on most of the marsh and so the only food available. These 2 insects are not especially restricted in other ways; they feed on various parts of the plants and in various degrees of exposure. The situation among the spiders and carnivorous insects has not been adequately investigated but there seem to be more species in these groups and perhaps more specialization in their niches.

There are 2 principal reasons why the salt marsh should have the less efficient alternative of the 2 paths to community stability. There is only one higher plant on the marsh and consequently a lack of variety of possible niches such as could be found in a forest at the same latitude. Possibly even more important is the restriction of biomass by the removal of much of the marsh production by the tidal currents. As Hutchinson (1959) has pointed out, if the total biomass is restricted, "then the rarer species in a community may be so rare that they do not exist." With these 2 limitations of the possible numbers of species that can survive in the marsh community, the only road to stability is the development of broad, unrestricted food habits such as is found in the marsh.

TABLE IX. Summary of salt marsh energetics

Input as light	600,000 kcal/m²/yr
Loss in photosynthesis	563,620 or 93.9%
Gross production	36,380 or 6.1% of light
Producer Respiration	28,175 or 77% of gross production
Net Production	8,205 kcal/m²/yr
Bacterial respiration	3,890 or 47% of net production
1° consumer respiration	596 or 7% of net production
2° consumer respiration	48 or 0.6% of net production
Total energy dissipation by consumers	4,534 or 55% of net production
Export	3,671 or 45% of net production

Table IX gives a summary of the energy flow for the system. The producers are the most important consumers in the marsh followed by the bacteria which degrade about 1/7 as much energy as the producers. The animals, both primary and secondary consumers, are a poor 3rd degrading only 1/7 as much energy as the bacteria. As far as the consumers are concerned, the situation in the salt marsh is not very different from that in other systems. But the high consumption by the producers is unusual. In a stable system such as a so-called "climax forest" consumption equals production and there is no accumulation of organic matter, but the trees are relatively unimportant consumers. Ovington (1957) gives data indicating that in mature pines respiration is something less than 10% of production. The fact that salt marsh *Spartina* respires over 70% of its production may be associated with existence in an osmotically difficult situation.

In spite of the high rate of producer respiration, net production in the salt marsh is 1.4% of incident light which is higher than in most systems studied (Teal 1957). Table IX shows that 45% of this production is lost to the estuarine waters. The fauna of the estuaries has not been quantitatively studied but the numbers of shrimp and crabs taken by the local fishery give evidence of their abundance. Since the waters of these estuaries are so turbid and well mixed that the phytoplankton spend most of their time in the dark and their net production is zero (Ragotzkie 1958), the estuarine animals must be living on the exported marsh production. There is about 1/2 as much estuarine as marsh area behind the sea islands and since 45% of the marsh production is exported to the estuaries, there can be 1.6 times as much consumer activity in the latter region as in the former.

The tides are of supreme importance in controlling the environment of the salt marshes. They limit the number of species that can occupy the system and so make it simple enough to be studied in the detail reported here. They are responsible for the high production of *Spartina,* as witnessed by the luxuriant growth along the tidal creeks as compared with that on the Short Spartina areas. At the same time the tides remove 45% of the production before the marsh consumers have a chance to use it and in so doing permit the estuaries to support an abundance of animals.

References

Burkholder, P. R., and G. H. Bornside. 1957. Decomposition of marsh grass by aerobic marine bacteria. Bull. Torrey Bot. Club **84**: 366-383.

Dunbar, M. J. 1960. The evolution of stability in marine environments. Natural selection at the level of the ecosystem. Amer. Nat. **94**: 129-136.

Hemmingsen, A. M. 1950. The relation of standard (basal) energy metabolism to total fresh weight of living organisms. Rep. Steno. Mem. Hosp. **4**: 1-58.

Hutchinson, G. E. 1959. Homage to Santa Rosalia or Why are there so many kinds of animals? Amer. Nat. **93**: 145-160.

Kimball, H. H. 1929. Amount of solar radiation that reaches the surface of the earth on land and on the sea, and methods by which it is measured. Mon. Weather Rev. **56**: 393-398.

Kuenzler, E. J. 1961. Structure and energy flow of a mussel population in a Georgia salt marsh. Limnol. and Oceanogr. **6**: 191-204.

Lindeman, R. L. 1942. The trophic-dynamic aspect of ecology. Ecology **23**: 399-418.

MacArthur, R. 1955. Fluctuations of animal populations and a measure of community stability. Ecology **36**: 533-536.

Marshall, S. M., and A. P. Orr. 1955. Experimental feeding of the copepod *Calanus finmarchicus* (Gunner) on phytoplankton cultures labelled with radioactive carbon. Pap. Mar. Biol. Oceanogr., Deep Sea Res. Suppl. **3**: 110-114.

Nielsen, C. O. 1949. Studies on the soil microfauna. II. The soil inhabiting nematodes. Natura Jutlandica **2**: 1-131.

Odum, E. P. 1961. Personal communication.

———**, and H. T. Odum.** 1959. Fundamentals of Ecology. Philadelphia: Saunders.

———**, and A. E. Smalley.** 1959. Comparison of population energy flow of a herbiverous and a deposit-feeding invertebrate in a salt marsh ecosystem. Proc. Nat. Acad. Sci.

Odum, H. T. 1957. Trophic structure and productivity of Silver Springs, Florida. Ecol. Monogr. **27**: 55-112.

———**, and E. P. Odum.** 1955. Trophic structure and productivity of a windward coral reef community on Eniwetok Atoll. Ecol. Monogr. **25**: 291-320.

Oney, J. 1954. Final report, Clapper rail survey and investigation study. Georgia Game Fish. Comm.

Ovington, J. D. 1957. Dry-matter production by *Pinus sylvestris* L. Ann. Bot. N. S. **21**: 287-314.

Patten, B. C. 1959. An introduction to the cybernetics of the ecosystem: the trophic-dynamic aspect. Ecology **40**: 221-231.

Pomeroy, L. R. 1959. Algal productivity in the salt marshes of Georgia. Limnol. and Oceanogr. **4**: 367-386.

Pringsheim, E. H. 1949. The relationship between bacteria and myxophyceae. Bact. Rev. **13**: 47-98.

Ragotzkie, R. A. 1959. Plankton productivity in estuarine waters of Georgia. Inst. Marine Sci. **6**: 146-158.

Richman, S. 1958. The transformation of energy by *Daphnia pulex*. Ecol. Monogr. **28**: 273-291.

Slobodkin, L. B. 1959. Energetics in *Daphnia pulex* populations. Ecology **40**: 232-243.

———. 1960. Ecological energy relationships at the population level. Amer. Nat. **94**: 213-236.

Smalley, A. E. 1959. The growth cycle of Spartina and its relation to the insect populations in the marsh. Proc. Salt Marsh Conf. Sapelo Island, Georgia.

———. 1960. Energy flow of a salt marsh grasshopper population. Ecology **41**: 672-677.

Teal, J. M. 1957. Community metabolism in a temperate cold spring. Ecol. Monogr. **27**: 283-302.

———. 1958. Distribution of fiddler crabs in Georgia salt marshes. Ecology **39**: 185-193.

———. 1959. Respiration of crabs in Georgia salt marshes and its relation to their ecology. Physiol. Zool. **32**: 1-14.

———**, and J. Kanwisher.** 1961. Gas exchange in a Georgia salt marsh. Limnol. and Oceanogr. **6**: 388-399.

———**, and W. Wieser.** 1961. Studies of the ecology and physiology of the nematodes in a Georgia salt marsh. Ms.

Wieser, W., and J. Kanwisher. 1961. Ecological and physiological studies on marine nematodes from a small salt marsh near Woods Hole, Massachusetts. Limnol. and Oceanogr. **6**: 262-270.

Zeuthen, E. 1953. Oxygen uptake as related to body size in organisms. Quart. Rev. Biol. **28**: 1-12.

CLIMATIC CHANGES IN SOUTHERN CONNECTICUT RECORDED BY POLLEN DEPOSITION AT ROGERS LAKE

MARGARET B. DAVIS

Department of Zoology and Great Lakes Research Division
University of Michigan, Ann Arbor, Mich.

(Accepted for publication November 2, 1968)

Abstract. Rates of deposition of pollen grains throughout late- and postglacial time were determined from the pollen concentration in radiocarbon-dated sediment. Changes by a factor of 5 or more for all except rare pollen types from one level to the next were considered significant indication of changes in the pollen input to the lake, reflecting changes in the pollen productivity of the surrounding vegetation.

Low pollen deposition rates in the oldest sediments reflect the prevalence of tundra vegetation between 14,000 and 12,000 years ago. An increase in the rate for tree pollen occurred 12,000 years ago, when boreal woodland became established. The rates continued to increase until a sudden sharp rise for white pine, hemlock, poplar, oak, and maple pollen 9,000 years ago marked the establishment of forest, similar perhaps to modern forests of the northern Great Lakes region. Pine pollen rates declined 8,000 years ago, and deciduous tree pollen became dominant. Ragweed pollen was deposited at relatively high rates 8,000 years ago, reflecting changes in the vegetation associated with the "prairie period" recorded in the Great Lakes region at this time. Subsequent changes in pollen deposition rates reflect the immigration of beech (6,500 years B.P.), hickory (5,500 years B.P.), and chestnut (2,000 B.P.) to southern Connecticut. During the past few hundred years pollen deposition rates reflect changes in the vegetation caused by disturbance by European settlers. Throughout much of postglacial time the pollen assemblages deposited at Rogers Lake are different from assemblages known from modern sediment. This makes climatic interpretation difficult and suggests that the forest associations of the region as they are recognized now are of quite recent origin.

INTRODUCTION

Fossil pollen grains from terrestrial plants occur abundantly and almost ubiquitously in Quaternary sediments. They form a continuous fossil record in the organic muds accumulating in lakes and peat bogs, providing one of the richest sources of information about the terrestrial environment of the past. Analyses of fossil pollen, presented as pollen diagrams, show the percentages of different pollen types changing characteristically from one stratigraphic level to another, defining various stratigraphic horizons, and thereby aiding in correlation of sediments. Many changes resemble differences between modern pollen assemblages associated with different vegetation types. Because these changes in pollen are widespread, occurring in contemporaneously deposited sediment of all types, they must reflect changes in ancient vegetation similar to the geographical differences in modern vegetation. Thus pollen diagrams provide evidence for a sequence of vegetation changes, induced by a series of changes in regional climate (von Post 1967).

In this paper I have sought to augment the information contained in the traditional pollen percentage diagram by estimating rates of deposition of pollen grains in the accumulating sediment. Deposition rates at different times in the past can be compared meaningfully, since the yearly influx of each pollen type is independent of the influx of other pollen types. Changes in deposition rates infer changes in species abundances of plants that cannot be inferred with certainty from pollen percentages. Furthermore, many fossil assemblages are different from pollen assemblages known in modern sediment and cannot be compared with the present. Interpretation of these pollen sequences can now be made by comparing pollen deposition rates as they change from level to level (Tsukada and Stuiver 1966). As more information becomes available on present pollen deposition rates and their relationship to modern vegetation, it may become possible to read pollen deposition rate diagrams directly as a record of density of vegetation on the landscape (Davis and Deevey 1964). Interpretations offered here have been more cautious because little is known about factors that control pollen input to and distribution within lakes (Davis 1967*a*, 1968). Nevertheless, the results have enlarged upon and strengthened previous interpretations of vegetation based on pollen percentages, providing new insights into the climatic history of southern Connecticut.

LOCATION OF SITE

Rogers Lake (Fig. 1) is a large (107 ha) lake in southern Connecticut (41° 22′ N, 72° 7′ W), 6 km east of the Connecticut River and 8 km north

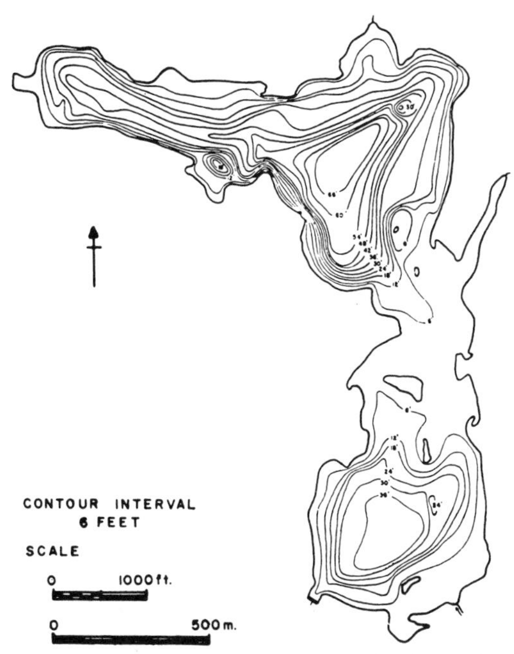

FIG. 1. Bathymetric map of Rogers Lake, Connecticut. Redrawn from Connecticut State Board of Fisheries and Game (1959).

of Long Island Sound. The lake has two basins, separated by a shoal only 2 m below the present water surface. The outlet at the south end of the lake has been dammed, raising the lake level about 1.3 m (Connecticut State Board of Fisheries and Game 1959).

Many cottages border Rogers Lake, but the surrounding landscape is heavily forested. The forest is almost entirely deciduous, with oaks making up almost half the basal area. Second to oak in abundance is red maple, followed in order by hickory, birch, and ash. Pine, beech, elm, and sugar maple are rare, and hemlock and alder still more infrequent (Davis 1965*b*, Davis and Goodlett, *unpublished data*). The forest is almost entirely second growth on abandoned farmland. Much of it has been further disturbed by cutting for cordwood and timber.

The topography of the region is irregular, with local relief of about 600 ft (ca. 200 m). Igneous bedrock is exposed at the surface on many hillsides, while other hills are covered with a mantle of glacial drift.

METHODS

Two cores of sediment, 5 cm in diameter, were collected from Rogers Lake by E. S. Deevey of Yale University and myself in September 1960, using a modified Livingstone piston corer (Deevey 1965) operated from a raft. An 11.5-m core was taken in 10.5-m water in the south basin. The upper 8 m of sediment was collected 1 or 2 m distant from the lower 4 m of sediment; correlation of the overlapping 1-m segment was confirmed by pollen analysis and radiocarbon dating. A second core was collected in 19.3-m water in the north basin of the lake. Here only 4.6 m of sediment was recovered before we encountered coarse sand.

The cores were stored in a cold room in the aluminum tubes in which they had been collected. Upon extrusion, samples were taken for pollen analysis by packing sediment into a 1-ml porcelain spatula. The remaining sediment was cut into segments, and 54 of these, varying in length from 5 to 20 cm, were submitted for age determination to Dr. Minze Stuiver of the Radiocarbon Laboratory at Yale University. Three 5-cm segments from the lowest meter of sediment collected in the north basin were also radiocarbon dated.

To assess the accuracy of the spatula method three series of five replicate samples were collected and weighed. The coefficients of variation of the mean wet weights were 11.5%, 3.4%, and 2.7%, respectively. Ninety-five per cent confidence intervals for mean wet weight per milliliter (assuming t-distribution) are $\pm 14.3\%$, $\pm 4.2\%$, and $\pm 3.4\%$, respectively. The 95% confidence interval of the pollen determinations was approximately $\pm 14\%$ of the estimated number in the sample for counts of 200 grains; where 1,000 grains were counted, it was approximately $\pm 6\%$ of the estimated number (Davis 1965*a*). The error in sample measurement is thus of the same magnitude as the error in pollen determination. If in future studies greater accuracy were desired, it would save time to measure the samples more accurately, perhaps by weighing them, rather than determining their pollen content more precisely by counting additional numbers of grains.

Samples were collected from the lowest meter of the Rogers Lake core by pushing a calibrated glass tube into the sediment. Samples collected with glass tubes were larger due to compaction than replicates collected with the spatula, and pollen counts from them were later corrected accordingly.

The 1-ml samples for pollen analysis were treated with KOH and acetolysis and, where necessary, with HF (Faegri and Iversen 1964). Assay of the pollen concentration was made by the aliquot slide method (Davis 1966). Control for loss of pollen during sample preparation was provided by adding a measured volume from a suspension of pure *Eucalyptus* pollen to the samples before preparation; the *Eucalyptus* pollen was counted together with the native pollen on the aliquot slides and the percentage recovery calculated. In only one case was the number of *Eucalyptus* pollen in prepared samples significantly different from the estimated number added. (The exception is the sample from 8.37-m depth.) The control was not in use at the time the late-glacial samples were prepared. In these samples the pollen concentration was determined using an older method (Davis 1965*a*) and cannot be accepted with equal confidence. From 200 to 1,000 grains total were counted on the aliquot slides; these counts were used both for estimation of the concentration of each type and for calculation of percentages.

Pollen from eastern white pine (*Pinus strobus*) was distinguished from pollen from the three other northeastern pine species (jack (*P. banksiana*), red (*P. resinosa*), and pitch (*P. rigida*), which were grouped as a single type) using the morphological criterion described by Ueno (1958). Microscope preparations from 11 samples were scanned under oil immersion and all pine grains were classified (Table 1); the results were extrapolated to the intervening levels. The mounting medium used was silicone fluid, so grains could be turned to improve visibility, but even so, almost one-third of the grains were torn or folded in such a way that critical identification could not be made. This proportion remained the same regardless of which type was dominant, however, indicating that failures to identify did not bias estimates of the relative abundances of the two types. Within the non-white pine species group, size will distinguish jack and/or red pine pollen from the much larger pollen of pitch pine (Whitehead 1964). Measurements were not made on the Rogers Lake material, but are available from stratigraphically equivalent levels at other sites.

TABLE 1. Percentage contribution of species to total pine pollen from selected levels in the Rogers Lake core (Values for other levels were obtained by interpolation)

Depth (m)	Estimated age of sample	*n*	White *P. strobus*	Other *P. banksiana*, *P. resinosa*, *P. rigida*	Unidentifiable
.075	+206	25	28	36	36
1.72	1,525	50	38	30	32
4.86	3,890	50	38	32	30
7.59	7,348	25	36	32	32
8.37	8,332	50	56	12	32
8.71	8,657	50	72	6	22
9.08	9,354	50	34	40	26
9.38	9,936	50	14	58	28
9.63	10,489	50	8	74	18
9.74	10,760	50	0	84	16
10.14	11,806	50	20	48	32

RESULTS AND DISCUSSION

Definition of terms

Pollen percentage is the percentage of any given pollen type among the total grains counted in a sample. Pollen percentages are plotted against depth or age of the sediment in the *pollen percentage diagram* (the traditional "pollen diagram").

Sediment matrix accumulation rate is the net thickness of sediment accumulated per unit time, after compaction and diagenesis. It may also be expressed as the amount of time per unit thickness of sediment; this has been termed *deposition time per centimeter* in a recent paper by Waddington (1969).

Pollen concentration is the number of grains per unit volume of wet sediment. Pollen concentration (per unit volume or weight of sediment) is often referred to in the literature as "absolute pollen frequency," or APF.

Pollen deposition rate, or *pollen accumulation rate,* is the net number of grains accumulated per unit area of sediment surface per unit time. It can also be called *pollen influx* per unit time. E. J. Cushing (*personal communication*) has suggested that this term is appropriate since we are measuring grains falling on an area. In the *pollen deposition rate diagram,* or *pollen influx diagram,* the number per square centimeter per year is plotted against the age of the sediment. This type of diagram has also been called an "absolute diagram," an undesirable term because it leaves unspecified whether the diagram shows pollen concentration or pollen deposition rate.

The general procedure in the study has been to multiply the pollen concentration at each level by the sediment matrix accumulation rate, in order to compute the pollen deposition rate.

Determination of sediment matrix accumulation rate

Matrix accumulation rate in the south basin of the lake was determined by plotting the radiocarbon ages of various levels in the core (Stuiver, Deevey, and Rouse 1963, Stuiver 1967) against depth. A smooth curve was fitted to the resulting points (Fig. 2); its slope represents the rate of accumulation. The deepest sediment, 11.5 m below the sediment surface, was 14,300 years old, indicating a fairly rapid average rate of accumulation of sediment during the lake's lifetime (about 0.08 cm per year). But marked fluctuations in accumulation rate of the sediment occurred from time to time (Fig. 2). Between 9,000 and 6,000 years ago, for example, the rate first increased to about ~~0.25~~ 0.12 cm per year, then dropped to a rate

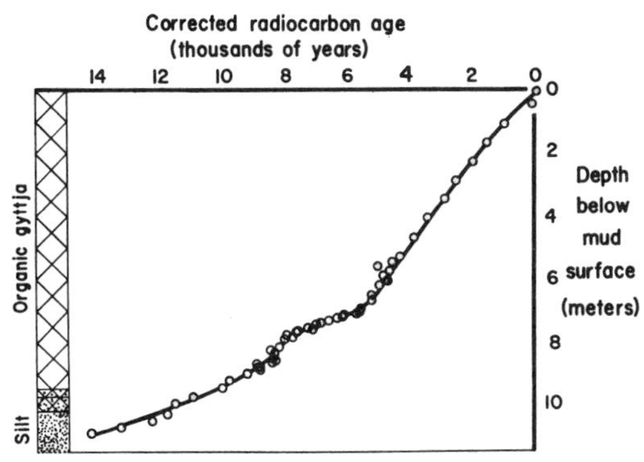

FIG. 2. Radiocarbon dates from the core from the south basin of Rogers Lake, plotted against the depth of the samples in the profile. The solid line is a composite curve drawn to express the general relationship between depth and age of the sediment.

one-fourth as rapid, and then increased again. There is no correlation between these changes and the type of sediment deposited (Fig. 2).

Matrix accumulation rate was also determined in the second core, from the north basin of the lake. The base of the core was 8,600 years old; the average rate of accumulation after that time was only 0.05 cm per year. Three radiocarbon dates (Y-1455, 1456, 1457) are available from the lowest 60-cm segment of the core, spanning the interval between 7,500 and 8,600 years B.P. (M. Stuiver, *personal communication*). Accumulation rate during this brief interval was about 0.06 cm per year. Rates in the south basin at that same time varied from 0.05 to 0.12 cm per year.

Two problems arise in evaluating estimates of matrix accumulation rates. One has its source in errors in the radiocarbon method itself. Because variations have occurred in the C^{14} concentration in the atmosphere, radiocarbon ages on material older than 2,500 years are slightly younger than the true ages of the samples, by an amount that increases proportionally with age (Damon, Long, and Greg 1966, Stuiver 1967). This anomaly causes the change in slope in the matrix accumulation curve at Rogers Lake at 2-m depth (2500 years B. P.) (Fig. 2) since samples below this level have C^{14} ages slightly younger than their true ages (Stuiver 1967). The difference in slope, however, is too small to be important in the present study.

Ancient carbon, apparently from ground water, is present in the water of Rogers Lake, giving an apparent age of 730 years (average of three samples) to the surface sediment. This amount has been subtracted from each age determination from the core, on the assumption that a fixed proportion of the carbon in the lake has always been C^{14}-deficient (Deevey et al. 1954, Broecker and Walton 1959, Stuiver et al. 1963). This assumption seems reasonable for the homogeneous postglacial sediment, although it might not hold for the older, silty sediment, deposited when the rate of carbon deposition was very low (Davis and Deevey 1964, Deevey 1964). Again, the change in slope resulting from a different correction would be too small to affect the results significantly.

The second problem in determining matrix accumulation rate is one of interpretation. The slope of the line fitted to the plot of radiocarbon dates vs. depth depends upon the shape of the curve used and on the closeness with which the curve has been fitted to the points. Some perspective on this problem is given by comparing the various curves that have been used for Rogers Lake, some using the less complete series of dates that were available earlier and others interpreting the existing data differently (Davis and Deevey 1964, Deevey 1964, 1968, Davis 1967*b,* Stuiver 1967). The differences between the resulting estimates of matrix accumulation rates, although larger than the differences mentioned above, are still relatively minor and should be considered a normal margin of error for this kind of study. Difficulties in interpretation arise because the size of the margin may not be predictable. The margin of error is influenced at any one site by the frequency and magnitude of fluctuations in accumulation rate and by the number of radiocarbon dates used to measure them.

Pollen concentration

The pollen concentration at various levels in the core has been plotted against radiocarbon age (not corrected for variation in atmospheric C^{14}, but corrected for C^{14} deficiency in the existing lake) (Fig. 3). Pollen concentrations vary from 20.000/ml for the oldest sediment to about 650,000/ml for sediment deposited 8,500 years ago. Samples from contiguous levels are generally similar in pollen concentration. An exception appears in sediment deposited about 10,200 years ago, where the pollen concentration suddenly falls to low levels, apparently because of dilution during an increase in matrix accumulation rate too brief to be detected by radiocarbon dating.

Deposition rate for total pollen

Pollen concentrations have been multiplied by the sediment accumulation rate (in centimeters per year) to obtain the pollen deposition rates (Fig. 3). The total influx of pollen to the sediment increased by a factor of 50 between 12,000 and 9,000 years ago, from 1,000 to 50,000 grains/cm^2 per year. Subsequently it fell to a rate of about

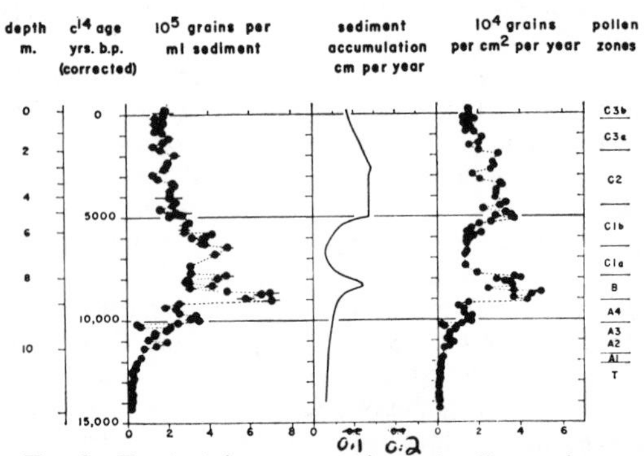

FIG. 3. To the left, concentrations of pollen grains per milliliter wet sediment plotted against the absolute age of the samples. Center column, accumulation rate for the sediment matrix, calculated from the data shown in Fig. 2. Right-hand column, deposition rates of total pollen from terrestrial plants.

$20{,}000/\text{cm}^2$ per year, a rate maintained for several thousand years until a maximum again occurred 5,000 years ago, followed by a gradual decrease until the present.

Some of these changes in pollen deposition are undoubtedly related to changes in the pollen input to the lake, while others represent changes without significance, within an expected margin of error. For example, the drop in pollen deposition rate 2,500 years ago is an artifact of the error in estimating matrix accumulation prior to that time (see above). The minimum rates between 5,000 and 8,000 years ago are also closely correlated with matrix accumulation rate. There is no clear explanation for this relationship, although one can speculate that a factor such as a pattern of water circulation that decreased the rate of sediment accumulation also discouraged the accumulation of pollen grains at the sampling site. Pollen grains in themselves occupy too little volume to affect the sediment accumulation rate directly. Another explanation is systematic error in radiocarbon dates from this part of the profile, caused by a temporary change in the atmospheric concentration of radiocarbon. The evidence Stuiver (1967) has presented argues against this possibility for this portion of postglacial time.

The magnitude of changes in pollen deposition rates that can be considered significant will emerge when measurements of pollen deposition at several sites in the same region can be compared for consistency. For the present, only a limited comparison is available between the pollen influx measured in the short, radiocarbon-dated segment of core from the north basin and the equivalent segment of the south basin core. The pollen concentration is similar in both cores. The sediment accumulation rates are also similar, although the north core rate was determined merely by a straight-line fit to three radiocarbon dates from closely spaced samples, while the variable accumulation rate in the south basin was measured by a smooth fit to many samples. The pollen deposition rates, which varied between 12,000 and 40,000, are similar within a factor of two at any level. A threefold decrease in deposition seen in the south basin, however, is not apparent at the other site. On this basis, changes in deposition rate less than a factor of three cannot be considered

TABLE 2. Pollen percentages not shown in Fig. 4 and 5 (Only samples in which rare types occur are included in the table)

Depth (m)	Thousands of radiocarbon years B.P.	Type and percentage
.08	+.21	Liguliflorae 0.2; *Zea* 0.7; Leguminosae 0.7
.32	0.17	Liguliflorae 0.4
.47	0.34	Gentianaceae 0.3
.58	0.45	*Plantago lanceolata* type 0.5; *Impatiens* 0.5
.79	0.67	Chenopodiaceae 0.2
.95	0.83	*Liriodendron* 0.4; Chenopodiaceae 0.4
1.00	0.88	*Plantago lanceolata* type 0.3
1.72	1.52	Leguminosae 0.3
1.98	1.76	Chenopodiaceae 0.3
2.22	1.95	*Impatiens* 0.5
2.70	2.32	Ranunculaceae 0.3
3.20	2.68	Betulaceae 0.2
3.84	3.14	*Plantago* sp. 0.4
4.24	3.43	Betulaceae 0.3
4.62	3.70	Scrophulariaceae 0.3; Leguminosae 0.3
4.86	3.89	Leguminosae 0.3
5.12	4.07	*Liriodendron* 0.6
5.66	4.47	Chenopodiaceae 0.2
6.06	4.76	*Liriodendron* 0.2
7.05	5.78	Chenopodiaceae 0.1
7.20	6.10	Betulaceae 0.3
7.24	6.22	Umbelliferae 0.2
7.25	6.23	*Impatiens* 0.1; Scrophulariaceae 0.2; Urticaceae 0.1; Leguminosae 0.1
7.34	6.65	*Smilax* 0.2; Chenopodiaceae 0.1
7.39	6.80	*Polygonum* 0.1; Chenopodiaceae 0.1
7.79	7.73	Chenopodiaceae 0.2; Umbelliferae 0.1
7.99	8.00	Chenopodiaceae 0.3
8.16	8.16	Chenopodiaceae 0.3
8.37	8.33	Chenopodiaceae 0.1
8.60	8.53	Leguminosae 0.2
8.66	8.59	Chenopodiaceae 0.2
8.70	8.66	*Eriocaulon* 0.1
8.87	9.00	Chenopodiaceae 0.2; Leguminosae 0.2
9.08	9.35	*Cornus* 0.2
9.15	9.49	Onagraceae 0.1
9.44	10.07	Chenopodiaceae 0.1
9.49	10.18	*Galium* 0.5
9.62	10.50	*Plantago* sp.
9.74	10.76	*Plantago* sp. 0.3
9.80	10.91	Umbelliferae 0.1
9.93	11.23	Liguliflorae 0.2; *Dryas* type 0.1; Betulaceae 0.2
9.96	11.32	*Plantago* sp. 0.4
10.33	12.36	Chenopodiaceae 0.1
10.38	12.50	Ranunculaceae 0.8
10.43	12.66	Liliaceae 0.3
10.49	12.84	*Polygonum* 0.2
10.53	12.97	*Galium* 0.6; Umbelliferae 0.3
10.63	13.29	Betulaceae 0.2; Leguminosae 0.2
10.78	13.79	Leguminosae 1.7
10.83	13.96	*Dryas* type 0.3

significant. (The margin of error is greater for rare pollen types whose concentration has not been determined as accurately.) Taking into account the additional error introduced at certain levels by problems in radiocarbon dating, changes in pollen deposition by a factor of five are probably the smallest attributable to variations in pollen input to the lake. Future studies will indicate whether this opinion is too conservative or too optimistic.

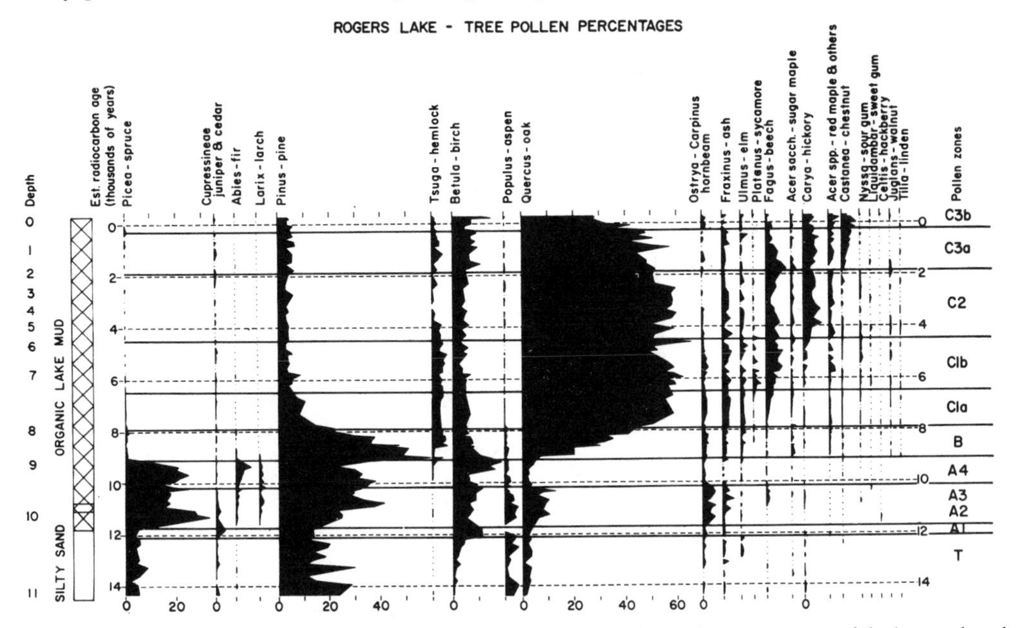

FIG. 4. Percentages for tree pollen types, calculated as percent total pollen from terrestrial plants, plotted against the absolute age of the sediment.

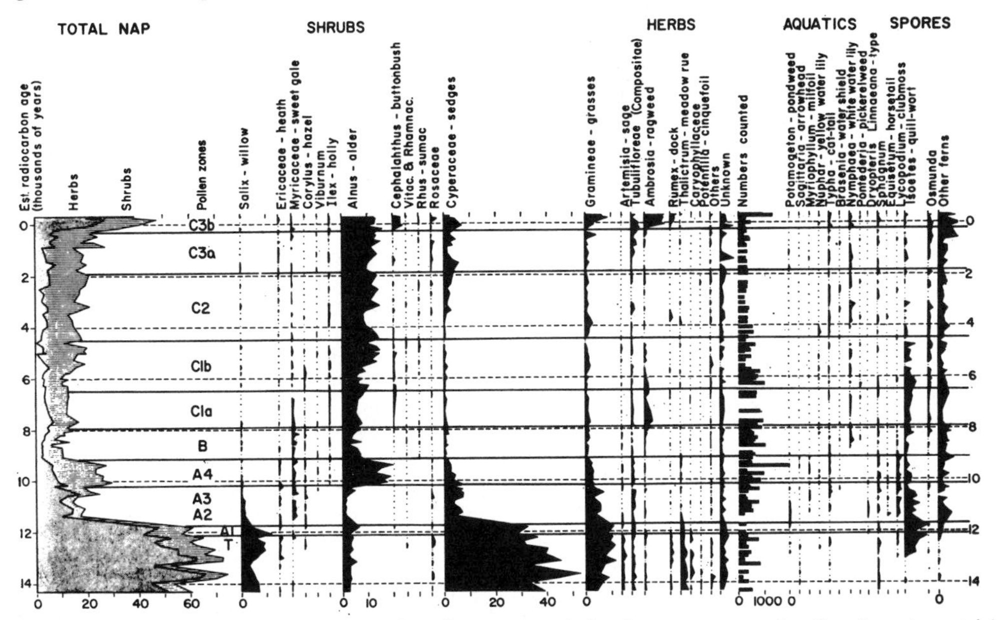

FIG. 5. Percentages for herb and shrub pollen types, calculated as percent total pollen from terrestrial plants, plotted against the absolute age of the sediment. Percentages for spores and aquatics, which are shown to the far right, are calculated as percentage of a sum equal to total terrestrial plant pollen.

Pollen stratigraphy and interpretation

The pollen percentage diagram from Rogers Lake (Fig. 4 and 5, Table 2) shows the stratigraphy described long ago for southern Connecticut (Deevey 1939, Leopold 1956, Beetham and Niering 1961). The new radiocarbon dates permit the use of absolute age, rather than depth, as the ordinate in the diagrams. The usual zonation of the pollen sequences for this region has been maintained here, except that zones C-1 and C-3 have been subdivided into rather different older (a) and younger (b) subzones. Zone B, however, is considered a single zone, rather than two subzones as at other sites (Davis 1958, 1960).

Deposition rate diagrams are given for all the major pollen types (Fig. 6 and 7). Several different scales have been used on the abscissae in order to magnify fluctuations in the rare pollen types. The taxa are arranged according to the stratigraphic level at which they reach maximum values.

Herb pollen zone (14,300–12,150 years ago)

The oldest, silty sediments are characterized by high percentages of pollen from herbaceous plants, especially sedges, grasses, *Artemisia, Rumex,* and *Thalictrum.* Willow is the most abundant shrub pollen, reaching 10% of the total, and pine, spruce, oak, and poplar are the most common tree pollen types.

The deposition rates (see especially Fig. 6C) show that all pollen types were in fact rare during this period. Even the herbs considered so characteristic of this pollen zone, such as grasses and *Artemisia,* are deposited at rates lower than those of the overlying spruce pollen zone. All trees are represented by very few pollen grains relative to the numbers deposited in sediment younger than 12,000 years; their prominence in the percentage diagram results from low pollen productivity by the herbaceous vegetation, which failed to dilute far-transported tree pollen in the sediment. Fluctuations in percentage values among the tree pollen types have been described from this zone at several sites in southern New England and have been attributed to climatic change (Leopold 1956, Deevey 1958, Leopold and Scott 1958, Ogden 1959, 1963). They involve small absolute numbers of grains, however, and are consequently of little or no significance as far as the local vegetation is concerned. Instead, they probably reflect chance fluctuations in the transport of pollen by wind from distant forests (Davis and Deevey 1964).

The vegetational and climatic meaning of the herb pollen zone in New England pollen diagrams has been the subject of much discussion. The presence of arctic or subarctic plant species in the region has been established from macrofossil evidence (Argus and Davis 1962). The percentages of non-arboreal pollen in the sediments resemble those in assemblages recovered from surface samples from arctic regions, very far from any forest, rather than from tundra along the forest margin (Davis 1967*c*). But it must be admitted that exact analogs to the fossil assemblages, which have appreciable percentages of pollen from oak and other temperate tree genera, have not yet been found in the modern Arctic. The very low deposition rates for all pollen types, however, are consistent with the low pollen productivity observed in modern Arctic vegetation (Aario 1940, Ritchie and Lichti-Federovich 1967).

Zone A-1, transition from herb to spruce zone (11,700–12,150 years ago)

This transitional pollen zone is recognizable in pollen diagrams from several sites in Connecticut and central Massachusetts (Leopold 1956, Davis 1958); it has a maximum of birch pollen, rising percentages of spruce pollen, and in some cases maximum poplar percentages. Earlier I had speculated (Davis 1958) that poplar might have formed the treeline of the advancing late-glacial forests. The deposition rates (Fig. 6B) show, however, that despite high percentage values in the lower part of the spruce zone, the deposition rates for poplar pollen are very low relative to rates later on, and probably do not represent an appreciable abundance of poplar trees. The maximum for birch pollen percentages in zone A-1 at Rogers Lake is also largely an artifact resulting from a decrease in deposition rates for sedge pollen (Fig. 6A and 6C).

Zone A-2-3, spruce-oak zone (10,200–11,700 years ago) and zone A-4, spruce-fir zone (9,100–10,200 years ago)

The pollen assemblage in zone A-2-3 is characterized by high percentages of spruce pollen and by a late-glacial percentage maximum (ca. 20%) for oak and other temperate deciduous trees such as hornbeam and ash. Pine percentages rise to a maximum near the upper boundary of the zone. Zone A-4 is characterized by decreased oak pollen percentages and maximal spruce, fir, and larch frequencies. Birch and alder pollen also increase.

Pollen deposition rates are much higher in the spruce zones than previously (Fig. 3 and 6B). Spruce rises steadily from fewer than 100 grains to maximum rates of 3,000 grains 10,000 years ago. Similar increases occur for fir, larch, and pine (Fig. 6B). The morphology of the pine

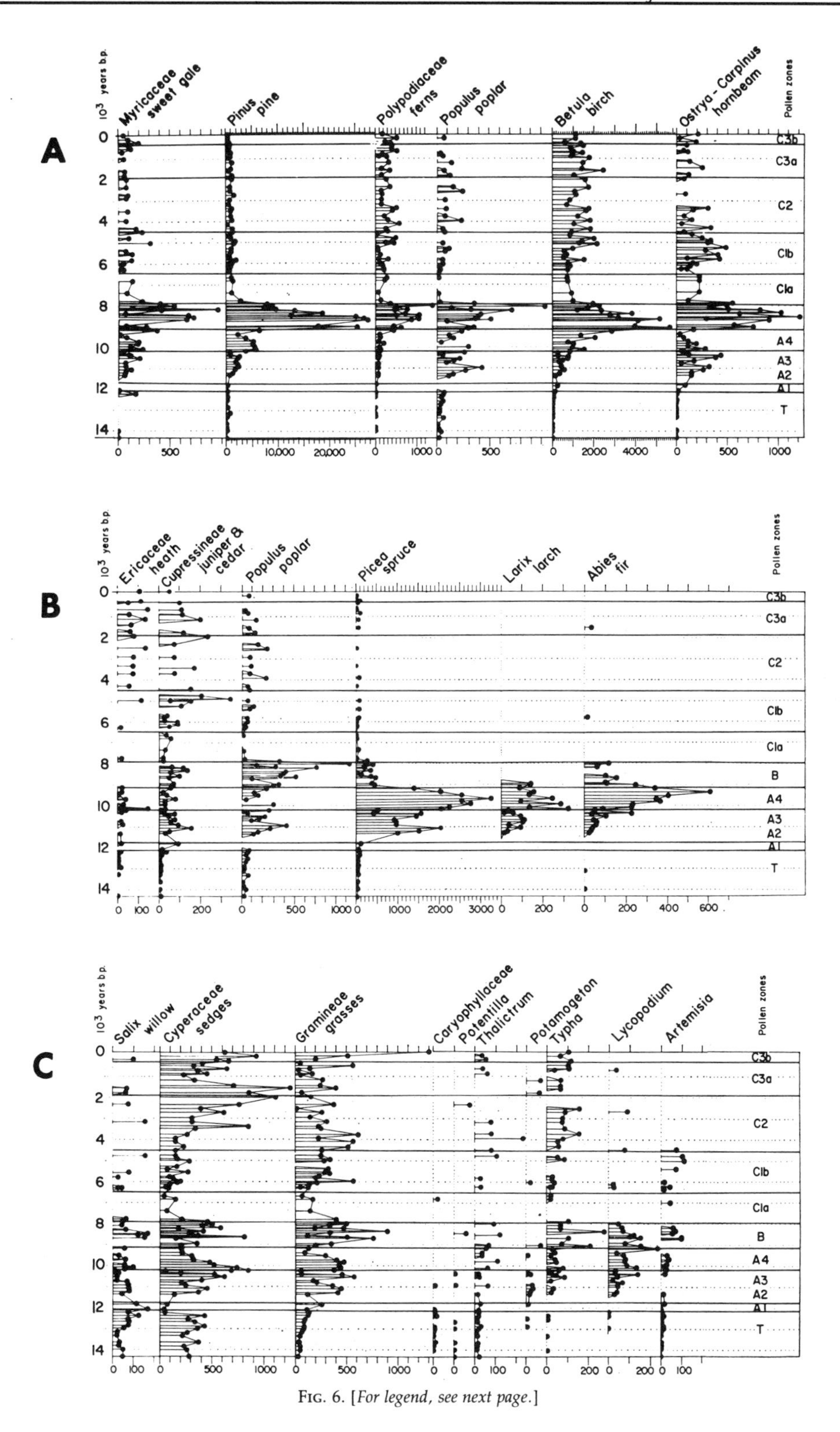

FIG. 6. [*For legend, see next page.*]

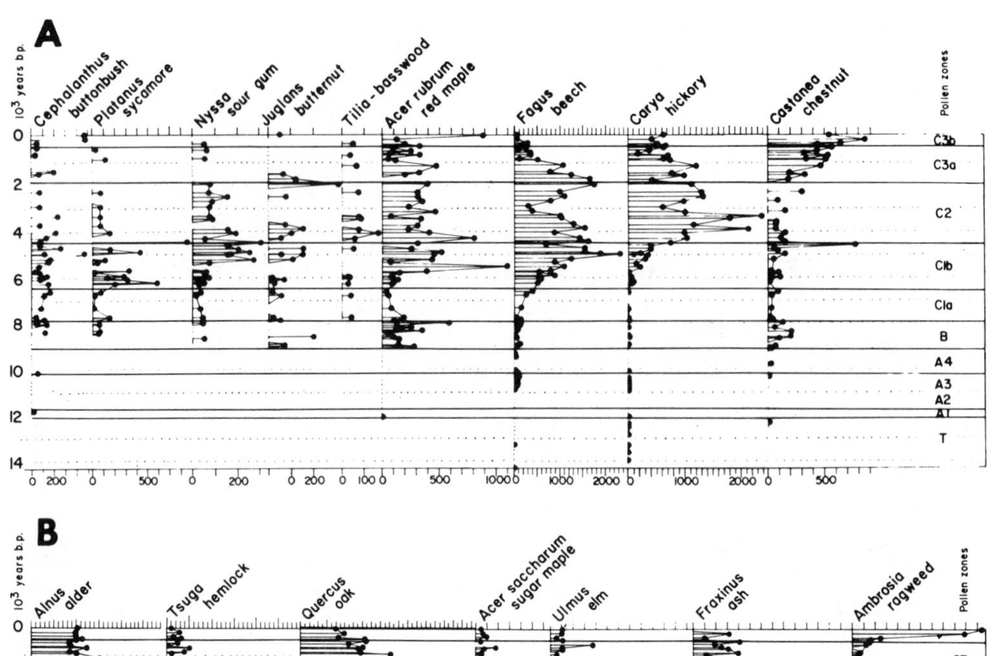

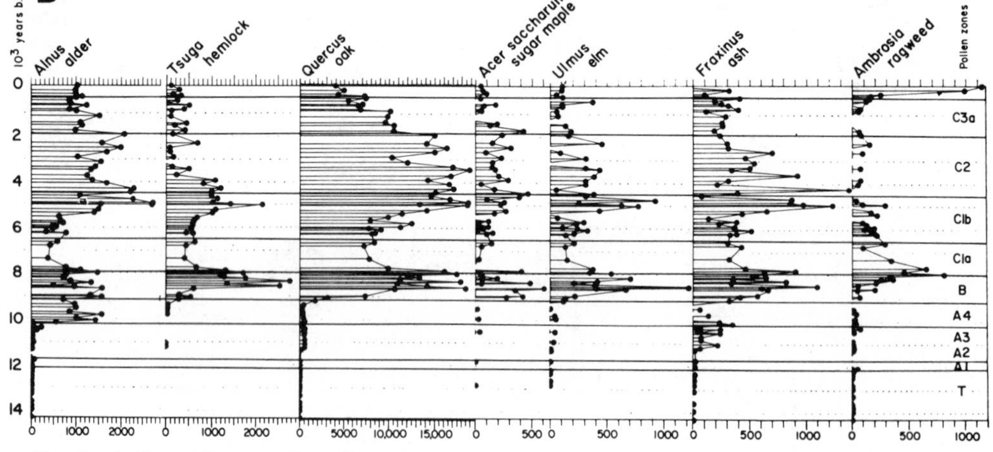

FIG. 7. A. Deposition rates for pollen types that display maximum rates in sediment younger than 6,000 years.

B. Deposition rates for pollen types that occur throughout the postglacial and for ragweed pollen.

pollen indicates that jack and/or red pine were the dominant species at this time (Fig. 8). (The small size of pollen measured at equivalent levels at other sites (Taunton, Mass., Davis, *unpublished data,* and Tom Swamp, Mass., Davis 1958) suggests that pitch pine was not present (Whitehead 1964)). Deciduous tree pollen deposition is significantly higher than in the preceding herb zone. The deposition rate for oak, however, is still very low (500) compared to the values attained in postglacial time (17,000) (Fig. 7B). Ash and hornbeam pollen influx (Fig. 6A, 7B), on the other hand, is not very different from the postglacial interval. Possibly the late-glacial vegetation of southern New England at the time of the spruce zone was similar in some respects to the contemporaneous vegetation of the Great Lakes region, which produced pollen assemblages dominated by spruce, in which ash pollen was an important component (Cushing 1965).

In zone A-4 the alder pollen deposition rate increased, while rates for the coniferous trees and

FIG. 6. A. Deposition rates for pollen types that reach maximum rates during the deposition of the pine zone (B). The different types of pine pollen are shown plotted separately in Fig. 8.

B. Deposition rates for pollen types that were abundant during the deposition of the spruce pollen zone (A).

C. Deposition rates for pollen types that occur in high percentages in the herb pollen zone (T).

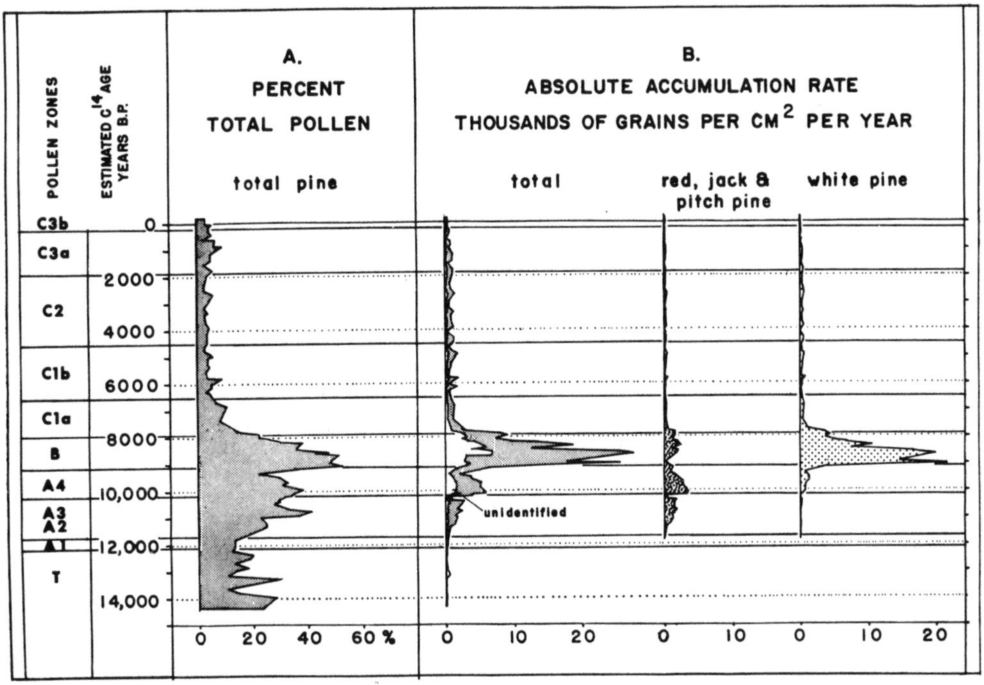

FIG. 8. Percentages and deposition rates for the various pine pollen types.

birch reached a maximum. Ash, hornbeam, and oak pollen deposition rates change slightly at this time, but the change is not quite large enough to be considered a significant indication of vegetation change.

The increased influx of tree pollen in the spruce zone, as well as the presence of macrofossils of trees at several other sites in the region (Davis 1965*b*) indicates that trees grew in southern New England at this time, increasing in abundance to maximum frequencies of boreal species in zone A-4. Whether the landscape was thickly forested, resembling the present-day boreal forest, remains in question, however. The pollen assemblages at most sites are unlike those deposited today within the boreal forest region. They resemble more strongly those from surface samples from a region north of the boreal forest in Quebec, where the prevailing vegetation type is open spruce woodland (Davis 1967*c*). At Rogers Lake, however, the comparison with Quebec is not satisfactory, as the ratio of pine to spruce pollen is higher than in the Quebec surface samples. High percentages for pine pollen are characteristic of zone A at all coastal sites in southern New England. Either spruce was less frequent at coastal sites and wind-blown pine pollen more common, or spruce was equally abundant at all sites while pine was more abundant near the coast. The pollen percentages in zone A-4 at Rogers are very similar to those of zone B-1 in central Massachusetts. Conceivably the two pollen zones could be the same age, although this is doubtful. The radiocarbon evidence is equivocal. (The only radiocarbon date from sites in Massachusetts is from the upper part of zone A-3. At 10,800 years, the date is a few hundred years older than equivalent levels at Rogers, but this is too small a difference to change the correlation for zone B-1, 30 cm higher in the section.) Percentages in zone B-1 in Massachusetts are similar to surface samples from southeastern Ontario, just south of the boreal forest. I am reluctant to interpret zone A-4 at Rogers as representing this kind of mixed coniferous-deciduous vegetation, however. White pine occurs in the vicinity of the surface sampling sites in Ontario (Rowe 1959), whereas its pollen is almost entirely absent from zone A-4 at Rogers.

Zone B, pine zone (7,900–8,100 years B.P.)

Pollen assemblages from zone B at Rogers are almost identical to those of zone B-2 in Massachusetts. White pine (*P. strobus*) dominates the assemblages, reaching percentages as high as 50% of total pollen. The deposition rate for white pine pollen increased rapidly at this time from a few hundred grains to 20,000 grains (Fig. 8). The increase was so rapid that pollen from several other genera, also increasing at this time, can hardly be noticed in the percentage diagram; the rate diagrams (Fig. 6A and 7B) show clearly, however, that pollen of the Myricaceae, poplar, birch, alder, hornbeam, oak, and hemlock, and fern spores reached a maximum. (The fern spores are from the Polypodiaceae, rather than *Pteridium,* which occurs characteristically in the pine zone in Minnesota (McAndrews 1966)). Also apparent (Fig. 6B) is the rapid decrease in deposition rates, as well as percentages, for boreal genera such as spruce, fir, and larch.

The climate must have changed dramatically at this time to produce such a striking change in the vegetation (Wright 1964, Ogden 1967). White pine, hemlock, and oak are distributed at present south of about 46° N lat in eastern Canada. Hemlock is the most restricted in range, occurring only to the south and east of Lake Superior. In the east white pine and hemlock range far southward along the Appalachian Mountain chain, but in the Midwest they drop out of the flora in southern Michigan, Wisconsin, and Minnesota (Fowell 1965). The climate of zone B time presumably approximated the modern climate somewhere within the geographical zone where these species ranges now coincide. Surface samples that precisely match the fossil percentages of zone B have not yet been found, but much of this area has not been sampled thoroughly. The high ratio of pine to birch pollen in zone B is similar to surface samples from central Canada northeast of the Great Lakes; it is in the Great Lakes region that analogs to the vegetation might occur.

Zones C-1, C-2, and C-3, oak zones (7,900 years ago to present)

The pollen diagrams show the postglacial stratigraphy described by Deevey (1939) many years ago: zone C-1, oak and hemlock, zone C-2, oak and hickory, and zone C-3, oak and chestnut. Additional details can now be seen: ragweed pollen occurs in the lower portion of zone C-1, designated here C-1a, while appreciable percentages of beech pollen characterize the upper portion (C-1b). In zone C-2 beech percentages (as well as hemlock) decline to a minimum, confirming the correlation of the beech minimum in Ohio with the hemlock minimum of zone C-2 in New England (Ogden 1966). In the upper portion of C-3 (C-3b) percentages of pollen from ragweed, *Rumex,* and other weeds increase.

The deposition rate for total pollen was relatively stable during this long period, and consequently the trends in rates for individual types are similar to the trends in percentages. Deposition rates for tree pollen seem to reach a minimum in zone C-1a, although the change is not large enough to be considered significant. At this time pollen from ragweed was deposited at increased rates. The maximum ragweed rate 8,000 years ago is approximately half as high as the modern rates, which record the response of this weedy species to forest clearance by European settlers. Further, the increased deposition of ragweed pollen in C-1a is a regional event: a similar maximum (of percentage) occurs in zone C-1 in a pollen diagram from Ohio (Ogden 1966). I believe that it indicates decreased density of forest, brought about by climatic changes associated with the expansion of prairie in the Great Lakes region 8,000 years ago (Cushing 1965). Wright (1968*b*), on the other hand, believes that the ragweed maximum in Northeastern profiles represents pollen blown from the expanded prairie in the Prairie Peninsula region of Illinois and Indiana. He points out that the warmest and driest part of postglacial time in the Northeast is generally considered to have occurred later, between 5,000 and 2,000 years ago. If the ragweed pollen were windblown, I would expect a ragweed maximum in postglacial sediment everywhere in New England. But although Vermont is no farther from the Prairie Peninsula than Connecticut, there is no ragweed maximum at Brownington Pond. There is, to the contrary, an increase in oak pollen percentages there at stratigraphically similar levels. In that region oak would increase in response to a drier climate (Davis 1965*b*). It is unfortunate that diagrams from Massachusetts (Davis 1958) are not detailed enough to show whether or not a ragweed maximum occurs in zone C-1.

An interesting feature in the Rogers Lake pollen sequence is the late arrival of several tree species (Fig. 7A, 7B). Red maple and hemlock first appeared 9,000 years ago (Figs. 7A, 7B), while hickory pollen was absent until 5,000 years ago, when its deposition rate suddenly increased to high levels. Chestnut pollen increased in abundance 2,000 years ago. Its pollen also appears sporadically throughout the profile. Presumably these late times of arrival and sudden increases to high frequencies are due to delayed migration from glacial refuges. This possibility has recently been emphasized by Wright (1964, 1968*a*). The absence of pollen from these important tree species in older sediment makes comparison with modern pollen assemblages and vegetation difficult. Surface samples from regions with a similar, although cooler, climate than southern Connecticut, for example central Massachusetts, contain about 20% pine pollen and 8% pollen from chestnut; none of the assemblages in Rogers Lake sediment is similar. In fact, only the uppermost levels can be compared with surface samples from anywhere in New England. The assemblages of C-1a might be compared with modern assemblages from the "Prairie Peninsula" region of the Midwest, since the high deposition rate for ragweed pollen is suggestive of forest with prairie openings. Furthermore, these regions are largely west of the present limit for beech, the pollen of which is rare in zone C-1a, while ash, hornbeam, and sycamore, the pollen of which is frequent in zone C-1a and C-1b, are characteristic trees of the prairie margin region in Indiana and Illinois. The comparison cannot be carried further, because surface sediments from that region rarely contain hemlock pollen and always contain hickory pollen, and thus they are very different from zone C-1a in Connecticut. This suggests that the forest in the vicinity of Rogers Lake during early postglacial time was different from any known today. Admittedly, much more intensive sampling of modern and presettlement sediments in the Midwest is necessary to prove that this view is correct. The soils and topography of southern New England are quite different from those of Michigan, Illinois, and Wisconsin. It seems probable that even in an identical climate, different abundances of species would prevail in the two regions.

The youngest pollen zones (C-3a and C-3b) date from the last two millenia; all the tree genera that are abundant in the modern forest surrounding Rogers Lake are present in the pollen assemblages. It should therefore be possible to make some comparisons between the fossil assemblages and the modern vegetation. In discussing the present distribution of tree species in southern Connecticut, Niering and Goodwin (1962) have emphasized the importance of fires set by prehistoric Indians. They believe that fire over a very long time interval has selected against fire-sensitive species, especially hemlock, while selecting for other more xerophytic types, such as oak, hickory, and chestnut. The pollen deposition rate diagram (Fig. 7B) shows that the frequency for hemlock has been low for the last 2,000 years; if fires were responsible, the effect must have started at around the time of the birth of Christ. Chestnut appears to have increased steadily in this interval; possibly its increase is related to disturbance, although climatic changes might also be responsible.

Later changes in the pollen content of the sediment (zone C-3b) are clearly related to forest clearance by European settlers. Ash and oak pollen percentages decline, and red maple, which can grow as a pioneer species in disturbed habitats, increases together with ragweed. *Rumex* and many other weedy herb pollen types increase in frequency. Buttonbush pollen increases (Fig. 5) perhaps as a result of changes in water level of the lake. The pollen changes are quite different from those that record agriculture in northwest Europe. In America ragweed is the prime indicator of agriculture, and plantain (*Plantago*) pollen, common in European deposits, is never abundant (Table 2). Cereal pollen is also extremely rare; only one *Zea* grain was found in the entire profile. In the uppermost samples (Fig. 4 and 5) a few very recent vegetation changes are recorded. A decline in total non-arboreal pollen corresponds with the very recent abandonment of many of the farms in the vicinity of Rogers Lake, and the two uppermost samples analyzed (which were not treated on an absolute basis) show decreased percentages for chestnut pollen, recording the chestnut blight, a disease that killed most of the trees of this species 30 to 45 years ago.

Conclusions

Application of a new method of pollen analysis to Rogers Lake sediments has confirmed and strengthened many of the previous interpretations of the pollen percentage diagrams from southern New England. New information has been obtained. Especially important is the sudden increase in pollen deposition rates for trees, from very low levels 12,000–14,000 years ago to high postglacial rates 9,000 years ago. The change in tree pollen fits well with the idea that a tundra vegetation, devoid of trees, was replaced first by woodland and then by forest as trees immigrated to the site and increased in frequency. Studies of modern pollen deposition have shown that although pollen productivity is much lower in tundra areas than in forest, the percentage composition of the pollen assemblages produced on either side of the northern tree line are very similar (Wright, Winter, and Patten 1963, Davis 1967*c*, Ritchie and Lichti-Federovich 1967). Consequently deposition rates are far more valuable than percentage diagrams as a pollen record of changes from tundra to ancient woodland or parkland.

Another interesting aspect of the late-glacial sequence at Rogers Lake is the sudden increase in the numbers of oak pollen grains deposited 11,500 years ago, a time just postdating the advance of the Valders ice sheet (Broecker and Farrand 1963). An increase in the abundance of oak somewhere within pollen dispersal distance is one possible explanation. Another is that the change reflects a change in prevailing wind direction from northwest to southwest correlated with the beginning of a time of very rapid glacial retreat.

The deposition rates throw new light on the nature of the vegetation 8,500–9,500 years ago, during the "pine period," when the pollen influx from white pine (*Pinus strobus*) increased very rapidly, implying a marked population increase for the species, followed by an equally rapid decline only a thousand years later. Many other pollen types, including poplar, maple, oak, and hemlock, also increase steeply 9,500 years ago, a change that is largely masked in the percentage diagram. Pollen influx for boreal species decreased, indicating that the change in the percentage pollen diagrams is not merely caused by dilution by increased pine pollen, but represents a real decline in numbers of boreal trees. The subsequent marked decline in deposition rate of pine pollen indicates a shift to almost purely deciduous forest. Perhaps the changes in species composition of the plankton in Rogers Lake at this time (Deevey 1968) are an additional reflection of changes in regional climate.

One of the new features of the Rogers Lake sequence is a maximum for ragweed pollen in zone C-1a, 7,000–8,000 years ago, a feature that has not been noticed in earlier analyses from the region, probably because emphasis was placed on tree pollen rather than upon herbs. Stratigraphically the ragweed maximum at Rogers Lake (Davis 1967*b*) is correlated with the Prairie Period of the Great Lakes region and reflects directly or indirectly vegetation changes associated with a xerothermic interval that occurred there between 5,000 and 8,000 years ago (Wright 1968*b*). The changes in pollen frequency that were originally thought to represent the zerothermic interval in southern New England (Deevey 1939) are in younger sediment, precisely dated now at 2,000–4,500 years B.P. These changes seem most easily interpreted now as the reflection of successive immigrations to Connecticut of first beech, then hickory, and then chestnut. Only during the last 2,000 years have the pollen assemblages in southern Connecticut been sufficiently similar to modern assemblages to permit comparison. This suggests that the forest communities there, as they are presently recognized, may be of very recent origin.

Acknowledgments

This paper is contribution 104 from the Great Lakes Research Division, University of Michigan. The work was initiated while the author was a research fellow at Yale University in 1960–61 and has been continued at the University of Michigan. Research support has been provided by the National Science Foundation, research grants G-19335, to Yale University, G-17830, GB-2377, and GB-5320 to the University of Michigan. I gratefully acknowledge the encouragement and cooperation of Edward S. Deevey, Jr. I also extend special thanks to Minze Stuiver, who has dated the sediment core from Rogers Lake.

Literature Cited

Aario, L. 1940. Waldgrenzen und subrezente Pollenspektren in Petsamo Lappland. Ann. Acad. Sci. Fenn., Ser. A, **54**(8): 1–120.

Argust, G. W., and M. B. Davis. 1962. Macrofossils from a late-glacial deposit at Cambridge, Massachusetts. Amer. Midland Natur. **67**: 106–117.

Beetham, Nellie, and W. A. Niering. 1961. A pollen diagram from southeastern Connecticut. Amer. J. Sci. **259**: 69–75.

Broecker, W. S., and W. R. Farrand. 1963. Radiocarbon age of the Two Creeks forest bed. Geol. Soc. Amer. Bull. **74**: 795–802.

Broecker, W. S., and Alan Walton. 1959. The geochemistry of C^{14} in fresh-water systems. Geochem. Cosmochim. Acta **16**: 15–38.

Connecticut State Board of Fisheries and Game. 1959. A fisheries survey of the lakes and ponds of Connecticut. Lake and Pond Surv. Unit Rep. 1. 395 p.

Cushing, E. J. 1965. Problems in the Quaternary phytogeography of the Great Lakes region, p. 403–416. *In* H. E. Wright and D. G. Frey [ed.] The Quaternary of the United States. Princeton Univ. Press, Princeton, N. J. 922 p.

Damon, P. E., A. Long, and D. C. Greg. 1966. Fluctuation of atmospheric C^{14} during the last six millennia. J. Geophys. Res. **71**: 1055–1063.

Davis, M. B. 1958. Three pollen diagrams from central Massachusetts. Amer. J. Sci. **256**: 540–570.

———. 1960. A late-glacial pollen diagram from Taunton, Massachusetts. Bull. Torrey Bot. Club. **87**: 258–270.

———. 1965*a*. A method for determination of absolute pollen frequency, p. 674–686. *In* B. Kummel and D. Raup [ed.] Handbook of paleontological techniques. W. H. Freeman, San Francisco, Calif. 852 p.

———. 1965*b*. Phytogeography and palynology of northeastern United States, p. 377–401. *In* H. E. Wright, Jr., and D. G. Frey [ed.] The Quaternary of the United States. Princeton Univ. Press, Princeton, N. J. 922 p.

———. 1966. Determination of absolute pollen frequency. Ecology **47**: 310–311.

———. 1967*a*. Pollen deposition in lakes as measured in sediment traps. Geol. Soc. Amer. Bull. **849**: 858.

———. 1967*b*. Pollen accumulation rates at Rogers Lake, Connecticut, during late- and postglacial time. Rev. Palaeobot. **2**: 219–230.

———. 1967*c*. Late-glacial climate in northern United States: a comparison of New England and the Great Lakes region, p. 11–43. *In* E. J. Cushing and H. E. Wright, Jr. [ed.] Quaternary paleoecology. Yale Univ. Press, New Haven, Conn. 425 p.

———. 1968. Pollen grains in lake sediments: redeposition caused by seasonal water circulation. Science **162**: 796–799.

Davis, M. B., and E. S. Deevey, Jr. 1964. Pollen accumulation rates: estimates from late-glacial sediment of Rogers Lake. Science **145**: 1293–1295.

Deevey, E. S., Jr. 1939. Studies on Connecticut lake sediments. I. A postglacial climatic chronology for southern New England. Amer. J. Sci. **237**: 691–724.

———. 1958. Radiocarbon-dated pollen sequences in eastern North America. Veröff. Geol. Inst. Rübel (Zurich) **34**: 30–37.

———. 1964. Preliminary account of fossilization of zooplankton in Rogers Lake. Proc. Intern. Ass. Theor. Appl. Limnol. **15**: 981–992.

———. 1965. Sampling lake sediments by use of the Livingstone sampler, p. 521–529. *In* B. Kummel and D. Raup [ed.] Handbook of paleontological techniques. W. H. Freeman, San Francisco, Calif. 852 p.

———. 1968. Cladoceran populations of Rogers Lake, Connecticut, during late- and postglacial time. 1st Inter. Conf. Paleolimnol. Proc. (1967), in press.

Deevey, E. S., Jr., M. S. Gross, G. E. Hutchinson, and H. L. Kraybill. 1954. The natural C^{14} contents of materials from hardwater lakes. Proc. Nat. Acad. Sci. **40**: 285–288.

Faegri, K., and J. Iversen. 1964. Textbook of pollen analysis. 2nd ed. Munksgaard, Copenhagen. 237 p.

Fowell, H. A. 1965. Silvics of forest trees of the United States. U. S. Dep. Agr. Handbook 271. 762 p.

Leopold, E. B. 1956. Two late-glacial deposits in southern Connecticut. Proc. Nat. Acad. Sci. **42**: 863–867.

Leopold, E. B., and R. A. Scott. 1958. Pollen and spores and their use in geology. Smithson. Inst. Rep., 1957: 303–323.

McAndrews, J. H. 1966. Postglacial history of prairie, savanna, and forest in northwestern Minnesota. Mem. Torrey Bot. Club **22**: 1–72.

Niering, W. A., and R. H. Goodwin. 1962. Ecological studies in the Connecticut Arboretum Natural Area. I. Introduction and a survey of vegetation types. Ecology **43**: 41–54.

Ogden. J. G., III. 1959. A late-glacial pollen sequence from Martha's Vineyard, Massachusetts. Amer. J. Sci. **257**: 366–381.

———. 1963. The Squibnocket cliff peat; radiocarbon dates and pollen stratigraphy. Amer. J. Sci. **261**: 344–353.

———. 1966. Forest history of Ohio. I. Radiocarbon dates and pollen stratigraphy of Silver Lake, Logan County, Ohio. Ohio J. Sci. **66**: 387–400.

———. 1967. Evidence for a sudden climatic change around 10,000 years ago, p. 117–127. *In* E. J. Cushing and H. E. Wright, Jr. [ed.] Quaternary paleoecology. Yale University Press, New Haven, Conn. 425 p.

von Post, L. 1967. Forest tree pollen in south Swedish peat bog deposits (transl.). Pollen et Spores **9**: 375–402.

Ritchie, J. C., and S. Lichti-Federovich. 1967. Pollen dispersal phenomena in arctic-subarctic Canada. Rev. Paleobot. Palynol. **3**: 255–266.

Rowe, J. S. 1959. Forest regions of Canada. Can. Dep. Northern Affairs and Natur. Res., Forest. Branch, Bull. 123. 71 p.

Stuiver, Minze. 1967. Origin and extent of atmospheric C^{14} variations during the past 10,000 years, p. 27–40. *In* Radio-active dating and methods of low-level counting. Int. Atom. Energy Agency, Vienna.

Stuiver, Minze, E. S. Deevey, Jr., and Irving Rouse. 1963. Yale natural radiocarbon measurements II. Radiocarbon **4**: 35–42.

Tsukada, M., and M. Stuiver. 1966. Man's influence on vegetation in Central Japan. Pollen et Spores **8**: 309–313.

Ueno, Jitsuro. 1958. Some palynological observations of Pinaceae. J. Inst. Polytechnics, Osaka City Univ., Ser. D **9**: 163–186.

Waddington, J. C. B. 1969. A stratigraphic record of the pollen influx to a lake in the Big Woods of Minnesota. Twelfth Conf. Great Lakes Res. Abstracts: 52–53.

Whitehead, D. R. 1964. Fossil pine pollen and full-glacial vegetation in southeastern North Carolina. Ecology **45**: 767–777.

Wright, H. E., Jr. 1964. Aspects of the early postglacial forest succession in the Great Lakes region. Ecology **45**: 439–448.

———. 1968*a*. History of the Prairie Peninsula, p. 78–88. *In* R. E. Bergstrom [ed.] The Quaternary of Illinois. Univ. Illinois Coll. Agr. Spec. Publ. 14.

———. 1968*b*. The roles of pine and spruce in the forest history of Minnesota and adjacent areas. Ecology **49**: 937–955.

Wright, H. E., Jr., T. C. Winter, and H. L. Patten. 1963. Two pollen diagrams from southeastern Minnesota: problems in the regional late-glacial and post-glacial vegetational history. Geol. Soc. Amer. Bull. **74**: 1371–1396.

PATTERN AND PROCESS IN THE PLANT COMMUNITY*

By ALEX. S. WATT, *Botany School, University of Cambridge*

(*With eleven Figures in the Text*)

CONTENTS

The plant community as a working mechanism

The plant community may be described from two points of view, for diagnosis and classification, and as a working mechanism. My primary concern is with the second of these. But inasmuch as the two aspects are not mutually exclusive, a contribution to our understanding of how a community is put together, and how it works, may contain something of value in description for diagnosis.

It is now half a century since the study of ecology was injected with the dynamic concept, yet in the vast output of literature stimulated by it there is no record of an attempt to apply dynamic principles to the elucidation of the plant community itself and to formulate laws according to which it maintains and regenerates itself. Pavillard's assessment of the dynamic behaviour of species comes very near it, but is essentially concerned with the 'influence (direct or indirect) of the species on the natural evolution of plant communities' (Braun-Blanquet & Pavillard, 1930). As things are, the current descriptions of plant communities provide information of some, but not critical, value to an understanding of them; how the individuals and the species are put together, what determines their relative proportions and their spatial and temporal relations to each other, are for the most part unknown. It is true that certain recent statistical work is stretching out towards that end, but the application of statistical technique, the formulation of laws and their expression in mathematical terms, will be facilitated if an acceptable qualitative statement of the nature of the relations between the components of the community is first presented. Such a statement is now made based on the study of seven communities in greater or less detail, for data of the kind required are seldom recorded.

The ultimate parts of the community are the individual plants, but a description of it in terms of the characters of these units and their spatial relations to each other is impracticable at the individual level. It is, however, feasible in terms of the aggregates of

* Presidential address to the British Ecological Society on 11 January 1947.

individuals and of species which form different kinds of patches; these patches form a mosaic and together constitute the community. Recognition of the patch is fundamental to an understanding of structure as analysed here.

In the subsequent analysis evidence is adduced to show that the patches (or phases, as I am calling them) are dynamically related to each other. Out of this arises that orderly change which accounts for the persistence of the pattern in the plant community. But there are also departures from this inherent tendency to orderliness caused by fortuitous obstacles to the normal time sequence. At any given time, therefore, structure is the resultant of causes which make for order and those that tend to upset it. Both sets of causes must be appreciated.

In describing the seven communities I propose in the first examples to emphasize those features which make for orderliness, in the later to content myself with little more than passing reference to these and to dwell specifically upon departures from it; in all examples to bring out special points for the illustration of which particular communities are well suited or for which data happen to be available.

For the present the field of inquiry is limited to the plant community, divorced from its context in the sere; all reference to relics from its antecedents and to invaders from the next state is omitted. I am assuming essential uniformity in the fundamental factors of the habitat and essential stability of the community over a reasonable period of time.

The evidence from seven communities

The regeneration complex

Regarded by some as an aggregate of communities, by others as one community, the regeneration complex fittingly serves as an introduction because its study emphasizes the underlying uniformity of the nature of vegetational processes. It consists of a mosaic of patches forming an intergrading series the members of which are readily enough assignable to a few types or phases. The samples of these phases are repeated again and again over the area; each is surrounded by samples of other, but not always the same phases.

As the work of Osvald (1923) and of Godwin & Conway (1939) shows, these phases are dynamically related to each other. For Tregaron Bog the sequence is briefly as follows. The open water of the pool is invaded by *Sphagnum cuspidatum* which in turn is invaded and then replaced first by *S. pulchrum*, then by *S. papillosum*, by whose peculiar growth a hummock is formed. This hummock is first crowned by *Calluna vulgaris*, *Erica tetralix*, *Eriophorum vaginatum* and *Scirpus caespitosus*, later by *Calluna vulgaris* with *Cladonia silvatica* forming a subsidiary layer. The proof of the time sequence lies in the vertical sequence of plant remains in the peat itself.

Each of the phases of the regeneration complex was at one time regarded as a community, and the whole as an aggregate of communities dynamically related to each other. Tansley (1939), for example, still calls them seral. The resemblance to a sere is close; each phase from the open water to the hummock with *Calluna* depends on its antecedents and is the forerunner of the next, the sequence depending on plant reaction.

The resemblance, however, is partial only, for a cycle of change is completed by the replacement of the hummock by the pool, a topographical change brought about by differential rates of rise in the different patches. That is, the immediate cause does not reside exclusively in the patch itself but in the spatial relation between patches and their relative changes in level. Thus the *Calluna* of the hummock dies or is killed and is followed in the place vacated by it by other species which have had no direct or indirect hand in its death. Thus in the full cycle we may distinguish an upgrade series and a downgrade.

Each patch in this space-time mosaic is dependent on its neighbours and develops under conditions partly imposed by them. The samples of a phase will in general develop under similar conditions but not necessarily the same, for the juxtaposition of phases will vary. This may be expected to affect the rate of development of the patches of a phase and their duration. But on the duration of the full cycle of change and its component phases there is inadequate information, although the *impression* is gained that in the upgrade series the net rate of production per annum is at first slow, then fast in the *Sphagnum papillosum* phase, then slow again in the final phases.

This brief summary presents the regeneration complex as a community of diverse phases forming a space-time pattern. Although there is change in time at a given place, the whole community remains essentially the same; the thing that persists unchanged is the process and its manifestation in the sequence of phases.

Dwarf Callunetum

It may well be argued that the regeneration complex is a special case. It is one of my objects to show that these dynamic phenomena are paralleled in a wide range of communities, all of which can hardly be set aside as special cases.

Occasionally in the regeneration complex there occur partial sequences in space which correspond with the time sequence. Complete correspondence between the space and time sequence is found, however, in certain communities under highly specialized conditions. Such conditions are found in the Arctic (Walton, 1922), where the prevailing winds determine the alinement of the plants in the community and their unidirectional vegetative spread. Similar phenomena are shown by the dwarf Callunetum of highly exposed places on the slopes of the Cairngorms.

The Callunetum consists of strips of *Calluna* separated by strips of bare wind-swept soil; all the *Calluna* plants lie side by side and in line, their apices spreading into the shelter created by the plant itself and their old parts dying away behind. In other places the strip of vegetation is double, consisting mainly of *Calluna* and *Arctostaphylos uva-ursi*; under the prevailing conditions both species have the same habit, but they differ so far in their specific make-up that *Arctostaphylos* is generally found to leeward of *Calluna* (Fig. 1). The spatial relation between these two species and their unidirectional growth suggest that as *Arctostaphylos* grows forward over the eroded soil, *Calluna* grows over the older parts of *Arctostaphylos*, suppressing the leaf-bearing shoots though not killing the old stems. This is, indeed, the case, and proof of the dynamic relation is found in the dead remains of *Arctostaphylos* below the *Calluna*.

The community just described is a three-phase (or more-phase depending on the refinement of the analysis) system with the phases arranged in linear series. In the dwarf Callunetum of more sheltered places unidirectional spread is replaced by centrifugal; each *Calluna* plant is free to spread until checked by other plants in its neighbourhood. The static pattern of the mature community is a background of *Calluna* with scattered patches of *Arctostaphylos*, *Cladonia silvatica* and bare soil in it. But the time sequence remains essentially the same, with four phases in the full cycle, which may be shortened to three or even to two phases.

4 *Pattern and process in the plant community*

Briefly summarized from details obtained by Dr G. Metcalfe (to whom I am also indebted for Fig. 1) during the Cambridge Botanical Expedition to the Cairngorms and kindly placed by him at my disposal, the history of the relations between the phases is as follows, the kind of evidence being the same (except for the positional relation) as that obtained from the linear series. The young vigorous shoots of *Calluna* are usually dense enough to exclude lichens; the older, fewer and less vigorous shoots are unable to

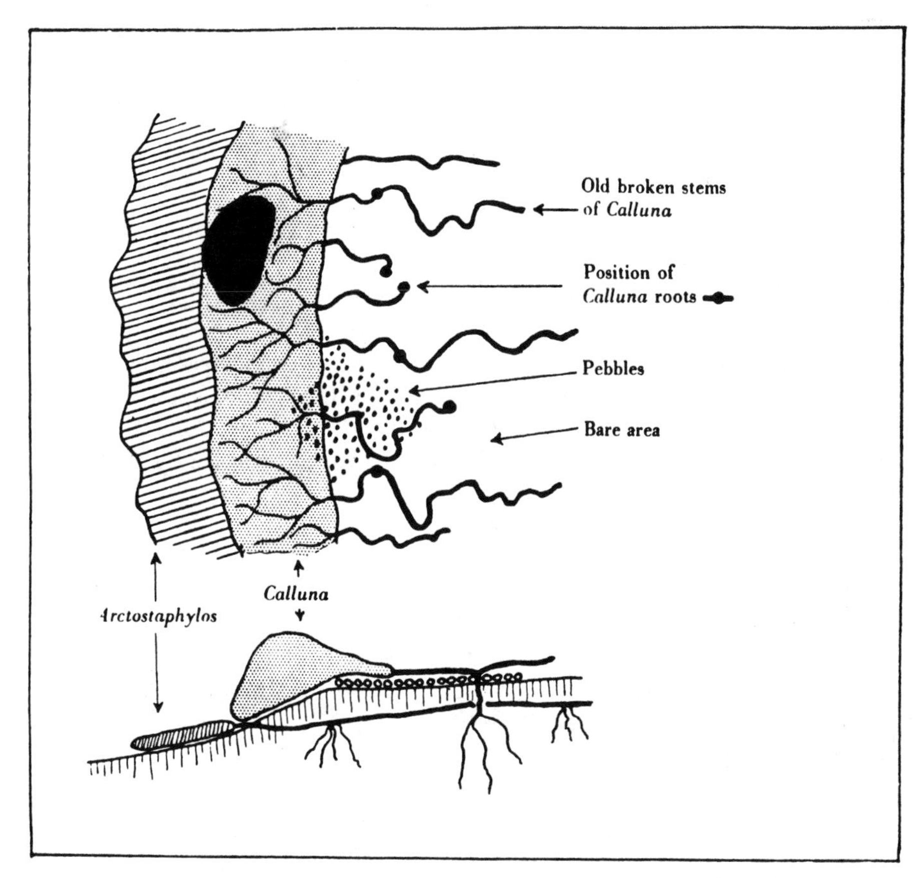

Fig. 1. Diagrammatic representation in plan and elevation to show the spatial relation between *Calluna* and *Arctostaphylos* in the 'double strips' separated by wind-swept bare soil in the Dwarf Callunetum of highly exposed places at approx. 2500 ft. on the northern slopes of the Cairngorms. The double strip moves forward in time; *Arctostaphylos* invades bare soil, and *Calluna* moves on and suppresses the adjacent *Arctostaphylos*.

do so and *Cladonia silvatica* is often abundant on the old parts of the plant. On its death *Cladonia* becomes dominant, anchored on the dead stems. (The counterpart of this phase in the linear series is the strip of dead *Calluna* stems behind the live; owing to the violence of the wind, *Cladonia*, although present, is unable to dominate.) In time the *Cladonia* mat disintegrates (in much the same way as it does in Breckland (Watt, 1937)) and bare soil is exposed, often with some remains of *Calluna* stems on it. If *Arctostaphylos* happens to be near such a gap (and a much ramified system of non-leaf bearing stems occurs under the *Calluna* mat) invasion is followed by complete occupation. In time by

ALEX. S. WATT 5

vegetative spread from the margin *Calluna* replaces the *Arctostaphylos*. The relations between the phases are indicated diagrammatically in Fig. 2, which also shows the short-circuiting in the absence of *Arctostaphylos*.

Clearly *Calluna* is the dominant plant; the locally dominant *Arctostaphylos* and *Cladonia* are allowed to occupy the ground which *Calluna* must temporarily vacate, *Arctostaphylos* as a phase in the upgrade series and *Cladonia* in the downgrade. Although no data are available it may be pointed out that from the annual rings of the woody stems of *Calluna* and *Arctostaphylos* some estimate of the duration of the phases is made possible.

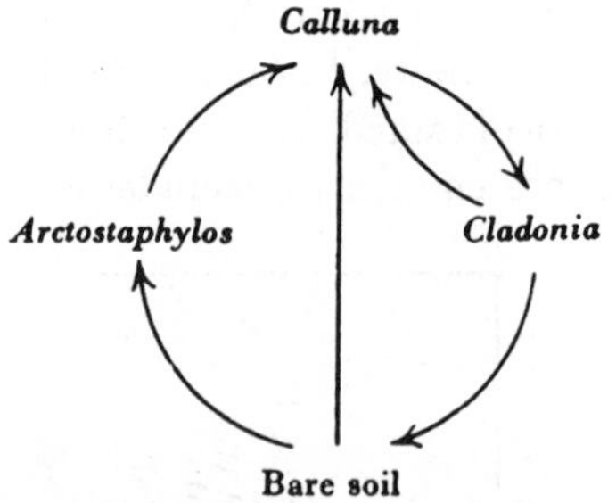

Calluna

Arctostaphylos

Cladonia

Bare soil

Fig. 2. Diagram illustrating the dynamic relations between the chief species in Dwarf Callunetum in less exposed places than those mentioned in Fig. 1. The arrows indicate the direction of change.

Eroded Rhacomitrietum

At somewhat higher altitudes on the Cairngorms an eroded community dominated by *Rhacomitrium lanuginosum* shows phenomena similar to those of the *Calluna* strips. For here, too, the direction of the prevailing wind, by determining the direction of spread of the plants, imposes on the patches of vegetation a space sequence which is also a time sequence.

The following very short account is based on details kindly supplied by Dr N. A. Burges, to whom I am also indebted for Fig. 3.

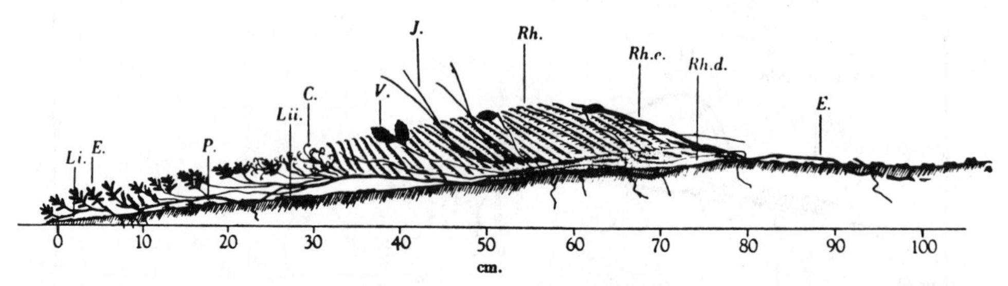

Fig. 3. Profile downwind or across a patch of vegetation in eroded Rhacomitrietum at approx. 3500 ft. on the northern slopes of Cairngorm. The spatial sequence is also a time sequence. *Li.*, *Lii.* and *P.* = bryophytes, *E.* = *Empetrum hermaphroditum*, *C.* = *Cladonia rangiferina*, *V.* = *Vaccinium myrtillus*, *J.* = *Juncus trifidus*, *Rh.* = *Rhacomitrium lanuginosum*, *Rh.e.* = *R. lanuginosum*, eroded face, *Rh.d.* = *R. lanuginosum*, dead.

The community consists of a network of patches of bare soil and of vegetation; each of the latter shows from the lee side to the exposed and eroded western side (Fig. 3) a series of phases characterized respectively by bryophytes, *Empetrum hermaphroditum*, a mixture of *Vaccinium myrtillus*, *V. uliginosum* and *Rhacomitrium lanuginosum*, and finally *Rhacomitrium* itself. Proof cf the temporal sequence is found in the vertical layering of peaty remains found under the last phase. The *Rhacomitrium* phase is eroded by the wind, the exposed accumulated humus dispersed and mineral soil once again laid bare. Since the stems of *Empetrum* stretch from end to end of the patch the number of rings enables an estimate to be made of the duration of the whole and of its phases. The rate of advance is approximately 1 m. in 50 years.

6 *Pattern and process in the plant community*

Bracken

The relation of *Arctostaphylos* to *Calluna* in the dwarf Callunetum of the less exposed places is deduced from internal evidence and supported by evidence provided by the study of their spatial and temporal relations in the specialized double strips. The Pteridietum (Fig. 4) which has been studied (Watt, 1947) provides a close parallel, for the area in which

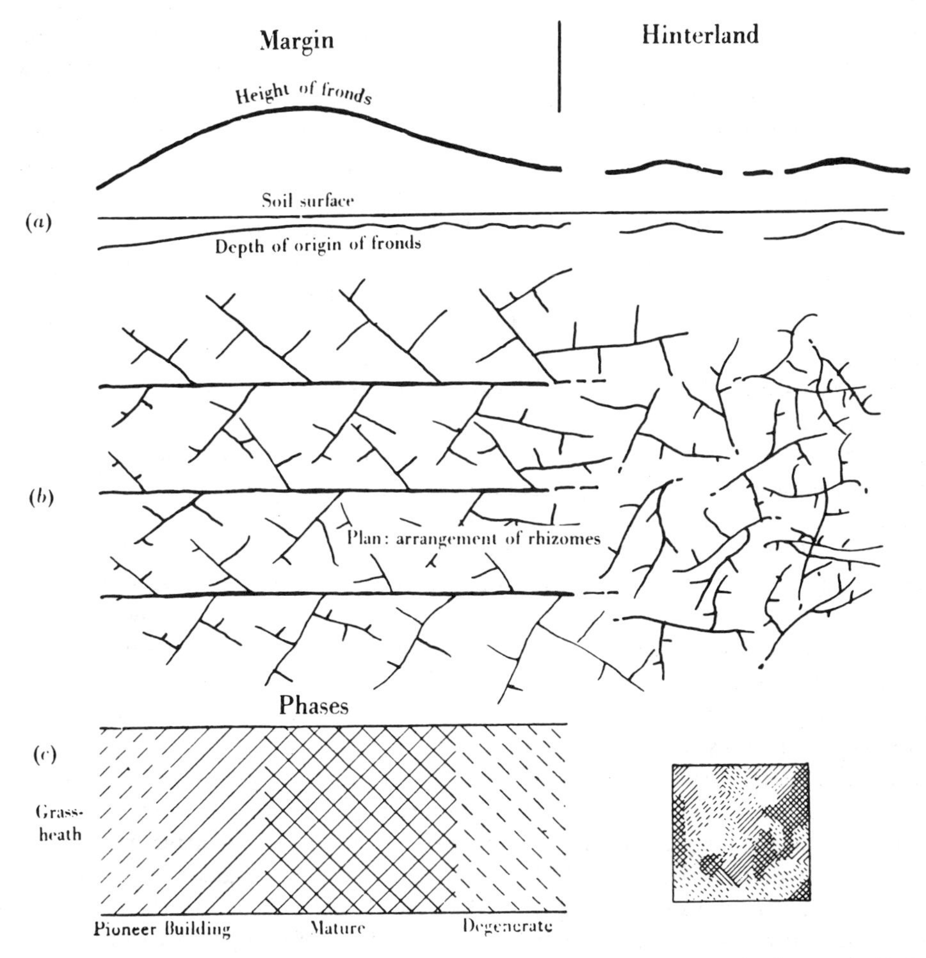

Fig. 4. Diagram to illustrate the spatial and temporal relations between the phases in the marginal belt and in the hinterland of a bracken community on Lakenheath Warren (Breckland). In (*a*) the change in height and in continuity of cover of the fronds are indicated: also the change in depth of origin of the fronds. In (*b*) the relative size and direction of growth of the main axes of the bracken plants are shown. In the marginal belt the axes are parallel, in the hinterland they form a network. In (*c*) the phases in the margin are in linear series: in the hinterland they are irregularly arranged.

it occurs consists of a hinterland (comparable with the continuous Callunetum) in which the fronds are patchily distributed and the axes of the bracken plants form a loose network; marginal to the hinterland is a belt of bracken invading grass-heath in which the individual plants lie side by side, with their main axes parallel to each other and their apices in line (cf. the strip of *Calluna*).

Such a marginal belt with unidirectional spread can be divided into a series of zones or phases by lines running parallel with the invading front. These phases can be distinguished by data from the frond and rhizome. The inherent circumstances make it clear that, from back to front, the space sequence of phases represents a sequence in time. Among the overdispersed fronds of the hinterland there are patches without fronds (the grass-heath phase) and patches with fronds which vary among themselves in features of frond and rhizome in much the same way as the sequence in the marginal belt; these are distinguished as the pioneer, building, mature and degenerate phases. These phases, together with the grass-heath phase, form a cycle of change in time at one place. Comparison of the two sets of data (Watt, 1945, 1947)—from the zones of the marginal belt and the patches of the hinterland—shows close agreement between the relations between the values within each set, an agreement all the more striking when allowance is made for the different circumstances. For in the marginal belt the phases are in linear series and bracken invades grass-heath free from bracken; in the hinterland the grass-heath phase is vacated by bracken but seldom completely before reinvasion takes place. There is thus some overlap between the beginning and end phases of the cyclic series.

The evidenee for cyclic change derived from the specialized marginal belt is supported by the internal evidence from the hinterland itself. For in the grass-heath phase without fronds there are abundant remains of decaying rhizome, and under the litter in the later phases with bracken are the recognizable remains of *Festuca* and *Agrostis* of the grass-heath phase. Phases with, and a phase without, fronds alternate in time.

The processes involved in the marginal belt and the hinterland are the same. In the marginal belt the natural tendency of the bracken to spread centrifugally is checked by lateral competition; spread is thus directed by ecological opportunity into the grass-heath which is free from bracken. Once established the sequence of phases depends on the development of the bracken itself and development as affected by its own reaction. In the hinterland, where the axes form a loose network, rejuvenation is similarly directed by ecological opportunity and essentially restricted to areas vacated by the death of the bracken. The phases which follow are a biological consequence of established invasion and are limited to the area of that invasion.

Running parallel with the change in the vegetation of the cycle, there is change in the factors of the habitat and in total habitat potential. The variable factors of microclimate and soil, e.g. shelter, light intensity, temperature, the amount of litter, the state of the humus and its distribution in the soil, are differentiae of the phases superposed on the original foundation of a uniform habitat. They are closely linked with the phases because they are the effects of the plants themselves; the effects of one phase become part cause of the next. Thus the spatial variation in the habitat is primarily caused by the variable vegetational cover. Further, taking a broad view, we may note that in the upgrade series there is an accumulation of plant material and an increase in habitat potential; in the downgrade both are dissipated.

The significance of the differential action of the phases on microclimate may be illustrated by the effect of frost during the winter 1939–40. In the grass-heath or pioneer phase rhizome apices were killed at a maximum depth of 20 cm.; in the mature phase, protected by a blanket of litter, the maximum depth of lethal damage was 6 cm. Action of this kind retards development and prolongs the duration of a phase, upsetting the smooth course of the time sequence.

8 *Pattern and process in the plant community*

Grassland A (Breckland)

The vegetation of Grassland A (Watt, 1940) is obviously patchy. In its hollows and hummocks the habitat presents a striking resemblance to the regeneration complex, and here, too, vegetational variation is linked with variation in microtopography and soil habitat.

In a reinvestigation of the community upon dynamic lines four phases are recognized: the hollow, building, mature, degenerate. There are intermediates, but little difficulty is experienced in assigning all parts of the typical community to one or other of these four phases.

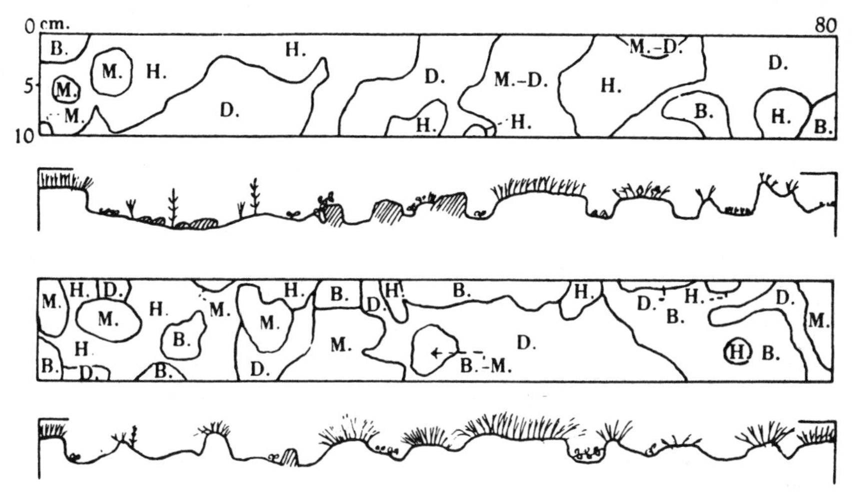

Fig. 5. The relative size and spatial relations of the phases in a plot of 160 × 10 cm. in Grassland A. The relation between the phases and the microtopography is seen in the profile taken along the upper edge of the plot.

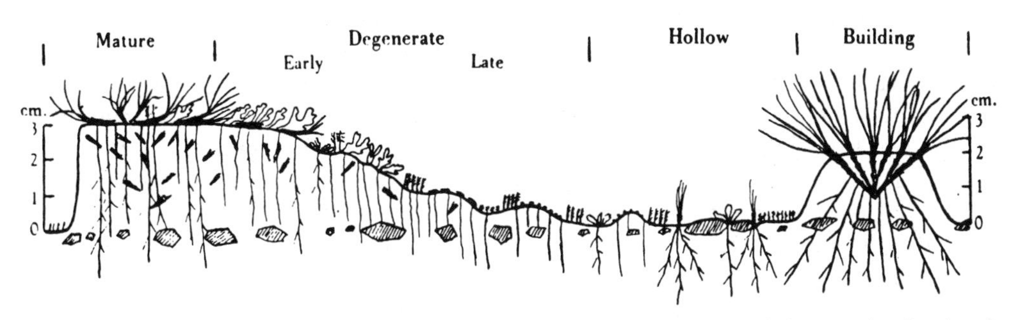

Fig. 6. Diagrammatic representation of the phases showing change in flora and habitat and indicating the 'fossil' shoot bases and detached roots of *Festuca ovina* in the soil.

Their spatial distribution is shown in Fig. 5; the patches are irregular in size and shape and their juxtaposition varies. These four phases form a time sequence, the evidence for which is emphasized in the following reconstruction (Fig. 6). The whole 'life' of the community centres round the reactions and life history of *Festuca ovina*, its growth and reproduction; it has every right to be called the dominant plant even although it occupies less than one-half of the total area (approximately 45%). The seedling becomes established among the stones which floor the hollow phase; as the plant grows and spreads the level of the mineral soil inside the tussock rises. This soil is free from stones and its accumulation is due to the activities of ants and earthworms, very probably also to wind-borne particles and particles water-borne in the splash of heavy rain. The young vigorous fescue with relatively long leaves and many inflorescences on relatively long stalks constitutes the building phase; in it the shoots of the plants are still attached to the parent stock. By further accumulation of soil the hummock increases in height, attaining a maximum of about 4 cm., and then carrying much less vigorous fescue, with shorter leaves and fewer inflorescences with shorter stalks. The original many-branched tussock is replaced by numerous small plants each lying horizontally and consisting of one or two shoots with vertically descending roots; they have arisen from the larger plant by the separation of branches through the death and decay of the parent stock, the evidence for this being the presence throughout the soil profile of the hummock of the bases of lateral branches which are resistant to decay and thus persist.

In the spaces between the individual plants in the mature phase, the fruticose lichens *Cladonia alcicornis* and *C. rangiformis* become established. They spread and ultimately form a mat below which the remains of fescue are found. This is the early degenerate phase. 'Fossil' shoots and unattached roots are found in the soil. In the late degenerate phase, characterized by the crustaceous lichens *Psora decipiens* and *Biatorina coeruleonigricans*, 'fossil' shoots and unattached roots are again found. In places these roots project above the soil surface and bear witness to the erosion (already suggested by the pitted surface) which gradually wears down the hummock. Some vestiges of the hummock survive into the hollow phase, but eventually these disappear and the erosion pavement of flint and chalk stones is once again exposed.

Every part of the surface of the typical community can be assigned to one or other of these four phases. A single set of them is fully representative of the whole and summarizes in itself the processes at work and their manifestations. The understanding of the community as a working mechanism is based on the elucidation of the relations of the phases to each other. They are its minimum representative (minimum area), its very core.

Now the contributions which the phases make to the community as a whole may vary among themselves and also from year to year. To eliminate one source of variability, and at the same time to take account of it as an ecological phenomenon of great importance, I propose to base the description of the community upon areas of equal size in the several phases (and to call the whole the unit pattern) and to make a separate assessment of the relative areas occupied by the phases.

Some of the data for the unit pattern of Grassland A are presented graphically in Fig. 7; the diversity as well as the unifying continuity of the phenomena are clearly shown. The cover percentage of fescue and 'bare soil and stones' are inversely related. The holophytic bryophytes are virtually excluded from the building and mature phases and increase to a maximum in the hollow, where competition from fescue is least. For the lichens, on the other hand, the maximum appropriately lies in the phase where the fungal partners can utilize the organic remains accumulated during previous phases.

The selective effect of the phasic microenvironments on the distribution of seedlings and their subsequent fate is shown in Table 1. The total number of seedlings varies directly with the 'bare soil and stones'. The distribution of the adults suggests failure of the seedlings to reach maturity in the mature and degenerate phases and shows restriction to the building and hollow phase and virtually to the hollow phase because the great bulk of

10 *Pattern and process in the plant community*

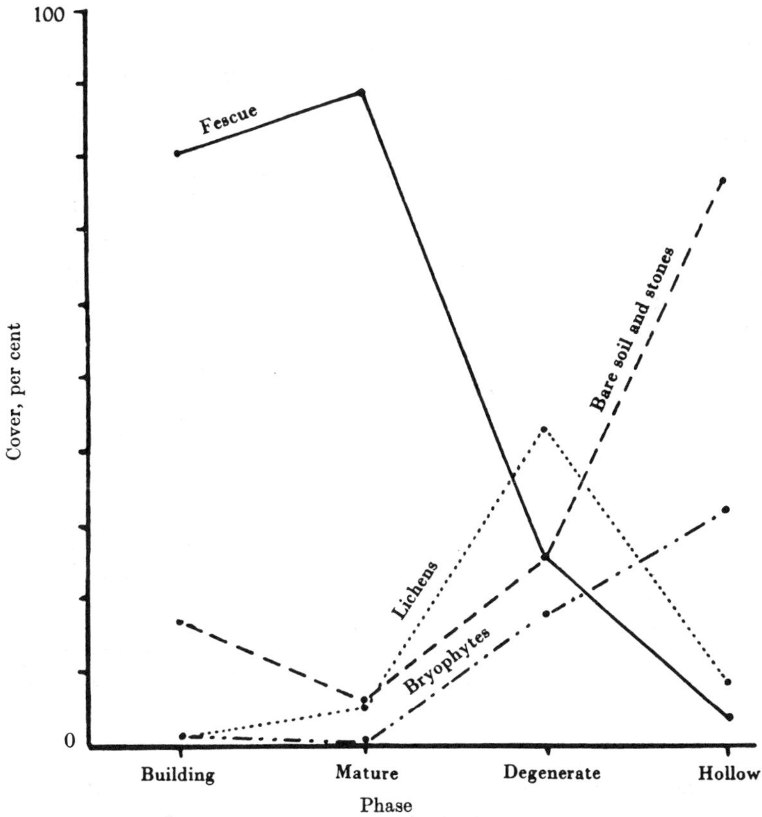

Fig. 7. Graphical presentation of some data for Grassland A. Note that most of the data for the building phase are from late stages.

Table 1. *Grassland A*

Species	Building	Mature	Degenerate	Hollow
	Phases			
	Total no. in 40 plots, each 5 × 5 cm.			
	Seedlings			
Arenaria serpyllifolia	0	0	0	1
Avena pratensis	0	0	0	2
Calamintha acinos	3	3	8	6
Cirsium lanceolatum	0	0	1	0
Crepis virens	24	7	10	21
Erigeron acre	25	17	31	68
Festuca ovina	2	0	10	66
Hieracium pilosella	9	5	6	7
Koeleria gracilis	0	0	1	0
Senecio jacobaea	0	0	0	1
Total	63	32	67	172
	Adults			
Arenaria serpyllifolia	0	0	0	1
Calamintha acinos	0	0	0	3
Cerastium semidecandrum	1	0	0	8
Crepis virens	1	0	0	0
Erigeron acre	3	0	1	3
Galium anglicum	11	0	0	7
Hieracium pilosella	2	0	0	4
Saxifraga tridactylites	1	0	0	0
Total	19	0	1	26

the adults recorded from the building phase are in the immediate surround of the fescue tussock and not in it. This restriction to the hollow phase in all probability holds for the surviving fescue seedlings as well. Thus at any given time the initiation of the cycle of change is restricted in space, and at any given place to the hollow phase in the time sequence.

Estimated by two different methods the relative areas of the phases hollow, building, mature, degenerate are, respectively, 25·7, 15·1, 16·3 and 42·9. As far as I can judge from the annual charting of two plots since 1936, but without actually noting the spatial limits of the phases, Grassland A is remarkably stable both in its unit pattern and the areal extent of its phases.

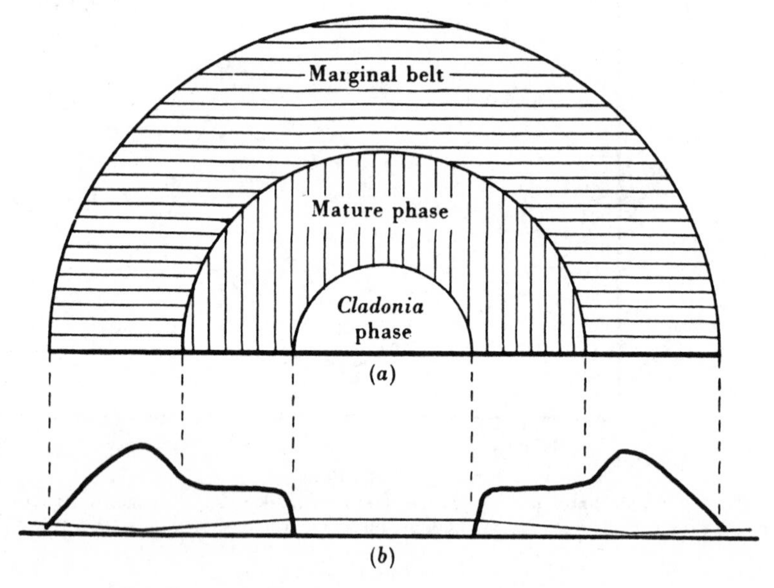

Fig. 8. Diagrammatic representation in plan (*a*) and profile (*b*) of an *Agrostis* ring set in a background of *Cladonia silvatica*. In (*b*) the darker line indicates the change in height (and/or number) of *Agrostis* shoots, the thinner line the change in thickness of the *Cladonia* mat.

Grass-heath on acid sands (Breckland)

The grass-heath on acid sands in Breckland has been sufficiently investigated to provide supporting evidence for the dynamic interpretation of the plant community and also to demonstrate the usefulness in description of separating the estimate of the relative areas of the phases from the unit pattern.

Fundamentally the community consists of rings (both solid and hollow) of *Agrostis tenuis* and *A. canina*, and occasional tussocks of *Festuca ovina* set in a background of *Cladonia silvatica*. (Further reference to *Festuca* is omitted because its relations to other species in the community are not fully elucidated.) Each ring is derived by vegetative spread from one plant and epitomizes in itself the spatial and temporal changes within the community. A typical ring (Fig. 8) of medium size consists of a peripheral zone of numerous, vigorous, vegetative and flowering shoots arising from large plants with rhizomes radiating outwards, an inner zone (mature phase) with fewer and shorter vegetative and flowering shoots, or no flowering shoots at all, on smaller plants whose rhizomes form a loose open network (cf. Bracken, p. 6). In the centre of the ring *Agrostis* as a live plant is absent, but its abundant dead remains are found below the mat of dominant *Cladonia silvatica* which is from 3 to 4 cm. thick; to this thickness it has gradually risen from a thickness of 1 cm. only in the middle of the peripheral zone.

Fuller investigation of the dynamic behaviour of the rings justifies the recognition of phases within the marginal belt and in a full cycle additional phases of degeneration and rebuilding following disruption of the lichen mat (Watt, 1938). But to keep the issue simple no more than the three phases need be further considered since they may complete a short cycle; samples from unit areas in each constitute the unit pattern.

The mutual relations between *Agrostis* and *Cladonia* have not been investigated to the degree necessary to decide whether the death of *Agrostis* is due to age, or in some measure to the influence of *Cladonia* or to both acting together. Similar problems arise at this stage of the cycle of change in all the communities examined; the problems are universal because they concern death and its causes. Their solution, although important, is not necessary to establish the fact of replacement.

The extension of the *Cladonia* phase at the expense of the mature phase (inner zone with *Agrostis*) is not, however, merely a question of time; it is primarily due to drought, which differentiates between the *Agrostis* plant of the peripheral zone (which survives) and that of the mature phase (which dies). As the result of a severe drought the mature phase passes wholesale and abruptly to the *Cladonia* phase, and the extent of the change will depend on the area of the phase capable of being affected by it. An area occupied by a large number of small patches of *Agrostis* will be relatively little affected by drought, while one which is almost entirely in the mature phase, through the spread and fusion of rings, will be almost wholly affected by it (Fig. 9). This explains the violent fluctuations met with in the populations of *Agrostis* in this community; in certain areas fluctuations have been so wide that estimates of abundance for one year are of no value as a diagnostic character. They do, however, reflect the vagaries in meteorological factors. On the other hand, data from the unit pattern would be much less affected.

Beechwood

The effect on the structure of a community of drought or other efficient cause may persist long after the cause has ceased to operate. In fact, at any given time, there may be no correspondence between structure and the current meteorological factors. The point is best illustrated by reference to communities with long-lived dominants. But first a brief note on phasic change in beechwoods.

The patchiness in some all-aged beechwoods on the Chilterns, reputedly managed on a selection system, is interpretable in terms of the temporal sequence of phases as revealed by the study of the life history of pure even-aged beechwoods of the same ecological type on the South Downs (Fig. 10) (Watt, 1925). To the three phases recognized—Bare, *Oxalis*, *Rubus*—there should be added a fourth, the gap phase, to which regeneration is confined because it is excluded from other phases. At any given place there is a cycle of change consisting of an upgrade series of phases in which there is a continual change in ecological structure, associated with changing age, rate of growth and density of the dominant trees and correlated with changes in the field layer, and a downgrade of the dying, dead and rotting stems and the vegetation of the gap.

For the time-productivity curve for the upgrade series we have the foresters' yield tables for managed woods. Although the data take no account of the yield from the subsidiary vegetation, we may assume this to be small in relation to the yield of the trees and without appreciable effect on the course of the curve. This is of the familiar growth type, rising slowly at first, then fast, then slowly again.

The phasic phenomena of woodland are on a scale to be immediately obvious. Again because of the scale, but also because we have the data, woodlands afford an excellent illustration of the cyclic relation between tree species and of causes having long-continued

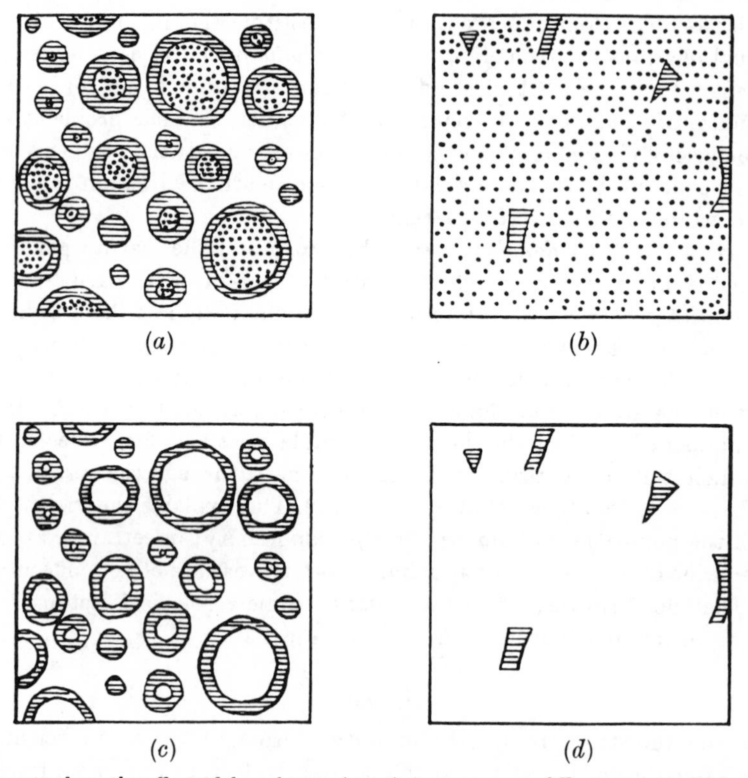

Fig. 9. Diagram to show the effect of drought on *Agrostis* in two areas of Festuco-Agrostidetum at different stages of development. In (*a*) there are many small (young) patches of *Agrostis*; in (*b*) the patches have grown and fused so that the bulk of the area is in the mature phase. In (*c*) and (*d*) drought has killed the *Agrostis* in the mature phase only of (*a*) and (*b*).

effects in structure. Both points have been dealt with recently by Jones (1945); brief reference is all that is necessary here.

Viewed against the ideal of sustained yield, aimed at most simply among foresters by allocating equal areas to each age class or group of age classes, the primeval wood by all accounts more often than not shows an unequal distribution of the areas occupied by its age classes. Aside from minor fluctuations over periods of time in, say, the number of deaths among old trees, there are exceptional factors of rare or sporadic occurrence, such as storms, fire, drought, epidemics, which create a gap phase of exceptional dimensions. If the whole gap phase is regenerated about the same time—and various circumstances, both inherent and external, like periodicity in seed years and suitable meteorological factors for seedling survival may help in that direction—then there is initiated an age class of abnormal area. This will persist like a tidal wave moving along the age classes until at least the death of the trees; it may even influence the structure of the next generation. In other words, the relative areas under the age classes (as a super refinement of phasic subdivision) need bear no relation to current meteorological factors but be explicable in terms of some past event which happened, it may be, 200 or 300 years ago.

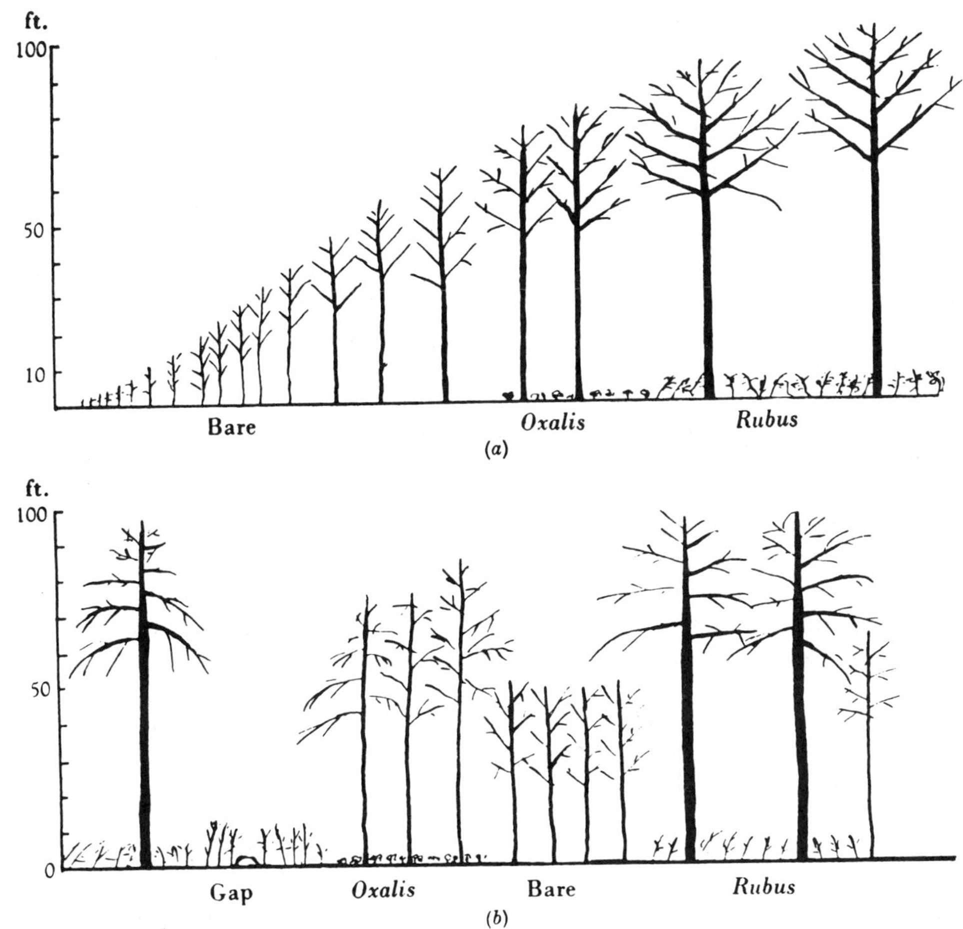

Fig. 10. Diagram to illustrate (*a*) the phasic change during the life history of an even-aged pure beechwood and (*b*) the distribution in space of the phases when the old wood is left to itself.

A series of sporadic exceptional events will obviously increase the difficulties of 'explaining' current relative areas, as also will the selective destruction of phases their juxtaposition in space. Fortuitous fluctuations of climate and other causes may thus bring about major departures from the normal or ideal wood.

The second point refers to the normal inclusion in the cycle of change of a phase dominated by another species of tree, in different kinds of beechwoods, for example, by ash, oak or birch. Such a phase (e.g. birch) may occupy small or large areas, but irrespective of the area it occupies, if the relation is cyclic and not seral (that is, one of alternating equilibrium in which successful regeneration of the beech is inhibited by itself but promoted by the conditions brought about by the other species of tree), then such a phase is part and parcel of the system. To recognize the birchwood and beechwood as separate entities may be convenient in classification, but successful practical management of such woods depends on their recognition as integral parts of a coordinated whole.

SUPPLEMENTARY EVIDENCE

Doubtless because of its universality, its familiarity and its apparent lack of significance the fact of the impermanence of the individual or group of individuals has seldom evoked comment. Butcher (1933) has, however, called attention to the changing pattern of river vegetation over a short period and suggests progressive impoverishment of the soil of the river bed as the cause of the disappearance of *Sium erectum* from a given place in it. Tutin (1938) has attributed the marked reduction in area occupied by *Zostera marina* primarily to deficiency in light intensity; this predisposed the plants to disease, which is a symptom of weakness rather than a cause of its virtual disappearance. Without contesting the accuracy of the immediate diagnosis, I would suggest that both phenomena may be more fully and satisfactorily explained against the background of cyclic change.

Phenomena similar to those described have been seen but not investigated in communities dominated respectively by *Polytrichum piliferum* (in Breckland), *Ammophila arenaria* and *Carex rigida-Rhacomitrium lanuginosum* of mountain tops. Poole (1937), from a study of mixed temperate rain forest in New Zealand, concludes that *Dacrydium cupressinum* on the one hand, and *Quintinia acutifolia* and *Weinmannia racemosa* on the other, alternate in time at one place.

The phenomena are not confined to temperate regions. In the high arctic the few species are represented by scattered individuals, and communities are distinguished by the quantitative relations of their chief components and/or by habitat factors (Griggs, 1934). Where there are few individuals competition is negligible or absent; factors of soil and climate are the primary selective agents. Griggs says there is not only change from place to place, but change in time at a given place (he refers to unpublished data from permanent quadrats as the basis for this latter statement). This distribution in place and in time, is, according to Griggs, erratic, but no details are given.

At the first remove in the low arctic a two-stage succession from open initial communities to the closed association of heath and bog shows the emergence of the organic environment into importance. Competition now is a factor limiting the distribution of species.

At the other extreme is tropical forest, the most complex of plant communities in which the biotic environment plays an outstanding part. It is not to be expected that such a complex community will be amenable to a simple dynamic interpretation. Yet presumptive evidence is given by Aubréville (1938) who has expressed views on the structure of the forest closely akin to those advocated here for a selection of communities. He finds that although there are in tropical forests very many species of tree the dominant or most abundant species are relatively few in number (about a score) their relative proportions varying from place to place, some dominating in one place, some in another.

But also at a given place there is a change in the dominant species with time, for, in the subsidiary layers the young trees (the dominants of the future) are often specifically different from the dominants of the uppermost layer. Aubréville, however, does not mention a regular cycle of change at a given place; in fact, as far as can be judged from his account the assemblage of species in a newly formed gap depends on the fortuitous occurrence of a number of factors, including distance of parents, mobility and abundance of seed, frequency of seed production, dispersal agents and their habits, and various factors like animals eating the seed, duration of life of the seed, light intensity, root competition and the kind of undergrowth, affecting the survival of the seed and the establishment of the seedling. Although the earlier stages of gap colonization may show diversity in floristic composition the later ones, according to Dr P. W. Richards (oral communication), are much more uniform.

If we assume with Aubréville that the general climatic and soil conditions are uniform over a large area then the differentiating biological characters of the many species assume primary importance, and variation in floristic composition from place to place is the resultant of the impact of these biological characters and environmental factors, largely, but not exclusively, residing in the plant community itself; that is, the community itself largely determines the distribution, density and gregariousness of its component parts.

Comparison and synthesis

The analysis of the seven communities, documented by observations over a period of time, by comparative studies and by appeal to the registration of the time change either in the plants themselves or in their subfossil remains, shows that there is an orderly sequence of phases in time at a given place and that the spatial relation between the phases can be interpreted in terms of their temporal relations. I now propose to correlate the observed phenomena and to offer a simple generalization and explanation.

The cycle of change in all the communities is divisible into two parts, an upgrade and a downgrade. The upgrade series is essentially a process or set of processes resulting in a building up of plant material and of habitat potential and the downgrade a dispersion of these mainly by fungi, bacteria, insects, etc., but also by inorganic agents. At each stage, of course, there is both a building up and a breaking down, but in the upgrade series the balance is positive, in the downgrade negative.

When bracken was being investigated the field phenomena evoked the spontaneous application of the names for the phases, pioneer, building, mature, degenerate. They undoubtedly conveyed impressions which later study hardened, for they are obviously applicable to several of the communities. The quantitative graphical expression of the change for the upgrade series for woodland and the qualitative description for other communities lead me to suggest that a curve of the growth or logistic type expresses the general course of change in total production in the upgrade series of each of the communities (Fig. 11 and appended correlation table). This is doubtless an oversimplification of the whole concrete situation. But there are data enough to show that we are dealing with something fundamental, with in fact, some biological law which operates in the plant community.

For the downgrade series a curve of similar form is tentatively suggested. There are many partial assessments of the course of the changes, but none to my knowledge dealing with the whole series in a comprehensive way.

The basis of these regular phenomena of change may be sought in the behaviour of populations and in the sequence of a plant succession on the restricted area of a patch. The community as we have seen consists of patches, each of limited area, and differentiated by floristic composition, age of dominant species and by habitat.

Alex. S. Watt 17

Patchiness can be readily enough detected in all the communities investigated and in the others referred to. In forest, the community about whose structural types more is known than about any other kind of community, patchiness is widespread. There is a form of 'irregular forest' that would appear to be an exception, although it is not known how far it is local or temporary, and until it has been examined by the use of criteria in addition to the distribution of crown classes, judgement must be suspended. A comparable structure may be found in other communities and at the moment no claim can be made that patchiness, as understood here, is universal. It certainly is widespread.

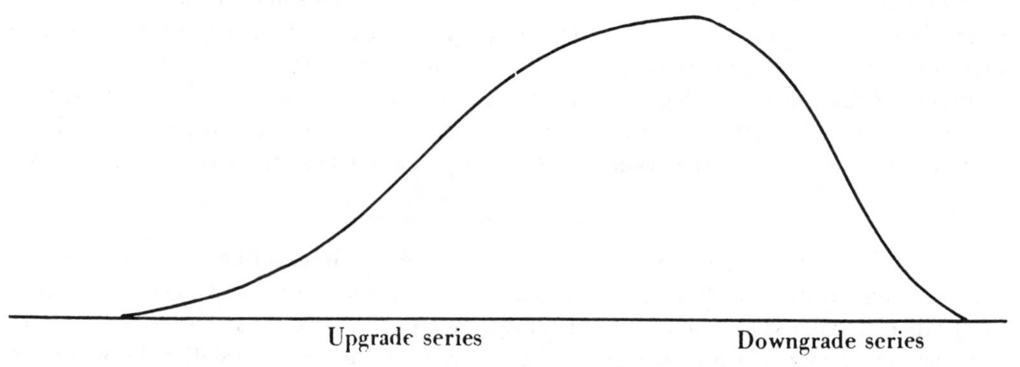

Fig. 11. The graphs illustrate in a general way the assumed course of change in the total productivity of the sequence of phases in the upgrade and downgrade portions of the cycle of change. The appended table correlates the phase sequences of the different communities.

Plant community				Phase		Hummock with:		
Regeneration complex	Pool	*Sphagnum cuspidatum*	*Sphagnum pulchrum*	*Sphagnum papillosum*		*Erica* *Scirpus*	*Calluna* *Cladonia*	Degenerate
Dwarf Callunetum	Bare soil	*Arctostaphylos*		*Calluna*		*Calluna*		*Cladonia*
Eroded Rhacomitrietum	Bare soil	Bryophytes *Empetrum*		*Rhacomitrium-Vaccinium*		*Rhacomitrium*		Eroding face
Pteridietum		Grass-heath	Pioneer	Building		Mature		Degenerate
Grassland A		Hollow		Building		Mature		Degenerate
Acid grass-heath		*Cladonia*	Pioneer	Building		Mature		*Cladonia*
Beechwood (sere 2)		Gap (late)	Bare stage		*Oxalis* stage	*Rubus* stage		Gap (early)

The widespread occurrence of patchiness is of interest in itself, but the persistence of the patch in time under continuing normal conditions is fundamental. Why, for example, starting with a set of patches, do we not get a general blending of the whole with the vegetation of every square foot a replica of every other?

May I first remind you that we are assuming uniformity in the fundamental soil and climatic factors of the community. Over and above this foundation we get, as we have seen, modifications of the soil and climate which are closely linked with the phases; these floating differentiae are in the first instance reactions of the phases and in the second, causes. We have, in fact, habitat variation in space as in time. (Also it should be noted that the habitat factors inherent in a phase modify and are themselves modified by the habitats of surrounding phases.) Seeds and propagules falling on a plant community therefore fall on a diversified environment (varying habitat and competitive power in the phases), and establishment is restricted to the phase where the plants survive; for many species this phase is the gap phase, or phase corresponding to it, formed by the death of the dominant in the mature phase or end-phase of the upgrade series. For plants spreading by rhizome the same holds good; in bracken, effective invasion is limited to the grass-heath phase, although it may begin in the mature phase and continue into the pioneer phase.

But not only is establishment of many species thus restricted in space, it is also restricted in time, to the period during which the gap phase is receptive. Checks to further colonization may be made by the invaders themselves, thus setting a time limit to the period within which establishment takes place. In this way, there is a tendency towards the production of even-aged or near-aged populations. Besides this restricted period during which colonization may take place, there are many other factors which tend to produce even-aged populations, some of which have already been alluded to and will not be further discussed.

Mainly because of competition, the plants established in the gap or corresponding phase are unable to invade and suppress surrounding phases. Thus prevented from spreading laterally, the patch becomes a microcosm of limited area with continuous but restricted food supply. Hence the population of that area behaves like a population of flies or human beings under similar restrictions of space and food. It is unable to spread, but it may develop.

This development in the upgrade series may take one of two forms, either as an even-aged population of a dominant species or as a series of dominants as in a plant succession. In the former the successive phases are distinguished by the species accompanying the dominant but primarily by the age and density of the dominant itself. The justification of the recognition of age in individuals or shoots as a basis for subdivision lies in the changing needs of the individual with age, its changing form (height, spread, root distribution, area of absorptive and anchoring surface, etc.) and its changing competitive power. The need for such recognition is obvious in a plant, whose gametophyte and sporophyte are separate and have different requirements, or in an insect with complete metamorphosis, in which the structure and habits of the larva differ so widely from those of the adult; but it holds also for higher plants because their requirements change during their life history.

The second type of change is found in communities where the successive phases (or some of them) have distinct dominants, that is, the phenomena are strictly comparable with those in a sere, where new dominants come in when the environment becomes suitable for them.

Now in both these, the even-aged population and the plant succession, the course of temporal change in total production per unit area is represented by a curve of the growth type (Pearl, 1926; Braun-Blanquet & Jenny, 1926). The reason for the general equation of corresponding phases now becomes obvious. In both, the phenomena of orderly change have their basis in the universal rhythmic phenomena of population development and of plant succession. It would thus appear that certain phenomena of the plant community can be brought within the scope of mathematical inquiry. This opens up an entirely new field in the study of natural communities with the possibility of putting the practical management of communities, as the forester has done the forest, upon a sound basis of natural law.

The description of the plant community is based on the unit pattern, which is the present expression of the continuous process of development and decline in a population broadly interpreted. The qualitative and quantitative assessment, therefore, of the plant community may be expressed in terms of the temporal changes in such a population.

Now it has been shown mathematically and confirmed experimentally that certain species with different ecological requirements live together and maintain stability only in a definite proportion (Gause, 1935). In like manner we may assume (although independent confirmation is necessary) that the plant community in a constant environment will show a steady state or a definite proportion between the constituent phases. This steady state we may call the phasic equilibrium, that is, the community is in harmony with itself and with its environment. Departures from this phasic equilibrium either in space or in time could then be measured and correlated with the changed factors of the environment.

The difficulties confronting this procedure are obvious. There are doubtless some communities (e.g. Grassland A) of which it would be possible to say with some assurance that they are in phasic equilibrium. But there are others like grass-heath on acid soil, and woodland, where, on account of the lag in response or repeated and fortuitous checks to normal orderliness, or the long delay between the incidence of the disturbing factor and return to a possible normal, anything like phasic equilibrium may rarely be achieved even although the tendency is always in that direction. In such circumstances we may copy the forester whose normal forest may be defined in terms of areas as one in which equal areas are occupied by a regular gradation of age classes. If we think in terms of natural groupings (phases) of unequal duration, then the areas would be proportional to the duration of the phases, the total equalling the duration of the cycle of change. The standard of comparison or normal would then be based on the unit pattern and duration of the phases. With such a standard, departures from it could then be assessed and comparisons made, but attempts to correlate such deviations with changes in the causal environmental factors would be confronted with great if not insurmountable difficulties. But once again there is brought into prominence the need for data on the duration of life of the plant in a variety of habitats and as affected by competition.

Relative areas of the phases is only one aspect of the general problem of the texture of the pattern or of the way in which the community is put together. The other two aspects are the size of the individual sample of the phase and the arrangement of the phases. Both these are important because the conditions of establishment and growth of plants vary with the size of the sample of the phase and with its surroundings.

SOME IMPLICATIONS

Animals and micro-organisms

I have thus far alluded almost solely to the higher plants and described the communities in terms of them, but it should be clear from the insistence on process that organisms other than higher plants play an important and sometimes a fundamental role, and that in the system of which the plant community is a part they ought to be investigated. On reading *Bioecology* (Clements & Shelford, 1939) I confess to a feeling of disappointment, because the data given to support the concept of the biome fall short of demonstrating that intimacy and integration between plants and animals which we are led to expect. Perhaps this is due to the lack of data for the communities described. The degree of intimacy varies, but all organisms whether loosely attached or occupying key positions in the system should be considered part of it.

20 *Pattern and process in the plant community*

Tragårdh (1923) has given a good illustration of the interactions between plants and insects in pinewoods in northern Sweden where the nun moth is a normal inhabitant, which never becomes epidemic as it does in pure pine or spruce woods in central Europe. This he attributes to the presence in the open pinewoods of *Calluna*, *Cynoglossum*, etc., which are the food plants of species of caterpillars which provide a constant supply of hosts for the polyphagous parasitic species of *Pimpla* which also attack the nun moth and so keep it in check. In the dense conifer forest of central Europe from which *Calluna* is excluded the population of *Pimpla* rises and falls with that of the nun moth but with a distinct lag; it also dies out with the extermination of its food supply. The reappearance of the nun moth is followed by rapid multiplication, and much damage is done before *Pimpla* arrives and increases in numbers to be an effective check to it. In the Swedish forest, on the other hand, a continuity of the population of the parasite is assured here because of the presence of alternative hosts feeding on *Calluna*, etc. Here we have interrelations between components of the tree layer and the field layer and the insects peculiar to each forming a complex web which makes for stability. This stability may be assured only so long as the food supply of the alternative hosts is present in sufficient amount. Further excellent examples of the intimacy and complexity of the relations between insects and plants are given by Beeson (1941).

This address would be too long if adequate reference were made to the variety of relationship between organisms associative, co-operative, antagonistic, symbiotic, etc.—and to the change in the nature of the food supply during the period covering the death of the plants in the mature phase and the complete disintegration of the accumulated organic matter. But that the change is regular is shown by the work of Blackman & Stage (1924) on the sequence of insects on living, dying, dead and decaying walnuts, of Graham (1925) on the earlier stages in the rotting of a log, of Saveley (1939) for insects and other invertebrates on dead pine and oak logs, and of Mohr (1943) on the changes in the species population during ageing and disintegration of cow droppings.

Comparable sequences are known among fungi. Parasites, facultative parasites and saprophytes succeed each other on dying and dead stems, and Garrett (1944) writes of the succession of micro-organisms on dead and decomposing roots. Generally on organic remains the sugar fungi develop first and are succeeded later by the cellulose fungi.

The part these organisms play is essential to the maintenance of the community; no assessment of the communal life can be complete without adequate appreciation of what is taking place.

The suggestion may be made that the life histories and food habits of micro-organisms in general, and insects and fungi, in particular those with complicated life histories, may be more profitably studied against the background of phasic change and that the parallelism between the change of food and habitat requirements during the life history of an organism (e.g. some insects with soil larvae) and the spatial change in the condition and floristic composition, or in age of dominant, of juxtaposed phases may be more than fanciful speculation about origins and survival.

The nature of the plant community

Tansley (1935) has coined the term ecosystem for the basic units in nature—units in which plants, animals and habitat factors of soil and climate interact in one system. The term is primarily used in reference to the larger geographical units. The words steppe, prairie, moor, bog apply to such systems and not to the plants alone. But the use of the term ecosystem is elastic; it may be applied to the smaller units like the 'plant community'.

The first step in the shattering of this unified system is the separation of the living plants and animals (the biome) from the non-living habitat, the components of the latter being regarded as factors affecting the former. The next step, which Tansley justifies partly on the ground of practical convenience, is to associate the animals with climate and soil as factors affecting plants. Further subdivision refers to the plants alone. According to Lippmaa (1933) the plant community is composed of synusiae or strata which he regards as the fundamental units of vegetation. The final step is Gleason's; to him the plant association is 'merely a fortuitous juxtaposition of plant individuals' (1936*b*).

In support of his view Gleason cites the randomicity of the component species of the association. This is by no means universal (Clapham, 1936; Blackman, 1935). The restriction of establishment in space and in time to certain phases in a diversified pattern tends towards the grouping of age classes (that is, there is overdispersed distribution in time), even although the species as a whole may be at random. Further, two species characteristic of a phase may be randomly distributed but show a high degree of association between them.

This assumed randomicity also enables Gleason to cut the community into small pieces, each of which he regards as representative of the whole. Such a proceeding ignores the significance of the diversity manifested in pattern; this pattern sets a lower limit to the representative minimum.

It is, however, important to realize that Gleason uses the word individual in the sense of an ecosystem of which the plant is the centre, that is, in which the plant and its environment form one interacting system. The individual plant (in the ordinary sense) is a part product of its environment and itself modifies the environment. Position therefore must be important.

Gleason himself cogently argues that by splitting an association into its synusiae, Lippmaa is separating synusiae which are 'united by relatively strong genetic and dynamic bonds' (1936*a*). The same argument is valid if it can be shown that in lateral spacing as well as in vertical there is dependence of one part on another, as is held to be shown by the evidence adduced here. His skeleton picture of a two-layered community consisting of a tree and a herb below it is divorced from its spatial moulding context and set in a moment in time. In time, both will change and their successors be influenced by their collective surroundings. In short Gleason has minimized the significance of the relations between the components of the community in horizontal space and in time. These relations constitute a primary bond in the maintenance of the integrity of the plant community; they give to it the unity of a co-ordinated system.

This co-ordinated unit may be studied as a plant community, the central object of study of plant sociology; round this unit knowledge is accumulated for use and classification. The functional relations of the community may be systematized in laws and hypotheses.

To draw a dividing line between the approach to the study of natural units through the plant community and through the ecosystem is difficult, if not impossible. But clearly it is one thing to study the plant community and assess the effect of factors which obviously and directly influence it, and another to study the interrelations of all the components of the ecosystem with an equal equipment in all branches of knowledge concerned. With a limited objective, whether it be climate, soil, animals or plants which are elevated into the central prejudiced position, much of interest and importance to the subordinate studies and perhaps even to the central study itself is set aside. To have the ultimate even if idealistic objective of fusing the shattered fragments into the original unity is of great scientific and practical importance; practical because so many problems in nature are problems of the ecosystem rather than of soil, animals or plants, and scientific because it is our primary business to understand.

What I want to say is what T. S. Eliot said of Shakespeare's work: we must know all of it in order to know any of it.

REFERENCES

Aubréville, A. (1938). La forêt coloniale. Les forêts de l'Afrique occidentale française. *Ann. Acad. Sci. colon., Paris,* **9.**

Beeson, C. F. C. (1941). *The Ecology and Control of the Forest Insects of India and the Neighbouring Countries.* India: Dehra Dun.

Blackman, G. E. (1935). A study by statistical methods of the distribution of species in grassland associations. *Ann. Bot., Lond.,* **49.**

Blackman, M. W. & Stage, H. W. (1924). On the succession of insects living in the bark and wood of dying, dead and decaying hickory. *N.Y. St. Coll. For. Tech. Publ.* no. 17. Syracuse, N.Y.

Braun-Blanquet, J. & Jenny, H. (1926). Vegetationsentwicklung und Bodenbildung in der alpinen Stufe der Zentralalpen. *N. Denkschr. schweiz. Ges. Naturfs.* **63.**

Braun-Blanquet, J. & Pavillard, J. (1930). *Vocabulary of Plant Sociology.* Translation by F. R. Bharucha. Cambridge.

Butcher, R. W. (1933). Studies in the ecology of rivers. *J. Ecol.* **21.**

Clapham, A. R. (1936). Overdispersion in grassland communities, and the use of statistical methods in plant ecology. *J. Ecol.* **24.**

Clements, F. E. & Shelford, V. E. (1939). *Bioecology.* London.

Garrett, S. D. (1944). *Root Disease Fungi.* Waltham, Mass., U.S.A.

Gause, G. F. (1935). Vérifications expérimentales de la théorie mathématique de la lutte pour la vie. *Act. Sci. et Ind.* no. 277.

Gleason, H. A. (1936*a*). Is the Synusia an Association? *Ecology,* **17.**

Gleason, H. A. (1936*b*). In *Mem. Brooklyn Bot. Gn,* no. 4. Brooklyn, N.Y.

Godwin, H. & Conway, V. M. (1939). The ecology of a raised bog near Tregaron, Cardiganshire. *J. Ecol.* **27.**

Graham, S. A. (1925). The felled tree trunk as an ecological unit. *Ecology,* **6.**

Griggs, R. F. (1934). The problem of arctic vegetation. *J. Wash. Acad. Sci.* **24.**

Jones, E. W. (1945). The structure and reproduction of the virgin forest of the North Temperate Zone. *New Phytol.* **44.**

Lippmaa, T. (1933). Aperçu general sur la vegetation autochtone du Lautaut (Hautes Alpes). *Acta Inst. bot. tartu,* **3.**

Mohr, Carl O. (1943). Cattle droppings as ecological units. *Ecol. Monogr.* no. 13.

Osvald, H. (1923). Die Vegetation des Hochmoores Komosse. *Akad. Abh. Uppsala.*

Pearl, R. (1926). *The Biology of Population.* London.

Poole, A. L. (1937). A brief ecological survey of the Pukekura State Forest, South Westland. *N.Z. J. For.* **4.**

Saveley, H. E. (1939). Ecological relations of certain animals in dead pine and oak logs. *Ecol. Monogr.* no. 9.

Tansley, A. G. (1935). The use and abuse of vegetational concepts and terms. *Ecology,* **16.**

Tansley, A. G. (1939). *The British Islands and their Vegetation.* Cambridge.

Tragårdh, I. (1923). Ziele und Wege in der Forstentomologie. *Medd. Stat. Skog.* **20.**

Tutin, T. G. (1938). The Autecology of *Zostera marina* in relation to its wasting disease. *New Phytol.* **37.**

Walton, J. (1922). A Spitzbergen salt marsh. *J. Ecol.* **10.**

Watt, A. S. (1925). On the ecology of British beechwoods with special reference to their regeneration. Part II. The development and structure of beech communities on the Sussex Downs. *J. Ecol.* **13.**

Watt, A. S. (1937). Studies in the ecology of Breckland. II. On the origin and development of blow-outs. *J. Ecol.* **25.**

Watt, A. S. (1938). Studies in the ecology of Breckland. III. The origin and development of the Festuco-Agrostidetum on eroded sand. *J. Ecol.* **26.**

Watt, A. S. (1940). Studies in the ecology of Breckland. IV. The Grass-heath. *J. Ecol.* **28.**

Watt, A. S. (1945). Contributions to the ecology of bracken. III. Frond types and the make up of the population. *New Phytol.* **44.**

Watt, A. S. (1947). Contributions to the ecology of bracken. IV. The structure of the community. *New Phytol.* **46.**

Wahrscheinlichkeit-Rechnung. Proc. Tokyo Mathem. Phys. Soc., 2nd Ser., **8**: 556-564.

Miyasita, K. 1955. Some consideration on the population fluctuation of the rice stem borer. (In Japanese.) Bull. Nat. Inst. Agr. Sci., Japan, C, **5**: 99-109.

Schwerdtfeger, F. 1935. Uber die Populationsdichte von *Bupalus piniarius, Panolis flammea, Dendrolimus pini, Sphinx pinastri* und ihren zeitlichen Wechsel. Z. Forst-u. Jagdwesen, **67**: 449-482, 513-540.

———. 1954. Grundsätzliches zur Populationsdynamik der Tiere, insbesondere der Insekten. Allgem. Forst-u. Jagdzeitung. **125**: 200-209.

Smith, H. S. 1935. The role of biotic factors in the determination of population densities. J. Econ. Ent., **28**: 873-898.

Tsutsui, K. et al. 1955. On the peculiar prevalence of rice stem borer in Tokai-Kinki district in 1954. (In Japanese.) Oyo-Kontyu, **11**: 53-58.

Utida, S. 1950. On the equilibrium state of the interacting population of an insect and its parasite. Ecol., **31**: 165-175.

———. 1950a. On the periodicity in the population fluctuations of some arctic mammals. (In Japanese.) Jap. J. Appl. Zool., **17**: 203-207.

———. 1954. "Logistic" growth tendency in the population fluctuation of the rice stem borer. (In Japanese.) Oyo-Kontyu, **10**: 3-10.

———. 1956. Effect of population density upon reproductive rate in relation to adult longevity. (In Japanese.) Jap. J. Ecol., **5**: 137-140.

———. 1957. "Logistic" growth tendency in the population fluctuation of the rice stem borer. (In Japanese.) Botyu-Kagaku, **22**: 57-63.

Varley, G. C. 1953. Ecological aspects of population regulation. Trans. 9th Int. Congr. Ent., **2**: 210-214.

POPULATION ECOLOGY OF SOME WARBLERS OF NORTHEASTERN CONIFEROUS FORESTS[1]

ROBERT H. MACARTHUR
Department of Zoology, University of Pennsylvania

INTRODUCTION

Five species of warbler, Cape May (*Dendroica tigrina*), myrtle (*D. coronata*), black-throated green (*D. virens*), blackburnian (*D. fusca*), and bay-breasted (*D. castanea*), are sometimes found together in the breeding season in relatively homogeneous mature boreal forests. These species are congeneric, have roughly similar sizes and shapes, and all are mainly insectivorous. They are so similar in general ecological preference, at least during years of abundant food supply, that ecologists studying them have concluded that any differences in the species' requirements must be quite obscure (Kendeigh, 1947; Stewart and Aldrich, 1952). Thus it appeared that these species might provide an interesting exception to the general rule that species either are limited by different factors or differ in habitat or range (Lack, 1954). Accordingly, this study was undertaken with the aim of determining the factors controlling the species' bundances and preventing all but one from being exterminated by competition.

LOGICAL NATURE OF POPULATION CONTROL

Animal populations may be regulated by two types of events. The first type occurs (but need not exert its effect) independently of the density of the population. Examples are catastrophes such as storms, severe winters, some predation, and some disease. The second type of event depends upon the density of the population for both its occurrence and strength. Examples are shortages of food and nesting holes. Both types seem to be important for all well-studied species. The first kind will be called density independent and the second density dependent. This is slightly different from the usual definitions of these terms which require the effects upon the population and not the occurrence to be density independent or dependent (Andrewartha and Birch 1954).

When density dependent events play a major role in regulating abundance, interspecific relations are also important, for the presence of an individual of another species may have some of the effects of an individual of the original density dependent species. This is clearly illustrated by the generalized habitats of the few species of passerine birds of Bermuda contrasted with their specialized habitats in continental North America where many additional species are also present (Bourne 1957).

If the species' requirements are sufficiently similar, the proposition of Volterra (1926) and Gause (1934), first enunciated by Grinnell (1922), suggests that only one will be able to persist, so that the existence of one species may even control the presence or absence of another. Because of this proposition it has become customary for ecologists to look for differences in food

[1] A Dissertation Presented to the Faculty of the Graduate School of Yale University in Candidacy for the Degree of Doctor of Philosophy, 1957.

or habitat of related species; such differences, if found, are then cited as the reason competition is not eliminating all but one of the species. Unfortunately, however, differences in food and space requirements are neither always necessary nor always sufficient to prevent competition and permit coexistence. Actually, to permit coexistence it seems necessary that each species, when very abundant, should inhibit its own further increase more than it inhibits the other's. This is illustrated in Figure 1. In this figure, the populations of the two species form the coordinates so that any point in the plane represents a population for each species. Each shaded area covers the points (*i.e.,* the sets of combined populations of the species) in which the species corresponding to the shading can increase, within a given environment. Thus, in the doubly-shaded area both species increase and in the unshaded area both species decrease. The arrows, representing the direction of population change, must then be as shown in the figure for these regions. In order that a stable equilibrium of the two species should exist, the arrows in the singly shaded regions obviously must also be as in the figure; an interchange of the species represented by the shading would reverse the directions of these arrows resulting in a situation in which only one species could persist. Thus, for stability, the boundaries of the shaded zones of increase must have the relative slope illustrated in the figure with each species inhibiting its own further increase more than the other's. The easiest way for this to happen would be to have each species' population limited by a slightly different factor. It is these different limiting factors which are the principal problem in an investigation of multispecific animal populations regulated by density dependent events.

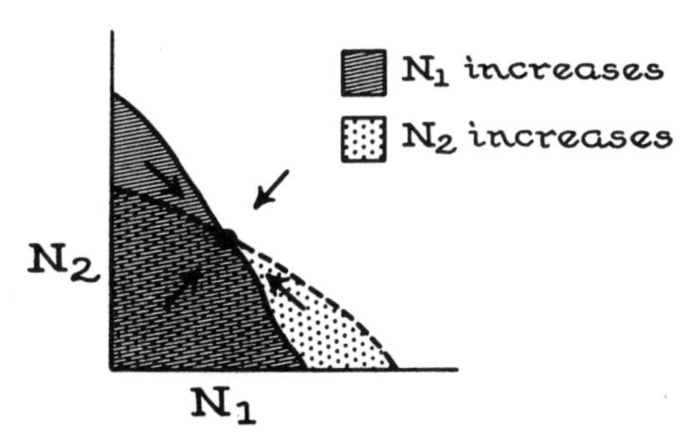

FIG. 1. The necessary conditions for a stable equilibrium of two species. The coordinate axes represent the populations of the species.

An example which has not received sufficient attention is competition in a heterogeneous environment. As has often been pointed out (Kluijver and Tinbergen 1953, Lack 1955, Hinde 1956) birds may emigrate or disperse from the most suitable areas where reproduction is successful into marginal habitats. Consider such a species which will be called A. Let B be a species that lives only in the area that is marginal for species A. Now, even if in an unlimited environment of this type, species B would eliminate species A by competition, in the heterogeneous environment species B may be eliminated from its own preferred habitat. For, if there is sufficient dispersal by species A, it may maintain, partly by immigration, such a high population in the marginal habitat that species B is forced to decrease. This process is probably very important in considering the environmental distributions of birds and implies that small areas of habitat typical for one species may not contain that species.

The study of limiting factors in nature is very difficult because ideally it requires changing the amount of the factor alone and observing whether this change affects the size of the population. Theoretically, if more than one factor changes, the analysis can still be performed, but in practice, if more changes of known nature occur, more of an unknown nature usually also occur. Limiting factors have been studied in two ways. The best way is artificially to modify single factors in the environment, observing the effect upon the birds. MacKenzie (1946) reviews some experiments of this type. The most notable was the increase from zero to abundant of pied flycatchers (*Muscicapa hypoleuca*) when nest boxes were introduced in the Forest of Dean. This showed conclusively that lack of nesting sites had limited the population. Such simple modifications are not always feasible. For instance, changing the food supply of an insectivorous bird is nearly impossible. The most feasible approach in such a case is to compare the bird populations in two regions which differ in the abundance of the factor being considered. Ideally, the two regions should differ only in this respect, but this is very improbable. A good example of this method of study is the work of Breckenridge (1956) which showed that the least flycatchers (*Empidonax minimus*) were more abundant in a given wood wherever the wood was more open.

The present study of the factors limiting warblers was conducted by the second approach. This is slightly less accurate than the first method, but permits studying more factors and requires less time. There are actually four parts to the study. First, it is shown that density dependent events play a large role in controlling the populations of the species. Second, a discussion of the general ecology of the species (food, feeding zones, feeding behavior, territoriality, predators, and mortality) is presented. The observations were made in the summers of 1956 and 1957. Third, the habits of the different species in different seasons are compared to see what aspects of the general ecology are invariant and hence characteristic of the species. Some observations on the species' morphology are discussed in the light of these characteristics. This was the project of the fall and winter of 1956 and the spring of 1957. Finally, a wood-to-wood comparison of species abundances, relative to the important constituents of their niches as determined in the earlier stages, is presented. This work was done in the summers of 1956 and 1957.

Density Dependence

It is the aim of this section to demonstrate that the five species of warbler are primarily regulated by density dependent events, that is, that they increase when rare and decrease when common (relative to the supply of a limiting factor). The strongest argument for this is the correlation of abundances with limiting factors discussed later. However, to avoid any risk of circularity, an independent partial demonstration will now be given.

If density independent events do not occur randomly but have a periodic recurrence, then a population controlled by these events could undergo a regular oscillation nearly indistinguishable in form from that of a population regulated by density dependent events. The distinction can be made, however, by observing the effect of the presence of an ecologically similar species. Here it will first be shown that increases and decreases are not random; then an argument will be given which renders the density independent explanation improbable.

If increases, I, and decreases, D, occur randomly, the sequence of observed I's and D's would have random order. A run of I's (or D's) is a sequence (perhaps consisting of one element) of adjacent I's (or D's) which cannot be lengthened; *i.e.*, the total number of runs is always one greater than the number of changes from I to D or D to I. If more runs of I's or D's are observed than would be expected in a random sequence, then an increase makes the following change more likely to be a decrease and conversely. This is what would be expected on the hypothesis of density dependent events.

There have been very few extensive censuses of any of the five species of warbler that are studied here. The longest are reproduced below from the data of Smith (1953) in Vermont, Williams (1950) in Ohio, and Cruickshank (1956) in Maine. The populations of myrtle, black-throated green, and blackburnian are listed; only those species are mentioned that are consistently present.

Myrtle													
Maine	7	5	7	7	6	7	8	10	9	8	10	10	10
			8	10	7	10							
Black-throated green													
Ohio	3	3	5	3	4	3	2	4	0	1	-	3	3
			3	4	3	3	4	3					
Vermont	7	7	7	6	8	5	6	6	7	8	8	6	8
			8	3	6	3							
Maine	8	9	11	10	10	10	11	11	11	10	11	9	8
			9	8	9								
Blackburnian													
Vermont	11	11	10	7	8	7	7	10	10	8	5	6	3
			3	6	6	5							
Maine	2	4	3	3	5	5	7	5	6	5	7	5	5
			7	6	6	3							

The increases, I, and decreases, D, of these censuses, in order are as follows:

Myrtle

D I D I I I D D I D I D I

Black-throated green

I D I D D I D I I I D I D

D I D I I I D I D I D

I I D I D I D D I D I

Blackburnian

D D I D I D D I D I D

I D I I D I D I D I D D

In this form the data from different censuses are perfectly compatible and, since censuses end or begin with I or D in no particular pattern, all the censuses for a given species may be attached:

Myrtle

D I D I I I D D I D I D I (10)

Black-throated green

I D I D D I D I I I D I D D I D

I I I D I D I D I I D I (27)

D I D D I D I

Blackburnian

D D I D I D D I D I D I D I I D I

D I D I D D (19)

Here the groups of letters underlined are runs and the number of runs is totalled in parentheses. From the tables of Swed and Eisenhart (1943), testing the one-sided hypothesis that there are no more runs than would be expected by chance, each of these shows a significantly large number of runs (the first less than 5% significance, the others less than 2.5%). That is, each species tends to decrease following an increase and to increase following a decrease, proving population control non-random. The mean periods of these fluctuations can easily be computed. For a run of increases followed by a run of decreases constitutes one oscillation. Thus, the periods of the oscillations of the three species are $13/5 = 2.6$, $70/27 = 2.6$, and $46/19 = 2.4$ years respectively. These fluctuations would require an unknown environmental cycle of period approximately 2.5 years if a regularly recurring density independent event were controlling the populations. Thus, from these data alone, it seems very probable that the three species (myrtle, black-throated green, blackburnian) are primarily regulated by density dependent events.

A species may be regulated by density dependent events and yet undergo dramatic changes in populations due to changes in the limiting factor itself. In this case tests by the theory of runs, used above, are likely to be useless. However, if a correlation can be made of the population with the environmental factor undergoing change, then not only can density dependence, *i. e.* existence of a limiting factor, be established, but also the nature of the limiting factor. For, if an increase in one environmental variable can be established, an experiment of the first type described above has been performed. That is, the habitat has been modified in one factor and a resulting change in bird population has been observed. Therefore, because the population changes, that one factor has been limiting.

This is apparently what happens in populations of Cape May and bay-breasted warblers. Kendeigh (1947), examining older material, established the fact that these species are abundant when there is an outbreak of *Choristoneura fumiferana* (Clem.), the spruce budworm. More recent information confirms this. To correlate with the fact (Greenbank 1956) that there have been continuously high budworm populations since 1909, there is the statement of Forbush (1929) that the Cape May warbler became more common about 1909, and the statement of Bond (pers. comm.) that the winter range of the species has been increasing in size. An outbreak of spruce budworms started in northern Maine in the late 1940's, and Stewart and Aldrich (1951) and Hensley and Cope (1951) studied the birds during 1950 and 1951. Cape May and bay-breasted warblers were among the commonest birds present, as in the earlier outbreaks, although both species were formerly not common in Maine (Knight 1908). The outbreak has continued through New Brunswick, where current bird studies (Cheshire 1954) indicate that bay-breasted is again the commonest bird, although for unknown reasons the Cape May has not been observed.

In conclusion, it appears that all five species are primarily regulated by density dependent events, and that a limiting factor is food supply for bay-breasted and Cape May warblers.

General Ecology

The density dependence tentatively concluded above implies that the presence of individuals of a species makes the environment less suitable for other individuals of that species. It would also be expected then that the presence of individuals of one species may make the environment less suitable for individuals of a different species. This is called interspecific competition. As mentioned above, this seems to mean that two sympatric species will have their populations limited by different factors so that each species inhibits its own population growth more than it inhibits that of the others. The factors inseparably bound to a species' persistence in a region are, then, its relation to other species and the presence of food, proper feeding zone, shelter from weather, and nesting sites (Andrewartha and Birch 1954, Grinnell 1914). In this section these factors as observed during the breeding season of the five species of warbler will be discussed.

The summer of 1956 was devoted to observations upon the four species, myrtle, black-throated green, blackburnian, and bay-breasted warblers, on their nesting grounds. The principal area studied was a 9.4 acre plot of mature white spruce (*Picea glauca*) on Bass Harbor Head, Mt. Desert Island, Hancock County, Maine. On 7 July 1956 the site of observations was changed to the town of Marlboro in Windham County, Vermont, where a red spruce (*P. rubens*) woodlot of comparable structure was studied. In the summer of 1957 more plots were studied. From 30 May until 5 June, eighteen plots of balsam fir (*Abies balsamea*), black spruce (*P. mariana*), and white spruce near Cross Lake, Long Lake, and Mud Lake in the vicinity of Guerette in Aroostook County, Maine, were studied. The remainder of the breeding season was spent on Mt. Desert Island, Maine, where five plots were censused. These will be described later.

Feeding Habits

Although food might be the factor for which birds compete, evidence presented later shows that differences in type of food between these closely related species result from differences in feeding behavior and position and that each species eats what food is obtainable within the characteristic feeding zone and by the characteristic manner of feeding. For this reason, differences between the species' feeding positions and behavior have been observed in detail.

For the purpose of describing the birds' feeding zone, the number of seconds each observed bird spent in each of 16 zones was recorded. (In the summer of 1956 the seconds were counted by saying "thousand and one, thousand and two, . . ." all subsequent timing was done by stop watch. When the stop watch became available, an attempt was made to calibrate the counted seconds. It was found that each counted second was approximately 1.25 true seconds.) The zones varied with height and position on branch as shown in Figure 2. The height zones were ten foot units measured from the top of the tree. Each branch could be divided into three zones, one of bare or lichen-covered base (B), a middle zone of old needles (M), and a terminal zone of new (less than 1.5 years old) needles or buds (T). Thus a measurement in zone T3 was an observation between 20 and 30 feet from the top of the tree and in the terminal part of the branch. Since most of the trees were 50 to 60 feet tall, a rough idea of the height above the ground can also be obtained from the measurements.

There are certain difficulties concerning these measurements. Since the forest was very dense, certain types of behavior rendered birds invisible. This resulted in all species being observed slightly disproportionately in the open zones of the trees. To combat this difficulty each bird was observed for as long as possible so that a brief excursion into an open but not often-frequented zone would be compensated for by the remaining part of the observation. I believe there is no serious error in this respect. Furthermore, the comparative aspect is independent of this error. A different difficulty arises from measurements of time spent in each zone. The error due to counting should not affect results which are comparative in nature. If a bird sits very still or sings, it might spend a large amount of time in one zone without actually requiring that zone for feeding. To alleviate this trouble, a record of activity, when not feeding, was kept. Because of these difficulties, non-parametric statistics have been used throughout the analysis of the study to avoid any *a priori* assumptions about distributions. One difficulty is of a different nature; because of the density of the vegetation and the activity of the warblers a large number of hours of watching result in disappointingly few seconds of worthwhile observations.

The results of these observations are illustrated in Figures 2-6 in which the species' feeding zones are indicated on diagrammatic spruce trees. While

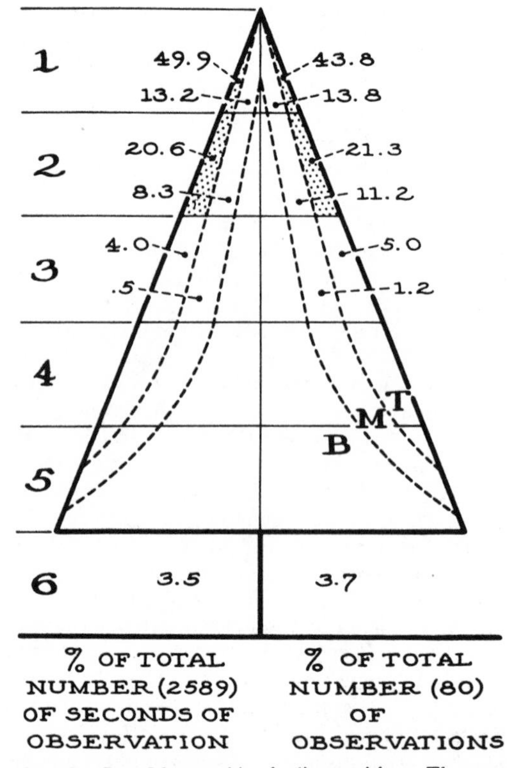

FIG. 2. Cape May warbler feeding position. The zones of most concentrated activity are shaded until at least 50% of the activity is in the stippled zones.

the base zone is always proximal to the trunk of the tree, as shown, the T zone surrounds the M, and is exterior to it but not always distal. For each species observed, the feeding zone is illustrated. The left side of each illustration is the percentage of the number of seconds of observations of the species in each zone. On the right hand side the percentage of the total number of times the species was observed in each zone is entered. The stippled area gives roughly the area in which the species is most likely to be found. More specifically, the zone with the highest percentage is stippled, then the zone with the second highest percentage, and so on until at least fifty percent of the observations or time lie within the stippled zone.

Early in the investigation it became apparent that there were differences between the species' feeding habits other than those of feeding zones. Subjectively, the black-throated green appeared "nervous," the bay-breasted slow and "deliberate." In an attempt to make these observations objective, the following measurements were taken on feeding birds. When a bird landed after a flight, a count

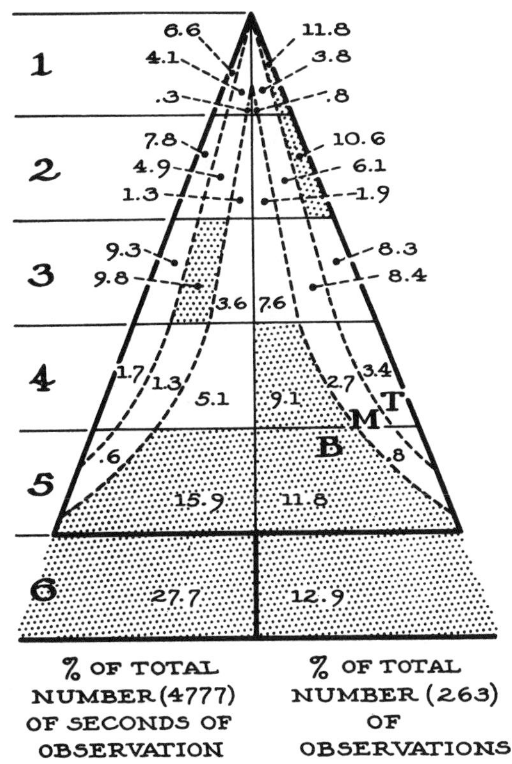

FIG. 3. Myrtle warbler feeding position. The zones of most concentrated activity are shaded until at least 50% of the activity is in the stippled zones.

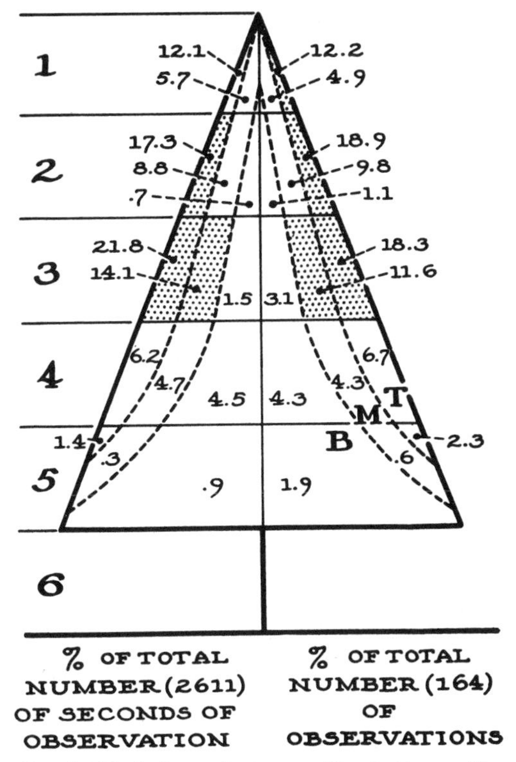

FIG. 4. Black-throated green warbler feeding position. The zones of most concentrated activity are shaded until at least 50% of the activity is in the stippled zones.

of seconds was begun and continued until the bird was lost from sight. The total number of flights (visible uses of the wing) during this period was recorded so that the mean interval between uses of the wing could be computed.

The results for 1956 are shown in Table I. The results for 1957 are shown in Table II. Except for the Cape May fewer observations were taken than in 1956.

By means of the sign test (Wilson, 1952), treating each observation irrespective of the number of flights as a single estimate of mean interval between flights, a test of the difference in activity can be performed. These data are summarized in the following inequality, where $\overset{95}{<}$ is interpreted to mean "has smaller mean interval between flights, with 95% certainty."

$$\text{Black-throated green} \overset{95}{<} \begin{Bmatrix} \text{Blackburnian} \\ \text{Myrtle} \end{Bmatrix} \overset{99}{<} \begin{Bmatrix} \text{Cape May} \\ \text{Bay-breasted} \end{Bmatrix}$$

The differences in feeding behavior of the warblers can be studied in another way. For, while all the species spend a substantial part of their time searching in the foliage for food, some appear to crawl along branches and others to hop across branches. To measure this the following procedure was adopted. All motions of a bird from place to place in a tree were resolved into components in three independent directions. The natural directions to use were vertical, radial, and tangential. When an observation was made in which all the motion was visible, the number of feet the bird moved in each of the three directions was noted. A surpringing degree of diversity was discovered in this way as is shown in Figure 7. Here, making use of the fact that the sum of the three perpendicular distances from an interior point to the sides of an equilateral triangle is independent of the position of the point, the proportion of motion in each direction is recorded within a triangle. Thus the Cape May moves predominantly in a vertical direction, black-throated green and myrtle in a tangential direction, bay-breasted and blackburnian in a radial direc-

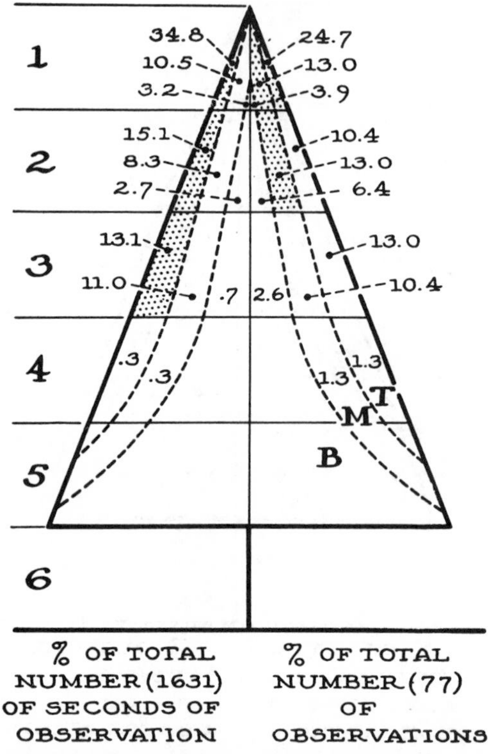

FIG. 5. Blackburnian warbler feeding position. The zones of most concentrated activity are shaded until at least 50% of the activity is in the stippled zones.

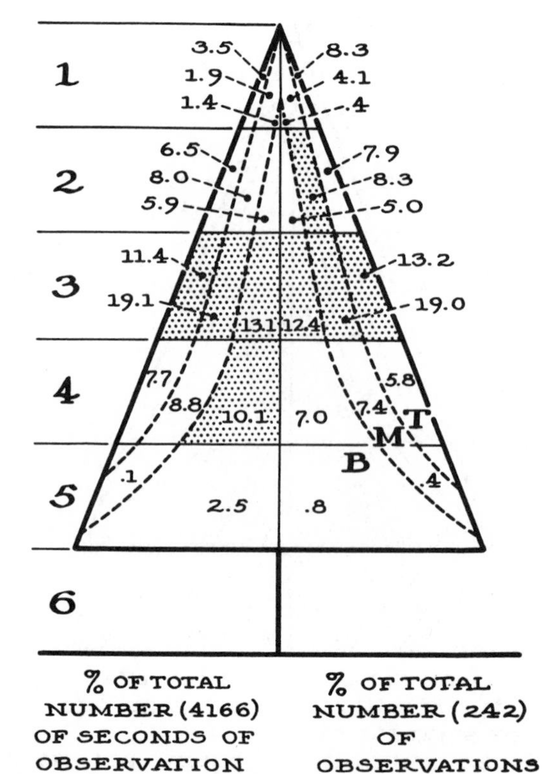

FIG. 6. Bay-breasted warbler feeding position. The zones of most concentrated activity are shaded until at least 50% of the activity is in the stippled zones.

tion. To give a nonparametric test of the significance of these differences Table III is required.

Each motion was classified according to the direction in which the bird moved farthest. Thus, in 47 bay-breasted warbler observations of this type, the bird moved predominantly in a radial direction 32 times. Applying a χ^2 test to these, bay-breasted and blackburnian are not different but all others are significantly ($P<.01$) different from one another and from bay-breasted and blackburnian.

There is one further quantitative comparison which can be made between species, providing additional evidence that during normal feeding behavior the species could become exposed to different types of food. During those observations of 1957 in which the bird was never lost from sight, occurrence of long flights, hawking, or hovering was recorded. A flight was called long if it went between different trees and was greater than an estimated 25 feet. Hawking is distinguished from hovering by the fact that in hawking a moving prey individual is sought in the air, while in hovering a nearly stationary prey individual is sought amid the foliage. This information is summarized in Table IV.

Both Cape May and myrtle hawk and undertake long flights significantly more often than any of the other species. Black-throated green hovers significantly more often than the others.

At this point it is possible to summarize differences in the species' feeding behavior in the breeding season. Unfortunately, there are very few original descriptions in the literature for comparison. The widely known writings of William Brewster (Griscom 1938), Ora Knight (1908), and S. C. Kendeigh (1947) include the best observations that have been published. Based upon the observations reported by these authors, the other scattered published observations, and the observations made during this study, the following comparison of the species' feeding behavior seems warranted.

Cape May Warbler. The foregoing data show that this species feeds more consistently near the top of the tree than any species expect blackburnian, from which it differs principally in type

TABLE I. The number of intervals between flights (I) recorded in 1956 and the total number of seconds (S) of observation counted

	Myrtle		Black-throated green		Black-burnian		Bay-breasted	
	I	S	I	S	I	S	I	S
	1	40	4	45	1	5	5	55
	4	32	8	20	13	77	3	22
	4	13	6	60	2	11	2	33
	4	17	5	35	5	18	1	7
	2	10	5	23	5	24	3	37
	1	10	3	7	3	16	2	20
	5	25	9	35	3	18	1	7
	5	10	8	25	3	15	4	10
	5	11	1	12	4	11	11	60
	6	30	7	20	3	12	5	41
	3	10	7	39	4	46	3	50
	1	5	13	25	2	26	1	17
	13	68	5	10			1	3
	1	5	5	12			1	49
	1	7	2	37			4	42
	4	26					5	60
							5	35
							5	26
							3	14
							4	38
							3	14
							1	11
							3	22
							2	29
Total..........	60	319	88	405	48	279	78	702
Total Adjusted to True seconds	60	399	88	506	48	349	78	876

TABLE II. Intervals between flights and seconds of observation in 1957

	Cape May		Myrtle		Black-throated green		Black-burnian		Bay-breasted	
	I	S	I	S	I	S	I	S	I	S
	3	47	4	18	3	18	1	22	7	110
	12	35	1	62	5	22	3	22	2	115
	5	50	5	47	3	15	9	110	1	18
	2	15	3	17	13	89	7	26	1	22
	1	15	1	5	23	40	3	8	2	19
	1	45	12	86					2	47
	1	20							3	27
	4	29							8	112
	11	129							7	46
	6	47								
	1	15								
	4	50								
	3	20								
	21	122								
	1	12								
	1	34								
	4	20								
	10	79								
Total................	91	782	26	235	47	184	24	188	33	576
Ad,usted 1956 Total...			60	399	88	506	48	349	78	876
Grand Total.........	91	782	86	644	125	690	72	537	111	1392
Mean Interval Between Flights.............		8.59		7.48		5.52		7.47		12.53

TABLE III. Number of times each species was observed to move predominantly in a particular direction. (Numbers ending in .5 result from ties)

Species	Radial	Tangential	Vertical	Total
Cape May.......	5.5	1.5	25	32
Myrtle..........	4.5	11.5	9	25
Black-throated green........	4	21	5	30
Blackburnian....	11	1	3	15
Bay-breasted....	32	7	8	47

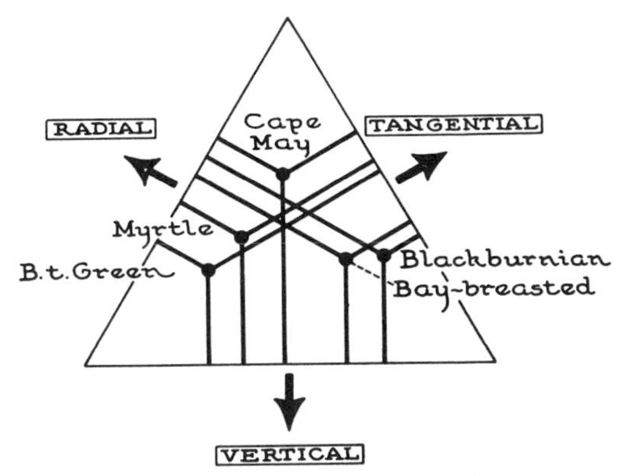

FIG. 7. Components of Motion. From the dot representing a species, lines are drawn to the sides of the triangle. The lengths of these lines are proportional to the total distance which the species moved in radial, tangential, and vertical directions, respectively.

TABLE IV. Classification of the flights observed for each species

Species	Long Flights	Hawking	Hovering	No. of Observ.
Cape May.........	35	12	0	53
Myrtle............	25	9	0	62
Black-throated green..........	1	0	7	42
Blackburnian......	0	1	0	35
Bay-breasted......	4	2	2	57

Applying χ^2 tests to pairs of species, the following conclusions emerge.

of feeding action. It not only hawks far more often than the blackburnian, but also moves vertically rather than radially in the tree, causing its feeding zone to be more restricted to the outer shell of the tree. Myrtle warblers when feeding in the tips of the trees nearly duplicate the feeding behavior of the Cape May. During rainy, windy, and cold weather Cape Mays were not found in the tree tops, but were instead foraging in the low willows (*Salix* sp.) and pin cherries (*Prunus pensylvanica*). Here they often fed among the flowers, for which their semitubular tongue (Gardner 1925) may be advantageous.

Because of this species' irregular breeding distribution, both in space and time, and its former rarity, there are very few published descriptions of its feeding behavior. Knight (1908), although he lived in Maine, had never seen one. Brewster (Griscom 1938) wrote that:

"It keeps invariably near the tops of the highest trees whence it occasionally darts out after passing insects. . . . In rainy or dark weather they came in numbers from the woods to feed among the thickets of low firs and spruces in the pastures. Here they spent much of their time hanging head downward at the extremity of the branches, often continuing in this position for nearly a minute at a time. They seemed to be picking minute insects from under the surface of the fir needles. They also resorted to a thicket of blossoming plum trees directly under the window, where we were always sure of finding several of them."

He also said that it was more active than the bay-breasted. Bond (1937) stated that all feeding was done more than twenty feet above the ground. Kendeigh (1947) said males tend to sing about seven feet from the tops of the trees, and that feeding is done at the same level. He also mentions that the birds sometimes hawk after passing insects. The rainy weather observations indicate behavior very much like that of winter and migration to be discussed later.

Myrtle Warbler. This species seems to have the most varied feeding habits of any species. Although it moves slightly more in a tangential direction than any except black-throated green, it is probably more correct to think of the myrtle as having the most nearly equal components (radial, tangential, and vertical) of any species. This is shown by its most nearly central location in Figure 7. It is also seen to have the most widely distributed feeding zone, although the ground feeding was nearly, but not completely, restricted to the gathering of emerging Tipulids for newly hatched young. Sometimes a substantial amount of this is hawking for flying insects; at other times it is largely by rapid peering (Grinnell 1921) amid the thick foliage near the tree tops. Myrtle, along with Cape May, makes a much higher proportion of flights to other trees than do the other species, often flying from one side of its territory to the other with no apparent provocation. The other three species tend to search one tree rather thoroughly before moving on. Further evidence of the plasticity of the myrtle warbler's feeding habits will be presented when the other seasons are considered. Grinnell and Storer (1924) stated that the Audubon's warbler (which often hybridizes with the myrtle and with it froms a superspecies (Mayr 1950)) also feeds in peripheral foliage and does a greater amount of hawking than other species. Kendeigh (1947) said that birds fed from ground to tops of trees, and also that two males covered two and four acres respectively in only a few minutes. Knight (1908) said "Many of the adult insects are taken on the wing, the warblers taking short springs and flights into the air for this purpose. The young for the first few days are fed on the softer sorts of insects secured by the parents, and later their fare is like that of the parents in every way."

Black-throated Green Warbler. Compared to the myrtle warbler this species is quite restricted in feeding habits. As seen in Figure 4, it tends to frequent the dense parts of the branches and the new buds, especially at mid elevations in the tree. Most of its motion is in a tangential direction, keeping the bird in foliage of a nearly constant type. It has the shortest interval between flights of any of the five species and thus appears the most active. Almost all feeding seems to be by the method of rapid peering, necessitating the frequent use of the wings which the observations indicate. The foliage on a white spruce is a thick, dense mat at the end of the branch, changing to bare branch rather more sharply than in the red spruce. The black-throated greens characteristically hop about very actively upon these mats, often, like the other species, looking down among the needles, and just as often, unlike the myrtle and bay-breasted, peering up into the next mat of foliage above. When food is located above, the bird springs into the air and hovers under the branch with its bill at the point whence the food is being extracted. While other species occasionally feed in this fashion, it is typical only of the black-throated green. After searching one branch, the black-throated green generally flies tangentially to an adjacent branch in the same tree or a neighboring one and continues the search. Only rarely, during feeding, does it make long flights. While it occasionally hawks for flying insects (missing a substantial proportion), this is not a typical behavior and the birds seldom sit motionless watching for flying insects in true hawking behavior. During its feeding, this species is very noisy, chipping almost incessantly, and, if it is a male and if it is early in the season, singing frequently. The other species are very quiet. A portion of this behavior can be confirmed from the literature. Knight (1908) said "Only rarely do they take their prey in the air, preferring to diligently seek it out among the branches and foliage" and Stanwood (Bent 1953) said "The bird is quick in its movements, but often spends periods of some length on one tree." Like the myrtle, this species enlarges its feeding zone while gathering food for its young. This is similar to the results of Betts (1955) indicating that the young tits eat different food from the adults.

Blackburnian Warbler. This species generally feeds high in the trees but is otherwise more or less intermediate between black-throated green and bay-breasted in its feeding behavior. This is true both of its flight frequency and its preferred feeding position on the limb (Figure 5). It is also intermediate in its method of hunting, usually moving out from the base to the tip of the branches looking down in the fashion of the bay-breasted and occasionally hopping about rapidly upon the mat of foliage at the branch tips looking both up and down for insects and even hovering occasionally. They seem to use the method of rapid peering, only occasionally hawking after a flying insect. As further evidence, Knight (1908) wrote "As a rule they feed by passing from limb to limb and examining the foliage and limbs of trees, more seldom catching anything in the air." Kendeigh (1945) said "It belongs to the treetops, singing and feeding at heights of 35 to 75 feet from the ground."

Bay-breasted Warbler. The usual feeding habits of this species are the most restricted of any of the species studied. All of the observations in the T1 zone and most in the T2 zone refer to singing males. This species uses its wings considerably less often than the other species, although it still appears to use the method of rapid peering in its hunting since it moves nearly continuously. These motions are, however, predominantly radial and seldom require the use of wings. The bird regularly works from the licheny base of the branches well out to the tip, although the largest part of the time is spent in the shady interior of the tree. It frequently stays in the same tree for long periods of time. This species very rarely hovers in the black-throated green fashion, and appears much less nimble in its actions about the tips of the branches, usually staying away from those buds which are at the edge of the mat of foliage. When it does feed at the edge of the mat, it is nearly always by hanging down rather than peering up. Other observers have emphasized the slowness. Brewster (Griscom 1938) called this warbler "slow and sluggish," and Kendeigh (1947) said "The birds do not move around much, but may sing and feed for long periods in the same tree." Forbush (1929) stated that it spends most of its time "moving about deliberately, after the manner of vireos."

Food

Two species may eat different food for only three reasons: 1. They may feed in different places or different times of day; 2. They may feed in such a manner as to find different foods; 3. They may accept different kinds of food from among those to which they are exposed. (Of course, a combination of these reasons is also possible.)

In the previous section it was shown that the warblers feed in different places and in a different manner, thus probably being exposed to different foods for the first and second reasons mentioned above. It is the aim of this section to show that the five warbler species have only small differences of the third kind. Theoretically, such differences, unaccompanied by morphological adaptations, would be disadvantageous, for, lacking the adaptations required to give greater efficiency in food collecting, and suffering a reduction in the number of acceptible food species, a bird would obtain food at a lesser rate. When the necessary adaptations are present, they usually consist of quite marked differences in bill structure such as those reported by Huxley (1942), Lack (1947), and Amadon (1950). As Table V shows, the mean bill measurements in millimeters of the five species of warbler considered in this study are quite similar. Twelve specimens of each species from the Peabody Museum of Natural History at Yale University were measured for each of the means given.

TABLE V. Mean dimensions of the bills of 12 specimens of each species

Species	Bill Length	Height at Nares	Width at Nares	Width 2.5mm from tip
Cape May	12.82	2.85	2.93	0.96
Myrtle	12.47	3.26	2.12	1.33
Black-throated green	12.58	3.38	3.15	1.34
Blackburnian	12.97	3.24	3.36	1.17
Bay-breasted	13.04	3.69	3.58	1.43

The Cape May alone has a noticeably different bill, it being more slender, especially at the tip. This bill houses a semi-tubular tongue as mentioned above, which is unique in the genus. These may be useful adaptations for their rainy weather flower feeding, but would seem ill-adapted for the characteristic flycatching of the breeding season (Gardner 1925). It is doubtless useful in other seasons, as will be discussed later. Aside from the Cape May, all other species differ in bill measurement by only a small fraction of a millimeter. Thus, for theoretical reasons, no pronounced differences of the third kind would be expected. Empirically, there is evidence to support this belief.

McAtee (1932) reported upon the analysis of eighty thousand bird stomachs, in an effort to disprove mimicry. Although his results were not conclusive, he claimed that insects appeared in bird stomachs about proportionately to their availability. Kendeigh (1945) agreed with this conclusion. Although McAtee (1926) said that no detailed studies of warbler food habits had been made, and no general ones seem to have appeared since, two very suggestive sets of analyses covering the five species have been published. Kendeigh (1947) reported upon the stomach contents of a collection made near Lake Nipigon, and Mitchell (1952) analyzed the stomachs of many birds taken during a budworm infestation in Maine. These data show, first, that most species of warbler eat all major orders of local arboreal arthropods. Furthermore, although there are differences in proportion of types of foods eaten by various species, these differences are most easily explained in terms of feeding zone. Thus, black-throated green and blackburnian which are morphologically the most similar of the five species have quite different foods. Kendeigh's table shows that black-throated green eats 4% Coleoptera, 31% Araneida, and 20% Homoptera, which blackburnian eats 22% Coleoptera, 2% Araneida, and 3% Homoptera. Dr. W. R. Henson has pointed out (pers. comm.) that Coleoptera can reasonably be assumed to come from inner parts of the tree where blackburnian has been shown to feed, whereas the Homoptera and most of the Araneida would be caught in the current year's growth where the black-throated green feeds more often, thus explaining the observed difference. Black-throated green and bay-breasted, with the most vireo-like bills (high at the nares), seem to eat more Lepidoptera larvae which are typical vireo food, but Mitchell's table shows that the other species too can eat predominantly Lepidoptera larvae when these are abundant. Otherwise, the food of the bay-breasted is more like that of the blackburnian, the feeding habits of which are similar. There are not sufficient data to analyze Cape May in this fashion. The myrtle warbler's feeding behavior is so flexible that no correlation between insects caught and a specific feeding zone or behavior is expected. This is shown by its having the most even distribution of food of any of the species considered. Thus, these correlations show that the differences in warbler food can be readily explained by morphological and zonal characteristics of the species.

Nest Location

The position of the nest is quite characteristic in warbler species. Nearly all species of the genus *Dendroica* nest off the ground. Figure 8 shows heights of the nests of the five species of warbler studied here. These data result from a combination of the records of Cruickshank (1956), the information in the egg collection of the American Museum of Natural History in New York, and that gathered in this study. Since the distributions are skewed and irregular, the median and confidence intervals for the median (Banerjee and Nair 1940) are appropriate measurements. As the figure shows, the Cape May, with 95% confidence interval for the median of 40-50 feet, and the blackburnian, whose interval is 30-50 feet, have quite similar nest heights, probably reflecting their tendency to feed at high elevations. The Cape May's nest is virtually always near the trunk in the uppermost dense cluster of branches in a spruce or occasionally a fir. The blackburnian may nest in a similar location or may nest farther out toward the branch tips. Myrtle and black-throated green have similar nesting heights, both species having 95% confidence interval for the median nest height of 15-20 feet. The black-throated green seems to prefer smaller trees for its nest, and is thus more likely to place its nest near the runk, but, in keeping with its other characteristics, the myrtle seems quite varied in this respect. Finally, the bay-breasted, which has the lowest feeding zone, has the lowest nest position, the median height being between 10 and 15 feet (95% confidence). Thus, the nest positions of the five species of warbler reflect their preferred feeding zones.

Territoriality

Defining territory as any defended area, warbler territories in the breeding season are of what Hinde (1956) called type A ("Large breeding area within which nesting, courtship, and mating and most food-seeking usually occur"). He pointed out that, since the behavioral mechanisms involved in defending a territory against others of the same species are the same as those involved in defending it against other species, this distinction need not be specified in the definition. From the ecological point of view, the distinction is of very great importance, however, for, as G. E. Hutchinson pointed out in conversation, if each species has its density (even locally) limited by a territorial behavior which ignores the other species, then there need be no further differences between the species to permit them to persist together. A weaker form of the same process, in which territories were compressible but only under pressure of a large population, would still be effective, along with small niche differences, in making each species inhibit its own population growth more than the others'—the

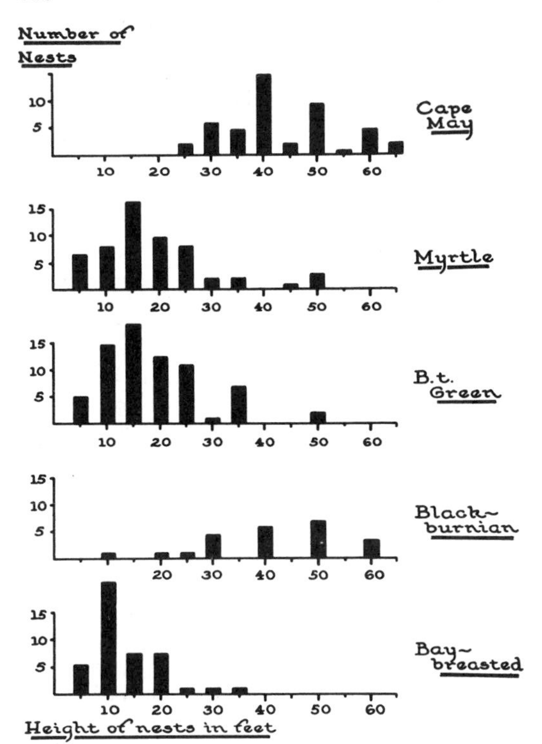

FIG. 8. Nesting heights of warblers.

necessary condition for the persistence of sympatric species.

Two further conditions make territory important for regulating populations. First, to have density dependent regulation, a species' regulating mechanism must have information of its own population density. Second, a predator ideally should keep its prey at that population level which permits the greatest rate of production. This means that the prey would not normally be particularly scarce. This, combined with the varied prey of the birds and the varied predators of the insects, would make food density a poorer criterion of a given bird species' density than size of territory. Thus, competition for food would be reduced from a "scramble" to a "contest" (Haldane 1955).

While the true nature of birds' territories has proved very elusive (Lack 1954, Hinde 1956), two separate lines of evidence suggest strongly that territories contribute to the regulation of local densities in warbler populations. Stewart and Aldrich (1951) and Hensley and Cope (1951) removed adult birds from their territories in 1949 and 1950 respectively, in a 40 acre plot in a budworm-infested area of Maine. The vacated territories were always filled by new pairs, the males singing the vigorous song of a bird setting up a territory. It seems nearly inescapable that these were part of a large floating population of birds only prevented from breeding by the absence of unoccupied territories. Since this was in a budworm outbreak, there seems little doubt that there would have been adequate food for a larger breeding population.

In a study over a series of years on the birds in New Brunswick, Cheshire (1954) recorded the populations and territory sizes of the various species as a budworm outbreak began and progressed. He showed that while the bay-breasted warbler (the commonest bird during the outbreak) underwent a five- to seven-fold increase as the outbreak began, their mean territory sizes remained constant instead of decreasing correspondingly. That is, there had been unoccupied interstices between territories initially; these were filled in by the incoming birds but territory sizes were left unchanged.

The facts suggest that the territory size is more or less fixed in this region (although, of course, it may vary from region to region) and that if territorial compression occurs during high population densities, it only does so during higher population densities than those observed. Of course, if high population densities persisted, natural selection might be expected to reduce territory size, but this is a different situation.

As for interspecific territoriality, there is no exclusion of the kind found in intraspecific territoriality, as is clearly shown in Kendeigh's (1947) territory maps. It is very difficult to distinguish a mild repulsion of other species by territorialism from a preference for slightly different habitats. Adequate information does not exist to make the distinction at present. However, it seems quite certain that interspecific territoriality is weaker than intraspecific and, therefore, that the effect of a large density of one species is greater on that species than on the others. It is thus probable that, in the warblers, territoriality helps reduce competition and acts as a stabilizing factor (as well as performing the well-known functions of pair formation and maintenance).

Natality and Mortality

In a population which has reached an equilibrium size, abundance is independent of birth and death rates. For species in equilibrium, then, a study of birth and death rates is not necessary to understand the control of the equilibrium abundance. However, as Darwin (1859) said, "A large number of eggs is of some importance to those species which depend upon a fluctuating amount of food, for it allows them rapidly to increase in numbers."

The five species of warbler studied here are very interesting in this respect. Table VI is a summary of the nesting data of the Museum of Comparative Zoology at Harvard, the American Museum of Natural History in New York, and the data of Harlow published by Street (1956).

TABLE VI. Mean clutch sizes for the 5 species

Species	No. of Nests	Clutch Size					Mean	St. Dev.
		3	4	5	6	7		
Cape May	48		4	11	24	9	5.792	.850
Myrtle	24	1	19	4			4.125	.449
Black-throated green	45	2	39	4			4.044	.366
Blackburnian	44	7	32	5			3.955	.526
Bay-breasted	49		5	21	20	3	5.429	.752

Cape May and bay-breasted warblers' nests were enough of a prize that it is quite certain that all found were kept and that the collections do not reflect any bias. There is a possibility of slight bias, collectors perhaps prizing larger clutches, in the other three species in the museum collections, but their clutch sizes are so constant that this seems improbable. The data of Harlow are not subject to his criticism, since he recorded all nests found.

While the sources of these collections vary in latitude from that of the Poconos of Pennsylvania to that of northern New Brunswick, there appears to be very little change in clutch size in this range of latitude. Thus the mean clutch size of 16 nests of the black-throated green in the Poconos is 4.06, while for the combined collections from Nova Scotia and New Brunswick (12 nests) the mean clutch is 4.17. The nests of the other species are from a narrow range of latitude and would not be expected to vary. Thus it was felt permissible to combine the data from different latitudes.

It is immediately apparent that Cape May and bay-breasted, the species which capitalize upon the periodic spruce budworm outbreaks, have considerably larger clutches than the other species, as Darwin would have predicted. It is of interest that the only other warbler regularly laying such large clutches is the Tennessee warbler (*Vermivora peregrina*) which is the other species regularly fluctuating with the budworms (Kendeigh 1947). Thus it seems that Darwin's statement provides an appropriate explanation for the larger clutches. It is also interesting that the standard deviation of the Cape May and bay-breasted warblers' clutch sizes is greater. This suggests a certain plasticity which can be verified, for the bay-breasted at least, as follows. If the time of the budworm outbreak in New Brunswick is taken as 1911-1920 (Swaine and and Craighead 1924), and other years from 1903 until 1938 are called non-budworm years, the bay-breasted warbler clutches from northeastern New Brunswick can be summarized as follows:

	Clutch Size			
	4	5	6	7
Budworm Years	1	5	15	3
Non-budworm Years	4	8	5	0

The U test (Hoel 1954) shows this to be significant at the .0024 level; that is, bay-breasted warblers lay significantly larger clutches during years of budworm outbreaks. There are not sufficient data to make a corresponding comparison for Cape May warblers. It is known (Wangersky and Cunningham 1956) that an increase in birth rate is likely to lead to instability. The easiest way to increase the stability, while still maintaining the large clutch which is desirable for the fluctuating food supply, is to have the clutch especially large when food is abundant. This is apparently the solution which the bay-breasted warbler, at least, has taken.

Mortality during the breeding season is more difficult to analyze. Disease is not normally important as a mortality factor in passerines (Lack, 1954) and this appeared to be the case for the warblers under observation. Predation may be important, however. Saw-whet owls (*Aegolius acadica*), Cooper's hawks (*Accipiter cooperii*), goshawks (*A. gentilis*), ravens (*Corvus corax*), crows (*C. brachyrhynchos*), and herring gulls (*Larus argentatus*) all occasionally were noted in the Maine woods, but no evidence was obtained of their preying upon the warblers. In fact, none of the established pairs of birds were broken up by predation of this type. Red squirrels (*Tamiasciurus hudsonicus*) were continually present in all plots and were frequently observed searching for nests. They certainly destroyed the nest of a black-throated green and of a brown creeper (*Certhia familiaris*) and were quite probably responsible for plundering one myrtle warbler nest which was robbed soon after eggs were laid. The most common evidence of mortality, however, was the frequent observation of parents feeding only one or two newly fledged young. Thus two pairs of myrtle warblers in 1956 and one in 1957 were observed the day the young left the nest feeding four young. One of the 1956 pairs succeeded in keeping all four young alive for at least three days, at which time they could no longer be followed. The remaining two pairs were only feeding two young on the day following the departure of the young from the nest. Similarly, of two black-throated green pairs (one in 1956, one in 1957) where young could be followed, one kept all four young alive and the second only raised two of the fledged four. It was difficult to determine the number of young the parents were feeding. It was also difficult to be at the nest site when the young left the nest to determine the number of fledged young. Consequently, no more observations suitable to report were made. When the young leave the nest, they fly to nearby trees quite independently of one another and apparently never return to the nest. The result is that within a few hours the young are widely scattered. In this condition they are very susceptible to predators and exposure, and should one fly when its parents were not nearby, it would rapidly starve. Normally, the young only fly or chatter loudly when a parent with food is calling nearby, and the parents seem remarkably good at remembering where the young have gone. At best, however, this is a very dangerous period. It is of some interest to note that adult warblers will feed not only young of other birds of their own species but also of other species. Skutch (1954) reviewed several published cases of this. Hence, when a wood is densely settled with warblers, the members of a large clutch might have a better chance of surviving, the straying young being fed by neighbors. This high density is, of course, the situation which obtains during a budworm outbreak when bay-breasted and Cape May warblers are so successful.

Time of Activities

So far, the nature and position of the species' activities during the breeding season have been compared. The time of these activities would also be a potential source of diversity. There could either be differences in the time of day in which feeding took place, or there could be differences in the dates during which eggs were laid and the young fed. The first type of difference (time of day) seems inherently improbable since, at least while feeding the fledged young, the parents are kept busy throughout the daylight hours gathering food. Record was taken of the time at which the various warbler species began singing in the morning of 19 June 1956. The results (Eastern Daylight Time) are: 0352, first warbler (magnolia, *D. magnolia*); 0357, first myrtle; 0400, both myrtle and magnolia singing regularly; 0401, first black-throated green; 0402, first parula warbler (*Parula americana*); 0403, first bay-breasted; 0405, all warblers singing regularly. Thus, within 13 minutes after the first warbler sang all species were singing regularly. The sequence of rising corresponds to the degree of exposure of the usual feeding zones for that date (see Figures 2-6), and therefore probably depends only upon the time at which the light reaches a certain intensity.

As for the breeding season, there is good evidence of differences in time of completion of clutches. Since date of completion can be expected to change from place to place, comparisons must be made at one fixed locality. Of 15 nests of black-throated green warbler found by Harlow in the Poconos of Pennsylvania, the mean date of clutch completion was June 3, and of 21 blackburnian the mean date of clutch completion was June 1. Thus, for this region, and there is no reason to think that the relative dates are different in other regions, there is little difference in time of nesting between blackburnian and black-throated green warblers. From the extensive collections of P. B. Philipp near Tabusintac, N. B., now in the American Museum of Natural History, and from a smaller number collected in the same region by R. C. Harlow, now in the Museum of Comparative Zoology, bay-breasted and Cape May warbler nest dates can be compared (Figure 9). It is quite clear that the bay-breasted with the median date of nest discovery of 25 June (95% confidence interval for the median 23-27 June) nest substantially later than the Cape May whose median date is 17 June (95% confidence interval for the median of 16 June-20 June). As the figure shows, the small number of nests of black-throated green and myrtle from the same region show a fairly wide spread but strongly suggest median dates intermediate between Cape May and bay-breasted. (The dates recorded by Palmer (1949) for Maine give a roughly similar sequence; myrtle, 30 May-6 June; black-throated green, 26 May-20 June; bay-breasted, after 7 June.)

It might be expected that the insects caught by the species which feed in the T zones and near the tree tops would reach peak abundance sooner thus making it desirable for those species to nest earlier. The sequence of nesting dates just presented seems to be consistent with this hypothesis.

Evidence from Winter Season

The five species of warbler migrate out of the coniferous forest through the deciduous forest and cultivated land in eastern North America and, mostly, into the West Indies and Central and South America. Therefore, any behavioral char-

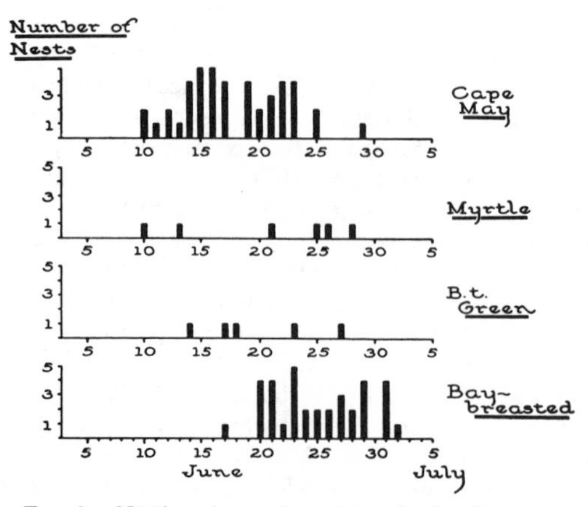

FIG. 9. Nesting dates of warblers in the Tabusintac, N. B., region.

acteristics that remain the same throughout the year must be nearly independent of the specific environment (at least within the range of environmental variation to which the bird is normally exposed). If any aspects of the breeding season behavior are retained throughout the year, these would be expected to be more fundamental than those aspects that varied with the local environment. This would be especially likely if the retained aspects of the behavior were controlled by morphological characteristics. The varying aspects would be interpreted as the result of interaction between the fundamental characteristics and the environment, as direct results of stimuli particular to that environment, or as seasonal aspects of birds' physiology. Thus it is of interest to compare behavior in different seasons.

Winter Distribution

Although Salomonsen (1954) said that species which breed in the same place tend to winter in different geographic regions, there is no evidence for this in the five warblers. More precisely, Salomonsen's statement suggests that a certain amount of competition might be avoided by having allopatric wintering grounds. Probably the most satisfactory way to test this is to determine whether the five warblers' ranges show less winter overlap than a randomly chosen group of five eastern warblers (Western warblers tend to winter in a different region and hence should not be included). To make definitions precise, two species of warbler were said to have a significant overlap in winter ranges if at least half of one species' winter range is included in the other's. From the winter range data of Bent (1953) the twenty-three species of warbler breeding in Maine (Palmer, 1949) show 253 significant overlaps, i.e., an average of 11 per species. Therefore, the probability that a randomly chosen pair of species of Maine warblers will show significant winter range overlap is 11/23 or .478. Considering the five species of warbler in the present study, Cape May overlaps with myrtle and possibly black-throated green; myrtle overlaps with Cape May, black-throated green, and blackburnian; black-throated green overlaps with Cape May (possibly), myrtle, and blackburnian; blackburnian overlaps with myrtle, black-throated green, and bay-breasted; and bay-breasted overlaps with blackburnian. There is thus a total of 10 certain overlaps among the 5 species, or 2 per species. There is thus a mean overlap per pair of species of 2/5 or .400 which is quite near the expected 0.478 suggesting that the five species overlap about randomly. It might be argued that the 23 species of Maine warblers themselves show a mutual repulsion in the winter ranges and hence are a poor standard of comparison. That this is not so can be seen as follows. The 23 species have a total of 315 significant summer overlaps in range, or 13.7 per species; *i.e.*, 13.7/23 or 0.596 is the probability that a randomly chosen pair of species will overlap in summer. As discussed above, the probability is 0.478 that a randomly chosen pair will overlap in winter. Therefore, if winter range is chosen independently of summer, $0.478 \times 0.596 = 0.285$ would be the expected probability of significant overlap in both winter and summer. Significant summer and winter overlaps were recorded in 164 cases, giving 7.1 per species, or a probability of $7.1/23 = 0.309$, which is even a very slightly higher figure than expected, showing a slight tendency for birds which summer together to winter together. Therefore, the Maine warblers do not repel one another in over-all winter range and they are therefore suitable for the comparison made earlier. It can be concluded that the five species show about the amount of overlapping of winter range that would be expected on a random basis.

This does not prove that the species occupy the same habitat in the winter, of course. Although there are no adequate data to investigate the problem, it is quite possible that because of habitat selection, wintering populations of the five species are isolated.

Winter Feeding Behavior

The period of 22 December 1956 until 9 January 1957 was spent in Costa Rica observing winter behavior of warblers. Although myrtle, black-throated green, and blackburnian warblers include Costa Rica in their winter range (Skutch, in Bent 1953, and pers. comm.), only black-throated green of the five were found during the author's study. However, many other species of warbler were present, and detailed notes were taken on them for comparison with summer behavior. Measurements of interval between flights were made for each species. These should be comparable with those made in other seasons. No strictly comparable measurement of feeding position could be made, however. In view of the great variety of tree heights in tropical forests, measurements of feeding height could not reasonably be made in terms of distance from the top of the tree. Instead, height above the ground was used, usually gauged by eye and occasionally checked by camera viewfinder. Zones such as base, middle, and tip of branch were not reasonable, but general reference to large limbs or leaves could be made. The actual behavior while gathering food is probably comparable with other seasons; it is fairly subjective, however, so that the comparison should be confirmed by the various other measurements. A general comparison of winter and summer behavior of warblers wintering in Costa Rica will be given first; this will be followed by a more detailed analysis of two species.

Thirteen species of Parulidae were observed in Costa Rica. Of these, nine, black and white (*Mniotilta varia*), Tennessee, golden-winged (*Vermivora chrysoptera*), yellow, black-throated green, sycamore (*Dendroica dominica albilora*)[2], chestnut-sided (*Dendroica pensylvanica*), Wilson's (*Wilsonia pusilla*), and redstart (*Setophaga ruticilla*), breed in northeastern United States and/or adjacent Canada. Their summer behavior has been observed, somewhat casually, by ornithologists for many years. These observations are summarized in a general way by Bent (1953) and are part of the common knowledge of most ornithologists. It is therefore of great interest that Skutch, who has made very careful observations of Central American birds, has stated (pers. comm.) that he thinks all warblers wintering in Costa Rica (except perhaps chestnut-sided which he feels spends more time in high trees in the winter) have the same general feeding behavior and feeding height in both seasons. Table VII summarizes general results of this study and helps confirm Skutch's impression.

Tennessee and black-throated green were observed in greater numbers than the others. Hence, a detailed comparison of their winter behavior in relation to their summer behavior is possible. Tennessee warblers often hopped along branches, while black-throated green more often hopped

[2] Rare in Costa Rica.

TABLE VII. Summary of observations in Costa Rica

Species	Costa Rica			Breeding Grounds	
	Approx. Number Observed	Principal Feeding Height	Principal Feeding Activities	Principal Feeding Height	Principal Feeding Activities
Black and White	5	0-35′	creeps on trunk and branches	same	same
Golden-winged	1	5-25′	hops	same	same
Tennessee	200	10-50′*	hops, esp. along branches*	0-40′	same*
Yellow	20	4-40′	hops	same	same
Black-th. green	50	10-50′*	hops, esp. across branches*	same*	same*
Sycamore	1	8-30′	creeps and hops	?	same
Chestnut-sided	15	0-50′	hops	0-30′	same
Wilson's	15	0-25′	hops	same	same
Amer. redstart	10	5-50′	hawks for flying insects	same	same

*See text for further information.

across branches. To measure this tendency, the following procedure was adopted. A count was made of the number of changes of feeding branch which required hopping or flying over a gap. The number of seconds of observation was also recorded. If Tennessees moved along branches, they should have had significantly fewer hops per second than black-throated greens which moved across branches. In the table below H stands for the number of hops or flights across an air gap in S seconds.

Black-throated green H	S	*Tennessee* H	S
		4	30
8	90	3	48
5	38	6	105
7	43	4	54
1	14	6	100
12	85	11	74
12	60	8	83
14	70	8	82
59	400	50	576

Mean No. of Seconds per hop 6.78 11.52

Black-throated greens thus hopped 59 times in 400 seconds for a mean number of seconds per hop of 6.78. Tennessees hopped 50 times in 576 seconds for a mean number of seconds per hop of 11.52. By an extended sign test (Dixon and Massey 1951) this difference is well within the 5.5% level of significance. Hence, it is clear that black-throated greens hop across branches more often than Tennessees, partially confirming the subjective impression described above.

Strictly comparable measurements cannot be made in the breeding season since black-throated green feeds in coniferous and Tennessee in deciduous trees. However, the black-throated green has been shown to move principally in a tangential direction in the summer, while of eleven Tennessee warbler observations in the summer of 1957 all showed the hopping along branches which characterized the winter feeding behavior.

From these data, it is evident that the general aspects of warbler behavior are nearly the same in winter and summer. For warblers wintering in the West Indies the same situation obtains; namely, the winter habitats bear no obvious similarity to those chosen in the breeding season, but the feeding behavior and height are roughly the same (Eaton 1953). Cape May and myrtle warblers are particularly interesting and a little atypical in this respect. Both Bond (1957) and Eaton mention that Cape May frequents gardens and plantations, where it is often near the ground. This behavior parallels that observed by Brewster in the summer and reported earlier. In the winter it spends much of its time feeding upon flowers, a fact which will be used later. Bond (1957) and Skutch (in Bent 1953) both mention that myrtle warblers, in their West Indian and Central American wintering grounds, are found from beaches to forests, frequenting open ground especially. This variety of feeding location combined with its enormous winter range confirms the summer observation of great flexibility of behavior.

These observations on the flexibility of myrtle warblers can be extended by including observations made on their wintering grounds in the United States. It is well known (Pettingill 1951) that myrtles winter in the northeast wherever there are extensive patches of bayberry (*Myrica pensylvanica*). Montauk Point on Long Island, which is such a place, was visited on January 26, 1957. Here myrtles were the commonest bird in the habitat in which the bayberry is abundant. The myrtles were moving about in flocks, frequently, as in the summer, making long flights. They fed principally while hopping upon the fallen leaves under bayberries. (There was no snow.) This behavior was very similar to that observed in the summer while they were catching emerging crane flies. Some were feeding upon the "wax" coat of the berries which is readily digestible and contains nodules which are rich sources of proteins and carbohydrates (Hausman 1927).

Population Control

Any factor that can control a local population has a space distribution. Examples of such factors are food, nesting sites, and predators. Thus all populations are limited by the amount of suitable space. The meaning of "suitable" for a given species is the interesting problem. Within a given environment each necessary activity requires a certain amount of space. That activity which requires the greatest amount of space is likely to be limiting. Thus, animals such as barnacles which wait for moving food to pass by and require very little space to catch it are likely to be limited by amount of surface on which to rest. Here "suitable space" means "space adequate for barnacle attachment." Similarly, for some insects the suitable space may be the space with sufficient food supply within easy dispersal distance; for some birds, suitable nest holes may be scarcer than adequate food and suitable space will mean proper nest hole. This section will be devoted to the nature of suitable space for the five species of warbler under consideration.

The five warblers do not seem to have special nesting space or nesting material requirements which would necessitate a larger amount of space than food gathering. Territory defence is probably the only activity of the warblers that requires an area of comparable size to that needed for food gathering. As discussed earlier, territoriality may exert a limiting effect upon populations under conditions of abundant food supply, thus acting as a stabilizing factor. However, if territory requirements limit populations under normal conditions, it may be inquired why natural selection has not reduced territory sizes thus permitting larger populations. Furthermore, variation in warbler population density from plot to plot suggests that more than an incompressible territory is responsible for population regulation. Therefore, like nesting space and nesting material requirements, territories probably require less space for warblers in normal years than does food gathering. Consequently, suitable space is probably the amount of space with an adequate food supply in which the bird is adapted to feed. Direct measurement of the food supply would require a very elaborate sampling scheme. However, measurement of the amount of foliage of the type in which the species has been shown to feed is quite feasible. If the density of breeding pairs of a species is proportional to the amount of foliage in a certain zone, then a census of a plot with twice the volume of this foliage should have twice the number of birds of corresponding species. Since the five species under consideration feed above 20 feet, only foliage above 20 feet was measured.

Five areas on Mt. Desert Island were censused and measured with this view in mind. All were in predominantly spruce forest, but except for this they were chosen to be as different as possible in order to exhibit as great a range of variation as possible. The volume of foliage above twenty feet was measured as follows. The volume of a cone is proportional to the product of its basal area with its height. The foliage of a spruce tree is roughly a hollow cone with walls of approximately constant thickness. The inner cone may be considered to begin ten feet below the outer. Finally, the basal area of the foliage is proportional to the trunk area. Consequently, the product of the height of foliage in the crown with the basal area of the trunks of the trees will give a figure proportional to the volume of the outer cone. A similar figure for the volume of the inner cone is calculated and the volume of foliage in the hollow shell is obtained by subtraction. The proportion of the volume lying above twenty feet of either the inner or the outer cone is the square of the proportion of the height lying above twenty feet, so that this adjustment is easily made. The number obtained is thus proportional to the desired volume. In practice, the height of foliage in the crown was measured with a rangefinder and the basal area was measured with a "Bitterlich reloscope" (Grosenbaugh 1952).

Plot A was a 3.8 acre section of an open sphagnum bog, the "Big Heath" on Mt. Desert Island. The only trees present were black spruce and tamarack (*Larix laricina*). Trees were very scattered and the highest were between 15 and 20 feet. While typical bog warblers were common, none of the five species considered here was present.

Plot B was a 4 acre strip along the edge of the bog. Red spruce largely replaced black in this plot, the trees occasionally reaching 50 feet in height. The strip was bordered on one side by a road and on the other by the open bog.

Plot C was a 9 acre area along the Hio Truck trail near Seawall on Mt. Desert Island. The forest here was quite mature, trees reaching 70 feet, and was predominantly composed of red spruce, white spruce, and balsam fir, with a higher proportion of white birch than occurred in the other plots.

Plot D was a dense, 4 acre stand of red spruce of moderate age near Southwest Harbor, Maine. Trees reached a height of 60 feet in some places, and had been thinned in part of the plot.

Plot E was the 9 acre stand of white spruce at Bass Harbor Head on Mt. Desert Island in which the observations of 1956 were made. Here the trees reached a height of 70 feet quite frequently, although the mean was nearer 60. This plot apparently originated as an old field stand with numerous, large, low branches. These have died and become covered with a layer of lichens, especially *Usnea*.

For each plot the composition of the warbler population and the volume of the foliage above 20 feet are indicated in figure 10. Here "others" refers to other tree-nesting warblers which fed above 20 feet. It is quite clear that the total tree-nesting warbler population is very nearly proportional to the foliage volume. The abundance of the myrtle warbler is evidently quite constant, only increasing slowly as the volume of foliage increases. This is probably due to its flycatching and ground feeding habits which help to make it less dependent upon the foliage.

Black-throated green warblers were the dominant birds in all mature spruce forest habitats on Mt. Desert Island. As the figure indicates, their abundance was nearly proportional to the foliage volume above 20 feet. Their abundance did fall off slightly in the two plots, C and E, in which very tall trees were present. Black-throated greens seldom use these tall tree tops for feeding, so that the amount of foliage suitable for black-throated greens should be slightly reduced from the amount calculated for plots C and E.

Blackburnians, which require foliage near the tops of the trees, were present in small numbers in both plots which had trees of height greater than 60 feet. A true understanding of their population control on Mt. Desert Island probably cannot be acquired without considering competition with other species. It is reasonable that the blackburnian can only persist where the forest is sufficiently old that the feeding zone of the dominant black-throated green stops well before the tree tops in which the blackburnian prefers to feed.

On Mt. Desert Island, bay-breasted warblers are only common where there are dense growths of lichen-covered lower branches of spruce in the shade of the forest crown. It is in this zone that a large part of their activity takes place and here their sluggish, radially moving, feeding behavior is well suited. This habitat appears when the forest becomes dense and has large trees. Consequently, the bay-breasted warblers only remained permanently in plots D and E. (Two set up territories in plot C but had apparently left by June 8.) Again it appears that this habitat is occupied by the bay-breasted partly because it is not occupied by the black-throated green. In the budworm infested spruce and balsam stands near Ft. Kent, Maine, the bay-breasted warbler was the dominant bird; here they occupied the dense young stands of conifers and black-throated greens were forced to occupy the ridges covered with mixed growths of hemlock and hardwoods. The type of competition in heterogeneous regions mentioned in the first section provides an appropriate explanation for the change in dominance; the forest composition of the whole region is more important than the very local conditions.

Although Cape May warblers did not occupy any of the census plots, observations on their feeding behavior suggest that they would be quite similar to the myrtle warbler as far as dependence on tree foliage is concerned. This is partially confirmed by the fact that of the 18 stations studied in northern Maine, Cape Mays were present in all the lowland ones and dominant only in the fairly open stands with mature trees—a habitat which is unsuitable for the bay-breasted but is quite satisfactory for both myrtle and Cape May.

Discussion and Conclusions

In this study competition has been viewed in the light of the statement that species can coexist only if each inhibits its own population more than the others'. This is probably equivalent to saying that species divide up the resources of a community in such a way that each species is limited by a different factor. If this is taken as a statement of the Volterra-Gause principle, there can be no exceptions to it. Ecological investigations of closely-related species then are looked upon as enumerations of the divers ways in which the resources of a community can be partitioned.

For the five species of warbler considered here, there are three quite distinct categories of "different factors" which could regulate populations. "Different factors" can mean different resources, the same resources at different places, or the same resources at different times. All three of these seem important for the warblers, especially if different places and times mean very different—different habitats and different years.

First, the observations show that there is every reason to believe that the birds behave in such a way as to be exposed to different kinds of food. They feed in different positions, indulge in hawking and hovering to different extents, move in different directions through the trees, vary from active to sluggish, and probably have the greatest need for food at different times corresponding to the different nesting dates. All of these differences are statistical, however; any two species show some overlapping in all of these activities. The species of food organisms which were widespread in the forest and had high dispersal rates would be preyed upon by all the warblers. Thus, competition for food is possible. The actual food eaten does indicate that the species have certain foods in common. The slight difference in habitat preference resulting from the species' different feeding zones is probably more important. This could permit each species to have its own center of dispersal to regions occupied by all species. Coexistence in one habitat, then, may be the result of each species being limited by the availability of a resource in different habitats. Even although the insects fed upon may be basically of the same type in the different habitats, it is improbable that the same individual insects should fly back and forth between distant woods; consequently, there would be no chance for competition. The habitat differences and, equivalently, the feeding zone differences, between blackburnian, black-throated green, and bay-breasted are sufficiently large that this explanation of coexistence is quite reasonable.

The myrtle warbler is present in many habitats in the summer but is never abundant. It has a very large summer and winter range, feeds from the tree tops to the forest floor, and by rapid peering or by hawking. It makes frequent long flights and defends a large territory. Probably it can be considered a marginal species which, by being less specialized and thus more flexible in its requirements, manages to maintain a constant, low population (Figure 10).

The Cape May warbler is in a different category, at least in the region near the southern limit of its range. For here it apparently depends upon the occasional outbreaks of superabundant food (usually spruce budworms) for its continued existence. The bay-breasted warbler, to a lesser degree, does the same thing. During budworm outbreaks, probably because of their extra large clutches, they are able to increase more rapidly than the other species, obtaining a temporary advantage. During the years between outbreaks they suffer reductions in numbers and may even be eliminated locally. Lack's hypothesis, that the clutch is adjusted so as to produce the maximum number of surviving offspring, provides a suitable explanation of the decrease during normal years of these large-clutched species. It may be asked why, if Lack's hypothesis is correct, natural selection favored large clutches in Cape May and bay-breasted. Cheshire's (1954) censuses suggest a tentative answer. During his years of censusing, increases in the bay-breasted warbler population reached a figure of over 300% per year. This probably far exceeds the maximum possible in-

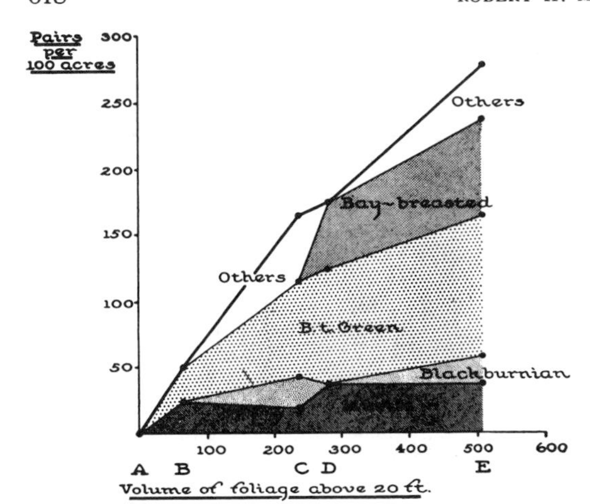

FIG. 10. Composition of the warbler population in plots A, B, C, D, and E. "Others" refers to other warbler species which feed at greater heights than 20 feet above the ground. The units of volume measurement are only proportional to the volume, but each unit roughly equals 1500 cubic feet per acre.

crease due to survival of nestlings raised in that place; probably immigration is the explanation. But if the species with large clutches search for areas in which food is superabundant and immigrate into these regions, then, for the species as a whole, the large clutch may be adapted to the maximum survival of offspring. Cape May and bay-breasted warblers may therefore be considered to be good examples of fugitive species (Hutchinson 1951).

Thus, of the five species, Cape May warblers and to a lesser degree bay-breasted warblers are dependent upon periods of superabundant food, while the remaining species maintain populations roughly proportional to the volume of foliage of the type in which they normally feed. There are differences of feeding position, behavior, and nesting date which reduce competition. These, combined with slight differences in habitat preference and perhaps a tendency for territoriality to have a stronger regulating effect upon the same species than upon others, permit the coexistence of the species.

ACKNOWLEDGMENTS

Prof. G. E. Hutchinson and Dr. S. D. Ripley have played indispensable roles in the development of this work, providing advice, encouragement, and support. The author had valuable discussion with Dr. A. F. Skutch, James Bond, and Paul Slud concerning birds on their wintering grounds. Dr. L. R. Holdridge provided invaluable help in Costa Rica, acting as naturalist and guide. Sincere thanks also go to the following persons. Miss Helen T. Mac Arthur prepared the illustrations. Drs. R. A. Paynter and Dean Amadon gave information about or provided access to the collections under their supervision. Drs. C. L. Remington, W. R. Henson, and P. B. Dowden provided entomological information. Dr. J. W. Mac Arthur helped with the observations. Finally, the author wishes to thank his wife who helped with observations, prepared the manuscript, and provided encouragement.

The work was supported by grants from the Peabody Museum of Natural History of Yale University and from the Chapman Memorial Fund of the American Museum of Natural History.

REFERENCES

Amadon, D. 1950. The Hawaiian honeycreepers (Aves, Drepaniidae). Bull. Amer. Mus. Nat. Hist. **95**: 157-262.

Andrewartha, H. G. and L. C. Birch. 1954. The distribution and abundance of animals. Chicago: Univ. Chicago Press.

Banerjee, S. K. and K. R. Nair. 1940. Tables of confidence intervals for the median in samples from any continuous population. Sankya **4**: 551-558.

Bent, A. C. 1953. Life histories of the North American wood warblers. U. S. Nat. Mus. Bull. **203**.

Betts, M. 1955. The food of titmice in an oak woodland. Jour. Anim. Ecol. **24**: 282-323.

Bond, J. 1937. The Cape May warbler in Maine. Auk **54**: 306-308.

———. 1957. North American wood warblers in the West Indies. Audubon Mag. **59**: 20-23.

Bourne, W. R. P. 1957. The breeding birds of Bermuda. Ibis **99**: 94-105.

Breckenridge, W. J. 1956. Measurements of the habitat niche of the least flycatcher. Wilson Bull. **68**: 47-51.

Cheshire, W. P. 1954. Bird populations and potential predation on the spruce budworm. Canada Dept. Agric. Sci. Serv., Annual Tech. Report, Green River Project 1953, Sect. **14**.

Cruickshank, A. D. 1956. Aud. Field Notes **10**: 431-432. (and earlier censuses of the same plot).

———. 1956a. Nesting heights of some woodland warblers in Maine. Wilson Bull. **68**: 157.

Darwin, C. R. 1859. The origin of species by means of natural selection or the preservation of favoured races in the struggle for life. London: Murray.

Dixon, W. J. and F. J. Massey. 1951. Introduction to statistical analysis. New York: McGraw-Hill.

Eaton, S. W. 1953. Wood warblers wintering in Cuba. Wilson Bull. **65**: 169-174.

Forbush, E. H. 1929. The birds of Massachusetts and other New England states. Vol. **III**. Boston: Mass. Dept. Agric.

Gardner, L. L. 1925. The adaptive modifications and the taxonomic value of the tongue in birds. Proc. U. S. Nat. Mus. 67, art. **19**: 1-49.

Gause, G. F. 1934. The struggle for existence. Baltimore: Williams and Wilkins.

Greenbank, D. O. 1956. The role of climate and dispersal in the initiation of outbreaks of the spruce budworm in New Brunswick. Can. Jour. Zool. **34:** 453-476.

Grinnell, J. 1914. Barriers to distribution as regards birds and mammals. Amer. Nat. **48:** 249-254.

———. 1921. The principle of rapid peering in birds. Univ. Calif. Chron. **23:** 392-396.

———. 1922. The trend of avian populations in California. Science **56:** 671-676.

——— **and T. Storer.** 1924. Animal life in the Yosemite. Berkeley: Univ. Calif. Press.

Griscom, L. 1938. The birds of Lake Umbagog region of Maine. Compiled from the diaries and journals of William Brewster. Bull. Mus. Comp. Zool. **66:** 525-620.

Grosenbaugh, L. R. 1952. Plotless timber estimates—new, fast, easy. Jour. Forestry **50:** 33-37.

Haldane, J. B. S. 1955. Review of Lack (1954). Ibis **97:** 375-377.

Hausman, L. A. 1927. On the winter food of the tree swallow (*Iridoprocne bicolor*) & the myrtle warbler (*Dendroica coronata*). Amer. Nat. **61:** 379-382.

Hensley, M. M. and J. B. Cope. 1951. Further data on removal and repopulation of the breeding birds in a spruce-fir forest community. Auk **68:** 483-493.

Hinde, R. A. 1956. The biological significance of territories of birds. Ibis **98:** 340-369.

Hoel, P. G. 1954. Introduction to mathematical statistics. New York: Wiley.

Hutchinson, G. E. 1951. Copepodology for the ornithologist. Ecology **32:** 571-577.

Huxley, J. 1942. Evolution, the modern synthesis. New York: Harper.

Kendeigh, S. C. 1945. Community selection birds on the Helderberg Plateau of New York. Auk **62:** 418-436.

———. 1947. Bird population studies in the coniferous forest biome during a spruce budworm outbreak. Biol. Bull. **1**, Ont. Dept. Lands and For.

Kluijver, H. N. and N. Tinbergen. 1953. Regulation of density in titmice. Arch. Ned. Zool. **10:** 265-289.

Knight, O. W. 1908. The birds of Maine. Bangor.

Lack, D. 1947. Darwin's finches. Cambridge: Cambridge Univ. Press.

———. 1954. The natural regulation of animal numbers. Oxford: Oxford Univ. Press.

———. 1955. The mortality factors affecting adult numbers. In Cragg, J. B. and N. W. Pirie 1955. The numbers of man and animals. London: Oliver and Boyd.

MacKenzie, J. M. D. 1946. Some factors influencing woodland birds. Quart. Jour. For. **40:** 82-88.

Mayr, E. 1950. Speciation in birds. Proc. Xth Int. Ornith. Cong. Uppsala.

McAtee, W. L. 1926. The relation of birds to woodlots in New York State. Roosevelt Wildlife Bull. **4:** 7-157.

———. 1932. Effectiveness in nature of the so-called protective adaptations in the animal kingdom, chiefly as illustrated by the food habits of Nearctic birds. Smithsonian Misc. Coll. **85(7):** 1-201.

Mitchell, R. T. 1952. Consumption of spruce budworms by birds in a Maine spruce-fir forest. Jour. For. **50:** 387-389.

Palmer, R. S. 1949. Maine birds. Bull. Mus. Comp. Zool. **102:** 1-656.

Pettingill, O. S. 1951. A guide to bird finding east of the Mississippi. New York: Oxford Univ. Press.

Salomonsen, F. 1954. Evolution and bird migration. Acta **XI** Cong. Int. Orn. Basil.

Skutch, A. F. 1954. Life histories of Central American birds. Pacific Coast Avifauna No. **31.**

Smith, W. P. 1953. Aud. Field Notes **7:** 337 (and earlier censuses of the same plot).

Stewart, R. E. and J. W. Aldrich. 1951. Removal and repopulation of breeding birds in a spruce-fir forest community. Auk **68:** 471-482.

———. 1952. Ecological studies of breeding bird populations in northern Maine. Ecol. **33:** 226-238.

Street, P. B. 1956. Birds of the Pocono Mountains, Pennsylvania. Delaware Valley Ornith. Club. Philadelphia.

Swaine, J. M. and F. C. Craighead. 1924. Studies on the spruce budworm (*Cacoecia fumiferana Clem.*) Part I. Dom. of Canada Dept. Agric. Bull. **37:** 1-27.

Swed, F. S. and C. Eisenhart. 1943. Tables for testing randomness of grouping in a sequence of alternatives. Ann. Math. Stat. **14:** 66-87.

Volterra, V. 1926. Variazione e fluttuazione del numero d'individiu in specie animali conviventi. Mem. Accad. Lincei **2:** 31-113. (Translated in Chapman, R. N. 1931. Animal ecology. New York: McGraw-Hill.)

Wangersky, P. J. and W. J. Cunningham. 1956. On time lags in equations of growth. Proc. Nat. Acad. Sci. **42:** 699-702.

Williams, A. B. 1950. Aud. Field Notes **4:** 297-298 (and earlier censuses of the same plot).

Wilson, E. B. 1952. An introduction to scientific research. New York: McGraw-Hill.

64. W. W. Wells and D. H. Neiderhiser, *J. Am. Chem. Soc.* **79**, 6569 (1957).
65. C. Djerassi, J. S. Mills, R. Villotti, *ibid.* **80**, 1005 (1958).
66. A. A. Kandutsch and A. E. Russell, *J. Biol. Chem.* **235**, 2253, 2256 (1960).
67. C. Djerassi, J. C. Knight, D. I. Wilkinson, *J. Am. Chem. Soc.* **85**, 835 (1963).
68. M. Slaytor and K. Bloch, unpublished.
69. J. A. Olson, M. Lindberg, K. Bloch, *J. Biol. Chem.* **226**, 941 (1957).
70. J. Pudles and K. Bloch, *ibid.* **235**, 12 (1960).
71. J. J. Britt, G. Scheuerbrandt, K. Bloch, unpublished.
72. M. Lindberg, F. Gautschi, K. Bloch, *J. Biol. Chem.* **238**, 1661 (1963).
73. W. M. Stokes and W. A. Fish, *ibid.* **235**, 2604 (1961).
74. J. Avigan, D. S. Goodman, D. Steinberg, *ibid.* **238**, 1283 (1963).
75. G. J. Schroepfer and I. D. Frantz, *ibid.* **236**, 3137 (1961); M. E. Dempsey, J. D. Seaton, G. J. Schroepfer, R. W. Trockmann, *ibid.* **239**, 1381 (1964).
76. N. L. R. Bucher, in *CIBA Foundation Symposium on the Biosynthesis of Terpenes and Sterols*, G. E. W. Wolstenholme and M. O'Connor, Eds. (Churchill, London, 1959), p. 46.
77. M. D. Siperstein and V. M. Fagan, *Advances in Enzyme Regulation* (Pergamon, Oxford, 1964), vol. 2, p. 249.
78. R. Y. Stanier and C. B. van Niel, *Arch. Mikrobiol.* **42**, 17 (1962).
79. The research described has been generously supported by grants in aid from the National Institutes of Health, the National Science Foundation, the Life Insurance Medical Research Fund, the Nutrition Foundation, and the Eugene Higgins Trust Fund of Harvard University.

Predation, Body Size, and Composition of Plankton

The effect of a marine planktivore on lake plankton illustrates theory of size, competition, and predation.

John Langdon Brooks and Stanley I. Dodson

During an examination of the distribution of the cladoceran *Daphnia* in the lakes of southern New England, it was noted that large *Daphnia*, although present in most of the lakes, could not be found among the plankton of several lakes near the eastern half of the Connecticut coast. The characteristic limnetic calanoid copepods of this region, *Epischura nordenskioldi* and *Diaptomus minutus*, and the cyclopoid *Mesocyclops edax* were also absent. Small zooplankters were abundant, especially the cladoceran *Bosmina longirostris* and the copepods *Cyclops bicuspidatus thomasi* and the small *Tropocyclops prasinus* (*1*).

All of these lakes lacking large zooplankters have sizable "landlocked" populations of the herring-like *Alosa pseudoharengus* (Wilson) = *Pomolobus pseudoharengus* (Fig. 1), known by several common names including "alewife" and "grayback" (*2*). This is originally an anadromous marine fish, breeding populations of which have become established in various bodies of fresh water, including Lake Cayuga, New York, and the Great Lakes (*3*). The marine populations live in the coastal waters of the western Atlantic, from the Gulf of St. Lawrence to North Carolina, and ascend rivers and streams to spawn in springtime. The young return to the sea in summer and autumn (*4*). The seven Connecticut lakes (Fig. 2) with self-perpetuating populations of alewives are within about 40 kilometers of the present coastline, and each is drained directly, by a small stream or river, or indirectly, through the estuaries of the larger Connecticut or Thames rivers, into Long Island Sound (Fig. 3). As such streams and rivers are normally ascended by marine alewives, it is assumed that the establishment of these self-sustaining populations in the lakes is natural.

The authors are affiliated with the Biology Department, Yale University, New Haven, Connecticut, where Dr. Brooks is associate professor of biology.

The "alewife lakes" are diverse in area and depth. Although we have not examined the food of the alewives in these Connecticut lakes (alewives are difficult to catch), studies in other lakes have revealed that planktonic copepods and Cladocera are the primary food. The indifference of alewives to nonfloating food is not surprising in view of the adaptation of the parent stock to feeding on zooplankton in the open waters of the sea (*5*).

The dominant crustaceans in the plankton of all the alewife lakes are the same small-sized species, *Bosmina longirostris* (or *Ceriodaphnia lacustris*) being most numerous; *Cyclops bicuspidatus thomasi* and *Tropocyclops prasinus*, present in varying ratios, are also numerous. By contrast, in the nonalewife lakes *Diaptomus* spp. and *Daphnia* spp. are always dominants, usually accompanied by the larger cyclopoids *Mesocyclops edax* and *Cyclops bicuspidatus thomasi*. The absence of all but one of these last-named larger zooplankters from the lakes inhabited by the planktivorous alewife may be due to differential predation by the alewives. The elimination of these pelagic zooplankters allows the primarily littoral species, such as *Bosmina longirostris*, to spread into the pelagic zone, from which, we conclude, they would otherwise be excluded by their larger competitors.

Changes in Crystal Lake Plankton

An opportunity to test this hypothesis was provided by the introduction into a lake in northern Connecticut of *Alosa aestivalis* (Mitchell), "glut herring," a species closely related, and very similar, to *Alosa pseudoharengus* (*6*). The plankton of Crystal Lake had been quantitatively sampled by one of us in 1942 before *Alosa* was introduced. At that time the zooplankton was dominated by the large forms (*Daphnia, Diaptomus, Mesocyclops*) expected in a lake of its size (see Table 1). Resampling 10 years after *Alosa* had become abundant should reveal plankton similar in composition to that common in the lakes with natural populations of *Alosa pseudoharengus* and unlike that characteristic of Crystal Lake before *Alosa* became abundant.

The plankton of the entire water column of Crystal Lake was sampled quantitatively on 30 June 1964 (*7*). All the crustacean zooplankters caught in the 1942 and 1964 collections (a total of 6623 specimens) were identified, and those belonging to each species were enumerated. As the nauplii of all species were lumped and merely enumerated as "nauplii," the figures for copepods given in Table 1 indicate only the percentages of adults and copepodids. The copepodid stages of some coexisting cyclopoids were so similar that differentiation of species was sometimes uncertain. For these species a lumped total is given in the table. The plankton of four lakes with natural populations of *Alosa* was sampled for comparison with that of Crystal Lake in 1964, and the plankton of four lakes, without *Alosa*, in the "alewife" region of southern Connecticut was sampled for comparison with the Crystal Lake plankton of 1942. All these lakes were sampled between 5 June and 7 July 1964 (*8*). The relative frequencies of those crustacean zooplankters which comprise 5 percent or more of the total are given in Table 1. The relatively large predaceous rotifer *Asplanchna priodonta*, the only noncrustacean recorded in Table 1, was remarkably numerous in some of the alewife lakes. The plankton of Crystal Lake in 1964, when *Alosa aestivalis* was abundant, was quite like that of the natural alewife lakes, and not at all like the plankton of Crystal Lake before *Alosa* was a significant element in the open-water community. Crystal Lake in 1942 resembled the lakes without alewives in that its plankton was dominated by *Diaptomus* and *Daphnia*. It might be added that resampling (*9*) of a majority of the other Connecticut lakes after the same 20-year interval has not revealed such a major change in the composition of the zooplankton anywhere else.

In order to examine more carefully the differences in body size between the dominants of alewife and those of non-alewife lakes, the size range of each species was determined. Body size was measured as body length, exclusive of terminal spines or setae. (The posterior limit of measurement for each genus is shown on the drawings of Fig. 4.) A summation of the numbers of each dominant that fell within each size interval yields a size-frequency diagram for the crustacean zooplankters. Although such diagrams were prepared for each lake, only those for Crystal Lake in 1942 and 1964 are presented here (Fig. 4). As size intervals represented by less than 1 percent of the population sample were left blank, the histogram does not indicate the presence of large but relatively rare forms. In Crystal Lake in 1942, specimens of the genera depicted occurred up to a length of 1.8 millimeters, and the predaceous *Leptodora kindtii* was represented by a few specimens between 5 and 10 millimeters. In 1964, by contrast, no zooplankters over 1 millimeter long could be found, although in other *Alosa* lakes there were occasional specimens up to 1.25 millimeters. The nauplii and metanauplii were counted but not measured, so that the totals of measured specimens are decreased by these amounts (see Table 1). The histograms (Fig. 4) show that the majority of the zooplankton are less than about 0.6 millimeter in length when *Alosa* is abundant, whereas the majority of specimens of the dominant species in the same lake before *Alosa* became abundant were over 0.5 millimeter long. The modal size in the presence of *Alosa* was 0.285 millimeter, whereas the modal size in the absence of *Alosa* was 0.785 millimeter. This seems clear evidence that predation by *Alosa* falls more heavily upon the larger plankters, eliminating those plankters more than about 1 millimeter in length.

Effects of Predation by Alosa

Whether or not a species will be eliminated (or reduced to extreme rarity) by *Alosa* predation will in good part depend upon the average size of the smallest instar of egg-producing females; a sufficient number of females must survive long enough to produce another generation. To assess the significance of this critical size, specimens approximating the average size of the smallest mature instar of the dominant species of both years were drawn to scale and appropriately placed on the size-frequency histograms. The smallest mature females of *Daphnia catawba, Mesocyclops edax, Epischura nordenskioldi*, and certainly *Leptodora kindtii* are too large to have a reasonable chance of surviving long enough to produce sufficient young. However, the elimination of *Diaptomus minutus* (depicted in Fig. 4) but not of *Cyclops bicuspidatus thomasi*, indicates that, for species maturing at a length between 0.6 and 1.0 millimeter, factors other than size (such as escape movements, spatial distribution) are also of significance. Aside from *Cyclops bicuspidatus thomasi*, all the characteristic zooplank-

Table 1. The relative frequency of planktonic Crustacea in lakes with and without *Alosa*. C, Cedar Pond; BA, Bashan Lake; BE, Beach Pond; G, Lake Gaillard; L, Linsley Pond; A, Amos Lake; Q, Lake Quonnipaug; R, Rogers Lake. Relative frequencies expressed in percentages.

Organism	Lakes without alewives				Crystal Lake		Natural alewife lakes			
	C	BA	BE	G	1942 (No *Alosa*)	1964 (*Alosa*)	L	A	Q	R
Cladocera										
Leptodora kindtii		*			*					
Holopedium spp.		*	6			*				5
Diaphanosoma spp.		*		*	5		*		*	
Daphnia galeata	59									
Daphnia catawba		*	7	11	14					
Ceriodaphnia lacustris						*	*	*	43	*
Bosmina coregoni			*	13						
Bosmina tubicens			*							5
Bosmina longirostris	*					34	39	44	16	10
Copepoda										
Epischura nordenskioldi		*	*	*	*					
Diaptomus minutus		84	76		52					
Diaptomus pygmaeus	*			50	*		*	6	*	5
Mesocyclops edax	21			}12						
Cyclops bicuspidatus	*					}29	34			35
Tropocyclops prasinus							*	11	9	20
Orthocyclops modestus									16	
Nauplii	11	13	10	12	11	28	7	30	8	18
Rotifera										
Asplanchna priodonta						*	17	6	*	
Dimensions										
Area (hectares)	9	112	158	440	80		9	42	45	107
Maximum depth (m)	5	14	19	20	14		14	14	14	20
Mean depth (m)	3	5	6		7		7	6	4	6

* Present, but comprising less than 4.5 percent.

ters of *Alosa* lakes mature at lengths of less than 0.6 millimeter.

Since *Alosa*, except during spawning, avoids the shores, its predation falls more heavily upon, and may eliminate, "lake species" of zooplankton that tend to avoid the shore, allowing littoral or "pond species" to thrive. Among the Cladocera, for example, a large *Daphnia* (*D. catawba, D. galeata*) usually occurs as a dominant in the zooplankton of lakes over 5 meters deep, while *Bosmina longirostris* is a common dominant in shallower bodies of water. When *Alosa* is present, any population of large *Daphnia* is eliminated or severely reduced, and *Bosmina longirostris*, relatively spared by its smaller size and the more littoral habits of a large part of its population, replaces *Daphnia* as the open-water dominant. The differential predation due to the tendency of *Alosa* to feed in open water away from the bottom (*5*) can also be seen among the medium-sized copepods. Of the species usually dominant in lake plankton, *Diaptomus minutus*, the smallest, is eliminated by *Alosa* in small lakes, and in those lakes the larger *Cyclops bicuspidatus thomasi* (see Fig. 4) achieves a dominance that it seldom enjoys otherwise. This differential predation is probably due to the fact that adult *Diaptomus minutus* are primarily epilimnetic in summer, while the adult *Cylops bicuspidatus thomasi* tend to be heavily concentrated in the inshore and bottom waters, only the immature being found in the open water. *Diaptomus pygmaeus*, intermediate between the above two species both in size at the onset of maturity and in spatial distribution of adults, often survives in *Alosa* lakes as a minor component of the plankton.

The alewife lakes of Connecticut are small. It is, therefore, of interest to examine the plankton of larger lakes into which *Alosa* has been introduced. *Alosa* has long been abundant in Lake Cayuga, the largest of the Finger Lakes of New York, with an area of 172 square kilometers and a maximum depth of 132 meters. The plankton of Cayuga is dominated by *Bosmina longirostris* (or *Ceriodaphnia* sp.). Large Cladocera, such as *Daphnia, Leptodora*, and *Polyphemus*, are never common, and *Diaptomus* is scarce. The calanoid *Senecella calanoides* (2.7 mm) is present only at depths below 80 meters. This numerical ascendancy of the smaller zooplankters in the upper waters is consistent with the concept of size-dependent predation by *Alosa*. The persistence of large zooplankters in the Laurentian Great Lakes indicates that *Alosa* has had a less dramatic effect on the plankton of these immense lakes (*10*).

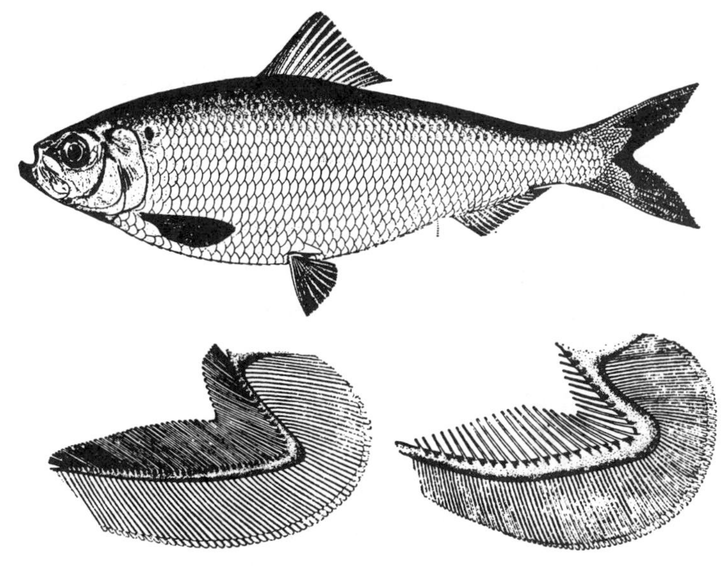

Fig. 1. *Alosa* (= *Pomolobus*) *pseudoharengus* (Wilson). Top, mature specimen, 300 mm long. Note that mouth opens obliquely. Bottom left, first branchial arch, with closely spaced gill rakers that act as a plankton sieve. Compare with (bottom right) the widely spaced gill rakers of *A. mediocris*, a species that feeds primarily upon small fish. [After Hildebrand (*4*), with the permission of the Sears Foundation for Marine Research, Yale University]

Size and Food Selectivity

To assess the significance of *Alosa* predation it is necessary to consider the importance of the size of food organisms throughout the open-water community. To simplify the discussion we shall consider the open-water community of a lake to comprise four trophic levels. Level four, the piscivores, consists chiefly of fish, even though their fry are planktivorous and thus belong to the third trophic level, planktivores. On the third level, also, fish are quantitatively most important (*11*). Level two consists of the herbivorous zooplankters. (Some zooplankters are predators and therefore on level three, but are quantitatively usually negligible.) The planktonic herbivores feed upon the microphytes—the larger algae (net phytoplankton) and the small algae (nannoplankton)—that comprise level one in the open waters, together with bacteria and a variety of nonliving organic particles. The nannoplankton, together with all the other particles that can pass through a 50-micron sieve, will be called nannoseston.

Animals choose their food on the basis of its size, abundance, and edibility, and the ease with which it is caught. However, there is a fundamental difference between food selection by herbivorous zooplankton and that by the predators of higher levels. For planktivores and piscivores, other things being equal, the least outlay of energy in relation to the reward is required if a smaller number of large prey, rather than a larger number of small ones, are taken (*12*). When the environment provides a choice, therefore, natural selection will tend to favor the predator that most consistently chooses the largest food morsel available. At the highest trophic levels, where the number of available prey is often low, a predator must often take either a small morsel or none at all. The variety (*13*) and abundance of the zooplankton, however, usually provide the planktivore with an array of sizes from which to choose. One would expect, therefore, that planktivores would prey upon larger organisms more consistently than do the piscivores. It should further be noted that visual discrimination is indispensable in the feeding of piscine planktivores, even of those such as the alewife, whose gill-rakers serve as plankton strainers (Fig. 1) (*14*).

In the selection of food by herbivorous zooplankters, on the other hand, visual discrimination plays a negligible role; indeed, the absolute dimensions of both herbivore and food particle may well determine the restricted role of vision (*15*). The lower limit of food-particle size for planktonic herbivores is determined by the mechanism that removes these particles from a water current flowing over a part of the body near the mouth. Studies of feeding in rotifers, cladocerans, and calanoid copepods have demonstrated that all can secure particles in the 1- to 15-micron range. This represents the entire range that can be taken by most herbivorous rotifers, while the upper size limit of particles that can also be taken by large cladocerans and calanoids probably accords roughly with the body size of the zooplankter, and commonly includes particles up to 50 microns (*16*). Among food particles of usable size, both rotifers and calanoids can exercise selection by rejecting individual particles, apparently on the basis of chemical or surface qualities, but the chief control that cladocerans exert is by varying the rate of the feeding movements (*17*).

Size-Efficiency Hypothesis

To differentiate between these two types of feeding, the planktivores and piscivores can be called "food selectors," because they continuously make choices, in large part on the basis of size. The herbivorous zooplankters, on the other hand, can be called "food collectors," because the size range of their food is more or less automatically determined. The ecological implications of size-dependent predation upon the array of planktonic food collectors are outlined in what we shall call the "size-efficiency hypothesis":

1) Planktonic herbivores all compete for the fine particulate matter (1 to 15 μ) of the open waters;

2) Larger zooplankters do so more efficiently and can also take larger particles;

3) Therefore, when predation is of low intensity the small planktonic herbivores will be competitively eliminated by large forms (dominance of large Cladocera and calanoid copepods).

4) But when predation is intense, size-dependent predation will eliminate the large forms, allowing the small zooplankters (rotifers, small Cladocera), that escape predation to become the dominants.

5) When predation is of moderate intensity, it will, by falling more heavily upon the larger species, keep the populations of these more effective herbivores sufficiently low so that slightly smaller competitors are not eliminated.

The data supporting this hypothesis are summarized below.

The view that the small particles present in open waters are the most important food element for all planktonic herbivores is supported by the following: Rotifers and large Cladocera (*Daphnia*) and calanoids (for example, *Eudiaptomus*) are all able to collect particles of the 1- to 15-micron range, as noted above. Particles in this range are heterogeneous (algae, bacteria, organic detritus, organic aggregates) and therefore constitute a relatively constant and demonstrably adequate source of food. Also, they are more digestible than many of the net phytoplankton (such as diatoms) which have a covering that impedes digestion and assimilation (*18*).

The competitive success of the larger planktonic herbivores is probably due to (i) greater effectiveness of food-collection; and (ii) relatively reduced metabolic demands per unit mass, permitting more assimilation to go into egg production.

The greater effectiveness of the larger zooplankters in collecting the nannoseston appears to be largely responsible for the replacement of small by larger species in nature, whenever circumstances permit (*19*). The probable basis for this greater effectiveness is the fact that in related species (with essentially identical food-collecting ap-

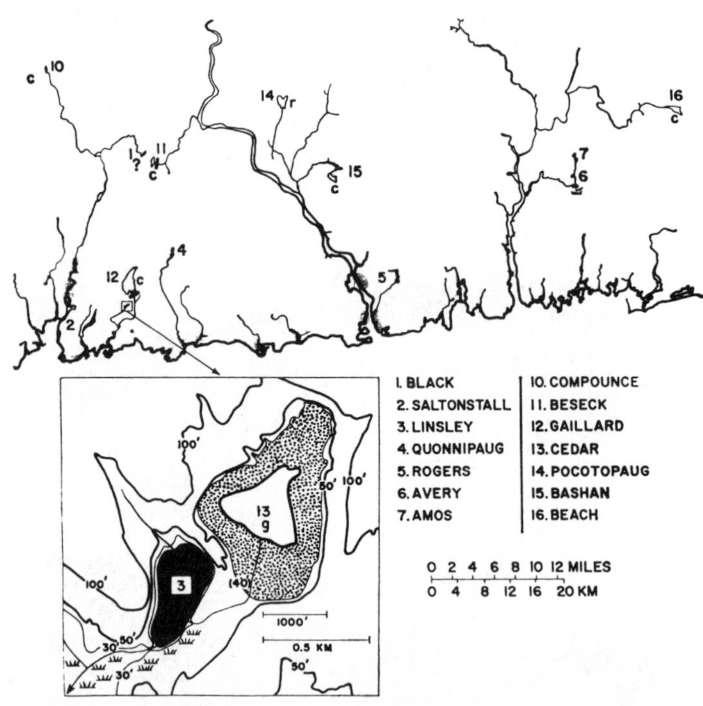

Fig. 2. The coastal strip of eastern Connecticut with the lakes (1–7) known to have natural "landlocked" populations of *A. pseudoharengus*. For the comparable lakes without *Alosa* (10–16), the species of openwater *Daphnia* present are indicated by the following symbols: *c, D. catawba; g, D. galeata; r, D. retrocurva*. Large *Daphnia* are missing in all the "alewife" lakes. The bars at the outlets of lakes 11 and 12 indicate that they have been dammed by man. Major intertidal marshes are cross-hatched. The query next to Black Pond (1) indicates that the plankton has not been studied. Inset shows details of Linsley and Cedar ponds. Stippled area around Cedar Pond is bog forest (see Fig. 3).

Fig. 3. Aerial view of Cedar and Linsley ponds (Branford, Connecticut). Cedar, in the foreground, lacks alewives, although they are common in Linsley, into which the outflow from Cedar drains through the surrounding bog forest (lighter in hue). Linsley in turn drains through a short meandering stream into Long Island Sound (Branford Harbor) which can be seen in the upper left corner (see Fig. 2). [Photograph by Truman Sherk]

paratus) the food-collecting surfaces are proportional to the square of some characteristic linear dimension, such as body length. In Crystal Lake, for example, the body length of *Daphnia catawba* is about four times that of *Bosmina longirostris*, so that the filtering area of the *Daphnia* will be about 16 times larger than that of *Bosmina*. Studies by Sushtchenia have shown that the relative rates at which *Daphnia* and *Bosmina* filter *Chlorella* are, indeed, proportional to the squares of their respective body lengths, suggesting that in Cladocera the area of the filtering surfaces is a major determinant of filtration rate. In addition to this greater ability to collect particles in the 1- to 15-micron range, the larger species can also exploit larger particles not available to smaller species; this appears to be especially significant in the greater competitive success of large calanoid copepods (*20*).

There is some indication that both basal metabolic rate and at least a part of the ordinary locomotor activity may be lower per unit mass in the larger than in the smaller of related species of zooplankters, although the depression of the basal metabolic rate may be slight (*21*). Locomotor activity is difficult to relate to body size, but it is likely that for herbivores a considerable proportion of such work is done in overcoming sinking. The rate of passive sinking of zooplankton up to the size of *Daphnia galeata*, with a carapace length of 1.5 millimeters, is proportional to the square of the body length. However, in *D. pulex*, which are about 1 millimeter longer, the rate is almost proportional to the body length itself (*22*). Therefore, locomotor activity probably increases with no more than the square of the length, and in larger forms shows an even lower rate of increase.

Thus both greater efficiency of food collecting and somewhat greater metabolic economy explain the demonstrably greater reproductive success of the larger of related species. This, together with the fact that generation time is but little greater in large cladocerans than in small ones, undoubtedly underlies the rapidity with which dominance can shift in this group (*23*).

The size-efficiency hypothesis predicts that whenever predation by planktivores is intense, the standing crop of small algae will be high because of relatively inefficient utilization by small panktonic herbivores, and that of large algae will also be high because these cannot be eaten by the small herbivores. Whenever the intensity of predation is diminished and large zooplankters predominate, the standing crops of both large (net phytoplankton) and small algae (nannoseston) should be relatively low, because of the greater efficiency of utilization of nannoseston and because some of the net phytoplankton can also be eaten.

When this prediction is tested, it is important that the biomass of the standing crops of large and small zooplankters be more or less equal, and the only data that meet these requirements are those obtained by Hrbáček *et al.* from Bohemian fish ponds with low and high fish stocks (*19*). In this excellent study, the authors made a qualitative and quantitative comparison between the zooplankton, net phytoplankton, and nannoseston of Poltruba Pond in 1957, when the fish stock was low, and the situation in the same pond in 1955, when the fish stock was high. In 1957, Poltruba Pond was also compared with a pond of roughly similar size (Procházka) with a large fish stock. Poltruba drains through a screened outlet. The biomasses (measured as organic nitrogen) of the zooplankton in the three situations were roughly equivalent, but in Poltruba in 1957, a large *Daphnia* comprised 80 percent of the zooplankton, whereas in the other two situations *Bosmina longirostris* was dominant and rotifers and ciliates were common. Where fish stocks were high and *Bosmina* was dominant, the net phytoplankton (especially diatoms and *Dinobryon*) was much more abundant than when *Daphnia* was dominant. Moreover, in both situations with *Bosmina* dominant the standing crop of nannoseston was two to three times greater than in the presence of *Daphnia*. This was true for both its organic nitrogen and its chlorophyll content. (Photosynthesis of the nannoseston in the presence of *Bosmina* was about five times greater than in the presence of *Daphnia*, a clear indication that the increase in the nannoseston in the presence of *Bosmina* was not merely an increase in the amount of slowly dying algae or detritus.) This result is precisely what the size-efficiency hypothesis predicts.

Size of Coexisting Congeners

In both aquatic and terrestrial habitats, pairs of closely related species of food selectors living in the same community and exploiting the same food source often apportion the available size array of food bits, in rough accord with their own divergent body sizes; that is, the larger species takes the larger bits and the smaller one the smaller bits. This is such a common and well-known phenomenon among congeneric birds (coexisting species of which may differ principally in body size, beak size, and size of food taken) that specific instances thereof and its evolutionary significance do not require discussion here (*24*). We wish only to emphasize that food apportioning according to body size is a path to stable coexistence seldom available to planktonic food collectors. Only in rare circumstances could it be advantageous for a species of large planktonic food collectors to abandon the 1- to 15-micron size range in favor of large par-

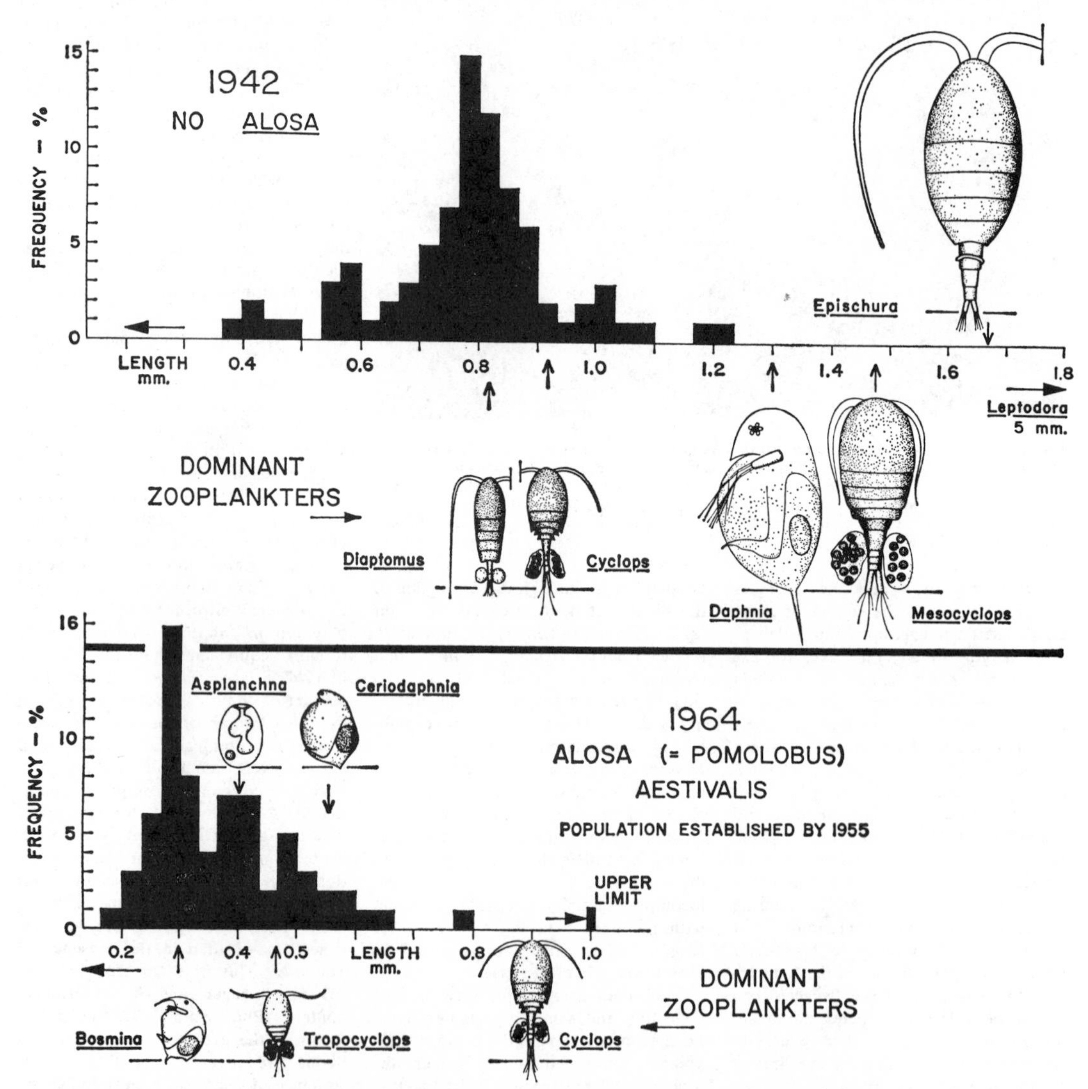

Fig. 4. The composition of the crustacean zooplankton of Crystal Lake (Stafford Springs, Connecticut) before (1942) and after (1964) a population of *Alosa aestivalis* had become well established. Each square of the histogram indicates that 1 percent of the total sample counted was within that size range. The larger zooplankters are not represented in the histograms because of the relative scarcity of mature specimens. The specimens depicted represent the mean size (length from posterior base lines to the anterior end) of the smallest mature instar. The arrows indicate the position of the smallest mature instar of each dominant species in relation to the histograms. The predaceous rotifer, *Asplanchna priodonta*, is the only noncrustacean species included; other rotifers were present but not included in this study.

ticles alone, leaving these small particles to small-sized congeneric competitors (*25*).

On the contrary, many coexisting congeneric zooplankters are of roughly similar size, and presumably—according to the size-efficiency hypothesis—of similar efficiency in food collecting (*26*). This tendency towards similarity in body size can be illustrated by the association in European lakes of *Daphnia galeata* and *D. cucullata*, the latter almost certainly derived from the former. This pair is well suited to our purpose because the various populations of *D. cucullata* exhibit a range of body size unusually large for *Daphnia*. During midsummer, some pond populations of *D. cucullata* mature when the carapace is only about 500 microns long, whereas in lake populations the body size at the onset of maturity may be as much as 900 microns. There is, of course, a complete array of intermediate body sizes in other populations. At the onset of maturity in midsummer, females of *Daphnia galeata* are slightly larger than the largest *D. cucullata*, usually at least 1 millimeter. As all these populations can be passively disseminated, clones of large, intermediate, and small forms of *D. cucullata* have almost certainly been introduced many times into each of the lakes in which *D. galeata* lives. But, as Wagler (*27*) has pointed out after examining 87 European populations of *D. cucullata*, it is only the clones with the largest body size that are found coexisting with *D. galeata*. The dwarf *D. cucullata* would be competitively excluded by *D. galeata* from lakes, just as Hrbáček observed it to be eliminated from ponds by larger species of *Daphnia*, whenever decreased predation allows the larger species to exist. In fact, clones of dwarf forms (carapace length in midsummer less than 550 μ) were found only in ponds where they were associated with *Bosmina longirostris*. It is also clear from Hrbáček's studies of fish ponds that dwarf *D. cucullata*, like *Bosmina longirostris*, can dominate only when predation by planktivores is intense (*28*).

Summary

In the predation by the normally marine clupeoid *Alosa pseudoharengus* ("alewife") upon lake zooplankton, the usual large-sized crustacean dominants (spp. of *Daphnia, Diaptomus*) are eliminated, and replaced by small-sized, basically littoral, species, especially *Bosmina longirostris*. The significance of size in food selection by planktivores as opposed to planktonic herbivores is examined, and it is proposed that all planktonic herbivores utilize small organic particles (1 to 15 μ). The large species, more efficient in collecting these small particles and capable of collecting larger particles as well, will competitively exclude their smaller relatives whenever size-dependent predation is of low intensity. Intense predation will eliminate the large species, and the relatively immune small species will predominate. These antagonistic demands of competition and predation are considered to determine the body size of the dominant herbivorous zooplankters.

References and Notes

1. For classification and distribution of North American *Daphnia*, see J. L. Brooks, *Mem. Conn. Acad. Arts Sci.* **13**, 1 (1957). Identifications and nomenclature of Cladocera, Calanoida, and Cyclopoida in this article accord with accounts by J. L. Brooks, M. S. Wilson, and H. C. Yeatman in *Fresh Water Biology*, W. T. Edmondson, Ed. (Wiley, New York, ed. 2, 1959).
2. C. W. Wilde, *A Fisheries Survey of the Lakes and Ponds of Connecticut* (Connecticut State Board of Fisheries and Game, Hartford, 1959). Although the term "land-locked" is commonly used to refer to self-sustaining freshwater populations of typically marine fish, not all of these populations are barred from returning to the sea, although low dams on most rivers probably isolate them from the anadromous marine stock.
   Opinions appear about equally divided between *Alosa* and *Pomolobus* as the generic name for the alewife. For example, *Pomolobus* is favored by S. F. Hildebrand, in "Fishes of the Western North Atlantic, Part III," *Mem. Sears Found. Marine Res.* **1**, 111 (1963); "*Alosa*" by G. A. Moore in *Vertebrates of the United States* (McGraw-Hill, New York, 1957). Opinion is similarly divided on the use of English names. Hildebrand favors "grayback" for *A.* (= *P.*) *pseudoharengus*, whereas the American Fisheries Society suggests "alewife." J. Hay [*The Run* (Doubleday, New York, ed. 2, 1965)] discusses the ecology of anadromous alewives.
3. Anadromous marine alewives are common, at least during the summer, in many coastal lakes in the Cape Cod area of Massachusetts and along the central portion of the Maine coast. [For references to pertinent surveys see J. L. Brooks and E. S. Deevey, Jr., in *Limnology in North America*, D. G. Frey, Ed. (Univ. of Wisconsin Press, Madison, 1963), pp. 117–62.] Permanent natural populations of *A. pseudoharengus* are known from several freshwaters of northeastern U.S. in addition to the Connecticut lakes. There are resident breeding populations of both this species and *A. aestivalis* (Mitchell) in the lower Mohawk River, New York, and in a few small lakes, different for each species, tributary thereto [*N.Y. State Dept. Conserv. Biol. Surv. Suppl. Ann. Rept.* (1935)]. There is a difference of opinion as to whether the *A. pseudoharengus* population of Lake Ontario is natural or has been artificially established [see R. R. Miller, *Trans. Am. Fisheries Soc.* **86**, 97 (1956)]. Regardless of its origin in Lake Ontario, it has spread into the other Great Lakes through man-made waterways. By the latter part of the 19th century, it had spread from a tributary of Lake Ontario via canals into Lake Cayuga and thence into the two adjoined Finger Lakes, Seneca and Keuka.
4. S. F. Hildebrand, "Fishes of the Western North Atlantic, Part III," *Mem. Sears Found. Marine Res.* **1**, 111 (1963).
5. For food habits in freshwater populations, see T. T. Odell, *Trans. Am. Fisheries Soc.* **64**, 118 (1934); A. L. Pritchard, *Univ. Toronto Biol. Ser. 38* (1929), p. 390. For a summary of food in sea, see S. F. Hildebrand (*4*).
6. *Alosa* (= *Pomolobus*) *aestivalis* (Mitchell) is referred to by several common names, among them "glut herring" which we shall use, and "blueback herring" or "blueback" which are preferred by Hildebrand (*4*). In a previous fisheries survey *Alosa* had not been found in Crystal Lake. See L. M. Thorpe, *Conn. State Geol. Nat. Hist. Surv. Bull. 63* (1942). For the 1955 survey that found *Alosa*, see C. W. Wilde (*2*). The Connecticut State Board of Fisheries and Game suspects that the glut herring was inadvertently introduced into Crystal Lake by the dumping of live bait, with which a few glut herring could have been intermixed, from the Connecticut River, some 15 miles away.
7. Sampled with Forest-Juday plankton trap of 10-liter capacity. In 1942 one sample was taken at each meter depth interval from the surface to just above the bottom. In 1964, two samples (2 meters apart horizontally) were taken at each depth interval.
8. Beach and Bashan lakes were sampled by the method used in sampling Crystal Lake in 1964 (*7*); the others were sampled by vertical tow-nettings, bottom to surface. All specimens taken in each trap series were counted. Subsamples from net collections were counted until a total of 350 to 600 specimens was reached for each lake.
9. J. L. Brooks, unpublished data.
10. E. B. Henson, A. S. Bradshaw, D. C. Chandler, *Mem. Cornell Univ. Agri. Exp. Sta. 378* (1961). Studies of Cayuga plankton: W. C. Muencher, *N.Y. State Dept. Conserv. Biol. Surv. Suppl. Ann. Rept.* (1928), p. 140; A. S. Bradshaw, *Proc. Intern. Assoc. Theor. Appl. Limnol.* **15**, 700 (1964). Bradshaw also gives data on the plankton of Lake Erie. For other references to plankton of the Laurentian Great Lakes, as well as a general summary, see A. M. Beeton and D. C. Chandler in *Limnology in North America*, D. G. Frey, Ed. (Univ. of Wisconsin Press, Madison, 1963), p. 535.
11. This is not to deny that predation by various invertebrates, such as phantom midge larvae (*Chaoborus* spp.) and true plankton predators (*Leptodora*, polyphemids) can be significant in some lakes. However, as predators they are almost certainly not as important as fish.
12. For the significance of size and number in biotic communities, see C. S. Elton, *Animal Ecology* (Sidgwick and Jackson, London, 1927). Analyses of the factors influencing food selection by various freshwater fish are given by V. S. Ivlev, *Experimental Ecology of the Feeding of Fishes*, D. Scott, Transl. (Yale Univ. Press, New Haven, 1961).
13. The ease with which zooplankters are passively dispersed makes it probable that most species present in any continental area will be introduced into a given lake within a reasonably short time (10 to 25 years).
14. M. Kozhoff, *Lake Baikal and Its Life* (Junk, The Hague, 1963); V. S. Ivlev (see *12*).
15. The aberrant cladoceran *Polyphemus pediculus*, although rare, is the smallest widespread member of the open-water community to have an eye sufficiently complex to form a distinct image. This species uses its relatively large, movable, many-lensed eye to locate prey that are 100 to 300 μ in length (or larger). While medium-sized herbivorous zooplankters are the same size as *Polyphemus*, the particulate food they collect is one-tenth the size of the food particles seen and seized by *Polyphemus*.
16. Information on food size in planktonic rotifers is summarized by W. T. Edmondson, *Ecol. Monographs* **35**, 61 (1965). Although *Polyarthra* (150 μ long) can take particles up to 35 μ, this does not vitiate the general statement that 15 μ is the characteristic upper size limit for the food of planktonic herbivorous rotifers.
   Sources of information on the size of food of Cladocera are too numerous to list here. Large *Daphnia*, among the largest of the cladoceran herbivores, are cultured on bac-

teria or algae (*Chlorella, Chlamydomonas* spp.) less than 10 μ in length. Most analyses reveal that most of the gut content is unidentifiable small particles, several microns in diameter. Over half the gut content of two large cladoceran species, *Daphnia galeata* and *D. catawba*, from a Maine lake was unidentifiable fine particulate material [D. W. Tappa, *Ecol. Monographs* **35**, 395 (1956)]. The identifiable algae present in the guts of both species were quantitatively similar and represented the forms, up to about 75 μ, that occurred in the water. The only times during the two summers included in the study by Tappa when this fine particulate material constituted less than half of the gut contents was during a *Dinobryon* bloom in July 1961, when the guts of both species were full of (the tests of) *Dinobryon*. *Dinobryon* sp. (single or in small clusters up to about 25 μ) was the most frequent identifiable alga in the guts of both species for the two years, and the diatom *Stephanodiscus* sp. (about 50 μ) was the second most abundant, being the dominant alga about one quarter as frequently as *Dinobryon*. Aside from the short period when the guts were full of *Dinobryon*, there were never more than 20 identifiable algal cells per gut; the mean for each species for all 26 dates during the two summers is less than 10 algal cells per *Daphnia*.

L. M. Sushtchenia [*Nauchn. Dokl. Vysshel Shkoly Biol. Nauki* **4**, 21 (1959)] notes that food particles about 3 μ are filtered more readily by *Bosmina, Diaphanosoma, Simocephalus* and *Daphnia* spp. than those 15 to 20 μ.

The great differences in the methods of food gathering by various species of cyclopoids makes their inclusion here impossible, but see G. Fryer, *Proc. Zool. Soc. London* **129**, 1 (1957); *J. Animal Ecol.* **26** (1957). The nutrition of *Diaptomus (Eudiaptomus) gracilis* Sars, a common Eurasian calanoid species of medium size (adults about 1.2 mm long), consists of small algae (nannoplankton), and G. Fryer [*Schweiz. Z. Hydrol.* **16**, 64 (1954)] also reported considerable amounts of fine detritus, apparently of vegetable origin, in the guts of this species in Lake Windemere. E. Nauwerck [*Arch. Hydrobiol.* **25**, 393 (1962)] demonstrated that this species in culture could ingest and derive nourishment from particles of animal detritus between 0.1 and 10 μ in diameter. We will consider this well-studied species to be characteristic of most freshwater calanoids, although, of course, larger species are expected to ingest larger particles as well.

17. For observations of selectivity in rotifers on the basis of qualities other than size see works of W. T. Edmondson, especially *Ecol. Monographs* **35**, 61 (1965). See also L. A. Erman, *Zool. Zh.* **41**, 34 (1962).

    Selectivity by *Diaptomus gracilis* in laboratory culture (see *16*) has been noted by A. G. Lowndes [*Proc. Zool. Soc. London* **1935**, 687 (1935)] and for cyclopoid copepods by G. Fryer (see *16*).

    L. M. Sushtchenia, (see *16*), provides data on the rates at which four species of planktonic Cladocera filter suspensions of various algae. The rate of food ingestion in *Daphnia magna* is controlled at various food concentrations by the rate of food collection and at high concentrations by rejection of the excess collected food from the food groove leading to the mouth [J. W. McMahon and F. H. Rigler, *Can. J. Zool.* **41**, 321 (1963); see biblography of that paper for other references to feeding rates].
18. See J. W. G. Lund, *Proc. Intern. Assoc. Theor. Appl. Limnol.* **14**, 147 (1961), for discussion of seasonal occurrence of microalgae in some English lakes. The quantitative importance of bacteria and detritus as food for planktonic herbivores has been mentioned too frequently for listing here. Detritus alone (although probably with some bacteria admixed) has been noted as food, but of poor quality, for *Diaptomus gracilis* (see Nauwerck, *16*) and for *Daphnia* [A. G. Rodina, *Zool. Zh.* **25**, 237 (1946)]. Although organic aggregates have been most thoroughly investigated to date in the sea [see G. A. Riley, *Limnology and Oceanography* **8**, 372 (1963)], their size range in the sea, 1 to 50 μ, is probably also characteristic of fresh waters where they certainly occur (G. A. Riley and P. Wangersky, private communications).

    The relative digestibility of flagellates, small green algae, and some bacteria as compared to *Scenedesmus, Raphidium*, and *Pediastrum* was noted by M. A. Kastal'skaia-Karzinkina [*Zool. Zh.* **21**, 153 (1942)]. M. Lefevre [*Bull. Biol. France Belg.* **76**, 250 (1942)], using egg production as a measure, assesses the nutritive value of 21 species of algae and shows that it is small size and fragile cell walls which make algae the most suitable as food.
19. For changes consequent upon alterations in fish stock in Bohemian ponds, see J. Hrbáček, M. Dvořakova, V. Kořínek, L. Procházkóva, *Proc. Intern. Assoc. Theor. Appl. Limnol.* **14**, 195 (1961); J. Hrbáček, *Cesk. Akad. Ved, Rada. Mat. Prirod. Ved* **72**, 1 (1962). W. Pennington [*J. Ecol.* **5**, 29 (1941)] details rapid replacement of rotifers by *Daphnia* in tub cultures of microalgae.
20. Sushtchenia (see *17*), in her Table 3, provides data on the filtration rates of *Bosmina longirostris, Diaphanosoma brachyurum, Simocephalus vetulus*, and *Daphnia magna*, at several temperatures and concentrations of *Chlorella*. When these results are reduced to the approximate values of 19°C with a standard concentration of *Chlorella* cells, the relative filtration rates of these four species are 1 : 1 : 10 : 18. The relative body lengths (from her Table 2) are 1 : 1.5 : 3 : 4, while the squares of the body lengths are as 1 : 2 : 9 : 16. Thus the filtration rates of these four cladoceran species are approximately proportional to the squares of their respective body lengths.

    Calanoid copepods may be quite unlike the Cladocera in the relationship between body length and rate of food collection, because of a compensatory decrease in the rate at which the head appendages beat. R. Schröder, *Arch. Hydrobiol. Suppl.* **25**, 348 (1961) has shown that in a *Diaptomus* species 2.4 mm long the rate of beating was one-half that in a species 1.2 mm long.
21. A general treatment of the realtionship between size and metabolism is given in E. P. Odum, *Fundamentals of Ecology*, (Saunders, Philadelphia, ed. 2, 1959), pp. 56–59. For zooplankton see G. C. Vinberg, *Zh. Obshch. Biol.* **11**, 367 (1950). Vinberg's "rule" states that decreasing the weight of a plankter by one decimal place increases its O_2 consumption per unit weight by 155 percent (transl. J. Hrbáček, see *19*). For summary of Crustacea in general, see H. P. Wolvekamp and T. H. Waterman in *Physiology of Crustacea*, T. H. Waterman, Ed. (Academic Press, New York, 1960), vol. 1.
22. J. L. Brooks and G. E. Hutchinson, *Proc. Nat. Acad. Sci. U.S.* **36**, 272 (1950).
23. At 20°C the eggs of large limnetic *Daphnia*, such as *D. galeata*, develop into neonates in 2.5 days and become egg layers (in nonturbulent culture) in a week; see D. J. Hall, *Ecology* **45**, 94 (1964). J. Green [*Proc. Zool. Soc. London* **126**, 173 (1956)] gives data relating body size to egg size and clutch size in *Daphnia* and other Cladocera. The longer generation time and the complexity of the life cycle of copepods may obscure the effects of body size upon competition.
24. T. W. Schoener [*Evolution* **19**, 189 (1965)] considers the ecological and evolutionary implications of this phenomenon in birds.
25. G. E. Hutchinson [*Ecology* **32**, 571 (1951)] based his suggestion of food-size selectivity in relation to body size in calanoid copepods on Lowndes's observation (see *17*) of selectivity in the feeding of *Diaptomus gracilis*. Although selectivity is clearly demonstrated by this copepod, it seems likely that attributes other than size of the algae were the basis of this selection. G. Fryer (see *16*) adduced his observation of differential food habits of *Diaptomus gracilis* and *Diaptomus laticeps* in support of such a food-partitioning hypothesis. During February, March, and April, when diatoms are abundant in Lake Windemere, *D. laticeps* (1.6 mm) had their guts full of *Melosira italica*, whereas *D. gracilis* had apparently been feeding primarily on nannoplankton. That it is the larger copepod that takes a somewhat larger food particle is not surprising, but it does not necessarily mean that *D. laticeps* is incapable of feeding on nannoseston under other conditions. Comparison with the food habits of the *Daphnia* in Aziscoos Lake, Maine (Tappa, see *16*), may be useful. One might assume that *Daphnia galeata* is a selective feeder on the basis of the July 1961 samples at which time the guts were crammed with *Dinobryon*. However, at all other times during the open water season of 1961 and all of 1962, over half of the gut was filled with ingested nannoseston. While copepods can be highly selective feeders, their discriminations do not appear to be based primarily upon size. To summarize, at certain times in certain ecosystems a population of a large-sized calanoid may ingest large food particles not utilizable by a coexisting, small-sized, congener. But it does not necessarily follow that the larger species always feeds exclusively on such large particles.
26. Such a stable association of *Daphnia galeata* and *Daphnia catawba* is not uncommon in the lakes of central Connecticut (J. L. Brooks, unpublished). The nature of the association between these two species in a Maine lake has been carefully examined by D. W. Tappa (see *16*). K. Patalas [*Roczniki Nauk Rolniczych Ser. B* **82**, 209 (1963)] presents data on the seasonal changes in the crustacean plankton of several Polish lakes and considers the relation of fish predation to the population size of the various competing zooplankters.
27. E. Wagler, *Intern. Rev. Ges. Hydrobiol. Hydrog.* **11**, 41, 262 (1923).
28. For a possible relationship between the seasonal incidence of predation and seasonal changes of size and form in *Daphnia* (cyclomorphosis), see J. L. Brooks, *Proc. Nat. Acad. Sci. U.S.* **53**, 119 (1965).
29. We thank the following for their assistance: D. W. Tappa for help with field work; Dr. S. Jacobson, New Haven Water Co., for permission to sample Lake Gaillard; J. Atz, American Museum of Natural History, for sharing his extensive knowledge of fish and their literature; G. E. Hutchinson and W. T. Edmondson for making translations of Erman and Sushtchenia available; and C. Goulden, G. E. Hutchinson, and G. A. Riley for critical perusal of the manuscript. S.I.D. was supported in part by the NSF Undergraduate Science Education Program at Yale University. The research of J.L.B. has been supported by grant GB1207 from the National Science Foundation.

PART

6 Experimental Manipulations in Lab and Field Systems

Manipulative Experiments as Tests of Ecological Theory

Jane Lubchenco and Leslie A. Real

An experiment, according to *Webster's Third New International Dictionary* (unabridged), is "an act or operation carried out under conditions determined by the experimenter (as in a laboratory) in order to discover some unknown principle or effect or to test, establish, or illustrate some suggested or known truth." An experiment then, by its very definition, can be distinguished from other methods of acquiring knowledge by its emphasis on purposeful control of conditions and directed aim of elucidating a principle or effect. By their nature, experiments cannot be conducted by either happenstance or serendipity. If a high school student accidentally combines two liquids that instantaneously polymerize, the result may be enormously important, but one cannot claim that any experiment was performed. Or when Fleming discovered the antibiotic effects of penicillin by observing rings of sterility in contaminated petri plates, he was not acting as an experimenter; rather, he was a keen observer whose insights saved thousands of lives. To say that some insight is the result of an experiment requires that the investigator establish appropriate conditions and controls.

An experiment is not the only approach to acquiring knowledge. If it were, then historians and paleontologists would have no claim to producing knowledge. A variety of methods for discovering general principles and truths certainly exists. Ecologists have traditionally used three different methods for discovering ecological principles. The first two we classify as "observational" methods, while the third is truly "experimental."

Observational methods have proven enormously valuable in ecological research and have dominated research programs in this area. Observations fall roughly into two classes: correlational and comparative. Both classes are primarily descriptive.

Correlational analysis attempts to explain how variation in a number of variables within a given system can be attributed to variation in some other set of variables. For example, variation in spatial location is associated with variation in temperature or soil moisture. The analysis is strictly associative, and (as we have all learned from basic statistics) no causal relations can be established through this method. Sometimes we can be easily mislead when we mistake an analysis of variation for an analysis of causation (see Lewontin 1974 for explicit treatment of this problem in population genetics), and any inference about the ecological process remains hypothetical when relying solely on correlational methods. The use of correlation methods have been especially valuable in examining geographic distribution. Associations between presence and absence of species and the presence or absence of resources or competitors (for example, Cody 1968; Pianka 1967; MacArthur 1958 [above, part 5]) comprise the majority of examples used in ecology textbooks. These classic analyses usually focus on a few individual species or on one particular system or geographic area. The alternative observational method—the comparative approach—emphasizes correlational patterns among many different species or multiple systems.

The comparative method has a venerable history and may have had its first modern expression in the catalogs of the French Encyclopedists. Darwin certainly used this method in constructing his theory for the formation of coral reefs (Darwin 1851). In this approach one compares many species in systems that are presumed to be at different stages or to represent different stages of a single process. Mayr (1963) provided an archetypal example of this method when he defined the different stages of geographic speciation. Evidence for the different stages was amassed by finding examples of species or relations among species that displayed particular features of the separate stages toward speciation, for example, species showing hybrid zones after secondary contact or partial reproductive isolation. The comparative method reconstructs processes by comparing sets of species in different stages of the process or subject to slight variations in selective environment. This method differs from the correlational approach in that no particular causal mechanisms are inferred or suggested by single data sets; rather, the hypothetical process is constructed by assembly of inferred stages. This method has been used most effectively to examine evolutionary processes (for example, see Partridge and Harvey 1988 for a recent example in life-history phenomena) but has also been used effectively in ecology, as in Olson's (1958) reconstruction of the process of succession on Michigan dunes or patterns of cyclical succession (Watt 1947, reprinted in part 5 of this volume).

A review by Diamond (1986) comparing different research strategies extols the merits of this kind of comparative approach. Diamond feels such methods lead to the greatest degree of generality, realism, and breadth of spatial and temporal scale. While such studies may possess these features, Diamond unfortunately considers this method to be experimental, producing what he calls "natural experiments." However, as we and others have observed (Hairston 1989), since there is no attempt at controlling or manipulating the conditions under which observations are made, the term "natural experiment" is a misnomer which only obscures the genuine contribution of the comparative method.

Truly experimental studies (either in the laboratory or in the field) must incorporate adequate controls and the intention to illustrate or test some principle. While ecology can boast of some stellar examples of experimental analyses, in general manipulative experiments have not been as common or as influential as in other fields of science. Ecology has primarily been a concept-driven rather than an experiment-driven science. There are at least two reasons why this has been the case.

The first, and by far the most important reason, is that ecological systems are tremendously variable. Individuals of a species may differ according to their genotype, sex, age, size, environment, or past history (Colwell 1984). Interactions among species may differ according to the identity of the individuals and the environment (Lubchenco 1986). The species composition of communities may vary from one location to the next; physical factors come in and out of play, and new structures appear as one moves across scales. Therefore, any particular experimental demonstration may have limited relevance to other systems, and the experiments in ecology have less generality. In other fields, such as physics or molecular biology, critical experiments are of enormous value and have shaped much of the thinking in those fields. For example, the Michaelson-Morely experiments definitively demonstrated the constancy of the speed of light in all directions, refuting the existence of an ether. Or from molecular biology, the Messelsohn-Stahl experiments definitively demonstrated the semiconservative replication of DNA. In each of these two examples, the object under scrutiny (the speed of light, the helical structure of DNA) is the same everywhere and at any time. This is certainly not the case with ecological systems. Part of the appeal of ecological research is the wide diversity of systems, but this diversity constrains our ability to accept experimental demonstrations as universally applicable.

Second, experiments are always best administered under conditions where the object under investigation can be treated in isolation—interaction effects are troublesome in every field of science. When physicists seek to investigate experimentally the properties of a gas, they make the box containing the gas large enough to ignore molecular interaction with the walls of the container. The molecules in the center of the box are essentially in isolation. Similarly, the biochemist isolates DNA from other cell components to explore its unique properties. Much of what ecologists find most interesting, however, resides in these interaction affects—the more complex the set of interactions, the less amenable the system will be to experimental study.

What experimental study does allow us (unlike any other method) is the critical determination of causal relationships that generate ecological pattern (Connell 1975; Hairston 1990). While the other methods allow us to infer or suggest causal mechanisms, only experiments allow for actual demonstration of causal mechanisms. Correlational and comparative approaches can support but they cannot test ecological hypotheses.

Despite the inherent challenge in conducting and interpreting them, ecological experiments are essential to the progress of our discipline and have played a unique role in leading to our current knowledge. In this section we have reprinted seven classic papers that explicitly document causal mechanisms in the generation of ecological pattern. These seven papers represent three categories: selection in natural systems, community structure and species interactions, and the ecology of scale.

Selection in Natural Systems: Ecological Genetics

Ecological genetics attempts to understand the forces determining patterns of genetic change across different geographic locations. Inevitably, as one moves from location to location, one encounters new species interactions, new environmental factors, and new sources of genetic variability. Changes in a population's average phenotype can occur quite dramatically over a small geographic range. For example, the maritime grass *Agrostis stolonifera* is a common perennial species in British dune systems. This grass occurs in two forms: a variety characterized by very short stolons, producing an individual rather clumped in appearance, and another variety with much longer stolons, producing a more sparse individual. The short-stolon form can colonize exposed areas of the dune, while the long-stolon form is restricted to protected sites. The transition between the two forms can be quite rapid, over only a few meters, as one moves from, say, the shelter provided by a large rock into more exposed areas (Aston and Bradshaw 1967). What accounts for this pronounced change in phenotype across space? Is the change a product of phenotypic plasticity or does it reflect genetic differentiation within populations on a very local level?

Experimental attempts to answer these kinds of questions were performed by three agricultural botanists, Jens Clausen, David Keck, and William Heisey from the Carnegie Institute of Washington's experimental station in California. In 1947, Clausen, Keck, and Heisey reported on the genetic basis for geographic variation in *Potentilla glandulosa.* This small perennial was experimentally manipulated at three elevational sites—coastal plain (Stanford), mid-montane (Mather), and high-montane (Timberline) at 11,000 feet. Clausen, Keck, and Heisey sowed individuals from seed from each of the three elevational sites and also raised F2 hybrids across the three elevational sites. Foreshadowing their later work, they concluded "at each station there is a great variation in vigor, but the plants that are vigorous in one environment may be weak in another, and vice versa. Physiological characteristics, such as earliness of flowering, frost resistance, and ability to survive in specific environments are genetically linked with morphological ones such as shape, size, and color of petals. Such linkages tend to keep the climatic races distinct even where they meet and intercross."

The basic conclusions from the *Potentilla* study were confirmed in their famous analysis of geographic genetic differentiation in the Yarrow *Achillea lanulosa* (Clausen, Keck, and Heisey 1948). This species' distribution is continuous along both the western and eastern slopes of the Sierra Nevada. At higher altitudes plants are smaller, and along the eastern slope they are subjected to more arid environments. The stunted growth form at high elevation could result from poor growth and thus may reflect plasticity in the phenotype. However, transplant experiments revealed that variation in plant stature was a product of genotypic differentiation across populations. The physiognomy of the parental lines was maintained when seeds were germinated and seedlings established under uniform environmental conditions. Also, seeds harvested from plants established in any one of three elevational sites (the same ones as in the *Potentilla* study) produced adult plants of the same phenotype as the originating population.

The 1948 study on *Achillea* is one of the most often cited but least read articles in ecological genetics. The most important results from the study are embedded in a very extensive analysis of plant species. Six volumes span thirty years of investigation, and the *Achillea* study is only one aspect of the study. Nonetheless, the importance of their work has long been recognized. Most researchers credit Clausen, Keck, and Heisey with the first documented experimental analysis of genetic differentiation in adjacent populations and with introducing transplant experiments as an appropriate technique for distinguishing genetic from environmental effects.

Transplant experiments suggesting genetic differentiation of local populations were used, however, as early as the 1920s by Turesson, who coined the term "ecotype" to describe genetic varieties within a single species (Turesson 1922, 1925, 1930). Concurrently, Clements and Goldsmith (1924) introduced the concept of a phytometer as a bioassay to assess the effects of local edaphic conditions on plant growth and development.

Perhaps the best known case of ecotype formation is geographic variation in the melanic forms of the pepper moth, *Biston betularia.* As Lewontin (1978) suggests, if anyone is in doubt about the existence of natural selection, they need only look at the case of industrial melanism to be convinced. Dobzhansky (1972) considered industrial melanism to be the "most striking example (of natural selection) outside the realm of microorganisms." Perhaps it is the most striking because both the mechanism of selection and the genetic basis of the varieties have been so well quantified experimentally.

The original form of this moth was a light grey morph peppered with black dots. The first record of a melanic, black variant dates to 1848, collected in Manchester, England. By 1898, 98% of the *Biston betularia* population in Manchester was black, and the melanic form quickly spread throughout the rest of industrialized England. By the mid-twentieth century, melanic forms were recorded in over 100 other species of moths in Britain, Europe, and North America.

The biological mechanisms accounting for this rapid evolutionary change were elucidated chiefly through the work of H. B. D. Kettlewell (1955 [reprinted here], 1956), C. A. Clark and P. M. Sheppard (1966), and E. B. Ford (1964, 1965). Kettlewell (1955, 1956) revealed the basic mechanism for natural selection through a series of elegant experiments.

In wooded areas subject to industrial pollution, the naturally occurring lichens that cover tree bark are killed and tree trunks are very darkly colored. In nonindustrial areas, lichen-covered tree trunks appear light grey in color. The melanic form of the pepper moth is cryptic on the dark trees but conspicuous on the light, lichen-covered trees, whereas the nonmelanic form is conspicuous on the dark trees but cryptic on the light. Natural selection favors the morph most protected from visual predators—mostly birds—which hunt by sight and can be observed picking moths from the tree's trunk.

To quantify the intensity of selection on the two morphs, Kettlewell (1955, 1956) released equal numbers of the melanic and nonmelanic variants in two different habitats, a polluted wood and a natural wood. In the polluted wood, where dark tree trunks were most frequent, 27.5% of the melanics but only 13.1% of the light forms were recovered. In the natural wood, where lichen-covered trees were most frequent, 12.5% of the light form but only 6.3% of the melanic variants were recaptured. The likelihood of recapture is related to the intensity of visual predation. In the polluted wood, Kettlewell observed 58 incidents of birds seizing released moths. Of these, 43 moths were light morphs, while fifteen were melanic despite their initial equal frequency in the habitat. In the natural, unpolluted wood, 86% of moths seized were melanic.

Breeding experiments using the two morphs have shown that the melanic form is most often a simple dominant, single gene difference from the light form. From the time of its first establishment, however, the melanic dominant has probably developed some heterotic advantage through gene modifiers acting on the major gene for melanism (Ford 1964, 1965).

Kettlewell's analysis certainly pertains to issues of genetic differentiation among local populations. However, what distinguishes Kettlewell's approach from Clausen, Keck, and Heisey's is his attention to quantifying the selective mechanism. In the plant study, we are convinced that populations are genetically distinct, but the selective mechanism maintaining divergence is only suggested. Kettlewell's focus is almost entirely on documenting selection. Differentiation is simply a by-product of selective mechanisms operating at the individual level.

Ecological genetics has experienced a rapid evolution since the pioneering efforts of Clausen, Keck, Heisey, and the English school represented by Kettlewell, Clark, Sheppard, and Ford. Early efforts were mostly directed at documenting selection on visible variations in natural populations. The pepper moth and land snail, *Cepaea nemoralis* (Cain and Sheppard 1954), represent the best examples of work of this type. Since this early phase, ecological genetics has developed two additional major foci (Via 1990).

By the mid 1960s, and especially beginning with the landmark paper of Lewontin and Hubby (1966), population geneticists recognized that natural populations harbored vast amounts of hidden variation if isozyme variation reflected the genome in general. What, if any, selection mechanism could account for the maintenance of polymorphisms at this level of hidden variation? This second focus, then, emphasizes the measurement of selection on hidden variation.

Selection on enzyme variants has now been well documented in natural systems. Koehn and Hilbish (1987) summarize research or natural selection on enzyme variants at the LAP locus in the common mussel *Mytilus edulis* along salinity gradients in Long Island Sound. The LAP94 allele, most common in areas of high salinity, steadily decreases in frequency as the mussels occupy more brackish and freshwater environments. The selective advantage of the LAP94 enzyme variant in highly saline waters (and its disadvantage in freshwater) can be related to the enzyme's function in regulating cellular osmotic pressure.

Demonstrating selection on hidden variation is important and interesting. However, many of the most important features of the organism's phenotype relating to evolutionary fitness are quantitative characters relating to life-history phenomena. A third focus in contemporary ecological genetics explores mechanisms of selection on quantitative characters in natural populations. By their nature, quantitative characters involve complicated sets of interactions among loci and with the interaction between genotypes and the environment. The general method relies on the relationship between the strength of selection response and the pattern of genetic variance and covariance among selected traits (Falconer 1981). While the earlier phases of ecological genetics may have been a little restrictive (there are, after all, only certain kinds of visible or hidden variations that are strictly qualitative in effect), in principle, any set of traits can be analyzed through quantitative techniques and appropriate experimental designs. The methods of quantitative genetics are often tedious and the statistical methods often daunting, yet the approach has considerable promise for opening the range of evolutionary phenomena that can be treated experimentally. Advances depending on quantitative genetics have already been recognized in exploring patterns of sexual selection (Arnold 1983), evolution of threshold characters (Lande 1978), plant-herbivore interactions (Via 1990; Gould 1983) and life-history phenomena (Lande 1982).

Such extended, complicated analyses would seem at best premature and at worst totally unjustified without the pioneering experimental demonstrations by Clausen, Keck, Heisey, Kettlewell, Ford, and Sheppard, which showed that natural selection can act quickly and decisively in shaping local populations. While the kinds of analysis eloquently championed by Kettlewell may seem outmoded today, they are still the studies which convince us that evolution by natural selection exists.

Community Structure and Species Interactions

"Community structure" refers to the patterns of distribution, abundance, size, diversity, and stability that characterize populations living together in a particular place. Community ecologists seek to understand the often complex interplay of biotic and abiotic factors that result in these patterns.

Experimental investigations of interactions among species have been particularly important in elucidating the complex nature of these interactions. Laboratory and field experiments on competition, predation, parasitism, and mutualism have provided critical tests of the processes involved and have also yielded a rich bonanza of unanticipated key insights altering the way ecologists understand the causes and consequences of species interactions.

Laboratory Studies of Competition

Competition has been of central interest to ecologists and has also been particularly amenable to experimental approaches both in the laboratory and in the field. The first published laboratory competition experiments were extensions of investigations of the population growth dynamics of easily cultured species. The Russian microbiologist G. F. Gause (1932, 1934; see also the discussion by Kingsland in the introduction to part 1 of this volume) extended his examinations of population trajectories of single species of yeast and protozoa to the dynamics of two-species cultures. His pioneering experiments were the first of many laboratory studies of competition. Gause's experiments and other subsequent laboratory investigations provided strong evidence that similar species could not coexist in simple laboratory settings. Subsequent investigations sought to define the conditions favoring one or the other species, to elucidate the mechanism by which one species won, and to examine the relationship between life history parameters and competitive ability.

Park (1948, reprinted here), reports results from laboratory experiments designed to examine interspecific competition between the grain beetles *Tribolium confusum* and *Tribolium castaneum*. Single species treatments were used to establish a baseline against which effects of interspecific competition could be measured. The results are particularly insightful because the experiments were designed with intimate knowledge of the life histories and natural history of the species, conducted over many generations, and replicated extensively. Thomas Park's experiments demonstrate convincingly that under the uniform, simple laboratory conditions specified: (1) competition between these two species invariably resulted in the demise of one species; and (2) the identity of the winner was predictable but not absolute. In this and later papers, Park examines some of the multiple factors determining the identity of the winner. Park's exploration of these factors foreshadowed many subsequent studies of competition.

Park's 1948 paper (also Park et al., 1965) is a powerful testimony to the importance of the interplay between competition and other biotic interactions. In his 1948 experiments, the presence or absence of a parasite, *Adelina tribolii*, determined the outcome of interspecific competition. Both *Tribolium* species could persist in single-species cultures either with or without the sporozoan parasite. However, in competition experiments, the presence or absence of *Adelina* could reverse the outcome of competition. If *Adelina* was absent, *T. confusum* was usually eliminated. If, on the other hand, *Adelina* was present, the opposite result usually occurred.

Competition in flour beetles was also affected by another, very different biotic interaction. Park suggested that reciprocal predation by adult and larval *Tribolium* on one another's eggs and pupae was the competition mechanism. This complex blend of competition and predation has been described for numerous species and is now termed "intraguild predation" (Polis and McCormick 1987). Recent empirical and theoretical studies have continued to explore the relationship between competition and other kinds of biotic interactions. Investigations of "predator-mediated coexistence" (Paine 1966, reprinted in this section) and intraguild predation (Polis et al., 1989) have significantly enhanced our understanding of processes affecting the distribution, abundance, and diversity of species.

Park's 1948 paper also anticipated many current investigations of how changes in the abiotic environment may affect competitive outcomes. His 1954 paper builds on the 1948 one, examining how some combinations of temperature and humidity provide environments favoring one species while other combinations favor a different species. Birch (1953) also reports that slight changes in temperature result in different competitive outcomes. In his experiments, one species of grain beetle, *Calandra oryzae*, has a higher innate capacity for increase (r_m) than does its competitor, *Rhizopertha dominica*, at one temperature (29.1°C), while the reverse is true at a slightly higher temperature (32.3°C).

Field Experiments of Competition

More than a decade before Gause's laboratory competition experiments were published, the British ecologist A. G. Tansley (1917) published field experiments examining competition between bedstraws (*Galium* spp). These experiments attempted to evaluate whether competition or the ability of each species to grow in different types of soils determined the observed distributions of these small herbaceous plants. Although his experimental design and execution were flawed, Tansley was ahead of his time in his attempts to use field experiments to distinguish among various hypothesized causes of the distribution of these species.

Studies of competition were primarily the domain of theoreticians and laboratory experimentalists until the 1960s. Although field studies involving correlations (for example, Hairston 1949; MacArthur 1958, reprinted in Part V; Kohn 1959) suggested that competition could play an important role in determining the distribution and abundance of species, experimental verification of these ideas was lacking. Connell's (1961*a*, reprinted in this section) classic investigation of competition between barnacle species on the seashore demonstrated the potential for conducting rigorous experiments in the field and testing unequivocally some of the ecological causes of species distribution patterns.

Having worked with difficult-to-trap wild rabbits for his master's thesis at the University of California at Berkeley, Connell was impressed with the immense potential of the barnacles he encountered on Scottish shores. Barnacles did not run away or hide, they were available in almost unlimited numbers, and they could be experimentally removed or transplanted.

Connell observed that although the larvae of *Chthamalus stellatus* settle in both the high and low portions of the shore, they did not survive lower down where *Balanus balanoides* lived. The timing of their demise and the complementary distribution of the two species' adults suggested to him that *Balanus* might be outcompeting *Chthamalus* at shore levels where *Balanus* thrived. Connell posed a very concise question: What is the cause of death of those *Chthamalus* settling lower on the shore? Possible answers included: (1) their inability to withstand longer immersion times; (2) competition from *Balanus;* and (3) predation from snails. Each of these possibilities was suggested because of correlations between each factor and the distribution of *Chthamalus.* Not content with correlations, Connell designed experiments to ascertain causality. He removed *Balanus* from some treatments but not others, transplanted stones with barnacles to different shore levels, and excluded predators, all along a bathymetric gradient. He then mapped and followed the fate of individual barnacles. These experiments demonstrated unequivocally that interspecific competition with *Balanus* prevents most *Chthamalus* from living lower on the shore. These elegant experiments provide insight into a variety of important aspects of competitive interactions.

1. Because the experiments were conducted along the shore-height gradient, Connell could examine the interaction between environment and competitive outcome. Unlike the uniform laboratory environment of Park's or Gause's experiments, in which one species invariably won, the intertidal gradient of Connell's experiments provides diverse environments and the opportunity for coexistence. The inferior competitor, *Chthamalus,* persists higher on the shore in a refuge zone where *Balanus* does not survive. Where both species settle, *Balanus* usually outcompetes *Chthamalus.* These experiments and other papers by Connell (1961*b,* 1970, 1972, 1974) laid the groundwork for many subsequent field experiments examining biotic interactions along physical gradients. These include studies not only of competition but also of predation and herbivory, and have been conducted along physical gradients such as desiccation, moisture, wave action, nutrients, depth or altitude (Connell 1961*a,* 1961*b;* 1975; Louda 1982, 1983; Lubchenco 1980, 1986; Menge 1976; Lubchenco and Menge 1978; Dayton 1971, 1975; Witman 1987; Tilman 1987, 1988). They provide strong evidence that biotic interactions change along physical gradients. These findings may be particularly useful in attempts to predict changes in species distributions and abundance in changing environments.

2. Connell's experiments provided graphic demonstration of the potential for direct interference as an important mechanism of competition. Individual *Balanus* were observed to smother, crush, or undercut individual *Chthamalus.* Still other mechanisms were observed in the plethora of field competition experiments that occurred in the 1970s and 1980s, prompting Schoener (1983) to propose the recognition of six mechanisms of competition in lieu of the traditional two. He suggests replacing the terms interference and exploitation with these six terms: consumptive, preemptive, overgrowth, chemical, territorial and encounter competition.

3. Connell's (1961*a*) *Chthamalus* paper was the first in a series of experimental investigations of the relation between competition and other factors affecting patterns of distribution and abundance. His study of *Balanus balanoides* on Scottish shores (1961*b*) posed many of the same questions as the *Chthamalus* investigation did, but it resulted in different answers. *Balanus's* distribution and abundance are determined primarily by predation by the snail *Thais lapillus.* A comparison of factors affecting the Scottish barnacles with factors influencing a different *Balanus, B. glandula,* along the west coast of the United States plus experimental field studies with predators (for example, Paine 1966, reprinted here), laid the groundwork for Connell's 1975 synthetic overview. This paper

asserted that ecologists needed to be more rigorous and critical in their interpretation of "evidence" supporting competition as *the* central mechanism in determining community structure. Connell suggested that competition was sometimes an important determinant of patterns of community structure, but that it was not as pervasive as had been commonly assumed. Connell (1975) called for a reexamination of "where competition occurs in nature and where it is prevented by predation or physical conditions." This call was partly responsible for an acceleration in experimental investigations of competition during the late 1970s and the 1980s.

This profusion of field experiments yielded several results. One was a more balanced view—based on more solid evidence—of factors determining patterns of community structure (Connell 1983; Hairston 1990; Menge and Sutherland 1976, 1987; Menge et al. 1986*a*, 1986*b*; Schoener 1983, 1985). Another was an increased appreciation of the complexities and subtleties of community interactions, and thus the difficulties in generalizing about community patterns (Colwell 1984; Lubchenco 1986). A third was a greater awareness of the importance of good experimental design and analysis (Hurlbert 1984; Hairston 1990).

Field Experiments on Predation

"It has long been recognized that predators can keep potential competitors so rare that they do not compete" (Connell 1975, p. 462). Although Darwin (1859) proposed this with his mown lawn experiments, and Gause (1934), Hutchinson (1961), and Slobodkin (1961) all suggested it could be important, the full implications of this concept were not realized until R. T. Paine's (1966) experiments on rocky intertidal coasts were reported and broadly appreciated.

Paine's experiments focused on the effects of predators on local patterns of species diversity to explain broader patterns of biogeographic diversity. Paine suggested that "local species diversity is directly related to the efficiency with which predators prevent the monopolization of the major environmental requisites by one species" (1966, p. 67). To test this hypothesis, Paine removed the top carnivore, the seastar *Pisaster ochraceus*, from an 8m × 2m stretch of shore on the outer coast of Washington and measured subsequent changes (relative to those on a control segment of the shore) in the number and kinds of macroscopic invertebrates and algae. His experiments demonstrated that removal of *Pisaster* resulted in a substantial decrease in local species diversity. The number of species dropped from fifteen to eight. The mechanism normally maintaining a higher level of species diversity was selective predation by *Pisaster* on a dominant competitor, the mussel, *Mytilus californianus*. In the absence of *Pisaster*, the mussel monopolized space on the shore.

This experiment demonstrated what Paine (1969) later termed the "keystone species" concept. "Keystone species are those which exert influences over other members of their ecological communities out of proportion to their abundances" (National Research Council 1986). The removal of a keystone species causes the community to change drastically.

Numerous species have been termed keystone predators, including sea otters (Estes et al. 1982), stream fish (Power et al. 1985), lake fish (Hall, Cooper, and Werner 1970; Carpenter et al. 1985), sea urchins (Paine and Vadas 1969), and snails (Lubchenco 1978). Not all communities have keystone predators, however. In the rocky intertidal community on the Pacific coast of Panama, for example, although a diverse suite of consumers has a dramatic effect on the distribution, abundance, and diversity of the community, no single species plays a keystone role (Menge et al. 1986*a*,*b*). A comparison of the types of communities with and without keystone species should provide new insights into factors affecting community structure and function.

Orians (National Research Council 1986, p. 6) suggests that the keystone species concept should be extended to include not only the keystone predators discussed above, but also keystone mutualists who "are essential for survival of other members of their communities, even though the amount of resources they consume is sometimes small. Well-known examples are pollinators and frugivores, ants that patrol and protect plants, and microorganisms associated with vascular plants (Alexander 1971; Gilbert 1975; Howe 1981; Janzen 1966)."

The keystone species concept and Paine's *Pisaster*-removal experiments have had a profound effect on the way ecologists think about communities. Paine's (1966; 1974; 1980) experiments laid the cornerstone for an experimental approach to community ecology. They graphically illustrated the idea introduced by Hairston, Smith, and Slobodkin (1960, reprinted above in part 3), that factors operating at one trophic level could determine patterns and processes at other levels. Dayton's (1971, 1975) elegant and innovative experiments in the same system provided new insight into the relative roles of predation and physical disturbance in determining community patterns and the roles played by different consumers. Paine's (1966) emphasis on incorporating food web dynamics into explanations of community and biogeographic patterns has been echoed in and reinforced by numerous studies (Paine 1980; Fretwell 1977; Oksanen et al. 1981; Carpenter and Kitchell 1988; Carpenter et al. 1985, 1987; Menge and Sutherland 1976, 1987).

Spatial Scale: Its Implications for Predation, Biogeography, and Ecosystems

Because ecologists are concerned with patterns and processes which range in size from molecules to the biosphere and in time from fractions of a second to centuries, the scale of events is important. Therefore, experiments need to be conducted over as wide a range of spatial and temporal scales as is feasible. This next section highlights the contributions of experiments to understanding some mechanisms which operate at a variety of spatial scales.

Spatial Heterogeneity and Predation

The experimental analysis of predation, like competition, began with G. F. Gause (1934). By the mid-1920s, a rigorous mathematical treatment of the stability of predator-prey populations had been established but had not been tested (Lotka 1925; Volterra 1926, reprinted above in part 2). The mathematical theory suggested that predators and their prey should show periodic oscillations over time and that the magnitude of the cycles would be determined by the initial conditions, that is, the starting densities of predator and prey. Gause undertook a laboratory study using a protozoan predator, *Didinium nasutum,* and a protozoan prey, *Paramecium caudatum,* cultured together in an oat medium. Unlike the predictions of the theory, when the two protozoans were cultured in the simple medium, the *Didinium* inevitably would exhaust and extinguish its *Paramecium* prey and eventually would become extinct in the culture itself. When Gause introduced a prey refuge into the medium (a sediment where *Didinium* was excluded), the *Paramecium* population flourished and the predatory *Didinium* went extinct. The *Didinium-Paramecium* system did not show the predicted oscillations through a variety of experimental manipulations including enlarging the vessel, restricting the initial density of *Didinium* to only a few individuals, and so forth.

In one experimental arrangement, however, Gause was able to promote oscillations at least through two cycles. Every third day of the experiment, he added one *Paramecium* and one *Didinium* to the culture flask. Such additions, in Gause's thinking, represented periodic immigrations into the system from outside sources. Over a sixteen-day period, the *Didinium-Paramecium* interactions mimicked the predictions of the Lotka-Volterra equations; the two populations cycled, with peaks in prey abundance shifting slightly to the left of the peaks in abundance of predators.

In 1958, C. B. Huffaker (reprinted here) extended Gause's original experimental approach and explored the simultaneous effects of migration and spatial heterogeneity in promoting predator-prey oscillations. Huffaker's experiments were the first empirical demonstrations of the importance of spatial heterogeneity and complexity in regulating and stabilizing predator-prey populations. His ingenious experiments focused on the predaceous mite, *Typhylodromus occidentalis,* and its prey, another mite, *Eotetranychus sexmaculatus,* which infests oranges.

At first, Huffaker was unable to stabilize the predatory-prey system. On single oranges, the predatory mite would quickly decimate the prey population and would then go extinct itself. Through an elaboration of his simple system, he manipulated the spatial environment in which the mites interacted. Huffaker altered the distance between the prey's food source (oranges), erected barriers to predator movement, and provided dispersal corridors for the prey. These manipulations effectively increased the patchiness of the environment and provided temporary refuge for the prey. Prey would locally go extinct, but would colonize new patches before the arrival of predators. This temporal escape in the patchy environment promoted sustained oscillations in both the predator and the prey for over 60 weeks. Huffaker provided the first convincing evidence that the spatial heterogeneity and patchiness of the environment could have direct effects on the relative rates of predator and prey dispersal and that differential capacity for dispersal could determine the stability of predator and prey populations.

The stability of predator-prey interactions in patchy environments is often enhanced when predators produce an "aggregation effect," that is, when predators are disproportionately drawn to areas or patches with high-prey densities, patches of low-prey densities experience a release from predation pressure. Aggregation effects essentially produce temporal-spatial refugia that exist while prey remain locally rare. Murdoch and Oaten (1975) discuss one example of such an aggregation effect in the behavior of starfish feeding on mussels off the coast of southern California. The starfish feed disportionately on high density patches of mussels, which eventually lead to the mussels' extinction. Small patches of mussels are not exploited by the starfish, however. The aggregative response of the starfish allows small clumps of mussels to grow and establish themselves. These patches, immune from predation for varying lengths of time, will stabilize the predator-prey interaction. Beddington et al. (1978) and Heads and Lawton (1983) explore the effects of aggregation response in insect predators and herbivores, and they reach the similar conclusion that heterogeneity and an aggregative response may be especially important in establishing stable low-density patterns of predator-prey populations.

While of obvious and documented importance, aggregation effects may not be essential for creating stable predator-prey interactions. Murdoch et al. (1984) found no evidence of an aggregation effect in the parasitoid control of the olive scale, *Parlatoria oleae.* Stabilization of this predatory-prey system seems to be associated with the stochastic temporal creation of new host patches through colonization, rather than through nonrandom search and patch exploitation. Nisbet and Gurney (1976) and Gurney and Nisbet (1978) had, in fact, already shown that in a purely deterministic predator-prey system, oscillations of the sort observed in the original Huffaker experiments could be produced simply as a result of the stochastic noise generated by the random colonization and extinction of patches.

Biogeography

From its origin, the study of biogeographic pattern was intrinsically a study in spatial pattern and scale. The very first studies in biogeography cataloged the locations of species groups and attempted to reach evolutionary conclusions based solely on patterns of cooccurence and association. Alfred Russell Wallace is unquestionably the father of evolutionary biogeography, and his three massive texts (1869, 1876, and 1880) still are exciting to read for their breadth of biological examples and insight. Brown and Gibson (1983) have since summarized the biogeographic principles advocated by Wallace. It is interesting and instructive to witness the large number of important principles shared by the nineteenth-century naturalist and more "modern" analysts. Wallace's emphasis on the importance of islands as models of biogeographic process and his emphasis on dispersal are particularly noteworthy.

The century following Wallace's initial contribution was dominated by the analysis of species spatial patterns and distribution. Little effort was directed at documenting the ecological processes that may have driven the observed patterns of abundance. Existing spatial associations among species and the composition of local flora and fauna were viewed as the result of evolutionary processes that had, by and large, already operated over extensive geological time scales. The constellation of species in any locale essentially represented the equilibrium configuration brought about by long evolutionary history. Current ecological processes were irrelevant except to the extent that they represented past forces which had contributed to forming biogeographic patterns (for example, Carlquist 1965; Udvardy 1969).

This view of biogeography is in sharp contrast to the revolutionary theoretical developments of MacArthur and Wilson (1963, 1967) and the emergence of the equilibrium theory of island biogeography. For some time, ecologists had recognized that large islands were high in species numbers while small islands characteristically had few species (Preston 1948, 1962). Distant and isolated islands were also characterized by few species and high degrees of endemism (Carlquist 1965, 1966). Traditionally, these patterns had been explained by speciation processes and chance historical events (Hamilton and Rubinoff 1963, 1964, 1967). MacArthur and Wilson, in a radical approach to explain these same patterns, suggested that insular patterns of species diversity were largely the result of ecological process operating on contemporary time scales. The number of species on an island was determined through the balancing of two processes: the immigration of new species to the island from outside sources; and the extinction of species on the island. MacArthur and Wilson showed that these dual processes, immigration and extinction, could account for most of the observed patterns of species diversity on islands. Since large islands will have higher immigration rates and lower extinction rates, they will have higher equilibrium numbers of species. Distant islands will have lower immigration rates and should have correspondingly lower species numbers. MacArthur and Wilson posed no explanation for patterns of endemism, which clearly must operate on a different time scale than the one they suggest which accounts for most of the variation in species richness across different locales.

The MacArthur-Wilson theory has two striking conclusions. First, an equilibrium number of species exists which characterizes any island of a given size and distance from its source of colonists. Second, the exact composition of species present on an island should change over time and depend on the historical processes of immigration and extinction. Since this equilibrium is established on an ecological time scale, the theory should be testable through defaunation experiments.

Daniel Simberloff, a graduate student of Wilson's, set out to test the equilibrium hypothesis in one of the most ambitious and successful large-scale experiments attempted in ecological research (Simberloff and Wilson 1969 [reprinted in this section], 1970; Wilson and Simberloff 1969; Simberloff 1969). In large tents on the Florida Keys Simberloff and Wilson enclosed islands of red mangrove (*Rhizophera mangle*). Methyl bromide gas was pumped into the enclosures to kill all of the island's animal inhabitants, a procedure that left vegetation unharmed. Recolonization of the defaunated islands was carefully monitored for several years following the initial perturbation. Recolonization by terrestrial arthropods was remarkably rapid. Within a year, all but the most distant islands had returned to their original species numbers. The recolonization curves seemed to approach a stable equilibrium with the exact number determined by size of island and distance from source islands in the Keys. The most convincing evidence for the equilibrium theory was the apparent rapid turnover in species composition once an equilibrium had been reached. Individual species colonized and disappeared, sometimes repeatedly, while the overall number of species on the island remained constant.

The MacArthur-Wilson equilibrium theory had been subjected to substantial criticism despite the strong support offered by the Simberloff-Wilson experiments. The equilibrium theory assumes that the dominant ecological processes determining species composition on islands are stochastic and equivalent across species. All species are essentially subject to the same random immigration and extinction rates. The theory does not account for any observed regularities in community organization, and MacArthur and Wilson were well aware of the important role of competition, predation, and evolution in structuring island communities. Islands are not simply random assemblages of species. The major focus of contemporary research in biogeography is the search for those processes, in addition to immigration and extinction, that account for overall community organization (Diamond 1975; Brown 1971, 1978; Brown and Gibson 1983). The major contribution of the MacArthur-Wilson-Simberloff approach was to direct attention away from historical accounts of species diversity that relied exclusively on correlational and comparative methods and toward ecological accounts that were open to manipulations and rigorous tests. With the advent of the equilibrium theory, biogeography became both an experimental and an ecological science.

Large-scale Manipulations of Ecosystems.

The analysis of ecosystem structure and function is by its very nature a problem of large scale. Ecosystems stretch across large and heterogenous geographic areas and consist of hundreds of interacting species varying in degrees of representation and importance. Prior to the emergence of an experimental approach to ecosystem science the structure and function of a large biotic system was restricted to monitoring the flows of energy and nutrients among ecosystem compartments, that is, primary producers, herbivores, and so on. The presumption was that changes in the ecosystem variable under study, for example the carbon or nitrogen level, between input and output from a compartment was the result of the biotic and abiotic interactions occurring within the compartment. The complete description of an ecosystem, then, would involve partitioning ecosystem structures into specific compartments and characterizing flows between compartments. The classic studies of H. T. Odum (1957) and J. M. Teal (1962, reprinted above in part 5) are of this nature.

Such descriptions of ecosystems are subject to two criticisms. First, this form of static representation is not useful in determining how ecosystems will respond to changes in basic compartmental structure. In order to ascertain the ecosystem response to perturbation, we must know the dynamics of processes occurring within compartments. Second, the static compartment model cannot be tested for completeness; that is, do we have all the important compartments included? The response to both of these criticisms demands the experimental manipulation of compartments on a large scale.

The large-scale manipulation of a forest ecosystem was undertaken by Gene Likens and Frank Bormann as part of the International Biological Program's terrestrial ecosystem site at Hubbard Brook Experimental Forest in the White Mountains of New Hampshire (Bormann and Likens 1979; Likens et al. 1967, 1977). The Likens-Bormann experiments represent one of the first planned manipulations of an ecosystem to ascertain the importance of vegetation in regulating biogeochemical cycles. From its inception in the 1960s, the Hubbard Brook project has continued to generate tremendous insights into ecosystem processes. In this volume, we have presented the first major report of the results of planned deforestation on nutrient dynamics (Likens et al. 1970).

The Hubbard Brook Forest is a second-growth hardwood forest maintained by the U.S. Forest Service. The forest is supported by hardrock substrate and is relatively impermeable to water. The area is subdivided into small watersheds each of which can act as an experimental unit. Likens and Bormann arranged for one 15.6 hectare watershed to be cut. No timber or vegetation was removed from the deforested area. Regrowth of vegetation was inhibited through the application of an aerial herbicide. After a two-year period, deforestation increased the average stream water concentrate of all major ions: nitrate concentration increased 56-fold, calcium 417%, magnesium 408%, potassium, 155%, and sodium 177%. Detritis and debris in stream outflow increased dramatically after deforestation, and annual runoff from the watershed was 41%, 28%, and 26% above the control over the three years following deforestation. Vegetation is apparently very important in influencing the processes governing the retention of essential nutrients in forest ecosystems. The biotic community is not simply a passive respondent to available abiotic resources, but instead seems instrumental in regulating that availability.

The effects of forest harvesting on nutrient availability have been explored in other forest ecosystems. Kimmins and Feller (1976) examined nutrient dynamics in a logged, coastal coniferous forest in British Columbia. In Sweden, Tamm et al. (1974) and Wiklander (1981) documented that forest cutting alters watershed hydrology and nutrient dynamics. Recent experimental studies have focused on the microbial mechanisms responsible for controlling rates of nutrient loss (Vitousek and Matson 1984, 1985), and the results of nutrient cycling experiments have been extended to tropical, as well as temperate, forest ecosystems (Vitousek and Denslow 1986; Matson et al. 1987).

Recently, aquatic ecosystems have been the focus of considerable experimental study. Lakes are particularly amenable to manipulations since they are largely contained structures whose boundaries are well defined. The Experimental Lakes Area in northwestern Ontario, Canada, was established in 1968 to evaluate experimentally the ecosystem processes involved in lake eutrophication. Large-scale experiments on eutrophication involve the addition of various combinations of phosphorous, nitrogen, and carbon to whole lakes (Schindler and Fee 1974; Schindler, 1974, 1975, 1988; Schindler et al. 1987). The results of these experiments explained why the trophic status of a lake responds primarily to increasing or decreasing imputs of phosphorous. These experiments illustrated the practical importance of removing human imputs of phosphorous when conducting restoration programs in lake management.

Schindler and his colleagues have also explored the implications of acidification in lakes on trophic structure and species composition (Schindler et al. 1989; Schindler et al. 1985). When the pH of experimental lakes was reduced from an initial pH of 6.5 to pH 5, the number of species in the aquatic community was reduced by 35%.

The Experimental Lakes Area projects have principally focused on the effects of nutrient dynamics on ecosystem function. Large-scale experimental manipulations of the biotic community in lakes have revealed the critical role of keystone species, and/or groups of species, in regulating species assemblages. Carpenter et al. (1987) undertook the removal of major fish predators from experimental lakes and documented substantial changes in both species composition and ecosystem function. For example, removal of fish predators increased the relative abundance of planktonic herbivores, thereby reducing the abundance of primary producers and decreasing lake primary productivity.

Large-scale experimental manipulations pose significant problems for the statistical analysis of results. Often, large-scale experiments are difficult if not impossible to replicate (due to the limitations of research effort, money, and available sites) or involve comparisons across large geographic areas that may subject control and experimental plots to different environments. Various approaches to the statistical analysis of large-scale experiments are being explored, including randomized intervention analysis (Carpenter et al. 1989) and time-series analysis (Matson and Carpenter 1991).

Literature Cited

Alexander, M. 1971. *Microbial ecology.* New York: John Wiley and Sons.

Arnold, S. J. 1983. Sexual selection: The interface of theory and empiricism. Pp. 67–107 in P. Bateson, ed., *Mate choice.* Cambridge: Cambridge University Press.

Aston, J., and A. D. Bradshaw. 1967. Evolution in closely adjacent plant populations. II. *Agrostis stolinifera* in maritime habitats. *Heredity* 21:649–64.

Beddington, J. R., C. A. Free, and J. H. Lawton. 1978. Modeling biological control: On the characteristics of successful natural enemies. *Nature* 273:513–19.

Birch, L. C. 1953. Experimental background to the study of the distribution and abundance of insects. III. The relations between innate capacity for increase and survival of different species of beetles living together on the same food. *Evolution* 7:136–44.

Bormann, F. H., and G. E. Likens. 1979. *Pattern and process in a forested ecosystem.* New York: Springer-Verlag.

Brown, J. H. 1971. Mammals on mountaintops: Non-equilibrium insular biogeography. *Amer. Nat.* 105: 467–78.

———. 1978. The theory of insular biogeography and the distribution of boreal birds and mammals. *Great Basin Nat. Mem.* 2:209–27.

Brown, J. H., and A. C. Gibson. 1983. *Biogeography.* St. Louis: Mosby.

Cain, A. J., and P. M. Sheppard. 1954. Natural selection in *Cepaea. Genetics* 39:89–116.

Carlquist, S. 1965. *Island life.* Garden City, N.Y.: Natural History Press.

———. 1966. The biota of long-distance dispersal. I. Principles of dispersal and evolution. *Quart. Rev. Biol.* 41:247–70.

Carpenter, S. R., T. M. Frost, D. Heisey, and T. K. Kratz. 1989. Randomized intervention analysis and the interpretation of whole-ecosystem experiments. *Ecology* 70:1142–1152.

Carpenter, S. R., and J. F. Kitchell. 1988. Consumer control of lake productivity. *Bioscience* 38:764–69.

Carpenter, S. R., J. F. Kitchell, and J. R. Hodgson. 1985. Cascading trophic interactions and lake productivity. *Bioscience* 35:634–39.

Carpenter, S. R., J. F. Kitchell, J. R. Hodgson, P. A. Cochran, J. J. Elser, M. M. Elser, D. M. Lodge, D. Kretchmen, X. He, and C. N. von Ende. 1987. Regulation of lake primary productivity by food web structure. *Ecology* 68:1863–1876.

Caswell, H. 1978. Predator-mediated coexistence: A non-equilibrium model. *Amer. Nat.* 112:127–54.

Clark, C. A., and P. M. Sheppard. 1966. A local survey of the distribution of industrial melanic forms in the moth *Biston betularia* and estimates of the selective values of these in an industrial environment. *Proc. Roy. Soc.*, B, 165:424–39.

Clausen, J., D. D. Keck, and W. M. Heisey. 1947. Heredity of geographically and ecologically isolated races. *Amer. Nat.* 81:114–33.

———. 1948. Experimental studies on the nature of species. III. Environmental responses of climatic races of *Achillea.* Publ. no. 581, pp. 1–129. Washington, D.C.: Carnegie Institution.

Clements, F. E., and G. W. Goldsmith. 1924. *The phytometer method in Ecology.* New York: Columbia University Press.

Cody, M. 1968. On the methods of resource division in grassland bird communities. *Amer. Nat.* 102: 107–48.

Colwell, R. K. 1984. What's new? Community ecology discovers biology. Pp. 387–96 in P. W. Price, C. N. Slobodchikoff, and W. S. Gaud, eds. *A new ecology: Novel approaches to interactive systems.* New York: John Wiley and Sons.

Connell, J. H. 1961*a*. The influence of interspecific competition and other factors on the distribution of the barnacle, *Chthamalus stellatus. Ecology* 42: 710–23.

———. 1961*b*. Effects of competition, predation by *Thais lapillus,* and other factors on natural populations of the barnacle *Balanus balanoides. Ecol. Monogr.* 31:61–104.

———. 1970. A predator-prey system in the marine intertidal region. I. *Balanus glandula* and several predatory species of *Thais. Ecol. Monogr.* 40:49–78.

———. 1972. Community interactions on marine rocky intertidal shores. *Ann. Rev. Ecol. Syst.* 3: 169–92.

———. 1974. Field experiments in marine ecology. In R. Mariscal, ed., *Experimental marine biology.* New York: Academic Press.

———. 1975. Some mechanisms producing structure in natural communities: A model and evidence from field experiments. Pp. 460–90, in M. L. Cody and J. M. Diamond, eds., *Ecology and Evolution of Communities.* Cambridge: Harvard University Press.

———. 1983. On the prevalence and relative importance of interspecific competition: Evidence from field experiments. *Amer. Nat.* 122:661–96.

Darwin, C. R. 1851. *Geological observations on coral reefs, volcanic islands, and on South America.* London: Smith, Elder.

———. 1859. *The origin of species by means of natural selection.* London: John Murray. Facsimile of first edition, Harvard University Press, 1964.

Dayton, P. K. 1971. Competition, disturbance and community organization: The provision and subsequent utilization of space in a rocky intertidal community. *Ecol. Monogr.* 41:351–89.

———. 1975. Experimental evaluation of ecological

dominance in a rocky intertidal algal community. *Ecol. Monogr.* 45:137–59.

Diamond, J. M. 1975. Assembly of species communities. Pp. 342–44 in M. L. Cody and J. M. Diamond, eds., *Ecology and evolution of communities.* Cambridge, Mass.: Belknap Press.

———. 1986. Laboratory experiments, field experiments, and natural experiments. Pp. 3–22 in J. Diamond and T. J. Case, eds., *Community ecology.* New York: Harper and Row.

Diamond, J. M., and R. M. May. 1976. Island biogeography and the design of natural reserves. Pp. 228–52 in R. M. May, *Theoretical ecology: Principles and applications.* Oxford: Blackwell.

Dobzhansky, Th. 1970. *Genetics of the evolutionary process.* New York: Columbia University Press.

Estes, J. A., R. J. Jameson, and E. B. Rhode. 1982. Activity and prey selection in the sea otter: Influence of population status on community structure. *Amer. Nat.* 120:242–58.

Falconer, D. S. 1981. *Introduction to quantitative genetics,* 2nd ed. London: Longman.

Ford, E. B. 1964. *Ecological genetics.* London: Methuen.

———. 1965. *Genetic polymorphism.* Cambridge, Mass.: MIT Press.

Fretwell, S. D. 1977. The regulation of plant communities by food chains exploiting them. *Perspect. Biol. Med.* 20:169–85.

Gause, G. F. 1932. Experimental studies on the struggle for existence. I. Mixed population of two species of yeast. *J. Exp. Biol.* 9:389–402.

———. 1934. *The struggle for existence.* Baltimore: Williams and Wilkins. Reprinted 1964 by Hafner, New York.

Gilbert, L. E. 1975. Ecological consequences of a coevolved mutualism between butterflies and plants. Pp. 210–40 in L. E. Gilbert and P. H. Raven, eds., *Coevolution of animals and plants.* Austin: University of Texas Press.

Gould, R. 1983. Genetics of plant herbivore systems: Interactions between applied and basic study. Pp. 599–654 in R. R. Denno and M. S. McClure, eds., *Variable plants and herbivores in natural and managed systems.* New York: Academic Press.

Gurney, W. S. C., and R. M. Nisbet. 1978. Predator-prey fluctuations in patchy environments. *J. Anim. Ecol.* 47:85–102.

Hairston, N. G. 1949. The local distribution and ecology of the plethodontid salamanders of the southern Appalachians. *Ecol. Monogr.* 19:47–73.

———. 1951. Interspecies competition and its probable influence upon the vertical distribution of Appalachian salamanders of the genus *Plethodon. Ecology* 32:266–74.

———. 1989. *Ecological experiments: Purpose, design, and execution.* Cambridge: Cambridge University Press.

Hairston, N. G., F. E. Smith, and C. B. Slobodkin. 1960. Community structure, population control, and competition. *Amer. Nat.* 94:421–25.

Hamilton,T. H., and I. Rubinoff. 1963. Isolation, endemism, and multiplication of species in the Darwin's finches. *Evolution* 17:388–403.

———. 1964. On models predicting abundance of species and endemics for the Darwin finches in the Galapagos Archipelago. *Evolution* 18:339–42.

———. 1967. On predicting insular variation in endemism and sympatry for the Darwin finches in the Galapagos Archipelago. *Amer. Nat.* 101: 161–71.

Hall, D. J., W. E. Cooper, and E. E. Werner. 1970. An experimental approach to the production dynamics and structure of freshwater animal communities. *Limn. and Ocean.* 15:839–928.

Heads, P. A., and J. H. Lawton. 1983. Studies on the natural enemy complex of the holly leaf-miner: The effects of scale on the detection of aggregative responses and the implications for biological control. *Oikos* 40:267–76.

Horn, H. S., and R. H. MacArthur. 1972. Competition among fugitive species in a harlequin environment. *Ecology* 53:749–52.

Howe, H. F. 1981. Removal of wild nutmeg (*Virola suranimensis*) crops by birds. *Ecology* 62:1093–1106.

Huffaker, C. B. 1958. Experimental studies on predation: Dispersion factors and predator-prey oscillations. *Hilgardia* 27:343–83.

Hurlbert, S. H. 1984. Pseudoreplication and the design of ecological field experiments. *Ecol. Monogr.* 54:187–211.

Hutchinson, G. E. 1961. The paradox of the plankton. *Amer. Nat.* 95:137–45.

Janzen, D. H. 1966. Coevolution of mutualism between ants and acacias in Central America. *Evolution* 20:249–75.

Kettlewell, H. B. D. 1955. Selection experiments on industrial melanism in the *Lepidoptera. Heredity* 9: 323–42.

———. 1956. Further selection experiments on industrial melanism in the *Lepidoptera. Heredity* 10: 287–302.

Kimmins, J. P., and M. C. Feller. 1976. Effects of clearcutting and broadcast slash-burning on nutrient budgets, streamwater chemistry, and productivity in Western Canada. *Proc. 16th. IUFRO World Congr.,* Oslo, Div. 1:186–97.

Koehn, R. K., T. J. Hilbish. 1987. The adaptive importance of genetic variation. *Amer. Sci.* 75:134–41.

Kohn, A. J. 1959. The ecology of *Conus* in Hawaii. *Ecol. Monogr.* 29:47–90.

Lande, R. 1978. Evolutionary mechanisms of limb loss in tetrapods. *Evolution* 32:73–92.

———. 1982. A quantitative genetic theory of life history evolution. *Ecology* 63:607–15.

Lewontin, R. C. 1974. The analysis of variance and the analysis of causes. *Amer. J. Hum. Genetics* 26: 400–411.

———. 1978. Adaptation. *Sci. Amer.* 239:212–30.

Lewontin, R. C., and J. V. Hubby. 1966. A molecular approach to the study of genic heterozygosity in natural populations. II. Amount of variation and degree of heterozygosity in natural populations of *Drosophila pseudoobsura. Genetics* 54:595–609.

Likens, G. E., F. H. Bormann, and N. M. Johnson. 1969. Nitrification: Importance to nutrient losses from a cutover forest ecosystem. *Science* 163: 1205–1206.

Likens, G. E., F. H. Bormann, N. M. Johnson, D. W. Fisher, and R. S. Pierce. 1970. Effects of forest cutting and herbicide treatment on nutrient budgets in the Hubbard Brook Watershed Ecosystem. *Ecol. Monogr.* 40:23–47.

Likens, G. E., F. H. Bormann, R. S. Pierce, J. S. Eaton, and N. M. Johnson. 1977. *Biochemistry of a forested ecosystem.* Berlin: Springer-Verlag.

Lotka, A. J. 1925. *Elements of physical biology.* Reprinted 1956. New York: Dover.

Louda, S. M. 1982. Distribution ecology: Variation in plant recruitment over a gradient in relation to insect seed predation. *Ecol. Monogr.* 52:25–41.

———. 1983. Seed predation and seedling mortality in the recruitment of a shrub, *Haplopappus venetus* Blake (Asteraceae), along a climate gradient. *Ecology* 64:511–21.

Lubchenco, J. 1978. Plant species diversity in a marine intertidal community: Importance of herbivore food preference and algal competitive abilities. *Amer. Nat.* 112:23–39.

———. 1980. Algal zonation in the New England Rocky intertidal community: An experimental analysis. *Ecology* 61:333–44.

———. 1986. Relative importance of competition and predation: Early colonization by seaweeds in New England. Pp. 537–55 in J. Diamond and T. Case, eds., *Community Ecology.* New York: Harper and Row.

Lubchenco, J., and B. A. Menge. 1978. Community Development and persistence in a low rocky intertidal zone. *Ecol. Monogr.* 48:67–94.

MacArthur, R. H. 1958. Population ecology of some warblers of northeastern coniferous forests. *Ecology* 39:599–619.

MacArthur, R. H., and E. R. Pianka. 1966. On optimal use of a patchy environment. *Amer. Nat.* 100:603–9.

MacArthur, R. H., and E. O. Wilson. 1963. An equilibrium theory of insular zoogeography. *Evolution* 17:373–87.

———. 1967. *The theory of island biogeography.* Princeton: Princeton University Press.

Matson, P. A., and S. R. Carpenter. 1990. Statistical analysis of ecological response to large-scale perturbations. *Ecology* 71:2037.

Matson, P. A., P. M. Vitousek, J. J. Ewel, M. J. Mazzarino, and G. P. Robertson. 1987. Nitrogen transformations following tropical forest felling and burning on a volcanic soil. *Ecology* 68:491–502.

Mayr, E. 1963. *Animal species and evolution.* Cambridge, Mass.: Harvard University Press.

Menge, B. A. 1976. Organization of the New England rocky intertidal community: Role of predation, competition, and environmental heterogeneity. *Ecol. Monogr.* 46:355–93.

Menge, B. A., J. Lubchenco, L. R. Ashkenas, and F. Ramsey. 1968*a*. Experimental separation of effects of consumers on sessile prey on a rocky shore in the Bay of Panama: Direct and indirect consequences of food web complexity. *J. Exp. Mar. Biol. Ecol.* 100:225–69.

Menge, B. A., J. Lubchenco, S. D. Gaines, and L. R. Ashkenas. 1986*b*. A test of the Menge-Sutherland model of community organization in a tropical rocky intertidal food web. *Oecologia* 71:75–89.

Menge, B. A., and J. P. Sutherland. 1976. Species diversity gradients: Synthesis of the roles of predation, competition, and temporal heterogeneity. *Amer. Nat.* 110:351–69.

———. 1987. Community regulation: Variation in disturbance, competition, and predation in relation to environmental stress and recruitment. *Amer. Nat.* 130:730–57.

Morin, P. J. 1983. Predation, competition, and the composition of larval anuran guilds. *Ecol. Monogr.* 53:119–38.

Murdoch, W. W., and A. Oaten. 1975. Predation and population stability. *Adv. Ecol. Res.* 9:1–131.

Murdoch, W. W., J. D. Reeve, C. B. Huffaker, and C. E. Kennett. 1984. Biological control of olive scale and its relevance to ecological theory. *Amer. Nat.* 123:371–92.

National Research Council. 1986. *Ecological knowledge and environmental problem-solving.* Washington, D.C.: National Academy Press.

Nisbet, R. M. and W. S. C. Gurney. 1976. A simple mechanism for population cycles. *Nature* 263: 319–20.

Odum, H. T. 1957. Trophic structure and productivity of Silver Springs, Florida. *Ecol. Monogr.* 27: 55–112.

Oksanen, L., S. D. Fretwell, J. Arruda, and P. Niemela. 1981. Exploitation ecosystems in gradients of primary productivity. *Amer. Nat.* 118: 240–61.

Olson, J. S. 1958. Rates of succession and soil changes on southern Lake Michigan sand dunes. *Bot. Gaz.* 119:125–69.

Paine, R. T. 1966. Food web complexity and species diversity. *Amer. Nat.* 100:65–75.

———. 1969. A note on trophic complexity and community stability. *Amer. Nat.* 103:91–93.

———. 1974. Intertidal community structure: Experimental studies on the relationship between a dominant competitor and its principal predator. *Oecologia* 15:93–120.

———. 1980. Food webs, linkage interaction strength,

and community infrastructure. *J. Anim. Ecol.* 49: 667–85.

Paine, R. T., and R. L. Vadas. 1969. The effects of grazing by sea urchins, *Strongylocentrotus*, on benthic algal populations. *Limn. and Ocean.* 14:710–19.

Park, T. 1948. Experimental studies of interspecies competition. I. Competition between populations of the flour beetles *Tribolium confusum* (Duvall) and *Tribolium castaneum* (Herst.) *Ecol. Monogr.* 18:267–307.

———. 1954. Experimental studies of interspecies competition. II. Temperature, humidity, and competition in two species of *Tribolim. Physiol. Zool.* 27:177–238.

Park, T., D. B. Mertz, W. Grodzinski, and T. Prus. 1965. Cannibalistic predation in populations of flour beetles. *Physiol. Zool.* 38:289–321.

Partridge, L., and P. H. Harvey. 1988. The ecological context of life history evolution. *Science* 241: 1449–55.

Pianka, E.R. 1967. On lizard species diversity: North American flatland deserts. *Ecology* 48:333–51.

Polis, G. A., and S. J. McCormick. 1987. Intraguild predation and competition among desert scorpions. *Ecology* 68:332–43.

Polis, G. A., C. A. Myers, and R. D. Holt. 1989. The ecology and evolution of intraguild predation: Potential competitors that eat each other. *Ann. Rev. Ecol. Syst.* 20:297–330.

Power, M. E., W. J. Matthews, and A. J. Steward. 1985. Grazing minnows, piscivorous bass and stream algae: Dynamics of a strong interaction. *Ecology* 66:1448–1456.

Preston, F. W. 1948. The commonness and rarity of species. *Ecology* 29:254–83.

———. 1962. The canonical distribution of commonness and rarity. *Ecology* 43:185–15, 410–32.

Quinn, J. F., and S. P. Harrison. 1988. Effects of habitat fragmentation and isolation on species richness and evidence from biogeographic patterns. *Oecologia* 75:132–40.

Rosenzweig, M. L., and R. H. MacArthur. 1963. Graphical representation and stability conditions of predator-prey interactions. *Amer. Nat.* 97:209–23.

Schindler, D. W. 1974, Eutrophication and recovery in experimental lakes: Implications for lake management. *Science* 195:260–62.

———. 1975. Whole-lake eutrophication experiments with phosphorous, nitrogen, and carbon. *Int. Ver. Theor. Angew. Limn. Verh.* 19:3221–3231.

———. 1988. Experimental studies of chemical stressors on whole lake ecosystems. *Verh. Internat. Verein. Limn.* 23:11–41.

Schindler, D. W., and E. J. Fee. 1974. The Experimental Lakes Area: Whole-lake experiments in eutrophication. *J. Fish. Res. BD. Can.* 31:937–53.

Schindler, D. W., R. H. Hesslein, and M. A. Turner. 1987. Exchange of nutrients between sediments and water after 15 years of experimental eutrophication. *Can. J. Fish. Aquatic Sci.* 44:26–33.

Schindler, D. W., S. E. M. Kasian, and R. H. Hesslein. 1989. Losses of biota from American aquatic communities due to acid rain. *Environ. Monit. Assess.* 12:269–85.

Schindler, D. W., K. H. Mills, D. F. Malley, D. L. Findlay, J. A. Shearer, I. J. Davies, M. A. Turner, G. A. Linsey, and D. R. Cruikshank. 1985. Long-term ecosystem stress: The effects of years of acidification on a small lake. *Science* 228:1395–1401.

Schoener, T. W. 1983. Field experiments on interspecies competition. *Amer. Nat.* 122:240–85.

———. 1985. Some comments on Connell's and my reviews of field experiments on interspecific competition. *Amer. Nat.* 125:730–40.

Simberloff, D. S. 1969. Experimental zoogeography of islands: A model for insular colonization. *Ecology* 50:296–314.

———. 1976. Experimental zoogeography of islands: Effects of island size. *Ecology* 57:629–48.

Simberloff, D. S., and L. G. Abele. 1976. Island biogeography theory and conservation practice. *Science* 191:285–86.

Simberloff, D. S., and E. O. Wilson. 1969. Experimental zoogeography of islands: The colonization of empty islands. *Ecology* 50:278–96.

———. 1970. Experimental zoogeography of islands: A two-year record of colonization. *Ecology* 51: 934–37.

Slobodkin, L. B. 1961. *Growth and regulation of animal populations.* New York: Holt, Rinehart and Winston.

Smith, F. E. 1952. Experimental methods in population dynamics: A critique. *Ecology* 33:441–50.

Soule, M. E. 1986. *Conservation biology: The science of scarcity and diversity.* Sunderland, Mass.: Sinauer Associates.

Soule, M. E., and B. A. Wilcox, eds. 1980. *Conservation biology: An evolutionary-ecological perspective.* Sunderland, Mass.: Sinauer Associates.

Tamm, C. O., H. Holmen, B. Popovic, and G. Wiklander. 1974. Leaching of plant nutrients from soils as a consequence of forestry operations. *Ambio* 3:211–21.

Tansley, A. G. 1917. On competition between *Galium saxatile* L. (*G. hercynicum* Weig.) and *Galium sylvestre* poll. (*G. asperum* Schreb.) on different types of soil. *J. Ecol.* 5:173–79.

Teal, J. M. 1962. Energy flow in the salt marsh ecosystem of Georgia. *Ecology* 43:614–24.

Tilman, D. 1987. Secondary succession and the pattern of plant dominance along experimental nitrogen gradients. *Ecol. Monogr.* 57:189–214.

———. 1988. *Plant strategies and the dynamics and structure of plant communities.* Princeton: Princeton University Press.

Turesson, G. 1922. The species and the variety as ecological units. *Hereditas* 3:100–113.

———. 1925. The plant species in relation to habitat and climate. *Hereditas* 6:147–236.

———. 1930. The selective effect of climate upon the plant species. *Hereditas* 14:99–152.

Udvardy, M. D. F. 1969. *Dynamic zoogeography.* New York: Van Nostrand Reinhold.

Via, S. 1990. Ecological genetics and host adaptation in herbivorous insects: The experimental study of evolution in natural and agricultural systems. *Ann. Rev. Entomol.* 35:421–46.

Vitousek, P. M., and J. S. Denslow. 1986. Nitrogen and phosphorous availability in treefall gaps in a lowland tropical rainforest. *J. Ecol.* 74:1167–1178.

Vitousek, P. M., and P. A. Matson. 1984. Mechanisms of nitrogen retention in forest ecosystems: A field experiment. *Science* 225:51–52.

Vitousek, P. M., and P. A. Matson. 1985. Disturbance, nitrogen availability, and nitrogen losses in an intensively managed loblolly pine plantation. *Ecology* 66:1360–1375.

Volterra, V. 1926. Fluctuations in the abundance of species considered mathematically. *Science* 118: 558–60.

Wallace, A. R. 1869. *The Malay Archipelago.* New York: Harper.

———. 1876. *The geographical distribution of animals.* 2 vols. London: MacMillan.

———. 1880. *Island life, On the phenomena and causes of insular faunas and floras.* London: MacMillan.

Watt, A. S. 1947. Pattern and process in the plant community. *J. Ecol.* 35:1–22.

Wiklander, G. 1981. Rapporteur's comment on clearcutting. *Ecol. Bull.* (Stockholm) 33:642–47.

Wilson, E. O., and D. S. Simberloff. 1969. Experimental zoogeography of islands: Defaunation and monitoring techniques. *Ecology* 50:267–78.

Witman, J. 1987. Subtidal coexistence: Storms, grazing, mutualism, and the zonation of kelps and mussels. *Ecol. Monogr.* 57:167–87.

PAPER 34

SELECTION EXPERIMENTS ON INDUSTRIAL MELANISM IN THE *LEPIDOPTERA*

Dr H. B. D. KETTLEWELL

Genetics Laboratory, Department of Zoology, University Museum, Oxford

Received 1.iv.55

I. THE SPREAD OF MELANISM

In view of the extremely rapid spread of melanic forms of at least fifty species of Lepidoptera in the industrial areas of Britain, experiments were carried out with the following objects. To ascertain: (i) Whether the black and pale forms are to an appreciable extent at an advantage on their appropriate backgrounds. (ii) If so, whether the difference can be evaluated. (iii) Whether birds and other visual predators search for and eat resting moths. (iv) If that be established, whether the insects are taken selectively, and in the order of conspicuousness that is registered by the human eye.

The importance of these questions will be appreciated from the following facts. The industrial melanism of the Lepidoptera is the most striking evolutionary change ever actually witnessed in any organism, animal or plant. Its current explanation (p. 2) assumes selective elimination by predators. Yet after more than twenty-five years of observation and constant enquiry, I have found no single instance in this country in which anyone has witnessed a bird detecting and eating a moth belonging to a protectively coloured (or "cryptic") species when sitting motionless on its correct background. Nevertheless, the effective concealing patterns found in great numbers of these insects (those affected by industrial melanism and others) seem explicable only on the assumption that predators hunting by sight are of serious danger to them. I determined, therefore, to subject this matter to experiment and precise observation on a scale not so far attempted. Since for this purpose it was necessary to use large numbers of specimens, I decided to concentrate upon a single species, which could act as a test case for this aspect of the theory of industrial melanism; and also in assessing the importance of predators as selective agents in the solution of the cryptic patterns of the Lepidoptera. The insect I chose for this work was the Peppered Moth, *Biston betularia* L., Selidosemidæ, the most famous example of industrial melanism, and the first to be detected. I accordingly reared this insect on a large scale in order to provide the experimental material that was required.

There is no question of dealing in this paper with other aspects of Industrial Melanism, such as the character and behaviour-differences associated with melanic and non-melanic insects, nor with air pollution. This latter topic involves a consideration of the degree of contamination of foliage in different areas of Britain, and its effect on larvæ, with the viability differences associated with the typical and the melanic forms. These subjects will be dealt with elsewhere.

2. THE EXPLANATION OF INDUSTRIAL MELANISM

The current explanation of industrial melanism was suggested by Ford (1937). Since a knowledge of it is essential to the purpose of this paper, it is necessary to summarise it for the benefit of those unacquainted with the subject. Briefly, it may be said that industrial melanism is ascribed to a change between selective advantage and disadvantage, due to industrialisation. It was noticed (i) that the species affected by this phenomenon rest fully exposed on tree trunks, walls or fences, apparently protected from predators by a concealing pattern. (ii) In nearly all of these, the melanics are unifactorial and not recessive (it is probable, and in some instances certain, that the quantity of melanin is greater in the homozygotes than in the heterozygotes). (iii) There was much somewhat anecdotal evidence, though derived from independent sources, that the successful melanics are more viable than the normal forms, which, nevertheless, until recently they have nowhere succeeded in displacing. This difference in viability was suggested by an excess of black varieties above expectation in segregating families. It should here be noticed in parenthesis that the type of metabolism involved in excess melanin production does not necessarily confer superior viability. Recessive melanics are known as rare varieties and these, though often as black as the darkest " dominants ", are much less hardy than the pale forms.

In view of these considerations, it was suggested that all genes conferring a physiological advantage had spread through the species, unless responsible also for some counter-balancing disadvantage, such as the destruction by excess melanin formation of the concealing pattern upon which the safety of the insect depends. Such melanics, though physiologically at an advantage, were therefore unable to establish themselves until the blackened vegetation of industrial areas provided an environment in which black colouring is no longer a handicap, perhaps even an asset, to concealment. Here, therefore, they have spread ; though the equally well protected, but relatively inviable, recessive melanics have not done so.

One aspect of this explanation, the superior viability of the industrial melanics, was soon established by experiment (Ford, 1940). The other, the selective action of predators in destroying the pale or blackish moths on inappropriate backgrounds, has remained unproved, and it is this which is studied in the present paper.

3. THE GENETICS OF MELANISM IN *BISTON BETULARIA*

In about fifty species showing industrial melanism, the black forms are more or less complete dominants (their heterozygotes cannot be distinguished from the homozygotes). Several have no dominance, and a few are multifactorial. *Biston betularia* represents the usual condition for such species in which melanism is unifactorial and dominant. There are, however, a few points of special interest in regard to its genetics.

This species is normally whitish grey with a sprinkling of black dots : a colour-pattern which conceals it extremely well on light bark, especially when lichens are present. It has two melanic forms. One, *carbonaria*, is completely black except for a small white dot at the base of each forewing and another where the antennæ approximate to the head. This is the famous industrial melanic of the species which has spread in such a spectacular fashion in industrial areas. It is now a complete dominant, but was probably not so when first found in the middle of last century. At that time there appears to have been in many of the heterozygotes (we do not know the appearance of the homozygotes) some trace of the pale markings. These are never seen now in districts where *carbonaria* constitute all but a small fraction of the population. It appears, therefore, that the gene-complex has been selected to produce the complete dominance of this form at the same time as it has become common during the last one hundred years. This matter is at present being subjected to genetic tests, which are not the subject of this paper. The other melanic form of *B. betularia* is less complete, being dark with a sprinkling of white scales. It is known as *insularia* and is much less common (for relative frequencies, see p. 334), though it has certainly spread in and outside industrial areas. It cannot be distinguished phenotypically in the presence of the *carbonaria* gene. It, too, seems to have become blacker during the last fifty years or so. It was long in doubt whether or not these forms were allelomorphs. I have, however, lately demonstrated that they are at separate loci, with no evidence of linkage.

4. THE NATURAL HISTORY OF THE MOTH

B. betularia has a long period of emergence, from early May to early August. It flies at night from late dusk till dawn. The male only comes to Mercury Vapour light freely, the female but rarely. Assembling to newly hatched females in nature takes place between 9 p.m. and midnight (G.M.T.), but will continue until dawn to caged females. By day the moth rests on tree trunks and boughs, on which it takes up its position at dawn. It hatches normally at dusk, thus avoiding visual predators on the first day of its imaginal life, before being free to choose its correct background.

5. METHOD OF SCORING THE MOTHS ON THEIR BACKGROUNDS

It must be emphasised that the relation of insects to backgrounds does not merely involve a simple colour-difference such as black on white or black on black. The complicated pattern of the typical cryptic insects melts into a surprising number of backgrounds, but only if they are light coloured or have a variegated pattern. For the purpose of analysis, the following system was devised :—

In the first instance, *at two yards distance*, the question had to be answered : is the insect conspicuous or inconspicuous on its background ? Surprisingly, in nearly every case this was found to be capable of immediate solution without difficulty. To obtain the degree of divergence as between insect and background, the following method was found satisfactory. After making the initial decision at two yards, if the insect was judged " conspicuous " one walked away from it until a point was reached where it became inseparable from its background, be it *carbonaria* on lichen covered oak trunk in Devon (50 yards), or a typical *betularia* on a Birmingham oak trunk (40 yards). In each case a value of −3 was given, and anything visible *over 30 yards* was awarded this. In trial experiments, with several observers taking part to test at which spot the moth disappeared into its background, a marked degree of agreement and uniformity was reached. The readings for degrees of conspicuousness were decided at the following distances :—

" *Conspicuous* " *insects* (*on incorrect background*)

30 yards and over −3
20-30 yards −2
10-20 yards −1
} Taken in the shade in average daylight.

Similarly, decisions in the direction of inconspicuousness were decided :—

" *Inconspicuous* " *insects* (*on correct background*)

2 yards +3
5 yards +2
10 yards +1
} Taken in the shade in average daylight.

In practice, it soon becomes unnecessary to pace the distance of each insect (a difficult procedure anyway in an aviary cage), and even after as short a schooling as one day, the independent scoring of others, taken at two yards, seldom diverged by more than plus or minus one unit.

6. RESULTS OF SCORING MOTHS ON THEIR BACKGROUNDS

The scorings of Dr R. A. Hinde and myself for sixteen insects assessed separately are given in table 1.

TABLE 1

Results obtained by two observers in scoring black and typical B. betularia *independently*

Phenotype *	Black (*carbonaria*) = C	Pale (*typical*) = T
Dr Hinde .	+2 +2 −3 +3 −2 −2 +3 −2 −1 +3	+3 −2 +3 −3 +2 +1
H. B. D. K. .	+2 +3 −3 +3 −3 −2 +3 −3 −2 +3	+3 −3 +2 −3 +1 +2

* Throughout this paper the three phenotypes of *B. betularia* : *carbonaria*, *typical* and *insularia* are designated by the letters C, T and I respectively.

This experimental design was decided on to give the least error in judging which of the six possible scores were applicable. The same method of scoring was used for both aviary and field experiments, the insects being released on trunks of different species of trees relative to their proportions within the wood. From the scoring alone (quite apart from what subsequently happened), it was possible to get some idea of the state of the average background.

Thus 651 male and female *betularia* were released in a circumscribed wood in the Birmingham district, where the melanic form comprised about 90 per cent. of the population. These consisted of 171 *typical*, 416 *carbonaria*, and 64 *insularia*. There were 33 release points, being the trunks and boughs of three birch trees and thirty oaks. The proportion of birch to oak being not more than 1 : 10 in this wood.

The following scoring was obtained :—

TABLE 2

Background scoring for betularia *and its melanics in a wood near Birmingham (males and females)*

Score (birch throughout in brackets)	+3	+2	+1	−1	−2	−3	No. of inconspicuous insects		No. of conspicuous insects		Total releases ♂s & ♀s	Average score per insect
Typical on oak .	3	5	9	7	44	86	17	11%	137	89%	154 } 171	−2·11
,, birch .	(10)	(6)	(1)	(0)	(0)	(0)	(17)	100%	(0)	0%	(17)	(+2·53)
Carbonaria on oak .	184	127	47	7	1	0	358	98%	8	2%	366 } 416	+2·33
,, on birch	(8)	(8)	(5)	(4)	(7)	(18)	(21)	42%	(29)	58%	50	(−0·54)
Insularia on oak .	21	5	10	8	9	3	36	64%	20	36%	56 } 64	+0·857
,, on birch .	(5)	(2)	(1)	(0)	(0)	(0)	(8)	100%	(0)	0%	(8)	(+2·50)

From table 2 it can be seen that the degree of crypsis (protection due to camouflage) as judged by the human eye varies greatly according to background and phenotype. On the oaks of Birmingham, the *carbonaria* are nearly always extremely well protected, but the reverse is shown for the *typical*, and the scoring of one is inversely proportional to that of the other. Birches, with their areas of black on light trunks, offer suitable crypsis to all phenotypes, which no oak trunk can possibly do. *Insularia*, though varying from light to dark, is seldom very conspicuous on any background, but is less perfectly concealed than *carbonaria* on dark trunks and than the *typical* form on light ones.

As a comparable experiment to this, I am able to quote one small release undertaken near Torcross, Devon, where all the trees and rocks are covered with lichen and algæ, and where no dark backgrounds similar to those found in the Midlands, were found. 128 *betularia* were released, 45 *typical* and 83 *carbonaria.* They gave the following scorings :—

TABLE 3

Background scoring for betularia *and melanic on Devon coast*

Score	+3	+2	+1	−1	−2	−3	Total conspicuous insects		Total in-conspicuous insects		Average score per insect
Typical	29	11	5	0	0	0	0	0%	45	100%	+2·53
Carbonaria	0	0	4	2	25	52	79	95%	4	5%	−2·46

This shows a reversal of the situation found in the Birmingham district, and in the location where I worked, there appeared to be no available resting sites for *carbonaria* other than those which were nearly always very conspicuous.

This method of scoring, then, appeared to be satisfactory as a means for our gauging background differences. It was, therefore, imperative to find whether bird predators, in the first instance, in confinement, appreciated the same set of values as decided by man, and finally to repeat, if possible, their reactions in the wild.

7. AVIARY EXPERIMENTS

(i) *General features of the aviary*

Due to the kindness of Dr Hinde and others, I was permitted to use the outdoor aviaries at the Research Station, Madingley, Cambridge. I chose a pair of nesting Great Tits (*Parus major*) which were in a wire netting frame approximately 18 yards by 6 yards, and 7 feet in height. The supports of the cage were made of dark larch and spruce trunks with the bark still present. There were 13 of these and in addition 20 other resting sites were introduced, making 33 in all (15 light and 18·dark).

There were also four horizontal poles. All the original construction trunks could be referred to as being lichen-free and with dark coloured bark, but among the introduced " furniture " birch and lichened trunks were included, and as variable an assortment of natural backgrounds as possible. The three forms of *betularia* were released on these.

(ii) *Method of introduction to birds*

The tits, having been driven to the far end of the cage, were screened off with a large sheet. The releasing was then carried out on selected trunks and boughs, each of which had a number, and the scoring was assessed in the manner previously described. The sheet was then taken down and the tits given the freedom of the aviary. From then on the investigation was conducted in two ways. Firstly, direct observation of the behaviour of the tits was kept from a distance through glasses. Secondly, at intervals the cage was entered and an inventory taken of the remaining moths.

(iii) *Results of the findings of aviary experiments*

In the first experiment, equal numbers of typical *betularia* and *carbonaria* were released on to the trunks and scored as being either on

TABLE 4

Order of predation by two great tits (Parus major) *in aviary experiment (A/16)*

Phenotype	Site No.	Score at commencement + −	Score after 20 minutes + −	Score after further 20 mins. + −	Score after 1½ hr.
Carbonaria .	birch, light	+3	present	present	taken
,, .	pine, dark	+3	,,	,,	,,
,, .	tarred post, dark	+3	,,	,,	,,
Typical . .	elder bough, light	+1	,,	,,	,,
,, . .	elm bough, light	+1	,,	,,	,,
Carbonaria .	elm bough, dark	+2	,,	taken	...
,, .	birch, light	−3	,,	,,	...
,, .	lichened birch, light	−2	,,	,,	...
,, .	birch, light	−3	,,	,,	...
,, .	lichened birch, light	−2	,,	,,	...
Typical . .	birch trunk, light	+3	,,	,,	...
,, . .	elm trunk, dark	−3	,,	,,	...
Carbonaria .	elm trunk, dark	+2	taken	...	...
,, .	birch trunk, light	−3	,,	...	...
Typical . .	post, dark	−3	,,	...	...
,, . .	birch, light	+2	,,	...	...
,, . .	tarred post, dark	−1	,,	...	...
No. of insects taken	...	...	5	7	5
No. of insects left	...	17	12	5	0
Total score left	...	+20 −20	+16 −13	+11 −0	0
Total score taken	...	...	+ 4 −7	+9 −20	+11 −0
Per cent. score taken	...	...	+25% −53·85%	+45% −100%	...

correct (+3) or incorrect (−3) backgrounds. Five were on correct and five on incorrect. The release was undertaken at 3 p.m. By 5 p.m. none had been taken. By 6 p.m. all those on incorrect backgrounds had been eaten as well as two scored as " correct " ; three remaining untouched, 2 *carbonaria* and 1 *typical*, all on correct backgrounds. The following day this was repeated using 18 *betularia*, 6 of each phenotype placed on backgrounds giving an equal score plus and minus for each phenotype. After a half hour, they had all been taken except two, one being a *typical* (+3), and one an *insularia* (+2). It was suggestive that the tits were becoming specialists on *betularia*, and subsequently they were seen to be searching each tree trunk eagerly one at a time immediately after admission, thereby defeating the object of the experiment. I decided, therefore, to try and widen their feeding interests and this was done by introducing into the cage at the same time as the *betularia* release, a number of other moths and insects in a proportion comparable to what was then occurring in nature, as judged by the local sample taken in my Mercury Vapour traps. These, of course, included individuals showing other types of colouration as well as cryptic. This proved successful, and in all further aviary experiments the introduction of other species of insects was carried out at the same time as that of *betularia*. In experiment A/16, 17 *betularia* (9 *carbonaria* and 8 *typical*) were released with the total score plus and minus in equality. The results are shown in table 4. For the purpose of statistical analysis it is accepted that it would have been more satisfactory if after each time the insect was taken it had been replaced by a similar phenotype. Experimentally this was impossible, nor in view of what happened was it necessary. It can be seen from table 4 that after two periods of 20 minutes each, when the tits were left alone, all 8 of the conspicuous insects had been taken with the score of—20, but in this time only 4 of those on correct background with a score of +9 had been eliminated, leaving five with a score of +11 untouched. A small series of similar experiments were conducted in each case with comparable results, except when pouring rain caused movement on the part of the insects and these experiments were abandoned.

An attempt was made to observe directly the order in which the *betularia* were taken by viewing from a distance through field glasses. The speed at which the two birds worked, however, in so limited a space made it impossible for accurate record to be kept. They would scan the trunks from nearby twigs, then fly and snatch the selected insect and frequently carry it down to the ground in order to eat it. At no time did I see a moth with warning colouration taken, but an Eyed Hawk, *S. ocellata* (showing " deflective coloration ") was eaten after one aborted attack.

It would appear then from these observations that birds, in this case, Great Tits, do in fact eat stationary cryptic insects including melanic forms, but in the first instance only after two hours of being in close proximity to them did they begin to do so. After an initial experience, however, they were quick to learn and took freely both light and black forms of *betularia*. Secondly, they took these in an order of conspicuousness similar to that gauged by the human eye. This is in accord with the Laboratory findings of Sumner (1934), who showed that predators failed to recognise camouflaged individuals in the same way as man. If the same could be shown in the field, it would provide the missing data needed to substantiate the current explanation of industrial melanism. Moreover, it would be possible to explain the stability in which complicated cryptic patterns are held, and at the same time to determine the relative selective values of light and black forms for crypsis in a given environment.

8. EXPERIMENTAL RELEASES IN AN INDUSTRIAL AREA

(i) *Choice of site*

B. betularia is not a colony insect, and in the course of its life the males, no doubt, and to a very much smaller extent the females, must frequently travel many miles from their origin. This is an entirely different state of affairs to that found in *Panaxia dominula*, the colony insect previously used by Fisher, Ford, Sheppard and others, for developing their technique for mark-release experiments (Fisher and Ford, 1947; Sheppard, 1955). Accordingly, it was of considerable interest to see whether their methods could be applied to this more usual state of affairs, that in which a species has a continuous range, and without having to resort to C. H. M. Jackson's "multiple square technique", as used by him in tsetse fly studies with all the labour it involves (C. H. M. Jackson, 1940). The site chosen, therefore, was of the greatest importance, and a wood with as many natural barriers as possible surrounding it, such as fields, rivers, or moorlands, was desirable. It is necessary to remember that the object of the experiments was to subject as many individuals as possible to maximum predation for as long a time as they could be exposed. For all these reasons, I chose a wood (or part of it) in the Christopher Cadbury Bird Reserve, near Rubery, Birmingham. This interesting Reserve is typical indigenous oak wood, with a fair sprinkling of large birch trees and an undergrowth of bracken, intersected by rides. No dead wood had been cut from the trees for years, hence this place boasts of a large bird population including the following trunk and bough feeders :—

Woodpeckers, *Picus viridis*, *Dryobates major*, *D. minor*.
Nuthatch, *Sitta affinis*.
Tree Creeper, *Certhia familiaris*.
Tits, *Parus caeruleus*, *P. major*, *P. ater*.
Flycatchers, *Muscicapa striata*, *M. hypoleuca*.

Nevertheless it must be judged an industrial area, being heavily polluted by smoke from the midlands.

I decided to use a peninsula of woodland, three sides of which were surrounded with fields and gardens, and which lay in a hollow in the hills. The fourth side continued into the main area of the reserve. The whole wood had sundry other copses almost exclusively consisting of birch, in its immediate neighbourhood. As stated, the proportion of birch to oak was less than 1 : 10, and none of the trunks or boughs had any lichen or algæ growing on them, having long since disappeared as the result of pollution (Eustace W. Jones, 1952). I decided, then, to undertake my releases in this peninsula of woodland and to surround the area as far as possible with assembling traps containing virgin females, as well as two Mercury Vapour lights. The assembling traps were of two kinds, one made of perforated zinc which allowed males to enter but not to escape, and the other were muslin cages at which I and two others collected assiduously throughout the night. Each trap contained one virgin female of each phenotype in case of scent differences. As the result of this,

TABLE 5

Release Experiment figures for Biston betularia (*males only*)
Rubery, near Birmingham, 1953

The letters C, T, I, stand for *carbonaria*, *typical* and *insularia* respectively throughout this paper

Date (1953)	Releases			Totals	Catches			Totals	Recaptures			Totals
	C	T	I		C	T	I		C	T	I	
25.6	10	12	10	*32*	8	0	1	*9*	...	...	...	...
26.6	0	0	0	*0*	127	15	7	*149*	3	1	1	*5*
27.6	33	11	15	*59*	34	5	1	*40*	1	0	1	*2*
28.6	37	21	5	*63*	23	3	3	*29*	2	0	2	*4*
29.6	0	0	0	*0*	55	10	1	*66*	5	4	0	*9*
30.6	68	26	8	*102*	37	3	2	*42*	1	0	1	*2*
1.7	90	21	3	*114*	76	9	2	*87*	19	2	2	*23*
2.7	74	21	3	*98*	75	13	4	*92*	28	6	0	*34*
3.7	68	15	0	*83*	77	11	10	*98*	25	3	1	*29*
4.7	67	10	2	*79*	66	9	2	*77*	23	2	0	*25*
5.7	0	0	0	*0*	73	3	5	*81*	16	0	0	*16*
Totals	**447**	**137**	**46**	**630**	**651**	**81**	**38**	**770**	**123**	**18**	**8**	**149**

	Catches			Totals
	C	T	I	
Wild Birmingham population	528	63	30	621
Per cent. phenotype	*85·03*	*10·14*	*4·83*	
Release after one day of self-determination	25	2	2	
Per cent. phenotype	*5·72*	*1·48*	*4·35*	
Per cent. return of releases	*27·5*	*13·0*	*17·4*	

the whole area must have been flooded with female scent from dusk to dawn, and this was probably responsible for holding the males in the wood. As previously stated, the three phenotypes were released on the trunks and boughs as early as possible in the day, scored, then visited later in the day and absent individuals recorded.

(ii) *Method of release*

In all the experiments, whether in the field or the aviary, boughs and trees selected for release-sites, were each given a number. In the wood, the proportion of such trees belonging to different species bore direct relationship to their estimated frequency in the locality.

Each insect, having been marked on its under-side with a small dot of quick drying cellulose paint with a colour-position for each day, was shaken from its box on to the bough or trunk. They generally wandered about for a few moments, and rapidly took up the optimum

TABLE 6

Comparative methods of collecting (with exception of first night). Mercury Vapour Light and Assembling

Phenotypes	26th June			27th June			28th June			29th June			30th June			1st July		
	C	T	I	C	T	I	C	T	I	C	T	I	C	T	I	C	T	I
M.V. Light— Totals	90	12	6	16	3	1	14	2	1	17	2	0	17	1	1	44	6	2
Recaptures	2	1	1	1	0	1	1	0	1	1	0	0	1	0	0	9	2	2
Assembling— Totals	37	3	1	18	2	0	9	1	2	38	8	1	20	2	1	32	3	0
Recaptures	1	0	0	0	0	0	1	0	1	4	4	0	0	0	1	10	0	0

Phenotypes	2nd July			3rd July			4th July			5th July			Totals	Percentage
	C	T	I	C	T	I	C	T	I	C	T	I	C T I	C T I
M.V. Light— Totals	45	7	4	41	6	7	33	8	1	45	0	1	433	*56·89*
Recaptures	16	3	0	13	2	1	10	2	0	5	0	0	75	*50·33*
Assembling— Totals	30	6	0	36	5	3	33	1	1	28	3	4	328	*43·11*
Recaptures	12	3	0	12	1	0	13	0	0	11	0	0	74	*49·67*

Phenotypes	Total Captures			Recaptures		
	C	T	I	C	T	I
M.V. Light— *Totals*	362	47	24	59	10	6
Percentage	*83·60*	*10·85*	*5·55*	*78·67*	*13·33*	*8·0*
Assembling— Totals	281	34	13	64	8	2
Percentage	*85·68*	*10·36*	*3·96*	*84·40*	*10·81*	*2·71*

position available. We have recently been able to show, experimentally, that black and light *betularia* are able to appreciate contrast differences between themselves and their backgrounds, and hence they tend to come to rest in the optimum positions offered by an available resting site (Kettlewell, 1955). Having done this, they did not move again. The position of the paint marks, being on the underside which is not exposed by day, excluded the possibility of an additional risk of attack by predators. It was important to see that the sun did not shine on the moths, and each one of them, having settled, was scored by the method described previously.

TABLE 6A

Captures at light and by means of assembling, showing the equivalence of the methods, for which $\chi^2_{(2)} = 1{\cdot}09$, *for which* $P = 0{\cdot}\text{—}0{\cdot}5$

(with the exception of first night)

	Captures		Total	χ^2
	at light	assembling		
carbonaria . . .	362	281	643	0·094
typical . . .	47	34	81	0·042
insularia . . .	24	13	37	0·957
	433	328	761	1·093

(iii) *Release results*

Six hundred and thirty *betularia* males (*carbonaria* 447, *typical* 137, *insularia* 46) were released. Out of a total of 770 caught (*carbonaria* 651, *typical* 81, *insularia* 38) 149 were recaptures (*carbonaria* 123, *typical* 18, *insularia* 8), HENCE THE RELATIVE RECOVERY VALUES FOR THE THREE PHENOTYPES OF MY RELEASES WERE *carbonaria* 27·5 per cent., *typical* 13·0 per cent., *insularia* 17·4 per cent. (see table 5).

Twenty-nine individuals were recaptured after forty-eight hours' absence, and therefore after at least twenty-four hours of self-determination. They gave survival values of *carbonaria* 5·72 per cent., *typical* 1·48 per cent., *insularia* 4·35 per cent. of the initial release.

Of the 770 caught, 621 were local insects not released by us. They comprised *carbonaria* 528, *typical* 63, *insularia* 30, giving a frequency of *carbonaria* 85·03 per cent., *typical* 10·14 per cent., *insularia* 4·83 per cent. for Rubery insects (see table 7).

In regard to the method of collecting, after the first night (there having been no release on the 25th), 433 males were taken at Mercury Vapour light, and 328 by assembling, with practically identical phenotype frequencies at both (see table 6A). This demonstrates the random nature of the samples collected. No one phenotype had behaviour differences as between attraction to light and assembling, always remembering that there were equal numbers of each female phenotype in all the assembling traps. The detailed results are shown in tables 5 and 6.

9. ECOLOGICAL OBSERVATIONS AND DATA

All three phenotypes of *Biston betularia* are cryptic. In attempting to decide their relative selective values in the Birmingham area we have three sets of figures to take into account :—

(i) Scoring values as gauged by human standards.
(ii) Direct observation as to what happened to the individuals so scored.
(iii) Recapture figures which provided data over longer periods.

TABLE 7

Daily totals of wild-caught Birmingham specimens

Date	C	Per cent.	T	Per cent.	I	Per cent.	Total
25.6.53	8	*88·9*	0	*0*	1	*11·1*	9
26.6.53	124	*86·1*	14	*9·7*	6	*4·2*	144
27.6.53	33	*86·84*	5	*13·6*	0	*0*	38
28.6.53	21	*84*	3	*12*	1	*4*	25
29.6.53	50	*87·7*	6	*10·5*	1	*1·8*	57
30.6.53	36	*90*	3	*7·5*	1	*2·5*	40
1.7.53	57	*89·06*	7	*10·94*	0	*0*	64
2.7.53	47	*81·04*	7	*12·07*	4	*6·89*	58
3.7.53	52	*75·36*	8	*11·59*	9	*13·05*	69
4.7.53	43	*82·69*	7	*13·40*	2	*3·85*	52
5.7.53	57	*87·69*	3	*4·62*	5	*7·69*	65
Totals	528	85·03	63	· 10·14	30	4·83	621

(i) *Scoring values*

The insects, having been released on to thirty-three trees, being a sample of the only trunks and boughs available in this wood, were then scored. Of 366 male and female *carbonaria* released on oak, 358 (97·83 per cent.) were gauged as inconspicuous, and 8 (2·17 per cent.) otherwise. Conversely, of 154 *typical* which took up their positions on oak, only 17 (11·1 per cent.) were judged inconspicuous as shown in tables 2 and 2A.

TABLE 2A

The concealment (or otherwise) of B. betularia *and its melanic* carbonaria *on oak trunks near Birmingham, as gauged by Man*

	Conspicuous	Inconspicuous	Total
carbonaria	8	358	366
typical	137	17	154
	145	375	520

The other melanic form, *insularia*, because of its small numbers, need not be considered in detail here. Releases of 50 *carbonaria* on to birch trees, which were in proportion to their frequency in the wood, gave this melanic conspicuous to inconspicuous in approximate equality. Thus it appeared that in this locality typical *betularia* were at an approximate disadvantage of 40 per cent.

(ii) *Direct observations*

After the first day of release it was noticed that some of the *betularia* were disappearing from one or two positions so observation of these sites was maintained continuously from a distance through field glasses. H. M. Kettlewell first observed a bird fly up out of the bracken, snatch a *betularia* and return to the ground, the whole incident being over in a flash. Subsequently we were able to watch this bird, a Hedge Sparrow (*Prunella modularis*) regularly at work and to score the order in which it took the phenotypes. At a somewhat later date moths began to disappear from another series of trees, and I and others witnessed a Robin (*Erithacus rubecula*) at work, particularly in the late afternoon. This bird was observed in a similar way to the Hedge Sparrow, flying on to the twigs and bracken near to the trees whence it viewed the trunks and branches, making occasional excursions to pick up *betularia*. This it frequently did on the wing, always returning to the ground to eat them, and there subsequently I found wings and remains as an additional check to what had happened. It came as a surprise to us to find the Robin and the Hedge Sparrow behaving in this way, and there were most certainly other birds at work, unseen by us. I give below (table 8) a record of the few occasions when there was no doubt of the order in which these birds took their insects.

TABLE 8

Direct observations of the predation of B. betularia *by two species of birds*

Robin	Oak	4th July 1953 Order of take	Robin	Oak	2nd July 1953 Order of take
typical	−3	1	*typical*	−3	1
carbonaria	+1	2	*typical*	−3	2
typical	−3	3	*carbonaria*	+3	3
typical	−2	4	*typical*	−3	4
carbonaria	+3	Not taken by 7 p.m.	*carbonaria*	+3	Not taken by 7 p.m.
carbonaria	+3		*carbonaria*	+2	

Hedge Sparrow	Oak	1st July 1953 Order of take
typical	−3	1
typical	−3	2
carbonaria	+3	3
typical	−1	4
carbonaria	+3	Not taken by 7 p. m.
carbonaria	+2	

From this it would appear that when a conspicuous insect had been found, it at once put other insects in the immediate vicinity at a disadvantage because of the birds' active searchings. This is corroborated by the increased predation which took place on those trees which harboured a moth with the score of −3 compared with trees where no such conspicuous individuals existed. Nevertheless, on oaks there were nearly always one or more *carbonaria* left which had been overlooked.

Total figures for male releases show that while 62·57 per cent. of *carbonaria* survived per day during observation (see table 9) only

TABLE 9

Observation release for 7 days (males only)

Score	*Carbonaria*						Total
	+3	+2	+1	−1	−2	−3	
25.6.53	3	3	2	0	0	0	8
27.6.53	11	15	5	1	1	0	33
30.6.53	17	20	15	6	2	0	60
1.7.53	43	28	6	0	0	4	81
2.7.53	47	14	3	1	0	6	71
3.7.53	24	18	12	0	2	3	59
4.7.53	29	14	7	0	1	3	54
Total release before midday (seven days)	174	112	50	8	6	16	366
Total missing (seven days) up to 5 p.m.)	65	35	20	3	3	11	137
Per cent. missing .	37·35	31·25	40	37·5	50	68·75	37·43
Per cent. survival .	*62·45*	*68·75*	*60*	*62·5*	*50*	*31·25*	*62·57*
Release escapes .	...	...	...	...	...	...	44
Per cent. escapes = activity	...	...	...	...	...	...	10·75

45·79 per cent. of the *typicals* did so (table 10). *Insularia* had a 57·14 per cent. survival rate for the comparatively few insects under observation (table 11). Furthermore, there is evidence that the birds took the individuals within each phenotype with regard to their degree of crypsis. Of the 508 males released for observation over seven days, 210 had disappeared at the end of the day for one reason or another.

(iii) *Recapture figures*

It is evident from table 5 that we recaptured more than twice as many *carbonaria* as *typical* relative to the number released. This can be accounted for in one or more ways.

(i) That the melanics were attracted to light more freely than the *typicals*.

(ii) That the *typicals* had a shorter span of life than the *carbonaria*.
(iii) That the *typicals* wander or migrate more than the melanics.
(iv) That there was a differential predation between the phenotypes.

TABLE 10

Observation release for seven days (males only)

Score	*Typical*						Total
	+3	+2	+1	−1	−2	−3	
25.6.53	1	0	0	2	2	2	7
27.6.53	1	1	0	0	3	6	11
30.6.53	2	2	2	5	6	7	24
1.7.53	2	0	2	0	3	11	18
2.7.53	2	1	0	3	8	11	25
3.7.53	3	0	1	0	2	7	13
4.7.53	0	0	0	0	3	6	9
Total release before midday	11	4	5	10	27	50	107
Total missing up to 5 p.m.	4	2	3	5	13	31	58
Per cent. missing	36·36	50	60	50	48·15	62	54·21
Per cent. survival	*53·64*	*50*	*40*	*50*	*51·83*	*38*	45·79
Release escapes	...	...	...	...	...	...	9
Per cent. escapes = activity	...	...	...	...	...	...	6·67

In regard to the first, we were able to show that an equal relative percentage of each form came both to light and assembling traps, which excludes behaviour differences in these directions. In regard to the life span, most wild caught insects were mark-released except for weak individuals, and when possible their numbers were subsidised by laboratory-bred specimens which were released at the same time. Many more of the *typicals* than *carbonaria* were bred ones and hence were in their first day of imaginal life, nor in regard to the wild individuals was there any evidence of a diminution of hatchings which would lead to the using of older moths, with a consequent shorter expectation of life (see table 7). Furthermore, the length of life of the two forms does not differ appreciably when bred and kept in the laboratory. The question of a different life span can therefore be excluded.

The possibility of a differential migration or dispersal rate taking place between the *typicals* and the melanics, must also be considered.

A comparison of the relative proportion of marked phenotypes taken at traps situated at the periphery of the release area to those caught at traps which were placed two hundred yards or more distant, shows

TABLE 11

Observation release for seven days (males only)

Score	*Insularia*						Total
	+3	+2	+1	−1	−2	−3	
25.6.53	2	0	1	3	0	0	6
27.6.53	3	1	4	1	3	1	13
30.6.53	4	3	0	0	1	0	8
1.7.53	1	0	1	0	0	1	3
2.7.53	3	0	0	0	0	0	3
3.7.53	0	0	0	0	0	0	0
4.7.53	1	1	0	0	0	0	2
Total release before midday	14	5	6	4	4	2	35
Total missing up to 5 p.m.	6	2	2	1	2	2	15
Per cent. missing .	33·33	50	33·33	25	50	100	42·36
Per cent. survival .	*66·67*	*50*	*66·67*	*75*	*60*	*0*	57·14
Release escapes .	...	...	...	...	...	...	6
Per cent. escapes = activity	...	...	...	...	...	...	17·39

no difference in the frequencies of the three forms. One must, therefore, accept that the figures represent selective differences. Moreover, having in mind the state of the tree trunks, the results of the observation-release and, lastly, the frequent witnessing of the selective nature of the predation undertaken by two species of birds, it is fair to assume that this is the correct interpretation.

10. DISCUSSION

On the, of course unverified, assumption that in the Manchester area *carbonaria* occupied 1 per cent. or less of the *betularia* population in 1848 and 99 per cent. or more of it in 1898, Haldane (1924) showed that *carbonaria* has a selective advantage of about 30 per cent. over the *typical*. This, of course, might not be entirely due to crypsis, in fact it almost certainly would not be so. That the wild *B. betularia* population of Rubery contained as much as 10·14 per cent. *typicals* is, in my opinion, due to the advantage of this form in the surrounding birch woods of which there were many in the neighbourhood. This is to some degree borne out by the fact that, within a mile and a half of the centre of Birmingham, where there are few birch trees and nearly all the tree trunks are black, the frequency of *carbonaria* is still higher, being 93·18 per cent., with *typical* 4·54 per cent., *insularia* 2·28 per cent., in a total of 176 (Bowater and others).

I am not aware of previous attempts to discover the selective advantages of melanic insects by mark-release recapture experiments. To the obvious criticism that the releases were not free to take up their own choice of resting site for the first day, I must answer that there were no other alternative backgrounds available for an insect that has to spend its days on trunks and boughs in this wood. I admit that, under their own choice, many would have taken up position higher in the trees, and that since the surface area of a tree increases proportional to the distance up the trunks and boughs, in so doing they would have avoided concentrations such as I produced. Tinbergen (1952), de Ruiter (1952) and others have shown the importance to cryptic insects of avoiding too high a density level, but this is no argument against the findings for the *relative* advantages of the three forms. It must be accepted, however, that, under natural conditions, predation, though selective, might take place at a lower tempo.

How low this may be under certain conditions may be seen from data provided by Heslop Harrison (1919-20) who kept daily observations on *Polia chi* (Agrotidæ), and its dark form, which took up positions in a state of nature on three types of wall in the Newcastle district. The frequency of the latter (= *olivacea*) was 10 per cent. in one locality and 50 per cent. in another. He examined " up to 300 examples daily " and " never was there any diminution of numbers in which more *olivacea* " (= the ' melanic ') " vanished than type *chi*. As a matter of fact, we used to consider it a marvellous thing if even a single one had disappeared." He quotes this " to demonstrate that the effect of natural selection is quite negligible as a factor in progressive melanism."

I am afraid I cannot agree with this. In the first place *olivacea* is not a melanic in the sense that the whole pattern has been obliterated, for this remains present in a darker olive-grey tint (= it is a " melanochroic " J. W. H. H.). For this reason, sitting on stone walls (and not tree trunks) as they do, both forms are inconspicuous against the same backgrounds. Secondly, in the absence of concentrations, predation may have been at a very low level, though highly selective, and however small an advantage (0·1 per cent. or less) selection is able to use that advantage and spread a gene through the population at a calculable rate (which is not a linear one) and which depends upon the population's size. Consequently there is, theoretically, no limit set to the smallness of an advantage which can be used in selection, and one involving one out of 300 individuals indicates quite a considerable selective influence (Fisher, 1930). Fortunately, in the present series of experiments the values are of a much higher order and the effects of natural selection on industrial melanics for crypsis in such areas can no longer be disputed.

11. SUMMARY

1. Industrial melanism in moths is the most striking evolutionary phenomenon ever actually witnessed in any organism, animal or plant.

2. In order to account for its success, a theory has been assumed in which an inter-relation exists between the superior viability of certain melanics and predation by birds which eliminate resting moths selectively.

3. It has been considered probable that those individuals are principally destroyed which do not effectively match their backgrounds: so favouring black forms in areas polluted by soot.

4. In general, predation by birds has been thought to be responsible for the evolution of concealing colour-patterns in moths.

5. In spite of exhaustive observations, however, no evidence of such a selective elimination has, up to now, ever been produced. This paper supplies it.

6. The work has been conducted on the pale cryptically coloured moth *Biston betularia*, and its two melanic forms : the extreme black *carbonaria*, and the less extreme and rarer *insularia*. Both are unifactorial and dominant. They are not allelomorphs.

7. To the human eye, *carbonaria* proved much the better concealed on the lichen-free tree trunks, blackened by pollution near Birmingham, but this advantage was reversed in favour of the pale form in unpolluted country. *Insularia* possessed an intermediate advantage in both places.

8. The three forms were exposed to the attacks of birds (*Parus major*) in an experimental aviary, where their selective elimination on incorrect backgrounds was first demonstrated.

9. Corresponding work was then carried out in a wood near the industrial area of Birmingham, where the proportions of the three forms occurring naturally are, *carbonaria* 85 per cent., *typical* 10 per cent., *insularia* 5 per cent.

10. 630 male *B. betularia* (447 *carbonaria*, 137 *typical* and 46 *insularia*) were released in this wood. The moths were each identified by a mark on the underside, invisible when the insect was at rest.

11. The released insects were recaptured by assembling to females and at light traps. Of 149 which were recovered, the proportions were *carbonaria* 27·5 per cent., *typical* 13·0 per cent., *insularia* 17·4 per cent. This indicates their relative survival rates, as other agencies affecting their recapture could be excluded.

12. Birds, *Erithracus rubecula* and *Prunella modularia*, were seen frequently to inspect the trees from the undergrowth and capture the resting moths.

13. They did so generally without alighting and with great rapidity. Consequently, the act can only be detected in large-scale planned experiments. It is for this reason that it has not been previously observed.

14. Three sets of figures, the camouflage efficiency as scored by man, the differential survival at the end of day (as the result of predation) and, mark-release results, each show that the more conspicuous moths (in the Birmingham district, the *typicals*) were principally taken. Consequently birds act as selective agents, as postulated by evolutionary theory.

Acknowledgments.—I wish to thank the Nuffield Foundation who have enabled me to undertake this work, also Dr E. B. Ford, F.R.S., for his constructive advice, and helpful criticism of this paper. I am indebted to Dr R. A. Hinde of the Ornithological Field Station, Madingley, Cambridge, for the excellent facilities he allowed me to make use of, and for his helpful suggestions. I have benefited greatly from the co-operation of Dr P. M. Sheppard, who has provided help throughout. I would like to make an acknowledgment to Mr and Mrs Christopher Cadbury and family who allowed us to make use of their private reserve and also helped us in certain observations in the experiments.

12. REFERENCES

DE RUITER, L. 1952. *Behaviour*, *4*, 3.
FISHER, R. A. 1930. *Genetical Theory of Natural Selection.*
FISHER, R. A., AND FORD, E. B. 1947. *Heredity*, *1*, 143-174.
FORD, E. B. 1937. *Biol. Rev.*, *12*, 461-503.
FORD, E. B. 1940. *Ann. Eugen.*, *10*, 227-252.
HALDANE, J. B. S. 1924. *Trans. Cam. Phil. Soc.*, *23*, 26.
HESLOP HARRISON, J. W. 1919-20. *J. Genet.*, *9*, 235.
JACKSON, C. H. N. 1940 *Ann. Eugen.*, *10*, 332-369.
JONES, E. W. 1952. *Rev.. Bry. et Lich.* T. xxi, 1-2.
KETTLEWELL, H. B. D. 1955. *Nature*, *175*, 943.
SHEPPARD, P. M. 1951. *Heredity*, *5*, 349-378.
SUMNER, F. B. 1934. *Proc. Nat. Acad. Sci. Wash.*, *20*, 559-564.
TINBERGEN, N. 1952. *Ibis*, *94*, 158-162.

PAPER **35**

EXPERIMENTAL STUDIES OF INTERSPECIES COMPETITION

1. COMPETITION BETWEEN POPULATIONS OF THE FLOUR BEETLES, *TRIBOLIUM CONFUSUM* DUVAL AND *TRIBOLIUM CASTANEUM* HERBST

INTRODUCTION

There is much speculation and some fact in ecological and evolutionary literature about competition between populations which are brought into tension with each other through their respective demands upon an environment shared by both but limited in its potentialities for exploitation. It is usually agreed by common consent that this problem is biologically significant; that it is complex in character and difficult to analyze, and that it merits active investigation in the field, the laboratory, and through stochastic processes.

A creditable start has been made along this line of ecological research. There is a considerable literature dealing with natural and experimental populations and with various theoretical aspects. Because of the recent publication (1947) of Crombie's valuable review devoted to the general topic, however, it is not necessary to discuss these papers here and the reader is referred to this source and papers by Elton (1946) and Williams (1947) for orientation.

The present study is designed to explore and describe what happens when two closely related species of flour beetles, *Tribolium confusum* Duval and *Tribolium castaneum* Herbst, are placed into direct competition in a shared environment to which both are individually well adapted. Tribolium seems particularly favorable for research of this type. Much is already known of its population behavior. It lends itself to laboratory manipulation and to accurate census. Its environment, flour, is not a synthetic but a natural environment. Further, there seems to be some theoretical meaning in using two species of the same genus instead of selecting forms of greater taxonomic divergence. The cogency of the last point was recognized by Charles Darwin who, in Chapter III of *The Origin of Species,* wrote, "As the species of the same genus usually have, though by no means invariably, much similarity in habits and constitution, and always in structure, the struggle will generally be more severe between them, if they come into competition with each other, than between the species of distinct genera."

The experiments shortly to be discussed have been rather broadly conceived and extensively carried out, both as to their duration and replication. It is considered desirable in any study of competition, whether field or laboratory, to observe the populations long enough so that a judgment can be reached about their eventual equilibria, and in sufficient numbers, so that variability within cultures of similar design and between cultures of dissimilar design can be statistically appraised. While these desiderata are not completely satisfied in every case reported here, as much has been done over the four-year period as time and facilities have permitted. Because the data raise more problems than they solve, this paper is to be considered primarily as descriptive in scope and not definitive. To signify the point that future research is needed and may be anticipated, a new series of publications is inaugurated herewith under the title "Experimental Studies of Interspecies Competition" of which this is the first contribution.

It is a privilege to acknowledge the assistance of a number of friends and colleagues who have aided me immeasurably in the conduct of this research and in the preparation of the manuscript. I am greatly indebted to Doctor Lowell J. Reed, Vice-President of the Johns Hopkins University and Professor of Biostatistics at that institution, for taking time from a busy schedule to discuss the data after the computations had been made. This conference was most stimulating and helpful. Professor Sewall Wright, of the University of Chicago, offered several cogent suggestions. Mr. L. C. Birch, of the University of Sydney, and Marian Burton Frank, of the University of Chicago, read the manuscript much to its profit. Doctor Clay G Huff, of the Naval Medical Research Center, generously assisted me with matters pertaining to the identification and life-cycle of the parasite *Adelina tribolii.* Doctor A. C. Crombie, of Cambridge University, loaned me a typescript copy of his important manuscript on interspecies competition in advance of its publication (1947). Mr. David Lack, of Oxford University, offered several stimulating suggestions when he was a visitor in this laboratory. Doctor Asher J. Finkel, of the University of Chicago, discussed with me certain segments of the data. Joan Barnes contributed greatly by her skillful and meticulous drafting of the figures. Finally, I am sincerely grateful to the following associates who at various times assisted untiringly in the exacting and laborious tasks of counting so many thousands of beetles and in keeping the laboratory routines in smooth working order: Marion Biggs Davis, Nina Leopold Elder, Marian Burton Frank, Virginia Miller Gregg, Shirley Horwitz, and Catharine Lutherman Kollros.

MATERIALS AND METHODS

Census Procedures

The technical procedures followed in this study were designed to satisfy three general requirements: to maintain the physical environment as constant as possible; to minimize and equalize handling of the beetles, and to supply the organisms with an optimal nutritive medium renewed at thirty-day intervals.

All populations were kept in dark incubators at a constant temperature of 29.5° C. (±0.5°) and at a relative humidity of 60 to 75 per cent. The medium consisted of 95 per cent whole-wheat flour sifted through No. 8 silk bolting-cloth and then fortified by the addition of dry brewer's yeast powder in the amount of five per cent by weight. To ensure homogeneity, the yeast and flour were thoroughly admixed in a mechanical rotating device. The populations were established in two sizes of glass containers and in three volumes of medium. The smallest volume (8 grams) was placed in shell vials 9.5 × 2.5 cm., and the larger volumes (40 and 80 grams) were placed in wide-mouthed bottles 10 × 7 cm.

These bottles containing the medium were then heated at 60° C., for three hours to kill any foreign organisms or parasitic cysts that might have been in the flour. Upon cooling, and before being used, they were stored for a week in an incubator in which they reach the requisite temperature and humidity equilibria of 29.5° and 60-75 per cent.

Census counts were taken at thirty-day intervals for every population regardless of its type. A census consists first in passing the infested medium through a No. 00 bolting cloth sieve that retains the imagoes, pupae, and larger larvae. These are gently brushed into a finger-bowl and, after dead forms and frass are removed, they are counted by means of a hand tally and the numbers recorded by stage categories. The finer siftings are then screened through a No. 2 bolting cloth that retains the larvae of intermediate sizes. These are placed on black paper, counted, and the number recorded. Most of the eggs are retained by this mesh (No. 2). They are carefully rolled off the black paper into the storage finger-bowl but not counted. The final sifting utilizes a No. 8 cloth that retains all eggs and the smallest larvae. The latter are counted and are placed, along with the larger stages and the eggs, into bottles containing the desired amount of fresh medium.

This method of census thus permits an accurate, *total* count to be made of all larvae, pupae, and imagoes, without any loss of eggs, and a re-establishment of the population in fresh medium for another thirty-day period. In the case of "experimental" or two-species cultures, each imago beetle is identified under the microscope as to species (i.e., either *Tribolium confusum* or *T. castaneum*) using the eye and antennal characters described and figured by Good (1936), and this essential datum is recorded.

Several comments are appropriately included here concerning fecundity, rate of metamorphosis, and imaginal longevity of the two species of flour beetles.

Fecundity

In two independent oviposition experiments, carried out several years apart in this laboratory at temperatures of 29° C. (±0.5), *T. castaneum* exhibited a significantly higher rate of fecundity than did *T. confusum.* These experiments were conducted under the relatively optimal conditions of one pair of young imago beetles in eight grams of medium with the latter renewed at regular intervals. The data merely show that a species difference in fecundity exists in a highly favorable environment. They tell nothing about oviposition under population conditions—conditions in which various degrees of crowding and competition obtain which greatly reduce egg production, as is known, and which perhaps reduce egg production differentially by species, as is not known. The last aspect has an obvious relevance for the present study and is now being investigated. Although, as indicated above, *T. castaneum* lays more eggs, there exists a rather large difference between the two species in the two assay experiments the reason for which is not understood. In one (the first conducted), *T. castaneum* had an oviposition rate, over a 60-day assay period, of 12.8 eggs per female per day as compared with 12.1 for *T. confusum*—a difference of approximately 6 per cent ($P = 0.0012$). Comparable rates for the second experiment were 14.9 for *T. castaneum* and 10.8 for *T. confusum*—a difference of approximately 28 per cent ($P < 0.0000$). Kollros (1944) studied the fecundity of *T. castaneum* using the same stocks and conditions and obtained a rate in close agreement with the last figure, namely: 13.5 eggs per female per day. Despite these divergencies it is reasonable to conclude that one species of flour beetle possesses a higher reproductive potential than the other. This datum is important for our present purposes.

Metamorphosis

Under the experimental conditions described earlier both *T. confusum* and *T. castaneum* require roughly about a month to pass through their respective post-embryonic periods which they do with negligible mortality at 29° C. *T. castaneum,* however, does this at a slightly faster pace largely by reducing the length of the egg and pupal stages. This can be summarized by reporting data presented by Kollros (1944) for *T. castaneum* and by Stanley (1946) for *T. confusum,* as follows:

	Egg stage (days)	*Larval period (days)*	*Pupal stage (days)*	*Total period (days)*
T. confusum	5.5	22.4	7.0	34.9
T. castaneum	3.8	22.8	6.2	32.8

Imago Longevity

Life tables for *Tribolium confusum* have been published by Pearl, Park, and Miner (1941), and some observations on mean life-duration of *T. castaneum* are included in the manuscript by Kollros. It seems clear that the latter species has a significantly shorter average life than does the former. Data, pertinent to this point and extracted from the two reports, are as follows (expressed as mean number of days):

	Males	Females
T. confusum	177.8 ± 2.8	198.5 ± 3.5
T. castaneum	115.6 ± 2.2	107.7 ± 2.4

These, again, are observations accumulated under favorable conditions of assay and do not necessarily represent events as they occur in populations.

Sufficient background has now been presented about the methods used in this study and about the ecological life-history and husbandry of the two species of Tribolium. Should the reader wish further orientation in such matters he is referred to the following papers: Boyce (1946), Chapman & Baird (1934), Dick (1937), Ford (1937), Good (1936), Holdaway (1932), Park (1934, 1941), and Park & Davis (1945).

THE EXPERIMENTAL DESIGN

The design of this study, patterned in part after that of Park, Gregg, and Lutherman (1941), can be clarified by reference to Table 1 from which the following points are to be noted:

(1) Three volumes of medium were used: 8 grams, designated hereafter as "I"; 40 grams, designated as "II"; and 80 grams, designated as "III."

(2) All populations were started by introducing imagoes in equal sex ratio in a density of one beetle per gram of medium. Thus, I had a total initial population of 4 males and 4 females; II, of 20 males and 20 females; and III, of 40 males and 40 females.

(3) There are "control" and "experimental" populations. The controls are single species cultures either of *T. confusum* or *T. castaneum*. The experimentals are mixed cultures in which interspecies competition is established within the ecosystem. Table 1 shows that the experimentals are started in one of three ways. First, *T. confusum* and *T. castaneum* adults are introduced in equal proportions (E-a). Second, *T. confusum* is introduced at a numerical advantage over *T. castaneum* in the proportion ¾ : ¼ (E-b). Third, *T. castaneum* is given a similar initial advantage over *T. confusum* (E-c).

(4) The table also lists total days of observation, number of replicated cases for each category,[1] and contains a code system by means of which each type of population can be readily identified. The reader is urged to memorize this code because it will be used repeatedly as the paper develops. In all instances the first notation is a Roman numeral that identifies the volume as explained above. The second notation identifies the culture as a control (C) or an experimental (E). For controls, the lower case "b" signifies a pure culture of *T. confusum;* "c" a pure culture of *T. castaneum.* For experimentals, "a" signifies that each species was introduced in a 50% ratio; "b" signifies that *T. confusum* comprised 75% of the total; and "c", that *T. castaneum* comprised 75% of the total.

(5) The bottom two rows of Table 1 refer to special populations grown under parasite-free conditions and designated "sterile" (S). These will be treated in detail later.

From the discussion of methods, and consideration of the experimental design, it should be clear that it is technically feasible to census, accurately and with negligible injury, two sorts of populations (single and competing species) cultured for long periods of time in a nutritious, regularly renewed environment of three different volumes under sufficiently constant conditions of temperature, moisture, and light.

[1] Because the taking of a census is time-consuming and laborious even for the volume I series, a practical limitation is necessarily imposed on number of replicates. Every effort has been made, however, to conduct this study utilizing as many cases as possible. It has taken approximately 40 hours a week over the four-year period to accumulate the data reported here.

TABLE 1. Design of the entire investigation.

Volume of Medium	Total Initial Imago Population	Initial Imago Density (per gram)	CONTROLS (single species)			EXPERIMENTALS (2 species in competition)			
			Code	Number of replicates at start	Total days of observation	Code	Initial ratios of imagoes at start	Number of replicates at start	Total days of observation
I (8 gm.)	8	1	I-C-b	20	1380	I-E-a	b = c(1/2:1/2)	15	780
						I-E-b	b > c(3/4:1/4)	15	750
			I-C-c	20	1380	I-E-c	b < c(1/4:3/4)	15	690
II (40 gm.)	40	1	II-C-b	17	1350	II-E-a	b = c(1/2:1/2)	9	1020
						II-E-b	b > c(3/4:1/4)	7	540
			II-C-c	18	1380	II-E-c	b < c(1/4:3/4)	7	930
III (80 gm.)	80	1	III-C-b	2	1410	III-E-a	b = c(1/2:1/2)	2	750
						III-E-b	b > c(3/4:1/4)	2	360
			III-C-c	2	.1410	III-E-c	b < c(1/4:3/4)	2	810
I STERILE	8	1	I-C-b-S	20	1140	I-E-a-S	b = c(1/2:1/2)	18	840
			I-C-c-S	20	1140				

SINGLE SPECIES POPULATIONS: CONTROLS

INTRODUCTORY

Although the primary concern of this study lies in the quantitative description of what happens when *Tribolium confusum* and *Tribolium castaneum* are cultured as competing populations, it is impossible to discuss this phenomenon with cogency until there is a clear understanding as to how each species behaves when grown by itself in the absence of such competition. An effect presumed to result from competitive pressure cannot be assumed unless it is shown that this particular effect is absent in the controls. Thus it is necessary to review in detail each type of control population described in Table 1; to compare these as to species noting differences and similarities, and to make some statement about their respective variabilities. This will be developed as follows. By means of tables and graphs, populations of *T. confusum* in the three volumes are discussed first. Then, comparable data for *T. castaneum* are presented. Next, the two control groups, designated in Table 1 as "steriles," are reviewed along with a discussion of their significance for the study as a whole. Finally, the species differences between *T. confusum* and *T. castaneum* are statistically analyzed for the total period of observation and conclusions, required in the interpretation of the experimental populations, are formulated.

The basic data for all eight types of control populations are presented in Tables 2-9. Each table is similar in construction and the entries are computed in the same way, as can be illustrated by an examination of Table 2. The table lists age of the populations by 30 day intervals, these corresponding, of course, to census periods. Then, the actual number of larvae and pupae, imagoes, and their sum totals for every age is noted. These figures are derived from all replicates and expressed as mean number per gram of medium. When expressed in this way it is possible to compare directly all three volumes. The tables also contain two columns in which the larvae and pupae, and the imagoes, are reported in percentages. As will be evident later, these percentages add information of interest when *T. confusum* controls are contrasted with *T. castaneum*. The last column *(n)* lists the number of individual populations existing at any particular age, or the number of cases from which the means were computed.

TRIBOLIUM CONFUSUM CONTROLS (I-C-B, II-C-B, III-C-B)

The data for control populations of *T. confusum* in the three volumes are tabulated in Tables 2, 3, and 4 and graphed in Figure 1. This figure, plotting mean number per gram as a logarithmic ordinate against time in days as the abscissa, presents the curves for the I, II, and III series. Each series is represented by three components: total numbers, number of imagoes, and number of larvae plus pupae (hereafter written "larvae-pupae").

From the graphs and tables it is possible to draw certain conclusions about populations of *T. confusum* kept under the particular conditions outlined in the preceding section.

In the first place the cultures maintain themselves successfully under these conditions. After 1410 days

TABLE 2. Larvae and pupae, imagoes, and sum totals for **I-C-b** over 1380 days of observation. Data are expressed as mean percentage of larvae and pupae, and imagoes, per 30-day census period and as mean number of individuals per gram of medium. (N.B. Slight discrepancies in the first decimal place exist for certain of the percentages in Tables 2 through 9 owing to the fact that these were computed before the numbers were reduced to "per gram" units).

Age (days)	Larvae and Pupae		Imagoes		Sum M/gm.	n
	M/gm.	Per cent	M/gm.	Per cent		
30...	24.5	90.6	2.5	9.4	27.0	20
60...	4.1	16.8	20.5	83.2	24.6	20
90...	1.5	6.8	20.7	93.2	22.2	20
120...	1.2	5.5	20.1	94.5	21.3	20
150...	1.0	4.8	19.2	95.2	20.2	20
180...	1.4	7.4	17.6	92.6	19.0	20
210...	1.5	9.2	14.7	90.8	16.2	20
240...	4.6	27.1	12.5	72.9	17.1	20
270...	4.4	30.4	10.1	69.6	14.5	20
300...	7.6	46.4	8.8	53.6	16.4	20
330...	7.3	45.3	8.9	54.7	16.2	20
360...	8.0	47.6	8.8	52.4	16.8	20
390...	7.0	44.3	8.8	55.7	15.8	20
420...	5.9	40.7	8.6	59.3	14.5	20
450...	7.6	49.5	7.7	50.5	15.3	20
480...	6.5	44.1	8.2	55.9	14.7	20
510...	6.0	44.1	7.6	55.9	13.6	20
540...	7.9	52.1	7.3	47.9	15.2	20
570...	8.4	52.2	7.7	47.8	16.1	20
600...	7.3	47.0	8.3	53.0	15.6	20
630...	7.3	48.9	7.7	51.1	15.0	20
660...	7.2	50.2	7.1	49.8	14.3	20
690...	11.8	63.7	6.8	36.3	18.6	20
720...	11.6	62.2	7.0	37.8	18.6	20
750...	14.7	66.8	7.3	33.2	22.0	19
780...	13.3	64.6	7.3	35.4	20.6	19
810...	13.2	64.9	7.1	35.1	20.3	19
840...	12.5	64.4	6.9	35.6	19.4	19
870...	11.8	64.5	6.5	35.5	18.3	19
900...	14.9	70.2	6.3	29.8	21.2	19
930...	14.6	67.0	7.2	33.0	21.8	19
960...	13.0	61.4	8.2	38.6	21.2	19
990...	11.8	60.2	7.8	39.8	19.6	19
1020...	16.4	68.5	7.6	31.5	24.0	19
1050...	15.2	67.5	7.3	32.5	22.5	19
1080...	18.6	72.7	7.0	27.3	25.6	19
1110...	14.9	67.3	7.2	32.7	22.1	19
1140...	15.8	70.0	6.8	30.0	22.6	19
1170...	13.6	66.2	6.9	33.8	20.5	19
1200...	16.0	71.3	6.4	28.7	22.4	10
1230...	14.6	68.6	6.7	31.4	21.3	10
1260...	13.5	69.0	6.1	31.0	19.6	10
1290...	12.5	67.4	5.9	32.6	18.4	10
1320...	13.0	68.7	5.9	31.3	18.9	10
1350...	16.0	72.2	6.2	27.8	22.2	10
1380...	15.8	71.5	6.3	28.5	22.1	10

the populations are as healthy, when judged by their size and equilibria, as they were when three (or more) years younger. The reduction in number of replicates shown in the *n*-column of Tables 2 and 3 was voluntary and does not signify culture mortality.

Secondly, it is clear from Figure 1 that volume of medium imposes no differences in the pattern of growth and maintenance. The curves for I, II, and III are remarkably similar to each other along the time axis as is immediately evident from inspection of the graphs. The curves for Series III are not as smooth as are those for the smaller volumes. This

TABLE 3. Larvae and pupae, imagoes and sum totals for **II-C-b** over 1350 days of observation. Data are expressed as mean percentages of larvae and pupae and imagoes per 30-day census period and as mean number of individuals per gram of medium.

Age (days)	Larvae and Pupae		Imagoes		Sum M/gm.	*n*
	M/gm.	Per cent	M/gm.	Per cent		
30...	23.4	85.9	3.8	14.1	27.2	17
60...	4.2	15.2	23.5	84.8	27.7	17
90...	1.4	5.7	23.4	94.3	24.8	17
120...	0.6	2.7	22.8	97.3	23.4	17
150...	0.8	3.2	21.5	96.8	22.3	17
180...	1.3	6.0	19.4	94.0	20.7	17
210...	2.2	11.6	17.1	88.4	19.3	17
240...	3.9	21.5	14.1	78.5	18.0	17
270...	5.2	31.3	11.4	68.7	16.6	17
300...	6.2	39.2	9.6	60.8	15.8	17
330...	7.6	47.4	8.3	52.6	15.9	17
360...	6.9	47.1	7.9	52.9	14.8	17
390...	7.8	52.8	7.0	47.2	14.8	17
420...	8.2	55.0	6.7	45.0	14.9	17
450...	8.8	56.1	6.9	43.9	15.7	17
480...	8.8	57.4	6.6	42.6	15.4	17
510...	9.3	59.7	6.3	40.3	15.6	17
540...	9.3	59.0	6.4	41.0	15.7	17
570...	8.9	57.6	6.6	42.4	15.5	17
600...	7.3	51.8	6.8	48.2	14.1	17
630...	7.8	55.2	6.4	44.8	14.2	17
660...	9.5	61.1	6.1	38.9	15.6	17
690...	9.8	63.3	5.6	36.7	15.4	17
720...	11.3	68.8	5.1	31.2	16.4	17
750...	12.8	73.3	4.7	26.7	17.5	17
780...	12.0	73.0	4.4	27.0	16.4	17
810...	10.7	69.2	4.7	30.8	15.4	17
840...	9.9	67.1	4.9	32.9	14.8	17
870...	10.3	67.0	5.1	33.0	15.4	17
900...	10.9	68.6	5.0	31.4	15.9	17
930...	11.7	66.3	6.0	33.7	17.7	17
960...	10.9	63.0	6.4	37.0	17.3	17
990...	10.9	62.8	6.5	37.2	17.4	17
1020...	10.0	59.1	6.9	40.9	16.9	17
1050...	11.1	61.1	7.0	38.9	18.1	17
1080...	11.5	63.5	6.6	36.5	18.1	17
1110...	12.7	66.4	6.4	33.6	19.1	17
1140...	10.7	63.3	6.2	36.7	16.9	17
1170...	11.6	66.9	5.7	33.1	17.3	17
1200...	10.9	67.7	5.2	32.3	16.1	10
1230...	11.7	70.7	4.8	29.3	16.5	10
1260...	10.3	67.6	5.0	32.4	15.3	10
1290...	10.9	68.8	4.9	31.2	15.8	10
1320...	10.1	63.8	5.7	36.2	15.8	10
1350...	11.4	65.1	6.1	34.9	17.5	10

TABLE 4. Larvae and pupae, imagoes, and sum totals for **III-C-b** over 1410 days of observation. Data are expressed as mean percentage of larvae and pupae and imagoes per 30-day census period and as mean number of individuals per gram of medium.

Age (days)	Larvae and Pupae		Imagoes		Sum M/gm.	*n*
	M/gm.	Per cent	M/gm.	Per cent		
30...	22.0	89.5	2.6	10.5	24.6	2
60...	2.2	9.4	21.1	90.6	23.3	2
90...	0.6	2.6	20.9	97.4	21.5	2
120...	0.7	3.2	20.6	96.8	21.3	2
150...	0.7	3.5	19.1	96.5	19.8	2
180...	1.9	10.6	16.3	89.4	18.2	2
210...	2.6	16.4	13.3	83.6	15.9	2
240...	4.3	30.3	9.8	69.7	14.1	2
270...	4.4	37.8	7.3	62.2	11.7	2
300...	5.8	52.3	5.3	47.7	11.1	2
330...	4.6	45.9	5.4	54.1	10.0	2
360...	9.2	55.8	7.3	44.2	16.5	2
390...	7.2	51.6	6.8	48.4	14.0	2
420...	11.2	66.3	5.7	33.7	16.9	2
450...	9.3	61.0	6.0	39.0	15.3	2
480...	6.7	55.7	5.3	44.3	12.0	2
510...	8.1	64.7	4.4	35.3	12.5	2
540...	5.8	60.9	3.7	39.1	9.5	2
570...	8.6	66.7	4.3	33.3	12.9	2
600...	6.9	55.1	5.6	44.9	12.5	2
630...	5.9	52.5	5.3	47.5	11.2	2
660...	7.5	61.9	4.6	38.1	12.1	2
690...	8.4	67.8	4.0	32.2	12.4	2
720...	8.2	66.5	4.1	33.5	12.3	2
750...	11.3	66.5	5.7	33.5	17.0	2
780...	11.1	69.9	4.8	30.1	15.9	2
810...	10.5	70.1	4.5	29.9	15.0	2
840...	11.6	74.9	3.9	25.1	15.5	2
870...	11.6	75.2	3.8	24.8	15.4	2
900...	15.2	82.3	3.3	17.7	18.5	2
930...	12.0	76.7	3.6	23.3	15.6	2
960...	11.4	73.1	4.2	26.9	15.6	2
990...	8.1	70.2	3.5	29.8	11.6	2
1020...	9.2	73.6	3.3	26.4	12.5	2
1050...	8.0	68.4	3.7	31.6	11.7	2
1080...	12.2	77.9	3.5	22.1	15.7	2
1110...	10.1	75.2	3.3	24.8	13.4	2
1140...	14.3	81.9	3.2	18.1	17.5	2
1170...	11.4	76.3	3.6	23.7	15.0	2
1200...	11.1	77.6	3.2	22.4	14.3	2
1230...	13.3	80.9	3.1	19.1	16.4	2
1260...	10.6	70.3	4.5	29.7	15.1	2
1290...	12.0	74.1	4.2	25.9	16.2	2
1320...	9.2	71.8	3.6	28.2	12.8	2
1350...	10.1	75.9	3.2	24.1	13.3	2
1380...	9.8	75.7	3.2	24.3	13.0	2
1410...	9.8	73.8	3.5	26.2	13.3	2

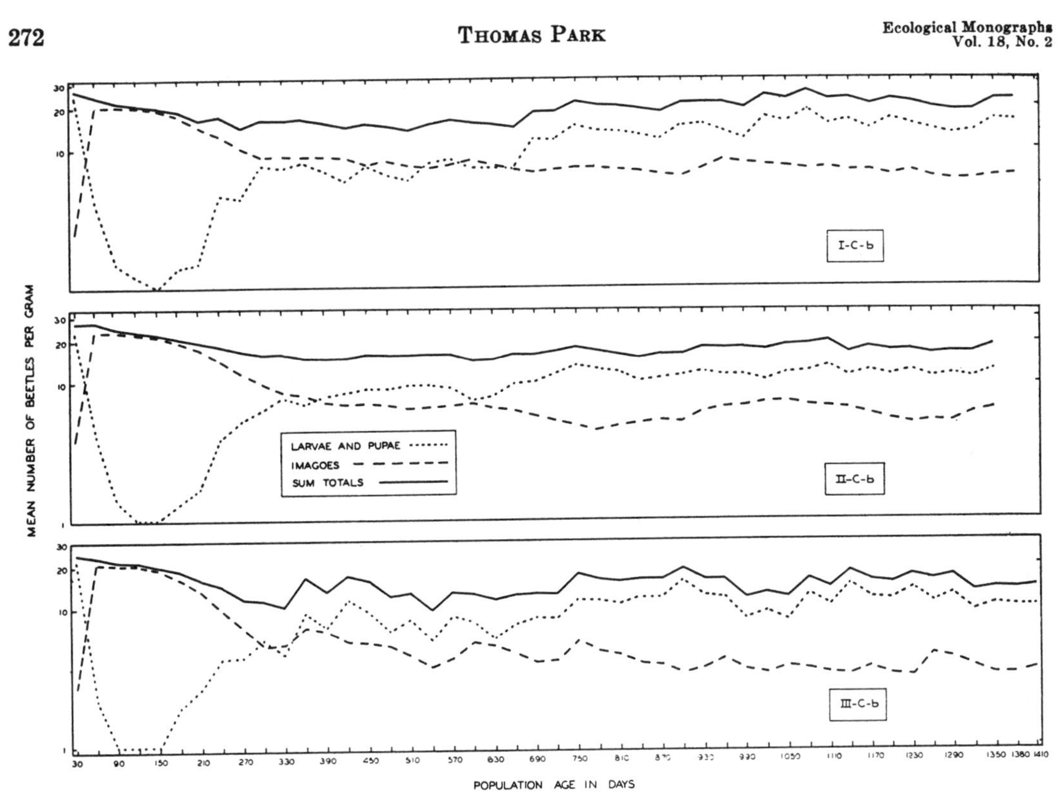

FIG. 1. Census history of I-C-b, II-C-b, and III-C-b populations *(Tribolium confusum)*. Mean number of beetles per gram as a logarithmic ordinate is plotted against age in days as the abscissa. Shown are larval-pupal, imaginal, and total population curves (see box).

undoubtedly reflects merely a small-sample effect—it was possible to run only two replicates in all the 80 gram cultures because of their great total size.

Although the pattern is similar between series, there is a tendency for mean total density to be inversely proportional to volume. This will be discussed later when statistical comparisons with *T. castaneum* and the "sterile" populations are introduced. Suffice it to say at this point that the mean total number per gram for Series I is 19.25, while for Series II and III these means are 17.36 and 14.94, respectively (see Table 10). The mean differences in relation to their respective probable errors are all highly significant.

When Figure I is examined with attention focussed on the first 300 days of growth, it is evident that there is exhibited by the larval-pupal and imaginal curves a characteristic "pre-equilibrium period" for all three volumes through which the populations pass before stabilization takes place. During this interval the total population declines somewhat but remains quite stable relative to its component curves. At the first census (30 days) the larval population is large, of course, because there has not been time to produce the first generation of imagoes. Only those imagoes introduced when the cultures were started and still alive are present at this count. By the second census (60 days) many imagoes have emerged and these uniformly young beetles constitute the bulk of the populations: Table 2 lists the number of larvae-pupae and imagoes to be 4.1 and 20.5, respectively, or 17 and 83%. A glance at Tables 3 and 4 shows that the figures are comparable for II-C-b and III-C-b. The number of adult Tribolium remains high during the next 90 days (i.e., to about day-150). These saturate their environment to a large extent and keep down the larval density probably through egg cannibalism, through crowding, and perhaps through other agencies. As these imagoes get older, their death-rate rises and they are reduced in actual numbers as well as in terms of percentages. This is clearly seen in Figure 1. There, the imago populations of all three series begin to decline after day-150 —a decline accompanied by a parallel increase of larvae. This sequence of events is perhaps somewhat oversimplified, but it will be extended shortly. From approximately day-300 to day-660 the larvae-pupae and imagoes are present in closely similar densities. This is more evident in I and II than in III. Thereafter, the larval-pupal curves rise above those for the adults and both remain essentially stabilized for the two year period remaining.

Finally, it is important for the reader to realize that, although the curves we have been discussing are based on averages, they do have validity as descriptions of events within *individual* populations. This is suggested, of course, by the similarity in pattern between the I, II, and III series. However, more direct evidence was accumulated by graphing single populations one-by-one and then observing that this assemblage of curves conforms closely with the means depicted in Figure 1. As a study of population variability, it would be interesting to publish these figures; but owing to space limitations, this is impracticable although something of the sort may appear at a later date.

TRIBOLIUM CASTANEUM CONTROLS (I-C-c, II-C-c, III-C-c)

The data for control populations of *T. castaneum*

TABLE 5. Larvae and pupae, imagoes, and sum totals for **I-C-c** over 1380 days of observation. Data are expressed as mean percentage of larvae and pupae and imagoes per 30-day census period and as mean number of individuals per gram of medium.

Age (days)	LARVAE AND PUPAE		IMAGOES		Sum M/gm.	*n*
	M/gm.	Per cent	M/gm.	Per cent		
30...	17.1	77.5	5.0	22.5	22.1	20
60...	9.3	36.0	16.6	64.0	25.9	20
90...	12.4	42.4	16.9	57.6	29.3	20
120...	12.3	43.4	16.1	56.6	28.4	20
150...	9.9	41.1	14.3	58.9	24.2	20
180...	6.8	36.2	12.1	63.8	18.9	20
210...	4.6	37.0	7.8	63.0	12.4	20
240...	7.5	59.8	5.1	40.2	12.6	20
270...	4.5	66.4	2.3	33.6	6.8	20
300...	6.2	79.8	1.6	20.2	7.8	19
330...	10.7	80.1	2.7	19.9	13.4	18
360...	6.5	66.2	3.3	33.8	9.8	18
390...	8.5	70.4	3.6	29.6	12.1	16
420...	7.9	67.8	3.8	32.2	11.7	16
450...	9.4	71.8	3.7	28.2	13.1	16
480...	7.5	69.2	3.4	30.8	10.9	15
510...	6.8	71.1	2.8	28.9	9.6	15
540...	7.0	68.3	3.2	31.7	10.2	14
570...	7.8	70.3	3.3	29.7	11.1	13
600...	6.9	69.9	3.0	30.1	9.9	13
630...	9.2	76.0	2.9	24.0	12.1	13
660...	7.5	73.1	2.8	26.9	10.3	13
690...	10.1	80.0	2.5	20.0	12.6	13
720...	10.1	77.2	3.0	22.8	13.1	13
750...	11.6	79.5	3.0	20.5	14.6	13
780...	11.0	79.4	2.9	20.6	13.9	13
810...	11.4	79.3	3.0	20.7	14.4	13
840...	10.5	78.6	2.9	21.4	13.4	13
870...	6.7	73.8	2.4	26.2	9.1	13
900...	8.0	79.6	2.1	20.4	10.1	13
930...	10.6	82.7	2.2	17.3	12.8	13
960...	10.6	78.4	2.9	21.6	13.5	13
990...	12.4	82.6	2.6	17.4	15.0	13
1020...	10.8	78.3	3.0	21.7	13.8	13
1050...	9.8	78.0	2.7	22.0	12.5	13
1080...	10.9	81.0	2.6	19.0	13.5	13
1110...	8.3	77.6	2.4	22.4	10.7	13
1140...	8.9	80.7	2.1	19.3	11.0	13
1170...	9.2	76.4	2.8	23.6	12.0	13
1200...	8.2	77.2	2.4	22.8	10.6	13
1230...	7.4	75.9	2.3	24.1	9.7	13
1260...	7.4	76.0	2.3	24.0	9.7	13
1290...	8.0	76.1	2.5	23.9	10.5	13
1320...	7.6	73.8	2.7	26.2	10.3	12
1350...	8.1	74.0	2.8	26.0	10.9	12
1380...	9.0	78.6	2.5	21.4	11.5	12

TABLE 6. Larvae and pupae, imagoes, and sum totals for **II-C-c** over 1380 days of observation. Data are expressed as mean percentage of larvae and pupae and imagoes per 30-day census period and as mean number of individuals per gram of medium.

Age (days)	LARVAE AND PUPAE		IMAGOES		Sum M/gm.	*n*
	M/gm.	Per cent	M/gm.	Per cent		
30...	13.0	55.5	10.4	44.5	23.4	18
60...	9.1	32.7	18.6	67.3	27.7	18
90...	11.4	38.2	18.6	61.8	30.0	18
120...	9.2	35.4	16.9	64.6	26.1	18
150...	8.2	38.5	13.2	61.5	21.4	18
180...	6.9	43.0	9.1	57.0	16.0	18
210...	3.2	42.6	4.4	57.4	7.6	18
240...	8.1	75.3	2.7	24.7	10.8	18
270...	5.3	76.8	1.6	23.2	6.9	18
300...	6.2	82.4	1.4	17.6	7.6	18
330...	7.7	82.2	1.7	17.8	9.4	18
360...	8.4	79.5	2.2	20.5	10.6	18
390...	8.6	78.9	2.3	21.1	10.9	18
420...	8.1	80.6	1.9	19.4	10.0	18
450...	8.6	81.6	1.9	18.4	10.5	18
480...	7.0	79.8	1.8	20.2	8.8	18
510...	7.0	80.7	1.7	19.3	8.7	18
540...	7.2	79.0	1.9	21.0	9.1	18
570...	6.7	77.9	1.9	22.1	8.6	18
600...	7.1	80.3	1.8	19.7	8.9	18
630...	6.9	81.0	1.6	19.0	8.5	18
660...	5.6	78.1	1.5	21.9	7.1	18
690...	6.9	84.0	1.3	16.0	8.2	18
720...	6.6	83.1	1.3	16.9	7.9	18
750...	7.7	83.9	1.5	16.1	9.2	18
780...	7.9	83.4	1.6	16.6	9.5	18
810...	7.1	82.6	1.5	17.4	8.6	18
840...	5.7	79.4	1.5	20.6	7.2	18
870...	6.2	77.9	1.8	22.1	8.0	18
900...	7.9	81.8	1.8	18.2	9.7	18
930...	9.1	80.7	2.1	19.3	11.2	18
960...	8.8	80.7	2.1	19.3	10.9	18
990...	7.5	81.4	1.7	18.6	9.2	18
1020...	8.2	82.8	1.7	17.2	9.9	18
1050...	7.6	82.3	1.6	17.7	9.2	18
1080...	7.3	82.5	1.6	17.5	8.9	18
1110...	6.7	82.0	1.5	18.0	8.2	18
1140...	6.6	81.8	1.5	18.2	8.1	18
1170...	7.9	82.2	1.7	17.8	9.6	18
1200...	8.3	81.9	1.8	18.1	10.1	18
1230...	8.4	79.6	2.2	20.4	10.6	18
1260...	8.4	78.9	2.2	21.1	10.6	18
1290...	8.8	80.2	2.2	19.8	11.0	18
1320...	9.0	81.1	2.1	18.9	11.1	18
1350...	8.1	80.7	2.0	19.3	10.1	18
1380...	9.1	83.1	1.8	16.9	10.9	18

grown in the three volumes are tabulated in Tables 5, 6, and 7 and graphed in Figure 2, using precisely the same methods of representation as those employed for *T. confusum.*

There is one difference between *T. castaneum* and *T. confusum* in terms of population survival that merits comment. As just reported, no cultures of *T. confusum* became extinct over the entire four-year period. This is equally true for the II and III series of *T. castaneum.* For the eight-gram series (I), however, there were the following eight exceptions out of twenty: one population became extinct at 300 days; one at 330 days; two at 390 days; one at 480 days; one at 540 days; and one at 1320 days. There can be no reasonable doubt but that the failure of these cultures was caused by an added mortality increment resulting from parasitic infection. This merely requires mention here because it will receive further attention shortly.

TABLE 7. Larvae and pupae, imagoes, and sum totals for **III-C-c** over 1410 days of observation. Data are expressed as mean percentage of larvae and pupae and imagoes per 30-day census period and as mean number of individuals per gram of medium.

Age (days)	LARVAE AND PUPAE		IMAGOES		Sum M/gm.	*n*
	M/gm.	Per cent	M/gm.	Per cent		
30...	14.9	63.3	8.6	36.7	23.5	2
60...	7.9	30.4	18.1	69.6	26.0	2
90...	12.9	42.3	17.6	57.7	30.5	2
120...	14.3	46.2	16.7	53.8	31.0	2
150...	12.6	45.5	15.1	54.5	27.7	2
180...	8.1	41.3	11.6	58.7	19.7	2
210...	6.7	48.2	7.2	51.8	13.9	2
240...	11.5	76.6	3.5	23.4	15.0	2
270...	6.1	74.4	2.1	25.6	8.2	2
300...	12.2	91.4	1.2	8.6	13.4	2
330...	12.1	83.7	2.3	16.3	14.4	2
360...	16.6	87.2	2.4	12.8	19.0	2
390...	12.0	80.9	2.8	19.1	14.8	2
420...	18.1	87.9	2.5	12.1	20.6	2
450...	14.8	84.2	2.8	15.8	17.6	2
480...	10.4	79.9	2.6	20.1	13.0	2
510...	16.9	87.4	2.4	12.6	19.3	2
540...	10.6	80.8	2.5	19.2	13.1	2
570...	12.4	84.4	2.3	15.6	14.7	2
600...	12.4	83.7	2.4	16.3	14.8	2
630...	9.6	81.7	2.2	18.3	11.8	2
660...	10.6	81.0	2.5	19.0	13.1	2
690...	11.6	83.6	2.3	16.4	13.9	2
720...	11.2	82.0	2.5	18.0	13.7	2
750...	10.0	80.7	2.4	19.3	12.4	2
780...	12.0	79.7	3.1	20.3	15.1	2
810...	14.6	79.1	3.9	20.9	18.5	2
840...	9.9	70.0	4.3	30.0	14.2	2
870...	9.4	68.0	4.4	32.0	13.8	2
900...	6.6	59.1	4.6	40.9	11.2	2
930...	7.2	60.7	4.7	39.3	11.9	2
960...	6.9	57.4	5.1	42.6	12.0	2
990...	3.7	42.9	5.0	57.1	8.7	2
1020...	9.0	66.7	4.5	33.3	13.5	2
1050...	10.4	69.6	4.5	30.4	14.9	2
1080...	12.6	77.1	3.7	22.9	16.3	2
1110...	9.0	75.8	2.9	24.2	11.9	2
1140...	7.9	79.0	2.1	21.0	10.0	2
1170...	7.4	82.4	1.6	17.6	9.0	2
1200...	3.5	81.6	0.8	18.4	4.3	2
1230...	6.5	88.8	0.8	11.2	7.3	2
1260...	11.2	79.6	2.9	20.4	14.1	2
1290...	9.1	63.5	5.2	36.5	14.3	2
1320...	9.0	65.7	4.7	34.3	13.7	2
1350...	9.2	66.0	4.7	34.0	13.9	2
1380...	10.3	69.4	4.5	30.6	14.8	2
1410...	10.1	71.3	4.1	28.7	14.2	2

Figure 2 makes a point similar to that made by Figure 1, namely: that volume does not affect the pattern of the curves. Under the conditions of this study, populations of *T. castaneum* describe a definite pattern of growth. This pattern is clearly similar for all three series, both in respect of performance by individual populations as well as for averages. The III-C-c curves, based on two replicates, are, again, more erratic than those of I-C-c and II-C-c. There are also relations between mean total density and volume. Over the entire period of observation these means are as follows: I, 13.30, II, 11.23, and III, 15.8 (Table 10). The relation between I and II is in the same direction as that reported for *T. confusum.* There is an exception in that, for *T. castaneum,* the 80 gram volumes (III) produce the largest total populations.

Tribolium castaneum populations also exhibit a pre-equilibrium period as is clearly seen in Figure 2. This period, although diverging in some respects from that shown in Figure 1, is as consistent and characteristic for *T. castaneum* as was the comparable phenomenon just described for *T. confusum.* In both cases, the duration of the period is about the same (approximately 300-330 days) and there are some similarities between the imago curves. There are real differences as well, and a brief recounting of these is appropriate. In the first place, the total population size for *T. castaneum* for all series rises to a maximum during the 90-120 day interval and then falls to assume relative stability from about day-330 on. The imago curves attain their maxima at day-60 but thereafter drop sharply until day-300 when they are lowest. At this time there are only 1.6, 1.4, and 1.2 adults per gram for I, II, and III, respectively. Analogous figures at the same census for *T. confusum* are 8.8, 9.6, and 5.3. The number of imagoes rises somewhat thereafter and remains quite stable over the three remaining years with the exception of a curious drop in III-C-c most pronounced at day-1200. The larval-pupal curves of Figure 2 do not decline so much as those graphed in Figure 1, and their eventual stabilization is at a higher actual, and percentage, level.

Probably the two major differences in terms of growth pattern between the two species when viewed over the entire period of observation lie in the facts (a) that *T. castaneum* maintains itself with a smaller

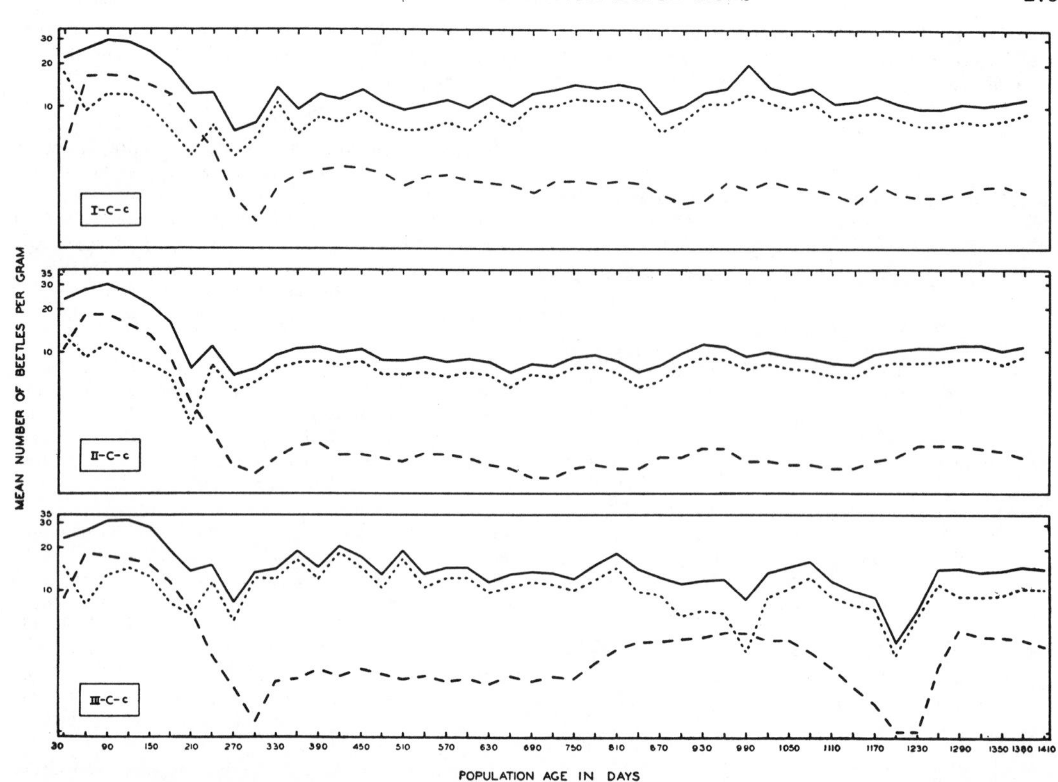

FIG. 2. Census history of I-C-c, II-C-c, and III-C-c populations *(Tribolium castaneum)*. Mean number of beetles per gram as a logarithmic ordinate is plotted against age in days as the abscissa. Shown are larval-pupal, imaginal, and total population curves (see box).

number of imagoes, both absolute and relative, and (b) that a depression in total numbers characteristically is seen at day-270 or day-300 which is not displayed by *T. confusum*.

TRIBOLIUM CONFUSUM AND T. CASTANEUM STERILE CONTROLS (I-C-B-S AND I-C-C-S)

In the section dealing with experimental design, and in the discussion of control cultures, the point has been made that both species of Tribolium are infected with a coccidian parasite, Adelina. Also, the inference has been advanced that this infection is related to the population behavior of the beetles and therefore must be considered in an interpretation of the findings. Chapman (1933) and Park, Gregg, and Lutherman (1941) concluded that Adelina infection could alter the course of growth and density of *T. confusum* populations. This, plus the finding of dead larvae and pupae in the siftings when censuses were taken, suggested that the phenomenon should be explored experimentally, thus leading to the establishment of the sterile control and experimental series reported upon in this paper.

Before discussing the actual data, however, it is in order to briefly review certain matters pertaining to this parasite. According to Wenyon (1926) the genus Adelina was founded by Hesse in 1911 to describe an organism parasitic within the oligochaete worm *Slavinia appendiculata*. In 1922 Riley & Krogh discussed a "new species" of coccidian that abundantly infected Tribolium but they did not identify it. In 1923 White mentioned that the fat bodies of both *T. confusum* and *T. castaneum* are invaded by a neosporidian parasite with the usual consequence that the larvae become "sick," distended, partially immobilized, and frequently die. He further noted that pupae and imagoes may also die from infection. White observed that mortality did not appear to any degree during the first month of a population's life-history but that "after a few months" deaths owing to this cause were plentiful. Chapman (1933) and Park (1934; on the basis of studies carried out with *T. confusum* in collaboration with Professor Clay G Huff) independently concluded that this parasite belonged to the genus Adelina. Bhatia (1937) worked with Adelina inhabiting *T. ferrugineum* (= *T. castaneum*) and described the species as *Adelina tribolii.* He also reported some data on its life-cycle, infectivity, and cellular morphology. Yarwood (1937) discusses *A. cryptocerci* whose host is the roach *Cryptocercus punctulatus.* Although her paper is not concerned with flour beetles, it does afford the best general account of the life-cycle of the genus and is useful for this reason. To date, it has not been critically determined, either by cytological evidence or cross-infection studies, whether *T. confusum* and *T. castaneum* are infected by the same or different species of Adelina. It is a reasonable working presumption that the parasite is taxonomically the same for both hosts. Studies designed to clarify the matter are now under way in this laboratory.

As well as it is understood, the life-cycle of Adelina in Tribolium follows this course. Infection is brought about when the larvae and imagoes ingest mature oöcysts. The oöcysts are taken into the gut as the beetles feed on flour containing them and/or as they feed on living or dead larvae, pupae, and imagoes that are themselves infected. In the mesenteron the oöcysts rupture and liberate motile sporozoites. These penetrate the epithelium passing through it to the haemocoels. The sporozoites invade the host cells, including those of the fat body, and undergo asexual reproduction or schizogony by which process merozoites are produced. These also are motile and further attack the tissues. Apparently schizogony may be repeated a number of times. Eventually sexual reproduction (gametogony) is initiated and gametoblasts differentiate into male and female sex cells. A zygote, formed upon fertilization, starts sporogony—the entire process occurring within a relatively heavy cyst wall. A number of sporoblasts differentiate within each cyst, and these later divide within themselves to form coiled, encased sporozoites. The end-result of the life-cycle is the production in several tissues of many mature oöcysts, each surrounded by a membrane and containing sporozoites. As mentioned above, these are the infective stages, but they do not become infective unless they get out of the first host and into another Tribolium. Since the oöcysts are inactive, they cannot escape through their own activity. Presumably they are liberated only when a beetle dies and breaks apart as typically happens in cultures, or, when such a beetle is cannibalized by its fellows. The nature of the pathology leading to host mortality is not understood. It is known that a beetle can harbor many oöcysts and still remain alive.

Adelina may be readily removed from Tribolium by taking advantage of the fact that the eggs of the latter are never internally parasitized. It is only necessary to separate the oöcysts from the egg membrane, allow the larvae to hatch, and then rear a stock in fresh medium sterilized by heat and prevented from becoming re-infected. By washing the eggs in soap solution to which mercuric chloride (1 : 1,000) is added, the oöcysts are either killed or slough off. This is followed by successive washings in sterile water, the eggs are permitted to dry, and then set aside in covered Petri dishes to await hatching. The young larvae are immediately transferred to the medium. Histological examination has shown this to be an effective method of ridding the beetles of the parasite (Park & Burrows, 1942). It is difficult to maintain Tribolium populations free of Adelina for long periods of time. Despite many tedious precautions (e.g., a separate laboratory and incubator, separate instruments and laboratory clothing, more complicated routines involving frequent heat sterilizations, etc.) some of the cultures eventually become infected. Fortunately, as will be seen later, this did not have serious consequences.

All of the populations designated as "sterile" were maintained in the smallest volume of medium (I); a procedure that reduces time and labor because the total counts relative to the other series are smaller. Since, as has already been seen for both species, volumes II and III diverge so little from I, such a limitation of the study seems legitimate as well as expedient.

The basic data for *Tribolium confusum* sterile con-

TABLE 8. Larvae and pupae, imagoes, and sum totals for **I-C-b-S** over 1140 days of observation. Data are expressed as mean percentage of larvae and pupae and imagoes per 30-day census period and as mean number of individuals per gram of medium.

Age (days)	LARVAE AND PUPAE		IMAGOES		Sum M/gm.	n
	M/gm.	Per cent	M/gm.	Per cent		
30...	25.0	95.6	1.1	4.4	26.1	20
60...	5.0	17.8	23.2	82.2	28.2	20
90...	1.5	6.1	23.6	93.9	25.1	20
120...	2.1	8.3	22.9	91.7	25.0	20
150...	1.2	5.1	21.7	94.9	22.9	20
180...	2.2	10.0	20.0	90.0	22.2	20
210...	2.0	10.6	16.7	89.4	18.7	20
240...	3.6	20.9	13.8	79.1	17.4	20
270...	3.2	22.9	10.8	77.1	14.0	20
300...	4.9	34.3	9.4	65.7	14.3	20
330...	4.8	33.3	9.6	66.7	14.4	20
360...	4.2	30.4	9.6	69.6	13.8	20
390...	7.2	42.5	9.7	57.5	16.9	17
420...	4.3	30.6	9.9	69.4	14.2	17
450...	5.4	36.6	9.5	63.4	14.9	17
480...	8.4	47.1	9.5	52.9	17.9	17
510...	10.8	50.4	10.5	49.6	21.3	17
540...	10.7	49.5	11.0	50.5	21.7	17
570...	7.8	34.8	14.5	65.2	22.3	17
600...	3.9	21.9	14.0	78.1	17.9	17
630...	5.4	30.8	12.0	69.2	17.4	17
660...	5.8	33.2	11.5	66.8	17.3	16
690...	7.2	39.6	11.0	60.4	18.2	16
720...	5.7	32.4	12.0	67.6	17.7	16
750...	12.0	47.2	13.3	52.8	25.3	16
780...	6.2	28.6	15.6	71.4	21.8	16
810...	4.4	23.2	14.6	76.8	19.0	16
840...	7.5	36.0	13.3	64.0	20.8	16
870...	5.3	29.3	12.6	70.7	17.9	16
900...	6.7	37.7	11.1	62.3	17.8	16
930...	5.6	34.6	10.5	65.4	16.1	15
960...	10.2	50.2	10.1	49.8	20.3	15
990...	3.5	21.9	12.4	78.1	15.9	15
1020...	6.1	36.9	10.6	63.1	16.7	15
1050...	6.2	37.0	10.5	63.0	16.7	15
1080...	6.0	36.5	10.5	63.5	16.5	15
1110...	6.2	37.2	10.5	62.8	16.7	15
1140...	7.9	42.6	10.7	57.4	18.6	15

trols (I-C-b-S) appear in Table 8 and are graphed as mean number per gram in Figure 3 in which the non-sterile control also is included for purposes of comparison. Figure 4 plots total size only for all series I groups. The *T. castaneum* data appear in Table 9 and Figure 5 and are similarly presented (I-C-c-S). The discussion will be developed by contrasting infected *T. confusum* with sterile *T. confusum* populations to be followed by a like treatment for populations of *T. castaneum*. Comparisons between the two species relative to the presence and absence of Adelina is then attempted.

It is possible to quickly summarize the differences and similarities between I-C-b and I-C-b-S by reference to Tables 2 and 8 and Figures 3 and 4.

In the first place, none of the individual cultures

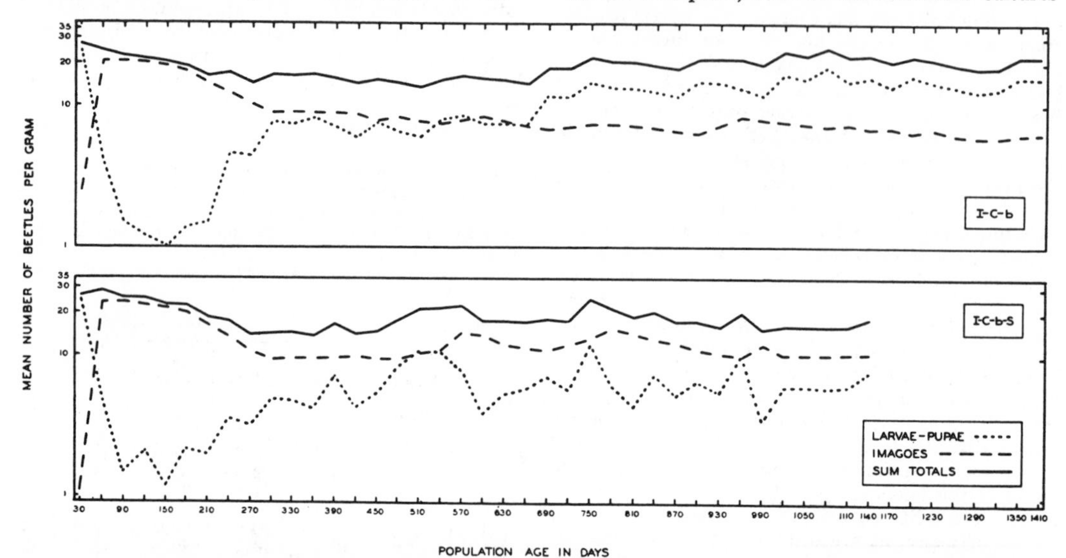

FIG. 3. Census history of *Tribolium confusum* infected controls (I-C-b) compared with *T. confusum* sterile controls (I-C-b-S). Shown are larval-pupal, imaginal and total population curves (see box). Logarithmic ordinate; arithmetical abscissa.

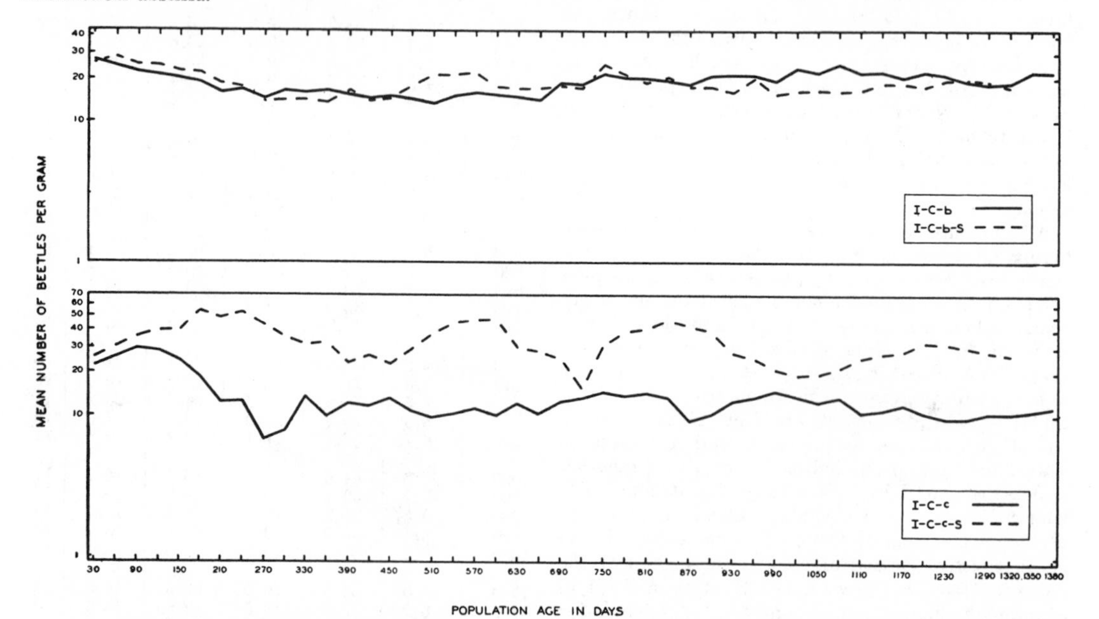

FIG. 4. Census history, in terms of total population size only, of *T. confusum* infected compared with *T. confusum* sterile (I-C-b cf. I-C-b-S) and *T. castaneum* infected compared with *T. castaneum* sterile (I-C-c cf. I-C-c-S). The reduced density of I-C-c-S shown at day-720 is caused by inadvertent low temperature as explained in text. Logarithmic ordinate; arithmetical abscissa.

became extinct over the entire period of observation. The reduction in the *n*-column of Table 8 subsequent to day-360 signifies inadvertent infection, not extinction.

In the second place, it is evident that total population size is unaffected by Adelina. This can be readily seen in Figures 3 and 4 or by comparison of the mean per gram columns of Tables 2 and 8. The divergence between the two curves along the time axis is accounted for by accidents of sampling. When the means for the entire period of observation are computed, they have the values 19.25 and 18.94 for I-C-b and I-C-b-S, respectively (Table 10). The mean difference is 0.31 with P., exceeding 50 per cent.

The third point is that a pre-equilibrium period is passed through by the sterile cultures also. This period is essentially similar to that displayed by the infected populations with the exception that the rise of the larval-pupal curve after day-210 is less pronounced.

The fourth point involves a fundamental difference between the two types of cultures rather than a similarity. The sterile *T. confusum* populations maintain themselves with a significantly greater number of imagoes, and a significantly smaller number of larvae-pupae, than do the infected populations both in terms of means per gram and percentages. This is especially evident after day-240. It leads to the speculation that something more than half a year is required before infection of *T. confusum* by Adelina, at least under these conditions of husbandry, reaches a degree sufficient to cause the infected cultures to deviate consistently from the steriles in terms of their composition by stages. The most probable interpretation of this deviation between the I-C-b and I-C-b-S series is, that in the absence of Adelina, there are fewer deaths in the late larval and pupal periods with the consequence that more imagoes emerge. It is interesting to note that the total size of the sterile populations, despite the fact that they contain more reproducers (imagoes), does not increase and exceed that of infected cultures (Fig. 4). As has been shown, this clearly does not happen. In explanation of this, one hypothesis is that some sort of balance exists between food supply and mean total population beyond the upper limit of which the cultures do not go. Another hypothesis is that the total *net*-reproduction of the sterile groups actually is not higher than that of the infected even though more adults are present. This heightened imago density could reduce the average rate of fecundity per female per unit of time, or increase the rate of egg cannibalism, or both. These hypotheses, and others that could be advanced, lend themselves to experimental analysis and verification—an analysis that should be forthcoming at a later date as the entire program on inter-species competition matures.

Up to this point, an impression of essential similarity between control cultures may have been established in the mind of the reader. Such similarities prevail in several directions: in terms of volume of medium, composition by stages, pattern of growth with time, equilibria, presence or absence of parasitization, and, to a considerable extent, even between species. This is not to say that important differences do not exist as well. They do, and certain of them have been described. But probably the general conception is more one of conformity than of divergence. When Adelina is removed from populations of *T. castaneum*, however, this conception is altered considerably. The I-C-c-S cultures, while consistent within themselves as a group of replicates, differ both quantitatively and qualitatively from all other controls. Because this phenomenon is interesting in its own right from the viewpoint of epidemiology, and

TABLE 9. Larvae and pupae, imagoes, and sum totals for **I-C-c-S** over 1140 days of observation. Data are expressed as mean percentage of larvae and pupae and imagoes per 30-day census period and as mean number of individuals per gram of medium.

Age (days)	LARVAE AND PUPAE		IMAGOES		Sum M/gm.	*n*
	M/gm.	Per cent	M/gm.	Per cent		
30...	22.2	86.8	3.4	13.2	25.6	20
60...	10.9	36.1	19.3	63.9	30.2	20
90...	15.6	44.3	19.6	55.7	35.2	20
120...	18.7	47.1	21.1	52.9	39.8	20
150...	18.0	45.2	21.8	54.8	39.8	20
180...	32.0	58.9	22.3	41.1	54.3	19
210...	20.8	42.8	27.7	57.2	48.5	18
240...	23.5	46.8	26.8	53.2	50.3	18
270...	18.6	43.4	24.3	56.6	42.9	18
300...	15.1	43.1	20.0	56.9	35.1	18
330...	14.7	46.6	16.9	53.4	31.6	18
360...	17.0	52.2	15.6	47.8	32.6	18
390...	12.5	53.6	10.8	46.4	23.3	18
420...	15.4	57.6	11.3	42.4	26.7	18
450...	11.5	49.9	11.5	50.1	23.0	18
480...	17.6	60.6	11.4	39.4	29.0	18
510...	25.3	66.9	12.6	33.1	37.9	18
540...	26.3	59.2	18.1	40.8	44.4	18
570...	25.8	55.7	20.6	44.3	46.4	18
600...	25.9	54.8	21.3	45.2	47.2	18
630...	12.2	40.5	17.9	59.5	30.1	17
660...	11.9	42.1	16.5	57.9	28.4	17
690...	10.6	41.9	14.7	58.1	25.3	17
720*..	6.0	39.1	9.4	60.9	15.4	17
750...	23.2	72.2	8.9	27.8	32.1	17
780...	22.9	60.1	15.3	39.9	38.2	17
810...	23.3	57.5	17.2	42.5	40.5	17
840...	27.6	58.4	19.6	41.6	47.2	16
870...	22.4	52.8	20.1	47.2	42.5	14
900...	20.2	50.8	19.6	49.2	39.8	13
930...	10.9	39.1	17.0	60.9	27.9	13
960...	11.9	46.5	13.7	53.5	25.6	11
990...	10.5	49.5	10.6	50.5	21.1	11
1020...	10.9	56.0	8.6	44.0	19.5	10
1050...	12.3	62.0	7.5	38.0	19.8	9
1080...	14.3	65.4	7.6	34.6	21.9	9
1110...	15.8	61.1	10.1	38.9	25.9	9
1140...	15.8	57.5	11.6	42.5	27.4	9

* See discussion of this census in text.

because it has an interpretative bearing on the problem of interspecies competition to be reviewed later, it is advisable to discuss the sterile *T. castaneum* controls in some detail, paying particular attention to them in comparison with infected populations of the same species.

From inspection of Figures 4 and 5 and certain of the tables, the following points are to be noted:

(1) None of the I-C-c-S cultures became extinct. As was true for I-C-b-S, some do acquire the parasite and these are then dropped from the computations. This accounts for the reduction in the *n*-column of Table 9; a greater reduction than that evident for *T. confusum*.

(2) In terms of total size, the sterile populations are larger than the infected at every point along the time axis. At the period when the two curves are closest together (720 days) the mean difference between these two points is 15.4 − 13.1 = 2.3.[2] Typically, the curves are much farther apart as is evident in the following tabulation showing differences (I-C-c-S minus I-C-c) arbitrarily selected at intervals of 150 days:

Day 90: 35.2 — 29.3 = 5.9 ± 1.74
Day 240: 50.3 — 12.6 = 37.7 ± 3.95
Day 390: 23.3 — 12.1 = 11.2 ± 1.54
Day 540: 44.4 — 10.2 = 34.2 ± 2.54
Day 690: 25.3 — 12.6 = 12.7 ± 1.82
Day 840: 47.2 — 13.4 = 33.8 ± 2.27
Day 990: 21.1 — 15.0 = 6.1 ± 1.64
Day 1140: 27.4 — 11.0 = 16.4 ± 1.42

These differences are all highly significant in relation to their probable errors.

For the entire period of study the mean range (maximum to minimum) for the steriles is 54.3 to 19.5; for the infected, 29.3 to 9.1. The actual differences between these ranges in terms of numbers per gram are 34.8 and 20.2. For the sterile cultures, the minimum is 36 per cent of the maximum; for the infected, 31% of the maximum.

Over 1140 days of observation for the steriles, and 1380 days of observation for the infected, the mean total population sizes are 33.48 and 13.30, respectively. This difference of 20.18 is highly significant statistically with P.<0.0000. In other words, the removal of infection by Adelina without any other known change in the populations' nutritional potential, physical environment, or husbandry, increases average size by two and a half times! This has significant implications. It suggests, under these experimental conditions of course, that a culture of *T. castaneum* is not limited by lack of food to that lower equilibrium maintained by I-C-c. It suggests that Adelina, when present in sufficient degree, has such potent adverse effects upon the beetles that lower densities result. Further, these effects must be specific for *T. castaneum* only. *T. confusum,* although its composition as to stages is influenced by parasitization, is not influenced so far as total numbers go.

(3) In addition to the fact just noted that the total size of *T. castaneum* sterile cultures is consistently greater than that of the infected cultures, there is also some divergence in the early trend of the two curves, as seen in Figure 4. The infected populations drop rather sharply from day-120 to day-270 —a drop not duplicated by the steriles. Presumably this drop reflects a build-up in infection pressure leading to subsequent mortality. In this connection, it is worth remembering that, although Adelina is present when the parasitized cultures are started, it is not present to any great extent because the few seeding imagoes are reared from *eggs* in clean medium. Thus, some time is required before the parasite gets well established. After some sort of numerical stabilization is reached between host and parasite populations, the curve of total numbers for I-C-c remains fairly level and the percentages of larvae-pupae to imagoes remain quite constant; the former comprising approximately 70 to 80 per cent of the total.

(4) As a preliminary to writing this manuscript, many study-graphs, assembled and plotted in different ways and designed to compare various phenomena, were constructed. Obviously, most of these cannot be reproduced owing to limitations of space and certain of them should not be incorporated into this report because they would obfuscate the major argument. Among these graphs, however, was one plotting mean total size of I-C-b, I-C-b-S, I-C-c, and I-C-c-S cultures. The co-ordinates of this figure were purposely exaggerated to emphasize variability in time within and between each of the four series by using an arithmetical, instead of a logarithmic, ordinate (mean per gram) and by contracting the abscissa (age in days) relative to the ordinate. Under these circumstances, an interesting feature emerges that can be seen in Figures 4 and 5 but that might be overlooked if one was not searching for it, namely: that the sterile *T. castaneum* populations pass through fluctuations of considerable amplitude consisting of

[2] This low figure of 15.4 has considerable physiological and ecological interest in its own right and merits further elaboration. Inadvertently, a failure in electrical power during the 690-720 day interval caused the sterile incubator (only) to drop to room temperature for some two to three weeks. It seems evident (a) that the low population density of *T. castaneum* at the 720 day census is caused by cold, and (b) that this depression does not obtain to any degree, if at all, for *T. confusum* cultures which were in the same chamber. The response was uniformly exhibited by *all* I-C-c-S replicates as the following protocol, comparing the total population size in number of beetles per eight grams of medium, clearly shows:

Replicate Number	Census age (days) 690	720	750
2	175	90	230
3	200	132	274
4	192	103	205
5	198	150	207
6	190	95	233
7	139	92	159
8	224	142	269
10	161	75	221
11	169	107	240
13	183	111	262
14	297	124	217
15	259	243	386
16	240	142	354
17	184	118	204
18	160	103	310
19	277	162	347
20	201	106	246

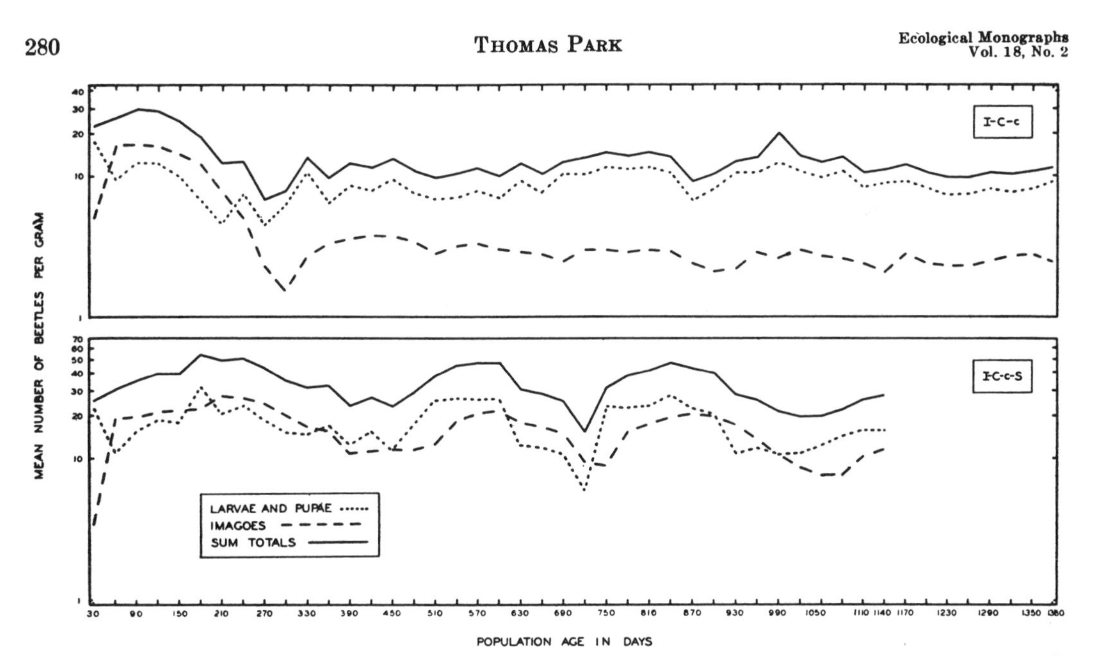

Fig. 5. Census history of *Tribolium castaneum* infected controls (I-C-c) compared with *T. castaneum* sterile controls (I-C-c-S). Shown are larval-pupal, imaginal, and total population curves (see box). Logarithmic ordinate; arithmetical abscissa. Low density of I-C-c-S at day 720 caused by cold (see text).

high peaks and low depressions. These are rather evenly spaced in time and are more characteristic of I-C-c-S than of the other groups. Peaks occur at 180, 600, 840, and 1200 days; and depressions, at 450, 720, and 1020 days.

It is of particular interest to report that these peaks and depressions are not primarily created by averaging the data and therefore do not represent artifacts in the sense that several extreme readings, either high or low, could produce them. Certainly something of this obtains, but, essentially, they are real phenomena exhibited at particular ages, or over several age ranges, by many of the individual populations.

(5) It has been shown for the I-C-b, I-C-b-S, and I-C-c cultures that there exists a pre-equilibrium period of some 300 days' duration during which the imagoes increase rapidly in numbers until they constitute well over half of the total population and then decline until a state of equilibrium, or, better, a state of relative stability, is reached. This varies in detail between the three types of controls. There is no comparable period in the sterile *T. castaneum* cultures as a glance at Figure 5 will show. After the sixtieth day, both larval-pupal and imaginal curves are close together and intercross. Thus, there is as much stability in composition in the early population history as there is later—perhaps even more.

(6) This crossing-over of the imaginal and larval-pupal curves of I-C-c-S characterizes the entire period and is not seen in I-C-c. In other words, the percentage composition of imagoes is higher in the absence of Adelina, an effect also noted for *T. confusum* and one to be anticipated because of the reduction of late larval and pupal mortality. The mean percentage of imagoes for the total period is 48.2 for the steriles and 29.5 for the infected ($P<0.0000$) (Table 12).

(7) In sum, it seems clear (a) that Adelina strikingly affects *T. castaneum* in terms of total population size while having no such effect on *T. confusum;* (b) that both species of beetles are similarly affected in the sense that the sterile cultures stabilize with a higher proportion of imagoes than do the infected cultures; (c) that removal of the parasite permits *T. castaneum* to exploit its environment in a remarkable way; and (d) that a characteristic pre-equilibrium period is absent in I-C-c-S.

By taking advantage of the misfortune that certain I-C-c-S cultures become infected with Adelina, it is possible to analyze further the relation of infection to total population size.

The data bearing on this were accumulated as follows. There are the sterile populations (I-C-c-S) that never became parasitized. There are populations initially of the same sort that go through a period free of Adelina only later to acquire the disease. There are the regular controls (I-C-c) that were always parasitized. These three groups can be compared with such comparisons made, not relative to a census age fixed for the entire sample, but relative to that time at which infection was first noted. If a certain I-C-c-S culture became parasitized, say, at day-630 and another at day-900, these were dealt with by tabulating, first, the total population counts one-by-one for each culture over 300 days preceding the infection (10 censuses) and, second, the counts for 240 days after the infection had been acquired (8 censuses). In order to have a legitimate control against which these "preinfection" and "infection" intervals could be tested, other always-sterile cultures were arbitrarily selected for every age represented in the sterile-becoming-infected sample. Mean total population size was computed for this sample; first, for the 300-day period, then, for the 240-day period. Each of these two I-C-c-S groups over the 540 days are represented by seven replicates. In addition, mean total size for seven I-C-c cultures was calculated for the last 240 days—the various census ages again being chosen in the same way. These means along with their probable errors are as follows:

I-C-c-S Cultures:

First interval (300 days)
- (a) Populations not infected...... 32.44 ± 0.9188
- (b) Populations not infected, but to become infected .... 34.61 ± 1.3702
- (c) (a) + (b) .... 33.53 ± 0.8208

Second interval (240 days)
- (d) Populations not infected (sample (a) above) .... 26.45 ± 0.8947
- (e) Populations infected (sample (b) above) .... 18.22 ± 0.7487

I-C-c Cultures:

- (f) Populations always infected (second interval) .... 10.75 ± 0.3397

In this table (a) and (b) do not differ significantly from each other ($P = 38.1\%$). The following differences are significant ($P < 5\%$): a minus d, b minus e, d minus e, and e minus f.

The most instructive point emerging from the table is, that after those populations originally sterile become parasitized (sample-e) their total size is significantly reduced below that of the cultures remaining sterile (sample-d), the mean difference being 8.23 ± 1.61. The sterile cultures also decline from an average of 32.44 for the first interval to one of 26.45 for the second interval—a decline obviously not induced by Adelina. It is worth noting that the I-C-c controls (sample-f) are considerably lower than is the mean for sample-e. Probably in the latter group there has not been enough time for the infection to attain maximum saturation.

CONCLUDING STATEMENT ABOUT ADELINA INFECTION

It is reasonable to conclude that Adelina influences *T. castaneum* populations through the agency of increased mortality directed especially to the immature stages. This conclusion is warranted from knowledge of the parasite and its life-cycle, from the observed census differences between infected and sterile populations, and from the "clinical" appearances of the two sorts of cultures at the time of counting when many dead larvae and pupae are present in the I-C-c replicates but essentially absent in the I-C-c-S. It does not necessarily follow, however, that culture growth is stimulated in the absence of Adelina only through reduction of the specific death-rate. Perhaps fecundity is heightened also; a point not yet analyzed. It should be recalled here that *T. castaneum,* under optimal conditions at least, possesses a reproductive capacity higher than that of *T. confusum.* This may be a factor favoring the demonstrated increase of I-C-c-S cultures.

There are other aspects to this entire matter of infection that are not yet clear. Why is it that *T. castaneum* steriles attain such a higher density than do those of *T. confusum?* Why, when the number of *T. confusum* imagoes increases as it does in the I-C-b-S series, does not total population size also increase? What limits, or, conversely stated, fails to stimulate, *T. confusum* in the absence of infection that works in the opposite direction for *T. castaneum?* Why is it that infection pressure can mount to a degree sufficient to cause certain of the I-C-c controls to become extinct when this never happens in the II and III volumes nor in any of the *T. confusum* populations? Is Adelina more infectious for *T. castaneum* than for *T. confusum* assuming only one species of parasite is incriminated, or, are both beetles equally susceptible with the infection resulting in greater pathology in *T. castaneum?* Are there any immune phenomena operating as part of this host-parasite interaction? Do the findings suggest a longer evolutionary association with better adaptation between *T. confusum* and Adelina than between *T. castaneum* and Adelina?

These are fascinating questions but ones that cannot be answered at this time. It is only possible to present the facts descriptively and postpone more final interpretations pending further study.

PERCENTAGE COMPOSITION OF CONTROL POPULATIONS

As an extension of the preceding discussion devoted largely to actual numbers per gram, it is helpful to examine and compare the eight types of control populations in terms of their percentage composition by stages. These percentages have already been reported in Tables 2-9, but the trends are more quickly evident in Figures 6 and 7.

A word about the construction of the two figures is appropriate. The ordinate is a percentage scale ranging from zero to 100 and the abscissa is a time scale reporting age by thirty-day census intervals. Each curve describes a particular type of population and is plotted from points computed as means of all replicates. The distance from the baseline to the curve is the percentage of larvae-pupae while the distance up from the curve to the horizontal line drawn at 100 per cent is the percentage of imagoes. Thus, when the curve is close to the abscissa, the proportion of imagoes is high; and, conversely, as the curve rises, the proportion of imagoes decreases and that of the larvae-pupae increases.

The figures are drawn, in each case, to compare two different control populations. Figure 6 compares I-C-b with I-C-c; II-C-b with II-C-c; III-C-b with III-C-c; and C-b with C-c, the three volumes being averaged. Figure 7, concerned entirely with the eight gram volume (I), compares I-C-b with I-C-b-S; I-C-c with I-C-c-S; and I-C-b-S with I-C-c-S.

In summary of percentage composition, the following points can be advanced:

(1) *Volume of medium.*—The curves of Figure 6 are consistent in pattern for each species, irrespec

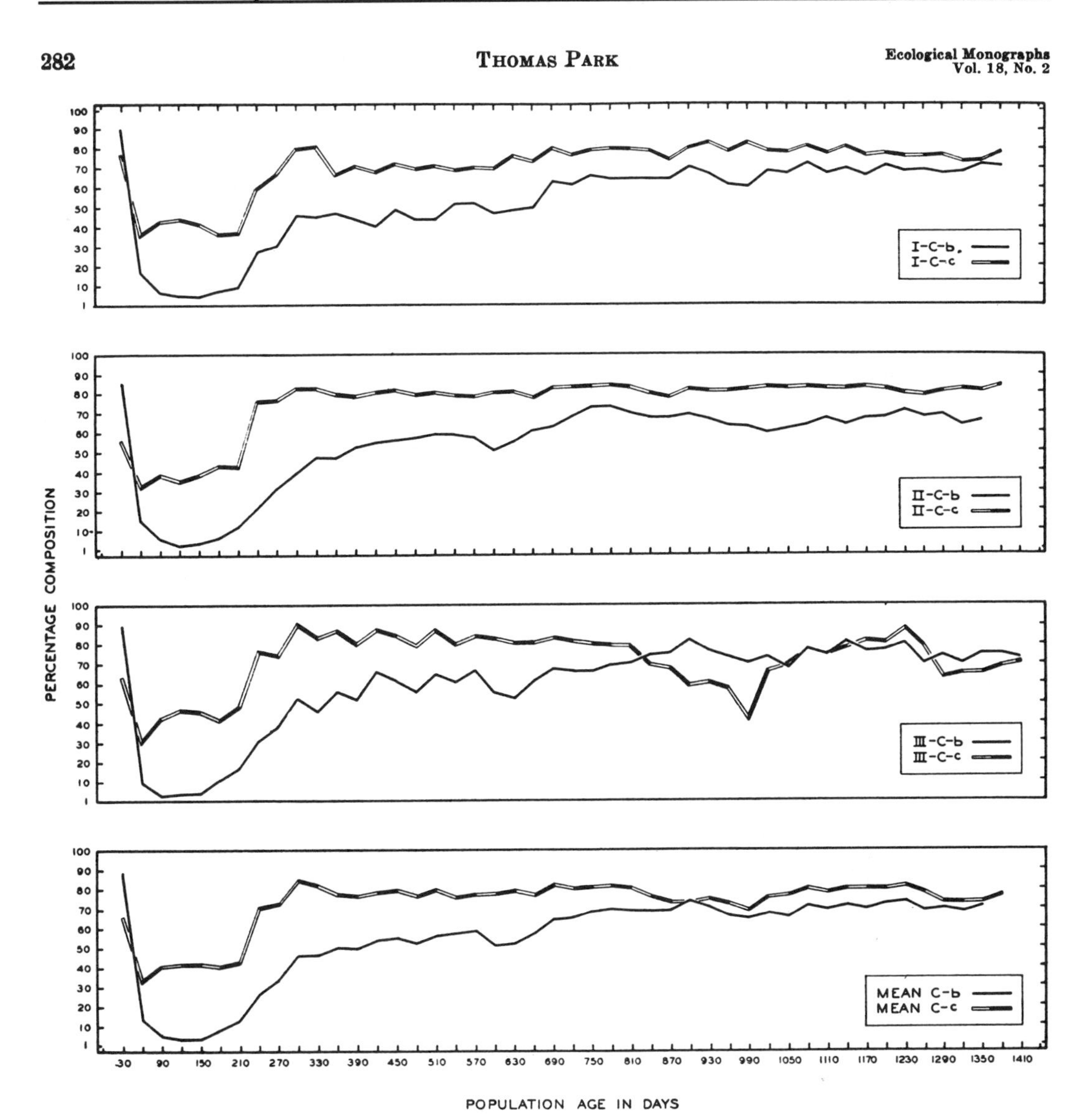

FIG. 6. Percentage composition curves for both species of Tribolium compared by volume of medium (upper three charts) and averaged together (lower chart). Distance from abscissa to curve is percentage of immature stages; from curve to 100 per cent line is percentage of adult beetles.

tive of total size (i.e., volume of medium). Some deviation from the usual pattern is exhibited by Series III between day-840 and day-1290, during which interval the lines overlap to a certain extent. It seems likely, however, that this, again, is a small-sample error as noted earlier.

(2) *Species differences, pre-equilibrium period.*—The percentage curves clearly describe the pre-equilibrium period for populations of both species of beetles and illustrate the species differences that obtain over this first 300 days. The proportion of larvae-pupae for *T. confusum* drops rapidly to a place where they comprise less than five per cent of the total population, then gradually rises until about half of the culture is so constituted. For *T. castaneum* this figure drops only to about 30-35 per cent and then increases until some three-fourths are immature and one-fourth are adult.

(3) *Species differences, equilibrium period.*—From approximately day-300 on, the percentage of *T. castaneum* stabilizes at something like 75-80% larvae-pupae, while *T. confusum* first stabilizes at about 50-60%. The latter curves then rise very slowly until they eventually come quite close to the proportions characteristic for *T. castaneum.*

(4) *Species differences, sterile and infected populations.*—The top graph of Figure 7, comparing I-C-b with I-C-b-S, extends the discussion already

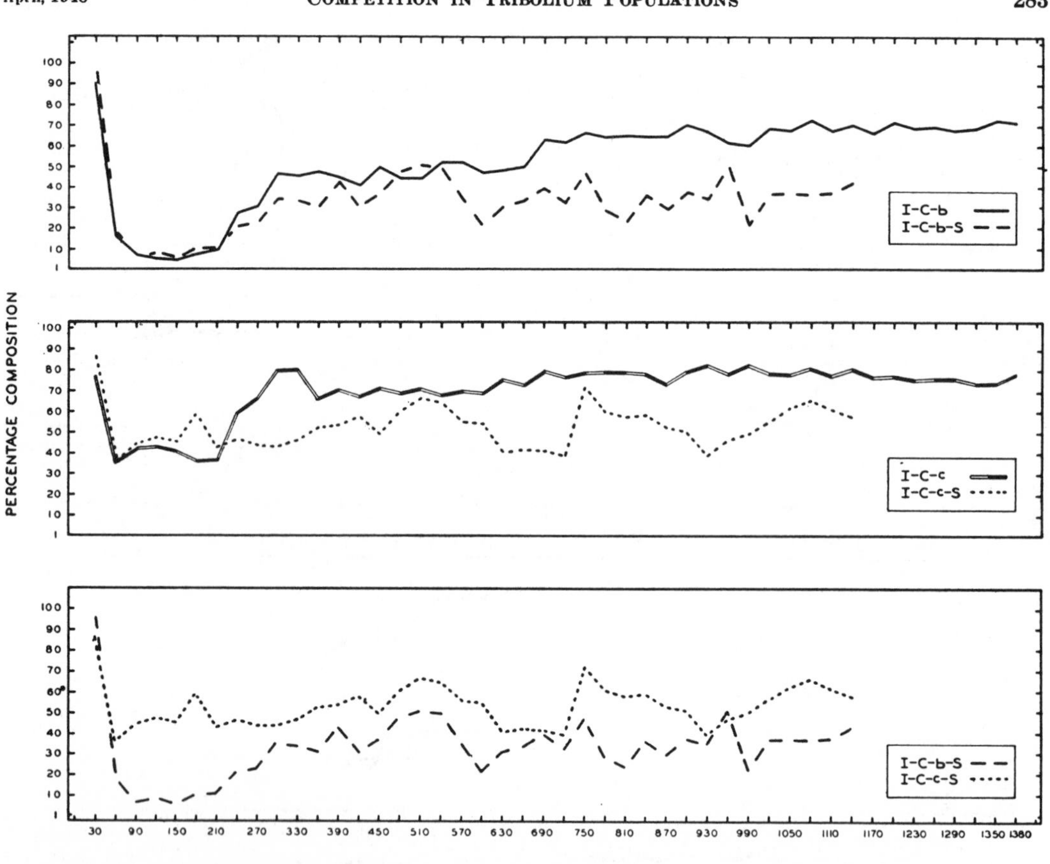

FIG. 7. Percentage composition curves for infected and sterile cultures of *T. confusum* and *T. castaneum* maintained in the eight-gram volume of medium. Respective comparisons indicated within the boxes.

presented based on actual numbers per gram. There is a close degree of confluence between the two curves until day-210, after which the sterile populations are characterized by a larger proportion of imagoes. According to our hypothesis, it requires about this much time for an Adelina infection to become sufficiently established so that mortality reduces the emergence of pupae from larvae, and the eclosion of imagoes from pupae. By day-1140 the adult percentages are 30.0 and 57.4 for infected and sterile, respectively.

In viewing the middle graph (I-C-c and I-C-c-S) it should be remembered that the sterile cultures are much larger both as totals as well as larvae-pupae and imagoes. From day-60 to day-210 the steriles exhibit a slightly lower percentage of imagoes than do the infected. After day-240, as even casual inspection will show, the sterile populations maintain themselves with a significantly higher proportion of imagoes: for example, at day-1140, I-C-c has 19.3% and I-C-c-S, 42.5%.

The bottom graph of Figure 7, contrasting the steriles of both species, suggests (a) that *T. confusum* still maintains a higher percentage of imagoes than does *T. castaneum* even though Adelina is absent, and (b) that the general effect of removing the parasite is to increase the proportion of adult beetles for both species of Tribolium to about the same degree.

CONTROL CULTURES CONSIDERED OVER THE TOTAL PERIOD: A SUMMARY

To afford a final summary, the eight types of single-species populations are now examined when averaged for the entire period of study—a period of nearly four years' duration. The topics considered are: statistics dealing with mean numbers per gram; statistics concerned with percentage composition; and statistics that describe variability.

Mean Number per Gram.—Table 10 lists for each type of control the range, mean, standard deviation, and coefficient of variability for larvae-pupae, imagoes, and sum totals. Table 11 records certain statistical comparisons between selected mean differences in relation to their probable errors. Figure 8 graphs the means for all types of cultures.

Our interest lies in summarizing very briefly four relationships, each considered from the point of view

TABLE 10. Statistics in terms of numbers per gram for total period of study.

Population	Stages	Mean Range		Mean ± P. E.	Standard Deviation	C. V.	n	Days Observed
		Maximum	Minimum					
I-C-b	L + P	18.6	0.96	10.38±0.52	5.25	50.6	46	1380
	I	20.6	5.9	8.87±0.41	4.15	46.8	46	1380
	Total	25.6	13.6	19.25±0.33	3.32	17.2	46	1380
I-C-b-S	L + P	12.0	1.2	6.21±0.43	3.98	64.1	38	1140
	I	23.6	9.4	12.73±0.53	4.82	37.9	38	1140
	Total	28.2	14.0	18.94±0.40	3.67	19.4	38	1140
II-C-b	L + P	12.8	0.6	8.97±0.40	3.94	43.9	45	1350
	I	23.5	4.4	8.39±0.53	5.31	63.3	45	1350
	Total	27.7	14.1	17.36±0.31	3.09	17.8	45	1350
III-C-b	L + P	15.2	0.6	8.66±0.40	4.09	47.2	47	1410
	I	21.1	3.2	6.28±0.49	4.99	79.5	47	1410
	Total	23.3	9.5	14.94±0.32	3.28	21.9	47	1410
I-C-c	L + P	12.4	4.4	8.98±0.23	2.27	25.3	46	1380
	I	16.9	1.6	4.32±0.39	3.95	91.4	46	1380
	Total	29.3	9.1	13.30±0.48	4.85	36.5	46	1380
I-C-c-S	L + P	32.0	6.0	17.63±0.66	6.00	34.0	38	1140
	I	27.7	7.5	15.85±0.62	5.63	35.5	38	1140
	Total	54.3	19.5	33.48±0.32	9.79	29.2	38	1140
II-C-c	L + P	11.4	3.2	7.72±0.15	1.53	19.8	46	1380
	I	18.6	1.3	3.51±0.44	4.48	127.6	46	1380
	Total	30.0	7.1	11.23±0.53	5.37	47.8	46	1380
III-C-c	L + P	18.1	3.5	10.48±0.31	3.13	29.9	47	1410
	I	18.1	0.8	4.70±0.41	4.17	88.7	47	1410
	Total	30.5	7.3	15.18±0.53	5.39	35.5	47	1410

of the total population, the number of imagoes, and the number of larvae-pupae. These relationships are:[3]

(1) Between volumes, within species (b in I, II, III; c in I, II, III)
(2) Between species, within volumes (b cf. c in I; b cf. c in II; b cf. c in III)
(3) Between parasitized and non-parasitized, within species and within volume (I-C-b cf. I-C-b-S; I-C-c cf. I-C-c-S)
(4) Between non-parasitized species, within volume (I-C-b-S cf. I-C-c-S)

Between Volumes, within Species.—It is clear from the tables and Figure 8 that the total population size of *T. confusum,* expressed as density of organisms per gram of medium, decreases significantly with increase of volume. Similar relations hold for larvae-pupae and adults, but only the difference between I-imagoes and III-imagoes is clearly significant. Because the trend is consistent, however, it is a reasonable conjecture that such an association of density with size of environment does exist.

The situation for *T. castaneum,* on the other hand, is not the same; namely, III>I>II with only the difference between III and II being decidedly significant (P values for III cf. I and I cf. II are borderline). While there is no indication of a trend between volumes so far as imagoes are concerned, it is obvious that the III populations maintain themselves with a greater number of immature stages and that I exceeds II in this respect.

One is then left with the conception (a) that volume of medium can be significantly correlated with density, and (b) that, while this correlation is quite consistent for *T. confusum* in a negative direction, it is not so clearcut for *T. castaneum.* An important point has been made earlier that volume, even though it affects density, does not appreciably alter the pattern or course of the populations in time.[4]

Between Species, within Volumes.—In terms of total population size, *T. confusum* in the 8 and 40 gram environments maintains itself at a significantly higher level than does *T. castaneum* with no appreciable difference displayed for the 80 gram environments. In other words, I-b>I-c by about 30%, II-b>II-c by about 35%, and III-b = III-c. Similar relations hold for the immature and adult stages of I and II considered independently although the number of *T. confusum* larvae-pupae in the smallest volumes does not differ significantly from that of *T.*

[3] In the statistical sense, two "interactions" are encompassed by this treatment: the interaction between species and volume, and that between species and parasitism.

[4] It is recognized that even though 47 entries, representing 47 consecutive censuses, comprise the III-C-b and III-C-c samples, actually these are based on only two replicates. Thus, their means do not have as much reliability as do those of the I and II series. This is unfortunate but, as pointed out earlier, quite unavoidable. We still place considerable general confidence in the samples, however, because they display a close similarity in pattern of growth to that shown by the two smaller volumes based on larger numbers. (Fig. 1. 2.)

TABLE 11. Statistical analysis of selected mean differences reported in Table 10 (by mean difference ÷ its probable error).

The Comparison	Stage	Mean difference	Largest member of pair	Probability
I-C-b and I-C-b-S	L and P	4.17	I-C-b	<0.0000
	I	3.86	I-C-b-S	<0.0000
	T	0.31	I-C-b	>0.5
I-C-b and II-C-b	L and P	1.41	I-C-b	0.1567
	I	0.48	I-C-b	>0.5
	T	1.89	I-C-b	0.0046
I-C-b and III-C-b	L and P	1.72	I-C-b	0.0795
	I	2.59	I-C-b	0.0070
	T	4.31	I-C-b	<0.0000
II-C-b and III-C-b	L and P	0.31	II-C-b	>0.5
	I	2.11	II-C-b	0.0505
	T	2.42	II-C-b	<0.0000
I-C-c and I-C-c-S	L and P	8.65	I-C-c-S	<0.0000
	I	11.53	I-C-c-S	<0.0000
	T	20.18	I-C-c-S	<0.0000
I-C-c and II-C-c	L and P	1.26	I-C-c	0.0019
	I	0.81	I-C-c	0.3450
	T	2.07	I-C-c	0.0505
I-C-c and III-C-c	L and P	1.50	III-C-c	0.0085
	I	0.38	III-C-c	>0.5
	T	1.88	III-C-c	0.0795
II-C-c and III-C-c	L and P	2.76	III-C-c	<0.0000
	I	1.19	III-C-c	0.1773
	T	3.95	III-C-c	<0.0000
I-C-b and I-C-c	L and P	1.40	I-C-b	0.0918
	I	4.55	I-C-b	<0.0000
	T	5.95	I-C-b	<0.0000
I-C-b-S and I-C-c-S	L and P	11.42	I-C-c-S	<0.0000
	I	3.12	I-C-c-S	0.0085
	T	14.54	I-C-c-S	<0.0000
II-C-b and II-C-c	L and P	1.25	II-C-b	0.0430
	I	4.88	II-C-b	<0.0000
	T	6.13	II-C-b	<0.0000
III-C-b and III-C-c	L and P	1.82	III-C-c	0.0152
	I	1.58	III-C-b	0.0918
	T	0.24	III-C-c	>0.5

castaneum (P = 9%). Within series-III the only statistically significant difference is that for the immature stages—the mean number being greater for *T. castaneum* than for *T. confusum* (P = 1.5%).

Between Parasitized and Non-parasitized, within Species, within Volume.—This relation has already been established and requires only brief summary here. The essential points are as follows:

(1) *T. confusum*-infected and *T. confusum*-sterile cultures maintain populations of equivalent total size. The latter group is characterized by more imagoes and fewer larvae-pupae. Both differences are highly significant (P<0.0000 in each instance). The matter of the higher imago densities in populations free of Adelina was discussed earlier.

(2) Non-infected *T. castaneum* cultures are decidedly larger than the infected over the entire period of observation. The mean differences are:

for total size,	33.48 — 13.30 = 20.18
for imagoes,	15.85 — 4.32 = 11.53
for larvae-pupae,	17.63 — 8.98 = 8.65

These differences all have a probability value of less than 0.0000.

Between Non-parasitized Species, within Volume.—When the sterile cultures grown in the same volume are compared as to species, it is immediately apparent that removal of Adelina from both allows *T. castaneum* to attain densities very much greater than those displayed by *T. confusum* whether parasitized or not. Under these conditions, *T. castaneum* has a mean total density of 33.48 organisms per gram as against 18.94 for *T. confusum* (P<0.0000); a mean imaginal density of 15.85 as against 12.73 (P = 0.0085); and a mean larval-pupal density of 17.63 as against 6.21 (P<0.0000). This becomes all the more striking when it is remembered that the population levels of *T. castaneum* in the presence of infection fall significantly below that of *T. confusum.*

Percentage Composition.—Statistics describing composition in percentages for the entire period are listed in Table 12, along with comparisons of selected mean differences. Only imagoes are treated because, obviously, the percentage of larvae-pupae is readily obtained for any case by subtracting a listed figure from 100.

The findings are so straightforward that they can be quickly summarized in the following statements:

(1) *T. confusum* populations are characterized by a higher proportion of imagoes (and a lower proportion of larvae-pupae) than are *T. castaneum* populations. This is true for all volumes and for parasitized and non-parasitized cultures alike. All such differences between the two species are highly significant statistically (P ranges from 0.0012 to <0.0000).

(2) There is extremely little divergence in percentage composition between the three volumes within the species. Thus, *T. confusum* displays 48.2 per cent imagoes in 8 grams of medium, 47.2 per cent in 40 grams, and 41.1 per cent in 80 grams. Comparable figures for *T. castaneum* are 29.5, 24.9, and 28.3. Any combination of differences within the two sets of percentages are completely insignificant (P ranges from 0.1208 to >0.5).

(3) For both beetles, removal of Adelina increases the imagoes relative to the larval-pupal component to a significant degree. *T. confusum* rises from 48.2 (I-C-b) to 69.0 (I-C-b-S) and *T. castaneum* from 29.5 (I-C-c) to 48.2 (I-C-c-S). Suggestively enough, the rise in each case is in the order of 20 per cent.

VARIABILITY WITHIN AND BETWEEN CONTROL POPULATIONS

This paper is not primarily a study of population variability. Its principal objective lies in making clear the major trends exhibited by single and mixed-species cultures in order that something can be advanced regarding the phenomenon of competition under controlled laboratory conditions. An interesting, but secondary, contribution could concern itself in a comprehensive way with such variability to which the design of the program and the data lend themselves. Possibly an extension in this direction will be forthcoming. The present responsibility, however, is to briefly summarize certain facts about populations in terms of their variability that seem relevant for the central argument.

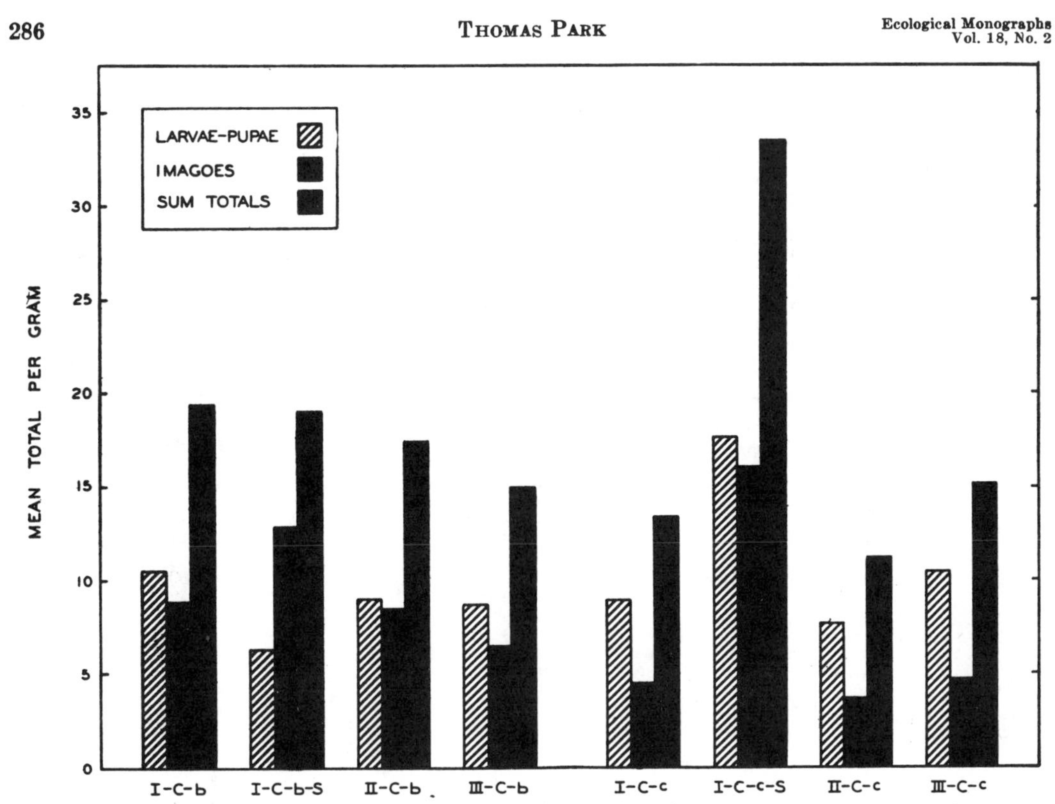

FIG. 8. Bar diagram showing the mean total population density of all eight types of control cultures averaged for the entire period of census.

This could be done in several ways. Perhaps the simplest, yet most meaningful, device would be to present graphs for the 119 control cultures whose census history has been followed and to discuss these with respect to similarities and differences. This method, however, would be too cumbersome for presentation here, although I have used it quite freely for my own edification. It would be possible also to treat the data and assay their variation by means of more elegant quantitative methods. While this may be attempted in the future, it has seemed sufficient for present purposes to rely upon such familiar statistics as the range, the standard deviation, and the coefficient of variability.

The variability displayed by each of the eight types of populations is assessed in two ways: first, for the entire period of study, and second, for eight census ages arbitrarily chosen at intervals of 150 days, beginning at day-90 (i.e., days 90, 240, 390, 540, 690, 840, 990, and 1140). The requisite figures appear in Table 10 for the grouped data, and in Table 13 for the data treated by intervals.

The Grouped Data.—Several interesting features are revealed by the ranges, standard deviations, and coefficients of variability of Table 10. First. no consistent trends are discernible for either species of beetle that can be related to volume of medium. Second, *T. confusum* in terms of total population size appears somewhat less variable, on both an absolute and relative basis, than does *T. castaneum.* This point is related, perhaps, to the greater effect exerted by Adelina on the latter beetle. Third, cultures of *T. confusum* exhibit more variation when the immature and adult components are examined individually than when the two are summed and considered together. Fourth, this same relation does not obtain with any measure of consistency for *T. castaneum.* Fifth, when the ranges are examined, it is clear that there is a greater spread between maximum and minimum values (for total population) for *T. castaneum* than for *T. confusum,* the former species thus attaining, for brief moments at least, greater and lower densities during the course of its census history. Sixth, removal of Adelina from the cultures appears to exert no systematic effect on variation beyond reducing the coefficient of variability for I-C-b-S imagoes, increasing (owing to the larger size of the cultures) the standard deviations for I-C-c-S. and lowering their coefficients.

The Data by Census Intervals.—The data are treated by census intervals for the primary purpose of seeing whether there are any marked trends of variability associated with age of cultures. This operation has been limited to mean total size and necessarily confined, because only two replicates exist

TABLE 12. Percentage imago composition for total period of study including selected statistical comparisons (30-day census excluded).

Population	Range		Mean ± P. E.	Standard Deviation	C. V.	n	Days Observed
	Maximum	Minimum					
I-C-b	95.2	27.8	48.2±2.05	20.4	42.3	45	1350
I-C-b-S	94.9	49.6	69.0±1.33	12.0	17.4	37	1110
II-C-b	97.3	26.7	47.2±2.10	20.7	43.8	44	1320
III-C-b	97.4	17.7	41.1±2.25	22.6	55.0	46	1380
I-C-c	64.0	17.3	29.5±1.33	13.2	44.7	45	1350
I-C-c-S	63.9	33.1	48.2±0.94	8.5	17.6	37	1110
II-C-c	67.3	16.0	24.9±1.46	14.5	58.2	45	1350
III-C-c	69.6	8.6	28.3±1.47	14.8	52.3	46	1380

The Comparison	Largest Number of pair	Mean Difference ± P. E.	Probability
I-C-b and I-C-b-S	I-C-b-S	20.8±2.44	<0.0000
I-C-b and II-C-b	I-C-b	1.0±2.93	>0.5
I-C-b and III-C-b	I-C-b	7.1±3.04	0.1208
II-C-b and III-C-b	II-C-b	6.1±3.08	0.1773
I-C-c and I-C-c-S	I-C-c-S	18.7±1.63	<0.0000
I-C-c and II-C-c	I-C-c	4.6±1.97	0.1208
I-C-c and III-C-c	I-C-c	1.2±1.98	>0.5
II-C-c and III-C-c	III-C-c	3.4±2.07	0.2805
I-C-b and I-C-c	I-C-b	18.7±2.44	<0.0000
I-C-b-S and I-C-c-S	I-C-b-S	20.8±1.63	<0.0000
II-C-b and II-C-c	II-C-b	22.3±2.56	<0.0000
III-C-b and III-C-c	III-C-b	12.8±2.69	0.0012

for the two series-III groups, to series-I and II. The results are presented in Table 13.

For study purposes, the standard deviations and coefficients of variability listed in the table were plotted against age for the six types of control cultures (I-C-b, I-C-b-S, II-C-b, I-C-c, I-C-c-S, and II-C-c). The curves lie rather closely together, frequently intercross, and essentially parallel the abscissa. The major point made by these graphs, however, is that the pattern of variability does not change in any appreciable way with time. This conclusion, although a negative one, is significant for our purposes because of its implication that a conception of progressive, temporal modification of variation can be excluded. The I-C-c-S populations afford something of an exception to this generalization. In these, the standard deviations fluctuate between high and low values. Such fluctuations are usually correlated with the magnitude of the mean and are related to a matter already noted, namely: that the sterile *T. castaneum* cultures tend to build up high densities, then undergo decline leading to lower densities, then build up again, and so on.

Other suggestions emerge from inspection of Table 13, but these are of relatively minor importance and need not be pursued.

MIXED-SPECIES POPULATIONS: EXPERIMENTALS

INTRODUCTORY

In the foregoing pages attention has centered on populations of *Tribolium confusum* and *Tribolium castaneum* reared apart as *intraspecies* systems. The emphasis now shifts to *interspecies* systems for which competition between the two forms can be detected, described, and partially analyzed. It is the relation of these competitions to the fate of both species of Tribolium maintained as mixed-species groups that we now wish to explore, making frequent use of the data and interpretations presented in the section just concluded.

It will be remembered that the experimental cultures differ from the controls only in the fact that *both* species were introduced into the vials when the populations were started. Apart from this, controls and experimentals are grown in the same medium allocated in the same three volumes (I, II, III), and maintained under similar conditions of physical environment, husbandry, and laboratory handling. Table 1 lists all types of experimental cultures and clarifies their respective designs. It need only be recalled at this point that the notation E-a signifies populations in which both beetles were introduced in equal ratio; E-b signifies that *T. confusum* possessed an initial numerical advantage over *T. castaneum;* and E-c, that *T. castaneum* had a similar advantage over *T. confusum.* In order to round out the total investigation, non-parasitized or "sterile" cultures were started, having the constitution I-E-a-S, and these will be considered later as a special case.

FINAL CONSEQUENCES OF COMPETITION

Before discussing in some detail each type of experimental culture, it is helpful to make clear just what is the final consequence, or end-result, of com-

TABLE 13. Variability of six control populations assayed for eight census periods at intervals of 150 days.

Age (days)	Population	Mean ± P. E. (per gram)	Standard Deviation (per gram)	Coefficient of Variability (per cent)	*n*
90	I-C-b	22.19± 0.9304	6.17	27.8	20
	I-C-b-S	25.1 ± 0.6439	4.27	17.0	20
	II-C-b	24.8 ± 0.3288	2.01	8.1	17
	I-C-c	29.3 ± 0.9938	6.59	22.5	20
	I-C-c-S	35.2 ± 0.7570	5.02	14.3	20
	II-C-c	30.0 ± 0.5295	3.33	11.1	18
240	I-C-b	17.1 ± 0.7600	5.04	29.5	20
	I-C-b-S	17.4 ± 0.4614	3.06	17.6	20
	II-C-b	18.0 ± 0.3697	2.26	12.5	17
	I-C-c	12.6 ± 0.7842	5.20	41.6	20
	I-C-c-S	50.3 ± 3.8716	24.35	48.4	18
	II-C-c	10.8 ± 0.6074	3.82	35.4	18
390	I-C-b	15.8 ± 0.8263	5.48	34.8	20
	I-C-b-S	16.9 ± 0.4451	2.64	15.6	17
	II-C-b	14.8 ± 0.5971	3.65	24.7	17
	I-C-c	12.1 ± 1.0622	6.30	52.1	16
	I-C-c-S	23.3 ± 1.1114	6.99	30.0	18
	II-C-c	10.9 ± 0.5535	3.48	31.9	18
540	I-C-b	15.2 ± 0.7102	4.71	30.9	20
	I-C-b-S	21.7 ± 0.7890	4.68	21.6	17
	II-C-b	15.7 ± 0.4368	2.67	17.0	17
	I-C-c	10.2 ± 0.6473	3.59	35.2	14
	I-C-c-S	44.4 ± 2.4629	15.49	34.9	18
	II-C-c	9.1 ± 0.2417	1.52	16.7	18
690	I-C-b	18.6 ± 1.2893	8.55	45.9	20
	I-C-b-S	18.2 ± 0.4231	2.51	13.8	16
	II-C-b	15.4 ± 0.4024	2.46	16.0	17
	I-C-c	12.6 ± 0.9467	5.06	40.5	13
	I-C-c-S	25.3 ± 0.8867	5.42	21.4	17
	II-C-c	8.2 ± 0.3482	2.19	26.7	18
840	I-C-b	19.4 ± 0.7193	4.65	23.9	19
	I-C-b-S	20.8 ± 0.6946	4.12	19.8	16
	II-C-b	14.8 ± 0.5955	3.64	24.6	17
	I-C-c	13.4 ± 0.7652	4.09	30.5	13
	I-C-c-S	47.2 ± 2.1345	12.66	26.8	16
	II-C-c	7.2 ± 0.2353	1.48	20.5	18
990	I-C-b	19.5 ± 0.7131	4.61	23.6	19
	I-C-b-S	15.9 ± 0.3344	1.92	12.1	15
	II-C-b	17.4 ± 0.3763	2.30	13.2	17
	I-C-c	15.0 ± 1.2311	6.58	44.2	13
	I-C-c-S	21.1 ± 1.0882	5.35	25.4	11
	II-C-c	9.2 ± 0.3482	2.19	23.8	18
1140	I-C-b	22.6 ± 0.8059	5.21	23.1	19
	I-C-b-S	18.6 ± 0.5870	3.37	18.1	15
	II-C-b	16.9 ± 0.3877	2.37	14.0	17
	I-C-c	11.0 ± 0.8270	4.42	40.2	13
	I-C-c-S	27.4 ± 1.1600	5.16	18.8	9
	II-C-c	8.1 ± 0.6471	4.07	50.2	18

petition as this relates to the fate of populations of both beetles. It has been shown that the control cultures, free of interspecies complications, eventually assume characteristic "equilibria" that are sustained for relatively long periods of time in terms of number of generations.

The major ecological finding of this entire study can be simply stated as follows: when *T. confusum* and *T. castaneum* are brought into competition, one of the two invariably becomes extinct with the other then assuming its "normal" (i.e., control) populational behavior. To put it differently, in the face of such stringent competition, it becomes impossible for both species to co-exist even though individually they are well adapted to their habitat and its surrounding physical environment.

Despite the fact that one species of Tribolium always becomes extinct in this two-species ecosystem, it is not always the *same* species that dies out. By examining what happens to each of the 92 experimental populations, it is a simple matter to tabulate which species dies along with the census age at which this occurs. These findings are systematically presented in Table 14. In certain instances, *T. confusum* became extinct in some populations, while *T. castaneum* disappeared in others. This is noted in the third column of the table under the heading "species becoming extinct." The last two columns list the mean and median ages of extinction.

From inspection of Table 14, the following points, all pertaining to species extinction, are to be noted:

(1) For the I-E-a populations consisting of 15 replicates, there are 11 cases of extinction of *T. castaneum* and 4 of *T. confusum.*

(2) For the I-E-c populations consisting, again, of 15 replicates, there are, again, 11 cases of extinction of *T. castaneum* and 4 of *T. confusum.*

(3) For the I-E-b populations, *T. castaneum* dies out in all of the 15 cultures.

(4) For the II and III series, totaling 29 populations, *T. castaneum* always becomes extinct whether the cultures are in the E-a, E-b, or E-c categories. Thus, 29 populations, originally established as mixed-species groups, eventually wound up as single-species cultures of *T. confusum.*

(5) When Adelina is removed and 18 sterile populations are established (I-E-a-S) the trend is reversed with *T. castaneum* typically winning over *T. confusum.* The former is successful in 12 out of 18 replicates.

The major conclusions emerging from the above summary are (a) that both species of beetles cannot live together for any extended period, and (b) that, in competition, *T. castaneum* characteristically disappears, with *T. confusum* remaining as the successful organism. The last event happens much more frequently than would be expected on the hypothesis that it is a matter of equal chance as to which species survives. Excluding for the moment the sterile cultures, Table 14 shows that from a group of 74 populations there are only 8 instances in which *T. confusum* becomes extinct, to be compared with 66 instances for which the reverse is true. This ratio differs significantly from the 37:37 distribution anticipated on an assumption of 50:50.

Accordingly, we are left with the not unreasonable conception that, in the presence of the parasite Adelina, a particular species of flour beetle characteristically survives in a competing system but that this does not invariably hold true, while, when the parasite is eliminated as a factor, the usual trend just described is reversed, although this, too, has its exceptions. In short, it is possible to predict the *most probable* outcome of interspecies competition for a particular experimental series, but this is not an all-or-none phenomenon since reversals sometime take place. These matters receive further attention later after each type of population has been reviewed in some detail.

The mixed-species populations.—The basic findings pertaining to competition between *Tribolium confusum* and *T. castaneum* will be presented by considering the following topics:

(1) Parasitized populations in which *T. castaneum* always becomes extinct.
(2) Extinction curves for *T. castaneum* and *T. confusum* compared.
(3) Extinction of *T. castaneum* and *T. confusum* in the absence of Adelina.

The data dealing with the mixed-species populations are presented as a series of tables and graphs similarly designed throughout. Two tables and graphs are required for each of the I-E-a, I-E-c, and I-E-a-S populations to record the fact that in these cultures certain replicates lost all their *T. castaneum* and became *T. confusum* controls, while in others the opposite occurred.

Each table reports census figures for both species of Tribolium as mean number per gram and as percentages of total size. The data are tabulated by 30 day periods along with which is shown, as the *n*-column, that number of replicates remaining as experimental cultures. For example, in Table 15 there are nine populations making up the sample at day-360 of which only six remain at day-390. In other words, three populations have become controls during the interval between censuses.

In all tables, the number of larvae-pupae is recorded both as a total and as species components of that total. While the former statistic is accurately determined by counting, the breakdown by species is necessarily based on computation rather than on census simple because the immature stages cannot be taxonomically identified. However, it seems logical to assume that the ratio between larvae-pupae and imagoes for each species in the experimentals is essentially the same as in the populations of single species (controls). Since this ratio is known for the controls it is possible to calculate the number of larvae-pupae in the experimentals that are probably *T. confusum* and those that are probably *T. castaneum*.[5]

The figures are graphed to stress competition between the species and hence, may be referred to as "competition curves." The ordinate plots the "total percentage of *T. castaneum*" (third column from the right in the tables) and the abscissa, "age in days." The distance from the abscissa to the curve denotes that percentage of the total population which is *T. castaneum;* the distance from the curve to the one-hundred per cent line denotes the percentage of *T. confusum.* Each figure shows at a glance the relative proportion of the two species at any census age. When the curve drops, *T. confusum* is increasing; and if the curve falls to the baseline, the populations have become *T. confusum* controls. When the curve reaches the top horizontal line, as happens in some of the Series I cultures, the populations have become exclusively *T. castaneum.*

In inspecting the various graphs, it should be kept in mind that, as the sample gets older, individual populations are dropping out through extinction of one of the two species. In other words, the points plotted in every figure are based on progressively fewer observations (cultures) until, eventually, only the life-history of one experimental culture remains to be depicted. From one point of view, this is an unfortunate method of representing the data in that reduction in number of cases obviously decreases in a measure the reliability of the curves during their later periods. On the other hand, the alternative method of graphing is less desirable. It is unrealistic to arbitrarily set as temporally equal for all replicates the census ages at which the expiring species was last counted, thus ignoring the chronology of the extinction, and then draw the curves backward in time from this assumed origin. Such a procedure was tried extensively in study graphs but forsaken because of the wide age-range within which species extinction can occur between different replicates. Also, the knowledge at our disposal about the trends exhibited by control cultures makes it clear that it is not legitimate to consider an experimental population of the age of, say, 180 days, to be ecologically equivalent with an-

[5] A note on the method is appropriate. For every census these data are available: total population size, number of *T. castaneum* imagoes, number of *T. confusum* imagoes, total number of larvae-pupae undifferentiated as to species, and percentages of larvae-pupae (and imagoes) in control cultures of both species for any desired age. These data can be utilized in differentiating the total number of larvae-pupae into its species components. Letting "b" symbolize *T. confusum* and "c" *T. castaneum*, the steps in the solution are as follows:

$$(1)\quad \frac{\text{c-imagoes}\left(\dfrac{\%\text{ larvae-pupae in c-controls}}{\%\text{ imagoes in c-controls}}\right)}{\text{b-imagoes}\left(\dfrac{\%\text{ larvae-pupae in b-controls}}{\%\text{ imagoes in b-controls}}\right)} = \frac{X}{Y}$$

$$(2)\quad \frac{X\text{ (observed total larvae-pupae)}}{(X+Y)} = \text{calculated larvae-pupae } (\textit{T. castaneum})$$

$$(3)\quad \frac{Y\text{ (observed total larvae-pupae)}}{(X+Y)} = \text{calculated larvae-pupae } (\textit{T. confusum})$$

Suppose this is applied to an actual case, e.g., the 300-day census for I-E-a reported in Table 15. There the total population is 13.0; number of b-imagoes is 3.7 and c-imagoes 0.8; total larvae-pupae, 8.5; per cent larvae-pupae at day-300 in I-C-b is 46% (from Table 2) and in I-C-c is 80 % (from Table 5).

Substituting,

$$(1)\quad \frac{0.8\left(\dfrac{80}{20}\right)}{3.7\left(\dfrac{46}{54}\right)} = \frac{3.2}{3.1}$$

$$(2)\quad \frac{(3.2)\ (8.5)}{(6.3)} = 4.3\ (\textit{T. castaneum}\text{ larvae-pupae})$$

$$(3)\quad \frac{(3.1)\ (8.5)}{(6.3)} = 4.2\ (\textit{T. confusum}\text{ larvae-pupae})$$

I am indebted to Mr. Thomas Burnett of the University of Chicago for originally suggesting this method and to Doctor Margaret Merrill of the Johns Hopkins University and Doctor L. C. Cole of Indiana University. for advising me as to its appropriateness.

TABLE 14. Age at extinction for each of the 92 mixed-species populations.

Experimental Type	Total Number of Replicates	Species Becoming Extinct		Age at Extinction of Individual Populations (Days)	Averages (Days)	
		Species	Frequency		Mean	Median
I-E-a	15	*T. castaneum*	11	270, 360, 390, 390, 390, 570, 600, 600, 630, 630, 780.	510±30.3	570±38.0
		T. confusum	4	510, 630, 690, 780.	652	660
II-E-a	9	*T. castaneum*	9	300, 360, 360, 390, 420, 420, 660, 690, 1020.	513±49.5	420±62.0
		T. confusum	0			
III-E-a	2	*T. castaneum*	2	840, 1470.		
		T. confusum	0			
I-E-a-S	18	*T. castaneum*	6	210, 270, 270, 360, 360, 450.	320±21.8	315±27.3
		T. confusum	12	510, 540, 570, 600, 600, 660, 660, 690, 690, 810, 810, 840.	665±20.2	660±25.3
I-E-b	15	*T. castaneum*	15	270, 300, 300, 300, 330, 360, 360, 360, 390, 390, 450, 600, 630, 720, 750.	434±26.8	360±33.4
		T. confusum	0			
II-E-b	7	*T. castaneum*	7	300, 330, 330, 360, 420, 450, 540.	390±20.1	360±25.2
		T. confusum	0			
III-E-b	2	*T. castaneum*	2	330, 360.	345	345
		T. confusum	0			
I-E-c	15	*T. castaneum*	11	330, 330, 360, 390, 390, 390, 450, 450, 510, 600, 600.	436±19.1	390±23.9
		T. confusum	4	180, 420, 540, 690.	458	480
II-E-c	7	*T. castaneum*	7	330, 390, 450, 480, 540, 690, 930.	549±46.9	480±58.8
		T. confusum	0			
III-E-c	2	*T. castaneum*	2	750, 810.	780	780
		T. confusum	0			

other population 840 days old, even though the two were started simultaneously under initially similar conditions.

Therefore, the number of cases used in computing the points has been sacrificed in order to maintain a strict chronology of events for both control and experimental cultures. This is probably not as serious as might be suspected at first glance because (1) the age of species extinction is accurately known for each culture, and (2) there is a general consistency between homologous figures that is quite reassuring.

Parasitized Populations in Which Tribolium castaneum Becomes Extinct

This category, involving the extinction of populations of *T. castaneum* when in competition with *T. confusum,* encompasses the majority (i.e., 66 out of 74) of the experimental cultures reported upon in this paper. The findings are presented by discussing the following sets of experiments in the order indicated:

(1) I-E-a, II-E-a, III-E-a
(2) I-E-b, II-E-b, III-E-b
(3) I-E-c, II-E-c, III-E-c
(4) General conclusions and comparisons between E-a, E-b, and E-c

The requisite data are contained in Tables 15 through 23, and selected competition curves are graphed as Figures 9, 10, and 11.

I-E-a, II-E-a, and III-E-a populations.—The feature that immediately strikes the eye when Figure 9 is examined is the essential conformity displayed by the three curves—curves summarizing experiments which differ in their design only in terms of volume of medium. *T. castaneum* attains its maximum abundance at day-90 when 80 per cent of the entire population is composed of this species. Thus, from an initial seeding of one-half *T. castaneum* to one-half *T. confusum,* the former increases at the expense of the latter until it occupies over three-fourths of the total habitat. This occurs early in the cultures' history despite the fact that it is *T. castaneum* that is definitely going to become extinct at a series of later dates! Reference to Tables 5, 6, and 7 will indicate that the 90-day census is also a large one for I-C-c, II-C-c, and III-C-c populations.

Returning to Figure 9, it is evident that the drop in percentage of *T. castaneum* starts at day-120, proceeds slowly for a time, and then accelerates. Between days-240 and 270 most of the cultures reach the 50 per cent line, and by day-360 the situation has so changed that now *T. confusum* constitutes something like 85 to 90 per cent of the total. It seems reasonable that much of this decline during the 90-360 day interval is conditioned, not primarily by competitive pressure exerted by *T. confusum,* but, rather, by the increase and spread of Adelina infection with its consequent and now well-known adverse effects upon *T. castaneum.* A decline is seen in the control cultures (I-C-c, II-C-c, and III-C-c) that is quite like that depicted in Figure 9, both in respect of magnitude and age interval of occurrence. The control cultures, of course, then proceed to recover[6] while *T. castaneum* in the experimental cultures does not.

[6] With the exception of the eight I-C-c populations discussed earlier in the section devoted to controls.

TABLE 15. I-E-a: *Tribolium castaneum* becoming extinct.

Age (days)	MEAN NUMBER PER GRAM									PERCENTAGE OF						n
	T. castaneum			*T. confusum*			Total			Larvae and Pupae		Imagoes		Total		
	L-P	Imag.	Sum	L-P	Imag.	Sum	L-P	Imag.	Sum	*T. cast.*	*T. conf.*	*T. cast.*	*T. conf.*	*T. cast.*	*T. conf.*	
30	3.9	1.6	5.5	6.6	0.9	7.5	10.5	2.5	13.0	30.0	50.8	12.3	6.9	42.3	57.7	11
60	4.9	9.4	14.3	1.1	5.2	6.3	6.0	14.6	20.6	23.8	5.3	45.6	25.3	69.4	30.6	11
90	14.3	9.2	23.5	0.9	4.9	5.8	15.2	14.1	29.3	48.8	3.1	31.4	16.7	80.2	19.8	11
120	7.7	8.1	15.8	0.4	4.5	4.9	8.1	12.6	20.7	37.2	1.9	39.1	21.8	76.3	23.7	11
150	6.5	7.2	13.7	0.3	4.3	4.6	6.8	11.5	18.3	35.5	1.6	39.4	23.5	74.9	25.1	11
180	5.9	5.7	11.6	0.6	3.9	4.5	6.5	9.6	16.1	36.6	3.7	35.4	24.3	72.0	28.0	11
210	4.8	3.4	8.2	1.0	3.7	4.7	5.8	7.1	12.9	37.2	7.7	26.4	28.7	63.6	36.4	11
240	6.5	2.4	8.9	2.4	3.4	5.8	8.9	5.8	14.7	44.2	16.3	16.4	23.1	60.6	39.4	11
270	4.9	1.6	6.5	2.2	3.3	5.5	7.1	4.9	12.0	40.8	18.3	13.4	27.5	54.2	45.8	10
300	4.3	0.8	5.1	4.2	3.7	7.9	8.5	4.5	13.0	33.1	32.3	6.1	28.5	39.2	60.8	10
330	3.6	0.6	4.2	5.7	4.7	10.4	9.3	5.3	14.6	24.7	39.0	4.1	32.2	28.8	71.2	10
360	1.1	0.5	1.6	6.1	5.8	11.9	7.2	6.3	13.5	8.1	45.2	3.8	42.9	11.9	88.1	9
390	2.4	0.8	3.2	4.9	4.9	9.8	7.3	5.7	13.0	18.5	37.7	6.1	37.7	24.6	75.4	6
420	2.2	0.7	2.9	5.5	5.3	10.8	7.7	6.0	13.7	16.1	40.1	5.1	38.7	21.2	78.8	6
450	1.0	0.3	1.3	7.2	5.6	12.8	8.2	5.9	14.1	7.1	51.1	2.1	39.7	9.2	90.8	6
480	1.7	0.6	2.3	6.1	5.8	11.9	7.8	6.4	14.2	12.0	43.0	4.2	40.8	16.2	83.8	6
510	1.3	0.5	1.8	5.4	6.3	11.7	6.7	6.8	13.5	9.6	40.0	3.7	46.7	13.3	86.7	6
540	1.0	0.4	1.4	8.1	6.0	14.1	9.1	6.4	15.5	6.4	52.3	2.6	38.7	9.0	91.0	6
570	0.8	0.3	1.1	7.2	6.1	13.3	8.0	6.4	14.4	5.5	50.0	2.1	42.4	7.6	92.4	5
600	1.2	0.3	1.5	8.1	5.3	13.4	9.3	5.6	14.9	8.0	54.4	2.1	35.5	10.1	89.9	3
630	2.5	0.5	3.0	5.8	3.8	9.6	8.3	4.3	12.6	19.8	46.0	4.0	30.2	23.8	76.2	1
660	0.9	0.2	1.1	6.5	3.8	10.3	7.4	4.0	11.4	7.9	57.0	1.8	33.3	9.7	90.3	1
690	1.0	0.1	1.1	9.8	4.0	13.8	10.8	4.1	14.9	6.7	65.8	0.7	26.8	7.4	92.6	1
720	0.5	0.1	0.6	12.6	4.9	17.5	13.1	5.0	18.1	2.8	69.6	0.5	27.1	3.3	96.7	1
750	0.5	0.1	0.6	14.5	5.9	20.4	15.0	6.0	21.0	2.4	69.0	0.5	28.1	2.9	97.1	1

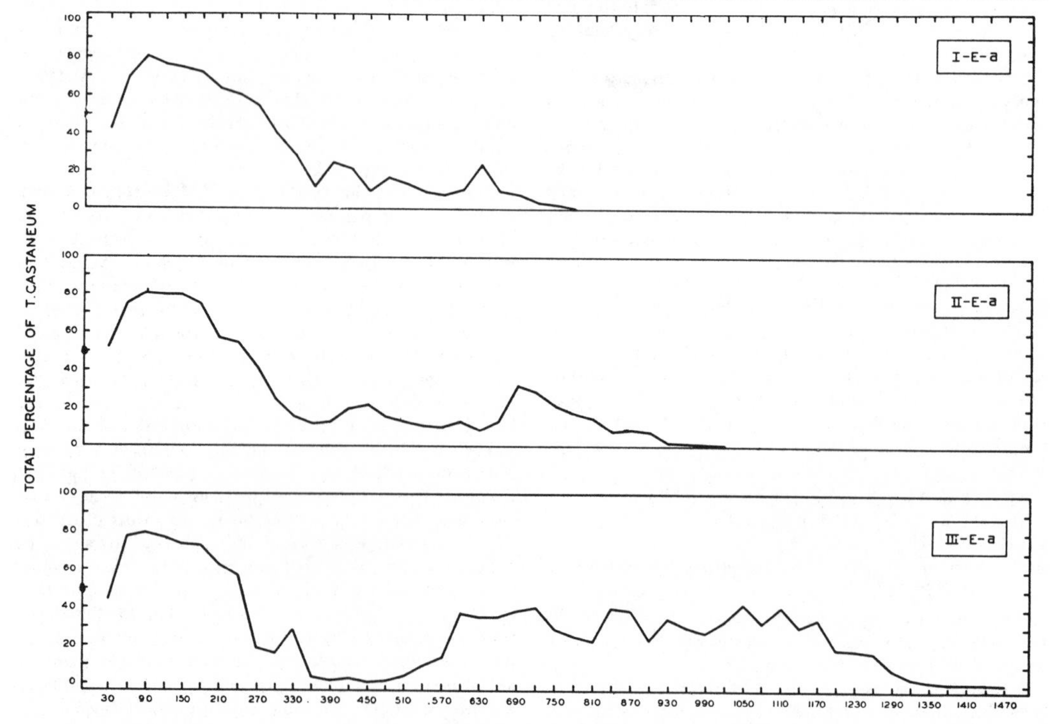

FIG. 9. Competition curves (percentage of *T. castaneum* against age in days) for I-E-a, II-E-a, and III-E-a populations in which *T. castaneum* eventually becomes extinct. (Distance from abscissa to curve is that percentage of the total count which is *T. castaneum*; from curve to 100 per cent line, the percentage of *T. confusum*.)

TABLE 16. II-E-a: *Tribolium castaneum* becoming extinct.

Age (days)	MEAN NUMBER PER GRAM									PERCENTAGE OF						n
	T. castaneum			*T. confusum*			Total			Larvae and Pupae		Imagoes		Total		
	L-P	Imag.	Sum	L-P	Imag.	Sum	L-P	Imag.	Sum	*T. cast.*	*T. conf.*	*T. cast.*	*T. conf.*	*T. cast.*	*T. conf.*	
30	7.2	3.8	11.0	9.2	1.0	10.2	16.4	4.8	21.2	34.0	43.4	17.9	4.7	51.9	48.1	9
60	6.7	11.3	18.0	1.1	5.0	6.1	7.8	16.3	24.1	27.8	4.6	46.9	20.7	74.7	25.3	9
90	9.0	11.2	20.2	0.4	4.6	5.0	9.4	15.8	25.2	35.7	1.6	44.5	18.2	80.2	19.8	9
120	7.6	10.1	17.7	0.1	4.5	4.6	7.7	14.6	22.3	34.1	0.4	45.3	20.2	79.4	20.6	9
150	7.6	8.4	16.0	0.1	4.1	4.2	7.7	12.5	20.2	37.6	0.5	41.6	20.3	79.2	20.8	9
180	6.1	5.3	11.4	0.3	3.6	3.9	6.4	8.9	15.3	39.9	2.0	34.6	23.5	74.5	25.5	9
210	4.0	1.8	5.8	1.2	3.2	4.4	5.2	5.0	10.2	39.2	11.8	17.7	31.3	56.9	43.1	9
240	5.8	0.9	6.7	2.2	3.5	5.7	8.0	4.4	12.4	46.8	17.7	7.2	28.3	54.0	46.0	9
270	5.0	0.5	5.5	3.9	3.9	7.8	8.9	4.4	13.3	37.6	29.3	3.8	29.3	41.4	58.6	9
300	2.9	0.4	3.3	5.3	5.1	10.4	8.2	5.5	13.7	21.2	38.7	2.9	37.2	24.1	75.9	8
330	1.9	0.3	2.2	6.8	5.6	12.4	8.7	5.9	14.6	13.0	46.6	2.1	38.3	15.1	84.9	8
360	1.1	0.3	1.4	4.7	5.9	10.6	5.8	6.2	12.0	9.2	39.2	2.5	49.1	11.7	88.3	6
390	1.4	0.3	1.7	6.9	4.7	11.6	8.3	5.0	13.3	10.5	51.9	2.3	35.3	12.8	87.2	5
420	1.8	0.4	2.2	5.1	3.9	9.0	6.9	4.3	11.2	16.1	45.5	3.5	34.9	19.6	80.4	3
450	2.5	0.5	3.0	6.5	4.6	11.1	9.0	5.1	14.1	17.7	46.1	3.6	32.6	21.3	78.7	3
480	1.7	0.5	2.2	6.5	5.7	12.2	8.2	6.2	14.4	11.8	45.1	3.5	39.6	15.3	84.7	3
510	1.2	0.4	1.6	5.8	5.4	11.2	7.0	5.8	12.8	9.4	45.3	3.1	42.2	12.5	87.5	3
540	1.0	0.3	1.3	6.4	5.1	11.5	7.4	5.4	12.8	7.8	50.0	2.4	39.8	10.2	89.8	3
570	0.9	0.3	1.2	5.9	5.2	11.1	6.8	5.5	12.3	7.3	50.0	2.5	40.2	9.8	90.2	3
600	1.3	0.3	1.6	5.9	5.1	11.0	7.2	5.4	12.6	10.3	46.8	2.4	40.5	12.7	87.3	3
630	0.8	0.2	1.0	6.2	5.3	11.5	7.0	5.5	12.5	6.4	49.6	1.6	42.4	8.0	92.0	3
660	1.3	0.5	1.8	6.6	5.6	12.2	7.9	6.1	14.0	9.3	47.1	3.6	40.0	12.9	87.1	2
690	3.2	0.8	4.0	4.8	3.7	8.5	8.0	4.5	12.5	25.6	38.4	6.4	29.6	32.0	68.0	1
720	3.2	0.8	4.0	6.5	3.6	10.1	9.7	4.4	14.1	22.7	46.1	5.7	25.5	28.4	71.6	1
750	2.3	0.6	2.9	7.1	3.7	10.8	9.4	4.3	13.7	16.8	51.8	4.4	27.0	21.2	78.8	1
780	2.1	0.5	2.6	8.9	3.7	12.6	11.0	4.2	15.2	13.8	58.5	3.3	24.4	17.1	82.9	1
810	1.9	0.4	2.3	9.4	4.5	13.9	11.3	4.9	16.2	11.7	58.0	2.5	27.8	14.2	85.8	1
840	0.9	0.2	1.1	8.9	4.0	12.9	9.8	4.2	14.0	6.4	63.6	1.5	28.5	7.9	92.1	1
870	0.6	0.3	0.9	5.3	4.6	9.9	5.9	4.9	10.8	5.6	49.1	2.7	42.6	8.3	91.7	1
900	1.0	0.2	1.2	10.7	4.2	14.9	11.7	4.4	16.1	6.2	66.5	1.3	26.0	7.5	92.5	1
930	0.2	0.05	0.25	10.3	5.95	16.25	10.5	6.0	16.5	1.2	62.4	0.3	36.1	1.5	98.5	1
960	0.1	0.05	0.15	8.4	6.72	15.12	8.5	6.77	15.27	0.6	55.0	0.4	44.0	1.0	99.0	1
990	0.1	0.02	0.12	9.2	6.55	15.75	9.3	6.57	15.87	0.6	58.0	0.2	41.2	0.8	99.2	1

Despite their fundamental similarity, there are also certain differences between the graphs of Figure 9—differences especially evident after the first year—and these require mention. While the I-E-a and II-E-a populations are themselves reasonably confluent, both diverge somewhat from III-E-a. For one thing, the III-E-a curve falls much lower at the 360 day reading, and the two replicates are maintained as experimental cultures only by virtue of the presence of a very few *T. castaneum* (0.5 to 3.8%). After day-540, however, this species stages so sharp a recovery that it returns to a level quite higher than that seen in the populations of smaller volumes. This is well documented by noting that, over the twenty census readings starting at day-600 and ending at day-1170, *T. castaneum* constitutes, on the average, 34% of the total. It is a possible conjecture that this increase is related to a temporary reduction in the extent and/or effectiveness of parasitization. Figure 9 shows also that the two III-E-a cultures display a longer life-duration as experimental populations than those of the I and II volumes. In fact, these are the most longevous of all the mixed species groups. The explanation of this, if not to be accounted for by accidents of sampling, is obscure.

I-E-b, II-E-b, and III-E-b populations.—Curves depicting these three sets of cultures are graphed in Figure 10 and are based on the information contained in Tables 18, 19, and 20. Here again, a general visual similarity is evident for the three curves, particularly during the first year of growth. Also, there are clear differences in pattern between them and those describing the E-a populations (Figure 9).

The early disparity existing between the E-a and E-b series is unquestionably related to the densities in which the two species of beetles were introduced when the experiments were started. In the E-a group, in which *T. confusum* and *T. castaneum* were associated in similar numbers, the latter attained greater abundance than the former during the first three months. In the E-b group, on the other hand, *T. confusum*, introduced at a numerical advantage of 50%, was just able to maintain itself on an equal basis with its competitor during this interval. For the E-c cultures, with *T. castaneum* constituting three-fourths of the initial population, *T. confusum* multiplied only slightly during the period and occupied less than ten per cent of the total habitat at day-90. In short, there is a positive relationship between initial concentration and size of population at the time of the third census, but *T. castaneum* takes greater advantage of this than does *T. confusum* in

TABLE 17. III-E-a: *Tribolium castaneum* becoming extinct.

Age (days)	MEAN NUMBER PER GRAM									PERCENTAGE OF						n
	T. castaneum			*T. confusum*			Total			Larvae and Pupae		Imagoes		Total		
	L-P	Imag.	Sum	L-P	Imag.	Sum	L-P	Imag.	Sum	*T. cast.*	*T. conf.*	*T. cast.*	*T. conf.*	*T. cast.*	*T. conf.*	
30......	6.7	2.9	9.6	11.0	0.9	11.9	17.7	3.8	21.5	31.2	51.2	13.5	4.1	44.7	55.3	2
60......	6.0	9.9	15.9	0.6	3.9	4.5	6.6	13.8	20.4	29.4	2.9	48.5	19.2	77.9	22.1	2
90......	8.3	9.6	17.9	0.1	4.5	4.6	8.4	14.1	22.5	36.9	0.4	42.7	20.0	79.6	20.4	2
120......	5.7	8.9	14.6	0.1	4.2	4.3	5.8	13.1	18.9	30.2	0.5	47.1	22.2	77.3	22.7	2
150......	4.4	6.8	11.2	0.2	3.8	4.0	4.6	10.6	15.2	28.9	1.3	44.8	25.0	73.7	26.3	2
180......	6.1	4.9	11.0	0.7	3.4	4.1	6.8	8.3	15.1	40.4	4.6	32.5	22.5	72.9	27.1	2
210......	4.7	2.2	6.9	1.2	2.9	4.1	5.9	5.1	11.0	42.7	10.9	20.0	26.4	62.7	37.3	2
240......	5:4	1.1	6.5	1.9	3.0	4.9	7.3	4.1	11.4	47.4	16.7	9.6	26.3	57.0	43.0	2
270......	1.9	0.3	2.2	5.8	4.0	9.8	7.7	4.3	12.0	15.8	48.3	2.5	33.4	18.3	81.7	2
300......	1.9	0.2	2.1	5.6	5.5	11.1	7.5	5.7	13.2	14.4	42.4	1.5	41.7	15.9	84.1	2
330......	3.4	0.8	4.2	4.4	6.3	10.7	7.8.	7.1	14.9	22.8	29.5	5.4	42.3	28.2	71.8	2
360......	0.59	0.06	0.65	10.91	5.78	16.69	11.50	5.84	17.34	3.4	62.9	0.4	33.3	3.8	96.2	2
390......	0.18	0.02	0.20	10.42	4.37	14.79	10.60	4.39	14.99	1.2	69.5	0.1	29.2	1.3	98.7	2
420......	0.42	0.02	0.44	15.67	3.79	19.46	16.09	3.81	19.90	2.1	78.7	0.1	19.1	2.2	97.8	2
450......	0.08	0.01	0.09	12.03	4.87	16.90	12.11	4.88	16.99	0.4	70.8	0.1	28.7	0.5	99.5	2
480......	0.14	0.03	0.17	10.67	6.12	16.79	10.81	6.15	16.96	0.8	62.9	0.2	36.1	1.0	99.0	2
510......	0.69	0.06	0.75	14.51	4.51	19.02	15.20	4.57	19.77	3.5	73.4	0.3	22.8	3.8	96.2	2
540......	1.3	0.2	1.5	10.5	4.1	14.6	11.8	4.3	16.1	8.1	65.2	1.2	25.5	9.3	90.7	2
570......	1.6	0.2	1.8	8.6	3.0	11.6	10.2	3.2	13.4	11.9	64.2	1.5	22.4	13.4	86.6	2
600......	4.5	0.4	4.9	6.0	2.3	8.3	10.5	2.7	13.2	34.1	45.4	3.0	17.5	37.1	62.9	2
630......	5.3	0.3	5.6	8.3	2.0	10.3	13.6	2.3	15.9	33.3	52.2	1.9	12.6	35.2	64.8	2
660......	4.2	0.6	4.8	6.3	2.4	8.7	10.5	3.0	13.5	31.1	46.7	4.5	17.7	35.6	64.4	2
690......	5.2	0.5	5.7	7.3	1.7	9.0	12.5	2.2	14.7	35.4	49.7	3.4	11.5	38.8	61.2	2
720......	4.5	0.5	5.0	5.8	1.5	7.3	10.3	2.0	12.3	36.6	47.1	4.1	12.2	40.7	59.3	2
750......	3.8	0.4	4.2	8.3	1.9	10.2	12.1	2.3	14.4	26.4	57.6	2.8	13.2	29.2	70.8	2
780......	3.7	0.3	4.0	10.3	1.4	11.7	14.0	1.7	15.7	23.6	65.6	1.9	8.9	25.5	74.5	2
810......	2.4	0.2	2.6	7.7	1.1	8.8	10.1	1.3	11.4	21.0	67.5	1.8	9.7	22.8	77.2	2
840......	3.5	0.7	4.2	5.3	0.8	6.1	8.8	1.5	10.3	34.0	51.5	6.8	7.7	40.8	59.2	1
870......	2.9	1.2	4.1	5.0	1.4	6.4	7.9	2.6	10.5	27.6	47.6	11.5	13.3	39.1	60.9	1
900......	2.7	0.8	3.5	10.2	1.0	11.2	12.9	1.8	14.7	18.4	69.4	5.4	6.8	23.8	76.2	1
930......	2.3	1.9	4.2	5.6	2.2	7.8	7.9	4.1	12.0	19.2	46.7	15.8	18.3	35.0	65.0	1
960......	4.6	1.1	5.7	11.6	1.4	13.0	16.2	2.5	18.7	24.6	62.0	5.9	7.5	30.5	69.5	1
990......	1.1	1.2	2.3	4.5	1.6	6.1	5.6	2.8	8.4	13.1	53.6	14.3	19.0	27.4	72.6	1
1020......	3.7	0.8	4.5	7.8	1.2	9.0	11.5	2.0	13.5	27.4	57.8	5.9	8.9	33.3	66.7	1
1050......	5.0	0.8	5.8	6.7	1.2	7.9	11.7	2.0	13.7	36.5	48.9	5.8	8.8	42.3	57.7	1
1080......	3.1	0.5	3.6	6.4	1.0	7.4	9.5	1.5	11.0	28.2	58.2	4.5	9.1	32.7	67.3	1
1110......	3.3	0.6	3.9	4.7	0.9	5.6	8.0	1.5	9.5	34.7	49.5	6.4	9.4	41.1	58.9	1
1140......	2.1	0.4	2.5	4.9	0.8	5.7	7.0	1.2	8.2	25.6	59.8	4.9	9.7	30.5	69.5	1
1170......	3.1	0.5	3.6	5.5	1.3	6.8	8.6	1.8	10.4	29.8	52.9	4.8	12.5	34.6	65.4	1
1200......	2.4	0.3	2.7	9.9	1.6	11.5	12.3	1.9	14.2	16.9	69.7	2.1	11.3	19.0	81.0	1
1230......	1.9	0.2	2.1	7.7	1.5	9.2	9.6	1.7	11.3	16.8	68.1	1.8	13.3	18.6	81.4	1
1260......	2.3	0.2	2.5	10.5	1.6	12.1	12.8	1.8	14.6	15.7	71.9	1.4	11.0	17.1	82.9	1
1290......	0.8	0.2	1.0	10.2	1.3	11.5	11.0	1.5	12.5	6.4	81.6	1.6	10.4	8.0	92.0	1
1320......	0.4	0.1	0.5	12.8	2.4	15.2	13.2	2.5	15.7	2.5	81.5	0.7	15.3	3.2	96.8	1
1350......	0.26	0.05	0.31	14.64	1.79	16.43	14.90	1.84	16.74	1.5	87.4	0.4	10.7	1.9	98.1	1
1380......	0.08	0.02	0.10	12.52	2.24	14.76	12.60	2.26	14.86	0.5	84.2	0.2	15.1	0.7	99.3	1
1410......	0.08	0.01	0.09	17.83	1.75	19.58	17.91	1.76	19.67	0.4	90.6	0.1	8.9	0.5	99.5	1
1440......	0.07	0.01	0.08	20.33	1.94	22.27	20.40	1.95	22.35	0.3	91.0	0.1	8.6	0.4	99.6	1

the sense that its relative increase in number of individuals is much greater. The point is readily documented by averaging together for the I, II, and III volumes those percentages listed in the appropriate tables for each of the three series (E-a, E-b, and E-c) at the 90-day census. These averages are as follows:

	T. castaneum %	*T. confusum* %
E-a	80.0	20.0
E-b	50.9	49.1
E-c	93.4	6.6

Thus, the proportional representation of *T. confusum* in E-a, E-b, and E-c is, respectively, 30.0, 25.9, and 18.4% *less* than would be expected on the assumption that, individual for individual, both species are evenly matched during this period of establishment. This interesting fact has no common relation to species *extinction,* however, for, as we have already seen, it is *T. castaneum* that eventually dies in all these 66 populations now under review. The major point reaffirmed here is that it *is* possible for *T. castaneum* to reproduce effectively in the presence of *T. confusum.*

The remarks above serve as a description of the early phase of growth of the E-b cultures. The period of decline for *T. castaneum* illustrated in Figure 10 should now be examined. This takes place, as it did for the E-a series, in all three volumes. The decline is of less magnitude because the maximal densities at day-90 were lower. But it is appreciable nevertheless and, occurring as it does over the same time interval, re-enforces the conclusion that it is caused primarily by an increasing infection pressure exerted by Adelina. This contraction of the *T. castaneum* component of the various experimental populations leads to the complete extinction of this species. By day-360, the two III-E-b cultures have become controls; the last I-E-b replicate is gone at day-750; and the last II-E-b, at day-540. The median extinction ages are 360, 360, and 345 days for volumes I, II, and III, respectively.

I-E-c, II-E-c, and III-E-c populations.—The performances of these three sets of experimental popu-

TABLE 18. I-E-b: *Tribolium castaneum* becoming extinct.

Age (days)	Mean Number Per Gram									Percentage of						n
	T. castaneum			*T. confusum*			Total			Larvae and Pupae		Imagoes		Total		
	L-P	Imag.	Sum	L-P	Imag.	Sum	L-P	Imag.	Sum	*T. cast.*	*T. conf.*	*T. cast.*	*T. conf.*	*T. cast.*	*T. conf.*	
30	5.8	1.1	6.9	15.0	1.0	16.0	20.8	2.1	22.9	25.3	65.5	4.8	4.4	30.1	69.9	15
60	3.1	5.1	8.2	2.2	10.3	12.5	5.3	15.4	20.7	15.0	10.6	24.6	49.8	39.6	60.4	15
90	5.3	5.1	10.4	1.1	10.1	11.2	6.4	15.2	21.6	24.5	5.1	23.7	46.7	48.2	51.8	15
120	3.5	4.1	7.6	0.7	9.8	10.5	4.2	13.9	18.1	19.3	3.9	22.7	54.1	42.0	58.0	15
150	4.3	3.6	7.9	0.9	9.2	10.1	5.2	12.8	18.0	23.9	5.0	20.0	51.1	43.9	56.1	15
180	4.1	3.0	7.1	1.5	8.5	10.0	5.6	11.5	17.1	24.0	8.8	17.5	49.7	41.5	58.5	15
210	2.7	2.0	4.7	1.6	7.6	9.2	4.3	9.6	13.9	19.4	11.5	14.4	54.7	33.8	66.2	15
240	2.6	1.5	4.1	2.8	6.6	9.4	5.4	8.1	13.5	19.3	20.7	11.1	48.9	30.4	69.6	15
270	3.3	1.1	4.4	4.2	6.4	10.6	7.5	7.5	15.0	22.0	28.0	7.3	42.7	29.3	70.7	14
300	3.7	0.9	4.6	5.5	6.2	11.7	9.2	7.1	16.3	22.7	33.7	5.5	38.1	28.2	71.8	11
330	2.9	0.8	3.7	4.9	6.7	11.6	7.8	7.5	15.3	18.9	32.0	5.3	43.8	24.2	75.8	10
360	1.4	0.8	2.2	5.0	6.3	11.3	6.4	7.1	13.5	10.4	37.0	5.9	46.7	16.3	83.7	7
390	3.1	0.9	4.0	6.0	5.1	11.1	9.1	6.0	15.1	20.5	39.7	6.0	33.8	26.5	73.5	5
420	2.1	0.6	2.7	6.6	5.8	12.4	8.7	6.4	15.1	13.9	43.7	4.0	38.4	17.9	82.1	5
450	2.4	0.7	3.1	7.8	5.9	13.7	10.2	6.6	16.8	14.3	46.4	4.2	35.1	18.5	81.5	4
480	1.2	0.5	1.7	6.5	7.2	13.7	7.7	7.7	15.4	7.8	42.2	3.2	46.8	11.0	89.0	4
510	1.2	0.4	1.6	7.4	8.0	15.4	8.6	8.4	17.0	7.1	43.5	2.3	47.1	9.4	90.6	4
540	0.6	0.3	0.9	8.9	8.6	17.5	9.5	8.9	18.4	3.3	48.4	1.6	46.7	4.9	95.1	4
570	0.4	0.2	0.6	7.8	9.3	17.1	8.2	9.5	17.7	2.3	44.1	1.1	52.5	3.4	96.6	4
600	0.3	0.2	0.5	4.8	10.4	15.2	5.1	10.6	15.7	1.9	30.6	1.3	66.2	3.2	96.8	3
630	0.2	0.1	0.3	6.8	10.3	17.1	7.0	10.4	17.4	1.1	39.1	0.6	59.2	1.7	98.3	2
660	0.2	0.1	0.3	7.0	9.1	16.1	7.2	9.2	16.4	1.2	42.7	0.6	55.5	1.8	98.2	2
690	0.3	0.1	0.4	11.6	8.6	20.2	11.9	8.7	20.6	1.5	56.3	0.4	41.8	1.9	98.1	2
720	0.2	0.1	0.3	10.2	11.6	21.8	10.4	11.7	22.1	0.9	46.1	0.5	52.5	1.4	98.6	1

TABLE 19. II-E-b: *Tribolium castaneum* becoming extinct.

Age (days)	Mean Number Per Gram									Percentage of						n
	T. castaneum			*T. confusum*			Total			Larvae and Pupae		Imagoes		Total		
	L-P	Imag.	Sum	L-P	Imag.	Sum	L-P	Imag.	Sum	*T. cast.*	*T. conf.*	*T. cast.*	*T. conf.*	*T. cast.*	*T. conf.*	
30	2.8	2.6	5.4	13.0	2.5	15.5	15.8	5.1	20.9	13.4	62.2	12.4	12.0	25.8	74.2	7
60	3.6	6.0	9.6	2.1	9.6	11.7	5.7	15.6	21.3	16.9	9.9	28.2	45.0	45.1	54.9	7
90	5.8	5.8	11.6	1.0	9.4	10.4	6.8	15.2	22.0	26.4	4.5	26.3	42.8	52.7	47.3	7
120	4.7	4.7	9.4	0.6	9.2	9.8	5.3	13.9	19.2	24.5	3.1	24.5	47.9	49.0	51.0	7
150	4.7	3.8	8.5	0.6	8.4	9.0	5.3	12.2	17.5	26.9	3.4	21.7	48.0	48.6	51.4	7
180	4.0	2.8	6.8	1.0	7.4	8.4	5.0	10.2	15.2	26.3	6.6	18.4	48.7	44.7	55.3	7
210	4.3	2.7	7.0	1.8	6.4	8.2	6.1	9.1	15.2	28.3	11.8	17.8	42.1	46.1	53.9	7
240	3.8	0.7	4.5	3.0	6.1	9.1	6.8	6.8	13.6	27.9	22.1	5.2	44.8	33.1	66.9	7
270	3.5	1.0	4.5	3.5	7.2	10.7	7.0	8.2	15.2	23.0	23.0	6.6	47.4	29.6	70.4	7
300	0.6	0.1	0.7	6.9	7.8	14.7	7.5	7.9	15.4	3.9	44.8	0.7	50.6	4.6	95.4	6
330	0.31	0.09	0.40	6.45	8.38	14.83	6.76	8.47	15.23	2.0	42.4	0.6	55.0	2.6	97.4	4
360	0.19	0.08	0.27	4.88	7.49	12.37	5.07	7.57	12.64	1.5	38.6	0.6	59.3	2.1	97.9	3
390	0.31	0.07	0.38	6.43	6.52	12.95	6.74	6.59	13.33	2.3	48.2	0.6	48.9	2.9	97.1	3
420	0.27	0.08	0.35	9.08	7.02	16.10	9.35	7.10	16.45	1.6	55.2	0.5	42.7	2.1	97.9	2
450	0.19	0.05	0.24	9.69	7.55	17.24	9.88	7.60	17.48	1.1	55.4	0.3	43.2	1.4	98.6	1
480	0.13	0.03	0.16	9.30	7.12	16.42	9.43	7.15	16.58	0.8	56.1	0.2	42.9	1.0	99.0	1
510	0.23	0.05	0.28	11.04	5.80	16.84	11.27	5.85	17.12	1.3	64.5	0.3	33.9	1.6	98.4	1

lations are recorded in Tables 21, 22, and 23 and graphed in Figure 11. The salient facts can be quickly summarized in the following five points:

(1) The three curves exhibit a general agreement between themselves and a general dissimilarity during the first year to those already discussed.

(2) Once again, the maximum percentage of *T. castaneum* is reached at day-90 at which time *T. confusum* constitutes 9.2 per cent of the total in the I volumes, 6.1% in the II volumes, and 4.5% in the III volumes. As suggested earlier, this very high proportion of *T. castaneum,* higher than that attained in any of the other cultures, is related to the fact that this beetle was introduced initially with a 50% advantage over its competitor plus the added and significant fact that, during this interval, it was able to multiply effectively.

(3) The decline in *T. castaneum,* starting about the time of the fourth or fifth census and so characteristic of E-a and E-b, is equally well displayed by the E-c populations. Actually, this decline is more dramatic for the last group because *T. castaneum* has a greater potentiality for contraction by virtue of its greater early abundance. The period of contraction is similar to that seen in the controls and the other two experimental series—a datum further incriminating Adelina as a causal agent.

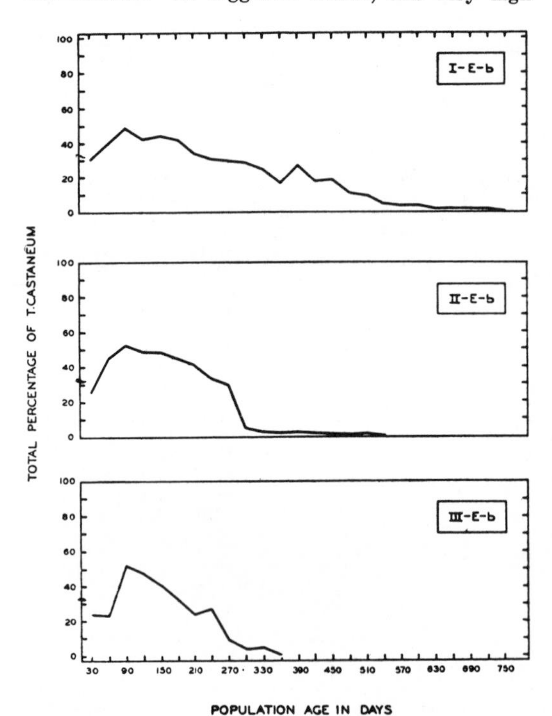

FIG. 10. Competition curves for I-E-b, II-E-b, and III-E-b populations in which *T. castaneum* becomes extinct.

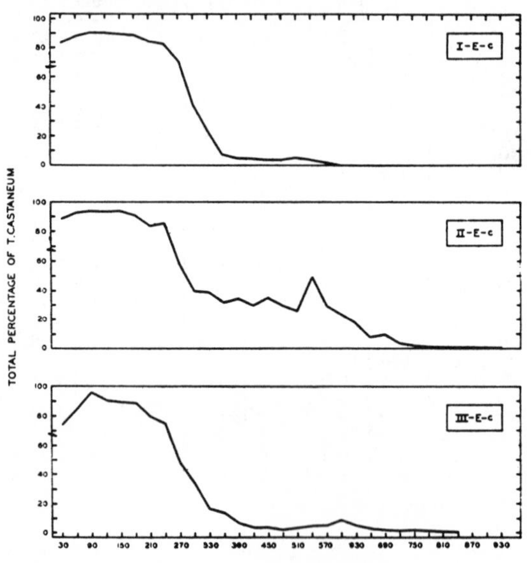

FIG. 11. Competition curves for I-E-c, II-E-c, and III-E-c populations in which *T. castaneum* becomes extinct.

TABLE 20. III-E-b: *Tribolium castaneum* becoming extinct.

Age (days)	MEAN NUMBER PER GRAM									PERCENTAGE OF						n
	T. castaneum			*T. confusum*			Total			Larvae and Pupae		Imagoes		Total		
	L-P	Imag.	Sum	L-P	Imag.	Sum	L-P	Imag.	Sum	*T. cast.*	*T. conf.*	*T. cast.*	*T. conf.*	*T. cast.*	*T. conf.*	
30.......	3.2	1.8	5.0	15.1	1.6	16.7	18.3	3.4	21.7	14.7	69.6	8.4	7.3	23.1	76.9	2
60.......	2.1	1.7	3.8	2.9	10.0	12.9	5.0	11.7	16.7	12.6	17.4	10.2	59.8	22.8	77.2	2
90.......	6.2	5.0	11.2	0.5	9.9	10.4	6.7	14.9	21.6	28.7	2.3	23.2	45.8	51.9	48.1	2
120.......	4.5	4.4	8.9	0.4	9.6	10.0	4.9	14.0	18.9	23.8	2.1	23.3	50.8	47.1	52.9	2
150.......	2.9	3.5	6.4	0.4	9.1	9.5	3.3	12.6	15.9	18.2	2.5	22.1	57.2	40.3	59.7	2
180.......	2.4	2.1	4.5	1.5	8.1	9.6	3.9	10.2	14.1	17.0	10.6	14.9	57.5	31.9	68.1	2
210.......	2.0	0.9	2.9	3.1	6.2	9.3	5.1	7.1	12.2	16.4	25.4	7.4	50.8	23.8	76.2	2
240.......	2.8	0.4	3.2	4.2	4.5	8.7	7.0	4.9	11.9	23.5	35.3	3.4	37.8	26.9	73.1	2
270.......	1.2	0.2	1.4	7.9	6.6	14.5	9.1	6.8	15.9	7.5	49.7	1.3	41.5	8.8	91.2	2
300.......	0.34	0.08	0.42	3.74	7.67	11.41	4.08	7.75	11.83	2.9	31.6	0.7	64.8	3.6	96.4	2
330.......	0.56	0.01	0.57	5.91	5.90	11.81	6.47	5.91	12.38	4.5	47.7	0.1	47.7	4.6	95.4	1

TABLE 21. I-E-c: *Tribolium castaneum* becoming extinct.

Age (days)	Mean Number Per Gram									Percentage of						n
	T. castaneum			*T. confusum*			Total			Larvae and Pupae		Imagoes		Total		
	L-P	Imag.	Sum	L-P	Imag.	Sum	L-P	Imag.	Sum	*T. cast.*	*T. conf.*	*T. cast.*	*T. conf.*	*T. cast.*	*T. conf.*	
30	14.8	5.9	20.7	3.6	0.5	4.1	18.4	6.4	24.8	59.7	14.5	23.8	2.0	83.5	16.5	11
60	6.7	15.0	21.7	0.4	2.4	2.8	7.1	17.4	24.5	27.3	1.6	61.3	9.8	88.6	11.4	11
90	8.5	15.2	23.7	0.2	2.2	2.4	8.7	17.4	26.1	32.6	0.8	58.2	8.4	90.8	9.2	11
120	7.6	13.6	21.2	0.1	2.2	2.3	7.7	15.8	23.5	32.3	0.4	57.9	9.4	90.2	9.8	11
150	7.7	10.7	18.4	0.1	2.0	2.1	7.8	12.7	20.5	37.6	0.5	52.2	9.7	89.8	10.2	11
180	6.4	7.9	14.3	0.1	1.7	1.8	6.5	9.6	16.1	39.7	0.6	49.1	10.6	88.8	11.2	11
210	5.6	5.0	10.6	0.4	1.6	2.0	6.0	6.6	12.6	44.4	3.2	39.7	12.7	84.1	15.9	11
240	7.2	2.9	10.1	0.8	1.4	2.2	8.0	4.3	12.3	58.8	6.5	23.3	11.4	82.1	17.9	11
270	6.2	1.4	7.6	1.6	1.6	3.2	7.8	3.0	10.8	57.4	14.8	13.0	14.8	70.4	29.6	11
300	3.7	0.8	4.5	3.2	3.3	6.5	6.9	4.1	11.0	33.6	29.1	7.3	30.0	40.9	59.1	11
330	2.6	0.6	3.2	5.1	5.7	10.8	7.7	6.3	14.0	18.6	36.4	4.3	40.7	22.9	77.1	9
360	0.7	0.4	1.1	6.7	7.9	14.6	7.4	8.3	15.7	4.5	42.7	2.5	50.3	7.0	93.0	8
390	0.5	0.2	0.7	6.5	8.5	15.0	7.0	8.7	15.7	3.2	41.4	1.3	54.1	4.5	95.5	5
420	0.5	0.2	0.7	7.1	8.5	15.6	7.6	8.7	16.3	3.1	43.6	1.2	52.1	4.3	95.7	5
450	0.3	0.2	0.5	4.7	8.6	13.3	5.0	8.8	13.8	2.2	34.1	1.4	62.3	3.6	96.4	3
480	0.3	0.2	0.5	5.4	8.1	13.5	5.7	8.3	14.0	2.1	38.6	1.5	57.8	3.6	96.4	3
510	0.5	0.2	0.7	6.1	7.1	13.2	6.6	7.3	13.9	3.6	43.9	1.4	51.1	5.0	95.0	2
540	0.4	0.2	0.6	7.8	7.3	15.1	8.2	7.5	15.7	2.5	49.7	1.3	46.5	3.8	96.2	2
570	0.2	0.1	0.3	9.5	8.4	17.9	9.7	8.5	18.2	1.1	52.2	0.6	46.1	1.7	98.3	2

TABLE 22. II-E-c: *Tribolium castaneum* becoming extinct.

Age (days)	Mean Number Per Gram									Percentage of						n
	T. castaneum			*T. confusum*			Total			Larvae and Pupae		Imagoes		Total		
	L-P	Imag.	Sum	L-P	Imag.	Sum	L-P	Imag.	Sum	*T. cast.*	*T. conf.*	*T. cast.*	*T. conf.*	*T. cast.*	*T. conf.*	
30	13.4	8.0	21.4	2.4	0.3	2.7	15.8	8.3	24.1	55.6	10.0	33.2	1.2	88.8	11.2	7
60	8.5	17.1	25.6	0.3	1.8	2.1	8.8	18.9	27.7	30.7	1.1	61.7	6.5	92.4	7.6	7
90	11.3	16.3	27.6	0.1	1.7	1.8	11.4	18.0	29.4	38.4	0.3	55.5	5.8	93.9	6.1	7
120	7.5	15.3	22.8	0.1	1.6	1.7	7.6	16.9	24.5	30.6	0.4	62.5	6.5	93.1	6.9	7
150	7.7	11.8	19.5	0.1	1.3	1.4	7.8	13.1	20.9	36.8	0.5	56.5	6.2	93.3	6.7	7
180	5.3	7.9	13.2	0.1	1.3	1.4	5.4	9.2	14.6	36.3	0.7	54.1	8.9	90.4	9.6	7
210	3.7	3.7	7.4	0.3	1.2	1.5	4.0	4.9	8.9	41.6	3.4	41.6	13.4	83.2	16.8	7
240	7.9	2.3	10.2	0.5	1.3	1.8	8.4	3.6	12.0	65.8	4.2	19.2	10.8	85.0	15.0	7
270	4.1	0.8	4.9	1.5	2.2	3.7	5.6	3.0	8.6	47.7	17.4	9.3	25.6	57.0	43.0	7
300	3.6	0.7	4.3	2.8	3.9	6.7	6.4	4.6	11.0	32.7	25.4	6.4	35.5	39.1	60.9	7
330	4.2	1.1	5.3	3.7	4.9	8.6	7.9	6.0	13.9	30.2	26.6	7.9	35.3	38.1	61.9	7
360	2.6	1.2	3.8	2.8	5.6	8.4	5.4	6.8	12.2	21.3	22.9	9.9	45.9	31.2	68.8	6
390	3.6	1.2	4.8	4.4	4.9	9.3	8.0	6.1	14.1	25.5	31.2	8.5	34.8	34.0	66.0	5
420	2.6	0.9	3.5	3.8	4.6	8.4	6.4	5.5	11.9	21.8	31.9	7.6	38.7	29.4	70.6	5
450	4.1	0.8	4.9	5.5	3.8	9.3	9.6	4.6	14.2	28.9	38.7	5.6	26.8	34.5	65.5	4
480	2.0	0.8	2.8	3.0	3.7	6.7	5.0	4.5	9.5	21.0	31.6	8.5	38.9	29.5	70.5	3
510	2.0	0.7	2.7	3.9	3.9	7.8	5.9	4.6	10.5	19.0	37.1	6.7	37.2	25.7	74.3	3
540	4.0	0.8	4.8	4.0	2.1	6.1	8.0	2.9	10.9	36.7	36.7	7.3	19.3	44.0	56.0	2
570	2.0	0.8	2.8	3.4	3.3	6.7	5.4	4.1	9.5	21.0	35.8	8.5	34.7	29.5	70.5	2
600	2.0	0.6	2.6	4.0	4.4	8.4	6.0	5.0	11.0	18.2	36.4	5.4	40.0	23.6	76.4	2
630	1.3	0.5	1.8	3.6	4.9	8.5	4.9	5.4	10.3	12.6	34.9	4.9	47.6	17.5	82.5	2
660	0.5	0.3	0.8	4.1	5.5	9.6	4.6	5.8	10.4	4.8	39.4	2.9	52.9	7.7	92.3	2
690	0.9	0.3	1.2	5.6	5.8	11.4	6.5	6.1	12.6	7.1	44.4	2.4	46.1	9.5	90.5	2
720	0.3	0.2	0.5	6.1	8.3	14.4	6.4	8.5	14.9	2.0	40.9	1.4	55.7	3.4	96.6	2
750	0.1	0.1	0.2	2.1	9.0	11.1	2.2	9.1	11.3	0.9	18.6	0.9	79.6	1.8	98.2	2
780	0.06	0.05	0.11	3.47	9.12	12.59	3.53	9.17	12.70	0.5	27.3	0.4	71.8	0.9	99.1	1
810	0.08	0.05	0.13	5.93	8.22	14.15	6.01	8.27	14.28	0.6	41.5	0.3	57.6	0.9	99.1	1
840	0.03	0.05	0.08	5.44	9.48	14.92	5.47	9.53	15.00	0.2	36.3	0.3	63.2	0.5	99.5	1
870	0.09	0.05	0.14	6.82	8.32	15.14	6.91	8.37	15.28	0.6	44.6	0.3	54.5	0.9	99.1	1
900	0.10	0.02	0.12	11.23	8.00	19.23	11.33	8.02	19.35	0.5	58.0	0.1	41.4	0.6	99.4	1

TABLE 23. III-E-c: *Tribolium castaneum* becoming extinct.

Age (days)	MEAN NUMBER PER GRAM									PERCENTAGE OF						n
	T. castaneum			*T. confusum*			Total			Larvae and Pupae		Imagoes		Total		
	L-P	Imag.	Sum	L-P	Imag.	Sum	L-P	Imag.	Sum	*T. cast.*	*T. conf.*	*T. cast.*	*T. conf.*	*T. cast.*	*T. conf.*	
30.......	11.4	5.5	16.9	5.5	0.5	6.0	16.9	6.0	22.9	49.8	24.0	24.0	2.2	73.8	26.2	2
60.......	6.3	13.6	19.9	2.0	1.8	3.8	8.3	15.4	23.7	26.6	8.4	57.4	7.6	84.0	16.0	2
90.......	12.3	11.2	23.5	0.1	1.0	1.1	12.4	12.2	24.6	50.0	0.4	45.5	4.1	95.5	4.5	2
120.......	9.8	12.5	22.3	0.1	2.3	2.4	9.9	14.8	24.7	39.7	0.4	50.6	9.3	90.3	9.7	2
150.......	6.3	10.2	16.5	0.1	1.9	2.0	6.4	12.1	18.5	34.0	0.5	55.2	10.3	89.2	10.8	2
180.......	7.5	6.8	14.3	0.3	1.5	1.8	7.8	8.3	16.1	46.6	1.9	42.2	9.3	88.8	11.2	2
210.......	4.5	2.7	7.2	0.5	1.4	1.9	5.0	4.1	9.1	49.4	5.5	29.7	15.4	79.1	20.9	2
240.......	5.8	1.2	7.0	0.9	1.5	2.4	6.7	2.7	9.4	61.7	9.6	12.8	15.9	74.5	25.5	2
270.......	4.6	0.6	5.2	3.5	2.2	5.7	8.1	2.8	10.9	42.2	32.1	5.5	20.2	47.7	52.3	2
300.......	4.3	0.5	4.8	4.6	5.0	9.6	8.9	5.5	14.4	29.9	31.9	3.4	34.8	33.3	66.7	2
330.......	1.8	0.4	2.2	4.8	6.4	11.2	6.6	6.8	13.4	13.4	35.8	3.0	47.8	16.4	83.6	2
360.......	1.1	0.4	1.5	3.3	6.6	9.9	4.4	7.0	11.4	9.6	28.9	3.6	57.9	13.2	86.8	2
390.......	0.6	0.2	0.8	5.5	6.3	11.8	6.1	6.5	12.6	4.8	43.6	1.6	50.0	6.4	93.6	2
420.......	0.4	0.1	0.5	7.6	6.2	13.8	8.0	6.3	14.3	2.8	53.1	0.7	43.4	3.5	96.5	2
450.......	0.49	0.09	0.58	9.55	6.76	16.31	10.04	6.85	16.89	2.9	32.9	0.5	63.7	3.4	96.6	2
480.......	0.20	0.07	0.27	6.56	6.64	13.20	6.76	6.71	13.47	1.5	48.7	0.5	49.3	2.0	98.0	2
510.......	0.45	0.08	0.53	10.62	6.04	16.66	11.07	6.12	17.19	2.6	61.8	0.5	35.1	3.1	96.9	2
540.......	0.5	0.2	0.7	7.7	7.4	15.1	8.2	7.6	15.8	3.2	48.7	1.2	46.9	4.4	95.6	2
570.......	0.4	0.2	0.6	5.6	6.9	12.5	6.0	7.1	13.1	3.0	42.7	1.6	52.7	4.6	95.4	2
600.......	0.9	0.2	1.1	6.3	6.1	12.4	7.2	6.3	13.5	6.7	46.7	1.5	45.1	8.2	91.8	2
630.......	0.6	0.1	0.7	9.6	5.6	15.2	10.2	5.7	15.9	3.8	60.4	0.6	35.2	4.4	95.6	2
660.......	0.28	0.09	0.37	8.08	6.16	14.24	8.36	6.25	14.61	1.9	55.3	0.6	42.2	2.5	97.5	2
690.......	0.26	0.06	0.32	11.22	5.64	16.86	11.48	5.70	17.18	1.5	65.3	0.4	32.8	1.9	98.1	2
720.......	0.21	0.05	0.26	12.90	6.64	19.54	13.11	6.69	19.80	1.1	65.1	0.2	33.6	1.3	98.7	2
750.......	0.27	0.05	0.32	16.00	5.49	21.49	16.27	5.54	21.81	1.2	73.4	0.3	25.1	1.5	98.5	1
780.......	0.11	0.02	0.13	12.44	7.42	19.86	12.55	7.44	19.99	0.5	62.2	0.2	37.1	0.7	99.3	1

(4) There is close confluence between the curves for I-E-c and III-E-c from day-240 until species extinction occurs. The percentage of *T. castaneum* in the II-E-c populations is higher for a number of census readings subsequent to day-270 and certain replicates persist as mixed-species cultures for quite a period.

(5) The temporal relations of extinction for the three experimental series are as follows:

	First extinction (days)	*Median (days)*	*Last survivor (days)*
I-E-c	330	390	600
II-E-c	360	480	930
III-E-c	750	780	810

General conclusions.—In summarizing this section devoted to the extinction of *T. castaneum* when in competition with *T. confusum,* it should be kept in mind that, while the former species characteristically dies out in the presence of the latter, there are exceptions which must be carefully examined in their own right. Such examination follows shortly. Our present responsibility is to discuss the extinction of *T. castaneum* and to offer certain broad suggestions as to how this may come about. These suggestions, advanced as first approximations only, must remain just that pending further research of a more analytical character. As pointed out earlier, the present study, despite its duration and scope, is essentially a *description* of events within populations established in several ways and as such makes no pretense at being anything like a final interpretation of these events. All interpretations in this paper, therefore, are to be regarded as exploratory and suggestive rather than as definitive.

Consideration of the following topics affords a brief summary of this section: (1) ages at which *T. castaneum* becomes extinct within and between the several experimental series; (2) increase of *T. castaneum* during the first ninety days of observation; (3) decrease of *T. castaneum* during the interval from day-90 to day-360 or thereabouts; (4) the final dying out of this species and conversion of the cultures to control populations of *T. confusum;* and (5) the rôle of parasitic infection and interspecies competition in the extinction process.

(1) It is possible to draw certain further conclusions regarding the time of extinction of *T. castaneum* from the data contained in Table 14. Since only two replicates exist for each of the three series III cultures, no statistical analysis is permissible. Something can be attempted for series I and II, however, despite the unavoidable fact that the numbers are admittedly inadequate. Mean ages at extinction have been compared between and within volumes, making use of a variant of Student's small-sample method that allows comparison of unpaired data for which the N is low. This method has been applied to the following mean differences, the appropriate probabilities being indicated within the parentheses:

I-E-a minus I-E-b (24%)
I-E-a minus I-E-c (21%)
I-E-c minus I-E-b (>50%)

II-E-a minus II-E-b (22%)
II-E-c minus II-E-a (>50%)
II-E-c minus II-E-b (8%)

II-E-a minus I-E-a (>50%)
I-E-b minus II-E-b (49%)
II-E-c minus I-E-c (13%)

From the above tabulation it is clear that no statistically significant differences exist between these figures. Differently put, and remembering the limitations imposed by the small size of the samples, no differential in mean survival time of *T. castaneum* populations in competition with *T. confusum* is demonstrated for series I and II. This holds true irrespective of the volume and of the initial ratios between the two species established when the experimentals were first started.

In some respects the median is a better index of survival time than is the mean because it is not so influenced by the long persistence of certain cultures as mixed-species populations and because most of the distributions are skewed, usually in a positive direction. Calculation of the significance of the difference between the various medians showed that only those comparisons of I-E-a with I-E-b and I-E-a with I-E-c exhibited a probability lower than one per cent, 210 ± 50.5 days and 180 ± 44.9 days, respectively. All other differences were in excess of the five per cent point.

(2) The preceding discussion has established the fact that *T. castaneum* increases disproportionately (relative to its associate) during the first ninety days irrespective of initial density. During this period the percentage of *T. confusum* is declining. This relationship holds even for the E-b populations in which *T. castaneum* is in competition with a large number of beetles of the other species, and it leads to the suggestion that the former species can multiply effectively when in association with *T. confusum*. At least three points can be advanced in explanation of this: (a) the higher reproductive potential of *T. castaneum;* (b) the slightly faster rate of development of this species; and (c) the fact that *T. castaneum* is not heavily infected with Adelina during these three months in consequence of which its rate of mortality is low. Of course, *T. confusum* also exerts some competitive effect upon *T. castaneum,* as is suggested perhaps by the observation that the experimental cultures do not attain as great densities of *T. castaneum* as those reported for comparable controls. This aspect of competition, however, is outweighed through the operation of the three factors enumerated above.

(3) All the experimental populations described up to this point are characterized by a drop in percentage of *T. castaneum* which is first seen at the fourth census. Were this decline, which typically reaches its nadir around day-360, confined to the mixed-species cultures, it could reasonably be assumed that the cause lay in competitive pressures directed against *T. castaneum* and established through the activities of the *T. confusum* populations. Something of this probably obtains for the latter species becomes increasingly more abundant as the first species contracts. But, since this reduction of *T. castaneum* is so characteristic of control populations when Adelina is present and since it essentially disappears when the parasite is removed, it seems clear that infection pressure *per se* is the primary cause.

(4) The final period during which *T. castaneum* actually becomes extinct can be visualized as setting in subsequent to the 300-360 day interval after this species has been reduced by parasitic infection. As we have seen (Table 14) extinction occurs soon thereafter for certain individual populations but it may be quite postponed for others. In one previously discussed case (III-E-a) a rather striking "rejuvenation" of *T. castaneum* occurred that resulted in an extended period of semi-equilibrium between the two species although even here the eventual outcome was the death of this form. Characteristically, the percentage of *T. castaneum* is low (usually well under 5% of the total) during the 90-150 days prior to extinction, although exceptions occur notably among the III-E-b replicates. This implies that the populations are not effectively reproducing during the end-period but surviving only until the last deaths take place. As the experimental cultures become single-species populations, equilibria, similar to those of the controls at the same ages, are established. Data supporting this are presented later in Table 28.

(5) Up to this point, our interest has centered on the extinction of *T. castaneum* when in competition with *T. confusum*. This, as has been described, is by far the most usual consequence of such competition in the sense that it happens much more frequently than would be expected purely on the basis of chance. Because the analysis has not yet gone far enough, a definitive explanation of the phenomenon cannot be advanced at this time. It is permissible, however, to order the known facts in a logically consistent pattern, this to be followed by broad generalizations of interpretative value. Two significant points have been established: (a) that *T. castaneum,* when not heavily parasitized, can multiply in association with *T. confusum,* and (b) that, as infection pressure mounts, the *T. castaneum* component of the ecosystem is reduced in much the same way as seen in the control or single species cultures. This leads to the conception that, after something like a year of existence in mixed-species populations, *T. castaneum* attains such low densities not primarily because of *T. confusum* but rather because of Adelina.

At this time the populations of *T. castaneum* are uniformly small, there is presumably a large reservoir of infection, and the relatively abundant *T. confusum,* in numerical terms at least, is the flourishing species. It is a reasonable supposition that these situations, operating together, are sufficient to cause the continuation of that observed decline of *T. castaneum* which leads to its eventual extinction. This view presupposes that *T. confusum* does exert some adverse effects upon its associate—effects especially important at this stage in the populations' life-history when *T. castaneum* is both sparse and stoutly infected and therefore highly vulnerable. The nature of this adverse competition is not identified, of course, by merely concluding that it exists, and the matter must be left in this state for the present time.

Earlier it was indicated that, for selected populations of I-E-a and I-E-c, *T. confusum* dies out instead of *T. castaneum*. This shows immediately that the events just outlined do not invariably occur in the combination and intensity required to drive out beetles of the latter species and logically introduces the next section devoted to this segment of the study.

EXTINCTION CURVES FOR T. CASTANEUM AND T. CONFUSUM COMPARED

Excluding the sterile, mixed-species cultures (I-E-a-S) shortly to be discussed, the extinction of *T. confusum* when competing with *T. castaneum* is an unusual event limited to eight series I replicates out of a total of 45. Four of these extinctions occur in I-E-a and four in I-E-c and these, as exceptions to the general rule, merit brief discussion. In passing, it is interesting to note that the I-E-b populations, started with a preponderance of *T. confusum* imagoes, and the series II and III cultures which together comprise a sample of 44, all wind up eventually as control populations of *T. confusum*.

Tables 15, 21, 24, and 25 deal with these two sets of I-E-a and I-E-c populations, and the appropriate competition curves are graphed in Figures 9, 11, and 12.

For convenience, those populations in which *T. castaneum* dies out will be referred to as "c-extinct" and those in which the opposite occurs as "b-extinct."

When the curves describing the extinction of *T. castaneum* (Figures 9 and 11) are compared with those for the extinction of *T. confusum* (Figure 12) dissimilarities are immediately evident.

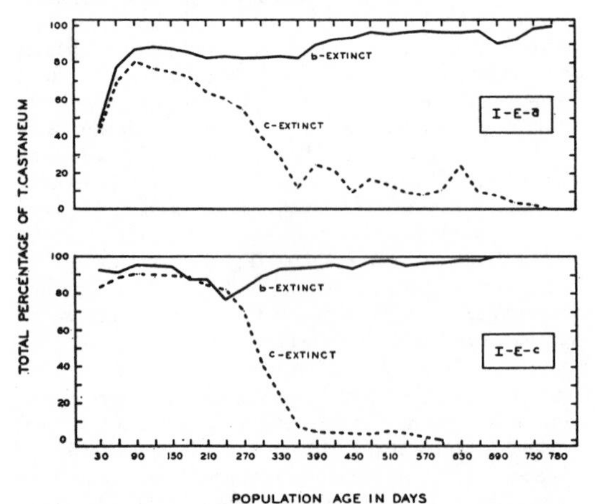

FIG. 12. Competition curves comparing the extinction of *T. castaneum* ("c-extinct") with *T. confusum* ("b-extinct") in I-E-a and I-E-c populations. (upper and lower charts, respectively)

The I-E-a, b-extinct cultures build up larger total populations of *T. castaneum* during the early period of growth than does the c-extinct group. This is true both in terms of percentages and actual numbers per gram. The percentage curves never intercross. For example, at day-90, 80.2% is *T. castaneum* in the one set of replicates (c-extinct) while 87.2% is *T. castaneum* in the other set (b-extinct). Comparable percentages for several succeeding censuses are as follows:

Age (days)	*(c-extinct)*	*(b-extinct)*
120	76.3	88.6
150	74.9	87.6
180	72.0	85.8
210	63.6	82.4
240	60.6	83.1

Expressed as actual numbers, there is, again no intercrossing between the two groups except at day-90. *T. castaneum* does exhibit its characteristic decline induced by Adelina, but this is not as marked for the b-extinct as it is for the c-extinct cultures.

Associated with these higher densities of *T. castaneum* are correspondingly lower densities of *T. confusum*. Apparently, the latter species, in the face of this increased competition, does not become as well established as in the other I-E-a populations (i.e., c-extinct). By the time *T. castenum* has passed through its major "infection crisis" in the b-extinct sample and has started to increase again as happens at day-360, *T. confusum* has undergone further reduction, is not able to control its now more sturdy associate, and, in consequence, declines gradually, to die out at days 510, 630, 690, and 780, respectively.

Occurrences within the I-E-c, b-extinct cultures are essentially similar to those just described, although not quite as clear-cut because *T. castaneum* reaches densities that are lower than in the c-extinct series at several censuses. There is once more, however, a decided rejuvenation of this species at day-270 strongly reminiscent of events in I-C-c—a rejuvenation not present when the beetle is to die out later.

A further discussion of these matters is best postponed until the sterile populations have been dealt with. With that accomplished, it is possible to attempt a synthesis of the various facts described thus far and to discuss in a general way the question of interspecies competition as it exists within populations of these two species of flour beetles.

EXTINCTION OF T. CASTANEUM AND T. CONFUSUM IN THE ABSENCE OF ADELINA

The data dealing with the sterile, mixed-species cultures are presented in Tables 26 and 27 and graphed in Figure 13. Figure 14 assembles in one place the trends of both *Tribolium confusum* and *Tribolium castaneum* in the four sets of Series I populations (I-E-a, c-extinct; I-E-a, b-extinct; I-E-a-S, c-extinct; and I-E-a-S, b-extinct) and uses an ordinate scale of mean number per gram.

As already mentioned, the most significant qualitative fact which emerges when Adelina is removed is that the typical outcome of competition, so charac-

TABLE 24. I-E-a: *Tribolium confusum* becoming extinct.

Age (days)	MEAN NUMBER PER GRAM									PERCENTAGE OF						n
	T. castaneum			*T. confusum*			Total			Larvae and Pupae		Imagoes		Total		
	L-P	Imag.	Sum	L-P	Imag.	Sum	L-P	Imag.	Sum	*T. cast.*	*T. conf.*	*T. cast.*	*T. conf.*	*T. cast.*	*T. conf.*	
30	8.2	1.1	9.3	10.8	0.5	11.3	19.0	1.6	20.6	39.8	52.4	5.3	2.5	45.1	54.9	4
60	6.1	10.4	16.5	1.7	3.0	4.7	7.8	13.4	21.2	28.8	8.0	49.0	14.2	77.8	22.2	4
90	10.5	10.6	21.1	0.3	2.8	3.1	10.8	13.4	24.2	43.4	1.2	43.8	11.6	87.2	12.8	4
120	14.8	10.2	25.0	0.4	2.8	3.2	15.2	13.0	28.2	52.5	1.4	36.1	10.0	88.6	11.4	4
150	8.3	9.4	17.7	0.1	2.4	2.5	8.4	11.8	20.2	41.1	0.5	46.5	11.9	87.6	12.4	4
180	7.0	8.1	15.1	0.3	2.2	2.5	7.3	10.3	17.6	39.8	1.7	46.0	12.5	85.8	14.2	4
210	5.0	5.8	10.8	0.3	2.0	2.3	5.3	7.8	13.1	38.2	2.3	44.2	15.3	82.4	17.6	4
240	7.9	3.9	11.8	0.8	1.6	2.4	8.7	5.5	14.2	55.6	5.6	27.5	11.3	83.1	16.9	4
270	7.9	2.9	10.8	0.8	1.5	2.3	8.7	4.4	13.1	60.3	6.1	22.1	11.5	82.4	17.6	4
300	7.4	2.2	9.6	0.8	1.2	2.0	8.2	3.4	11.6	63.8	6.9	19.0	10.3	82.8	17.2	4
330	5.9	2.0	7.9	0.6	1.0	1.6	6.5	3.0	9.5	62.1	6.3	21.1	10.5	83.2	16.8	4
360	12.1	2.3	14.4	2.1	0.9	3.0	14.2	3.2	17.4	69.5	12.1	13.3	5.1	82.8	17.2	4
390	10.6	3.6	14.2	0.8	0.8	1.6	11.4	4.4	15.8	67.1	5.1	22.8	5.0	89.9	10.1	4
420	9.9	3.7	13.6	0.5	0.6	1.1	10.4	4.3	14.7	67.3	3.4	25.2	4.1	92.5	7.5	4
450	12.9	3.9	16.8	0.6	0.5	1.1	13.5	4.4	17.9	72.1	3.3	21.7	2.9	93.8	6.2	4
480	8.8	4.1	12.9	0.2	0.3	0.5	9.0	4.4	13.4	65.7	1.5	30.6	2.2	96.3	3.7	4
510	10.3	4.1	14.4	0.3	0.4	0.7	10.6	4.5	15.1	68.2	2.0	27.2	2.6	95.4	4.6	3
540	10.9	4.0	14.9	0.3	0.2	0.5	11.2	4.2	15.4	70.8	1.9	25.9	1.4	96.7	3.3	3
570	10.0	3.9	13.9	0.2	0.2	0.4	10.2	4.1	14.3	69.9	1.4	27.3	1.4	97.2	2.8	3
600	10.2	3.3	13.5	0.3	0.2	0.5	10.5	3.5	14.0	72.9	2.1	23.5	1.5	96.4	3.6	3
630	8.8	3.3	12.1	0.2	0.2	0.4	9.0	3.5	12.5	70.4	1.6	26.4	1.6	96.8	3.2	2
660	10.2	3.4	13.6	0.2	0.2	0.4	10.4	3.6	14.0	72.9	1.4	24.2	1.5	97.1	2.9	2
690	17.1	2.5	19.6	0.7	0.2	0.9	17.8	2.7	20.5	83.4	3.4	12.2	1.0	95.6	4.4	1
720	13.3	4.7	18.0	0.3	0.2	0.5	13.6	4.9	18.5	71.9	1.6	25.4	1.1	97.3	2.7	1
750	10.7	4.4	15.1	0.1	0.1	0.2	10.8	4.5	15.3	69.9	0.6	29.0	0.5	98.9	1.1	1

TABLE 25. I-E-c: *Tribolium confusum* becoming extinct.

Age (days)	MEAN NUMBER PER GRAM									PERCENTAGE OF						n
	T. castaneum			*T. confusum*			Total			Larvae and Pupae		Imagoes		Total		
	L-P	Imag.	Sum	L-P	Imag.	Sum	L-P	Imag.	Sum	*T. cast.*	*T. conf.*	*T. cast.*	*T. conf.*	*T. cast.*	*T. conf.*	
30	14.2	7.9	22.1	1.5	0.3	1.8	15.7	8.2	23.9	59.4	6.3	33.1	1.2	92.5	7.5	4
60	6.4	15.9	22.3	0.3	1.8	2.1	6.7	17.7	24.4	26.2	1.2	65.2	7.4	91.4	8.6	4
90	8.8	15.9	24.7	0.1	1.1	1.2	8.9	17.0	25.9	34.0	0.4	61.4	4.2	95.4	4.6	4
120	8.5	15.2	23.7	0.1	1.1	1.2	8.6	16.3	24.9	34.1	0.4	61.1	4.4	95.2	4.8	4
150	7.9	12.7	20.6	0.1	1.0	1.1	8.0	13.7	21.7	36.4	0.5	58.5	4.6	94.9	5.1	4
180	3.2	7.0	10.2	0.1	1.3	1.4	3.3	8.3	11.6	27.6	0.9	60.3	11.2	87.9	12.1	3
210	6.1	3.0	9.1	0.2	1.1	1.3	6.3	4.1	10.4	58.6	1.9	28.9	10.6	87.5	12.5	3
240	4.6	1.7	6.3	0.8	1.1	1.9	5.4	2.8	8.2	56.1	9.8	20.7	13.4	76.8	23.2	3
270	9.9	1.5	11.4	1.4	1.0	2.4	11.3	2.5	13.8	71.7	10.1	10.9	7.3	82.6	17.4	3
300	8.7	2.2	10.9	0.6	0.7	1.3	9.3	2.9	12.2	71.3	4.9	18.0	5.8	89.3	10.7	3
330	8.2	2.9	11.1	0.3	0.5	0.8	8.5	3.4	11.9	68.9	2.5	24.5	4.1	93.4	6.6	3
360	9.0	3.0	12.0	0.5	0.3	0.8	9.5	3.3	12.8	70.3	3.9	23.4	2.4	93.7	6.3	3
390	7.3	2.2	9.5	0.3	0.3	0.6	7.6	2.5	10.1	72.3	3.0	21.8	2.9	94.1	5.9	3
420	6.7	1.8	8.5	0.2	0.2	0.4	6.9	2.0	8.9	75.3	2.2	20.2	2.3	95.5	4.5	2
450	9.2	1.4	10.6	0.5	0.2	0.7	9.7	1.6	11.3	81.4	4.4	12.4	1.8	93.8	6.2	2
480	8.9	2.4	11.3	0.2	0.1	0.3	9.1	2.5	11.6	76.7	1.7	20.7	0.9	97.4	2.6	2
510	7.1	1.9	9.0	0.1	0.1	0.2	7.2	2.0	9.2	77.2	1.1	20.6	1.1	97.8	2.2	2
540	2.7	1.1	3.8	0.1	0.1	0.2	2.8	1.2	4.0	67.5	2.5	27.5	2.5	95.0	5.0	1
570	6.2	1.6	7.8	0.2	0.1	0.3	6.4	1.7	8.1	76.5	2.5	19.8	1.2	96.3	3.7	1
600	7.8	1.4	9.2	0.2	0.1	0.3	8.0	1.5	9.5	82.1	2.1	14.7	1.1	96.8	3.2	1
630	7.9	1.9	9.8	0.1	0.1	0.2	8.0	2.0	10.0	79.0	1.0	19.0	1.0	98.0	2.0	1
660	9.7	1.6	11.3	0.2	0.1	0.3	9.9	1.7	11.6	83.6	1.7	13.8	0.9	97.4	2.6	1

TABLE 26. I-E-a-S: *Tribolium castaneum* becoming extinct.

Age (days)	MEAN NUMBER PER GRAM									PERCENTAGE OF						n
	T. castaneum			*T. confusum*			Total			Larvae and Pupae		Imagoes		Total		
	L-P	Imag.	Sum	L-P	Imag.	Sum	L-P	Imag.	Sum	*T. cast.*	*T. conf.*	*T. cast.*	*T. conf.*	*T. cast.*	*T. conf.*	
30.......	3.5	1.7	5.2	16.0	2.2	18.2	19.5	3.9	23.4	15.0	68.4	7.2	9.4	22.2	77.8	6
60.......	3.1	4.5	7.6	3.5	13.4	16.9	6.6	17.9	24.5	12.6	14.3	18.4	54.7	31.0	69.0	6
90.......	4.0	4.5	8.5	1.0	13.6	14.6	5.0	18.1	23.1	17.3	4.3	19.5	58.9	36.8	63.2	6
120.......	6.0	3.9	9.9	0.2	13.1	13.3	6.2	17.0	23.2	25.9	0.9	16.8	56.4	42.7	57.3	6
150.......	2.5	3.0	5.5	0.7	12.5	13.2	3.2	15.5	18.7	13.4	3.7	16.0	66.9	29.4	70.6	6
180.......	3.3	2.2	5.5	1.4	12.1	13.5	4.7	14.3	19.0	17.4	7.4	11.6	63.6	29.0	71.0	6
210.......	2.4	1.3	3.7	2.7	8.6	11.3	5.1	9.9	15.0	16.0	18.0	8.7	57.3	24.7	75.3	5
240.......	1.8	0.8	2.6	5.2	7.8	13.0	7.0	8.6	15.6	11.5	33.3	5.2	50.0	16.7	83.3	5
270.......	0.4	0.6	1.0	2.8	8.4	11.2	3.2	9.0	12.2	3.3	22.9	4.9	68.9	8.2	91.8	3
300.......	0.3	0.3	0.6	7.0	8.4	15.4	7.3	8.7	16.0	1.9	43.7	1.9	52.5	3.8	96.2	3
330.......	0.3	0.2	0.5	6.5	10.4	16.9	6.8	10.6	17.4	1.7	37.4	1.2	59.7	2.9	97.1	3
360.......	0.2	0.2	0.4	5.4	10.2	15.6	5.6	10.4	16.0	1.2	33.7	1.3	63.8	2.5	97.5	1
390.......	0.2	0.1	0.3	12.0	9.1	21.1	12.2	9.2	21.4	0.9	56.1	0.5	42.5	1.4	98.6	1
420.......	0.2	0.1	0.3	7.9	12.0	19.9	8.1	12.1	20.2	1.0	39.1	0.5	59.4	1.5	98.5	1

TABLE 27. I-E-a-S: *Tribolium confusum* becoming extinct.

Age (days)	MEAN NUMBER PER GRAM									PERCENTAGE OF						n
	T. castaneum			*T. confusum*			Total			Larvae and Pupae		Imagoes		Total		
	L-P	Imag.	Sum	L-P	Imag.	Sum	L-P	Imag.	Sum	*T. cast.*	*T. conf.*	*T. cast.*	*T. conf.*	*T. cast.*	*T. conf.*	
30.......	6.1	4.6	10.7	7.6	1.6	9.2	13.7	6.2	19.9	30.6	38.2	23.2	8.0	53.8	46.2	12
60.......	6.3	8.3	14.6	1.7	6.0	7.7	8.0	14.3	22.3	28.2	7.6	37.3	26.9	65.5	34.5	12
90.......	10.4	8.4	18.8	0.6	6.0	6.6	11.0	14.4	25.4	40.9	2.4	33.1	23.6	74.0	26.0	12
120.......	13.5	8.0	21.5	0.9	5.8	6.7	14.4	13.8	28.2	47.9	3.2	28.3	20.6	76.2	23.8	12
150.......	10.0	7.3	17.3	0.5	5.5	6.0	10.5	12.8	23.3	42.9	2.1	31.3	23.7	74.2	25.8	12
180.......	14.6	6.9	21.5	0.9	5.3	6.2	15.5	12.2	27.7	52.7	3.2	24.9	19.2	77.6	22.4	12
210.......	11.8	7.8	19.6	1.2	4.8	6.0	13.0	12.6	25.6	46.1	4.7	30.5	18.7	76.6	23.4	12
240.......	14.3	8.7	23.0	2.2	4.4	6.6	16.5	13.1	29.6	48.3	7.4	29.4	14.9	77.7	22.3	12
270.......	12.9	9.0	21.9	2.3	4.1	6.4	15.2	13.1	28.3	45.6	8.1	31.8	14.5	77.4	22.6	12
300.......	11.2	9.0	20.2	3.0	3.5	6.5	14.2	12.5	26.7	41.9	11.2	33.7	13.2	75.6	24.4	12
330.......	13.5	9.7	23.2	2.4	3.0	5.4	15.9	12.7	28.6	47.2	8.4	33.9	10.5	81.1	18.9	12
360.......	21.1	11.8	32.9	1.8	2.5	4.3	22.9	14.3	37.2	56.7	4.8	31.7	6.8	88.4	11.6	12
390.......	13.1	12.2	25.3	1.2	1.8	3.0	14.3	14.0	28.3	46.3	4.2	43.1	6.4	89.4	10.6	12
420.......	18.3	12.7	31.0	0.7	1.5	2.2	19.0	14.2	33.2	55.1	2.1	38.3	4.5	93.4	6.6	12
450.......	15.3	14.1	29.4	0.8	1.2	2.0	16.1	15.3	31.4	48.7	2.5	44.9	3.9	93.6	6.4	12
480.......	22.7	14.7	37.4	1.0	1.1	2.1	23.7	15.8	39.5	57.5	2.5	37.2	2.8	94.7	5.3	12
510.......	25.1	16.6	41.7	0.7	0.9	1.6	25.8	17.5	43.3	58.0	1.6	38.3	2.1	96.3	3.7	11
540.......	25.8	22.2	48.0	0.6	0.8	1.4	26.4	23.0	49.4	52.2	1.2	45.0	1.6	97.2	2.8	10
570.......	24.8	21.5	46.3	0.3	0.6	0.9	25.1	22.1	47.2	52.5	0.6	45.6	1.3	98.1	1.9	9
600.......	21.5	22.0	43.5	0.1	0.4	0.5	21.6	22.4	44.0	48.9	0.2	50.0	0.9	98.9	1.1	7
630.......	12.3	16.5	28.8	0.1	0.3	0.4	12.4	16.8	29.2	42.1	0.3	56.5	1.1	98.6	1.4	7
660.......	11.0	16.1	27.1	0.2	0.4	0.6	11.2	16.5	27.7	39.7	0.7	58.1	1.5	97.8	2.2	5
690.......	10.7	11.8	22.5	0.3	0.3	0.6	11.0	12.1	23.1	46.4	1.3	51.0	1.3	97.4	2.6	3
720.......	5.5	7.4	12.9	0.1	0.3	0.4	5.6	7.7	13.3	41.3	0.7	55.7	2.3	97.0	3.0	3
750.......	19.0	8.2	27.2	0.2	0.2	0.4	19.2	8.4	27.6	68.8	0.7	29.7	0.8	98.5	1.5	3
780.......	17.6	12.8	30.4	0.1	0.2	0.3	17.7	13.0	30.7	57.3	0.3	41.7	0.7	99.0	1.0	3
810.......	23.3	13.1	36.4	0.1	0.1	0.2	23.4	13.2	36.6	63.7	0.3	25.7	0.3	99.4	0.6	1

teristic of infected cultures, is reversed. In other words, *T. castaneum* now wins more often then does *T. confusum*—an event taking place in 12 out of 18 replicates. This suggests immediately that, in the absence of infection, the usual occurrences within the populations may be, and in fact frequently are, markedly altered. The matter cannot be interpreted exclusively on this basis, however, because *T. confusum* does sometimes drive out *T. castaneum* in sterile cultures as just noted.

This section dealing with the non-infected, competition cultures can be logically developed as follows:

(1) I-E-a compared with I-E-a-S; *T. confusum* becoming extinct.

(2) I-E-a compared with I-E-a-S; *T. castaneum* becoming extinct.

(3) I-E-a-S (b-extinct) compared with I-E-a-S (c-extinct).

(1) *I-E-a compared with I-E-a-S (b-extinct).*—When I-E-a and I-E-a-S populations in which *T. confusum* becomes extinct are contrasted, certain differences are apparent despite the fact that the eventual outcome is similar. These differences require brief discussion. In the first place, the number of *T. confusum* is greater for the sterile than for the infected group during the first year of the cultures' history. This suggests perhaps that removal of

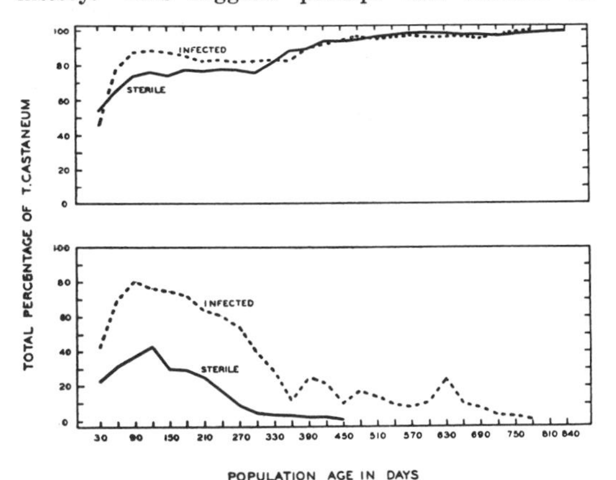

FIG. 13. Competition curves showing the extinction of *T. confusum* (upper chart) and the extinction of *T. castaneum* (lower chart) in I-E-a (marked "infected") and I-E-a-S (marked "sterile") populations.

Adelina has some stimulatory or beneficial effect upon this beetle even though the degree of such stimulation is much less than that seen for *T. castaneum.* The control cultures support this suggestion only to the extent that *T. confusum,* when not parasitized, maintained itself over the total period of observation with a higher percentage of imagoes than it did when parasitized. It is also noteworthy that *T. castaneum* in the sterile series starts its exaggerated growth after day-300 and does not ever exhibit any sign of the depression so very characteristic when Adelina is present and so very important in inducing that decline which, taken in conjunction with competition, results in final extinction. The fact that the numbers of *T. castaneum* are consistently less during the first 150 days for the sterile than for the infected group, also deserves mention. This may well result from an increase in competition pressure caused by the presence of larger populations of *T. confusum* and is supported by the fact that the sterile *controls* of *T. castaneum* (I-C-c-S) are larger than the infected *controls* (I-C-c) even at the first three censuses. Eventually, however, the populations of *T. castaneum* in the sterile cultures become so large that *T. confusum* is completely crowded out. The extinction of *T. confusum* in the I-E-a cultures has already been discussed, and this unusual event was related to the observation that, in these four instances, *T. castaneum* does not attain as low a density for the 270-330 day interval as is usually the case.

(2) *I-E-a compared with I-E-a-S (c-extinct).*—Extinction of *T. castaneum* is the typical end-result of competition in parasitized cultures and the atypical

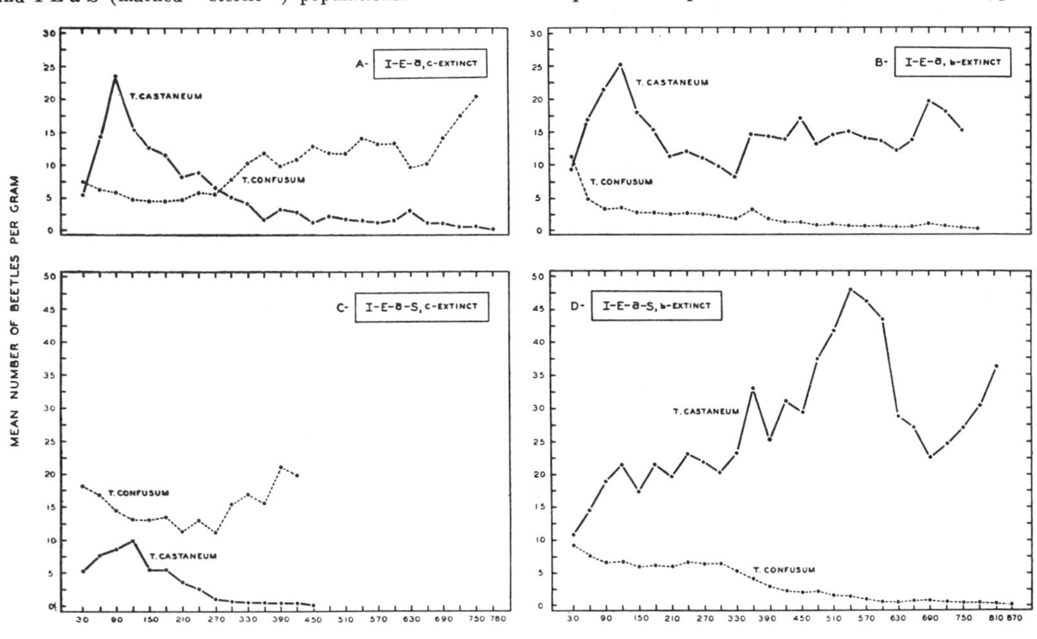

FIG. 14. The actual population trends in mean number per gram of both *T. confusum* and *T. castaneum* in the following four categories of cultures: I-E-a, c-extinct (chart "A"); I-E-a, b-extinct ("B"); I-E-a-S, c-extinct ("C"); and I-E-a-S, b-extinct ("D").

result in non-parasitized cultures. The pattern of events, having been sufficiently discussed for the former group, needs no further elaboration. The pattern for the latter group (sterile) is dissimilar to any thus far seen and requires brief description. These six, I-E-a-S populations behave very differently from their infected and sterile counterparts. The significant feature is that *T. confusum,* rather than *T. castaneum,* is the abundant form from the time of the first census, a dominance which is maintained until *T. castaneum* completely disappears. Furthermore, this disappearance occurs sooner than that seen for any other cultures of any experiment: the mean time is 320 days, the median, 315 days, and the range, 210-450 days. It seems evident that the factors operating to produce a result of this sort are different, in intensity and combination at least, from those already outlined.

With Adelina gone, *T. castaneum* normally would be expected to multiply abundantly as it does in I-C-c-S and, as it does after day-150, in I-E-a-S (b-extinct). Clearly, this does not happen, the reason for which is obscure. Is *T. castaneum* in these replicates inferior in some way even though it is not burdened with parasitization and does this "inferiority" then allow *T. confusum* to reproduce very effectively from the beginning on? Or, does removal of Adelina afford, in these instances, just enough of an added stimulus for *T. confusum* so that it goes ahead to out-compete its associate? The first alternative seems unlikely because it occurs too often to be mere happenstance and because no evidence of such "inferiority," apart from that induced by infection, has been unearthed. The second alternative has something to support it in that, as we have just seen, *T. castaneum* does show some gain during its early period of growth in the b-extinct, I-E-a-S group, and, it might be argued, this gain is sufficient to boost the species into its position of dominance. But this too is not a satisfying explanation for it does not identify causal factors nor does it indicate why the respective densities of the two species are reversed relative to each other after just thirty days of co-existence. This last point really states the crux of the problem and, if answered, would go a long way toward clarifying the whole matter.

About all that can be said at this time beyond merely describing the observed population trends, is to indicate (1) that *T. confusum* when uninfected drives out *T. castaneum* about one-third of the time and in the process builds up relatively large populations at an early date; (2) that this extinction pattern is of a different configuration than that seen for the parasitized populations which require a low density induced by Adelina at about day-300 before the species dies out; and (3) that, taken as a group, the extinctions occur sooner than those characteristic of the other experimental cultures.

(3) *I-E-a-S (b-extinct) compared with I-E-a-S (c-extinct).*—These two groups of sterile cultures, one involving the extinction of *T. confusum* in two-thirds of the cases and the other involving the extinction of *T. castaneum* in one-third of the cases, have been discussed sufficiently in the preceding sections so that only adumbration is required here. When the data for the two groups are graphed in terms of mean number per gram, the two following observations are worth noting:

(a) The density of *T. castaneum* in the b-extinct sample is always greater than in the c-extinct sample. For the former, this species starts its exceptional growth at day-300 and, without ever passing through a period of depression, soon attains the massive populations already described for the sterile controls. *T. confusum* starts out higher (by day-60) than it does in infected cultures, an interesting point stressed earlier, and maintains something of an equilibrium until the three-hundredth day, at which time it starts a decline that coincides exactly with the growth of its competitor. The curves for these two species (b-extinct group) suggest (1) that *T. confusum* is more abundant during the first year than it would be if Adelina were present; (2) that *T. castaneum* is not quite as abundant over the same interval as it would be if *T. confusum* did not display such a heightened density; and (3) that *T. castaneum,* unencumbered by the parasite, eventually achieves such large populations despite the presence of its competitor that the latter disappears from the cultures.

(b) The curious situation in the c-extinct, sterile sample was dealt with in the last section. In sum, this is characterized by the immediately large populations of *T. confusum,* the essential persistence and exaggeration of this condition, the early reduction in numbers of *T. castaneum,* and the relatively rapid dying-out of this species. The data show that *T. castaneum* does increase slightly between day-30 and day-120 and that during this interval *T. confusum* declines somewhat. This substantiates a general principle emerging from the mixed-species cultures, namely: that when one species trends upward, the other species trends downward, and *vice-versa.*

CONVERSION OF MIXED-SPECIES POPULATIONS INTO SINGLE-SPECIES CONTROLS

It is valuable to compare the total size of those cultures that had been experimental populations with the total size of counterpart controls of the same age. These comparisons are presented in Table 28 for all groups except series III. The last are omitted because there are only two replicates for each sample. The age selected for analysis was arbitrarily chosen to include the maximum number of cases for each group. The table lists the mean numbers of beetles per gram and, when N is large enough, their respective probable errors.

The findings are straightforward. There are no significant differences in terms of total size between the populations that had recently been in competition with another species of Tribolium (formerly, experimentals) and those that had never experienced such an association (controls). To state the point differently, such competition imposes no lasting effect upon the species that survives. The latter goes ahead

TABLE 28. Mean total number of beetles per gram from cultures originally of both species after these become single species population, compared with appropriate controls of the same age.

Populations:	Age at Comparison	Mean per Gram	Mean Difference Between Controls and Experimentals	n
I-E's (c extinct)....	840	18.6± 0.6303	0.8± 0.95	30
I-C-b...............	840	19.4± 0.7193	....	19
I-E's (b extinct)....	840	12.4	1.0	5
I-C-c...............	840	13.4± 0.7652	....	13
I-E-a-S (c extinct)....	840	21.9	....	6
I-C-b-S.............	840	20.8± 0.6946	1.1	16
I-E-a-S (b extinct)....	840	45.4± 4.41	1.8± 4.8	11
I-C-c-S.............	840	47.2± 2.13	....	16
II-E s (c extinct)....	1140	15.3± 0.4458	1.6± 1.8	22
II-C-b...............	1140	16.9± 0.3877	....	17

to assume the normal pattern that typifies it under these conditions of husbandry.

DISCUSSION

The studies of single-species control cultures of *Tribolium confusum* and *Tribolium castaneum* grown apart under similar and essentially constant environmental conditions have shown that these two flour beetles maintained themselves successfully, eventually established equilibria characteristic for each, and exhibited no more variability, and perhaps less, than would be expected for ecological data of this sort. In addition, the sporozoan parasite, *Adelina tribolii,* was demonstrated to have a dramatic effect especially upon populations of *T. castaneum.* When Adelina had attained the requisite saturation within cultures of this beetle, the host dropped in numbers, and, after partial recovery and as if in adjustment to the parasite, then carried on at reduced density. Removal of the parasite also completely removed this low depression and permitted the beetle to attain an abundance considerably over twice that displayed by counterpart, but infected, populations. Adelina did not influence the behavior of *T. confusum* in anything like the same way. When parasitized, this beetle does not exhibit the depression mentioned above and, even when cultured under parasite-free conditions, does not produce larger total populations. It was shown for both Tribolium species that the relative abundance of imagoes, as indexed by percentages, increased something in the order of twenty per cent in complete absence of infection. Many other facts about the control populations were established earlier, but the above, sketched broadly as they are, are sufficient for present purposes.

When *T. confusum* and *T. castaneum* are placed together in competition, one of the two always becomes extinct. This may occur as early as 180 days after starting the experiment or, rarely, as late as 1470 days; but occur it does. Furthermore, it is not a particular species which invariably dies out as has been made abundantly clear. After extinction takes place, the population assumes "control behavior" by which is meant that it does not differ significantly from cultures of the same age and species that have never had the experience of competition in their past history.

The conclusion thus seems clearly established that competition between the two species always ensues when they are brought together as components of the same ecosystem. This is to be expected on the basis of their ecology. They inhabit the same medium to which both are well adapted individually; they exploit the same source of food; their life-histories are fundamentally quite similar; they are not divergent as to size, although *T. confusum* is somewhat larger; their patterns of reproduction, and so far as known, their mortality, are generally alike; and they are undoubtedly closely related in the taxonomic sense. With these facts in mind, it could be forecast *a priori* that stringent competitive pressures would arise between the two populations and that these would probably lead to definite end-results in respect of species survival (Crombie 1945, 1946, 1947). Such end-results have been classified and, in a large way, some of the factors, or factor-complexes at least, have been identified.

The story is complicated, although not rendered insolvable, by the finding that competition does not invariably result in only one outcome. This is not regarded as surprising—in fact, it might well be expected on the basis of general knowledge of population ecology as well as on the basis of the fact that *T. confusum* and *T. castaneum* considered individually as species are both successful groups with wide zoögeographic ranges (see Elton 1946, Lack 1937, Williams 1947).

The disclosure that either species may die in the presence of the other (although with different frequencies depending upon the experimental design) suggests that, to occur, the extinction process requires an antecedent, reticulum of related events which, for a specific type of population, typically unfold in a particular way but which, in exceptional cases, may be so altered as to produce the dying out of the other organism.

At least four categories of factors are operating within these mixed-species populations. These are:

(1) Competition exerted against *T. confusum* by *T. castaneum.*
(2) Competition exerted against *T. castaneum* by *T. confusum.*
(3) The presence or absence of Adelina in relation to *T. confusum.*
(4) The presence or absence of Adelina in relation to *T. castaneum.*

Obviously, these four factors interact as a nexus, their intensity and effectiveness varying with time as a density-dependent function (Nicholson 1933, Schwerdtfeger 1942, Smith 1935, Thompson 1939, Varley 1947). Further, it should be possible to dissect them experimentally. Competition means nothing in the analytical sense unless the operations comprising it are identified and quantitatively evaluated. Parasitization, if present, has maximum meaning when the degree and pathogenicity of infection pressure is assayed relative to the differential susceptibilty of both species and to the age and history of the ecosystem considered in its entirety. As stressed several times in this report, answers to such matters cannot be presented here but depend on a program of further research, a segment of which is now in progress.

It is evident that the two species of flour beetles compete with each other and in this way alter their respective futures as populations. With the parasite present, and during the earlier stages of growth, *T. castaneum,* perhaps in part because of its higher reproductive capacity, increases, while *T. confusum* contracts. *T. castaneum* then suffers a considerable decline owing to the heightened mortality consequent upon the build-up of Adelina after which its associate gradually starts to multiply. The former species eventually attains a critically low density and is driven from the cultures when in this vulnerable state. If, as occasionally happens, the parasite does not bring about a minimum density sufficiently within the critical limits, *T. castaneum* stages a recovery of the sort characteristic of control cultures, and this leads to the extinction of *T. confusum.* Removal of Adelina reverses the frequency of occurrence of these events largely (1) by eliminating that depression in numbers exhibited by infected controls and experimentals at about the three-hundredth day, and (2) by enabling this species to establish populations of much larger size. The last phenomenon, of course, further increases the competition pressure exerted against *T. confusum.* However, as we have seen, *T. confusum* also appears to profit to some extent from the absence of the parasite and about a third of the time, is able to inhibit *T. castaneum* enough so that its early decline and extinctions is initiated. This particular sequence is not understood at the moment.

Before concluding this discussion, it is germane for purposes of orientation to speculate briefly about certain types of factors that may prove to be involved in the phenomenon of interspecies competition explored in this study. As general background, it should be remembered that both flour beetle populations are competing for space and for food in a limited environment in which, as single species, they maintain indefinitely an equilibrium so long as the medium is regularly renewed and the physical environment remains favorable.

Following are some possible ways that the two species of Tribolium could influence each other when grown together:

(1) They may exploit in differing degrees the quantity and/or nutritive quality of the available food supply. This could be brought about by interactions ("coactions") between the beetles and induce a change in their physiology, or, it could alter the milieu directly ("conditioning"). (Park 1941).[7]

(2) They may be differentially tolerant to simple crowding both inside the medium and on its surface. This could impose an effect upon their fecundity, fertility, rate and success of post-embryonic development, imaginal mortality, and age-distribution. (Park 1933, Boyce 1946).

(3) The presence of a competitor could modify the intraspecies relations in a way that these would diverge markedly from those characteristic of control cultures.

(4) There may exist actual behavior coactions between the two forms that inhibit one, inhibit both, or favor one and which undoubtedly would vary with the density and ecological age of the ecosystem. For example, such coactions could involve,

(a) alteration of the temperature, moisture content, and oxygen concentration *within* the medium ("microclimate") from that obtaining in the controlled incubators in such a way as to affect, differentially by species, the natality and/or mortality rates;

(b) differential predation in which, say, the larvae and imagoes of one species might consume the eggs and pupae of the other at a rate statistically higher than that directed against their own kind;

(c) differential interference with copulation or fecundation; and

(d) differential conditioning of the medium (Crombie 1945). This might concern the reduction of available food, the elaboration of environmental poisons, such as excrement and carbon dioxide, or comminution of the flour itself.

Much remains to be done before it can be said that interspecies competition in the genus Tribolium is well understood. But it is hoped that a start has been made and that future analytical research, wherever conducted, will contribute to this general problem which, in the author's opinion at least, is an important one for population ecology and evolution.

SUMMARY

(1) The census history of 211 laboratory populations of the flour beetles, *Tribolium confusum* and *Tribolium castaneum,* maintained in three volumes of favorable medium was accurately followed for a period of about four years with counts made every thirty days. Some of these populations consisted only of *T. confusum* ("controls"), others were exclusively *T. castaneum* ("controls"), and still others contained both species brought together in competition in a shared environment ("experimentals"). The medium was regularly renewed at each census and all cultures were kept in constant temperature incubators. In addition, certain control and experimental populations ("steriles") were rendered free of a sporozoan parasite, *Adelina tribolii,* that inhabits with consequent pathogenicity the haemocoels and other tissues of the beetles and is distributed as

[7] For *T. castaneum* at least, there is a suggestion that limitation of food supply does not play a significant rôle in controlling the equilibrium density achieved by these populations. The point was made explicit in the section on single-species populations and re-affirmed in the section on mixed-species populations. It will be remembered from the discussion of the sterile controls of *T. castaneum* (I-C-c-S) that the density increased nearly three times when parasitization was eliminated. This occurred without any alteration in the total amount or nutritive quality of the flour-yeast medium.

inactive, but infectious, oöcysts throughout the flour. Since this parasite is a potent source of mortality, especially for *T. castaneum,* it was necessary to evaluate the statistical effect of its presence and absence in the ecosystem before the findings could be interpreted.

(2) Single-species, control populations of *T. confusum* and *T. castaneum* grew successfully and eventually established equilibrium densities that were maintained for the duration of the study. *T. confusum* stabilized at a higher total density than did *T. castaneum.* The latter exhibited a marked depression in numbers at about the three-hundredth day from which it typically recovered. This depression is attributed to increased mortality resulting from parasitic infection. Evidence is advanced, for *T. confusum* particularly, that density is inversely proportional to the total volume of medium.

(3) The absence of the parasite (Adelina) influenced the populations of the two beetles in different ways. *T. castaneum,* when uninfected, attained equilibria nearly three times greater than those characteristic of counterpart, infected cultures. Along with this went a complete removal of the depression mentioned in the paragraph above. On the other hand, *T. confusum,* when uninfected, maintained populations that did not differ significantly in terms of total size from the parasitized controls. Sterile cultures of both species possessed about twenty per cent more imagoes than did infected controls.

(4) Data are presented for all single-species cultures that analyze the variability within and between census counts both in respect of the total period of observation and for selected census intervals.

(5) Data are presented that describe the percentage composition of immature relative to mature stages for the various cultures, and these are discussed with reference to the design of the program.

(6) For experimental cultures in which interspecies competition is established, it is shown that one species of beetle *always* becomes extinct under these conditions, after which the surviving form then assumes a census trend statistically identical with comparable controls. However, not always the same species dies when faced with the competitive pressures of its associate. Typically, but with exceptions which are carefully dealt with, *T. castaneum* becomes extinct when Adelina is present, and *T. confusum* becomes extinct when Adelina is absent.

(7) The phenomenon of interspecies competition in Tribolium is discussed and the known facts are ordered into a preliminary hypothesis intended to partially collate the observed results.

(8) Future areas of research, analytical in character, are broadly sketched.

LITERATURE CITED

Bhatia, M. L. 1937. On Adelina tribolii, a coccidian parasite of Tribolium ferrugineum F. Parasitology **29:** 239-246.

Boyce, Janet Mabry. 1946. The influence of fecundity and egg mortality on the population growth of Tribolium confusum Duval. Ecology **27:** 290-302.

Chapman, R. N. 1933. The causes of fluctuations of populations of insects. Proc. Haw. Ent. Soc. **8:** 279-297.

Chapman, R. N., & Lillian Baird. 1934. The biotic constants of Tribolium confusum Duval. J. Exp. Zool. **68:** 293-304.

Crombie, A. C. 1945. On competition between different species of graminivorous insects. Proc. Roy. Soc. (B) **132:** 362-395.
1946. Further experiments on insect competition. Ibid. **133:** 76-109.
1947. Interspecific competition. J. Anim. Ecol. **16:** 44-73.

Dick, John. 1937. Oviposition in certain coleoptera. Ann. Applied Biol. **24:** 762-796.

Elton, Charles. 1946. Competition and the structure of ecological communities. J. Anim. Ecol. **15:** 54-68.

Ford, John. 1937. Research on populations of Tribolium confusum and its bearing on ecological theory: a summary. J. Anim. Ecol. **6:** 1-14.

Good, N. E. 1936. The flour beetles of the genus Tribolium. U. S. Dept. Agr. Technical Bull. **498:** 1-57.

Hesse, E. 1911. Sur le genre Adelea à propos d'une nouvelle coccidie des oligochètes. Arch. Zool. Exp. Gén. **7:** xv-xx.

Holdaway, F. G. 1932. An experimental study of the growth of populations of the "flour beetle," Tribolium confusum Duval, as affected by atmospheric moisture. Ecol. Monogr. **2:** 261-304.

Kollros, Catharine Lutherman. 1944. A study of the gene, pearl, in populations of Tribolium castaneum Herbst. Doctorate dissertation: University of Chicago Libraries.

Lack, David. 1947. Darwin's finches. Cambridge: 1-208.

Nicholson, A. J. 1933. The balance of animal populations. J. Anim. Ecol. **2:** 132-178.

Park, Thomas. 1933. Studies in population physiology. II. Factors regulating initial growth of Tribolium confusum populations. J. Exp. Zool. **65:** 17-42.

Park, Thomas. 1934. Observations on the general biology of the flour beetle, Tribolium confusum Duval. Quart. Rev. Biol., **9:** 36-54.
1941. The laboratory population as a test of a comprehensive ecological system. Ibid., **16:** 274-293, 440-461.

Park, Thomas, & William Burrows. 1942. The reproduction of Tribolium confusum Duval in a semisynthetic wood-dust medium. Physiol. Zool. **15:** 476-484.

Park, Thomas, & Marion Biggs Davis. 1945. Further analysis of fecundity in the flour beetles, Tribolium confusum Duval and Tribolium castaneum Herbst. Ann. Ent. Soc. Amer. **38:** 237-244.

Park, Thomas, E. V. Gregg, & C. Z. Lutherman. 1941. Studies in population physiology. X. Inter-specific competition in populations of granary beetles. Physiol. Zool. **14:** 395-430.

Pearl, Raymond, Thomas Park, and J. R. Miner. 1941.

Experimental studies on the duration of life. XVI. Life tables for the flour beetle, Tribolium confusum Duval. Amer. Nat. **75**: 5-19.

Riley, W. A., & Laurence Krogh. 1922. A coelomic coccidian of Tribolium. Anat. Rec. **23**: 121. (abstract)

Schwerdtfeger, F. 1942. Über die Uraschen des massenwechsels der Insekten. Zeit. f. angewandte entomologie **28**: 254-303.

Smith, H. S. 1935. The rôle of biotic factors in the determination of population densities. J. Econ. Ent. **28**: 873-898.

Stanley, John. 1946. The environmental index, a new parameter, as applied to Tribolium. Ecology **27**: 303-314.

Thompson, W. R. 1939. Biological control and the theories of interactions of populations. Parasitology **31**: 299-388.

Varley, G. C. 1947. The natural control of population balance in the knapweed gall-fly (Urophora jaceana). J. Anim. Ecol. **16**: 139-187.

Wenyon, C. M. 1926. Protozoology. N. Y. 1-1563 (2 vols.)

White, G. F. 1923. On a neosporidian infection in flour beetles. Anat. Rec. **26**: 359. (abstract)

Williams, C. B. 1947. The generic relations of species in small ecological communities. J. Anim. Ecol. **16**: 11-18.

Yarwood, E. A. 1937. The life cycle of Adelina cryptocerci Sp. Nov., a coccidian parasite of the roach Cryptocercus punctulatus. Parasitology **29**: 370-390.

HILGARDIA

A Journal of Agricultural Science Published by the California Agricultural Experiment Station

Vol. 27 AUGUST, 1958 No. 14

EXPERIMENTAL STUDIES ON PREDATION: DISPERSION FACTORS AND PREDATOR-PREY OSCILLATIONS[1,2]

C. B. HUFFAKER[3]

INTRODUCTION

THIS PAPER is the second covering a series of experiments designed to shed light upon the fundamental nature of predator-prey interaction, in particular, and the interrelations of this coaction with other important parameters of population changes, in general. In the first of this series (Huffaker and Kennett, 1956),[4] a study was made of the predatory mites, *Typhlodromus cucumeris* Oudemans[5] and *Typhlodromus reticulatus* Oudemans, and their prey species, *Tarsonemus pallidus* Banks, the cyclamen mite which attacks strawberries. In that paper the authors discussed in a broad way the need for detailed studies of this kind and the implications of such results for theories of population dynamics, particularly the role of predation—which role has been minimized by a number of researchers (e.g., Uvarov, 1931; Errington, 1937, 1946; Leopold, 1954).

A significant result of the experiments of Huffaker and Kennett (1956) was the demonstration of two types of fluctuations in density. Where predators were excluded there was a regularized pattern of fluctuations of **decreasing** amplitude, a result of reciprocal density-dependent interaction of the phytophagous mite and its host plant. The other, sharply contrasting type of regularized fluctuation occurred as a primary result of predation on the phytophagous mite by the predatory form. The interacting reciprocal dependence of the prey and predator populations resulted in greatly reduced densities and amplitude of fluctuations, comparing this with the status when predators were absent.

In the present effort a considerable body of quantitative data is presented. Also, the outlines of an experimental method and design sufficiently flexible for use in studying some of the principles of population dynamics are

[1] Received for publication August 16, 1957.

[2] These results were obtained during a period of sabbatical leave in 1955. The generous assistance of C. E. Kennett and F. E. Skinner in the preparation of the illustrations is gratefully acknowledged.

[3] Entomologist in Biological Control in the Experiment Station, Berkeley.

[4] See "References" on page 383.

[5] H. Womersley and C. E. Kennett now consider that this predator is really *Typhlodromus bellinus* Womersley.

delineated. The specific results are discussed with respect to the much-debated question of whether the predator-prey relation is inherently self-annihilative, and the bearing on this of the type of dispersion and hazards of searching. Certain trends are exhibited; if these are further verified by later experiments they may be theoretically significant, but such possibilities will be covered only after the accumulation of additional data from this continuing series of studies.

The immediate objective in the present effort has been the establishment of an ecosystem in which a predatory and a prey species could continue living together so that the phenomena associated with their interactions could be studied in detail. Once conditions are established giving a measure of assurance against overexploitation, various other features could be introduced to study their relations to the periods and amplitudes of such oscillations in density as are demonstrated. This could include such factors as differences in temperature, humidity, or physical terrain, for example. Also, the effects could be studied of using two or more predatory species competing for the one prey species, or the simultaneous employment of two species of prey acceptable to the one predator. Many variations along these lines could be expected to furnish valuable information, and the present data represent only a beginning.

Some of the many questions that could ultimately be answered include:

1. Are such oscillations inherently of **increasing** amplitude?

2. Even if so, are there commonly present forces which act to cancel this tendency, and if so, what are these forces?

3. Is the predator-prey relation adequately described by the Gause theory of overexploitation and auto-annihilation except under conditions involving immigration from other ecosystems?

4. Does the presence of other significant species in addition to the two primary or original coactors introduce a stabilizing or disturbing effect?

5. What may be the effect of changes in the physical conditions upon the degree of stability or permanence of the predator-prey relation?

6. Can evidence be obtained supporting or refuting the concept that the prey, as well as the predators, benefits from the relation?

7. What is the order of influence on stability of population density of such parameters as shelter (from physical adversity of environment), food, disease, and natural enemies of other kinds?

There are no published accounts wherein the predator-prey relation has been followed under controlled conditions beyond a single wave or "oscillation" in density. Authorities differ as to whether this relation is inherently disoperative, leading inevitably to annihilation of either the predatory species alone or both the predator and its prey in the given universe or microcosm employed. In this controversy there is confusion as to what constitutes a **suitable** experimental microcosm. Published examples of such studies have been contradictory or inconclusive.

In the classic experiments of Gause (1934) and Gause *et al.* (1936), the predator and prey species survived together only under quite arbitrary conditions—either when a portion of the prey population was protected by a "privileged sanctuary" or when reintroductions were made at intervals. Gause concluded that such systems are self-annihilative, that predators characteristically overexploit their prey, and that in nature immigrants must repopulate the local environments where this has occurred. He argued against the theory that repeated waves or oscillations conforming to mathematical formulae have an inherent meaning in the absence of immigrations.

Nicholson (1933, 1954) advocates the contrary view, and he and Winsor (1934) criticized Gause's experiments on the grounds that the universes or microcosms he employed were too small to even approximate a **qualitative,** to say nothing of a **quantitative,** conformity to theory.

DeBach and Smith (1941) conducted a stimulating experiment with a special type of predator (an entomophagous "parasite") in which they tested the biological parameter of searching capacity against Nicholson's formula. The results conformed to theory very neatly. Ecologists consider the results as based upon too arbitrary assignments or omission of other biological parameters—such as length of a generation, fecundity, undercrowding phenomena at very low densities, et cetera. However, their work remains a strikingly successful pioneer endeavor in this field, and their method of isolating the variables other than **searching** was productive.

In the present study an effort was made to learn if an adequately large and complex laboratory environment could be set up in which the predator-prey relation would not be self-exterminating, and in which all the biological parameters are left to the free play of the interacting forces inherent in the experiment, once established. Consequently, the procedure was to introduce the prey species and the predator species only at the initiation of an experiment and to follow population trends afterward without any further introductions or manipulations. No assignments of biological parameters were made. Furthermore, no areas restrictive to the predators were furnished. Food "conditioning" is the only complicating variable and this disturbance was minimized by the methods used. However, as experiments were terminated because of the annihilative force of predation in the initial, limited universes employed, the conditions set up for subsequent experiments were made progressively more complex in nature and the areas larger.

EXPERIMENTAL DESIGN AND PROCEDURE

General Aspects

The six-spotted mite, *Eotetranychus sexmaculatus* (Riley), was selected as the prey species and the predatory mite *Typhlodromus occidentalis* Nesbitt as the predator. These species were selected because successful methods of rearing them in the insectary were already known, and because earlier observations had revealed this *Typhlodromus* as a voracious enemy of the six-spotted mite. It was known to develop in great numbers on oranges infested with the prey species, to destroy essentially the entire infestation, and then to die *en masse.* At this author's suggestion, Waters (1955) had studied the detailed biology of both species and had followed population trends on individual

Fig. 1. Orange wrapped with paper and edges sealed, ready for use with sample areas delineated. (Photograph by F. E. Skinner.)

oranges as a problem in predator-prey dynamics.[6] This work was valuable in the conduct of the present research.

Uniformity in certain characteristics was maintained throughout the course of these experiments. Temperature was maintained constant at 83°F. Relative humidity varied some but was not allowed to fall below 55 per cent. There were no means of dehumidifying the room, but automatic humidity controls assured against the damaging action of low humidity. The room was kept dark.

Uniformity in total areas of the universes was achieved by utilizing various combinations of oranges and rubber balls equivalent to them in size (see figs. 1, 2, 3, and 4, for examples). This made it possible to change either or both the total primary food substrate (orange surface) and the degree of dispersion of that substrate without altering the total area of surfaces in the universes or the general distribution of units of surface in the systems. The object was to make it possible to vary the surface of orange utilized and its

[6] The design of Waters' experiments, however, was not such as to answer some of the questions posed by this study. His universes were restricted and simple, with no possibility for return to oranges by individuals leaving them by "dropping off." His results were similar to those of Gause, but he drew several conclusions which are more generally applicable than some of Gause' generalizations.

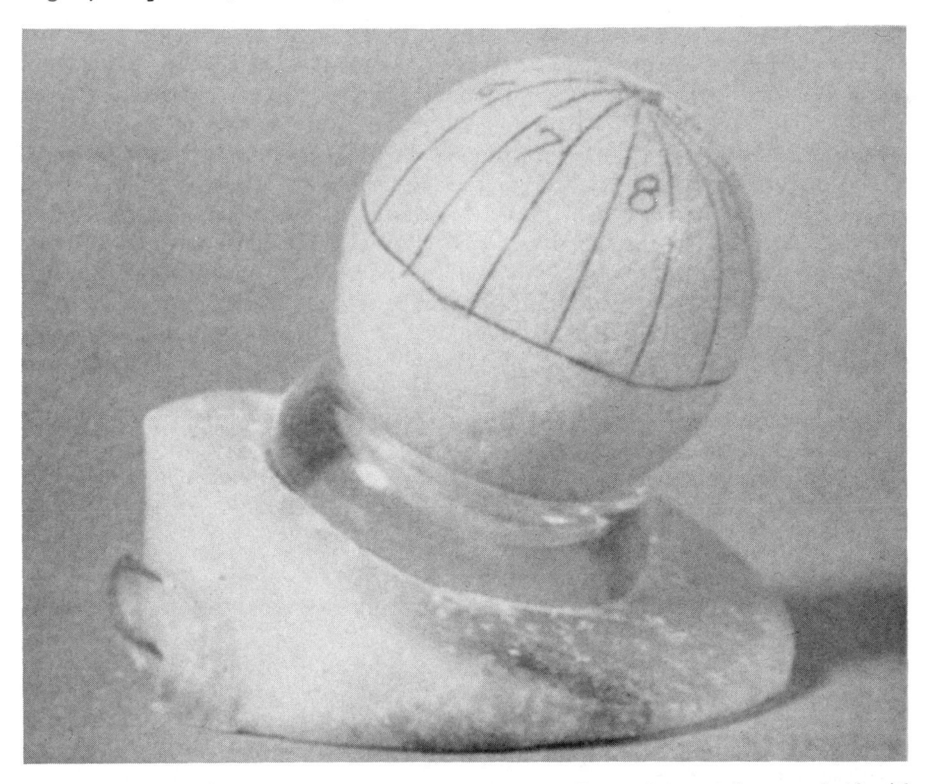

Fig. 2. Orange with lower half covered with paraffin and exposed upper half with sample areas delineated. Fuzzy surface is due to lint used. Paraffin base serves to bring all areas into focus under the microscope (see text). (Photograph by F. E. Skinner.)

Fig. 3. Four oranges, each with half-surfaces exposed (see fig. 2), grouped and joined with a wire loop, remainder of positions occupied by waxed, linted rubber balls, a 2-orange feeding area on a 4-orange dispersion, grouped. (Photograph by F. E. Skinner.)

distribution in order to complicate the search for food by both the prey and predator. Thus, a simple environment where all the food was concentrated to a maximum degree (fig. 3) could be compared with one in which the food was dispersed according to arbitrary degrees (fig. 4) throughout the system (the oranges being arranged or randomly dispersed among the rubber balls). Also, the quantity of food as well as the nature of the dispersion were varied by covering the oranges with paper to whatever degree desired, the paper being wrapped tightly, twisted and tied, and with circular holes then cut to expose the required areas of orange surface. The edges of the holes and the twisted ends were then sealed with paraffin to exclude the mites from gaining entrance to the covered surfaces. An example of an orange ready for use is shown in figure 1.

Fig. 4. Four oranges, each with half-surfaces exposed (see fig. 2), randomized among the 40 positions, remainder of positions occupied by waxed, linted rubber balls—a 2-orange feeding area on a 4-orange dispersion, widely dispersed. (Photograph by F. E. Skinner.)

Considerable difficulty was encountered in arriving at a proper means of limiting the feeding area on a given orange. The first method tried was to dip the naked oranges in hot paraffin, leaving the desired areas clean. Considerable time was lost during the operation of the first two series of experiments because the oranges which were almost completely coated with paraffin or, later, even those covered with polyethylene bags, rotted before results could be obtained. It was only subsequent to this difficulty that a good grade of typing paper was tried. The paper proved to be an excellent material but somewhat difficult to form to an adequately smooth, mite-excluding covering. This fault was corrected by very slightly dampening the paper before wrapping and by using paraffin as a sealing material.

A primary difficulty foreseen in this study was that of replenishing the food material as it is used or becomes conditioned. Under the conditions employed, oranges last from 60 to 90 days as suitable food for the prey species if not fed upon to the extent of conditioning. However, a heavy infestation can deplete an orange of suitable nourishment and thus condition it within three to five days. It is relatively impracticable to remove the food factor as a limiting feature. Also, localized food depletion by the prey species just as much as food depletion by the predator species is inherent to the natural scene. Yet, it was hoped that depletion of food could be evaluated and reduced to a minor position in limiting the populations. It was desirable to build into the design a schedule of removal of old oranges, whether or not conditioned, and their replacement by fresh ones. This would make possible a continuing system which would not automatically end if and when the original oranges became too old or conditioned. Also, by the use of careful notes on conditioning and by comparing universes where predators were introduced with universes which had none, conclusions could be drawn as to the relative importance of any interference occasioned by the conditioning of the oranges. The schedule consisted of removing ¼ of the oranges (the oldest or obviously most unsuitable ones) at intervals of 11 days. This gave a complete change of oranges every 44 days—a period amply in advance of general unsuitableness because of age alone. One restriction was imposed. No **significant last** of a population in a subsection was removed—any such orange otherwise "due" being held another 11 days.

Fig. 5. 120 oranges, each with 1/20 orange-area exposed (method of fig. 1), occupying all positions in a 3-tray universe with partial-barriers of vaseline and wooden posts supplied—a 6-orange feeding area on a 120-orange dispersion with a complex maze of impediments (see text). Trays are broadly joined by use of paper bridges. (Photograph by F. E. Skinner.)

An estimate was made from general observations and the results of Waters (1955) that an orange area equivalent to that of two oranges, each 2½ inches in diameter, would be adequate, as a beginning, to study the predator-prey relation. With this premise, the smallest working basis for this design would utilize four oranges with each orange half-covered. This is because it was not desired to change more than ¼ of the food surfaces at a given time; that is, one of the four oranges used. This made it impossible to go to the ultimate in simplicity of searching and greatest concentration of the orange surfaces by utilizing only two whole, uncovered oranges.

Each universe in the earlier experiments conformed to this pattern, but certain other changes were made later. Each universe consisted of a flat metal tray, 40 inches long and 16 inches wide, with a side wall 1 inch in height and with 40 Syracuse watch glasses on each of which rested either an orange or a rubber ball (see figs. 3, 4, and 5). The positions were arbitrarily arranged in rows to conform to the dimensions of the trays—10 oranges and/or balls along the long dimension and four along the short dimension. This gave a center to center distance of 4 inches. The upper rim of the bordering wall of each tray was kept coated with white petroleum jelly to prevent movement of mites into or out of the trays. The predators and prey were therefore free to move onto or leave oranges or rubber balls but were not permitted to leave or enter the universes.

Lint-covered oranges as used by Finney (1953) to culture the six-spotted mite were used in all the experiments. The lint gives an ideal physical environment for propagation of the species employed. It produces an orange surface similar to the covering of fine hairs and setae on the surfaces of many plant leaves. However, it also adds to the complexity of the searching problem for the predator and increases the maximal potentials in populations of both species. In addition, it was noted that populations on well-linted oranges were less subject to the adverse effects of low humidity.

Initially, experiments were arranged in duplicate; check units having the prey species alone were carried along with those having both the prey and predator species. However, as the experiments developed, certain universes automatically terminated, and new ones not synchronized chronologically or exactly comparable in other respects were substituted. With each new universe employed, some change designed to give a better chance for perpetuation of the predator and prey in the system was incorporated. These changes were based on deductive thought and trial and error processes. During this study, the author had the good fortune to have Dr. A. J. Nicholson of Australia, one of the world's leading theoreticians on population dynamics, examine the experiments, and he confirmed the view that these changes must largely be decided upon by trial and error processes. Concerning some points it was not known whether a given change would detract or add to the chances of perpetuation of the predator-prey relations—such is the reciprocal interdependence of some actions.

The exposed area of an orange was in the form of a circle which was stamped on the orange by use of a shell vial of the proper size and an inking pad. The space between the outer edge of this line was then covered with hot paraffin and joined with the edge of the paper surrounding the hole. Several layers of hot paraffin were laid down as a seal by use of a small camel's hair brush. The oranges were then placed in a refrigerator for about 1 hour to chill the surfaces. Upon removal to the laboratory, only two or three oranges at a time, condensation on the surfaces was sufficient for dampening the point of an indelible pencil as lines were drawn on the surfaces of the exposed circular areas. Diameter lines were drawn in this way, dividing the area into 16 or more sampling sections with each section numbered. This greatly facilitated the counting of the populations. The counting had to be done under a stereoscopic microscope.

When the populations reached very low levels, total populations were counted, but normally the populations were counted on only ¼ or ⅛ of the total surfaces. The sample areas were taken in each case so as to be distributed evenly around the face of the "clock," but the first section to be counted was always taken at random.

Considering the universes employing oranges with all or one half their surfaces exposed, much difficulty was initially encountered in manipulating the oranges under the microscope so that the populations in the sample areas could be viewed fully, and without disturbance of the populations. However, the device shown in figure 2 solved this problem. It consists of a paraffin block cut so that one side at its highest point is 1½ inches and is tapered to the other side which is only ⅛ inch high. An orange placed on the Syracuse watch glass, which sits loosely in a depression in the block, can then be turned by rotating the watch glass so that any desired sample area can be brought into focus without touching or awkwardly manipulating the orange.

Sampling Procedure

Sampling of partial areas and populations in the present study was necessary in order to reduce the time required in counting large populations. Hence, the populations in most cases were sampled. Statistical analyses of test examples furnished estimates of the loss in confidence occasioned by such sampling.

It was found that a population estimate based upon a subsample of a given size on an orange is more reproducible if composed of two or more noncontiguous areas evenly distributed among the position (see fig. 1). This method was used in all sampling.

TABLE 1

ANALYSIS OF VARIANCE FOR DATA OF APRIL 26 IN FIGURE 8.

Orange no.	1	2	3	4	5	6	7	8	9	10	11	12	13	14	15	16	17	18	19	20
Subsample																				
1	20	2	1	1	2	2	8	1	20	15	17	0	16	7	6	2	1	0	2	0
2	17	2	5	1	2	4	6	5	30	8	9	4	5	2	7	8	0	0	0	4
3	21	2	9	3	0	3	0	0	32	4	8	7	1	0	8	1	0	0	0	1
4	23	1	1	5	0	2	4	3	24	2	15	5	2	4	7	1	0	0	1	1
Sums	81	7	16	10	4	11	18	9	106	29	49	16	24	13	28	12	1	0	3	6

Source V.	d.f.	S.S.	M.S.
Total	79	4,160	
Between oranges	19	3,679	194*
Within oranges	60	481	8

Standard error general mean, $S\bar{x}_g = 0.81$, or 14.6% of $\bar{x}_g$.

Standard error between oranges, $S\bar{x}_o = 6.9$, or 31.2% of $\bar{x}_o$.

Using this method of assignment of the positions of the subsamples on an orange, examples of data were analyzed to establish whether the within-orange variance which is associated with subsampling is significantly greater than the variance between the oranges. The data of Table 1 for April 26 illustrate the nature of the variance, and show that the within-orange variance, which is associated with the subsamples, is very small, and that if greater accuracy were required, it could best be achieved by counting populations on more oranges rather than by altering the technique of subsampling on a given orange.

It was therefore decided that the samples include every orange, but the subsamples on each orange would be varied with the approximate densities of the populations encountered, the usual proportion being ½ or ¼ the total exposed area on each orange.

Early in the study in deciding upon the technique of sampling, the **entire** population of the prey species was counted on a representative group of 44 orange units, with each unit exposing $1/20$ of an orange area. This population was thus finite and known. The data used were for July 5 from the experiment illustrated in figure 18. The mean, x, for the 44 items was 6.95 mites per exposed area, with a standard error of ± 1.09, which is 15.7 per cent of the mean. This standard error reflects the variance inherent to the particular type of conglomerate distribution exhibited by such populations.

In order to determine if the subsamples could be used to estimate the total populations, half-area counts were first used. Series of six half-area random lots of the component items were drawn from the aforementioned total population on the 44 representative oranges. The means, standard errors, and the coefficient of variation were then compared with the values based upon the total known population. These statistical parameters were little changed: the standard errors being ± 0.52, ± 0.68, ± 0.65, ± 0.63, ± 0.56, and ± 0.64, respectively, as compared with a half-value of ± 0.55 for the total population; the coefficient of variation varying from 16.2 per cent to 19.3 per cent compared with 15.7 per cent (the coefficient of variation of the whole population); and the means, as estimates of the mean of the total population, averaging only 4.7 per cent higher or lower than the corresponding half-value for the total population—the range being from 3.3 to 7.8.

A test was also made to determine if a further reduction in the counting (to ¼ the total area) would give adequate estimates of the population when large universes or high populations were encountered. Both the predator and prey populations on August 1 (see fig. 18) were analyzed for this purpose.

It should be noted that some basic change had occurred between July 5 and August 1 contributing to greater skewness of distribution. This is revealed by an increase in the coefficient of variation from 15.7 per cent for the whole population count of July 5 (analysis just discussed) to 22.3 per cent for the large sample count on August 1. The six subsamples of reduced size taken on August 1 closely approximated the large sample in coefficient of variation, these being 23.3 per cent, 23.9 per cent, 23.3 per cent, 22.8 per cent, 23.3 per cent, and 23.5 per cent, respectively.

There are two probable reasons for this change in the nature of the variation. In the former instance, the predators had not yet been introduced while they were significantly active on August 1. The predator-prey relation characteristically contributes to skewness, colonial, or conglomerate distribution. Also, the prey were introduced into the universe in equal numbers on all oranges at the initiation of the experiment, and some time is necessary for the typical conglomerate distribution to become manifested, even disregarding predation. Therefore, the larger, although uniform coefficient of variation of the samples on August 1 are not the result of inadequate sampling technique but express the nature of the distribution of the population.

Comparing the prey populations of the ¼-area samples with the composite "total" or ½-area sample on August 1, the standard errors were little changed: these being ± 2.43, ± 2.63, ± 2.70, ± 2.67, ± 2.60, and ± 2.48, respectively, for the six ¼-area samplings, as compared with a half-value of ± 2.39 for the ½-area sample; the coefficient of variation, previously listed, varied only from 22.8 per cent to 23.9 per cent, compared with 22.3 per cent for the sample of double size; and the means, as estimates of the mean of the population present on twice the area, averaging only 4.7 per cent higher or lower than the corresponding half-value for the larger sample—the range being from 1.6 to 9.2.

Predator populations were more variable than were the prey populations. The standard error for total counts of the large (½ area) sample was ± 0.254, with a half-value of ± 0.127, while the values for the counts made on six ¼-area samples were ± 0.117, ± 0.150, ± 0.155, ± 0.167, ± 0.153, and ± 0.151. The coefficient of variation also varied more than the corresponding values for the prey population, these being 24.6 per cent, 34.6 per cent, 32.6 per cent, 29.0 per cent, 32.8 per cent, and 27.2 per cent, compared with 26.7 per cent for the sample of double size. The same was true for the means, these values, as estimates of the mean of the population on the larger area, averaging 5.7 per cent higher or lower, but having a range from 0.0 to 12.1.

When such ranges of error relating to the various observed points in the illustrations are considered, comparing positions of high and low densities, it is obvious that there is adequate accuracy in the estimates to establish the validity of the major trends or patterns of population change exhibited with respect to the predators and their prey in the various experiments. Yet, obviously, some of the minor, inconsistent changes following no general trend in time may be the result of inadequacy of sampling, and hence have a random character independent of predation.

RESULTS

A group of eight universes was started on February 4 and February 10. These were duplicates of an earlier group—which, as previously stated, had to be discarded because the oranges were rotting—except that the covering used on the oranges was part polyethelene material and part paraffin, rather than paraffin alone. This group also had to be discarded except for certain universes which utilized four oranges each, and these were half-covered with paraffin. Oranges only half-covered with paraffin proved satisfactory, and those units were retained.

A basic idea in this study has been the comparison of results when the plant food (oranges) is readily accessible (massed in one location in the universes) with other examples having the food widely dispersed, with the problems of dispersal and searching thus made more difficult for both the predators and the prey. The control universes which reveal the approximate levels of density of the prey species in the **absence** of predation, thus limited by the availability of food, were followed under several conditions of dispersion of the food material. These are considered representative of densities permitted by the respective levels of availability of food; and the degree to which the prey fail to reach these densities under the pressure of predation in the other universes is a measure of the effect of that predation.

The specific designs of experiments which differ from the general methods and procedures discussed previously will be covered, along with the results obtained, under each type of universe employed.

I. Densities of Prey and Fluctuations in the Absence of Predation

The following three universes were used as a measure of the population dynamics of the prey species in the absence of predators.

A. Predators Absent, Simplest Universe, Four Large Areas of Food, Grouped at Adjacent, Joined Positions. In this universe a 2-orange feeding area on a 4-orange dispersion was employed, and the unit was started February 10 and ended July 1 (see figs. 3 and 6). The initial colony was established by placing 20 female six-spotted mites, *E. sexmaculatus* (the prey species), on one of the four oranges. Movement to the other oranges was delayed until the period between March 4 and March 8 at which time the orange originally colonized was beginning to become conditioned and migrants had started moving. Thus, on a feeding area as large as a half orange, overpopulation may be delayed for about three weeks. This is significant with respect to attempts to establish self-sustained existence of both the predator and its prey in a universe. If the prey do not move readily or at least are not moving from some arenas rather readily most of the time, the predators only have to locate a colony arena and stay with it until it is overexploited, resulting in its own extinction and possibly that of its prey as well. On a given orange, this predator commonly overexploits such a colony in much less than the three week period required for conditioning pressure under these conditions (see figs. 9, 10, and 11). This question will be discussed further in relation to the data of Subsections F, G, H, and I of Section II, and it led to the arrangements used in those universes.

Regarding densities, the approximate mean population reached in this universe (fig. 6) was 9,400 *E. sexmaculatus* (all stages), or 4,700 per orange-area. Two major peaks above that level, once population growth had progressed that far, and two subsequent, resultant depressions below it occurred. These indicate a trend of a somewhat "oscillatory" nature due to occurrence of waves of maximal or excessive utilization of the food, followed by inadequate food to support the high levels. This trend may be only a carry-over result of the arbitrary unnaturally high abundance of entirely "unconditioned" or unutilized food at the initiation of the experiment, in interaction with the pattern of orange replacement. This example was not continued long enough to learn if the degree of such fluctuations would continue undiminished in amplitude or whether an inherent oscillation associated with factors of dispersal and population density under related conditions is a real feature of well-established, long-term system—i.e., ones which have reached internal balance, or relative calm.

These results do establish that a relatively high mean population is characteristic of this experimental arrangement, contrasted to that which results when predators are present. Perhaps a sizeable part of the large fluctuations may be the result of variations in the nutritional qualities of the

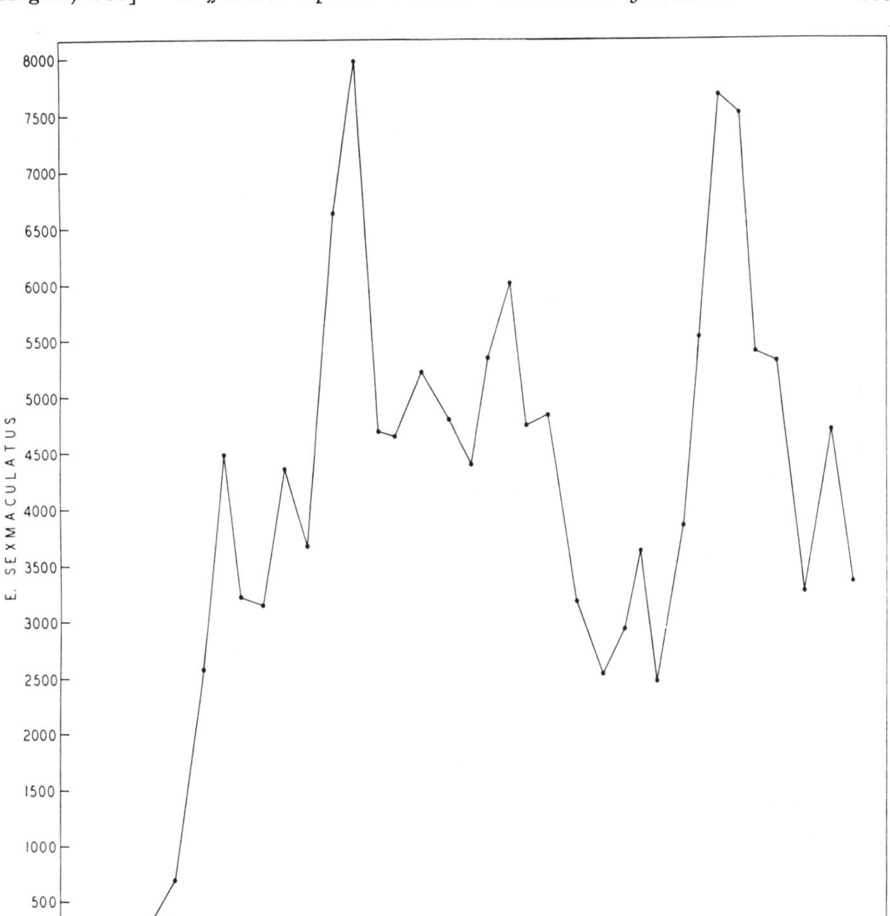

Fig. 6. Densities per orange-area of *Eotetranychus sexmaculatus* in the absence of predators in the simplest universe used, with four large areas of food (orange surface) grouped at adjacent, joined positions—a 2-orange feeding area on a 4-orange dispersion (see fig. 3 and text, Subsection A, Section I of "Results").

oranges supplied during the course of the experiment. It is known that oranges do vary in nutritional value for this mite. Beginning with March 30, at which time the population had first attained maximal utilization of the food, the oranges afterward removed in the replacement scheme were invariably fully utilized or conditioned. This full utilization is probably the most important reason why the mean level of this population was higher than that in the next two universes discussed. In fact, it was characteristic that the populations on these oranges reached conditioning levels well in advance of the dates for removal of the respective oranges, and such conditioning pressure, sometimes from two oranges at once, accomplished very prompt natural

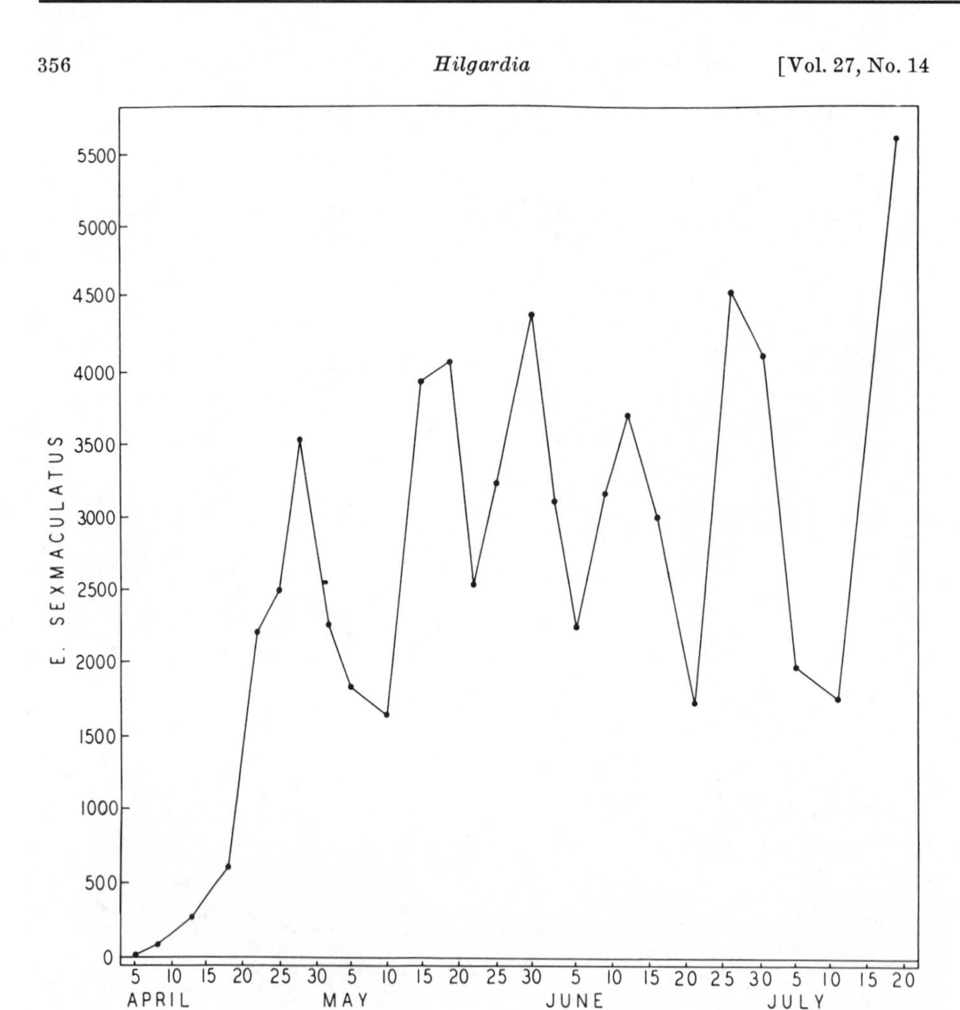

Fig. 7. Densities per orange-area of *Eotetranychus sexmaculatus* in the absence of predators, with four large areas of food (orange surface) widely dispersed among 36 foodless positions—a 2-orange feeding area on a 4-orange dispersion (see fig. 4 and text, Subsection B, Section I of "Results").

colonization of each new orange added, and thus little loss of time in utilization of the food was occasioned. Also, even after such prompt conditioning, the oranges continued to support for some time a much reduced but sizeable population of mites, and these factors support the position that in this universe the more complete utilization accounted for the higher mean level of density, comparing this universe with those of Subsections B, figure 7, and C, figure 8, of this section.

B. Predators Absent, Four Large Areas of Food Widely Dispersed. In this universe, as in the last, a 2-orange feeding area on a 4-orange dispersion was employed (in this case, not grouped), and the unit was started on April 5 and ended July 18 (see figs. 4 and 7).

The mean level of density of *E. sexmaculatus* subsequent to the initial period prior to May 20 was 7,000, or 3,500 per orange-area. It is seen, there-

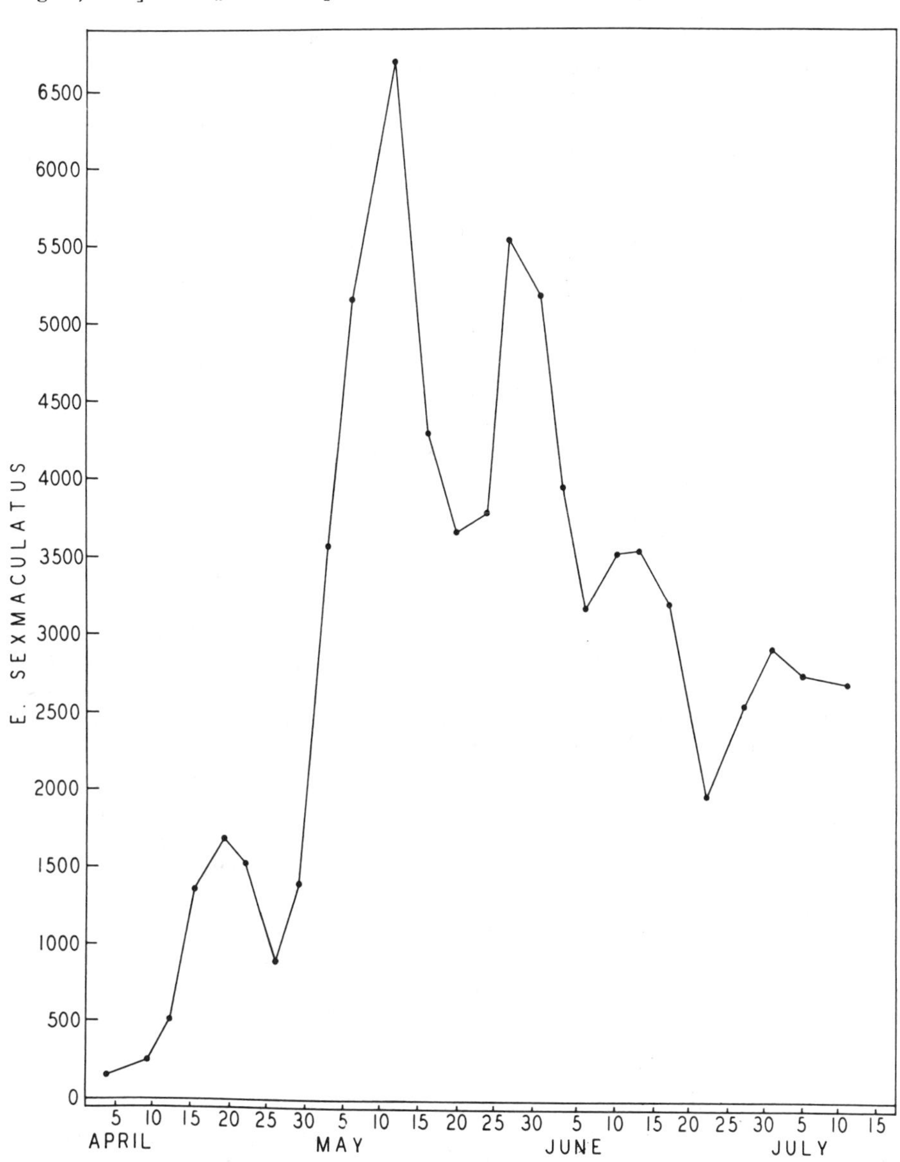

Fig. 8. Densities per orange-area of *Eotetranychus sexmaculatus* in the absence of predators with 20 small areas of food alternating with 20 foodless positions occupied by rubber balls—a 2-orange feeding area on a 20-orange dispersion (no photograph of this exact arrangement, but see text, Subsection C, Section I of "Results").

fore, that although the same quantity of food was supplied, the utilization was somewhat lower due to the difficulty the mites had in quickly locating the new orange units and the resulting loss in population in doing so, with reduced utilization during the time the oranges were in the universe.

There was also a different pattern of fluctuation in numbers, the changes being somewhat more regular in occurrence, and less marked maxima exhibited, than was so when the four oranges were placed adjacent and joined at one arena in the universe. The interaction between the difficulty of dispersal and the supply of food appears to have had a slightly stabilizing effect, compared with the condition when the supply of food alone was the primary feature and problems of dispersal minimal in effects. Unquestionably, the known variation in the quality of the oranges used is a source of error—but doubtfully sufficient to nullify this indication.

Tracing this population, 20 female mites were placed on a single orange and the population growth occurred almost entirely on the single originally stocked orange until just prior to April 28 at which time conditioning of that orange had forced migrants to search for food. By May 2 they had located the other three oranges, but not in sufficient numbers to forestall the decline in the general population resulting from conditioning of the orange originally stocked. The second ascent followed an expected course and was also short-circuited by conditioning on one of the first naturally colonized oranges on which the population first got well under way, at a time before the numbers which had located the other two oranges had increased enough to offset the decline. Subsequent events were similar; the distances between positions with food, and the difficulty the mites had in locating them were such that usually only one orange was highly productive at a time and occasionally an orange came due for removal prior to full utilization. However, none was removed which harbored a major portion of the total population. This would account for the mean density being somewhat lower than that which was experienced in the universe of Subsection A, figure 6, of this section, in which relatively full utilization was experienced.

Unusually high levels of density were the result of a partial chance occurrence of simultaneous productivity on two or more of the four oranges, but the resultant steep ascents such as those occurring during the last half of June and again between July 12 and July 18 are certain to be followed swiftly by corresponding declines in density.

C. Predators Absent, 20 Small Areas of Food Alternating with 20 Positions with No Food. In this universe, also, a 2-orange feeding area was employed, but this was segmented into 20 parts of 1/10 orange-area each, with one part on each of 20 oranges. The 20 oranges were placed in alternate positions with 20 rubber balls. The universe was started on March 31 and ended on July 11, but the first count was made on April 4 (see fig. 8).

It should be noted that although the food material was segmented into 20 parts and dispersed over 20 oranges (equalizing to a much greater degree the variation in orange quality) and 20 positions in the universe, each orange was only one position distant from another and the positions having no food (the rubber balls) were much fewer in number. Consequently, although the food was widely dispersed it was readily accessible and sources of migrants were close at hand to any new orange added. In this respect the unit more nearly resembled those universes in which all the food was massed at one arena in the universe and on adjacent oranges.

The mean level of density of mites subsequent to the initiation period prior to May 20 was 6,600, or 3,300 per orange-area. This is slightly lower than the results of the previously discussed universe. It is likely that the differences in the actual levels of the means in these last two instances do not have real meaning, but the patterns of change in density are probably meaningful of an inherent relation to the types of dispersion employed. Yet, if the initial extreme high in density, occurring during the middle of May, is **included,** the mean level would be approximately 3,800 per orange-area—still somewhat below the mean level of the universe in which the oranges were grouped and joined (fig. 6), but slightly higher than that exhibited in the universe just discussed.

Tracing the history of this population, it was initiated by placing 10 female *E. sexmaculatus* on each of two of the 20 oranges. The first count on April 4 showed a mean population of 152 per orange-area, and these were still located only on the two colonized, or stocked, oranges. Not until April 19, at which time the population on these two oranges was first noted as causing conditioning and under competitive pressure, had migrants generally moved to other oranges; only one other orange had a few mites prior to that time. That date also marked the decline of the population, and this decline was due to the conditioning on the two oranges stocked originally, before the other 18 oranges had been located and population growth on them gotten under way. In this universe nearly all the oranges were found at this same initial period of migrants and the subsequent very steep population increase was the result of simultaneous utilization of the unused oranges in the entire universe. The second depression in the middle of May was due to rather general conditioning of many of those oranges, and the next increase was made possible by the replacement of utilized oranges by new ones according to the predetermined schedule.

There was in this case a strong indication that the period of initiation and establishment of a balance between density and the schedule of supplying food is prolonged much beyond a period of 45 days. The large amplitude of fluctuations in the early stages of the experiment, considered in relation to the successive **steady decrease** in this amplitude, and in view of the much reduced probable source of error associated with variations in orange quality, makes it likely that a position around 5,500, or 2,750 mites per orange-area, is nearer to a true equilibrium, and that the wide fluctuations and high densities which persisted during the early course of the experiment were adjustments prior to establishment of a semblance of such balance.

In contrast to this, the data illustrated in figure 7 probably represent a meaningful difference in patterns of population change. In that case the large fluctuations did **not** diminish with time. In the universe illustrated in figure 7, a position with food is found with great difficulty, but each such position has a supply five-fold in quantity. In the present example, figure 8, there are five times as many positions with food, but each has only 1/5 the quantity. The positions with food are more readily located but each is depleted more rapidly. Additional replicates would need to be run and continued over a longer period of time in order to answer the questions raised.

II. Population Changes under Predator and Prey Interactions

A. Predators Present, Simplest Universe, Four Large Areas of Food, Grouped at Adjacent Joined Positions. In this universe a 2-orange feeding area on a 4-orange dispersion was employed. The unit was started February 4 by stocking with 20 female six-spotted mites. It ended April 5 (fig. 9). The

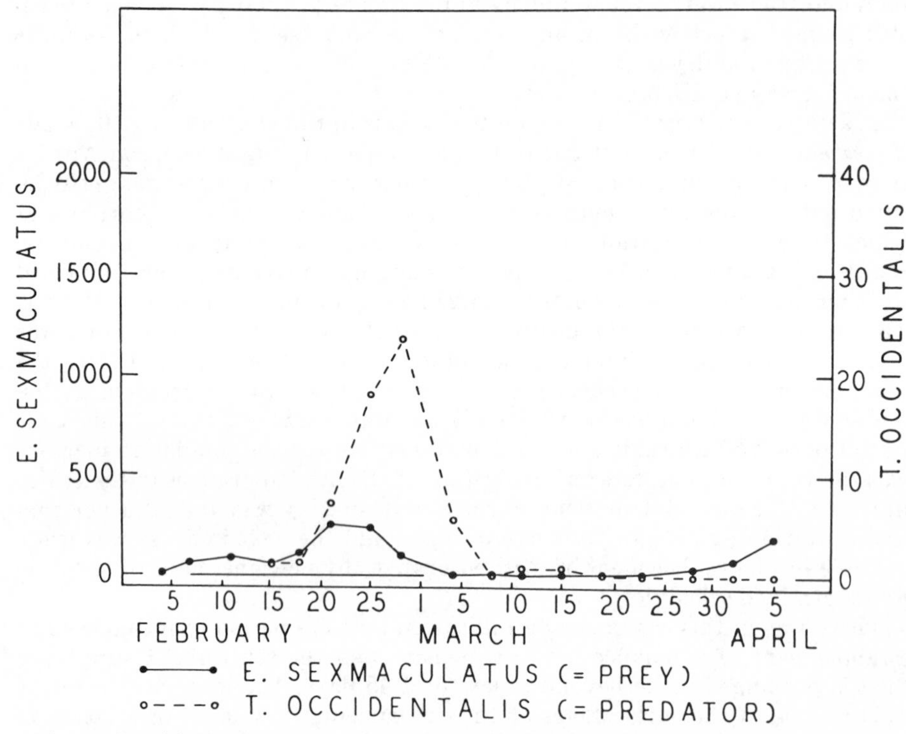

Fig. 9. Densities per orange-area of the prey, *Eotetranychus sexmaculatus,* and the predator, *Typhlodromus occidentalis,* with 4 large areas of food for the prey (orange surface) grouped at adjacent, joined positions—a 2-orange feeding area on a 4-orange dispersion (see fig. 3 and text, Subsection A, Section II of "Results").

arrangement was the same as the control universe of Subjection A of Section I (see also figs. 3 and 6) except that in this universe the predatory species was present. Two female predators were introduced 11 days after the introduction of the six-spotted mites. Both predators were placed on a single orange. This scheme was followed with all the universes except as otherwise stated.

The stocked orange spoiled between February 7 and February 11, and, consequently, the prey then declined in numbers and only increased after moving to the adjacent oranges. The prey then increased to a level of 500,

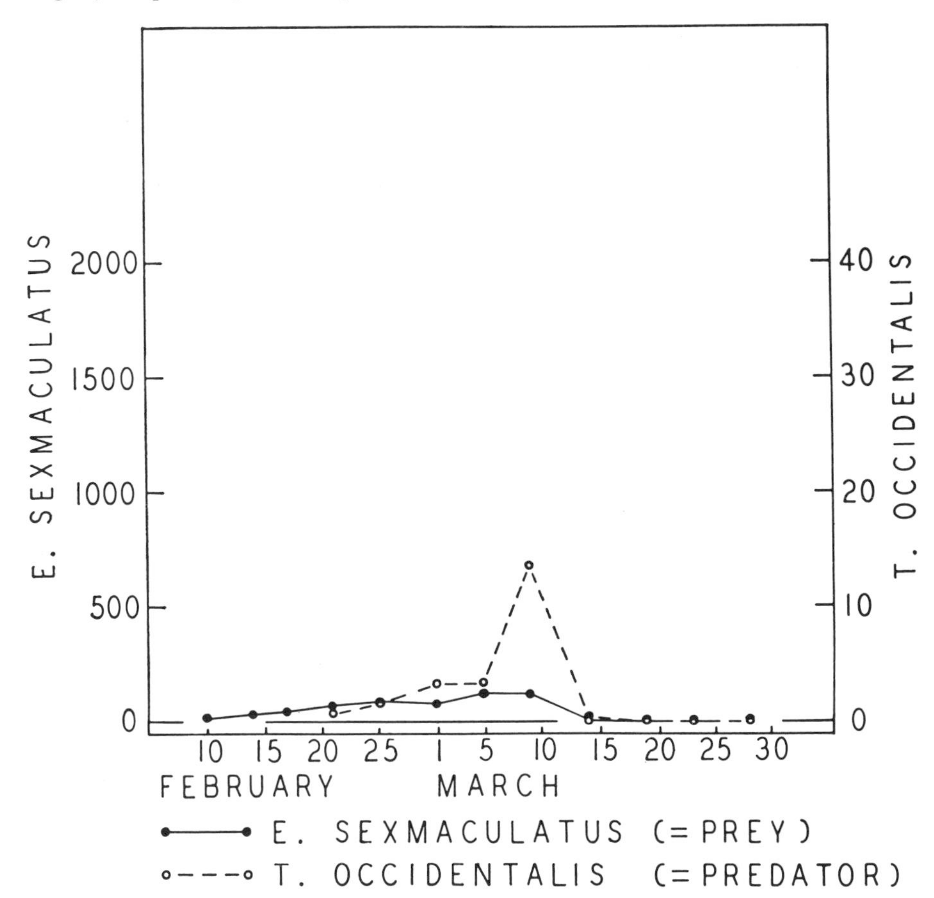

Fig. 10. Densities per orange-area of the prey, *Eotetranychus sexmaculatus,* and the predator, *Typhlodromus occidentalis,* with 8 large areas of food for the prey (orange surface) grouped at adjacent, joined positions—a 4-orange feeding area on an 8-orange dispersion (no photograph of this exact arrangement, but it was similar to that of figure 3 except that 8 oranges were used; see also text, Subsection B, Section II of "Results").

or 250 per orange-area, at which time it was preyed upon so severely that it was reduced to a nil density within 10 days and all the predators subsequently starved. The consequent characteristic, very gradual increase in the numbers of the prey was then prolonged for about 15 days before there was attained a state of vigorous population growth (see "Discussion").

B. Predators Present, Eight Large Areas of Food, Grouped at Adjacent Joined Positions. In this universe a 4-orange feeding area on an 8-orange dispersion was employed, and the unit was started February 10 and ended March 28 (fig. 10). The eight oranges were grouped in one end of the tray and joined with wire loops. In this case, because of the larger quantity of food supplied, 40 female six-spotted mites, or prey, were colonized initially, 20 on each of two of the eight oranges. Two female predators were added 11 days later.

The notes taken during the early days of this universe reveal that the female six-spotted mites used for colonizing were old and not from the usual stock colony of vigorous young females. These females died quickly without producing the usual quota of eggs after colonizing. Hence, the population could not increase at the usual rate until the first daughters became fecund. By the time that occurred and the normal population increase would other-

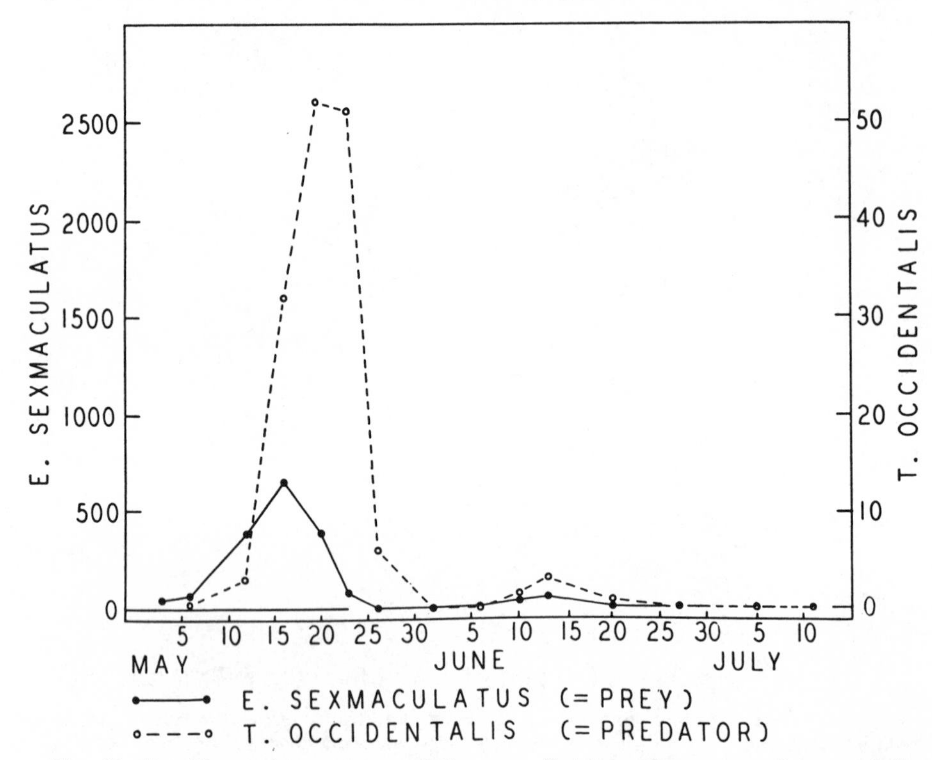

Fig. 11. Densities per orange-area of the prey, *Eotetranychus sexmaculatus,* and the predator, *Typhlodromus occidentalis,* with 6 large areas of food for the prey (orange surface) grouped at adjacent joined positions—a 6-orange feeding area on a 6-orange dispersion (no photograph of this exact arrangement, but it was similar to that of figure 3 except that 6 whole oranges were used; see text, Subsection C, Section II of "Results").

wise have ensued, the predators had become sufficiently abundant that the increase never occurred at all, even though migrants had moved to and populated at least six of the eight oranges. Hence, the population reached a maximal level of only 451 mites, or 113 per orange-area.

In this universe the predators overexploited the prey by March 14 to the extent that not only did they annihilate themselves but they also annihilated the prey species, even though the latter had dispersed successfully throughout the universe.

C. Predators Present, Six Whole Oranges as Food, Grouped at Adjacent Joined Positions. In this universe a 6-orange feeding area on a 6-orange dispersion was employed. The unit was started with 20 female six-spotted mites on April 26 and ended July 11 (fig. 11). The prey were introduced on two oranges, the two female predators on only one of them.

The prey temporarily thus escaped severe predator action on one of the oranges and a few migrants moved onto some of the other oranges, but this was not until about May 15, and before these were able to effect an appreci-

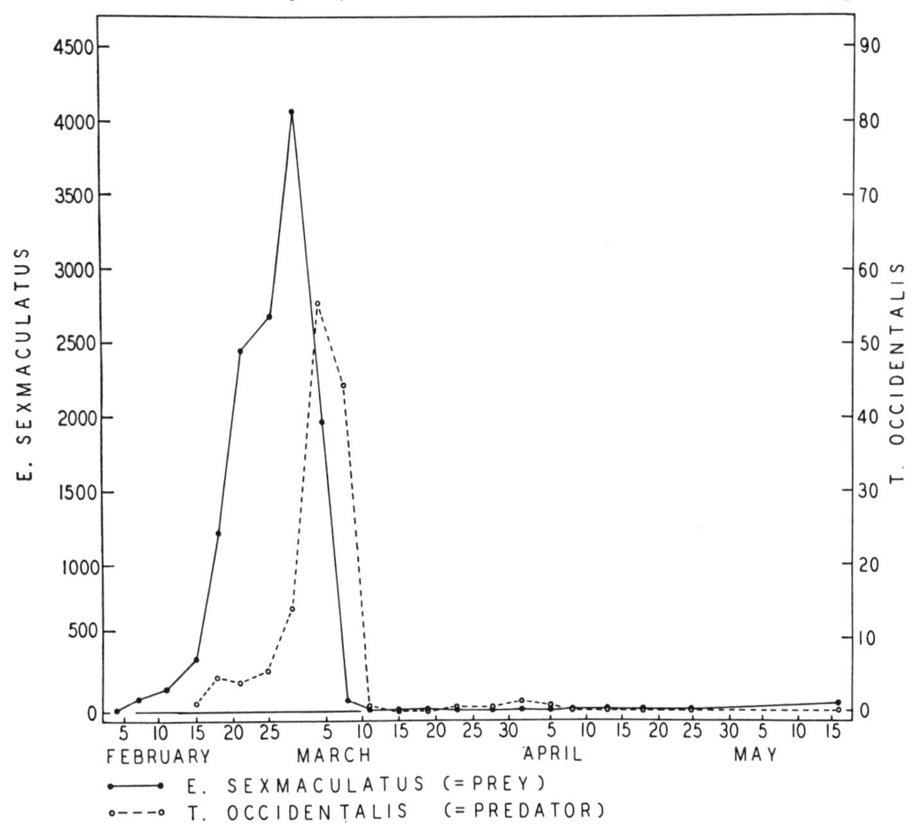

Fig. 12. Densities per orange-area of the prey, *Eotetranychus sexmaculatus,* and the predator, *Typhlodromus occidentalis,* with 4 large areas of food for the prey (orange surface) widely dispersed among 36 foodless positions—a 2-orange feeding area on a 4-orange dispersion (see fig. 4 and text, Subsection D, Section II of "Results").

able general population growth, the predators reached all the infested oranges. The peak population reached was 3,900 mites, or 650 per orange-area. After May 18 to 20, the prey suffered the characteristic crash effect. Contrary to what happened in most similar universes, there was limited temporary survival of the prey **and** the predator species. The prey increased very slightly from June 5 to June 13, and this was followed by a corresponding increase in the predators, after which time the predators quickly annihilated their prey and thus themselves.

D. Predators Present, Four Large Areas of Food Widely Dispersed. In this universe a 2-orange feeding area on a 4-orange dispersion was used (see figs. 4 and 12). The oranges were placed at randomized positions among rubber balls as shown in figure 4. Twenty female six-spotted mites were colonized on one orange on February 4, and two female predators were put on the same orange on February 11. The universe was ended May 17.

The wide dispersal of the food among the 40 positions presented an obstacle to movement of both the prey and the predators. In fact, neither species reached the unstocked oranges until March 28 when both did and, thus, densities on the other oranges were never substantial. The colonized orange was

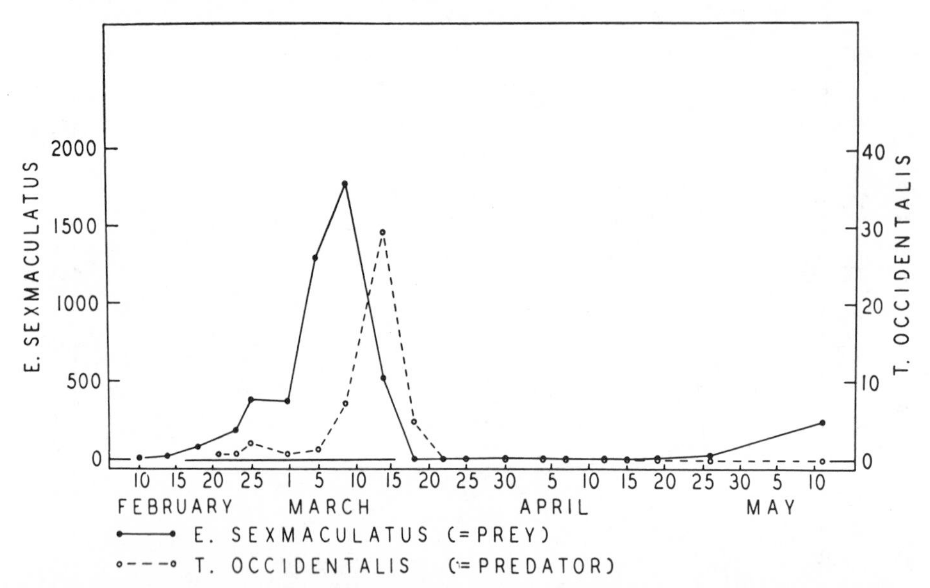

Fig. 13. Densities per orange-area of the prey, *Eotetranychus sexmaculatus,* and the predator, *Typhlodromus occidentalis,* with 8 large areas of food for the prey (orange surface) widely dispersed among 32 foodless positions—a 4-orange feeding area on an 8-orange dispersion (no photograph of this arrangement, but it was similar to that of figure 4 except that 8 oranges were used; see also text, Subsection E, Section II of "Results").

apparently phenomenal in nutritional quality for on it the prey reproduced at a very high rate, so much so that the predators did not quickly overtake it even though the latter were present on that orange from the eleventh day. The population of predators did not increase rapidly at first, although at the low density at that time it is probable that the numbers missed in the counting may have been enough to explain a part or most of this retarded increase in the midst of an abundance of prey.

At any rate, the prey population reached the high level of 8,113, or 4,056 per orange-area. This level could not be maintained on the single orange longer than a few days even in the absence of predation; thus, both conditioning of that orange and intense predation jointly accounted for the very abrupt crash which followed. Nearly all the predators then starved but a very few survived and prevented any resurgence of the prey until after April 8 at which time the last predator died and the prey began a very gradual increase in numbers (see "Discussion").

E. Predators Present, Eight Large Areas of Food Widely Dispersed. In

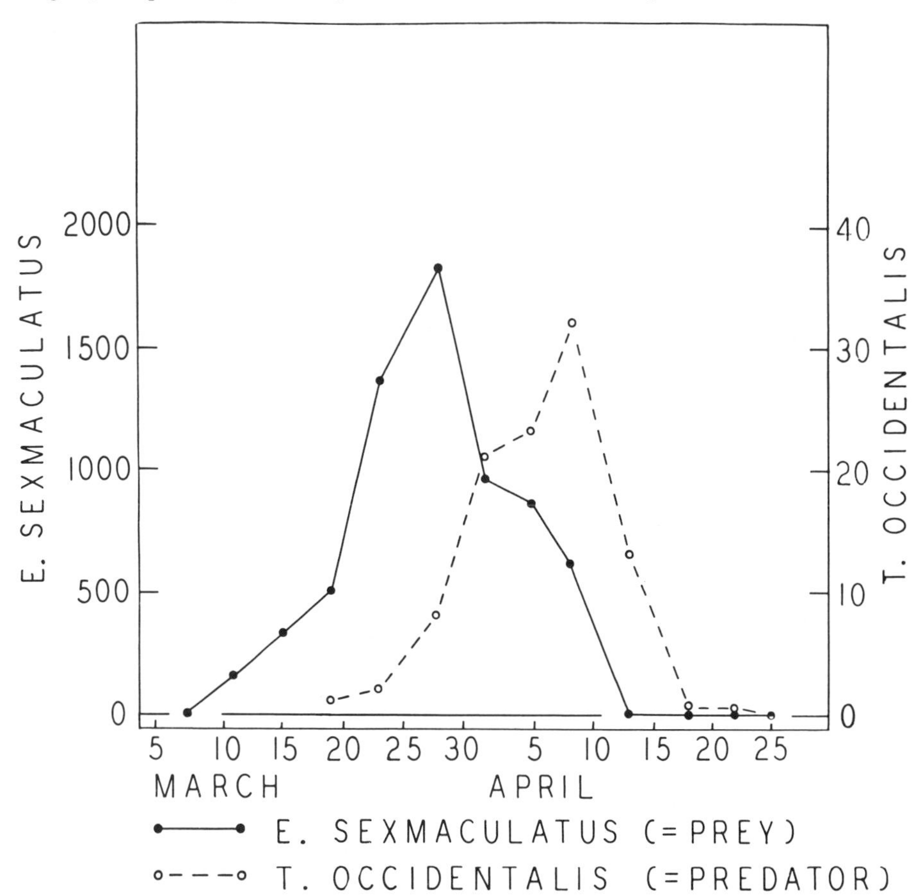

Fig. 14. Densities per orange-area of the prey, *Eotetranychus sexmaculatus*, and the predator, *Typhlodromus occidentalis*, with 20 small areas of food for the prey (orange surface) alternating with 20 foodless positions—a 2-orange feeding area on a 20-orange dispersion (no photograph of this exact arrangement, but see text, Subsection F, Section II of "Results").

this universe a 4-orange feeding area on an 8-orange dispersion was utilized, the remainder of the 40 positions being occupied by rubber balls. The unit was started February 10 and ended May 11 (fig. 13). Twenty female six-spotted mites were colonized on each of two of the eight oranges, whereas the two female predators were introduced 11 days later on one of the oranges colonized with the prey species.

There was a logical delay of about 14 days between the ascent in the prey population and the ascent in the predator population. Thus, the prey could increase unabated on the oranges which did not receive predators until such time as the latter moved onto them. The dispersed condition of the oranges made more likely such a lag in general predator action. In the examples of the universes otherwise comparable except that the food was grouped in one area and joined (figs. 9, 10, and 11), the ascents in predator densities fol-

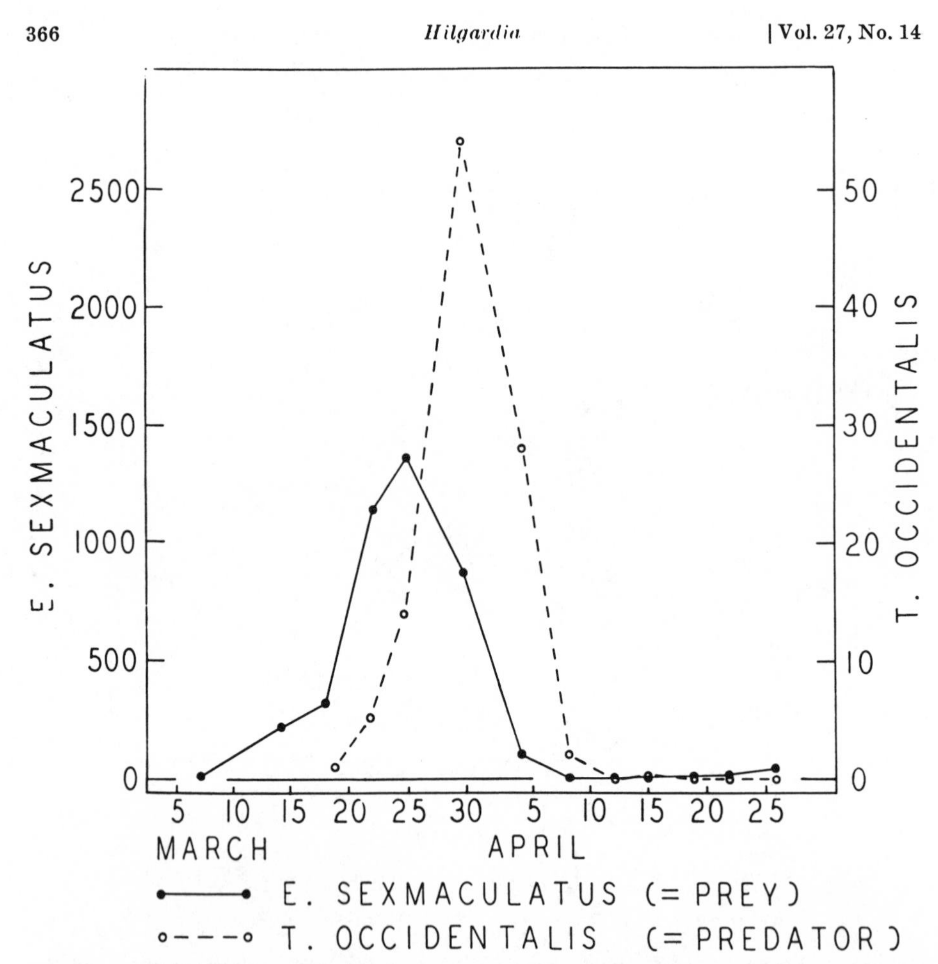

Fig. 15. Densities per orange-area of the prey, *Eotetranychus sexmaculatus*, and the predator, *Typhlodromus occidentalis*, with 20 small areas of food for the prey (orange surface) alternating with 20 foodless positions—a 2-orange feeding area on a 20-orange dispersion (no photograph of this exact arrangement, but see text, Subsection F, Section II of "Results").

lowed the respective ascents in prey densities within intervals of two or three days. Also, in those universes the prey never reached such high levels.

The lag period in this universe was sufficient for the general prey population to reach a level of 7,046, or 1,761 per orange-area before the predators moved through the universe and reduced the prey to a very low level, after which all the predators starved. The very gradual subsequent increase in the numbers of the prey up to May 25 is obvious in figure 13. After that date the undercrowding effects from the intense predator action had been overcome and a *substantial* increase followed.

F. Predators Present, 20 Small Areas of Food Alternating with 20 Foodless Positions. Two universes of this arrangement were used. In each, a 2-orange feeding area on a 20-orange dispersion was used and both were started on March 7. One was ended on April 25, the other on April 26 (figs.

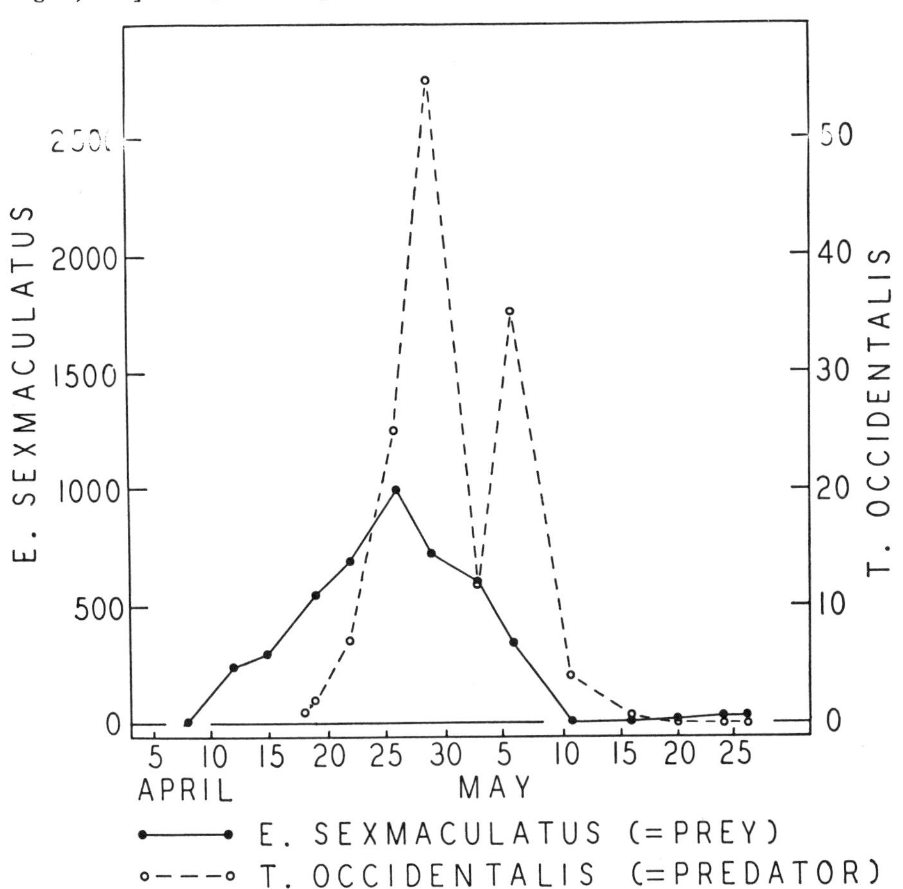

Fig. 16. Densities per orange-area of the prey, *Eotetranychus sexmaculatus,* and the predator, *Typhlodromus occidentalis,* with 40 small areas of food for the prey (orange surface) occupying all 40 positions—a 2-orange feeding area on a 40-orange dispersion, but with units of food thus adjacent (no photograph of this exact arrangement, but it was similar to ⅓ of the universe shown in figure 5 except that the wooden posts were not used and the maze of vaseline partial-barriers was much less complex; see also text, Subsection G, Section II of "Results").

14 and 15). In each universe, 10 female six-spotted mites were introduced onto each of two of the oranges, and in each, two female predators were introduced onto one of the two oranges 11 days later.

In these universes the feeding area employed on a given orange was reduced to a $\frac{1}{10}$-orange area. Large areas support the prey species for long periods of time. Dispersal pressure, or overpopulation—which is now known to cause practically all movement from orange to orange—is delayed. It was felt that by decreasing the feeding surface at each orange position—thus having more positions—the prey would be kept on the move from more individual sources so that following a localized crash from predation, there would occur sooner a subsequent population pressure which would cause more rapid dispersal and resultant repopulation on a broader spatial basis in the universe. This seemed to offer a better possibility for achieving perpetuation of the predators and prey than would wider dispersion of the smaller number of larger areas of food.

With one of the examples, the period of lag in the increase in predators was just as long, and, with the other, it was three to four days shorter than it was for the units just previously discussed. However, the prey populations did not reach levels quite so high. The higher level, however, occurred in the universe where there was the longer period of lag in the predator response to prey increase (see fig. 14). In this universe the predators exterminated the prey by April 18 and then died themselves. In the other example (see fig. 15), the prey was almost, but not entirely, exterminated by April 8. The predators quickly starved after that date and subsequently the prey gradually increased in numbers. Thus, even when 20 positions of food were used, the prey was exterminated in one instance although it survived in the other (see "Discussion").

G. Predators Present, 40 Small Areas of Food Occupying All Positions. In this universe a 2-orange area was again utilized, but the feeding area on each orange was further reduced to $\frac{1}{20}$-orange area; thus, all 40 positions were occupied by oranges (a 40-orange dispersion)—i.e., no rubber balls were used. The unit was started April 8 and ended May 26 (fig. 16). Ten female six-spotted mites were colonized on each of two of the oranges and 10 days later two female predators were colonized on one of them.

The tray used was also divided into three areas, mostly, but not entirely, separated by vaseline barriers as an impediment but not an exclusion to movement. The barrier pattern was not as complicated as that used in the later experiments such as those shown in figure 19. It was felt that the presence of the barriers would introduce greater difficulty for the predators in contacting all general positions of prey at a time, and the smaller areas of food would insure quicker movement of the prey and repopulation of depopulated areas by migrants from areas missed at the time of greatest predator abundance and pressure.

Subsequent to the expected initial increase, there was a sharp decline in the predators between April 29 and May 3 to a level below that which could be supported by the prey population in the total universe at the time. There was then a second sharp increase in the predators as they moved into one of the areas where they had not previously made contact with the prey. The second decline in predators and prey was general throughout the universe and the predators then starved since the prey reached a level which would not support a single predator. The prey population then began a gradual increase in numbers.

In this universe it became obvious that the history of events could not be properly illustrated by use of a simple line graph plotting densities, and this became increasingly true when still greater complexity was introduced. The counts made on the individual oranges revealed, unquestionably, that the sharp drop in predator abundance between April 29 and May 3, and the immediately subsequent sharp increase in numbers, were reliable reflections of the changes in the total populations in the universe. Only a pictorial record

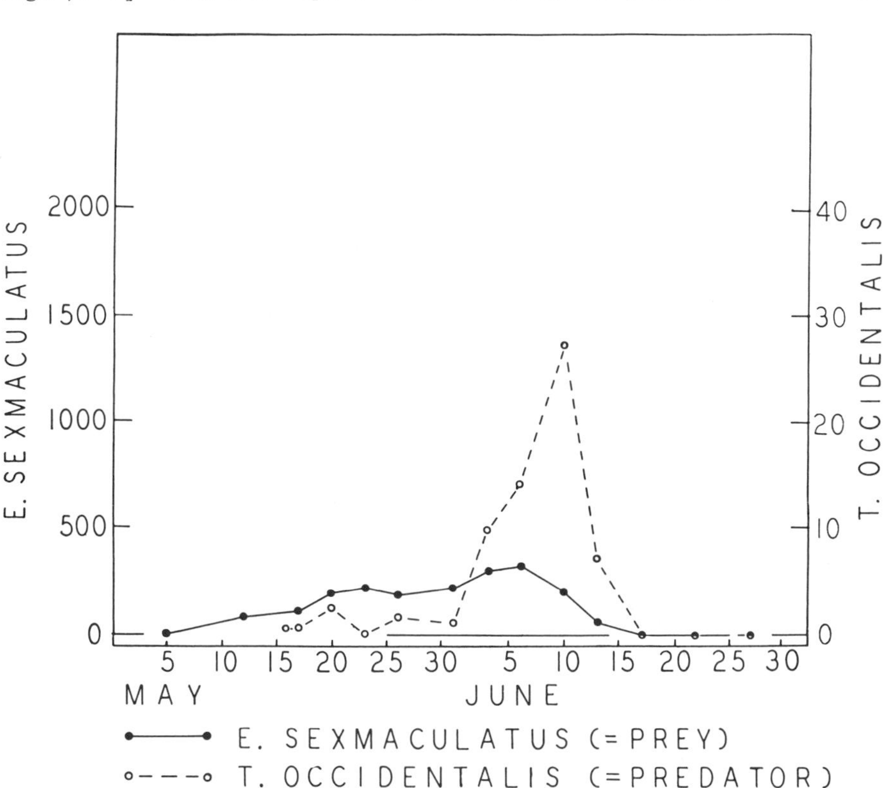

Fig. 17. Densities per orange-area of the prey, *Eotetranychus sexmaculatus,* and the predator, *Typhlodromus occidentalis,* with 120 small areas of food for the prey (orange surface) occupying all 120 positions in a 3-tray universe—a 6-orange feeding area on a 120-orange dispersion, with a simple maze of vaseline partial-barriers utilized (no wooden posts), but with the stocking done in a very restricted manner (see fig. 5 and text, Subsection H, Section II of "Results").

of the densities of the predators and the prey in the specific geographic areas in which the predators were active, and in which they were not active, could be expected accurately to portray the situation. Otherwise, the data reveal a sharp decline in predators, followed by a rapid increase, and then followed again by a rapid decline—all taking place synchronously with a rather steady general decline in density of the prey, if the data for the whole universe are plotted as a single unit as in figure 16.

Such events are contrary to the known fact of specific dependence of this predator upon this prey in these universes and, as well, contrary to the fact of the predator's rapid response to changes in the numbers of its prey by corresponding changes in its own numbers if it contacts its prey. Such a pictorial record was constructed for the data of the most important of these experiments (see fig. 18), but the time required to do this for each universe would be excessive. In any event, this does not appear to be necessary when simple universes are used (see "Discussion").

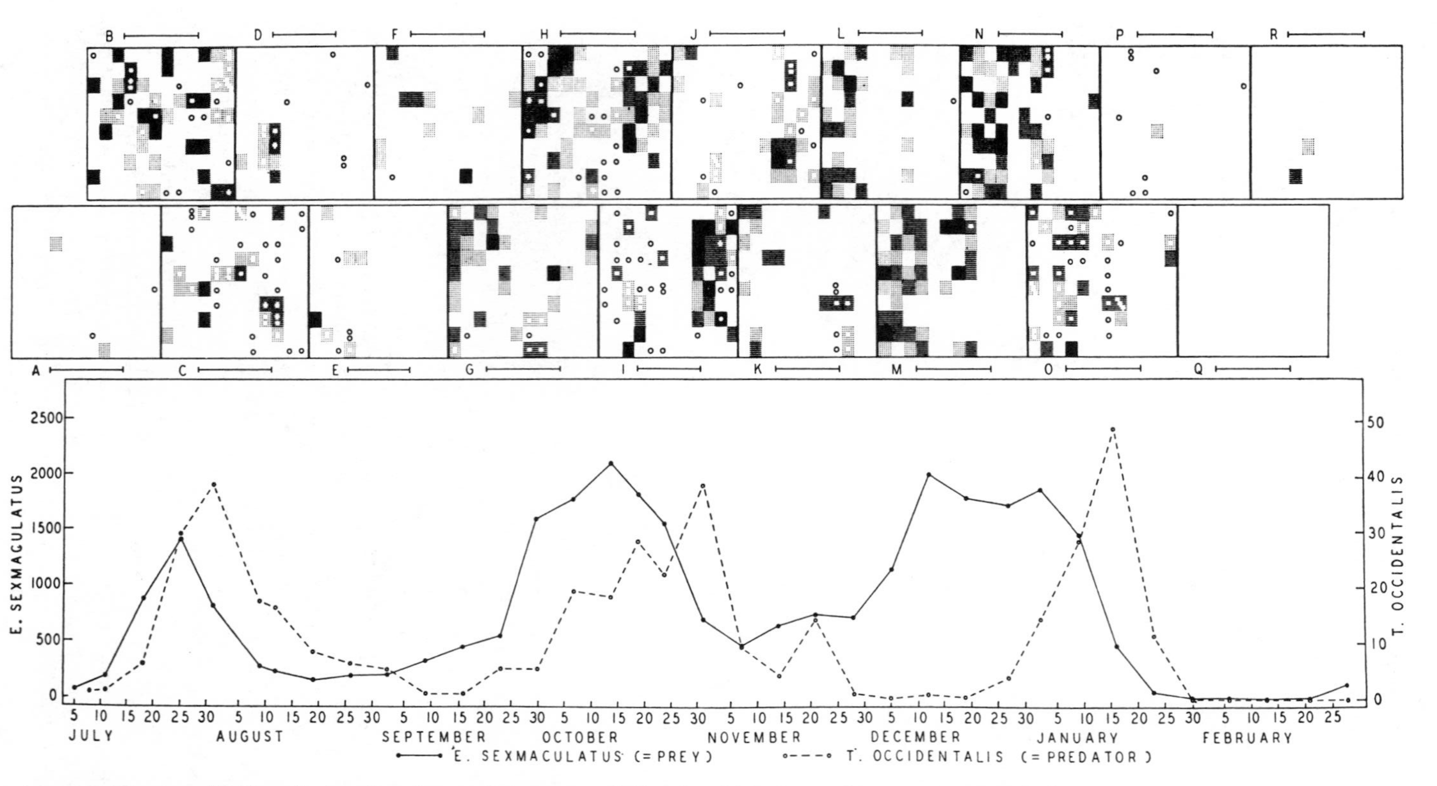

Fig. 18. Three oscillations in density of a predator-prey relation in which the predatory mite, ***Typhlodromus occidentalis***, preyed upon the orange feeding six-spotted mite, ***Eotetranychus sexmaculatus***.

The graphic record below shows the sequence of densities per orange-area, while the pictorial record, charts A to R, above, shows both *densities* and *positions* within the universe. The horizontal line by each letter "A," "B," et cetera, shows the period on the time scale represented by each chart. A photograph of the arrangement of this universe is shown in figure 5 and a sketch of the complex maze of vaseline partial-barriers in figure 19—a 6-orange feeding area on a 120-orange dispersion (see text, Subsection I, Section II of "Results").

H. Predators Present, 120 Small Areas of Food Occupying All 120 Positions, a 6-Orange Area. In this universe the area of orange exposed at a position was the same as that used in the universe just discussed and the food occupied all positions (no rubber balls were used), but the need for an increase in the total potentials and complexity had become obvious. Hence, in this unit, the food potential and the total areas required to be covered in searching were trebled, i.e., a 6-orange area was used on a 120-orange dispersion. The universe consisted of three of the trays (previously used singly) joined together (see fig. 5). An arrangement of vaseline partial-barriers was again used (see fig. 19). The universe was started with 10 female six-spotted

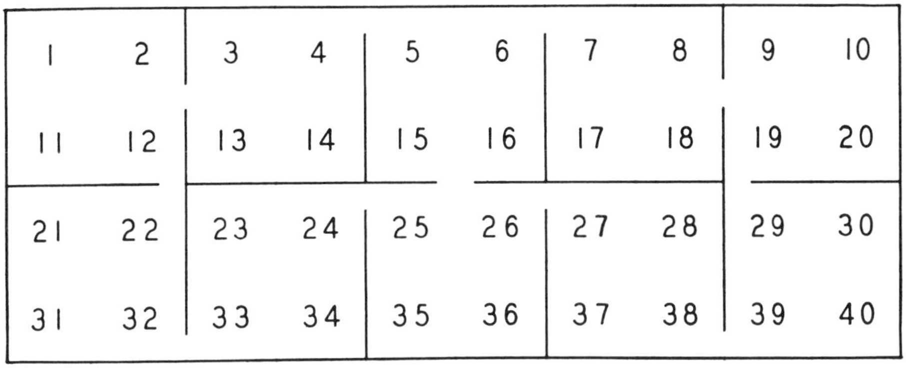

1	2	3	4	5	6	7	8	9	10
11	12	13	14	15	16	17	18	19	20
21	22	23	24	25	26	27	28	29	30
31	32	33	34	35	36	37	38	39	40

Fig. 19. Diagram of a tray used in the complex 3-tray universes (see figs. 17 and 18) with the positions of vaseline barriers shown by black lines.

mites, placed on each of two oranges in one of the trays on May 5. Two female predators were added on one of these oranges on May 16. The universe was ended June 27 (fig. 17).

In this universe the predator action was again delayed. However, movement of the prey from the one tray in which the initial colonies of both species were introduced never occurred, and it thus became obvious that with the use of such universes a wider arbitrary spread of the initial stock should be employed. Otherwise, the data of this universe added nothing new.

I. Predator-Prey Oscillations, 120 Small Areas of Food Occupying all 120 Positions, a 6-orange Area. This universe was basically like that of the last discussed but greater complexity and a different scheme of introducing the initial colonizing stock were employed. The universe was a three-tray arrangement, and a 6-orange feeding area on a 120-orange dispersion was again used. The partial barriers of vaseline were also used (see fig. 19). Contrary to previous procedure, 120 female six-spotted mites were introduced on June 30 (the graph of fig. 18 shows the first count date, July 5). That is, one mite was placed on each of the 120 oranges. In this universe also, the predators were added only five days later so as to permit them to become effective prior to general conditioning of the stocked oranges by the prey. Also contrary to the previous procedure, 27 female predators were introduced, and these were distributed, one on each of 27 oranges, these being representative of all major sections of the universe. This scheme assured that the populations would not become annihilated prior to dispersal to all parts of the universe—such as happened in the unit previously discussed. This experiment was ended March 27, although because of engraving difficulties, the data were plotted only to February 28 (see fig. 18).

In addition, small wooden posts were placed in upright positions in each of the major sections of the universe. This was to give the prey species a maximal opportunity to disperse over the vaseline barriers, thus utilizing its adaptive ability to drop by silken strands and be carried by air currents to new locations. An electric fan was used to create a mild air movement in the room. Although the predators have superior dispersal ability within a **limited** environment where movement by wind is not involved, they do not utilize this method of movement. On the other hand, by virtue of such adaptations the prey species has very superior abilities to disperse over greater distances to entirely new areas and environments. Therefore, a restricted environment or universe of the kind used in these experiments utilizes the superior dispersal power of the predator within local areas without giving chance for expression of the equally important superior dispersal power of the prey across greater distances and obstacles. The wooden posts were introduced in an effort to partially correct this condition in these experiments. They did not prove entirely satisfactory, and a more elaborate arrangement to accomplish this purpose should be employed.

Although they were joined into a single universe, the three trays may be looked upon as adjacent microenvironments, and so may the smaller subdivisions within the trays. By this scheme the changes in various areas or the geographic waves in distribution (see charts A to R of fig. 18), as well as the general density changes in the whole universe were followed. It should be noted that the horizontal dimensions of these charts were reduced (relative to the vertical dimensions) because of difficulties in reproduction on a single page. The horizontal dimension of each universe was 50 inches, the dimension shown vertically was 40 inches.

It is obvious in the charts of figure 18 that the divisions between trays, although not covered with a barrier, were an impediment to movement of somewhat greater effect than that caused by the vaseline barriers used **within** each tray.

The charts, A to R of figure 18 represent a compromise with the ideal of showing the exact locations in the universe and the densities of the entire population **on each** date of sampling. In order to have the charts on the same page and running synchronously with the linear graphs of density on the same time scale, the data for each pair of sampling dates were combined. The horizontal lines by each letter "A," "B," "C," et cetera indicate the corresponding time period of the two dates of counting. Note that the chart for the first period, "A," is below, the second, above, the third, below, the fourth, above, et cetera. The classes of density used for the predators and for the prey were limited in number.

In some instances a few prey were present but not enough to be shown. In some instances of predators being shown in areas where no prey are shown, that is a true condition; in others, it may mean only that although the prey were present, there were too few to justify shading. In either instance, such rarity of the prey means that the predators there would be doomed shortly to starvation, and the charts reveal this fact. Note also that the predators are shown in such "white" areas by white circles (ringed), whereas within the shaded areas indicating prey densities, they are shown by white circles also, but no bordering rings are used.

Considering the trays individually, or any sections of the trays individually, the predators either moved away or died in every case just as was true in Gause's experiments with protozoa and with cheese mites, or has been so in all the universes previously discussed herein. However, by utilizing the large and more complex environment so as to make less likely the predators' contact with the prey at all positions at once, or essentially so, it was possible to produce three waves or oscillations in density of predators and prey. That these waves represent a direct and reciprocal predator-prey dependence is obvious.

The maximal density of prey during the first oscillation was 8,550, or 1,425 per orange-area. The predator population responded quickly to the increase in abundance of the prey since the two species occupied the same arenas by virtue of the manual distribution in stocking of the oranges initially (see "Discussion").

The second peak density of prey occurred about October 15 and was somewhat higher than the first at 12,624, or 2,104 per orange-area. The higher level was an automatic effect of the greater lag in response by the predator population during the initial period of this oscillation—i.e., from September 10 to September 30 or somewhat beyond. This lag resulted from the predator's lack of contact with the main masses of the prey which were present in two of the three trays (left third and right third—see charts F and G, fig. 18), although they were in substantial contact with the prey of the center tray (lower center area of chart G) and slightly so with the much larger population in the left tray, or section. Thus the prey were able to sustain a marked increase in density in two of the three trays, as is shown by the progress seen from chart F to chart H and as seen in the graph for this period. More general contact with the prey was eventually achieved but the pattern of achievement reveals the reason for the rather erratic changes in density of the prey during the process (from October 7 to October 31). On October 7 the predators had reached a moderate density in the universe of 19 per orange-area, but most of these were located in the lower section of the middle tray (chart G), the aftereffect of which is shown in chart H in the elimination of most of the prey in that area and the numbers of predators still present in that area but without food. This chart also shows that their general movement had been partly onto oranges where their prey was abundant and partly onto ones where they found little or no food. As shown in this chart, also, the predators had made substantial contact with the main mass of prey in the left tray but still had not done so in the right tray in which the prey were now rapidly increasing in numbers. It was not until later (chart I) that the right tray also was reached and general population decline of both predators and prey ensued. Thus, localized discontinuity in contact accounts for the zig-zag pattern of increase in the density of the predators during the period involved. A smoothed curve of densities of the predators would correspond to the usual pattern of a predator-prey relation.

During the third major increase in the population of the prey, the maximal density reached was 11,956, or 1,993 per orange-area. In this instance, the prey had escaped substantial predation for a long period. Using chart K of figure 18 to represent the end of the second oscillatory wave, it is seen that the predators survived only in the lower right area of the universe where only a minor portion of the prey was present. With the near annihilation of this localized center of population by the predators, the latter then starved as is shown in the subsequent chart L, although one female predator had wandered off into an area where there was no food but from which position it later moved to the left and located the edge of the main mass of prey (chart M). During this time (charts K to M), the prey increased greatly in the absence of predators in the large area it inhabited. Considering the universes having predators, conditioning of the oranges for the first time became a dominant depressive feature for the prey population.

Shortage of food was the principal reason why the population leveled off at a high density of approximately 1,800 mites per orange-area between December 12 and January 2. Except for the fact that the main masses of the population encountered a shortage of food at that time because of the predators' loss of contact with it (charts L, M, and N of fig. 18), the numbers almost certainly would have increased to a position approximating at least 2,500 per orange-area. It is probable that such an increase would have contributed to a compensating, slightly earlier rise in the predator population and, consequently, a slightly earlier decline in the prey population. The resultant crash recorded just subsequent to January 2 was largely an effect of predation, although, as stated, the level from which it was initiated would have been higher (and the resultant aftereffect correspondingly more drastic) except for the ameliorating effect of the shortage of food for the prey just prior to that time. During this time and just subsequently, the approximate proportion of conditioned oranges among those which had reached the age for replacement was 38 per cent (see also "Discussion").

Also, during the peak period of the second oscillation there was a substantial but not principal contribution toward leveling off of the prey population at the approximate position of 1,700 mites per orange-area. In this instance, the proportion of conditioned oranges among those removed from the universe was 25 per cent, but the period of this influence was of shorter duration, and the predators earlier achieved more significant contact with the main masses of the prey, thus preventing a greater degree of food conditioning by the prey. In this instance, it is doubtful whether the prey would have increased to a significantly higher level even if the food had not become limiting to this degree. Thus, 75 per cent of the oranges remained unused to a damaging degree for the 56 days they were present in the universe, and predator action was the principal reason for this.

During the initial oscillatory wave, shortage of food did not enter as a contributive factor. Only six of the 120 oranges removed from the universe were conditioned, i.e., 5 per cent, 95 per cent remaining unconditioned for the 56 days (in this experiment only) each was present in the universe.

The data of this universe with relation to certain points will be covered further under the section on "Discussion."

DISCUSSION

The Experimental Data

Discussed in this section are certain topics pertaining to the data of the various universes (figs. 1 to 18), collectively, and, as well, the significance of these results as exemplifying the role of dispersion in the predator-prey relation.

Since the universe illustrated in figure 18 approaches in result one of the main objectives of this study, those data will be compared with the results in various of the other universes. The arbitrary imposition of wide distribution of both interacting species throughout all sections of the universe in the initial stocking of this universe had several effects which bear a relation to the subsequent events in this universe and to those occurring in other universes otherwise similar: 1) Both species found favorable quantities of food readily at hand for population growth, for during the initial period neither the predator nor the prey faced impediments. 2) The increase in density of the predator in response to increase in density of its prey was immediate. 3) There was very little conditioning of the oranges by the prey during this early phase, for the predators increased their action swiftly and precluded this. 4) The changes in general density of both species during this phase represent rather smooth curvilinear regressions, for the changes in density were relatively simultaneous throughout the universe, with few localized departures from the general pattern (the data thus support the contention that the actual error of sampling is small—see subsection on "Sampling Procedures"). Some of these interrelated points require clarification.

Since in this first oscillation it was not necessary that the predators overcome substantial impediments (with consequent lag effects and losses in numbers) in locating sources of prey, this oscillation is perhaps typical of one where dispersion of food and habitat and the hazards associated with finding them are minimal. In general, the simple universes previously used where the food was massed in one area gave similar results.

On the other hand, the lag effect of predator action exhibited in most of the complex universes discussed earlier, where the colonizing stock of predators was introduced at only a single or very restricted number of positions, is typical of the **second** and **third** oscillatory waves in density as shown for this universe (see fig. 18). Furthermore, the gradual, progressive change in the nature of the distributions (from one oscillation to another) in this universe was such that introductions of the predators into more than a limited number of arenas or introductions of the prey into all sections or arenas of universes would appear to create a condition of distribution not at all natural to such interactions of predators and prey. In this still-too-restricted universe, the predators survived the two critical, post-crash periods only at a single arena and in extremely small numbers, perhaps only a single female in each instance, certainly so in the second. During the third critical, post-crash phase, all the predators perished. Thus, the time required for the interacting populations to adjust to patterns of spatial and quantitative distribution more characteristic of a predator-prey relation which has come closer to **internal** balance may be the principal reason why the lag effect was accumulative from oscillation to oscillation.

Regarding the results of the universes illustrated in figures 14 and 15, the significant fact is that the exact course which may be taken locally at such very low levels of density is a product of chance events, the course of which could be best expressed as a probability of occurrence under various stipulated conditions. In the one instance, the prey were annihilated by the predtors, whereas in the other identical universe the predators starved before all the prey were dead, and the prey population then gradually recovered.

Thus, generally, as to whether the participants survive the critical phase and thus make possible the second oscillation is locally a matter of chance, but as the universe considered is increased in complexity and total potentials, the probability that the participants **will** survive is increased.

In this connection it is obvious that the prey must survive the exploitation by the predator as a prerequisite to any possibility of the predator's survival. Thus, the first object is to devise an ecosystem in which there is a near certainty that the prey will survive. In the first universes employed, this condition was not even approached, but the larger, more complex universe employed (see fig. 18) comes closer to this requirement (but, considering the position of the predator, is still far from adequate).

Obviously, for given conditions, the probability that three or four successive oscillations will occur is progressively more remote and is the **product** of the separate probabilities of survival of both participants through **each** component critical phase. This would be true even disregarding the view of Nicholson (1933, 1954) that the amplitude of such oscillations will increase with time. If his view is correct and its tenets **not** modified by damping features, the probability of such a relation continuing for a successively longer number of oscillations would be correspondingly even more reduced. Yet, this cannot be interpreted as contrary to the principle that as density of the prey decreases, the pressure of predator action on it will also decrease.

In this connection, Huffaker and Kennett (1956), as previously stated, demonstrated that biotic interaction between a phytophagous form and its plant host, (with examples which feed in a way as to cause **reaction** by the plant in a manner as to alter the food potential produced subsequently) may be such that the oscillations in the absence of predation may be of **decreasing** amplitude, due to progressive weakening of the plants. Franz (in press) also showed that such interaction may predetermine in a rather subtle way the potentials of subsequent populations of plant feeding forms and, correspondingly, the natural enemies which attack them. Such mechanisms tend to reduce the amplitude of predator-prey oscillations as interactions occur between predator actions and nutritional limitations in time and place.

Shortage of food was also discussed in Subsection I of Section II in relation to damping of the amplitude in a predator-prey universe. It should be noted also that in the universe illustrated in figure 12, there was substantial shortage of food for the prey in local arenas, such was the interaction of problems of dispersal (in this universe where the oranges were widely dispersed) and predator action (see also Subsection D of Section II). While 1/4 of the orange supply was fully utilized at its replacement, the predators prevented the utilization of the other 3/4 of the food.

Thus, it is obvious that even though action of a predator may be **locally** insignificant at a given time and compensatory in nature (only a substitute for food conditioning which would surely limit the density **there** anyway), the predation may be **far more significant** throughout the larger sphere which would be reached by migrants from the nutritionally overpopulated area. That is, such migrants could proceed to overpopulate the new areas as well but for the predators which preclude the possibility in an example such as this.

This type of control by predation, associated or not with shortage of food for the prey in local arenas, is generally illustrated by the data of this study. The degree of lag in appearance or introduction of the predators into the ecosystem or local arenas is the critical feature of how much of the plant food may be depleted prior to effective curtailment of the plant feeding form. Significantly, in the presence of an effective predator, overpopulation by the plant feeding form in one arena is to a marked degree an assurance against such overpopulation in other arenas. Thus, the common contention by biological control specialists that the farmer should be willing to accept some crop injury is theoretically sound and has been practically demonstrated many times.

In the universes of Subsections A, B, and C of Section II (illustrations of figures 9, 10, and 11), the food of the prey was readily accessible, joined and grouped (a minimum of dispersion). In those universes the predators readily found their prey, responded more quickly to changes in density of the prey and were able quickly to destroy them. This condition appears to offer greater likelihood that **both** the predators and the prey will be annihilated, although in one of the three examples the prey escaped that end.

The occurrence of an almost imperceptible second wave or increase in the universe in which six whole oranges were used (see fig. 11) does not justify the conclusion that simple increase in the area or quantity of food used necessarily greatly increases the chances of creating a self-perpetuating predator-prey system. It is logical to assume that increased complexity is a more important element of the prerequisites than increased area or quantity of food for the prey. Such complexity creates greater relative refuge or protection against the prey's being overexploited, and also reduces its effective reproduction, but it is significant that refuges **restrictive** to the predators such as envisaged by Gause (1934) and Gause, *et al.* (1936) are not implied as essential.

Comparatively, increase of a prey population recovering from the effects of extreme predation is much less rapid than that which results when an original colony is started with an equally low number of colonizing individuals. This has been a characteristic feature in these experiments. Examples of this may be seen in figures 9, 12, 15, 16, and 18 by comparing the steepness of the curves, at the initiation of the universes, with the obviously very gradual increases which resulted from the small numbers which escaped the predators at the end of the crashes in the populations and subsequent to starvation of all the predators.

The reason seems to lie with certain undercrowding phenomena and with the fact that very few females escape the predators, and these are often unmated. Unless copulation occurs later they produce only male offspring. If the female survives long enough and remains in the area, promoting likelihood of contact, she may then copulate with one of her own sons or perhaps another male, and female progeny would result. Two or more generations may be required for the population to attain a favorable proportion of fertilized females and, thus, vigorous population growth, even when predators are no longer present.

Another partial explanation is the observed fact that the females which survive are more commonly found on partially or heavily conditioned oranges. On these oranges the presence of a much greater quantity of webbing, cast skins, and bodies of dead mites affords a relative sheltering effect and thus reduces the probability of the predators' destroying the last survivors. These heavily conditioned oranges are very poor sources of food for population increase; hence, a slow recovery results from the survivors.

CONCLUSIONS

The aforestated considerations suggest that the most satisfactory universe employed (see fig. 18) is still far too restricted, and that, for a perpetuating system, sufficient potentialities must be incorporated to assure several or many such arenas of "last survivors" of predators. This system would leave little probability that all such "last survivors" will simultaneously starve and none find new arenas inhabited by the prey. Thus, there is envisaged in such a system many intergrading, larger, nearly self-sustaining subuniverses or ecosystems, each one as adequate or more adequate than the one illustrated in figure 18.

A major difficulty in demonstrating the existence of reciprocal predator-prey oscillations in nature is associated with the patchy or wavelike occurrence of the predation in time and place, particularly true with examples which are wingless such as the mites and which may have limited extensive dispersal power over distances or from tree to tree, for example. In such studies an inherent oscillatory relation would be confused if the sample area taken to reveal the dynamics of a population unit is too large and, obviously, if it is too small.

Nearly any field entomologist who has studied the action of natural enemies of insect pests has noted that the pattern of action is often patchy in occurrence, proceeding in irregular waves from one or more centers. It is obvious that in one local arena the predator-prey relation may be in one phase of an oscillation while in an adjacent arena it may be in a diametrically opposed phase. Therefore, any combining of two such populations into one would not give a reliable picture of the inherent oscillatory nature of the relation.

Thus, in selecting an environmental area it must be large enough to permit the continued existence of both the predator and its prey, yet not so large that the populations in its several sections may proceed asynchronously, due to too limited interchange of the biotic participants.

It is thus more philosophical than factual to discuss whether or not the predator-prey relation is "inherently" disoperative or self-exterminative in arbitrarily restricted environments. To use an extreme example, the end result would be certain if a small universe or enclosure were employed in which only one pair of mountain lions was confined with only one pair of mule deer. Although in the case of this predatory mite and its prey, an orange is a far more nearly adequate base for a suitable ecosystem, yet, on a single orange the predator has invariably overexploited its prey and become exterminated as a local population. With many examples, the prey has been exterminated also, but with others the prey has been able to recover after the starvation of all the predators. Obviously, if the area is sufficiently small and arbitrarily simple, the biological parameters which have been present during the long evolutionary origins of the relations involving the participants are absent, and capture is so simple that the coaction is disoperative. It is necessary that a system be adequate to assure a high probability that some prey will be missed and that somewhere reasonably accessible, but not too readily so, there are local populations of prey which are thriving and sending emigrants to repopulate the depopulated areas. Also, this predator cannot survive on very low populations (although it requires many fewer prey than does the beetle, *Stethorus* spp., which feeds on the same prey—see also Kuenen, 1945), but must contact a fair density of prey at least at small micro-arenas in order to reproduce and survive.

That self-sustaining predator-prey coactions cannot be maintained without "migration" is self-evident. In this type of study the distinction between migration and any movement at all becomes rather ephemeral. The author disagrees that these migrations must be from beyond the limits of a reasonably adequate system. They may be a result of normal movements within the system—if the system is adequate to give expression to the inherent balance in the biological relations of the predator, its prey, and their coinhabited environment. In an unpublished study, the author and C. E. Kennett have demonstrated that a single strawberry plant is an adequate universe during its life span to sustain a predator-prey coaction. No smaller universe utilizing strawberries is conceivable since a single leaf or flower is not a self-perpetuating living unit.

The speed of local-arena extermination by a predator does not define the period of an oscillation. In fact, it appears to bear little relation to that period. Local extermination of the prey on an orange exposing only $\frac{1}{20}$ orange area has often occurred within three days of the entry of a female predator in that area. Even if the density of the prey population is high, with several hundreds on such an area, they are often exterminated within a period of five or six days. If the environment considered is increased to a single half-orange unit, the time required for self-extermination has been, on the average, longer. In most of the more complicated environments, and involving at least a 2-orange area, the time required to produce the drastic decline in population sufficiently general to jeopardize the predator's existence or cause its extermination has been greatly extended—20 to 40 days for a complicated arrangement involving a single-tray universe and a 2-orange feeding area widely dispersed. The same interval was increased to 30 to 60 days for the most complicated system employing a 6-orange feeding area dispersed over 120 oranges and including a maze-effect of vaseline partial-barriers.

It seems, therefore, that the complexity of the dispersal and searching relationships, combined with the period of time required for the prey species to recover in local arenas from the effects of severe predation and accomplish general repopulation, is more important in determining the period of oscillation than is the intensity of predation once contact is made with local arenas of prey. The rapid recovery of the prey is essential to maintain the predator unless there are arenas of high population which are missed.

It is thought that the existence of barriers increases the chances that the prey species will survive at a level conducive to its rapid recovery. However, it is recognized that the barriers may act as a double-edge sword and defeat the purpose of their use. They do increase the incompleteness of contact and cause marked delay in predator increase. This delay also subsequently causes a greater predator population, which then tends to offset the purpose of the barriers during the crash period. Only further experimentation would really prove whether the barrier feature of this experiment has been a deterrent or an aid to continuance of the coaction. Theoretically, the greater violence of oscillation caused by the barriers would be disoperative in nature, but, on the other hand, they do create the partial asynchrony in geographic position and promote earlier population recovery of the prey species. These features are essential to survival of the predator. They may be more than enough to offset the disoperative pressure created by the higher populations achieved at the crests of population densities. Perhaps, also, the partial ameliorating effect from conditioning of the food in local arenas tends to cancel the greater amplitude otherwise occasioned by the use of the barriers.

The complexity of the environment being searched by both predators and prey lends to the relations a marked inconstancy of hazards from micro-area to micro-area. The idea of a constant area of discovery for the predator or of a constancy in dispersal effectiveness for the prey is difficult to visualize in this environment or in nature. There is not only inconstancy, but nothing resembling a progressive gradation in the hazards. The area of discovery, or that area effectively covered and in which all prey are destroyed by a predator of this species, would vary with its hunger, the density of the prey population independently of hunger of the predator (due to the greater webbing, the added cast skins, debris, et cetera present), the complexity of the general environment with respect to the variability of physical barriers or restrictions of all kinds, and the degree of synchrony in responses, preferences, and tolerances between predator and the prey to such conditions as gravity, light, temperature, moisture, the physical surfaces, air movements, et cetera. The idea of a constant area of discovery has theoretical meaning, particularly where simple, uniform areas are involved to which both predators and prey are rigidly restricted and no chance afforded them to express the broad or narrow ranges of asynchrony in behavior and ecology.

The author feels that the balance or stability observed in nature is characteristic of the total environments in which the evolution of a relation occurred, and forms related to one another in a manner notable for the lack of stability in the community would tend to be replaced by others whose relations are more stable and, thus, the assets of the environment more efficiently utilized. The same effect would be achieved if they were forced by better adjusted competitors to occupy progressively less significant niches within which adequate stability does prevail. It cannot be overstressed that what happens with one predator-prey relation, in one ecosystem, or under a given environmental complex, as to seasons or period of years, for instance, does not necessarily apply to others.

While these data indicate that, other things being equal, simplified monocultures of crops are likely to have greater problems with insect pests than are diversified plantings, it is, nevertheless, known that a single species of introduced natural enemy has in many cases throughout the world permanently solved the most severe problems relating to such pests of monocultures. In this connection, it is interesting to note that Taylor (1955) expressed the opinion that because of the variety and mosaic of small plantings in Britain, the complex of forces for solid natural control are more favorable than in regions where extensive acreages of monocultures are the rule—the reasons for which he considered are yet unknown.

SUMMARY

An experimental study of the role of dispersion in the predator-prey relation was made, using the predatory mite, *Typhlodromus occidentalis,* and the phytophagous mite, *Eotetranychus sexmaculatus,* as the prey. Earlier experimental work by G. F. Gause and associates had led to some acceptance among ecologists of the view that the predator-prey relation is inherently self-annihilative and that continuation of this relation or coaction is dependent upon either: 1) immigrations into the depopulated areas from without, or 2) the existence of definite refuges restrictive to the predators.

In this study, a wide variety of different arrangements in dispersion of plant food (and microhabitat) was tested experimentally. In all the simple universes employed the conclusions of Gause with respect to the predator, but not to the prey, seemed to apply. The unacceptability of that view was demonstrated by the use of a larger, much more complex universe utilizing wide dispersion and incorporating also partial barriers, thus increasing still further the relative dispersion while still not incorporating restrictive refuges. By this method, predator-prey coaction was maintained for three successive oscillations. It is thus quite probable that a controlled, experimental ecosystem can be established in which the predator-prey coaction would not be inherently self-annihilative. It is believed also that various damping mechanisms would come into play which would serve to ameliorate the theoretically sound concept that oscillations arising from this coaction are inherently of increasing severity in amplitude.

The whole controversy becomes rather more philosophical than factual, considering that the earlier view incorporated the purely relative concept of immigration of new stock from without, and any distinction between immigration or emigration and any movement at all on the part of the participants can hardly be upheld. The suggestion seems more appropriate that artificial universes are inadequate if they do not give possibility of expression of the major parameters intrinsic to the specific predator-prey coaction in the natural habitat, and that conclusions drawn from such data as to principles have limited value. The success we have in sustaining such a coaction under experimental conditions is probably a measure of the degree to which we have duplicated the inherent essentials.

In this study, arbitrary selection of different degrees of dispersion and segmentation of the units of food for the prey was accomplished without altering the total surfaces to be searched, and, when desired, without altering the total amounts of food used. This was done by covering oranges to various degrees, leaving known exposed portions, and dispersing them as desired among waxed rubber balls of the same size. The technique offers possibilities of elaborate and varied studies along these lines. For example, further modifications could make it possible to study the predator-prey relation with greater assurance against overexploitation, and, thus, various other features, such as the introduction of a competing predatory species or a competing prey species, could be introduced in order to study their relations to the periods and amplitudes of the oscillations. By elaboration along these lines it should be possible to establish empirically whether employment of quite diversified agricultures may offer prospects of relief from insect pests—in comparison with extensive cultivations of single crops.

ERRATA:

Page 370: The legend for fig. 18 was omitted; should read as follows:

Prey: 0–5 nil density (white); 6–25 low density (light stipple); 26–75 medium density (horizontal lines); 76 or over, high density (solid black).

Predator: 1–8 (one white circle).

Page 373, line 35, last word "prey" should read "predator."

HILGARDIA 27-14

REFERENCES

DeBach, P., and H. S. Smith
1941. Are population oscillations inherent in the host-parasite relations? Ecology **22**: 363–69.

Errington, P.
1937. What is the meaning of predation? Smithsn. Inst. Ann. Rpt. **1936**:243–52.
1946. Predation and vertebrate populations. Quart. Rev. Biol. **21**:144–77.

Finney, G. L.
1953. A technique for mass-culture of the six-spotted mite. Jour. Econ. Ent. **46**:712–13.

Franz, J.
In Press. The effectiveness of predators and food as factors limiting gradations of *Adelges* (*Dreyfusia*) *piceae* (Ratz.) in Europe. Tenth Inter. Cong. Ent., 1956.

Gause, G. F.
1934. The struggle for existence. (163 pp.) Williams & Wilkins, Baltimore. Md.

Gause, G. F., N. P. Smaragdova, and A. A. Witt
1936. Further studies of interaction between predators and prey. Jour. Anim. Ecol. **5**:1–18.

Huffaker, C. B., and C. E. Kennett
1956. Experimental studies on predation: Predation and cyclamen-mite populations on strawberries in California. Hilgardia **26**(4):191–222.

Kuenen, D. J.
1945. On the ecological significance of *Metatetranychus ulmi* C. L. Koch (Acari, Tetranychidae). Tijdschr. v. Ent. **88**:303–12.

Leopold, A. S.
1954. The predator in wildlife management. Sierra Club Bul. **39**:34–38.

Nicholson, A. J.
1933. The balance of animal populations. Jour. Anim. Ecol. **2**, Supp.:132–78.
1954. An outline of the dynamics of animal populations. Austral. Jour. Zool. **2**:9–65.

Taylor, T. H. C.
1955. Biological control of insect pests. Ann. Appl. Biol. **42**:190–96.

Uvarov, B. P.
1931. Insects and climate. Ent. Soc. London, Trans. **78**:1–247.

Waters, N. D.
1955. Biological and ecological studies of *Typhlodromus* mites as predators of the six-spotted mite. (Unpublished Ph.D. dissertation, University of California, Berkeley.)

Winsor, C. P.
1934. Mathematical analysis of growth of mixed populations. Cold Spring Harbor Symposia on Quant. Biol. **2**:181–89.

THE INFLUENCE OF INTERSPECIFIC COMPETITION AND OTHER FACTORS ON THE DISTRIBUTION OF THE BARNACLE *CHTHAMALUS STELLATUS*

Joseph H. Connell

Department of Biology, University of California, Santa Barbara, Goleta, California

Introduction

Most of the evidence for the occurrence of interspecific competition in animals has been gained from laboratory populations. Because of the small amount of direct evidence for its occurrence in nature, competition has sometimes been assigned a minor role in determining the composition of animal communities.

Indirect evidence exists, however, which suggests that competition may sometimes be responsible for the distribution of animals in nature. The range of distribution of a species may be decreased in the presence of another species with similar requirements (Beauchamp and Ullyott 1932, Endean, Kenny and Stephenson 1956). Uniform distribution is space is usually attributed to intraspecies competition (Holme 1950, Clark and Evans 1954). When animals with' similar requirements, such as 2 or more closely related species, are found coexisting in the same area, careful analysis usually indicates that they are not actually competing with each other (Lack 1954, MacArthur 1958).

In the course of an investigation of the animals of an intertidal rocky shore I noticed that the adults of 2 species of barnacles occupied 2 separate horizontal zones with a small area of overlap, whereas the young of the species from the upper zone were found in much of the lower zone. The upper species, *Chthamalus stellatus* (Poli) thus settled but did not survive in the lower zone. It seemed probable that this species was eliminated by the lower one, *Balanus balanoides* (L), in a struggle for a common requisite which was in short supply. In the rocky intertidal region, space for attachment and growth is often extremely limited. This paper is an account of some observations and experiments designed to test the hypothesis that the absence in the lower zone of adults of *Chthamalus* was due to interspecific competition with *Balanus* for space. Other factors which may have influenced the distribution were also studied. The study was made at Millport, Isle of Cumbrae, Scotland.

I would like to thank Prof. C. M. Yonge and the staff of the Marine Station, Millport, for their help, discussions and encouragement during the course of this work. Thanks are due to the following for their critical reading of the manuscript: C. S. Elton, P. W. Frank, G. Hardin, N. G. Hairston, E. Orias, T. Park and his students, and my wife.

Distribution of the species of barnacles

The upper species, *Chthamalus stellatus*, has its center of distribution in the Mediterranean; it reaches its northern limit in the Shetland Islands, north of Scotland. At Millport, adults of this species occur between the levels of mean high water of neap and spring tides (M.H.W.N. and M.H.W.S.: see Figure 5 and Table I). In southwest England and Ireland, adult *Chtham-*

alus occur at moderate population densities throughout the intertidal zone, more abundantly when *Balanus balanoides* is sparse or absent (Southward and Crisp 1954, 1956). At Millport the larvae settle from the plankton onto the shore mainly in September and October; some additional settlement may occur until December. The settlement is most abundant between M.H.W.S. and mean tide level (M.T.L.), in patches of rock surface left bare as a result of the mortality of *Balanus*, limpets, and other sedentary organisms. Few of the *Chthamalus* that settle below M.H. W.N. survive, so that adults are found only occasionally at these levels.

Balanus balanoides is a boreal-arctic species, reaching its southern limit in northern Spain. At Millport it occupies almost the entire intertidal region, from mean low water of spring tides (M.L.W.S.) up to the region between M.H.W.N. and M.H.W.S. Above M.H.W.N. it occurs intermingled with *Chthamalus* for a short distance. *Balanus* settles on the shore in April and May, often in very dense concentrations (see Table IV).

The main purpose of this study was to determine the cause of death of those *Chthamalus* that settled below M.H.W.N. A study which was being carried on at this time had revealed that physical conditions, competition for space, and predation by the snail *Thais lapillus* L. were among the most important causes of mortality of *Balanus balanoides*. Therefore, the observations and experiments in the present study were designed to detect the effects of these factors on the survival of *Chthamalus*.

Methods

Intertidal barnacles are very nearly ideal for the study of survival under natural conditions. Their sessile habit allows direct observation of the survival of individuals in a group whose positions have been mapped. Their small size and dense concentrations on rocks exposed at intervals make experimentation feasible. In addition, they may be handled and transplanted without injury on pieces of rock, since their opercular plates remain closed when exposed to air.

The experimental area was located on the Isle of Cumbrae in the Firth of Clyde, Scotland. Farland Point, where the study was made, comprises the southeast tip of the island; it is exposed to moderate wave action. The shore rock consists mainly of old red sandstone, arranged in a series of ridges, from 2 to 6 ft high, oriented at right angles to the shoreline. A more detailed description is given by Connell (1961). The other barnacle species present were *Balanus crenatus* Brug and *Verruca stroemia* (O. F. Muller), both found in small numbers only at and below M.L.W.S.

To measure the survival of *Chthamalus*, the positions of all individuals in a patch were mapped. Any barnacles which were empty or missing at the next examination of this patch must have died in the interval, since emigration is impossible. The mapping was done by placing thin glass plates (lantern slide cover glasses, 10.7 $\times$ 8.2 cm, area 87.7 cm^2) over a patch of barnacles and marking the position of each *Chthamalus* on it with glass-marking ink. The positions of the corners of the plate were marked by drilling small holes in the rock. Observations made in subsequent censuses were noted on a paper copy of the glass map.

The study areas were chosen by searching for patches of *Chthamalus* below M.H.W.N. in a stretch of shore about 50 ft long. When 8 patches had been found, no more were looked for. The only basis for rejection of an area in this search was that it contained fewer than 50 *Chthamalus* in an area of about 1/10 m^2. Each numbered area consisted of one or more glass maps located in the 1/10 m^2. They were mapped in March and April, 1954, before the main settlement of *Balanus* began in late April.

Very few *Chthamalus* were found to have settled below mid-tide level. Therefore pieces of rock bearing *Chthamalus* were removed from levels above M.H.W.N. and transplanted to and below M.T.L. A hole was drilled through each piece; it was then fastened to the rock by a stainless steel screw driven into a plastic screw anchor fitted into a hole drilled into the rock. A hole ¼" in diameter and 1" deep was found to be satisfactory. The screw could be removed and replaced repeatedly and only one stone was lost in the entire period.

For censusing, the stones were removed during a low tide period, brought to the laboratory for examination, and returned before the tide rose again. The locations and arrangements of each area are given in Table I; the transplanted stones are represented by areas 11 to 15.

The effect of competition for space on the survival of *Chthamalus* was studied in the following manner: After the settlement of *Balanus* had stopped in early June, having reached densities of 49/cm^2 on the experimental areas (Table I) a census of the surviving *Chthamalus* was made on each area (see Figure 1). Each map was then divided so that about half of the number of

TABLE I. Description of experimental areas*

Area no.	Height in ft from M.T.L.	% of time submerged	Population Density: no./cm² in June, 1954 — Chthamalus, autumn 1953 settlement: Undisturbed portion	Population Density: no./cm² in June, 1954 — Chthamalus, autumn 1953 settlement: Portion without *Balanus*	Population Density: no./cm² in June, 1954 — All barnacles, undisturbed portion	Remarks
MHWS	+4.9	4	—	—	—	—
1	+4.2	9	2.2	—	19.2	Vertical, partly protected
2	+3.5	16	5.2	4.2	—	Vertical, wave beaten
MHWN	+3.1	21	—	—	—	—
3a	+2.2	30	0.6	0.6	30.9	Horizontal, wave beaten
3b	"	"	0.5	0.7	29.2	" " "
4	+1.4	38	1.9	0.6	—	30° to vertical, partly protected
5	+1.4	"	2.4	1.2	—	" " " " "
6	+1.0	42	1.1	1.9	38.2	Horizontal, top of a boulder, partly protected
7a	+0.7	44	1.3	2.0	49.3	Vertical, protected
7b	"	"	2.3	2.0	51.7	" "
11a	0.0	50	1.0	0.6	32.0	Vertical, protected
11b	"	"	0.2	0.3	—	" "
12a	0.0	100	1.2	1.2	18.8	Horizontal, immersed in tide pool
12b	"	100	0.8	0.9	—	" " " " "
13a	−1.0	58	4.9	4.1	29.5	Vertical, wave beaten
13b	"	"	3.1	2.4	—	" " "
14a	−2.5	71	0.7	1.1	—	45° angle, wave beaten
14b	"	"	1.0	1.0	—	" " " "
MLWN	−3.0	77	—	—	—	—
MLWS	−5.1	96	—	—	—	—
15	+1.0	42	32.0	—	—	Chthamalus of autumn, 1954 settlement; densities of Oct., 1954.
7b	+0.7	44	5.5	3.7	—	(same as above)

*The letter "a" following an area number indicates that this area was enclosed by a cage; "b" refers to a closely adjacent area which was not enclosed. All areas faced either east or south except 7a and 7b, which faced north.

Chthamalus were in each portion. One portion was chosen (by flipping a coin), and those *Balanus* which were touching or immediately surrounding each *Chthamalus* were carefully removed with a needle; the other portion was left untouched. In this way it was possible to measure the effect on the survival of *Chthamalus* both of intraspecific competition alone and of competition with *Balanus*. It was not possible to have the numbers or population densities of *Chthamalus* exactly equal on the 2 portions of each area. This was due to the fact that, since *Chthamalus* often occurred in groups, the *Balanus* had to be removed from around all the members of a group to ensure that no crowding by *Balanus* occurred. The densities of *Chthamalus* were very low, however, so that the slight differences in density between the 2 portions of each area can probably be disregarded; intraspecific crowding was very seldom observed. Censuses of the *Chthamalus* were made at intervals of 4-6 weeks during the next year; notes were made at each census of factors such as crowding, undercutting or smothering which had taken place since the last examination. When necessary, *Balanus* which had grown until they threatened to touch the *Chthamalus* were removed in later examinations.

To study the effects of different degrees of immersion, the areas were located throughout the tidal range, either *in situ* or on transplanted stones, as shown in Table I. Area 1 had been under observation for 1½ years previously. The effects of different degrees of wave shock could not be studied adequately in such a small area

FIG. 1. Area 7b. In the first photograph the large barnacles are *Balanus,* the small ones scattered in the bare patch, *Chthamalus.* The white line on the second photograph divides the undisturbed portion (right) from the portion from which *Balanus* were removed (left). A limpet, *Patella vulgata,* occurs on the left, and predatory snails, *Thais lapillus,* are visible.

of shore but such differences as existed are listed in Table I.

The effects of the predatory snail, *Thais lapillus,* (synonymous with *Nucella* or *Purpura,* Clench 1947), were studied as follows: Cages of stainless steel wire netting, 8 meshes per inch, were attached over some of the areas. This mesh has an open area of 60% and previous work (Connell 1961) had shown that it did not inhibit growth or survival of the barnacles. The cages were about 4 × 6 inches, the roof was about an inch above the barnacles and the sides were fitted to the irregularities of the rock. They were held in place in the same manner as the transplanted stones. The transplanted stones were attached in pairs, one of each pair being enclosed in a cage (Table I).

These cages were effective in excluding all but the smallest *Thais.* Occasionally small *Thais,* ½ to 1 cm in length, entered the cages through gaps at the line of juncture of netting and rock surface. In the concurrent study of *Balanus* (Connell 1961), small *Thais* were estimated to have occurred inside the cages about 3% of the time.

All the areas and stones were established before the settlement of *Balanus* began in late April, 1954. Thus the *Chthamalus* which had settled naturally on the shore were then of the 1953 year class and all about 7 months old. Some *Chthamalus* which settled in the autumn of 1954 were followed until the study was ended in June, 1955. In addition some adults which, judging from their large size and the great erosion of their shells, must have settled in 1952 or earlier, were present on the transplanted stones. Thus records were made of at least 3 year-classes of *Chthamalus.*

RESULTS

The effects of physical factors

In Figures 2 and 3, the dashed line indicates the survival of *Chthamalus* growing without contact with *Balanus.* The suffix "a" indicates that the area was protected from *Thais* by a cage.

In the absence of *Balanus* and *Thais*, and protected by the cages from damage by water-borne objects, the survival of *Chthamalus* was good at all levels. For those which had settled normally on the shore (Fig. 2), the poorest survival was on the lowest area, 7a. On the transplanted stones (Fig. 3, area 12), constant immersion in a tide pool resulted in the poorest survival. The reasons for the trend toward slightly greater mortality as the degree of immersion increased are unknown. The amount of attached algae on the stones in the tide pool was much greater than on the other areas. This may have reduced the flow of water and food or have interfered directly with feeding movements. Another possible indirect effect of increased immersion is the increase in predation by the snail, *Thais lapillus*, at lower levels.

Chthamalus is tolerant of a much greater degree of immersion than it normally encounters. This is shown by the survival for a year on area 12 in a tide pool, together with the findings of Fischer (1928) and Barnes (1956a), who found that *Chthamalus* withstood submersion for 12 and 22 months, respectively. Its absence below M.T.L. can probably be ascribed either to a lack of initial settlement or to poor survival of newly settled larvae. Lewis and Powell (1960) have suggested that the survival of *Chthamalus* may be favored by increased light or warmth during emersion in its early life on the shore. These conditions would tend to occur higher on the shore in Scotland than in southern England.

The effects of wave action on the survival of *Chthamalus* are difficult to assess. Like the degree of immersion, the effects of wave action may act indirectly. The areas 7 and 12, where relatively poor survival was found, were also the areas of least wave action. Although *Chthamalus* is usually abundant on wave beaten areas and absent from sheltered bays in Scotland, Lewis and Powell (1960) have shown that in certain sheltered bays it may be very abundant. Hatton (1938) found that in northern France, settlement and growth rates were greater in wave-beaten areas at M.T.L., but, at M.H.W.N., greater in sheltered areas.

At the upper shore margins of distribution *Chthamalus* evidently can exist higher than *Balanus* mainly as a result of its greater tolerance to heat and/or desiccation. The evidence for this was gained during the spring of 1955. Records from a tide and wave guage operating at this time about one-half mile north of the study area showed that a period of neap tides had coincided with an unusual period of warm calm weather in April so that for several days no water, not even waves, reached the level of Area 1. In the period

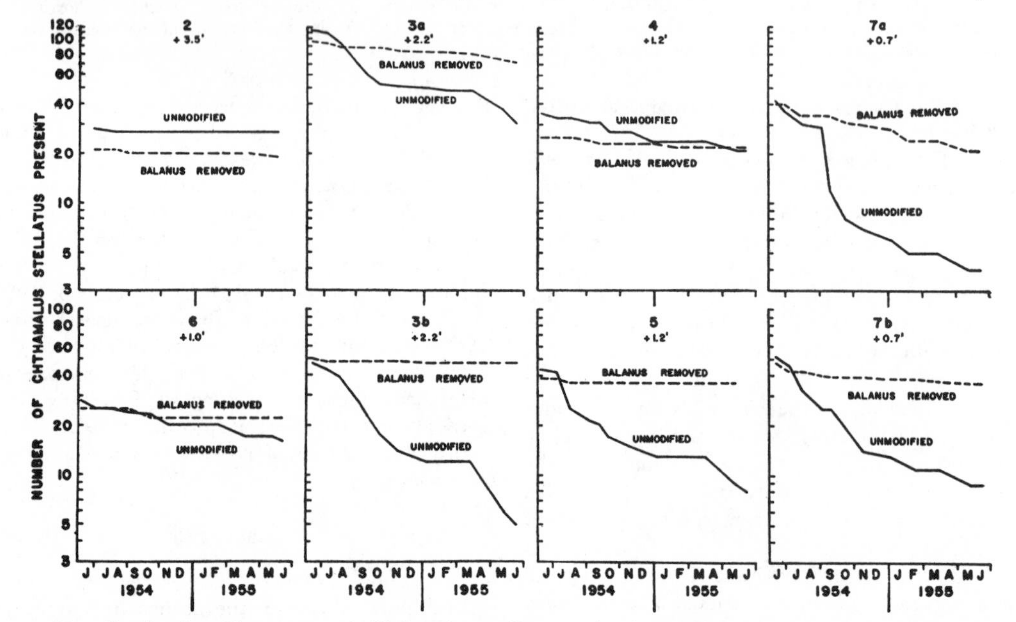

FIG. 2. Survivorship curves of *Chthamalus stellatus* which had settled naturally on the shore in the autumn of 1953. Areas designated "a" were protected from predation by cages. In each area the survival of *Chthamalus* growing without contact with *Balanus* is compared to that in the undisturbed area. For each area the vertical distance in feet from M.T.L. is shown.

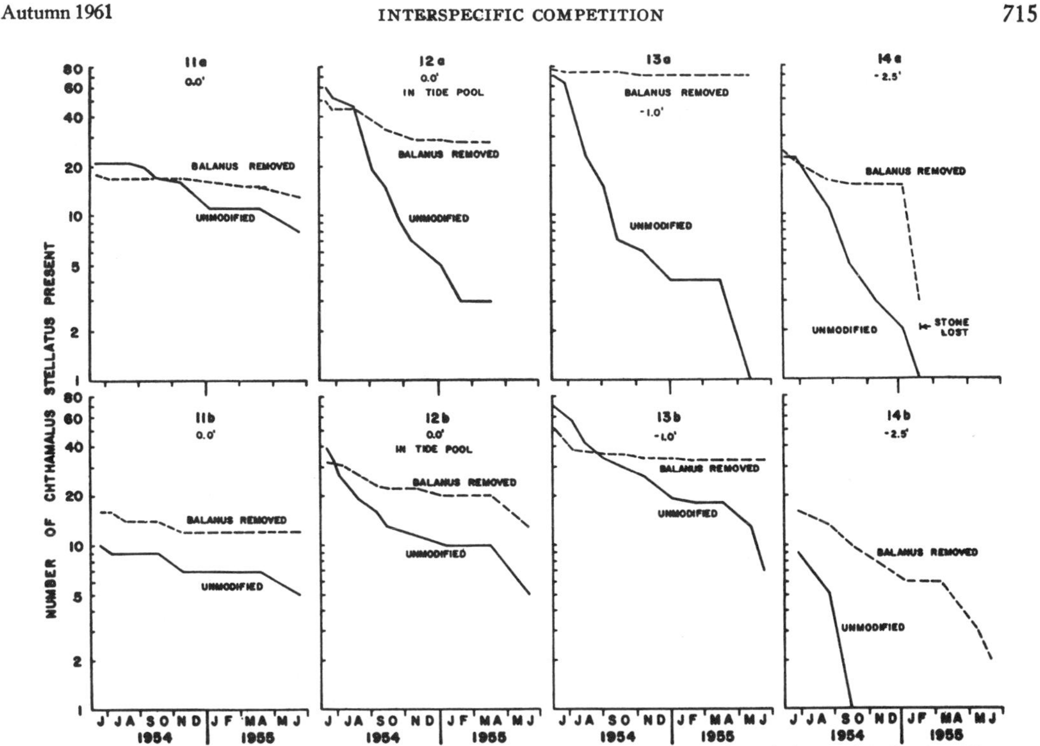

FIG. 3. Survivorship curves of *Chthamalus stellatus* on stones transplanted from high levels. These had settled in the autumn of 1953; the arrangement is the same as that of Figure 2.

between the censuses of February and May, *Balanus* aged one year suffered a mortality of 92%, those 2 years and older, 51%. Over the same period the mortality of *Chthamalus* aged 7 months was 62%, those 1½ years and older, 2%. Records of the survival of *Balanus* at several levels below this showed that only those *Balanus* in the top quarter of the intertidal region suffered high mortality during this time (Connell 1961).

Competition for space

At each census notes were made for individual barnacles of any crowding which had occurred since the last census. Thus when one barnacle started to grow up over another this fact was noted and at the next census 4-6 weeks later the progress of this process was noted. In this way a detailed description was built up of these gradually occurring events.

Intraspecific competition leading to mortality in *Chthamalus* was a rare event. For areas 2 to 7, on the portions from which *Balanus* had been removed, 167 deaths were recorded in a year. Of these, only 6 could be ascribed to crowding between individuals of *Chthamalus*. On the undisturbed portions no such crowding was observed. This accords with Hatton's (1938) observation that he never saw crowding between individuals of *Chthamalus* as contrasted to its frequent occurrence between individuals of *Balanus*.

Interspecific competition between *Balanus* and *Chthamalus* was, on the other hand, a most important cause of death of *Chthamalus*. This is shown both by the direct observations of the process of crowding at each census and by the differences between the survival curves of *Chthamalus* with and without *Balanus*. From the periodic observations it was noted that after the first month on the undisturbed portions of areas 3 to 7 about 10% of the *Chthamalus* were being covered as *Balanus* grew over them; about 3% were being undercut and lifted by growing *Balanus;* a few had died without crowding. By the end of the 2nd month about 20% of the *Chthamalus* were either wholly or partly covered by *Balanus;* about 4% had been undercut; others were surrounded by tall *Balanus*. These processes continued at a lower rate in the autumn and almost ceased during the later winter. In the spring *Balanus* resumed growth and more crowding was observed.

In Table II, these observations are summarized for the undistributed portions of all the areas. Above M.T.L., the *Balanus* tended to overgrow the *Chthamalus,* whereas at the lower levels, undercutting was more common. This same trend was evident within each group of areas, undercutting being more prevalent on area 7 than on area 3, for example. The faster growth of *Balanus* at lower levels (Hatton 1938, Barnes and Powell 1953) may have resulted in more undercutting. When *Chthamalus* was completely covered by *Balanus* it was recorded as dead; even though death may not have occurred immediately, the buried barnacle was obviously not a functioning member of the population.

TABLE II. The causes of mortality of *Chthamalus stellatus* of the 1953 year group on the undisturbed portions of each area

Area no.	Height in ft from M.T.L.	No. at start	No. of deaths in the next year	Percentage of Deaths Resulting From:			
				Smothering by *Balanus*	Undercutting by *Balanus*	Other crowding by *Balanus*	Unknown causes
2..........	+3.5	28	1	0	0	0	100
3a..........	+2.2	111	81	61	6	10	23
3b..........	"	47	42	57	5	2	36
4..........	+1.4	34	14	21	14	.0	65
5..........	+1.4	43	35	11	11	3	75
6..........	+1.0	27	11	9	0	0	91
7a..........	+0.7	42	38	21	16	53	10
7b..........	"	51	42	24	10	10	56
11a..........	0.0	21	13	54	8	0	38
11b..........	"	10	5	40	0	0	60
12a..........	0.0	60	57	19	33	7	41
12b..........	"	39	34	9	18	3	70
13a..........	−1.0	71	70	19	24	3	54
13b..........	"	69	62	18	8	3	71
14a..........	−2.5	22	21	24	42	10	24
14b..........	"	9	9	0	0	0	100
Total, 2- 7..	—	383	264	37	9	16	38
Total, 11-14..	—	301	271	19	21	4	56

In Table II under the term "other crowding" have been placed all instances where *Chthamalus* were crushed laterally between 2 or more *Balanus,* or where *Chthamalus* disappeared in an interval during which a dense population of *Balanus* grew rapidly. For example, in area 7a the *Balanus,* which were at the high population density of 48 per cm², had no room to expand except upward and the barnacles very quickly grew into the form of tall cylinders or cones with the diameter of the opercular opening greater than that of the base. It was obvious that extreme crowding occurred under these circumstances, but the exact cause of the mortality of the *Chthamalus* caught in this crush was difficult to ascertain.

In comparing the survival curves of Figs. 2 and 3 within each area it is evident that *Chthamalus* kept free of *Balanus* survived better than those in the adjacent undisturbed areas on all but areas 2 and 14a. Area 2 was in the zone where adults of *Balanus* and *Chthamalus* were normally mixed; at this high level *Balanus* evidently has no influence on the survival of *Chthamalus.* On Stone 14a, the survival of *Chthamalus* without *Balanus* was much better until January when a starfish, *Asterias rubens* L., entered the cage and ate the barnacles.

Much variation occurred on the other 14 areas. When the *Chthamalus* growing without contact with *Balanus* are compared with those on the adjacent undisturbed portion of the area, the survival was very much better on 10 areas and moderately better on 4. In all areas, some *Chthamalus* in the undisturbed portions escaped severe crowding. Sometimes no *Balanus* happened to settle close to a *Chthamalus,* or sometimes those which did died soon after settlement. In some instances, *Chthamalus* which were being undercut by *Balanus* attached themselves to the *Balanus* and so survived. Some *Chthamalus* were partly covered by *Balanus* but still survived. It seems probable that in the 4 areas, nos. 4, 6, 11a, and 11b, where *Chthamalus* survived well in the presence of *Balanus,* a higher proportion of the *Chthamalus* escaped death in one of these ways.

The fate of very young *Chthamalus* which settled in the autumn of 1954 was followed in detail in 2 instances, on stone 15 and area 7b. The *Chthamalus* on stone 15 had settled in an irregular space surrounded by large *Balanus.* Most of the mortality occurred around the edges of the space as the *Balanus* undercut and lifted the small *Chthamalus* nearby. The following is a tabulation of all the deaths of young *Chthamalus* between Sept. 30, 1954 and Feb. 14, 1955, on Stone 15, with the associated situations:

Lifted by *Balanus*	: 29
Crushed by *Balanus*	: 4
Smothered by *Balanus* and *Chthamalus*	: 2
Crushed between *Balanus and Chthamalus*	: 1
Lifted by *Chthamalus*	: 1
Crushed between two other *Chthamalus*	: 1
Unknown	: 3

This list shows that crowding of newly settled *Chthamalus* by older *Balanus* in the autumn mainly takes the form of undercutting, rather than of smothering as was the case in the spring. The reason for this difference is probably that the *Chthamalus* are more firmly attached in the spring so that the fast growing young *Balanus* grow up over them when they make contact. In the autumn the reverse is the case, the *Balanus* being firmly attached, the *Chthamalus* weakly so.

Although the settlement of *Chthamalus* on Stone 15 in the autumn of 1954 was very dense, 32/cm^2, so that most of them were touching another, only 2 of the 41 deaths were caused by intraspecific crowding among the *Chthamalus*. This is in accord with the findings from the 1953 settlement of *Chthamalus*.

The mortality rates for the young *Chthamalus* on area 7b showed seasonal variations. Between October 10, 1954 and May 15, 1955 the relative mortality rate per day $\times$ 100 was 0.14 on the undisturbed area and 0.13 where *Balanus* had been removed. Over the next month, the rate increased to 1.49 on the undisturbed area and 0.22 where *Balanus* was absent. Thus the increase in mortality of young *Chthamalus* in late spring was also associated with the presence of *Balanus*.

Some of the stones transplanted from high to low levels in the spring of 1954 bore adult *Chthamalus*. On 3 stones, records were kept of the survival of these adults, which had settled in the autumn of 1952 or in previous years and were at least 20 months old at the start of the experiment. Their mortality is shown in Table III; it was always much greater when *Balanus* was not removed. On 2 of the 3 stones this mortality rate was almost as high as that of the younger group. These results suggest that any *Chthamalus* that managed to survive the competition for space with *Balanus* during the first year would probably be eliminated in the 2nd year.

Censuses of *Balanus* were not made on the experimental areas. However, on many other areas in the same stretch of shore the survival of *Balanus* was being studied during the same period (Connell 1961). In Table IV some mortality rates measured in that study are listed; the *Balanus* were members of the 1954 settlement at population densities and shore levels similar to those of the present study. The mortality rates of *Balanus* were about the same as those of *Chthamalus* in similar situations except at the highest level, area 1, where *Balanus* suffered much greater mortality than *Chthamalus*. Much of this mortality was caused by intraspecific crowding at all levels below area 1.

TABLE III. Comparison of the mortality rates of young and older *Chthamalus stellatus* on transplanted stones

Stone No.	Shore level	Treatment	Number of *Chthamalus* present in June, 1954		% mortality over one year (or for 6 months for 14a) of *Chthamalus*	
			1953 year group	1952 or older year groups	1953 year group	1952 or older year groups
13b	1.0 ft below MTL	*Balanus* removed	51	3	35	0
		Undisturbed	69	16	90	31
12a	MTL, in a tide pool, caged	*Balanus* removed	50	41	44	37
		Undisturbed	60	31	95	71
14a	2.5 ft below MTL, caged	*Balanus* removed	25	45	40	36
		Undisturbed	22	8	86	75

TABLE IV. Comparison of annual mortality rates of *Chthamalus stellatus* and *Balanus balanoides**

Area no.	Height in ft from M.T.L.	Population density: no./cm^2 June, 1954	% mortality in the next year
	Chthamalus stellatus, autumn 1953 settlement		
1	+4.2	21	17
3a	+2.2	31	72
3b	"	29	89
6	+1.0	38	41
7a	+0.7	49	90
7b	"	52	82
11a	0.0	32	62
13a	−1.0	29	99
12a	(tide pool)	19	95
	Balanus balanoides, spring 1954 settlement		
1 (top)	+4.2	21	99
1:Middle Cage 1	+2.1	85	92
1:Middle Cage 2	"	25	77
1:Low Cage 1	+1.5	26	88
Stone 1	−0.9	26	86
Stone 2	"	68	94

* Population density includes both species. The mortality rates of *Chthamalus* refer to those on the undisturbed portions of each area. The data and area designations for *Balanus* were taken from Connell (1961); the present area 1 is the same as that designated 1 (top) in that paper.

In the observations made at each census it appeared that *Balanus* was growing faster than *Chthamalus*. Measurements of growth rates of the 2 species were made from photographs of the areas taken in June and November, 1954. Barnacles growing free of contact with each other were measured; the results are given in Table V. The growth rate of *Balanus* was greater than that of *Chthamalus* in the experimental areas; this agrees with the findings of Hatton (1938) on the shore in France and of Barnes (1956a) for continual submergence on a raft at Millport.

TABLE V. Growth rates of *Chthamalus stellatus* and *Balanus balanoides.* Measurements were made of uncrowded individuals on photographs of areas 3a, 3b and 7b. Those of *Chthamalus* were made on the same individuals on both dates; of *Balanus,* representative samples were chosen

	CHTHAMALUS		BALANUS	
	No. measured	Average size, mm.	No. measured	Average size, mm.
June 11, 1954	25	2.49	39	1.87
November 3, 1954	25	4.24	27	4.83
Average size in the interval	3.36		3.35	
Absolute growth rate per day x 100	1.21		2.04	

After a year of crowding the average population densities of *Balanus* and *Chthamalus* remained in the same relative proportion as they had been at the start, since the mortality rates were about the same. However, because of its faster growth, *Balanus* occupied a relatively greater area and, presumably, possessed a greater biomass relative to that of *Chthamalus* after a year.

The faster growth of *Balanus* probably accounts for the manner in which *Chthamalus* were crowded by *Balanus.* It also accounts for the sinuosity of the survival curves of *Chthamalus* growing in contact with *Balanus.* The mortality rate of these *Chthamalus,* as indicated by the slope of the curves in Figs. 2 and 3, was greatest in summer, decreased in winter and increased again in spring. The survival curves of *Chthamalus* growing without contact with *Balanus* do not show these seasonal variations which, therefore, cannot be the result of the direct action of physical factors such as temperature, wave action or rain.

Seasonal variations in growth rate of *Balanus* correspond to these changes in mortality rate of *Chthamalus.* In Figure 4 the growth of *Balanus* throughout the year as studied on an intertidal panel at Millport by Barnes and Powell (1953), is compared to the survival of *Chthamalus* at about the same intertidal level in the present study. The increased mortality of *Chthamalus* was found to occur in the same seasons as the increases in the growth rate of *Balanus.* The correlation was tested using the Spearman rank correlation coefficient. The absolute increase in diameter of *Balanus* in each month, read from the curve of growth, was compared to the percentage mortality of *Chthamalus* in the same month. For the 13 months in which data for *Chthamalus* was available, the correlation was highly significant, P = .01.

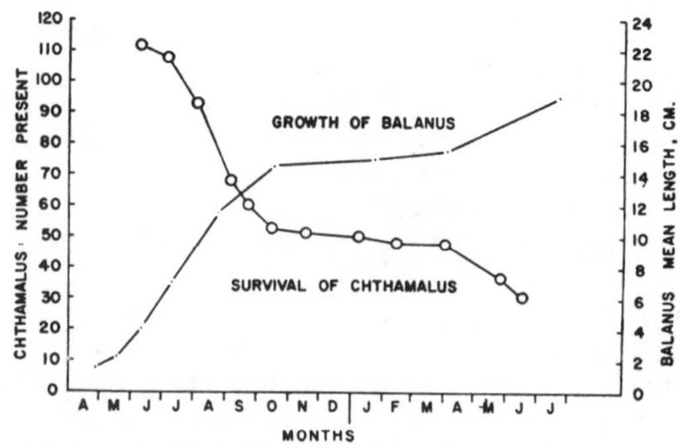

FIG. 4. A comparison of the seasonal changes in the growth of *Balanus balanoides* and in the survival of *Chthamalus stellatus* being crowded by *Balanus.* The growth of *Balanus* was that of panel 3, Barnes and Powell (1953), just above M.T.L. on Keppel Pier, Millport, during 1951-52. The *Chthamalus* were on area 3a of the present study, one-half mile south of Keppell Pier, during 1954-55.

From all these observations it appears that the poor survival of *Chthamalus* below M.H.W.N. is a result mainly of crowding by dense populations of faster growing *Balanus.*

At the end of the experiment in June, 1955, the surviving *Chthamalus* were collected from 5 of the areas. As shown in Table VI, the average size was greater in the *Chthamalus* which had grown free of contact with *Balanus;* in every case the difference was significant ($P < .01$, Mann-Whitney U. test, Siegel 1956). The survivors on the undisturbed areas were often misshapen, in some cases as a result of being lifted on to the side of an undercutting *Balanus.* Thus the smaller size of these barnacles may have been due to disturbances in the normal pattern of growth while they were being crowded.

These *Chthamalus* were examined for the presence of developing larvae in their mantle cavities. As shown in Table VI, in every area the proportion of the uncrowded *Chthamalus* with larvae was equal to or more often slightly greater than on the crowded areas. The reason for this may be related to the smaller size of the crowded *Chthamalus.* It is not due to separation, since *Chthamalus* can self-fertilize (Barnes and Crisp

TABLE VI. The effect of crowding on the size and presence of larvae in *Chthamalus stellatus*, collected in June, 1955

Area	Treatment	Level, feet above MTL	Number of *Chthamalus*	DIAMETER IN MM		% of individuals which had larvae in mantle cavity
				Average	Range	
3a......	Undisturbed	2.2	18	3.5	2.7-4.6	61
"......	*Balanus* removed	"	50	4.1	3.0-5.5	65
4.......	Undisturbed	1.4	16	2.3	1.8 3.2	81
".......	*Balanus* removed	"	37	3.7	2.5-5 1	100
5.......	Undisturbed	1.4	7	3.3	2.8-3.7	70
".......	*Balanus* removed	"	13	4.0	3.5-4.5	100
6.......	Undisturbed	1.0	13	2.8	2.1-3.9	100
".......	*Balanus* removed	"	14	4.1	3.0-5.2	100
7a & b..	Undisturbed	0.7	10	3.5	2.7-4.5	70
" ..	*Balanus* removed	"	23	4.3	3.0-6.3	81

1956). Moore (1935) and Barnes (1953) have shown that the number of larvae in an individual of *Balanus balanoides* increases with increase in volume of the parent. Comparison of the cube of the diameter, which is proportional to the volume, of *Chthamalus* with and without *Balanus* shows that the volume may be decreased to ¼ normal size when crowding occurs. Assuming that the relation between larval numbers and volume in *Chthamalus* is similar to that of *Balanus,* a decrease in both frequency of occurrence and abundance of larvae in *Chthamalus* results from competition with *Balanus.* Thus the process described in this paper satisfies both aspects of interspecific competition as defined by Elton and Miller (1954): "in which one species affects the population of another by a process of interference, i.e., by reducing the reproductive efficiency or increasing the mortality of its competitor."

The effect of predation by Thais

Cages which excluded *Thais* had been attached on 6 areas (indicated by the letter "a" following the number of the area). Area 14 was not included in the following analysis since many starfish were observed feeding on the barnacles at this level; one entered the cage in January, 1955, and ate most of the barnacles.

Thais were common in this locality, feeding on barnacles and mussels, and reaching average population densities of 200/m² below M.T.L. (Connell 1961). The mortality rates for *Chthamalus* in cages and on adjacent areas outside cages (indicated by the letter "b" after the number) are shown on Table VII.

If the mortality rates of *Chthamalus* growing without contact with *Balanus* are compared in and out of the cages, it can be seen that at the upper levels mortality is greater inside the cages,

TABLE VII. The effect of predation by *Thais lapillus* on the annual mortality rate of *Chthamalus stellatus* in the experimental areas*

Area	Height in ft from M.T.L.	% mortality of *Chthamalus* over a year (The initial numbers are given in parentheses)					
		a: Protected from predation by a cage			b: Unprotected, open to predation		
		With *Balanus*	Without *Balanus*	Difference	With *Balanus*	Without *Balanus*	Difference
Area 3..	+2.2	73 (112)	25 (96)	48	89 (47)	6 (50)	83
Area 7..	+0.7	90 (42)	47 (40)	43	82 (51)	23 (47)	59
Area 11..	0	62 (21)	28 (18)	34	50 (10)	25 (16)	25
Area 12 .	0†	100 (60)	53 (50)	47	87 (39)	59 (32)	28
Area 13..	−1.0	98 (72)	9 (77)	89	90 (69)	35 (51)	55

*The records for 12a extend over only 10 months; for purposes of comparison the mortality rate for 12a has been multiplied by 1.2.
†Tide pool.

at lower levels greater outside. Densities of *Thais* tend to be greater at and below M.T.L. so that this trend in the mortality rates of *Chthamalus* may be ascribed to an increase in predation by *Thais* at lower levels.

Mortality of *Chthamalus* in the absence of *Balanus* was appreciably greater outside than inside the cage only on area 13. In the other 4 areas it seems evident that few *Chthamalus* were being eaten by *Thais.* In a concurrent study of the behavior of *Thais* in feeding on *Balanus balanoides,* it was found that *Thais* selected the larger individuals as prey (Connell 1961). Since *Balanus* after a few month's growth was usually larger than *Chthamalus,* it might be expected that *Thais* would feed on *Balanus* in preference to *Chthamalus.* In a later study (unpublished) made at Santa Barbara, California, *Thais emarginata* Deshayes were enclosed in cages on the shore with mixed populations of *Balanus glandula* Darwin and *Chthamalus fissus* Darwin. These species were each of the same size range as the corresponding species at Millport. It was found that *Thais emarginata* fed on *Balanus glandula* in preference to *Chthamalus fissus.*

As has been indicated, much of the mortality of *Chthamalus* growing naturally intermingled with *Balanus* was a result of direct crowding by *Balanus.* It therefore seemed reasonable to take the difference between the mortality rates of *Chthamalus* with and without *Balanus* as an index of the degree of competition between the species. This difference was calculated for each area and is included in Table VII. If these differences are compared between each pair of adjacent areas in and out of a cage, it appears that the difference, and therefore the degree of competition, was greater outside the cages at the upper shore levels and less outside the cages at the lower levels.

Thus as predation increased at lower levels, the degree of competition decreased. This result would have been expected if *Thais* had fed upon *Balanus* in preference to *Chthamalus*. The general effect of predation by *Thais* seems to have been to lessen the interspecific competition below M.T.L.

Discussion

"Although animal communities appear qualitatively to be constructed as if competition were regulating their structure, even in the best studied cases there are nearly always difficulties and unexplored possibilities" (Hutchinson 1957).

In the present study direct observations at intervals showed that competition was occurring under natural conditions. In addition, the evidence is strong that the observed competition with *Balanus* was the principal factor determining the local distribution of *Chthamalus*. *Chthamalus* thrived at lower levels when it was not growing in contact with *Balanus*.

However, there remain unexplored possibilities. The elimination of *Chthamalus* requires a dense population of *Balanus*, yet the settlement of *Balanus* varied from year to year. At Millport, the settlement density of *Balanus balanoides* was measured for 9 years between 1944 and 1958 (Barnes 1956b, Connell 1961). Settlement was light in 2 years, 1946 and 1958. In the 3 seasons of *Balanus* settlement studied in detail, 1953-55, there was a vast oversupply of larvae ready for settlement. It thus seems probable that most of the *Chthamalus* which survived in a year of poor settlement of *Balanus* would be killed in competition with a normal settlement the following year. A succession of years with poor settlements of *Balanus* is a possible, but improbable occurrence at Millport, judging from the past record. A very light settlement is probably the result of a chance combination of unfavorable weather circumstances during the planktonic period (Barnes 1956b). Also, after a light settlement, survival on the shore is improved, owing principally to the reduction in intraspecific crowding (Connell 1961); this would tend to favor a normal settlement the following year, since barnacles are stimulated to settle by the presence of members of their own species already attached on the surface (Knight-Jones 1953).

The fate of those *Chthamalus* which had survived a year on the undisturbed areas is not known since the experiment ended at that time. It is probable, however, that most of them would have been eliminated within 6 months; the mortality rate had increased in the spring (Figs. 2 and 3), and these survivors were often misshapen and smaller than those which had not been crowded (Table VI). Adults on the transplanted stones had suffered high mortality in the previous year (Table III).

Another difficulty was that *Chthamalus* was rarely found to have settled below mid tide level at Millport. The reasons for this are unknown; it survived well if transplanted below this level, in the absence of *Balanus*. In other areas of the British Isles (in southwest England and Ireland, for example) it occurs below mid tide level.

The possibility that *Chthamalus* might affect *Balanus* deleteriously remains to be considered. It is unlikely that *Chthamalus* could cause much mortality of *Balanus* by direct crowding; its growth is much slower, and crowding between individuals of *Chthamalus* seldom resulted in death. A dense population of *Chthamalus* might deprive larvae of *Balanus* of space for settlement. Also, *Chthamalus* might feed on the planktonic larvae of *Balanus;* however, this would occur in March and April when both the sea water temperature and rate of cirral activity (presumably correlated with feeding activity), would be near their minima (Southward 1955).

The indication from the caging experiments that predation decreased interspecific competition suggests that the action of such additional factors tends to reduce the intensity of such interactions in natural conditions. An additional suggestion in this regard may be made concerning parasitism. Crisp (1960) found that the growth rate of *Balanus balanoides* was decreased if individuals were infected with the isopod parasite *Hemioniscus balani* (Spence Bate). In Britain this parasite has not been reported from *Chthamalus stellatus*. Thus if this parasite were present, both the growth rate of *Balanus*, and its ability to eliminate *Chthamalus* would be decreased, with a corresponding lessening of the degree of competition between the species.

The causes of zonation

The evidence presented in this paper indicates that the lower limit of the intertidal zone of *Chthamalus stellatus* at Millport was determined by interspecific competition for space with *Balanus balanoides*. *Balanus*, by virtue of its greater population density and faster growth, eliminated most of the *Chthamalus* by directing crowding.

At the upper limits of the zones of these species no interaction was observed. *Chthamalus* evidently can exist higher on the shore than *Balanus* mainly as a result of its greater tolerance to heat and/or desiccation.

The upper limits of most intertidal animals are probably determined by physical factors such as these. Since growth rates usually decrease with increasing height on the shore, it would be less likely that a sessile species occupying a higher zone could, by competition for space, prevent a lower one from extending upwards. Likewise, there has been, as far as the author is aware, no study made which shows that predation by land species determines the upper limit of an intertidal animal. In one of the most thorough of such studies, Drinnan (1957) indicated that intense predation by birds accounted for an annual mortality of 22% of cockles (*Cardium edule* L.) in sand flats where their total mortality was 74% per year.

In regard to the lower limits of an animal's zone, it is evident that physical factors may act directly to determine this boundary. For example, some active amphipods from the upper levels of sandy beaches die if kept submerged. However, evidence is accumulating that the lower limits of distribution of intertidal animals are determined mainly by biotic factors.

Connell (1961) found that the shorter length of life of *Balanus balanoides* at low shore levels could be accounted for by selective predation by *Thais lapillus* and increased intraspecific competition for space. The results of the experiments in the present study confirm the suggestions of other authors that lower limits may be due to interspecific competition for space. Knox (1954) suggested that competition determined the distribution of 2 species of barnacles in New Zealand. Endean, Kenny and Stephenson (1956) gave indirect evidence that competition with a colonial polychaete worm, (*Galeolaria*) may have determined the lower limit of a barnacle (*Tetraclita*) in Queensland, Australia. In turn the lower limit of *Galeolaria* appeared to be determined by competition with a tunicate, *Pyura*, or with dense algal mats.

With regard to the 2 species of barnacles in the present paper, some interesting observations have been made concerning changes in their abundance in Britain. Moore (1936) found that in southwest England in 1934, *Chthamalus stellatus* was most dense at M.H.W.N., decreasing in numbers toward M.T.L. while *Balanus balanoides* increased in numbers below M.H.W.N. At the same localities in 1951, Southward and Crisp (1954) found that *Balanus* had almost disappeared and that *Chthamalus* had increased both above and below M.H.W.N. *Chthamalus* had not reached the former densities of *Balanus* except at one locality, Brixham. After 1951, *Balanus* began to return in numbers, although by 1954 it had not reached the densities of 1934; *Chthamalus* had declined, but again not to its former densities (Southward and Crisp 1956).

Since *Chthamalus* increased in abundance at the lower levels vacated by *Balanus*, it may previously have been excluded by competition with *Balanus*. The growth rate of *Balanus* is greater than *Chthamalus* both north and south (Hatton 1938) of this location, so that *Balanus* would be likely to win in competition with *Chthamalus*. However, changes in other environmental factors such as temperature may have influenced the abundance of these species in a reciprocal manner. In its return to southwest England after 1951, the maximum density of settlement of *Balanus* was 12 per cm^2; competition of the degree observed at Millport would not be expected to occur at this density. At a higher population density, *Balanus* in southern England would probably eliminate *Chthamalus* at low shore levels in the same manner as it did at Millport.

In Loch Sween, on the Argyll Peninsula, Scotland, Lewis and Powell (1960) have described an unusual pattern of zonation of *Chthamalus stellatus*. On the outer coast of the Argyll Peninsula *Chthamalus* has a distribution similar to that at Millport. In the more sheltered waters of Loch Sween, however, *Chthamalus* occurs from above M.H.W.S. to about M.T.L., judging the distribution by its relationship to other organisms. *Balanus balanoides* is scarce above M.T.L. in Loch Sween, so that there appears to be no possibility of competition with *Chthamalus*, such as that occurring at Millport, between the levels of M.T.L. and M.H.W.N.

In Figure 5 an attempt has been made to summarize the distribution of adults and newly settled larvae in relation to the main factors which appear to determine this distribution. For *Balanus* the estimates were based on the findings of a previous study (Connell 1961); intraspecific competition was severe at the lower levels during the first year, after which predation increased in importance. With *Chthamalus*, it appears that avoidance of settlement or early mortality of those larvae which settled at levels below M.T.L., and elimination by competition with *Balanus* of those which settled between M.T.L. and M.H.W.N., were the principal causes for the absence of adults below M.H.W.N. at Millport. This distribution appears to be typical for much of western Scotland.

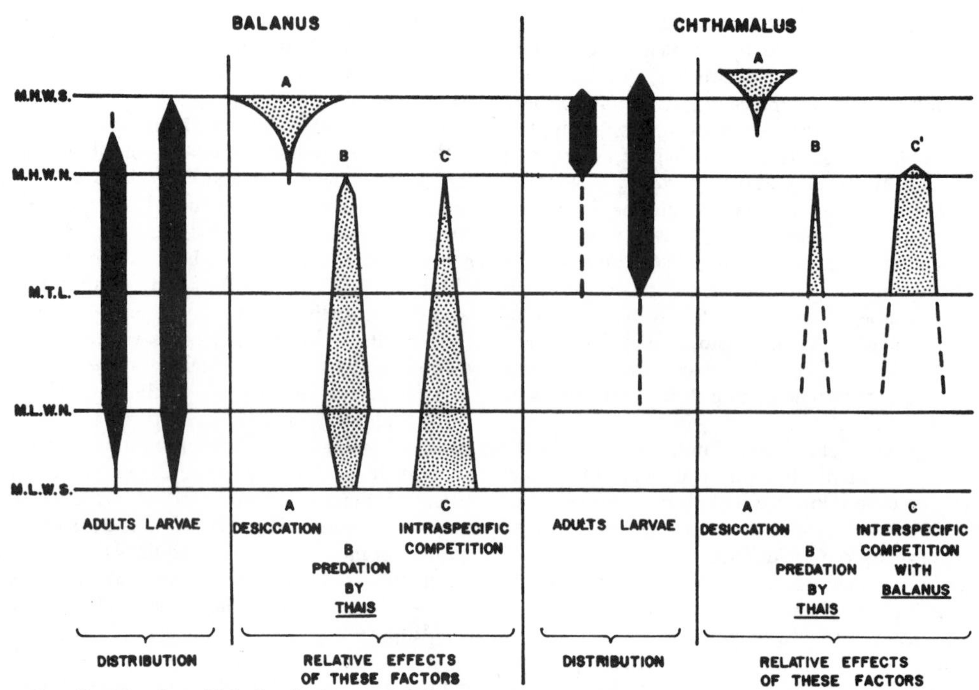

FIG. 5. The intertidal distribution of adults and newly settled larvae of *Balanus balanoides* and *Chthamalus stellatus* at Millport, with a diagrammatic representation of the relative effects of the principal limiting factors.

SUMMARY

Adults of *Chthamalus stellatus* occur in the marine intertidal in a zone above that of another barnacle, *Balanus balanoides*. Young *Chthamalus* settle in the *Balanus* zone but evidently seldom survive, since few adults are found there.

The survival of *Chthamalus* which had settled at various levels in the *Balanus* zone was followed for a year by successive censuses of mapped individuals. Some *Chthamalus* were kept free of contact with *Balanus*. These survived very well at all intertidal levels, indicating that increased time of submergence was not the factor responsible for elimination of *Chthamalus* at low shore levels. Comparison of the survival of unprotected populations with others, protected by enclosure in cages from predation by the snail, *Thais lapillus,* showed that *Thais* was not greatly affecting the survival of *Chthamalus*.

Comparison of the survival of undisturbed populations of *Chthamalus* with those kept free of contact with *Balanus* indicated that *Balanus* could cause great mortality of *Chthamalus*. *Balanus* settled in greater population densities and grew faster than *Chthamalus*. Direct observations at each census showed that *Balanus* smothered, undercut, or crushed the *Chthamalus;* the greatest mortality of *Chthamalus* occurred during the seasons of most rapid growth of *Balanus*. Even older *Chthamalus* transplanted to low levels were killed by *Balanus* in this way. Predation by *Thais* tended to decrease the severity of this interspecific competition.

Survivors of *Chthamalus* after a year of crowding by *Balanus* were smaller than uncrowded ones. Since smaller barnacles produce fewer offspring, competition tended to reduce reproductive efficiency in addition to increasing mortality.

Mortality as a result of intraspecies competition for space between individuals of *Chthamalus* was only rarely observed.

The evidence of this and other studies indicates that the lower limit of distribution of intertidal organisms is mainly determined by the action of biotic factors such as competition for space or predation. The upper limit is probably more often set by physical factors.

References

Barnes, H. 1953. Size variations in the cyprids of some common barnacles. J. Mar. Biol. Ass. U. K. **32:** 297-304.

———. 1956a. The growth rate of *Chthamalus stellatus* (Poli). J. Mar. Biol. Ass. U. K. **35**: 355-361.

———. 1956b. *Balanus balanoides* (L.) in the Firth of Clyde: The development and annual variation of the larval population, and the causative factors. J. Anim. Ecol. **25**: 72-84.

——— **and H. T. Powell.** 1953. The growth of *Balanus balanoides* (L.) and *B. crenatus* Brug. under varying conditions of submersion. J. Mar. Biol. Ass. U. K. **32**: 107-128.

——— **and D. J. Crisp.** 1956. Evidence of self-fertilization in certain species of barnacles. J. Mar. Biol. Ass. U. K. **35**: 631-639.

Beauchamp, R. S. A. and P. Ullyott. 1932. Competitive relationships between certain species of freshwater Triclads. J. Ecol. **20**: 200-208.

Clark, P. J. and F. C. Evans. 1954. Distance to nearest neighbor as a measure of spatial relationships in populations. -Ecology **35**: 445-453.

Clench, W. J. 1947. The genera *Purpura* and *Thais* in the western Atlantic. Johnsonia **2**, No. 23: 61-92.

Connell, J. H. 1961. The effects of competition, predation by *Thais lapillus,* and other factors on natural populations of the barnacle, *Balanus balanoides.* Ecol. Mon. **31**: 61-104.

Crisp, D. J. 1960. Factors influencing growth-rate in *Balanus balanoides.* J. Anim. Ecol. **29**: 95-116.

Drinnan, R. E. 1957. The winter feeding of the oystercatcher (*Haematopus ostralegus*) on the edible cockle (*Cardium edule*). J. Anim. Ecol. **26**: 441-469.

Elton, Charles and R. S. Miller. 1954. The ecological survey of animal communities: with a practical scheme of classifying habitats by structural characters. J. Ecol. **42**: 460-496.

Endean, R., R. Kenny and W. Stephenson. 1956. The ecology and distribution of intertidal organisms on the rocky shores of the Queensland mainland. Aust. J. mar. freshw. Res. **7**: 88-146.

Fischer, E. 1928. Sur la distribution geographique de quelques organismes de rocher, le long des cotes de la Manche. Trav. Lab. Mus. Hist. Nat. St.-Servan **2**: 1-16.

Hatton, H. 1938. Essais de bionomie explicative sur quelques especes intercotidales d'algues et d'animaux. Ann. Inst. Oceanogr. Monaco **17**: 241-348.

Holme, N. A. 1950. Population-dispersion in *Tellina tenuis* Da Costa. J. Mar. Biol. Ass. U. K. **29**: 267-280.

Hutchinson, G. E. 1957. Concluding remarks. Cold Spring Harbor Symposium on Quant. Biol. **22**: 415-427.

Knight-Jones, E. W. 1953. Laboratory experiments on gregariousness during setting in *Balanus balanoides* and other barnacles. J. Exp. Biol. **30**: 584-598.

Knox, G. A. 1954. The intertidal flora and fauna of the Chatham Islands. Nature Lond. **174**: 871-873.

Lack, D. 1954. The natural regulation of animal numbers. Oxford, Clarendon Press.

Lewis, J. R. and H. T. Powell. 1960. Aspects of the intertidal ecology of rocky shores in Argyll, Scotland. I. General description of the area. II. The distribution of *Chthamalus stellatus* and *Balanus balanoides* in Kintyre. Trans. Roy. Soc. Edin. **64**: 45-100.

MacArthur, R. H. 1958. Population ecology of some warblers of northeastern coniferous forests. Ecology **39**: 599-619.

Moore, H. B. 1935. The biology of *Balnus balanoides.* III. The soft parts. J. Mar. Biol. Ass. U. K. **20**: 263-277.

———. 1936. The biology of *Balanus balanoides.* V. Distribution in the Plymouth area. J. Mar. Biol. Ass. U. K. **20**: 701-716.

Siegel, S. 1956. Nonparametric statistics. New York, McGraw Hill.

Southward, A. J. 1955. On the behavior of barnacles. I. The relation of cirral and other activities to temperature. J. Mar. Biol. Ass. U. K. **34**: 403-422.

——— **and D. J. Crisp.** 1954. Recent changes in the distribution of the intertidal barnacles *Chthamalus stellatus* Poli and *Balanus balanoides* L. in the British Isles. J. Anim. Ecol. **23**: 163-177.

———. 1956. Fluctuations in the distribution and abundance of intertidal barnacles. J. Mar. Biol. Ass. U. K. **35**: 211-229.

Vol. 100, No. 910 The American Naturalist January–February, 1966

FOOD WEB COMPLEXITY AND SPECIES DIVERSITY

ROBERT T. PAINE

Department of Zoology, University of Washington, Seattle, Washington

Though longitudinal or latitudinal gradients in species diversity tend to be well described in a zoogeographic sense, they also are poorly understood phenomena of major ecological interest. Their importance lies in the derived implication that biological processes may be fundamentally different in the tropics, typically the pinnacle of most gradients, than in temperate or arctic regions. The various hypotheses attempting to explain gradients have recently been reviewed by Fischer (1960), Simpson (1964), and Connell and Orias (1964), the latter authors additionally proposing a model which can account for the production and regulation of diversity in ecological systems. Understanding of the phenomenon suffers from both a specific lack of synecological data applied to particular, local situations and from the difficulty of inferring the underlying mechanism(s) solely from descriptions and comparisons of faunas on a zoogeographic scale. The positions taken in this paper are that an ultimate understanding of the underlying causal processes can only be arrived at by study of local situations, for instance the promising approach of MacArthur and MacArthur (1961), and that biological interactions such as those suggested by Hutchinson (1959) appear to constitute the most logical possibilities.

The hypothesis offered herein applies to local diversity patterns of rocky intertidal marine organisms, though it conceivably has wider applications. It may be stated as: "Local species diversity is directly related to the efficiency with which predators prevent the monopolization of the major environmental requisites by one species." The potential impact of this process is firmly based in ecological theory and practice. Gause (1934), Lack (1949), and Slobodkin (1961) among others have postulated that predation (or parasitism) is capable of preventing extinctions in competitive situations, and Slobodkin (1964) has demonstrated this experimentally. In the field, predation is known to ameliorate the intensity of competition for space by barnacles (Connell, 1961b), and, in the present study, predator removal has led to local extinctions of certain benthic invertebrates and algae. In addition, as a predictable extension of the hypothesis, the proportion of predatory species is known to be relatively greater in certain diverse situations. This is true for tropical vs. temperate fish faunas (Hiatt and Strasburg, 1960; Bakus, 1964), and is seen especially clearly in the comparison of shelf water zooplankton populations (81 species, 16% of which are carnivores) with those of the presumably less productive though more stable Sargasso Sea (268 species, 39% carnivores) (Grice and Hart, 1962).

In the discussion that follows no quantitative measures of local diversity are given, though they may be approximated by the number of species represented in Figs. 1 to 3. No distinctions have been drawn between species within certain food categories. Thus I have assumed that the probability of, say, a bivalve being eaten is proportional to its abundance, and that predators exercise no preference in their choice of any "bivalve" prey. This procedure simplifies the data presentation though it dodges the problem of taxonomic complexity. Wherever possible the data are presented as both number observed being eaten and their caloric equivalent. The latter is based on prey size recorded in the field and was converted by determining the caloric content of Mukkaw Bay material of the same or equivalent species. These caloric data will be given in greater detail elsewhere. The numbers in the food webs, unfortunately, cannot be related to rates of energy flow, although when viewed as calories they undoubtedly accurately suggest which pathways are emphasized.

Dr. Rudolf Stohler kindly identified the gastropod species. A. J. Kohn, J. H. Connell, C. E. King, and E. R. Pianka have provided invaluable criticism. The University of Washington, through the offices of the Organization for Tropical Studies, financed the trip to Costa Rica. The field work in Baja California, Mexico, and at Mukkaw Bay was supported by the National Science Foundation (GB-341).

THE STRUCTURE OF SELECTED FOOD WEBS

I have claimed that one of the more recognizable and workable units within the community nexus are subwebs, groups of organisms capped by a terminal carnivore and trophically interrelated in such a way that at higher levels there is little transfer of energy to co-occurring subwebs (Paine, 1963). In the marine rocky intertidal zone both the subwebs and their top carnivores appear to be particularly distinct, at least where macroscopic species are involved; and observations in the natural setting can be made on the quantity and composition of the component species' diets. Furthermore, the rocky intertidal zone is perhaps unique in that the major limiting factor of the majority of its primary consumers is living space, which can be directly observed, as the elegant studies on interspecific competition of Connell (1961a,b) have shown. The data given below were obtained by examining individual carnivores exposed by low tide, and recording prey, predator, their respective lengths, and any other relevant properties of the interaction.

A north temperate subweb

On rocky shores of the Pacific Coast of North America the community is dominated by a remarkably constant association of mussels, barnacles, and one starfish. Fig. 1 indicates the trophic relationships of this portion of the community as observed at Mukkaw Bay, near Neah Bay, Washington (ca. 49° N latitude). The data, presented as both numbers and total calories consumed by the two carnivorous species in the subweb, *Pisaster ochraceus*,

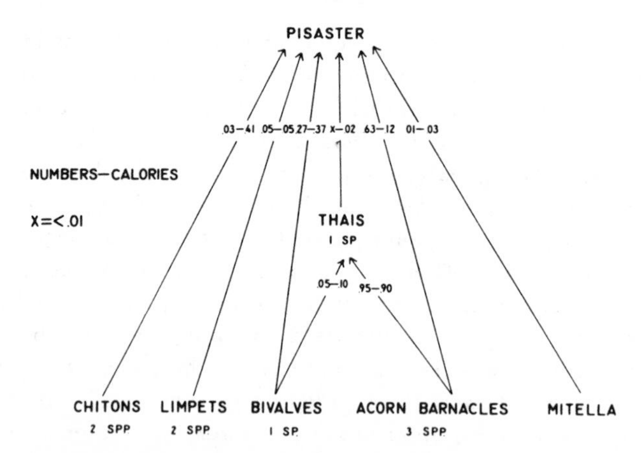

FIG. 1. The feeding relationships by numbers and calories of the *Pisaster* dominated subweb at Mukkaw Bay. *Pisaster*, N = 1049; *Thais*, N = 287. N is the number of food items observed eaten by the predators. The specific composition of each predator's diet is given as a pair of fractions; numbers on the left, calories on the right.

a starfish, and *Thais emarginata*, a small muricid gastropod, include the observational period November, 1963, to November, 1964. The composition of this subweb is limited to organisms which are normally intertidal in distribution and confined to a hard rock substrate. The diet of *Pisaster* is restricted in the sense that not all available local food types are eaten, although of six local starfish it is the most catholic in its tastes. Numerically its diet varies little from that reported by Feder (1959) for *Pisaster* observed along the central California coastline, especially since the gastropod *Tegula*, living on a softer bottom unsuitable to barnacles, has been omitted. *Thais* feeds primarily on the barnacle *Balanus glandula*, as also noted by Connell (1961b).

This food web (Fig. 1) appears to revolve on a barnacle economy with both major predators consuming them in quantity. However, note that on a nutritional (calorie) basis, barnacles are only about one-third as important to *Pisaster* as either *Mytilus californianus*, a bivalve, or the browsing chiton *Katherina tunicata*. Both these prey species dominate their respective food categories. The ratio of carnivore species to total species is 0.18. If *Tegula* and an additional bivalve are included on the basis that they are the most important sources of nourishment in adjacent areas, the ratio becomes 0.15. This number agrees closely with a ratio of 0.14 based on *Pisaster*, plus all prey species eaten more than once, in Feder's (1959) general compilation.

A subtropical subweb

In the Northern Gulf of California (ca. 31° N.) a subweb analogous to the one just described exists. Its top carnivore is a starfish (*Heliaster kubiniji*), the next two trophic levels are dominated by carnivorous gastropods, and the main prey are herbivorous gastropods, bivalves, and barnacles. I have collected there only in March or April of 1962–1964, but on both sides of the Gulf at San Felipe, Puertecitos, and Puerta Penasco. The resultant trophic arrangements (Fig. 2), though representative of springtime conditions and indicative of a much more stratified and complex community, are basically similar to those at Mukkaw Bay. Numerically the major food item in the diets of *Heliaster* and *Muricanthus nigritus* (a muricid gastropod), the two top-ranking carnivores, is barnacles; the major portion of these predators' nutrition is derived from other members of the community, primarily herbivorous mollusks. The increased trophic complexity presents certain graphical problems. If increased trophic height is indicated by a decreasing percentage of primary consumers in a species diet, *Acanthina tuberculata* is the highest carnivore due to its specialization on *A. angelica*, although it in turn is consumed by two other species. Because of this, and ignoring the percentages, both *Heliaster* and *Muricanthus* have been placed above *A. tuberculata*. Two species, *Hexaplex* and *Muricanthus* eventually become too large to be eaten by *Heliaster*, and thus through growth join it as top predators in the system. The taxonomically-difficult gastropod family Columbellidae, including both herbivorous and carnivorous species (Marcus and Marcus, 1962) have been placed in an intermediate position.

The Gulf of California situation is interesting on a number of counts. A new trophic level which has no counterpart at Mukkaw Bay is apparent, interposed between the top carnivore and the primary carnivore level. If higher level predation contributes materially to the maintenance of diversity, these species will have an effect on the community composition out of proportion to their abundance. In one of these species, *Muricanthus*,

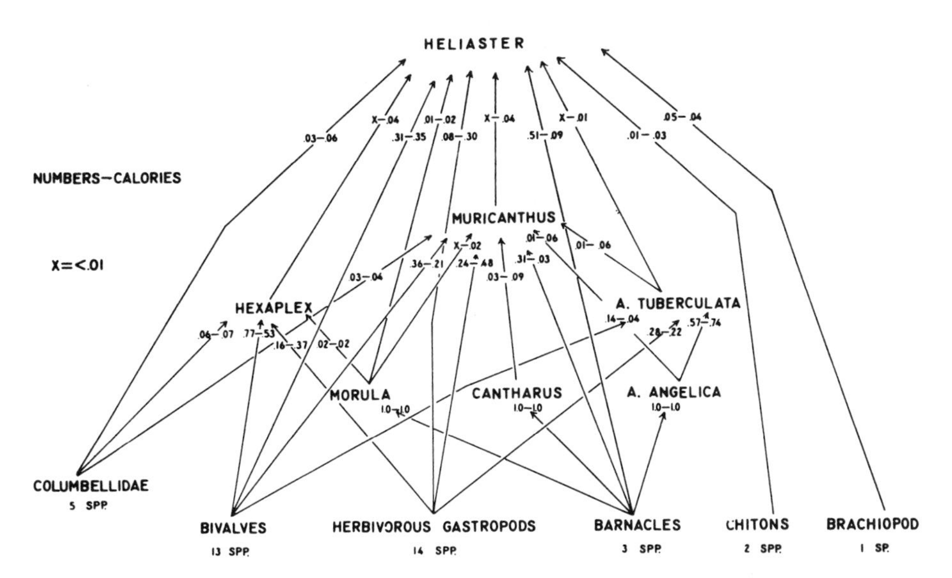

FIG. 2. The feeding relationships by numbers and calories of the *Heliaster* dominated subweb in the northern Gulf of California. *Heliaster*, N = 2245; *Muricanthus*, N = 113; *Hexaplex*, N = 62; *A. tuberculata*, N = 14; *A. angelica*, N = 432; *Morula*, N = 39; *Cantharus*, N = 8.

the larger members belong to a higher level than immature specimens (Paine, unpublished), a process tending to blur the food web but also potentially increasing diversity (Hutchinson, 1959). Finally, if predation operates to reduce competitive stresses, evidence for this reduction can be drawn by comparing the extent of niche diversification as a function of trophic level in a typical Eltonian pyramid. *Heliaster* consumes all other members of this subweb, and as such appears to have no major competitors of comparable status. The three large gastropods forming the subterminal level all may be distinguished by their major sources of nutrition: *Hexaplex*—bivalves (53%), *Muricanthus*—herbivorous gastropods (48%), and *A. tuberculata*—carnivorous gastropods (74%). No such obvious distinction characterizes the next level composed of three barnacle-feeding specialists which additionally share their resource with *Muricanthus* and *Heliaster*. Whether these species are more specialized (Klopfer and MacArthur, 1960) or whether they tolerate greater niche overlap (Klopfer and MacArthur, 1961) cannot be stated. The extent of niche diversification is subtle and trophic overlap is extensive.

The ratio of carnivore species to total species in Fig. 2 is 0.24 when the category Columbellidae is considered to be principally composed of one herbivorous (*Columbella*) and four carnivorous (*Pyrene*, *Anachis*, *Mitella*) species, based on the work of Marcus and Marcus (1962).

A tropical subweb

Results of five days of observation near Mate de Limon in the Golfo de Nocoya on the Pacific shore of Costa Rica (approx. 10° N.) are presented in Fig. 3. No secondary carnivore was present; rather the environmental resources were shared by two small muricid gastropods, *Acanthina brevidentata* and *Thais biserialis*. The fauna of this local area was relatively simple and completely dominated by a small mytilid and barnacles. The co-occupiers of the top level show relatively little trophic overlap despite the broad nutritional base of *Thais* which includes carrion and cannibalism. The relatively low number of feeding observations (187) precludes an accurate appraisal of the carnivore species to total web membership ratio.

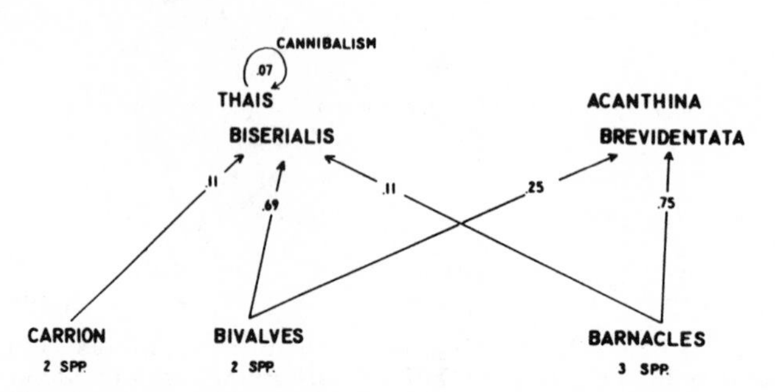

FIG. 3. The feeding relationship by numbers of a comparable food web in Costa Rica. *Thais*, N = 99; *Acanthina*, N = 80.

CHANGES RESULTING FROM THE REMOVAL OF THE TOP CARNIVORE

Since June, 1963, a "typical" piece of shoreline at Mukkaw Bay about eight meters long and two meters in vertical extent has been kept free of *Pisaster.* An adjacent control area has been allowed to pursue its natural course of events. Line transects across both areas have been taken irregularly and the number and density of resident macroinvertebrate and benthic algal species measured. The appearance of the control area has not altered. Adult *Mytilus californianus*, *Balanus cariosus*, and *Mitella polymerus* (a goose-necked barnacle) form a conspicuous band in the middle intertidal. The relatively stable position of the band is maintained by *Pisaster* predation (Paris, 1960; Paine, unpublished). At lower tidal levels the diversity increases abruptly and the macrofauna includes immature individuals of the above, *B. glandula* as scattered clumps, a few anemones of one species, two chiton species (browsers), two abundant limpets (browsers), four macroscopic benthic algae (*Porphyra*-an epiphyte, *Endocladia*, *Rhodomela*, and *Corallina*), and the sponge *Haliclona*, often browsed upon by *Anisodoris*, a nudibranch.

Following the removal of *Pisaster*, *B. glandula* set successfully throughout much of the area and by September had occupied from 60 to 80% of the available space. By the following June the *Balanus* themselves were being crowded out by small, rapidly growing *Mytilus* and *Mitella.* This process of successive replacement by more efficient occupiers of space is continuing, and eventually the experimental area will be dominated by *Mytilus*, its epifauna, and scattered clumps of adult *Mitella.* The benthic algae either have or are in the process of disappearing with the exception of the epiphyte, due to lack of appropriate space; the chitons and larger limpets have also emigrated, due to the absence of space and lack of appropriate food.

Despite the likelihood that many of these organisms are extremely long-lived and that these events have not reached an equilibrium, certain statements can be made. The removal of *Pisaster* has resulted in a pronounced *decrease* in diversity, as measured simply by counting species inhabiting this area, whether consumed by *Pisaster* or not, from a 15 to an eight-species system. The standing crop has been increased by this removal, and should continue to increase until the *Mytilus* achieve their maximum size. In general the area has become trophically simpler. With *Pisaster* artificially removed, the sponge-nudibranch food chain has been displaced, and the anemone population reduced in density. Neither of these carnivores nor the sponge is eaten by *Pisaster*, indicating that the number of food chains initiated on this limited space is strongly influenced by *Pisaster*, but by an indirect process. In contrast to Margalef's (1958) generalization about the tendency, with higher successional status towards "an ecosystem of more complex structure," these removal experiments demonstrate the opposite trend: in the absence of a complicating factor (predation), there is a "winner" in the competition for space, and the local system tends toward simplicity. Predation by this interpretation interrupts the successional process and, on a local basis, tends to increase local diversity.

No data are available on the microfaunal changes accompanying the gradual alteration of the substrate from a patchy algal mat to one comprised of the byssal threads of *Mytilus*.

INTERPRETATION

The differences in relative diversity of the subwebs diagrammed in Figs. 1–3 may be represented as Baja California (45 spp.) >> Mukkaw Bay (11 spp.) > Costa Rica (8 sp.), the number indicating the actual membership of the subwebs and not the number of local species. All three areas are characterized by systems in which one or two species are capable of monopolizing much of the space, a circumstance realized in nature only in Costa Rica. In the other two areas a top predator that derives its nourishment from other sources feeds in such a fashion that no space-consuming monopolies are formed. *Pisaster* and *Heliaster* eat masses of barnacles, and in so doing enhance the ability of other species to inhabit the area by keeping space open. When the top predator is artificially removed or naturally absent (i.e., predator removal area and Costa Rica, respectively), the systems converge toward simplicity. When space is available, other organisms settle or move in, and these, for instance chitons at Mukkaw Bay and herbivorous gastropods and pelecypods in Baja California, form the major portions of the predator's nutrition. Furthermore, *in situ* primary production is enhanced by the provision of space. This event makes the grazing moiety less dependent on the vagaries of phytoplankton production or distribution and lends stability to the association.

At the local level it appears that carnivorous gastropods which can penetrate only one barnacle at a time, although they might consume a few more per tidal interval, do not have the same effect as a starfish removing 20 to 60 barnacles simultaneously. Little compensation seems to be gained from snail density increases because snails do not clear large patches of space, and because the "husks" of barnacles remain after the animal portion has been consumed. In the predator removal area at Mukkaw Bay, the density of *Thais* increased 10- to 20-fold, with no apparent effect on diversity although the rate of *Mytilus* domination of the area was undoubtedly slowed. Clusters (density of 75–125/m^2) of *Thais* and *Acanthina* characterize certain rocks in Costa Rica, and diversity is still low. And, as a generality, wherever acorn barnacles or other space-utilizing forms potentially dominate the shore, diversity is reduced unless some predator can prevent the space monopoly. This occurs in Washington State where the shoreline, in the absence of *Pisaster*, is dominated by barnacles, a few mussels, and often two species of *Thais*. The same monopolistic tendencies characterize Connell's (1961a,b) study area in Scotland, the rocky intertidal of northern Japan (Hoshiai, 1960, 1961), and shell bags suitable for sponge settlement in North Carolina (Wells, Wells, and Gray, 1964).

Local diversity on intertidal rocky bottoms, then, appears directly related to predation intensity, though other potential factors are mentioned below. If one accepts the generalizations of Hedgpeth (1957) and Hall

(1964) that ambient temperature is the single most important factor influencing distribution or reproduction of marine invertebrates, then the potential role of climatic stability as measured by seasonal variations in water temperature can be examined. At Neah Bay the maximum range of annual values are 5.9 to 13.3 C (Rigg and Miller, 1949); in the northern Gulf of California, Roden and Groves (1959) recorded an annual range of 14.9 to 31.2 C; and in Costa Rica the maximum annual range is 26.1 to 31.7 C (Anon., 1952). Clearly the greatest benthic diversity, and one claimed by Parker (1963) on a regional basis to be among the most diverse known, is associated with the most variable (least stable) temperature regimen. Another influence on diversity could be exercised by environmental heterogeneity (Hutchinson, 1959). Subjectively, it appeared that both the Mukkaw Bay and Costa Rica stations were topographically more distorted than the northern Gulf localities. In any event, no topographic features were evident that could correlate with the pronounced differences in faunal diversity. Finally, Connell and Orias (1964) have developed a model for the organic enrichment of regions that depends to a great extent on the absolute amount of primary production and/or nutrient import, and hence energy flowing through the community web. Unfortunately, no productivity data are available for the two southern communities, and comparisons cannot yet be made.

PREDATION AND DIVERSITY GRADIENTS

To examine predation as a diversity-causing mechanism correlated with latitude, we must know why one environment contains higher order carnivores and why these are absent from others. These negative situations can be laid to three possibilities: (1) that through historical accident no higher carnivores have evolved in the region; (2) that the sample area cannot be occupied due to a particular combination of *local* hostile physiological effects; (3) that the system cannot support carnivores because the rate of energy transfer to a higher level is insufficient to sustain that higher level. The first possibility is unapproachable, the second will not apply on a geographic scale, and thus only the last would seem to have reality. Connell and Orias (1964) have based their hypothesis of the establishment and maintenance of diversity on varying rates of energy transfer, which are determined by various limiting factors and environmental stability. Without disagreeing with their model, two aspects of primary production deserve further consideration. The animal diversity of a given system will probably be higher if the production is apportioned more uniformly throughout the year rather than occurring as a single major bloom, because tendencies towards competitive displacement can be ameliorated by specialization on varying proportions of the resources (MacArthur and Levins, 1964). Both the predictability of production on a sustained annual basis and the causation of resource heterogeneity by predation will facilitate this mechanism. Thus, per production unit, greater stability of production should be correlated with greater diversity, other things being equal.

The realization of this potential, however, depends on more than simply the annual stability of carbon fixation. Rate of production and subsequent transfer to higher levels must also be important. Thus trophic structure of a community depends in part on the physical extent of the area (Darlington, 1957), or, in computer simulation models, on the amount of protoplasm in the system (Garfinkel and Sack, 1964). On the other hand, enriched aquatic environments often are characterized by decreased diversity. Williams (1964) has found that regions of high productivity are dominated by few diatom species. Less productive areas tended to have more species of equivalent rank, and hence a greater diversity. Obviously, the gross amount of energy fixed by itself is incapable of explaining diversity; and extrinsic factors probably are involved.

Given sufficient evolutionary time for increases in faunal complexity to occur, two independent mechanisms should work in a complementary fashion. When predation is capable of preventing resource monopolies, diversity should increase by positive feedback processes until some limit is reached. The argument of Fryer (1965) that predation facilitates speciation is germane here. The upper limit to local diversity, or, in the present context, the maximum number of species in a given subweb, is probably set by the combined stability and rate of primary production, which thus influences the number and variety of non-primary consumers in the subweb. Two aspects of predation must be evaluated before a generalized hypothesis based on predation effects can contribute to an understanding of differences in diversity between *any* comparable regions or faunistic groups. We must know if resource monopolies are actually less frequent in the diverse area than in comparable systems elsewhere, and, if so, why this is so. And we must learn something about the multiplicity of energy pathways in diverse systems, since predation-induced diversity could arise either from the presence of a variety of subwebs of equivalent rank, or from domination by one major one. The predation hypothesis readily predicts the apparent absence of monopolies in tropical (diverse) areas, a situation classically represented as "many species of reduced individual abundance." It also is in accord with the disproportionate increase in the number of carnivorous species that seems to accompany regional increases in animal diversity. In the present case in the two adequately sampled, structurally analagous, subwebs, general membership increases from 13 at Mukkaw Bay to 45 in the Gulf of California, a factor of 3.5, whereas the carnivore species increased from 2 to 11, a factor of 5.5.

SUMMARY

It is suggested that local animal species diversity is related to the number of predators in the system and their efficiency in preventing single species from monopolizing some important, limiting, requisite. In the marine rocky intertidal this requisite usually is space. Where predators capable of preventing monopolies are missing, or are experimentally removed, the systems become less diverse. On a local scale, no relationship between latitude (10° to 49° N.) and diversity was found. On a geographic scale, an increased stability of annual production may lead to an increased capacity for systems to support higher-level carnivores. Hence tropical, or other, ecosystems are more diverse, and are characterized by disproportionately more carnivores.

LITERATURE CITED

Anon. 1952. Surface water temperatures at tide stations. Pacific coast North and South America. Spec. Pub. No. 280: p. 1-59. U. S. Coast and Geodetic Survey.

Bakus, G. J. 1964. The effects of fish-grazing on invertebrate evolution in shallow tropical waters. Allan Hancock Found. Pub. 27: 1-29.

Connell, J. H. 1961a. Effect of competition, predation by *Thais lapillus*, and other factors on natural populations of the barnacle *Balanus balanoides*. Ecol. Monogr. 31: 61-104.

______. 1961b. The influence of interspecific competition and other factors on the distribution of the barnacle *Chthamalus stellatus*. Ecology 42: 710-723.

Connell, J. H., and E. Orias. 1964. The ecological regulation of species diversity. Amer. Natur. 98: 399-414.

Darlington, P. J. 1957. Zoogeography. Wiley, New York.

Feder, H. M. 1959. The food of the starfish, *Pisaster ochraceus*, along the California coast. Ecology 40: 721-724.

Fischer, A. G. 1960. Latitudinal variations in organic diversity. Evolution 14: 64-81.

Fryer, G. 1965. Predation and its effects on migration and speciation in African fishes: a comment. Proc. Zool. Soc. London 144: 301-310.

Garfinkel, D., and R. Sack. 1964. Digital computer simulation of an ecological system, based on a modified mass action law. Ecology 45: 502-507.

Gause, G. F. 1934. The struggle for existence. Williams and Wilkins Co., Baltimore.

Grice, G. D., and A. D. Hart. 1962. The abundance, seasonal occurrence, and distribution of the epizooplankton between New York and Bermuda. Ecol. Monogr. 32: 287-309.

Hall, C. A., Jr. 1964. Shallow-water marine climates and molluscan provinces. Ecology 45: 226-234.

Hedgpeth, J. W. 1957. Marine biogeography. Geol. Soc. Amer. Mem. 67, 1: 359-382.

Hiatt, R. W., and D. W. Strasburg. 1960. Ecological relationships of the fish fauna on coral reefs of the Marshall Islands. Ecol. Monogr. 30: 65-127.

Hoshiai, T. 1960. Synecological study on intertidal communities III. An analysis of interrelation among sedentary organisms on the artificially denuded rock surface. Bull. Marine Biol. Sta. Asamushi. 10: 49-56.

______. 1961. Synecological study on intertidal communities. IV. An ecological investigation on the zonation in Matsushima Bay concerning the so-called covering phenomenon. Bull. Marine Biol. Sta. Asamushi. 10: 203-211.

Hutchinson, G. E. 1959. Homage to Santa Rosalia or why are there so many kinds of animals? Amer. Natur. 93: 145–159.

Klopfer, P. H., and R. H. MacArthur. 1960. Niche size and faunal diversity. Amer. Natur. 94: 293–300.

______. 1961. On the causes of tropical species diversity: niche overlap. Amer. Natur. 95: 223–226.

Lack, D. 1949. The significance of ecological isolation, p. 299–308. *In* G. L. Jepsen, G. G. Simpson, and E. Mayr [eds.], Genetics, paleontology and evolution. Princeton Univ. Press, Princeton.

MacArthur, R., and R. Levins. 1964. Competition, habitat selection, and character displacement in a patchy environment. Proc. Nat. Acad. Sci. 51: 1207-1210.

MacArthur, R. H., and J. W. MacArthur. 1961. On bird species diversity. Ecology 42: 594–598.

Marcus, E., and E. Marcus. 1962. Studies on Columbellidae. Bol. Fac. Cienc. Letr. Univ. Sao Paulo 261: 335–402.

Margalef, R. 1958. Mode of evolution of species in relation to their place in ecological succession. XVth Int. Congr. Zool. Sect. 10, paper 17.

Paine, R. T. 1963. Trophic relationships of 8 sympatric predatory gastropods. Ecology 44: 63–73.

Paris, O. H. 1960. Some quantitative aspects of predation by muricid snails on mussels in Washington Sound. Veliger 2: 41–47.

Parker, R. H. 1963. Zoogeography and ecology of some macro-invertebrates, particularly mollusca in the Gulf of California and the continental slope off Mexico. Vidensk. Medd. Dansk. Natur. Foren., Copenh. 126: 1-178.

Rigg, G. B., and R. C. Miller. 1949. Intertidal plant and animal zonation in the vicinity of Neah Bay, Washington. Proc. Calif. Acad. Sci. 26: 323–351.

Roden, G. I., and G. W. Groves. 1959. Recent oceanographic investigations in the Gulf of California. J. Marine Res. 18: 10–35.

Simpson, G. G. 1964. Species density of North American recent mammals. Syst. Zool. 13: 57-73.

Slobodkin, L. B. 1961. Growth and regulation of Animal Populations. Holt, Rinehart, and Winston, New York.

______. 1964. Ecological populations of Hydrida. J. Anim. Ecol. 33 (Suppl.): 131-148.

Wells, H. W., M. J. Wells, and I. E. Gray. 1964. Ecology of sponges in Hatteras Harbor, North Carolina. Ecology 45: 752–767.

Williams, L. G. 1964. Possible relationships between plankton-diatom species numbers and water-quality estimates. Ecology 45: 809–823.

EXPERIMENTAL ZOOGEOGRAPHY OF ISLANDS: THE COLONIZATION OF EMPTY ISLANDS

DANIEL S. SIMBERLOFF[1] AND EDWARD O. WILSON
The Biological Laboratories, Harvard University, Cambridge, Massachusetts 02138

(Accepted for publication December 16, 1968)

Abstract. We report here the first evidence of faunistic equilibrium obtained through controlled, replicated experiments, together with an analysis of the immigration and extinction processes of animal species based on direct observations.

The colonization of six small mangrove islands in Florida Bay by terrestrial arthropods was monitored at frequent intervals for 1 year after removal of the original fauna by methyl bromide fumigation. Both the observed data and climatic considerations imply that seasonality had little effect upon the basic shape of the colonization curves of species present vs. time. By 250 days after defaunation, the faunas of all the islands except the most distant one ("E1") had regained species numbers and composition similar to those of untreated islands even though population densities were still abnormally low. Although early colonists included both weak and strong fliers, the former, particularly psocopterans, were usually the first to produce large populations. Among these same early invaders were the taxa displaying both the highest extinction rates and the greatest variability in species composition on the different islands. Ants, the ecological dominants of mangrove islands, were among the last to colonize, but they did so with the highest degree of predictability.

The colonization curves plus static observations on untreated islands indicate strongly that a dynamic equilibrium number of species exists for any island. We believe the curves are produced by colonization involving little if any interaction, then a gradual decline as interaction becomes important, and finally, a lasting dynamic equilibrium. Equations are given for the early immigration, extinction, and colonization curves.

Dispersal to these islands is predominantly through aerial transport, both active and passive. Extinction of the earliest colonists is probably caused chiefly by such physical factors as drowning or lack of suitable breeding sites and less commonly by competition and predation.

[1] Present address: Department of Biological Science, Florida State University, Tallahassee, Florida 32306

As population sizes increase it is expected that competition and predation will become more important. Observed turnover rates showed wide variance, with most values between 0.05 and 0.50 species/day. True turnover rates are probably much higher, with 0.67 species/day the extreme lower limit on any island. This very high value is at least roughly consistent with the turnover equation derived from the MacArthur-Wilson equilibrium model, which predicts turnover rates on the order of 0.1–1.0 species/day on the experimental islands.

Introduction

In the first article of this series (Wilson and Simberloff 1969) we showed how the recent formulation of mathematical biogeographic theory has both intensified the need for studies of the entire colonization process and defined the measurements required for such studies. The idea was conceived of approaching the problem experimentally by the removal of entire arthropod faunas from series of small islands. Six very small mangrove islands of the Florida Keys, each consisting of only one to several *Rhizophora mangle* trees standing in shallow water, were selected. Elimination of the faunas ("defaunation") was achieved through fumigation with methyl bromide; and techniques were worked out for censusing the arthropod species during the recolonization process.

In the present article we discuss the criteria we used for counting species and describe the recolonization process on all six islands, from the moment of defaunation to the reattainment of equilibrial numbers of species less than a year later.

Species Counts

For the species counts and discussion which follow these definitions will be used:

- Propagule: the minimum number of individuals of a species capable of breeding and population increase under ideal conditions for that species (unlimited food supply and proper habitat, no predators, etc.)
- Colonization: the existence of at least one propagule of a species on an island
- Extinction: the disappearance of a species from an island
- Invasion: the arrival of one or more propagules on an island
- Immigration: the arrival of a propagule on an island unoccupied by the species

The distinction between invasion and immigration should be noted. It is incorrect to speak of an immigration rate for one species, since an immigration rate for an island is in units of species/time. A species can have an invasion rate on a given island, however; this is simply the number of propagules of that species landing per unit time.

In analyzing species counts made at discrete intervals, as in our monitorings, every species for which at least one propagule exists is designated a colonist. It is also designated an immigrant if it was not a colonist at the preceding count. Every immigrant is also, by definition, a colonist. This definition says nothing about whether food and a breeding site exist; a species whose propagule lands on one of our islands is a colonist even if it is doomed to quick extinction for purely physical reasons (e.g., the absence of a suitable nest site in the *Rhizophora* for a given species of ant).

Because of the relative nearness of our experimental islands to source areas and, to a lesser extent, their small size and ecological simplicity, ambiguities concerning the state of colonization exist that would not arise if we were dealing with truly distant and larger islands. Except for a few birds, any animal species for which a propagule is recorded either avoids leaving the boundary of the island or (much more rarely) leaves and perishes in the sea. The island, therefore, is not simply an extension in some sense of the mainland.

In our experiment a small percentage of the animal species, less than 10% of all species sighted, behave as though the distances to the experimental islands are not qualitatively different from the same distances overland. We wish to discount these species in our calculations, unless insularity becomes important in particular instances. Two classes of species can be recognized in this connection. Several kinds of insects (cicadas, odonates, foraging bees and wasps) treat small mangrove islands as part of a fine-grained foraging area, traveling readily and frequently among several islands and adjacent shore regions. A species of the wasp genus *Polistes* forages regularly over small mangrove islands but rarely nests there. When nesting does occur the wasps apparently restrict their foraging largely to the nesting island. Only an extant nest qualifies the *Polistes* as a colonist. In similar cases actual breeding, rather than just the presence of sufficient animals to breed under the most favorable conditions, was employed as the criterion for colonization. Transient adult butterflies, particularly *Ascia monuste* and *Phoebis agarithe* (Pieridae), migrating over and beyond the experimental islands, occasionally skim briefly through but do not stop to breed. These will not be considered colonists for reasons similar to those used to discount fine-grained foragers.

A second difficulty associated with the nearness of our islands concerns intermittent breeding by strong-flying insects combined with continuous foraging by adults. Females of the moth *Automeris io* (Saturniidae) fly frequently onto small mangrove islands and occasionally deposit eggs. The life cycle from egg to adult of *Automeris* on *Rhizophora* is about 2½ months, and if (as may have happened on E3) adults breed on an island at intervals greater than that, extinction rates would appear to be high. For after a brood matures, the survivors generally all disperse from the island and no new adults may breed there for a period. This is obviously not extinction in its classical sense—it is not caused by competition for food or space, predation, climatic catastrophe, etc. —but it does accord with our strict definition given above. This situation was fortunately rare, almost entirely restricted to a few lepidopterans. Species of this type will be considered colonists, with one immigration only and no extinction, until a definitive and extended absence is recorded. The behavior of the very few nesting birds would place them in this category but they were nevertheless discounted in our analysis.

A few arthropods live in and among *Rhizophora* roots at or below the water level, some foraging on mud at low tide. These include the isopod *Ligia exotica*, an unidentified amphipod, and three insects: *Trochopus plumbeus* (Hemiptera: Veliidae), *Axelsonia littoralis* (Collembola: Isotomidae), and *Anurida maritima* (Collembola: Poduridae). These will be excluded from the species counts because they are essentially part of a surface marine community and apparently do not interact significantly with the arboreal mangrove fauna. All are ubiquitous around small mangrove islands and cannot be eradicated with certainty.

The impermanent mudbanks on E8 and E9 (Wilson and Simberloff 1969, Fig. 2) that remain wet but above water for several weeks in calm weather harbor a characteristic marine arthropod community, listed in Table 1. Most of the species are concentrated in washed-up and wet debris and algae. That all but the earwig *Labidura riparia* are virtually marine and do not breed on the islands is indicated first by their never having been collected on *Rhizophora* (even when the mud is submerged and debris washed away by wind-driven high tide), second by most species having been observed swimming from one patch of mud or debris to another, and finally by the swift recolonization by large populations of most species observed after extinction caused by extended flooding. *L. riparia* is the only species found on the islands proper, and it is probably the only one whose energetic interaction with the mangrove and its arboreal fauna is significant. All the others apparently feed on seaweed or washed-up detritus, or else prey upon those which do. Consequently *L. riparia* alone is considered a colonist.

TABLE 1. Arthropod community of intermittently submerged mudbanks on small mangrove islands

Taxon	Species
INSECTS	
Collembola	
Poduridae:	gen. sp.
Dermaptera	
Labiduridae:	*Labidura riparia*
Coleoptera	
Carabidae:	*Bembidion* sp. nr. *contractum*
	Tachys occulator
Corylophidae:	*Anisomeristes* sp.
Ptiliidae:	*Actinopteryx fucicola*
Staphylinidae:	gen. sp.
Hemiptera	
Saldidae:	*Pentacora sphacelata*
OTHER	
Acarina	
Veigaiaidae:	*Veigaia* sp.
Isopoda	
Oniscoidea:	*Ligia exotica*

The tree snail *Littorina angulifera* and tree crab *Aratus pisonii* inhabit all but the upper canopy of small mangrove islands but will not be counted for the following two reasons. Neither can be removed—*Littorina* is unaffected by 50 kg/1000 m^3 of methyl bromide for 3 hr, and *Aratus* simply drops to the water and may swim under the tent—and both have planktonic larvae. Again, the interaction of these species with the remainder of the arboreal community appears superficially not to be significant.

Our definition of a propagule dictates that animals with zero reproductive value (e.g., a male ant landing on an island after a nuptial flight) not be considered colonists.

Although birds were not counted here, bird parasites which establish breeding populations on the islands rather than wholly on the birds were listed as colonists. Specifically, the hippoboscid flies *Olfersia sordida* and *Lynchia albipennis* (whose puparia are commonly found in tree crevices) and the tick *Argas radiatus* (all stages of which live under dead mangrove bark) were counted.

Finally, except for the rare larvae and pupae, all Diptera were excluded. Monitoring of flies proved too difficult to warrant faith in the accuracy of species counts, and the extreme scarcity of immature stages indicates that small mangrove islands rarely support breeding dipteran populations.

Deep-boring beetles were deemed valid colonists

in this series. The uncertainty of extinction discussed in our first report (Wilson and Simberloff 1969) notwithstanding, the data imply that the few initially surviving cerambycid larvae were destroyed by a delayed effect, and that the weevils may have succumbed in the same way. In any event there are but four species involved, and rarely were more than two found on a single island.

Acceptable colonists were counted conservatively. In all instances of uncertainty about the number of species of a given taxon present, the minimum possible number is taken. For example, occasionally the records from one census revealed two thrips, *Neurothrips magnafemoralis* and *Liothrips* sp. (both Tubulifera), as well as larval thrips identifiable only to the suborder Tubulifera. Since the larvae could conceivably be ascribed to one of the two species known present, only two species are recorded for this period. Similarly the observations on E1 (May 29, 1967) of a moth caterpillar *Bema ?ydda* (Phycitidae), and a small adult moth similar to *Bema* but seen too briefly to be so recorded with certainty yield a species count of one only, since both individuals could belong to one species.

The presence of an immature animal need not imply breeding on an island; spiderlings balloon more readily than adults, and any insect larva could be blown or rafted to an island, although for some, of course, the probability of this is quite low. Similarly, an adult female does not constitute a propagule or part of one if she has not been fertilized, is not of a parthenogenetic species, and no male is present. Nevertheless, we assume here that an adult female, an adult of indeterminate sex, and an immature animal each imply the presence of a propagule. Adult males are not so counted.

In the Appendix are given the complete recorded histories of all of the colonists on E1, E2, E3, ST2, E7, and E9, the six islands whose entire faunas were removed by fumigation (see Wilson and Simberloff 1969). These records are based on direct observation in over 90% of the cases. In certain instances (hatched bars) animals were assumed present through one or more monitoring cycles when not actually observed. To ensure consistency in such interpolations the rules given in Table 2 were followed.

Table 2. General rules applied in interpolation of colonists

Animal	Number of cycles interpolated	Justification
Ants: workers seen first	2 (previous)	Conservative on physiological grounds
Ants: queen seen first	indefinite	Obvious for short periods. Data of E9 support for longer periods.
Deep-boring beetles	2+	Present in relatively low density and only 10% of twigs broken at each monitoring.
Small leaf dwellers	1-2	Inconspicuous, but densities usually increase rapidly.
Bark-dwellers	1	Often become dense quickly, and habitat examined completely.
Araneids	1	Spiderlings may be minute, but webs are conspicuous.
Tetragnatha	2	Position of webs makes less conspicuous than araneids.
Salticids	2-3	Furtive, often inconspicuous, and usually present in low densities.
Anyphaenids	1	Build up relatively high densities quickly.
Small crawlers	2	Usually low densities; difficult to record.
Fliers	2	Frequently conspicuous, but may be inactive because of weather.
Caterpillars	1 (usually)	No generalization possible; large ones are conspicuous.

Seasonality

We must first discuss whether any aspects of the colonization depicted in the Appendix (and Figs. 1–3) are artifacts of the particular season at which defaunation was performed. That is, if all the islands were fumigated in September instead of March would the tables and derived curves be qualitatively different?

This would obviously be so in much of the United States, where dispersal stages of most insects and spiders occur at short, distinct periods (usually in the summer). The Florida Keys are subtropical, however, with the mean temperature of the coldest month (20.9°C) only 7.7°C lower than that of the warmest month. Frost has never been recorded. The mean humidity of the driest month is but 8% lower than that of the most humid. Rainfall is less homogeneous, September–October averaging about 150 mm and December–January only 38 mm. The precise amount for all months is quite variable, however, and there are obviously no extreme dry or wet seasons. Wind is also relatively constant over the year in both speed and direction: it averages about 18 km/hr from the eastern quadrant. It is not surprising, therefore, that we found most of the mangrove arthropod species active throughout the year. All life stages, including dispersal forms, of many and probably most species of insects and spiders were present during every month. We can make no quantitative assertions, but it seemed to us that there were no striking seasonal decreases among the more abundant mangrove inhabitants in either population size or activity, including flight—except for mosquitoes, which were much more numerous in the summer.

We had the good fortune to be present February 26, 1967, when the temperature at Key West fell briefly to 9.5°C, the lowest reading in several years. Our surveys during that cold spell revealed no apparent mortality or even great lessening of activity of mangrove inhabitants other than a decrease in flight activity quite normal for the prevailing wind speed.

Finally, and most importantly, the conclusion that propagules were constantly hitting our islands is incontrovertible from the data summarized in the Appendix. Moreover, these data provide no clear indication of seasonality in the dispersal of any taxon. The colonization of E7, defaunated approximately half a year before the other islands, should certainly have manifested any strong seasonal component. Yet in both form and specifics (Fig. 2 and Appendix) it is consistent with the colonization of the other islands.

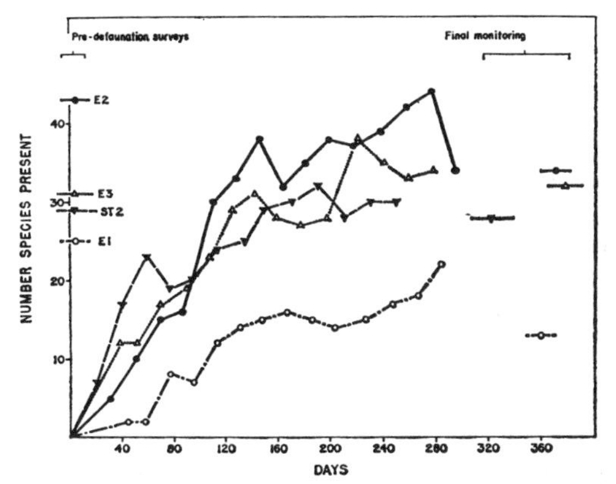

FIG. 1. The colonization curves of the experimental islands in series 1. In each curve the last and next-to-last census points are not connected by a line, only in order to stress the greater period of elapsed time compared with the times separating earlier censuses.

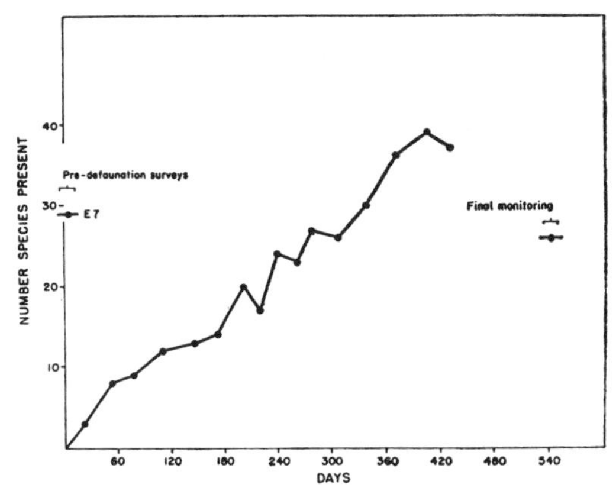

FIG. 2. The colonization curve of island E7.

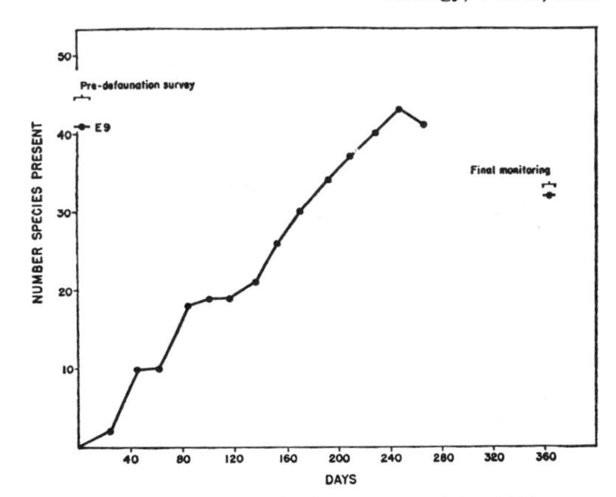

FIG. 3. The colonization curve of island E9.

PATTERNS OF COLONIZATION

As would be expected in a system involving but one plant, succession in the usual sense, a progression of discrete and relatively stable communities, did not occur. This does not imply a lack of order in the time course of colonization; indeed, the invasion of species was remarkably regular. But it was not accompanied by wholesale extinction of distinct animal associations.

Several broad patterns are nevertheless evident. First, although the earliest immigrants on all islands included both strong fliers (especially moths and wasps) and weak fliers or nonfliers (particularly psocopterans, chrysopids, and spiders), the latter built up large populations more rapidly and became numerically dominant. That wind should transport many of the early invaders is not surprising. The first animal recorded on Krakatau after its eruption was a spiderling (Cotteau 1885), while psocopterans (including wingless nymphs) and spiders are prominent in aerial plankton samples (Glick 1939). But their success in colonization deserves further comment. The food supply for psocopterans, algal and lichen growth on the mangrove itself, is evidently sufficient to allow much larger populations than one normally finds on undisturbed *Rhizophora* islands. This implies that on untreated islands there may be predation by animals not present on recently defaunated islands. In fact, we have observed the ants *Pseudomyrmex elongatus* and *Pseudomyrmex "flavidula"* (both of which colonized later) carrying apparently freshly killed psocopterans to their nests. Running and jumping spiders can, of course, eat the psocopterans. *Maevia vittata* (Salticidae) was seen catching a *Psocidus texanus* adult on E7, while the webs of even small araneid spiderlings commonly trapped transient flies, especially tipulids and ceratopogonids, in the same size range as psocopterans.

Although it involves only a relatively small subset of the entire Florida Keys fauna, colonization by these early weak fliers was more variable than by other classes of arthropods, both in time of arrival on individual islands and species composition at any given time among islands. The psocopterans were particularly unpredictable; from the 2d month after defaunation there were usually 1–4 species on each island at any census period, but over the course of the experiment a total of 24 species were involved. Although certain psocopterans, especially species of *Psocidus* and *Peripsocus,* were generally more prominent than others, there was little correlation among the sets of psocopterans found on different islands. Furthermore, as can be seen from the Corrodentia (= Psocoptera) sections of the Appendix, many psocopterans persisted for less than a month. As a group they invaded and multiplied readily, and became extinct almost as readily. On E7, only 5 species of the 15 colonists remained as long as 2 months.

Spiders as a group were less variable than psocopterans in their colonization pattern but much more so than that of most later colonists. Although the majority of the 36 spider species which colonized the islands followed the pattern just described for the psocopterans—that is, they immigrated readily and were extinguished quickly, and occurred on but 1 or 2 islands—a few species behaved quite differently. *Eustala* sp., *Tetragnatha* sp., *Leucauge venusta, Hentzia palmarum,* and *Aysha velox* in particular, colonized most of the islands and usually persisted for at least several months. These include most of the spiders found on the islands before defaunation.

Wasps present a similarly heterogeneous picture of colonization, many species appearing on one or two islands and vanishing rapidly while a few, notably *Pachodynerus nasidens, Scleroderma macrogaster,* and *Calliephialtes ferrugineus,* colonized many islands and persisted for long intervals. Mites did not invade as early as did wasps and spiders, but displayed the same pattern of many short-lived species, a few persisting and recurring. Most of the 20-odd species of Acarina were recorded from one or two islands only and disappeared within a month, while *Amblyseius* sp. and *Galumna* sp. were omnipresent and their populations long-lived. An apparent correlation exists in the spiders, mites, and wasps between mean length of persistence and number of islands colonized. This relation, however, may be artifactual, for the following reason. If two species, the first with a very high, the second with a very low initial probability of extinction (long expected persistence time), invaded all six islands with equal frequency, we would expect to see the former on one or two islands only and the latter on most or all of them.

Thrips, lepidopterans, orthopterans and ants display a regularity of colonization in sharp contradistinction to the relatively unordered patterns of colonization previously described. The last three groups, particularly the ants, mount the largest populations in undisturbed mangrove animal communities.

Only five species of thrips colonized, four of which were widespread among the islands. Almost all invasions occurred 4–5 months after defaunation. Most thysanopteran colonizations endured for at least 3 months. Large populations were occasionally produced but rarely persisted—the late colonizing ants may have attacked thrips. There was no consistent order of colonization: all species commonly immigrated about the same time.

Only 8 lepidopteran species colonized the experimental islands, of about 30 species known from mangrove and a few hundred from the general Keys fauna. (This figure does not count the 2 or 3 fine-grained foragers discussed earlier.) All eight were recorded more than once, and six were widespread. Most colonizations were sustained and many involved sizable populations. There was a somewhat predictable order of invasion, with *Phocides batabano, Bema ?ydda,* and *Ecdytolopha* sp. usually the first arrivals, *Nemapogon* sp. appearing somewhat later, and *Alarodia slossoniae* and *Automeris io* usually not seen until about 200 days.

The orthopteroids colonized still more predictably. Of approximately 25 species that occur in mangrove swamps and 60 or more that occur in the Keys as a whole, only 9 invaded the experimental islands. If the two very near islands (E2 and E7) are discounted, only four species were involved and all have multiple records. These four were rarely extinguished; two, the green tree cricket *Cyrtoxipha confusa* and roach *Latiblattella* n. sp., produced large populations. All appeared capable of early invasion.

The ants displayed the most orderly pattern of colonization. These insects are also numerically,

TABLE 3. Colonization of experimental islands by ants

Island	E1	E2	E3	E7	E9	ST2
Ant species before defaunation[a]	5 6 11	3 8 9 11 13 14	5 6 9 11 13	1 2 9 11 14	4 8 9 11 12 13	5 6 9 11 13 14
Ant colonists in order of colonization	6	{6, 14} 11 1	6 11 14 {8, 10} 9	6 16[b] 1 {10, 11, 14} 12	{12, 15[b]} {1, 6} 11 14	{6, 7[b], 11} 14 1 13 9

[a]Species are coded as follows:
1 = *Camponotus floridanus*
2 = *Camponotus planatus*
3 = *Camponotus tortuganus*
4 = *Camponotus* sp.
5 = *Camponotus* (*Colobopsis*) sp.
6 = *Crematogaster ashmeadi*
7 = *Crematogaster atkinsoni*
8 = *Monomorium floricola*
9 = *Paracryptocerus varians*
10 = *Paratrechina bourbonica*
11 = *Pseudomyrmex elongatus*
12 = *Pseudomyrmex "flavidula"*
13 = *Tapinoma littorale*
14 = *Xenomyrmex floridanus*
15 = *Brachymyrmex* sp.
16 = *Hypoponera opacior*
[b] indicates later extinction.

and probably energetically, the dominant animals on all small mangrove islands. Of the more than 50 species found in the Keys, 20 species inhabit red mangrove swamps and about 12 of these normally occur on small islands. The pre-defaunation surveys revealed a highly ordered fall-off of ant species in two directions. First, on islands of equal size but varying distance from source area, the most distant islands contain *Crematogaster ashmeadi;* those somewhat nearer, both the *Crematogaster* and *Pseudomyrmex elongatus;* those nearer still, these two species plus *Paracryptocerus varians, Tapinoma littorale,* and *Camponotus* (*Colobopsis*) sp.; and on islands near shore, most or all of the above species plus one or more species of *Camponotus, Pseudomyrmex "flavidula," Monomorium floricola,* and *Xenomyrmex floridanus.*

If instead one fixes a distance (usually small) from the source area and examines islands of increasing size, he generally finds on the smallest bush (ca. 1 m high) *Crematogaster ashmeadi;* on slightly larger bushes, *C. ashmeadi* and/or *Pseudomyrmex elongatus;* on small trees these two with perhaps two species drawn from among *Paracryptocerus, Tapinoma, Colobopsis, Xenomyrmex,* and *Monomorium;* and on islands the size of our experimental ones, the full complement expected on an island of the appropriate distance from source. In short, the ability to colonize increasingly smaller islands parallels closely the ability to colonize increasingly distant ones.

The order of colonization by ants of the experimental islands (Table 3) provides a curious zoogeographic analog of Haeckel's biogenetic law. In almost every instance *Crematogaster ashmeadi* was the first colonist, even on the two near islands (E2 and E7) where it was not present before defaunation. On all but the most distant island (E1) *Pseudomyrmex elongatus* was an early and prominent colonist. Moreover, the small subset of the mangrove ants which comprised the remainder of the colonists was almost identical to that found on small islands as one moves nearer to the source areas. Only 3 extinctions were observed, and 2 were of species believed unable to nest in red mangrove forests. As a group, the ants colonized later than most other taxa and at the close of the first phase of our study (April 1968) few species had built up populations numerically similar to those on untreated islands.

COLONIZATION CURVES

Figures 1–3 show the original numbers of species present and the colonization curves (number of species present vs. time) for the experimental islands.

Estimation of the number of species present before defaunation and after regular monitoring had ended is complicated by two factors. First, as can be seen from the data presented in the Appendix, a few colonists are inferred present at each regular post-defaunation monitoring period without actually having been observed because they were recorded at both preceding and subsequent periods. Since there was but one pre-defaunation survey, its total species number must be increased by the mean number of species inferred present without having been observed for all regular post-defaunation monitorings after an approximate equilibrium $\hat{S}$ had been reached. The last four regular monitorings were used for this purpose, since the population structure then was probably closest to that before defaunation. A similar correction must of course be made for the final census, after regular monitorings had ended. In addition to this correction, the pre-defaunation surveys of E1, E2, E3, E6, and E7 were believed deficient for leaf fauna. Since this habitat normally harbors 3 to 5 species on all islands, it was assumed that a total of 4 such species were present on each of these islands, though the actual number seen was 1 to 4. Figures 1–3 do not include a correction for unseen short-lived species, estimated at about 2 species per monitoring period but with very high variance (Simberloff 1969).

The numbers of species recorded on the control islands before and after the experiment were:

	before	after
E6	30	28*
E10	20	23

* plus 9 spp. in bird nests only

These cannot be corrected for unseen colonists, but show conclusively that no important long-term effect was operating in the Keys that might have distorted the results from the defaunated islands. Although the numbers of species on the control islands did not change significantly, the species composition varied considerably, implying that the number of species, S, approaches a dynamic equilibrium value, $\hat{S}$.

The most apparent implication of the colonization curves and other information presented so far is that this equilibrium $\hat{S}$ does exist. Three lines of evidence are relevant. First is the fact, just mentioned, that S on the control islands did not change greatly from the beginning to the end of the year-long period in which the nearby experimental islands were being colonized. Second is the static observation that untreated islands with similar area and distance from source have similar $\hat{S}$ (note E3 and ST2 before defaunation).

Perhaps the most convincing argument for an equilibrium $\hat{S}$, however, is the increase of species present on all our islands to approximately the same number as before defaunation, and then rough oscillation about this number. This $\hat{S}$ may be only a quasi-equilibrium—that is, the curve of S versus time after $\hat{S}$ is reached may not be truly stationary—because of two long-term processes. The first is that when an approximate $\hat{S}$ is first reached the population structure of the particular set of species on the island is still changing rapidly; some species are represented by few individuals and may have large r (intrinsic rate of increase). Others may have abnormally large populations. Population sizes are generally fluctuating much more rapidly than on untreated islands. We have already indicated one manifestation of this process in our experiment: psocopterans colonized early and built up immense populations before presumed ant predators appeared in number. The effect of this extreme population fluctuation on the colonization curve is that species may be eliminated and added at a rate systematically different from that on an untreated island. The concept of a dual dynamic equilibrium—of species number and population structure—will be mentioned briefly in this paper and described more completely and formally by Simberloff (1969).

The joint evolution of the particular constellation of species on an island ought logically to raise $\hat{S}$ systematically over very long periods of time (Wilson and Taylor 1967), but we will neglect this effect because the time course of such a change is obviously beyond that of this experiment. Also, the invasion rates for most species on our islands are probably too high to allow significant genetic alteration to occur in populations on individual islands; they are not isolated in an evolutionary sense.

The curves of Figs. 1–3, except for that of E1, are believed best explained by the following equation, based on a model devised by W. H. Bossert and P. N. Holland:

$$\mathbf{E}[S(t)] = \sum_{\alpha=1}^{P} \frac{i_\alpha}{i_\alpha + e_\alpha}\left(1 - e^{-(i_\alpha + e_\alpha)t}\right)$$

where $\mathbf{E}[S(t)]$ = the expected number of colonists present at time t

P = number of species in the pool

i_α = invasion rate of species α

e_α = intrinsic probability of extinction for species α

The derivation of this equation and a discussion of the concepts involved will be presented elsewhere (Simberloff 1969). The important aspects of this theory for our immediate purpose, however, are as follows:

The variance of $S(t)$ is high:

$$\mathbf{var}\,[S(t)] = \sum_{\alpha=1}^{P} \mathbf{E}[S_\alpha(t)]\,\{1 - \mathbf{E}\,[S_\alpha(t)]\}$$

S_α is a species indicator variable, which equals 1 when species α is present and 0 when species α is absent. Also,

$$\mathbf{E}[S_\alpha(t)] = \frac{i_\alpha}{i_\alpha + e_\alpha}\left(1 - e^{-(i_\alpha + e_\alpha)t}\right).$$

Once the number of species on these particular islands is between about 75% and 90% of the equilibrium value of

$$\mathbf{E}[S(t)] \simeq \sum_{\alpha=1}^{P} \frac{i_\alpha}{i_\alpha + e_\alpha} = \hat{S}$$

the major premise of this stochastic version of the MacArthur-Wilson equilibrium theory, namely non-interaction of species, would be invalidated. From this point onward $S(t)$ declines slightly, at a slow rate which cannot yet be predicted well stochastically. It equilibrates ultimately at an enduring $\hat{S}$ partly determined by interaction and close to the number that existed before defaunation. The decrease on all islands in number of species present for the final census (after regular monitoring had ended) is believed to be a manifestation of this decline.

E1 was so distant from its presumed source area that a few early invaders were able to build up large populations before their probable competitors arrived. Interactions thus became important before even a small fraction of the non-interaction $\hat{S}$ was achieved. We predict that on E1 the colonization curve will ascend slowly and irregularly to an equilibrium near the pre-defaunation $\hat{S}$. This enduring equilibrium will probably be the same one to which the colonization curve of E1 would ultimately have descended had interactions not become important until a large fraction of $\hat{S}$ had been achieved, as on the nearer islands.

As the distance of the island from the faunal source increases, the non-interaction $\hat{S}$ should decrease because of decreases in the i_α. Furthermore, the time necessary to reach any given percentage of the non-interaction $\hat{S}$, though not readily expressed mathematically, can be shown to increase with increasing distance from source. If this time is sufficiently long the few early colonists are able to produce large enough populations to interact significantly with later immigrants. This in fact is what happened on E1.

The colonization curve of E7 (Fig. 2) must be considered in light of the fact that 85% of the tree was killed by the fumigation (Wilson and Simberloff 1969). The dead portion did not disappear, but rather deteriorated until by the end of one year the wood was brown and dry and much of the bark was peeling. Thus the island was not a constant factor, and the relative proportions of the various microhabitats changed drastically until ultimately there was far more dead wood and bark and far less leafy canopy than on an untreated island. If a consistent measure of area existed for these islands it would probably have remained unchanged on E7, but the expected number of species would not because the different microhabitats normally support different numbers of species. In particular, dead bark shelters numerous species while mangrove leaves rarely support more than four species on a single island. Before its gradual decline, the colonization curve of E7 rose to a far higher percentage of the original $\hat{S}$ (135%) than did that of any other island. In addition, it was still rising a year after defaunation, when those of all other islands but E1 had leveled off. Arachnids (excluding orb-weavers) and psocopterans were the animals largely responsible for the higher S near the end of the experiment. Both groups are primarily bark dwellers.

All the above considerations imply that one of the determinants of the shape of the colonization curve on this island alone was variation of the island habitat, and that the enduring $\hat{S}$ which will ultimately be achieved on E7 may be very different from the pre-defaunation figure.

Dispersal

The agents of dispersal for specific immigrations can rarely be given with assurance, but the evidence implies that aerial transport, passive and active, is the major mode of invasion.

For the several parasites of vertebrates that breed on the islands proper, zoochorous transport is certain. These arthropods include the hippoboscid flies *Olfersia sordida* and *Lynchia albipennis* and tick *Argas radiatus* on E1 and E9. All three species parasitize the cormorants and pelicans which roost on these islands. A number of mite colonists could also have arrived on transient larger animals. This appears to be so for *Entonyssus* sp. on E2; it is an obligate parasite in snake lungs, and a *Natrix* was recorded sloughing on E2. The invasions of E1, E3, and E7 by the mite *Ornithonyssus bursa,* an avian parasite, were probably ornithochorous.

A more interesting possibility involves phoresy on birds, particularly by some of the psocopterans prominent among early immigrants. Mockford (1967) has hypothesized that phoresy may be a more efficient means of insular invasion than wind transport, especially for smaller species. Among four species found on Asian birds he lists *Ectopsocopsis cryptomeriae,* which has also colonized E3 and E7. Whether phoresy is a significant phenomenon in the Keys remains to be determined.

Other arthropods that may utilize phoresy are the pseudoscorpion *Tyrannochelifer* sp. found on E7, and free-living mites on all islands. These could have been transported by larger insects as well as by birds.

A related transport mechanism utilizes nesting material carried by birds. We suspect without proof that several mites in our experiment arrived in this fashion. Two beetle larvae were found in deep excavations in Green Heron nest twigs soon after the nests were built: *Chrysobothris tranquebarica* on E3 and ?*Sapintus fulvipes* on ST2 (the latter in a twig not of *Rhizophora*). It seems certain that both were in the twigs when the birds constructed their nests. Meyerriecks (1960) has observed green herons using twigs from other trees and from the nest tree itself. It is our impression that most nesting material that goes into new nests on small mangrove islands is brought from elsewhere. This was assuredly so for the non-mangrove twig on ST2.

Hydrochorous transport, either free or on rafts, was not as important as aerial transport in the invasion of our islands. Many mangrove colonists remain afloat almost indefinitely in salt water, but only the orthopteroids seem able to achieve significant independent, oriented motion. Furthermore, most floating insects and spiders of all sizes are rapidly devoured by fingerling fishes which are immensely numerous about all mangrove islands. Even actively swimming crickets and earwigs frequently meet this fate, attracting more attackers by the very vigor of their efforts.

Nevertheless, at least one invasion (though not immigration, since the species was already present) occurred this way, and there are other pertinent observations. An adult female *Cyrtoxipha* cricket was seen floating from Upper Snipe Key to E2, a distance of only about 2 m. She successfully climbed a root and disappeared into the lower canopy. That a winged individual fully capable of oriented flight should actively choose to disperse by water is dubious, but at least over this short a distance the method is feasible in case the animal accidentally lands in the water.

Four *Automeris io* caterpillars were seen to float from one end of E9 to the other, a distance of about 8 m. Three of them lodged on roots and eventually climbed up out of the water. The weakening of leaves attacked simultaneously by several large *Automeris* caterpillars (a common occurrence) might cause them to fall into water occasionally. Also, the caterpillars release their hold readily when the branch is shaken, apparently as a defensive maneuver. It seems doubtful, however, that a high proportion of those that land in the water could successfully travel between islands. Aside from fish attack, they would be plagued by their inability to direct their motion and to climb readily from the water.

Occasionally earwigs (*Labidura riparia*) were seen swimming from one root to another within an island. We do not know how well this species flies, but its swimming ability seems adequate for aquatic travel of considerable distance, and it readily climbs out of the water. Fishes would still be a hazard, however.

Our experiments resulted in no unequivocal evidence of invasions by rafting. In fact, several considerations imply that rafting must play a minor role in mangrove colonization. There is rarely land on which rafts can lodge, since *Rhizophora* islands generally have no supratidal ground. Drifting wood usually hits an island, gets trapped temporarily among roots, and eventually floats away. Even more importantly, green *Rhizophora* wood sinks immediately and many dead twigs also fail to float, thus precluding rafting by much of the wood-boring fraction of the fauna. Finally, except during hurricanes, there is very little floating debris in Florida Bay—far less than the amount found at river mouths.

The preceding considerations lead, by elimination, to the inference that aerial transport must be an important means of dispersal to our islands. Several observations of actively flying propagules support this hypothesis. Buprestid and cerambycid beetles, lepidopterans, wasps, and a lacewing have all been seen to land on the islands after flight from an outside source. Many other mangrove colonists are evidently carried passively by wind, especially psocopterans, thrips, neuropterans, and most spiders. Many of these more or less passively dispersed organisms are minute, and direct evidence on their invasion method is therefore scarce. Occasionally spiderlings and psocopterans were found in the air around our islands. In general, they are usually important components of aerial plankton (Glick 1939), and it is known from numerous anecdotal records (e.g., Bristowe 1958) that ballooning for the distances involved in our experiment is regularly achieved by some species of spiders. So far, the small sizes of most of these animals has made it impossible to follow a flight visually from source to island. One suggestive record, however, is that of a spider dragline stretching the 2 m between Upper Snipe Key and E2.

An attempt to correlate Figures 1–3 with hourly wind data of the U. S. Weather Bureau station at Key West was inconclusive, but this piece of negative evidence is hardly damaging to the thesis of dominance of aerial transport. For such data give only timed readings at one nearby point, while dispersal depends largely on specific gusts at odd times in highly circumscribed areas about the islands. Furthermore the published wind data are from a single fixed anemometer, while the entire wind profile would be necessary to describe dispersal.

Immigration and Extinction Rates

The intermittent nature of the monitoring had the unavoidable result that many immigrations and extinctions (perhaps two-thirds) were not observed, the species involved being obligate or near-obligate transients which immigrate and are extinguished all within one interval between two monitoring periods. For this reason absolute immigration and extinction curves cannot be derived from the observational data.

The observed immigration and extinction curves for all islands were highly variable, with no apparent pattern except for generally higher immigration rates on nearer islands during the first 150 days. Rates were usually between 0.05 and 0.50 species/day. Employing a statistical method devised by Simberloff (1969) it was determined that the very least the expected error in immigration and extinction rates could be is 0.67 species/day, and furthermore that the variance about the "expected" immigration and extinction curves is very high. The observed curves are consequently of limited interest, since they yield only an extreme lower limit for turnover rates and cannot be used to intuit shapes of the "expected" curves. On the other hand, the values of the turnover rates are of considerable interest even when they can be approximated only to the nearest order of magnitude. They are of course surprisingly high, in the vicinity of 1% of the equilibrial species number per day or higher. Yet this is at least roughly consistent with the MacArthur-Wilson (1967) model, which predicts that the turnover (= extinction) rate at equilibrium is 1.15 (mean $\hat{S}/t_{0.90}$, where mean $\hat{S}$ is the average equilibrial species number and $t_{0.90}$ is the time (in days) required to reach 90% of the equilibrial number. According to this formula, which is based on the simplest non-interactive version of the model, the turnover rates in our experimental islands should fall somewhere between 0.1 and 1.0 species/day. The relation between this version of the model and the more precise stochastic form of the model will be treated later at length by Simberloff (1969). The MacArthur-Wilson formulation is a special case of the many cases covered by the stochastic version and it has the advantage of permitting this first rough (and approximately correct) prediction of turnover rates.

In testing such predictions with measurements in the field there is reason to expect that the invasions not observed will occur by different means than those which are recorded. The assumption that most propagules arrived by air in the experimental keys is therefore probably valid. The evidence against a major seasonal component of dispersal was given earlier. In sum, we have a large body of information which implies that i_α (invasion rate of species α) is nearly constant through time for all α. What we lack at present is quantitative information on the sizes and distribution of the i_α.

Extinction rates, at least during most of the rise of the colonization curve from 0 to a large fraction of $\hat{S}$ are adequately represented by the unchanging, species-characteristic e_α, without an additional S-dependent or density-dependent factor included. The main arguments behind this assertion are:

i) Most of the species in the Florida Keys pool are obligate transients on these small mangrove islands. For a variety of reasons not directly related to their own densities or to other species they are doomed to swift extinction.
ii) Population sizes during most of the rise of S from 0 to near $\hat{S}$ are uniformly low.

The observed data from this experiment provide rough quantitative information on both the distribution and sizes of the e_α unlike with the i_α. Whatever the i_α and e_α, the expected curve of immigration rate vs. time is represented during the rise of S from 0 to a large fraction of the non-interaction $\hat{S}$ by:

$$\mathbf{E}[I(t)] = \sum_{\alpha=1}^{P} i_\alpha - \frac{i_\alpha^2}{i_\alpha + e_\alpha}\left(1 - e^{-(i_\alpha + e_\alpha)t}\right)$$

$$= \sum_{\alpha=1}^{P} i_\alpha - i_\alpha \mathbf{E}[S_\alpha(t)]$$

while the expected curve of extinction rate vs. time during the same period is:

$$\mathbf{E}[E(t)] = \sum_{\alpha=1}^{P} \frac{i_\alpha e_\alpha}{i_\alpha + e_\alpha}\left(1 - e^{-(i_\alpha + e_\alpha)t}\right)$$

$$= \sum_{\alpha=1}^{P} e_\alpha \mathbf{E}[S_\alpha(t)]$$

Beyond this point accurately predicted curves are impossible. It is nevertheless clear that during the slight decline of S to an enduring $\hat{S}$, $E(t)$ must be, on the average, slightly greater than $I(t)$, while after the equilibrium is reached the two must remain approximately equal. On any real island, of course, the two curves would cross and recross indefinitely. It also seems reasonable that after interactions become important, the $E(t)$ and $I(t)$ curves still do not change much, since the contribution from their common major component, the transients, does not change with time.

Whereas evidence on the specific agents of dispersal, and hence on immigration, has been plentiful during this study, observations on the causes of extinction have been meager. Obviously the probability of witnessing the death or disappearance of the last member of a population is exceedingly low. Some inferences can be drawn from observed means of population decrease.

Population decline should be most apparent when associated with interaction, especially predation, yet the small sizes of most populations during our experiment reduced interactions enormously. A few cases of predation have already been mentioned. Insectivorous birds, particularly warblers and red-winged blackbirds, were frequently observed eating numerous insects of many species. Wasps, both parasitic and nonparasitic, were seen destroying several insects and spiders. Some of these attacks may have led directly to extinction, when the prey populations were small. Examples include the parasitism on *Automeris io* caterpillars by *Apanteles hemileucae* (Braconidae) and the destruction of the salticids *Hentzia palmarum* and *Stoidis aurata* by *Trypoxylon collinum* (Sphecidae).

Exclusion can also provide indirect evidence of extinction through interaction. A possible instance is the apparent predation of crickets on E9 by a large population of the centipede *Orphnaeus brasilianus*. E8, with no centipedes, had immense populations of four cricket species. After defaunation of E9 removed *Orphnaeus,* a large population of the cricket *Cyrtoxipha confusa* was rapidly established.

From observation of the pre-defaunation distribution of ant species' numbers and population sizes on the various islands, it seems probable that when one ant species is able to build up large populations before other species invade, it can exclude one or more other species. However, direct evidence of aggressive behavior among the mangrove ants is lacking; the observed exclusion may have been the result of nest site pre-emption.

Most extinctions were probably not the result of interactions, at least during the initial rise of the colonization curves, but rather resulted from the inability of most species in the Florida Keys pool to colonize these tiny mangrove islands under any conditions. Lack of proper food or nest site and hostile physical conditions are probably common causes. An example was the observation of a dealate queen of *Brachymyrmex* sp. on E9. Since this species nests in soil, which is lacking on the experimental islands, it is not surprising that no workers were subsequently observed, despite the fact that the queen landed in a totally ant-free environment.

Even species which can survive on small mangrove islands have high probabilities of extinction not related to interaction. A number of animals have been found drowned during this experiment, including entire small ant colonies, numerous lepidopterous larvae, beetles, and psocopterans. It seems probable that during the hurricanes which periodically buffet these islands such deaths would be commonplace.

Similarly, numerous mangrove colonists inhabit hollow twigs and even under normal conditions many of these fall into the water. Certainly such events are multiplied during storms. The ant *Camponotus floridanus* and the oedemerid beetle *Oxycopis* sp. seem particularly vulnerable in this respect: they typically inhabit low, weakly anchored, hollow roots.

Acknowledgments

We wish to thank Joseph A. Beatty and Edward L. Mockford for identification of specimens and advice on experimental technique. Robert E. Silberglied performed the survey of Keys arthropods and aided in monitoring. S. Peck, H. Nelson, and R. B. Root participated in island censuses. John Ogden cultured many arthropod colonists. The following specialists gave generously of their time to identify animals encountered during the experiment: D. M. Anderson, E. W. Baker, J. R. Barron, J. C. Bequaert, L. Berner, B. D. Burks, J. A. Chemsak, K. A. Christiansen, W. J. Clench, C. M. Clifford, B. Conde, R. E. Crabill, P. J. Darlington, G. W. Dekle, H. Dybas, H. E. Evans, J. G. Franclemont, R. C. Froeschner, A. B. Gurney, J. L. Herring, W. Herrnkind, R. W. Hodges, R. L. Hoffman, D. G. Kissinger, J. P. Kramer, J. F. Lawrence, H. W. Levi, J. E. Lloyd, E. G. MacLeod, P. M. Marsh, R. Matthews, W. B. Muchmore, S. B. Mulaik, L. Pinter, C. C. Porter, E. S. Ross, L. M. Roth, L. M. Russell, R. L. Smiley, T. J. Spilman, L. J. Stannard, R. L. Usinger, B. D. Valentine, G. B. Vogt, T. J. Walker, L. M. Walkley, R. E. Warner, D. M. Weisman, F. G. Werner, R. E. White, S. L. Wood, R. E. Woodruff, P. W. Wygodzinsky. The work was supported by NSF grant GB-5867.

Literature Cited

Bristowe, W. S. 1958. The world of spiders. Collins, London. 308 p.

Cotteau, E. 1885. A week at Krakatau. Proc. Geogr. Soc. Australia, N.S.W. and Victorian branches, II.

Glick, P. A. 1939. The distribution of insects, spiders, and mites in the air. U.S.D.A. Tech. Bull. 673. 151 p.

MacArthur, R. H., and E. O. Wilson. 1967. The theory of island biogeography. Princeton University Press. 203 p.

Meyerriecks, A. J. 1960. Comparative breeding behavior of four species of North American herons. Publ. Nuttall Ornithological Club, No. 2. 158 p.

Mockford, E. L. 1967. Some Psocoptera from plumage of birds. Proc. Entomol. Soc. Wash. **69:** 307–309.

Simberloff, D. 1969. Experimental zoogeography of islands. III. A theory of insular colonization. Ecology (in press).

Wilson, E. O., and D. Simberloff. 1969. Experimental zoogeography of islands. Defaunation and monitoring techniques. Ecology **50:** 267–278.

Wilson, E. O., and R. W. Taylor. 1967. An estimate of the potential evolutionary increase in species density in the Polynesian ant fauna. Evolution **21:** 1–10.

APPENDIX. The colonists of six experimentally defaunated islands. Headings identify the island and give the number of days after defaunation for each census; "Pre" is the pre-defaunation census. Solid entries indicate that a species was seen; shaded, that it was inferred to be present from other evidence; open, that it was not seen and inferred to be absent.

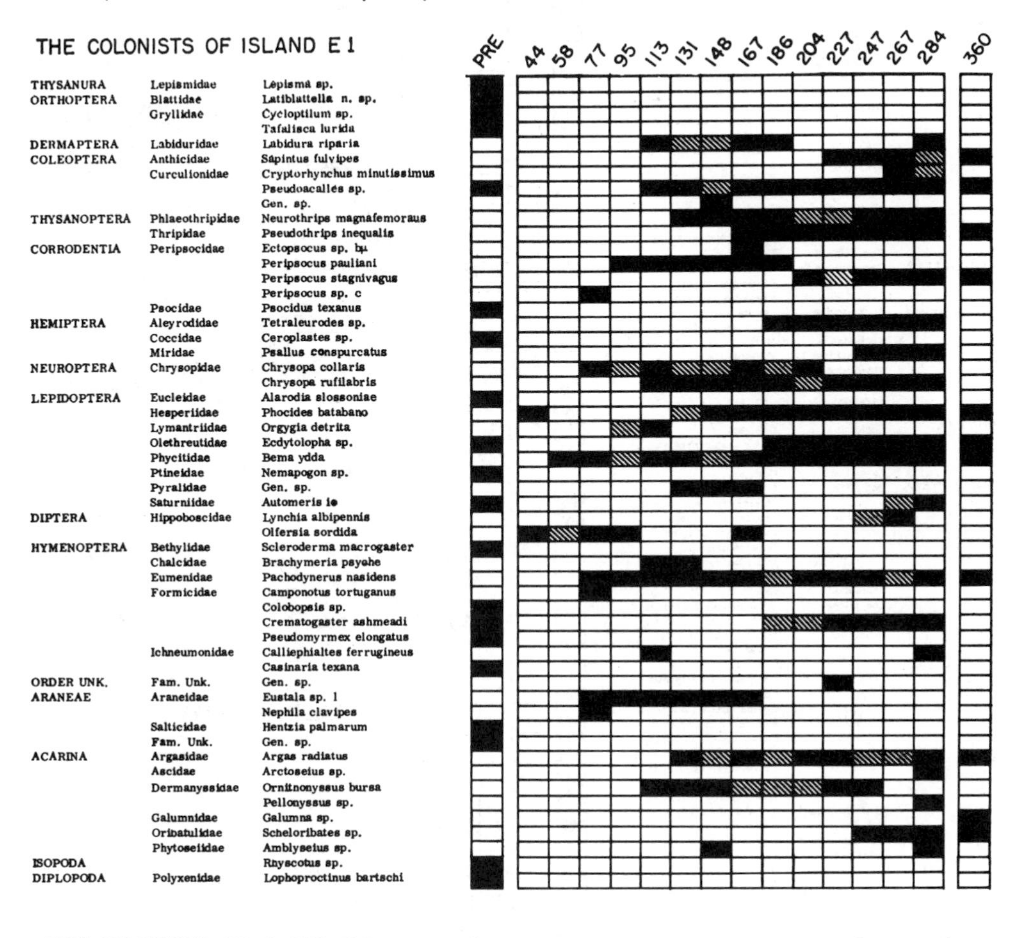

THE COLONISTS OF ISLAND E2

PRE 31 51 69 86 111 128 146 164 181 191 218 238 258 276 295 371

EMBIOPTERA	Teratembiidae	Gen. sp.
ORTHOPTERA	Blattidae	Aglaopteryx sp.
		Latiblattella n. sp.
		Latiblattella rehni
	Gryllidae	Cycloptilum sp.
		Cyrtoxipha sp.
		Orocharis gryllodes
		Tafalisca lurida
	Tettigoniidae	Turpilia rostrata
ISOPTERA	Kalotermitidae	Kalotermes jouteli
		Neotermes castaneus
COLEOPTERA	Anobiidae	Tricorynus sp.
	Anthicidae	Sapintus fulvipes
	Buprestidae	Actenodes auronotata
		Chrysobothris sexfasciatus
		Chrysobothris tranquebarica
	Cerambycidae	Ataxia sp.
		Styloleptus biustus
	Chrysomelidae	Gen. sp.
	Cucujidae	Gen. sp.
	Curculionidae	Cryptorhynchus minutissimus
		Pseudoacalles sp.
	Lampyridae	Micronaspis floridana
	Oedemeridae	Copidita suturalis
		Oxacis sp.
		Oxycopis sp.
	Scolytidae	Poecilips rhizophorae
		Trischidias atoma
	Fam. Unk.	Gen. sp.

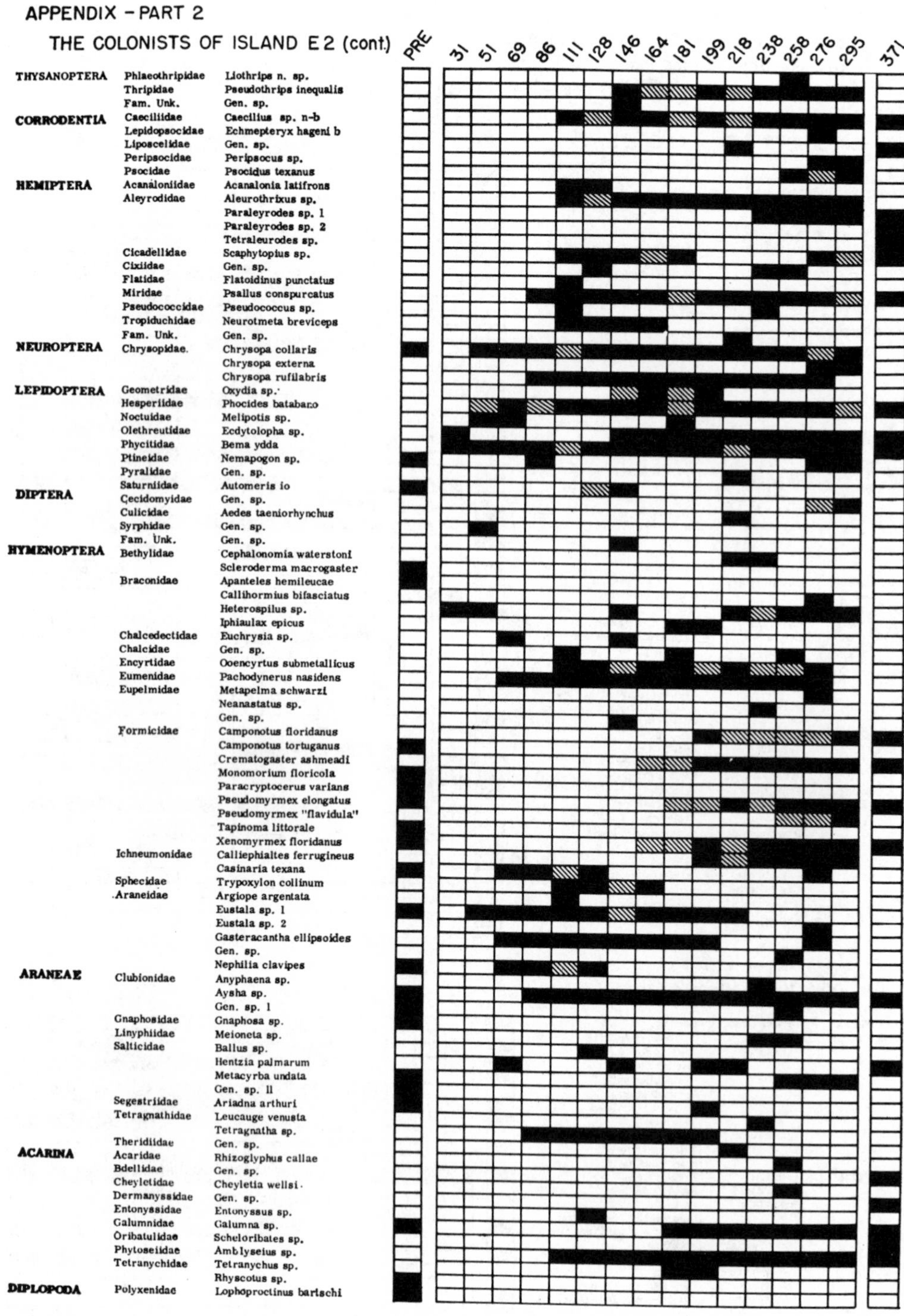
APPENDIX - PART 2
THE COLONISTS OF ISLAND E2 (cont.)
PRE
31 51 69 86 111 128 146 164 181 199 218 238 258 276 295
371
THYSANOPTERA Phlaeothripidae Liothrips n. sp.
Thripidae Pseudothrips inequalis
Fam. Unk. Gen. sp.
CORRODENTIA Caeciliidae Caecilius sp. n-b
Lepidopsocidae Echmepteryx hageni b
Liposcelidae Gen. sp.
Peripsocidae Peripsocus sp.
Psocidae Psocidus texanus
HEMIPTERA Acanaloniidae Acanalonia latifrons
Aleyrodidae Aleurothrixus sp.
Paraleyrodes sp. 1
Paraleyrodes sp. 2
Tetraleurodes sp.
Cicadellidae Scaphytopius sp.
Cixiidae Gen. sp.
Flatidae Flatoidinus punctatus
Miridae Psallus conspurcatus
Pseudococcidae Pseudococcus sp.
Tropiduchidae Neurotmeta breviceps
Fam. Unk. Gen. sp.
NEUROPTERA Chrysopidae Chrysopa collaris
Chrysopa externa
Chrysopa rufilabris
LEPIDOPTERA Geometridae Oxydia sp.
Hesperiidae Phocides batabano
Noctuidae Melipotis sp.
Olethreutidae Ecdytolopha sp.
Phycitidae Bema ydda
Ptineidae Nemapogon sp.
Pyralidae Gen. sp.
Saturniidae Automeris io
DIPTERA Cecidomyidae Gen. sp.
Culicidae Aedes taeniorhynchus
Syrphidae Gen. sp.
Fam. Unk. Gen. sp.
HYMENOPTERA Bethylidae Cephalonomia waterstoni
Scleroderma macrogaster
Braconidae Apanteles hemileucae
Callihormius bifasciatus
Heterospilus sp.
Iphiaulax epicus
Chalcedectidae Euchrysia sp.
Chalcidae Gen. sp.
Encyrtidae Ooencyrtus submetallicus
Eumenidae Pachodynerus nasidens
Eupelmidae Metapelma schwarzi
Neanastatus sp.
Gen. sp.
Formicidae Camponotus floridanus
Camponotus tortuganus
Crematogaster ashmeadi
Monomorium floricola
Paracryptocerus varians
Pseudomyrmex elongatus
Pseudomyrmex "flavidula"
Tapinoma littorale
Xenomyrmex floridanus
Ichneumonidae Calliephialtes ferrugineus
Casinaria texana
Sphecidae Trypoxylon collinum
Araneidae Argiope argentata
Eustala sp. 1
Eustala sp. 2
Gasteracantha ellipsoides
Gen. sp.
Nephilia clavipes
ARANEAE Clubionidae Anyphaena sp.
Aysha sp.
Gen. sp. 1
Gnaphosidae Gnaphosa sp.
Linyphiidae Meioneta sp.
Salticidae Ballus sp.
Hentzia palmarum
Metacyrba undata
Gen. sp. II
Segestriidae Ariadna arthuri
Tetragnathidae Leucauge venusta
Tetragnatha sp.
Theridiidae Gen. sp.
ACARINA Acaridae Rhizoglyphus callae
Bdellidae Gen. sp.
Cheyletidae Cheyletia wellsi
Dermanyssidae Gen. sp.
Entonyssidae Entonyssus sp.
Galumnidae Galumna sp.
Oribatulidae Scheloribates sp.
Phytoseiidae Amblyseius sp.
Tetranychidae Tetranychus sp.
Rhyscotus sp.
DIPLOPODA Polyxenidae Lophoproctinus bartschi

APPENDIX – PART 3

THE COLONISTS OF ISLAND E3

Census columns: PRE, 38, 52, 69, 89, 107, 125, 142, 159, 177, 197, 221, 241, 259, 277, 379

Order	Family	Species
COLLEMBOLA	Poduridae	Gen. sp.
EMBIOPTERA	Teratembiidae	Diradius caribbeana
ORTHOPTERA	Blattidae	Latiblattella n. sp.
	Gryllidae	Cycloptilum spectabile
		Tafalisca lurida
ISOPTERA	Kalotermitidae	Kalotermes jouteli
COLEOPTERA	Anobiidae	Tricorynus sp.
	Anthicidae	Sapintus fulvipes
	Buprestidae	Actenodes auronotata
		Chrysobothris tranquebarica
	Cerambycidae	Styloleptus biustus
	Curculionidae	Cryptorhynchus minutissimus
		Pseudoacalles sp.
	Lathridiidae	Melanophthalma floridana
	Oedemeridae	Oxacis sp.
		Oxycopis sp.
	Scolytidae	Trischidias atoma
THYSANOPTERA	Phlaeothripidae	Haplothrips flavipes
		Liothrips n. sp.
		Neurothrips magnafemoralis
	Thripidae	Pseudothrips inequalis
	Fam. Unk.	Gen. sp.
CORRODENTIA	Liposcelidae	Liposcelis bostrychophilus
	Peripsocidae	Ectopsocopsis cryptomeriae
		Ectopsocus sp. c
		Peripsocus pauliani
	Psocidae	Psocidus texanus
HEMIPTERA	Aleyrodidae	Tetraleurodes sp.
	Anthocoridae	Dufouriellus afer
		Orius sp.
	Cicadellidae	Scaphytopius sp.
	Lygaeidae	Gen. sp. 2
	Membracidae	Gen. sp.
	Miridae	Psallus conspurcatus
	Nabidae	Carthasis decoratus
NEUROPTERA	Chrysopidae	Chrysopa collaris
		Chrysopa externa
		Chrysopa rufilabris
LEPIDOPTERA	Eucleidae	Alarodia slossoniae
	Geometridae	Oxydia sp.
	Hesperiidae	Phocides batabano
	Olethreutidae	Ecdytolopha sp.
	Phycitidae	Bema ydda
	Ptineidae	Nemapogon sp.
	Saturniidae	Automeris io
HYMENOPTERA	Bethylidae	Nesepyris floridanus
		Scleroderma macrogaster
		Gen. sp.
	Braconidae	Heterospilus sp.
		Iphiaulax epicus
		Macrocentrus sp.
	Chalcidae	Brachymeria psyche
		Gen. sp.
	Encyrtidae	Gen. sp.
	Eulophidae	Entedontini sp.
		Melittobia chalybii
	Eumenidae	Pachodynerus nasidens
	Eupelmidae	Gen. sp. 1
		Gen. sp. 2
	Formicidae	Colobopsis sp.
		Crematogaster ashmeadi
		Monomorium floricola
		Paracryptocerus varians
		Paratrechina bourbonica
		Pseudomyrmex elongatus
		Tapinoma littorale
		Xenomyrmex floridanus
	Ichneumonidae	Calliephialtes ferrugineus
		Casinaria texana
	Scelionidae	Probaryconus sp.
		Telonemus sp.
	Sphecidae	Trypoxylon collinum
	Vespidae	Gen. sp.
ARANEAE	Araneidae	Eustala sp. 1
		Gasteracantha ellipsoides
	Clubionidae	Anyphaena sp.
		Aysha sp.
	Gnaphosidae	Sergiolus sp.
	Salticidae	Hentzia palmarum
	Segestriidae	Ariadna arthuri
	Tetragnathidae	Leucauge venusta
		Tetragnatha antillana
	Theridiidae	Gen. sp.
ACARINA	Acaridae	Tyrophagus putrescentiae
	Ascidae	Asca sp.
	Dermanyssidae	Gen. sp.
	Eupodidae	Eupodes sp. nr. fusifer
	Galumnidae	Galumna sp.
	Oribatulidae	Scheloribates sp.
	Phytoseiidae	Amblyseius sp.
DIPLOPODA	Polyxenidae	Lophoproctinus bartschi
ISOPODA		Rhyscotus sp.

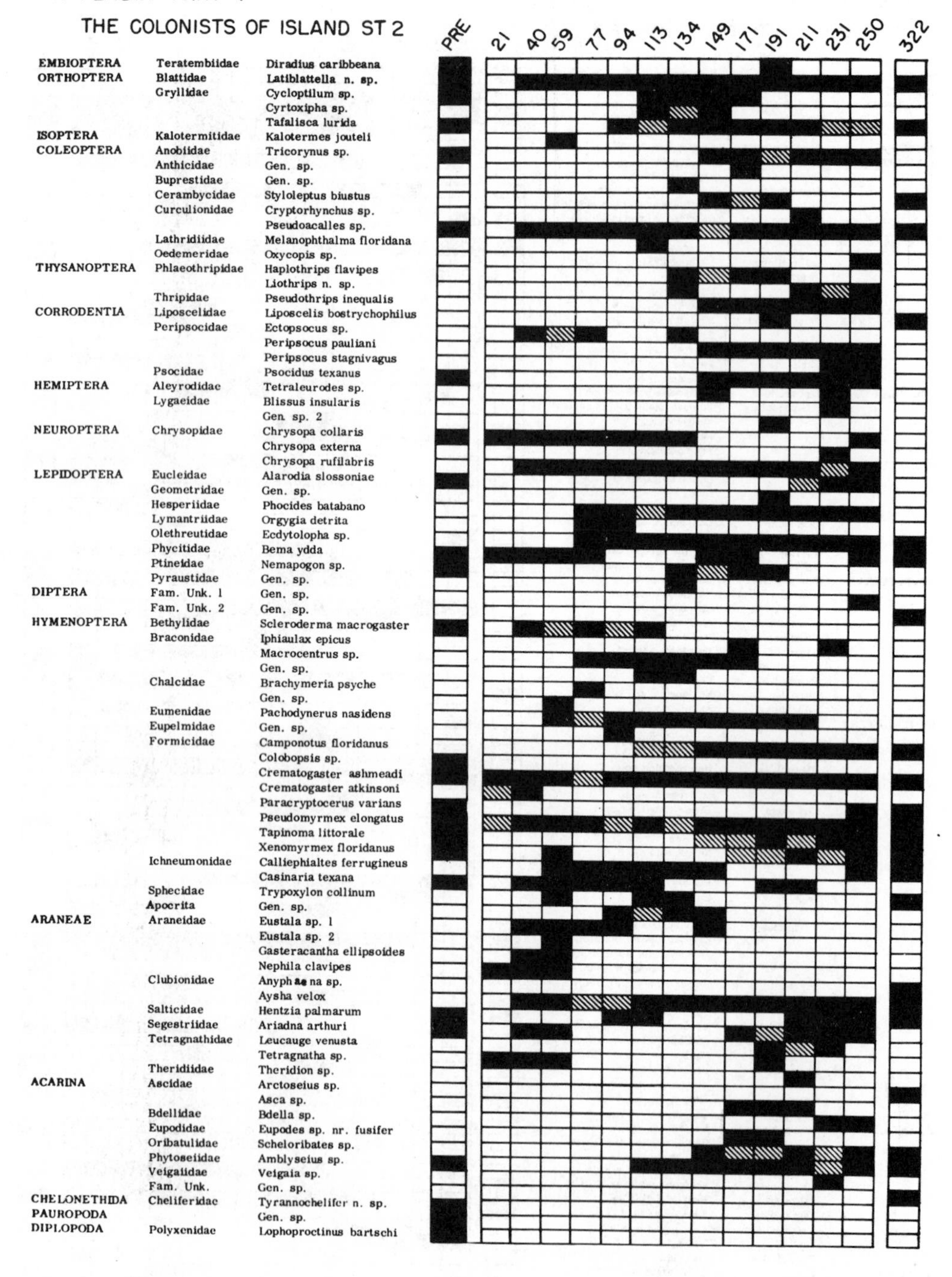
APPENDIX - PART 4
THE COLONISTS OF ISLAND ST 2
PRE
21
40
59
77
94
113
134
149
171
191
211
231
250
322
EMBIOPTERA Teratembiidae Diradius caribbeana
ORTHOPTERA Blattidae Latiblattella n. sp.
Gryllidae Cycloptilum sp.
Cyrtoxipha sp.
Tafalisca lurida
ISOPTERA Kalotermitidae Kalotermes jouteli
COLEOPTERA Anobiidae Tricorynus sp.
Anthicidae Gen. sp.
Buprestidae Gen. sp.
Cerambycidae Styloleptus biustus
Curculionidae Cryptorhynchus sp.
Pseudoacalles sp.
Lathridiidae Melanophthalma floridana
Oedemeridae Oxycopis sp.
THYSANOPTERA Phlaeothripidae Haplothrips flavipes
Liothrips n. sp.
Thripidae Pseudothrips inequalis
CORRODENTIA Liposcelidae Liposcelis bostrychophilus
Peripsocidae Ectopsocus sp.
Peripsocus pauliani
Peripsocus stagnivagus
Psocidae Psocidus texanus
HEMIPTERA Aleyrodidae Tetraleurodes sp.
Lygaeidae Blissus insularis
Gen. sp. 2
NEUROPTERA Chrysopidae Chrysopa collaris
Chrysopa externa
Chrysopa rufilabris
LEPIDOPTERA Eucleidae Alarodia slossoniae
Geometridae Gen. sp.
Hesperiidae Phocides batabano
Lymantriidae Orgygia detrita
Olethreutidae Ecdytolopha sp.
Phycitidae Bema ydda
Ptineidae Nemapogon sp.
Pyraustidae Gen. sp.
DIPTERA Fam. Unk. 1 Gen. sp.
Fam. Unk. 2 Gen. sp.
HYMENOPTERA Bethylidae Scleroderma macrogaster
Braconidae Iphiaulax epicus
Macrocentrus sp.
Gen. sp.
Chalcidae Brachymeria psyche
Gen. sp.
Eumenidae Pachodynerus nasidens
Eupelmidae Gen. sp.
Formicidae Camponotus floridanus
Colobopsis sp.
Crematogaster ashmeadi
Crematogaster atkinsoni
Paracryptocerus varians
Pseudomyrmex elongatus
Tapinoma littorale
Xenomyrmex floridanus
Ichneumonidae Calliephialtes ferrugineus
Casinaria texana
Sphecidae Trypoxylon collinum
Apocrita Gen. sp.
ARANEAE Araneidae Eustala sp. 1
Eustala sp. 2
Gasteracantha ellipsoides
Nephila clavipes
Clubionidae Anyphaena sp.
Aysha velox
Salticidae Hentzia palmarum
Segestriidae Ariadna arthuri
Tetragnathidae Leucauge venusta
Tetragnatha sp.
Theridiidae Theridion sp.
ACARINA Ascidae Arctoseius sp.
Asca sp.
Bdellidae Bdella sp.
Eupodidae Eupodes sp. nr. fusifer
Oribatulidae Scheloribates sp.
Phytoseiidae Amblyseius sp.
Veigaiidae Veigaia sp.
Fam. Unk. Gen. sp.
CHELONETHIDA Cheliferidae Tyrannochelifer n. sp.
PAUROPODA Gen. sp.
DIPLOPODA Polyxenidae Lophoproctinus bartschi

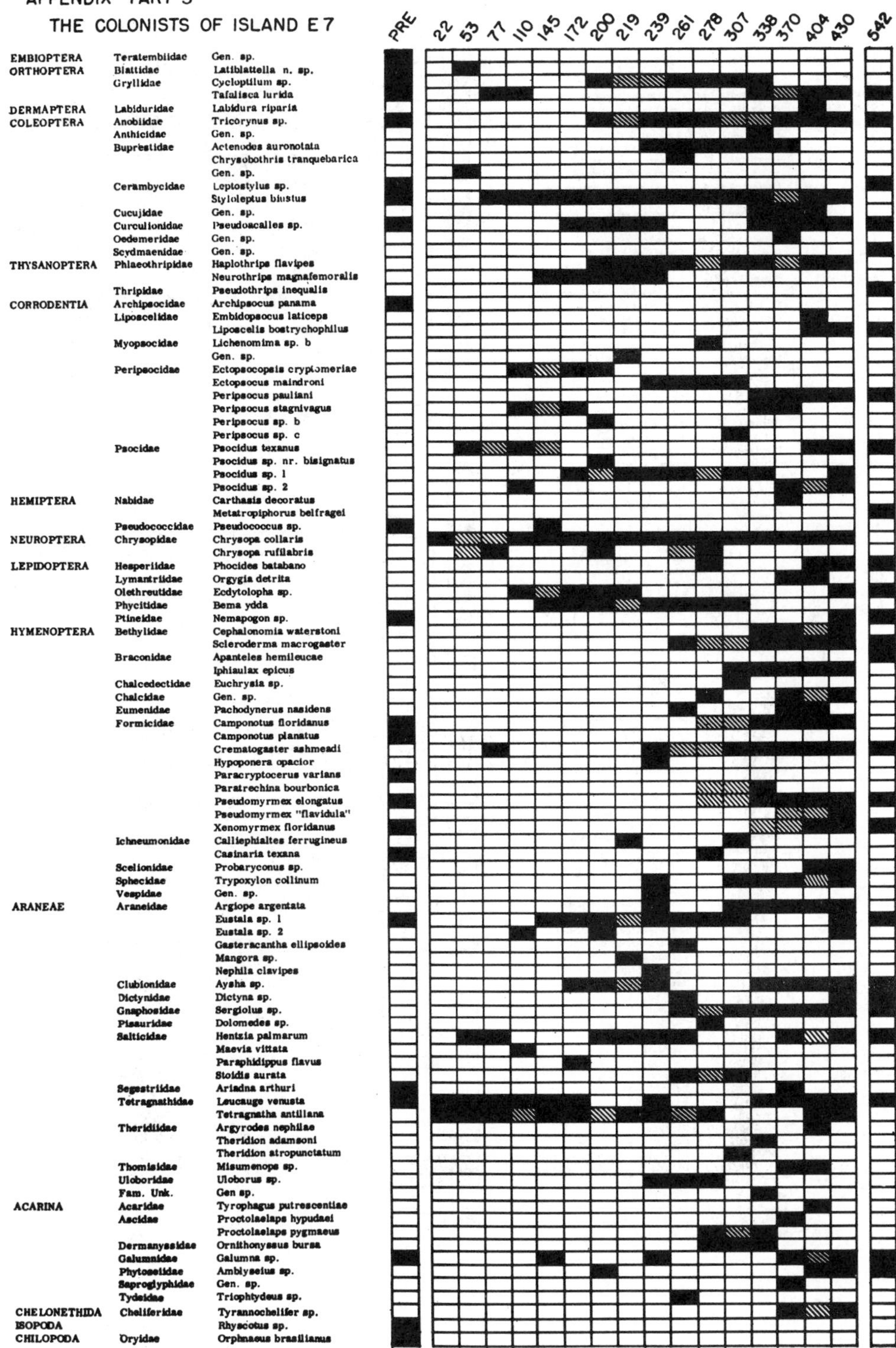
APPENDIX – PART 5
THE COLONISTS OF ISLAND E7
PRE
22
53
77
110
145
172
200
219
239
261
278
307
338
370
404
430
542
EMBIOPTERA Teratembiidae Gen. sp.
ORTHOPTERA Blattidae Latiblattella n. sp.
Gryllidae Cycloptilum sp.
Tafalisca lurida
DERMAPTERA Labiduridae Labidura riparia
COLEOPTERA Anobiidae Tricorynus sp.
Anthicidae Gen. sp.
Buprestidae Actenodes auronotata
Chrysobothris tranquebarica
Gen. sp.
Cerambycidae Leptostylus sp.
Styloleptus biustus
Cucujidae Gen. sp.
Curculionidae Pseudoacalles sp.
Oedemeridae Gen. sp.
Scydmaenidae Gen. sp.
THYSANOPTERA Phlaeothripidae Haplothrips flavipes
Neurothrips magnafemoralis
Thripidae Pseudothrips inequalis
CORRODENTIA Archipsocidae Archipsocus panama
Liposcelidae Embidopsocus laticeps
Liposcelis bostrychophilus
Myopsocidae Lichenomima sp. b
Gen. sp.
Peripsocidae Ectopsocopsis cryptomeriae
Ectopsocus maindroni
Peripsocus pauliani
Peripsocus stagnivagus
Peripsocus sp. b
Peripsocus sp. c
Psocidae Psocidus texanus
Psocidus sp. nr. bisignatus
Psocidus sp. 1
Psocidus sp. 2
HEMIPTERA Nabidae Carthasis decoratus
Metatropiphorus belfragei
Pseudococcidae Pseudococcus sp.
NEUROPTERA Chrysopidae Chrysopa collaris
Chrysopa rufilabris
LEPIDOPTERA Hesperiidae Phocides batabano
Lymantriidae Orgyia detrita
Olethreutidae Ecdytolopha sp.
Phycitidae Bema ydda
Ptineidae Nemapogon sp.
HYMENOPTERA Bethylidae Cephalonomia waterstoni
Scleroderma macrogaster
Braconidae Apanteles hemileucae
Iphiaulax epicus
Chalcedectidae Euchrysia sp.
Chalcidae Gen. sp.
Eumenidae Pachodynerus nasidens
Formicidae Camponotus floridanus
Camponotus planatus
Crematogaster ashmeadi
Hypoponera opacior
Paracryptocerus varians
Paratrechina bourbonica
Pseudomyrmex elongatus
Pseudomyrmex "flavidula"
Xenomyrmex floridanus
Ichneumonidae Calliephialtes ferrugineus
Casinaria texana
Scelionidae Probaryconus sp.
Sphecidae Trypoxylon collinum
Vespidae Gen. sp.
ARANEAE Araneidae Argiope argentata
Eustala sp. 1
Eustala sp. 2
Gasteracantha ellipsoides
Mangora sp.
Nephila clavipes
Clubionidae Aysha sp.
Dictynidae Dictyna sp.
Gnaphosidae Sergiolus sp.
Pisauridae Dolomedes sp.
Salticidae Hentzia palmarum
Maevia vittata
Paraphidippus flavus
Stoidis aurata
Segestriidae Ariadna arthuri
Tetragnathidae Leucauge venusta
Tetragnatha antillana
Theridiidae Argyrodes nephilae
Theridion adamsoni
Theridion atropunctatum
Thomisidae Misumenops sp.
Uloboridae Uloborus sp.
Fam. Unk. Gen sp.
ACARINA Acaridae Tyrophagus putrescentiae
Ascidae Proctolaelaps hypudaei
Proctolaelaps pygmaeus
Dermanyssidae Ornithonyssus bursa
Galumnidae Galumna sp.
Phytoseiidae Amblyseius sp.
Saproglyphidae Gen. sp.
Tydeidae Triophtydeus sp.
CHELONETHIDA Cheliferidae Tyrannochelifer sp.
ISOPODA Rhyscotus sp.
CHILOPODA Oryidae Orphnaeus brasilianus

APPENDIX - PART 6

THE COLONISTS OF ISLAND E9

PRE 24 45 62 84 101 117 136 153 171 193 210 229 247 266 364

Order	Family	Species
ORTHOPTERA	Gryllidae	Cycloptilum sp.
		Cyrtoxipha confusa
		Orocharis sp.
DERMAPTERA	Labiduridae	Labidura riparia
COLEOPTERA	Anobiidae	Cryptorama minutum
		Tricorynus sp.
	Anthicidae	Sapintus fulvipes
		Vacusus vicinus
	Buprestidae	Actenodes auronotata
		Chrysobothris tranquebarica
	Cantharidae	Chauliognathus marginatus
	Cerambycidae	Styloleptus biustus
	Curculionidae	Cryptorhynchus minutissimus
		Pseudoacalles sp.
	Lathridiidae	Holoparamecus sp.
	Oedemeridae	Oxacis sp.
	Fam. Unk.	Gen. sp.
THYSANOPTERA	Phlaeothripidae	Haplothrips flavipes
		Neurothrips magnafemoralis
	Thripidae	Pseudothrips inequalis
CORRODENTIA	Caeciliidae	Caecilius sp. np.
	Lachesillidae	Lachesilla n. sp.
	Lepidopsocidae	Echmepteryx hageni b
	Liposcelidae	Belaphotroctes okalensis
		Embidopsocus laticeps
		Liposcelis sp. not bostrychophilus
	Peripsocidae	Ectopsocus sp. bμ
		Peripsocus stagnivagus
	Psocidae	Psocidus texanus
		Psocidus sp. 1
	Trogiomorpha	Gen. sp.
HEMIPTERA	Anthocoridae	Dufouriellus afer
	Cixiidae	Oliarus sp.
	Miridae	Psallus conspurcatus
	Pentatomidae	Oebalus pugnax
	Fam. Unk.	Gen. sp.
NEUROPTERA	Chrysopidae	Chrysopa collaris
		Chrysopa externa
		Chrysopa rufilabris
LEPIDOPTERA	Eucleidae	Alarodia slossoniae
	Olethreutidae	Ecdytolopha sp.
	Phycitidae	Bema ydda
	Psychidae	Oiketicus abbottii
	Ptineidae	Nemapogon sp.
	Pyralidae	Tholeria reversalis
	Saturniidae	Automeris io
	Fam. Unk.	Gen. sp.
DIPTERA	Hippoboscidae	Olfersia sordida
	Fam. Unk.	Gen. sp.
HYMENOPTERA	Braconidae	Apanteles hemileucae
		Apanteles marginiventris
		Callihormius bifasciatus
		Ecphylus n. sp. nr. chramesi
		Iphiaulax epicus
	Chalcidae	Gen. sp. 1
		Gen. sp. 2
		Gen. sp. 3
		Gen. sp. 4
	Eulophidae	Euderus sp.
	Eumenidae	Pachodynerus nasidens
	Eupelmidae	Gen. sp.
	Formicidae	Brachymyrmex sp.
		Camponotus floridanus
		Camponotus sp.
		Crematogaster ashmeadi
		Monomorium floricola
		Paracryptocerus varians
		Pseudomyrmex elongatus
		Pseudomyrmex "flavidula"
		Tapinoma littorale
		Xenomyrmex floridanus
		Gen. sp.
	Ichneumonidae	Calliephialtes ferrugineus
		Casinaria texana
	Pteromalidae	Urolepis rufipes
	Sphecidae	Trypoxylon collinum
	Vespidae	Polistes sp.
ARANEAE	Araneidae	Argiope argentata
		Eriophora sp.
		Eustala sp.
		Gasteracantha ellipsoides
		Metepeira labyrinthea
		Nephila clavipes
	Clubionidae	Aysha sp.
	Dictynidae	Dictyna sp.
	Gnaphosidae	Sergiolus sp.
	Linyphiidae	Meioneta sp.
	Lycosidae	Pirata sp.
	Salticidae	Hentzia palmarum
	Scytodidae	Scytodes sp.

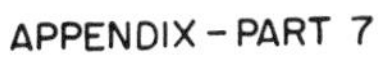

THE COLONISTS OF ISLAND E9 (cont.)

Order	Family	Species
	Tetragnathidae	Leucauge venusta
		Tetragnatha antillana
		Tetragnatha sp. 2
	Theridiidae	Gen. sp.
	Uloboridae	Uloborus sp.
ACARINA	Argasidae	Argas radiatus
	Ascidae	Lasioseius sp.
		Melichares sp.
	Bdellidae	Bdella sp.
	Carpoglyphidae	Carpoglyphus lactis
	Erythraeidae	Sphaerolophus sp.
	Galumnidae	Galumna sp.
	Phytoseiidae	Amblyseius sp.
ISOPODA		Rhyscotus sp.
CHILOPODA	Oryidae	Orphnaeus brasilianus

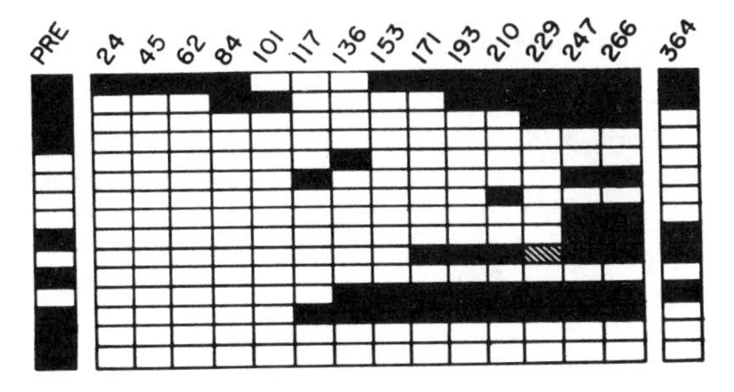

EFFECTS OF FOREST CUTTING AND HERBICIDE TREATMENT ON NUTRIENT BUDGETS IN THE HUBBARD BROOK WATERSHED-ECOSYSTEM[1]

GENE E. LIKENS[2]

Department of Biological Sciences
Dartmouth College
Hanover, New Hampshire

F. HERBERT BORMANN

School of Forestry
Yale University
New Haven, Connecticut

NOYE M. JOHNSON

Department of Earth Sciences
Dartmouth College
Hanover, New Hampshire

D. W. FISHER

U. S. Geological Survey
Washington, D. C.

ROBERT S. PIERCE

Northeastern Forest Experiment Station
Forest Service, U. S. Department of Agriculture
Durham, New Hampshire

(Accepted for publication July 31, 1969)

Abstract. All vegetation on Watershed 2 of the Hubbard Brook Experimental Forest was cut during November and December of 1965, and vegetation regrowth was inhibited for two years by periodic application of herbicides. Annual stream-flow was increased 33 cm or 39% the first year and 27 cm or 28% the second year above the values expected if the watershed were not deforested.

Large increases in streamwater concentration were observed for all major ions, except NH_4^+, $SO_4^{=}$ and HCO_3^-, approximately five months after the deforestation. Nitrate concentrations were 41-fold higher than the undisturbed condition the first year and 56-fold higher the second. The nitrate concentration in stream water has exceeded, almost continuously, the health levels recommended for drinking water. Sulfate was the only major ion in stream water that decreased in concentration after deforestation. An inverse relationship between sulfate and nitrate concentrations in stream water was observed in both undisturbed and deforested situations. Average streamwater concentrations increased by 417% for Ca^{++}, 408% for Mg^{++}, 1558% for K^+ and 177% for Na^+ during the two years subsequent to deforestation. Budgetary net losses from Watershed 2 in kg/ha-yr were about 142 for NO_3-N, 90 for Ca^{++}, 36 for K^+, 32 for SiO_2-Si, 24 for Al^{+++}, 18 for Mg^{++}, 17 for Na^+, 4 for Cl^-, and 0 for SO_4-S during 1967-68; whereas for an adjacent, undisturbed watershed (W6) net losses were 9.2 for Ca^{++}, 1.6 for K^+, 17 for SiO_2-Si, 3.1 for Al^{+++}, 2.6 for Mg^{++}, 7.0 for Na^+, 0.1 for Cl^-, and 3.3 for SO_4-S. Input of nitrate-nitrogen in precipitation normally exceeds the output in drainage water in the undisturbed ecosystems, and ammonium-nitrogen likewise accumulates in both the undisturbed and deforested ecosystems. Total gross export of dissolved solids, exclusive of organic matter, was about 75 metric tons/km² in 1966–67, and 97 metric tons/km² in 1967–68, or about 6 to 8 times greater than would be expected for an undisturbed watershed.

The greatly increased export of dissolved nutrients from the deforested ecosystem was due to an alteration of the nitrogen cycle within the ecosystem.

The drainage streams tributary to Hubbard Brook are normally acid, and as a result of deforestation the hydrogen ion content increased by 5-fold (from pH 5.1 to 4.3).

Streamwater temperatures after deforestation were higher than the undisturbed condition during both summer and winter. Also in contrast to the relatively constant temperature in the undisturbed streams, streamwater temperature after deforestation fluctuated 3–4°C during the day in summer.

Electrical conductivity increased about 6-fold in the stream water after deforestation and was much more variable.

Increased streamwater turbidity as a result of the deforestation was negligible, however the particulate matter output was increased about 4-fold. Whereas the particulate matter is normally 50% inorganic materials, after deforestation preliminary estimates indicate that the proportion of inorganic materials increased to 76% of the total particulates.

Supersaturation of dissolved oxygen in stream water from the experimental watersheds is common in all seasons except summer when stream discharge is low. The percent saturation is dependent upon flow rate in the streams.

Sulfate, hydrogen ion and nitrate are major constituents in the precipitation. It is suggested that the increase in average nitrate concentration in precipitation compared to data from 1955–56, as well as the consistent annual increase observed from 1964 to 1968, may be some measure of a general increase in air pollution.

TABLE OF CONTENTS

INTRODUCTION

Management of forest resources is a worldwide consideration. Approximately one-third of the surface of the earth is forested and much of this is managed or deforested by one means or another.

Forests may be temporarily or permanently reduced by wind, insects, fire, and disease or by human activities such as harvesting or management utilizing physical or chemical techniques. Management goals range from simple harvest of wood and wood products, to increased water yields, to military stratagems involving defoliation of extensive forested areas.

Despite the importance of the forest resource, there is very little quantitative information at the ecosystem level of understanding on the biogeochemical interactions and implications resulting from large-scale changes in habitat or vegetation. This gap in our understanding results because it is particularly difficult to get quantitative ecological information that allows predictions about the entire ecosystem. The goal of the Hubbard Brook Ecosystem study is to understand the energy and biogeochemical relationships of northern hardwood forest watershed-ecosystems as completely as possible in order to propose sound land management procedures.

The small watershed approach to the study of hydrologic-nutrient cycle interaction used in our investigations of the Hubbard Brook Experimental Forest (Bormann and Likens 1967) provides an opportunity to deal with complex problems of the ecosystem on an experimental basis. The Hubbard Brook Experimental Forest, maintained and operated by the U.S. Forest Service, is especially well-suited to this approach since ecosystems can be defined as discrete watersheds with similar northern hardwood forest vegetation and a homogeneous bedrock, which forms an impermeable base (Bormann and Likens, 1967; Likens, *et al.,* 1967). Thus, the six small watersheds we have used at Hubbard Brook provide a replicated experimental design for manipulations at the ecosystem level of organization.

All vegetation on Watershed 2 (W2) was cut during the late fall and winter of 1965, and subsequently treated with herbicides in an experiment designed to determine the effect on 1) the quantity of stream water flowing out of the

[1] This is Contribution No. 14 of the Hubbard Brook Ecosystem Study. Financial support for this work was provided by NSF Grants GB 1144, GB 4169, GB 6757, and GB 6742. The senior author acknowledges the use of excellent facilities and resources at the Brookhaven National Laboratory during the preparation of part of this manuscript. Also, we thank J. S. Eaton for special technical assistance, and W. A. Reiners and R. C. Reynolds for critical comments and suggestions. Published as a contribution to the U. S. Program of the International Hydrological Decade, and the International Biological Program.

[2] Present address: Division of Biological Sciences, Cornell University, Ithaca, New York 14850.

watershed, and 2) fundamental chemical relationships within the forest ecosystem, including nutrient relationships and eutrophication of stream water. In effect this experiment was designed to test the homeostatic capacity of the ecosystem to adjust to cutting of the vegetation and herbicide treatment. This paper will discuss the results of this experimental manipulation in comparison to adjacent, undisturbed watershed-ecosystems.

THE HUBBARD BROOK ECOSYSTEM

The hydrology, climate, geology, and topography of the Hubbard Brook Experimental Forest have been reported in detail elsewhere (Likens, *et al.,* 1967).

The climate of this region is dominantly continental. Annual precipitation is about 123 cm (Table 1), of which about one-third to one-fourth is snow. Although precipitation is evenly distributed throughout the year, stream flow is not. Summer and early autumn stream flow is usually low; whereas the peak flows occur in April and November. Loss of water due to deep seepage appears to be minimal in the Hubbard Brook area (Likens, *et al.,* 1967). The bedrock of the area is a medium to coarse-grained sillimanite-zone gneiss of the Littleton Formation and consists of quartz, plagioclase and biotite with lesser amounts of sillimanite. The mantle of till is relatively shallow and has a similar mineral and chemical composition to the bedrock. The soils are podzolic with a pH less than 7. Despite extremely cold winter air temperatures, soil frost seldom forms since insulation is provided by several centimeters of humus and a continuous winter snow cover (Hart *et al.,* 1962).

TABLE 1. Average annual water budgets for Watersheds 1 through 6 of the Hubbard Brook Experimental Forest. Watershed 2 has been excluded from the averages for 1965–68; 1967–68 is based on Watersheds 1, 3, and 6 only

Water Year (1 June-31 May)	Precipitation (P) (cm)	Runoff (R) (cm)	P-R (Evaporation and Transpiration) (cm)
1963-64	117.1	67.7	49.4
1964-65	94.9	48.8	46.1
1965-66	124.5	72.7	51.8
1966-67	132.5	80.6	51.9
1967-68	141.8	89.4	52.4
1963-68	122.2	71.8	50.4
1955-68	122.8	71.9	50.9

METHODS AND PROCEDURES

Precipitation is measured in the experimental watershed with a network of precipitation gauges, approximately 1 for every 12.9 hectares of watershed. Streamflow is measured continuously at stream-gauging stations, which include a V-notch weir or a combination of V-notch weir and San Dimas flume anchored to the bedrock at the base of each watershed.

Weekly samples of precipitation and stream water were obtained from the experimental areas for chemical analysis. Rain and snow were collected in two types of plastic containers, 1) those continuously uncovered or 2) those uncovered only during periods of rain or snow. One-liter samples of stream water were collected in clean polyethylene bottles approximately 10 m above the weir in both the deforested and undisturbed watersheds. Chemical concentrations characterizing a period of time are reported as weighted averages, computed from the total amount of precipitation or streamflow and the total calculated chemical content during the period. Details concerning the methods used in collecting samples of precipitation and stream water, analytical procedures, and measurement of various physical characteristics have been given by Bormann and Likens (1967), Likens, *et al.* (1967), and Fisher, *et al.* (1968).

During November and December of 1965 all trees, saplings and shrubs of W2 (15.6 ha) were cut, dropped in place, and limbed so that no slash was more than 1.5 m above the ground. No roads were made on the watershed and great care was taken to minimize erosion. No timber or other vegetation was removed from the watershed. Regrowth of vegetation was inhibited by aerial application of the herbicide, Bromacil ($C_9H_{13}Br N_2O_2$), at 28 kg/ha on 23 June 1966. Approximately 80% of the mixture applied was Bromacil and 20% was largely inert carrier (H. J. Thome, personal communication). Also, during the summer of 1967, approximately 87 liters of an ester of 2, 4, 5-trichlorophenoxyacetic acid (2, 4, 5-T) was individually applied to scattered regrowths of stump sprouts.

The results reported cover the period immediately following the cutting of the vegetation on W2, 1 January 1966 through 1 June 1968.

HYDROLOGIC PARAMETERS

The annual hydrologic regime at Hubbard Brook has varied greatly since we began our study in 1963 (Table 1). The 1964-65 water-year was exceptionally dry, and 1967–68 was very wet. These fortuitous extremes have provided a wide range of hydrologic conditions for our study of the hydrologic-nutrient cycle interactions.

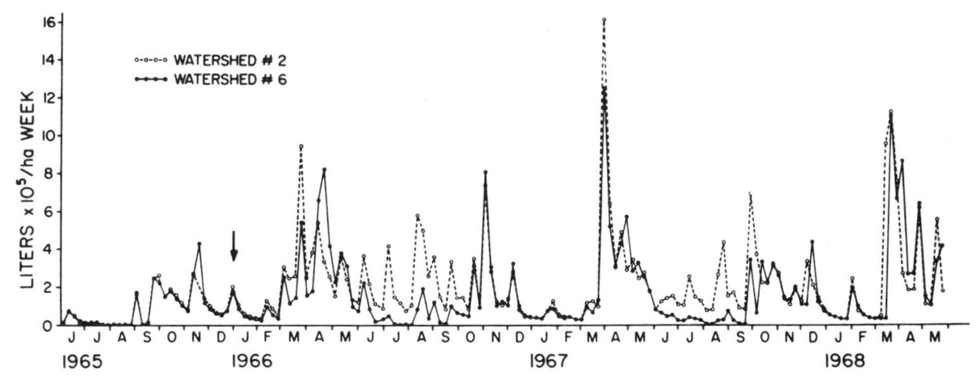

FIG. 1. Average weekly stream water discharge from Watersheds 2 and 6 during 1965–68. The vegetation on Watershed 2 was cut during November and December 1965. The arrow indicates the completion of the cutting.

Chemical input into a watershed from meteorological sources is based in part on the volume of precipitation, thus it is important to determine whether precipitation is distributed randomly throughout the watersheds. Our studies in 1963–64 and 1964–65 indicated no significant difference between rain gauges at different elevations (Likens *et al.,* 1967). Subsequent data indicate that there is generally very little difference in the precipitation pattern with elevation, but one or two storms of high intensity may significantly alter the spatial distribution for total annual precipitation within the area. Since the overall precipitation pattern for 1966–67 and 1967–68 was more variable than previous years, we have calculated the precipitation input for each watershed, for these two years, on the basis of Thiessen averages established for the area (see Thiessen, 1923). In spite of the variation in precipitation and runoff, the amount of water lost by evaporation and transpiration from undisturbed watersheds remained about the same each year during 1963 to 1968 (Table 1).

Cutting the vegetation of W2 produced a significant effect on the distribution of water loss from the watershed. These changes are reported in detail elsewhere (Hornbeck, *et al.*).

The annual runoff in 1967–68 and 1968–69 from the deforested ecosystem increased by 39% and 28%,[3] respectively, over the values expected had the watershed not been cut (Table 2). The greatest difference occurred during June through September, when runoff values were 414% (1966–67) and 380% (1967–68) greater than expected.

[3] The slight discrepancy from data presented by Hornbeck, *et al.,* is due to rounding errors incurred in the conversion of English to metric units.

The increased streamflow during summer is directly attributable to the removal of transpiring surface. In addition to the increased yield of

TABLE 2. Runoff and precipitation for Watershed 2 (cut-over) of the Hubbard Brook Experimental Forest, expressed in cm of water. The watershed was clear-cut in November and December of 1965. Predicted values for runoff were calculated from regression analyses established during a pretreatment calibration period. (After Hornbeck, *et al.,* 1970)

	1965-66	1966-67	1967-68
Precipitation	124.6	132.1	138.7
Predicted runoff	(77.7)	(86.5)	(96.7)
Measured runoff	79.9	119.9	123.4
Evaporation and Transpiration	44.7	12.2	15.4

drainage water from the watershed, the snowmelt was advanced a few days and was more rapid, particularly during 1967-68 (Fig. 1; Federer, 1969; Hornbeck and Pierce, 1969).

In various experiments throughout the world, with 100% reduction in forest cover by clear-cutting or chemical treatment, stream flow increases averaged about 20 cm the first year after treatment (Hibbert, 1967). Detailed studies at the Coweeta Hydrologic Laboratory, Southeastern Forest Experiment Station in North Carolina showed maximum increases in water yield of about 41 cm during the first year following the complete removal of the hardwood forest vegetation on small watersheds (Hoover, 1944).

PRECIPITATION CHEMISTRY

Sulfate and hydrogen ions are the most abundant constituents (in terms of chemical equivalents) in precipitation falling on the watersheds

TABLE 3. Average weighted ion content of bulk precipitation collected within the Hubbard Brook Experimental Forest from 1 June to 31 May expressed in mg/liter.

	1965-66	1966-67	1967-68
calcium	0.22	0.21	0.20
magnesium	0.04	0.03	0.05
potassium	0.05	0.05	0.05
sodium	0.16	0.10	0.12
aluminum	***	0.1	***
ammonium	0.21	0.18	0.22
hydrogen ion	0.07	0.08	0.07
nitrate	1.41	1.49	1.56
sulfate	3.3	3.1	3.3
chloride	0.21+	0.50##	0.35
bicarbonate	#	#	#
dissolved silica	***	***	***

***Not determined; probably less than 0.1
#Virtually absent
##Based on samples for 9 months. Samples collected from 1 September through 30 November were discarded because plastic screens were used on the collection apparatus (Juang and Johnson 1967; Fisher, *et al.* 1968).
+Calculated for the period September through August (Juang and Johnson, 1967).

at Hubbard Brook. The pH of rain and snow samples is frequently less than 4.0. Nitrate is next in abundance, and significant amounts of ammonium, chloride, sodium and calcium are usually present. Lesser amounts of magnesium, potassium and aluminum are also found (Table 3).

Since precipitation samples are composited over weekly intervals, it is difficult to identify the origin of chemical impurities in individual air masses or storms.

Our chemical input data are based on bulk precipitation, i.e., a mixture of rain or snow and dry fallout (Whitehead and Feth, 1964). We have been concerned with the contributions from dry fallout since the beginning of the study. Juang and Johnson (1967) suggested that dry fallout was a source of chloride in the Hubbard Brook ecosystem. However, based upon comparisons between precipitation collectors that were continuous open (Likens, *et al.*, 1967) and those that opened only during periods of rain or snow (Wong Laboratories, Mark IV), it is clear that the bulk of chemical input to the ecosystem comes in rain and snow. Our attempts to quantify the much smaller contributions from dry fallout have been thus far inconclusive.

Average values for the ion content of precipitation, interpolated from isopleth maps for central New Hampshire (Junge, 1958; Junge and Werby, 1958), do not agree closely with the average weighted concentration of ions measured in precipitation at Hubbard Brook (Table 3). Considering the yearly variation in precipitation chemistry (Likens, *et al.*, 1967; Fisher, *et al.*, 1968; Table 3), and that our samples were taken in 1965–68, it is not surprising that there is not better agreement between our values and those obtained in 1955–56 by Junge and Werby. Our cation values are nearly 50% lower than those reported by Junge (1958) and Junge and Werby (1958); whereas our anion values, with the exception of chloride, are somewhat higher.

Soil and road dust are among the principal sources of base metal ions in local precipitation (*e.g.*, Gambel and Fisher, 1966). In the Hubbard Brook area, the almost total forest cover strongly minimizes the generation of soil dust, thereby reducing the concentration of these ions in the local precipitation.

The weighted concentration of nitrate in precipitation at Hubbard Brook is considerably higher than the concentrations reported by Junge (1958) for New England, and in fact, for most regions of the U. S. Since air pollution can affect the concentration of nitrate in precipitation (*e.g.*, Junge, 1963), our data might reflect some measure of increased air pollution between 1955–56 and 1965–66, and its resultant effect on nutrient budgets in rural as well as urban areas. In this regard it is interesting that our average values for nitrate in precipitation (rain and snow plus dry fallout) has increased each year since 1964; whereas most of the other chemical constituents remained constant or fluctuated slightly (Table 3).

CHEMICAL INPUT THROUGH HERBICIDE APPLICATION

Our nutrient budgets are based on the difference between chemical input in precipitation and output in stream water from the watershed-ecosystems (Bormann and Likens, 1967). Thus, it is important to account for any extraneous chemical inputs to the ecosystem, such as an application of herbicide. Approximately 3650 liters of Bromacil solution were sprayed on the cutover watershed in June 1966. Based on a chemical analysis of the herbicide solution, and assuming complete decomposition, 0.04 Ca^{++}/ha, 0.09 kg Mg^{++}/ha, 0.01 kg K^{+}/ha, 0.11 kg Na^{+}/ha, and 10.6 kg NO_3^{-}/ha were added to the watershed. These are relatively insignificant inputs to the budgets for the deforested watershed. Bromide is equivalent to chloride with the analytical method we use for chloride (Iwasuki *et al.*, 1952). Thus we have calculated a maximum possible input of 3.0 kg/ha to the chloride budget from Bromacil.

Approximately 87 liters of an ester (propylene glycol butyl ether) of 2, 4, 5-T were sprayed on the cutover watershed during the summer of 1967. This herbicide was mixed with water from the drainage stream of the watershed to produce the spray solution. Chemical analyses of the solution showed that negligible amounts of cations were added to the watershed-ecosystem from this application of herbicide. However we calculated a maximum input of 0.7 kg Cl^-/ha to the watershed from the herbicide solution. Inputs for other anions were negligible.

STREAMWATER PARAMETERS

Temperature

The drainage streams in the undisturbed watersheds during summer are in deep shade beneath the forest canopy; in winter they are under a deep snow pack. Thus, for most of the year, daily stream water temperatures are relatively constant and the annual temperature range is only about 16°C (Figs. 2 and 3).

The altitudinal temperature gradient in stream water varies somewhat in the undisturbed watersheds, and in Watershed 4 during midsummer it is about 5–7°C. At times it may be as much as 12°C, however the steepest portion of this thermal gradient is confined largely to the upper 50 m of the stream (McConnochie and Likens, 1969). Macan (1958) and others have shown that small streams usually warm and reach equilibrium temperatures rapidly, and that the average water temperature approximates the average air temperature. In winter, the altitudinal temperature gradient in streams within the watersheds is not more than 1 or 2°C.

Streamwater temperatures in the deforested watershed (W2) were higher than in the undisturbed watersheds during both summer and winter. In the absence of shade, temperature varied by 3–4°C during the day in summer; the maximum temperature occurred about 1500 hours and the minimum at 0700–0800 hours (Fig. 3). The annual variation was about 18 to 20°C (Fig. 2).

Even though the mean January air temperature is −9°C (U. S. Forest Service, 1964), the streams do not freeze solid. In fact, they are relatively warm below the thick snow pack in winter (Figs. 2 and 3). A thicker snow depth in the stream channel of the deforested watershed could provide more insulation and may account for the somewhat higher streamwater temperature. As mentioned earlier, frost is uncommon in the soils of the watersheds.

The streams seem to have discrete summer and winter temperature regimes with rapid seasonal transitions. Warm-up during the spring is very rapid, occurring mostly in May in the undisturbed situation (Fig. 2).

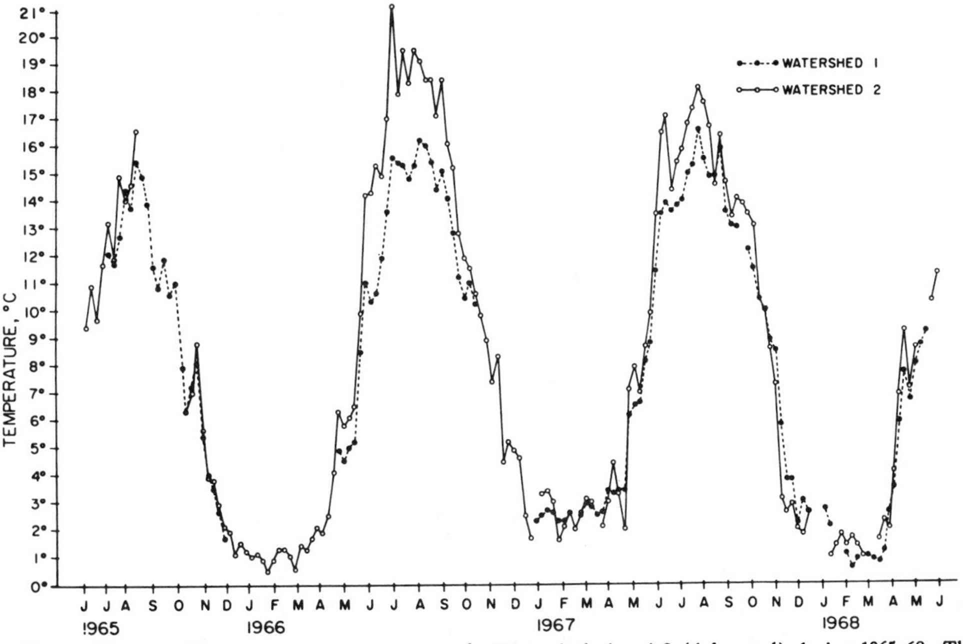

FIG. 2. Mean weekly stream water temperatures in Watersheds 1 and 2 (deforested) during 1965–68. The recording thermometers were located above the weirs in each stream. The points were not connected during periods when the stream discharge was negligible or when the thermograph functioned improperly.

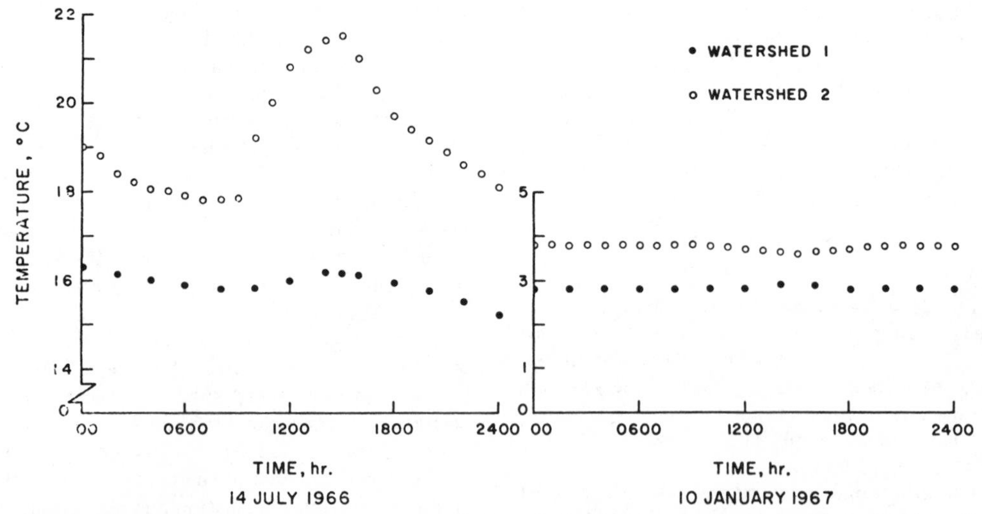

FIG. 3. Mean hourly stream water temperatures during one day in the summer and one day in the winter in Watersheds 1 and 2 (deforested).

DISSOLVED OXYGEN

The stream water from the Hubbard Brook watersheds is normally saturated or slightly supersaturated with dissolved oxygen, except during periods of very low flow, *e.g.,* late summer and early autumn (Fig. 4). There is no apparent seasonal pattern separate from this dependence on discharge. The dissolved oxygen values are adjusted for altitude and streamwater temperature to calculate the percent saturation. Ruttner (1953) indicates that the dissolved oxygen in streams, even in cascading water, should not exceed the saturation equilibrium with air. However, supersaturation of dissolved oxygen has been found by several other workers in natural streams and rivers (*e.g.,* Järnefelt, 1949; Harvey and Cooper, 1962; Minckley, 1963; Woods, 1960). The explanation of the supersaturated condition in streams tributary to Hubbard Brook is not readily apparent, although it may be "forced" to supersaturation by turbulence (Lindroth, 1957; Harvey and Cooper, 1962) or may be a function of the altitudinal temperature gradient in the streams. Photosynthetic organisms are not abundant in these tributary streams, and diel variations in dissolved oxygen are not apparent.

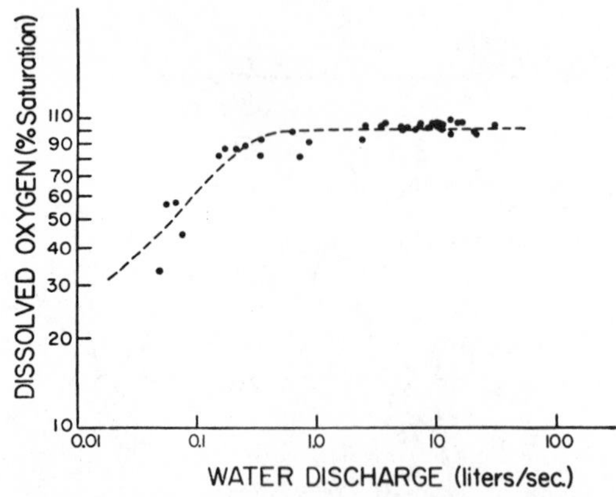

FIG. 4. Relationship between dissolved oxygen and stream water discharge in undisturbed Watershed 3 during 1965–66. The equation for the dashed line is $C = (1 - e^{-(D)(0.0693)})\ 79.26 + 22.59$, where C = dissolved oxygen in % saturation and D = stream discharge in liters/sec. The constants in this equation were determined by linear regression analysis and the F-ratio for the regression line is very highly significant (>0.001).

At low flows the water mostly seeps from one small pool to the next through gravel and organic debris dams. The biological oxygen demand from decaying leaves and other organic debris in the stream is apparently high enough to reduce significantly the dissolved oxygen concentration when flow and turbulence are low. Rapid depletion of dissolved oxygen resulting from decomposition of leaves in small streams has been demonstrated by several workers (see Minckley, 1963).

Increased water yield from the deforested watershed, particularly in the summer months (Fig. 1), results in greater discharge, more turbulence and a constant high level of dissolved oxygen in this stream. Thus, during the summer and autumn large differences in dissolved oxygen concentration may be anticipated between the streams

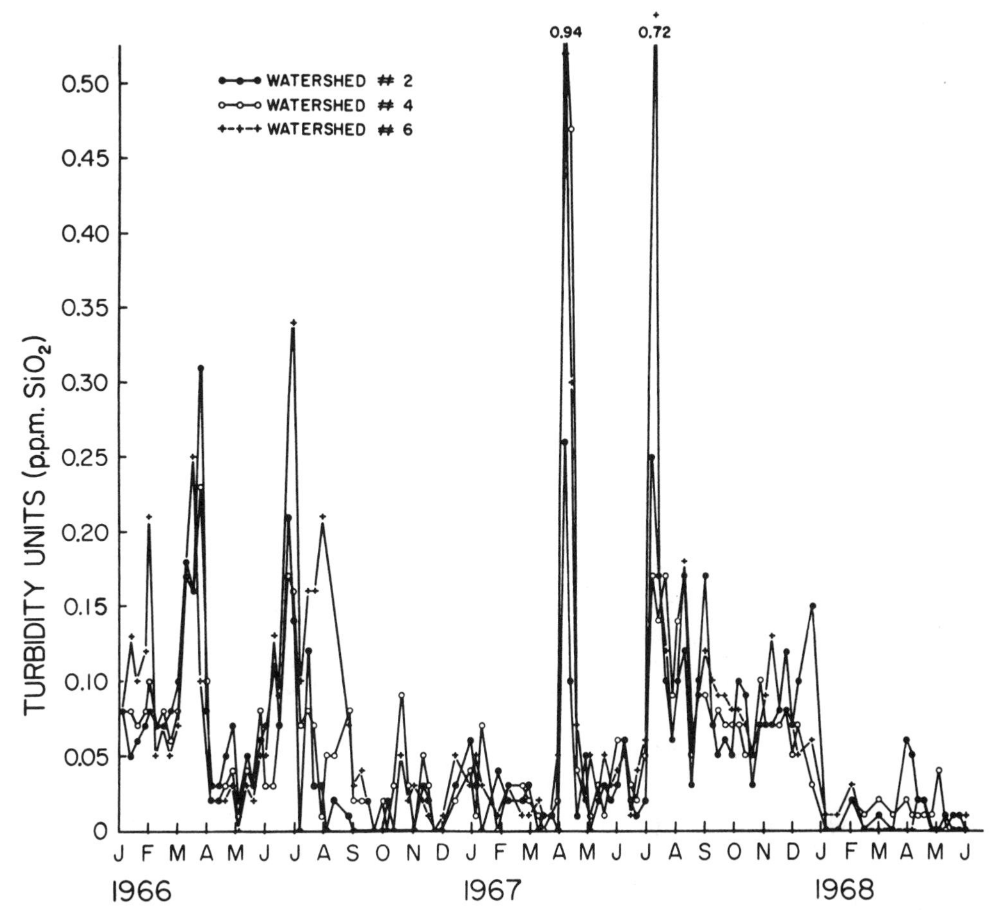

FIG. 5. Turbidity values for stream water from Watersheds 2 (deforested), 4 and 6 during 1966–1968.

in the deforested and undisturbed watersheds. With increased water temperature, sunlight and high concentrations of dissolved oxygen, more rapid decomposition of the organic debris would be expected in the stream of W2.

TURBIDITY

Following disturbance of the vegetation and litter in forested watersheds, drainage waters may become quite turbid as the result of erosion and transport of inorganic and organic matter from the watershed (*e.g.*, Lieberman and Hoover, 1948a; Tebo, 1955). For example, this frequently occurs as a consequence of unregulated commercial logging operations. However, with care and planning, turbidity and sedimentation in drainage streams may be minimized following commercial logging (Hewlett and Hibbert, 1961; Lieberman and Hoover, 1948b; Trimble and Sartz, 1957).

Stream water draining the Hubbard Brook Experimental Forest is very clear, and no obvious differences were noted in the turbidity of the stream water from W2 following the cutting and herbicide treatment of the vegetation (Fig. 5). In fact, the peak turbidity values seemed to be depressed in comparison with values for streams in the undisturbed watersheds. Of the three watersheds compared, W6 showed the greatest extremes in turbidity, and these extreme values were also slightly out-of-phase with changes in the other watersheds (Fig. 5). High values were not always correlated with high runoff values, *e.g.*, during June 1966 and July 1967. All in all, the measurements of turbidity were of little value in assessing the changes in water quality of the stream water of these forested and deforested ecosystems.

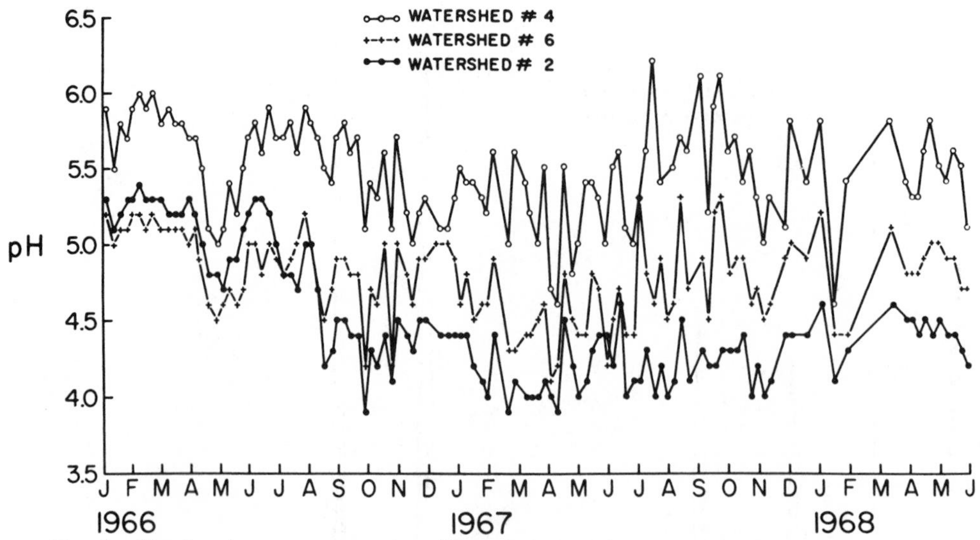

FIG. 6. pH values for stream water from Watersheds 2 (deforested), 4 and 6 during 1966–1968.

PARTICULATE MATTER

Undisturbed forest ecosystems lose relatively little organic and inorganic particulate matter. Average annual losses are about 25 kg/ha-yr., and are about equally divided between organic and inorganic particulate matter (Bormann, *et al.*, 1969). These minor particulate matter losses, about one-sixth of the dissolved substance losses, are attributed to the operation of biotic factors, which 1) decrease the erodability of the ecosystem, 2) decrease the amount of runoff and 3) tend to damp the frequency of high discharge rates.

The influence of these biotic factors has been severely limited by removal of the living vegetation in Watershed 2. Compared to undisturbed Watershed 6, the new conditions in Watershed 2 are reflected by a 4-fold increase in particulate matter in the settling basin above the V-notch weir for the period May 1966 to May 1968. Also, particulate matter output from Watershed 2 is becoming increasingly inorganic in content (76% inorganic, May 1966 to May 1968).

The increased output in particulate matter, and particularly the increased proportion of inorganic materials, from W2 occurred primarily from the unraveling of the stream channel. In many places the banks have been eroded and many of the small debris dams, composed of leaves, twigs, etc., that were common in the undisturbed stream, have worn away, with a subsequent release of trapped inorganic and organic materials. Because these dams no longer exist, and without the annual replenishment of leaves and other organic debris to build new dams, stream water is able to transport much more particulate material downslope. Also, stream channel erosion is increased since the binding action of roots has been reduced, and leaves, which provide a protective cover on the banks, have been removed and not replenished.

pH

Acidic water characterizes the drainage streams from the undisturbed watersheds. This is typical for many streams in New England with podzolic soils (*e.g.*, Anderson and Hawkes, 1958). The pH values are variable throughout the year, but consistently show the same relationship between watersheds (Fig. 6). There was particularly good agreement in pattern between the undisturbed watersheds, W4 and W6, although stream water from W4 was relatively less acidic.

As a result of deforestation, and concurrently with other chemical changes in the drainage waters during June 1966, the hydrogen ion content (calculated from pH measurements) in stream water from W2 increased by 5-fold (Fig. 6). That is, the weighted average pH, decreased from 5.1 during 1965–66 to 4.3 during 1966–67 and 1967–68 (Likens, *et al.*, 1969). During the same period the weighted pH value remained relatively unchanged for W4 and W6.

ELECTRICAL CONDUCTIVITY

Electrical conductivity averages about 20 μmhos/cm² at 25°C in stream water of the undis-

TABLE 4. Weighted average concentration in stream water from watersheds 2, 4, and 6 of the Hubbard Brook Experimental Forest, expressed in mg/liter

	W2			W4#		W6		
Ion	1965-66	1966-67	1967-68	1965-66	1966-67	1965-66	1966-67	1967-68
Ca^{++}	1.81*	6.45	7.55	1.82*	1.80	1.36*	1.27	1.28
Mg^{++}	0.37	1.35	1.51	0.41	0.40	0.36	0.35	0.36
K^{+}	0.19	1.92	2.96	0.22	0.24	0.18	0.20	0.26
Na^{+}	0.87	1.51	1.54	1.13	1.10	0.83	0.80	0.93
Al^{+++}	0.22	1.5	2.0	0.12	0.12	0.32	0.33	0.32
NH_4^{+}	0.14	0.07	0.05	0.12	0.06	0.12	0.05	0.02
NO_3^{-}	0.94	38.4	52.9	0.86	0.88	0.85	0.69	1.30
$SO_4^{=}$	6.8	3.8	3.7	6.4	6.2	6.2	6.0	6.1
Cl^{-}	0.54	0.89	0.75	0.56**	0.58	0.57***	0.55	0.56
HCO_3^{-}	0.8	0.1	0	1.6	2.0	0.1	0.2	0.3
SiO_2—aq	4.1	5.6	5.7	5.4	5.5	4.1	4.4	3.8
Total	16.8	61.6	78.7	18.6	18.9	15.0	14.8	15.2

*Values for the initial 7 months of the water-year have been increased by a factor of 1.6 to compensate for analytical interferences (Likens, *et al.*, 1967; Johnson, *et al.*, 1968)
**Juang and Johnson (1967) calculated a value of 0.51 for the period September through August.
***Juang and Johnson calculated a value of 0.50 for the period September through August.
#Data not available for 1967-68.

turbed watersheds and changes very little either on a daily or seasonal basis. This would be expected from the relative constancy of cation and anion concentrations in stream water at Hubbard Brook (Likens, *et al.,* 1967; Fisher, *et al.,* 1968; Johnson, *et al.,* 1969).

In contrast, the conductivity of stream water from the deforested watershed is quite variable, ranging from 65 and 160 μmhos/cm² at 25°C. Usually the conductivity decreased during rain storms. However occasionally the conductivity increased or was unaffected. The hydrogen ion concentration of rain water added to the stream water is important in this regard, as well as other variables such as amount of rainfall, its duration, streamwater discharge, and water content of the soil.

IONS

Ammonium and Nitrate

The ammonium ion occurs in very low concentration in stream water from undisturbed watersheds of the Hubbard Brook Experimental Forest (Fisher, *et al.,* 1968). Essentially no change was observed in the concentration of this ion in drainage water from the deforested watershed. In contrast the nitrate concentration increased from an average weighted value of 0.9 mg/liter prior to cutting of the vegetation, to 53 mg/liter, two years later (Table 4). Measured concentrations soared to 82 mg/liter in October 1967 (Fig. 7). It should be noted that the initial increase (June 1966) in streamwater nitrate concentration in W2 occurred 16 days before the application of the Bromacil and at the same time the nitrate concentration in the stream water from the undisturbed watershed showed the normal late-spring decline (Fig. 7; Bormann, *et al.,* 1968).

Nitrate concentrations in stream water from the undisturbed watersheds, show a pronounced, recurring seasonal pattern (Bormann, *et al.,* 1968; Fisher, *et al.,* 1968). Average monthly concentrations are low (<0.1 mg/l) throughout the summer growing season, increase in November, and reach values as high as 2 mg/liter during the spring (April) thaw (Fig. 8). This variation may be explained by a combination of mechanical and biological effects (Johnson, *et al.,* 1969).

The decline of nitrate concentrations during May and the low concentrations throughout the summer correlate with heavy nutrient demands by the vegetation and increased heterotrophic activity associated with warming of the soil. The winter pattern of NO_3^- concentration may be explained in strictly physical terms, since the input of nitrate in precipitation from November through May largely accounts for nitrate lost in stream water during this period (Bormann, *et al.,* 1968). Also, sublimation and evaporation of water from the snow pack could account for some 15–20% increase in concentration of NO_3^- in the stream water in the spring. Concentration of nitrate in stream water after deforestation show a pattern that is nearly the reciprocal of the undisturbed situation (Fig. 8).

Since yearly input of nitrate-nitrogen in precipitation exceeds losses in stream water in the undisturbed ecosystems (Table 5), the concentration of nitrate in stream water provides no conclusive evidence for nitrification in these un-

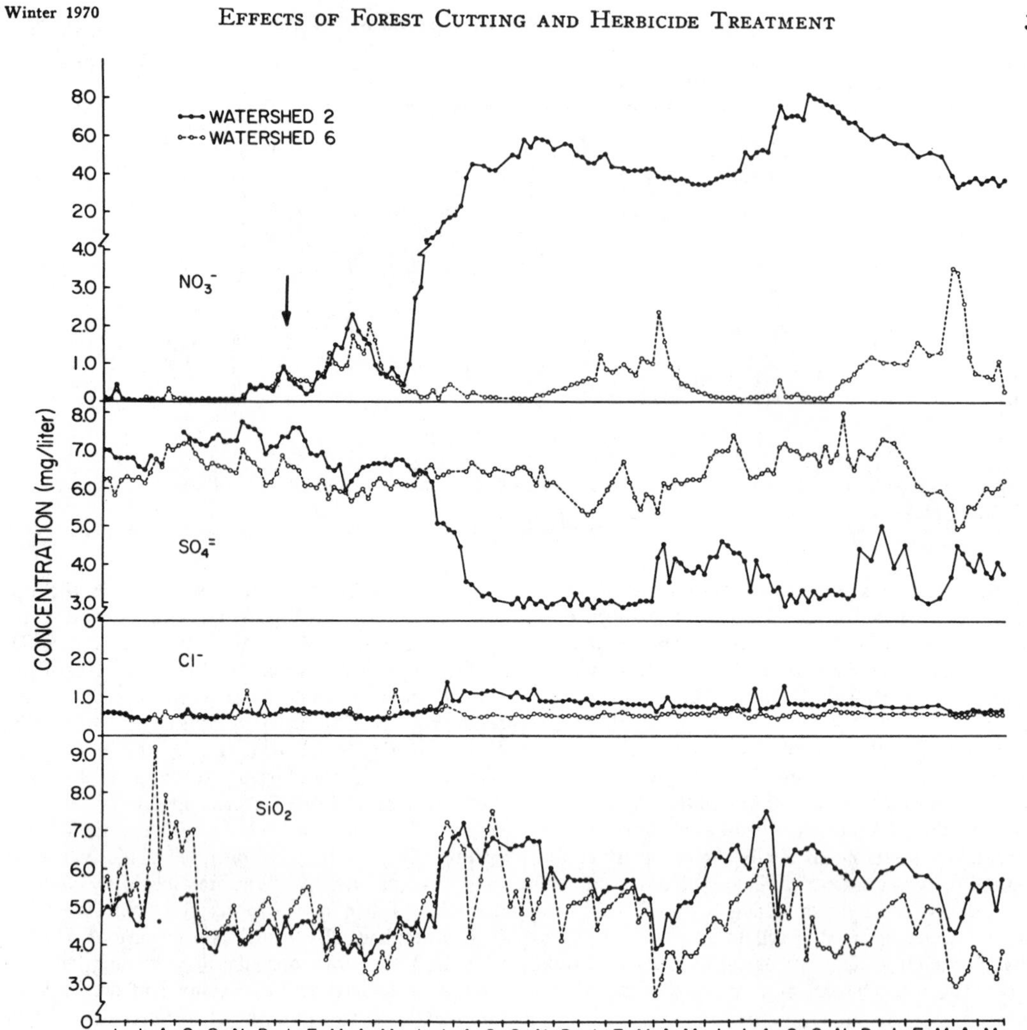

FIG. 7. Measured streamwater concentrations for nitrate, sulfate, chloride and dissolved silica in Watersheds 2 (deforested) and 6. Note the change in scale for the nitrate concentration. The arrow indicates the completion of cutting in W2.

disturbed acid soils. The low levels of ammonia and nitrate in the drainage water of the undisturbed ecosystem (W6) may attest to the efficiency of the oxidation of ammonia to nitrate, and to the efficiency of the vegetation in utilizing nitrate. However, Nye and Greenland (1960) state that growing, acidifying vegetation represses nitrification; thus the vegetation may draw directly on the ammonium pool, and little nitrate may be produced within the undisturbed ecosystem. However, in the absence of forest vegetation, the microflora of the deforested watershed apparently oxidize ammonia to nitrate, and the nitrate is rapidly flushed from the watershed-ecosystem (Bormann, *et al.,* 1968; Likens, *et al.,* 1969).

Many microorganisms convert organic nitrogen into ammonia. The ammonia then may be oxidized to nitrite by bacteria of the genus *Nitrosomonas.* The nitrite may be further oxidized to nitrate by bacteria of the genus *Nitrobacter.* The important end products of these reactions in terms of nutrient losses from the watershed-ecosystem are the increased production of nitrate and hydrogen ions. Increased biological nitrification apparently is the principal factor responsible for

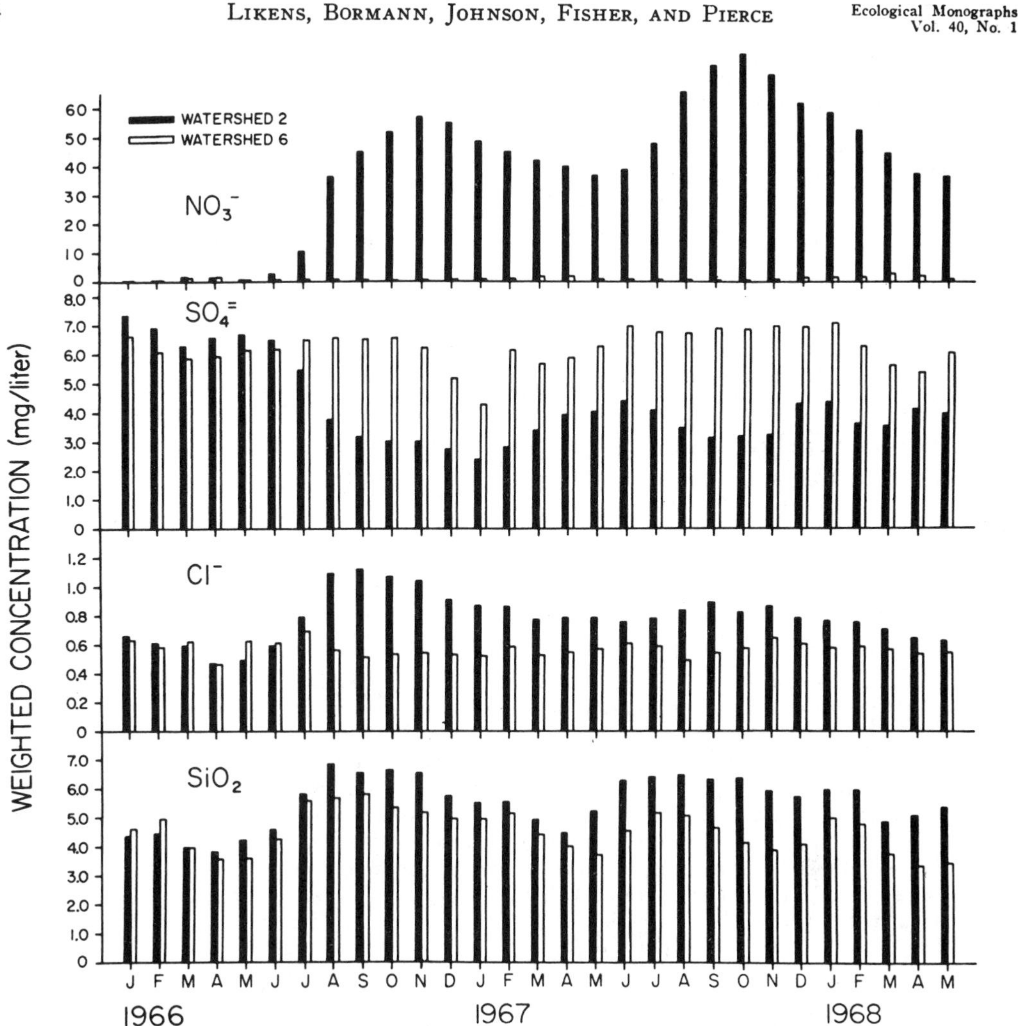

FIG. 8. Weighted average monthly concentrations of nitrate, sulfate, chloride and dissolved silica in stream water from Watersheds 2 (deforested) and 6.

the total flush of ions from the deforested watershed. The increase in the milliequivalent value for nitrate in the water years subsequent to cutting, balances (within 10–15% error limits for the watershed systems) the increased net losses of cations and the decreased net losses of anions (Likens, *et al.,* 1969). This will be discussed in more detail in a later section.

Alexander (1967) indicates that nitrification usually decreases greatly below a pH of 6.0 and becomes negligible at a pH of 5.0. Our data would indicate that this is not the case for our deforested watershed as shown by the increased concentration of nitrate in the drainage water during 1966–68 (Fig. 7), and the low pH of stream waters (Fig. 6) and soils. Also, in comparison with soils under undisturbed forest, bacteria of the genera *Nitrosomonas* and *Nitrobacter* have increased 18-fold and 34-fold, respectively, in the soil of Watershed 2 (Smith, *et al.,* 1968). Other workers have shown nitrification at similarly low pH's (Boswell, 1955; Weber and Gainey, 1962). It may be that we are dealing with relatively little known species of nitrifying bacteria adapted to more acid conditions (*e.g.,* Alexander, 1967).

Sulfate

Sulfate concentration in stream water from the undisturbed watersheds is relatively constant in relation to the highly variable stream discharge

TABLE 5. Nutrient budgets for undisturbed (W6) and cutover (W2) watersheds of the Hubbard Brook Experimental Forest. Values are expressed in kg/ha for the period 1 June to 31 May

Element	W6 Input	W6 Output	W6 Net loss or gain	W2 Input	W2 Output	W2 Net loss or gain
1966-67						
Ca	2.4	10.7	−8.3	2.3	77.3	−75.0
Mg	0.4	2.9	−2.5	0.5	16.2	−15.7
K	0.6	1.7	−1.1	0.5	23.0	−22.5
Na	1.3	6.8	−5.5	1.3	18.1	−16.8
Al	1.4	2.8	−1.4	1.3	18.2	−16.9
NH_4−N	1.9	0.3	+1.6	1.8	0.7	+ 1.1
NO_3−N	4.6	1.3	+3.3	6.8	104	−97.2
SO_4−S	14.4	17.1	−2.7	13.6	15.3	− 1.7
Cl	6.9**	4.6	+2.3	9.5**	10.6	− 1.1
HCO_3−C	*	0.4	−0.4	*	0.2	− 0.2
SiO_2−Si	*	17.2	−17	*	31.3	−31
1967-68						
Ca	3.0	12.2	−9.2	2.7	93.1	−90.4
Mg	0.8	3.4	−2.6	0.7	18.6	−17.9
K	0.8	2.4	−1.6	0.7	36.5	−35.8
Na	1.8	8.8	−7.0	1.7	19.0	−17.3
Al	*	3.1	−3.1	*	24.5	−24
NH_4−N	2.6	0.2	+2.4	2.4	0.5	+ 1.9
NO_3−N	5.2	2.8	+2.4	4.9	147	−142
SO_4−S	16.0	19.3	−3.3	15.16	15.15	0
Cl	5.2	5.3	−0.1	5.5	9.2	− 3.7
HCO_3−C	*	0.5	−0.5	*	0	0
SiO_2−Si	*	17.0	−17	*	32.6	−32

*Not determined, but very low.
**Based on data for 9 months.

(Figs. 7 and 8; Fisher, *et al.,* 1968); and on close examination the sulfate concentration shows a very small and irregular volume concentration effect (Johnson, *et al.,* 1969). In addition streamwater concentrations of sulfate seem to show some general, recurring seasonal patterns. There is a gradual rise to maximum values in the autumn with lower values during the late winter and early spring. This pattern is nearly the reciprocal of the nitrate pattern (Figs. 7 and 8).

The weighted concentration of $SO_4^{=}$ in drainage water from the deforested watershed decreased by about 45% in the first year subsequent to cutting and herbicide treatment (Table 4). The decrease in sulfate concentration was somewhat delayed in relation to the increase in nitrate concentration (Fig. 7). The explanation for this decrease in sulfate concentration is complicated and will be elaborated in the section on Nutrient Budgets.

Chloride

Since the quantity of chloride bearing rocks in the Hubbard Brook watershed-ecosystems is small (Juang and Johnson, 1967), and since biological immobilization is small, we expected the weighted concentration of Cl^- in precipitation after correction for evapotranspiration, to balance the concentration in drainage water from undisturbed ecosystems. However, during 1965–66 a higher concentration was found in the stream water of the undisturbed watersheds than could be accounted for by precipitation input (Juang and Johnson, 1967). The excess Cl^- was attributed to the accumulation of Cl^- in the ecosystem by dry removal of aerosols through impaction on the forest canopy. More recent comparisons of the chemical results from precipitation collectors that are continuously open and those that open only during periods of rain or snow are inconclusive, but do not support the dry fallout hypothesis. Data for years subsequent to 1965–66 show that the long-term, mean chloride concentration in stream water is about the same as the average concentration in precipitation, after adjustment for water loss by evapotranspiration (Johnson, *et al.,* 1969). We feel that the problems encountered previously in the chemical analysis and collection of chloride samples (*e.g.,* Fisher *et al.,* 1968) were greatly minimized during 1967–68, and significantly, the chloride input in precipitation balanced the output in stream water in undisturbed W6 (Table 5).

After deforestation of W2, the streamwater concentration of chloride increased, but the increase was somewhat delayed relative to most of the other ions (Fig. 7). The weighted concentration of chloride in drainage waters from W2

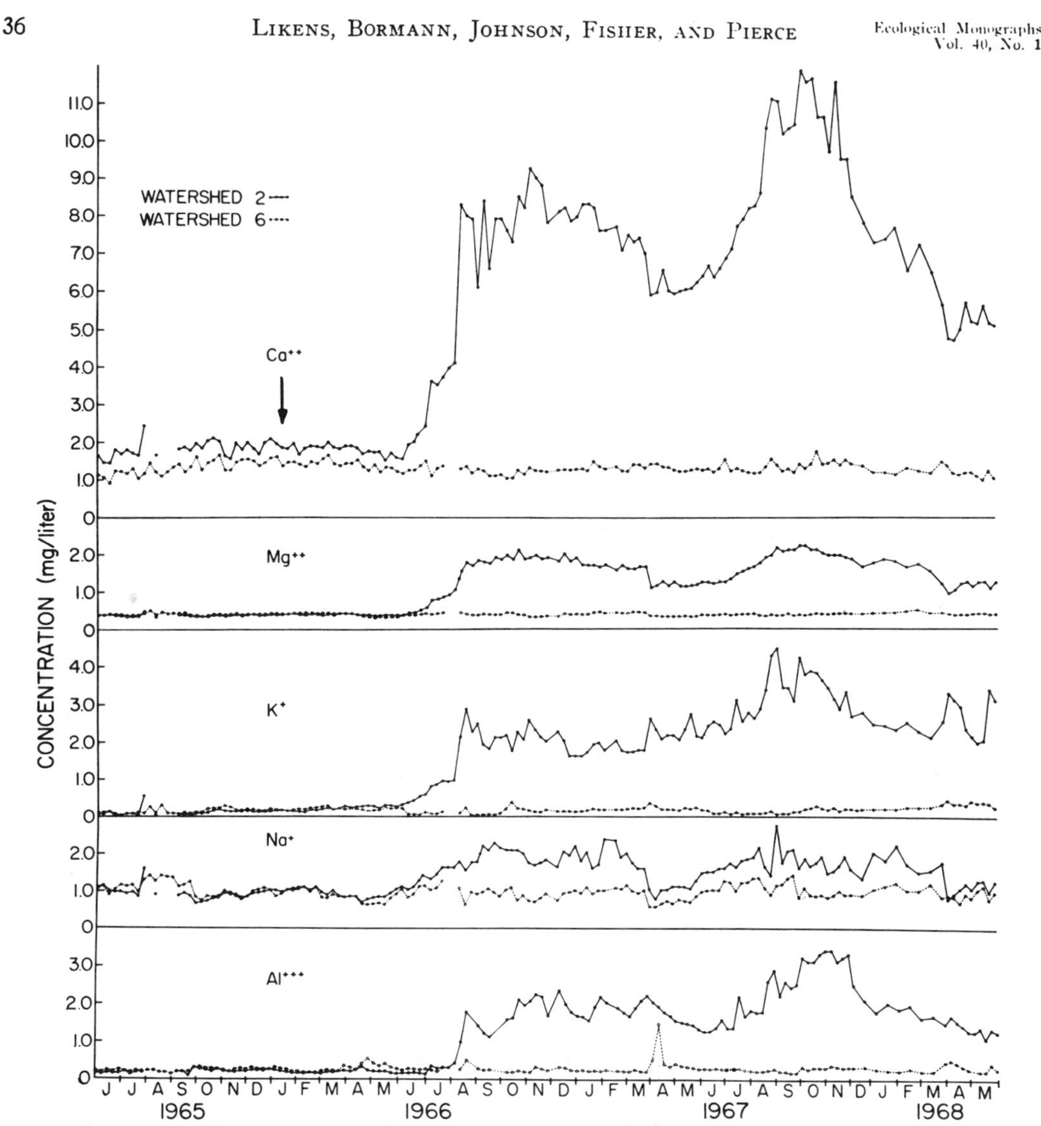

FIG. 9. Measured streamwater concentrations for Ca++, Mg++, K+, Na+, and Al+++ in Watersheds 2 (deforested) and 6. The points were not connected during periods when the stream discharge was negligible. The arrow indicates the completion of cutting in W2.

increased about 65% in the first year (Table 4), indicating the existence of a small chloride reservoir within the ecosystem. A maximum of about 70% of this increased chloride concentration could be explained by the addition of the Bromacil solution to the watershed, however not all of the Bromacil was lost from the watershed during the first year (Pierce, 1969). A continuing but relatively smaller increased chloride concentration also was observed in stream water from W2 during the second year, 1967–68 (Table 4; Fig. 7). Some 30% of the increase in streamwater concentration during 1967–68 may be attributed to the addition of the 2, 4, 5-T solution to the watershed.

Calcium, Magnesium, Potassium and Sodium

The concentration of these cations characteristically has been relatively constant in stream water of the undisturbed watersheds despite the highly variable discharge of water (Likens, *et al.*, 1967). The constancy of magnesium is phenomenal in this regard. The relatively small variations that do occur may be explained by appropriate equations for dilution and concentration, based on

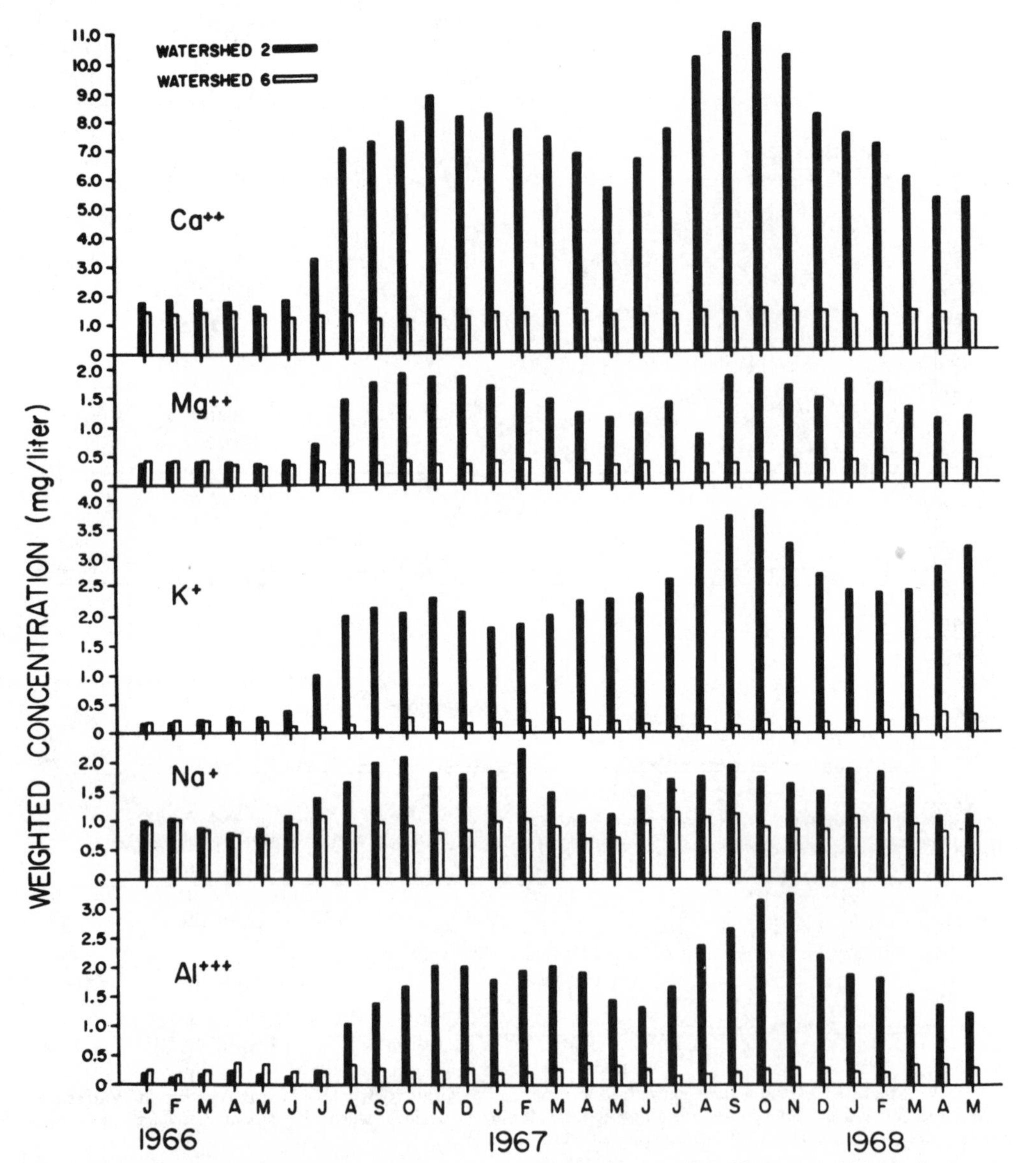

FIG. 10. Weighted average monthly concentrations of Ca^{++}, Mg^{++}, K^{+}, Na^{+}, and Al^{+++} in stream water from Watersheds 2 (deforested) and 6.

water discharge (Johnson, *et al.*, 1969). In the deforested watershed the volume-concentration model still applies but is less important as a controlling factor than the effect of nitrification.

Remarkable increases in average stream water concentration (417% for Ca^{++}, 408% for Mg^{++}, 1558% for K^{+} and 177% for Na^{+}) were observed during the two years subsequent to deforestation (Figs. 9 and 10; Table 4). The increased concentration and resultant net losses of these cations in drainage waters is dependent upon the increased concentration of nitrate and hydrogen ions brought about by nitrification (Likens, *et al.*, 1969). With decreasing pH and available nitrate these cations are more readily mobilized and leached from the watershed. This may occur 1) as the hydrogen ions replace the cations on the humic-clay ion exchange complexes of the soil, 2) as organic compounds are decomposed and 3) as chemical decomposition of bedrock and till is accelerated. However, since nitrification is occurring in the humic layers of the soil, the excess ions are most likely released from the decay of humic substances (Likens, *et al.*, 1969). This is best shown by the changes in the ratio of Ca:Na in the drainage water of Watershed 2 before and after deforestation. Prior to cutting, the Ca:Na ratio in the stream water of Watershed 2 was 1.6/1.0. This ratio is consistent with that observed in adjacent undisturbed watersheds, and has been attributed to the steady-state chemical weathering of bedrock and till (Johnson, *et al.*, 1968). However, after deforestation, the ratio climbed to an average of 4.8/1.0 for the period 1966–68. More significantly, the Ca:Na ratio for the "excess" ions of the stream (that is, the amount added above the undisturbed condition) is 7.4/1.0. Thus, the net chemical effect of the deforestation was to differentially produce soluble calcium within the ecosystem. Hardwood forest litter is rich in calcium relative to sodium (Ca:Na is >20/1.0; Scott 1955) in contrast to average bedrock for the Hubbard Brook watershed-ecosystems (Ca:Na is 0.9/1.0; Johnson, *et al.*, 1968), suggesting that most of the "excess" ions produced are derived from the bulk decomposition of humic materials (Likens, *et al.*, 1969).

Only a small portion of the 6- to 22-fold increase in net losses of calcium, magnesium and potassium were apparently derived from an accelerated rate of chemical weathering (Table 5). Based on the change in the sodium budget of Watershed 2, and attributing net sodium losses solely to chemical weathering (Johnson, *et al.*, 1968), the chemical decomposition of bedrock and till within Watershed 2 has apparently accelerated no more than 3-fold.

Aluminum

Aluminum concentrations in stream water from the undisturbed watersheds show seasonal pattern and differences between watersheds (Fisher, *et al.*, 1968). Concentrations of aluminum are highest in April during peak runoff (Figs. 9 and 10). In fact, there is a direct relationship between aluminum concentration and stream discharge (Johnson, *et al.*, 1969). Concentration of aluminum in stream water varies with watershed, with W4 lower than the other streams (Table 4). The average stream water concentration for W4 is about 1/3 of the value for W6.

Deforestation resulted in an increased concentration of aluminum in stream water from W2, but the increase was delayed relative to other ions (Figs. 9 and 10). Aluminum concentration increased by 10-fold in two years following deforestation. The primary source for aluminum in the system is the bedrock and till. Hence, the increase in aluminum concentration indicates the importance of decreasing pH on the solubility of common minerals containing aluminum in the soils, especially the clay minerals such as kaolin and vermiculite. The decrease in pH from 5.2 to 4.3, we observed as a result of the deforestation, is critical to the solubility of alumina (*e.g.*, Mason, 1966, p. 167).

Dissolved Silica

Seasonal variations in the concentration of dissolved silica in stream water are inversely correlated with runoff (Johnson, *et al.*, 1969). Maximum streamwater concentrations occur in late summer, whereas minimum concentrations occur during peak flow periods associated with spring thaw. There is apparently a yearly recurring slump in the dissolved silica concentration in August preceding the seasonal decline in September (Fig. 7) for which the explanation is not clear.

After forest cutting the average weighted concentration of dissolved silica increased by about 37%. This increase probably reflects the increased solution or weathering of the geologic substrate, or both.

The relatively new finding that the solubility of amorphorus silica is inversely related to pH between about pH 3 and 7 (see Marshall, 1964, p. 81) partially explains the increase in dissolved silica in stream water from the deforested watershed.

Bicarbonate

Although the bicarbonate ion is normally the most abundant in freshwaters, it is a relatively minor constituent of the acidic runoff water from the undisturbed Hubbard Brook watersheds (Fisher, *et al.*, 1968; Johnson, *et al.*, 1968; Likens, *et al.*, 1969; and Table 4). In fact, it is barely detectable with routine methods in the stream water from W6.

In W2 the measured concentration dropped to nearly zero by the middle of the first year after cutting (1966–67), and concurrently with the decrease in streamwater pH. Thus the weighted value for 1966–67 (Table 4) reflects the conditions during the first part of the water-year and is somewhat misleading as a temporal average for the year. The streamwater concentration is zero for 1967–68 as expected since the average pH was 4.3.

TABLE 6. Dependence of chemical concentration on the volume of water discharged from Watersheds 2 (cut-over) and 6 during 1967–68. Regression lines were fitted to the relationship $y = ax + b$, where $y =$ chemical concentration in mg/liter, $a =$ slope, $x =$ a function of discharge $\left(\frac{1}{1+\beta D}\right)$, β is a proportionality constant, and D is stream discharge in liters/ha-day, and $b = y$ intercept (Johnson, *et al.*, 1969). The F-ratio for the values given for slope are significant at the 0.01 level

		W2	W6
calcium	slope Y-intercept	ns* —	ns —
magnesium	slope Y-intercept	ns —	ns —
sodium	slope Y-intercept	1.08 1.23	0.66 0.69
potassium	slope Y-intercept	-1.79 4.40	-0.36 0.50
aluminum	slope Y-intercept	ns —	-0.34 0.57
hydrogen	slope Y-intercept	ns —	ns —
nitrate	slope Y-intercept	ns —	-1.63 1.49
sulfate	slope Y-intercept	ns —	ns —
chloride	slope Y-intercept	ns —	ns —
dissolved silica	slope Y-intercept	1.85 5.31	3.24 2.78

*Not significant

EFFECT OF NITRIFICATION ON CATION LOSSES

In the undisturbed watersheds, the relatively small variations in cation and anion chemistry may be explained almost entirely by the effects of dilution and concentration as brought about by changes in stream discharge and biological activity (Johnson, *et al.*, 1969). Initially we thought that the large oscillations in streamwater chemistry after deforestation (Figs. 7, 8, 9 and 10) should be explained by this same model. However a regression analysis showed that only sodium, potassium, and dissolved silica were significantly (<0.01) related to discharge of water after deforestation (Table 6).[4] Moreover, the slope of the regression lines and the Y-intercepts for these three elements were changed greatly, relative to the undisturbed situation (Table 6; Johnson, *et al.*, 1969). The correlation coefficients for these regression lines were all less than 0.57, which indicates a great deal of scatter in the points and suggests either experimental error or other important controlling factors. In addition to large changes in the slope and Y-intercept for the solutes that show volume or concentration effects with changes in stream discharge after deforestation (Johnson, *et al.*, 1969), aluminum and nitrate, which usually were related to discharge, no longer show any significant relationship (Table 6). Thus, the increased nitrification after deforestation has swamped the rather small volume and concentration effects of discharge, characteristic of these ions in the undisturbed watersheds.

Regression analyses for each ion on nitrate concentration in stream water showed very high significance (<0.001) for all ions except hydrogen, ammonium and dissolved silica (Table 7). The more important cations (Ca^{++}, Mg^{++}, and Al^{+++}), in terms of milliequivalent values, had extremely high correlation coefficients in the regression analysis; whereas the correlation coefficients were lower for Na^+ and K^+ (Table 7; Figs. 11 and 12). Thus, these data show the quantitative importance of nitrification as a major controlling factor in determining the quantity and quality of dissolved materials flushed from the cutover watershed in drainage waters. Thus, in the undisturbed watershed, dilution and concentration effects on water discharge were the principal mechanisms controlling ionic concentrations, while in the deforested watershed nitrification was the more important controlling factor.

TABLE 7. Relationship between the nitrate concentration and other ions in stream water from the cut-over watershed (W2) during 1967–68. Regression lines fitted to the relationship $y = ax + b$, where $y =$ chemical concentration in mg/liter, $a =$ slope, $x =$ concentration of nitrate in mg/liter, and $b = y$ intercept. An F-ratio > 12.6 for the slope is significant at the 0.001 level for these regression analyses

	Slope	Correlation coefficient	F-ratio for slope
magnesium	0.024	0.97	673
calcium	0.137	0.95	424
aluminum	0.044	0.93	279
potassium	0.023	0.60	23.9
sodium	0.014	0.57	20.4
sulfate	-0.023	-0.01	30.4
chloride	0.005	-0.28	21.5
dissolved silica	0.017	0.37	6.8
ammonia	0.0006	-0.28	1.0
hydrogen	0.0003	0.08	0.3

[4] The same pattern of results was obtained for 1966–67, however, the following discussion is based entirely on 1967–68 because of the complexities associated with the sharp chemical transition period that occurred during the water-year of 1966–67 (Figs. 7 and 9).

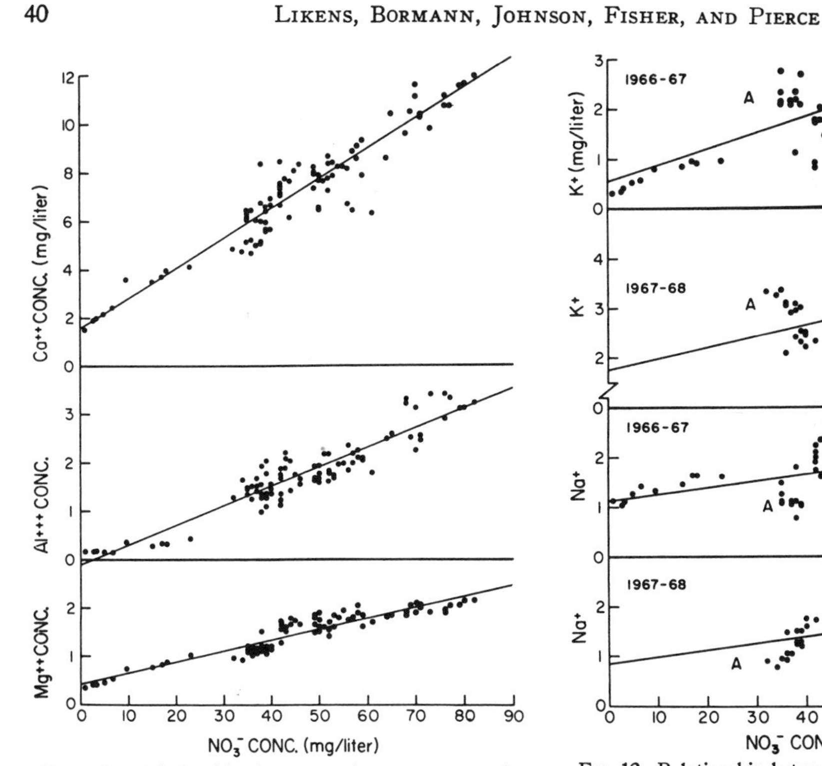

FIG. 11. Relationship between nitrate concentration and calcium, magnesium and aluminum concentrations in stream water from Watershed 2 during 1 June 1966 to 31 May 1968.

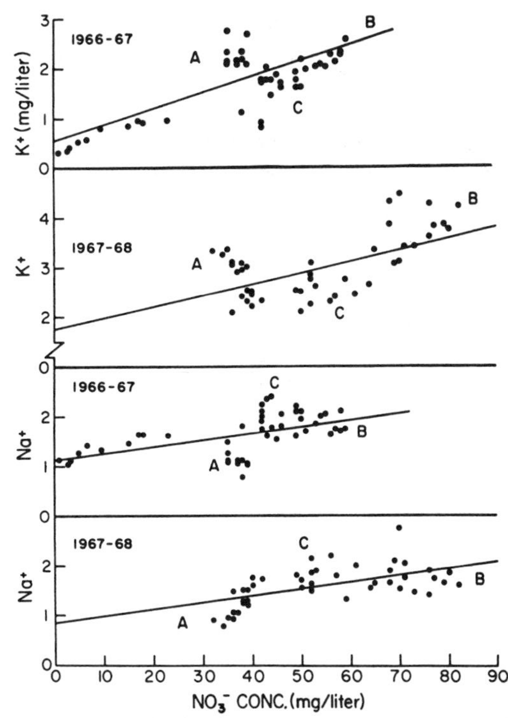

FIG. 12. Relationship between nitrate concentration and sodium and potassium concentrations in stream water from W2 during 1 June 1966 to 31 May 1968. All points in Area A were observed during the period of high spring runoff; points in Area B were observed in relatively high runoff periods in November; and Area C is a mixture of summer and winter points with lesser runoff. The points for nitrate concentrations <25 mg/liter represent the chemical transitional period between undisturbed and deforested conditions (1 June 1966 through 31 July 1966). See text for full explanation.

Plots of the actual points in the regression of Na^+ and K^+ on NO_3^- show an interesting relationship and indicate the relative importance of 1) nitrate concentration and 2) discharge in regulating ionic concentration of stream water (Fig. 12). That is, if the regression line is taken as the reference, points above the line represent concentration of Na^+ or K^+ and points below the line represent dilution of Na^+ or K^+ over that expected for the relationship between each of these cations and nitrate. An analysis of the temporal distribution for these points reveals that this is just the case. The points above the line for K^+ vs. NO_3^- (Area A) are all associated with periods of high spring runoff; Area B is associated with higher runoff periods in November; and Area C is a mixture of summer and winter points with lower runoff values. According to the volume-concentration model for undisturbed watersheds, K^+ concentrations are directly related to volume of water discharge (Table 6; Johnson, *et al.*, 1969).

The reverse is shown for the relationship for Na^+ vs. NO_3^-. Area A, represents the spring runoff period; B is the November period; and C is the summer and winter periods. Sodium is inversely related to the volume of discharge in the undisturbed watersheds (Johnson, *et al.*, 1969; Table 6), and in the cutover watershed (Table 6).

The sodium and potassium were normalized for volume and concentration effects during 1967-68 according to the equations presented by Johnson, *et al.* (1969). As a result, the F-ratio and correlation coefficients for a regression analysis of nitrate and normalized sodium and potassium concentrations were improved. The residual scatter of points about this latter regression line indicates the presence of other minor controlling factors.

The significant slope but poor correlation coefficient for NO_3^- vs. $SO_4^=$ (Table 7) is the result of the marked curvilinear relationship after deforestation shown in Fig. 14.

NUTRIENT BUDGETS

Nutrient budgets for dissolved ions and dissolved silica for the Hubbard Brook watershed-ecosystems were determined from the difference between the meteorologic input per hectare and the geologic output per hectare (Bormann and Likens, 1967). Input was calculated from the product of the ionic concentration (mg/liter) and the volume (liters) of water as precipitation (Likens, *et al.*, 1967 and Fisher, *et al.*, 1968). Additional input from applications of herbicides was added to the precipitation input. Output was calculated as the product of the volume (liters) of water draining from the watershed-ecosystems and its ionic concentration (mg/liter). Budgets for all of the ions and substances measured are given in Table 5.

Net losses were greatly increased after deforestation and herbicide treatment for all ions except ammonium, sulfate, and bicarbonate. Two factors are involved in the removal of nutrients from the deforested watershed: 1) increased runoff and 2) increased ionic concentrations in stream water. If the concentrations had not increased from the undisturbed condition, increased runoff would have accounted for a 39% increase in gross export the first year and a 28% increase the second year after deforestation. However, the gross outputs for 1967–68 were greater than the undisturbed watershed, W6 (Table 5), by 7.6-fold for Ca^{++}, 5.5-fold for Mg^{++}, 15.2-fold for K^{+}, 2.2-fold for Na^{+}, 46-fold for NO_3-N, 1.8-fold for Cl^{-}, 7.9-fold for Al^{+++} and 1.9-fold for SiO_2-Si, clearly indicating that increased stream water concentrations are primarily responsible for the increased nutrient loss from the ecosystem.

Nitrogen losses from W2 after deforestation, although very large already, do not take into account volatilization. Allison (1955) reported volatilization losses averaging 12 percent of the total nitrogen losses from 106 fallow soils. However, denitrification is an anaerobic process and requires a nitrate substrate generated aerobically (Jansson, 1958); consequently, for substantial denitrification to occur in fields, aerobic and anaerobic conditions must exist in close proximity. The large increases in subsurface flow of water from the deforested watershed suggests that such conditions may have been more common than in the undisturbed ecosystem. Moreover, Alexander (1967) points out, "When ammonium oxidation takes place at a pH lower than 5.0 to 5.5, or where the acidity produced in nitrification increases the hydrogen ions to an equivalent extent, the formation of nitrite can lead to a significant chemical volatilization of nitrogen."

Net losses of SO_4-S from the deforested ecosystem were about 40% lower in 1966–67, and 100% lower in 1967–68 than from undisturbed watersheds. In fact, the 1967–68 budget for SO_4-S in W2 was balanced in contrast to the undisturbed situation (Table 5). Precipitation is by far the major source of sulfate for the undisturbed watersheds (Fisher, *et al.*, 1968). Although the amount of sulfate added by precipitation in 1967–68 was increased slightly relative to previous years, the net export of sulfate was zero, with sulfate input in precipitation exactly balancing streamwater export (Table 5). The decreases in streamwater sulfate concentration and gross export from the ecosystem occurred concurrently with the increases in streamwater nitrate concentration and gross export after forest cutting (Figs. 7 and 8).

Average sulfate concentrations in stream water were 3.8 and 3.7 mg/liter during 1966–67 and 1967–68 (Table 4), far below the 6.4 and 6.8 mg/liter values recorded in 1964–65 and 1965–66 before cutting (Fisher, *et al.*, 1968). Much of this change can be explained by two facts, 1) the 39 to 28% increase in streamwater discharge from 1966 to 1968, which resulted from the elimination of transpiration by deforestation, and 2) the elimination of sulfate generation by sources internal to the ecosystem. If the decreases in sulfate concentrations were wholly due to increased runoff after deforestation, concentrations calculated on the basis of expected runoff (*i.e.*, normal for the undisturbed system, Table 2) and measured gross sulfate lost from W2 during 1966–67 and 1967–68 (Table 5), should approximate the weighted streamwater concentrations for the undisturbed period, 1964–66. However, these calculated concentrations (5.3 and 4.7 mg/liter) equal only 79 and 70% respectively of the average weighted concentrations for 1964–66. These differences in concentration may be due to some year-to-year variation, but are largely explained by a sharp reduction in the internal release of sulfate from the ecosystem, which we earlier attributed to chemical weathering and biological activity (Fisher, *et al.*, 1968, Likens, *ct al.*, 1969). The average annual internal release of sulfate (i.e., an amount equivalent to net loss) supplies about 10 kg/ha in the undisturbed watersheds (Table 5). Removal of this source of sulfate would account for the lower than expected adjusted sulfate concentrations mentioned above.

Thus, apparently, the normal, relatively small release of $SO_4^{=}$ from the ecosystem by chemical weathering and microbial activity (Table 5) probably became negligible following forest cutting. There are at least two possible mechanisms, operating simultaneously or separately, which may account for this:

i) There may be decreased oxidation of various sulfur compounds to $SO_4^{=}$. Waksman (1932) has suggested that a high concentration of nitrate is very toxic to sulfur oxidizing bacteria, such as *Thiobacillus thiooxidans*. This species may be important in sulfate oxidation in the deforested watershed since *T. thiooxidans* is capable of active growth at low pH (Alexander 1967). In the undisturbed watersheds we have observed a highly significant inverse linear relationship between the concentration of nitrate and sulfate in drainage water. This relationship is particularly clear in plots of sulfate concentrations against nitrate concentration using data from November through April, when the vegetation is dormant (Fig. 13). The inverse relationship between NO_3^- and

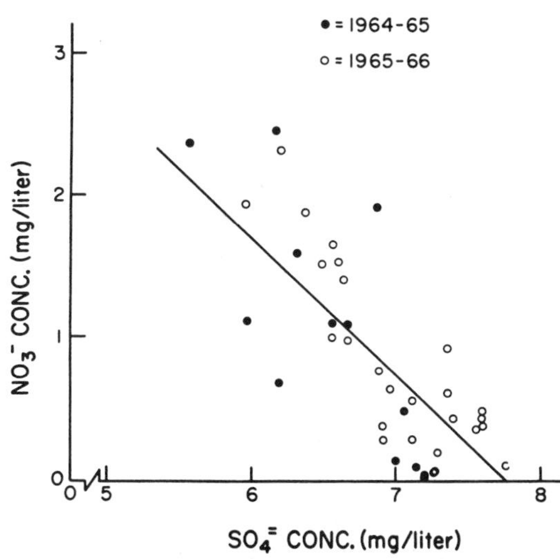

FIG. 13. Relationship between nitrate and sulfate concentrations in stream water from Watershed 2. Data were obtained during November through April of 1964–65 and 1965–66, which was prior to the increase in nitrate concentration resulting from clearing of the forest vegetation (Fig. 7). The F-ratio for this regression line is very highly significant ($p < 0.001$) and the correlation coefficient is 0.79.

$SO_4^{=}$ concentrations is very obvious in the first water-year after deforestation, 1966–67, when nitrate concentrations in stream water increased from normal (undisturbed) values to very high concentrations (Fig. 14). During the second wa-

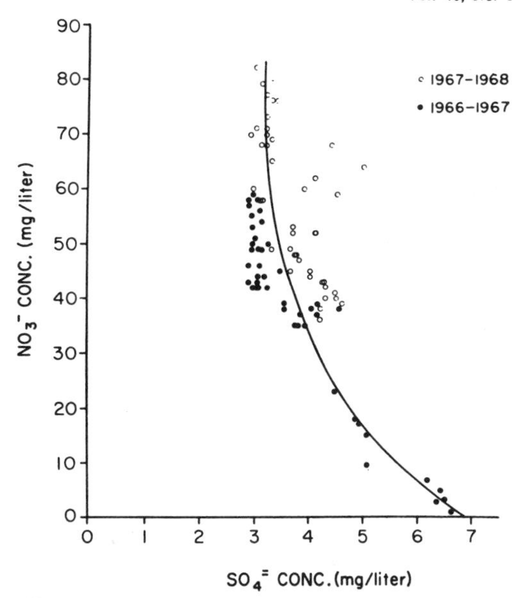

FIG. 14. Relationship between nitrate and sulfate concentrations in stream water from Watershed 2 during 1966–67 and 1967–68. Nitrate values less than 25 mg/liter indicate the chemical transition period (1 June 1966 through 31 July 1966, Fig. 7) between undisturbed and deforested conditions.

ter-year after cutting, 1967–68, the nitrate concentration in stream water from W2 increased even more, whereas the sulfate concentration decreased very little and coincided with the concentration of sulfate in precipitation after adjustment for water loss by evaporation. Perhaps there is an intricate feedback mechanism between the toxicity of the nitrate concentrations and microbial oxidation of sulfur compounds within the soil. Another possibility is that the number of sulfur oxidizing bacteria have been selectively reduced by the herbicides in the deforested watershed.

ii) Although somewhat unlikely, there may be increased sulfate reduction brought about by more anaerobic conditions, particularly in the lower more inorganic horizons of the soil (*e.g.*, Waksman, 1932). That is, an increased zone of water saturation in the deeper layers and in topographic lows on the cutover watershed probably has less free oxygen than in the undisturbed situation, promoting sulfur reduction. One difficulty, however, is that the growth of the most important sulfur reducing bacteria (*Desulfovibrio* spp.) is greatly retarded by acid conditions (Alexander, 1967). Also, molecular hydrogen released by anaerobic bacterial decomposition of organic matter may be used for the reduction of sulfate (Postgate, 1949; Rankama and Sahama, 1950).

The chloride budget for the undisturbed watershed during 1966–67 showed that input in precipitation exceeded the gross output, whereas the budget was essentially balanced during the 1967–68 water-year (Table 5). However after deforestation, significant net losses of chloride were observed (Table 5). The application of Bromacil in 1966–67 potentially added the equivalent of 3.0 kg Cl/ha or about 50% of the chloride input as precipitation. From the pattern of chloride changes in stream water following the addition of this herbicide (Fig. 7), it would appear that the herbicide and/or its degradation products were lost from the watershed quite gradually throughout the year. Measurements of Bromacil in stream water seemed to confirm this (Pierce, 1969). Since the Bromacil (1966–67) and possibly 2, 4, 5–T (1967–68) were not all flushed from the ecosystem within a year, then the internal release of chloride from the ecosystem probably represented an even greater percentage of the gross annual output (Table 5). Based upon streamwater concentrations in W2 and W6 (Fig. 7), it would appear that the internal reservoir (plus external inputs from herbicides) of chloride within the ecosystem has been essentially exhausted in two years following deforestation.

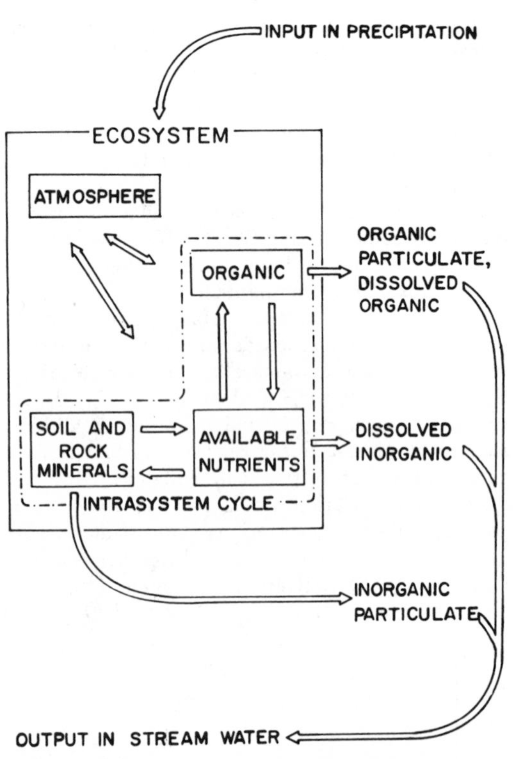

FIG. 15. Diagrammatic model for sites of accumulation and pathways of nutrients in the Hubbard Brook ecosystem (after Bormann, *et al.*, 1969).

GENERAL DISCUSSION AND SIGNIFICANCE

The intrasystem cycle of a terrestrial ecosystem links the organic, available nutrient, and soil and rock mineral compartments through rate processes including decomposition of organic matter, leaching and exudate from the biota, nutrient uptake by the biota, weathering of primary minerals, and formation of new secondary minerals (Fig. 15). The deforestation experiment was designed to test the effects of blockage on a major ecosystem pathway, *i.e.*, nutrient and water uptake by vegetation, on other components of the intrasystem cycle and on the export behavior of the system as a whole. The block was imposed by cutting all of the forest vegetation and subsequently preventing regrowth with herbicides. We hoped that this experimental procedure would provide information about the nature of the homeostatic capacity of the ecosystem. The deforested condition has been maintained since 1 January 1966.

Forest clearing and herbicide treatment had a profound effect on the hydrologic and nutrient relationships of our northern hardwood ecosystem. Annual runoff (water export) increased by some 33 cm or 39% in the first year and 27 cm or 28% in the second year over that expected. Moreover, the discharge pattern was altered so that sustained, higher flows occurred in the summer months and the snow pack melted earlier in the spring. This overall increase in stream runoff is large compared to the average increase (about 20 cm) found for other such experiments throughout the world (Hibbert, 1967), but is less than the maximum increase of 41 cm found for clear-cut watersheds in North Carolina (Hoover, 1944).

No previous comprehensive measurements have been made of the homeostatic ability of a watershed-ecosystem to retain nutrients despite major shifts in the hydrologic cycle, including increased discharge, following deforestation (Odum, 1969).

Our results showed that cation and anion export did not change for the first 5 months (winter and spring) after deforestation, but then the ionic concentrations increased spectacularly, and remained at high levels for the 2 years of observation (Tables 4 and 5, Figs. 7, 8, 9 and 10). Annual net losses in kg/ha amounted to about 142 for nitrate-nitrogen, 90 for calcium, 36 for potassium, 32 for dissolved silica, 24 for aluminum, 18 for magnesium, 17 for sodium, and 4 for chlorine during 1967–68. These losses are much

TABLE 8. Comparative net gains or losses of dissolved solids in runoff following clear-cutting of Watershed 2 in the Hubbard Brook Experimental Forest for the period 1 June to 31 May. In metric tons/km²-yr

	1966-67		1967-68	
	W2	*W6*	*W2*	*W6*
Ca	−7.5	−0.8	−9.0	−0.9
K	−2.3	−0.1	−3.6	−0.2
Al	−1.7	−0.1	−2.4	−0.3
Mg	−1.6	−0.3	−1.8	−0.3
Na	−1.7	−0.6	−1.7	−0.7
NH_4	+0.1	+0.2	+0.2	+0.3
NO_3	−43.0	+1.5	−62.8	+1.1
SO_4	−0.5	−0.8	0	−1.0
HCO_3	−0.1	−0.2	0	−0.3
Cl	−0.1	+0.2	−0.4	0
SiO_2−aq	−6.6	−3.6	−6.9	−3.6
Total	−65.0	−4.6	−88.4	−5.9

greater than for adjacent undisturbed ecosystems (Table 5). Ammonium-nitrogen was essentially unchanged relative to the undisturbed condition during this period and showed an annual net gain of about 1 to 2 kg/ha. In comparison with the undisturbed watershed-ecosystems the greatest changes occurred in nitrate-nitrogen and potassium export. Nitrate-nitrogen is normally accumulated in the undisturbed ecosystem in contrast to this very large export, and the net potassium output increased about 18-fold. The total net export of dissolved inorganic substances from the deforested ecosystem is 14–15 times greater than from the undisturbed ecosystem (Table 8).

The terrestrial ecosystem is one of the ultimate sources of dissolved substances in surface water. The contribution of dissolved solids (gross export) by our undisturbed forest ecosystems, 13.9 metric tons/km² (Bormann, *et al.,* 1969) is only about 25% of the dissolved load predicted by Langbein and Dawdy (1964) for regions with 75 cm of annual runoff. Their estimates were based on data from watersheds of the north Atlantic slope, which probably include areas disturbed by agriculture or logging. The difference between our undisturbed forest ecosystems and the regional prediction is credited in part to the operation of various regulating biotic factors associated with mature undisturbed forest (Bormann, *et al.,* 1969).

Deforestation markedly altered the ecosystem's contribution of dissolved solids to the drainage waters. Total gross export, exclusive of dissolved organic matter, was about 75 metric tons/km² in 1966–67 and 97 metric tons/km² in 1967–68. These figures exceed the regional prediction of Langbein and Dawdy (1964). However, it should be noted that the accelerated export of dissolved substances results primarily from mining the nutrient capital of the ecosystem and cannot be sustained indefinitely.

Surprisingly, the net export of dissolved inorganic substances from the cutover watershed is about double the annual value estimated for particulate matter removed by debris avalanches in the White Mountains (Bormann, *et al.,* 1969). Thus, the effects of deforestation may have almost twice the importance of avalanches in short-term catastrophic transport of inorganic materials downslope in the White Mountains.

Coupled with this increase in gross and net export of dissolved substances, there has been at least a 4-fold increase in the export of inorganic and organic particulate matter from the deforested ecosystem. This increase indicates that the biotic mechanisms that normally minimize erosion and transport (Bormann, *et al.,* 1969) are also becoming less effective.

The greatly increased export of nutrients from the deforested ecosystem resulted primarily from an alteration of the nitrogen cycle within the ecosystem. Whereas in the undisturbed system, nitrogen is cycled conservatively between the living and decaying organic components, in the deforested watershed, nitrate produced by microbial nitrification from decaying organic matter, is rapidly flushed from the system in drainage waters. In fact, the increased nitrate output accounts for the net increase in total cation and anion export from the ecosystem (Likens, *et al.,* 1969). With the increased availability of nitrate ions and hydrogen ions from nitrification, cations are readily leached from the system. Cations are mobilized as hydrogen ions replace them on the various exchange complexes of the soil and as organic and inorganic materials decompose. Based upon the increased output for sodium, chemical decomposition of inorganic materials in the deforested ecosystem is also accelerated by about 3-fold.

If the streamwater concentrations had remained constant after deforestation, increased water output alone would have accounted for 39% of the increased nutrient export the first year and 28% of the increased nutrient export the second year. However, the very large increase in annual export of dissolved solids from the deforested ecosystem occurred primarily because the streamwater concentrations were vastly increased, mostly as a direct result of the increased nitrification. The increased output of nutrients originated predominantly from the organic compartment of the watershed-ecosystem (Fig. 15).

Our study shows that the retention of nutrients within the ecosystem is dependent on constant and efficient cycling between the various components of the intrasystem cycle, *i.e.*, organic, available nutrients, and soil and rock mineral compartments (Fig. 15). Blocking of the pathway of nutrient uptake by destruction of one subcomponent of the organic compartment, *i.e.*, vegetation, leads to greatly accelerated export of the nutrient capital of the ecosystem. From this we may conclude that one aspect of homeostatis of the ecosystem, *i.e.*, maintenance of nutrient capital, is dependent upon the undisturbed functioning of the intrasystem nutrient cycle, and that in this ecosystem no mechanism acts to greatly delay loss of nutrients following sustained destruction of the vegetation.

The increased output of water from the deforested watershed was readily visible during the summer months, however the increased ion and particulate matter concentrations were not. The stream water from the deforested watershed appeared to be just as clear and potable as that from adjacent, undisturbed watersheds. However this was not the case. By August, 1966, the nitrate concentration in stream water exceeded (at times almost doubled) the concentration recommended for drinking water (Public Health Service, 1962).

The high nutrient concentrations, plus the increased amount of solar radiation (absence of forest canopy) and higher temperature in the stream, resulted in significant eutrophication. A dense bloom of *Ulothrix zonata* (Weber and Mohr) Kütz, has been observed during the summers of 1966 and 1967 in the stream of W2. In contrast the undisturbed watershed streams are essentially devoid of algae of any kind. This represents a good example of how an overt change in one component of an ecosystem, alters the structure and function, often unexpectedly, in another part of the same ecosystem or in another interrelated ecosystem. Unless these ecological interrelationships are understood, naive management practices can produce unexpected and possibly widespread deleterious results.

CONCLUSIONS

1. The quantity and quality of drainage waters were significantly altered subsequent to deforestation of a northern hardwoods watershed-ecosystem. All vegetation on Watershed 2 of the Hubbard Brook Experimental Forest was cut, but not removed, during November and December of 1965; and vegetation regrowth was inhibited by periodic application of herbicides.

2. Annual runoff of water exceeded the expected value, if the watershed were undisturbed, by 33 cm or 39% during the first water-year after deforestation and 27 cm or 28% during the second water-year. The greatest increase in water discharge, relative to an undisturbed situation, occurred during June through September, when runoff was 414% (1966–67) and 380% (1967–68) greater than the estimate for the untreated condition.

3. Deforestation resulted in large increases in streamwater concentrations of all major ions except NH_4^+, $SO_4^=$ and HCO_3^-. The increases did not occur until 5 months after the deforestation. The greatest increase in streamwater ionic concentration after deforestation was observed for nitrate, which increased by 41-fold the first year and 56-fold the second year above the undisturbed condition.

4. Sulfate was the only major ion in stream water from Watershed 2 that decreased in concentration after deforestation. The 45% decrease the first year (1966–67) resulted mostly from increased runoff of water and by eliminating the generation of sulfate within the ecosystem. The concentration of sulfate in stream water during 1967–68 equalled the concentration in precipitation after adjustment for water loss by evaporation. Sulfate concentrations were inversely related to nitrate concentrations in stream water in both undisturbed and deforested watersheds.

5. In the undisturbed watersheds the stream water can be characterized as a very dilute solution of sulfuric acid (pH about 5.1 for W2); whereas after deforestation the stream water from Watershed 2 became a relatively stronger nitric acid solution (pH 4.3), considerably enriched in metallic ions and dissolved silica.

6. The increase in average nitrate concentration in precipitation for the Hubbard Brook area compared to data from 1955–56, as well as the consistent annual increase observed from 1964–1968, may be some measure of a general increase in air pollution.

7. The greatly increased export of dissolved nutrients from the deforested ecosystem was due to an alteration of the nitrogen cycle within the ecosystem. Whereas nitrogen is normally conserved in the undisturbed ecosystem, in the deforested ecosystem nitrate is rapidly flushed from the system in drainage water. The mobilization of nitrate from decaying organic matter, presumably by increased microbial nitrification, quantitatively accounted for the net increase in

total cation and anion export from the deforested ecosystem.

8. Increased availability of nitrate and hydrogen ions resulted from nitrification. Cations were mobilized as hydrogen ions replaced them on various exchange complexes of the soil and as organic and inorganic materials were decomposed. Chemical decomposition of inorganic materials in the deforested ecosystem was accelerated about 3-fold. However, the bulk of the nutrient export from the deforested watershed originated from the organic compartment of the ecosystem.

9. The total net export of dissolved inorganic substances from the deforested ecosystem was 14–15 times greater than from undisturbed ecosystems. The increased export occurred because the streamwater concentrations were vastly increased, primarily as a direct result of the increased nitrification, and to a much lesser extent because the amount of stream water was increased.

10. The deforestation experiment resulted in significant pollution of the drainage stream from the ecosystem. Since August, 1966, the nitrate concentration in stream water has exceeded, almost continuously, the maximum concentration recommended for drinking water. As a result of the increased temperature, light and nutrient concentrations, and in sharp contrast to the undisturbed watersheds, a dense bloom of algae has appeared each year during the summer in the stream from Watershed 2.

11. Nutrient cycling is closely geared to all components of the ecosystem; decomposition is adjusted to nutrient uptake, uptake is adjusted to decomposition, and both influence chemical weathering. Conservation of nutrients within the ecosystem depends upon a functional balance within the intrasystem cycle of the ecosystem. The uptake of water and nutrients by vegetation is critical to this balance.

LITERATURE CITED

Alexander, M. 1967. Introduction to Soil Microbiology. John Wiley and Sons, Inc., New York, 472 pp.

Allison, F. E. 1955. The enigma of soil nitrogen balance sheets. Advan. Agron. **7**: 213–250.

Anderson, D. H., and H. E. Hawkes. 1958. Relative mobility of the common elements in weathering of some schist and granite areas. Geochim. Cosmochim. Acta **14**(3): 204–210.

Bormann, F. H., and G. E. Likens. 1967. Nutrient cycling. Science **155**(3761): 424–429.

Bormann, F. H., G. E. Likens, D. W. Fisher, and R. S. Pierce. 1968. Nutrient loss accelerated by clear-cutting of a forest ecosystem. Science **159**: 882–884.

Bormann, F. H., G. E. Likens, and J. S. Eaton. 1969. Biotic regulation of particulate and solution losses from a forest ecosystem. BioScience **19**(7): 600–610.

Boswell, J. G. 1955. The microbiology of acid soils. IV. Selected sites in Northern England and Southern Scotland. New Phytol. **54**(2): 311–319.

Federer, C. A. 1969. Radiation and snowmelt on a clear-cut watershed. E. Snow Conf. Proc., Boston, Mass. (**1968**) pp. 28–41.

Fisher, D. W., A. W. Gambell, G. E. Likens, and F. H. Bormann. 1968. Atmospheric contributions to water quality of streams in the Hubbard Brook Experimental Forest, New Hampshire. Water Resources Res. **4**(5): 1115–1126.

Gambell, A. W., and D. W. Fisher. 1966. Chemical composition of rainfall, eastern North Carolina and southeastern Virginia, U.S. Geol. Survey Water-Supply Paper **1535K**: 1–41.

Harvey, H. H., and A. C. Cooper. 1962. Origin and treatment of a supersaturated river water. Internat. Pacific Salmon Fish. Comm., Prog. Rept. No. **9**: 1–19.

Hart, G., R. E. Leonard, and R. S. Pierce. 1962. Leaf fall, humus depth, and soil frost in a northern hardwood forest. Forest Res. Note **131**, Northeastern For. Exp. Sta., Durham, N. H.

Hibbert, A. R. 1967. Forest treatment effects on water yield. pp. 527–543. *In*: Proc. Internat. Symposium on Forest Hydrology, ed. by W. E. Sopper and H. W. Lull, Pergamon Press, N. Y.

Hoover, M. D. 1944. Effect of removal of forest vegetation upon water yields. Trans. Amer. Geophys. Union, Part **6**: 969–975.

Hornbeck, J. W., and R. S. Pierce. 1969. Changes in snowmelt run-off after forest clearing on a New England watershed. E. Snow Conf. Proc., Portland, Maine (1969). (In press).

Hornbeck, J. W., R. S. Pierce, and C. A. Federer. Streamflow changes after forest clearing in New England. (In preparation).

Iwasaki, I., S. Utsumi, and T. Ozawa. 1952. Determination of chloride with mercuric thiocyanate and ferric ions. Chem. Soc. Japan Bull. **25**: 226.

Hewlett, J. D., and A. R. Hibbert. 1961. Increases in water yield after several types of forest cutting. Quart. Bull. Internatl. Assoc. Sci. Hydrol. Louvain, Belgium, pp. 5–17.

Jansson, S. L. 1958. Tracer studies on nitrogen transformations in soil with special attention to mineralisation-immobilization relationships. Kungl. Lantbrukshögskolans Annaler. **24**: 105–361.

Järnefelt, H. 1949. Der Einfluss der Stromschnellen auf den Sauerstoff-und Kohlensäuregehalt und das pH des Wassers im Flusse Vuoksi. Verh. Internat. Ver. Limnol. **10**: 210–215.

Johnson, N. M., G. E. Likens, F. H. Bormann, and R. S. Pierce. 1968. Rate of chemical weathering of silicate minerals in New Hampshire. Geochim. Cosmochim. Acta. **32**: 531–545.

Johnson, N. M., G. E. Likens, F. H. Bormann, D. W. Fisher, and R. S. Pierce. 1969. A working model for the variation in streamwater chemistry at the Hubbard Brook Experimental Forest, New Hampshire. Water Resources Res. 5(6): 1353–1363.

Juang, F. H. F., and N. M. Johnson. 1967. Cycling of chlorine through a forested watershed in New England. J. Geophys. Research **72**(22): 5641–5647.

Junge, C. E. 1958. The distribution of ammonia and nitrate in rain water over the United States. Trans.

Amer. Geophys. Union **39:** 241–248.

———. 1963. Air Chemistry and Radioactivity. Academic Press, N. Y. 382 pp.

———, **and R. T. Werby.** 1958. The concentration of chloride, sodium, potassium, calcium and sulphate in rain water over the United States. J. Meteorol. **15:** 417–425.

Langbein, W. B., and D. R. Dawdy. 1964. Occurrence of dissolved solids in surface waters in the United States. U. S. Geol. Survey Prof. Paper **501–D:** D115–D117.

Lieberman, J. A., and M. D. Hoover. 1948a. The effect of uncontrolled logging on stream turbidity. Water and Sewage Works **95**(7): 255–258.

———. 1948b. Protecting quality of stream flow by better logging. Southern Lumberman: 236–240.

Likens, G. E., F. H. Bormann, N. M. Johnson, and R. S. Pierce. 1967. The calcium, magnesium, potassium, and sodium budgets for a small forested ecosystem. Ecology **48**(5): 772–785.

Likens, G. E., F. H. Bormann and N. M. Johnson. 1969. Nitrification: Importance to nutrient losses from a cutover forested ecosystem. Science **163**(3872): 1205–1206.

Lindroth, A. 1957. Abiogenic gas supersaturation of river water. Arch. für Hydrobiol. **53:** 589–597.

Macan, T. T. 1958. The temperature of a small stony stream. Hydrobiologia **12:** 89–106.

Marshall, C. E. 1964. The physical chemistry and minerology of soils. Vol. I. Soil materials. Wiley and Sons, N. Y. 388 pp.

Mason, B. 1966. Principles of geochemistry. 3rd ed. Wiley and Sons, N. Y. 329 pp.

McConnochie, K., and G. E. Likens. 1969. Some Trichoptera of the Hubbard Brook Experimental Forest in central New Hampshire. Canadian Field Naturalist. **83**(2): 147–154.

Minckley, W. L. 1963. The ecology of a spring stream, Doe Run, Meade County, Kentucky. Wildlife Monogr. **11:** 1-124.

Nye, R. H., and D. J. Greenland. 1960. The soil under shifting cultivation. Commonwealth Bureau of Soils, Harpenden, England, Tech. Bull. No. **51,** 156 pp.

Odum, E. P. 1969. The strategy of ecosystem development. Science **164**(3877): 262–270.

Pierce, R. S. 1969. Forest transpiration reduction by clearcutting and chemical treatment. Proc. Northeastern Weed Control Conference. **23:** 344–349.

Postgate, J. R. 1949. Competitive inhibition of sulphate reduction by selenate. Nature (**4172**): 670–671.

Rankama, K., and T. G. Sahama. 1950. Geochemistry. Chicago Univ. Press, 912 pp.

Ruttner, F. 1953. Fundamentals of Limnology. Univ. of Toronto Press. (Transl. by D. G. Frey and F. E. J. Fry). 242 pp.

Scott, D. R. M. 1955. Amount and chemical composition of the organic matter contributed by overstory and understory vegetation to forest soil. Yale Univ. School of Forestry. Bull. No. **62,** 73 pp.

Smith, W., F. H. Bormann, and G. E. Likens. 1968. Response of chemoautotrophic nitrifiers to forest cutting. Soil Science **106**(6): 471–473.

Tebo, L. D. 1955. Effects of siltation, resulting from improper logging, on the bottom fauna of a small trout stream in the southern Appalachians. Prog. Fish Culturist **12**(2): 64–70.

Thiessen, A. H. 1923. Precipitation for large areas. Monthly Weather Review **51:** 348–353.

Trimble, G. R., and R. S. Sartz. 1957. How far from a stream should a logging road be located? J. Forestry **55**(5): 339–341.

U. S. Forest Service. Northeastern Forest Experiment Station. 1964. Hubbard Brook Experimental Forest. Northeast. For. Exp. Sta., Upper Darby, Penn. 13 pp.

U. S. Public Health Service. 1962. Drinking water standard. U.S. Public Health Service Publ. **956.** Washington, D. C.

Waksman, S. A. 1932. Principles of soil microbiology. 2nd ed. Williams and Wilkins Co., Baltimore. 894 pp.

Weber, D. F., and P. L. Gainey. 1962. Relative sensitivity of nitrifying organisms to hydrogen ions in soils and in solutions. Soil Sci. **94:** 138–145.

Whitehead, H. C., and J. G. Feth. 1964. Chemical composition of rain, dry fallout, and bulk precipitation at Menlo Park, California, 1957-1959. J. Geophys. Res. **69**(16): 3319–3333.

Woods, W. J. 1960. An ecological study of Stony Brook, New Jersey. Ph.D. Thesis, Rutgers Univ. New Brunswick. 307 pp.

Contributors

James H. Brown
Department of Biology
University of New Mexico
Albuquerque, New Mexico 87131

Sharon E. Kingsland
Department of the History of Science
Johns Hopkins University
Baltimore, Maryland 21218

Joel G. Kingsolver
Department of Zoology NJ-15
University of Washington
Seattle, Washington 98195

Simon A. Levin
Section of Ecology and Systematics
Cornell University
Ithaca, New York 14853

Jane Lubchenco
Department of Zoology
Oregon State University
Corvallis, Oregon 97331-2914

Robert T. Paine
Department of Zoology
University of Washington
Seattle, Washington 98195

Robert K. Peet
Department of Biology
Campus Box 3280
University of North Carolina
Chapel Hill, North Carolina 27599-3280

Leslie A. Real
Department of Biology
Campus Box 3280
University of North Carolina
Chapel Hill, North Carolina 27599-3280